Handbook of Experimental Pharmacology

Volume 86

The Cholinergic Synapse

Contributors

D. V. Agoston · T. Bartfai · E. Borroni · A. N. Davison · I. Ducis
F. Fonnum · E. Giacobini · I. Hanin · J. Järv · D. A. Johnson
H. Kilbinger · W.-D. Krenz · M. E. Kriebel · K. Krnjević · A. Maelicke
J. Massoulié · E. G. McGeer · P. L. McGeer · G. R. Pilar · P. E. Potter
G. P. Richardson · P. J. Richardson · D. W. Schmid · H. Stadler
J.-P. Toutant · K. V. Toyka · S. Tuček · M. Tytell · K. Wächtler
V. P. Whittaker · H. Zimmermann

Editor

V. P. Whittaker

Springer-Verlag
Berlin Heidelberg New York
London Paris Tokyo

Professor Dr. VICTOR P. WHITTAKER
Arbeitsgruppe Neurochemie
Max-Planck-Institut für biophysikalische Chemie
Postfach 2841
D-3400 Göttingen

With 98 Figures

ISBN 3-540-18613-1 Springer-Verlag Berlin Heidelberg New York
ISBN 0-387-18613-1 Springer-Verlag New York Berlin Heidelberg

Library of Congress Cataloging-in-Publication Data. The Cholinergic synapse / contributors, D. V. Agoston ... [et al.]; editor, V. P. Whittaker. p. cm. -- (Handbook of experimental pharmacology; v. 86) Includes bibliographies and index. ISBN 0-387-18613-1 (U.S.) 1. Cholinergic mechanisms. 2. Synapses. I. Agoston, D. V. II. Whittaker, V. P. (Victor Percy), 1919- . III. Series. [DNLM: 1. Cholinergic Fibers. 2. Synapses. W1 HA51L v. 86 / WL 102.8 C547] QP905.H3 vol. 86 [QP368.7] 615'.1 s-dc 19 [599'.0188] DNLM/DLC for Library of Congress 88-4468

Typesetting: K+V Fotosatz GmbH, Beerfelden
Printing and bookbinding: Brühlsche Universitätsdruckerei, Giessen
2122/3130-543210 - Printed on acid-free paper

List of Contributors

D. V. AGOSTON, Laboratory of Cell Biology, National Institute of Mental Health, Building 36, Room 3A17, Bethesda, MD 20892, USA

T. BARTFAI, Department of Biochemistry, Arrhenius Laboratory, University of Stockholm, S-10691 Stockholm

E. BORRONI, Arbeitsgruppe Neurochemie, Max-Planck-Institut für biophysikalische Chemie, Karl-Friedrich-Bonhoeffer-Institut, Postfach 2841, D-3400 Göttingen

A.N. DAVISON, Department of Neurochemistry, Institute of Neurology, National Hospital, Queen Square, London WC1N 3BG, Great Britain

I. DUCIS, Department of Pathology (D-33), Division of Neuropathology, University of Miami School of Medicine, P.O. Box 016960, Miami, FL 33101, USA

F. FONNUM, Division for Environmental Toxicology, Norwegian Defence Research Establishment, P.O. Box 25, N-2007 Kjeller

E. GIACOBINI, Department of Pharmacology, Southern Illinois University School of Medicine, P.O. Box 3926, Springfield, IL 62708, USA

I. HANIN, Department of Pharmacology and Experimental Therapeutics, Loyola University Stritch School of Medicine, 2160 South First Avenue, Maywood, IL 60153, USA

J. JÄRV, Department of Organic Chemistry, Tartu University, Tartu, Estonian SSR, USSR

D.A. JOHNSON, The Biological Science Group, College of Liberal Arts and Science, University of Connecticut, Storrs, CT 06268, USA

H. KILBINGER, Pharmakologisches Institut der Universität Mainz, Abteilung für Neuropharmakologie, Obere Zahlbacher Straße 67, D-6500 Mainz

W.-D. KRENZ, Zoologisches Institut, Universität Basel, Rheinsprung 9, CH-4051 Basel

M.E. KRIEBEL, Department of Physiology, State University of New York, College of Medicine, Health Science Center, 750 East Adams Street, Syracuse, NY 13210, USA

K. KRNJEVIĆ, Departments of Anaesthesia Research and Physiology, McGill University, McIntyre Building, Room 1208, 3655 Drummond Street, Montreal PQ H3G 1Y6, Canada

A. MAELICKE, Max-Planck-Institut für Ernährungsphysiologie, Rheinlanddamm 201, D-4600 Dortmund 1

J. MASSOULIÉ, Laboratoire de Neurobiologie, École Normale Supérieure, 46 rue d'Ulm, F-75230 Paris Cedex 05

E.G. MCGEER, Kinsmen Laboratory of Neurological Research, Department of Psychiatry, University of British Columbia, 2255 Wesbrook Mall, Vancouver, BC V6T 1W5, Canada

P.L. MCGEER, Kinsmen Laboratory of Neurological Research, Department of Psychiatry, University of British Columbia, 2255 Wesbrook Mall, Vancouver BC V6T 1W5, Canada

G.R. PILAR, The Biological Sciences Group, College of Liberal Arts and Sciences, University of Connecticut, U-42, Room 416, 75 N. Eagleville Rd., Storrs, CT 06268, USA

P.E. POTTER, Department of Pharmacology and Experimental Therapeutics, Loyola University Stritch School of Medicine, 2160 South First Avenue, Maywood, IL 60153, USA

G.P. RICHARDSON, Ethology and Neurophysiology Group, School of Biology, University of Sussex, Brighton, Sussex, Great Britain

P.J. RICHARDSON, Department of Clinical Biochemistry, Addenbrooke's Hospital, Hills Road, Cambridge CB2 2QR, Great Britain

D.W. SCHMID, Laboratoire de Neurobiologie et Physiologie Comparées, Université de Bordeaux I, Place du Docteur Bertrand Peyneau, F-33120 Arcachon

H. STADLER, Arbeitsgruppe Neurochemie, Max-Planck-Institut für biophysikalische Chemie, Karl-Friedrich-Bonhoeffer-Institut, Postfach 2841, D-3400 Göttingen

J.-P. TOUTANT, Laboratoire de Neurobiologie, École Normale Supérieure, 46 rue d'Ulm, F-75230 Paris Cedex 05

K.V. TOYKA, Neurologische Klinik der Universität Düsseldorf, Moorenstraße 5, D-4000 Düsseldorf 1

S. TUČEK, Institute of Physiology, Czechoslovak Academy of Sciences, Vídeňská 1083, CS-14220, Prague, CSSR

M. TYTELL, Department of Anatomy, Wake Forest University, Medical Center, Bowman School of Medicine, Winston-Salem, NC 27103, USA

K. WÄCHTLER, Institut für Zoologie, Tierärztliche Hochschule, Bünteweg 17, D-3000 Hannover 71

V.P. WHITTAKER, Arbeitsgruppe Neurochemie, Max-Planck-Institut für biophysikalische Chemie, Postfach 2841, D-3400 Göttingen

H. ZIMMERMANN, Arbeitskreis Neurochemie, Zoologisches Institut der Johann Wolfgang Goethe-Universität, Siesmayerstraße 70, D-6000 Frankfurt/Main 11

Preface

One of the most impressive works of scholarship in the field of experimental pharmacology has been the Heffter-Heubner *Handbuch der experimentellen Pharmakologie*, internationalized some years ago under the title *Handbook of Experimental Pharmacology* and kept up to date by a series of numbered *Ergänzungswerke* or supplementary volumes which have now replaced in importance the original *Handbuch*. These volumes constitute a valuable and continuously updated multiauthor review series of topics important in modern pharmacology and allied sciences.

The Editorial Board of the *Handbook* invited me 2 years ago to undertake, as subeditor, the preparation of a new volume entitled *The Cholinergic Synapse*. A previous volume in this series, vol. 15, *Cholinesterases and Anticholinesterase Agents*, edited by GEORGE KOELLE, was published in 1963 and was far wider in scope than its title suggested: it was, in fact an authoritative summing up of the whole subject of cholinergic function and still has some value today as an account of the state of the art as it was at that time. Since then another excellent review, of a specific cholinergic synapse, has appeared in this series: this was vol. 42, *Neuromuscular Junction*, edited by ELEANOR ZAIMIS and published in 1976. A third volume, vol. 53, *Pharmacology of Ganglionic Transmission*, which appeared in 1980 and was edited by D.A. KHARKEVICH, includes important aspects of autonomic cholinergic function. The Editors felt that the time was ripe for a fourth volume which would summarize the extensive new literature on cholinergic function in general which has appeared in the late 1970s and early 1980s. In accordance with the increasing interest being shown in the cell- and molecular-biological aspects of synaptic function it was decided to emphasize these. To give adequate coverage and to reduce the burden on individual contributors it was decided to have a fairly large number of shorter chapters rather than a few longer reviews, but care has been taken to coordinate the subject matter and, to some extent, the style of each and to avoid too many gaps and overlaps: with what success, we leave the reader to judge.

The volume is divided into seven parts. It begins with three introductory chapters which draw attention to the main model systems from the intensive study of which much of what we know about cholinergic mechanisms – especially at the cell- and molecular-biological level – has been derived. The second part, consisting of Chaps. 4 and 5, is concerned with biological aspects of the cholinergic system, its phylogeny and development. In the third part, a series of chapters (Chaps. 6–10) then examines the main molecular components of the cholinergic system, including the postsynaptic nicotinic and muscarinic receptors (Chaps. 9

and 10). The two long and detailed chapters on the cholinesterases (Chaps. 8a and b) provide a fitting link with G. B. KOELLEs predecessor volume. Consideration is then given to the cellular organization of the cholinergic neuron beginning with that key organelle, the synaptic vesicle, the main storage site of the transmitter in the nerve terminal (Chap. 11). Six further chapters follow (Chaps. 12–17) covering several other aspects of cellular organization, among them, axonal transport, the high-affinity choline uptake system, cholinergic-specific antigens and cholinergic co-transmitters. The division of topics between parts III and IV is somewhat arbitrary, and one important topic – the Ca^{2+} channel – is excluded, on the grounds that it is not peculiar to cholinergic endings but is a ubiquitous synaptic component. The same consideration applies to the cellular mechanisms of vesicle exo- and endocytosis. In the fifth part of the book, "Peripheral Cholinergic Systems" (Chaps. 18–20), and a companion sixth, "Central Cholinergic Systems" (Chaps. 21–23), the main types of peripheral and central cholinergic synapses are examined from the physiological, biochemical, pharmacological and anatomical points of view. A final section covers, in three chapters, the neuropathology, experimental and clinical, of the cholinergic system, including the effect of aging.

A book such as this, with its many hundreds of references, often arranged for convenience in summary tables, cannot be brought to completion without the devoted work of many people. Many thanks go in the first place to my fellow authors, whose authoritative contributions I deeply appreciate and whose forbearance and prompt responses during the editorial process, when diverse contents, styles and treatments in the individual chapters must be harmonized, I gratefully acknowledge. I am much indebted to Dr. MARGARET DICKINS for devoted help during the press editing. My warm thanks also goes to Mrs. ELSE VETTER and Mrs. IRENE FRIED for their skilled secretarial work.

Regarding the final form of the book, my responsibilities effectively ended, for reasons beyond my control, with the submission of the edited typescripts to the publisher. The quality, accuracy and consistency of the printed text and index then became the responsibility of the publisher, assisted by the authors.

Göttingen V. P. WHITTAKER

Contents

Part I. Model Cholinergic Systems

CHAPTER 9

Structure and Function of the Nicotinic Acetylcholine Receptor

CHAPTER 10

Muscarinic Acetylcholine Receptors

Part IV. Cellular Organization of the Cholinergic System

CHAPTER 11

Cholinergic Synaptic Vesicles

CHAPTER 14

The High-Affinity Choline Uptake System

CHAPTER 15

Cholinergic-Specific Antigens

CHAPTER 16

Cholinergic False Transmitters

CHAPTER 17

Cholinergic Co-transmitters

Part V. Peripheral Cholinergic Synapses

CHAPTER 18

The Neuromuscular Junction

Part VI. Central Cholinergic Systems

CHAPTER 21

Central Cholinergic Pathways: The Biochemical Evidence

CHAPTER 22

Central Cholinergic Pathways: The Histochemical Evidence

CHAPTER 23

Central Cholinergic Transmission: The Physiological Evidence

Part VII. Neuropathology of Cholinergic Transmission

CHAPTER 24

The Cholinergic System in Aging

CHAPTER 25

Disorders of Cholinergic Synapses in the Peripheral Nervous System

List of Abbreviations

AcCoA	acetylcoenzyme A
ACh	acetylcholine
AChE	acetylcholinesterase
AChR	acetylcholine receptor
ADP	adenosine diphosphate
AH5183	vesamicol
AHP	after-hyperpolarization
ALS	amyotrophic lateral sclerosis
AMP	adenosine monophosphate
anti-ChE	anticholinesterase
4-AP	4-aminopyridine
ATP	adenosine triphosphate
ATPase	adenosine triphosphatase
BLP	bombesin-like peptides
α-BTX	α-bungarotoxin
β-BTX	β-bungarotoxin
BuChE	butyrylcholinesterase
BW284C51	bis(4-allyldimethylammoniumphenyl) pentan-3-one dibromide
BWSV	black widow spider venom
cAMP	cyclic adenosine monophosphate
CCK	cholecystokinin
cGMP	cyclic guanosine monophosphate
CGRP	calcitonin gene-related peptide
ChAT	choline acetyltransferase
ChE	cholinesterase
ChK	choline kinase
CMA	choline mustard aziridinium ion
CNS	central nervous system
CNTF	ciliary neuronotrophic factor
DAF	decay acceleration factor
3,4-DAMP	3,4-diaminopyridine
DFP	diisopropylphosphorofluoridate
DHP	dihydropyridine
DMP	dimethylphenylpiperazinium

EAMG	experimental autoimmune myasthenia gravis
ECMA	monoethylcholine mustard aziridinium ion
EDTA	ethylenediaminetetra-acetic acid
EGTA	ethylene glycol tetra-acetic acid
EOD	electric organ discharge
EpC	electroplaque current
EpP	electroplaque potential
EPP	end-plate potential
EPSP	excitatory postsynaptic potential
GABA	γ-aminobutyric acid
GAL	galanin
GC	gas chromatography
GCMS	gas chromatography-mass spectrometry
GTP	guanosine triphosphate
HACU	high-affinity choline uptake
HC	homocholine
HC-3	hemicholinium-3
HPLC	high-performance liquid chromatography
HPLC-ED	high-performance liquid chromatography with electrochemical detection
HRP	horseradish peroxidase
HSPG	heparansulphate proteoglycan
IgG	immunoglobulin G
IPSP	inhibitory postsynaptic potential
LACU	low-affinity choline uptake
LDC	large dense-cored vesicle
LEMS	Lambert-Eaton myasthenic syndrome
LHRH	luteinizing hormone releasing hormone
mAChR	muscarinic acetylcholine receptor
MBTA	4-(*N*-maleimido)benzyltrimethylammonium
MEC	monoethylcholine
mEpC	miniature electroplaque current
mEPC	miniature end-plate current
mEPP	miniature end-plate potential
MG	myasthenia gravis
MPLM	myenteric plexus-longitudinal muscle
nAChR	nicotinic acetylcholine receptor
NBD-5-acylcholine	*N*-7-(4-nitrobenzo-2-oxa-1,3-diazole)-5-aminopentanoic acid β-(*N*-trimethyl-ammonium)ethyl ester
NCI	noncompetitive inhibitor
NGF	nerve growth factor

NMJ	neuromuscular junction
NMR	nuclear magnetic resonance
NPY	neuropeptide Y
p38	synaptophysin
2-PAM	1-methyl-2-hydroxyiminomethylpyridinium
PC	phosphatidylcholine
PE	phosphatidylethanolamine
PI	phosphatidylinositol
PIPLC	phosphatidylinositol-specific phospholipase C
PNS	peripheral nervous system
PS	phosphatidylserine
PSPM	presynaptic plasma membrane
PyC	pyrrolcholine
$[^3H]$ 3-QNB	tritium-labelled 3-quinuclidine benzylate
RER	rough endoplasmatic reticulum
RMP	resting membrane potential
SCG	superior cervical ganglion
SDS-PAGE	sodium dodecylsulphate polyacrylamide gel electrophoresis
SEL	small electrolucent vesicle
SOM	somatostatin
SP	substance P
SRP	signal recognition particle
TEC	triethylcholine
TID	3-trifluoromethyl-3(*m*-iodophenyl)-diazirine
TPMP	triphenylmethylphosphonium
TR	turnover rate
TTX	tetrodotoxin
UTP	uridine triphosphate
VIP	vasoactive intestinal polypeptide
VSCC	voltage-sensitive Ca^{2+} channel
VSG	variant surface glycoprotein
WGA	wheat germ agglutinin

Part I Model Cholinergic Systems

CHAPTER 1

Model Cholinergic Systems: An Overview

V. P. WHITTAKER

A. The Value of Model Systems

Progress in biology is frequently dependent upon and stimulated by the intensive study of particular instances of a more general phenomenon. Such instances are often referred to as experimental model systems. Their usefulness is dependent on such attributes as accessibility, homogeneity and availability in quantity. Our understanding of cholinergic function, in particular, has benefitted from the study of model systems. In this chapter several model cholinergic synapses will be described that have been found particularly useful and will often be mentioned again in later chapters of this book. All have considerable potential for future work. Four of them – the electromotor synapse, the myenteric plexus, the ciliary ganglion and the diaphragm – are peripheral synapses and two are closely related in that the electromotor synapse is a modified neuromuscular junction (NMJ). The two others are central nervous system preparations relatively rich in cholinergic synapses. Of course, many other preparations, such as the perfused heart and the salivary gland might have been included and will also be referred to in later chapters.

Work with model systems gives hints of what to look for, or expect, in systems less easily probed. This advantage must be set against the risk that the model, in one way or another, is not typical of its class. Thus, one must always seek to test the general validity of any findings with a model system by comparison with other, less easily explored members of the same class. Examples of such comparisons will also be found in the chapters that follow.

B. Pros and Cons of Selected Model Systems

I. Cephalopod Optic Lobe

Table 1 lists, in order of acetylcholine (ACh) content, several tissues from various species that have been either intensively studied or offer considerable potential for future work. All have their pros and cons.

One of the richest sources of ACh known is the optic lobe of the head ganglion of cephalopods (octopus, squid). The afferents from the well-developed eyes of these animals are cholinergic and the optic lobes contain a rich neuropil. Since the cephalopod retina consists essentially of a single layer of photoreceptors, much of the processing which occurs in the mammalian retina is performed, in the cephalopods, in the optic lobes. The main cholinergic contribution comes

Table 1. Some cholinergic model systems

Species	Tissue	ACh content (nmol g^{-1})	Tissue mass per animal (g)	Purely cholinergic	Reference
Squid (*Loligo*)	Optic lobe	2100 ±220 (4)	0.42 ± 0.06 (13)	No	1
Electric ray (*T. marmorata*)	Electric organ	920 ± 90 (24)	120 ±10 (23)	Yes	2
Guinea pig	MPLM	120 ± 15 (8)	1.1 ± 0.2 (8)	No	3
Cat	SCG	100	0.015	No	4
Fowl	Ciliary ganglion	56 ± 10	0.001	Yes	5
Guinea pig	Caudate nucleus	20 ± 3 (5)	2.8 ± 0.1 (5)	No	6
	Diaphragm	0.9± 0.1 (13)	1.2 ± 0.1 (30)	Yes	7

Values are means±SEM (no. of determinations in parentheses).
Abbreviations: MPLM, myenteric plexus-longitudinal muscle; SCG, superior cervical ganglion.
References: 1, DOWDALL and WHITTAKER (1973); 2, ZIMMERMANN and WHITTAKER (1974); 3, DOWE et al. (1980); 4, MACINTOSH and COLLIER (1976); 5, PILAR and TUTTLE (1982); 6, LAVERTY et al. (1963); 7, V. P. WHITTAKER, unpublished data

from the input from the photoreceptors, which terminates in large carrot-shaped endings. The phylogenetic significance of the cholinergic character of sensory nerves in arthropods and cephalopods is discussed in Chap. 4.

ACh was first isolated from nervous tissue in chemically characterized amounts using the head ganglia of octopus as the source (BACQ and MAZZA 1935) and this tissue is also rich in choline acetyltransferase (ChAT) (NACHMANSOHN and WEISS; NACHMANSOHN 1963). Cholinergic-rich synaptosome and synaptic vesicle preparations can be readily prepared from squid optic lobes (DOWDALL and WHITTAKER 1973) and the high affinity choline uptake (HACU) system was first discovered in such synaptosomes (WHITTAKER 1972; DOWDALL and SIMON 1973). The amount of tissue per specimen is quite sufficient for such work; six to eight squid give as much nervous tissue as a guinea-pig cortex. Perhaps because squid are difficult to transport to inland laboratories, very little work has been done on cholinergic transmission in this species.

II. Electric Organs

Another tissue rich in cholinergic synapses is the electric organ of *Torpedo.* Two species have been mainly used. *T. californica* and *T. marmorata.* This tissue too, has an ACh content which – if lower than that of cephalopod optic lobes – is almost an order of magnitude greater than the highest found in mammalian tissues. Each specimen of *T. marmorata,* the commonly available European species, furnishes over 100 g, in a larger fish almost 0.5 kg of electric organ, an amount sufficient for the isolation, in bulk, of synaptosomes, synaptic vesicles and receptor-rich membranes. Much of what we now know of presynaptic transmitter pools (Chap. 6), vesicle composition and recycling (Chap. 11), the HACU system (Chap. 14), transmitter release (Chap. 16), cholinergic-specific surface antigens (Chap. 15), and the nicotinic ACh receptor (nAChR) (Chap. 9) has been obtained from work with this system, and its importance merits the treatment in greater detail it receives in Chap. 2. Electric organs are found in several distinct families of fish and another frequently used species has been the unrelated freshwater electric eel of the Amazon and other South American rivers, *Electrophorus electricus.* The electric organs of this species are capable of giving much larger shocks than that of *Torpedo* but the synaptic region on each electrocyte is more restricted and the ACh and nAChR content is about one-tenth that of *Torpedo* per unit weight. Another electric fish which has considerable potential for further work is the Nile catfish *Malapterurus.* In this species the entire cholinergic innervation of the electric organs is supplied by only two giant cells, one on each side of the head, which allows scope for microinjection experiments.

An advantage of electric organ over other tissues for studying cholinergic function is that its innervation is *purely* cholinergic, though the presence of adrenergic vasomotor fibres has not been rigorously excluded. The usefulness of the model has recently been extended by the discovery of a form of vasoactive intestinal polypeptide (VIP) (AGOSTON and CONLON 1986) and a bombesin-like neuropeptide (AGOSTON and SHAW 1987) in the electromotor neurons of *Torpedo* (Chap. 17).

III. Myenteric Plexus-Longitudinal Muscle Preparation

Among mammalian tissues the myenteric plexus-longitudinal muscle (MPLM) preparation of guinea-pig ileum introduced by PATON and ZAR (1968) has one of the highest ACh contents and richest cholinergic innervations of any mammalian tissue. Two types of cholinergic synapse are present: ganglionic and neuromuscular. The former are blocked by typical 'nicotinic' blocking agents, the latter are classed as muscarinic and the muscle innervated is smooth muscle. An ACh agonist like butyrylcholine which has primarily a nicotinic action generates only partial contractions of the ileum since its action is via ganglia. A comparison of the effect of the two esters on the contraction of the ileum shows that muscarinic synapses predominate.

Separation of the longitudinal muscle with its adherent myenteric plexus from underlying muscle layers (Fig. 1 a) is readily achieved by drawing the everted ileum onto a glass rod of suitable diameter and gently peeling the MPLM tissue off. Other putative transmitters such as 5-hydroxytryptamine and noradrenaline are mainly left behind (Table 2); nevertheless, though the MPLM is highly enriched in cholinergic synapses, it is not purely cholinergic. In particular, several pharmacologically active neuropeptides are also present in relatively high concentration (Table 2).

The MPLM preparation readily lends itself to subcellular fractionation. Fairly pure (10% – 20%) cholinergic synaptic vesicles have been isolated by centrifugal density gradient separation in a zonal rotor from homogenates made by the application of moderately intense liquid shear to the tissue (Dowe et al. 1980). Three other types of vesicles were separated at the same time (Fig. 1 b): a second larger and denser ACh-containing vesicle which also stores vasoactive intestinal polypeptide (VIP) and two others which store somatostatin and substance P (SP)

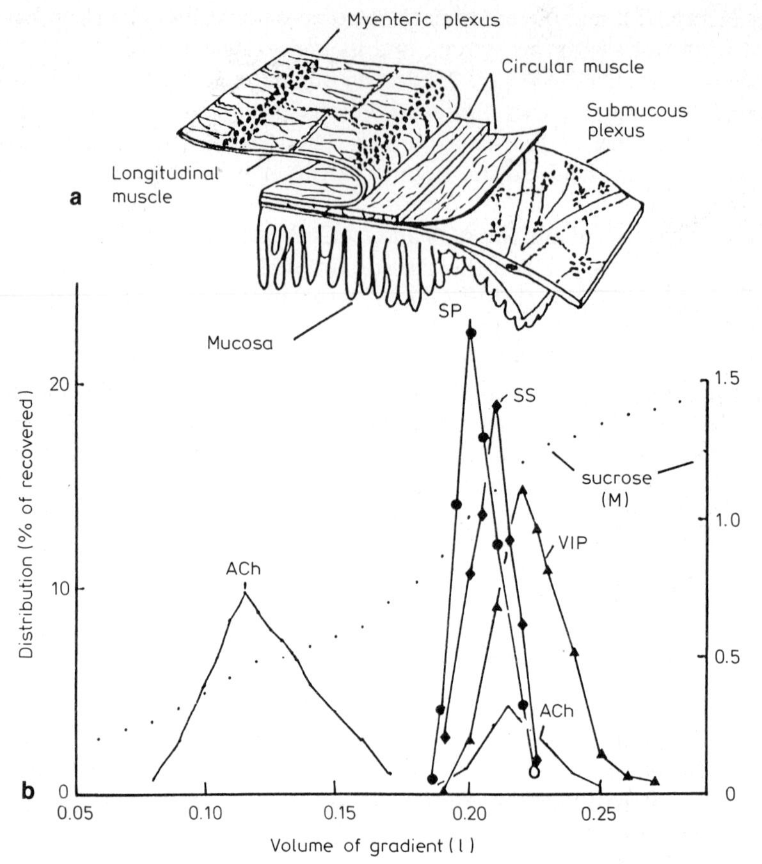

Fig. 1. a MPLM preparation: **b** Separation of vesicles containing ACh and neuropeptides from a MPLM homogenate using a sucrose density gradient in a zonal rotor. Note that ACh is bimodally distributed with a second peak coincident with that of vasoactive intestinal polypeptide (*VIP*) and that the three neuropeptides substance P (*SP*), somatostatin (*SS*) and VIP are separately packaged in granules of increasing density

respectively (ÁGOSTON et al. 1985b, Table 2; Chap. 17, this volume). The cholinergic nerve terminals in this preparation are labile and are not readily pinched off to form synaptosomes; however, these can be obtained in low yield from homogenates prepared under conditions of mild shear (JONAKAIT et al. 1979; DOWE et al. 1980).

IV. Ganglion Preparations

1. Superior Cervical Ganglion

Among ganglion preparations the superior cervical ganglion (SCG) of the cat and the avian ciliary ganglion are both classical model systems for studying choliner-

Table 2. Neuroactive components in MPLM and characteristics of their storage particles

Component	Amount in MPLM		Characteristics of storage particles		
	As % of total[a]	Units[b] (g^{-1} wet wt)	Diameter[c] (nm)	Density[d] ($g\,m^{-1}$)	Enrichment factor[d] (relative to MPLM)
ACh					
Total	79	129 ± 16 (6)	–	–	–
Bound	77	63 ± 6 (6)	61 ± 6	1.066 ± 0.006	44 ± 6
Substance P	80	475 ± 50 (8)	65 ± 3	1.123 ± 0.007	62 ± 8
Somatostatin	43	171 ± 14 (6)	87 ± 3	1.138 ± 0.008	52 ± 6
VIP[e]	40	4.8 ± 0.6 (8)	110 ± 6	1.148 ± 0.008	39 ± 5
Noradrenaline	52	7.2 ± 1.2 (4)	–	–	–
Dopamine	31	2.1 ± 0.4 (4)	–	–	–
5-Hydroxytryptamine	2	0.5 ± 0.2 (4)	–	–	–

The table contains results of Dowe et al. (1980) and Ágoston et al. (1985b)
[a] Refers to gut wall before separation of MPLM
[b] Units are (ACh, noradrenaline, dopamine, 5-hydroxytryptamine) nmol, remainder, pmol. Values are means ± SD of number of experiments shown in parentheses
[c] Equatorial diameter; values are means ± SD derived from over 1000 profiles
[d] Means ± SD of 4 experiments
[e] VIP storage granule also contained ACh with an enrichment factor of 18 ± 3

gic synaptic function. The ACh content is almost of the same order as the MPLM preparation. Much use has been made of the perfused SCG by MacIntosh and coworkers (MacIntosh and Collier 1976). This led to the recognition that ACh is present in the cholinergic preganglionic nerve terminals in more than one pool, the 'readily releasable' now equated with part or all of the vesicular pool, the 'reserve' or cytoplasmic and the 'surplus', the extra ACh accumulating in the cytoplasmic pool in the presence of an anticholinesterase (anti-ChE) agent.

The perfused SCG has proved to be a highly suitable preparation for the study of the pharmacology of cholinergic ganglionic synapses and the effect of drugs on the presynaptic pools of transmitter, but because of their small mass and connective tissue-rich structure, SCGs are not the easiest material for subcellular fractionation. Nevertheless synaptic vesicles rich in ACh have been isolated from them and an estimate of their ACh content obtained (1600 molecules per vesicle; Wilson et al. 1973, Table 3).

2. Avian Ciliary Ganglion

The avian ciliary ganglion preparation as developed especially by Pilar and coworkers (reviewed by Pilar and Tuttle 1982; see also Chap. 3, this volume), is a purely cholinergic system at the level of chemical transmission, but the synapses also display a degree of electrical coupling. The ganglionic synapses are axosomatic, but the ciliary ganglion cells are themselves cholinergic as well as cholinoceptive, and terminate in the iris to form NMJs; thus the iris preparation, too, provides a useful model cholinergic system. The amount of tissue from a

Table 3. Estimates of quantal size and vesicle content

Synapse/ junction	Species	No. of molecules	Comments	Reference
I. *Electrophysiological: quantal size*				
NMJ		900	1	MACINTOSH (1959)
NMJ	Frog	4000	2	KATZ and MILEDI (1972)
NMJ	Snake	6000	3	KUFFLER and YOSHIKAMA (1975)
NMJ	Rat	100000	4	KRNJEVIĆ and MITCHELL (1961)
NMJ	Rat	30000 (7500)	5	POTTER (1970)
NMJ	Frog	13000	5	MILEDI et al. (1983)
NMJ	Rat	6000	5	FLETCHER and FORRESTER (1975)
EMS	*Torpedo*	7000		DUNANT and MULLER (1986)
II. *Morphological: for core of 31 (49) nm diameter*				
For isotonic (0.16 M) ACh		1500 (6000)		WHITTAKER (1966)
For solid ACh		50000		WHITTAKER (1966)
III. *Biochemical: from isolated vesicles*				
EMS	*Torpedo*	200000	6	OHSAWA et al. (1979)
		20000	7	GIOMPRES and WHITTAKER (1986)
SCG	Ox	1600		WILSON et al. (1973)
Cortex	Guinea pig	300	8	WHITTAKER and SHERIDAN (1964)
		2000	9	
		6000	10	

Comments: (1) lowest reported estimate; (2) based on ratio of magnitudes of mEPP and 'unit potential' generated by a single nAChR channel opening, which requires two ACh molecules; (3) estimated by direct iontophoretic application of ACh to an exposed postsynaptic membrane and probably the most reliable; (4) calculated from resting output of ACh in eserinized diaphragm preparation, later recognized to include a large non-quantized component; (5) from evoked ACh release; (6) for 'reserve' vesicles; (7) for 'recycling' vesicles with 10% loading; (8) value for total vesicle preparation and (9, 10) assuming (9) 15%, (10) 5% of total cortical vesicle population is cholinergic (the latter is more likely). Abbreviations: NMJ, neuromuscular junction; EMS, electromotor synapse; SCG, superior cervical ganglion

single animal is extremely small, and thus unsuitable for tissue fractionation techniques, but the preparation has been extensively used for other types of biochemical study: for studying cholinergic neuronal differentiation, synaptogenesis, the role of trophic factors in development and axonal transport. The preparation is thus, like the *Torpedo* electric organ, of sufficient importance and future potential to merit separate treatment (Chap. 3).

V. Mammalian Central Nervous System Preparations

Because man is a mammal, mammalian preparations made from the common laboratory animals have inevitably appealed more to biomedical research workers than avian, lower vertebrate or invertebrate preparations, in spite of their often greater complexity and much lower cholinergic involvement. One aim of these first three chapters, by high-lighting non-mammalian model systems, is partially to redress this balance, since the cholinergic neuron, as pointed out in Chap. 4, is highly conserved in evolution. Nevertheless, completeness and comprehensiveness demand an adequate mention of mammalian preparations, particularly those derived from the central nervous system (CNS), useful in the investigation of the cholinergic synapse. Later chapters (Chaps. 21 to 23) describe the biochemical, anatomical and physiological evidence for the existence of cholinergic neurons and cholinergic synaptic transmission in the mammalian CNS and from this emerges the paradox that while the biochemical and cytochemical evidence for a cholinergic representation in the mammalian CNS whose characteristics closely resemble those of peripheral cholinergic systems is good, the electrophysiological evidence for cholinergic transmission is not as strong as the biochemical and cytochemical results would imply. More work with isolated preparations – explants or dissociated neurons in culture, brain slices, isolated neurons and nerve terminal (synaptosomal) preparations – might help to resolve this paradox.

The distribution of cholinergic neurons, tracts and terminals in the CNS is ‘patchy’ but the various brain regions provide the investigator with workable amounts of tissue and concentrations of cholinergic components. The caudate nucleus, hippocampus and other forebrain regions are used, often as slices or synaptosome preparations. The caudate nucleus contains a large number of cholinergic interneurons and has one of the highest ACh and ChAT contents of any brain area (Table 1). On homogenization in eserinized sucrose to preserve free ACh a high proportion of these cholinergic markers is recovered in the synaptosome fraction (LAVERTY et al. 1963) showing that the cell bodies contain little ChAT or free ACh. The hippocampus is often used as slices from which electrophysiological recordings can be made. This brain region has an extremely well defined septal cholinergic input which can be interrupted by cutting the fimbria. After 7–14 days all cholinergic markers – ACh, ChAT, HACU, AChE – progressively disappear as the cholinergic input degenerates (KUHAR et al. 1973). A good reproducibility between animals is obtained if intermediate levels are expressed as a percentage of the non-operated contralateral control values.

Other cortical areas can be studied in a variety of ways, separately or in combination: as intact systems, as slices or after subcellular fractionation. The Oborin cup technique, in which a plastic sleeve is placed on the surface of the cortex through a hole in the skull and filled with Locke's solution, may be used to study the release of ACh from cholinergic nerve terminals in the underlying cortical tissue (MITCHELL 1963; CHAKRIN et al. 1972; CASAMENTI et al. 1980). A resting release of ACh (approximately 0.5 pmol $cm^{-2} min^{-1}$) is increased by stimulation. Labelled ACh is released if labelled choline is applied to the cortex via the cup and the underlying labelled tissue may be subjected to subcellular fractionation at the end of the collection periods. This technique has proved very useful for establishing the subcellular compartment from which ACh is released (VON SCHWARZENFELD 1979; Chap. 16, this volume). In a similar way, experiments may be performed on slices which may then be subjected to subcellular fractionation in order to learn more about the subcellular compartments involved (cf. BOKSA and COLLIER 1980).

A convenient technique for studying transmitter function in synaptosomes is in the form of a 'synaptosome bed', i.e. a layer of synaptosomes formed by pouring a suspension onto a filter bed under slight pressure (DE BELLEROCHE and BRADFORD 1972). The bed can be superfused or treated as a tissue slice. The uptake of transmitter or transmitter precursor and the release of transmitter on stimulation, electrically or by raising the K^+ concentration of the superfusion liquid, also the action of pharmacological agents on these processes, can be conveniently studied in this preparation. With brain synaptosomes the applicability of the technique is of course not limited to cholinergic nerve terminals.

Another useful technique is the use, with synaptosomes in suspension under metabolizing conditions, of fluorescent probes which signal the magnitude of the resting potential across the synaptosomal plasma membrane (BLAUSTEIN and GOLDRING 1975). This technique has been applied to the purely cholinergic synaptosome preparations derived from electromotor nerve terminals (MEUNIER 1984; for critique, see Chap. 12, this volume). It has given useful information about the relationship between depolarization, Ca^{2+} influx and transmitter release.

VI. Amphibian and Mammalian Voluntary Muscle

1. Quantized Release of Transmitter

The frog and mammalian NMJs have been the subject of intense electrophysiological studies since the early 1950s. From such studies have emerged two basic findings which must be accommodated in any theoretical model of the cholinergic synapse: the phenomenon of the quantized release of transmitter (reviewed in KATZ 1966) and the detection of micropotentials from which the channel-open time of single nAChR molecules can be deduced (KATZ and MILEDI 1972; ANDERSON and STEVENS 1973).

2. Vesicular Storage and Release as the Physical Basis for Quantized Release

The classical work of KATZ and coworkers (reviewed in KATZ 1966) showed that the miniature endplate potentials (mEPPs) observed to occur in a random, spontaneous fashion at the frog NMJ could not possibly be accounted for by the molecular diffusion of ACh from the terminal but implied the release of the transmitter in presynaptically generated packets of several thousand molecules per event. How could such a 'quantized' release be envisaged? The answer lay to hand in the then soon to be discovered synaptic vesicles. Calculations based on the size of the vesicle core and the number of molecules of ACh that would be present if the vesicle core were filled with a somewhat hyperosmotic ACh solution gave values of the same order of magnitude as the best estimates then available of the number of ACh molecules needed to generate a mEPP (Table 3, blocks I and II). The successful isolation of similarly sized synaptic vesicles from brain using the then relatively recently developed techniques of subcellular fractionation (WHIT-

TAKER et al. 1964) and the demonstration that approximately the requisite number of ACh molecules was indeed present (WHITTAKER and SHERIDAN 1964) greatly strengthened the identification of these organelles as the physical basis for quantal release (Table 3, block III). Until recently the very low nerve terminal content of muscle and its resistance to homogenization has precluded the isolation of vesicles from this source but with new techniques this is now becoming possible (VOLKNANDT and ZIMMERMANN 1986). The SCG and electric tissue are further examples of peripheral tissue with a purely or predominantly cholinergic innervation from which cholinergic synaptic vesicles have been isolated. Thus, in the discussion which follows information derived from synapses other than the NMJ has been taken into account.

Critics (e.g. MACINTOSH and COLLIER 1976) have pointed to a number of weak links in the identification of quanta with vesicle-bound transmitter: the vesicles were not isolated from NMJs; the calculations regarding transmitter content involve a number of assumptions; newly synthesized transmitter is preferentially released, yet the total population of vesicles does not contain much newly synthesized transmitter; it is relatively easy to deplete transmitter stores, but not synaptic vesicles, by repetitive stimulation. However, most of these criticisms have been met and it may be said that in the 30 years that have elapsed since the 'vesicle hypothesis' was first formulated, the evidence for it has grown steadily stronger.

In its original form the vesicle hypothesis was undoubtedly an oversimplification and had a narrow experimental base; both vesicles and mEPPs were thought to be more uniform than they are now known to be. The existence of a subclass of recycling, partially loaded and slightly smaller and denser vesicles than the main, reserve population (BARKER et al. 1972; ZIMMERMANN and DENSTON 1977; ÁGOSTON et al. 1985a; Chap. 11, this volume), was not suspected and mEPPs were regarded as extremely uniform, totally random events. As Chap. 18 shows, later work has revealed the presence at several NMJs of both 'giant' mEPPs and subminiature endplate potentials (sub-mEPPs), which might be accounted for in a number of ways. Giant potentials may represent the synchronized discharge of mEPPs due to intracellular vesicle fusion or compound ('piggy-back') exocytosis; sub-mEPPs may result from nerve damage by the electrode, the recording of mEPPs occurring at some distance from the electrode or partial refilling of recycling vesicles. That sub-mEPPs, like mEPPs, are vesicle-linked phenomena is shown by the fact that their occurrence, like that of mEPPs, ceases after vesicle depletion (KRIEBEL and FLOREY 1983). mEPP frequency, usually thought of as that of spontaneous random events, can be manipulated in both mouse neuromuscular and *Torpedo* electromotor junctions by mechanical pressure from the electrodes. The number of junctions with the morphology typical for chemical synapses for which quantal release has been electrophysiologically demonstrated is rather small, but it has been shown to occur at sympathetic ganglia (BLACKMAN et al. 1963), the fowl ciliary ganglion (MARTIN and PILAR 1964), the electromotor synapse (Chap. 19), glutamergic (DUDEL and KUFFLER 1961) and adrenergic (BURNSTOCK and HOLMAN 1962) junctions and in the CNS (KATZ and MILEDI 1963; KUNO 1964; KRENZ 1985). It thus seems to be a property common to vesicle-containing, chemically transmitting synapses, irrespective of the nature of the transmitter used.

A further possible complication for the vesicle hypothesis has been the discovery that at least some cholinergic nerve terminals contain co-transmitters. Thus in electromotor vesicles and some mammalian vesicles ACh and ATP are packaged together (Chap. 11). The effect of ATP on the postsynaptic action of ACh is at present rather ill-defined but ought to be taken into account when the electrogenic effect of ACh is being measured. Neuropeptides, especially VIP, are also present in some cholinergic nerve terminals (Chap. 17). They are usually packaged in vesicles larger than the classical synaptic vesicles, but, at least in the myenteric plexus, there is some evidence that ACh is stored in VIP-containing vesicles as well as in classical synaptic vesicles (ÁGOSTON et al. 1985b). Possibly some of the deviant mEPPs now being recognized involve such vesicles and the effect of ACh is modified by the simultaneous release of the neuropeptide during such events.

3. Electrogenic Effect of ACh and Size of the Quantum

In attempting to identify quanta with vesicles, we need to know how many molecules of ACh form a quantum. Three approaches have been used: attempts have been made to simulate mEPPs with known doses of ACh ejected from a micropipette; the amount of ACh liberated from a nerve terminal over a period of time has been measured by bioassay or mass fragmentography (Chap. 6) and divided by the number of quanta observed during the same period; and the magnitude of the micropotential believed to represent a unit event, i.e. the combination of two molecules of ACh with a molecule of receptor, has been compared with the magnitude of a mEPP. The first of these methods suffers from the difficulty of approaching the postsynaptic membrane sufficiently closely with the tip of the pipette; the most reliable measurements have been made by KUFFLER and YOSHIKAMA (1975) who injected ACh into an oil droplet covering a postsynaptic muscle membrane from which the presynaptic nerve terminal had been removed (Table 3, entry 3). This is in fair agreement with an estimate based on micropotentials (Table 3, entry 2).

In making estimates of the size of the quantum by dividing the amount of assayable ACh released in a given time by the number of quanta observed during the same time, conditions must be carefully selected and there are a number of sources of error. Estimates from measurements at rest are virtually impossible to make because most of the ACh released is non-quantized; it has been estimated (MITCHELL and SILVER 1963; FLETCHER and FORRESTER 1975; MILEDI et al. 1983) that the non-quantized release is 20–100 times greater than the quantized (as detected by mEPPs). Failure to appreciate this at the time led to the pioneer, but erroneously high estimate of quantal size by KRNJEVIĆ and MITCHELL (1961) (Table 3, entry 4). Fortunately, non-quantized release is not increased (MILEDI et al. 1983) or increased only to a small extent (VIZI and VYSKOČIL 1979) by stimulation whereas quantal release is greatly increased; thus the release of ACh evoked by stimulation can be corrected for by non-quantal release measured at rest. When this has been done, the ACh release evoked by partial depolarization is proportional to the number of quanta released (MILEDI et al. 1983). The amount of stimulation required to generate enough assayable ACh varies according to the sensitivity of the assay and difficulties may arise in estimating the number of quanta released if the rate of release is too high (HAIMANN et al. 1985).

In careful work by FLETCHER and FORRESTER (1975) with rat diaphragm, the amount of ACh release per stimulus declined with increasing rates of stimulation, apparently because of progressive anoxic block of nerve conduction to some branches of the motor nerve. A frequency of 1 Hz was found to be optimal. The evoked release for 3600 impulses was estimated to be 144 pmol; assuming the quantal content per impulse at 1 Hz to be 400 and the number of endplates activated per impulse to be 10^4, the quantal size was calculated to be $(144\times6\times10^{11})/(3600\times400\times10^4)$ or 6000, in excellent agreement with KUFFLER and YOSHIKAMA's (1975) estimate. In an earlier study POTTER (1970) had estimated that the evoked release was between 3.5 and 7.0 amol per pulse per endplate which is close to FLETCHER and FORRESTER's figure of 4 amol [i.e. $(144\times10^{-12})/(3600\times10^4)$ mol], but he assumed a much lower quantal content per impulse (100) giving a mean of 30000 molecules per quantum. This is almost certainly too high; if FLETCHER and FORRESTER's assumptions are applied, the figure would be 7500.

Another rather high estimate was obtained by MILEDI et al. (1983). Using frog sartorius muscle soaked in trimethylammonium, which greatly increased the number of quanta released per stimulus (from 400 to 80000), the evoked release of ACh by 10 stimuli was estimated to be 17.6 pmol. Assuming for this muscle that 10^3 endplates were activated per stimulus the quantal content was thought to be $(17.6\times10^{-12}\times6\times10^{23})/(8\times10^4\times10\times10^3)$ or about 13000 molecules of ACh, in reasonable agreement with a morphological estimate of vesicle content (see next section) of 11000 molecules per vesicle (MILEDI et al. 1982).

4. Morphological Evidence for Identifying Quantal Release with Vesicle Exocytosis

A reasonable prediction of the vesicular release hypothesis is, that on stimulating a junction to exhaustion, vesicle numbers should be depleted. In fact, without assistance from agents (the Ca^{2+} ionophore A 23187, 5-aminopyridine, α-latrotoxin or La^{3+} ions) that bypass the easily fatiguable Ca^{2+} channels or others (ouabain, cooling) that block vesicle retrieval, depletion requires strong stimulation and is never complete. Failure to suppress vesicle recycling will cause discrepancies in depletion experiments between the number of mEPPs released and the number of vesicles disappearing and will give spuriously high values for the amount of ACh per vesicle when the total amount of ACh is divided by the number of vesicles disappearing. However, high frequency (10 Hz) stimulation alone does cause vesicle depletion at the NMJ (KORNELIUSSEN 1972; CECCARELLI et al. 1972, 1973); HEUSER and REESE 1973; CECCARELLI and HURLBUT 1975) and the electromotor synapse (ZIMMERMANN and WHITTAKER 1974a). Vesicle retrieval occurs very rapidly – at the electromotor synapse faster than refilling (ZIMMERMANN and WHITTAKER 1974b; SUSZKIW 1980) – and is complete in a variety of synapses, including the NMJ, in 1 h (for literature review, see ZIMMERMANN 1979). At lower rates of stimulus (0.2 Hz) vesicle retrieval keeps pace with exocytosis but in the Mauthner's fibre-giant fibre synapse of the hatchet fish retrieval can be reversibly blocked by cooling (MODEL et al. 1975), and this quickly leads to vesicle depletion and transmission block. Using pharmacological means to bypass the Ca^{2+} channel or block vesicle retrieval, there is a massive activation of vesicular exocytosis and a cascade of mEPPs occurs which ceases when the stock of vesicles is used up.

In an exemplary study of the ouabain-treated frog cutaneous pectoris NMJ HAIMANN et al. (1985) estimated that the release of between 7.4×10^5 and 8.8×10^5 quanta were accompanied by the disappearance of 75% of the total stock of 5×10^5 vesicles. At the same time, the surface area of the presynaptic plasma membrane (axolemma) expanded to form infoldings that penetrated deeply into the terminal cytoplasm; in confirmation of earlier work with NMJs (HEUSER and REESE 1973) and torpedine electromotor synapses (BOYNE et al. 1975) the plasma membrane expansion balanced vesicle loss, i.e. total membrane area was conserved. No detectable uptake of horseradish peroxidase into vesicles or cisternae took place and no further release of quanta occurred when α-latrotoxin was applied at the end of the experiment.

In another study with frog sartorius muscle (MILEDI et al. 1980, 1982) $LaCl_3$ was used to bring about total vesicle depletion and the amount of ACh released was measured. Attributing this to the discharge of 10^6 vesicles from all of 10^3 endplates gave an estimate of the number of molecules of ACh per vesicle of 11000, in satisfactory agreement with the same authors' estimate of the size of a quantum (MILEDI et al. 1983).

Another prediction of the hypothesis is that the time-course of the occurrence of morphologically detected presynaptic exocytotic events will correlate with the rise-time of a single post-junctional potential. To investigate this correlation experimentally is technically demanding since, using normal methods, the rate at which a post-junctional potential can be recorded greatly exceeds the rate at which a motor nerve terminal can be fixed for electron microscopy. However, TORRI-TARELLI et al. (1985) have largely overcome these difficulties by making use of the very thin frog cutaneous pectoris NMJ preparation, cryosubstitution of the frozen tissue and an improved design of quick-freezing apparatus consisting of a copper plate cooled in liquid helium and protected until a fraction of a millisecond before impact by a vacuum maintained by a sealing shutter. In this way the build-up of a thin insulating layer of ice on the cold plate is prevented. Under these conditions an excellent correlation between the rate of formation of exocytotic figures and the rising phase of the post-junctional potential has been obtained. Cryosubstitution followed by thin section was found to be preferable to freeze fracture for evaluating exocytotic events since the later favours the exposure of E faces of the nerve terminal plasma membrane and not the requisite P faces in the narrow zone of well-preserved NMJs. Evidence to the contrary obtained with small blocks of electric tissue (GARCIA-SEGURA et al. 1986) may have resulted from the use of a less rapid freezing technique and freeze fracture.

A third prediction of the hypothesis is that the core of the vesicle should be able to accommodate a quantum. Block II of Table 3 gives the number of molecules of ACh that could be accommodated in a core of 31 nm in diameter on the assumption that the core is filled with iso-osmotic (entry 9) or solid (entry 10) ACh. Physicochemical studies with *Torpedo* electromotor vesicles suggest that they contain iso-osmotic ACh (GIOMPRES and WHITTAKER 1984, 1986). However it should be pointed out that the value calculated is proportional to volume and thus sensitive to errors in diameter; the value of 31 nm is a rather low estimate for the diameter of the core of a typical synaptic vesicle. If a correction factor is made to the mean profile diameter of vesicles in section (47 nm) to obtain the true diameter (57 nm) and the vesicle membrane is assumed to be 4 nm thick instead of 8 nm, then the core is 49 nm in diameter and its volume almost four times greater, giving a value for the number of molecules of ACh in that volume of iso-osmotic ACh as 6000 – the same as KUFFLER and YASHIKAMA's electrophysiological estimate.

5. Vesicle Metabolic Heterogeneity

Historically, one difficulty for the vesicle hypothesis was the finding in brain cortex that when the various pools of ACh were labelled by the introduction of a radioactively labelled precursor, the classical fraction of monodispersed synaptic vesicles (D band of WHITTAKER et al. 1964) was but little labelled, yet the released ACh was highly labelled (CHAKRIN et al. 1972). This selective release of newly synthesized or newly taken up transmitter has been demonstrated for a variety of synapses, including adrenergic ones. The difficulty was resolved when it was realised that vesicles are not metabolically homogeneous, but that there exists, in stimulated terminals and probably to a smaller extent in resting terminals, a subpopulation of recycling vesicles which rapidly takes up labelled ACh formed from labelled precursor by cytoplasmic ChAT and releases it on continued stimulation (Chaps. 11, 16). This metabolic heterogeneity of the synaptic vesicle pool was first shown in mammalian brain (BARKER et al. 1972) but has also been demonstrated in *Torpedo* electromotor terminals (ZIMMERMANN and DENSTON 1977) and cholinergic terminals of the myenteric plexus (ÁGOSTON et al. 1985a). An extension to the NMJ may soon be technically feasible.

The use of false transmitters (Chap. 16) has enabled the cytoplasmic and recycling vesicular pools in brain and electromotor terminals to be unequivocally distinguished and in all cases, the mixture of transmitters released by stimulation from labelled pools always has the same composition as that in the recycling vesicle pool irrespective of that in the cytoplasmic pool.

6. ACh Content of Isolated Vesicles

Synaptic vesicles have been successfully isolated from brain cortex, electromotor terminals and the SCG (Table 3, block III), and their ACh content determined. This involves counting the number of vesicles in a given volume and this has been done either by tagging the preparation with a known number of polystyrene beads and then determining the ratio, in electron micrographs of negatively stained preparations, of vesicles to beads, or sectioning beds of vesicles of known surface area and depth and using stereological analysis to deduce the number of vesicles per unit area of section. With brain tissue a large correction factor has to be applied for the dilution of cholinergic vesicles by those originating from non-cholinergic terminals. A major uncertainty is the extent to which vesicles retain their ACh during isolation. However, values for the ACh content of isolated cortical vesicles agree well with estimates based on synaptosomes and the average number of vesicles per synaptosome, and the best estimate (entry 16) agrees well with the best estimates of quantal size (Table 3, block I, entries 3 and 7); furthermore, it is consistent with the maintenance of isotonicity across the vesicle membrane.

The much larger (90 nm) electromotor synaptic vesicles have been estimated to contain several times the amount of ACh than the smaller vesicles characteristic of most nerve terminals; this may be expected from their greater capacity and the higher osmolarity of the *Torpedo* body fluids. However, actively recycling vesicles are estimated to contain much less ACh, perhaps only one-tenth of that present in reserve vesicles (GIOMPRES and WHITTAKER 1986). Even so, the packet is three to four times larger than that in mammalian vesicles. This large packet may be needed to compensate for the low impedance of the electrocyte membrane, since mEPPs recorded at the electromotor synapse are similar in magnitude to those of the frog NMJ (Chap. 19).

Vesicles may be expected to contain more ACh than quanta, since the released ACh has to pass a barrier of AChE to reach the receptors. However at a concentration of 10 m*M* or more, ACh is an effective AChE inhibitor. The puff of iso-osmotic ACh directed at the postsynaptic membrane which results from vesicular release may be essential to bring about this temporary inhibition of AChE. It is difficult to estimate the losses due to diffusion and hydrolysis which may reduce the efficiency of this vesicular discharge, and they are no doubt influenced by variations in the concentration of AChE and the geometry of the cleft in different synapses. Probably in NMJs losses are small, and below the limit of accuracy of estimates of vesicular content on the one hand and the size of the quantum on the other. At the electromotor synapse, by contrast, they may be larger, since there is a very high AChE concentration in the electric organ. In general we can conclude that such discrepancies as do exist between estimates of vesicle transmitter content whether morphological or based on isolated vesicle preparations and those of the size of quantum are not decisive.

7. Cytoplasmic ACh and its Significance

There is little doubt that a small cytoplasmic pool of transmitter exists in cholinergic nerve terminals: this follows from the cytoplasmic location of ChAT. In frog sartorius muscle most of the ACh is intraneuronal; however, a small but significant pool of ACh exists in the non-neuronal portion of muscles where it is probably synthesized by extraneuronal carnitine acetyltransferase (MILEDI et al. 1980). Further small quantities of ACh exist in nerve axons and Schwann cells. Of the intraterminal ACh most is located in synaptic vesicles. MILEDI et al. (1980) used 2 m*M* $LaCl_3$ to bring about depletion of synaptic vesicles, denervation to destroy intraneural ACh and homogenization in the presence and absence of an anti-ChE, diethyldimethylpyrophosphonate, to estimate the sizes of the various free ACh compartments. They concluded that of the 43 pmol of ACh in the muscle 25% was extraneural, 11% was in the nerves and 64% in the nerve terminals; of this less than 10% was extravesicular, i.e. cytoplasmic. Estimates of the size of the cytoplasmic pool in electromotor terminals vary from 6% to 22% depending on techniques and conditions (BABEL-GUÉRIN 1979; SUSZKIW 1980; WEILER et al.

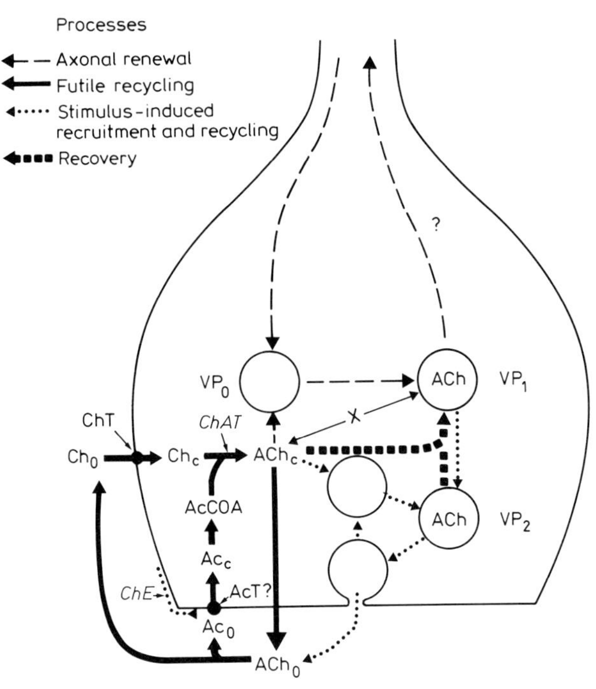

Fig. 2. The functional organization of the cholinergic terminal. As shown by immunocytochemistry and covalent ^{35}S-labelling, synaptic vesicles are formed in the cell body and transported – largely empty – to the terminal (VP_0). Here they fill with cytoplasmic acetylcholine (ACh_c) and enter the pool of reserve vesicles (VP_1). On stimulation, a proportion of the VP_1 vesicles are recruited into the recycling (VP_2) pool. Such vesicles only partially refill from the cytoplasm and undergo partial osmotic dehydration, becoming smaller and denser. At rest, the recycled pool slowly takes up more acetylcholine and reacquires the biophysical properties of the reserve pool. By contrast, there is little or no direct exchange between reserve vesicles and the cytoplasm. Cytoplasmic acetylcholine is subject to 'futile recycling' (*arrows*). Transmitter leaking out of the terminal (ACh_0) is rapidly hydrolysed by acetylcholinesterase (ChE) in the cleft and the acetate (Ac_0) and choline (Ch_0) are salvaged; in both cases uptake is facilitated by carriers (*AcT, ChT*). Cytoplasmic choline (Ch_c) and acetate (Ac_c) are resynthesized to cytoplasmic acetylcholine by the soluble enzyme choline acetyltransferase (*ChAT*); acetate must however be first converted to acetylcoenzyme A (*AcCoA*). (Not shown in the diagram are the endogenous pools of choline and AcCoA which could be called upon to restore cytoplasmic acetylcholine when this is depleted by stimulation.) The cytoplasmic and VP_2 pools of transmitter can be specifically labelled by means of false transmitters and thus cytoplasmic release and that brought about by vesicle recycling can be readily distinguished. A similar cycle exists for synaptic vesicle ATP (see Chap. 11). There is no evidence as yet for antidromic axonal transport of 'worn out' vesicles, although by analogy with other systems this is likely to occur

1982; Ágoston et al. 1986). It is also clear that this cytoplasmic ACh is continuously leaking out (Katz and Miledi 1977); in the absence of an anti-ChE, it is immediately hydrolysed and the products (choline and acetate) are taken up again by specific transporters. This futile recycling probably indicates as in other metabolic systems a regulating function of the cytoplasmic pool and accounts for the ease with which this pool can be labelled. It plays a central role in one scheme for the organization of the cholinergic terminal (WHITTAKER 1986; and Fig. 2). EDWARDS et al. (1985) have suggested that leakage is facilitated by a fraction of the vesicular ACh carrier that has become incorporated into the plasma membrane and has escaped retrieval by the recycling mechanism. However such a carrier would have the same specificity for choline esters as the vesicular carrier and would therefore be expected to release false transmitters in the same ratio as they are present in recycling vesicles. This is not what has been found (Chap. 16); if false transmitters are selected whose ratio in recycling vesicles is different from that in which they are present in cytoplasm, the ratio in which they are released at rest (non-quantized release) conforms to their cytoplasmic ratio but changes on stimulation to that of the recycling vesicles (LUQMANI et al. 1980).

Although most authors accept that cytoplasmic ACh is the main source of the non-quantized release referred to earlier, some have postulated that this release is quantized and that it is, indeed, the pool from which evoked ACh release originates. Most of the evidence is against this. Thus MILEDI et al. (1982) found that when vesicles were depleted from frog sartorius motor nerve terminals by La^{3+}, mEPPs were no longer observed, yet cytoplasmic ACh actually rose. This ACh could not be released by stimulation or by the Ca^{2+} ionophore A 23187. DUNANT et al. (1977) have also noted, in *Torpedo* electric organ, that cytoplasmic ACh can vary through wide limits without affecting mEPP frequency. It is indeed intrinsically difficult to imagine what type of channel, gate or transporter inserted into the presynaptic plasma membrane could impose quantal release on the cytoplasmic transmitter pool especially as the size of the quantum is known to be independent of the cytoplasmic transmitter concentration. Even if the opening time of a gate were inversely proportional to the cytoplasmic concentration of transmitter – which implies an unusual sensing mechanism – this would have effects on the duration of the mEPP which have not been observed. On the other hand, if the operative mechanism is a carrier, it would have to bind, and release, several thousand molecules of transmitter simultaneously, which postulates a uniquely large number of binding sites for a carrier molecule and an affinity constant low enough to ensure saturation at cytoplasmic concentrations a few percent of normal. Both mechanisms imply that the release of ACh is electrogenic, which is contrary to the experimental evidence.

References

Agoston DV, Conlon JM (1986) Presence of vasoactive intestinal polypeptide-like immunoreactivity in the cholinergic electromotor system of *Torpedo marmorata*. J Neurochem 47:445–453

Agoston DV, Shaw C (1988) Presence of GRP-like immunoreactivity in the cholinergic electromotor system of *Torpedo marmorata* (in preparation)

Ágoston DV, Kosh JW, Lisziewicz J, Whittaker VP (1985a) Separation of recycling and reserve synaptic vesicles from the cholinergic nerve terminals of the myenteric plexus of guinea pig ileum. J Neurochem 44:299–305

Ágoston DV, Ballmann M, Conlon JM, Dowe GHC, Whittaker VP (1985b) Isolation of neuropeptide-containing vesicles from the guinea pig ileum. J Neurochem 45:398–406

Ágoston DV, Dowe GHC, Fiedler W, Giompres PE, Roed IS, Walker JH, Whittaker VP, Yamaguchi T (1986) The use of synaptic vesicle proteoglycan as a stable marker in kinetic studies of vesicle recycling. J Neurochem 47:1580–1592

Anderson CR, Stevens CF (1973) Voltage clamp analysis of acetylcholine produced end-plate current fluctuations at frog neuromuscular junction. J Physiol (Lond) 235:655–691

Babel-Guérin E (1974) Métabolisme du calcium et libération de l'acétylcholine dans l'organe électrique de la torpille. J Neurochem 23:525–532

Bacq ZM, Mazza FP (1935) Recherches sur la physiologie et la pharmacologie du système nerveux autonome. XVIII. Isolement de chloroaurate d'acétylcholine à partir d'un extrait de cellules nerveuses d'*Octopus vulgaris*. Arch Int Physiol 42:43–46

Barker LA, Dowdall MJ, Whittaker VP (1972) Choline metabolism in the cerebral cortex of guinea pig. Biochem J 130:1063–1075

Belleroche JD de, Bradford HF (1972) Metabolism of beds of mammalian cortical synaptosomes: response to depolarizing influences. J Neurochem 19:585–602

Blackman JC, Ginsborg BL, Ray Y (1963) On the quantal release of the transmitter at a sympathetic synapse. J Physiol (Lond) 167:402–415

Blaustein MP, Goldring JM (1975) Membrane potentials in pinched-off presynaptic nerve terminals monitored with a fluorescent probe: evidence that synaptosomes have potassium diffusion potentials. J Physiol (Lond) 247:589–615

Boksa P, Collier B (1980) Spontaneous and evoked release of acetylcholine and a cholinergic false transmitter from brain slices: comparison to true and false transmitter in subcellular stores. Neuroscience 5:1517–1532

Boyne AF, Bohan TP, Williams TH (1975) Changes in cholinergic synaptic vesicle populations and the ultrastructure of the nerve terminal membranes of *Narcine brasiliensis* electric organ stimulated to fatigue *in vivo.* J Cell Biol 67:814–825

Burnstock G, Holman ME (1962) Spontaneous potentials at sympathetic nerve endings in smooth muscle. J Physiol 160:446–460

Casamenti F, Pedata F, Corradetti R, Pepeu G (1980) Acetylcholine output from the cerebral cortex, choline uptake and muscarinic receptors in morphine-dependent, freely-moving rats. Neuropharmacology 19:597–605

Ceccarelli B, Hurlbut WP (1975) The effects of prolonged repetitive stimulation in hemicholinium on the frog neuromuscular junction. J Physiol (Lond) 247:163–168

Ceccarelli B, Hurlbut WP, Mauro A (1972) Depletion of vesicles from frog neuromuscular junctions by prolonged tetanic stimulation. J Cell Biol 54:30–38

Ceccarelli B, Hurlbut VP, Mauro A (1973) Turnover of transmitter and synaptic vesicles at the frog neuromuscular junction. J Cell Biol 57:499–524

Chakrin LW, Marchbanks RM, Mitchell JF, Whittaker VP (1972) Origin of acetylcholine released from the surface of the cortex. J Neurochem 19:2727–2736

Dowdall MJ, Simon EJ (1973) Comparative studies on synaptosomes: uptake of [*N-Me-*^{3}H]choline by synaptosomes from squid optic lobes. J Neurochem 21:969–982

Dowdall MJ, Whittaker VP (1973) Comparative studies on synaptosome formation: the preparation of synaptosomes from the head ganglion of the squid *Loligo pealii.* J Neurochem 20:921–935

Dowe GHC, Kilbinger H, Whittaker VP (1980) Isolation of cholinergic synaptic vesicles from the myenteric plexus of guinea-pig small intestine. J Neurochem 35:993–1003

Dudel J, Kuffler SW (1961) The quantal nature of transmission and spontaneous miniature potentials at the crayfish neuromuscular junction. J Physiol (Lond) 155:514–529

Dunant Y, Muller D (1986) Quantal release of acetylcholine evoked by depolarization at the *Torpedo* nerve-electroplaque junction. J Physiol (Lond) 373:461–478

Dunant Y, Israël M, Lesbats B, Manaranche R (1977) Oscillation of acetylcholine during nerve activity in the *Torpedo* electric organ. Brain Res 125:123–140

Edwards C, Doležal V, Tuček S, Zemková H, Vyskočil F (1985) Is an acetylcholine transport system responsible for nonquantal release of acetylcholine at the rodent myoneural junction? Proc Natl Acad Sci USA 82:3514–3518

Fletcher P, Forrester T (1975) The effect of curare on the release of acetylcholine from mammalian motor nerve terminals and an estimate of quantum content. J Physiol (Lond) 251:131–144

Garcia-Segura LM, Muller D, Dunant Y (1986) Increase in the number of presynaptic large intramembrane particles during synaptic transmission at the *Torpedo* electroplaque junction. Neuroscience 19:63–79

Giompres PE, Whittaker VP (1984) Differences in the osmotic fragility of recycling and reserve synaptic vesicles from the cholinergic electromotor nerve terminals of *Torpedo* and their possible significance for vesicle recycling. Biochim Biophys Acta 770: 166–170

Giompres PE, Whittaker VP (1986) The density and free water of cholinergic synaptic vesicles as a function of osmotic pressure. Biochim Biophys Acta 882:398–409

Haimann C, Torri-Tarelli F, Fesce R, Ceccarelli B (1985) Measurement of quantal secretion induced by ouabain and its correlation with depletion of synaptic vesicles. J Cell Biol 101:1953–1965

Heuser JE, Reese TS (1973) Evidence for recycling of synaptic vesicle membranes during transmitter release at the frog neuromuscular junction. J Cell Biol 57:315–344

Jonakait GM, Gintzler AR, Gershon MD (1979) Isolation of axonal varicosities (autonomic synaptosomes) from the enteric nervous system. J Neurochem 32: 1387–1400

Katz B (1966) Nerve, muscle and synapse. McGraw-Hill, New York

Katz B, Miledi R (1963) A study of spontaneous miniature potentials in spinal motoneurones. J Physiol (Lond) 168:389–422

Katz B, Miledi R (1972) The statistical nature of the acetylcholine potential and its molecular components. J Physiol (Lond) 224:665–699

Katz B, Miledi R (1977) Transmitter leakage from motor nerve endings. Proc R Soc Lond B196:59–72

Korneliussen H (1972) Ultrastructure of normal and stimulated motor endplates. Z Zellforsch 130:28–57

Krenz WD (1985) Morphological and electrophysiological properties of the electromotoneurones of the electric ray *Torpedo marmorata* in *in vivo* and *in vitro* brain slices. Comp Biochem Physiol 82A:59–65

Kriebel ME, Florey E (1983) Effect of lanthanum ions on the amplitude distributions of miniature endplate potentials and on synaptic vesicles in frog neuromuscular junctions. Neuroscience 9:535–547

Krnjević K, Mitchell JF (1961) The release of acetylcholine in the isolated rat diaphragm. J Physiol (Lond) 155:246–262

Kuffler SW, Yoshikama D (1975) The number of transmitter molecules in a quantum: an estimate from iontophoretic application of acetylcholine at the neuromuscular synapse. J Physiol (Lond) 251:465–482

Kuhar MJ, Sethy VH, Roth RH, Aghajanian GK (1973) Choline: selective accumulation by central cholinergic neurons. J Neurochem 20:581–593

Kuno M (1964) Quantal components of excitatory synaptic potentials in spinal motoneurones. J Physiol (Lond) 175:81–99

Laverty R, Michaelson IA, Sharman DF, Whittaker VP (1963) The subcellular localization of dopamine and acetylcholine in the dog caudate nucleus. Br J Pharmacol 21:482–490

Luqmani YA, Sudlow G, Whittaker VP (1980) Homocholine and acetylhomocholine: false transmitters in the cholinergic electromotor system of *Torpedo.* Neuroscience 5:153–160

MacIntosh FC (1959) Formation, storage and release of acetylcholine at nerve endings. Can J Biochem 37:343–356

MacIntosh FC, Collier B (1976) Neurochemistry of cholinergic nerve terminals. In: Zaimis E (ed) Neuromuscular junction. Springer, Berlin Heidelberg New York, pp 99–228 (Handbook of experimental pharmacology, vol 42)

Martin AR, Pilar G (1964) Quantal components of the synaptic potential in the ciliary ganglion of the chick. J Physiol (Lond) 175:1–16

Meunier FM (1984) Relationship between presynaptic membrane potential and acetylcholine release in synaptosomes from *Torpedo* electric organ. J Physiol (Lond) 354:121–137

Miledi R, Molenaar PC, Polak RL (1980) The effect of lanthanum ions on acetylcholine in frog muscle. J Physiol (Lond) 309:199–214
Miledi R, Molenaar PC, Polak RL (1982) Free and bound acetylcholine in frog muscle. J Physiol (Lond) 333:189–199
Miledi R, Molenaar PC, Polak RL (1983) Electrophysiological and chemical determination of acetylcholine release at the frog neuromuscular junction. J Physiol (Lond) 334:245–259
Mitchell JF (1963) The spontaneous and evoked release of acetylcholine from the cerebral cortex. J Physiol (Lond) 165:98–116
Mitchell JF, Silver A (1963) The spontaneous release of acetylcholine from the denervated hemidiaphragm of the rat. J Physiol (Lond) 165:117–129
Model PG, Highstein SM, Bennett MVL (1975) Depletion of vesicles and fatigue of transmission at a vertebrate central synapse. Brain Res 98:209–228
Nachmansohn D (1963) Choline acetylase. In: Koelle GB (ed) Cholinesterases and anticholinesterase agents. Springer, Berlin Göttingen Heidelberg, pp 40–54 (Handbook of experimental pharmacology, vol 15)
Nachmansohn D, Weiss MS (1948) Studies on choline acetylase. IV Effect of citric acid. J Biol Chem 172:677–697
Ohsawa K, Dowe GHC, Morris SJ, Whittaker VP (1979) The lipid and protein content of cholinergic synaptic vesicles from the electric organ of *Torpedo marmorata* purified to constant composition: implications for vesicle structure. Brain Res 161:447–457
Paton WDM, Zar MA (1968) The origin of acetylcholine released from guinea-pig intestine and longitudinal muscle strips. J Physiol (Lond) 194:13–33
Pilar G, Tuttle JB (1982) A simple neuronal system with a range of uses: the avian ciliary ganglion. In: Hanin I, Goldberg AM (eds) Progress in cholinergic biology: model cholinergic synapses. Raven, New York, pp 213–247
Potter LT (1970) Synthesis, storage and release of [^{14}C]acetylcholine in isolated rat diaphragm muscles. J Physiol (Lond) 246:145–166
Schwarzenfeld I von (1979) Origin of transmitters released by electrical stimulation from a small, metabolically very active vesicular pool of cholinergic synapses in guinea-pig cerebral cortex. Neuroscience 4:477–493
Suszkiw JB (1980) Kinetics of acetylcholine recovery in *Torpedo* electromotor synapses depleted of synaptic vesicles. Neuroscience 5:1341–1349
Torri-Tarelli F, Grohovaz F, Fesce R, Ceccarelli B (1985) Temporal coincidence between synaptic vesicle fusion and quantal secretion of acetylcholine. J Cell Biol 101:1386–1399
Vizi ES, Vyskočil F (1979) Changes in total and quantal release of acetylcholine in the mouse diaphragm during activation and inhibition of membrane ATPase. J Physiol (Lond) 286:1–14
Volknandt W, Zimmermann H (1986) Acetylcholine, ATP and proteoglycan are common to synaptic vesicles isolated from the electric organs of electric eel and electric catfish as well as from rat diaphragm. J Neurochem 47:1449–1462
Weiler M, Roed IS, Whittaker VP (1982) The kinetics of acetylcholine turnover in a resting cholinergic nerve terminal and the magnitude of the cytoplasmic compartment. J Neurochem 38:1187–1191
Whittaker VP (1966) The binding of acetylcholine by brain particles *in vitro*. In: von Euler VS, Rosell S, Uvnas B (eds) Mechanism of release of biogenic amines. Pergamon, Oxford, pp 147–164
Whittaker VP (1972) The storage and release of acetylcholine. Biochem J 128:73–74P
Whittaker VP (1986) Acetylcholine storage and release. Trends Pharmacol Sci 7:312–315
Whittaker VP, Sheridan MN (1964) The morphology and acetylcholine content of isolated cerebral cortical synaptic vesicles. J Neurochem 12:363–372
Whittaker VP, Michaelson IA, Kirkland RJA (1964) The separation of synaptic vesicles from nerve ending particles ('synaptosomes'). Biochem J 90:293–303
Wilson WS, Schulz RA, Cooper JR (1973) The isolation of cholinergic synaptic vesicles from bovine superior cervical ganglion and estimation of their acetylcholine content. J Neurochem 20:659–667

Zimmermann H (1979) Vesicle recycling and transmitter release. Neuroscience 4: 1773–1804

Zimmermann H, Denston CR (1977) Separation of synaptic vesicles of different functional states from the cholinergic synapses of the *Torpedo* electric organ. Neuroscience 2:715–730

Zimmermann H, Whittaker VP (1974a) Effect of electrical stimulation on the yield and composition of synaptic vesicles from the cholinergic synapses of the electric organ of *Torpedo*: a combined biochemical, electrophysiological and morphological study. J Neurochem 22:435–450

Zimmermann H, Whittaker VP (1974b) Different recovery rates of the electrophysiological, biochemical and morphological parameters in the cholinergic synapses of the *Torpedo* electric organ after stimulation. J Neurochem 22:1109–1114

CHAPTER 2

Model Cholinergic Systems: The Electromotor System of *Torpedo*

V. P. WHITTAKER and D. W. SCHMID

A. Embryology

In the fifth week of embryonic development of the electric ray, *Torpedo marmorata*, normal-looking, vertically orientated myotubes in the branchial region of the embryo deviate from the expected course of development of such cells and turn into electrocytes (FRITSCH 1890; FOX and RICHARDSON 1978, 1979) (Fig. 1). The lower pole of the myotube dilates, its myofibrils disintegrate, its nuclei migrate into the horizontal plane and the cell takes on a pear-shaped appearance. A few days, later the narrow upper part of the cell also loses its myofibrils and is drawn down. The embryonic electrocytes now flattened in the horizontal plane and arrange themselves in closely packed stacks which have a honeycomb appearance when viewed from above.

This unusual course of myotube development appears to be genetically preprogrammed since it also takes place *in vitro* in cultured explants of the embryonic electric organ *Anlage* in the absence of nervous tissue (RICHARDSON et al. 1981).

The still non-innervated electrocytes rapidly acquire a high density of nicotinic acetylcholine receptor (nAChR) on their ventral surface and, with a short lag, a high density of acetylcholinesterase (AChE) (KRENZ et al. 1980; WITZEMANN et al. 1983a) (Fig. 2). Only after differentiation of the ventral membrane do the axons of the electromotor neurons begin to branch and send neurites terminating in growth-cones into the stacks of electrocytes. Synaptogenesis begins at an embryo length of 50 mm (about 7 weeks) but functional synapses are not detectable until 90 mm. By birth (120–130 mm) the ventral surfaces of the electroytes are 80% covered with synapses. However, choline acetyltransferase (ChAT) activity at this stage is only 50% of the adult value and this low capacity to synthesize transmitter may explain the rapid fatiguability of the neonatal organ (KRENZ et al. 1980).

The *Torpedo* is ovoviviparous, that is, the fish does not lay eggs but carries the fertilized ova in a uterus-like expansion of the genital tract. Up to 16 embryos may be found in a gravid female; gestation is 6–8 months. This slow development separates events by weeks which may take place in the avian egg or the embryos of smaller mammals in hours; on the other hand work with *Torpedo* embryos is made difficult by their limited seasonal and geographic availability and the usual problems of working with wild material. However, explants of embryonic material have been successfully grown (RICHARDSON et al. 1981) and observations of potential importance made (RICHARDSON et al. 1985) (Figs. 3, 4).

Evidence for the existence in electric organ extracts of two factors – a heat-stable dialysable neuronotrophic factor needed to stabilize electromotor neurons

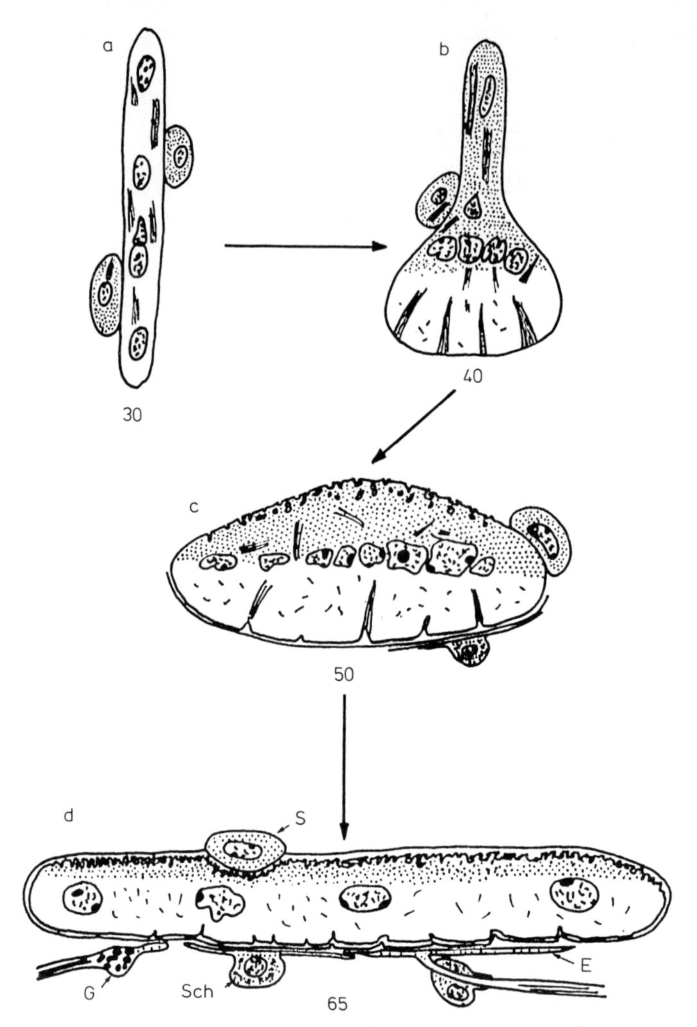

Fig. 1a–d. Transformation of myotubes (**a**) in the branchial region of the embryonic *Torpedo* into flattened electrocytes (**d**). At 40 mm embryo length the vertically orientated myotubes become pear-shaped (**b**); the nuclei migrate into the horizontal plane and the bulging lower portions of the cells lose their myofibrils. By 50 mm the upper pole of the cells also lose their myofibrils and are drawn down (**c**). Satellite cells (*S*) fuse with the nascent electrocyte to enlarge it, a process well under way by 65 mm. Meanwhile growth cones (*G*) have penetrated the stacks of electrocytes and synaptogenesis (*E*) takes place. Schwann cells (*Sch*) cover the axons

in culture and a heat-labile, non-dialysable cholinotrophic factor needed to evoke the expression of cholinergic function in the electromotor neurons – was obtained by comparing the survival and ChAT content of explants of electric lobe from 50-mm long embryos in media to which extracts of electric organ were added with those of control explants. As seen in Fig. 3, such explants (blocks 1, 2) do not survive after 7 days in unsupplemented culture; however, if supplemented with

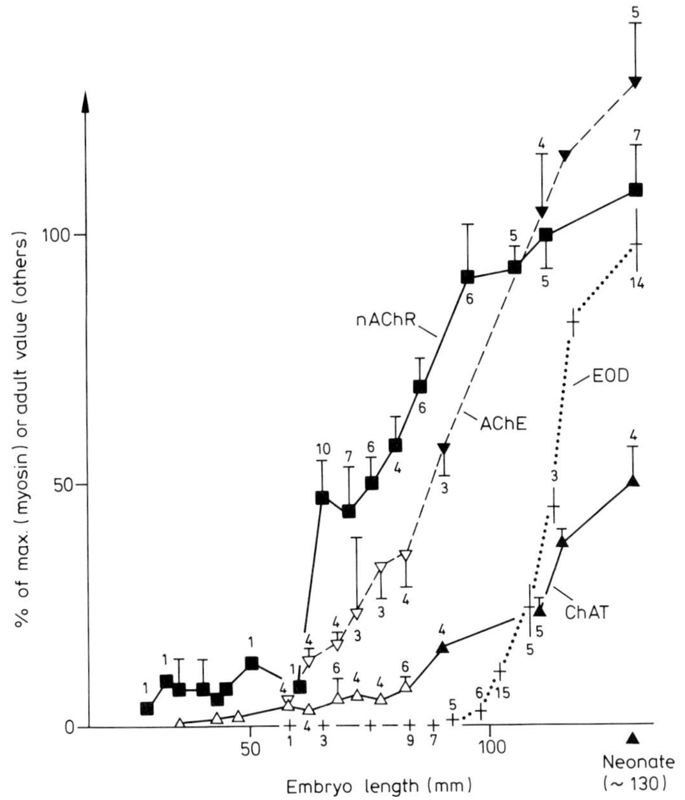

Fig. 2. Time course of increases in nAChR, AChE, ChAT and the electrical discharge in response to single stimuli via the nerve (*EOD*) accompanying differentiation of the electric organ. Unless otherwise indicated by the *small figures, points* are means of three observations. *Bars* indicate SEMs

an extract of late embryonic or neonatal electric organ, they survive (block 3) and produce ChAT (block 4). The factor needed for survival is heat- (block 5) and cold- (block 6) stable but is lost on dialysis (block 7); it is probably a relatively low molecular mass substance. By contrast, heat-treated extracts partially lose their ability to induce ChAT formation (compare block 8 with block 4), but the factor involved is cold-stable (compare block 9 with block 4). The fact that dialysis inactivates the cholinotrophic as well as the neuronotrophic factor (compare block 10 with block 4) is not conclusive evidence that both factors have a small molecular mass since the cholinotrophic factor clearly cannot exert its effect unless the explant survives.

Figure 4a shows the results of assaying extracts of electric organs at different stages of development for the two factors. There is a rapid rise in titre between 60 and 80 mm. This corresponds to the end of a second wave of cell death in the electric lobe *in vivo* (Fig. 4b). This wave affects the whole organ (Fox and Richardson 1982), in contrast to the first wave which is confined to the caudal region of the lobe and is attributed to the degeneration of a vestigial accessory

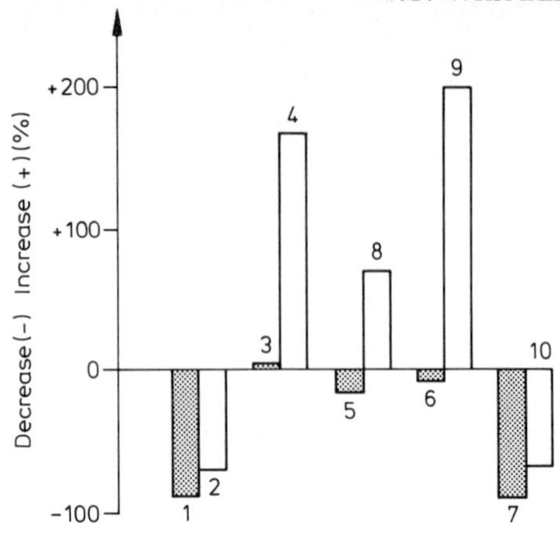

Fig. 3. Changes in neuronal cell count (*stippled blocks*) and ChAT (*white blocks*) in 50–60 mm explants of *Torpedo* electric lobe after 7 days in culture without supplement (*blocks 1, 2*) or supplemented with untreated extracts of electric organ from 105-mm long embryos (*blocks 3, 4*) or with similar extracts heat treated (*blocks 5, 8*), kept in the cold (*blocks 6, 9*) or dialysed (*blocks 7, 10*). For explanation see text

electric organ which survives in the related species *Narcine brasiliensis* with, however, inverse polarity (FOX et al. 1985).

B. Components of the Electromotor System

I. Electric Lobe

1. General Features

The cell bodies of the electromotor nerves lie in paired nuclei on the dorsal surface of the brain-stem just behind the cerebellum (Fig. 5). These cell bodies are among the largest motor neuron cell bodies found in vertebrates, being 125–135 μm in diameter with nuclei of about 40 μm. Their cytoplasm (Fig. 6) abounds in endoplasmic reticulum, Golgi membranes, polysomes, vesicles, mitochondria and lipofuscin granules, a picture denoting intense metabolic activity. This is not surprising when one considers that each cell body has to service at least 30 times its own volume of nerve terminal. Each lobe contains 60000 to 70000 neuron cell bodies comprising about one-half the weight of the tissue (about 400 mg in a large fish); the rest consists of dendrites, axons, nerve terminals, glia and blood vessels.

Each neuron cell body has up to 20 dendrites which are the targets for the axons of neurons in the oval nucleus located in the brain stem (KRENZ 1985). The synapses formed by these axons are purely axodendritic and they are the only type present in the lobe, since there are no recurrent collaterals of the efferent motor fibres, interneurons or inhibitory axosomatic terminals as in the motoneuron pools of the spinal cord. The oval or command nucleus receives inputs from the viscera and sensory organs and determines if and when the electromotor neurons fire in response to hunger or nociception. The transmitter at these lobe synapses is not ACh and is probably glutamate (W. D. KRENZ, personal communication).

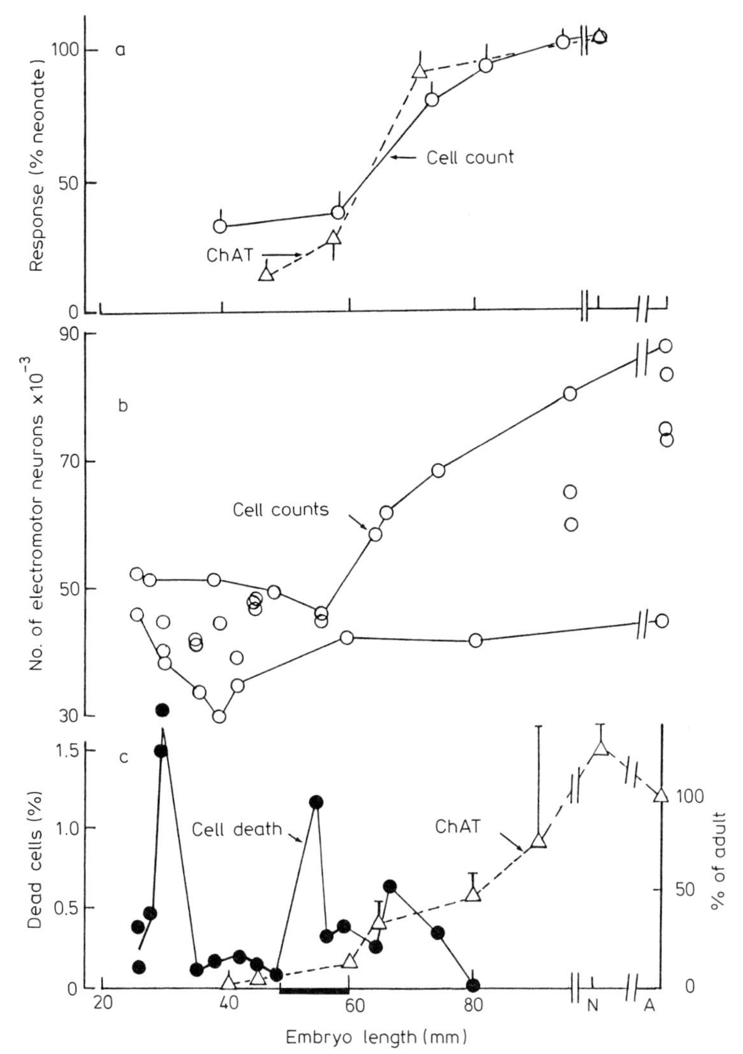

Fig. 4. a Assay of neuronotrophic and cholinotrophic factors in extracts of electric organ from embryos 40 mm in length to neonates (120–130 mm in length). The assays are based on the response of 50–60 mm explants of *Torpedo* electric lobe after 7 days to the extracts (as shown in Fig. 3, blocks 3 and 4 for extracts from 105-mm long embryos), expressed as percentages of that to extracts of neonatal electric organ. The parameters measured were ChAT activity (*triangles*) and proportion of neuronal cell bodies surviving (*circles*) after 7 days. Results are replotted from those of RICHARDSON et al. (1985). **b** Number of neuronal cell bodies in the electric lobes at different embryo lengths (*circles*); *lines* connect lowest and highest counts. **c** Percentage of dead cells (*filled circles*) and ChAT activity (as % of adult) (*triangles*). Note that the end of the second wave of cell death and the rise in ChAT activity corresponds to a rise in the content of trophic factors in the electric organ as graphed in **a**

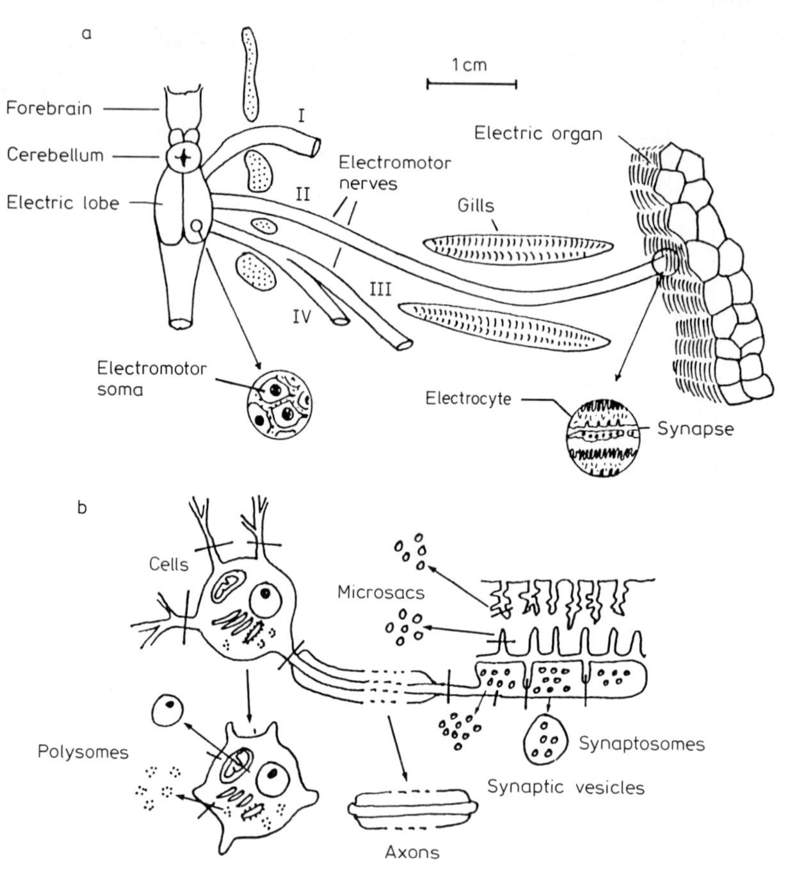

Fig. 5a, b. Diagram showing **a** the electromotor system with the cell bodies of the electromotor neurons in the electric lobes and their terminals in the electric organ, and **b** the types of preparation obtainable by tissue fractionation

2. Tissue Fractionation

Gentle homogenization releases the electromotor cell bodies from the tissue and detaches them from their axons and dendrites (Fig. 5) (DOWDALL et al. 1976). Centrifuging the homogenate into a simple step gradient separates the cells from dendritic, axonal and glial fragments and erythrocytes. The yield is over 70%. Such cell preparations are an excellent starting point for the isolation of mRNA (SCHMID et al. 1982). As might be expected from the perikaryal morphology and the vast terminal expansion, good yields of highly translatable poly(A) mRNA may be obtained either from the isolated cell preparation or the whole lobe tissue. This mRNA and its complementary DNA (cDNA) provide a means of analysing the portion of the genome coding for cholinergic function that has so far been little exploited; the work that has been done up to now is summarized in the next section.

Homogenization of lobe tissue at a somewhat higher rate of shear generates synaptosomes from the axodendritic nerve terminals; these have been partially purified (V. P. WHITTAKER, unpublished) and constitute a homogeneous prepa-

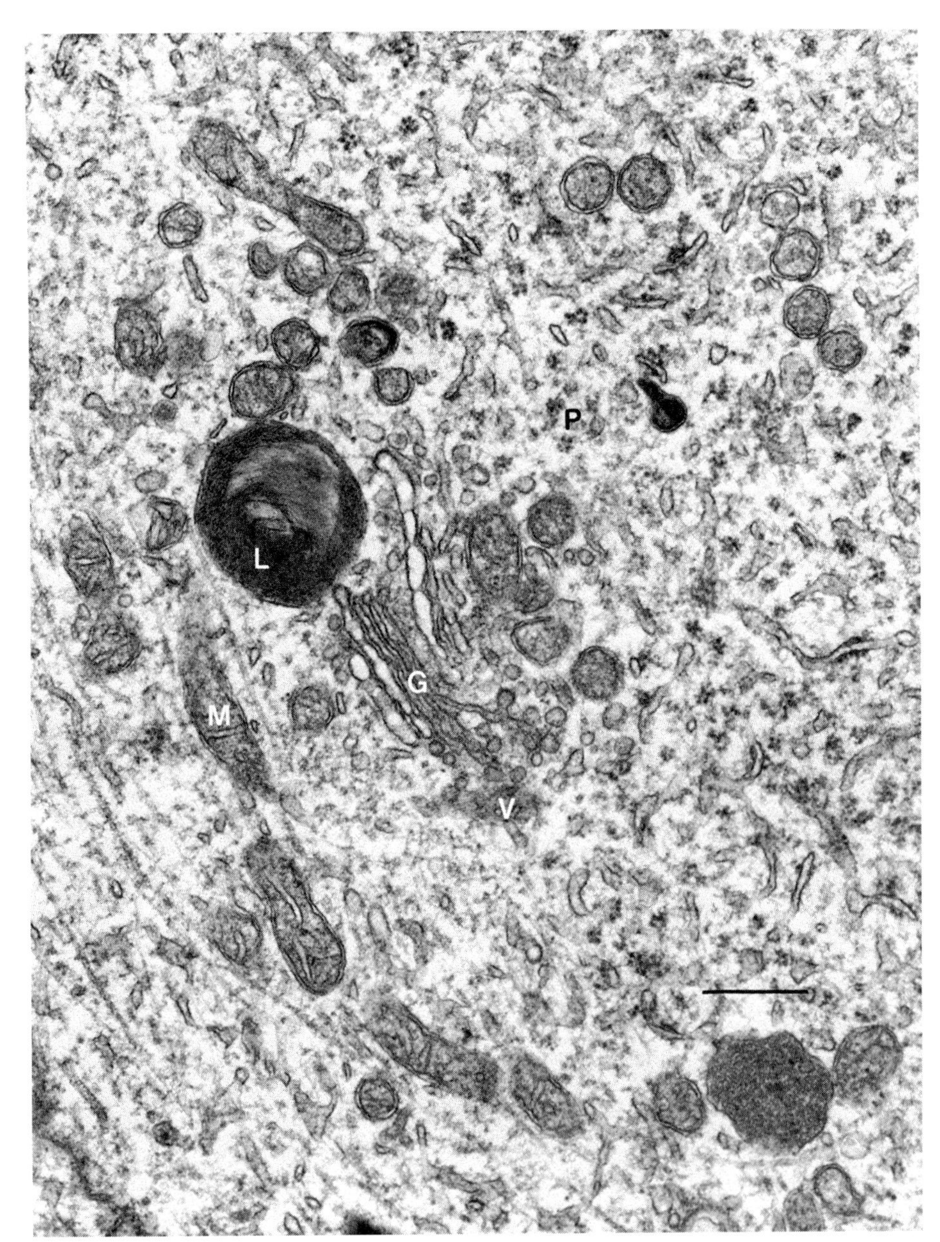

Fig. 6. Cytoplasm of electromotor neuron cell bodies. Note Golgi membranes (*G*), vesicles (*V*), mitochondria (*M*), lipofuscin granules (*L*) and polysomes (*P*). *Bar* represents 0.5 μm

ration of non-cholinergic (probably glutamergic) synaptosomes with considerable potential for further work.

These homogenates should also provide a rich source of membranes in which the biochemical mechanisms of synaptic vesicle formation could be studied *in vitro.*

3. Molecular Genetics

When poly(A) mRNA is translated in a cell-free rabbit reticulocyte system, some 150 major translation products can be separated by two-dimensional gel electrophoresis. By comparison with the translation products obtained with cerebellar mRNA, many of these can be designated lobe-specific and so putatively cholinergic specific. Of these several are identifiable as nerve terminal proteins including some synaptic vesicle proteins (SCHMID et al. 1982) (Fig. 7a).

cDNA has been successfully prepared from the lobe mRNA inserted into the pBR 322 vector and cloned in *E. coli* (SCHMID and GIROU 1987). Radioactive cDNA was made from lobe, cerebellum and muscle for use as probes in an effort to identify clones containing cDNA specific for electromotor neurons. Cerebellum was chosen as a region of the nervous tissue low in cholinergic neurons and muscle as a representative non-nervous tissue. It was expected that many clones would react to all three probes indicating the presence of cDNA coding for pro-

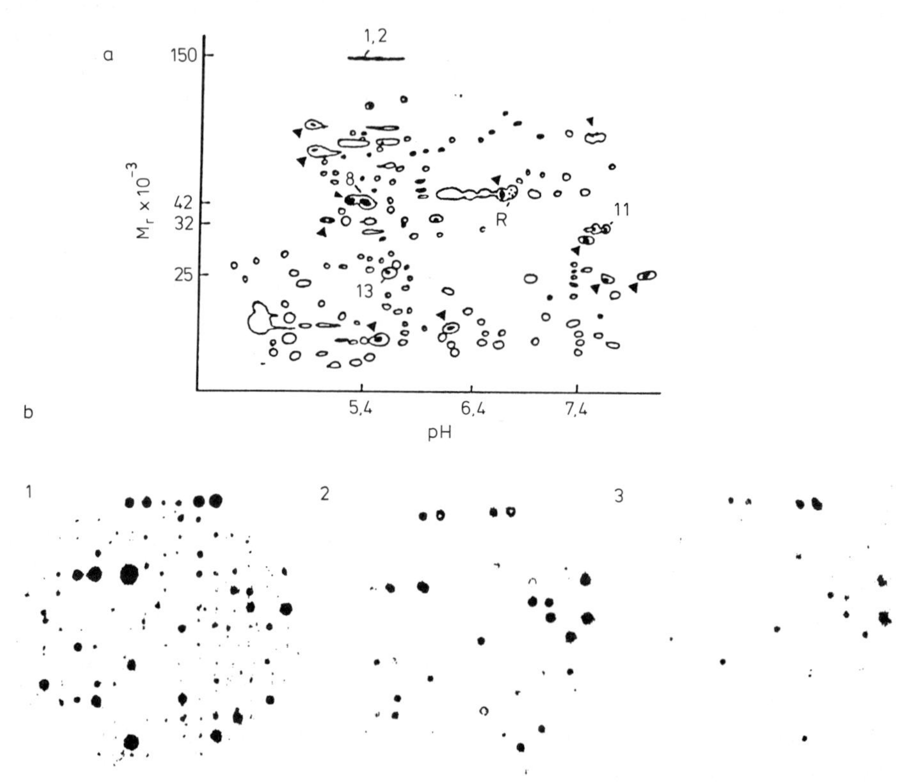

Fig. 7. a Two-dimensional gel electrophoresis of translation products of electric lobe poly(A) mRNA. *Outlines* designate position of radioactive products; *black spots*, Coomassie-blue synaptosomal proteins; *numbered spots*, synaptic vesicle components; *arrows*, other synaptic proteins. Several are coincident with radioactive translation products, others (e.g. *1, 2*) evidently require post-translational modification and are not represented among the direct translation products. **b** Colony screen of clones derived from lobe cDNA. Probing with cDNA derived from (*1*) lobe, (*2*) cerebellum and (*3*) muscle shows that relatively few clones contain sequences also expressed in cerebellum and/or muscle

teins common to most cells. A smaller number would react only to the two probes derived from nervous tissue. Such clones might well contain cDNA coding from proteins unique to the nervous tissue but not transmitter-related. Among these would be proteins of glial cells and those concerned with aspects of neuronal function common to all nerve cells; axonal transport and the mechanism of exocytosis, to name but two. A final group of clones might fail to react to either the muscle- or cerebellum-derived probes but hybridize to the lobe probe. Such clones could be provisionally designated cholinergic-specific. These expectations were realized: out of some 3000 clones tested in two rounds of screening some 50 proved to be lobe-specific (Fig. 7b). Northern blot hybridization showed that lobe-specific poly(A) RNA inserts ranged in length from 1700 to 3700 nucleotides. We do not know how many proteins are needed to designate a neuron as cholinergic, but apart from ChAT and AChE there is the transporter mediating high affinity choline uptake (HACU), the proteins of the synaptic vesicle including the acetylcholine (ACh) and ATP transporters and the Ca^{2+}-activated and proton-translocating ATPase (Chap. 11) and the enzymes needed to synthesize the cholinergic-specific surface markers (Chap. 15). In principle, given antibodies to these proteins and the use of expression vectors to construct a cDNA library from the lobe mRNA, their sequences could eventually be established. This will be an important field for the future.

4. Vesicle Synthesis

Vesicles of the size characteristic of synaptic vesicles are numerous in the electromotor perikaryon. When an antiserum raised against the proteoglycan of cholinergic synaptic vesicles is applied to electric lobe sections, the vesicular antigen is seen to be distributed in punctate fashion, consistent with localization in vesicles. The highest concentration of antigen is in the axon hillock region of the neuronal cell body, as though awaiting export to the terminal by axonal transport (JONES et al. 1982). Electromotor neurons are able, albeit rather slowly and inefficiently, to incorporate ^{35}S, administered as $^{35}SO_4^{2-}$ into vesicle proteoglycan; as described in greater detail in Chap. 13, this fraction of covalently bound ^{35}S detaches itself from the main pool of radioactivity and is transported to the terminal at the 'fast' rate expected for synaptic vesicles (STADLER et al. 1986; KIENE 1986).

These observations will serve to show the potential of this relatively large mass of purely cholinergic motor neuron cell bodies for further studies of synaptic vesicle synthesis.

II. Axons

The axons of the electromotor neurons are conveyed in eight nerve trunks, four on each side of the rostrocaudal axis, to the electric organs. There are on average about 15000–16000 heavily myelinated axons in each trunk but some trunks have more axons, and innervate a larger territory, than others. Since up to 5 cm of each of the 2–5 mm thick nerves can easily be obtained per specimen, and fish tolerate the surgical intervention involved in ligating the nerve trunks well, these nerves provide excellent preparations with which to study axonal transport *in vivo* and *in vitro.* The axonal transport of vesicular ACh and ATP at the fast rate, and of ChAT at the slow rate has been studied *inter alia* by DAVIES et al. (1977) and DAVIES (1978) (for further details see Chap. 13).

III. Electric Organ

1. General Features

Viewed from above, in a specimen from which the skin has been removed, each electric organ appears as a lobe-shaped gelatinous mass with a honeycomb appearance, each cell of which is in reality a stack of flattened electrocytes. Seen in scanning electron microscopy, each electrocyte is copiously innervated on its ventral surface by fine branches of the electromotor axons; usually three axons enter the stack from different corners and branch about seven times to form a dense interdigitating pattern of nerve terminals (G. Q. FOX and S. HONMA, personal communication; FOX et al. 1980).

The electrocytes grow throughout life; this is achieved by fusion with a gradually decreasing population of satellite cells. Their plasma membranes are highly invaginated. On the dorsal non-innervated face the invaginations expand and diminish to form a dense reticular pattern; on the ventral face they are simpler in structure and resemble the similar invaginations of the postsynaptic membranes of muscle. The membranes have a high content of nAChR (Chap. 9); negatively stained fragments show the characteristic five-sided rosette with a central hole, the presumptive ion channel. Other, so-called *v* proteins are also present (Hamilton et al. 1979). These include actin (WITZEMANN et al. 1983a, b) and a membrane-bound, 'low salt soluble' creatine phosphokinase (WITZEMANN 1985).

The plasma membranes of the electrocyte are covered, into their deepest invaginations, by a basal lamina. In the synaptic region a high concentration of AChE is inserted into this (Chap. 8).

The presynaptic nerve terminals resemble motor nerve terminals in skeletal muscle, except that their synaptic vesicles are larger (90 nm in diameter). This size difference between the vesicles of electromotor and motor nerve terminals holds for the *Torpedo* also.

The electrocyte cytoplasm is rich in intracellular filaments including desmin filaments. These may be attached to the cytoplasmic face of the plasma membrane by calelectrin, a protein of molecular mass 36000 kDa which binds to membrane lipids in a Ca^{2+}-regulated manner (WALKER 1982; FIEDLER and WALKER 1985). Electric organ is relatively rich in this protein and antisera to it cross-react with similar proteins in a wide variety of tissues, including liver, brain, leucocytes, adrenal medulla and spermatozoa (SÜDHOF et al. 1984). It is now recognized that these constitute a new family of Ca^{2+}-regulated proteins which possess hydrophobic repeat sequences in common (GEISOW et al. 1986).

2. Preparations of the Electric Organ

For many purposes, including release studies, it is convenient to work with blocks of intact electric tissue of varying sizes *in vitro*. Innervated blocks of 30 g or more, perfused at 20 °C through cannulae inserted into the accompanying blood vessels, may be kept in a functional state for up to 24 h or more and provide ample tissue for the isolation and separation of synaptic vesicles at the end of the incubation period (ZIMMERMANN and DENSTON 1977a; SUSZKIW et al. 1978; GIOMPRES et al. 1981a; ÁGOSTON et al. 1986). If 20 μ*M* eserine or other anti-ChE is added to the perfusion medium, released ACh or false transmitters may be recovered in 70%–80% yield. Stimulation may be applied via the attached electromotor nerves or by field stimulation.

Smaller blocks (approximately 2×2 cm in area) may be perfused by inserting a stiff, fine nylon cannula into the middle of the block or simply incubated in oxygenated *Torpedo* Ringer-Locke solution (SUSZKIW 1980; WEILER et al. 1982; KOSH and WHITTAKER 1985). They may also be field stimulated (LUQMANI et al. 1980).

Very small blocks have proved highly convenient for electrophysiological work. Normal electrocyte resting potentials have been recorded with intracellular microelectrodes inserted into the cut electrocytes, showing that electrocytes seal up after being cut (M.E. KRIEBEL, personal communication). Such preparations are convenient for detecting the electromotor equivalent of miniature endplate potentials, i.e. the miniature electroplaque potentials, and for studying the quantal content of the electric organ discharge (ERDÉLYI and KRENZ 1984; Chap. 20, this volume). It has not, however, proved possible with *Torpedo* electric tissue to dissociate columns of electrocytes into a suspension of single cells as is possible with some other electric organs, notably those in the tails of certain species of skate.

3. Subcellular Fractionation of the Electric Organ

a) Isolation of Synaptic Vesicles

The gelatinous, collagen-rich electric organ is difficult to comminute at 0 °C in a conventional homogenizer but becomes brittle and easily crushed to a coarse powder after freezing in liquid nitrogen (WHITTAKER et al. 1972; OHSAWA et al. 1979). This unconventional method of comminution does little damage to the tissue at the fine-structural level but tears open the nerve terminals (SOIFER and WHITTAKER 1972), allowing the extraction of synaptic vesicles and cytoplasmic protein from the softened tissue when this is suspended in ice-cold iso-osmotic sucrose, NaCl or glycine.

The synaptic vesicles may be separated from soluble cytoplasmic proteins and coarser membrane fragments by centrifuging the cytoplasmic extract into a continous sucrose-NaCl density gradient in a zonal rotor. A preliminary fractionation on a step gradient may precede the continuous density gradient centrifugation to reduce contamination of synaptic vesicle fractions by cytoplasmic proteins (TASHIRO and STADLER 1978). The limiting concentration of ACh is about 6 nmol mg^{-1} protein. Using such vesicles a detailed knowledge of synaptic vesicle structure has been obtained (Chap. 11). A second, denser type of vesicle, containing a form of vasoactive intestinal polypeptide (VIP), has recently been isolated and partially purified (D. V. ÁGOSTON, unpublished; Chap. 17, this volume).

b) Separation of Recycling from Reserve Vesicles

When synaptic vesicles are isolated from electric organs that have been stimulated just prior to vesicle isolation, the distribution of vesicular ACh and ATP in the density gradient is broader than when unstimulated tissue is used. The distribution is stimulus- and time-dependent, and under appropriate conditions may be bimodal with one peak (VP_1) coinciding with the vesicle peak obtained with unstimulated tissue as the starting material and the second (VP_2) corresponding to a denser region of the gradient. After several hours' recovery, the unimodal, rather sharply localized distribution of these two vesicle markers is reestablished (ZIMMERMANN and DENSTON 1977a; SUSZKIW et al. 1978; GIOMPRES et al. 1981a, b; ÁGOSTON et al. 1986).

A clue as to the significance of this stimulus-induced heterogeneity in vesicle density was provided by experiments with blocks of stimulated electric tissue through which *Torpedo* Ringer solution containing isotopically labelled ACh or ATP (ZIMMERMANN 1978; ZIMMERMANN and BOKOR 1979) precursors had been perfused. The isotopically labelled (i.e. newly synthesized) ACh or ATP was subsequently found mainly in the denser (VP_2) region of the gradient. This suggested that the dense vesicles are those undergoing cycles of exo- and endocytosis with concomitant release of pre-existing ACh and ATP and reloading with the newly synthesized substances.

This identification of the VP_2 fraction with the pool of recycling vesicles has been confirmed in various ways. If a marker of appropriate molecular mass and electron density (dextran has been used) is perfused through the stimulated block, dextran particles enter the recycling pool. After the pool has been generated and labelled, transmitter or co-transmitter released on stimulation has the same isotopic ratio (specific radioactivity) as that in the recycling pool. The increase in density (and the decrease in vesicular diameter that accompanies it; ZIMMERMANN and DENSTON 1977a, b; GIOMPRES et al. 1981a, b) has been satisfactorily accounted for by osmotic dehydration attendant upon partial reloading during vescicle recycling (GIOMPRES and WHITTAKER 1984, 1986). The optimum stimulus parameters for the formation of a large pool of recycling vesicles and the kinetics of their recovery have been studied (ÁGOSTON et al. 1986). This topic is further discussed in Chaps. 11 and 16.

In addition to the VP_1 and VP_2 synaptic vesicle pools, there is now considerable evidence (STADLER et al. 1986; KIENE 1986) for the existence in electromotor axons and terminals of a population of newly formed, immature vesicles which contain a proteoglycan of higher molecular mass than that of the two mature populations and little or no ACh or ATP. This pool has been designated the VP_0 pool and can be detected in density gradients by its low mean density and its content of stable vesicle markers. For further details the reader is referred to Chaps. 11 and 13.

c) Synaptosomes

Comminution of tissue minces by forcing them through grids of diminishing mesh size and submitting the resultant slurry to centrifugal fractionation on a step gradient enables sealed portions of the electromotor nerve terminals to be isolated in rather low yields. These synaptosomes may be of smaller or larger size depending on the comminution protocol, but there is little difference in their properties (Chap. 12). They take up choline by the HACU system, release ACh on depolarization and, as work with fluorescent probes which report membrane potentials indicates, a proportion of them (perhaps those with internal mitochondria) have normal membrane potentials (MEUNIER 1984); however their average ability to pump ions and presumably their general metabolic performance is rather low compared to mammalian brain synaptosomes even when care is taken to prepare them in such a way that their internal Na^+ and K^+ are initially close to their physiological values (RICHARDSON and WHITTAKER 1981). *Torpedo* electromotor synaptosome preparations differ from mammalian brain synaptosomes in the following ways:

- they are purely cholinergic
- they are on average much less active metabolically
- they have few, if any postsynaptic attachments
- they are generated in much lower yields

Perhaps their greatest value is as a source of purely cholinergic presynaptic plasma membranes.

d) Presynaptic Plasma Membranes

To prepare these, synaptosomes are lysed in water and the lysate refractionated on a step gradient (STADLER and TASHIRO 1979). They are useful as a source of cholinergic-specific surface antigens (Chap. 15) and, as vesiculated membrane fragments energized by an inwardly directed Na^+ and an outwardly directed K^+ gradient, for studying the ion dependency of the HACU system (Chap. 14) uncomplicated by metabolism-linked energy sources. In this application the parent synaptosome preparation is best isolated on a Ficoll gradient. The HACU system has been reconstituted in artificial membrane vesicles (DUCIS and WHITTAKER 1985).

e) Microsacs

Microsacs are vesiculated fragments of electrocyte plasma membranes formed by homogenizing comminuted electric tissue and are of two types: those derived from invaginations of the ventral (postsynaptic) membrane and those formed by pinching off the canaliculi of the dorsal membrane.

The ventral microsacs are extremely rich in nAChR and can be separated from other particles by reason of their relatively great density. When loaded with ^{23}Na, they have provided a useful system with which to study the ionophore properties of the receptor (KASAI and CHANGEUX 1971; COHEN et al. 1972). The kinetics of channel opening and desensitization have been studied with fluorescent ligands in such preparations (Chap. 9).

The dorsal microsacs are rich in Na^+, K^+-ATPase. Relatively little use has been made of them. Some parts of the reticulated invaginations may pinch off to form vacuolated bodies which may contaminate some synaptosome preparations (SHERIDAN et al. 1966).

f) Cytoskeleton and Basal Lamina

Triton treatment of frozen and crushed electric organ in the presence of Mg^{2+} ions removes membrane lipids leaving a 'cytoskeletal fraction' containing characteristic components of the cytoskeleton such as desmin, actin, fodrin, neurofilaments and tubulin (WALKER et al. 1985). More extensive detergent treatment of tissue fragments leaves pieces of basal lamina from which a receptor-aggregating factor has been extracted (GODFREY et al. 1984). Tissue slices similarly extracted leave a cage-like structure of basal lamina (G. P. RICHARDSON, personal communication). These preparations have been relatively little exploited.

C. Conclusions

It will be clear from the foregoing account and many of the further chapters of this book that cholinergic neurobiology has enormously benefitted from the intensive study of the electromotor system of *Torpedo* and other fish. The electric organ of *Torpedo* has a higher content of neural elements in it than other electric organs and the electrocytes are electrically inexcitable; they respond only to acetylcholine (reviewed in GRUNDFEST 1957). These electric organs have contributed to our understanding of the cholinergic synapse in somewhat the same way as the intensive study of the adrenal medulla and the chromaffin granule has to the adrenergic synapse but the model is a better one in that both pre- and postsynaptic elements are present.

The nAChR of *Torpedo* electric organ was the first pharmacological receptor to be isolated and sequenced. The electromotor synaptic vesicle is among the most completely understood of storage granules. But, as this chapter shows, the usefulness of the electromotor system as a model is far from being exhausted; synaptic vesicle genesis, the mechanism of vesicular exo- and endocytosis, the role of VIP and other peptides in cholinergic neurons and the gene expression of the cholinergic neuron are among the problems that could be usefully studied with this model in the future.

References

Ágoston DV, Dowe GHC, Fiedler W, Giompres PE, Roed IS, Walker JH, Whittaker VP, Yamaguchi T (1986) The use of synaptic vesicle proteoglycan as a stable marker in kinetic studies of vesicle recycling. J Neurochem 47:1580–1592
Cohen JB, Weber M, Huchet M, Changeux JP (1972) Purification from *Torpedo marmorata* electric tissue of membrane fragments particularly rich in cholinergic receptor. FEBS Lett 26:43–47
Davies LP (1978) ATP in cholinergic nerves: evidence for the axonal transport of a stable pool. Exp Brain Res 33:149–157

Davies LP, Whittaker VP, Zimmermann H (1977) Axonal transport in the electromotor nerves of *Torpedo marmorata*. Exp Brain Res 30:493–510

Dowdall MJ, Fox GQ, Wächtler K, Whittaker VP, Zimmermann H (1976) Recent studies on the comparative biochemistry of the cholinergic neuron. Cold Spring Harbor Symp Quant Biol 40:65–81

Ducis I, Whittaker VP (1985) High-affinity sodium-gradient-dependent transport of choline into vesiculated presynaptic plasma membrane fragments from the electric organ of *Torpedo marmorata* and reconstitution of the solubilized transporter into liposomes. Biochim Biophys Acta 815:109–127

Erdélyi L, Krenz WD (1984) Electrophysiological aspects of synaptic transmission at the electromotor junctions of *Torpedo marmorata*. Comp Biochem Physiol 79A:505–511

Fiedler W, Walker JH (1985) Electron microscopic localization of calelectrin, a M_r 36000 calcium-regulated protein, at the cholinergic electromotor synapse of *Torpedo*. Eur J Cell Biol 38:34–41

Fox GQ, Richardson GP (1978) The developmental morphology of *Torpedo marmorata*: electric organ – myogenic phase. J Comp Neurol 179:677–698

Fox GQ, Richardson GP (1979) The developmental morphology of *Torpedo marmorata*: electric organ – electrogenic phase. J Comp Neurol 185:293–316

Fox GQ, Richardson GP (1982) The developmental morphology of *Torpedo marmorata*: electric lobe – electromotoneuron proliferation and cell death. J Comp Neurol 207:183–190

Fox GQ, Jones RT, Honma S (1980) Preparation of electric tissue for scanning electron microscopy with special emphasis on presynaptic structure. Proceedings VIIth European congress on electron microscopy. The Hague 2:800–801

Fox GQ, Richardson GP, Kirk C (1985) *Torpedo* electromotor system development: neuronal cell death and electric organ development in the fourth branchial arch. J Comp Neurol 236:274–281

Fritsch G (1890) Die elektrischen Fische. II, Die Torpedineen. Von Veit, Leipzig

Geisow MJ, Fritsche U, Hexham JM, Dash B, Johnson T (1986) A consensus amino-acid sequence repeat in *Torpedo* and mammalian Ca^{2+}-dependent membrane-binding proteins. Nature (Lond) 320:636–638

Giompres PE, Whittaker VP (1984) Differences in the osmotic fragility of recycling and reserve synaptic vesicles from the cholinergic electromotor nerve terminals of *Torpedo* and their possible significance for vesicle recycling. Biochim Biophys Acta 770:166–170

Giompres PE, Whittaker VP (1986) The density and free water of cholinergic synaptic vesicles as a function of osmotic pressure. Biochim Biophys Acta 882:398–409

Giompres PE, Zimmermann H, Whittaker VP (1981a) Purification of small dense vesicles from stimulated *Torpedo* electric tissue by glass bead column chromatography. Neuroscience 6:765–774

Giompres PE, Zimmermann H, Whittaker VP (1981b) Changes in the biochemical and biophysical parameters of cholinergic synaptic vesicles on transmitter release and during a subsequent period of rest. Neuroscience 6:775–785

Godfrey EW, Nitkin RM, Wallance BG, Rubin LL, McMahan UJ (1984) Components of *Torpedo* electric organ and muscle that cause aggregation of acetylcholine receptors on cultured muscle cells. J Cell Biol 99:615–627

Grundfest H (1957) The mechanism of discharge of the electric organs in relation to general and comparative electrophysiology. Prog Biophys Biophys Chem 7:1–85

Hamilton SL, McLaughlin M, Karlin A (1979) Formation of disulfide-linked oligomers of acetylcholine receptor in membrane from *Torpedo* electric organ. Biochemistry 18:155–163

Jones RT, Walker JH, Stadler H, Whittaker VP (1982) Immunohistochemical localization of a synaptic-vesicle antigen in a cholinergic neuron under conditions of stimulation and rest. Cell Tissue Res 223:117–126

Kasai M, Changeux JP (1971) In vitro excitation of purified membrane fragments by cholinergic agonists. I. Pharmacological properties of the excitable membrane fragments. J Membr Biol 6:1–23

Kiene ML (1986) Charakterisierung eines Proteoglykans acetylcholinspeichernder synaptischer Vesikel und Beiträge zur Dynamik der Vesikel in der Nervenendigung. Dissertation, Göttingen

Kosh JW, Whittaker VP (1985) Is propionylcholine present in or synthesized by electric organ? J Neurochem 45:1148–1153

Krenz WD (1985) Morphological and electrophysiological properties of the electromotoneurones of the electric ray *Torpedo marmorata in vivo* and *in vitro* brain. Comp Biochem Pharmacol 82A:59–65

Krenz WD, Tashiro T, Wächtler K, Whittaker VP, Witzemann V (1980) Aspects of the chemical embryology of the electromotor system of *Torpedo marmorata* with special reference to synaptogenesis. Neuroscience 5:617–624

Luqmani YA, Sudlow G, Whittaker VP (1980) Homocholine and acetylhomocholine: false transmitters in the cholinergic electromotor system of *Torpedo.* Neuroscience 5:153–160

Meunier FM (1984) Relationship between presynaptic membrane potential and acetylcholine release in synaptosomes from *Torpedo* electric organ. J Physiol (Lond) 354:121–137

Ohsawa K, Dowe GHC, Morris SJ, Whittaker VP (1979) The lipid and protein content of cholinergic synaptic vesicles from the electric organ of *Torpedo marmorata* purified to constant composition: implications for vesicle structure. Brain Res 161:447–457

Richardson GP, Fox GQ (1982) The developmental morphology of *Torpedo marmorata*: the electric lobes. J Comp Neurol 211:331–352

Richardson PJ, Whittaker VP (1981) The Na^+ and K^+ content of isolated *Torpedo* synaptosomes and its effect on choline uptake. J Neurochem 36:1536–1542

Richardson GP, Krenz WD, Kirk C, Fox GQ (1981) Organotypic culture of embryonic electromotor system tissues from *Torpedo marmorata.* Neuroscience 6:1181–1200

Richardson GP, Rinschen B, Fox GQ (1985) *Torpedo* electromotor system development: developmentally regulated neuronotrophic activities of electric organ tissue. J Comp Neurol 231:339–352

Schmid DW, Giroux C (1987) Cloning of cDNA derived from mRNA of the electric lobe of *Torpedo marmorata* and selection of putative cholinergic-specific sequences. J Neurochem 48:307–312

Schmid DW, Stadler H, Whittaker VP (1982) The isolation, from electromotor neurone perikarya, of messenger RNAs coding for synaptic proteins. Eur J Biochem 1272:633–639

Sheridan MN, Whittaker VP, Israël M (1966) The subcellular fractionation of the electric organ of *Torpedo.* Z Zellforsch 74:291–307

Soifer D, Whittaker VP (1982) Morphology of subcellular fractions derived from the electric organ of *Torpedo.* Biochem J 128:845–846

Stadler H, Tashiro T (1979) Isolation of synaptosomal plasma membranes from cholinergic nerve terminals and a comparison of their proteins with those of synaptic vesicles. Eur J Biochem 101:171–178

Stadler H, Kiene ML, Harlos P, Welscher U (1986) Structure and function of cholinergic synaptic vesicles. In: Hamprecht B, Neuhoff V (eds) Colloquium Mosbach 1986: neurobiochemistry, selected topics. Springer, Berlin Heidelberg New York, pp 55–65

Südhof TC, Ebbecke M, Walker JH, Fritsche U, Boustead C (1984) Isolation of mammalian calelectrins: a new class of ubiquitous Ca^{2+}-regulated proteins. Biochemistry 23:1103–1109

Suszkiw JB (1980) Kinetics of acetylcholine recovery in *Torpedo* electromotor synapses depleted of synaptic vesicles. Neuroscience 5:1341–1349

Suszkiw JB, Zimmermann H, Whittaker VP (1978) Vesicular storage and release of acetylcholine in *Torpedo* electroplaque synapses. J Neurochem 30:1269–1280

Tashiro T, Stadler H (1978) Chemical composition of cholinergic synaptic vesicles from *Torpedo marmorata* based on improved purification. Eur J Biochem 90:479–487

Walker JH (1982) The isolation from cholinergic synapses of a protein that binds to membranes in a calcium-dependent manner. J Neurochem 39:815–823

Walker JH, Boustead CM, Witzemann V, Shaw G, Weber K, Osborn M (1985) Desmin is a major protein of the *Torpedo* electrocyte: distribution of desmin, actin, fodrin, neurofilaments, and tubulin at the cholinergic synapse. Eur J Cell Biol 38:123–133

Weiler M, Roed IS, Whittaker VP (1982) The kinetics of acetylcholine turnover in a resting cholinergic nerve terminal and the magnitude of the cytoplasmic compartment. J Neurochem 38:1187–1191

Whittaker VP, Essman WB, Dowe GHC (1972) The isolation of pure synaptic vesicles from the electric organs of elasmobranch fish of the family Torpedinidae. Biochem J 128:833–845

Witzemann V (1985) Creatine phosphokinase: isoenzymes in *Torpedo marmorata*. Eur J Biochem 150:201–210

Witzemann V, Richardson GP, Boustead C (1983a) Characterization and distribution of acetylcholine receptors and acetylcholinesterase during electric organ development in *Torpedo marmorata*. Neuroscience 8:333–349

Witzemann V, Schmid D, Boustead C (1983b) Differentiation-dependent changes of nicotinic synapse-associated proteins. Eur J Biochem 1371:235–245

Zimmermann H (1978) Turnover of adenine nucleotides in cholinergic synaptic vesicles of the *Torpedo* electric organ. Neuroscience 3:827–836

Zimmermann H, Bokor JT (1979) [Adenosine]5′-phosphate recycles independently of acetylcholine in cholinergic synaptic vesicles. Neurosci Lett 13:319–324

Zimmermann H, Denston CR (1977a) Recycling of synaptic vesicles in the cholinergic synapses of the *Torpedo* electric organ during induced transmitter release. Neuroscience 2:695–714

Zimmermann H, Denston CR (1977b) Separation of synaptic vesicles of different functional states from the cholinergic synapses of the *Torpedo* electric organ. Neuroscience 2:715–730

CHAPTER 3

Model Cholinergic Systems: The Avian Ciliary Ganglion

G. R. PILAR and D. A. JOHNSON

A. Introduction

The avian ciliary ganglion, especially that of the domestic fowl ('chick'), has emerged in recent years as a model system for the study of ganglionic structure and function. Its accessibility both for *in vivo* manipulation and *in vitro* experimentation has made it useful for studies examining biochemical, morphological, electrophysiological and tissue culture aspects of neuronal interaction. Furthermore, the ciliary ganglion is a preparation amenable to manipulation and study *in ovo*, a feature that has made it particularly useful as a developmental model system (Chap. 5). However, what was once heralded as a simply organized, purely cholinergic system has been found to be somewhat more complex: recent work has demonstrated the presence of transmitters and neuromodulators other than ACh.

This overview aims not only to introduce this model system and to summarize its structural and functional organization, but also to review recent developments suggesting an increased degree of synaptic complexity which now challenges those who are familiar with this preparation.

B. Anatomy

The ciliary ganglion (of domestic fowl unless otherwise stated) comprises the cells mediating the parasympathetic functions of the third cranial (N. III) or oculomotor nerve. The efferent input to the ganglion arises from the ipsilateral accessory oculomotor nucleus, a midbrain locus analogous in position and function to the Edinger-Westphal nucleus of mammals (COWAN and WENGER 1968; NARAYANAN and NARAYANAN 1976; REINER et al. 1983) and distinct from the mammalian accessory oculomotor nucleus, a structure believed to be involved in eye movement integration (CARPENTER and PETER 1970).

The ganglion is located intraorbitally, just lateral to the exit of the optic nerve from the globe. Preganglionic fibres run in a branch of the oculomotor nerve which also carries other fibres to N. III targets inferiorly (inferior and medial rectus and inferior oblique). Exiting distally are typically two groups of fibres, the ciliary nerves and the choroid nerves, which enter the sclera to innervate respectively the iris-ciliary body complex and the choroid smooth muscle coat (MERINEY and PILAR 1987). Several millimetres distal to the emergence from the ganglion of the postganglionic nerves the ciliary nerve is joined by a nerve ramus of the ophthalmic branch of the trigeminal nerve. This carries sensory afferents which do not enter the ganglia. There is no evidence for axonal fibres from any source which pass through the ciliary ganglion without making synaptic contact.

Virtually all elements in this preparation are accessible to surgical manipulation (PILAR and TUTTLE 1982). Presynaptic denervation may be accomplished either by an intracranial approach or by intraorbital nerve section (LANDMESSER and PILAR 1970). The postganglionic nerves to the internal eye targets may be cut either intraocularly via a small scleral incision or extraocularly by section of the nerves while directly viewing the ganglion and its postganglionic branches (MARWITT et al. 1971).

Similar procedures, though more challenging because of the smaller size of the preparation, can be achieved with embryonic preparations. Furthermore, removal of entire target structures, such as the optic vesicle, can be undertaken *in ovo* (COWAN and WENGER 1968; LANDMESSER and PILAR 1974a).

Ganglion cells take their origin from the cranial neural crest cells (NARAYANAN and NARAYANAN 1978); these migrate to their intraorbital location at approximately day 2 of incubation which corresponds to stage 9 of the HAMBURGER and HAMILTON (1951) staging classification. By stage 25 (4 days) a distinct ganglion is recognizable and cell proliferation has ceased.

However, the number of cells at that time, approximately 6500, is roughly twice the number ultimately found in mature ciliary ganglia (LANDMESSER and PILAR 1974b). Over a 4-day period beginning at stage 35 (9 days of incubation) about one-half the ganglion cells die such that, by stage 42, approximately 3200 cells remain. This process has been shown to be a competitive phenomenon highly dependent on interaction of the neurons with the target tissue (LANDMESSER and PILAR 1974a, b, 1978; PILAR et al. 1980).

The cells remaining after the period of cell death comprises two morphologically and functionally distinct cell populations, the ciliary cells (approximately 45%) and the choroid cells (approximately 55%) (Fig. 1; and MARWITT et al. 1971). The two cell types tend to be segregated within the ganglion and each has identifiable postganglionic nerves leaving the ganglion at distinct sites.

The ciliary cell population, although slightly fewer in number, is made up of cells of larger size and therefore occupies a greater proportion of ganglionic volume.

Individual ciliary cells receive innervation from only a single presynaptic axon. These oculomotor nerve fibres are of higher conduction velocity and lower threshold than those innervating choroid cells (MARWITT et al. 1971) and end upon ciliary cells in large cup-like endings termed calyces (DE LORENZO 1960). The calyx is a transient phenomenon: beginning during the second week of life, the calyciform ending slowly breaks up into more typical bouton-like endings which are concentrated at the hilar region of the cell (DE LORENZO 1960; CANTINO and MUGNAINI 1975). A unique view of the extent of the calyciform ending can be seen using cobalt filling of preganglionic neurons (D.A. JOHNSON, M. ICHIKI and G. PILAR, in preparation). It was found that following cobalt application, in isolated preparations, to the oculomotor nerve (anterograde) or to the postganglionic nerves (retrograde), the tracer was localized in the respective presynaptic or postsynaptic element. Furthermore, in ciliary but not choroid cells, cobalt apparently crossed the synaptic cleft via the gap junctions known to exist at this

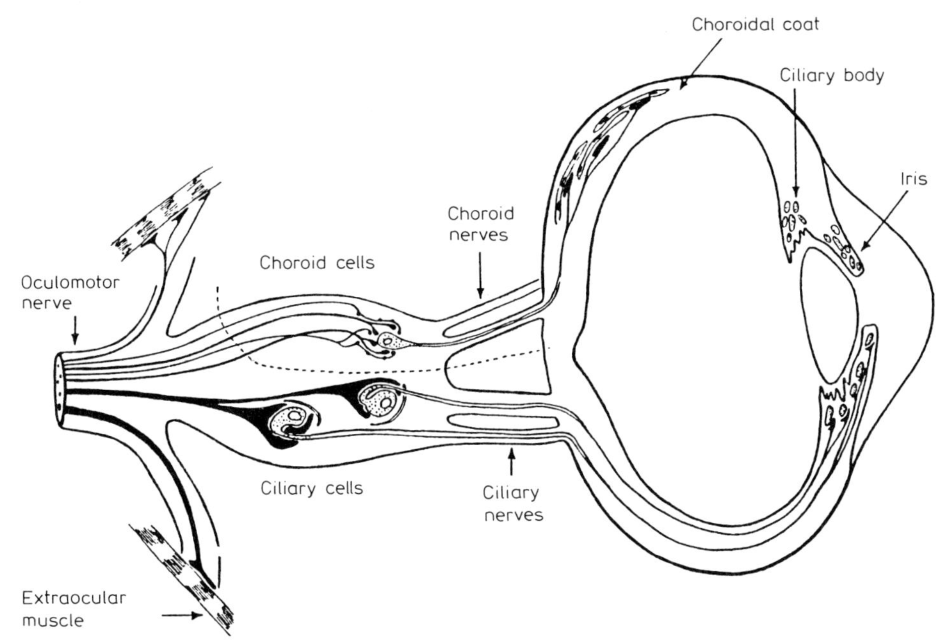

Fig. 1. Ciliary ganglion and its target organs in the eye. The preganglionic fibres run together in the oculomotor nerve with sympathetic fibres and motor axons for the extraocular muscles: inferior rectus, inferior oblique, medial rectus and superior rectus. Ciliary cells with calyciform synapses send axons to form the ciliary nerves to striated muscle in the iris and ciliary muscles. The multiply innervated choroid cells send axons in the choroid nerves to innervate vascular smooth muscle in the choroidal coat. Adrenergic varicosities course near terminals and choroid cells. (LANDMESSER and PILAR 1974a)

synapse (DE LORENZO 1966; CANTINO and MUGNAINI 1975) and labelled the synaptic partner. Techniques known to uncouple gap junctions in other preparations (exposure to dinitrophenol or incubation at temperatures below 6 °C; POLITOFF et al. 1969) blocked transfer of cobalt ions across low resistance junctions in the ciliary ganglia. Thus, under these conditions, application of cobalt in the anterograde direction labelled presynaptic calyces while leaving the postsynaptic elements unstained. The configuration of these endings is of significance in that the calyx is of sufficient size to be impaled by a microelectrode for intracellular recording (MARTIN and PILAR 1963).

Ciliary cells themselves are somewhat ovoid in shape and are, on average, 30–40 μm in diameter with an eccentric nucleus. The entire cell body is enveloped by a thin myelin sheath in hatched animals (HESS 1965). A single axon emerges from the hillock region of the cell. In many cells, this region faces proximally such that the axon initially courses toward the preganglionic pole of the ganglion for a short distance before turning back toward the periphery. Ciliary cell axons pierce the sclera and can be seen to course subsclerally to anterior intraocular targets.

Electron-microscopic analysis has confirmed these light microscopic observations (DE LORENZO 1960, 1966; CANTINO and MUGNAINI 1975). Ciliary cells are

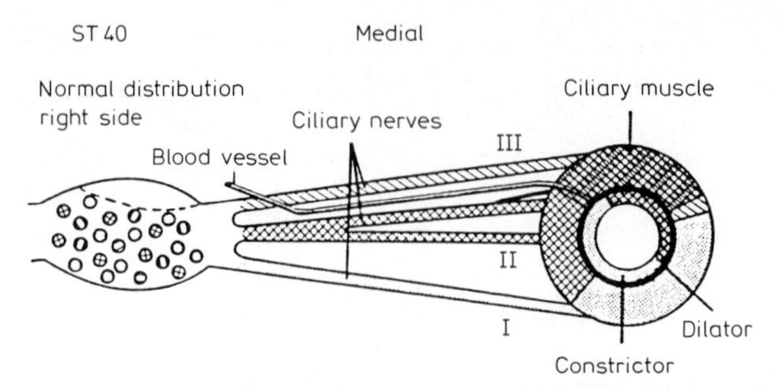

Fig. 2. Peripheral innervation pattern of the ciliary population. Each of the main ciliary nerve branches emanating from the ganglion innervates different parts of the peripheral target. Branch I (*stippled*) innervates the iris constrictor and the lateral portions of the ciliary muscle. Branch III (*hatched*) innervates only the medial portion of the ciliary muscle, whereas branch II (*cross-hatched*) innervates intermediate areas of the ciliary muscle as well as the iris constrictor muscle. A small branch (*solid*) emerging from branch II innervates the iris dilator. The cells that project out of each of these nerves are not localized but are distributed throughout the ciliary portion of the ganglion. (PILAR et al. 1980)

seen to form short evaginations and small pseudo-dendrites, many of which serve as loci of synaptic junctions (TAKAHASHI and HAMA 1965; TAKAHASHI 1967). Typical gap junctions, although sparse, can be clearly identified forming connections between presynaptic terminals and ciliary cell bodies. Freeze-fracture techniques have demonstrated that these gap junctions possess the pattern of membrane particle aggregates typical of gap junctional structures seen in other preparations (CANTINO and MUGNAINI 1975). Consistently with the dual chemical and electrical nature of synaptic transmission at these synapses, electron micrographs occasionally reveal synaptic vesicles and gap junctions at the same region of the synaptic interface.

Although ciliary cells form a morphologically homogeneous cell population, it has been shown that there exist distinct subpopulations, each innervating a different portion of the target structures (PILAR et al. 1980) (Fig. 2). This segregation appears to be localized at the level of the postganglionic nerves, each of which innervate a different portion of the periphery; retrograde labelling of individual nerves showed that the somata of those nerve fibres were widely distributed throughout the ganglion.

The remainder of the principal cells of the ciliary ganglion are the choroid cells. Many of these are multiply innervated, receiving input from as few as one to as many as four or more preganglionic axons (DRYER and CHIAPINELLI 1987). These synaptic endings are typical *boutons terminaux*.

Choroid cell axons emerge from the ganglion in the choroid nerve branches; these enter the globe to innervate a smooth muscle layer of the choroid coat. While definitive evidence is lacking, it is believed that activity among choroid neurons plays a role in subretinal vasoregulatory processes. These cells tend to be smaller than ciliary cells (20–25 μm versus 30–40 μm), are innervated by preganglionic axons of lesser velocity and higher threshold and occupy a lesser proportion of ganglionic volume, being largely restricted to the dorsomedial and distal portion of the ganglion (PILAR et al. 1980). Choroid nuclei are typically central and no myelin sheath surrounds the cells. The electron-microscopical appearance of choroid cells is typical of cells in other autonomic ganglia.

There are no interneurons in the fowl ciliary ganglion. There are, however, fine varicose adrenergic terminals which are widely dispersed throughout the ganglion but are more concentrated in the region of the choroid cells (EHINGER 1967; CANTINO and MUGNAINI 1974). These fibre endings do not appear to be related to the vasculature and, in favourable settings, seem to form pericellular baskets of seemingly synaptic character around ganglion cells. The somata of origin of these terminals are not intraganglionic and are presumed to reside in the cranial cervical ganglion, the avian analogue of the mammalian superior cervical ganglion. Their functional significance is not known.

C. Electrophysiology and Biochemistry

I. Chemical and Electrical Components in Synaptic Transmission

The distinction between the two cell populations of the ciliary ganglia, apparent from their anatomic dissimilarities, is further manifested by their electrophysiological differences. This section will briefly discuss the classical synaptic organization of the ciliary ganglion and then consider some previously unpublished work suggesting that the ciliary ganglion is more than a mere way-station for nervous impulses directed from the brain to target tissue.

MARTIN and PILAR (1963, 1964) first reported intracellular recordings from chick ciliary ganglion cells that revealed two populations of cells. In one, synaptic transmission resembled that seen in other autonomic ganglia and at the neuromuscular junction, i.e. presynaptic excitation elicited an action potential in the postsynaptic cell, which, when rendered subthreshold by hyperpolarization of the membrane potential, revealed an underlying excitatory postsynaptic potential (EPSP). These cells were subsequently shown to comprise the choroid cell population (MARWITT et al. 1971).

The other cell population also responded to orthodromic stimulation with a suprathreshold action potential. However, blockage of the action potential by hyperpolarization revealed an EPSP preceded by a short latency coupling potential which was largely independent of membrane potential and presumed to represent direct current flow from presynaptic to postsynaptic neural elements. This group of cells was ultimately shown to represent the ciliary cell population.

Pharmacological experiments at that time established the nicotinic nature of transmission through the ciliary ganglia (MARTIN and PILAR 1963a). D-Tubocurarine effectively abolished the chemical component of synaptic transmission (i.e. the EPSP) but had no effect on the amplitude or time course of the electrical coupling potential.

The extracellularly recorded responses to orthodromic and antidromic stimulation also reflected this dual mode of synaptic transmission. In response to preganglionic stimulation of the oculomotor nerve, potential changes recorded in postganglionic ciliary nerves were bimodal. Presumably the initial peak represented those axons whose somata were driven to superthreshold levels by electrical coupling while the peak of longer latency comprises those axons whose cell bodies were excited by chemical synaptic coupling. This hypothesis was confirmed pharmacologically when, in the presence of curare or high Mg^{2+} and low Ca^{2+}, the later peak of activity was abolished while the first remained largely unaltered.

Other features of synaptic transmission in this preparation are revealed in extracellular records. With supramaximal stimulation conditions, the area under each peak represents roughly the proportion of postganglionic cells excited by each mode of synaptic transmission. Thus, experimental manipulations designed to alter either electrical or chemical synaptic transmission may be screened for their effects on one type of transmission relative to the other. For example, it is well known that elimination of Ca^{2+} from experimental solutions will block chemical synaptic transmission. As expected, lowering of the external Ca^{2+} concentration ($[Ca^{2+}]_O$) was seen to reduce the chemical component of the extracellular action potential. Conversely, elevated $[Ca^{2+}]_O$ enhanced chemical transmission. These effects are probably related to the Ca^{2+} dependence of acetylcholine (ACh) release. The size of the peak arising from electrically excited cells behaved in the opposite fashion, increasing in size in low $[Ca^{2+}]_O$, decreasing in high $[Ca^{2+}]_O$. This effect is reversible and readily reproduced with multiple changes of $[Ca^{2+}]_O$ in superfusion solutions. Just why this occurs is not entirely clear but it may be secondary to effects of Ca^{2+} ions on membrane surface charges, such that elevation of $[Ca^{2+}]_O$ screens fixed negative surface charges on the membranes, thereby altering the transmembrane potential gradient and effectively raising the threshold for firing (GILBERT and EHRENSTEIN 1969).

The relative excitability of an individual ciliary cell in terms of chemical *versus* electrical firing is not fixed. Rather, cells are capable of being excited by either mode under certain conditions. This is shown by recordings in which the electrical coupling potential is subthreshold and the cell generates an action potential in response to an EPSP. However, with slight depolarization, conditions are attained such that the electrical coupling potential has become suprathreshold. Thus, changes in the synaptic elements or their environment may play a role in determining the mode (chemical or electrical) by which a nerve cell is excited.

That gap-junctional structures may underlie the electric synaptic coupling was first suggested by DE LORENZO (1966) who described tight junctions between ciliary cells and their presynaptic contacts. Using freeze-fracture techniques, CANTINO and MUGNAINI (1975) further examined the organization of gap junctions and concluded that, although they occupy only a small proportion (0.17%) of the synaptic interface, they are present in sufficient numbers and locations to mediate electrical transmission.

It should be noted that electrical coupling occurs in the antidromic as well as the orthodromic direction, although not with similar efficiency (MARTIN and PILAR 1963a, b; 1964). Thus, antidromic stimulation of the ciliary nerves of post-hatched fowl typically gives rise to action potentials which can be recorded extracellularly in the oculomotor nerve.

As already mentioned, the presynaptic nerve endings form large calyces into which microelectrodes may be placed. When an electrode is so placed, stimulation of the oculomotor nerve evokes an action potential of short latency which, when blocked by membrane hyperpolarization by current injection, reveals a small electrotonic potential with decay parameters identical to those of the membrane itself (as opposed to the prolonged potential decay typically following an EPSP). In those cells in which there is antidromic electrical coupling, stimulation of the postsynaptic ciliary nerves evokes activity in the nerve terminal, be it a small electrotonic potential or a full action potential.

The above constitutes a brief overview of the classical synaptic organization of the fowl ciliary ganglion. However, recent investigations based on earlier work on this simple ganglion now suggest a more complicated functional arrangement. Three areas will be discussed: (a) the presence of substance P (SP), enkephalins and morphine activities in the ciliary ganglion; (b) the presence of presynaptic ACh receptors (AChR) on ciliary nerve terminals; and (c) the existence of two populations of postsynaptic receptor types, muscarinic (mAChR) and nicotinic (nAChR).

II. Presence of Neuropeptides and its Implications

Following earlier investigations of SP-like and leu-enkephalin-like immunoreactivities in several regions of the avian nervous system, ERICHSEN et al. (1982a, b) examined the pigeon ciliary ganglion for similar substances. They found that SP-like and enkephalin-like immunoreactivity was localized in preganglionic terminals in the ciliary ganglion, using an anti-SP monoclonal antibody and an anti-leu-enkephalin antiserum, respectively, and probing for these reactivities with either indirect fluorescence or peroxidase-antiperoxidase methods (ERICHSEN et al. 1982a). These substances were restricted to the presynaptic nerve terminals and electron-microscopic analysis showed that the reaction product was limited to small dense-core vesicles. Both calyciform and *bouton terminal*-type endings were labelled, suggesting that presynaptic endings upon both choroid and ciliary cells contained SP and leu-enkephalin. Not only did this work demonstrate the existence of putative neurotransmitters other than ACh, it distinguished between the two anatomically defined nerve terminal types. It is of interest to note the striking similarity between the appearance of presynaptic calyces when stained with the immunoprecipitation techniques used by ERICHSEN et al. (1982a, b) and when labelled with cobalt precipitate.

The role, if any, of SP and/or leu-enkephalin in ganglionic transmission is not understood. ERICHSEN et al. (1982a, b) noted that the labelled dense-core vesicles were not located among small clear vesicles (presumably containing ACh) at synaptic release sites (active zones), but were somewhat removed from the portion of the nerve terminal abutting the synaptic cleft. More recently DRYER and CHIAPINELLI (1987) have shown that non-cholinergic slow excitatory postsynaptic potentials recorded from choroid cells can be mimicked by the application of SP. While generation of slow potentials among choroid cells may be ascribed to SP release from presynaptic terminals, DRYER and CHIAPINELLI (1987) did not observe any effects of SP on ciliary neurons. Thus, the role of SP in calyciform endings, as well as leu-enkephalin localized in both types of terminals, awaits further clarification.

III. Presence of Presynaptic ACh Receptors

Our understanding of the synaptic organization of the fowl ciliary ganglion has been further complicated by evidence demonstrating presynaptic AChRs. The concept that receptors for neurotransmitters may be localized on the presynaptic nerve terminal as well as postsynaptically, is not new (see review in STARKE et al. 1977); indeed, that pre-junctional adrenoreceptors exist and play a role in the feedback modulation of noradrenergic transmission has met with general acceptance (LANGER 1980).

The ciliary ganglion of the fowl is an ideal preparation for the study of putative presynaptic receptors in that not only is it composed of a population of relatively homogeneous cell types, a feature that facilitates biochemical analysis, but also, as previously mentioned, presynaptic terminals ending upon ciliary cells are of such size that intracellular recording is feasible. In the presence of 3-μ*M* atropine, the release of ACh evoked by an increase in the external K^+ concentration, depolarization or orthodromic electrical stimulation of the oculomotor nerve was enhanced; the result was interpreted as demonstrating an inhibitory feedback of ACh on further transmitter release such that, when blocked, release was enhanced.

While recording intracellularly from presynaptic terminals, exogenously applied cholinergic agonists (ACh and carbachol) produced a depolarization of the presynaptic membrane potential, and a resultant decrease in presynaptic action potential amplitude. Concomitantly there was a decrease in nerve terminal input resistance, indicative of increased conductance. This effect could be blocked with atropine. Application of cholinergic agonists to superfusion solutions while the presynaptic membrane potential was being monitored at high sensitivity revealed increases in membrane potential fluctuations similar to those ascribed by KATZ and MILEDI (1970, 1971) to ACh-induced voltage noise. During continuous exposure to ACh, the quantal content of the individual synaptic stimuli measured in postsynaptic cells was seen to be decreased. This effect was shown to be independent of the process of desensitization of postsynaptic receptors in response to such maintained ACh exposure. Thus, considerable evidence exists in support of the hypothesis that presynaptic receptors are present and play a role in synaptic transmission. It should be noted that this population of ciliary ganglion cells is, in the mature domestic fowl, chemically and electrically synaptically coupled. Thus, even a marked reduction in transmitter released during stimulation may not render the synaptic events subthreshold in the light of the fact that in many cells, the electrical coupling potential, which is unaffected by membrane potential, is suprathreshold.

The mechanism underlying these effects is not known. Neither the observation that transmitter release is altered during prolonged depolarization by elevated K^+ concentration nor that the quantal content is reduced during ACh exposure can be explained by a mechanism involving presynaptic depolarization whereby the presynaptic action potential amplitude is reduced, so limiting the amount of transmitter released. The experimental findings are however compatible with an alternative mechanism whereby phosphorylation of a specific protein localized in the synaptic region may partially uncouple the excitation-secretion coupling mechanism (MICHAELSON et al. 1979).

Thus, while many details of the presynaptic mAChR hypothesis, not the least of which is a satisfactory demonstration of a feedback modulatory role in normal ganglionic transmission, remain to be fully elucidated, the ciliary ganglion of the fowl has proved to be a useful preparation for demonstrating that such receptors exist and respond to agonists in a fashion similar to postsynaptic AChRs.

IV. Postsynaptic Receptor Heterogeneity

The third feature to be discussed concerns postsynaptic heterogeneity, i.e., the presence of both nAChRs and mAChRs at this classical nicotinic synapse.

AChRs have historically been distinguished by their pharmacological characteristics as either nicotinic or muscarinic, the former typically eliciting a fast synaptic depolarization at skeletal muscle and ganglionic sites whereas the latter is associated with slow depolarization or hyperpolarizations, often related to changes in cGMP in tissues such as smooth or cardiac muscle (EL-FAKAHANY and RICHELSON 1981). Each receptor category has type-specific agonists and antagonists.

Chemical synaptic transmission in the ciliary ganglion was described as nicotinic, based upon the abolition of the chemically induced EPSP by the application of D-tubocurarine, a known nicotinic antagonist (MARTIN and PILAR 1964). Indeed, attempts to demonstrate an effect of muscarinic blockade, by atropine, on ganglionic transmission induced by orthodromic electrical stimulation revealed no effect. However, it was noted that in response to exogenous application of nicotine or ACh the frequency and duration of action potentials recorded from small nerve filaments (containing an electrophysiologically distinct number of nerve fibres) was increased over control during the presence of eserine, an anticholinesterase (anti-ChE). Furthermore, this potentiation of drug effect persisted after denervation, localizing the effect to the postsynaptic portion of the synapse. The effect of eserine was completely blocked by atropine. These experiments raise the possibility that both nAChRs and mAChRs are present postsynaptically and both may be present on the same cell. However, the atropine-sensitive element appears to play no role in normal ganglionic transmission, being demonstrable only by pharmacological manipulation. Perhaps this receptor class is functionally or structurally extrasynaptic, such that, while being capable of inducing action potentials under pharmacological constraints, it does not play an integral role in normal synaptic transmission. What is further perplexing is the ability of atropine to alter a nicotine-induced effect. Perhaps this results from cross-sensitivities between nicotinic and muscarinic agonists, antagonists and receptors or possibly arises from a secondary release of ACh from another site, possibly the cell soma (JOHNSON and PILAR 1980), that under normal circumstances is not exposed to significant levels of neurotransmitter.

As can be seen by the above three examples, what was once believed to be a purely cholinergic, relatively simply organized parasympathetic ganglion, is in fact a system of great complexity. However, it remains a preparation ideally suited for electrophysiological analysis and may well serve to unravel the complexities of neurotransmitter interactions and modulations.

V. Regulation of ACh Synthesis

The usefulness of the avian ciliary ganglion extends to the study of the biochemistry as well as the electrophysiology of the synapse. In particular, the ciliary nerve-iris muscle has proved suitable for many studies of the regulation of transmitter synthesis.

As mentioned previously, this preparation, in accordance with general principles of autonomic nervous system organization, utilizes ACh as the preganglionic neurotransmitter; indeed, PILAR et al. (1973) determined that most of the ganglionic ACh was localized in the presynaptic nerve terminals. They did note, however, that roughly one-third of the total ACh could be attributed to ganglionic cell sites. Later work showed that, not only could this cell body pool of ACh be labelled by incubation in [^{3}H]ACh, but that, in denervated preparations in which all nerve terminals had degenerated, stimulation of the cell bodies evoked the release of [^{3}H]ACh (JOHNSON and PILAR 1980).

Furthermore, release of [^{3}H]ACh by stimulation (elevated external K^+ cholinergic agonists or antidromic electrical stimulation of the ciliary nerves) was shown to be Ca^{2+}-dependent. The existence of this cellular pool of ACh complicates the interpretation of electrophysiological and biochemical studies of transmitter dynamics – as does the presence, already referred to – of non-cholinergic neuroactive substances, such as leu-enkephalin and SP. Nevertheless, the preparation has proved ideal for the study of enzyme activity levels in relationship to developmental events. Thus, CHIAPINELLI et al. (1976) showed that increases in activity of acetylcholinesterase (AChE) and choline acetyltransferase (ChAT) were temporally correlated with electrophysiological and anatomical indicators of synapse formation. Presumably, contact between neuron and target triggers the enzyme induction necessary for the maintenance of adequate neurotransmitter stores and the termination of transmitter action.

Even more information has been gleaned from the ciliary nerve-iris muscle preparation. This preparation has several advantages for biochemical studies of synaptic transmission: (a) the neurons innervating the iris form a homogeneous population of cholinergic cells that are accessible for *in vivo* manipulation; (b) the iris muscle in birds is striated muscle and synaptic junctions are very similar to the vertebrate neuromuscular junction; (c) the preparation can be isolated very rapidly (within minutes); (d) the densely innervated iris has an extracellular space that equilibrates rapidly with superfusion solutions, thus facilitating accurate measurements of tracer fluxes into and out of the tissue (SUSZKIW et al. 1976; BEACH et al. 1980).

SUSZKIW and PILAR (1976) showed that ciliary nerve terminals in the iris possessed two kinetically distinct choline uptake processes, a finding that had previously been described in brain synaptosomes (HAGA and NODA 1973; DOWDALL and SIMON 1973; YAMAMURA and SNYDER 1973). One is a low-affinity system, which is Na^+-independent and saturable in the millimolar range, and the other, a high-affinity, Na^+-dependent choline uptake (HACU) system that operates in the electromotor range and is tightly coupled to ACh synthesis.

The HACU system has been further characterized and its relationship to membrane depolarization and ACh synthesis has been detailed (VACA and PILAR

1979; PILAR and VACA 1979; BEACH et al. 1980). Conditions that depolarize the presynaptic membrane potential inhibit HACU and the resultant ACh synthesis, indicating that electrochemical gradients for both Na^+ and the choline ion provide an appropriate driving force for uptake into terminals. However, following depolarization, uptake of choline and ACh synthesis is accelerated, a phenomenon also found in synaptosomes (MURRIN and KUHAR 1976).

This and other work provide evidence that supports a model of ACh synthesis whereby nerve terminal depolarization sets in motion a series of events beginning with an influx of Ca^{2+} which, in turn, mediates ACh release, presumably from a vesicular pool. Such release drives the translocation of cytoplasmic ACh into the vesicular pool, stimulating, by mass action, further intraterminal ACh synthesis. This synthesis requires choline; this enters the terminal in a Na^+-dependent fashion down its electrochemical gradient which in turn is dependent upon membrane polarization and Na, K-ATPase activity.

D. Cell and Tissue Culture

The suitability of ciliary ganglion cells for study of neuronal function is not limited to *in vivo* studies and examination of isolated, intact ganglia but extends also to partially and fully dissociated preparations.

BRENNER and MARTIN (1976) examined electrophysiological characteristics and neuronal sensitivity to ACh in normal and axotomized cells in partially dissociated ciliary ganglia. After brief incubation with collagenase, small clusters of cells could be teased apart with fine needles and viewed with NOMARSKI interference contrast optics for intracellular recording and iontophoretic ACh application.

In the late 1970s, a new-found ability to maintain isolated ciliary ganglion cells in culture stimulated the use of such preparations to answer a number of questions concerning development in autonomic ganglia in general and parasympathetic ganglia in particular (NISHI and BERG 1977, 1979; Chap. 5, this volume). Numerous studies helped to define the criteria for successful long-term survival in culture and described refinements in harvesting techniques (TUTTLE et al. 1980a; NISHI and BERG 1979; COLLINS 1978a, b). Furthermore, techniques already utilized in the study of intact ganglia, such as intracellular recording, electron microscopy, and biochemical analyses, were found to be applicable to cultured cells.

Ciliary ganglion cells in culture not only make structural contact with other neurons and muscle cells but also form functional contacts (MARGIOTTA and BERG 1982; NISHI and BERG 1977). Such studies have given new information on the formation of synaptic connections and, by observing events that do not normally occur *in vivo* (e.g. the formation of interneuronal synaptic connections), provide opportunities for speculation concerning those constraints *in vivo* which prohibit interganglionic connections. Interestingly, cultured neurons differentiate sufficiently in appropriate media to be capable of releasing ACh from growth cones in the absence of target myotubes (HUME et al. 1983).

An area of great interest concerns possible cholinergic neuronotrophic factors. For the survival of sympathetic ganglion cells in culture, nerve growth factor

(NGF) is an essential constituent of the growth medium. The successful culture of ciliary ganglion cells provides a bioassay for detecting a cholinergic analogue of NGF.

In their initial report, NISHI and BERG (1979) reported the increased survival of cultured cells when embryo extract supplemented the culture medium. Without this, ganglion cells required co-culture with myotubes for survival, suggesting that some trophic factor existed which promoted prolonged survival. Several such factors have been reported, the first by ADLER et al. (1979), who found an intraocular substance which promoted survival. TUTTLE et al. (1980a) reported the purification of a protein fraction of molecular mass 12 kDa that supported long-term cell survival.

Other workers have isolated survival factors from ox heart (BONYHADY et al. 1982) and from embryonic eye extracts in which two factors affect neuronal function, one of molecular mass 20 kDa which stimulated growth while a second of molecular mass 50 kDa appears to promote enzyme synthesis (NISHI and BERG 1981). Indeed, there appears to be something unique about the target tissue that not only permits extended survival but allows further differentiation toward the mature functional state. Thus, TUTTLE et al. (1983) found that although neurons grow in myotube-conditioned medium or with fibroblasts, conditions that promote survival, it is only in the presence of myotubes themselves that ciliary neurons appear to be specifically induced toward maturity as measured by enhanced synthesis of ACh from [^{3}H]choline and enhanced release of neurotransmitter. For further discussion of cholinergic synaptogenesis the reader is referred to Chap. 5.

E. Concluding Remarks

This review has emphasized some of the features of the ciliary ganglion that have made it an attractive preparation for studies of neuronal development, morphology, synaptic biochemistry, electrophysiology and isolated cells in tissue culture.

As mentioned in the introduction, this model neuronal system, presumably purely cholinergic, has recently been shown to be considerably more complex, with new features of neuronal development and synaptic function to be resolved. Yet it remains a system that readily lends itself to multidisciplinary analysis of experimental problems with substantial previous information available as background for new ideas and novel approaches.

References

Adler R, Landa K, Manthorpe M, Varon S (1979) Cholinergic neuronotropic factors: II. Intraocular distribution of trophic activity for ciliary neurons. Science 204:1434–1436
Beach RL, Vaca K, Pilar G (1980) Ionic and metabolic requirements for high-affinity choline uptake and acetylcholine synthesis in nerve terminals at a neuromuscular junction. J Neurochem 34:1387–1398
Bonyhady RE, Hendry IA, Hill CE (1982) Reversible dissociation of a bovine cardiac factor that supports survival of avian ciliary ganglionic neurones. J Neurosci Res 7:11–21
Brenner HR, Martin AR (1976) Reduction in acetylcholine sensitivity of axotomized ciliary ganglion cells. J Physiol 260:159–175

Cantino D, Mugnaini E (1974) Adrenergic innervation of the parasympathetic ciliary ganglion in the chick. Science 185:279–281
Cantino D, Mugnaini E (1975) The structural basis for electrotonic coupling in the avian ciliary ganglion: a study with thin sectioning and freeze-fracturing. J Neurocytol 4:505–536
Carpenter MB, Peter P (1970) Accessory oculomotor nuclei in the monkey. J Hirnforsch 12:405–418
Chiapinelli V, Giacobini E, Pilar G, Uchimura H (1976) Induction of cholinergic enzymes in chick ciliary ganglion and iris muscle cells during synapse formation. J Physiol 257:749–766
Collins F (1978a) Axon initiation by ciliary neurons in culture. Dev Biol 65:50–57
Collins F (1978b) Induction of neurite outgrowth by a conditioned-medium factor bound to the culture substratum. Proc Natl Acad Sci USA 75:5210–5213
Cowan MW, Wenger E (1968) Degeneration in the nucleus of origin of the preganglionic fibers to the chick ciliary ganglion following early removal of the optic vesicle. J Exp Zool 168:105–124
de Lorenzo AJD (1960) The fine structure of synapses in the ciliary ganglion of the chick. J Biophys Biochem Cytol 7:31–36
de Lorenzo AJD (1966) Electron microscopy: tight junctions in synapses of the ciliary ganglion. Science 152:76–78
Dowdall MJ, Simon EJ (1973) Comparative studies on synaptosomes: uptake of [*N-Me*-^{3}H]choline by synaptosomes from squid optic lobes. J Neurochem 21:969–982
Dryer SF, Chiapinelli VA (1987) Properties of choroid and ciliary neurons in the avian ciliary ganglion and evidence for subtance P as a neurotransmitter. J Neurosci (in press)
Ehinger B (1967) Adrenergic nerves in the avian eye and ciliary ganglion. Z Zellforsch 82:577–588
El-Fakahani E, Richelson E (1981) Effect of lanthanides on muscarinic acetylcholine receptor function. Mol Pharmacol 19:282–290
Erichsen JT, Karten HJ, Eldred WD, Brecha NC (1982a) Localization of substance P-like and enkephalin-like immunoreactivity within preganglionic terminals of the avian ciliary ganglion: light and electron microscopy. J Neurosci 2:994–1003
Erichsen JT, Reiner A, Karten HJ (1982b) Co-occurrence of substance P-like and leu-enkephalin-like immunoreactivities in neurones and fibres of avian nervous system. Nature 295:407–410
Gilbert DL, Ehrenstein G (1969) Effect of divalent cations on potassium conductance of squid axons: determination of surface charge. Biophys J 9:447–463
Haga T, Noda H (1973) Choline uptake systems of rat brain synaptosomes. Biochim Biophys Acta 291:564–575
Hamburger V, Hamilton HL (1951) A series of normal stages in the development of the chick embryo. J Morphol 86:49–92
Hess A (1965) Developmental changes in the structure of the synapse on the myelinated cell bodies of the chick ciliary ganglion. J Cell Biol 25:1–19
Hume RI, Role LW, Fischbach GD (1983) Acetylcholine release from growth cones detected with patches of acetylcholine receptor-rich membranes. Nature 305:632–634
Johnson DA, Pilar G (1980) The release of acetylcholine from post-ganglionic cell bodies in response to depolarization. J Physiol 299:605–619
Katz B, Miledi R (1970) Membrane noise produced by acetylcholine. Nature 226:962–963
Katz B, Miledi R (1971) Further observations on acetylcholine noise. Nature (New Biology) 232:124–126
Landmesser L, Pilar G (1970) Selective reinnervation of two cell populations in the adult pigeon ciliary ganglion. J Physiol 211:203–216
Landmesser L, Pilar G (1974a) Synapse formation during embryogenesis on ganglion cells lacking a periphery. J Physiol 241:715–736
Landmesser L, Pilar G (1974b) Synaptic transmission and cell death during normal ganglionic development. J Physiol 241:738–749
Landmesser L, Pilar G (1978) Interactions between neurons and their targets during *in vivo* synaptogenesis. Fed Proc 37:2016–2022
Langer SZ (1980) Presynaptic regulation of the release of catecholamines. Pharmacol Rev 32:337–362

Margiotta JF, Berg DK (1982) Functional synapses are established between ciliary ganglion neurones in dissociated cell culture. Nature 296:152–154

Martin AR, Pilar G (1963) Dual mode of synaptic transmission in the avian ciliary ganglion. J Physiol 168:443–463

Martin AR, Pilar G (1964) Presynaptic and post-synaptic events during post-tetanic potentiation and facilitation in the avian ciliary ganglion. J Physiol 175:17–32

Marwitt R, Pilar G, Weakly JN (1971) Characterization of two ganglion cell populations in avian ciliary ganglia. Brain Res 25:317–334

Meriney SD, Pilar G (1987) Cholinergic innvervation of the smooth muscle in the choroid coat of the chick eye and its development. J Neurosci 7:3827–3839

Michaelson DM, Avissar S, Kloog Y, Sokolovsky M (1979) Mechanism of acetylcholine release: possible involvement of presynaptic muscarinic receptors in regulation of acetylcholine release and protein phosphorylation. Proc Natl Acad Sci USA 76:6336–6340

Murrin LC, Kuhar MJ (1976) Activation of high-affinity choline uptake *in vitro* by depolarizing agents. Mol Pharmacol 12:1082–1090

Narayanan CH, Narayanan Y (1976) An experimental inquiry into the central source of preganglionic fibers to the chick ciliary ganglion. J Comp Neurol 166:101–110

Narayanan CH, Narayanan Y (1978) On the origin of the ciliary ganglion in birds studied by the method of interspecific transplantation of embryonic brain regions between quail and chick. J Embryol Exp Morphol 47:137–148

Nishi R, Berg D (1977) Dissociated ciliary ganglion neurons *in vitro*: survival and synapse formation. Proc Natl Acad Sci USA 74:5171–5175

Nishi R, Berg DK (1979) Survival and development of ciliary ganglion neurones grown alone in cell culture. Nature 277:232–234

Nishi R, Berg DK (1981) Two components from eye tissue that differentially stimulate the growth and development of ciliary ganglion neurons in cell cultue. J Neurosci 1:505–513

Pilar G, Tuttle JB (1982) A similar neuronal system with a wide range of uses: the avian ciliary ganglion. In: Hanin I, Goldberg AM (eds) Progress in cholinergic biology: model cholinergic synapses. Raven, New York, pp 213–247

Pilar G, Vaca K (1979) Regulation of acetylcholine synthesis in cholinergic nerve terminals. Brain Res 49:97–106

Pilar G, Jenden DJ, Campbell B (1973) Distribution of acetylcholine in the normal and denervated pigeon ciliary ganglion. Brain Res 49:245–256

Pilar G, Landmesser L, Burstein L (1980) Competition for survival among developing ciliary ganglion cells. J Neurophysiol 43:233–254

Politoff AL, Socolar JJ, Loewenstein WR (1969) Permeability of a cell membrane junction. Dependence on energy metabolism. J Gen Physiol 53:498–515

Suszkiw JB, Pilar G (1976) Selective localization of a high affinity choline uptake system and its role in ACh formation in cholinergic nerve terminals. J Neurochem 26:1133–1138

Suszkiw JB, Beach RL, Pilar GR (1976) Choline uptake by cholinergic neuron cell somas. J Neurochem 26:1123–1131

Takahashi K (1967) Special somatic spine synapses in the ciliary ganglion of the chick. Z Zellforsch 83:70–75

Takahashi K, Hama K (1965) Some observations on the fine structure of the synaptic area in the ciliary ganglion of the chick. Z Zellforsch 67:174–184

Tuttle JB, Greene C, Pilar G, Lucas-Lenard J (1980a) Neuronal survival protein (NSP): a substance supporting survival of parasympathetic neurons in cell culture. Soc Neurosci Abstr 6:377

Tuttle JB, Suszkiw JB, Ard M (1980b) Long-term survival and development of dissociated parasympathetic neurons in culture. Brain Res 183:161–180

Tuttle JB, Vaca K, Pilar G (1983) Target influence on ^{3}H-ACh synthesis and release by ciliary ganglion neurons *in vitro*. Dev Biol 97:255–263

Vaca K, Pilar G (1979) Mechanisms controlling choline transport and acetylcholine synthesis in motor nerve terminals during electrical stimulation. J Gen Physiol 73:605–628

Yamamura HI, Snyder SH (1973) High affinity transport of choline into synaptosomes of rat brain. J Neurochem 21:1355–1374

Part II Biology of the Cholinergic Synapse

CHAPTER 4

Phylogeny of the Cholinergic Synapse *

K. WÄCHTLER

A. Introduction

Contacts between neurons or between neurons and effector cells with structures typical of chemical synapses have been described in nervous systems of all phylogenetic levels including the nerve nets of sedentary coelenterates (WESTFALL et al. 1971). From numerous studies of model systems such as vertebrate or arthropod muscle endplates or identified neurons of invertebrates (squid, snail, leech, locust) and vertebrates (lamprey, fish, domestic fowl, rat) it can be concluded that chemical synaptic transmission is basically similar in all nervous systems (AKERT and WASER 1969; PAPPAS and PURPURA 1972; KANDEL and SCHWARTZ 1985). Several transmitter substances in various combinations are consistently involved. Their list at present is probably far from being complete (COOPER et al. 1974; MCGEER et al. 1978; BUIJS et al. 1982). All comparative observations suggest that there is no nervous system specialized on the production and use of only one transmitter. Even at the most fundamental level of neural organization nerve cells with different chemical specializations coexist (KOOPOWITZ and KEENAN 1982), so that the well known transmitter heterogeneity found in all advanced nervous systems is likely to date back to the very origin of neurons (FLOREY and MICHELSON 1973; SAKHAROV 1974).

Acetylcholine (ACh) is one of the transmitter substances with a phylogenetically long tradition (GERSCHENFELD 1973; KOOPOWITZ and KEENAN 1982) (Fig. 1).

Although present as a byproduct of choline metabolism in bacteria, higher plants and protozoa (WHITTAKER 1963a, b; FLOREY and MICHELSON 1973; SMALLMAN and MANECKJEE 1981), it is unlikely that ACh was already present in the earliest chemical synapses. Coelenterates, notably sedentary ones like *Hydra* and other polyps, give an idea of how the first nervous systems might have been organized, which is why the nerve nets of these animals have attracted so many investigators (BULLOCK and HORRIDGE 1965; LENTZ 1968; SCHALLER and GIERER 1973). Electron microscope studies reveal that in the nervous system of *Hydra* all the typical features of chemical synapses can be recognized (WESTFALL et al. 1971). It is unlikely however that these synapses are cholinergic. Although there are some reports of cholinesterase activity in coelenterate tissue homogenates (BULLOCK and NACHMANSOHN 1942; ERZEN and BRZIN 1978) all

* Dedicated to the memory of two inspiring academic teachers at the University of Kiel: A. Remane (1898–1976) and W. Bargmann (1906–1978).

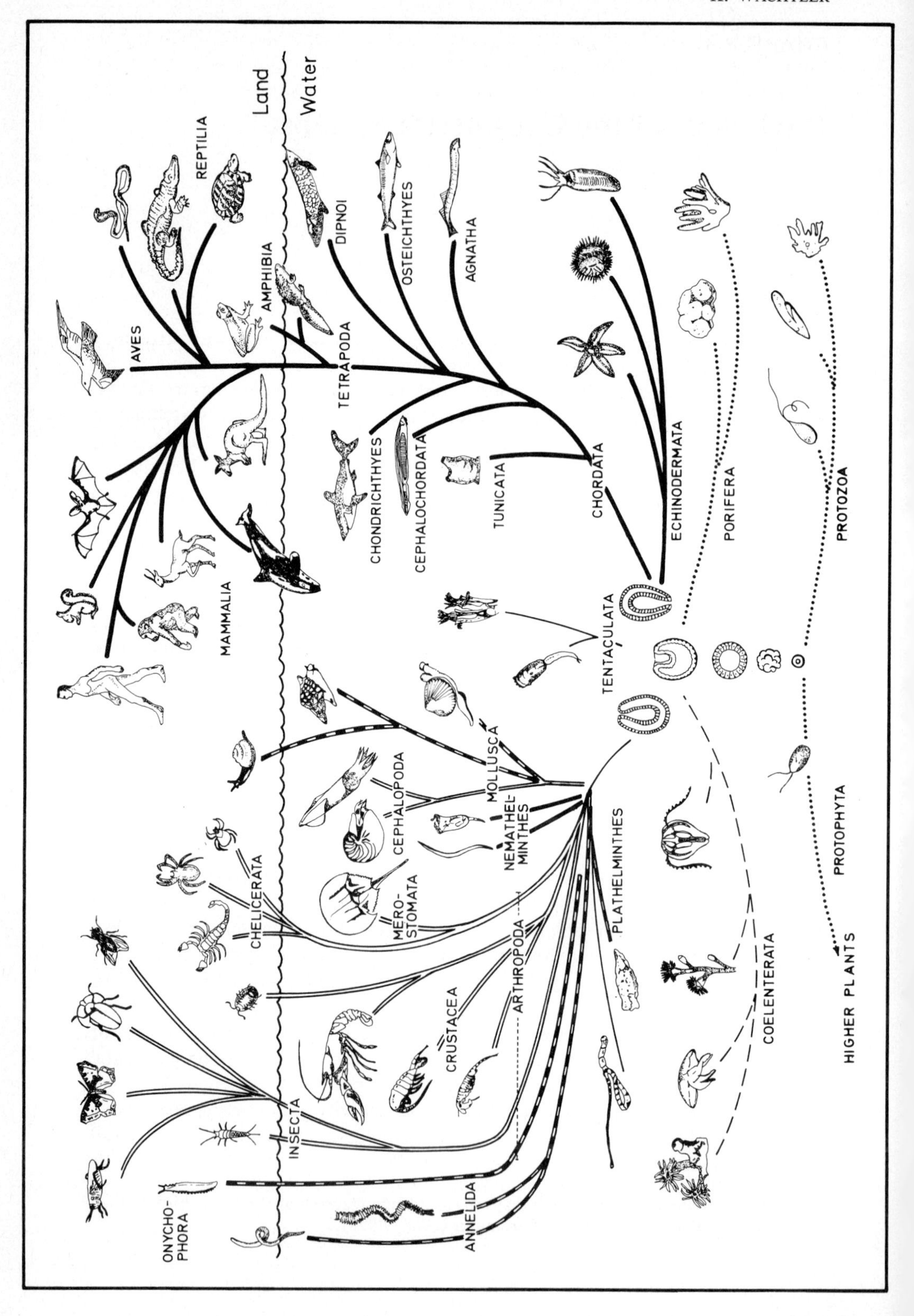
Land
Water
REPTILIA
AVES
AMPHIBIA
DIPNOI
OSTEICHTHYES
AGNATHA
TETRAPODA
CHONDRICHTHYES
CEPHALOCHORDATA
TUNICATA
CHORDATA
ECHINODERMATA
PORIFERA
PROTOZOA
MAMMALIA
TENTACULATA
MOLLUSCA
CEPHALOPODA
NEMATHEL-
MINTHES
PROTOPHYTA
CHELICERATA
MERO-
STOMATA
PLATHELMINTHES
ARTHROPODA
CRUSTACEA
COELENTERATA
HIGHER PLANTS
INSECTA
ONYCHO-
PHORA
ANNELIDA

attempts to extract ACh have been negative, at least in *Hydra* (K. WÄCHTLER unpublished; and C.H. SCHALLER personal communication). For the more advanced nervous systems of free-swimming coelenterates, the medusae, there are no positive reports either (FISCHER 1971; MICHELSON 1973a).

In the nervous systems of Bilateria, however, cholinergic neurons are present in all phyla studied (Bilateria are all those animals in Fig. 1 above the Coelenterata and the Porifera). Details are given in several comparative reviews (WHITTAKER 1963b; TREHERNE 1966; FLOREY 1967a, b; FISCHER 1971; MICHELSON 1973a, b; FLOREY and MICHELSON 1973; GERSCHENFELD 1973; SATTELLE 1980; BURNSTOCK 1981; WALKER 1982). The most primitive Bilateria to possess cholinergic innervation are the echinoderms. The longitudinal muscles of the body walls of sea cucumbers (e.g. *Stichopus regalis*, WHITTAKER 1963b) are extremely sensitive to applied ACh and the effect is potentiated by eserine. The muscles produce an ACh-like substance on stimulation (BACQ 1939). The oesophagus of the sea urchin, *Strongylocentrotus drobachiensis*, is also sensitive to ACh (MCLENNAN 1959). The flat worms are the most primitive animals to show cephalization of the nervous system: *Planaria maculata* has a relatively high concentration of ACh in its anterior nervous system; this diminishes caudally (WELSH 1946). The dorsal muscle of the leech contracts in response to very low doses of ACh and provides a classical preparation for the assay of ACh (SZERB 1961). Recent observations on cholinergic components in the nervous system of nematodes (ROZHKOVA et al. 1980) and rotifers (NOGRADY and ALAI 1983) fit well into this general picture.

The object of this chapter is to examine whether the obvious evolutionary changes in the structure and complexity of nervous systems are parallelled by changes in the cholinergic synapse and its distribution pattern.

B. Variation at the Synaptic Level

I. Parameters Considered

The cholinergic synapse has all the cytological specializations found in most other chemical synapses (PETERS et al. 1970; PAPPAS and PURPURA 1972; KANDEL and SCHWARTZ 1985). Some variability in morphology, chemical characteristics and physiology may be observed both from neuron to neuron and between species (LANDAU et al. 1981), so that it is difficult to detect changes of possible evolutionary significance. Species differences reported for cholinergic neurons mainly relate to ACh synthesis, the termination of the action of ACh, and the characteristics of receptors for ACh.

◄———

Fig. 1. Simplified phylogenetic diagram of the animal kingdom (based on one by STÖRMER and HEINTZ, Museum of Palaeontology, Oslo). The main trends in the use of ACh as a neurotransmitter are indicated: ..., organisms without neurons; – – –, no cholinergic neurons; ———, ACh system poorly investigated; ▬▬, ACh system predominantly in the motor system; ═══, ACh system mainly sensory; ▭▭, ACh used in both motor and sensory system

II. Synthesis of ACh

ACh is a simple molecule synthesized in one step from choline and acetylcoenzyme A by the enzyme choline acetyltransferase (ChAT). Since the introduction of reliable microassays (HILDEBRAND et al. 1974; FONNUM 1974) this enzyme has been measured in nervous tissue from a variety of invertebrates and vertebrates (HILDEBRAND et al. 1974; MCCAMAN and MCCAMAN 1976; SUGDEN and NEWSHOLME 1977; OSBORNE 1978; BREER 1981a, b; WÄCHTLER 1981, 1983a, b). Comparative analyses, notably those using immunocytochemical techniques with monoclonal antibodies (LEVEY and WAINER 1982), show that the enzyme is basically similar in nervous tissues from widely separated phyla. According to these authors some differences in antibody response can however be observed among mammalian species. Differences in the sensitivity towards inhibitors of ChAT from invertebrates and vertebrates have also been reported (ROSSIER 1977).

As ChAT is consistently found in cholinergic neurons it has been much used as a marker enzyme. Where ACh occurs, ChAT and acetylcholinesterase (AChE) are present too, though AChE alone is not always a reliable marker. The quantitative correlation of these substances can however vary considerably, so that the tissue level of one of them can scarcely be predicted from that of the others. Discrepancies between ChAT and ACh levels have been observed, *inter alia*, in gastropods (MCCAMAN and MCCAMAN 1976), arthropods (HILDEBRAND et al. 1974; BREER 1981b; SUKUMAR et al. 1983) and vertebrates (WÄCHTLER 1981; MARCHI et al. 1981; HOEHMANN et al. 1983).

The degree of fluctuation in vertebrates was studied in 27 species selected in an attempt to cover the main evolutionary levels from cyclostomes to mammals (WÄCHTLER 1981). The ChAT activity determined in brain homogenates varied from the species with the highest activity to that with the lowest by 160:1 whereas the corresponding ratio for AChE activity was 23:1 and that for the ACh content 15:1. Large fluctuations in all three cholinergic components were observed in both unrelated and closely related species so that no correlation allowing an evolutionary interpretation of low or high values is found in a given species.

The production of ACh is consistently linked to a high-affinity choline uptake (HACU) system (DOWDALL and SIMON 1973; YAMAMURA and SNYDER 1973; SIMON and KUHAR 1976; KUHAR and MURRIN 1978; JOPE 1979; BEACH et al. 1980; see also Chap. 19, this volume). This has been used as an additional cholinergic marker (TAKANO et al. 1980). Good correlation of HACU with ACh and ChAT levels has been found in the nervous systems of various species (DOWDALL and SIMON 1973; DOWDALL et al. 1975; WHITTAKER and DOWDALL 1975; TAKANO et al. 1980; BREER 1982). Here, again, no evolutionary trends can be detected.

III. Termination of the Action of ACh

The mechanisms by which released ACh can be removed from the synaptic cleft have been studied from the comparative point of view by SAKHAROV and TURPAEV (1967) and TURPAEV and SAKHAROV (1973). Based mainly on observations on molluscs, these authors describe three alternative removal mechanisms for ACh which they interpret as evolutionary steps. These are:

- Diffusion of ACh to extrajunctional spaces as is the case with other transmitter substances (BUIJS et al. 1982; KANDEL and SCHWARTZ 1985).
- Desensitization of ACh receptors by an ACh-dependent release of antagonists.
- The degradation of ACh by the action of AChE.

According to TURPAEV and SAKHAROV (1973) the most primitive stage of development is characterized by (a) low ChE activity in the neuropil augmented by some activity in the body fluids and (b) long lasting postsynaptic potentials largely unchanged by ChE blockers. This stage of evolution is exemplified by some bivalves and gastropods.

A further stage is observed in the cholinergic innervation of the heart of bivalves (*Anodonta*). It involves the desensitization of the ACh receptors by an ACh-induced release of ATP; this exerts an antagonistic effect by shortening the lifetime of the receptor-transmitter interaction (SAKHAROV and NISTRATOVA 1963). A similar mechanism has been described in the frog heart (PUTINSEVA and TURPAEV 1959; MAGLEBY and PALLOTTA 1981). In cephalopods the inactivation of ACh is achieved by considerable AChE activities so that the third evolutionary level of molluscan cholinergic synaptic organization would correspond to the situation found in the vertebrate nervous system. So far there is no information on similar trends in synaptic perfection to be observed in animal phyla other than molluscs. A recent observation made in the author's laboratory (K.H. KÖRTJE and K. WÄCHTLER unpublished) may be relevant to the above observations on the removal of ACh from the synaptic cleft by diffusion, namely that larval lampreys (ammocoetes) show no disturbance in their swimming activity if kept in solutions of diisopropylphosphorofluoridate (DFP) concentrated enough to inhibit AChE activity completely. As the AChE activity in lampreys is lower than that of most other vertebrates (WÄCHTLER 1974) it remains to be seen whether AChE in lampreys has the same essential function as in the muscular innervation of higher vertebrates.

IV. Characteristics of ACh Receptors

The effect produced by a cholinergic terminal on the postsynaptic cell depends on the interaction of released ACh with its receptors. From this interaction a broad spectrum of postsynaptic effects may result leading finally to either an excitatory or an inhibitory message. Depending on the type of receptor, the influence may be a direct one on ion channels (ionotropic) or indirect by changing the metabolic activities of the postsynaptic cell (metabotropic, MCGEER et al. 1978; KANDEL and SCHWARTZ 1985; or, more correctly, metabolotropic, WHITTAKER 1979).

The existence of several ACh receptors in both post- and presynaptic locations (ROSSINI 1981; KOKETSU and YAMADA 1982), the fact that one receptor type may control different ion channels (KANDEL and SCHWARTZ 1985) and the observation that two terminals of the same neuron may innervate a postsynaptic cell via two kinds of ACh receptors (KANDEL and GARDNER 1972; HORN and DODD

1983) well illustrate the functional versatility of cholinergic synapses found in both vertebrates and invertebrates. The diversity of possible receptor activities may not be found in any one species; it represents, rather, the range of different types encountered in phylogeny. It is, therefore, impossible to classify any one feature of an ACh receptor as phylogenetically 'basic' or 'derived'.

On the basis of their drug selectivity nicotinic (nAChR) and muscarinic (mAChR) ACh receptors (DALE 1914) can be distinguished in neurons of both vertebrates and invertebrates. Their pharmacological properties and chemical structure have been analysed in detail (KERKUT et al. 1970; YAMAMURA et al. 1984; BIRDSALL et al. 1980; SATTELLE 1980; WAMSLEY et al. 1980; PEPER et al. 1982; LINDSTROM 1986; see also Chaps. 9 and 10, this volume). The characteristics of both receptor types seem to be largely unchanged through evolution. Thus VENTER et al. (1984) found very similar reactions against monoclonal antibodies in preparations of mAChRs from insect heads and human central nervous system (CNS).

Comparative analysis of nAChRs of different sources also reveals much similarity in both properties and molecular structure. Its four component polypeptide subunits have been found in vertebrates (RAFTERY et al. 1980) and insects (BREER et al. 1985). From the sequence homology of the four subunits in *Torpedo* electrocytes it has been concluded (NODA et al. 1983a, b) that these subunits originated from an ancestral gene (possibly coding for a simple ion channel unresponsive to ACh) which duplicated twice, giving rise to the genes responsible for the α-, β-, γ- and δ-subunits. The structure of these subunits has been conserved over long periods of evolutionary history (CONTITROCONTI et al. 1982; STROUD and FINER-MOORE 1985). Sequence studies of the subunits indicate that the α-unit shows less evolutionary changes than the others, which is why it is thought to be the most important part of the nAChR (STROUD and FINER-MOORE 1985).

At which phylogenetic level mAChRs and nAChRs became differentiated is not known; according to MICHELSON (1973b) observations on echinoderms (*Holothuria*) suggest that cholinergic receptors might have gone through a 'polyvalent' stage which responded to both nicotinic and muscarinic drugs and that such a stage might have been succeeded by one in which receptor selectivity increased to a level typical for highly evolved nervous systems. However, observations with other invertebrates like sea-squirts (*Ascidia*) are inconsistent with this view. Here the innervation of the body wall is mediated by nAChRs in one species (*Ciona intestinalis*) and at least in part by mAChRs in another (*Tethyum auranticum*) (FLOREY 1967b; MICHELSON 1973b). Within the vertebrates the purely nicotinic innervation of the heart in the lamprey is replaced by a purely muscarinic one in all other vertebrates (CAMPBELL and BURNSTOCK 1973). However, before classifying this change as an evolutionary progression one has to note that muscarinic activities have also been found in the hearts of tunicates (KRIEBEL 1968) and crustacea (FLOREY and RATHMEYER 1980).

Information on ACh receptors in the gastrointestinal tract of invertebrates is not very detailed so far. mAChRs appear to play a role in this system early on in evolution. They are reported from annelids (ROZHKOVA 1973) and crustacea (MARDER 1977) and occur in visceral muscles of vertebrates from cyclostomes to mammals (CAMPBELL and BURNSTOCK 1973). The innervation of body wall and appendages is mediated by nicotinic ACh receptors (nAChRs) in almost all phyla with the interesting exception of arthropods (see below) (FISCHER 1971; ROZHKOVA et al. 1980; MARSDEN et al. 1981).

In the CNS ACh receptors have been studied intensively in annelids (MULLER et al. 1981), molluscs (LEVITAN and TAUC 1972; GERSCHENFELD 1973), insects (MEYER and EDWARDS 1980; KEHOE and MARDER 1976; SATTELLE 1980) and vertebrates (YAMAMURA et al. 1984; BIRDSALL et al. 1978, 1980; BARLOW et al. 1980; BARNARD and DOLLY 1982). The two types of ACh receptor vary considerably in relative abundance in the nervous systems of different species. In insects central nAChRs are estimated to be ten times more numerous than mAChRs (DUDAI 1978, 1979; SATTELLE 1980; BREER 1981 a, b; AGUILAR and LUNT 1984). This ratio is reversed in the vertebrate brain (YAMAMURA et al. 1974; SEGAL et al. 1978; MCGEER et al. 1978; SALVATERRA and FODERS 1979). Difficulties in the interpretation of comparative data may result from the fact that in some invertebrates, at least in molluscs and insects, a third type of ACh receptor has been described in addition to muscarinic and nicotinic ones which has no counterpart in vertebrates (KEHOE 1972a, b; MICHELSON 1973b; SATTELLE 1980; ROSSINI 1981).

As MICHELSON (1973b) points out, it can by and large be stated that nAChRs prevail in systems where rapid response is needed as in many striated muscles, whereas muscarinic receptors play an important role at neurons with modulatory functions as in the innervation of the heart and viscera and in the CNS notably at associative sites. But little can in fact be said about evolutionary changes in ACh receptors. Most of the comparative data on receptor function lead to the conclusion that cholinergic and cholinoceptive neurons fully equipped with several types of ACh receptors and specific ion channels emerge early in evolution, and that there is no obvious arrangement typical for a certain group of animals or a certain evolutionary level (CAMPBELL and BURNSTOCK 1973; TRIGGLE and TRIGGLE 1976).

C. Qualitative and Quantitative Changes in Cholinergic Innervation

I. Qualitative Changes

In all nervous systems, where they are present, cholinergic neurons constitute a relatively small proportion of the total neuron population with characteristic distribution patterns (WHITTAKER 1965; FIBINGER 1982). The contribution of these neurons to nervous function can vary from one animal phylum to another and within a given nervous system. Both qualitative and quantitative changes may be observed in comparative studies. Cholinergic neurons may dominate the input into the nervous system in some phyla but be replaced in another by neurons utilizing another transmitter. On the other hand, once present in a given system in a particular role it may basically keep this but vary its contribution by quantitative changes.

The most striking case of a qualitative change in function (summarized in Fig. 1) is the observation that the cholinergic system is involved in the innervation of the motor system in many invertebrates and all vertebrates but not in arthropods where the skeletal motor system is activated by glutamate and inhibited by GABA (FLOREY 1967a; GERSCHENFELD 1973). This is a particularly striking example of phylogenetic discontinuity in that arthropods are related to annelids and onychophora both of which use ACh as the transmitter in the motor system (FLOREY and FLOREY 1965; MULLER et al. 1981). What makes the arthropods different from other phyla and so successful is the development of a complicated exoskeleton with unusual tegumental appendages but there is no plausible reason why cholinergic innervation is no longer suited to the new motor requirements implied by this structural progress.

In arthropods and some other invertebrates ACh is the main transmitter at synapses mediating the sensory input from the periphery (BARKER et al. 1972). This, in contrast to the motor innervation, is what arthropods do inherit from the annelids which use ACh both in the sensory and the motor system (FLOREY and MICHELSON 1973; MARSDEN et al. 1981; GARDNER and WALKER 1982). This leads particularly in species with pronounced visual systems like flies or cephalopods to high concentrations of cholinergic components in the brain centres responsible for processing the visual input (LEWIS 1953; JONES et al. 1981; DOWDALL and WHITTAKER 1973; DOWDALL 1974). Thus the optic lobe of the octopus has one of the highest ACh contents of any neural tissue ($2\ \mu mol\, g^{-1}$ tissue). Interestingly, neuronal ACh was first isolated and unequivocally chemically identified from this tissue by BACQ and MAZZA in 1935.

In the vertebrate visual system a cholinergic contribution to retino-tectal projections is confined to lower forms, such as amphibia and teleosts (OSWALD et al. 1979; OSWALD and FREEMAN 1980). In mammals the cholinergic components found in mesencephalic, thalamic and cortical visual centres are independent of the retinal input (FIBINGER 1982; BIGL and SCHOBER 1977). This indicates a qualitative shift of transmitter use within the vertebrates for which no functional explanation can be given. Although there are cholinergic afferents to the olfactory bulb and olfactory epithelium (MACRIDES et al. 1981; HEDLUND and SHEPHARD 1983), there are no cholinergic synapses at the entrance of the olfactory nerve. This is also true for the auditory nerve in vertebrates.

In their comparative review on the vertebrate peripheral nervous system (PNS) CAMPBELL and BURNSTOCK (1973) stress the overall similarity of cholinergic innervation patterns in vertebrates. On the other hand they provide interesting examples of qualitative changes that have taken place in the course of evolution. These include variations in the innervation of the gastrointestinal tract, gall bladder, lungs, urinary bladder and reproductive tract. Whereas cholinergic innervation of visceral smooth muscle in lower vertebrates is provided by the spinal autonomic outflow it originates in reptiles, birds and mammals mainly in the cranial and sacral parasympathic. The reason for this shift of cholinergic function from the sympathetic to the parasympathetic system is not known. An increasing separation of a sacral component in the parasympathetic system is another example of a qualitative change in the cholinergic system of vertebrates.

II. Quantitative Changes

As systematic investigations of a larger number of species with well-defined phylogenetic positions are lacking in invertebrates, comparative observations on quantitative changes in the distribution of cholinergic components are confined to vertebrates.

In the vertebrate brain evolutionary progress relates to all subsystems. As nearly all phylogenetic levels from lampreys to primates are represented in the contemporary fauna (Fig. 2) the study of structural changes in the common ground-plan has fascinated many anatomists with an interest in evolution (KUHLENBECK 1927, 1967–1978; ARIENS KAPPERS et al. 1936; STARCK 1962; KARTEN 1969; EBBESON 1980; NORTHCUTT 1981). As SAKHAROV (1974) has remarked, comparative neurochemistry can provide information on the regional preferences of transmitter systems in the course of evolution.

Neurons producing ACh are present at all levels in relative small quantities and in an uneven distribution. By comparing ChAT levels in the main brain parts HEBB and RATKOVIĆ (1964) found a shift of cholinergic activities from caudal to rostral with increasing phylogenetic rank. If a larger number of species is investigated this trend is less consistent (WÄCHTLER 1980, 1981). A clearly descending caudorostral gradient in ACh distribution can, however, be seen in some ancient vertebrates, such as the primitive fish *Petromyzon*, *Polypterus* and *Lepidosiren* and the reverse distribution pattern in some mammals, for instance the mouse (Fig. 3) (WÄCHTLER 1980; KÖRTJE 1987).

In spinal cord and medulla oblongata the ACh system is typically located in motor neurons (Fig. 4). In the cerebellum it is poorly represented in most vertebrates. In a few species, the electric ray (*Torpedo*), *Lepidosiren* and lizard (*Lacerta*) the ACh content is relatively high (Fig. 3). Whether this is associated with an increase of cholinergic cerebellar afferents is not known. The mesencephalic tectum has high ACh values in most vertebrates. In several teleosts, amphibia and reptiles with well developed visual systems the cholinergic activities in this region exceed those of the medulla oblongata. In the diencephalon the contribution of the cholinergic system varies considerably. It is very high in the hagfish (*Myxine*), elasmobranchs and some teleosts but also in birds and some mammals so that there are no hints for changes that could be correlated with phylogeny.

The only brain region where something like an evolutionary trend can be made out is the telencephalon. Here with the exception of *Myxine* and elasmobranchs low values for all cholinergic components are found in lower vertebrates while in the telencephalon of reptiles, birds and mammals a quantitative increase in the ACh system is observed (WÄCHTLER 1978, 1980, 1982, 1983a, b) (Fig. 5). To some extent this late appearance of cholinergic components in the telencephalon has its counterpart in the ontogenesis of vertebrates (SHEN et al. 1955; SILVER 1974; PENG LOH 1976; SCHLESINGER 1981).

The study of the cholinergic system in homologous brain structures of the telencephalon of higher vertebrates encounters some difficulties. This is due to the fact that the more or less tube-shaped hemisphere of ancient vertebrates evolved on two divergent lines: one leading to the birds (globiform development)

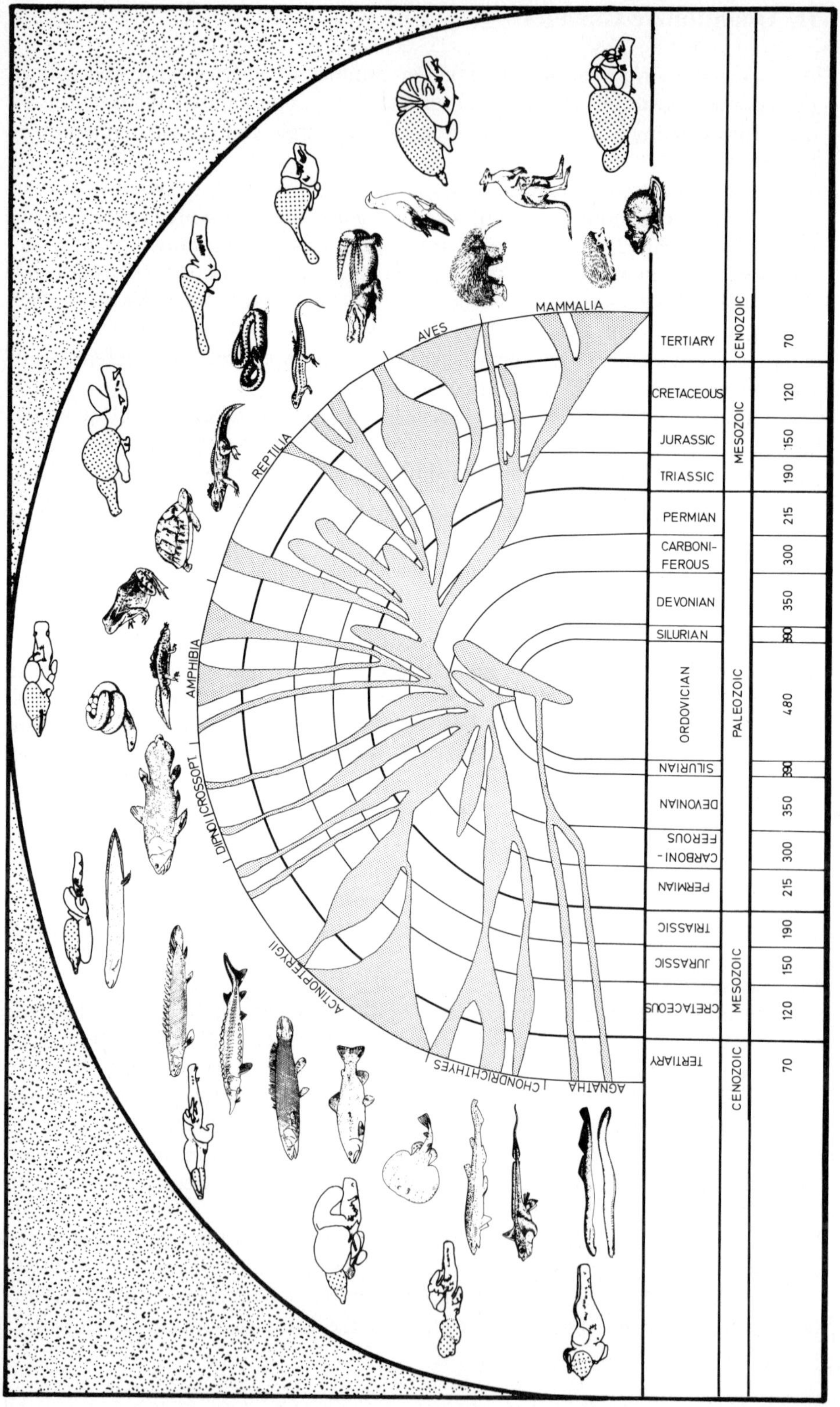

MAMMALIA
AVES
REPTILIA
AMPHIBIA
DIPNOI
CROSSOPT
ACTINOPTERYGII
CHONDRICHTHYES
AGNATHA
TERTIARY
CENOZOIC
70
CRETACEOUS
MESOZOIC
120
JURASSIC
150
TRIASSIC
190
PERMIAN
215
CARBONI-
FEROUS
300
DEVONIAN
350
SILURIAN
390
ORDOVICIAN
PALEOZOIC
480
SILURIAN
390
DEVONIAN
350
CARBONI-
FEROUS
300
PERMIAN
215
TRIASSIC
190
JURASSIC
150
MESOZOIC
CRETACEOUS
120
TERTIARY
CENOZOIC
70

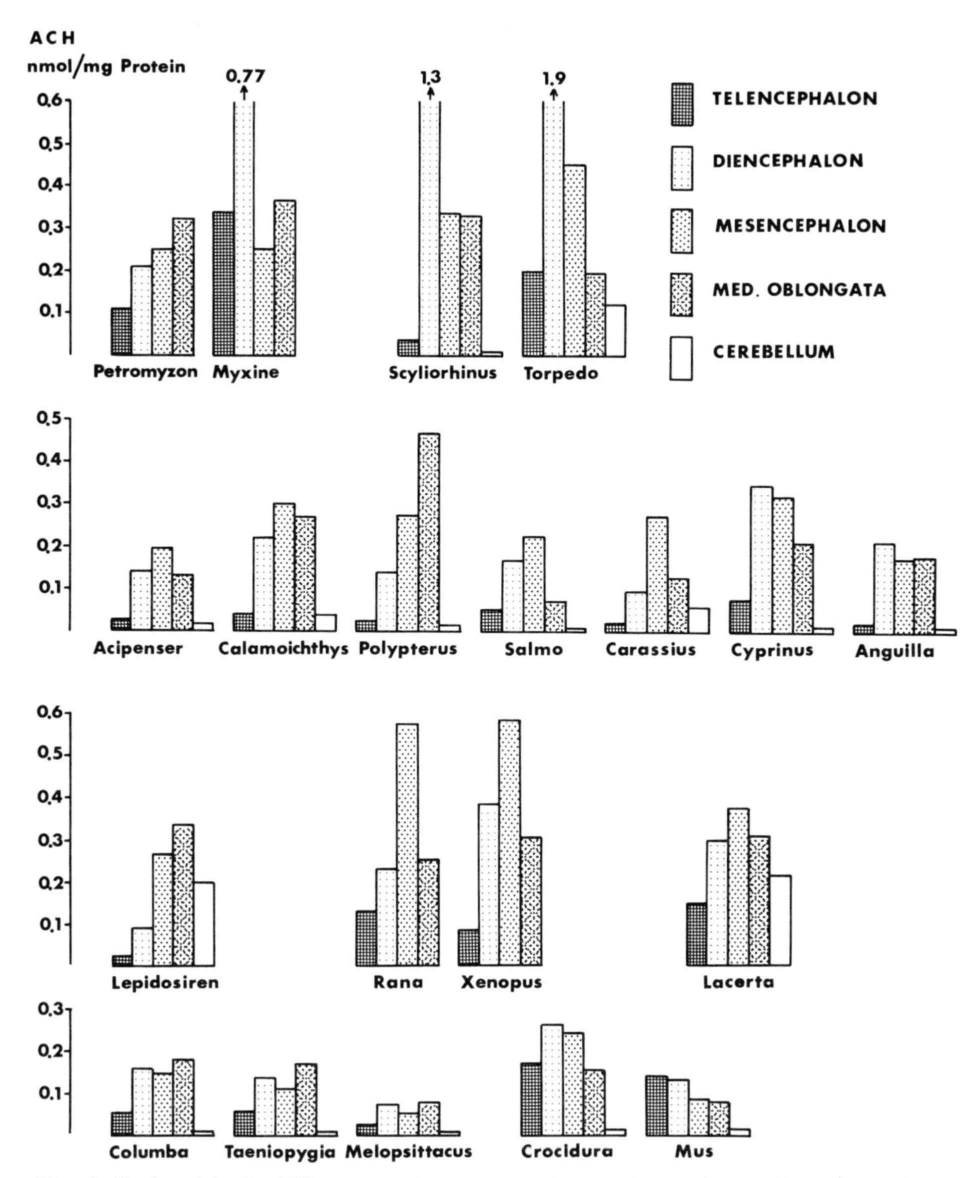

Fig. 3. Regional brain ACh content in representative vertebrates in nmol mg^{-1} protein. (From WÄCHTLER 1980)

Fig. 2. Time course (in millions of years) and radiation in the evolution of vertebrates. Increases (e.g. bony fish, birds, mammals) and decreases (e.g. crocodiles) in the number of species from their origin to the present day are indicated by the varying widths of the evolutionary lines. The basic design of the brain has been conserved over a period of about 480 million years. The telencephalon (*stippled*) has undergone the most obvious quantitative and qualitative changes

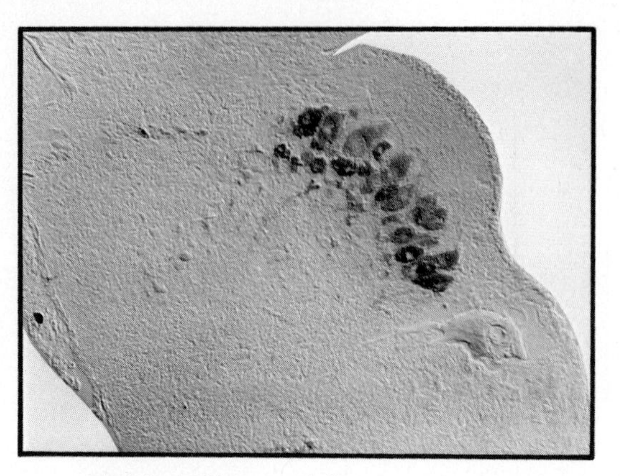

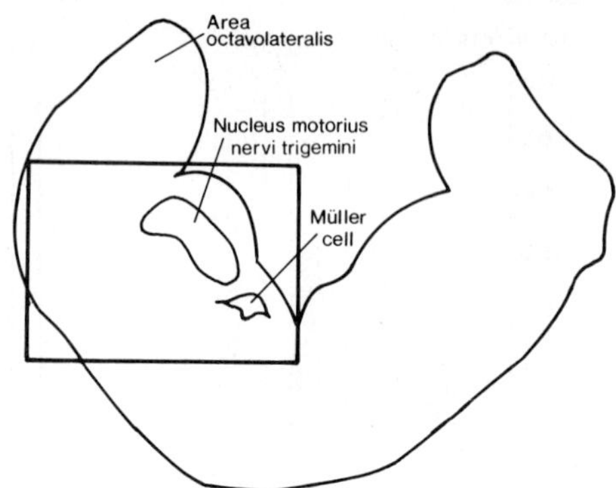

Fig. 4. AChE-containing motor neurons stained by the thiocholine method as described by KARNOVSKY and ROOTS (1964) in the nucleus motorius V (innervating the sucker) of a lamprey after 8-h recovery from DFP treatment *in vitro* and AChE-free Müller neuron. (From KÖRTJE and WÄCHTLER 1985)

the other to the mammals (laminar development) (KARTEN 1969; KIRSCHE 1972; BENOWITZ 1980; NORTHCUTT 1981). In birds a dorsolateral thickening of the hemisphere increases in volume and bulges centrally, whereas in mammals parts of the dorsal hemisphere give rise to a cortical sheet. This cell arrangement becomes the prevailing surface structure and increasingly covers the whole telencephalon and other brain parts (KARTEN 1969; NAUTA and KARTEN 1970; KARTEN and DUBBLEDAM 1973; BENOWITZ 1980).

Most of the compact cell masses in the avian telencephalon have been considered expansions of the striatum. This concept is still preserved in the terminology of avian telencephalic nuclei. It has been shown however that only the basal parts, the palaeostriatum augmentatum and palaeostriatum primitivum correspond to the mammalian striatum. This interpretation was strongly supported by histochemical AChE studies (NAUTA and KARTEN 1970; PARENT and OLIVIER 1970; WÄCHTLER 1973). The so-called neostriatum in birds can, on the basis of its relative low ACh, ChAT, AChE and mAChR content, be clearly separated from those components derived from the basolateral portion of the hemisphere (WÄCHTLER 1985) (Fig. 6).

The neostriatum in birds is like the dorsal-ventricular ridge in reptiles and the cortex in mammals of pallial and not of basal origin. Thus the avian telencephalon

→

Fig. 5. Representative vertebrates (*right*) and their brains (*left*). The protein content (*dashed horizontal line*) and cholinergic components of the telencephalon (*stippled*) are given as a percentage of the corresponding whole brain values (*middle*). Only in mammals does the percentage of telencephalic ACh, ChAT and AChE exceed that of telencephalic protein

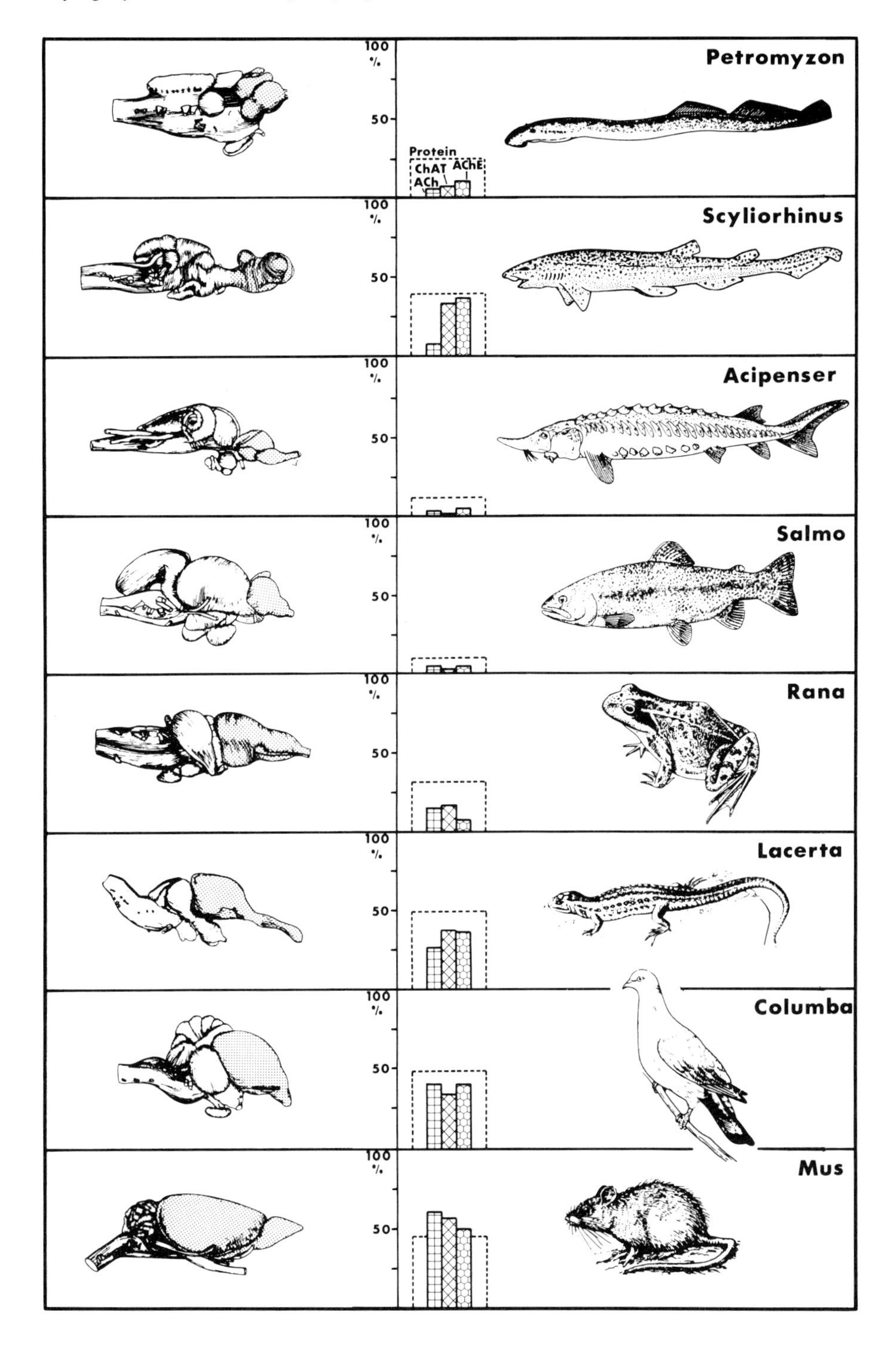
Petromyzon
Scyliorhinus
Acipenser
Salmo
Rana
Lacerta
Columba
Mus
100
%
50
Protein
ChAT
AChE
ACh

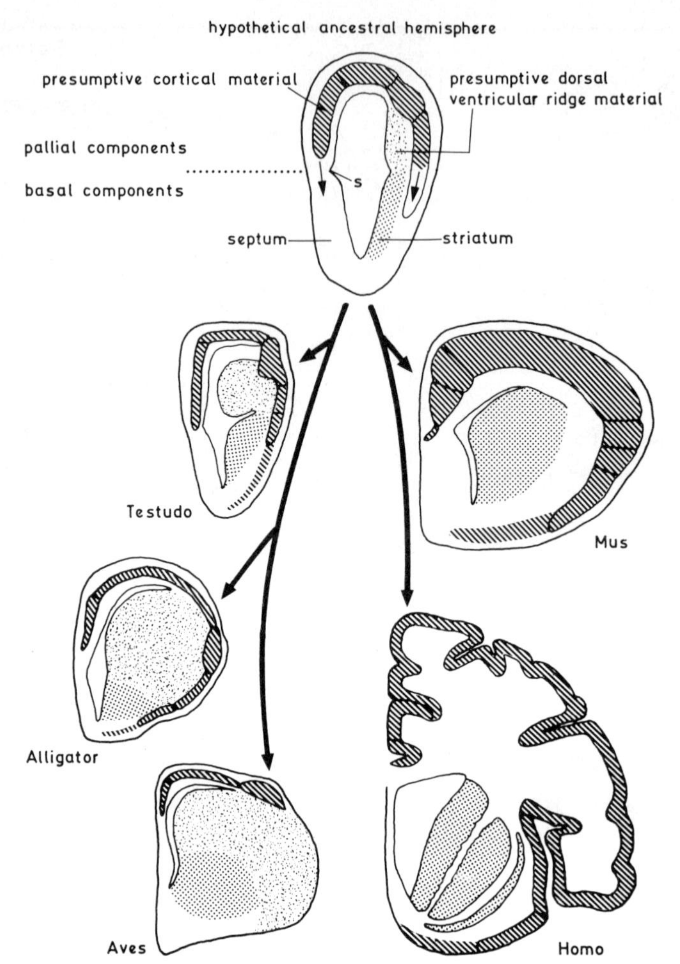

Fig. 6. Two lines of telencephalon evolution in vertebrates (modified from KIRSCHE 1972) shown in transverse sections. The ventricular sulcus (*S*) separates the telencephalic cell masses into pallial and basal components. The pallial matrix gives rise either to various cortical structures (*hatched*) or to a thickening of the dorsolateral hemisphere, the dorsal ventricular ridge (*stippled*). The basal components form septum and striatum (*dotted*). Thus alligators and birds are not characterized by an expansion of the basal ganglia but mainly by an expansion of the dorsal ventricular ridge

is not characterized as often stated by a tremendous increase of the basal ganglia but as in mammals by the proliferation of pallial cell masses which receive increasing sensory input. The main difference is that in birds there is very little cortex formation.

As to the cholinergic distribution pattern within the telencephalon (Figs. 7 and 8) there is with few exceptions (*Lacerta*, *Crocidura*) little cholinergic activity in the olfactory bulb (Fig. 7). This at least in mammals is likely to be entirely due to cholinergic afferents most of which come from the diagonal band (FIBINGER 1982). In the hemisphere basal telencephalic nuclei contain more ACh, ChAT and AChE

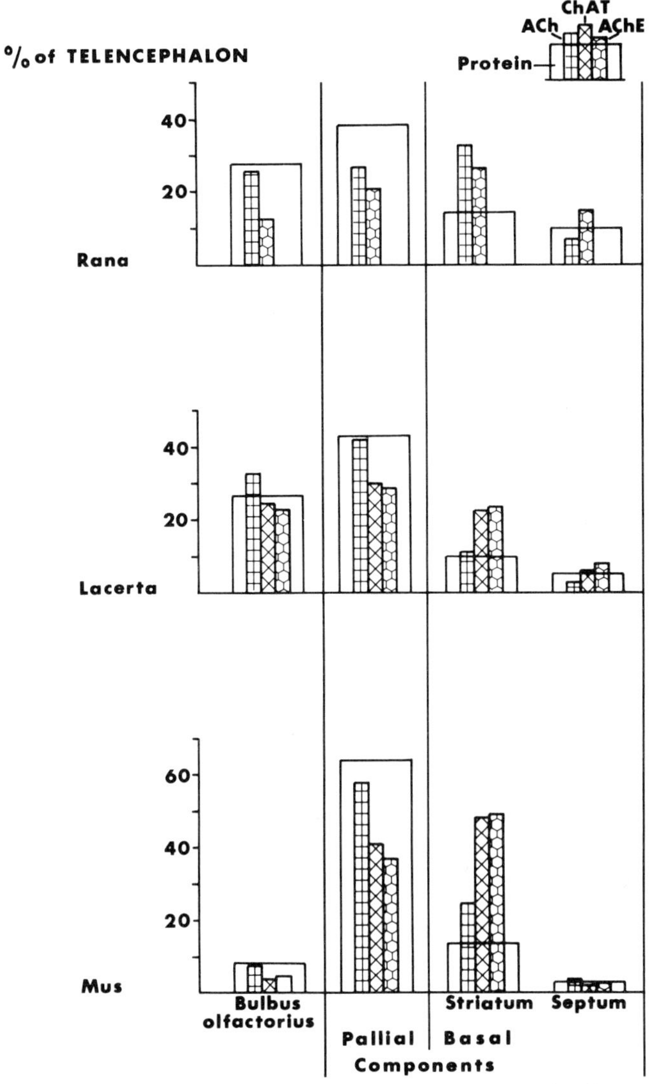

Fig. 7. Protein content, ACh, ChAT and AChE in the telencephalon in representatives of amphibia (frog), reptiles (lizard) and mammals (mouse). The distribution of the components in the main subdivisions is very similar in the three species. The high concentration relative to protein of cholinergic components in septum and striatum can be noticed at all evolutionary levels. (From KÖRTJE 1980)

than pallial components (PARENT and OLIVIER 1970; WÄCHTLER 1973; NORTHCUTT 1974, 1978; BRAUTH and KITT 1980; and Fig. 7). In autoradiographs of mAChRs this contrast is less obvious (Fig. 8).

In the mammalian striatum the population of cholinergic neurons is estimated to be about 1% of the total. They are large intrinsic cells with short processes. The cholinergic neurons in the septum, on the other hand, are projecting cells;

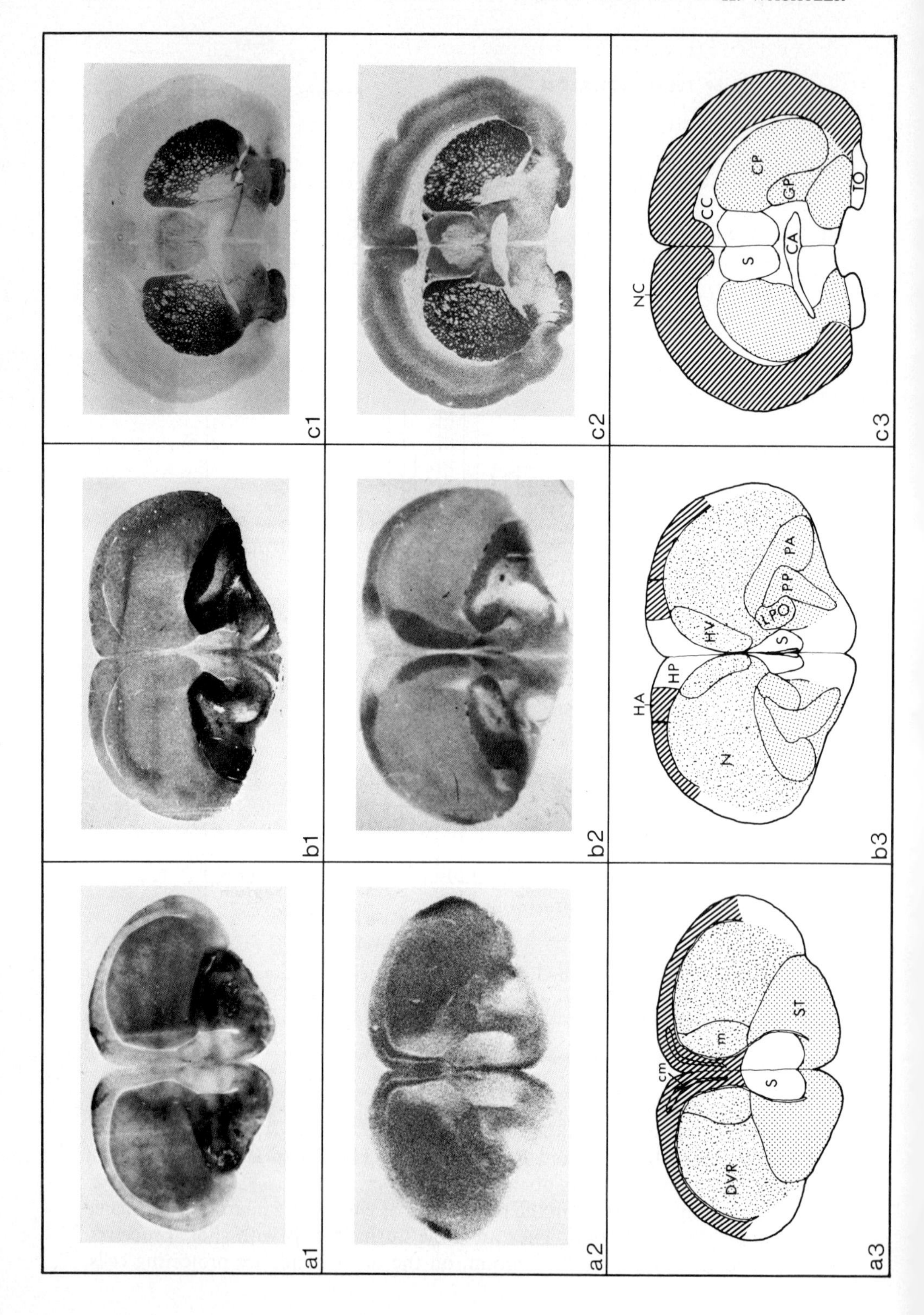
a1
a2
a3
b1
b2
b3
c1
c2
c3
DVR
cm
m
S
ST
HA
HP
HV
N
S
LPO
PP
PA
NC
CC
CP
GP
TO
S
CA

they are the main source of cholinergic afferents to the hippocampus. The knowledge about the origin of cholinergic projections to the cerebral cortex is still incomplete. A substantial contribution of the nucleus basalis of MEYNERT has been demonstrated in the brain of rats and primates (WENK et al. 1980; MESULAM et al. 1983; NIEUWENHUYS 1985).

It is likely that the relatively small number of cholinergic neurons along the base of the forebrain, which according to FIBINGER (1982) is responsible for most of the cholinergic activity in the telencephalon, is a consistent component of the vertebrate brain (HOEHMANN et al. 1983). The relative increase of ACh, ChAT and AChE in the telencephalon of higher vertebrates (Fig. 5) probably results from the progressive histological differentiation of these neurons. Increase in ramification and thus in synaptic surface of nerve cells in the telencephalon from frog to mammals has been demonstrated in comparative Golgi studies (POLIAKOV 1964; RAMON-MOLINER 1969).

The observed increase in the proportion of total brain cholinergic markers associated with the telencephalon (Fig. 5) is thus not due to the selective development of structures such as the dorsal-ventricular ridge, cortex or basal ganglia that might be thought to be especially rich in cholinergic elements but is accounted for by a general process of telencephalization in which all parts of the telencephalon remain in approximately the same relationship to each other (KÖRTJE 1980). Figure 7 shows that the contributions of bulbus olfactorius, pallium, septum and striatum to the total telencephalic activity remain approximately the same in frog, lizard and mouse.

D. Conclusions

The observations summarized above may be somewhat disappointing for those who expect clearly defined evolutionary progress manifested in stepwise development of the cholinergic synapse and an increase in its importance in the functioning of nervous systems. Although in a number of animals the nervous system is too small to provide the necessary experimental evidence, the comparative data as a whole strongly suggest that the functional repertoire of a cholinergic neuron is much the same throughout phylogeny (CAMPBELL and BURNSTOCK 1973;

Fig. 8a–c. Frontal sections through the telencephalon of **a** a reptile (*Iguana*); **b** a bird (*Columba*); **c** a mammal (*Mus*). Distribution of AChE (upper series: **a1, b1, c1**) and mAChRs (middle series: **a2, b2, c2**; autoradiographs of [^{3}H]QNB-binding; for methods see WÄCHTLER 1985). The main brain structures are indicated in outline drawings (lower series: **a3, b3, c3**). The good agreement of activities in the striatal region and its contrast to the pallium can be seen (with the interesting exception of the mAChR distribution in the iguana, **a2**). The basic similarity of dorsal ventricular ridge (*DVR*) derivatives in reptile and bird is obvious (with the exception of *QNB* binding in the hyperstriatum ventrale (*HV*) in **b2**). *CA*, commissura anterior; *CC*, corpus callosum; *cm*, medial cortex; *CP*, caudato putamen; *DVR*, dorsal ventricular ridge; *GP*, globus pallidus; *HA*, hyperstriatum accessorium; *HP*, hippocampus; *HV*, hyperstriatum ventrale; *LPO*, lobus parolfactorius; *m*, medial part of DVR (sensory projection); *N*, neostriatum; *NC*, neocortex; *PA*, palaeostriatum augmentatum; *PP*, palaeostriatum primitivum; *S*, septum; *ST*, striatum; *TO*, tuberculum olfactorium

CUELLO and SOFRONIEW 1984; TRIGGLE and TRIGGLE 1976), and that the whole well-documented sequence of activities from the synthesis of ACh to its degradation and the reuptake of its hydrolysis products is well conserved.

Although some degree of variability at several points in this sequence, from species to species (LANDAU et al. 1981), from neuron to neuron within a species or in the same neuron at different physiological or developmental states (MORIN and WASTERLAIN 1980; MARCHI et al. 1981; STEINBACH 1981) may occur, the basic features of the cholinergic synapse, after it first appeared early in the evolution of the nervous system, remained largely unchanged throughout the course of evolution (CAMPBELL and BURNSTOCK 1973; TRIGGLE and TRIGGLE 1976). Just as the basis for the evolution of nervous systems is an increase in the number of neurons rather than qualitative changes in them, the basis for the evolution of the cholinergic system is an increase in the number and diversity of cholinergic synapses rather than their improved functioning.

As comparative data on cholinergic neurons become more plentiful, the statement made by TURPAEV and SAKHAROV (1973) seems more and more doubtful:

> In any line of animal evolution, the higher representatives have more elaborated and effective synaptic transmission than lower ones. The junction itself seems to undergo evolution – chemical, structural or both.

For the cholinergic synapse at least the evidence for its basic structural and functional stability is more compelling than that for progressive change and improvement. One would rather agree with the conclusions KOOPOWITZ and KEENAN (1982) draw from their neurochemical work on the turbellarian brain, where they found cholinergic neurons coexisting with those using other identified transmitters, when they say:

> The lesson we have learned from this group is fascinating. In the evolution of nervous systems nearly all of the physiological functions and arrangements must have emerged early on. The brain possesses a variety of cell types, including interneurons that process multimodality information. There is considerable plasticity. Perhaps in more advanced animals it is the addition of greater numbers of neurons that produce novel emergent properties while the basic cellular physiological code has remained relatively unchanged through the millenia of organismic evolution.

Acknowledgements. I wish to thank Mrs. G. Jarosch for skilful technical assistance over the years and for providing the illustrations, Mrs. M. Betka for her patience in preparing the manuscript and the Deutsche Forschungsgemeinschaft (DFG) for financial support.

References

Aguilar JS, Lunt GG (1984) Cholinergic binding sites with muscarinic properties on membranes from the supraesophageal ganglion of the locust (*Schistocerca gregaria*). Neurochem Int 6:501–507

Akert K, Waser PG (eds) (1969) Mechanisms of synaptic transmission. Progress in brain research, vol 31. Elsevier, Amsterdam

Ariens Kappers CU, Huber GC, Crosby EC (1936) The comparative anatomy of the nervous system of vertebrates including man. Macmillan, New York

Bacq ZM (1939) Un test marin pour l'acétylcholine. Arch Int Physiol 49:20–22

Bacq ZM, Mazza FP (1935) Isolement de chloroaurate d'acétylcholine à partir d'un extrait de cellules nerveuses d'*Octopus vulgaris*. Arch Int Physiol 42:43–46
Barker DL, Herbert E, Hildebrand JG, Kravitz EA (1972) Acetylcholine and lobster sensory neurones. J Physiol 226:205–229
Barlow RB, Burston KN, Vis A (1980) Three types of muscarinic receptors? Br J Pharmacol 68:141–142
Barnard EA, Dolly JO (1982) Peripheral and central nicotinic ACh receptors: how similar are they? Trends Neurosci 5:325–327
Beach RL, Vaca K, Pilar G (1980) Ionic and metabolic requirements for high-affinity choline uptake and ACh synthesis in nerve terminals at a neuromuscular junction. J Neurochem 34:1387–1398
Benowitz L (1980) Functional organization of the avian telencephalon. In: Ebbeson SO (ed) Comparative neurology of the telencephalon. Plenum, New York, pp 389–419
Bigl V, Schober W (1977) Cholinergic transmission in subcortical and cortical visual centers of rats: no evidence for the involvement of primary optic system. Exp Brain Res 27:211–219
Birdsall NJM, Burgen ASV, Hulme EC (1978) Binding of agonists to brain muscarinic receptors. Mol Pharmacol 14:723–736
Birdsall NJM, Hulme EC, Burgen A (1980) Character of the muscarinic receptors in different regions of the rat brain. Proc R Soc Lond [Biol] 207:1–12
Brauth SAE, Kitt CA (1980) The palaeostriatal system of caiman crocodiles. J Comp Neurol 189:437–466
Breer H (1981 a) Isolation and characterization of synaptosomes from the central nervous system of *Locusta migratoria*. Neurochem Int 3:155–163
Breer H (1981 b) Comparative studies on cholinergic activities in the central nervous system of *Locusta migratoria*. J Comp Physiol 141:271–276
Breer H (1982) Uptake of [*N-Me*-^{3}H] choline by synaptosomes from the central nervous system of *Locusta migratoria*. J Neurobiol 13:107–117
Breer H, Kleene R, Hinz G (1985) Molecular forms and subunit structure of the acetylcholine receptor in the central nervous system of insects. J Neurosci 5:3386–3392
Buijs RM, Pevet P, Saab DF (eds) (1982) Chemical transmission in the brain: the role of amines, amino acids and peptides. Progress in brain research, vol 55. Elsevier, Amsterdam
Bullock TH, Horridge GH (1965) Structure and function in the nervous systems of invertebrates. Freeman, San Francisco
Bullock TH, Nachmansohn D (1942) Cholinesterase in primitive nervous systems. J Comp Physiol 20:239–242
Burnstock G (1981) Neurotransmitters and trophic factors in the autonomic nervous system. J Physiol (Lond) 313:1–36
Campbell G, Burnstock G (1973) Pharmacology of cholinergic systems in lower vertebrates. In: Michelson MJ (ed) International encyclopedia of pharmacology and therapeutics 85/1, Pergamon, Oxford, pp 43–79
Conti-Troconti BM, Hunkapiller MW, Lindstrom JM, Raftery MA (1982) Subunit structure of the acetylcholine receptor from *Electrophorus electricus*. Proc Natl Acad Sci USA 79:6489–6493
Cooper JR, Bloom FE, Roth RH (1974) The biochemical basis of neuropharmacology, 2nd edn. University Press, Oxford
Cuello AC, Sofroniew MV (1984) The anatomy of the CNS cholinergic neurons. Trends Neurosci 7:74–78
Dale HH (1914) The action of certain esters and ethers of choline and their relation to muscarine. J Pharmacol 6:147–196
Dowdall MJ (1974) Biochemical properties of synaptosomes from the optic lobe of squid (*Loligo pealii*). Biochem Soc Trans 2:270–272
Dowdall MJ, Simon EJ (1973) Comparative studies on synaptosomes: uptake of [*N-Me*-^{3}H] choline by synaptosomes from squid optic lobes. J Neurochem 21:969–982
Dowdall MJ, Whittaker VP (1973) Comparative studies on synaptosomes formation: the preparation of synaptosomes from the head ganglion of the squid, *Loligo pealii*. J Neurochem 20:921–935

Dowdall MJ, Wächtler K, Henderson F (1975) Comparative studies on the high-affinity choline uptake system. (Abstr.) Vth Meeting International Society of Neurochemistry, Barcelona, p 122

Dudai Y (1978) Properties of an α-bungarotoxin binding cholinergic nicotinic receptor from *Drosophila melanogaster.* Biochim Biophys Acta 539:505–517

Dudai Y (1979) Cholinergic receptors in insects. Trends Biochem Sci 4:40–44

Ebbeson SO (ed) (1980) Comparative neurology of the telencephalon. Plenum, New York

Erzen J, Brzin M (1978) Cholinergic mechanisms in Hydra. Comp Biochem Physiol 59C:39–43

Fibinger HC (1982) The organization and some projections of cholinergic neurons of the mammalian forebrain. Brain Res Rev 4:327–388

Fischer H (1971) Vergleichende Pharmakologie der Überträgersubstanzen in tiersystematischer Darstellung. (Handbook of experimental pharmacology, vol 26) Springer, Berlin Heidelberg New York

Florey E (1967a) Neurotransmitters and modulators in the animal kingdom. Fed Proc 26:1164–1178

Florey E (1967b) Cholinergic neurons in tunicates: an appraisal of the evidence. Comp Biochem Physiol 22:617–628

Florey E, Florey E (1965) Cholinergic neurons in the onychophora: a comparative study. Comp Biochem Physiol 15:125–136

Florey E, Michelson MJ (1973) Occurrence, pharmacology, and significance of cholinergic mechanisms in the animal kingdom. In: Michelson MJ (ed) International encyclopedia of pharmacology and therapeutics 85/1. Pergamon, Oxford, pp 11–41

Florey E, Rathmeyer M (1980) Pharmacological characterization of cholinoreceptors of cardiac ganglion cells of crustaceans. Gen Pharmacol 11:47–54

Fonnum F (1974) A rapid radiochemical method for the determination of choline acetyltransferase. J Neurochem 24:407–409

Gardner CR, Walker RJ (1982) The roles of putative neurotransmitters and neuromodulators in annelids and related invertebrates. Prog Neurobiol 18:81–120

Gerschenfeld HM (1973) Chemical transmission in invertebrate central nervous systems and neuromuscular junctions. Physiol Rev 53:1–119

Hebb C, Ratković D (1964) Choline acetylase in the evolution of the brain in vertebrates. In: Richter D (ed) Comparative neurochemistry. Pergamon, Oxford, pp 347–354

Hedlund B, Shephard GM (1983) Biochemical studies on muscarinic receptors in the salamander olfactory epithelium. FEBS Lett 162:428–431

Hildebrand JG, Townsel JG, Kravitz EA (1974) Distribution of acetylcholine, choline and choline acetyltransferase in regions and single identified axons of the lobster nervous system. J Neurochem 23:951–963

Hoehmann CF, Carroll PT, Ebner FF (1983) Acetylcholine levels and choline acetyltransferase activity in turtle cortex. Brain Res 258:120–122

Horn JP, Dodd J (1983) Inhibitory cholinergic synapses in autonomic ganglia. Trends Neurosci 6:180–184

Jones SW, Sundershan P, O'Brien RD (1981) Alpha-bungarotoxin binding in housefly heads and *Torpedo* electroplax. J Neurochem 36:447–453

Jope RS (1979) High affinity choline transport and acetylCoA production in brain and their roles in the regulation of acetylcholine synthesis. Brain Res Rev 1:313–344

Kandel ER, Gardner D (1972) The synaptic actions mediated by the different branches of a single neuron. In: Kopien IJ (ed) Neurotransmitters. Res Publ Assoc Res Nerv Ment Dis 50:91–146

Kandel ER, Schwartz JH (eds) (1985) Principles of neural science, 2nd edn. Elsevier, New York

Karnovsky MJ, Roots L (1964) A 'direct-coloring' thiocholine method for cholinesterase. J Histochem Cytochem 12:219–221

Karten H (1969) The organization of the avian telencephalon and some speculations on the phylogeny of the amniote telencephalon. Ann NY Acad Sci 167:164–185

Karten H, Dubbledam JC (1973) The organization and projections of the palaeostriatal complex in the pigeon (*Columba livia*). J Comp Neurol 148:61–90

Kehoe JS (1972a) Three acetylcholine receptors in Aplysia neurons. J Physiol (Lond) 225:115–146
Kehoe JS (1972b) The physiological role of three acetylcholine receptors in *Aplysia* neurons. J Physiol (Lond) 225:147–172
Kehoe JS, Marder E (1976) Identification and effects of neural transmitters in invertebrates. Ann Rev Pharmacol 16:245–268
Kerkut GA, Newton A, Pitman RM, Walker RJ, Woodruff GN (1970) Acetylcholine receptors of invertebrate neurons. Br J Pharmacol 40:586–587
Kirsche W (1972) Die Entwicklung des Telencephalons der Reptilien und deren Beziehungen zur Hirn-Bauplanlehre. Nova Acta Leopoldina 37:9–78
Körtje KH (1980) Über das Vorkommen von Acetylcholin, Cholinacetyltransferase und Acetylcholinesterase in homologen Vorderhirnregionen einiger Wirbeltiere. Thesis, Hannover
Körtje KH (1987) Untersuchungen zum Acetylcholin-System im zentralen Nervensystem des Flußneunauges (*Lampetra fluziatilis* L.). Dissertation, Hannover
Körtje KH, Wächtler K (1985) The localization of cholinergic neurons in the lamprey in tissue sections and whole mounts after DFP treatment. Mikroskopie 42:305–309
Koketsu K, Yamada M (1982) Presynaptic muscarinic receptors inhibiting active acetylcholine release in the bullfrog sympathetic ganglion. Br J Pharmacol 77:75–82
Koopowitz H, Keenan L (1982) The primitive brains of plathyhelminthes. Trends Neurosci 5:77–79
Kriebel M (1968) Cholinoceptive and adrenoceptive properties of the tunicate heart pacemakers. Biol Bull Woods Hole 135:174–180
Kuhar MJ, Murrin LC (1978) Sodium-dependent high affinity choline uptake. J Neurochem 30:15–21
Kuhlenbeck H (1927) Vorlesungen über das Zentralnervensystem der Wirbeltiere. Eine Einführung in die Gehirnanatomie auf vergleichender Grundlage. Fischer, Jena
Kuhlenbeck H (1967–1978) The central nervous system of vertebrates (in 5 vols). Karger, Basel
Landau EM, Gavish B, Nachshen DA, Lotan I (1981) pH dependence of the acetylcholine receptor channel: a species variation. J Gen Physiol 77:647–666
Lentz TL (1968) Primitive nervous systems. Yale University Press, New Haven
Levey AJ, Wainer B (1982) Cross-species and inter-species reactivities of monoclonal antibodies against choline acetyltransferase. Brain Res 234:469–473
Levitan H, Tauc L (1972) Acetylcholine receptors: topographic distribution and pharmacological properties of two receptor types on a single molluscan neuron. J Physiol (Lond) 222:537–558
Lewis SE (1953) Acetylcholine in blowflies. Nature (Lond) 172:1004–1005
Lindstrom J (1986) Probing nicotinic acetylcholine receptors with monoclonal antibodies. Trends Neurosci 9:401–407
Macrides F, Davis BJ, Youngs WM, Nadi NS, Margolis FL (1981) Cholinergic and catecholaminergic afferents to the olfactory bulb in the hamster: a neuroanatomical, biochemical and histochemical investigation. J Comp Neurol 203:495–514
Magleby KL, Pallotta BS (1981) A study of desensitization of acetylcholine receptors using nerve-released transmitter in the frog. J Physiol (Lond) 316:225–250
Marchi M, Hoffmann DW, Giacobini E, Volle R (1981) Development and aging of cholinergic synapses. VI. Mechanisms of acetylcholine biosynthesis in the chick iris. Dev Neurosci 4:442–450
Marder E (1977) Pharmacological analysis of transmitter effects in the crustacean stomatogastric ganglion. Proc Int Union Physiol Sci 13:477
Marsden JR, Bsata N, Cain H (1981) Evidence for a cerebral cholinergic system and suggested pharmacological patterns of neural organization in the prostomium of the polychaete *Nereis virens*. Tissue Cell 13:255–268
McCaman RE, McCaman MW (1976) Biology of individual cholinergic neurons in the invertebrate CNS. In: Goldberg A, Hanin J (eds) Biology of cholinergic function. Raven, New York, pp 485–513
McGeer PL, Eccles JC, McGeer EG (1978) Molecular neurobiology of the mammalian brain. Plenum, New York

McLennan H (1959) The identification of one active component from brain extracts containing factor I. J Physiol 146:358–368
Mesulam MM, Mufson EJ, Levey AI, Wainer BH (1983) Cholinergic innervation of cortex by the basal forebrain: cytochemistry and cortical connections of the septal area diagonal band nuclei, nucleus basalis (substantia innominata) and hypothalamus in the Rhesus monkey. J Comp Neurol 214:170–197
Meyer MR, Edwards JS (1980) Muscarinic cholinergic binding sites in an orthopteran central nervous system. J Neurobiol 11:215–220
Michelson MJ (1973a) Pharmacology of cholinergic systems in some other phyla. In: Michelson MJ (ed) International encyclopedia of pharmacology and therapeutics 85/1. Pergamon, Oxford, pp191–228
Michelson MJ (1973b) Structure and mutual disposition of cholinreceptors and changes in their disposition in the course of evolution. In: Michelson MJ (ed) International encyclopedia of pharmacology and therapeutics, 85/1. Pergamon, Oxford, pp 357–390
Morin AM, Wasterlain CG (1980) Aging and ratbrain muscarinic receptors as measured by quinuclidinyl benzilate binding. Neurochem Res 5:301–308
Muller K, Nicholls J, Stent G (1981) Neurobiology of the leech. Cold Spring Harbor Laboratory, New York
Nauta WJH, Karten HJ (1970) A general profile of vertebrate brain, with sidelights on the ancestry of the cerebral cortex. In: Schmitt FO (ed) The neurosciences: second study program. Rockefeller University Press, New York, pp 7–26
Nieuwenhuys R (1985) Chemoarchitecture of the brain. Springer, Berlin Heidelberg New York, pp 7–11
Noda M, Furutami Y, Takahashi H, Toyosato M, Tanabe T, Shimizu S, Kikyotani S, Kayano T, Hirose T, Inayama S, Numa S (1983a) Cloning and sequence analysis of calf cDNA and human genomic DNA encoding α-subunit precursors of muscle acetylcholine receptors. Nature 305:818–823
Noda M, Takahashi H, Tanabe T, Toyosato M, Kikyotani S, Furutani Y, Hirose T, Takashima H, Inayama S, Miyara T, Numa S (1983b) Structural homology of *Torpedo californica* acetylcholine receptor subunits. Nature 302:528–532
Nogrady T, Alai M (1983) Cholinergic neurotransmission in rotifers. In: Peiler B, Starkweather R, Nogrady T (eds) Biology of rotifers. Proceedings of the third international rotifer symposium, Uppsala, 1982. Hydrobiologia 104:149–153
Northcutt RG (1974) Some histochemical observations on the telencephalon of the bullfrog, *Rana catesbeiana* (AChE, MAO, SDH). J Comp Neurol 157:379–390
Northcutt RG (1978) Forebrain and midbrain organization in lizards and its phylogenetic significance. In: Greenberg N, McLean PD (eds) National Institute of Mental Health, Rockville, pp 11–63
Northcutt RG (1981) Evolution of the telencephalon in nonmammals. Ann Rev Neurosci 4:301–350
Osborne N (ed) (1978) Biochemistry of characterised neurons. Pergamon, Oxford
Oswald RE, Freeman JA (1980) Optic nerve transmitters in lower vertebrate species. Life Sci 27:527–533
Oswald RE, Schmidt DE, Freeman JA (1979) Assessment of acetylcholine as an optic nerve transmitter in *Bufo marinus*. Neuroscience 4:1129–1136
Pappas G, Purpura D (eds) (1972) Structure and function of synapses. Raven, New York
Parent A, Olivier A (1970) Comparative histochemical study of the corpus striatum. J Hirnforsch 12:73–81
Peng Loh Y (1976) Developmental changes in activity of cholinacetyltransferase, acetylcholinesterase and glutamic acid decarboxylase in the central nervous system of the toad *Xenopus laevis*. J Neurochem 26:1303–1305
Peper K, Bradley RJ, Dreyer F (1982) The acetylcholine receptor at the neuromuscular junction. Physiol Rev 62:1271–1333
Peters H, Palay SL, Webster H (1970) The fine structure of the nervous system. Harper and Row, New York
Poliakov GJ (1964) Development and complication of the cortical part of the coupling mechanism in the evolution of vertebrates. J Hirnforsch 7:253–273

Putinseva TG, Turpaev TM (1959) On a cardiac stimulant secreted by the frog heart's ventricle under the influence of acetylcholine (in Russian). Dokl Akad Nauk SSSR 129:1442–1444

Raftery MA, Hunkapiller MW, Strader CD, Hood LE (1980) Acetylcholine receptor: complex of homologous subunits. Science 208:1454–1457

Ramon-Moliner E (1969) The leptodendric neuron: its distribution and significance. Ann NY Acad Sci 167:65–70

Rossier J (1977) Choline acetyltransferase: a review with special reference to its cellular and subcellular localization. Int Rev Neurobiol 20:283–337

Rossini L (1981) Classifying cholinergic receptors. Trends Parmacol Sci 2:1–4

Rozhkova EK (1973) Pharmacology of cholinergic systems in annelids. In: Michelson MJ (ed) International encyclopedia of pharmacology and therapeutics 85/1. Pergamon, Oxford, pp 169–190

Rozhkova EK, Malyntina TA, Shishov BA (1980) Pharmacological characteristics of cholinoreception in somatic muscles of the nematode, *Ascaris suum*. Gen Pharmacol 11:141–146

Sakharov DA (1974) Evolutionary aspects of transmitter heterogeneity. J Neural Transm [Suppl] 11:43–59

Sakharov DA, Nistratova SN (1963) Peculiarities of response in the heart of *Anodonta* (in Russian). Fiziol Zh SSSR 49:1475–1480

Sakharov DA, Turpaev TM (1967) Evolution of cholinergic transmission. In: Salanki J (ed) Neurobiology of invertebrates. Akademiai Kiado, Budapest, pp 305–314

Salvaterra PM, Foders RM (1979) ^{125}I-α-bungarotoxin and [^{3}H] quinuclidinylbenzilate binding in the central nervous system of different species. J Neurochem 32:1509–1517

Sattelle DB (1980) Acetylcholine receptors of insects. Adv Insect Physiol 15:215–315

Schaller CH, Gierer H (1973) Distribution of the head activating substance in hydra and its localisation in membranous particles in nerve cells. J Embryol Exp Morphol 29:39–52

Schlesinger C (1981) Zur Ontogenese des Acetylcholinsystems im Gehirn des südafrikanischen Krallenfrosches (*Xenopus laevis* Daudin). J Hirnfosch 22:543–553

Segal M, Dudai Y, Amsterdam A (1978) Distribution of an alpha-bungarotoxin binding cholinergic receptor in rat brain. Brain Res 148:105–119

Shen SC, Greenfield P, Boell EJ (1955) The distribution of cholinesterases in the frog brain. J Comp Neurol 102:717–743

Silver A (1974) The biology of cholinesterases. North-Holland, Amsterdam

Simon JR, Kuhar MJ (1976) High affinity choline uptake: ionic and energy requirements. J Neurochem 27:93–99

Smallman BN, Maneckjee A (1981) The synthesis of acetylcholine by plants. Biochem J 194:361–364

Starck D (1962) Die Evolution des Säugetiergehirns. Steiner, Wiesbaden

Steinbach JH (1981) Developmental changes in acetylcholine receptor aggregates at rat skeletal neuromuscular junctions. Dev Biol 84:267–276

Stroud RM, Finer-Moore J (1985) Acetylcholine receptor: structure function and evolution. Ann Rev Cell Biol 1:317–351

Sugden PH, Newsholme EA (1977) Activities of choline acetyltransferase, acetylcholinesterase, glutamate decarboxylase, 4-aminobutyrate aminotransferase in nervous tissue from some vertebrates and invertebrates. Comp Biochem Physiol 56C:89–94

Sukumar R, Ivy M, Townsel JG (1983) The distribution of cholinergic components in a marine arthropod. Comp Biochem Physiol 74C:201–209

Szerb JC (1961) The estimation of acetylcholine, using leech muscle in a microbath. J Physiol 158:8–9

Takano Y, Kohjimoto Y, Uchimura K, Kamiya AO (1980) Mapping of the distribution of high-affinity choline uptake and ChAT in the striatum. Brain Res 194:583–587

Treherne JE (1966) Neurochemistry of arthropods. Cambridge University Press, Cambridge

Triggle DJ, Triggle CR (1976) Chemical pharmacology of the synapse. Academic, London

Turpaev TM, Sakharov DA (1973) Comparative physiology of the mechanisms terminating the synaptic action of acetylcholine. In: Michelson MJ (ed) International encyclopedia of pharmacology and therapeutics 85/1. Pergamon, Oxford, pp 345–356

Venter JC, Eddy B, Hall LM, Fraser CM (1984) Monoclonal antibodies detect the conservation of muscarinic cholinergic receptor structure from *Drosophila* to human brain and detect possible structural homology with α_1-adrenergic receptors. Proc Natl Acad Sci USA 81:272–276

Wächtler K (1973) Vergleichend-histochemische Untersuchungen zur Acetylcholinesterase-Verteilung im Telencephalon der Wirbeltiere. Habilitationsschrift, Hannover

Wächtler K (1974) The distribution of acetylcholinesterase in the cyclostome brain. I. *Lampetra planeri* (L.). Cell Tissue Res 152:259–270

Wächtler K (1978) The regional distribution of choline acetyltransferase and acetylcholinesterase in the brain of the lamprey (*Petromyzon fluviatilis*). Comp Biochem Physiol 61C:411–413

Wächtler K (1980) The regional production of acetylcholine in the brain of lower and higher vertebrates. Comp Biochem Physiol 65C:1–16

Wächtler K (1981) The regional distribution of acetylcholine, cholineacetyltransferase and acetylcholinesterase in vertebrate brains of different phylogenetic levels. In: Pepeu G, Ladinsky H (eds) Cholinergic mechanisms. Plenum, New York, pp 59–72

Wächtler K (1982) Observations on the evolution of the cholinergic system in the telencephalon of vertebrates. Comp Biochem Physiol 72C:357–361

Wächtler K (1983a) The regional distribution of cholinergic components in the brain of the ratfish (*Chimaera monstrosa*). Comp Biochem Physiol 74C:481–484

Wächtler K (1983b) The acetylcholine system in the brain of cyclostomes with special reference to the telencephalon. J Hirnforsch 24:63–70

Wächtler K (1985) The regional distribution of muscarinic acetylcholine receptors in the telencephalon of the pigeon. J Hirnforsch 26:85–89

Walker RJ (1982) Current trends in invertebrate neuropharmacology. Verh Dtsch Zool Ges 31–59

Wamsley JK, Zarbin M, Birdsall N, Kuhar MJ (1980) Muscarinic cholinergic receptors: autoradiographic localization of high and low affinity agonist binding sites. Brain Res 200:1–12

Welsh JH (1946) Acetylcholine in *Planaria*. Anat Rec 95:79

Wenk H, Bigl V, Meyer V (1980) Cholinergic projections from magnocellular nuclei of the basal forebrain to cortical areas in rats. Brain Res Rev 2:295–316

Westfall JA, Yamataka S, Enos PD (1971) Ultrastructural evidence of polarized synapses in the nerve net of *Hydra*. J Cell Biol 51:318–323

Whittaker VP (1963a) Identification of acetylcholine and related esters of biological origin. In: Koelle GB (ed) Cholinesterases and anticholinesterase agents. Springer, Berlin Heidelberg New York, pp 1–39 (Handbook of experimental pharmacology, vol 15)

Whittaker VP (1963b) Cholinergic neurohormones. In: von Euler US, Heller H (eds) Comparative endocrinology, vol 2. Academic, New York, pp 182–208

Whittaker VP (1965) The application of subcellular fractionation techniques to the study of brain function. Prog Biophys Mol Biol 15:39–91

Whittaker VP (1979) Re: Metabotropic neurotransmission. Trends Neurosci 2:7

Whittaker VP, Dowdall MJ (1975) Current state of research on cholinergic synapses. In: Waser PG (ed) Cholinergic mechanisms. Raven, New York, pp 23–42

Yamamura HJ, Snyder SH (1973) High affinity transport of choline into synaptosomes of rat brain. J Neurochem 21:1355–1374

Yamamura HJ, Kuhar MJ, Greenberg D, Snyder SH (1974) Muscarinic cholinergic receptor binding: regional distribution in monkey brain. Brain Res 66:541–546

Yamamura HJ, Enna SJ, Kuhar MJ (1984) Neurotransmitter receptor binding, 2nd edn. Raven, New York

CHAPTER 5

Development of the Cholinergic Synapse: Role of Trophic Factors

G. P. RICHARDSON *

A. Introduction

Considerable advances have been made in the last decade in the search for the trophic factors that might be operational during the development of cholinergic systems. The search has been inspired by the discovery of nerve growth factor (NGF), a protein essential for the development and maintenance of both sympathetic and sensory neurons, and various lines of experimental evidence suggesting the necessity of trophic factors for cholinergic neuron development. The use of tissue culture models has provided the technical means of approaching the problem and has now led to the purification of several factors that are putatively involved in various aspects of the development of cholinergic neurons *in vivo*. This chapter will focus on the work that has been directed to purifying the factors required for the survival and development of the cholinergic motor neurons of the parasympathetic nervous system and the spinal cord *in vitro* and, in view of its potential role in the expression of the cholinergic phenotype, on the factor purified from heart-cell conditioned medium that induces cholinergic characteristics in sympathetic neurons.

B. Parasympathetic Ciliary Neurons

I. Ciliary Ganglion as a Model System

The avian ciliary ganglion has provided an ideal model system for studying the development of the cholinergic neurons of the parasympathetic nervous system. As a tightly defined, easily accessible group of purely cholinergic neurons well separated from both the central nervous system and its localized synaptic targets it has been the focus of considerable study. In addition, the development of the ganglion *in vivo* has been thoroughly described; thus the relevance of *in vitro* results to normal development can be critically evaluated. Naturally occurring neuronal cell death reduces the size of the domestic fowl ('chick') ciliary neuron population by approximately 50% between stages 34 and 40 (days 8–14) of development (LANDMESSER and PILAR 1974a), and the amount of cell death can be increased to around 90% by removal of the eye (LANDMESSER and PILAR 1974b) or decreased by up to 30% by the addition of a supernumerary eye (NARAYANAN

---

* From School of Biology, University of Sussex, Falmer, Sussex, UK (current address see list of contributors).

and NARAYANAN 1978). This is the primary piece of evidence, suggesting that the target tissue is providing a limited amount of some trophic agent that is necessary for the survival of ciliary neurons.

II. Necessity of Trophic Factors for the Survival and Growth of Ciliary Neurons *In Vitro*

In 1976 it was reported that dissociated ciliary ganglion cells from 8-day chick embryos (stages 33–34) would survive *in vitro* for at least 48 h on a polyornithine (PORN) substrate providing the medium had been previously conditioned by chick heart cells and, in addition to surviving, would grow extensive neuritic processes (HELFAND et al. 1976). No neurons survived 48 h in unconditioned medium and the activity necessary for neuron survival in heart-cell conditioned medium (HCM) was retained on dialysis, did not bind to the substrate and was not inhibited by anti-sera to NGF. NGF itself had no effect on ciliary neuron survival. Subsequently NISHI and BERG (1977, 1979) showed that all the ciliary neurons recovered by the dissociation procedure at stage 32 (i.e. before the onset of naturally occurring cell death) would survive at least 25 days (i.e. well beyond the normal period of cell death) when co-cultured with skeletal myotubes or when cultured on PORN in either HCM or medium containing 5% chick embryo extract (CEE). COLLINS (1978) then demonstrated that HCM contained a trypsin-sensitive factor that attaches to a PORN substrate and induces rapid neurite outgrowth from ciliary neurons under conditions in which even short term survival (24 h or greater) was not possible (i.e. in unconditioned medium or in medium containing no embryo extract). VARON et al. (1979) analysed neuron survival, neurite outgrowth and choline acetyltransferase (ChAT) activity in cultures of ciliary ganglion cells; these were grown on either PORN or an adhesive collagen substrate in serum-containing media supplemented with either HCM or CEE. Both sources (HCM and CEE) contained activities that would support neuron survival and stimulate neurite outgrowth, but their effects depended considerably on the choice of substrate. This initial study first suggested that neuron survival, neurite outgrowth and ChAT activity in ciliary ganglion cell cultures were all subject to independent regulation.

At the same epoch a large number of other studies demonstrated that either co-culturing chick ciliary neurons with other cells or the use of tissue extracts and fluids as media supplements or conditioned media from a variety of sources were necessary for ciliary neuron survival *in vitro*. These studies can be found listed in a recent review by MANTHORPE and VARON (1985). The extracts and fluids have been obtained from a variety of different species and tissues – for example extracts of rat skeletal muscle (HILL and BENNETT 1983), rat brain (NIETO-SAMPEDRO et al. 1983; MANTHORPE et al. 1983c), ox heart (BONYHADY et al. 1980) and embryonic *Torpedo* electric organ (RICHARDSON et al. 1985) and human serum (TOUZEAU and KATO 1983) have all been shown to support the short term survival of chick ciliary neurons *in vitro*.

III. Developmental Regulation of Survival Activity for Ciliary Neurons

Using quantitative assays for ciliary neuron survival based on dissociated cell cultures obtained from the 8-day chick embryos, it was shown that nearly 33% of the total activity necessary for the *in vitro* survival of ciliary neurons present in 12-day CEE was contained in the eye, that the specific activity was 7-fold higher in eye extracts than in whole embryo extracts and that, within the eye itself, trophic activity was highest in the choroid and iris (ADLER et al. 1979). Furthermore it was shown that the specific trophic activity in the eye increases rapidly between days 11 and 13 (stages 37 and 39), at a time when naturally occurring cell death in the ciliary ganglion is normally ending (LANDA et al. 1980). Specific activity begins to plateau in the eye between days 13 and 15 (stages 39 and 41). In contrast, the specific trophic activity of chick heart extracts necessary for the *in vitro* survival of ciliary neurons increases most rapidly between days 19 and 20 and reaches a plateau at birth (HILL et al. 1981). However the maximum specific trophic activity of chick heart extracts is 40 trophic units (TU) per milligram of protein (by hatching at day 21), whereas that of extracts of the combined ciliary body, choroid, iris and pigment epithelium (CIPE) of the 15-day eye ranges from 12000 – 20000 TU mg^{-1} protein (MANTHORPE et al. 1980). The rapid increase in specific trophic activity in the chick heart occurs at a time when parasympathetic innervation of the heart is taking place. It has also been reported that 18-day chick heart explants evoke maximum stimulation of neurite outgrowth from ciliary ganglia pieces in a co-culture system (EBENDAL 1979).

IV. Characterization of Ciliary Neuron Survival Factors

Using the selected intra-ocular tissues of the 15-day eye (CIPE) the activity necessary for the 24-h survival of chick ciliary neurons *in vitro* (the ciliary neuronotrophic factor, CNTF) has been purified approximately 400-fold to apparent homogeneity (BARBIN et al. 1984; MANTHORPE et al. 1982a; MANTHORPE et al. 1980) by a sequence of DEAE ion-exchange chromatography, ultrafiltration-concentration, sucrose-gradient centrifugation and preparative sodium dodecylsulphate polyacrylamide gel electrophoresis (SDS-PAGE). Purified chick CNTF has a M_r of 20400 (both on sucrose gradients under non-denaturing conditions and in SDS-PAGE) and is an acidic molecule focusing in iso-electric gels in a broad region between pI 4.8 and 5.2. Using gel filtration on Sephadex G100 and G200, earlier studies (MANTHORPE et al. 1980) had shown the activity eluted with an apparent M_r of 35000 – 40000 and also that it focused sharply in iso-electric gels in a region at pH 5.0. It has been suggested (BARBIN et al. 1984) that these differences between purified CNTF and cruder fractions may result from the dissociation of a dimer into monomers and also be due to the action of lytic enzymes generating multiple charge forms.

Using a similar assay system BONYHADY et al. (1980) reported that the activity in ox heart extracts supporting the *in vitro* survival of ciliary neurons behaved on gel-filtration columns as a complex with a M_r greater than 50000 that dissociated into lower molecular mass components with a M_r of 20000. Dissociation was later shown to be reversible (BONYHADY et al. 1982). The activity in ox heart extracts binds to DEAE columns at neutral pH, is eluted from them by 0.11–0.16 *M* NaCl and has a pI of 6.2. Chick eye CNTF also binds to DEAE at neutral pH and elutes between 0.07 and 0.25 *M* NaCl but is more acidic (pI 5.0). However other studies have shown that the activities in crude extracts of both chick heart and eye have iso-electric points between 4.5 and 5.5 (HILL et al. 1981). Using ciliary ganglion explants and evaluating fibre outgrowth as a means of detecting activity, EBENDAL et al. (1979) reported that chick heart extracts contained activity which had a M_r of 40000 on G200 gel-filtration columns. More recent studies have shown that extracts of rat peripheral nerves and spinal nerve roots contain a CNTF for chick cells at a specific activity equal to or greater than that found in extracts of the combined, selected intraocular tissues of the chick eye (WILLIAMS et al. 1984). The activities present in mouse sciatic nerve, chick eye, chick heart and chick extra-embryonic membranes can all be recovered from SDS gels of crude extracts (COLLINS 1985) and all have almost identical molecular masses (21.5–22.0 kDa).

The purified eye CNTF is not specific for the cholinergic neurons of the ciliary ganglion and will also support, for example, the survival of 10-day chick dorsal root ganglion cells and 11-day chick spinal ganglion cells (BARBIN et al. 1984), both of which are traditional targets of NGF. The 8-day chick ciliary ganglion is not, however, responsive to NGF. The brain-derived growth factor purified by BARDE et al. (1982) from pig brain also does not support the survival of 8-day chick ciliary neurons (VARON et al. 1983).

V. Factors Affecting the Growth and Development of Ciliary Neurons *In Vitro*

Elevating K^+ levels in medium has been shown to increase neuronal survival in a number of neuronal culture systems (human dorsal root ganglion, SCOTT 1977; cerebellar neurons, LASHER and ZAGON 1972; sympathetic neurons, PHILLIPSON and SANDLER 1975), and the general assumption has been that high K^+ causes depolarization and mimics, to a certain degree, the effects of neuronal activity. High extracellular K^+ (25 m*M*) allows both the short- and long-term survival of chick ciliary neurons *in vitro* in standard non-supplemented culture media in both dissociated and explant culture systems (BENNETT and WHITE 1979; NISHI and BERG 1981 b; KATO and RAY 1982).

By exploiting this possibility of maintaining ciliary neurons in long-term culture by elevating extracellular K^+, NISHI and BERG (1981 a) demonstrated that two components could be resolved in extracts of 17-day chick embryo eyes after gel filtration on Bio-Gel P-200 columns which, independently from cell survival, stimulated growth. One component had an apparent M_r of 20000 and stimulated growth (increase in somal size and lactate dehydrogenase content) without affecting ChAT content per neuron and the other had an apparent M_r of 50000 and stimulated ChAT activity without affecting neuronal growth. These two activities were called growth promoting activity (GPA, 20 kDa) and ChAT stimulating activity (CSA, 50 kDa), although it should be noted that both the CSA and GPA fractions of eye extract also supported ciliary neuron survival in the absence of high K^+. In the presence of 25 mM K^+ saturating levels of CSA and GPA caused 2-fold increases in the ChAT content per neuron and lactate dehydrogenase (LDH) content per neuron respectively, relative to control cultures after 9 days *in vitro.*

Unfortunately an exact comparison between the rate of ChAT accumulation *in vitro* and *in vivo* is difficult to make due to the presence of pre-synaptic cholinergic terminals in the intact ciliary ganglion. Total ChAT activity per ganglion increases nearly 30-fold during the 7-day period between day 11 and day 18 (stages 37–44) (CHIAPPINELLI et al. 1976). TUTTLE et al. (1980) have calculated that ChAT activity per ciliary neuron increases 25-fold over the same period. Starting with cultures from 7.5 day embryos (stages 32–33), the results of NISHI and BERG (1981 a) show that ciliary ganglion cells increase their ChAT content by 6-fold over the 7-day culture period between days 3 and 10 *in vitro* (equivalent to embryonic days 11–18). The ChAT activity per neuron after 10 days *in vitro* was 0.3 pmol per neuron. This can be compared to an *in ovo* estimate of 5 pmol per neuron at day 17 (TUTTLE et al. 1980). While such comparisons should be treated with caution, they do suggest that ChAT accumulation is slow *in vitro* even under conditions that are considered to be optimal for both growth and survival. In this respect it should be noted that TUTTLE et al. (1980) reported that HCM, although toxic in the long-term for the survival of ciliary neurons in their hands, could cause a very rapid increase in ChAT activity per neuron (up to 100-fold) during the first 8 days *in vitro.* This result is interesting in view of the increases in the ChAT content of cultured sympathetic neurons caused by either co-culture with ganglionic non-neuronal cells or the use of HCM (PATTERSON and CHUN 1977 a, b; see below, Sect. D). This suggests the possibility that factors similar to those in HCM but not present in eye extracts are required for the development of normal ChAT levels in ciliary neurons *in vitro.* However KATO and RAY (1982) have examined the effects of co-culturing ciliary neurons with myotubes and ganglionic non-neuronal cells in the presence of eye extract and with normal extracellular K^+ and have found no additional stimulation of either acetylcholine (ACh) synthesis or accumulation.

The addition of 25 mM K^+ to ciliary neuron cultures grown in saturating levels of 17-day eye extract on a substrate of lysed fibroblasts also causes a 1.5- to 2.0-fold increase in ChAT and LDH activity, choline uptake, and both ACh and protein synthesis relative to controls grown in normal extracellular K^+ after a period of 11 days *in vitro*, without affecting neuronal survival (NISHI and BERG 1981 b).

VI. Factors Stimulating Neurite Growth in Ciliary Neuron Cultures

COLLINS (1978) first demonstrated that HCM contained macromolecules that would bind to a PORN substrate and promote the outgrowth of neurites from ciliary neurons. Segregation of the neurite-promoting activity and the activity necessary for ciliary neuron survival present in HCM that had originally been described by HELFAND et al. (1976) was achieved by ADLER and VARON (1980). Overnight incubation of HCM in PORN-coated tissue culture plastic dishes yields two products: an HCM-primed PORN substrate that will promote neurite production but is unable on its own to support 24-h survival and an absorbed HCM that will not promote neurite outgrowth on a PORN substrate but is capable of providing 24-h support of ciliary neurons on either a PORN or a collagen substrate. Using a similar procedure but looking at the rate of neurite elongation as opposed to neuron survival COLLINS and DAWSON (1982) also demonstrated the presence of two separate activities in HCM: one that binds to PORN and promotes neurite production and another that is soluble, does not bind to PORN and can be shown to increase the rate of neurite elongation in a survival-independent assay. The two activities could also be separated by filtration of HCM through an Amicon membrane with a 100-kDa cut-off; the PORN-binding neurite-promoting material was retained and the filtrate contained the soluble activity that stimulated the rate of neurite elongation. PORN-binding neurite-promoting factors (PNPFs) to which ciliary neurons are responsive can be found in media conditioned by a wide variety of chick primary cell cultures and cell lines from mouse, rat and human tissues (ADLER et al. 1981). Little species specificity is observed and the effect is not specific for cholinergic neurons. Interestingly, PNPFs cannot be detected in media conditioned by certain chick primary cultures, such as ganglionic non-neuronal cell cultures and pure neuronal cell cultures.

The extracellular matrix protein laminin has been shown to have neurite-promoting activity when bound either to tissue culture plastic or PORN, but neurite-promoting activity can be detected at lower dilutions when a PORN substrate is utilized (MANTHORPE et al. 1983a). Half-maximal neurite promotion can be obtained by coating substrata with laminin at concentrations of 50 ng ml^{-1}, which represents a concentration of approximately 40 p*M*. This study also showed that an antiserum to rat laminin would not block the neurite-promoting activity of a rat PNPF partially purified from a rat RN22 Schwannoma-conditioned medium though it would block the activity of both purified rat and mouse laminins. However, in a more recent study of conditioned media from six different sources, all of which contained PNPF(s) for sympathetic neurons, it was shown that while anti-laminin antibodies failed to block the neurite-promoting activity bound to the substrate from five out of six media, they would immunoprecipitate the activity from all of the media (LANDER et al. 1985). While this suggests that laminin may be the active factor present in all conditioned media, it also raises the possibility that laminins from different sources may have slight differences perhaps either in their structure or in material associated with them.

Using a system to assay for factors stimulating the rate of neurite elongation as opposed to factors promoting the formation of neurites COLLINS and DAWSON (1983) have shown that mouse 2.5-S NGF will stimulate the rate of neurite elongation in cultures of chick ciliary neurons. This effect of NGF was seen at concentrations between 0.1 and 10 ng ml^{-1}, the concentration range over which NGF normally exerts its biological effects on sympathetic and sensory neurons. Mouse submaxillary NGF has no effect on ciliary neuron survival *in vitro* (BARBIN et al. 1984; see also HELFAND et al. 1976, 1978), although it is retrogradely transported by ciliary nerves *in vivo* (MAX et al. 1978).

C. Spinal Cord Neurons

I. Greater Complexity of the System

The establishment of assays for studying the survival requirements of spinal motor neurons has proven to be more complex than it is for ciliary neurons. This is mainly due to the large number of different cell types present in the spinal cord and an apparent ability of the neurons to survive *in vitro* with no special media supplements. The experimental evidence for the existence of a growth factor for motor neurons has been recently reviewed by SLACK et al. (1983) and the experimental evidence suggesting the operation of a motor neuron survival factor during development is similar to that described for the ciliary neurons. During development of the lateral motor columns an excess of motor neurons is produced and the population size decreases by nearly 40% between days 5 and 9 (HAMBURGER 1975). This naturally occurring motor neuron cell death can be increased by limb bud removal (HAMBURGER 1934) and reduced by the addition of supernumerary limbs (HOLLYDAY and HAMBURGER 1976). This evidence can be interpreted as suggesting that motor neuron survival is dependent on a supply of some trophic agent from the limb.

II. Survival Factors for Spinal Cord Neurons

Early studies with dissociated cell cultures were prepared from mice on embryonic days (ED) 12–14 or 4-day chick embryo spinal cords showed that spinal cord cells had no absolute requirement for the presence of muscle cells for their survival *in vitro* (GILLER et al. 1973; GILLER et al. 1977; BERG 1978). However, cultures were prepared at high surface density (2500 mm^{-2} for chick cultures and 1400 mm^{-2} for mouse cells); they contained a large number of different cell types, and no positive identification of motor neurons was made. BENNETT et al. (1980), using dissociated cell cultures from 9-day chick spinal cords containing motor neurons prelabelled with horseradish peroxidase (HRP) that had been retrogradely transported from the limb muscles, were able to show that 48-h survival of the HRP-labelled motorneurons *in vitro* on a polylysine substrate required the presence of serum containing skeletal muscle-conditioned medium. Serum-containing media conditioned by smooth muscle cells or kidney cells were ineffective in providing 48-h support. Plating densities were not quoted in this study but were probably considerably lower than in earlier ones. In 1981 SCHNAAR and SCHAFFNER used metrizamide gradients to fractionate dissociated 6-day chick spinal cord preparations into three fractions. Cultures prepared from the fraction with the lowest buoyant density, which contained most of the large cells present in the dissociate and 84% of the ChAT activity recovered from the gradient, required both a polylysine substrate and muscle-conditioned medium for 6-day survival *in vitro*. The cells were plated at a low surface density (150 cells mm^{-2}) in this study and a combination of muscle cell-conditioned medium with a collagen substrate was insufficient for 6-day survival. Cells from the lower two fractions did not require muscle cell-conditioned medium for 6-day survival *in vitro*.

Subsequently LONGO et al. (1982), using dissociates from the 4-day chick lumbar spinal cord and plating cells at a surface density of 25 cell mm^{-2} in serum-free medium, reported that spinal cord cell neuronotrophic factors could be detected in serum-free conditioned media from chick heart and muscle primaries, mouse Schwann cell primaries and rat RN22 Schwannoma cells. Partially purified chick CNTF and mouse NGF were both ineffective in supporting spinal cell survival on a PORN substrate. The trophic titres of the conditioned media were low, ranging from 4–16 TU ml^{-1} conditioned medium. If 4-day lumbar cord cells are plated at higher surface density (250 cells mm^{-2}) in serum-free medium, no exogenous factors are required for 24-h survival and medium conditioned by high-density lumbar cord cell-cultures will support low density cultures (MANTHORPE et al. 1983b). The putative spinal cord cell neuronotrophic factors present in conditioned media necessary for 24-h survival are dialysable and heat- and trypsin-insensitive (MANTHORPE et al. 1982b). One of the heat-stable, trypsin-insensitive, dialysable components of astroglial-conditioned medium that supports the 24-h survival of avian and rodent central neurons *in vitro* (VARON et al. 1984) has been shown to be pyruvate (SELAK et al. 1985). This is not a component of all basal media and this observation has led to the suggestion (SELAK et al. 1985) that certain classes of neurons may display a dependence on an exogenous supply of pyruvate for their survival at certain stages of their development. However, even in low density cultures of cell sorter-purified, 6-day chick cord motor neurons maintained in a serum-free, pyruvate-containing medium the percentage of healthy cells with neurites rapidly declines after 2 days *in vitro* in the absence of serum-free muscle cell-conditioned medium (CALOF and REICHARDT 1984). More recently DOHRMANN et al. (1986) have used a procedure similar to that described by SCHNAAR and SCHAFFNER (1981) to purify prelabelled motoneurons and have shown that motoneuron survival for periods of longer than 3–4 days requires the presence of embryonic muscle tissue extracts in the medium.

High titres of heat- and trypsin-sensitive, non-dialysable activities necessary for the short-term survival of low-density lumbar spinal cord cell cultures can be found in peripheral and central nervous system wound fluids (MANTHORPE et al. 1982b). The trophic titres of these wound fluids range from 100–12000 TU ml^{-1}. Long-term *in vitro* survival of ED 13–14 rat-cord dissociated cell cultures grown at high surface density (550 cells mm^{-2}) in the presence of pyruvate-containing basal medium requires the presence of serum and, after gel filtration of serum at low pH, activity necessary for long term survival was recovered in a fraction with a molecular mass of 55 kDa (KAUFMAN and BARRETT 1983). Iso-electric focusing of this fraction showed that the activity focused at a point between pH 5 and 6. Extracts of denervated adult rat and mouse muscles (NURCOMBE et al. 1984; SLACK and POCKETT 1982) and CEE (SLACK and POCKETT 1982) have also been shown to contain activity that will support the survival of identified chick motor neurons *in vitro* but these activities have not been characterized.

III. Factors Stimulating ChAT Activity in Spinal Cord Cultures

The studies of GILLER et al. (1973, 1977) showed that the ChAT activities of dissociated spinal cord cell cultures from 12–14 ED mice could be increased by up to 20-fold after 2 weeks *in vitro* relative to control cultures by co-culturing them with either mouse myotubes or by growing them in medium conditioned by mouse myotubes. ChAT accumulation rate was very slow (almost negligible) in the absence of myotubes or myotube-conditioned medium, whereas it rose steadily with time in the presence of myotubes increasing 10-fold over a 12-day period from days 2–14 *in vitro*. AChE, creatine phosphokinase, phosphoglucomutase, myokinase and phosphorylase activities were not affected by the presence of myotubes, nor was cell survival. Co-culture also caused a sharp decrease in glutamic acid decarboxylase activity after 7 days *in vitro*, but this effect could not be mimicked by conditioned medium. The active agent in conditioned medium was non-dialysable and had a molecular mass greater than 50 kDa. Media conditioned by both rat and mouse muscle cells stimulate the development of ChAT activity in mouse spinal cord cell cultures, but medium conditioned by chick myotubes does not (GODFREY et al. 1980). Media conditioned by mouse heart, liver and kidney primary cell cultures also stimulate development of ChAT activity and therefore, although the effect is relatively species specific, the active agent can be produced by a variety of cell types. BROOKES et al. (1980) repeated the experiments of GILLER et al. (1977) and found that co-culturing, although it stimulated ChAT activity, did not cause significant increase in the ACh content of the cultures, although this may have been due to a non-saturating choline concentration in the medium. Using 4-day chick spinal cord cells BERG (1978) was able only to stimulate a small increase in ACh synthesis by co-culturing the cells with myotubes. Only a 50% increase in ACh synthesis and storage and a 60% increase in ChAT activity was found after co-culture and this was only evident after the first week *in vitro*. However, the ability to synthesize and store ACh did increase 10-fold during the 12-day culture period between days 3 and 14 *in vitro* in co-cultures, suggesting that the main difference between this study and that of GILLER et al. (1977) lies in the ability of chick spinal neurons to accumulate ChAT in the absence of myotubes (but in the presence of embryo extract). However it should be noted that ChAT activity per spinal cord of the chick increases at least 60-fold between days 6 and 9 (MANTHORPE et al. 1983b) whereas the increase in ACh synthesis and storage in the cultures of BERG (1978) is less than 5-fold over the equivalent period *in vitro*. MEYER et al. (1979) examined ACh synthesis in both explant and dissociated cell cultures of chick and mouse spinal cords grown in either the presence or absence of muscle cells from the same species. Although motor neuron cell curvival was not evaluated, the presence of muscle cells caused a stimulation of ACh synthesis with dissociated spinal cord cells from both chick and mouse, but no stimulation was observed in explant cultures. ACh synthesis in co-cultured dissociated spinal cord cell cultures was nearly 3-fold higher than in controls. The amount of ACh synthesis observed in spinal cord explants was similar to that observed in muscle-stimulated dissociated cord cell cultures suggesting that intrinsic influences within the cord are as potent a stimulus as that provided by the muscle cells. These experiments do show that co-culturing can effect ACh synthesis in dissociated chick cord cell cultures as well as in mouse cord cell cultures. Unfortunately the plating density was not quoted in the experiments of MEYER et al. (1979) so that a direct comparison with the study of BERG (1978) cannot be made.

ISHIDA and DEGUCHI (1983a) reported that co-culturing ED 13–14 mouse spinal cord dissociates with heart cells caused a 30-fold stimulation in ChAT activity relative to controls after 11 days *in vitro*. These authors also reported that the neurotransmitter L-noradrenaline (L-norepinephrine) could cause up to 8-fold increases in ChAT activity of dissociated spinal cord cell cultures 8 days after its first addition to the medium. This effect was probably mediated by the proliferating non-neuronal cells of the cultures as it was largely absent when non-neuronal cell numbers were reduced by treatment of the cultures with anti-mitotics. Similar effects to those produced by noradrenaline could also be caused by dibutyryl cAMP and 8-bromo-cAMP, and it was suggested that noradrenaline was acting indirectly on the neuronal cells via the release of cAMP from the non-neuronal cells. BRENNEMAN and WARREN (1983) also reported that dibutyryl cAMP and 8-bromo-cAMP would cause 2- to 3-fold stimulation in the ChAT activity of 12–14 ED dissociated mouse cord cell cultures but did not consider the possibility that the effect was mediated by the non-neuronal satellite cells. These results illustrate some of the dangers of working with mixed populations of cells. Thyrotropin-releasing hormone has also been reported to cause a 16-fold increase in the ChAT content of 13–15 ED rat spinal cord explant cultures relative to controls after 3–5 weeks *in vitro* (SCHMIDT-ACHERT et al. 1984), but cell survival was not evaluated in these experiments.

ACh synthesis and ChAT activity in anti-mitotic-treated, dissociated 14–15 ED rat ventral cord cell cultures grown in pyruvate containing basal medium with horse serum at a surface density of 500 cells mm^{-2} can be significantly increased after 4 days *in vitro* by a heat- and trypsin-labile factor present in a soluble extract of neonatal rat muscle (SMITH and APPEL 1983). Extracts from cardiac muscle and cerebral cortex from neonatal rats were effective in this system, but extracts from spleen, liver, lung, skin or kidney had little effect on ACh synthesis in the cultures. The presence of a low molecular weight (1300–1500 Da), trypsin-insensitive peptide in 1–5 day post-natal rat muscle extracts that stimulates ACh synthesis in this culture system has also been reported (MCMANAMAN et al. 1985). Using high surface density 14-ED rat cord cell cultures (550 cells mm^{-2}), maintained in medium containing the serum fraction necessary for long-term cord cell survival (KAUFMAN and BARRETT 1983; see above) to detect activity stimulating growth, KAUFMAN et al. (1985) have used high-performance liquid chromatography (HPLC) to characterize the active components of muscle- and lung-conditioned media and neonatal muscle extracts. Activity from both the conditioned media and the muscle extracts elutes in a fraction with a molecular mass of 40 kDa, and it increases ACh synthesis, ChAT activity, protein synthesis and LDH content of the cord cell cultures but depresses γ-aminobutyrate (GABA) accumulation. Crude muscle extracts did not depress GABA accumulation but conditioned media did. The active fraction caused 2- to 3-fold increases in ChAT activity after a 10-day culture period relative to controls over a concentration range of 50–100 μg protein ml^{-1} culture medium. GEISS and WEBER (1984) have also examined the effects of muscle-conditioned medium on 14-ED rat cord cell dissociates and have shown that the activity stimulating ACh synthesis co-purifies with the activity in muscle-conditioned medium that stimulates ChAT accumulation and ACh synthesis in sympathetic superior cervical ganglion cell cultures (see below, Sect. D), suggesting that the two activities reside in the same molecule. Conditioned medium also stimulated AChE synthesis in these cultures. However the response of the superior cervical ganglion cultures to the conditioned medium factor(s) was always at least one order of magnitude greater than the response of the spinal cord cells although the amount of ACh storage and synthesis in untreated cultures was equivalent. The number of motor neurons and superior cervical ganglion cells in the two types of culture have not been compared.

As with ciliary ganglion cells, depolarizing agents (high extracellular K^+ and veratridine) also cause an increase in the ChAT levels and ACh synthesis in foetal mouse spinal cord cell cultures without significantly effecting protein synthesis (ISHIDA and DEGUCHI 1983b). AChE levels are depressed by these agents. Calcium antagonists (Mg^{2+} and vermapil) blocked the effects of the depolarizing agents, suggesting that calcium influx was mediating the effect.

IV. Factors Stimulating Neurite Growth in Spinal Cord Cultures

In 1980 DRIBIN and BARRETT reported that media conditioned by primary cultures of embryonic and neonatal skeletal muscle, fibroblasts or lung cells cause a dramatic increase in the amount of neurite outgrowth obtained from 15–16 ED rat spinal cord explants when they are grown on a polylysine-coated collagen substrate. The factor had a molecular mass greater than 100 kDa and was stable at 58 °C for 30 min. Subsequent studies showed that the effect could be observed on both collagen and polylysine substrates and that little species specificity was encountered: chick, rat and mouse primary cell culture conditioned media all stimulated neuritic outgrowth from rat, chick and mouse spinal cord explants although media from mammalian primaries were better for all cord types (DRIBIN 1982).

Media conditioned by primary cultures of heart, skeletal muscle and Schwann cells and rat RN22 Schwannoma cells, and extracts of muscle tissue (HENDERSON et al. 1981; LONGO et al. 1982; HENDERSON et al. 1983, 1984; CALOF and REICHARDT 1984; NURCOMBE et al. 1984) will stimulate neurite outgrowth from spinal cord cells *in vitro*. Activity can be detected using plastic, collagen, polylysine and polyornithine substrates. DRIBIN and BARRETT (1982b) have reported that there are two distinct fractions in both muscle- and fibroblast-conditioned media, one with a molecular mass of approximately 50 kDa which must be continuously present in the medium to exert its effect and a second with a molecular mass greater than 300 kDa which is active when bound to the substrate. A combination of the two components is required for the full expression of the outgrowth-promoting potential of unfractionated conditioned medium. The situation appears to be very similar to that described above for the ciliary neurons (Sect. B. VI) and it seems most likely that the high molecular mass, substrate-bound activity is laminin or has laminin associated with it. DOHRMANN et al. (1986) have shown that the neurite-promoting activity present in embryonic skeletal muscle extracts that operates on purified motor neurons can be immunoprecipitated by anti-laminin serum. Some further characterization of the

50-kDa component present in conditioned medium has been made (DRIBIN and BARRETT 1982a), but it has not as yet been purified.

D. Sympathetic Neurons

I. Evidence for a Cholinergic Subpopulation

The majority of sympathetic neurons in the mammal are adrenergic and require NGF for their survival and for the maintenance of differentiated function. However a small proportion of these neurons are cholinergic, and a recent study (LANDIS et al. 1985) has shown that treatment of neonatal rats with antiserum to NGF eliminates the cholinergic sympathetic innervation of rat sweat glands. The cholinergic neurons that innervate the sweat glands are therefore probably dependent on NGF for their survival *in vivo*, although they may be rather unusual and express adrenergic phenotype at an early stage in their development before becoming cholinergic. Nonetheless, cultures of rat superior cervical ganglia can be switched to a cholinergic phenotype under appropriate experimental conditions (see below) and even when expressing cholinergic characteristics still maintain a dependence on NGF (CHUN and PATTERSON 1977). The situation in the chick appears to be different, where a subpopulation of sympathetic neurons from the paravertebral ganglia which have a high ChAT content and a low tyrosine hydroxylase content appear to require a factor from HCM for their survival *in vitro* and not NGF (EDGAR et al. 1981).

II. Induction of Cholinergic Transmitter Status in Sympathetic Neurons

In 1974 PATTERSON and CHUN showed that primary, dissociated cultures of sympathetic neurons from the rat superior cervical ganglion which normally synthesize considerable amounts of dopamine and noradrenaline and little ACh (100- to 1200-fold less for cultures grown in buffered L-15 medium in air) developed the capacity to synthesize large amounts of ACh when grown in the presence of non-neuronal satellite cells from the ganglion. By measuring the net synthesis of [^{3}H]ACh from [^{3}H]choline over a 4-h period in the presence of eserine sulphate it was shown that ACh synthesis was increased 10- to 30-fold by the presence of non-neuronal cells relative to sister cultures grown under identical conditions. As had been previously shown by MAINS and PATTERSON (1973), catecholamine (CA) synthesis was little affected by the presence or absence of ganglionic non-neuronal cells. Subsequent studies (PATTERSON et al. 1975; PATTERSON and CHUN 1977a) showed that ChAT activity was from 50- to 1000-fold higher in homogenates of mixed cultures of non-neuronal and neuronal cells than in either neuronal or non-neuronal cells alone, that ACh synthesis and ChAT activity could be stimulated by HCM and that such treatment had no effect on neuronal survival or growth: total neuron numbers or the amount of phospholipid or protein per neuron were unaffected by the presence of conditioned medium. Mixing homogenates from pure neuronal cultures and mixed cultures provided evidence that the mixed cultures did not contain a ChAT activator, and that the pure neuronal cell cultures did not contain a ChAT inhibitor (PATTERSON and CHUN 1977a). These experiments therefore demonstrated that certain non-neuronal cell cultures secreted a diffusible trophic agent into the medium that could greatly stimulate the production of ACh and ChAT in rat sympathetic neurons that were normally adrenergic.

These initial studies also showed that the effect of HCM was dose-dependent; ACh production after 20 days *in vitro* rose 40-fold over the 0%–62% conditioned-medium concentration range studied. Over the same concentration range of HCM there was, in these experiments, a 25-fold decrease in CA production. The ability to synthesize and store CA was inversely related to the percentage of HCM in the medium. Experiments performed on isolated, single sympathetic neurons (REICHARDT and PATTERSON 1977) grown under control conditions or conditions that stimulate ACh production in mass culture suggested that most neurons synthesized either CA or ACh depending on the condition, and that neither very few dual function neurons synthesizing both CA and ACh nor silent neurons synthesizing neither transmitter were present. This suggested that the neurons responded in a 'flip-flop' fashion to the cholinergic-specifying factor present in conditioned medium, either becoming purely cholinergic in its presence or purely adrenergic in its absence. However, a high-affinity uptake system for noradrenaline is retained by sympathetic neurons producing ACh (REICHARDT and PATTERSON 1977; WAKSHULL et al. 1978), and, in a study of the enzymes responsible for CA production and both the synthesis and breakdown of ACh, SWERTS et al. (1983) found that in cultures of sympathetic neurons grown for 20 days in muscle-conditioned medium, tyrosine hydroxylase, dopamine decarboxylase, dopamine β-hydroxylase and AChE activities were only decreased 3-fold while ChAT activity was increased 50-fold. CA production decreased from 2- to 3-fold and ACh production was increased 27-fold. Evidently the decrease in enzymes involved in CA production occurs at a much slower rate than the increase in ChAT activity. Complete loss of adrenergic properties depends on the concentration and source of the conditioning stimulus, the duration of the culture period and the culture conditions. For example, using a completely different set of culture conditions it can be shown that ChAT and dopamine decarboxylase activities can increase *pari passu in vitro* (JOHNSON et al. 1980). Increases in ChAT activity and ACh production caused by HCM are accompanied by a gradual change in transmitter vesicle type; dense-core vesicles become less numerous and clear vesicles increase in number (LANDIS 1976; JOHNSON et al. 1976).

III. Characterization of the Cholinergic-Specifying Factor

The partial purification and characterization of the cholinergic-specifying factor present in media conditioned by rat C6 glioma cells or primary heart cells in the presence of 10% fetal calf serum was described by WEBER (1981). The active factor was purified approximately 1500-fold by a sequence of ammonium sulphate precipitation, DEAE- and CM-cellulose ion exchange chromatography and gel filtration on Sephadex G-100. Activity eluted from the Sephadex columns as a symmetrical peak with an apparent molecular mass of 40–45 kDa, and this fraction caused a 20-fold increase in the ACh:CA ratio at a concentration of 1 μg ml^{-1} in the culture medium. The activity does not bind to DEAE columns at neutral pH, indicating that it is a basic molecule, and that the activity of the fraction not binding to DEAE is not inactivated by treatment with reducing agents, denaturants or a combination of both treatments. The activity eluting from the CM column with 0.2 *M* NaCl was lost after periodate treatment and reduced by 30%–35% after treatment with immobilized pronase. In a further study WEBER et al. (1985) used muscle cell-conditioned medium and included an additional initial purification step with hydroxyapatite to which the activity does not bind. Molecular sieving on Ultragel AcA 44 also showed that the activity behaved as if it had a Stokes radius similar to that of ovalbumin. However using sucrose-gradient centrifugation the partial specific volume was found to be considerably lower than most non-glycosylated globular proteins (0.68 ml g^{-1} versus 0.7–0.75 ml g^{-1}) and the sedimentation coefficient was 2.1 *S*. From this data a molecular mass of 21 kDa and a frictional ratio of 1.56 were determined. This partially purified factor was found to be heat stable, losing only 13% of its activity after treatment at 100 °C for 10 min, but crude fractions were sometimes completely inactivated by this treatment.

Further progress in the purification of this factor has been made by FUKADA (1985) using hormone-supplemented, serum-free conditioned medium (FUKADA 1980) as the starting material, the same sequence of purification steps as WEBER (1981), followed by SDS-PAGE. Activity was eluted from a gel slice with a molecular mass corresponding to 45 kDa. The 45-kDa band could be converted in a stepwise manner by treatment with endoglycosidase F to a 22-kDa component, again suggesting that the cholinergic-specifying factor is heavily glycosylated. Endoglycosidase F treatment does not, however, lead to a loss of biological activity. It was estimated in this study that the cholinergic-specifying factor had been purified over 100000 times by this method.

E. Conclusions

In the last decade several factors have been shown to have the effects *in vitro* that are expected of trophic agents that might be operational during the development of cholinergic neurons *in vivo*. CNTF, NGF, the cholinergic-specifying factor of PATTERSON and colleagues, and laminin have all been shown or suggested to have effects on various aspects of cholinergic motor neuron development *in vitro*. These factors are listed in Table 1. Of these agents the best demonstration of any significant role *in vivo* exists only for NGF – with the demonstration that treatment of rat pups with antiserum to NGF causes elimination of cholinergic innervation of the sweat glands (LANDIS et al. 1985). CNTF is only the third neuronotrophic factor to be purified to homogeneity and it is operational on chick ciliary neurons *in vitro* at molar concentrations equal to or even 10-fold lower than those at which mouse NGF is active on chick dorsal root ganglion cells (BARBIN et al. 1984). However CNTF is present in very high concentrations in the eye even in early development, suggesting that the ciliary neurons may be constantly saturated with CNTF unless its release and availability are tightly regulated.

Table 1. Trophic factors required for the survival and development of cholinergic neurons

Factor	Molecular mass (daltons)	Isoelectric point (pI)	Sedimentation coefficient (S)	Property
CNTF	20400	4.8 – 5.2	–	Supports the survival of ciliary neurons *in vitro*
β-NGF	26518 (dimer)	9.3	2.5	Stimulates rate of axon elongation in ciliary neuron cultures; increases ChAT in central neurons
Cholinergic specifying factor	21000	basic	2.1	Increases ChAT in spinal cord cultures; stimulates expression of cholinergic phenotype in sympathetic neurons
Laminin	$\simeq 1 \times 10^6$ (3 or 4 subunits)	–	–	Neurite-promoting agent

The development of neutralizing antibodies to CNTF should enable its role *in vivo* to be critically evaluated. However, the ability to purify CNTF will now allow its effects and those of NGF and the cholinergic-specifying factor on ChAT accumulation in ciliary neurons *in vitro* to be examined more carefully. Only in the sympathetic nervous system may the survival of cholinergic neurons be dependent on NGF, but NGF does stimulate neurite growth in the parasympathetic system (Collins and Dawson 1983), and more recent studies are suggesting a role for NGF in the control of ChAT levels in central neurons (Gnahn et al. 1983; Mobley et al. 1985; Martinez et al. 1985). None of the four factors cited above show any strict specificity for cholinergic neurons but there is no *a priori* reason to expect growth factors to recognize the boundaries set up by transmitter choice unless they are determining it; they may simply operate *in vivo* as a function of their distribution within both target- and supporting-tissues.

At present, the purification of the macromolecules necessary for spinal motor neuron survival, growth and development provides one of the most intriguing challenges in neurobiology. The multitude of different culture and assay systems that have been used by different investigators makes progress hard to define although it appears that serum factors, low molecular mass agents, muscle-derived factors and intrinsic, cord-derived factors all might play some role in spinal motor neuron development. The use of purified motor neuron preparations (Schnaar and Schaffner 1981; Dohrmann et al. 1986) should enable more progress to be made in this field.

References

Adler R, Varon S (1980) Cholinergic neuronotrophic factors: V. Segregation of survival- and neurite-promoting activities in heart-conditioned media. Brain Res 188:437–448

Adler R, Landa KB, Manthorpe M, Varon S (1979) Cholinergic neuronotrophic factors: intraocular distribution of trophic activity for ciliary neurones. Science 204:1434–1436

Adler R, Manthorpe M, Skaper SD, Varon S (1981) Polyornithine-attached neurite-promoting factors (PNPFs): culture sources and responsive neurons. Brain Res 206:129–144

Barbin G, Manthorpe M, Varon S (1984) Purification of the chick eye ciliary neuronotrophic factor. J Neurochem 43:1468–1478

Barde Y-A, Edgar D, Thoenen H (1982) Purification of a new neuronotrophic factor from mammalian brain. EMBO J 1:548–549

Bennett MR, White W (1979) The survival and development of cholinergic neurons in potassium-enriched media. Brain Res 173:549–553

Bennett MR, Lai K, Nurcombe V (1980) Identification of embryonic motoneurons *in vitro*: their survival is dependent on skeletal muscle. Brain Res 190:537–542

Berg D (1978) Acetylcholine synthesis by chick spinal cord neurons in dissociated cell culture. Dev Biol 66:500–512

Bonyhady RE, Hendry IA, Hill CE, McLennan IS (1980) Characterization of a cardiac muscle factor required for the survival of cultured parasympathetic neurones. Neurosci Lett 18:197–201

Bonyhady RE, Hendry IA, Hill CE (1982) Reversible dissociation of a bovine cardiac factor that supports survival of avian ciliary ganglionic neurones. J Neurosci Res 7:11–21

Brenneman DE, Warren D (1983) Induction of cholinergic expression in developing spinal cord cultures. J Neurochem 41:1349–1356

Brookes N, Burt DR, Goldberg AM, Bierkamper GG (1980) The influence of muscle-conditioned medium on cholinergic maturation in spinal cord cell cultures. Brain Res 186:474–479

Calof AL, Reichardt LF (1984) Motoneurons purified by cell sorting respond to two distinct activities in myotube-conditioned medium. Dev Biol 106:194–201

Chiappinelli V, Giacobini E, Pilar G, Uchimura H (1976) Induction of cholinergic enzymes in chick ciliary ganglion and iris muscle cells during synapse formation. J Physiol (Lond) 257:749–766

Chun LLY, Patterson PH (1977) Role of nerve growth factor in the development of rat sympathetic neurons *in vitro*. III. Effect on acetycholine production. J Cell Biol 75:712–718

Collins F (1978) Induction of neurite outgrowth by a conditioned-medium factor bound to the culture substratum. Proc Natl Acad Sci USA 75:5210–5213

Collins F (1985) Electrophoretic similarity of the ciliary ganglion survival factors from different tissues and species. Dev Biol 109:255–258

Collins F, Dawson A (1982) Conditioned medium increases the rate of neurite elongation: separation of this activity from the substratum-bound inducer of neurite outgrowth. J Neurosci 2:1005–1010

Collins F, Dawson A (1983) An effect of nerve growth factor on parasympathetic neurite outgrowth. Proc Natl Acad Sci USA 80:2091–2094

Dohrmann U, Edgar D, Sendtner M, Thoenen H (1986) Muscle-derived factors that support survival and promote fiber outgrowth from embryonic chick spinal motor neurons in culture. Dev Biol 118:209–221

Dribin LB (1982) On the species and substrate specificity of conditioned medium enhancement of neuritic outgrowth from spinal cord explants. Dev Brain Res 3:300–304

Dribin LB, Barrett JN (1980) Conditioned medium enhances neuritic outgrowth from rat spinal cord explants. Dev Biol 74:184–195

Dribin LB, Barrett JN (1982a) Characterization of neuritic outgrowth-promoting activity of conditioned medium on spinal cord explants. Exp Brain Res 4:435–441

Dribin LB, Barrett JN (1982b) Two components of conditioned medium increase neuritic outgrowth from rat spinal cord explants. J Neurosci Res 8:271–280

Ebendal T (1979) Stage dependent stimulation of neurite outgrowth exerted by nerve growth factor and chick heart in cultured embryonic ganglia. Dev Biol 72:276–290

Ebendal T, Belew M, Jacobson CO, Porath J (1979) Neurite outgrowth elicited by embryonic chick heart: Partial purification of the active factor. Neurosci Lett 14:91–95

Edgar D, Barde Y-A, Thoenen H (1981) Subpopulations of cultured chick neurones differ in their requirements for survival factors. Nature 289:294–295

Fukada K (1980) Hormonal control of neurotransmitter choice in sympathetic neurone cultures. Nature 287:553–555
Fukada K (1985) Purification and partial characterization of a cholinergic neuronal differentiation factor. Proc Natl Acad Sci USA 82:8795–8799
Giess MC, Weber MJ (1984) Acetylcholine metabolism in rat spinal cord cultures: regulation by a factor involved in the determination of the neurotransmitter phenotype of sympathetic neurons. J Neurosci 4:1442–1452
Giller EL, Schrier BK, Shainberg A, Fisk HR, Nelson PG (1973) Choline acetyltransferase activity is increased in combined cultures of spinal cord and muscle cells from mice. Science 182:588–589
Giller EL, Neale JH, Bullock PN, Schrier BK, Nelson PG (1977) Choline acetyltransferase activity of spinal cord cell cultures increased by co-culture with muscle and by muscle-conditioned medium. J Cell Biol 74:16–29
Gnahn H, Hefti F, Heumann R, Schwab ME, Thoenen H (1983) NGF-mediated increase of choline acetyltransferase (ChAT) in the neonatal rat forebrain: Evidence for a physiological role of NGF in the brain? Dev Brain Res 9:45–52
Godfrey EW, Schrier BK, Nelson PG (1980) Source and target cell specificities of a conditioned medium factor that increases choline acetyltransferase activity in cultured spinal cord cells. Dev Biol 77:403–418
Hamburger V (1934) The effects of limb bud extirpation on the development of the central nervous system in the chick spinal cord. J Exp Zool 68:449–494
Hamburger V (1975) Cell death in the development of the lateral motor column of the chick embryo. J Comp Neurol 160:535–546
Helfand SL, Smith GA, Wessells NK (1976) Survival and development in culture of dissociated parasympathetic neurons from ciliary ganglia. Dev Biol 50:541–547
Helfand SL, Riopelle RJ, Wessells NK (1978) Non-equivalence of conditioned medium and nerve growth factor for sympathetic, parasympathetic, and sensory neurons. Exp Cell Res 113:39–45
Henderson CE, Huchet M, Changeux JP (1981) Neurite outgrowth from embryonic chicken spinal neurons is promoted by media conditioned by muscle cells. Proc Natl Acad Sci USA 78:2625–2629
Henderson CE, Huchet M, Changeux JP (1983) Denervation increases a neurite-promoting activity in extracts of skeletal muscle. Nature 302:609–611
Henderson CE, Huchet M, Changeux JP (1984) Neurite-promoting activities for embryonic spinal neurons and their developmental changes in the chick. Dev Biol 104:336–347
Hill CE, Hendry IA, Bonyhady RE (1981) Avian parasympathetic neurotrophic factors: Age-related increases and lack of regional specificity. Dev Biol 85:258–261
Hill MA, Bennett MR (1983) Cholinergic growth factor from skeletal muscle elevated following denervation. Neurosci Lett 35:31–35
Hollyday M, Hamburger V (1976) Reduction of naturally occurring motorneuron loss by enlargement of the periphery. J Comp Neurol 170:311–320
Ishida I, Deguchi T (1983a) Regulation of acetyltransferase in primary cell cultures of spinal cord by neurotransmitter l-Norepinephrine. Dev Brain Res 7:13–23
Ishida I, Deguchi T (1983b) Effect of depolarising agents on choline acetyltransferase and acetylcholinesterase activities in primary cell cultures of spinal cord. J Neurosci 3:1818–1823
Johnson M, Ross D, Meyers M, Rees R, Bunge R, Vakshull E, Burton H (1976) Synaptic vesicle cytochemistry changes when cultured sympathetic neurones develop cholinergic interactions. Nature 262:308–310
Johnson MI, Ross CD, Meyers M, Spitznagel EL, Bunge RP (1980) Morphological and biochemical studies on the development of cholinergic properties in cultured sympathetic neurons. I. Cholinergic properties. J Cell Biol 84:680–691
Kato AC, Ray MJ (1982) Chick ciliary ganglion in dissociated cell culture. I. Cholinergic properties. Dev Biol 94:121–130
Kaufman LM, Barrett JN (1983) Serum factor supporting long-term survival of rat central neurons in culture. Science 220:1394–1396

Kaufman LM, Barry SR, Barrett JN (1985) Characterization of tissue-derived macromolecules affecting transmitter synthesis in rat spinal cord neurons. J Neurosci 5:160–166

Landa KB, Adler R, Manthorpe M, Varon S (1980) Cholinergic neuronotrophic factors. III. Developmental increase of trophic activity for chick embryo ciliary ganglion neurons in their intraocular target tissues. Dev Biol 74:401–408

Lander AD, Fujii DK, Reichardt LF (1985) Laminin is associated with the 'neurite outgrowth-promoting factors' found in conditioned media. Proc Natl Acad Sci USA 82:2183–2187

Landis SC (1976) Rat sympathetic neurons and cardiac myocytes developing in microcultures: correlation of the fine structure of endings with neurotransmitter function in single neurons. Proc Natl Acad Sci USA 73:4220–4224

Landis SC, Fredieu JR, Yodlowski M (1985) Neonatal treatment with nerve growth factor antiserum eliminates cholinergic sympathetic innervation of rat sweat glands. Dev Biol 112:222–229

Landmesser L, Pilar G (1974a) Synaptic transmission and cell death during normal ganglion development. J Physiol (Lond) 241:737–749

Landmesser L, Pilar G (1974b) Synapse formation during embryogenesis on ganglion cells lacking a periphery. J Physiol (Lond) 241:715–736

Lasher J, Zagon R (1972) The effect of potassium on neuronal differentiation in cultures of dissociated newborn rat cerebellum. Brain Res 41:482–488

Longo FM, Manthorpe M, Varon S (1982) Spinal cord neuronotrophic factors (SCNTF's): I. Bioassay of Schwannoma and other conditioned media. Dev Brain Res 3:277–294

Mains RE, Patterson PH (1973) Primary cultures of dissociated sympathetic neurons. I. Establishment of long-term growth in culture and studies of differentiated properties. J Cell Biol 59:329–345

Manthorpe M, Varon S (1985) Regulation of neuronal survival and neuritic growth in the avian ciliary ganglion by trophic factors. In: Guroff G (ed) Growth and maturation factors, vol III. Wiley, New York, p 77

Manthorpe M, Skaper S, Adler R, Landa K, Varon S (1980) Cholinergic neuronotrophic factors: fractionation properties of an extract from selected chick embryonic eye tissues. J Neurochem 24:69–75

Manthorpe M, Barbin G, Varon S (1982a) Isoelectric focusing of the chick eye ciliary neuronotrophic factor. J Neurosci Res 8:233–239

Manthorpe M, Longo FM, Varon S (1982b) Comparative features of spinal neuronotrophic factors in fluids collected *in vitro* and *in vivo*. J Neurosci Res 8:241–250

Manthorpe M, Engvall E, Ruoslahti E, Longo FM, Davis GE, Varon S (1983a) Laminin promotes neuritic regeneration from cultured peripheral and central neurons. J Cell Biol 97:1882–1890

Manthorpe M, Luyten W, Longo FM, Varon S (1983b) Endogenous and exogenous factors support neuronal survival and choline acetyltransferase activity in embryonic spinal cord cultures. Brain Res 267:57–66

Manthorpe M, Nieto-Sampedro M, Skaper SD, Barbin G, Longo FM, Varon S (1983c) Neuronotrophic activity in brain wounds in the developing rat: correlation with implant survival in the wound cavity. Brain Res 267:47–56

Martinez HJ, Dreyfus CF, Miller Jonakait G, Black IB (1985) Nerve growth factor promotes cholinergic development in brain striatal cultures. Proc Natl Acad Sci USA 82:7777–7781

Max SR, Schwab M, Dumas M, Thoenen H (1978) Retrograde transport of nerve growth factor in the ciliary ganglion of the chick and rat. Brain Res 159:411–415

McManaman JL, Smith RG, Appel SH (1985) Low-molecular-weight peptide stimulates cholinergic development in ventral spinal cord cultures. Dev Biol 112:248–252

Meyer T, Burkart W, Jockusch H (1979) Choline acetyltransferase induction in cultured neurons: Dissociated spinal cord cells are dependent on muscle cells, organotypic explants are not. Neurosci Lett 11:59–62

Mobley WC, Rutkowski JL, Tennekoon GI, Buchanan K, Johnston MV (1985) Choline acetyltransferase activity in striatum of neonatal rats increased by nerve growth factor. Science 229:284–287

Narayanan CH, Narayanan Y (1978) Neuronal adjustments in developing nuclear centers of the chick embryo following transplantation of an additional optic primordium. J Embryol Exp Morphol 44:53–70

Nieto-Sampedro M, Manthorpe M, Barbin G, Varon S, Cotman CW (1983) Injury-induced neuronotrophic activity in adult rat brain. Correlation with survival of delayed implants in wound cavity. J Neurosci 3:2219–2229

Nishi R, Berg DK (1977) Dissociated ciliary ganglion neurons *in vitro*: survival and synapse formation. Proc Natl Acad Sci USA 74:5171–5175

Nishi R, Berg DK (1979) Survival and development of ciliary ganglion neurones grown alone in cell culture. Nature 277:232–234

Nishi R, Berg DK (1981 a) Two components from the eye tissue that differentially stimulate the growth and development of ciliary ganglion neurons in cell culture. J Neurosci 1:505–512

Nishi R, Berg DK (1981 b) Effects of high K^+ concentrations on the growth and development of ciliary ganglion in cell culture. Dev Biol 87:301–307

Nurcombe V, Hill MA, Eagleson KL, Bennett MR (1984) Motor neuron survival and neuritic extension from spinal cord explants induced by factors released from denervated muscle. Brain Res 291:19–28

Patterson PH, Chun LLY (1974) The influence of non-neuronal cells on catecholamine and acetylcholine synthesis and accumulation in cultures of dissociated sympathetic neurons. Proc Natl Acad Sci USA 71:3607–3610

Patterson PH, Chun LLY (1977a) The induction of acetylcholine synthesis in primary cultures of dissociated rat sympathetic neurons. I. Effects of conditioned medium. Dev Biol 56:263–280

Patterson PH, Chun LLY (1977b) The induction of acetylcholine synthesis in primary cultures of dissociated rat sympathetic neurons. II. Developmental aspects. Dev Biol 60:473–481

Patterson PH, Reichardt LF, Chun LLY (1975) Biochemical studies on the development of primary sympathic neurons in culture. Cold Spring Harbor Symp Quant Biol 40:389–397

Phillipson OT, Sandler M (1975) The influence of nerve growth factor, potassium depolarisation and dibutyryl (cyclic) adenosine 3′,5′-monophosphate on explant cultures of chick embryo sympathetic ganglia. Brain Res 90:273–281

Reichardt LF, Patterson PH (1977) Neurotransmitter synthesis and uptake by isolated sympathetic neurones in microcultures. Nature 270:147–151

Richardson GP, Rinschen B, Fox GQ (1985) Torpedo electromotor system development: developmentally regulated neurotrophic activities of electric organ tissue. J Comp Neurol 231:339–352

Schmidt-Achert KM, Askanas V, Engel WK (1984) Thyrotropin-releasing hormone enhances choline acetyltransferase and creatine kinase in cultured spinal ventral horn neurons. J Neurochem 43:586–589

Schnaar RL, Schaffner AE (1981) Separation of cell types from embryonic chicken and rat spinal cord: characterization of motoneuron-enriched fractions. J Neurosci 1:204–217

Scott BS (1977) The effect of elevated potassium on the time course of neuron survival in cultures of dissociated dorsal root ganglia. J Cell Physiol 91:305–316

Selak I, Skaper SD, Varon S (1985) Pyruvate participation in the low molecular weight activity for central nervous system neurons in glia-conditioned media. J Neurosci 5:23–28

Slack JR, Pockett S (1982) Motor neurotrophic factor in denervated adult skeletal muscle. Brain Res 247:138–140

Slack JR, Hopkins WG, Pockett S (1983) Evidence for a motor nerve growth factor. Muscle Nerve 6:243–252

Smith RG, Appel SH (1983) Extracts of skeletal muscle increase neurite outgrowth and cholinergic activity of fetal rat spinal motor neurons. Science 291:1079–1081

Swerts JP, Thai A Le Van, Vigny A, Weber MJ (1983) Regulation of enzymes responsible for neurotransmitter synthesis and degradation in cultured rat sympathetic neurons. I. Effects of muscle-conditioned medium. Dev Biol 100:1–11

Touzeau G, Kato AC (1983) Effects of myotrophic lateral sclerosis sera on cultured cholinergic neurones. Neurology 33:317–322
Tuttle JB, Suszkiw JB, Ard M (1980) Long-term survival and development of dissociated parasympathetic neurons in culture. Brain Res 183:161–180
Varon S, Manthorpe M, Adler R (1979) Cholinergic neuronotrophic factors: I. Survival, neurite outgrowth and choline acetyltransferase activity in monolayer cultures from chick embryo ciliary ganglia. Brain Res 173:29–45
Varon S, Manthorpe M, Selak I, Skaper SD (1983) Humoral agents modulating neuronal survival *in vitro*. In: Biggio G, Spanno PF, Toffano G, Gessa GL (eds) Symposia in neuroscience, 3rd Capo Boi conference on neuroscience, Sardinia, Italy. Pergamon, Oxford
Varon S, Skaper SD, Barbin G, Selak I, Manthorpe M (1984) Low molecular weight agents support survival of cultured neurons from the central nervous system. J Neurosci 4:654–658
Wakshull E, Johnson MI, Burton H (1978) Persistance of an amine uptake system in cultured rat sympathetic neurons which use acetylcholine as their transmitter. J Cell Biol 79:121–131
Weber M (1981) A diffusible factor responsible for the determination of cholinergic functions in cultured sympathetic neurons. J Biol Chem 256:3447–3453
Weber MJ, Raynaud B, Delteil C (1985) Molecular properties of a cholinergic differentiation factor from muscle-conditioned medium. J Neurochem 45:1541–1547
Williams LR, Manthorpe M, Barbin G, Nieto-Sampedro M, Cotman CW, Varon S (1984) High ciliary neuronotrophic specific activity in rat peripheral nerve. Int J Dev Neurosci 2:177–180

Part III Molecular Components of the Cholinergic System

CHAPTER 6

Estimation of Acetylcholine and Choline and Analysis of Acetylcholine Turnover Rates *In Vivo*

P. E. POTTER * and I. HANIN *

A. Introduction

The development within the last 15–20 years of sensitive and specific new methods for the measurement of acetylcholine (ACh) in biological tissues has sparked an explosion of interest in the dynamics of the cholinergic system. Evidence for cholinergic dysfunction in conditions such as Alzheimer's disease, depression, schizophrenia, Huntington's chorea, Gilles de la Tourette's syndrome and tardive dyskinesia have furthered the interest in this system and emphasized the need for a greater understanding of cholinergic function.

Our knowledge of the cholinergic system has lagged behind that of other classical neurotransmitters, e.g. the biogenic amines, partly because of the lack, until recently, of reliable methods for the assay of ACh other than bioassay. In addition, ACh is hydrolysed extremely rapidly *in vivo*, thus making it essential that *post-mortem* degradation is kept to a minimum prior to the extraction of ACh, otherwise one would not be able to determine ACh content accurately *in vivo*. Finally, factors which influence cholinergic neurons will not necessarily affect ACh levels; instead the rates of formation, release, and/or degradation of ACh may change in order to compensate for increased or decreased activity of the neurons. For this reason it is important to study not only the steady-state levels of ACh, but also the *dynamics* of the cholinergic system. The dynamic state of the system is represented by *turnover rate*, which estimates the rate of renewal of a substance (ACh) from its precursor (choline) under steady-state conditions (ZILVERSMIT 1960).

This chapter will describe briefly several of the techniques that are used for measuring ACh and choline. Many of these have already been reviewed extensively (e.g. HANIN 1974; HANIN and GOLDBERG 1976; JENDEN 1978; HANIN 1982). The principles involved in estimating ACh turnover rate will also be reviewed in this chapter, and the usefulness of the various methods for determining ACh turnover will be evaluated. For the sake of clarity, emphasis will be placed on the measurement of ACh and choline and the estimation of ACh turnover in the brains of experimental animals, especially the mouse. However, these techniques are equally applicable to other tissues.

* From Loyola University of Chicago Stritch School of Medicine (current addresses see list of contributors).

B. Preparation of Samples for ACh and Choline Analysis

I. General Requirements for Measurement of ACh and Choline

In all methods for the measurement of ACh and choline certain requirements and conditions for the extraction and preservation of these compounds must be satisfied in order to ensure their optimal and accurate evaluation in tissue extracts. These requirements are listed and detailed below.

Ideally, any method for measurement of ACh and choline should fulfil the following criteria:

1. *post-mortem* changes in tissue ACh and choline content should be kept to a minimum
2. ACh and choline should be extracted quantitatively, with sample concentration made possible during the procedure
3. an appropriate internal standard should be used to estimate recovery
4. there should be no interference in the assay from drugs, ACh- or choline-like compounds, or other chemicals
5. the method should have absolute specificity for ACh and choline
6. the method should provide good separation and estimation of ACh and choline in the same sample
7. there should be sufficient sensitivity to permit accuracy of the analysis
8. the method should be simple and should use readily available equipment
9. the cost per sample should be low.

II. Extraction of ACh and Choline from Tissues, Subcellular Fractions and Perfusates

Most extraction procedures commonly used for ACh and choline take advantage of the stability of ACh between pH 3.8 and 4.5 (WHITTAKER 1963). Dilute acids (0.1 – 0.4 *N* perchloric; 10% trichloroacetic; citric acid-phosphate, pH 4) sometimes combined with an organic solvent (15% formic acid, 85% acetone or 0.2% glacial acetic acid made up in 95% ethanol) are generally used for homogenization of tissues for ACh and choline extraction (see JENDEN and CAMPBELL 1971). For suspensions of subcellular fractions (e.g. synaptosomes, synaptic vesicles) and perfusates, where the protein concentration is low, simply acidifying the ice-cold sample to pH 4.0, heating it at 100 °C for 10 min and chilling it again to 0 °C releases any bound ACh and stabilizes it (HEBB and WHITTAKER 1958).

In many procedures ACh and choline are further extracted, e.g. with sodium tetraphenylboron (FONNUM 1969); by precipitation either with ammonium reineckate (BENTLEY and SHAW 1952) as the periodide salt (SCHMIDT et al. 1969) or by ion exchange/column chromatography (STAVINOHA and WEINTRAUB 1974a; GILBERSTADT and RUSSELL 1984), then concentrated by evaporation of the solvents utilized.

III. Preservation of ACh and Choline *Post-Mortem*

Following decapitation marked and rapid changes occur (within seconds) in the levels of both ACh and choline in the brain. ACh levels are reduced through enzymatic degradation. Simultaneously the breakdown of ACh as well as of phosphorylcholine and/or glycerylphosphorylcholine leads to a dramatic increase in choline levels (DROSS and KEWITZ 1972; KEWITZ et al. 1973; DROSS 1975). In order to avoid these *post-mortem* changes, near-freezing techniques were first developed (TAKAHASHI and APRISON 1964; STAVINOHA et al. 1967). However, since it has been estimated that 10.4 s are required to chill rat brain to almost 0 °C (TAKAHASHI and APRISON 1964), the near-freezing techniques may be too slow to eliminate hydrolysis of ACh and choline-containing phospholipids *in situ*. Therefore, rapid freezing (brain blowing) techniques have also been developed (VEECH et al. 1973; VEECH and HAWKINS 1974), which freeze the animal's brain within seconds. Although this approach is useful in the analysis of whole brain, dissection of the frozen tissue into discrete brain areas is impossible.

These problems have been eliminated by the development of techniques for killing animals by head-focused microwave irradiation (STAVINOHA et al. 1973), which rapidly inactivates *in situ* the enzymes involved in ACh metabolism and minimizes *post-mortem* changes in measured levels of ACh and choline (SCHMIDT et al. 1972; STAVINOHA et al. 1974; GUIDOTTI et al. 1974; SCHMIDT 1976). Microwave irradiation has now become a standard technique for preparing tissues for estimation of ACh and choline.

C. Analysis of ACh and Choline

Numerous approaches have been used for the analysis of ACh and choline; from simple bioassay to complex gas chromatography-mass spectrometry (GCMS). These methods will be discussed in general terms in the narrative which follows. Commentary will be provided as to how well each meets the criteria mentioned above, particularly those of specificity, sensitivity and simplicity.

I. Bioassay

The earliest methods for the measurement of ACh utilized bioassay. Tissues which have been used include the frog rectus abdominis, leech dorsal muscle, clam heart, and guinea-pig ileum (CHANG and GADDUM 1933; MACINTOSH and PERRY 1950; RYALL et al. 1964; PATON and VIZI 1969; DUDAR and SZERB 1970; OHSAWA 1980). The effect of ACh on the contractility of these tissues has been used as an index of ACh content. Blood pressure changes in anaesthetized cat and rat following administration of ACh have also been used to measure ACh levels in tissue extracts (WHITTAKER 1963).

The most convenient and sensitive bioassays for ACh are the eserinized morphinized guinea-pig ileum and the eserinized dorsal muscle of the leech. The principles of bioassay and the practical details of setting up the preparations have been outlined by WHITTAKER and BARKER (1972). With both preparations ACh is washed out before contraction is complete; this has the effect of causing the preparation to relax quickly to its baseline state, ready for the next dose. The ileum can be used on a cycle as short as 1.5 min (10 s contraction, 80 s relaxation), but for the leech a 3.5 – 4 min cycle is necessary (1 – 1.5 min contraction, 2.5 min

Table 1. Chemical methods for ACh determination (developed before 1972)

Authors	Detection	Principle	Sensitive	Specific
AMMON (1933)	CO_2, Warburg's apparatus	Enzyme liberates acetic acid, which reacts with sodium bicarbonate to produce CO_2	No	Yes
BEATTIE (1936)	Colorimetric	Detection of ammonium reineckate precipitated with ACh/choline and dissolved in acetone	No	No
SHAW (1938)	Colorimetric	Modification of BEATTIE'S method, choline separated from ACh	No	Yes
HESTRIN (1949)	Colorimetric	$RCOOR' + NH_2OH \rightarrow RCONHOH + R'OH$ $RCONHOH + \frac{1}{3}Fe^{3+} \rightarrow$ [RC(=O)–NH–O chelate]$Fe_{1/3}$	Yes	No
MASLOVA (1964)	Polarographic	Analysis of Fe^{3+} ions obtained by HESTRIN'S method	Yes	No
VAN GULIK (1956)	Photometric	Detection on paper chromatograph treated by HESTRIN'S method	Yes	No
SHUMACHER and EHL (1970)	Photometric	Analysis of Fe hydroxamic acid analogue of ACh	No	No
ECKSBORG and PERSSON (1971)	Photometric	Analysis of dipicrylamine ion pair analogue of ACh	No	No
MITCHELL and CLARK (1952)	Spectrophotometry	Complex with Bromophenol Blue in alkaline medium	Yes	No
SINSHEIMER and HONG (1965)	Amperometric	Titration of ACh with tetraphenylboron, using graphite electrode	No	Yes
KUNEC-VAGIC and WEBER (1967)	Chemiluminescence	Luminol, indole, benzidine and O-dianisidine/ACh in water-acetone solution in presence of H_2O_2 or perchlorate salt	No	Yes
COOPER (1964)	Fluorimetry	Detection of NADH ACh → choline + acetate Acetate → Ethanol → acetaldehyde (NAD → NADH)	Yes	No
FELLMAN (1969)	Fluorimetry	Acetylhydrazyl salicylhydrazone formed from ACh	No	Yes
O'NEILL and SAKAMOTO (1970)	Fluorimetry	Detection of NADH, formed as acetate from ACh is converted to citrate, in malate-oxalacetate-linked reaction	No	Yes
BROWNING (1972)	Fluorimetry	Detection of NADH generated during formation of pyruvate from phosphoenolpyruvate and ADP formed during phosphorylation of choline	No	Yes

relaxation). The ileum is suspended in a 1 ml bath, the leech muscle (a $10 \times 0.25 \times 0.25$ mm slip of the dorsal muscle) in a 0.1 ml bath. Contractions are recorded with a linear motion transducer and recorded. Thresholds are $1-2$ pmol ml^{-1} (ileum) or $0.2-2$ pmol ml^{-1} (leech), and the range is from $2-10$ times threshold.

Bioassay provides a rapid, sensitive, convenient, inexpensive and quite accurate ($\pm 5\%$) way of estimating ACh, especially when numerous fractions (e.g. superfusates, chromatographic column effluents or density gradient fractions from synaptosome or synaptic vesicle isolations) must be assayed. In one day $60-100$ samples can be assayed routinely. Bioassay is, however, quite sensitive to interfering substances and is thus contraindicated when drugs are present in the samples to be assayed, unless these are first removed by separation techniques such as high performance liquid chromatography (HPLC; see VIZI et al. 1985 for a description of bioassay combined with HPLC). Choline may be assayed by acetylating it and assaying the ACh produced (BLIGH 1952), but this is somewhat inconvenient.

II. Chemical Techniques Before 1972, Excluding Radioenzymatic and Gas Chromatographic (GC) Assays

Numerous simple and ingenious chemical techniques, each based on specific properties of the ACh or choline molecule, were developed prior to 1972. In each, attempts were made to devise approaches that combined simplicity with sensitivity and specificity. Although all of these assays are generally simple and inexpensive, none offers both sensitivity and specificity, and they have therefore not been widely used. As these assays have been previously described in detail (HANIN 1968, 1974, 1982), they will not be discussed here; instead, they are summarized in Table 1.

Several techniques which utilize radioenzymatic assays, GC or GCMS were also developed prior to 1972. They have withstood the test of time, and are used widely today. These techniques, therefore, are discussed more extensively in the next two sections.

III. Radioenzymatic Assays

Two general approaches were developed independently for the radioenzymatic analysis of ACh (and choline). These assays are based on the hydrolysis of ACh to choline, and subsequent acetylation or phosphorylation of choline to form a radiolabelled product, which is then detected by liquid scintillation spectrometry. Acetylation is catalysed by ChAT and phosphorylation by choline kinase (ChK). The enzyme-catalysed reactions are:

$$\text{choline} + [^{14}\text{C}]\ \text{AcCoA} \xrightarrow{\text{ChAT}} [^{14}\text{C}]\ \text{ACh} + \text{CoA}$$

or

$$\text{choline} + [^{32}\text{P}]\ \text{ATP} \xrightarrow{\text{ChK}} [^{32}\text{P}]\ \text{phosphorylcholine} + \text{ADP}\ .$$

Because these methods specifically measure choline, ACh and choline must first be separated; ACh must then be hydrolysed to choline for analysis. ACh and choline are separated either by paper electrophoresis (SAELENS et al. 1970; FEIGENSON and SAELENS 1969; REID et al. 1971; SCHUBERTH and SUNDWALL 1971) or by addition of ChK before the radiolabelled substrate, so that endogenous choline is converted to nonradioactive phosphorylcholine (SHEA and APRISON 1973; GOLDBERG and MCCAMAN 1973). Radioactive substrate is then added, ACh is hydrolysed to choline by the addition of AChE or elevated pH, and the choline formed is converted to a radioactive product. Detailed descriptions of these assays can be found in the handbook of methods for analysis of choline and ACh, edited by HANIN (1974).

The radioenzymic assays are very sensitive (as little as 2 pmol can be measured by some of these methods), can be carried out using routine laboratory equipment and are especially useful for processing a large number of samples. However, because of the need for radioactive substrates, they are more expensive than the classical chemical assays. In addition, they are not entirely chemically specific, and it is possible that the enzymes might be affected by drugs or chemicals in the medium. Another drawback is that samples must be split and run in duplicate in order to assay both ACh and choline.

Moreover, in studies of turnover dual-labelling analyses must in general be conducted in order to discriminate radiolabelled products of the radioenzymatic assay from the radiolabelled precursors that were administered to the animals (see below). For this reason, although these methods have been used extensively and reliably for the determination of ACh and choline content, they have not been used as frequently for the estimation of ACh turnover.

IV. GC and GCMS Analyses

A significant advance in the study of cholinergic function was the application of GC to measure ACh and choline. An essential step in the method is conversion of the quaternary amines to volatile derivatives.

The first GC method which was described measured the production of ethanol by reduction of ACh with borohydride (STAVINOHA et al. 1964; STAVINOHA and RYAN 1965). Others measured acetic acid (CRANMER 1968) or olefin (MITCHNER and JENNINGS 1967) formation after the appropriate chemical conversion. These methods are not entirely specific since they measure a by-product of a reaction with ACh rather than a derivative of the original compound.

In subsequent GC methods volatile compounds specific to each choline ester were produced by esterification of choline, followed by *N*-demethylation (e.g. dimethylaminoethyl acetate was generated from ACh) resulting either from reaction with sodium benzenethiolate (JENDEN et al. 1968; HANIN and JENDEN 1969; HANIN et al. 1970; JENDEN 1970) or pyrolysis (SZILAGYI et al. 1968; SCHMIDT et al. 1969; SZILAGYI et al. 1972; STAVINOHA and WEINTRAUB 1974b; SCHMIDT and SPETH 1975). The compounds were identified by their GC retention time and measured by flame ionization or nitrogen detection.

The coupling of mass spectrometry to gas chromatography (GCMS) greatly extended the sensitivity and specificity of this method (JENDEN 1973; JENDEN et al. 1974; KARLÉN et al. 1974; POLAK and MOLENAAR 1974; STAVINOHA and WEINTRAUB 1974b; FREEMAN et al. 1975). Fragments are now obtained either by electron impact ionization, which produces a base peak at m/e of 58 from all choline esters (HAMMAR et al. 1968); or by chemical ionization, which produces a peak with m/e specific for each starting compound (HANIN and SKINNER 1975).

GCMS is a very powerful and versatile method for the estimation of ACh and choline. The ability to detect stable (deuterium-labelled) isotopes of choline and ACh in the same sample has made it extremely useful for the study of ACh turnover (see below). Deuterium-labelled analogues of choline and ACh can also be used as internal standards. The method has gained wide acceptance and has now been used reliably for over 15 years. Moreover, there have been numerous modifications of the original procedures (e.g. POLAK and MOLENAAR 1979; MARUYAMA et al. 1979; KOSH et al. 1979; KHANDELWAL et al. 1981). For a detailed discussion of the uses and potential of GCMS, see HANIN and SHIH (1980). The disadvantages of GCMS are that the equipment required is expensive and not generally available. Moreover, rigorous sample preparation is required.

V. New Developments Since 1978

In the past 7 years a number of novel methods for ACh have been developed. Two sophisticated new approaches using *mass spectrometry* have been designed (LEHMANN et al. 1978; UNGER et al. 1981) for the direct analysis of underivatized ACh in tissue extracts. These approaches, as with most of those in the present section, have not been used extensively as yet.

Two *radioimmunoassay* methods have been introduced (SPECTOR et al. 1978; KAWASHIMA et al. 1980). In the first, immunogen is formed from the choline salt of 6-aminocaproic acid conjugated with bovine serum albumin (SPECTOR et al. 1978). In the other method, the immunogen is formed from a conjugate of glutaric anhydride, choline and bovine serum albumin (KAWASHIMA et al. 1980). These methods are very specific and sensitive (<2 pmol). They do, however, require production of antibody over a period of months, and it is essential that the antibody exhibits little cross-reactivity with compounds other than ACh. Choline cannot be measured with this method.

A *radioreceptor* assay, based on the displacement of a specific radiolabelled muscarinic agonist has also been described (EHLERT et al. 1982). The sensitivity of the method is comparable to that of other highly sensitive methods (<1 pmol). The authors note however that as with the radioenzymic assay, a major disadvantage of this method is interference from drugs and ions, particularly in concentrated solutions or if the samples have been exposed to muscarinic agonists. An advantage of this assay is that samples with a large (>1000-fold) excess of choline can be assayed. This method cannot be used for the estimation of choline.

ISRAËL and LESBATS (1980, 1981 a) have developed a *chemiluminescent* method for choline and ACh, based on the conversion of choline to betaine and hydrogen peroxide (H_2O_2) by the enzyme choline oxidase (IKUTA et al. 1977). The H_2O_2 generated during this reaction produces chemiluminescence in the presence of luminol and horseradish peroxidase. Chemiluminescence can be detected with a photomultiplier tube, which is commercially available (Aminco). As this method detects choline, samples are first added to the mixture of choline oxidase, luminol, and peroxidase and the light response to choline is allowed to decay, then acetylcholinesterase (AChE) is added to hydrolyse ACh, and the light response is again measured. The sensitivity of the assay is determined by the purity of the AChE preparation used; the authors report a sensitivity of approximately 10 pmol. This method is especially useful for monitoring the time course of the depolarization-induced release of ACh from isolated tissues (ISRAËL and LESBATS 1981b).

The oxidation of choline by choline oxidase to produce betaine and H_2O_2 has also been utilized in the most recent method developed for the analysis of ACh (POTTER et al. 1983), which employs HPLC with electrochemical detection (HPLC-ED). Choline and ACh were separated on a reverse phase column and, as they eluted from the column, were mixed with the enzymes choline oxidase and AChE in a post-column reaction coil (POTTER et al. 1983). The H_2O_2 produced by the reaction with choline oxidase was monitored by electrochemical detection at +0.5 V across a platinum electrode. Subsequently, there have been numerous modifications of this technique, one of which involves the immobilization of choline oxidase and AChE on various types of columns (EVA et al. 1984; DAMSMA et al. 1985; HONDA et al. 1985; KANEDA et al. 1985). Although the immobilized enzyme has been a great improvement, the chromatographic procedures employed (e.g. EVA et al. 1984) have not been satisfactory. In general, the best separation of ACh and choline is performed on reverse phase columns (BYMASTER et al. 1985; KANEDA et al. 1985). Use of an Altex ODS-5 column with 0.025 *M* Tris-maleate, pH 7.0, containing 3 ml l^{-1} sodium octanesulphonic acid and 300 mg l^{-1} at a flow rate of 1 ml min^{-1}, with the immobilized enzyme of EVA et al. (1984) offers excellent separation and sensitivity of detection of ACh and choline.

There are numerous advantages to this system: the sensitivity ($\leqq 1-2$ pmol) is equivalent to that of the best ACh methods; specificity for ACh and choline are provided by the HPLC column and the enzyme, for which there are few substrates (POTTER et al. 1983); ACh and choline are separated and measured in the same sample; an internal standard can be used to correct for recovery; sample preparation is relatively uncomplicated, with no need for derivatization or prior separation; the method uses equipment which is available in most neurochemistry laboratories; and it is relatively simple to use. Because ACh and choline are measured in the same sample, this method can be easily adapted to estimate ACh turnover (POTTER et al. 1984; EVA et al. 1985; LUND et al. 1986).

D. Methods for Measuring ACh Turnover

I. General Considerations

The turnover of a substance is defined as the synthesis or renewal of that substance at steady state (ZILVERSMIT 1960). The dynamics of this renewal process are estimated by the *turnover rate in vivo*. Estimates of turnover rate are of importance because the activity of a group of neurons may not necessarily be reflected accurately in their *content* of neurotransmitter. For example, it has been shown that in an area such as the hippocampus, there is a low-to-moderate ACh content, while the ACh turnover rate is very high, indeed much higher than in other brain nuclei with as great or greater levels of ACh (RACAGNI et al. 1974). Furthermore, when the activity of the hippocampus is increased by stimulation of the afferent neurons from the septum, ACh turnover rate is increased, but ACh content remains unchanged (MORONI et al. 1978). Similarly, when the afferent input to hippocampus is removed by cutting the fimbria, thus reducing the activity of hippocampal neurons, the ACh turnover rate decreases dramatically, whereas there is only a slight increase in ACh content (MORONI et al. 1978).

Thus, it is evident that one cannot make assumptions about the activity of cholinergic neurons based on ACh content, for it is likely that the fraction of ACh which is actually being released and turning over is a small portion of the total ACh in the neuron (for a discussion of the significance of pools of ACh on neuronal ACh turnover, see WEILER et al. 1982). In the intact animal the effect of a treatment which might greatly enhance or decrease ACh function might be missed completely if ACh turnover was not measured.

In catecholamine-containing neurons one can estimate the turnover rate from the content of the transmitter, its precursor, and its metabolites. This is not the case in cholinergic neurons where choline, in addition to its function in phospholipid metabolism, is both the precursor and the metabolite of ACh. This multiple role of choline makes the estimation of ACh turnover rate rather complex. However, a number of approaches have been designed to estimate the turnover rate of brain ACh *in vivo.* Before attempting to explain these approaches, certain terms used in estimating ACh turnover should be defined.

1. *Steady state:* This refers to the content of the substances to be measured (ACh and choline). The content must be constant throughout the experimental period, i.e. the rate of disappearance must be equalled by the rate of synthesis.
2. *Specific activity:* In isotopic studies, this refers to the amount of label per unit of concentration of a given substance (e.g. ACh or choline). In turnover studies the specific activity is described at particular times after injection of labelled precursor.
3. *Fractional rate constant* k_{ACh}: The turnover rate is determined as a function of a rate constant defined as k_{ACh}. This value reflects the proportion of the compartment that is metabolized during a given time period and is independent of the size of the compartment.
4. *Turnover rate* TR_{ACh}: The amount of ACh which is formed or resynthesized per unit of time. TR_{ACh} is calculated from the product of the fractional rate constant and the steady state concentration of substance, i.e. $TR_{\mathrm{ACh}} = k_{\mathrm{ACh}}\,[\mathrm{ACh}]$.

Approaches that have been used to estimate TR_{ACh} can be divided into two groups: non-isotopic methods and isotopic methods.

II. Non-isotopic Methods

These methods all perturb the steady state of ACh and then measure either the rate of return to pre-existing conditions or the rate of change of the system after such perturbation.

In the earliest attempt to determine TR_{ACh}, RICHTER and CROSSLAND (1949) measured the *rate at which ACh levels returned to normal after depletion by electroshock*. From this initial rate of ACh re-synthesis they calculated a TR_{ACh} of 38 nmol g^{-1} min^{-1} in whole rat brain. We now know that this value is considerably higher than that calculated with most other methods (for comparison of values obtained with various TR_{ACh} methods, see HANIN and COSTA 1976; NORDBERG 1977).

Another approach has been to measure the *rate of decline in ACh content after inhibition of ACh synthesis* by intraventricular injection of hemicholinium-3 (HC-3; DOMINO and WILSON 1972). The rate of depletion should equal the rate of efflux of ACh, and can be used to estimate the fractional rate constant which, when multiplied by the steady-state ACh content, would give TR_{ACh}. Theoretically this approach stems from those applied earlier to measurement of turnover of biogenic amines (BRODIE et al. 1966; NEFF et al. 1969). However, when applied to the study of TR_{ACh}, it has been criticized on the following points: (a) it has not been shown that ACh synthesis is completely inhibited at zero time by HC-3, and the mechanism of action of HC-3 is not entirely clear; (b) inhibition of ACh synthesis by HC-3 occurs through competition with choline for high-affinity uptake into nerve endings (BIRKS and MACINTOSH 1961; GARDINER 1961; BUTERBAUGH and SPRATT 1968), but may also include direct suppression of ChAT (MACINTOSH 1963; COLLIER and LANG 1969; RODRIGUEZ DE LORES ANAIZ et al. 1970) and/or diversion of choline from ACh synthesis to the metabolism of phospholipids (GOMEZ et al. 1970); and (c) the decline of ACh reaches a plateau within 30 min of HC-3 administration, suggesting that ACh synthesis is not totally inhibited under these conditions. Turnover rates calculated with this method (1.2 nmol g^{-1} min^{-1} for whole mouse brain) are somewhat lower than with most estimates using other methods.

STAVINOHA et al. (1976) have calculated the *in vivo accumulation of ACh following inhibition of AChE* with Dichlorvos (15 mg kg^{-1} intravenously). The accumulation of ACh should reflect the rate of ACh synthesis and can be used to calculate the fractional rate constant. Similar considerations apply to this method, however, as to that using HC-3. Specifically, cholinesterase (ChE) should be completely inactivated at zero time without effects on other systems. Furthermore, since a significant portion of the choline used for ACh synthesis is derived from ACh breakdown, it is possible that availability of choline could eventually be compromised due to endogenous feedback mechanisms *in vivo*. Despite these concerns the values for TR_{ACh} in various brain areas determined by this method are in the same ranges are those calculated by other methods (e.g. RACAGNI et al. 1974).

If ACh content is maintained at a constant level, the amount of ACh released must be equalled by the amount synthesized. Thus, measurement of ACh output should provide an estimate of ACh synthesis rate and turnover when a constant ACh content has been established in a particular experimental situation. SZERB et al. (1970) estimated TR_{ACh} by measuring the ACh released into a cortical cup from the exposed, perfused cortex of non-anaesthetized cats. Pre-perfusion with echothiolate was used to inhibit AChE, HC-3 was then added and the rate of decline in ACh output was determined. Studies were also conducted in the presence of atropine, which greatly enhances ACh output. The TR_{ACh} calculated in these experiments was 0.15 nmol g^{-1} min^{-1} in the absence and 0.92 nmol g^{-1} min^{-1} in the presence of atropine. These values are low in comparison to most estimates. This study, while imaginative, also is subject to the concerns raised *vis-á-vis* the methods listed above, using HC-3 and ChE inhibitors as tools for the estimation of TR_{ACh} *in vivo*.

The non-isotopic methods, although each innovative, have been largely overshadowed by the isotopic methods for measuring TR_{ACh} *in vivo*, which were developed at about the same time (early 1970s) and have been used far more extensively to date.

III. Isotopic Methods

1. General Principles

There are a number of isotopic methods for estimating TR_{ACh} in brain (for reviews, see HANIN and COSTA 1976; JENDEN 1976; CHENEY and COSTA 1977, 1978; NORDBERG 1977). The methods differ in the labelled precursor used, the mode of administration of precursor, the technique used to measure ACh and choline, and the kinetic analysis of the data. However, they all share one basic property: each attempts to measure the *rate at which a labelled precursor is incorporated into ACh in vivo.*

Experimentally, in all of these methods the specific activities and the steady-state levels of ACh and choline must be determined.

2. Methodology

In the early isotopic studies of TR_{ACh} intravenous injections of [*Me*-^{3}H] choline (SCHUBERTH et al. 1969, 1970) or of [*Me*-^{14}C] choline (SAELENS et al. 1974) were employed. These studies, however, were criticized at the time because the amounts of choline administered were high (0.4–0.66 μmol per mouse), owing to the low specific activity of the radiolabelled choline then available. Brain choline has been shown to be increased, at least temporarily, after elevation of plasma choline (COHEN et al. 1975; RACAGNI et al. 1975). Thus, injection of these non-tracer amounts of choline may have perturbed brain choline levels, violating the principles of steady-state kinetics. Moreover, it has been shown that the parameters measured are not consistent over a wide range of precursor concentrations, and thus the estimate of TR_{ACh} may vary with the dose of radiolabelled choline administered (KARLÉN et al. 1982).

To circumvent this difficulty HANIN and co-workers (HANIN et al. 1973; CHENEY et al. 1975a) injected [*Me*-^{14}C]phosphorylcholine intravenously. This substance is hydrolysed rapidly *in vivo* to choline (HANIN et al. 1972) without affecting either plasma or brain levels of choline. The pattern of appearance of radioactivity into choline and ACh is similar to that seen after administration of choline.

Pulse labelling resulted in sufficient incorporation of label for measurement of TR_{ACh} in whole brain or in active areas, such as striatum. It was not, however, as reliable for the measurement of TR_{ACh} in small, less active areas. Therefore, an infusion technique was adopted. Initially, ^{14}C-labelled phosphorylcholine was infused intravenously at a rate of 0.012 μmol min^{-1} per mouse for 8 min, a dose previously shown not to perturb the steady-state levels of ACh or choline (RACAGNI et al. 1974, 1975; CHENEY et al. 1975b). Subsequently, infusion of deuterium-labelled phosphorylcholine was employed with a similar positive outcome (ZSILLA et al. 1977).

One advantage of the infusion technique was that substantial labelling was achieved, thus allowing the measurement of TR_{ACh} in very small brain areas (RACAGNI et al. 1974). In addition, the amount of labelled choline increased steadily throughout the experimental period. This reduced to a minimum the contribution to ACh synthesis of recycled labelled choline. The disadvantage of this technique is that animals must be restrained throughout the infusion period.

With the development of radiolabelled choline of very high specific activity, the need for infusion techniques has diminished. SCHUBERTH et al. (1969), in their pioneering experiments had to inject 0.5 μmol of [*Me*-^{3}H]choline with a specific activity of 0.2 Ci $mmol^{-1}$. More recently, far smaller amounts have been used, e.g. 15 nmol per mouse (specific activity 16.5 Ci $mmol^{-1}$; NORDBERG et al. 1977), 3 nmol per mouse (specific activity 70 Ci $mmol^{-1}$; VOCCI et al. 1979), and 50 pmol per mouse (specific activity 80 Ci $mmol^{-1}$; POTTER et al. 1984). These represent true tracer amounts of choline.

To date the following four approaches, combining some of the principles mentioned above, have been employed for the determination of steady-state levels of ACh and choline, as well as TR_{ACh}.

1. Labelled ACh and choline extracted from tissues of animals injected with a radiolabelled precursor have been separated by paper electrophoresis and steady-state levels of ACh and choline determined in separate samples by radioenzymatic assay (SCHUBERTH et al. 1969, 1970; SAELENS et al. 1974; NORDBERG and SUNDWALL 1977; VOCCI et al. 1979).
2. After extraction from tissue (treated as above) ACh and choline have been separated and measured by GC, the effluent samples trapped, and the radioactivity of each peak determined using liquid scintillation spectrometry (HANIN et al. 1973; CHENEY et al. 1975a; RACAGNI et al. 1974; RACAGNI et al. 1975).
3. Deuterium-labelled precursors have been injected and the levels of endogenous and deuterium-labelled ACh and choline determined by GCMS (CHOI et al. 1973; JENDEN et al. 1974; KOSLOW et al. 1974; CHENEY et al. 1975b; ZSILLA et al. 1976). This is the most convenient method for the measurement of TR_{ACh}, since all of the parameters can be determined simultaneously.
4. After extraction from tissues of animals pre-treated with radiolabelled choline, ACh and choline have been separated and measured by HPLC-ED. The effluent under each peak has been collected and assayed for radioactivity (POTTER et al. 1984; EVA et al. 1985; LUND et al. 1986).

3. Kinetic Analysis of ACh Turnover Data

Five different approaches have been developed for the determination of TR_{ACh} from the experimental data. These approaches are based generally on the principles of precursor-product relationships outlined by ZILVERSMIT (1960). Each approach assumes that:

1. Steady-state levels of labelled ACh and choline are maintained throughout the time of the analysis;
2. Labelled compounds behave identically to endogenous ones;
3. Choline and ACh mix rapidly and completely *in vivo*;
4. Choline is not recycled, or, if it does, it mixes completely with free choline;
5. ACh is contained in a single compartment; this is probably not true and one approach does even account for two compartments.

The equations that have been derived for the estimation of TR_{ACh} as well as values calculated for whole mouse brain are summarized in Table 2. The first of these approaches that developed by SCHUBERTH et al. (1969), measures the peak ratio between labelled ACh and choline and multiplies this value by the endogenous choline content. Thus the value for TR_{ACh} can be affected greatly by changes in choline content, which may be marked if microwave irradiation is not used. SAELENS et al. (1974) modified this method by calculating the pool of choline available for synthesis of ACh using the following equation:

$$[\text{choline}] = (\text{Ch}^*/\text{ACh}^*)_{\text{peak}} \times [\text{ACh}] \tag{1}$$

Values calculated with equation (1) are often higher than those calculated with other methods (for definition of symbols see Table 2).

In the second approach, developed by JENDEN and co-workers (CHOI et al. 1973; JENDEN et al. 1974), $d\text{ACh}^*/dt$ is determined by non-linear regression analysis of the experimental data and plotted against $(\text{Ch}^*/[\text{Ch}] - \text{ACh}^*/[\text{ACh}])$. The slope of the line measures the turnover rate of ACh. If ACh is contained in two compartments, the plot will yield two lines with different slopes (JENDEN et al. 1974).

In the third approach the experimental data points are fitted using non-linear regression analysis (HANIN et al. 1973; CHENEY et al. 1975a). From the curves calculated to fit the experimental points the rate constant k_{NC} can be determined. This method was developed for studies using pulse injections of phosphorylcholine.

The fourth approach accounts for the entry of labelled choline from slowly infused phosphorylcholine into the choline pool involved in endogenous ACh synthesis (RACAGNI et al. 1974; CHENEY et al. 1975b). The rate constant, k_{A}, representing disappearance of choline, can be calculated from the specific activities of choline at various time points. The rate constant for the formation of ACh, k_{B}, can be calculated once k_{A} is known. This approach assumes that a single compartment of ACh exists and cannot be used to determine values for a two-compartment model. By assuming that k_{A} does not change under experimental conditions, RACAGNI et al. (1974) developed a mathematical model which allows TR_{ACh} to be determined at a single time point. This reduces the number of analyses required to estimate TR_{ACh}. However, it should always be verified that k_{A} is not changed in different groups of animals (e.g. control versus drug treated).

The finite difference approach (the fifth approach) was originally developed by NEFF et al. (1971) for the analysis of catecholamine turnover. More recently,

Table 2. Approaches for calculation of TR_{ACh} using isotopic methods

		TR_{ACh} (whole mouse brain)	References
(1)	$TR_{ACh} = \frac{ACh^*}{Ch^*} \times [Ch]$	50 10 6.5 22, 7	SCHUBERTH et al. (1969) SCHUBERTH et al. (1970) SAELENS et al. (1974) VOCCI et al. (1979)
(2)	$\frac{dACh^*}{dt} = TR_{ACh}\left(\frac{Ch^*}{[Ch]} - \frac{ACh^*}{[ACh]}\right)$	7.9, 1.4	JENDEN et al. (1974)
(3)	$TR_{ACh} = k_{NC}[ACh]$ k_{NC} is calculated from $S_{NC} = \frac{k_{NB}k_{NC}}{k_{NB}-1} - \frac{e^{-t/r} - e^{-k}NC^{t/r}}{k_{NC}-1} - \frac{e^{-k_{NB}t/r} - e^{-k_{NC}t/r}}{k_{NC}-k_{NB}}$	6.0	CHENEY et al. (1975a)
(4)	$TR_{ACh} = k_B[ACh]$ where $k_B = \frac{-1}{t_1} \ln \left[1 - \frac{k_b}{k_A} + \frac{k_B}{k_A} e^{-k_A t_1} + \frac{1}{m}\left(\frac{k_B}{k_A} - 1\right)(1 - e^{-k_A t_1})\right]$ and $k_A = \frac{-1}{t_2} \ln \left[1 - \frac{S_{Ch(t_2)}}{S_{Ch(t_1)}}(1 - e^{-k_A t_1})\right]$ where $m = \frac{S_{Ch(t_1)}}{S_{ACh(t_1)}}$		RACAGNI et al. (1974) CHENEY et al. (1975b)
(5)	$TR_{ACh} = k_B[ACh]$ where $k_B = \frac{2[S_{ACh(t_2)} - S_{ACh(t_1)}]}{(t_2 - t_1)[S_{Ch(t_1)} - S_{ACh(t_1)} + S_{Ch(t_2)} - s_{ACh(t_2)}]}$	22, 5.9 4.5	VOCCI et al. (1979) CHENEY et al. (1975a)

Ch, ACh, endogeneous choline, ACh; Ch*, ACh*, labelled choline, ACh; S_{Ch}, specific activity of choline; S_{ACh}, specific activity of ACh; k_A, rate constant for choline; k_B, rate constant for ACh; k_{NC}, rate constant

it has been applied to the analysis of TR_{ACh} *in vivo* (CHENEY et al. 1975a; NORDBERG 1977; SHIH and HANIN 1978; VOCCI et al. 1979; BLUTH et al. 1984; POTTER et al. 1984; BROWN et al. 1986). This approach also assumes a one-compartment model. Mathematically it is simpler than approaches 2, 3 or 4. The value for k_B, and thus TR_{ACh}, can be calculated from the experimental data in a straightforward fashion.

TR_{ACh} values for the *same data* have been calculated and compared by several authors, using both the finite differences method and one of the other equations described above. CHENEY et al. (1975a) found that turnover rates calculated by approaches 3 and 5 were similar. NORDBERG (1977) found that values for TR_{ACh} calculated by approach 1 were higher than with other approaches, but those determined with 2 and 5 were comparable. The values obtained depended upon whether S_{ACh} had reached its peak value. BLUTH et al. (1984) also found identical values using approaches 2 and 5. VOCCI et al. (1979) calculated similar values with approaches 1 and 5.

In surveying the literature on *in vivo* TR_{ACh} determination in brain, one finds a diversity of values. This probably results from the different methodologies used in different laboratories. It should be remembered, however, that TR_{ACh} is not an absolute value but an *estimate* of ACh turnover. The relative changes in TR_{ACh} in response to various treatments (i.e. increase or decrease) should thus be the same regardless of the method used.

Of the various approaches used to measure TR_{ACh}, the authors favour the finite differences method, since it is relatively simple and has already been used, with comparable results, in a number of laboratories (e.g. CHENEY et al. 1975a; NORDBERG 1977; SHIH and HANIN 1978; VOCCI et al. 1979; BLUTH et al. 1984; POTTER et al. 1984; BROWN et al. 1985).

E. Concluding Remarks

The literature to date regarding the development of techniques for the determination of ACh and choline, as well as TR_{ACh}, has been surveyed in this chapter. Clearly, there are a variety of techniques available. Each technique has its advantages and disadvantages, and the selection of which to use depends on one's specific requirements and preferences.

In evaluating ACh methodology the test of time is also important. Some methods have gained wide acceptance because they are easy to use, sensitive, and reliable; others have not. At present GCMS, radioenzymatic assays and HPLC appear to be the most commonly used methods for the estimation of ACh, although bioassay is still very useful in particular applications.

The situation regarding estimation of TR_{ACh} is less clear cut. Each development has been part of a progression to better and more effective methods; indeed, sufficient tools and methodology are now available to answer key questions concerning the regulation of cholinergic function. With time it is almost certain that these approaches will continue to be improved and streamlined. Hopefully, some conformity in the approaches for estimating TR_{ACh} will then emerge, which will facilitate comparison of data obtained in different laboratories.

Acknowledgement. This study was supported by grants no. MH 26320 and MH 34893 from the National Institute of Mental Health.

References

Ammon R (1933) Die fermentitative Spaltung des Acetylcholins. Pflugers Arch Ges Physiol 233:486–491
Beattie FJR (1936) CCXVIII: a colorimetric method for the determination of choline and acetylcholine in small amounts. Biochem J 30:1554–1559
Bentley GA, Shaw FH (1952) The separation and assay of acetylcholine in tissue extracts. J Pharmacol Exp Ther 106:193–199
Birks R, MacIntosh FC (1961) Acetylcholine metabolism of a sympathetic ganglion. Can J Biochem Physiol 39:787–827
Bligh J (1952) The level of free choline in plasma. J Physiol (Lond) 117:234–240
Bluth R, Langnickel R, Morgenstern R, Oelszner W (1984) Comparison of two methods for estimating the acetylcholine turnover in discrete rat brain structures. Pharmacol Biochem Behav 20:169–174
Brodie BB, Costa E, Dlabac A, Neff NH, Smookler HH (1966) Application of steady-state kinetics to the estimation of synthesis rate and turnover time of tissue catecholamines. J Pharmacol Exp Ther 154:493–498
Brown OM, Salata JJ, Graziani LA (1986) Parasympathetic innervation of the heart: acetylcholine turnover *in vivo*. In: Hanin I (ed) Dynamics of cholinergic function. Plenum, New York, pp 1071–1080
Browning ET (1972) Fluorometric enzyme assay for choline and acetylcholine. Anal Biochem 46:624–638
Buterbaugh GC, Spratt JL (1968) Effect of hemicholinium on choline uptake in the isolated perfused rabbit heart. J Pharmacol Exp Ther 159:255–260
Bymaster FP, Perry KW, Wang DT (1985) Measurement of acetylcholine and choline in brain by HPLC with electrochemical detection. Life Sci 37:1775–1781
Chang HC, Gaddum JH (1933) Choline esters in tissue extracts. J Physiol (Lond) 79:255–285
Cheney DL, Costa E (1977) Pharmacological implications of brain acetylcholine turnover measurements in rat brain nuclei. Ann Rev Pharmacol Toxicol 17:369–386
Cheney DL, Costa E (1978) Biochemical pharmacology of cholinergic neurons. In: Lipton MA, DiMascio A, Killam KF (eds) Psychopharmacology: a generation of progress. Raven, New York, pp 283–291
Cheney DL, Costa E, Hanin I, Trabucchi M, Wang CT (1975a) Application of principles of steady-state kinetics to the *in vivo* estimation of acetylcholine turnover rate in mouse brain. J Pharmacol Exp Ther 192:288–296
Cheney DL, Trabucchi M, Racagni G, Wang C, Costa E (1975b) Effects of acute and chronic morphine on regional rat brain acetylcholine turnover rate. Life Sci 15:1977–1990
Choi RL, Roch M, Jenden DJ (1973) A regional study of acetylcholine turnover in rat brain and the effect of oxotremorine. Proc West Pharmacol Soc 16:188–190
Cohen EL, Kwok M-L, Wurtman RJ (1975) Increase in brain acetylcholine levels following administration of choline to rats. Fed Proc 34:2933
Collier B, Lang C (1969) The metabolism of choline by a sympathetic ganglion. Can J Physiol Pharmacol 47:119–126
Cooper JR (1964) The fluorometric determination of acetylcholine. Biochem Pharmacol 13:795–797
Cranmer MF (1968) Estimation of the acetylcholine levels in brain tissue by gas chromatography of acetic acid. Life Sci 7:995–1000
Damsma G, Westerink BHC, Horn AS (1985) A simple, sensitive, and economic assay for choline and acetylcholine using HPLC, and enzyme reactor, and an electrochemical detector. J Neurochem 45:1649–1659

Domino EF, Wilson AE (1972) Psychotropic drug influences on brain acetylcholine utilization. Psychopharmacologia (Berlin) 25:291–298
Dross K (1975) Effects of di-isopropylfluorophosphate on the metabolism of choline and phosphatidylcholine in rat brain. J Neurochem 24:701–706
Dross K, Kewitz H (1972) Concentration and origin of choline in the rat brain. Naunyn Schmiedebergs Arch Pharmacol 274:91–106
Dross VK, Kewitz H (1966) Der Einbau von i. v. zugeführtem Cholin in das Acetylcholin des Gehirns. Naunyn Schmiedebergs Arch Pharmacol 255:10
Dudar JD, Szerb JC (1970) Bioassay of acetylcholine on leech muscle suspended in a microbath. Exp Physiol Biochem 3:341–350
Ecksborg S, Persson BA (1971) Photometric determination of acetylcholine in rat brain after selective isolation by ion pair extraction and micro column separation. Acta Pharm Suec 8:205–216
Ehlert FJ, Roeske WR, Yamamura HI (1982) A simple and sensitive radioreceptor assay for the estimation of acetylcholine. Life Sci 31:347–354
Eva C, Hadjiconstantinou M, Neff NH, Meek JL (1984) Acetylcholine measurement by high-performance liquid chromatography using an enzyme-loaded postcolumn reactor. Anal Biochem 143:320–324
Eva C, Meek JL, Costa E (1985) Vasoactive intestinal peptide which coexists with acetylcholine decreases acetylcholine turnover in mouse salivary glands. J Pharm Exp Ther 232:670–674
Feigenson MF, Saelens JK (1969) An enzyme assay for acetylcholine. Biochem Pharmacol 18:1479–1486
Fellman JH (1969) A chemical method for the determination of acetylcholine: its application in a study of presynaptic release and a choline acetyltransferase assay. J Neurochem 16:135–143
Fonnum F (1969) Isolation of choline esters from aqueous solutions by extraction with sodium tetraphenylboron in organic solvents. Biochem J 113:291–298
Freeman JJ, Choi RL, Jenden DJ (1975) The effect of hemicholinium on behavior and on brain acetylcholine and choline in the rat. Psychopharmacol Commun 1:15–27
Gardiner JE (1961) The inhibition of acetylcholine synthesis in brain by a hemicholinium. Biochem J 81:297–303
Gilberstadt ML, Russell JA (1984) Determination of picomole quantities of acetylcholine and choline in physiologic salt solutions. Anal Biochem 138:78–85
Goldberg AM, McCaman RE (1973) The determination of picomole amounts of acetylcholine in mammalian brain. J Neurochem 20:1–8
Gomez M, Domino EF, Sellinger OZ (1970) Effect of hemicholinium on choline distribution *in vivo* in the canine caudate nucleus. Biochem Pharmacol 19:1753–1760
Guidotti A, Cheney DL, Trabucchi M, Doteuchi M, Wang C, Hawkins RA (1974) Focussed microwave radiation: a technique to minimize *post mortem* changes of cyclic nucleotides, DOPA and choline, and to preserve brain morphology. Neuropharmacology 13:1115–1122
Hammar C-G, Hanin I, Holmstedt B, Kitz RJ, Jenden DJ, Karlén B (1968) Identification of acetylcholine in fresh rat brain by combined gas chromatography-mass spectrometry. Nature 220:915–917
Hanin I (1968) A new gas chromatographic procedure for the microestimation of acetylcholine in tissue extracts. PhD thesis, Department of Pharmacology, UCLA School of Medicine, Los Angeles
Hanin I (ed) (1974) Choline and acetylcholine. Handbook of chemical assay methods. Raven, New York
Hanin I (1982) Methods for the analysis and measurement of acetylcholine: an overview. In: Spector S, Back N (eds) Modern methods in pharmacology. Liss, New York, pp 29–38
Hanin I, Costa E (1976) Approaches used to estimate brain acetylcholine turnover rate *in vivo*: effects of drugs on brain acetylcholine turnover rate. In: Goldberg AM, Hanin I (eds) Biology of cholinergic function. Raven, New York, pp 355–377

Hanin I, Goldberg AM (1976) Appendix I: quantitative assay methodology for choline, acetylcholine, choline acetyltransferase, and acetylcholinesterase. In: Goldberg AM, Hanin I (eds) Biology of cholinergic function. Raven, New York, pp 647–654

Hanin I, Jenden DJ (1969) Estimation of choline esters in brain by a new gas chromatographic procedure. Biochem Pharmacol 18:837–845

Hanin I, Shih T-M (1980) Gas chromatography, mass spectrometry, and combined gas chromatography-mass spectrometry. In: Hanin I, Koslow SH (eds) Physico-chemical methodologies in psychiatric research. Raven, New York, pp 111–154

Hanin I, Skinner RF (1975) Analysis of microquantities of choline and its esters utilizing gas chromatography-chemical ionization mass spectrometry. Anal Biochem 66:568–583

Hanin I, Massarelli R, Costa E (1970) Environmental and technical preconditions influencing choline and acetylcholine concentrations in rat brain. In: Heilbronn E, Winter A (eds) Drugs and cholinergic mechanisms in the CNS. Research Institute of National Defense, Stockholm, pp 33–54

Hanin I, Cheney DL, Trabucchi M, Massarelli R, Wang CT, Costa E (1973) Application of principles of steady-state kinetics to measure acetylcholine turnover rate in rat salivary glands: effects of deafferentation and duct ligation. J Pharmacol Exp Ther 187:68–77

Hebb CO, Whittaker VP (1958) Intracellular distribution of acetylcholine and choline acetylase. J Physiol 142:187–196

Hestrin S (1949) The reaction of acetylcholine and other carboxylic acid derivatives with hydroxylamine, and its analytical application. J Biol Chem 180:249–267

Honda K, Miyaguchi K, Nishino H, Tanaka H, Yao T, Imai K (1986) High performance liquid chromatography followed by peroxyoxalate chemiluminescence detection of acetylcholine and choline using immobilized enzymes. Anal Biochem 153:50–53

Ikuta S, Imamura S, Misabi H, Horiuti Y (1977) Purification and characterization of cholineoxidase from arthrobacter globiformus. J Biochem 82:1714–1749

Israël M, Lesbats B (1980) Détection continue de la liberation d'acetylcholine de l'organe électrique de la Torpille à l'aide d'une réaction de chimiluminescence. CR Acad Sci [D] (Paris) 291:713–716

Israël M, Lesbats B (1981 a) Chemiluminescent determination of acetylcholine and continuous detection of its release from *Torpedo* electric organ synapses and synaptosomes. Neurochem Int 3:81–90

Israël M, Lesbats B (1981 b) Continuous determination by a chemiluminescent method of acetylcholine release and compartmentation in *Torpedo* electric organ synaptosomes. J Neurochem 37:1475–1483

Jenden DJ (1970) Recent developments in the determination of acetylcholine. In: Heilbronn E, Winter A (eds) Drugs and cholinergic mechanisms in the CNS. Research Institute of National Defense, Stockholm, pp 3–13

Jenden DJ (1973) The use of gas chromatography-mass spectrometry to measure tissue levels and turnover of acetylcholine. In: Costa E, Holmstedt B (eds) Advances in biochemical psychopharmacology, vol 7. Raven, New York, pp 69–81

Jenden DJ (1976) Approaches to the study of acetylcholine turnover in brain. Proc West Pharmacol Soc 19:1–7

Jenden DJ (1978) Estimation of acetylcholine and the dynamics of its metabolism. In: Jenden DJ (ed) Cholinergic mechanisms and psychopharmacology. Plenum, New York, pp 139–162

Jenden DJ, Campbell LB (1971) Measurement of choline esters. In: Glick D (ed) Analysis of biogenic amines and their related enzymes, vol 19. Interscience, New York, pp 183–216

Jenden DJ, Hanin I, Lamb SI (1968) Gas chromatographic microestimation of acetylcholine and related compounds. Anal Chem 48:125–128

Jenden DJ, Choi L, Silverman RW, Steinborn JA, Roch M, Booth RA (1974) Acetylcholine turnover estimation in brain by gas chromatography/mass spectrometry. Life Sci 14:55–63

Kaneda N, Noro Y, Nagatsu T (1985) Highly sensitive assay for acetylcholinesterase activity by high performance liquid chromatography with electrochemical detection. J Chromatogr 344:93–100

Karlén B, Lundgren G, Nordgren I, Holmstedt B (1974) Ion pair extraction in combination with gas phase analysis of acetylcholine. In: Hanin I (ed) Choline and acetylcholine: handbook of chemical assay methods. Raven, New York, pp 163–174

Karlén B, Lundgren G, Lundin J, Holmstedt B (1982) On the turnover of acetylcholine in mouse brain: influence of dose size of deuterium labelled choline given as precursor. Biochem Pharmacol 31:2867–2872

Kawashima K, Ishikawa H, Mochizuki M (1980) Radioimmunoassay for acetylcholine in the rat brain. J Pharmacol Methods 3:115–123

Kewitz H, Dross K, Pleul O (1973) Choline and its metabolic successors in brain. In: Genazzani E, Herken H (eds) Central nervous system – studies on metabolic regulation and function. Springer, Berlin Heidelberg New York, pp 21–32

Khandelwal JK, Szilagyi PIA, Barker LA, Green JP (1981) Simultaneous measurement of acetylcholine and choline in brain by pyrolysis-gas chromatography-mass spectrometry. Eur J Pharmacol 76:145–156

Kosh JW, Smith MB, Sowell JW, Freeman JJ (1979) Improvements in the gas chromatographic analysis of acetylcholine and choline. J Chromatogr 163:206–211

Koslow S, Racagni G, Costa E (1974) Mass fragmentographic measurement of norepinephrine, dopamine, serotonin and acetylcholine in seven discrete nuclei of the rat telediencephalon. Neuropharmacology 13:1123–1130

Kunec-Vagic E, Weber K (1967) Acetylcholine as an activator of oxidation reactions. Experientia 23:432–433

Lehmann WD, Schulten H-R, Schroder N (1978) Determination of choline and acetylcholine in distinct rat brain regions by stable isotope dilution and field desorption mass spectrometry. Biomed Mass Spectrom 5:591–595

Lund DD, Oda RP, Pardini BJ, Schmid PG (1986) Vagus nerve stimulation alters regional acetylcholine turnover in rat heart. Circ Res 58:373–377

MacIntosh FC (1963) Synthesis and storage of acetylcholine in nervous tissue. Can J Biochem Physiol 41:2555–2571

MacIntosh FC, Perry WLM (1950) Biological estimation of acetylcholine. Methods Med Res 3:78–92

Maruyama Y, Kusaka M, Mori J, Horikawa A, Hasegawa Y (1979) Simple method for the determination of choline and acetylcholine by pyrolysis gas chromatography. J Chromatogr 164:121–127

Maslova AF (1964) Quantitative determination of acetylcholine in biological materials by polarographic analysis. Vopr Med Khim 10:311–316

Mitchell R, Clark BB (1952) Determination of quaternary ammonium compounds including acetylcholine, tetraethylammonium and hexamethonium. Proc Soc Exp Biol Med 81:105–109

Mitchner H, Jennings EC (1967) Modified Hoffman degradation for the analysis of *N*-alkyl benzyldimethylammonium chlorides by gas chromatography. II. Benzalkonium chloride. J Pharm Sci 56:1595–1598

Moroni F, Malthe-Sorenssen D, Cheney DL, Costa E (1978) Modulation of ACh turnover in the septal-hippocampal pathway by electrical stimulation and lesioning. Brain Res 150:333–341

Neff NH, Lin RC, Ngai SH, Costa E (1969) Turnover rate measurement of brain serotonin in unanesthetized rats. In: Costa E, Greengard P (eds) Advances in biochemical psychopharmacology, vol 1. Raven, New York, pp 91–110

Neff NH, Spano PF, Groppetti A, Wang CT, Costa E (1971) A simple procedure for calculating the synthesis rate of norepinephrine, dopamine and serotonin in rat brain. J Pharmacol Exp Ther 176:701–710

Nordberg A (1977) Apparent regional turnover of acetylcholine in mouse brain: methodological and functional aspects. Acta Physiol Scand [Suppl] 445:1–51

Nordberg A, Sundwall A (1977) Effect of sodium pentobarbital on the apparent turnover of acetylcholine in different brain regions. Acta Physiol Scand 99:336–344

Ohsawa K (1980) Fluid-potentiometer and acetylcholine bioassay with clam entire heart. Life Sci 26:111–115
O'Neill JJ, Sakamoto T (1970) Enzymatic fluorometric determination of acetylcholine in biological extracts. J Neurochem 17:1451–1460
Paton WDM, Vizi ES (1969) The inhibitory action of noradrenaline and adrenaline on acetylcholine output by guinea-pig longitudinal muscle strip. Br J Pharmacol 35:10–28
Polak RL, Molenaar PC (1974) Pitfalls in determination of acetylcholine from brain by pyrolysis-gas chromatography/mass spectrometry. J Neurochem 23:1295–1297
Polak RL, Molenaar PC (1979) A method for determination of acetylcholine by slow pyrolysis combined with mass fragmentography on a packed capillary column. J Neurochem 32:407–412
Potter PE, Meek JL, Neff NH (1983) Acetylcholine and choline in neuronal tissue measured by HPLC with electrochemical detection. J Neurochem 41:188–194
Potter PE, Hadjiconstantinou M, Meek JL, Neff NH (1984) Measurement of acetylcholine turnover rate in brain: an adjunct to a simple HPLC method for choline and acetylcholine. J Neurochem 43:288–290
Racagni G, Cheney DL, Trabucchi M, Wang C, Costa E (1974) Measurement of acetylcholine turnover rate in discrete areas of rat brain. Life Sci 15:1961–1975
Racagni G, Trabucchi M, Cheney DL (1975) Steady-state concentration of choline and acetylcholine in rat brain parts during a constant rate infusion of deuterated choline. Naunyn Schmiedebergs Arch Pharmacol 290:99–105
Reid WD, Haubrich DR, Krishna G (1971) Enzymatic radioassay for acetylcholine and choline in brain. Anal Biochem 42:390–397
Richter D, Crossland J (1949) Variation in acetylcholine content of the brain with physiological state. Am J Physiol 159:247–255
Rodriguez de Lores Arnaiz G, Zieher LM, deRobertis E (1970) Neurochemical and structural studies on the mechanism of action of hemicholinium-3 in central cholinergic synapses. J Neurochem 17:221–229
Ryall W, Stone N, Watkins JC (1964) The cholinomimetic activity of extracts of brain nerve terminals. J Neurochem 11:621–637
Saelens JK, Allen MP, Simke JP (1970) Determination of acetylcholine and choline by an enzymatic assay. Arch Int Pharmacodyn 186:279–286
Saelens JK, Simke JP, Schuman J, Allen MP (1974) Studies with agents which influence acetylcholine metabolism in mouse brain. Arch Int Pharmacodyn 209:250–258
Schmidt DE (1976) Regional levels of choline and acetylcholine in rat brain following head focussed microwave sacrifice: effect of (+)amphetamine and (±)parachloramphetamine. Neuropharmacology 15:77–84
Schmidt DE, Speth RC (1975) Simultaneous analysis of choline and acetylcholine in rat brain by pyrolysis gas chromatography. Anal Biochem 67:353–357
Schmidt DE, Szilagyi PIA, Alkon DL, Green JP (1969) Acetylcholine: release from neural tissue and identification by pyrolysis-gas chromatography. Science 165:1370–1371
Schmidt DE, Speth RC, Welsch F, Schmidt MJ (1972) The use of microwave radiation in the determination of acetylcholine in the rat brain. Brain Res 38:377–389
Schuberth J, Sundwall A (1971) A method for the determination of choline in biological materials. Acta Pharmacol Toxicol 29 [Suppl 4]: S1
Schuberth J, Sparf B, Sundwall A (1969) A technique for the study of acetylcholine turnover in mouse *in vivo.* J Neurochem 16:695–700
Schuberth J, Sparf B, Sundwall A (1970) On the turnover of acetylcholine in nerve endings of mouse brain *in vivo.* J Neurochem 17:461–468
Schumacher H, Ehl R (1970) Der Einfluß von Diethyl-p-nitro-phenyl-phosphat auf den Acetylcholingehalt des Kammerwassers. Arzneimittelforsch 20:1476–1479
Shaw FH (1938) CXXXV. The estimation of choline and acetylcholine. Biochem J 32:1002–1007
Shea PA, Aprison MH (1973) An enzymatic method for measuring picomole quantities of acetylcholine and choline in CNS tissue. Anal Biochem 56:165–177

Shih T-M, Hanin I (1978) Effects of chronic lead exposure on levels of acetylcholine and choline and on acetylcholine turnover rate in rat brain areas *in vivo*. Psychopharmacology 58:263–269

Sinsheimer JE, Hong D (1965) Recorded amperometric tetraphenylborate titrations of amines, quaternary nitrogen compounds, and potassium salts. J Pharma Sci 54:805–806

Spector S, Felix AM, Finberg JPM (1978) Radioimmunoassay for acetylcholine. In: Jenden DJ (ed) Cholinergic mechanisms and psychopharmacology. Plenum, New York, pp 163–168

Stavinoha WB, Ryan LC (1965) Estimation of the acetylcholine content of rat brain by gas chromatography. J Pharmacol Exp Ther 150:231–235

Stavinoha WB, Weintraub ST (1974a) Estimation of choline and acetylcholine in tissue by pyrolysis gas chromatography. Anal Chem 46:757–760

Stavinoha WB, Weintraub ST (1974b) Choline content of rat brain. Science 183:964–965

Stavinoha WB, Ryan LC, Treat EL (1964) Estimation of acetylcholine by gas chromatography. Life Sci 3:689–693

Stavinoha WB, Endecott BR, Ryan LC (1967) Freezing techniques to prepare brain tissue for analysis of acetylcholine. Pharmacologist 9:252

Stavinoha WB, Weintraub ST, Modak AT (1973) The use of microwave heating to inactivate cholinesterase in the rat brain prior to analysis for acetylcholine. J Neurochem 20:361–371

Stavinoha WB, Weintraub ST, Modak AT (1974) Regional concentrations of choline and acetylcholine in the rat brain. J Neurochem 23:885–886

Stavinoha WB, Modak AT, Weintraub ST (1976) Rate of accumulation of acetylcholine in discrete regions of the rat brain after dichlorvos treatment. J Neurochem 27:1375–1378

Szerb JC, Malik H, Hunter EC (1970) Relationship between acetylcholine content and release in the cat's cerebral cortex. Can J Physiol Pharmacol 48:780–790

Szilagyi PIA, Schmidt DE, Green JP (1968) Microanalytical determination of acetylcholine, other choline esters and choline by pyrolysis-gas chromatography. Anal Chem 40:2009–2013

Szilagyi PIA, Green JP, Brown OM, Margolis S (1972) The measurement of nanogram amounts of acetylcholine in tissues by pyrolysis gas chromatography. J Neurochem 19:2555–2566

Takahashi R, Aprison MH (1964) Acetylcholine content of discrete areas of the brain obtained by a near-freezing method. J Neurochem 11:887–898

Toru M, Aprison MH (1966) Brain acetylcholine studies: a new extraction procedure. J Neurochem 13:1533–1544

Unger SE, Ryan TM, Cooks RG (1981) The identification of choline and its esters by secondary ion mass spectrometry. SIA Surface Interface Analysis 3:12–15

Van Gulik WJ (1956) Small-scale chomatography and the direct photometric determination of choline esters on paper. Acta Physiol Pharmacol Neerl 5:222

Veech RL, Hawkins RA (1974) Brain blowing: a technique for *in vivo* study of brain metabolism. In: Marks N, Rodnight R (eds) Research methods in neurochemistry, vol 2. Plenum, New York, pp 171–182

Veech RL, Harris RL, Veloso D, Veech EH (1973) Freeze blowing: a new technique for the study of brain *in vivo*. J Neurochem 20:183–188

Vizi ES, Hársing LG Jr, Duncalf D, Nagashima H, Potter PE, Foldes FF (1985) A simple and sensitive method of acetylcholine identification and assay: bioassay combined with mini-column gel filtration or high performance liquid chromatography. J Pharmacol Methods 13:201–211

Vocci FJ Jr, Karbowski MJ, Dewey WL (1979) Apparent *in vivo* acetylcholine turnover rate in whole mouse brain: evidence for a two compartment model by two independent kinetic analysis. J Neurochem 32:1417–1422

Weiler M, Roed IS, Whittaker VP (1982) The kinetics of acetylcholine turnover in a resting cholinergic nerve terminal and the magnitude of the cytoplasmic compartment. J Neurochem 38:1187–1191

Whittaker VP (1963) Identification of acetylcholine and related esters of biological origin. Springer, Berlin Heidelberg New York, pp 1–39 (Handbook of experimental pharmacology, vol 15)
Whittaker VP, Barker LA (1972) The subcellular fractionation of brain tissue with special reference to the preparation of synaptosomes and their component organelles. In: Fried R (ed) Methods of neurochemistry, vol 2. Dekker, New York, pp 1–52
Zilversmit DB (1960) The design and analysis of isotope experiments. Am J Med 29:832–848
Zsilla G, Cheney DL, Racagni G, Costa E (1976) Correlation between analgesia and the decrease of acetylcholine turnover rate in cortex and hippocampus elicited by morphine, meperidine, viminol R_2 and azidomorphine. J Pharm Exp Ther 199:662–668
Zsilla G, Racagni G, Cheney DL, Costa E (1977) Constant rate infusion of deuterated phosphorylcholine to measure the effects of morphine on acetylcholine turnover rate in specific nuclei of rat brain. Neuropharmacology 16:25–30

CHAPTER 7

Choline Acetyltransferase and the Synthesis of Acetylcholine

S. TUČEK

A. Introduction

Soon after the discovery that acetylcholine (ACh) mediates synaptic transmission in sympathetic ganglia, BROWN and FELDBERG (1936) noted that the amount of ACh released from perfused stimulated ganglia during their experiments was several-fold higher than the amount of ACh that had been present in the ganglia at the start of the experiments. It became apparent from this and other observations that ACh is synthesized in the nerve terminals which use it as their transmitter and that the rate of its synthesis varies so as to keep the stores of ACh in the nerve terminals at a constant level under most physiological conditions. Systematic studies of the process of ACh synthesis were started by QUASTEL and MANN (QUASTEL et al. 1936, MANN et al. 1938, 1939) and by STEDMAN and STEDMAN (1937). In a few years, NACHMANSOHN and MACHADO (1943) proved able to demonstrate the synthesis of ACh in cell-free tissue extracts with added ATP, acetate and choline; they ascribed the synthesis to the activity of a new enzyme which they called 'choline acetylase'. It has been clarified in the following work that at least two enzymes must have been active in the system used by NACHMANSOHN and MACHADO (1943), one enzyme producing acetylcoenzyme A (AcCoA) from acetate, coenzyme A (CoA) and ATP (AcCoA synthetase by the present nomenclature), and another enzyme producing ACh from AcCoA and choline. The name of choline acetylase was preserved for the latter enzyme and was modified to choline acetyltransferase (acetyl-CoA:choline-*O*-acetyltransferase, EC 2.3.1.6) (ChAT) in the EC-IUB enzyme nomenclature.

The synthesis of ACh proceeds according to the equation (see Fig. 1 for formulas):

$$\text{Choline} + \text{AcCoA} \xrightleftharpoons{\text{ChAT}} \text{ACh} + \text{CoA} .$$

The reaction is reversible but the equilibrium is shifted to the right, the equilibrium constant K_{eq} being close to 13 (12.3 according to PIEKLIK and GUYNN 1975; 13.3 according to HERSH 1982).

An attempt is made in this chapter to provide a general outline of the present understanding of how ACh is synthesized in nerve cells. The main areas reviewed are the basic properties of ChAT, the origin and supply of substrates for the synthesis, and the question of how the synthesis is controlled. Considerably more is known about the synthesis of ACh in the nervous system of mammals than in that of other animals; this is why data obtained on mammalian brain and sympathetic neurons prevail throughout the chapter. Limited space does not permit going into

acetyl-CoA + choline ⇌ CoA + acetylcholine

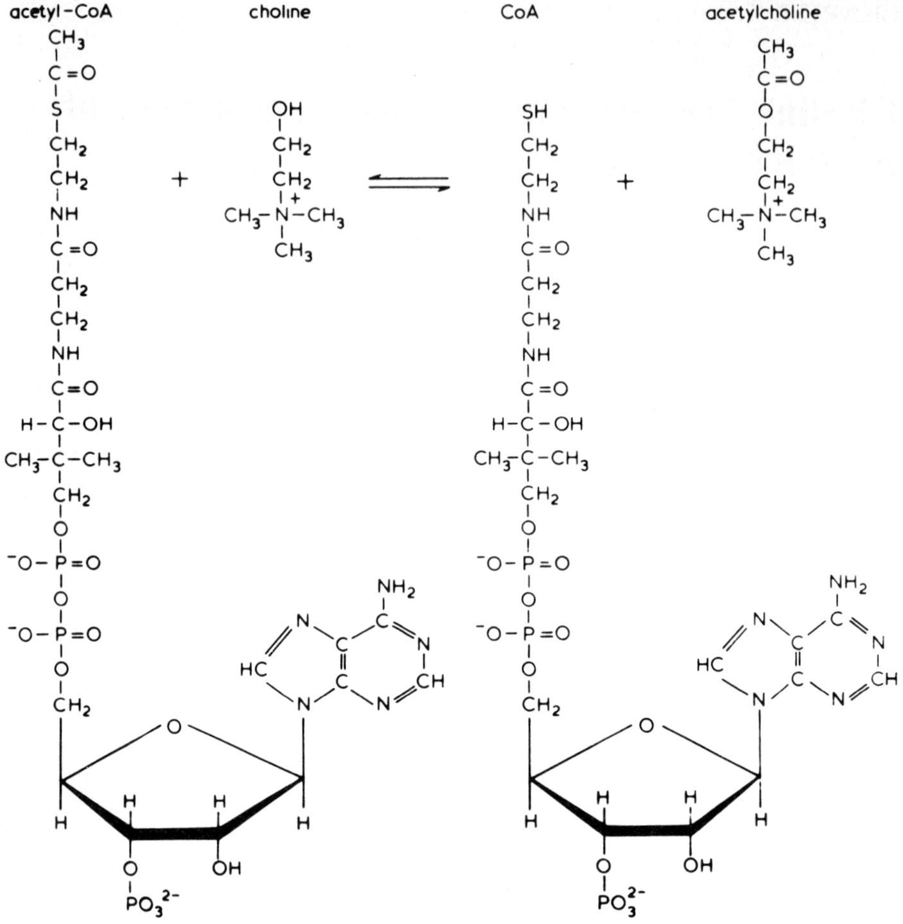

Fig. 1. Reaction for the biosynthesis of ACh

details and many important aspects have had to be omitted. Interested readers are advised to seek additional information in specialized books (GOLDBERG and HANIN 1976; TUČEK 1978, 1979; HANIN and GOLDBERG 1982) and review articles (MACINTOSH and COLLIER 1976; ROSSIER 1977a; JOPE 1979; MALTHE-SØRENSSEN 1979; MAUTNER 1980, 1986; TUČEK 1983a, b, 1984, 1985; MCGEER et al. 1984).

B. Choline Acetyltransferase

I. The Enzyme Molecule

ChAT is a globular protein with a radius estimated at 3.39 nm (ROSSIER 1976) and with a comparatively high positive surface charge (FONNUM 1970), which is probably due to the high content of glutamine (BRUCE et al. 1985). Data on its relative molecular mass (M_r) vary within a wide range (see Table 1 in TUČEK 1983a), but most investigators found it close to 68000 (ROSSIER 1976; ECKENSTEIN and THOENEN 1982; CRAWFORD et al. 1982a; BRUCE et al. 1985). HERSH et al. (1984) have shown that limited proteolysis of ChAT (not accompanied by a loss of activity) is likely to occur during tissue homogenization and isolation of the enzyme, and it appears that the true molecular mass of native ChAT is somewhat higher (73000 based on the work of HERSH et al. 1984).

The purification of ChAT proved to be difficult, mainly in view of its low content in tissues and of the presence of other proteins of similar mass, shape and charge. In addition to conventional protein separation methods, chromatographic procedures taking advantage of certain distinctive properties of ChAT were successfully applied, such as chromatography on hexylamine-Sepharose (ECKENSTEIN et al. 1981), octyl-Sepharose (SLEMMON et al. 1982), blue dextran-Sepharose (ROSKOSKI et al. 1975) with AcCoA (MALTHE-SØRENSSEN et al. 1978; HERSH et al. 1978a; DIETZ and SALVATERRA 1980) or CoA (ECKENSTEIN et al. 1981) used for elution, decamethylene-naphthylvinylpyridine-Sepharose (COZZARI and HARTMAN 1983a), CoA-Sepharose (DRISKELL et al. 1978; RYAN and MCCLURE 1979) or CoA-hexane-Agarose (STRAUSS and NIRENBERG 1985). Successful production of monospecific antibodies against ChAT opened the door for its purification by immunoabsorption (ECKENSTEIN and THOENEN 1982; PENG et al. 1983; BRUCE et al. 1985; STRAUSS and NIRENBERG 1985; JOHNSON and EPSTEIN 1986). References to several of the more successful purifications of ChAT are listed in Table 1; for a more complete list, see Table II in TUČEK (1983a) and Table I in MCGEER et al. (1984).

Many data accumulated in the literature after 1972 indicating that multiple molecular forms of ChAT (either of different electric charge or of different molecular mass) exist in the nervous tissue of most animal species (for reviews, see MALTHE-SØRENSSEN 1979; TUČEK 1983a; for more recent references, HERSH et al. 1984). The existence of multiple electric charge or molecular mass variants was regarded as the result of differences in posttranslational processing of the enzyme. It appears more likely at present, however, that the observed heterogeneity was due to a limited proteolysis of the enzyme during homogenization and purification (HERSH et al. 1984). The findings of isoelectric variants with virtually identical molecular mass may have been caused by various degrees of *post-mortem* deamidation of glutamine and asparagine residues in the enzyme. A new type of molecular heterogeneity of ChAT has been described recently, based on differences in its hydrophilic versus amphiphilic properties (EDER-COLLI et al. 1986). Relevant evidence will be mentioned in connection with the question of the intracellular localization of ChAT.

During sodium dodecyl sulphate-polyacrylamide gel electrophoresis (SDS-PAGE) ChAT moves as one single band or as 2–3 bands of very similar mobility (e.g. BRANDON and WU 1978; DRISKELL et al. 1978; DIETZ and SALVATERRA 1980; ECKENSTEIN et al. 1981; BRUCE et al. 1985; PENG et al. 1986; BRUCE and HERSH 1987) and the bulk of available evidence supports the view that the enzyme (or its molecular mass isoforms, if they exist *in vivo*) is a simple protein, not composed of non-covalently associated subunits. Observations contradicting this view have been published by SU et al. (1980) and BADAMCHIAN and

Table 1. Selected references to ChAT purification

Source of enzyme		Final activity (μmol min^{-1} mg^{-1} protein)	Estimated molecular mass (kDa)	References
Species	Tissue			
Drosophila	Head Ganglia	500	67 (67 + 54)	Slemmon et al. (1982)
Squid	Optic lobes	67	120; 125; 200	Husain and Mautner (1973)
Fowl	Optic lobes	50	64	Johnson and Epstein (1986)
Rat	Brain	77	67	Dietz and Salvaterra (1980)
	Brain	29	–	Strauss and Nirenberg (1985)
Ox	Caudate nuclei	28	69	Malthe-Sørenssen et al. (1978)
	Caudate nuclei	59	–	Ryan and McClure (1979)
	Caudate nuclei	142	72; 76	Cozzari and Hartmann (1983a)
Pig	Brain	135	67	Eckenstein et al. (1981)
Human	Placenta	93	68	Hersh et al. (1978a)
	Placenta	45	68	Bruce et al. (1985)
	Striatum (washed membranes)	37	67; 62	Peng et al. (1986)

CARROLL (1985). Experiments with *Drosophila* ChAT (SLEMMON et al. 1984; SALVATERRA and MCCAMAN 1985) suggest that a proportion of the enzyme which has an M_r of 67000 undergoes a limited posttranslational proteolytic cleavage yielding two peptides with M_r values of 54000 and 13000, and that the two subunits (of which the heavier one is probably enzymatically active) remain non-covalently associated in the cells.

ChAT is a weak immunogen and attempts to produce monospecific antibodies against it were not particularly successful until the end of the last decade. The situation changed in the present decade, and the successful production of antibodies against ChAT, both polyclonal (ECKENSTEIN et al. 1981; COZZARI and HARTMAN 1980) and monoclonal (LEVEY et al. 1981, 1983; ECKENSTEIN and THOENEN 1982; CRAWFORD et al. 1982a, b; ROSS et al. 1983; NAGAI et al. 1983; ICHIKAWA et al. 1983; STRAUSS and NIRENBERG 1985; JOHNSON and EPSTEIN 1986) has been reported from many laboratories. Their utilization was of crucial importance for the progress in the immunocytochemical identification of cholinergic neurons and pathways in the central nervous system (Chap. 22; HOUSER et al. 1983; WAINER et al. 1984; MCGEER et al. 1984; KÁSA 1986). Although most of the monoclonal antibodies do not bind to the active centre of the enzyme, some of them do. Data obtained by CRAWFORD et al. (1982a) suggest that antibody determinants are clustered in a localized region on the enzyme's surface. The antibodies differ strongly in their species specificity or non-specificity.

Substantial progress in the unravelling of the chemical structure of ChAT is likely to result from the utilization of the methods of molecular genetics. The localization of the gene responsible for the synthesis of ChAT has been determined in *Drosophila* chromosomes by GREENSPAN (1980) and successful cloning of complementary DNA for *Drosophila* ChAT has been achieved by ITOH et al. (1986); its analysis suggests that *Drosophila* ChAT is derived from a much larger precursor. Promising for further progress is the finding by GUNDERSEN et al. (1985) that the synthesis of ChAT can be induced in *Xenopus* oöcytes by injecting them with the appropriate mRNA.

II. Catalytic Action

Several explanations of the kinetic mechanism of the ChAT-catalysed reaction were proposed in earlier studies (see TUČEK 1983a) but the most probable one seems to be that based on recent investigations performed in HERSH's laboratory. Analysis of measurements of initial reaction velocities at different concentrations of substrates and products associated with the use of dead-end inhibitors and of alternate substrates (HERSH and PEET 1977; HERSH 1980) and with an investigation of isotope exchange at equilibrium (HERSH 1982) has led to the conclusion that the ChAT-catalysed reaction follows a random Theorell-Chance mechanism in which a low but finite amount of ternary complex exists. Although the binding of substrates and the subsequent release of products are random, the predominant pathway is that in which AcCoA binds to the enzyme before choline and ACh dissociates from the enzyme before CoA (Fig. 2). Within the range of physiological values of ionic strength the rate of the reaction depends on the interconversion

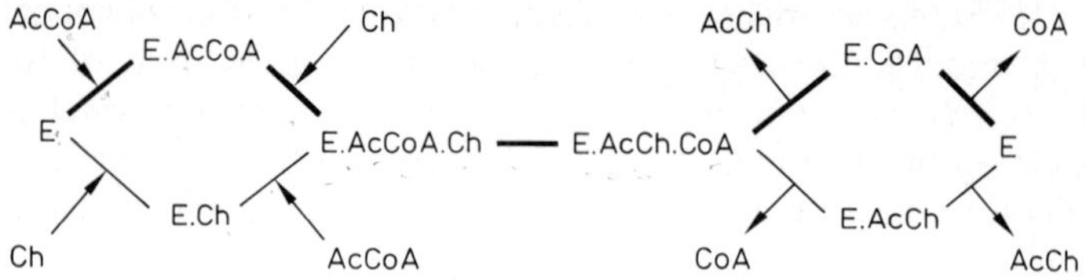

Fig. 2. Mechanism of the reaction catalysed by ChAT (according to HERSH 1982). *AcCh*, acetylcholine; *AcCoA*, acetyl coenzyme A; *CoA*, coenzyme A; *Ch*, choline; *E*, enzyme

rate of the ternary complex; when the reaction proceeds at a low ionic strength, the dissociation of CoA becomes the rate-limiting step (HERSH et al. 1978b).

Data concerning the arrangement of the active sites of ChAT and its requirements regarding the structure of substrates have been reviewed by MAUTNER (1986) and TUČEK (1983a). Choline binds with its cationic head to an anionic binding site and its binding is very little affected after one of its three methyl groups had been structurally modified; on the other hand, replacements of more than one of the methyl groups or modifications of the hydroxyethyl chain of choline may have strong effects on the binding.

Both coulombic and hydrophobic forces play a role in the binding of AcCoA, which seems to be mainly effected through the adenosine part of its molecule (BANNS et al. 1977; MALTHE-SØRENSSEN et al. 1978; ROSSIER 1977b) and through the 3′-phosphate on the ribose part (CURRIER and MAUTNER 1977; ROSSIER 1977b; WEBER et al. 1979). The 3′-phosphate probably binds to an arginine residue in the active centre (MAUTNER et al. 1981b).

Very important for the catalytic function of ChAT is the imidazole group of a histidine residue within the active centre (CURRIER and MAUTNER 1974; ROSKOSKI 1974; MALTHE-SØRENSSEN 1976). It probably acts through general-base catalysis and enables the hydroxyl group of choline to attack the thiol ester grouping of AcCoA (MAUTNER 1980); a less likely possibility is that the imidazole group ensures a nucleophilic attack on acetylcoenzyme A, a transient formation of acetylated histidine and the transfer of the acetyl group to choline (MALTHE-SØRENSSEN 1979).

ChAT contains free thiol groups (SEVERIN and ARTENIE 1967) and can be inhibited by thiol reagents (BERMAN-REISBERG 1957). In spite of an earlier claim to the contrary (ROSKOSKI 1973), a consensus had been reached that the thiol groups do not play a role in the binding of substrates and are not essential for enzyme activity. At least one thiol group is present in the active site and reagents binding to it inhibit ChAT activity by steric hindrance (MALTHE-SØRENSSEN 1976; HERSH et al. 1979).

ChAT binds easily to blue dextran columns (ROSKOSKI et al. 1975) from which it can be eluted by AcCoA or CoA (MALTHE-SØRENSSEN et al. 1978; HERSH et al. 1978a; DIETZ and SALVATERRA 1980; ECKENSTEIN et al. 1981). Reactive blue 2 (the chromophore of blue dextran) and blue dextran act as competitive inhibitors with regard to AcCoA (ROSKOSKI et al. 1975; ROSSIER 1977b; MAUTNER et al. 1981a). It was suggested that the high affinity for reactive blue 2 indicates the presence on various enzyme molecules of a 'dinucleotide fold', which is thought to be a structure mediating the binding of the adenosyl diphosphoryl part of CoA and other cofactors. Although this generalization is not always fully justified (MAUTNER et al. 1981a), data on the binding and inhibitory properties of reactive blue 2 and other aromatic dyes with regard to ChAT may be taken as an indication that there is a hydrophobic site involved in the attachment of AcCoA to ChAT (see MAUTNER 1986). It has not been fully excluded that the binding of ChAT to nonspecific hydrophobic columns observed during purification (MAUTNER 1977; MALTHE-SØRENSSEN et al. 1978; PENG et al. 1980; ECKENSTEIN et al. 1981; SLEMMON et al. 1982) occurred via this specific AcCoA-binding hydrophobic site, but it seems rather more likely that other hydrophobic areas on the surface of the enzyme were involved (DIETZ and SALVATERRA 1980).

A schematic drawing by MAUTNER (1986) suggesting the possible arrangement of ChAT binding sites and of the attached substrates is reproduced in Fig. 3.

Apparent K_m values of mammalian ChAT are usually found in the range of 5–20 μM for AcCoA and of 0.3–1.2 mM for choline. Examples of values actually reported are given in Table 2. Both the K_m and V_{max} of ChAT are strongly influenced by the ionic strength of the medium; they increase with increasing concentrations of monovalent ions and reach a maximum at salt concentrations of about 0.3 M. The activating effect of monovalent ions is basically nonspecific (MORRIS and TUČEK 1966; POTTER et al. 1968; HERSH 1979, 1980). Divalent cations increase the activity of ChAT at low and inhibit at high concentrations

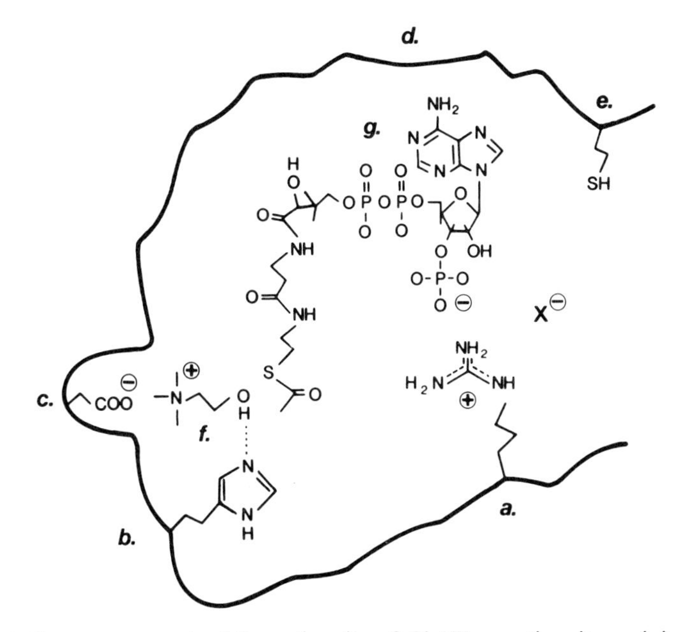

Fig. 3. Tentative arrangement of the active site of ChAT; *a*, active-site arginine; *b*, active-site histidine; *c*, anionic binding site; *d*, hydrophobic binding site; *e*, cysteine; *f*, choline; *g*, acetyl coenzyme A. Reprinted, with permission, from MAUTNER (1986)

Table 2. Examples of apparent K_m values of choline acetyltransferase for its natural substrates

Source of enzyme		K_m values for				References
Species	Tissue	Choline (m*M*)	Acetyl-coenzyme A (μ*M*)	Acetyl-choline (m*M*)	Coenzyme A (μ*M*)	
Drosophila	Head ganglia	0.047	90	–	–	DRISKELL et al. (1978)
	Head ganglia	6.8	110	–	–	SALVATERRA and MCCAMAN (1985)
Squid	Ganglia	1.9	47	–	–	CURRIER and MAUTNER (1977)
Torpedo	Electric organ[a]	0.72	22.6	–	–	EDER-COLLI et al. (1986)
	Electric organ[b]	1.08	16.6	–	–	EDER-COLLI et al. (1986)
Rat	Brain	0.41	18	–	–	WHITE and WU (1973a)
	Brain	1.01	46.5	–	–	MALTHE-SØRENSSEN (1976)
Ox	Striatum	0.75	10	–	–	GLOVER and POTTER (1971)
Human	Caudate nucleus	0.6	6.6	–	–	WHITE and WU (1973a)
	Placenta	0.41	11.9	1.3	8.8	HERSH and PEET (1977)

[a] Hydrophilic; [b] amphiphilic enzyme

(MORRIS and TUČEK 1966; POTTER et al. 1968; HERSH and PEET 1978). The increased V_{max} at high ionic strength is caused by an increase in the rate at which CoA (or AcCoA, in the case of reverse reaction) dissociates from the enzyme (HERSH et al. 1978b; HERSH 1979).

Most product inhibition studies agree that ACh acts as a competitive inhibitor with respect to choline and a non-competitive inhibitor with respect to AcCoA while CoA inhibits competitively with respect to AcCoA and non-competitively with respect to choline. RYAN and MCCLURE (1980) calculated apparent K_i values of 45.5 mM for ACh and 37.7 μM for CoA. WHITE and WU (1973a) derived K_i values of the same orders of magnitude but considerably lower values were calculated by HERSH and PEET (1977).

Data concerning the pH optimum of ChAT vary. Examples of reported values are 7.4 (MORRIS and TUČEK 1966), 7.5–10.0 (GLOVER and POTTER 1971), 8.5 (HERSH and PEET 1978) and 8.0–10.5 (RYAN and MCCLURE 1980). The pH-activity profiles depend on ionic strength (HERSH and PEET 1978).

The activity of ChAT can be more or less specifically inhibited by several groups of compounds. Information on ChAT inhibitors has not been included here because of limitation in space, and the reader is referred to reviews by HAUBRICH (1976), CHAO (1980) and MAUTNER (1986). A peptide which might act as an endogenous inhibitor of ChAT has been described by COZZARI and HARTMAN (1983b).

III. Cellular and Subcellular Localization

The principal role of ACh in the body is that of a synaptic transmitter and it is therefore not surprising that the enzyme producing it is mainly found in nerve cells. Among these the presence of ChAT seems to be specific for those which actually utilize ACh as their transmitter, i.e. for cholinergic neurons. The presence of ChAT, however, was also described in cells other than neurons (including some bacteria and plant cells) and the topic of non-neural ChAT has been reviewed by SASTRY and SADAVONGVIVAD (1979).

The most striking example of non-neuronal localization of ChAT in mammals is its presence in the placenta (a non-innervated tissue in which the functional role of ACh is not known). Human placenta is a particularly rich source of ChAT and there was much concern whether the neural and placental enzymes are identical. The answer is 'yes'; the enzymes from both sources have an identical molecular mass (ROSKOSKI et al. 1975; BRUCE et al. 1985), are recognized by the same monoclonal (LEVEY et al. 1982) and polyclonal (BRUCE et al. 1985) antibodies, are identically affected by a number of reagents (ROSKOSKI et al. 1975), yield identical peptide maps after tryptic digestion (BRUCE et al. 1985), have the same affinities for their substrates (Table 2) and utilize the same reaction mechanism (HERSH and PEET 1977; HERSH 1980).

It has been shown that the Schwann cells in the end-plate area of denervated frog muscles release ACh (BIRKS et al. 1960) and TUČEK et al. (1978) demonstrated the presence of ChAT in degenerated frog nerves composed virtually exclusively of Schwann cells; it appeared that these cells started to produce ChAT only after the nerve had been transected. As far as Schwann cells in mammalian nerves are concerned, it is unlikely that they produce ChAT even after nerve degeneration since the activity of ChAT in degenerating nerve trunks falls close to zero values (TUČEK 1974). On the other hand, ChAT is likely to be present in the Schwann cells of normal squid nerves (HEUMANN et al. 1981). The reason for these species variations is not clear.

The presence of ChAT in normal glial cells is unlikely. This view is mainly based on the complete absence of ChAT from the optic nerve (HEBB and SILVER 1956), which contains glial cells. ChAT was not found in cultured glia from rat striatum (SINGH and VAN ALSTYNE 1978) and in cultures of non-neuronal flat cells from chick embryonic brain (PETERSON et al. 1973). The enzyme was discovered, however, in altered glial cells, such as some gliomas (HEBB and SILVER 1970) and non-neuronal cell lines obtained from chemically induced brain tumours (SCHUBERT et al. 1974).

Recent immunocytochemical observations suggest that ChAT is also present in a small proportion of capillary endothelial cells (PARNAVELAS et al. 1985).

It is well known that ACh is regularly found in denervated skeletal muscles and in non-innervated portions of normal muscles (BHATNAGAR and MACINTOSH 1960; HEBB 1962; KRIEBEL et al. 1978; POLAK et al. 1981; MILEDI et al. 1977, 1982a, b; DOLEŽAL and TUČEK 1983) and released from them during incubations (MITCHELL and SILVER 1963; KRNJEVIĆ and STRAUGHAN 1964; POLAK et al. 1981; MILEDI et al. 1982b; DOLEŽAL and TUČEK 1983). However, the enzyme responsible for the synthesis of ACh in skeletal and cardiac muscle fibres is carnitine acetyltransferase and not ChAT (TUČEK 1982; SLAVÍKOVÁ and TUČEK 1982). Carnitine acetyltransferase can acetylate choline instead of carnitine, its natural substrate, because both compounds are similar (WHITE and WU 1973b). The functional significance of the ACh that is synthesized by carnitine acetyltransferase in the muscle fibres is not known. The synthesis of ACh by carnitine acetyltransferase explains earlier claims that ChAT is present in spermatozoa (GOODMAN et al. 1984).

Histochemical and immunocytochemical evidence indicates that ChAT is distributed throughout all parts of cholinergic neurons, i.e. in the perikarya, dendrites and neurites (KÁSA et al. 1970; HOUSER et al. 1983), but its distribution in subcellular fractions of brain homogenates shows that the enzyme is preferentially localized in the nerve terminals, i.e. close to the sites of ACh release (HEBB and WHITTAKER 1958; WHITTAKER 1965; TUČEK 1967a). FONNUM et al. (1973) estimated that in cholinergic neurons of the hypoglossal nucleus the perikarya contain 2%, the axons in the main nerve trunks 42%, and the preterminal axons and their endings 56% of the total ChAT content of these neurons.

To investigate the localization of ChAT inside the synaptosomes, they were disrupted by hypo-osmotic treatment and fractionated. It was found that the behaviour of the enzyme differed according to species. While most of the enzyme appeared in the soluble fraction in some species (WHITTAKER et al. 1964), a considerable proportion of it remained associated with membrane-containing particulate fractions in other species (MCCAMAN et al. 1965; TUČEK 1966a). Most of the enzyme is released from the membranes into solution, however, when the ionic strength of the medium is increased (FONNUM 1966, 1967, 1968; POTTER et al. 1968), and there is little doubt that under conditions of physiological ionic strength and pH the major part of ChAT is in a soluble state in the nerve terminal cytoplasm. The view that the enzyme is specifically associated with synaptic vesicles (DE ROBERTIS 1964) was not substantiated in later studies (TUČEK 1966b; EDER-COLLI and AMATO 1985).

It has been pointed out recently that a proportion of ChAT behaves as if it were rather firmly bound to nerve terminal membranes. Three varieties of the enzyme have been distinguished in the nerve terminals, based on experiments with differential solubilization of ChAT from brain (BENISHIN and CARROLL 1983) and electric organ (EDER-COLLI and AMATO 1985) synaptosomes: (a) the predominant form of soluble ChAT; (b) the ionically membrane-bound (high-salt-soluble) ChAT, extractable with 0.5 *M* NaCl; and (c) the non-ionically membrane-bound (detergent-soluble) ChAT. During extraction with Triton X-114 (a detergent which permits the separation of hydrophilic from amphiphilic proteins) about 12%–13% of ChAT in the electric organ behaved as an amphiphilic protein (EDER-COLLI et al. 1986) and appeared to be identical with the non-ionically membrane-bound (detergent-soluble) ChAT observed after differential solubilization (EDER-COLLI and AMATO 1985). The soluble and non-ionically membrane-bound forms of ChAT have been reported to have identical or very similar molecular mass, to behave similarly during SDS-PAGE (BADAMCHIAN and CARROLL 1985), to be recognized by the same antibody (PENG et al. 1986) and to have the same affinities for choline and AcCoA (BENISHIN and CARROLL 1983; EDER-COLLI and AMATO 1985; EDER-COLLI et al. 1986). On the other hand, the non-ionically membrane-bound enzyme has been reported to differ from the soluble enzyme in that it is capable of acetylating homocholine (BENISHIN and CARROLL 1983) and is less sensitive to inhibition by ACh and CoA (EDER-COLLI et al. 1986; see also SMITH and CARROLL 1980; BENISHIN and CARROLL 1981a, b).

While the observation that a proportion of ChAT in the electric organ is amphiphilic and differs from the bulk of the enzyme (EDER-COLLI et al. 1986) appears unequivocal, there is a strong possibility that the 'membrane-bound' enzyme found in studies with differential solubilization of brain ChAT was the soluble enzyme occluded in undisrupted synaptosomes or membrane vesicles forming during homogenization and hyposmotic treatment. Hyposmotic treatment of synaptosomes does not make all ChAT accessible to its substrates (TUČEK 1967d) and BRUCE and HERSH (1987) have indeed found that the solubilization of 'membrane-bound' brain ChAT with a detergent was accompanied by the solubilization of a corresponding proportion of lactate dehydrogenase, which is a marker for cytoplasmic enzymes. The soluble and detergent-released ChAT appeared identical by several criteria.

It seems likely that the amphiphilic ChAT in the electric organ (EDER-COLLI et al. 1986) represents the same molecule as the hydrophilic ChAT, but that its properties have been altered by an association with the membrane. Conceivably, some enzyme molecules become either non-covalently attached to an integral membrane protein, serving as an anchor, or covalently attached to a membrane lipid (e.g. phosphatidylinositol; see LOW et al. 1986). The attachment to the membrane would be expected to occur after the enzyme had been transported to the nerve terminals since very little ChAT was found associated with the membranes in *Torpedo* nerve trunk (EDER-COLLI and AMATO 1985).

Information about the transport of ChAT in the axons has been summarized by DAHLSTRÖM (1983); most data support the view that the transport is slow and unidirectional in mammals (e.g. TUČEK 1975) and *Torpedo* (DAVIES et al. 1977).

C. Supply of Substrates

I. Acetylcoenzyme A

AcCoA is an obligatory intermediary product in the metabolism of carbohydrates, fatty acids and several amino acids and is therefore present in all types of animal cells. Published data on its concentration in the brain vary (see Table 2 in TUČEK 1984), but most findings have been in the range of 1.9–6.0 nmol g^{-1} wet weight of brain tissue (GUYNN 1976: 5.1; ŘÍČNÝ and TUČEK 1980a: 1.9; MITZEN and KOEPPEN 1984: 1.9 nmol g^{-1}). Since 1 g of the brain is likely to contain about 550 μl of intracellular water, the mean concentration of AcCoA in the intracellular water would be expected to be in the range of 3.5–10.9 μM. The concentration is probably higher in the intramitochondrial water and lower in the cytosol but it is difficult to make a realistic estimate. In isolated adipocytes about one-half of their AcCoA seems to be in the cytosol and one-half in the mitochondria; the cytosol was found to contain 36% of total AcCoA after incubation without insulin and 72% after incubation with insulin (PAETZKE-BRUNNER et al. 1978). No data are available on the content of AcCoA in the electric organ of electric fish (for discussion see CORTHAY et al. 1985).

Many investigations have devoted themselves to the question of the origin of the acetyl groups in the particular pool of AcCoA that is used for the synthesis of ACh (reviews QUASTEL 1978; JOPE 1979; TUČEK 1978, 1983b, 1984). As far as the mammalian brain is concerned, there is little doubt that in the long run the acetyl groups of ACh are derived from glucose and pyruvate. This conclusion is based on *in vitro* and *in vivo* experiments. In brain slices incubated with labelled glucose or pyruvate all ACh synthesized during incubation was found to originate from the labelled substrates in the medium, with no contribution by endogenous precursors (BROWNING and SCHULMAN 1968; LEFRESNE et al. 1973, 1977; WILLOUGHBY et al. 1986). After intracisternal injections of a number of labelled substrates to rats *in vivo*, glucose, pyruvate and lactate were found to make the largest quantitative contribution to the synthesis of brain ACh (TUČEK and CHENG 1970, 1974).

Since AcCoA is produced from pyruvate (and from glucose and lactate, via pyruvate) in a reaction catalysed by the pyruvate dehydrogenase complex and since this complex is localized in the mitochondrial matrix, the question arose as to how the intramitochondrially produced, pyruvate-generated AcCoA passes mitochondrial membranes before being utilized for the extramitochondrial synthesis of ACh. The highly permeable outer mitochondrial membrane appears to be no major obstacle. On the other hand, the inner mitochondrial membrane is generally assumed to be impermeable to AcCoA and the view prevails that the acetyl groups from the intramitochondrial AcCoA can only be transferred across it after they have been incorporated into some other compound which serves as an 'acetyl carrier' (GREVILLE 1969). After crossing the membrane, the 'acetyl carrier' is transformed back to AcCoA; among several substances considered for this role, citrate appears particularly important in many cells.

Studies with isotopically labelled substrates designed to clarify the way in which the pyruvate-generated acetyl groups are transferred from brain mitochondria to the extramitochondrial ChAT have not, however, yielded a clear-cut and unequivocal picture. It would seem that labelled atoms from a compound acting as an 'acetyl carrier' for the synthesis of ACh should be incorporated into ACh at least as efficiently as labelled atoms from glucose or pyruvate, but no substrate fulfilling this condition has been found among the substances considered as potential 'acetyl carriers'. Labelled atoms from citrate (NAKAMURA et al. 1970; TUČEK and CHENG 1970, 1974; LEFRESNE et al. 1977; CHENG and BRUNNER 1978; BENJAMIN and QUASTEL 1981; DOLEŽAL and TUČEK 1981), acetate (NAKAMURA et al. 1970; TUČEK and CHENG 1970, 1974; LEFRESNE et al. 1977; DOLEŽAL and TUČEK 1981), glutamate (NAKAMURA et al. 1970; TUČEK and CHENG 1970, 1974) and *N*-acetylaspartate (S. TUČEK and S.C. CHENG, unpublished experiments) are comparatively little incorporated into brain ACh in experiments with brain slices and synaptosomes or after *in vivo* administration. The only compound which has been found to be relatively efficiently utilized (only 0.6 times less in comparison with glucose) for the synthesis of ACh was acetylcarnitine (DOLEŽAL and TUČEK 1981). Acetate was efficiently incorporated into ACh in perfused sympathetic ganglia but, surprisingly, its utilization for the synthesis of ACh was diminished rather than increased by stimulation (KWOK and COLLIER 1982).

In spite of the low incorporation of label from citrate into ACh, other evidence makes it nearly certain that citrate is responsible for the supply of approximately one-fourth to one-third of the acetyl groups in brain ACh. Citrate is formed from oxaloacetate and AcCoA by citrate synthase and transported from the mitochondria by the tricarboxylate carrier in the inner mitochondrial membrane. AcCoA can again be produced from citrate and ATP by ATP citrate lyase, a cytoplasmic enzyme (TUČEK 1967b). ATP citrate lyase is specifically inhibited by (−)-hydroxycitrate, and its apparently complete inhibition has been found to diminish the synthesis of ACh from glucose in brain slices or synaptosomes by 20%−40% (STERLING and O'NEILL 1978; GIBSON and SHIMADA 1980; TUČEK et al. 1981; STERLING et al. 1981; SZUTOWICZ et al. 1981; ŘÍČNÝ and TUČEK 1982). A similar degree of inhibition of ACh synthesis was also observed in the perfused rat diaphragm (ENDEMANN and BRUNENGRABER 1980). The contribution of the citrate pathway to the synthesis of ACh may be higher than the estimated one-third in the cell bodies and lower in the nerve terminals (GIBSON and PETERSON 1983).

It has been concluded from experiments with double-labelled glucose (^{14}C and ^{3}H in position 6) that all acetyl groups entering ACh are supplied via citrate (SOLLENBERG and SÖRBO 1970; STERLING and O'NEILL 1978) because the ratio of $^{3}H/^{14}C$ subsequently found in ACh was about 0.7 times lower than that in the glucose from which ACh had been synthesized. The loss of one-third of ^{3}H (relative to ^{14}C) atoms is expected to occur when AcCoA and oxaloacetate are joined to form citrate. However, there are other ways in which ^{3}H atoms may be lost. An exchange of ^3H between AcCoA and water is catalysed by both citrate synthase (EGGERER 1965; SRERE 1967) and ATP citrate lyase (DAS and SRERE 1972), and a big loss of ^3H would be expected if the metabolites were recycled in the Krebs cycle. The relative loss of ^3H was smaller in similar experiments performed by CLARK et al. (1982). It would be difficult to reconcile the idea that citrate supplies all or most acetyl groups for ACh with the already mentioned observations that the labelled atoms from citrate are very little incorporated into ACh and that complete inhibition of ATP citrate lyase decreases the synthesis of ACh by no more than 20%–40%.

The role of ATP citrate lyase in the provision of a proportion of AcCoA for the synthesis of ACh is supported by the observed correlations between the activities of ChAT and ATP citrate lyase in different brain areas (SZUTOWICZ and ŁYSIAK 1980; SZUTOWICZ et al. 1982b, 1983b; but see HARVEY et al. 1982, for contradicting data) and neurons (HAYASHI and KATO 1978) and in different populations of neuroblastoma cells (SZUTOWICZ et al. 1983a). Cholinergic denervation of the hippocampus was followed by decreases in the activities of ChAT and ATP citrate lyase (SZUTOWICZ et al. 1982a), a finding which suggests a preferential localization of ATP citrate lyase in the cholinergic nerve endings. This is at variance with the data obtained after cholinergic denervation of the interpeduncularis nucleus (STERRI and FONNUM 1980; see below).

It is not certain what other ways are used, in parallel with the citrate pathway, for the supply of acetyl groups for the synthesis of brain ACh. The following two appear most likely:

(1) Transfer via acetylcarnitine. In this pathway, carnitine acetyltransferase must ensure the transfer of acetyl groups from AcCoA to carnitine in the inner mitochondrial space and back to CoA in the outer mitochondrial space (TZAGOLOFF 1982). STERRI and FONNUM (1980) observed that a cholinergic denervation of nucleus interpeduncularis was followed by a sharp decrease in its ChAT and a moderate decrease in its pyruvate dehydrogenase and carnitine acetyltransferase activities, while the activities of two other AcCoA-producing enzymes, ATP citrate synthase and AcCoA synthetase, were little diminished. This finding suggested that carnitine acetyltransferase is preferentially located in the cholinergic nerve terminals in the interpeduncular nucleus, but it is not clear why the situation appears different in the hippocampus (SZUTOWICZ et al. 1982a; see above). DOLEŽAL and TUČEK (1981) compared the utilization of various labelled precursors of acetyl groups for the synthesis of ACh in striatal slices with the utilization of glucose and found that the efficiency of the utilization of acetylcarnitine was comparatively high (0.6 times that of glucose).

(2) Direct passage of AcCoA through Ca^{2+}-induced hydrophilic pores (or channels) in the inner mitochondrial membrane. It has been observed that mitochondria release AcCoA into the incubation medium in the presence of Ca^{2+} ions. While the concentrations of Ca^{2+} used in the earlier work were of the order of 1 m*M* (TUČEK 1967c; BENJAMIN and QUASTEL 1981), it has been shown by BENJAMIM et al. (1983) that 10 μ*M* Ca^{2+} was sufficient to produce a five-fold increase in the output of AcCoA from liver mitochondria, and RÍČNÝ and TUČEK (1983) discovered in experiments with Ca^{2+}-EGTA buffers and brain mitochondria that the output of AcCoA was 41% higher at an estimated concen-

tration of 0.1 μM as compared to 10 nM Ca^{2+}. If Ca^{2+} had the same effect *in vivo*, its influx into the presynaptic nerve terminals during synaptic activity might promote the supply of mitochondrial AcCoA for the synthesis of ACh. The mechanism in which Ca^{2+} affects the permeability of the inner mitochondrial membrane for AcCoA has not been clarified. Its effect was reversible on liver mitochondria (BENJAMIN et al. 1983). HUNTER and coworkers (HUNTER et al. 1976; HUNTER and HAWORTH 1979a, b; HAWORTH and HUNTER 1979, 1980) observed that Ca^{2+} brings about a generalized increase in the permeability of inner mitochondrial membranes for substances with a molecular mass of up to 1000 Da and ascribed it to the formation of Ca^{2+}-induced hydrophilic channels. The reversibility and kinetics of the permeabilization evoked by Ca^{2+} have been analysed by AL-NASSER and CROMPTON (1986).

While glucose is the usual source of the acetyl groups of ACh in the brain, it seems likely that it might be replaced by acetoacetate or 3-hydroxybutyrate if its utilization were impaired and acetoacetate and 3-hydroxybutyrate were present at high concentrations. The synthesis of ACh from acetoacetate and 3-hydroxybutyrate has been demonstrated both *in vitro* (ITOH and QUASTEL 1970; GIBSON and BLASS 1979; GIBSON and SHIMADA 1980; STERLING et al. 1981; PETERSON and GOLDMAN 1986) and *in vivo* (TUČEK and CHENG 1970, 1974). Acetoacetate and 3-hydroxybutyrate are more important in the energy metabolism of the developing compared to the adult brain (DRAHOTA et al. 1965), and this may hold true for their role in the metabolism of ACh as well (ITOH and QUASTEL 1970; GIBSON and BLASS 1979; PETERSON and GOLDMAN 1986).

When the origin of the acetyl groups of ACh was investigated in the electric organ of *Torpedo*, the results were in sharp contrast to those obtained with the brain (ISRAËL and TUČEK 1974; MOREL 1975; MOREL et al. 1977): acetate was found to be utilized very efficiently, while glucose very little. Pyruvate was utilized with a similar efficiency to acetate when it was present as the only substrate, but the synthesis of ACh proceeded mainly from acetate when both pyruvate and acetate were present simultaneously (ISRAËL and TUČEK 1974; CORTHAY et al. 1985). High utilization of acetate for the synthesis of ACh has also been reported for lobster nerves (CHENG and NAKAMURA 1970), the ganglia of *Aplysia* (BAUX et al. 1979), cholinergic neuroblastoma cells (KATO et al. 1977), rabbit cornea (FITZGERALD and COOPER 1967) and rat skeletal muscle (DREYFUS 1975). It would be interesting to clarify whether there is a more general metabolic background to the reported species and tissue differences in the utilization of either glucose-derived or acetate-derived AcCoA for the synthesis of ACh. The transport of acetate into the nerve terminals in the electric organ has been analysed by MOREL (1975) and O'REGAN (1983, 1984).

II. Choline

1. Biosynthesis

The biosynthesis of choline proceeds via a metabolic pathway in which phosphatidylaminoethanol is gradually *N*-methylated to phosphatidylmonomethylaminoethanol, phosphatidyldimethylaminoethanol and phosphatidylcholine (PC) (for references see TUČEK 1978). Free choline can be released from PC in one of four ways: (a) by its deacylation to glycerylphosphorylcholine and subsequent cleavage of glycerylphosphorylcholine to choline and glycerophosphoric acid (catalysed by glycerylphosphorylcholine glycerophosphohydrolase; see MANN 1975); this way is the most important quantitatively; (b) alternatively, glycerylphosphorylcholine may be cleaved to glycerol and phosphorylcholine by glycerylphosphorylcholine cholinephosphohydrolase (ABRA and QUINN 1975) and choline released from phosphorylcholine by alkaline phosphatase; (c) by direct cleavage of PC to choline and diacylphosphatidate (catalysed by phospholipase D; see HATTORI and KANFER 1985; WITTER and KANFER 1985); (d) by exchange for serine or ethanolamine (catalysed by enzymes that have not been definitely identified; see GAITI et al. 1974; BUCHANAN and KANFER 1980). Consequently, PC is the main store and source of free choline in the body. Its synthesis occurs mainly in the liver, from which it is supplied via the blood to other organs.

The brain itself is a major producer of free unesterified choline. The blood leaving the brain contains more free choline than the blood entering it (DROSS and KEWITZ 1972; CHOI et al. 1975; AQUILONIUS et al. 1975; SPANNER et al. 1976), and it was calculated from the negative arterio-venous difference in rats that the rate of cerebral free choline formation equals 7 nmol min^{-1} g^{-1} (DROSS and KEWITZ 1972). There is no doubt that free choline is taken up from the blood and utilized in the brain, but it is apparent that more free choline is returned back to the blood. The question of the origin of this additional free choline has not been satisfactorily resolved (reviews ANSELL and SPANNER 1982; BLUSZTAJN and WURTMAN 1983). It is virtually certain that the additional free choline must be produced by the hydrolysis of choline phospholipids in the brain, but it is uncertain whether practically all of these phospholipids are supplied to the brain by the blood (and, in that case, in what form), or whether an important amount of choline is produced by the hydrolysis of PC which is synthesized directly in the brain (including *de novo* synthesis of its choline groups in the brain).

It was believed for a long time that the choline moiety of PC cannot be produced in the brain because of apparent inability of the brain tissue to methylate phosphatidylethanolamine (PE). Although the presence of PE methyltransferases in the brain or cultured brain cells has been firmly established in recent years (SKURDAL and CORNATZER 1975; BLUSZTAJN et al. 1979; ZEISEL et al. 1979; MOZZI and PORCELLATI 1979; CREWS et al. 1980; MOZZI et al. 1981; BLUSZTAJN and WURTMAN 1981; DAINOUS et al. 1981, 1982; PERCY et al. 1982; HATTORI et al. 1984; TACCONI and WURTMAN 1985), the activities discovered are very low; thus, the rate of methylation calculated from the report by CREWS et al. (1980) corresponds to only 61 pmol g^{-1} min^{-1} and from that by BLUSZTAJN and WURTMAN (1981) to 6 pmol g^{-1} min^{-1}, i. e. to a minute fraction of the rate of the net cerebral free choline production. Therefore, it appears necessary to assume that much choline is being supplied to the brain by the blood in the form of choline phospholipids. The incorporation of labelled atoms from intravenously administered lysophosphatidylcholine into several compounds in the brain (including ACh) has been described by ILLINGWORTH and PORTMAN (1972), while comparable direct observations with PC (which would seem to be a more likely transporter), started by MANN and BENNETT (1979), need completion.

2. Concentrations in Body Fluids

Reports on the content of free choline in the blood plasma vary (see Table 5.1 in TUČEK 1978), but most data are close to or slightly above 10 μ*M* (for man, AQUILONIUS et al. 1970; SCHUBERTH and SUNDWALL 1971; ECKERNÄS 1977; for dog, BLIGH 1952; SCHUBERTH and SUNDWALL 1971; for rat, FREEMAN et al. 1975; for rabbit, SCHUBERTH and JENDEN 1975). In newborn rats the plasma concentration of choline is higher: from 65 μ*M* at birth it decreases to about 10 μ*M* at 20 days (ZEISEL and WURTMAN 1981).

The concentration of choline in the cerebrospinal fluid is lower than that in blood plasma. Data listed by TUČEK (1978, Table 5.2) are in the range of 0.4–5.1 μ*M* and 2.5 μ*M* appears to be a reasonable rounded-up mean value. It seems important to note that lower concentrations of choline in the cerebrospinal fluid than in the blood plasma are also regularly reported in studies in which both values have been determined in parallel; the difference is therefore unlikely to depend on the method of assay used. Thus AQUILONIUS et al. (1970) found 2.2 μ*M* choline in the cerebrospinal fluid and 10.9 μ*M* in the plasma of man, SCHUBERTH and JENDEN (1975) reported 5.1 μ*M* choline in the cerebrospinal fluid and 10.1 μ*M* in the plasma of rabbit, and GROWDON et al. (1977) discovered 1.8 μ*M* choline in the cerebrospinal fluid and 13.6 μ*M* in the serum of human patients with Huntington's disease.

3. Concentration in the Brain

Determinations of the content of choline in the brain are complicated by the fact that a rapid *post-mortem* increase of free choline occurs in cerebral tissues (SCHUBERTH et al. 1970; DROSS and KEWITZ 1972; HANIN and SCHUBERTH 1974; ZEISEL 1985); rather high values were therefore reported before this was realized. DROSS and KEWITZ (1972) measured the concentrations of choline in the brain at various intervals after death and by interpolation to the moment of decapitation obtained the value of 27.5 nmol g^{-1} as the likely cerebral concentration of choline in rats *in vivo*. Arresting brain metabolism with microwaves has led to values of 24–26 nmol g^{-1} (STAVINOHA and WEINTRAUB 1974a, b), while MANN and HEBB (1977) arrived at lower values by 'freeze-blowing' the brain. Considerably higher concentrations of choline are regularly found in experiments with brain tissue *in vitro*.

The concentration of choline in the intercellular water of the brain has not been directly measured. As discussed by TUČEK (1984), the absence of major diffusional barriers between the cerebrospinal fluid and the interstitial space of the brain (DAVSON 1970; WRIGHT 1978; BRADBURY 1979) makes it likely that the concentration of choline in brain intercellular water is identical with (or very close to) its concentration in the cerebrospinal fluid. Although the movement of choline from the blood plasma to the brain tissue occurs rapidly (DROSS and KEWITZ 1972; AQUILONIUS et al. 1973), it is severely restricted by the low capacity

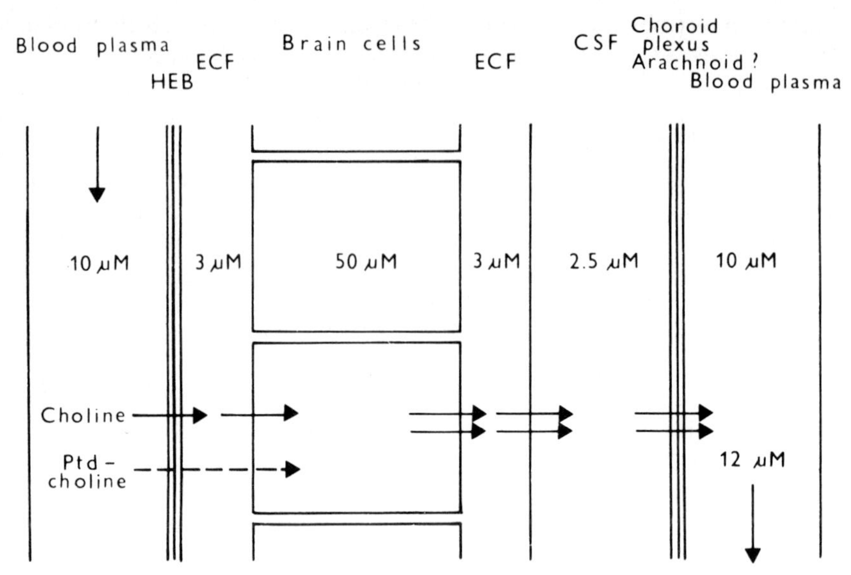

Fig. 4. Schematic representation of the movement of choline between the blood, the interstitial fluid (*ECF*) and intracellular fluid of the brain, and the cerebrospinal fluid (*CSF*). The concentrations of choline indicated for the blood plasma and CSF are from measured data in the literature, whereas those in the interstitial and intracellular fluids are predicted values (see text). *Ptd-choline*, choline phospholipids. Reprinted, with permission, from TUČEK (1985)

of choline carriers in the haematoencephalic barrier (CORNFORD et al. 1978; PARDRIDGE et al. 1979; BRAUN et al. 1980). An equilibration of choline concentrations in the blood plasma on the one side and in the intercellular water and cerebrospinal fluid on the other side is apparently prevented by continuous pumping of choline out of the cerebrospinal fluid back into the blood by carriers located in the choroid plexuses (TOCHINO and SCHANKER 1965, 1966; AQUILONIUS and WINDBLADH 1972; LANMAN and SCHANKER 1980) and (in the bullfrog) in the arachnoid (WRIGHT 1978; EHRLICH and WRIGHT 1982). A schematic representation of the presumed movement of choline between the blood plasma, intercellular and cerebrospinal fluids and the cells of the brain is given in Fig. 4.

Choline is taken up from the intercellular fluid by the high- and low-affinity carriers in cell membranes and the concentration of choline in the cells is expected to be higher than in the extracellular fluid in view of the electronegativity of the cells' interior. An estimate of the intracellular concentration of choline can be made from data on its total content in the brain, assuming that it is equally distributed between all types and parts of brain cells (TUČEK 1984). Out of the total content of 27.5 nmol choline g^{-1} brain tissue (DROSS and KEWITZ 1972), only 0.5 nmol g^{-1} brain is likely to be in the intravascular water (50 μl at 10 μM concentration) and 0.45 nmol g^{-1} brain in the interstitial water (150 μl at 3 μM concentration); the remaining 26.55 nmol g^{-1} brain are bound to be in the intracellular water. Taking that there are 550 μl of intracellular water per gram of brain, the concentration of choline in the intracellular water is likely to be close to 48 μ*M*.

Although it seems natural to admit that there may be short-term deviations from an electrochemical equilibrium in the distribution of choline on both sides of cell membranes, it appears likely that concentrations will stabilize close to an equilibrium in the long run. An estimate of the ratio of intracellular to extracellular concentrations of choline at electrochemical equilibrium can be made using the Nernst equation. If

$$E = \frac{RT}{ZF} \times 2.303 \log_{10} \frac{[\text{choline}]_i}{[\text{choline}]_e}\,, \tag{1}$$

the measurements are performed at 38 °C and the 'mean' membrane potential of the brain cells is taken to be −0.075 V, then

$$\log_{10} \frac{[\text{choline}]_i}{[\text{choline}]_e} = \frac{0.075 \times (+1) \times 96500}{8.314 \times 311 \times 2.303} = 1.215 \tag{2}$$

and

$$\frac{[\text{choline}]_i}{[\text{choline}]_e} = 16.4\ . \tag{3}$$

Consequently, in the case of electrochemical equilibrium in the distribution of choline across cell membranes at a membrane potential of −0.075 V, the intracellular concentration of choline is expected to be 16.4 times higher than the extracellular one; if the extracellular concentration in the brain is 3 μ*M*, then the predicted intracellular concentration is 49.2 μ*M*. The close agreement between this estimate and the 48 μ*M* concentration calculated for intracellular choline in the preceding paragraph supports the assumption that, indeed, electrochemical equilibrium prevails in the overall distribution of choline between the intra- and extracellular spaces in the brain.

It is not known if the concentration of choline in the cholinergic nerve endings is the same as in other parts of cholinergic neurons and in other cells of the nervous system. It is conceivable that it is higher. While the electrochemical gradient in the distribution of choline probably acts as the sole driving force for the transport of choline into the bulk of brain cells, recent observations on proteoliposomes (VYAS and O'REGAN 1985; O'REGAN and VYAS 1986) and membraneous vesicles (DUCIS and WHITTAKER 1985; BREER and KNIPPER 1985) together with some of the earlier work on synaptosomes (for reviews see TUČEK 1985; Chap. 19, this volume) suggest that the gradient in the distribution of Na^+ may act as an additional driving force for the high-affinity carrier-mediated transport of choline into cholinergic nerve terminals. This is a point of utmost importance, since the utilization of the Na^+ gradient (i. e. co-transport of Na^+ with choline) would allow an increase in the intraterminal concentration of choline far above that predicted from the Nernst equation (KOTYK 1983); in this case, however, the low-affinity carriers would be expected to counteract and carry choline away from the terminals. A gradient of Na^+ concentrations was clearly unimportant for activity-stimulated choline uptake in sympathetic ganglia in experiments described by BIRKS (1985).

III. Source of Choline for ACh Synthesis

A large body of evidence indicates that, under most experimental conditions and particularly under conditions close to the physiological, most of the choline that is used for the synthesis of ACh in cholinergic neurons originates from the extracellular fluid surrounding them. Choline is transported into neurons by carriers in their membranes (Chap. 19; TUČEK 1978, 1984; KUHAR 1979; JOPE 1979; SPETH and YAMAMURA 1979; MURRIN 1980); the high-affinity carriers are mainly responsible for the transport into presynaptic nerve endings, while the low-affinity carriers mediate the transport into other parts of cholinergic neurons (neuronal perikarya, SUSZKIW et al. 1976). There is no proof, however, that the choline which is taken up from the extracellular fluid is the *exclusive* source of choline for the synthesis of ACh, and the available evidence is compatible with the view that the choline which is transported into the nerve terminals becomes mixed with the choline already present there, after which the synthesis, although dependent on the supply of additional choline from the outside, proceeds from the mixed pool.

The contention that extracellular choline is crucial for the supply of precursor for the synthesis of ACh is a generalization based on numerous experiments, including the following:

(1) Nearly all unlabelled ACh in sympathetic ganglia (COLLIER and MACINTOSH 1969) and the diaphragm (POTTER 1970) can be replaced with labelled ACh when the ganglia or the phrenic nerve are stimulated while labelled choline is present in the perfusion or incubation fluid.

(2) The presence of choline in the perfusate is necessary to prevent depletion of ACh stores in perfused sympathetic ganglia during preganglionic nerve stimulation (BIRKS and MACINTOSH 1961). Addition of choline to the incubation media or superfusion fluids increases the stimulated release of ACh from brain prisms or slices (CARROLL and GOLDBERG 1975; RICHTER 1976; MILLINGTON and GOLDBERG 1982; WEILER et al. 1983) and eserinized sympathetic ganglia (SACCHI et al. 1978) and the synthesis of ACh in diluted suspensions of brain synaptosomes (GUYENET et al. 1973).

(3) Labelled choline injected intravenously is rapidly incorporated into brain ACh (SCHUBERTH et al. 1970; DROSS and KEWITZ 1972; AQUILONIUS et al. 1973). Intense incorporation of labelled choline from the medium into ACh also proceeds during *in vitro* experiments with stimulated nervous tissue (POLAK et al. 1977; HIGGINS and NEAL 1982a, b; WEILER et al. 1983).

(4) The synthesis of ACh in minced brain tissue (BHATNAGAR and MACINTOSH 1967), synaptosomes (GUYENET et al. 1973; MEYER et al. 1982), diaphragm (POTTER 1970), sympathetic ganglia (BIRKS and MACINTOSH 1961), stimulated ileum (SOMOGYI et al. 1977), vagus nerve terminals in the heart (LEWARTOWSKI and BIELECKI 1963; LINDMAR et al. 1983) and other preparations is inhibited by hemicholinium-3 (HC-3), an inhibitor of choline transport across cell membranes, and the inhibition can usually be overcome by increasing the concentration of choline in the incubation medium. Within 30 min after an injection of HC-3 into the caudate nuclei *in vivo*, their ACh content is diminished to 12% of control values (HEBB et al. 1964). Changes in ACh synthesis in synaptosomes

are directly proportional to changes in the activity of choline carriers caused by HC-3 (GUYENET et al. 1973) or by Triton X-100 (LEFRESNE et al. 1975).

(5) The synthesis of ACh can be lowered by omitting Na^+ from the incubation media (BIRKS 1963; PATON et al. 1971; COLLIER et al. 1972; GREWAAL and QUASTEL 1973; LINDMAR et al. 1977). The effect of low Na^+ on the synthesis of ACh can be overcome by increasing the concentration of choline in the extracellular fluid and is best explained by an inhibition of choline transport into neurons.

(6) Most of the choline taken up into synaptosomes by carriers in their membranes is converted to ACh (YAMAMURA and SNYDER 1973; HAGA and NODA 1973; MARCHBANKS et al. 1981; WEILER et al. 1981; RICHARDSON 1986).

(7) In perfused hearts the increase in the release of ACh by nerve stimulation is followed by a proportional decrease in the continuous efflux of choline from the heart; the decrease is caused by choline uptake into the cells (probably into the nerve terminals resynthesizing ACh) and can be abolished by HC-3 (LINDMAR et al. 1980).

The dependence of the production of ACh on the presence of choline in the incubation or perfusion media varies in experiments on different tissues. While the synthesis of ACh in perfused sympathetic ganglia (BIRKS and MACINTOSH 1961) and isolated diaphragms (POTTER 1970) is highly dependent on exogenous choline, the resting synthesis of ACh in brain slices does not require the addition of choline to the medium (LEFRESNE et al. 1973; MILLINGTON and GOLDBERG 1982). It is well known that brain tissue produces choline during incubations (BHATNAGAR and MACINTOSH 1967; COLLIER et al. 1972; RICHTER 1976; CEDER and SCHUBERTH 1977; WEILER et al. 1978, 1983; DOLEŽAL and TUČEK 1982; ZEISEL 1985), and this choline is apparently taken up from the extracellular space and utilized for the synthesis of ACh in the nerve terminals. When the uptake of choline is blocked by the absence of Na^+, brain slices lose most of their ACh, but they recover their ACh stores after the addition of Na^+ even in an incubation medium to which no choline had been added (COLLIER et al. 1972).

It is not clear which types of brain cells participate in the release of choline to the extracellular fluid and to what extent. Glia may be very important in view of the observation that the rate of synthesis of PC is higher in cultured glia than neurons (DAINOUS et al. 1982; see also WALUM 1981, for interesting data indicating possible role of glia in maintaining the homeostasis of choline in the brain). WURTMAN et al. (1985) hypothesized that the hydrolysis of PC in cholinergic neurons may be the source of choline for ACh synthesis under emergency conditions. Much free choline is produced directly in the intercellular space by the hydrolysis of synaptically released ACh and can be re-utilized after re-uptake into nerve terminals. In sympathetic ganglia, the re-utilization of choline produced by the hydrolysis of released ACh accounts for 50%–60% of the total supply of choline for new ACh synthesis (COLLIER and MACINTOSH 1969; COLLIER and KATZ 1974); choline produced from previously released ACh competes for re-entry into the nerve terminals with other choline present in the extracellular fluid (COLLIER and KATZ 1974). The release of choline to the extracellular fluid is stimulated by the activation of muscarinic receptors (LADINSKY et al. 1974; LUNDGREN et al. 1977; CORRADETTI et al. 1982, 1983; DOLEŽAL and TUČEK

1984; BREHM et al. 1985), probably causing the hydrolysis of PC by an activation of phospholipase D (LINDMAR et al. 1986).

D. Regulation of ACh Synthesis

I. Kinetic Considerations

The concentration of ACh is very stable in the mammalian nervous system and tissues with cholinergic innervation (BIRKS and MACINTOSH 1961; POTTER 1970); when lowered by intense stimulation, it rapidly returns to pre-stimulation values (RICHTER and CROSSLAND 1949). The stability of the level of ACh appears lower in the electric organ of electric fish (DUNANT et al. 1974). The rate of ACh synthesis adapts to the rate of ACh release so as to keep the concentration of ACh in the nerve endings constant (COLLIER and MACINTOSH 1969), and any explanation of the mechanism by which the synthesis of ACh is controlled has to explain the way in which the levels of ACh in the nerve terminals exert their feedback effect on the rate of synthesis.

A scheme of the control of ACh synthesis that would accommodate all known experimental observations has not yet been devised. In spite of this, it seems possible to give a reasonably safe outline of the basic principle of the control and of the main factor involved. The role of additional, as yet unidentified factors is suspected, and the true importance which individual factors have under physiological circumstances is difficult to evaluate with the incomplete information available.

It is very likely that the concentration of ACh in the compartment of its synthesis in the nerve terminals at rest is close to the concentration expected according to the law of mass action when the reaction catalysed by ChAT is at equilibrium. Factors such as the hydrolysis of ACh by cholinesterases or the removal of ACh into synaptic vesicles do not operate with an intensity sufficient to produce a permanent displacement of the concentration of ACh away from equilibrium; if they did, rapid and large alterations in the concentration of ACh should be evoked by their pharmacological elimination or by synaptic activity, which is not the case. Therefore, it seems justified to expect that the concentration of ACh in the compartment of its synthesis obeys the equation

$$[\mathrm{ACh}] = K_{\mathrm{eq}}\,[\mathrm{choline}]\,\frac{[\mathrm{AcCoA}]}{[\mathrm{CoA}]} \tag{4}$$

and depends on the concentration of choline and on the degree of acetylation of CoA. As long as the ratio of [AcCoA]/[CoA] is constant, the ratio [ACh]/[choline] is also expected to be constant. Taking that $K_{\mathrm{eq}} = 12.3$ (PIEKLIK and GUYNN 1975) and that the concentrations of AcCoA and CoA in the cholinergic nerve endings in the brain are the same as in the whole brain (5.1 and 17.9 $\mathrm{nmol\,g^{-1}}$, respectively; GUYNN 1976), we can calculate that in cerebral cholinergic nerve terminals

$$\frac{[\mathrm{ACh}]}{[\mathrm{choline}]} = 12.3 \times \frac{5.1}{17.9} = 3.5\ . \tag{5}$$

Based on the assumptions of equation (5), one expects that 78% (= 3.5/4.5) of the choline entering the nerve terminals becomes acetylated and 22% remains unacetylated. In accordance with this, 60%–75% of the choline that had been taken up by the high-affinity carriers was found to become acetylated in synaptosomes (HAGA and NODA 1973; YAMAMURA and SNYDER 1973; MARCHBANKS et al. 1981; RICHARDSON 1986).

In line with assumptions used in the preceding paragraph are the findings by McGEE et al. (1978) that in hybrid neuroblastoma x glioma cells the intracellular levels of ACh change in parallel with the extracellular levels of choline and those of EISENSTADT and SCHWARTZ (1975) that intracellular injections of increasing amounts of choline into a cholinergic neuron bring about the synthesis of increasing amounts of ACh and that the conversion of injected AcCoA to ACh is greatly increased when choline is injected into the cell simultaneously. Various lines of evidence supporting the equilibrium model of the regulation of ACh synthesis (originally proposed in POTTER et al. 1968 and GLOVER and POTTER 1971) have been discussed earlier (TUČEK 1978, 1984, 1985).

II. Regulatory Factors

The following factors appear particularly important in determining the concentration of ACh in the nerve terminals and ensuring the adjustment of ACh synthesis to its release.

1. Concentration of ACh in the Compartment of Synthesis

Its decrease during synaptic activity is likely to have two consequences: (a) a shift in the equilibrium of the reaction catalysed by ChAT, which automatically leads to the synthesis of new ACh; and (b) a disinhibition of the high-affinity choline carriers, for which ACh is an inhibitor (POTTER 1968; YAMAMURA and SNYDER 1973; MARCHBANKS et al. 1981); this will increase the supply of choline into the nerve endings. The concentration of ACh in the compartment of synthesis is not known. If the concentration of choline is 48 μ*M* as calculated in Sect. C.II.3 and the concentration of ACh is 3.5 times higher according to equation (5), one would expect 168 μ*M* ACh to be present, which is much more than the concentrations of ACh found to inhibit the high-affinity uptake of choline into synaptosomes by at least 50% (1 μ*M*, YAMAMURA and SNYDER 1973; 12 μ*M*, SIMON et al. 1975). Although it is not known if ACh is equally effective in causing the *trans*-inhibition of uptake *in vivo* as it is in producing the *cis*-inhibition *in vitro*, it seems necessary to take into account that its inhibitory action on the high-affinity carriers may be quite strong.

2. Uptake of Choline

An increase in the rate of the uptake of choline induced by synaptic activity has been demonstrated under many circumstances:

1. in synaptosomes prepared from the brains of animals that had been injected with agents increasing the turnover of ACh shortly before death (SIMON and

KUHAR 1975; ATWEH et al. 1975; SIMON et al. 1976; JENDEN et al. 1976; MURRIN et al. 1977; RICHTER et al. 1982);
2. in synaptosomes prepared from brain slices that had been exposed to electrical stimulation (ANTONELLI et al. 1981);
3. in brain slices or synaptosomes that had been pre-incubated under depolarizing conditions or with veratridine (GREWAAL and QUASTEL 1973; POLAK et al. 1977; MURRIN and KUHAR 1976; ROSKOSKI 1978; WEILER et al. 1978, 1981; MARCHBANKS et al. 1981; MARCHBANKS 1982);
4. in sympathetic ganglia *in vitro* (HIGGINS and NEAL 1982a, b), isolated iris (VACA and PILAR 1979) and heart atria (WETZEL and BROWN 1983) after a period of depolarization;
5. in perfused sympathetic ganglia during or after a period of electrical stimulation of the preganglionic fibres (COLLIER and ILSON 1977; O'REGAN and COLLIER 1981; O'REGAN et al. 1982; COLLIER et al. 1983), mostly utilizing analogues of choline for measurements of the activity of choline carriers (reviewed in COLLIER 1986);
6. in perfused hearts following a period of vagal stimulation (LINDMAR et al. 1980).

The results obtained have been discussed in more detail by TUČEK (1984). It seems apparent that in most experiments utilizing synaptosomes or brain slices the release of ACh was the main (perhaps the sole) factor responsible for the increase of choline uptake. The effect of previous depolarization was usually (but not always – BARKER 1976; MURRIN et al. 1977) dependent on the presence of Ca^{2+} during depolarization and thus probably on the release of ACh (MURRIN and KUHAR 1976; POLAK et al. 1977; ROSKOSKI 1978; WEILER et al. 1978; MARCHBANKS et al. 1981) and could be prevented by a blocker of calcium channels (MURRIN and KUHAR 1976; MURRIN et al. 1977). The content of ACh in synaptosomes was measured in several studies and was inversely correlated with the observed rate of choline uptake (JENDEN et al. 1976; WEILER et al. 1978; ROSKOSKI 1978; ANTONELLI et al. 1981).

In addition to the release of ACh, a post-stimulation increase in the activity of Na^+, K^+-ATPase and the resulting hyperpolarization was probably responsible for the post-stimulation activation of choline uptake in isolated iris (VACA and PILAR 1979) since the activation could be diminished by tetrodotoxin or omission of Na^+ from the incubation medium. On perfused sympathetic ganglia, the increased uptake during stimulation appeared to depend on the release of ACh under most (but not all) conditions. The increase in uptake which occurs during the post-stimulation period and depends on the presence of Na^+ during that period seems to be best explained by assuming post-stimulation activation of electrogenic Na^+, K^+-ATPase resulting in a hyperpolarization of the nerve terminals. O'REGAN and COLLIER (1981) point out, however, that there are differences between the conditions of their experiments and the conditions known to produce after-hyperpolarization in the work of others (e. g. replaceability of Ca^{2+} by Ba^{2+}). It remains to be established if some other factor was involved in their experiments, in addition to the release of ACh and hyperpolarization.

3. Post-stimulation Hyperpolarization

The entry of Na^+ into the nerve terminals during nerve impulses activates electrogenic Na^+, K^+-ATPase and the ensuing hyperpolarization is likely to increase the uptake of choline. An increase of the membrane potential from 0.075 to 0.085 V would increase the intracellular concentration of choline at electrochemical equilibrium by 45% (see Table 1 in TUČEK 1985). Experiments in which after-hyperpolarization may have played a role have been mentioned in the previous paragraph. An activation of Na^+, K^+-ATPase was found to increase the high-affinity uptake of choline and the synthesis and stores of ACh in sympathetic ganglia by BIRKS (1983, 1985), who suggests direct association between the uptake of choline and the sodium pump; it seems difficult to exclude the role of hyperpolarization in his experiments.

4. Effect of Ca^{2+} on the Supply of AcCoA

The increase in the concentration of Ca^{2+} in the nerve terminals during synaptic activity might promote the supply of AcCoA from mitochondria in two ways: (a) by increasing the permeability of the inner mitochondrial membrane (see Sect. C); or (b) by stimulating the activity of pyruvate dehydrogenase phosphate phosphatase and increasing the activity of pyruvate dehydrogenase (WIELAND 1983; HANSFORD and CASTRO 1985; MCCORMACK 1985). Hypothetically, Ca^{2+}-induced increases in the concentration of AcCoA in the cytosol might be responsible for the otherwise difficult to explain increase in the concentration of ACh in synaptosomes or nerve terminals induced by depolarization or stimulation under conditions when the release of ACh was blocked (WONNACOTT and MARCHBANKS 1976; COLLIER et al. 1986).

5. Metabolic Changes in the Availability of Choline

Although the concentration of choline in the blood plasma is very stable and the store of choline in the extracellular fluid surrounding the nerve terminals is seemingly inexhaustible, there are grounds to suspect that the stores of immediately available choline may become rate-limiting for ACh synthesis during periods of high synaptic activity in the brain. The content of ACh can be diminished in certain brain areas by drugs increasing its turnover and the effect of these drugs on brain ACh content can be prevented or attenuated by pre-loading the organism with large doses of choline (WECKER and SCHMIDT 1980; SCHMIDT and WECKER 1981; TROMMER et al. 1982; JOPE 1982), as if the supply of choline had been rate-limiting.

6. Metabolic Changes in the Availability of AcCoA

The synthesis of ACh in brain slices exposed to media with a low concentration of glucose (ŘÍČNÝ and TUČEK 1980b) or with metabolic inhibitors (ŘÍČNÝ and TUČEK 1981) has been found to decrease in parallel with the content of AcCoA in slices, which provides an explanation of earlier data on the effect of metabolic inhibitors (GIBSON et al. 1975; JOPE et al. 1978) or starvation (KUNTSCHEROVÁ and VLK 1968) on the synthesis and content of ACh in the brain. The depletion of striatal ACh induced by atropine was lower in rats pretreated with a large dose of glucose (DOLEŽAL and TUČEK 1982), suggesting that the supply of AcCoA may have been rate-limiting during intense ACh release.

7. Activity of ChAT

The stability or rapid restoration of ACh stores in mammalian tissues during and after stimulation indicates that the amount of ChAT in the nerve terminals is sufficient to prevent a depletion of ACh during physiological rates of synaptic activity; conditions may be different in the electric organ of electric fish (DUNANT et al. 1974) and in the head ganglia of *Drosophila* (SALVATERRA and MCCAMAN 1985). When measured in detergent-treated homogenates in the presence of saturating concentrations of substrates, the activity of mammalian ChAT is always several times higher than the maximum rates of ACh release from corresponding tissues. It is likely, however, that the activity of ChAT *in situ* is considerably lower than its activity with an excess of substrates *in vitro*. The concentrations of substrates likely to be present in the cells (50 μ*M* choline and 5 μ*M* AcCoA) are lower than the corresponding K_m values of ChAT (e. g. 400 μ*M* for choline and 10 μ*M* for AcCoA, see Table 1), and it can be computed (TUČEK 1984) or extrapolated from published kinetic data (WHITE and WU 1973a; RYAN and MCCLURE 1980) that, at the concentrations and K_m values considered above, the activity of ChAT will be less than 1/20 of its V_{max}. The maximum rates of ACh release during intense stimulation may therefore be close to the maximum activity that ChAT can display at the low intracellular concentrations of substrates. The activity of ChAT was probably rate-limiting in experiments in which brain ACh became depleted under a combined influence of naphthylvinylpyridine, an inhibitor of ChAT, and of swimming stress (KÁSA et al. 1982; BUDAI et al. 1986).

It is difficult to evaluate whether the amphiphilic and membrane-bound ChAT play any special role in the synthesis of ACh in the nerve endings, before the contradictions mentioned above are resolved. The amphiphilic enzyme is probably localized close to the space occupied by the releasable pool of ACh (corresponding to just 15% of total ACh in a sympathetic ganglion – COLLIER et al. 1986) and to the sites of choline uptake and ACh release, and its preferential involvement in the synthesis of releasable ACh appears likely. This does not modify the concept of the control of ACh synthesis in any substantial way. It is clear that the catalytic sites of ChAT are in the interior of the nerve endings and choline must be transported before being acetylated; evidence indicating the independence of choline transport and acetylation has been reviewed earlier (TUČEK 1984). The low sensitivity of the amphiphilic (EDER-COLLI et al. 1986) and membrane-associated (BENISHIN and CARROLL 1981b) enzyme to inhibition by CoA may improve conditions for the synthesis of ACh.

It is not yet known if the peptide inhibiting ChAT and described by COZZARI and HARTMAN (1983b) is in contact with ChAT in intact neurons and has any function.

Genetically determined variations in the activity of ChAT have been studied on temperature-sensitive ChAT strutural gene mutants of *Drosophila melanogaster* (SALVATERRA and MCCAMAN 1985). The activities of ChAT and the concentrations of ACh in *Drosophila* heads were found to vary in parallel after transitions from one temperature to another, as if the activity of ChAT determined the concentration of ACh. This observation is difficult to interpret in terms of the equilibrium model of the control of ACh synthesis, unless one assumes that the changes in the activity of ChAT were brought about by sudden switchings on or off of its production in individual neurons, rather than by gradual increases or decreases.

An important question still awaiting clarification concerns the mechanism responsible for the oscillations in the content of ACh and ATP in the electric organ of electric fish observed during electric stimulations (DUNANT et al. 1974, 1977; ISRAËL et al. 1977; CORTHAY et al. 1985).

References

Abra RM, Quinn PJ (1975) A novel pathway for phosphatidylcholine catabolism in rat brain homogenates. Biochim Biophys Acta 380:436–441

Al-Nasser I, Crompton M (1986) The reversible Ca^{2+}-induced permeabilization of rat liver mitochondria. Biochem J 239:19–29

Ansell GB, Spanner S (1982) Choline transport and metabolism in the brain. In: Horrocks LA, Ansell GB, Porcellati G (eds) Phospholipids in the nervous system, vol. I: Metabolism. Raven, New York, pp 137–144

Antonelli T, Beani L, Bianchi C, Pedata F, Pepeu G (1981) Changes in synaptosomal high affinity choline uptake following electrical stimulation of guinea-pig cortical slices: effect of atropine and physostigmine. Br J Pharmacol 74:525–531

Aquilonius S-M, Windbladh B (1972) Cerebrospinal fluid clearance of choline and some other amines. Acta Physiol Scand 85:78–90

Aquilonius S-M, Schuberth J, Sundwall A (1970) Choline in the cerebrospinal fluid as a marker for the release of acetylcholine. In: Heilbronn E, Winter A (eds) Drugs and cholinergic mechanisms in the CNS. Almquist and Wiksell, Stockholm, pp 399–410

Aquilonius S-M, Flentge F, Schuberth J, Sparf B, Sundwall A (1973) Synthesis of acetylcholine in different compartments of brain nerve terminals *in vivo* as studied by the incorporation of choline from plasma and the effects of pentobarbital on this process. J Neurochem 20:1509–1521

Aquilonius S-M, Ceder G, Lying-Tunell U, Malmlund HO, Schuberth J (1975) The arteriovenous difference of choline across the brain of man. Brain Res 99:430–433

Atweh S, Simon JR, Kuhar MJ (1975) Utilization of sodium-dependent high affinity choline uptake *in vitro* as a measure of the activity of cholinergic neurons *in vivo*. Life Sci 17:1535–1544

Badamchian M, Carroll PT (1985) Molecular weight determinations of soluble and membrane-bound fractions of choline-O-acetyltransferase in rat brain. J Neurosci 5:1955–1964

Banns H, Hebb C, Mann SP (1977) The synthesis of ACh from acetyl-CoA, acetyl-dephospho-CoA and acetylpantetheine phosphate by choline acetyltransferase. J Neurochem 29:433–437

Barker LA (1976) Modulation of synaptosomal high affinity choline transport. Life Sci 17:1535–1544

Baux G, Simonneau M, Tauc L (1979) Transmitter release: ruthenium red used to demonstrate a possible role of sialic acid containing substrates. J Physiol (Lond) 291:161–178

Benishin CG, Carroll PT (1981 a) Acetylation of choline and homocholine by membrane-bound choline-O-acetyltransferase in mouse forebrain nerve endings. J Neurochem 36:732–740
Benishin CG, Carroll PT (1981 b) Differential sensitivity of soluble and membrane-bound forms of choline-O-acetyltransferase to inhibition by coenzyme A. Biochem Pharmacol 30:2483–2484
Benishin CG, Carroll PT (1983) Multiple forms of choline-O-acetyltransferase in mouse and rat brain: solubilization and characterization. J Neurochem 41:1030–1039
Benjamin AM, Quastel JH (1981) Acetylcholine synthesis in synaptosomes: mode of transfer of mitochondrial acetyl coenzyme A. Science 213:1495–1496
Benjamin AM, Murphy CRK, Quastel JH (1983) Calcium-dependent release of acetyl-coenzyme A from liver mitochondria. Can J Physiol Pharmacol 61:154–158
Berman-Reisberg R (1957) Properties and biological significance of choline acetylase. Yale J Biol Med 29:403–435
Bhatnagar SP, MacIntosh FC (1960) Acetylcholine content of striated muscle. Proc Can Fed Biol Soc 3:12–13
Bhatnagar SP, MacIntosh FC (1967) Effects of quaternary bases and inorganic cations on acetylcholine synthesis in nervous tissue. Can J Physiol Pharmacol 45:249–268
Birks RI (1963) The role of sodium ions in the metabolism of acetylcholine. Can J Biochem Physiol 41:2573–2597
Birks RI (1983) Activation of feline acetylcholine synthesis in the absence of release: dependence on sodium, calcium and the sodium pump. J Physiol (Lond) 344:347–357
Birks RI (1985) Activation of acetylcholine synthesis in cat sympathetic ganglia: dependence on external choline and sodium-pump rate. J Physiol (Lond) 367:401–417
Birks RI, MacIntosh FC (1961) Acetylcholine metabolism of a sympathetic ganglion. Can J Biochem Physiol 39:787–827
Birks R, Katz B, Miledi R (1960) Physiological and structural changes at the amphibian myoneural junction, in the course of nerve degeneration. J Physiol (Lond) 150:145–168
Bligh J (1952) The level of free choline in plasma. J Physiol (Lond) 117:234–240
Blusztajn JK, Wurtman RJ (1981) Choline biosynthesis by a preparation enriched in synaptosomes from rat brain. Nature 290:417–418
Blusztajn JK, Wurtman RJ (1983) Choline and cholinergic neurons. Science 221:614–620
Blusztajn JK, Zeisel SH, Wurtman RJ (1979) Synthesis of lecithin (phosphatidylcholine) from phosphatidylethanolamine in bovine brain. Brain Res 179:319–327
Bradbury M (1979) The concept of a blood brain barrier. Wiley, Chichester
Brandon C, Wu J-Y (1978) Purification and properties of choline acetyltransferase from *Torpedo californica*. J Neurochem 30:791–797
Braun LD, Cornford EM, Oldendorf WH (1980) Newborn rabbit blood-brain barrier is selectively permeable and differs substantially from the adult. J Neurochem 34:147–152
Breer H, Knipper M (1985) Choline fluxes in synaptosomal membrane vesicles. Cell Mol Neurobiol 5:285–295
Brehm R, Corradetti R, Krahn V, Löffelholz K, Pepeu G (1985) Muscarinic mobilization of choline in rat cerebral cortex does not involve alterations of blood-brain barrier. Brain Res 345:306–314
Brown GL, Feldberg W (1936) The acetylcholine metabolism of a sympathetic ganglion. J Physiol (Lond) 88:265–283
Browning ET, Schulman MP (1968) (^{14}C)Acetylcholine synthesis by cortex slices of rat brain. J Neurochem 15:1391–1405
Bruce G, Hersh LB (1987) Studies on detergent released choline acetyltransferase from membrane fractions of rat and human brain. Neurochem Res 12:1059–1066
Bruce G, Wainer BH, Hersh LB (1985) Immunoaffinity purification of human choline acetyltransferase: comparison of the brain and placental enzymes. J Neurochem 45:611–620
Buchanan AG, Kanfer JN (1980) Topographical distribution of base exchange activities in rat brain subcellular fractions. J Neurochem 34:720–725

Budai D, Říčný J, Kása P, Tuček S (1986) 4-(1-Naphthylvinyl)pyridine decreases brain acetylcholine *in vivo*, but does not alter the level of acetyl-CoA. J Neurochem 46:990–992

Carroll PT, Goldberg AM (1975) Relative importance of choline transport to spontaneous and potassium depolarized release of ACh. J Neurochem 25:523–527

Ceder G, Schuberth J (1977) In vivo formation and post-mortem changes of choline and acetylcholine in the brain of mice. Brain Res 128:580–584

Chao L-P (1980) Choline acetyltansferase: purification and characterization. J Neurosci Res 5:85–115

Cheng S-C, Brunner EA (1978) Alteration of tricarboxylic acid cycle metabolism in rat brain slices by halothane. J Neurochem 30:1421–1430

Cheng S-C, Nakamura R (1970) A study on the tricarboxylic acid cycle and the synthesis of acetylcholine in the lobster nerve. Biochem J 118:451–455

Choi RL, Freeman JJ, Jenden DJ (1975) Kinetics of plasma choline in relation to turnover of brain choline and formation of acetylcholine. J Neurochem 24:735–741

Clark JB, Booth RFG, Harvey SAK, Leong SF, Patel TB (1982) Compartmentation of the supply of the acetyl moiety for acetylcholine synthesis. In: Bradford HF (ed) Neurotransmitter interaction and compartmentation. Plenum, New York, pp 431–460

Collier B (1986) Choline analogues: their use in studies of acetylcholine synthesis, storage and release. Can J Physiol Pharmacol 64:341–346

Collier B, Ilson D (1977) The effect of preganglionic nerve stimulation on the accumulation of certain analogues of choline by a sympathetic ganglion. J Physiol (Lond) 264:489–509

Collier B, Katz HS (1974) Acetylcholine synthesis from recaptured choline by a sympathetic ganglion. J Physiol (Lond) 238:639–655

Collier B, MacIntosh FC (1969) The source of choline for acetylcholine synthesis in a sympathetic ganglion. Can J Physiol Pharmacol 47:127–135

Collier B, Poon P, Salehmoghaddam S (1972) The formation of choline and of acetylcholine by brain *in vitro*. J Neurochem 19:51–60

Collier B, Kwok YN, Welner SA (1983) Increased acetylcholine synthesis and release following presynaptic activity in a sympathetic ganglion. J Neurochem 40:91–98

Collier B, Welner SA, Říčný J, Araujo DM (1986) Acetylcholine synthesis and release by a sympathetic ganglion in the presence of 2-(4-phenylpiperidino)cyclohexanol (AH 5183). J Neurochem 46:822–830

Cornford EM, Braun LD, Oldendorf WH (1978) Carrier mediated bloodbrain barrier transport of choline and certain choline analogs. J Neurochem 30:299–308

Corradetti R, Lindmar R, Löffelholz K (1982) Physostigmine facilitates choline efflux from isolated heart and cortex *in vivo*. Eur J Pharmacol 85:123–124

Corradetti R, Lindmar R, Löffelholz K (1983) Mobilization of cellular choline by stimulation of muscarinic receptors in isolated chicken heart and rat cortex *in vivo*. J Pharmacol Exp Ther 226:826–832

Corthay J, Dunant Y, Eder L, Loctin F (1985) Incorporation of acetate into acetylcholine, acetylcarnitine, and amino acids in the *Torpedo* electric organ. J Neurochem 45:1809–1819

Cozzari C, Hartman BK (1980) Preparation of antibodies specific to choline acetyltransferase from bovine caudate nucleus and immunohistochemical localization of the enzyme. Proc Natl Acad Sci USA 77:7453–7457

Cozzari C, Hartman BK (1983a) Choline acetyltransferase. Purification procedure and factors affecting chromatographic properties and enzyme stability. J Biol Chem 258:10010–10013

Cozzari C, Hartman BK (1983b) An endogenous inhibitory factor for choline acetyltransferase. Brain Res 276:109–118

Crawford GD, Correa L, Salvaterra PM (1982a) Interaction of monoclonal antibodies with mammalian choline acetyltransferase. Proc Natl Acad Sci USA 79:7031–7035

Crawford GO, Slemmon JR, Salvaterra PM (1982b) Monoclonal antibodies selective for *Drosophila melanogaster* choline acetyltransferase. J Biol Chem 257:3853–3856

Crews FT, Hirata F, Axelrod J (1980) Identification and properties of methyltransferases that synthesize phosphatidylcholine in rat brain synaptosomes. J Neurochem 34:1491–1498
Currier SF, Mautner HG (1974) On the mechanism of action of choline acetyltransferase. Proc Natl Acad Sci USA 71:3355–3358
Currier SF, Mautner HG (1977) Interaction of analogues of coenzyme A with choline acetyltransferase. Biochemistry 16:1944–1948
Dahlström A (1983) Presence, metabolism, and axonal transport of transmitters in peripheral mammalian axons. In: Lajtha A (ed) Handbook of neurochemistry, vol 5. Plenum, New York, pp 405–441
Dainous F, Avola R, Hoffmann D, Louis J-C, Freysz L, Dreyfus H, Vincendon G, Massarelli R (1981) Synthèse de la choline dans les neurones. C R Acad Sci (Paris) [Série III] 293:781–783
Dainous F, Freysz L, Mozzi R, Dreyfus H, Louis JC, Porcellati G, Massarelli R (1982) Synthesis of choline phospholipids in neuronal and glial cell cultures by the methylation pathway. FEBS Lett 146:221–223
Das N, Srere PA (1972) Exchange of methyl protons of acetyl coenzyme A catalysed by adenosine triphosphate citrate lyase. Biochemistry 11:1534–1537
Davies LP, Whittaker VP, Zimmermann H (1987) Axonal transport in the electromotor nerves of *Torpedo marmorata*. Exp Brain Res 30:493–510
Davson H (1970) A textbook of general physiology. Churchill, London
De Robertis EDP (1964) Histophysiology of synapses and neurosecretion. Pergamon, Oxford
Dietz GW, Salvaterra PM (1980) Purification and peptide mapping of rat brain choline acetyltransferase. J Biol Chem 255:10612–10617
Doležal V, Tuček S (1981) Utilization of citrate, acetylcarnitine, acetate, pyruvate and glucose for the synthesis of acetylcholine in rat brain slices. J Neurochem 36:1323–1330
Doležal V, Tuček S (1982) Effects of choline and glucose on atropine-induced alteration of acetylcholine synthesis and content in the brain of rats. Brain Res 240:285–293
Doležal V, Tuček S (1983) The synthesis and release of acetylcholine in normal and denervated rat diaphragms during incubation *in vitro*. J Physiol 334:461–474
Doležal V, Tuček S (1984) Activation of muscarinic receptors stimulates the release of choline from brain slices. Biochem Biophys Res Commun 120:1002–1007
Drahota Z, Hahn P, Mourek J, Trojanová M (1965) The effect of acetoacetate on oxygen consumption of brain slices from infant and adult rats. Physiol Bohemoslov 14:134–136
Dreyfus P (1975) Identification de l'acétate comme précurseur du radical acétyle de l'acétylcholine des jonctions neuromusculaires du rat. C R Acad Sci (Paris) D280:1893–1894
Driskell WJ, Weber BH, Roberts E (1978) Purification of choline acetyltransferase from *Drosophila melanogaster*. J Neurochem 30:1135–1141
Dross K, Kewitz H (1972) Concentration and origin of choline in the rat brain. Naunyn Schmiedebergs Arch Pharmacol 274:91–106
Ducis I, Whittaker VP (1985) High affinity, sodium-gradient-dependent transport of choline into vesiculated presynaptic plasma membrane fragments from the electric organ of *Torpedo marmorata* and reconstitution of the solubilized transporter into liposomes. Biochim Biophys Acta 815:109–127
Dunant Y, Gautron J, Israël M, Lesbats B, Manaranche R (1974) Evolution de la décharge de l'organe électrique de la Torpille et variations simultanées de l'acétylcholine au cours de la stimulation. J Neurochem 23:635–643
Dunant Y, Israël M, Lesbats B, Manaranche R (1977) Oscillation of acetylcholine during nerve activity in the *Torpedo* electric organ. Brain Res 125:123–140
Eckenstein F, Thoenen H (1982) Production of specific antisera and monoclonal antibodies to choline acetyltransferase: characterization and use for identification of cholinergic neurons. EMBO J 1:363–368

Eckenstein F, Barde Y-A, Thoenen H (1981) Production of specific antibodies to choline acetyltransferase purified from pig brain. Neuroscience 6:993–1000
Eckernäs S-A (1977) Plasma choline and cholinergic mechanisms in the brain. Acta Physiol Scand [Suppl] 449:1–62
Eder-Colli L, Amato S (1985) Membrane-bound choline acetyltransferase in *Torpedo* electric organ: a marker for synaptosomal plasma membranes? Neuroscience 15:577–589
Eder-Colli L, Amato S, Froment Y (1986) Amphiphilic and hydrophilic forms of choline-O-acetyltransferase in cholinergic nerve endings of the *Torpedo*. Neuroscience 19:275–288
Eggerer H (1965) Zum Mechanismus der biologischen Umwandlung von Citronensäure. VI. Citrat-synthase ist eine Acetyl-CoA-Enolase. Biochem Z 343:111–138
Ehrlich BE, Wright EM (1982) Choline and PAH transport across blood-CSF barriers: the effect of lithium. Brain Res 250:245–249
Eisenstadt ML, Schwartz JH (1975) Metabolism of acetylcholine in the nervous system of *Aplysia californica*. III. Studies of an identified cholinergic neuron. J Gen Physiol 65:293–313
Endemann G, Brunengraber H (1980) The source of acetyl coenzyme A for acetylcholine synthesis in the perfused rat phrenic nerve-hemidiaphragm. J Biol Chem 255:11091–11093
Fitzgerald GG, Cooper JR (1967) Studies on acetylcholine in the corneal epithelium. Fed Proc Fed Am Soc Exp Biol 26:651
Fonnum F (1966) Is choline acetyltransferase present in synaptic vesicles? Biochem Pharmacol 15:1641–1643
Fonnum F (1967) The 'compartmentation' of choline acetyltransferase within the synaptosome. Biochem J 103:262–270
Fonnum F (1968) Choline acetyltransferase binding to and release from membranes. Biochem J 109:389–398
Fonnum F (1970) Surface charge of choline acetyltransferase from different species. J Neurochem 17:1095–1100
Fonnum F, Frizell M, Sjöstrand J (1973) Transport, turnover and distribution of choline acetyltransferase and acetylcholinesterase in the vagus and hypoglossal nerves of the rabbit. J Neurochem 21:1109–1120
Freeman JJ, Choi RL, Jenden DJ (1975) Plasma choline: its turnover and exchange with brain choline. J Neurochem 24:729–734
Gaiti A, de Medio GE, Brunetti M, Amaducci L, Procellati G (1974) Properties and function of the calcium-dependent incorporation of choline, ethanolamine and serine into the phospholipids of isolated rat brain microsomes. J Neurochem 23:1153–1159
Glover VAS, Potter LT (1971) Purification and properties of choline acetyltransferase from ox brain striate nuclei. J Neurochem 18:571–580
Gibson GE, Blass JP (1979) Proportional inhibition of acetylcholine synthesis accompanying impairment of 3-hydrobutyrate oxidation in rat brain. Biochem Pharmacol 28:133–139
Gibson GE, Peterson C (1983) Acetylcholine and oxidative metabolism in septum and hippocampus *in vitro*. J Biol Chem 258:1142–1171
Gibson GE, Shimada M (1980) Studies on the metabolic pathway of the acetyl group for acetylcholine synthesis. Biochem Pharmacol 29:167–174
Gibson GE, Jope R, Blass JP (1975) Decreased synthesis of acetylcholine accompanying impaired oxidation of pyruvic acid in rat brain minces. Biochem J 148:17–23
Goldberg AM, Hanin I (eds) (1976) Biology of cholinergic function. Raven, New York
Goodman DR, Adatsi FK, Harbison RD (1984) Evidence for the extreme overestimation of choline acetyltransferase in human sperm, human seminal plasma and rat heart: a case of mistaking carnitine acetyltransferase for choline acetyltransferase. Chem Biol Interact 49:39–53
Greenspan RJ (1980) Mutations of choline acetyltransferase and associated neural defects in *Drosophila melanogaster*. J Comp Physiol 137:83–92

Greville GD (1969) Intracellular compartmentation and the citric acid cycle. In: Lowenstein JM (ed) Citric acid cycle. Dekker, New York, pp 1–136
Grewaal DS, Quastel JH (1973) Control of synthesis and release of radioactive acetylcholine in brain slices from the rat. Biochem J 132:1–14
Growdon JH, Cohen EL, Wurtman RJ (1977) Effects of oral choline administration on serum and CSF choline levels in patients with Huntington's disease. J Neurochem 28:229–231
Gundersen CB, Jenden DJ, Miledi R (1985) Choline acetyltransferase and acetylcholine in Xenopus oocytes injected with mRNA from the electric lobe of *Torpedo.* Proc Natl Acad Sci USA 82:608–611
Guyenet P, Lefresne P, Rossier J, Beaujouan JC, Glowinski J (1973) Inhibition by hemicholinium-3 of (^{14}C)acetylcholine synthesis and (^{3}H)choline high-affinity uptake in rat striatal synaptosomes. Mol Pharmacol 9:630–639
Guynn RW (1976) Effect of ethanol on brain CoA and acetyl-CoA. J Neurochem 27:303–304
Haga T, Noda H (1973) Choline uptake systems of rat brain synaptosomes. Biochim Biophys Acta 291:564–575
Hanin I, Goldberg AM (eds) (1982) Progress in modern cholinergic biology; model cholinergic synapses. Raven, New York
Hanin I, Schuberth J (1974) Labelling of acetylcholine in the brain of mice fed on a diet containing deuterium labelled choline: studies utilizing gas chromatography – mass spectrometry. J Neurochem 23:819–824
Hansford RG, Castro F (1985) Role of Ca^{2+} in pyruvate dehydrogenase interconversion in brain mitochondria and synaptosomes. Biochem J 227:129–136
Harvey SAK, Long SF, Clark JB (1982) Noncorrelation of choline kinase or ATP citrate lyase with cholinergic activity in rat brain. J Neurochem 39:1481–1484
Hattori H, Kanfer JN (1985) Synaptosomal phospholipase D potential role in providing choline for acetylcholine synthesis. J Neurochem 45:1578–1584
Hattori H, Bansal VS, Orihel D, Kanfer JN (1984) Presence of phospholipid-*N*-methyltransferases and base-exchange enzymes in rat central nervous system axolemma-enriched fractions. J Neurochem 43:1018–1024
Haubrich DR (1976) Choline acetyltransferase and its inhibitors. In: Goldberg AM, Hanin I (eds) Biology of cholinergic function. Raven, New York, pp 239–268
Haworth RA, Hunter DR (1979) The Ca^{2+}-induced membrane transition in mitochondria. II. Nature of the Ca^{2+} trigger site. Arch Biochem Biophys 195:460–467
Haworth RA, Hunter DR (1980) Allosteric inhibition of the Ca^{2+}-activated hydrophilic channel of the mitochondrial membrane by nucleotides. J Membr Biol 54:231–236
Hayashi H, Kato T (1978) Acetyl-CoA synthesizing enzyme activities in single nerve cell bodies of rabbit. J Neurochem 31:861–869
Hebb CO (1962) Acetylcholine content of the rabbit plantaris muscle after denervation. J Physiol (Lond) 163:294–306
Hebb CO, Silver A (1956) Choline acetylase in the central nervous system of man and some other mammals. J Physiol (Lond) 134:718–728
Hebb CO, Silver A (1970) Biochemical parameters of central cholinergic nerves, mapping the distribution of acetylcholine and the enzymes which control its metabolism in the brain. In: Tedeschi CG (ed) Neuropathology: methods and diagnosis. Little, Brown, Boston, pp 665–690
Hebb CO, Whittaker VP (1958) Intracellular distribution of acetylcholine and choline acetylase. J Physiol (Lond) 142:187–196
Hebb CO, Ling GM, McGeer EG, McGeer PL, Perkins D (1964) Effect of locally applied hemicholinium on the acetylcholine content of the caudate nucleus. Nature (Lond) 204:1309–1311
Hersh LB (1979) The lack of specificity towards salts in the activation of choline acetyltransferase from human placenta. J Neurochem 32:991–996
Hersh LB (1980) Studies on the kinetic mechanism and salt activation of bovine brain choline acetyltransferase. J Neurochem 34:1077–1081

Hersh LB (1982) Kinetic studies of the choline acetyltransferase reaction using isotope exchange at equilibrium. J Biol Chem 257:12820–12834
Hersh LB, Peet M (1977) Re-evaluation of the kinetic mechanism of the choline acetyltransferase reaction. J Biol Chem 252:4796–4802
Hersh LB, Peet M (1978) Effect of salts on the physical and kinetic properties of human placental choline acetyltransferase. J Neurochem 30:1087–1093
Hersh LB, Coe B, Casey L (1978a) A fluorimetric assay for choline acetyltransferase and its use in the purification of the enzyme from human placenta. J Neurochem 30:1077–1085
Hersh LB, Barker LA, Rush B (1978b) Effect of sodium chloride on changing the rate limiting step in the human placental choline acetyltransferase reaction. J Biol Chem 253:4966–4970
Hersh LB, Nair RV, Smith DJ (1979) The reaction of choline acetyltransferase with sulfhydryl reagents. Methoxycarbonyl-CoA disulphide as an active site-directed reagent. J Biol Chem 254:11988–11992
Hersh LB, Wainer BH, Andrews LP (1984) Multiple isoelectric and molecular weight variants of choline acetyltransferase. Artifact or real? J Biol Chem 259:1253–1264
Heumann R, Villegas J, Herzfeld DW (1981) Acetylcholine synthesis in the Schwann cell and axon in the giant nerve fiber of the squid. J Neurochem 36:765–768
Higgins AJ, Neal MJ (1982a) Potassium activation of (^{3}H)choline accumulation by isolated sympathetic ganglia of the rat. Br J Pharmacol 77:573–580
Higgins AJ, Neal MJ (1982b) Effect of potassium depolarization and pregaglionic nerve stimulation on the metabolism of (^{3}H)choline in rat isolated sympathetic ganglia. Br J Pharmacol 77:581–590
Houser CR, Crawford GD, Barber RP, Salvaterra PM, Vaughn JE (1983) Organization and morphological characteristics of cholinergic neurons: an immunocytochemical study with a monoclonal antibody to choline acetyltransferase. Brain Res 266:97–120
Hunter DR, Haworth RA (1979a) The Ca^{2+}-induced membrane transition in mitochondria. I. The protective mechanisms. Arch Biochem Biophys 195:453–459
Hunter DR, Haworth RA (1979b) The Ca^{2+}-induced membrane transition in mitochondria. III. Transitional Ca^{2+}-release. Arch Biochem Biophys 195:468–477
Hunter DR, Haworth RA, Southard JH (1976) Relationship between configuration, function and permeability of calcium-treated mitochondria. J Biol Chem 251:5069–5077
Husain SS, Mautner HG (1973) The purification of choline acetyltransferase of squid-head ganglia. Proc Natl Acad Sci USA 70:3749–3753
Ichikawa T, Ishida I, Deguchi T (1983) Monoclonal antibodies to choline acetyltransferase of rat brain. FEBS Lett 155:306–310
Illingworth DR, Portman OW (1972) The uptake and metabolism of plasma lysophosphatidylcholine *in vivo* by the brain of squirrel monkeys. Biochem J 130:557–567
Israël M, Tuček S (1974) Utilization of acetate and pyruvate for the synthesis of "total", "bound" and "free" acetylcholine in the electric organ of *Torpedo*. J Neurochem 22:487–491
Israël M, Lesbats B, Manaranche R, Marsal J, Mastour-Frachon P, Meunier M (1977) Related changes in amounts of ACh and ATP in resting and active *Torpedo* nerve electroplaque synapses. J Neurochem 28:1259–1267
Itoh T, Quastel JH (1970) Acetoacetate metabolism in infant and adult rat brain *in vitro*. Biochem J 116:641–655
Itoh N, Slemmon JR, Hawke DH, Williamson R, Morita E, Itakura K, Roberts E, Shively JE, Crawford GD, Salvaterra PM (1986) Cloning of Drosophila choline acetyltransferase cDNA. Proc Natl Acad Sci USA 83:4081–4085
Jenden DJ, Jope RS, Weiler MH (1976) Regulation of acetylcholine synthesis: does cytoplasmic acetylcholine control high affinity choline uptake? Science 194:635–637
Johnson CD, Epstein ML (1986) Monoclonal antibodies and polyvalent antiserum to chicken choline acetyltransferase. J Neurochem 46:968–976
Jope RS (1979) High affinity choline transport and acetyl-CoA production in brain and their roles in the regulation of acetylcholine synthesis. Brain Res Rev 1:313–344

Jope RS (1982) Effects of phosphatidylcholine administration to rats on choline in blood and choline and acetylcholine in brain. J Pharmacol Exp Ther 220:322–328
Jope RS, Weiler MH, Jenden DJ (1978) Regulation of acetylcholine synthesis: control of choline transport and acetylation in synaptosomes. J Neurochem 30:949–954
Kása P (1986) The cholinergic systems in brain and spinal cord. Prog Neurobiol 26:211–272
Kása P, Mann SP, Hebb C (1970) Localization of choline acetyltransferase. Nature (Lond) 226:812–816
Kása P, Szepesy G, Gulya K, Bánsághy K, Rakonczay Z (1982) The effect of 4-(1-naphthylvinyl)-pyridine on the acetylcholine and on the number of synaptic vesicles in the central nervous system of the rat. Neurochem Int 4:185–193
Kato AC, Lefresne P, Berwald-Netter Y, Beaujouan JC, Glowinski J, Gros F (1977) Choline stimulates the synthesis and accumulation of acetate in a cholinergic neuroblastoma clone. Biochem Biophys Res Commun 78:350–356
Kotyk A (1983) Coupling of secondary active transport with $\Delta\tilde{\mu}_{H^+}$. J Bioenerg Biomembr 15:307–319
Kriebel ME, Matterson DR, Pappas GD (1978) Acetylcholine content in physiologically fatigued frog nerve-muscle preparations and in denervated muscle. Gen Pharmacol 9:229–234
Krnjević K, Straughan DW (1964) The release of acetylcholine from the denervated rat diaphragm. J Physiol (Lond) 170:371–378
Kuhar MJ (1979) Sodium-dependent high affinity choline uptake. In: Tuček S (ed) The cholinergic synapse. Prog Brain Res 49:71–76
Kuntscherová J, Vlk J (1968) Über die Bedeutung der Cholin- und Glukosezufuhr für die Normalisierung der durch Hungern herabgesetzten Acetylcholinvorräte in den Geweben. Plzen Lek Sb 31:5–12
Kwok YN, Collier B (1982) Synthesis of acetylcholine from acetate in sympathetic ganglion. J Neurochem 39:16–26
Ladinsky H, Consolo S, Peri G (1974) Effect of oxotremorine and physostigmine on choline levels in mouse whole brain, spleen and cerebellum. Biochem Pharmacol 23:1187–1193
Lanman RC, Schanker LS (1980) Transport of choline out of the cranial cerebrospinal fluid spaces of the rabbit. J Pharmacol Exp Ther 215:563–568
Lefresne PL, Guyenet P, Glowinski J (1973) Acetylcholine synthesis from (2-^{14}C)pyruvate in rat striatal slices. J Neurochem 20:1083–1097
Lefresne PL, Guyenet P, Beaujouan JC, Glowinski J (1975) The subcellular localization of ACh synthesis in rat striatal synaptosomes investigated with the use of Triton X-100. J Neurochem 25:415–422
Lefresne P, Hamon M, Beaujouan JC, Glowinski J (1977) Origin of the acetyl moiety of acetylcholine synthesized in rat striatal synaptosomes. Biochimie 59:197–215
Levey AI, Aoki M, Fitch FW, Wainer BH (1981) The production of monoclonal antibodies reactive with bovine choline acetyltransferase. Brain Res 218:383–387
Levey AI, Rye DB, Wainer BH (1982) Immunochemical studies of bovine and human choline-O-acetyltransferase using monoclonal antibodies. J Neurochem 39:1652–1659
Levey AI, Armstrong DM, Atweh SF, Terry RD, Wainer BH (1983) Monoclonal antibodies to choline acetyltransferase: production, specificity, and immunohistochemistry. J Neurosci 3:1–9
Lewartowski B, Bielecki K (1963) The influence of hemicholinium No. 3 and vagal stimulation on acetylcholine content of rabbit atria. J Pharmacol Exp Ther 142:24–30
Lindmar R, Löffelholz K, Pompetzki H (1977) Acetylcholine overflow during infusion of a high potassium-low sodium solution into the perfused chicken heart in the absence and presence of physostigmine. Naunyn Schmiedebergs Arch Pharmacol 229:17–21
Lindmar R, Löffelholz K, Weide W, Witzke J (1980) Neuronal uptake of choline following release of acetylcholine in the perfused heart. J Pharmacol Exp Ther 215:710–715
Lindmar R, Löffelholz K, Weide W, Weis S (1983) Evidence for bilateral vagal innervation of postganglionic parasympathetic neurons in chicken heart. J Neural Transm 56:239–247

Lindmar R, Löffelholz K, Sandmann J (1986) Characterization of choline efflux from the perfused heart at rest and after muscarine receptor activation. Naunyn Schmiedebergs Arch Pharmacol 332:224–229

Low MG, Ferguson AJ, Futerman AH, Silman I (1986) Covalently attached phosphatidylinostitol as a hydrophobic anchor for membrane proteins. Trends Biochem Sci 11:212–215

Lundgren G, Karlén B, Holmstedt B (1977) Acetylcholine and choline in mouse brain – influence of peripherally acting cholinergic drugs. Biochem Pharmacol 26:1607–1612

MacIntosh FC, Collier B (1976) Neurochemistry of cholinergic terminals. In: Zaimis E (ed) Neuromuscular junction. Springer, Berlin Heidelberg New York, pp 99–228 (Handbook of experimental pharmacology, vol 42)

Malthe-Sørenssen D (1976) Choline acetyltansferase – evidence for acetyl transfer by a histidine residue. J Neurochem 27:873–881

Malthe-Sørenssen D (1979) Recent progress in the biochemistry of choline acetyltransferase. In: Tuček S (ed) The cholinergic synapse. Prog Brain Res 49:45–58

Malthe-Sørenssen D, Lea T, Fonnum F, Eskeland T (1978) Molecular characterization of choline acetyltransferase from bovine brain caudate nucleus and some immunological properties of the highly purified enzyme. J Neurochem 30:35–46

Mann PJG, Tennenbaum M, Quastel JH (1938) On the mechanism of acetylcholine formation in brain *in vitro*. Biochem J 32:243–261

Mann PJG, Tennenbaum M, Quastel JH (1939) Acetylcholine metabolism in the central nervous system: the effects of potassium and other cations on acetylcholine liberation. Biochem J 33:822–835

Mann SP (1975) Distribution of glycerophosphorylcholine diesterase in rat brain. Experientia 31:1256–1258

Mann SP, Bennett RC (1979) The fate of choline in the circulating plasma of the rat. Experientia 35:211–212

Mann SP, Hebb C (1977) Free choline in the brain of the rat. J Neurochem 28:241–244

Marchbanks RM (1982) The activation of presynaptic choline uptake by acetylcholine release. J Physiol (Paris) 78:373–378

Marchbanks RM, Wonnacott S, Rubio MA (1981) The effect of acetylcholine release on choline fluxes in isolated synaptic terminals. J Neurochem 36:379–393

Mautner HG (1977) Choline acetyltransferase. CRC Crit Rev Biochem 4:341–370

Mautner HG (1980) Acetylcholine and choline acetyltransferase. Neurochem Int 2:123–134

Mautner HG (1986) Choline acetyltransferase. In: Boulton AA, Baker GB, Yu PH (eds) Neurotransmitter enzymes. Humana, Clifton, pp 273–317 (Neuromethods, vol 5)

Mautner HG, Merrill RE, Currier SF, Harvey G (1981 a) Interaction of aromatic dyes with the coenzyme A binding site of choline acetyltransferase. J Med Chem 24:1534–1537

Mautner HG, Pakula AA, Merrill RE (1981 b) Evidence for presence of an arginine residue in the coenzyme A binding site of choline acetyltransferase. Proc Natl Acad Sci USA 78:7449–7452

McCaman RE, Rodríguez de Lores Arnaiz G, de Robertis E (1965) Species differences in subcellular distribution of choline acetylase in the CNS. J Neurochem 12:927–935

McCormack JG (1985) Characterization of the effects of Ca^{2+} on the intramitochondrial Ca^{2+}-sensitive enzymes from rat liver and within intact rat liver mitochondria. Biochem J 231:581–595

McGee R, Simpson P, Christian C, Mata M, Nelson P, Nirenberg M (1978) Regulation of acetylcholine release from neuroblastoma x glioma hybrid cells. Proc Natl Acad Sci USA 75:1314–1318

McGeer PL, McGeer EG, Peng JH (1984) Choline acetyltransferase: purification and immunohistochemical localization. Life Sci 34:2319–2338

Meyer EM Jr, Engel DA, Cooper JR (1982) Acetylation and phosphorylation of choline following high or low affinity uptake by rat cortical synaptosomes. Neurochem Res 7:749–759

Miledi R, Molenaar PC, Polak RL (1977) An analysis of acetylcholine in frog muscle by mass fragmentography. Proc R Soc Lond 197:285–297

Miledi R, Molenaar PC, Polak RL (1982a) Free and bound acetylcholine in frog muscle. J Physiol 333:189–199
Miledi R, Molenaar PC, Polak RL, Tas JWM, van der Laaken T (1982b) Neural and non-neural acetylcholine in the rat diaphragm. Proc R Soc Lond B 214:153–168
Millington WR, Goldberg AM (1982) Precursor dependence of acetylcholine release from rat brain *in vitro*. Brain Res 243:263–270
Mitchell JF, Silver A (1963) The spontaneous release of acetylcholine from the denervated hemidiaphragm of the rat. J Physiol (Lond) 165:117–129
Mitzen EJ, Koeppen AH (1984) Malonate, malonyl-coenzyme A, and acetyl-coenzyme A in developing rat brain. J Neurochem 43:499–506
Morel N (1975) Incorporation d'acétate dans l'acétylcholine de l'organe électrique de Torpille: effets des concentrations d'acétate et de choline. C R Acad Sci D 280:999–1001
Morel N, Israël M, Manaranche R, Masour-Frachon P (1977) Isolation of pure cholinergic nerve endings from *Torpedo* electric organ. J Cell Biol 75:43–55
Morris D, Tuček S (1966) The influence of inorganic cations on the activities of choline acetyltransferase, phosphate acetyltransferase and acetylphosphatase. J Neurochem 13:333–345
Mozzi R, Porcellati G (1979) Conversion of phosphatidylethanolamine to phosphatidylcholine in rat brain by the methylation pathway. FEBS Lett 100:363–366
Mozzi R, Siepi D, Andreoli V, Piccinin GL, Porcellati G (1981) *N*-methylation of phosphatidylethanolamine in different brain areas of the rat. Bull Mol Biol Med 6:6–15
Murrin LC (1980) High affinity transport of choline in neuronal tissue. Pharmacology 21:132–140
Murrin LC, Kuhar MJ (1976) Activation of high-affinity choline uptake *in vitro* by depolarizing agents. Mol Pharmacol 12:1082–1090
Murrin LC, DeHaven RN, Kuhar MJ (1977) On the relationship between (^{3}H) choline uptake activation and (^{3}H)acetylcholine release. J Neurochem 29:681–687
Nachmansohn D, Machado AL (1943) The formation of acetylcholine. A new enzyme: 'choline acetylase'. J Neurophysiol 6:397–403
Nagai T, Pearson T, Peng F, McGeer EG, McGeer PL (1983) Immunohistochemical staining of the human forebrain with monoclonal antibody to human choline acetyltransferase. Brain Res 265:300–306
Nakamura R, Cheng S-C, Naruse H (1970) A study on the precursors of the acetyl moiety of acetylcholine in brain slices. Biochem J 118:443–450
O'Regan S (1983) Uptake of acetate and propionate by isolated nerve endings from the electric organ of *Torpedo marmorata* and their incorporation into choline esters. J Neurochem 41:1596–1601
O'Regan S (1984) Evaluation of acetate uptake and its relative conversion to acetylcholine in *Torpedo* electric organ synaptosomes under different ionic and metabolic conditions. Neurochem Int 6:339–346
O'Regan S, Collier B (1981) Factors affecting choline transport by the superior cervical ganglion during and following stimulation, and the relationship between choline uptake and acetylcholine synthesis. Neuroscience 6:511–520
O'Regan S, Vyas S (1986) Modifications in choline transport activity as a function of membrane potential and the sodium gradient. J Physiol (Paris) 81:325–331
O'Regan S, Collier B, Israël M (1982) Studies on presynaptic cholinergic mechanisms using analogues of choline and acetate. J Physiol (Paris) 78:454–460
Paetzke-Brunner I, Schön H, Wieland OH (1978) Insulin activates pyruvate dehydrogenase by lowering the mitochondrial acetyl-CoA/CoA ratio as evidenced by digitonin fractionation of isolated fat cells. FEBS Lett 93:307–311
Pardridge WM, Cornford EM, Braun LD, Oldendorf WH (1979) Transport of choline and choline analogues through the blood-brain barrier. In: Barbeau A, Growdon JH, Wurtman RJ (eds) Nutrition and the brain, vol 5. Raven, New York, pp 25–34
Parnavelas JG, Kelly W, Burnstock G (1985) Ultrastructural localization of choline acetyltransferase in vascular endothelial cells in rat brain. Nature 316:724–725

Paton WDM, Vizi ES, Zar MA (1971) The mechanism of acetylcholine release from parasympathetic nerves. J Physiol 215:819–848
Peng JH, McGeer PL, Kimura H, Sung SC, McGeer EG (1980) Purification and immunochemical properties of choline acetyltransferase from human brain. Neurochem Res 5:943–962
Peng JH, McGeer PL, McGeer EG (1983) Anti-human choline acetyltransferase fragments: antigen binding (Fab) Sepharose chromatography for enzyme purification. Neurochem Res 8:1481–1486
Peng JH, McGeer PL, McGeer EG (1986) Membrane-bound choline acetyltransferase from human brain: purification and properties. Neurochem Res 11:959–971
Percy AK, Moore JF, Waechtler CJ (1982) Properties of particulate and detergent-solubilized phospholipid *N*-methyltransferase activity from calf brain. J Neurochem 38:1404–1412
Peterson C, Goldman JE (1986) In vitro acetylcholine synthesis and oxidative metabolism during development of normal and brindled mouse brain. Dev Brain Res 29:153–159
Peterson GR, Webster GW, Shuster L (1973) Characteristics of choline acetyltransferase and cholinesterases in two types of cultured cells from embryonic chick brain. Dev Biol 34:119–134
Pieklik JR, Guynn RW (1975) Equilibrium constants of the reactions of choline acetyltransferase, carnitine acetyltransferase, and acetylcholinesterase under physiological conditions. J Biol Chem 250:4445–4450
Polak RL, Molenaar PC, van Gelder M (1977) Acetylcholine metabolism and choline uptake in cortical slices. J Neurochem 29:477–485
Polak RL, Sellin LC, Thesleff S (1981) Acetylcholine content and release in denervated or botulinum poisoned rat skeletal muscle. J Physiol 319:253–259
Potter LT (1968) Uptake of choline by nerve endings isolated from the rat cerebral cortex. In: Campbell PN (ed) The interaction of drugs and subcellular components of animal cells. Churchill, London, pp 293–304
Potter LT (1970) Synthesis, storage and release of (^{14}C)acetylcholine in isolated rat diaphragm muscles. J Physiol 206:145–166
Potter LT, Glover VAS, Saelens JK (1968) Choline acetyltransferase from rat brain. J Biol Chem 243:3864–3870
Quastel JH (1978) Source of the acetyl group in acetylcholine. In: Jenden DJ (ed) Cholinergic mechanisms and psychopharmacology. Plenum, New York, pp 411–430
Quastel JH, Tennenbaum M, Wheatley AHM (1936) Choline ester formation in, and choline esterase activities of, tissues *in vitro*. Biochem J 30:1668–1681
Richardson PJ (1986) Choline uptake and metabolism in affinity-purified cholinergic nerve terminals from rat brain. J Neurochem 46:1251–1255
Richter D, Crossland J (1949) Variation in acetylcholine content of the brain with physiological state. Am J Physiol 159:247–255
Richter JA (1976) Characteristics of acetylcholine release by superfused slices of rat brain. J Neurochem 26:791–797
Richter JA, Gormley JM, Holtman JR, Simon JR (1982) High-affinity choline uptake in the hippocampus: its relationship to the physiological state produced by administration of barbiturates. J Neurochem 39:1440–1445
Říčný J, Tuček S (1980a) Acetylcoenzyme A in the brain: radio-enzymatic determination, use of microwaves, and postmortem changes. Anal Biochem 103:369–376
Říčný J, Tuček S (1980b) Relation between the content of acetylcoenzyme A and acetylcholine in brain slices. Biochem J 188:683–688
Říčný J, Tuček S (1981) Acetylcoenzyme A and acetylcholine in slices of rat caudate nuclei incubated in the presence of metabolic inhibitors. J Biol Chem 256:4919–4923
Říčný J, Tuček S (1982) Acetylcoenzyme A and acetylcholine in slices of rat caudate nuclei incubated with (−)-hydroxycitrate, citrate and EGTA. J Neurochem 39:668–673
Říčný J, Tuček S (1983) Ca^{2+} ions and the output of acetylcoenzyme A from brain mitochondria. Gen Physiol Biophys 2:27–37
Roskoski R (1973) Choline acetyltransferase. Evidence for an acetylenzyme reaction intermediate. Biochemistry 12:3709–3714

Roskoski R (1974) Choline acetyltransferase and acetylcholinesterase: evidence for essential histidine residues. Biochemistry 13:5141–5144
Roskoski R (1978) Acceleration of choline uptake after depolarization-induced acetylcholine release in rat cortical synaptosomes. J Neurochem 30:1357–1361
Roskoski R, Lim C-T, Martinsek Roskoski L (1975) Human brain and placental choline acetyltransferase purification and properties. Biochemistry 14:5105–5110
Ross ME, Park DH, Teitelman G, Pickel VM, Reis DJ, Joh TH (1983) Immunohistochemical localization of choline acetyltransferase using a monoclonal antibody: a radioautographic method. Neuroscience 10:907–922
Rossier J (1976) Biophysical properties of rat brain choline acetyltransferase. J Neurochem 26:555–559
Rossier J (1977a) Choline acetyltransferase: a review with special reference to its cellular and subcellular localization. Int Rev Neurobiol 20:283–337
Rossier J (1977b) Acetyl-coenzyme A and coenzyme A analogues: their effects on rat brain choline acetyltransferase. Biochem J 165:321–326
Ryan RL, McClure WO (1979) Purification of choline acetyltransferase from rat and cow brain. Biochemistry 18:5357–5365
Ryan RL, McClure WO (1980) Physical and kinetic properties of choline acetyltransferase from rat and bovine brain. J Neurochem 34:395–403
Sacchi O, Consolo S, Peri G, Prigioni I, Ladinsky H, Perri V (1978) Storage and release of acetylcholine in the isolated superior cervical ganglion of the rat. Brain Res 151:443–456
Salvaterra PM, McCaman RE (1985) Choline acetyltransferase and acetylcholine levels in *Drosophila melanogaster*: a study using two temperature-sensitive mutants. J Neurosci 5:903–910
Sastry BVR, Sadavongvivad C (1979) Cholinergic systems in non-nervous tissues. Pharmacol Rev 30:65–132
Schmidt DE, Wecker L (1981) CNS effects of choline administration: evidence for temporal dependence. Neuropharmacology 20:535–539
Schubert D, Heinemann S, Carlisle W, Tarikas H, Kimes B, Patrick J, Steinbach JH, Culp W, Brandt BL (1974) Clonal cell lines from the rat central nervous system. Nature (Lond) 249:224–227
Schuberth J, Jenden DJ (1975) Transport of choline from plasma to cerebrospinal fluid in the rabbit with reference to the origin of choline and to acetylcholine metabolism in brain. Brain Res 84:245–256
Schuberth J, Sundwall A (1971) A method for the determination of choline in biological materials. Acta Pharmacol 29[Suppl 4]:51
Schuberth J, Sparf B, Sundwall A (1970) On the turnover of acetylcholine in nerve endings of mouse brain *in vivo*. J Neurochem 17:461–468
Severin SE, Artenie V (1967) Extraction, partial purification and some properties of choline acetyltransferase from rabbit brain (in Russian). Biokhimiia 32:125–132
Simon JR, Kuhar MJ (1975) Impulse-flow regulation of high affinity choline uptake in brain cholinergic nerve terminals. Nature 255:162–163
Simon JR, Mittag TW, Kuhar ML (1975) Inhibition of synaptosomal uptake of choline by various choline analogs. Biochem Pharmacol 24:1139–1142
Simon JR, Atweh S, Kuhar MJ (1976) Sodium-dependent high affinity choline uptake: a regulatory step in the synthesis of acetylcholine. J Neurochem 26:909–922
Singh VK, van Alstyne D (1978) Glial cells from normal adult rat brain established in continous culture. Brain Res 155:418–421
Skurdal DN, Cornatzer WE (1975) Choline phosphotransferase and phosphatidyl ethanolamine methyltransferase activities. Int J Biochem 13:887–892
Slavíková J, Tuček S (1982) Choline acetyltransferase in the heart of adult rats. Pflugers Arch 392:225–229
Slemmon JR, Salvaterra PM, Crawford GD, Roberts E (1982) Purification of choline acetyltransferase from Drosophila melanogaster. J Biol Chem 257:3847–3852
Slemmon JR, Salvaterra PM, Roberts E (1984) Molecular characterization of choline acetyltransferase from *Drosophila melanogaster*. Neurochem Int 6:519–526

Smith PC, Carroll PT (1980) A comparison of solubilized and membrane bound forms of choline-O-acetyltransferase (EC 2.3.1.6) in mouse brain nerve endings. Brain Res 185:363–371

Sollenberg J, Sörbo B (1970) On the origin of the acetyl moiety of acetylcholine in brain studied with a differential labelling technique using ^{3}H-^{14}C-mixed labelled glucose and acetate. J Neurochem 17:201–207

Somogyi GT, Vizi ES, Knoll J (1977) Effect of hemicholinium-3 on the release and net synthesis of acetylcholine in Auerbach's plexus of guinea pig ileum. Neuroscience 2:791–796

Spanner S, Hall RC, Ansell B (1976) Arterio-venous differences of choline and choline lipids across the brain of rat and rabbit. Biochem J 154:133–140

Speth RC, Yamamura HI (1979) Sodium dependent high affinity neuronal choline uptake. In: Barbeau A, Growdon JH, Wurtman RJ (eds) Nutrition and the brain, vol 5. Raven, New York, pp 129–139

Srere PA (1967) A magnetic resonance study of the citrate synthase reaction. Biochem Biophys Res Commun 26:609–614

Stavinoha WB, Weintraub ST (1974a) Estimation of choline and acetylcholine in tissue by pyrolysis gas chromatography. Anal Chem 46:757–760

Stavinoha WB, Weintraub ST (1974b) Choline content of rat brain. Science 183:964–965

Stedman E, Stedman E (1937) The mechanism of the biological synthesis of acetylcholine. I. The isolation of acetylcholine produced by brain tissue in vitro. Biochem J 31:817–827

Sterling GH, O'Neill JJ (1978) Citrate as the precursor of the acetyl moiety of acetylcholine. J Neurochem 31:525–530

Sterling GH, McCafferty MR, O'Neill JJ (1981) β-Hydroxybutyrate as a precursor to the acetyl moiety of acetylcholine. J Neurochem 37:1250–1259

Sterri SH, Fonnum F (1980) Acetyl-CoA synthesizing enzymes in cholinergic nerve terminals. J Neurochem 35:249–254

Strauss WL, Nirenberg M (1985) Inhibition of choline acetyltransferase by monoclonal antibodies. J Neurosci 5:175–180

Su YYT, Wu J-Y, Lam DMK (1980) Purification and some properties of choline acetyltransferase from catfish brain. J Neurochem 34:438–445

Suszkiw JB, Beach RL, Pilar GR (1976) Choline uptake by cholinergic neuron cell somas. J Neurochem 26:1123–1131

Szutowicz A, Łysiak W (1980) Regional and subcellular distribution of ATP-citrate lyase and other enzymes of acetyl-CoA metabolism in rat brain. J Neurochem 35:775–785

Szutowicz A, Bielarcyk H, Łysiak W (1981) The role of citrate derived from glucose in the acetylcholine synthesis in rat brain synaptosomes. Int J Biochem 13:887–892

Szutowicz A, Srere PA, Allen CN, Crawford IL (1982a) Effects of septal lesions on enzymes of acetyl-CoA metabolism in the cholinergic system of the rat hippocampus. J Neurochem 39:458–463

Szutowicz A, Stepień M, Bielarczyk H, Kabata J, Łysiak W (1982b) ATP citrate lyase in cholinergic nerve endings. Neurochem Res 7:799–810

Szutowicz A, Morrison MR, Srere PA (1983a) The enzymes of acetyl-CoA metabolism in differentiating cholinergic (S-20) and noncholinergic (NIE-115) neuroblastoma cells. J Neurochem 40:1664–1670

Szutowicz A, Harris NF, Srere PA, Crawford IL (1983b) ATP-citrate lyase and other enzymes of acetyl-CoA metabolism in fractions of small and large synaptosomes from rat brain hippocampus and cerebellum. J Neurochem 41:1502–1505

Tacconi M, Wurtman RJ (1985) Phosphatidylcholine produced in rat synaptosomes by N-methylation is enriched in polyunsaturated fatty acids. Proc Natl Acad Sci USA 82:4828–4831

Tochino Y, Schanker LS (1965) Active transport of quaternary ammonium compounds by the choroid plexus *in vitro*. Am J Physiol 208:666–673

Tochino Y, Schanker LS (1966) Serum and tissue factors that inhibit amine transport by the choroid plexus *in vitro*. Am J Physiol 210:1229–1233

Trommer BA, Schmidt DE, Wecker L (1982) Exogenous choline enhances the synthesis of acetylcholine only under conditions of increased cholinergic neuronal activity. J Neurochem 39:1704–1709

Tuček S (1966a) On subcellular localization and binding of choline acetyltransferase in the cholinergic nerve endings of the brain. J Neurochem 13:1317–1327
Tuček S (1966b) On the question of the localization of choline acetyltransferase in synaptic vesicles. J Neurochem 13:1329–1332
Tuček S (1967a) Observations on the subcellular distribution of choline acetyltransferase in the brain tissue of mammals and comparisons of acetylcholine synthesis from acetate and citrate in homogenates and nerve-endings fractions. J Neurochem 14:519–529
Tuček S (1967b) Subcellular distribution of acetyl-CoA synthetase, ATP citrate lyase, citrate synthase, choline acetyltransferase, fumarate hydratase and lactate dehydrogenase in mammalian brain tissue. J Neurochem 14:531–545
Tuček S (1967c) The use of choline acetyltransferase for measuring the synthesis of acetyl-coenzyme A and its release from brain mitochondria. Biochem J 104:749–756
Tuček S (1967d) Activation of choline acetyltransferase in subcellular fractions of the brain by hyposmotic treatment and by ether. Biochem Pharmacol 16:109–113
Tuček S (1974) Transport and changes of activity of choline acetyltransferase in the peripheral stump of an interrupted nerve. Brain Res 82:249–261
Tuček S (1975) Transport of choline acetyltransferase and acetylcholinesterase in the central stump and isolated segments of a peripheral nerve. Brain Res 86:259–270
Tuček S (1978) Acetylcholine synthesis in neurons. Chapman and Hall, London
Tuček S (ed) (1979) The cholinergic synapse. Prog Brain Res 49
Tuček S (1982) The synthesis of acetylcholine in skeletal muscles of the rat. J Physiol 322:53–69
Tuček S (1983a) The synthesis of acetylcholine. In: Lajtha A (ed) Enzymes in the nervous system. Plenum, New York, pp 219–249 (Handbook of neurochemistry, 2nd edn, vol 4)
Tuček S (1983b) Acetylcoenzyme A and the synthesis of acetylcholine in neurons: review of recent progress. Gen Physiol Biophys 2:313–324
Tuček S (1984) Problems in the organization and control of acetylcholine synthesis in brain neurons. Prog Biophys Mol Biol 44:1–46
Tuček S (1985) Regulation of acetylcholine synthesis in the brain. J Neurochem 44:10–24
Tuček S, Cheng S-C (1970) Precursors of acetyl groups in acetylcholine in the brain *in vivo*. Biochim Biophys Acta 208:538–540
Tuček S, Cheng S-C (1974) Provenance of the acetyl group of acetylcholine and compartmentation of acetyl-CoA and Krebs cycle intermediates in the brain *in vivo*. J Neurochem 22:893–914
Tuček S, Zelená J, Ge I, Vyskočil F (1978) Choline acetyltransferase in transected nerves, denervated muscles and Schwann cells of the frog: correlation of biochemical, electron microscopical and electrophysiological observations. Neuroscience 3:709–724
Tuček S, Doležal V, Sullivan AC (1981) Inhibition of the synthesis of acetylcholine in rat brain slices by (−)-hydroxycitrate and citrate. J Neurochem 36:1331–1337
Tzagoloff A (1982) Mitochondria. Plenum, New York
Vaca K, Pilar G (1979) Mechanisms controlling choline transport and acetylcholine synthesis in motor nerve terminals during electric stimulation. J Gen Physiol 73:605–628
Vyas S, O'Regan S (1985) Reconstitution of carrier-mediated choline transport in proteoliposomes prepared from presynaptic membranes of *Torpedo* electric organ, and its internal and external ionic requirements. J Membr Biol 85:111–119
Wainer BH, Levey AI, Mufson EJ, Mesulam M-M (1984) Cholinergic systems in mammalian brain identified with antibodies against choline acetyltransferase. Neurochem Int 6:163–182
Walum E (1981) Uptake and release of choline in cultures of human glioma cells. Cell Mol Neurobiol 1:389–400
Weber BH, Hariri M, Martin D, Driskell WJ (1979) Enzymic properties of choline acetyltransferase from heads of *Drosophila melanogaster*. J Neurochem 32:1597–1598
Wecker L, Schmidt DE (1980) Neuropharmacological consequences of choline administration. Brain Res 184:234–238

Weiler MH, Jope RS, Jenden DJ (1978) Effect of pretreatment under various cationic conditions on acetylcholine content and choline transport in rat whole brain synaptosomes. J Neurochem 31:789–796

Weiler MH, Gundersen CB, Jenden DJ (1981) Choline uptake and acetylcholine synthesis in synaptosomes: investigations using two different labelled variants of choline. J Neurochem 36:1802–1812

Weiler MH, Bak IJ, Jenden DJ (1983) Choline and acetylcholine metabolism in rat neostriatal slices. J Neurochem 41:473–480

Wetzel GT, Brown JH (1983) Relationships between choline uptake, acetylcholine synthesis and acetylcholine release in isolated rat atria. J Pharmacol Exp Ther 226:343–348

White HL, Wu JC (1973a) Kinetics of choline acetyltransferase (EC 2.3.1.6) from human and other mammalian central and peripheral nervous tissues. J Neurochem 20:297–307

White HL, Wu JC (1973b) Choline and carnitine acetyltransferase of heart. Biochemistry 12:841–846

Whittaker VP (1965) The application of subcellular fractionation techniques to the study of brain function. Prog Biophys Mol Biol 15:39–96

Whittaker VP, Michaelson IA, Kirkland RJA (1964) The separation of synaptic vesicles from nerve-ending particles ('synaptosomes'). Biochem J 90:293–303

Wieland OH (1983) The mammalian pyruvate dehydrogenase complex: structure and regulation. Rev Physiol Biochem Pharmacol 96:123–170

Willoughby J, Harvey SAK, Clark JB (1986) Compartmentation and regulation of acetylcholine synthesis at the synapse. Biochem J 235:215–223

Witter B, Kanfer JN (1985) Hydrolysis of endogenous phospholipids by rat brain microsomes. J Neurochem 44:155–162

Wonnacott S, Marchbanks RM (1976) Inhibition by botulinum toxin of depolarization-evoked release of [^{14}C]acetylcholine from synaptosomes *in vitro*. Biochem J 156:701–712

Wright EM (1978) Transport processes in the formation of the cerebrospinal fluid. Rev Physiol Biochem Pharmacol 83:1–34

Wurtman RJ, Blusztajn JK, Maire J-C (1985) 'Autocannibalism' of choline-containing membrane phospholipids in the pathogenesis of Alzheimer's disease – a hypothesis. Neurochem Int 7:369–372

Yamamura HI, Snyder SH (1973) High affinity transport of choline into synaptosomes of rat brain. J Neurochem 21:1355–1374

Zeisel SH (1981) Dietary choline: biochemistry, physiology, and pharmacology. Rev Nutr 1:95–121

Zeisel SH (1985) Formation of unesterified choline by rat brain. Biochim Biophys Acta 835:331–343

Zeisel SH, Wurtman RJ (1981) Developmental changes in rat blood choline concentration. Biochem J 198:565–570

Zeisel SH, Blusztajn JK, Wurtman RJ (1979) Brain lecithin biosynthesis: evidence that bovine brain can make choline molecules. In: Barbeau A, Growdon JH, Wurtman RJ (eds) Choline and lecithin in brain disorders. Raven, New York, pp 47–55 (Nutrition and the brain, vol 5)

CHAPTER 8a

Vertebrate Cholinesterases: Structure and Types of Interaction

J. MASSOULIÉ and J.-P. TOUTANT

A. Introduction

Cholinesterases (ChEs) hydrolyse choline esters specifically and rapidly and thus play an essential role in cholinergic transmission, e. g. at the neuromuscular junctions of vertebrates. Vertebrates possess two distinct ChEs, acetylcholinesterase (AChE, EC 3.1.1.7) and butyrylcholinesterase, also called pseudocholinesterase or non-specific cholinesterase (BuChE, EC 3.1.1.8). Both these enzymes present multiple molecular variants which possess identical catalytic activity but differ in their molecular structure and interactions. The molecular variants can be differentiated by various analytical procedures. We find it useful first to classify them as molecular forms, according to their hydrodynamic parameters (sedimentation coefficient, Stokes radius), corresponding to different quaternary structures.

All vertebrates appear to possess homologous sets of molecular forms for each of the two enzymes, AChE and BuChE; in particular, it is possible to distinguish two classes of molecules, globular and asymmetric forms, according to the presence of a collagen-like 'tail' associated with the catalytic subunits.

The molecular forms of AChE from electric organs of the electric eel *Electrophorus* and the electric ray *Torpedo* have been analysed in detail using sedimentation and sodium dodecylsulphate polyacrylamide gel electrophoresis (SDS-PAGE): in the case of *Electrophorus*, the quaternary associations correspond to the simple scheme illustrated in Fig. 1; the situation is more complex in the case of *Torpedo*. We will therefore successively describe the molecular forms of AChE in *Electrophorus*, in *Torpedo* and in other torpedinid electric rays, and examine how the conclusions may be extended to other species. We discuss particularly the characteristic interactions of the molecular forms: ionic interactions of asymmetric forms, and hydrophobic or non-hydrophobic character of globular forms. We then examine how each molecular form, corresponding to a given quaternary assembly, may be further subdivided by applying other analytical procedures, such as non-denaturing electrophoresis and the binding of lectins. We also discuss the specificity of immunological recognition between AChE and BuChE, among enzymes of different species and between enzyme fractions within a given organism.

Finally we examine the recent contribution of molecular genetics to our understanding of the generation of the molecular diversity of these enzymes.

We will cover with particular attention the more recent literature, since these subjects have been previously reviewed by MASSOULIÉ and BON (1982), BRIMI-

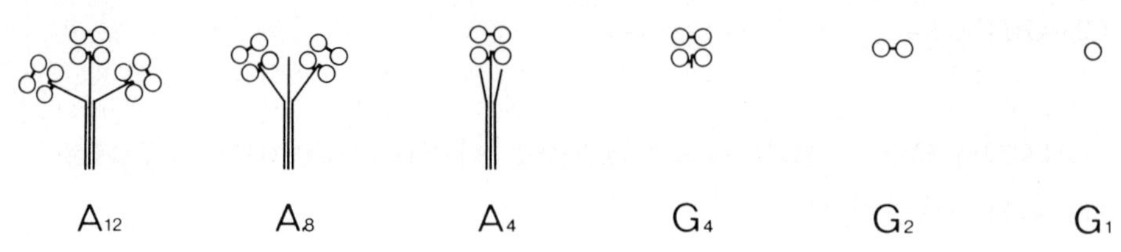

Fig. 1. Schematic representation of the quaternary structure of asymmetric and globular forms of cholinesterases. The quaternary structure of the different forms has been elucidated in the case of AChE from *Electrophorus* electric organs, in which all subunits appear to be identical (BON et al. 1976; ROSENBERRY and RICHARDSON 1977; ANGLISTER and SILMAN 1978). Homologous sets of molecular forms exist in all vertebrates, including mammals and birds (BON et al. 1979) and the primitive Agnatha, *Petromyzon marinus* (lamprey) (PEZZEMENTI et al. 1987). It is useful to apply the same nomenclature for all species, to express homologies between corresponding molecules, for both AChE and BuChE. (After MASSOULIÉ and BON 1982)

JOIN (1983), BRZIN et al. (1983), OTT (1985), ROSENBERRY (1985), RAKONCZAY (1986), BRODBECK (1986), BRIMIJOIN and RAKONCZAY (1986), TAYLOR et al. (1986), and TOUTANT and MASSOULIÉ (1987). A large part of a recent international meeting on ChEs has also been devoted to these problems (BRZIN et al. 1984).

B. Duality of ChEs: AChE and BuChE

ChEs are serine hydrolases, which hydrolyse ester bonds by an acylation-deacylation mechanism. They are therefore sensitive to irreversible alkylating inhibitors such as organophosphates and may be specifically labelled by radioactive diisopropylphosphorofluoridate (DFP) (see reviews in KOELLE 1963; SILVER 1974; for a review on the catalytic mechanism of AChE, see ROSENBERRY 1975).

The catalytic centre of ChEs consists of an esteratic subsite containing the active serine and an ionic subsite which binds the quaternary ammonium group of choline and of reversible inhibitors such as edrophonium. In addition, AChE possesses one or several peripheral anionic sites which preferentially bind some ligands like propidium (MOOSER and SIGMAN 1974; TAYLOR and LAPPI 1975; for a review, see ROSENBERRY 1975). The distance between catalytic and peripheral anionic sites has been evaluated as 1·4 nm from the binding of bisquaternary ligands and 1·9–3·7 nm from the interaction between fluorescent ligands (BERMAN et al. 1980); the difference between the two values possibly indicates conformational changes in the protein. Substrate inhibition, a phenomenon that exists for AChE but not for BuChE, is often considered to arise from ligand binding to peripheral sites, but in several cases it has been shown to be due to the binding of ligands to the active site of the acetylated enzyme intermediate (ROSENBERRY 1975a).

ChEs are defined by their preferential hydrolysis of choline esters as opposed to other esterases (aliesterases, aromatic esterases). However they also hydrolyse other esters. For example the uncharged analogues of choline esters, 3,3-dimethylbutyl esters, are rapidly hydrolysed, at about one-third of the rate of the corresponding choline esters (ADAMS and WHITTAKER 1949), a striking example of the principle of the complementariness of conformation between an enzyme and its substrate. A wide range of aliphatic esters are hydrolysed at a rate determined by the closeness of their structure to that of the preferred choline esters (ADAMS and WHITTAKER 1949; HASAN et al. 1980, 1981).

ChEs are characteristically inhibited by the quaternary carbamate, eserine (physostigmine). Two ChEs which differ in their substrate specificity apparently coexist in all vertebrates. AChE hydrolyses acetylcholine (ACh) faster than other choline esters and does not hydrolyse butyrylcholine at a significant rate. BuChE is characterized by the fact that, in addition to ACh, it hydrolyses other choline esters including butyrylcholine. Another major difference is that AChE is inhibited by high concentrations of ACh (>3 mM) whereas BuChE is not, as mentioned above.

The substrate preference of BuChE for ACh and its higher analogues, propionylcholine and butyrylcholine, is more variable than that of AChE: in most organisms, butyrylcholine is the most rapidly hydrolysed substrate; in some cases (e.g. in rat) it is propionylcholine, and in *Torpedo*, a relatively primitive vertebrate, it is ACh (TOUTANT et al. 1985). For this reason, this enzyme has also been called propionylcholinesterase, non-specific cholinesterase or pseudocholinesterase. These terms also express the notion that AChE, being predominant in brain and muscle, is a more noble, synaptic enzyme, whereas BuChE, typically found in serum, is a second-rate, scavenging enzyme. It is true that BuChE hydrolyses esters like benzoylcholine, which are not substrates for AChE; the analogue acetyl-β-methylcholine is however hydrolysed by AChE, but not by BuChE (MENDEL et al. 1943). In addition, the structural polymorphism of the two enzymes is equivalent, and it is possible that both ChEs play partially distinct but important physiological roles, as discussed in Chap. 8b.

Besides their relative rates of substrate hydrolysis, AChE and BuChE may also be distinguished by their sensitivity to specific inhibitors (see SILVER 1974). For example, AChE is inhibited by low concentrations of the reversible bisquaternary inhibitor 1,5-bis(4-allyldimethylammoniumphenyl)-pentan-3-one dibromide (BW 284C51; 50% inhibition at around 10^{-7} M) while BuChE is inhibited only at much higher concentrations. Conversely BuChE is generally inhibited by low concentrations of ethopropazine, and of the organophosphate tetra-mono-*iso*propyl-pyrophosphortetramide (*iso*OMPA), but this is not the case for *Torpedo* BuChE. In each organism, the characteristics of inhibition of AChE and BuChE should therefore be established, and never used *a priori* as a definition of these two enzymes. The two enzymes also differ in their thermal inactivation rate (VIGNY et al. 1978; EDWARDS and BRIMIJOIN 1983) and in their immunoreactivity.

It has been reported that human brain and meningiomas contain an enzyme which has a low activity on butyrylthiocholine and is sensitive to BW 284C51 but also to *iso*OMPA under conditions that normally do not affect AChE (RAZON et al. 1984). This might represent a variant of AChE or an intermediate between AChE and BuChE. The existence of intermediate enzymes in the brain of the domestic fowl was also suggested in a study which combined analyses of reactivity towards organophosphorus inhibitors and hydrolysis of acetyl- and butyrylthiocholine (CHEMNITIUS et al. 1983). Differences in the kinetics (substrate inhibition) and stability of AChE from embryonic and adult rat cortex have been described (INESTROSA and RUIZ 1985) and interpreted as suggesting the sequential expression of two different enzymes during development.

Although such results suggest the existence of catalytic differences between ChE molecules, most available data indicate that there is no ambiguity in the definition of AChE and BuChE. In the rest of this review, we adopt the perspective that the two ChEs represent distinct enzymes, each of which consists of several catalytically equivalent molecular variants.

C. Molecular Forms of ChEs: The Value and Limitations of a General Scheme and a Nomenclature for Their Quaternary Structure

Within a given organism AChE and BuChE each present a multiplicity of molecules which possess the same catalytic activity (substrate and inhibitor specificity, kinetic parameters) but differ in a variety of molecular properties.

In order to explore the complexity of the molecular variants of these enzymes, it is necessary to apply well-defined operational criteria in a systematic, ordered manner. We define molecular forms first corresponding to general types of quaternary structures by their hydrodynamic properties.

The asymmetric, or collagen-tailed, molecules present characteristic properties (large Stokes radius, low salt aggregation, sensitivity to collagenase) and consist of one, two or three tetramers of catalytic subunits (in *Torpedo* a combination of catalytic and non-catalytic subunits) attached to the strands of the triple helical tail (A_4, A_8, A_{12}). The globular forms consist of monomers (G_1), dimers (G_2) and tetramers (G_4) of the catalytic subunits (BON et al. 1979).

It is useful to adopt the same nomenclature for the molecular forms of AChE and BuChE in all vertebrate species because there is a clear correspondence between the sets of molecules in spite of considerable variations in their individual characteristics, e. g. in the size of the catalytic subunit and in the sedimentation coefficients (Table 1). For example, the sedimentation coefficient of the A_{12} form of AChE varies from 16–17 *S* in mammals, 17.4 *S* in *Torpedo*, 18.4 *S* in *Electrophorus*, 19.2 *S* in domestic fowl (MASSOULIÉ and BON 1982) and 20 *S* in quail (ROTUNDO 1984a) to 20.8 *S* in pigeon (CHATONNET and BACOU 1983). The *S* values of homologous forms of AChE and BuChE are clearly distinct (*Torpedo*, TOUTANT et al. 1985; fowl, SILMAN et al. 1979; ALLEMAND et al. 1981; rat, VIGNY et al. 1978). It should be emphasized that our terminology only expresses the overall size of the molecules and the presence or absence of a collagen-like tail but does not claim to describe the detailed structure of the molecules. It proposes a convenient, operational means of identifying similar forms in a variety of situations, in spite of their interspecific and intraspecific variability.

In addition to the fact that the structure of some polymeric forms is more complex than suggested by the scheme of Fig. 1 (e. g. *Torpedo* A forms and ox caudate nucleus G_4 form, see below), there are two aspects of the polymorphism of ChEs that are not taken into account by the simple nomenclature that we propose.

Table 1. Sedimentation coefficients and Stokes radii of cholinesterase molecular forms. The sedimentation coefficients of globular forms were usually determined in the presence of Triton X-100 in the sucrose gradients; asterisks indicate values obtained in the absence of detergent. Detergents did not influence the sedimentation of asymmetric forms. Data in parentheses refer to lytic forms. Globular lytic forms of *Electrophorus* and G_4 lytic form of *Torpedo* AChE were proteolytic products of native A forms. G_2 and G_1 lytic forms of *Torpedo* AChE were obtained after digestion of the native G_2 form by pronase or proteinase K. In domestic fowl the lytic forms of AChE and BuChE were obtained by trypsin treatment of the corresponding native forms. From PEZZEMENTI et al. (1987) for *Petromyzon*, BON and MASSOULIÉ (1980) for *Torpedo*, MASSOULIÉ et al. (1980) for *Electrophorus*, BON et al. (1979) and ALLEMAND et al. (1981) for fowl and mammalian ChEs

		AChE					BuChE
		Petromyzon	*Torpedo*	*Electrophorus*	Domestic fowl	Mammals	Domestic fowl
G_1	(a)	3.7	(5)	(5.3)	5.2*	3.5 to 4 4.8*	4.9* (4.7*)
	(b)	–	(4.6)	(3.7)	5.6	3.7	5.6 5.4
G_2	(a)	6.1	6.3 (7.1)	(7.7)	7.9* (7.2*)	6 7.8*	7.1*
	(b)	–	8.1 (5.9)	(5.9)	7.7 (5.8)	6.1	7.1
G_4	(a)	9.6	(10.9)	(11.8)	11.8* (10.9*)	10.5*	11.1 (11*)
	(b)	–	(8.1)	(8.7)	10.2 (7.9)	8.1	9.7 9.2
A_4	(a)	8.6	9.1	9.1	(absent)	8.7	(absent)
	(b)	–	13.1	12.4		13	
A_8	(a)	12.6	13.4	14.2	14.8	13	–
	(b)	–	16.2	14.4	15.5	14.4	–
A_{12}	(a)	16.0	17.4	18.4	19.2 to 20	16.7 to 17.1	18.6
	(b)	–	17.2	15.6	17.5 to 18.1	15.5 to 16.2	17.3

(a) Sedimentation coefficient in Svedberg units
(b) Stokes radius in nm

Firstly, each molecular form is not homogeneous but may consist, in fact, of variants that can be distinguished by various analytical procedures. It is, for example, important to distinguish amphiphilic and non-amphiphilic globular molecules, according to their interactions with non-denaturing detergents. The presence of different variants may complicate the sedimentation profiles, e. g. amphiphilic and non-amphiphilic G_4 forms in soluble extracts from ox brain (GRASSI et al. 1982) or fowl embryonic muscle (TOUTANT 1986). In addition, the binding of antibodies or lectins also leads to further subdivisions within each molecular form.

Secondly, polymeric assemblies of catalytic, and in some cases non-catalytic subunits, that do not correspond to the scheme of Fig. 1, although marginal, have been shown to exist in some cases: for example, soluble extracts from *Torpedo* electric organs contain amphiphilic AChE dimers loosely associated with non-catalytic proteins (BON and MASSOULIÉ 1980). Another exception is that of a minor 15-*S* AChE form that exists in primary cultures of murine neurons (M. LAZAR, M. VIGNY and Y. NETTER, unpublished results) and also transiently in extracts from early embryonic human and rat brains (MULLER et al. 1985). This form was insensitive to collagenase and was not readily dissociated into smaller molecules by reduction (unpublished data). The low proportion of this form prevented a detailed analysis. It may be analogous to the well-defined, stable aggregates that were obtained by LANDAUER et al. (1984) when the detergent was removed from Triton X-100-solubilized G_4 AChE from human brain. It is not certain whether such unusual polymers occur naturally, or whether they only arise from the rearrangement of solubilized molecules.

The existence of multiple variants of similar quaternary structure, within each molecular form, and the occurrence of additional polymers or aggregates may explain reports describing a multiplicity of sedimentation components of undefined structure (DREYFUS et al. 1984). In any case, it is important to note that asymmetric forms cannot be identified solely on the basis of their sedimentation coefficient because of the existence of fast-sedimenting associations of globular molecules.

In the following sections we describe the structure and interaction properties of the asymmetric and globular forms. Asymmetric forms have been studied in detail in the case of AChE from *Electrophorus* and *Torpedo* electric organs, and globular forms in the case of AChE from *Torpedo* electric organs and human erythrocytes, as well as BuChE from human serum.

D. Electric Organs of Fish: A Privileged Source of AChE for Biochemical Studies

Fish electric organs (Chap. 2) have been a favourite model for biochemists of cholinergic transmission ever since NACHMANSOHN first reported the exceptionally high cholinesterase activity of *Torpedo* electric tissue in 1938 (MARNAY and NACHMANSOHN 1938).

A large part of our knowledge of the molecular basis of cholinergic transmission, and particularly of the biochemical structure of cholinesterase molecular forms, has been obtained by analysing the electric organs of the fresh-water electric eel, the gymnotid *Electrophorus electricus* (osteichthyes, teleosts) and various species of marine electric rays of the torpedinid family (chondrichthyes, elasmobranchs). Various species of electric rays are used by different laboratories, depending on their local accessibility. The most widely used are *Torpedo marmorata* (Atlantic coast of Europe and western part of the Mediterranean), *T. torpedo*, also known as *T. ocellata* (eastern Mediterranean), *T. californica* (California), *Narke japonica* (Japan), *Discopyge tschudii* (Pacific coast of Peru and Chile, Atlantic coast of Brazil and Argentina).

The electric organs of these fishes appear to be very similar morphologically, physiologically and biochemically (ABE et al. 1983; MÉNDEZ et al. 1984; SAKAI et al. 1985a, b). They are phylogenetically derived from muscles which have lost their contractile apparatus. They constitute an extremely useful material for the study of ACh receptors and AChE. In addition, the anatomy of the electromotor system of torpedinid electric rays also makes it possible to analyse the ChE contents of the motoneurons, whose cell bodies are located in a specific nucleus of the myelencephalon, the electric lobes and their axons, which form the voluminous electric nerves. Moreover, *T. marmorata* is a viviparous animal from which it is possible to obtain embryos of different stages and thus analyse the development of the electromotor system both morphologically (MELLINGER et al. 1978; FOX and RICHARDSON 1978, 1979) and biochemically (KRENZ et al. 1980; BON 1982; WITZEMANN and BOUSTEAD 1982) *in vivo* as well as in cultures *in vitro* (RICHARDSON and WITZEMANN 1986).

E. Quaternary Structure and Interactions of Asymmetric Forms

I. *Electrophorus*

Electrophorus electric organs constitute a rather simple system since they mostly contain asymmetric AChE forms, A_{12}, A_8 and A_4. Proteolytic digestion or sonication may convert these molecules into lytic globular forms, mostly tretramers (G_4) but also dimers (G_2) and monomers (G_1).

The asymmetric, or collagen-tailed forms, consist of one, two or three tetramers (A_4, A_8, A_{12}), associated with the three strands of the triple helical 'tail'. Figure 1 gives a schematic representation of the quaternary structure of the molecular forms, derived from analyses of the enzyme of *Electrophorus* electric organs.

Because of their particular structures, the asymmetric forms possess very characteristic properties, i. e. a large Stokes radius ($Re > 9$ nm) and a specific sensitivity to collagenase digestion. Bacterial collagenase cleaves the tail of these molecules in a temperature-dependent fashion (BON and MASSOULIÉ 1978). Intermediate products in which the distal part of the tail is removed present a lower Stokes radius, but their sedimentation coefficients are increased by $2-3$ *S*. Extensive digestion by collagenase or by trypsin produces isolated tetramers. In addi-

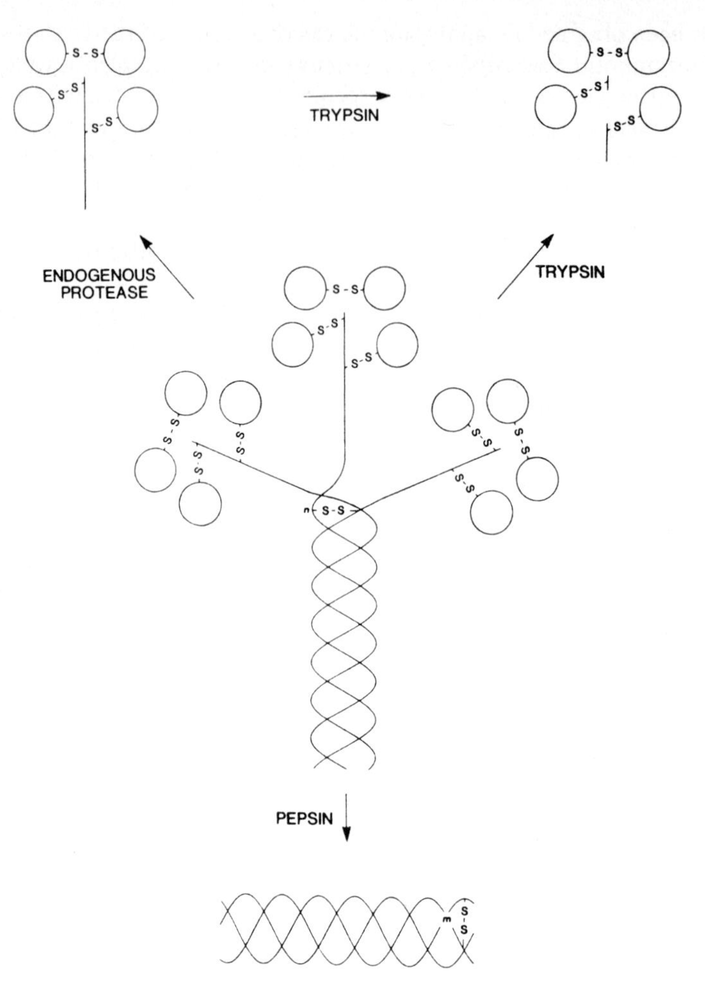

Fig. 2. Quaternary structure of *Electrophorus* A_{12} AChE and of its degradation products. Each catalytic subunit is linked by one interchain disulphide bridge, either to another subunit to form a peripheral dimer or to one of the three tail peptides. Reduction of the native enzyme cleaves the interchain disulphides but not the intrachain disulphide bonds. In non-reducing conditions the external dimers are separated by SDS from the central core, containing the tail peptides (ROSENBERRY and RICHARDSON 1977; ANGLISTER and SILMAN 1978). Proteases such as trypsin or an endogenous protease release active tetramers by cleaving the tail peptides. Pepsin degrades the catalytic subunit and the non-collagen-like domain of the tail (after ROSENBERRY et al. 1980). Bacterial collagenase cleaves the helical part of the tail; it produces intermediate molecules that sediment faster than the native enzyme and eventually releases tetramers

tion, the asymmetric forms present characteristic ionic interactions, as discussed below, and generally aggregate in low-salt conditions.

The arrangement of catalytic and tail subunits in *Electrophorus* A_{12} AChE is schematically represented in Fig. 2 (ROSENBERRY et al. 1980; ROSENBERRY 1985). It is remarkable that this structure closely resembles electron-microscopic images of stored A_{12} AChE (CARTAUD et al. 1975).

The 12 catalytic subunits appear to be identical. Half form disulphide linked dimers, and the others are linked by pairs to the non-collagenous ends of the tail peptides. The tail peptides are themselves linked by several disulphide bonds. Thus, denaturation of the molecule without reduction separates the external dimers from a core structure containing the tail and the internal dimers. It is noteworthy that reduction of the disulphide bonds in the native enzyme does not dissociate this structure (BON and MASSOULIÉ 1976).

Reduction in the absence of denaturation was found to be very selective for interchain disulphide bonds. By labelling the sulphydryl groups with [^{14}C]*N*-ethylmaleimide, ROSENBERRY showed that there are seven cysteine residues in each subunit, but that only one may be titrated after reduction of the native enzyme (ROSENBERRY 1975a, 1985; T.L. ROSENBERRY and P. BARNETT, unpublished results). If all subunits are assumed to be identical, the same cysteine residue is responsible for the linkage of catalytic subunits as dimers and in their association to the tail peptides.

The tail consists of a pepsin-sensitive, non-collagenic domain and a pepsin-resistant domain which presents the spectroscopic properties of triple-helical collagen (ROSENBERRY et al. 1980). The collagenic domain of the tail peptides contains hydroxyproline, hydroxylysine and a high content of glycine. Its mass has been evaluated as 24 kDa (MAYS and ROSENBERRY 1981); 8-kDa peptides, originating from the non-collagenic domain, remain attached to isolated tetramers after cleavage by an endogenous protease (BARNETT and ROSENBERRY 1978). The mass of the tail subunit is evaluated as about 32 kDa, but because of the migration properties of collagen in SDS-PAGE its apparent mass according to non-collagenous standards is 41 kDa (ROSENBERRY et al. 1980, 1982; ROSENBERRY 1985).

Most of the hydroxylysine residues of the tail peptides are linked to 2-O-α-D-glucopyranosyl-O-β-D-galactopyranose (ROSENBERRY et al. 1980) which is characteristic of basement membrane collagens. No cross-reactivity was observed between asymmetric AChE forms from *Electrophorus* and from ox superior cervical ganglia, with various antisera directed against different collagens including the basement membrane collagens types IV and V (GRASSI et al. 1983). In addition, the A_{12} enzymes were not found to interact with laminin or fibronectin (GRASSI et al. 1983). Thus the collagen-like tail appears to constitute a novel, hitherto unclassified type of collagen. Monoclonal antibodies to the tail are now available (LAPPIN and RUBIN 1984, 1987; SAKAI et al. 1986), and it will be possible to examine whether the asymmetric forms of AChE and BuChE, where they coexist, possess identical tails, whether this collagen occurs exclusively in the context of ChEs and whether it is also associated with other proteins or exists in a free form.

II. *Torpedo*

The structure of the *Torpedo* asymmetric AChE forms is somewhat more complex than that of *Electrophorus*. TAYLOR and his colleagues have found that the enzyme of *T. californica* contain not only tail peptides (with an apparent mass of 55 kDa each) and 68-kDa catalytic subunits but also non-catalytic 100-kDa sub-

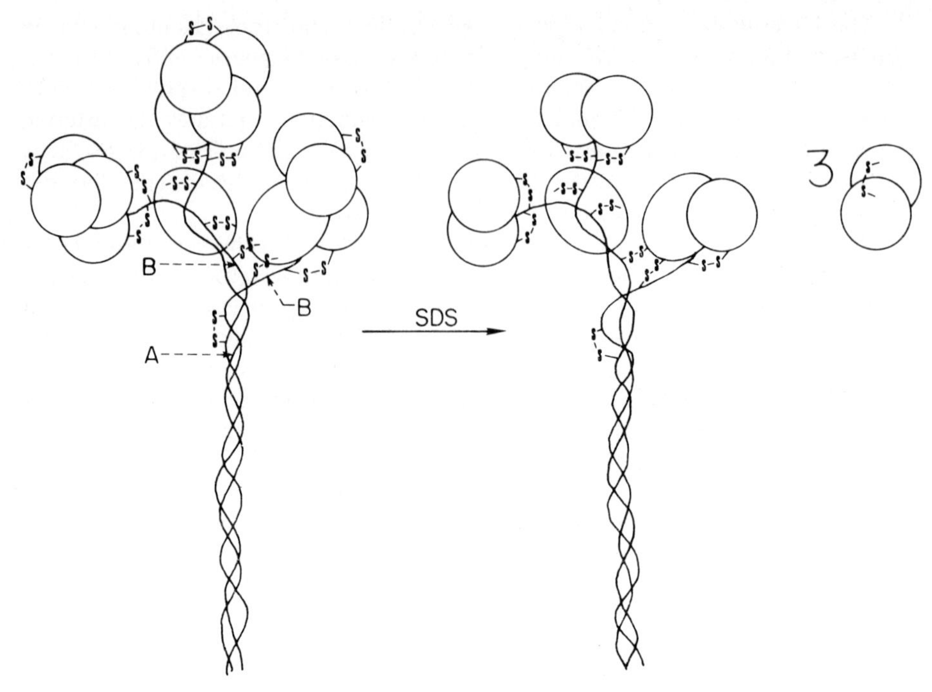

Fig. 3. Quaternary structure of *Torpedo* A_{12} form. This structure contains a variable proportion of non-catalytic globular subunits (100 kDa) that are disulphide-linked to the tail peptides. Denaturation by SDS under non-reducing conditions separates dimers and core structures which differ in their mass according to the proportion of 100-kDa subunits present. (After LEE and TAYLOR 1982)

units (LEE and TAYLOR 1982; LEE et al. 1982a; LAPPIN et al. 1987). The presence of an additional 100-kDa subunit has also been demonstrated in the case of *T. marmorata* (VERDENHALVEN and HUCHO 1985) and in very low proportion in *N. japonica* (SAKAI et al. 1985b).

The 100-kDa structural subunits are linked by disulphide bonds to the tail subunits in addition to, or in the place of, a variable fraction of internal catalytic subunits (Fig. 3). As a consequence, when the enzyme is analysed by SDS-PAGE without reduction, its core structure forms multiple bands, differing in their proportions of 100-kDa and 68-kDa peptides (LEE et al. 1982a). It should be noted that this molecular heterogeneity does not result in a marked polydispersity of the sedimentation coefficients (the 100-kDa subunits replace about 15%–30% of the tail-linked catalytic subunits; see LEE et al. 1982a).

The 100-kDa subunit is not labelled with DFP and therefore does not contain an active site. It does not react with antibodies raised against the catalytic subunits, and it presents a distinct peptide map pattern (LEE et al. 1982a). It therefore does not correspond to a precursor of the catalytic subunit but possibly to a procollagen or non-collagenous component of the basal lamina, which would be progressively replaced by catalytic subunits during the biosynthetic assembly of the asymmetric forms (LEE et al. 1982a).

III. Asymmetric AChE Forms from Other Species, BuChE Asymmetric Forms, Hybrid AChE-BuChE Asymmetric Forms

Asymmetric forms have been characterized by their hydrodynamic properties, low-salt aggregation and collagenase sensitivity in all classes of vertebrates, including mammals and birds (BON et al. 1979), even the very primitive agnatha, as shown in the case of the lamprey (*Petromyzon marinus*) by PEZZEMENTI et al. (1987).

The presence of a non-collagenous, non-catalytic subunit has so far been observed only in *Torpedo.*

BuChE asymmetric forms are in general much less abundant than their AChE counterparts. They have been unambiguously characterized however in *Torpedo* heart and electric lobes (TOUTANT et al. 1985), in fowl muscle (SILMAN et al. 1979; ALLEMAND et al. 1981) and nerve (DI GIAMBERARDINO and COURAUD 1978), and in rat superior cervical ganglion (SCG) (VIGNY et al. 1978) and sciatic nerve (DI GIAMBERARDINO and COURAUD 1978).

Since the complex structure of the asymmetric forms is readily dissociated by proteases which convert them into globular forms (usually G_4), their proportion in a tissue extract may be underestimated as a result of endogenous proteolysis. For this reason it is advisable to solubilize these molecules in the presence of anti-proteolytic agents (SILMAN et al. 1978; LAI et al. 1986).

TSIM et al. (1988) recently reported the existence of hybrid AChE-BuChE A_{12} forms in pectoralis muscle of 1-day-old fowl. They found that this asymmetric ChE could be immunoprecipitated by two monoclonal antibodies, ACB-2 and 7D-11, which recognize, respectively, the AChE and BuChE catalytic subunits (RANDALL et al. 1987). The purified enzyme contained 110-kDa, 72-kDa and 58-kDa polypeptide chains. The 58-kDa one was destroyed by collagenase and therefore represented the tail subunit. The 110-kDa and 72-kDa components could be labelled by DFP, and this was blocked respectively by BW284C51 and by *iso*OMPA, indicating that they represent catalytic subunits of AChE and BuChE. The two kinds of catalytic subunits were present in equivalent numbers, as judged by the amount of protein in the two bands and by the ratio of AChE and BuChE activity in the preparation, the turnover number of BuChE active sites being 1/20 that of AChE active sites.

In the case of this hybrid AChE-BuChE A_{12} form, the index 12 indicates the total number of catalytic subunits of both types. The detailed structure of these hybrids raise a number of questions concerning the arrangement of AChE and BuChE subunits. It is unlikely that they could be associated randomly: their quaternary association surfaces are certainly not equivalent since no hybrid globular forms have been observed in rat (VIGNY et al. 1978) or in *Torpedo* (TOUTANT et al. 1985). It is possible that the catalytic subunits form homogeneous tetramers attached to different strands of the tail. However, the fact that the two types were found in equal numbers suggests that they might be organized in a regular manner with, for example, BuChE subunits forming internal dimers attached to the tail peptides and AChE subunits forming external dimers (TSIM et al. 1988). In any case, the fact that the AChE and BuChE profiles are shifted by a small but clear difference of about 1 *S* in the A_{12} region of gradients, when both activities were assayed in the same gradient fractions (ALLEMAND et al. 1981), implies that different types of hybrid molecules may exist, or that hybrid molecules coexist with AChE A_{12}. In the adult the hybrid form is replaced by the classical AChE A_{12} form (E. BARNARD, personal communication).

IV. Ionic Interactions

In general the asymmetric forms of ChEs are not soluble in low ionic strength buffers, but may be solubilized in the presence of salt, usually 1 *M* NaCl. In the case of *Torpedo*, Mg^{2+} was found to be more efficient than Na^{+} at the same ionic strength (BON 1982). Recently SKETELJ and BRZIN (1985) reported than 1 *M* LiCl solubilized the enzyme from rat muscles with a better efficiency that 1 *M* $MgCl_2$ or 1 *M* NaCl.

The asymmetric forms precipitate or produce polydisperse aggregates in low-salt conditions (BON et al. 1979). This aggregation may be prevented by acylating the enzyme with maleic anhydride (DUDAI and SILMAN 1973) or acetic anhydride (CARTAUD et al. 1978; BON et al. 1979), suggesting that it reflects ionic interactions involving positively charged NH_2 groups. The aggregation of *Electrophorus* A_{12} AChE was lost after an extensive purification of the enzyme but could be restored by adding a variety of natural or synthetic polyanions or a non-proteic fraction of electric organ which contained acidic glycosaminoglycans (BON et al. 1978). When observed in the electron microscope, the aggregates appeared as bundles of molecules in which the tails were packed together, with the globular parts at both extremities (CARTAUD et al. 1978). The aggregation of the asymmetric enzyme was abolished after collagenase digestion, even if only a fraction of the tail had been removed (BON and MASSOULIÉ 1978). Thus the aggregation appears to involve only the collagenous domain of the tail. These interaction properties are probably responsible for the *in situ* anchoring of the collagen-tailed molecules, e. g. in the basal lamina of neuromuscular junctions.

INESTROSA and his colleagues have recently shown that heparin can solubilize the A_{12} and A_8 forms but not the globular forms of AChE from rat neuromuscular junctions (TORRES and INESTROSA 1983) and from sciatic nerves (TORRES et al. 1983). They also found that the asymmetric forms from rat muscle and from the electric organ of *Discopyge* were quantitatively and selectively retained on a heparin-Agarose column, and that the binding was abolished after collagenase digestion (BRANDAN and INESTROSA 1984). The bound enzyme could be released by NaCl (0.55 *M*) or by heparin at low ionic strength. Chondroitin sulphate and hyaluronic acid appear unable to solubilize the asymmetric forms from tissues or to displace them from the heparin column. Heparin was found to solubilize both class I and class II (see below) asymmetric forms from fowl retina and skeletal muscle at low ionic strength as well-defined 24-*S* aggregates but only class I forms in the presence of salt (BARAT et al. 1986). The polyanion that binds the enzyme *in situ* appears to be an heparan sulphate proteoglycan (BRANDAN et al. 1985; BRANDAN and INESTROSA 1986), in agreement with previous findings of VIGNY et al. (1983), or possibly dermatan sulphate proteoglycan (BRANDAN and INESTROSA 1987).

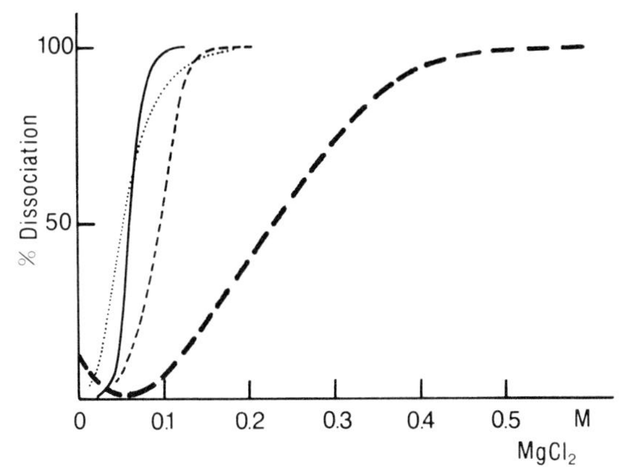

Fig. 4. Solubilization of asymmetric forms as a function of salt concentration. AChE asymmetric forms from muscle and nervous tissues of various species, as well as *Electrophorus* electric organs, are generally solubilized below 0.2 *M* $MgCl_2$ (or 1 *M* NaCl). In the case of *Torpedo* electric organs the concentration required to solubilize the enzyme is higher. A significant fraction of asymmetric AChE is solubilized in the absence of divalent ions but not in the presence of 50 m*M* $MgCl_2$ (After MASSOULIÉ et al. 1980; BON and MASSOULIÉ 1980). *Unbroken line*, *Electrophorus* electrical organ; *heavy broken line*, *Torpedo* electric organ; *light broken line*, fowl muscle; *dotted line*, ox superior cervical ganglion

The solubilization properties of the asymmetric forms are largely variable in different tissues. This is illustrated in *Torpedo* by the electric organs and the electric lobes. The asymmetric forms appear to be identical in both tissues, but in contrast with electric organs (BON and MASSOULIÉ 1980) they are largely solubilized from electric lobes at low ionic strength in the form of small aggregates (BON 1982). This is consistent with the hypothesis that the A forms are anchored in the basal lamina in electric organs, whereas in the electric lobes they are essentially intracellular.

In *Torpedo* electric organs a small proportion of asymmetric forms is solubilized in a low-salt buffer together with globular forms (see below). Although high concentrations of Mg^{2+} efficiently solubilize the A forms (Fig. 4; BON and MASSOULIÉ 1980), a low concentration of Mg^{2+} (50 m*M*) prevents their low ionic strength solubilization, making it possible to obtain only globular forms in solution. It is important to take this effect into account to obtain a satisfactory separation of the two classes of molecules by sequential solubilization.

RAMÍREZ and his colleagues have reported that fowl retina contains a significant proportion of A_{12} AChE but this enzyme can be solubilized only in the presence of a chelating agent, EDTA (GÓMEZ-BARRIOCANAL et al. 1981). Studying the solubilization of AChE asymmetric forms in other tissues, such as rat muscle and *Torpedo* electric organ (RAMÍREZ et al. 1984), these authors found that a fraction of the A forms could be solubilized in high salt (1 *M* NaCl) without EDTA, and that a second pool could then be solubilized in the presence of EDTA. They defined these two pools as class I and II forms (RAMÍREZ et al. 1984; BARAT et al. 1984). They determined the content of class I and II forms in a number of tissues (retina, brain, skeletal muscle and smooth muscle) of various

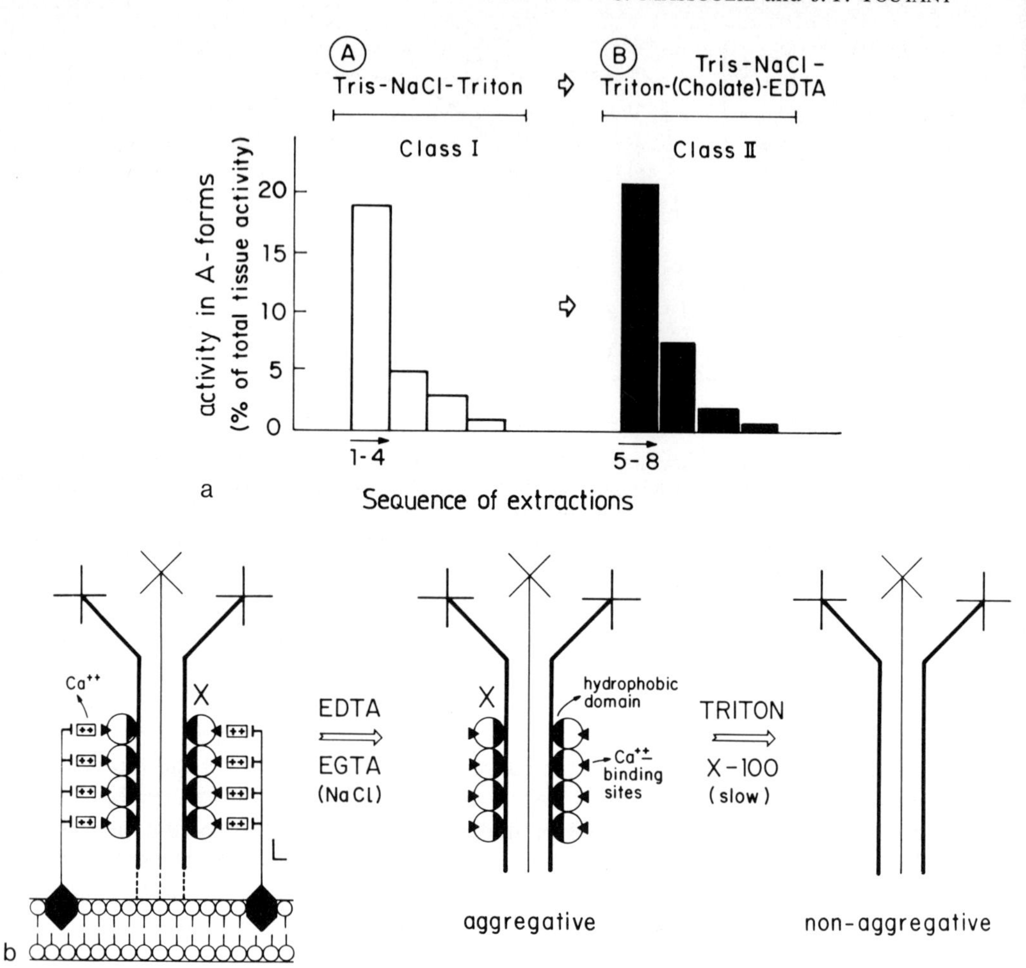

Fig. 5a, b. Existence of two fractions of asymmetric forms that may be solubilized in the absence or in the presence of chelating agents. **a** Solubilization of asymmetric AChE from fowl skeletal muscle; activity recovered in 100000-g supernatants from successive extracts in 10 m*M* Tris-HCl pH 7, 1 *M* NaCl, 0.5% Triton X-100 followed by extracts in the same buffer (*A*) plus 3 m*M* EDTA (*B*). The enzyme solubilized in the first and second extraction media are called, respectively, class I and class II A forms. (After BARAT et al. 1984). **b** Proposed scheme of attachment of class II AChE A forms to membranes. In this model the tail interacts with unidentified *X* molecules, which allow its attachment to an anionic component (*L*) of the membrane or of the extracellular matrix through Ca^{2+}. The enzyme is solubilized with the attached *X* molecules by chelating agents. These complexes aggregate in the presence of Ca^{2+}. The *X* components may be removed by detergents (Triton X-100) generating a non-aggregative AChE A form. (After RAMÍREZ et al. 1984)

species (trout, frog, fowl, rat and rabbit) and found their proportions to be very variable.

Ca^{2+} and less efficiently Mg^{2+} caused a reversible aggregation of the class II, EDTA-soluble A forms and this effect was abolished in the presence of Triton X-100. RAMÍREZ and collagues propose a model in which the tail of the enzyme is linked to cellular membranes via two hypothetical components, X and L, and divalent ions (Fig. 5). In their view, the tail would interact with X through hydrophobic interactions. Thus the enzyme might be solubilized as complexes with X in the absence of detergent, and X would be removed by the detergent. Since the sedimentation of the solubilized enzyme appears identical in the presence or in the absence of detergent, it seems that the two classes of forms, as defined by their solubilization requirements, correspond to different interactions with their environment *in situ* rather than to intrinsic differences in the structure of the enzyme molecules.

Non-denaturing detergents have often been noted to facilitate the solubilization of asymmetric forms (DUDAI and SILMAN 1974). This effect has been taken as indicating that they are engaged in hydrophobic interactions (DREYFUS et al. 1983; NICOLET et al. 1984). These forms however have never been found to interact directly with detergents. It is likely that the effect of detergents is due to the disruption of subcellular structures, e. g. closed membrane vesicles, in which the intracellular enzyme is entrapped.

Several authors have demonstrated that asymmetric AChE forms bind to certain types of phospholipid liposomes. Although OCHOA (1978) observed a binding of the enzyme from *Discopyge* to lecithin liposomes, other studies showed that this effect is highly specific for sphingomyelin (WATKINS et al. 1977; KAUFMANN and SILMAN 1980). The binding required the integrity of the collagen tail and was resistant to high salt (1 *M* NaCl). COHEN and BARENHOLZ (1984) recently confirmed these results and studied the interactions in detail. They showed that the binding does not depend on the physical state of the sphingomyelin bilayer or on the nature of its acyl chain but decreases if a proportion of phosphatidylcholine (PC) is introduced. It does not seem to involve hydrophobic or electrostatic interactions but possibly hydrogen bonds between the OH groups of hydroxylysine and hydroxyproline and the C2 amide group or C3 hydroxyl group of sphingosine. It is not specific for collagen-tailed AChE but is also observed with other collagens.

V. Asymmetric AChE Forms and the Basal Lamina

The now classical experiments of MCMAHAN and his colleagues (MARSHALL et al. 1977; MCMAHAN et al. 1978) have shown that AChE subsists in the basal lamina of frog pectoral muscle after complete degeneration of both nerve and muscle elements. The asymmetric forms are obvious candidates for this localization because of their ionic interactions and because they may be removed by collagenase from the end-plate regions of muscles (HALL and KELLY 1971; HALL 1973). Attempts to analyse directly the molecular forms that remained in the basal lamina preparation in the degeneration experiments have, however, unexpectedly shown the presence of both asymmetric and globular forms (NICOLET et al. 1986) or even only globular forms (ANGLISTER and MCMAHAN 1983). Because the type of interactions that might anchor globular molecules in this structure is not clear, and because these experiments apply to a very unphysiological situation, it is not possible to conclude that globular AChE forms are normally associated with basal lamina. Only the use of specific probes, for example monoclonal antibodies directed against distinct epitopes of each type of molecular form (e. g. against the tail), will make it possible to directly identify molecular forms in intact basal lamina *in situ*.

Studying the molecular forms of AChE in the rat diaphragm, YOUNKIN et al. (1982) found that a large fraction of extracellular asymmetric forms at the endplates cannot be extracted by conventional methods but only by collagenase digestion. It is possible that this enzyme is covalently cross-linked to basal lamina components.

F. Physicochemical Properties of Globular Forms

I. Solubility and Hydrophobicity: Operational Definition of Amphiphilic Molecules

The globular forms of ChEs may be classified according to their solubility in the presence or in the absence of detergents and according to their interactions with non-denaturing detergents. It must be strongly emphasized that the two properties do not coincide, because (a) molecules interacting with detergent are frequently found to be soluble in the absence of detergent (BON and MASSOULIÉ 1980; GRASSI et al. 1982; TOUTANT and MASSOULIÉ 1987) and (b) reciprocally, nonhydrophobic molecules may be entrapped in vesicles, and released by detergents, as has already been suggested in the case of the asymmetric forms (BON 1982).

It is therefore essential to distinguish clearly between the solubility and the interaction properties of the molecules. Amphiphilic molecules must be defined operationally by their interactions with non-denaturing detergents, as evinced by physicochemical properties. The binding of detergent micelles modifies the sedimentation coefficient, the Stokes radius and, depending on the ionic nature of the detergent, the electrophoretic mobility of the enzyme (MASSOULIÉ et al. 1984; STIEGER and BRODBECK 1985; TOUTANT 1986).

II. Solubility of ChE Globular Forms: Soluble and Detergent-Soluble Fractions

In general, quantitative solubilization of ChEs from a tissue cannot be achieved in an aqueous buffer in the absence of detergent, but, as already mentioned, a significant fraction of the activity is usually solubilized under these conditions.

In the case of *Torpedo* we have found it useful to define a multistep solubilization procedure (BON and MASSOULIÉ 1980). The tissue was first extracted without detergent in low-salt, yielding a low-salt soluble fraction (LSS). We found it necessary to include 50 m*M* $MgCl_2$ at this stage in order to prevent the solubilization of a small proportion of asymmetric forms. This extraction could be followed by a low-salt detergent extraction in the presence of Triton X-100 (detergent-soluble fraction, DS) and then by a high-salt extraction, which allowed the solubilization of asymmetric forms in a high-salt soluble fraction (HSS).

In these conditions the three fractions contain approximately 10%–20% (LSS), 50%–60% (DS) and 30%–40% (HSS) of the total activity. Repeated extractions in the detergent-free buffer did not significantly increase the LSS activity, so that these fractions really correspond to distinct pools of enzyme (S. BON, unpublished results).

Both LSS and DS fractions contain amphiphilic G_2 dimers. In the LSS fraction these dimers are solubilized in association with hydrophobic proteins of variable size, forming 8- to 9-*S* complexes with a Stokes radius of 5.5 – 9.0 nm (BON and MASSOULIÉ 1980). These associations appear to be rather loose: they are dissociated by dilution so that the enzyme then sediments at 7 *S* (S. BON, unpublished results).

We applied such fractionation procedures to the ox caudate nucleus (GRASSI et al. 1982). This allowed us to extract most of the globular forms in the first steps (LSS and DS), making it possible to demonstrate the presence of trace amounts of asymmetric A_{12} and A_8 forms in a subsequent HSS fraction. The soluble fraction (LSS) represented about 20% of the total activity and contained mostly amphiphilic G_1, together with amphiphilic G_2 and both amphiphilic and hydrophilic G_4. In a LSS extract of rabbit semi-membranosus muscle we also demonstrated the presence of an amphiphilic G_1 AChE form, together with a small proportion of detergent-insensitive G_4 form (see TOUTANT and MASSOULIÉ 1987).

A similar proportion of soluble activity and a similar composition of G_1 and G_4 molecular forms was observed in a number of mammalian brains: rat (MULLER et al. 1985), human (SØRENSEN et al. 1982; MULLER et al. 1985; GENNARI and BRODBECK 1985; SØRENSEN et al. 1985; ATACK et al. 1986), dog, horse, swine (SØRENSEN et al. 1985) and rabbit (RACKONCZAY and BRIMIJOIN 1985). The detergent interactions of the soluble G_1 form of ox were demonstrated by the change of sedimentation coefficient (from 5 to 4 *S*) and Stokes radius (from about 4.0 nm to 6.5 nm) that occurred when Triton X-100 was added (GRASSI et al. 1982). In this case the molecule appears to be solubilized in an isolated form even without detergent.

When discussing the results of a stepwise solubilization of ChE, it is obvious that the significance of the different fractions depends on the order in which they are obtained. If, for example, *Torpedo* electric organ is first extracted in high salt (STIEGER and BRODBECK 1985), the resulting extract will contain both the LSS and HSS enzyme. In addition, the composition of solubility fractions will depend on the precise operational conditions, composition of buffer, buffer/tissue ratio, homogenization procedure, temperature, etc. Solubility alone cannot therefore be used to characterize molecular species.

III. Non-amphiphilic or Hydrophilic Globular Forms

Non-amphiphilic or hydrophilic forms are operationally defined in a negative manner, in opposition to the detergent-interacting amphiphilic molecules. Hydrophilic globular forms are readily obtained as lytic products by dissociation of asymmetric forms or by digestion of amphiphilic forms by proteases or phosphatidylinositol-specific phospholipase C (PIPLC) (see Sect. F. IV. 1).

Native hydrophilic globular forms occur in the serum of most vertebrates. The best-studied example is that of human serum BuChE (LOCKRIDGE et al. 1979;

MASSON 1979). The predominant G_4 form of this enzyme has been shown to consist of two dimers of identical subunits linked in pairs by a single disulphide bond. Reduction of these interchain disulphide bonds is easily achieved in the native enzyme, in contrast to intrachain disulphide bonds, and does not disrupt the quaternary structure of the molecule (LOCKRIDGE et al. 1979). This arrangement of disulphide bonds appears, therefore, very similar to that observed in the external dimeric units of *Electrophorus* asymmetric forms.

A transient high level of soluble AChE, possibly originating from differentiating nervous tissue, has been observed in the plasma of mammalian foetuses (B. P. DOCTOR, personal communication). Foetal calf serum is thus a convenient source for the purification of soluble G_4 AChE (about 1 $\mu g\,ml^{-1}$)(RALSTON et al. 1985). Soluble AChE persists at the adult stage in the serum of some species such as the rabbit, which contains a G_4 form (RAKONCZAY and BRIMIJOIN 1985). Human serum contains two forms of AChE, probably G_2 and G_4 (SØRENSEN et al. 1986).

Hydrophilic forms of AChE have also been observed in the cerebrospinal fluid of rabbits (HODGSON and CHUBB 1983). It occurs in human cerebrospinal fluid and in amniotic fluid, which also contains G_4 BuChE (F. MULLER, unpublished results). Hydrophilic G_4 AChE has also been characterized together with its amphiphilic analogue in the secretion products of T28 cells (LAZAR et al. 1984).

G_1 forms are usually amphiphilic, but a G_1 non-amphiphilic form has been described in soluble extracts of rabbit enterocytes (which also possess a membrane-bound G_2 form) (SINE and COLAS 1987).

Hydrophilic forms of ChE also occur as a minor component in muscle and nervous tissue. These forms are totally recovered together with amphiphilic forms in detergent-free extracts ('soluble' fraction). For example, detergent-free extracts from ox caudate nucleus were found to contain approximately equivalent proportions of hydrophilic and amphiphilic G_4 AChE, together with amphiphilic G_2 and G_1 forms (GRASSI et al. 1982) and similar results were obtained in the case of human brain (GENNARI and BRODBECK 1985) and fowl muscles (TOUTANT 1986). Hydrophilic G_4 BuChE was characterized in *Torpedo* heart (TOUTANT et al. 1985).

HODGSON and CHUBB (1983) have reported that an edrophonium-Sepharose affinity column retains a specific form of soluble G_4 AChE in ox serum as well as in soluble fractions of ox adrenal medulla and rabbit brain. Other AChE components from the same fractions were not bound to the column. This selectivity is surprising since edrophonium is considered a catalytic site-specific ligand. The bound form was identified by non-denaturing electrophoresis and apparently corresponds to a secreted hydrophilic form, although this was not fully demonstrated. If this is confirmed, it will be interesting to investigate the relationship between the binding to the columns and the hydrophilic character of the protein.

IV. Membrane-Bound, Intracellular and Extracellular Amphiphilic Forms

Most native globular ChE forms are amphiphilic, i. e. they interact with non-denaturing detergent in a manner similar to that described for the G_2 forms from *Torpedo* electric organs and human erythrocytes. These enzymes may be converted into hydrophilic forms by limited proteolytic digestion, as shown for the amphiphilic G_4 form from human brain (GENNARI and BRODBECK 1985) and for the G_2 and G_4 forms from *Torpedo* muscle and spinal cord (BON et al. 1988b)

Amphiphilic BuChE forms have been characterized in the heart (G_1, G_2), nerve, and spinal cord (G_4) of *Torpedo* (TOUTANT et al. 1985), but amphiphilic AChE forms have been studied more systematically. Erythrocytes from ox (MÉFLAH et al. 1984, RAKONCZAY and BRIMIJOIN 1985), rabbit (RAKONCZAY and BRIMIJOIN 1985), fowl (LYLES et al. 1982) and *Torpedo* (BON et al. 1988a, b) contain a G_2 form, like human erythrocytes. AChE from rat erythrocyte membranes has been described as a G_1 form (RIEGER et al. 1976; FERNANDEZ et al. 1979; RAKONCZAY and BRIMIJOIN 1985) or as a G_2 form (FUTERMAN et al. 1985c).

Ox lymphocytes also contain G_2 AChE. Mouse neural cells in culture (T28) present a G_4 form on their outer membrane surface (LAZAR and VIGNY 1980) and this is also predominant in membrane preparations from mammalian brain.

The existence of intracellular amphiphilic molecules has been demonstrated by numerous studies on the subcellular localization of AChE forms; a well documented example is that of the G_1 form of T28 cells (LAZAR and VIGNY 1980).

Extracellular, secreted forms may also be amphiphilic. In culture the T28 cells actively secrete G_1 and G_4 AChE, with a minor proportion of G_2, at a rate of 1.4% of their cellular activity per hour. G_1, G_2 and a fraction of the released G_4 forms clearly interacted with Triton X-100 and must be classified as amphiphilic. For example, G_1 sedimented at 4.8 *S* in the absence of detergent and at 3.8 *S* in the presence of Triton X-100 – a small but significant difference. It should be noted, however, that this form differed from the cellular form, sedimenting at 3.4 *S* in the same conditions (LAZAR et al. 1984; MASSOULIÉ et al. 1985). An amphiphilic G_2 AChE form has been characterized in the plasma of *Torpedo*, indicating that the release of such molecules may occur in physiological conditions (BON et al. 1988a). Thus, amphiphilic G_1, G_2 and G_4 forms may exist intracellularly in association with plasma membranes and even as extracellular soluble proteins.

V. Interactions with Detergent Micelles

The detergent interaction properties of amphiphilic forms of AChE have been studied in detail in the case of the G_2 forms of *Torpedo* electric organ (BON and MASSOULIÉ 1980; VIRATELLE and BERNHARD 1980; STIEGER et al. 1984) and of the G_2 forms of human erythrocyte membranes (OTT and BRODBECK 1978; WIEDMER et al. 1979; WEITZ et al. 1984; DUTTA-CHOUDHURY and ROSENBERRY 1984; ROSENBERRY and SCOGGIN 1984). The two enzymes are disulphide-linked dimers (LEE et al. 1982b; ROSENBERRY and SCOGGIN 1984) and possess very similar hydrophobic interactions.

In the presence of non-denaturing detergents, such as Triton X-100, they form well-defined complexes, in which one detergent micelle is associated with each

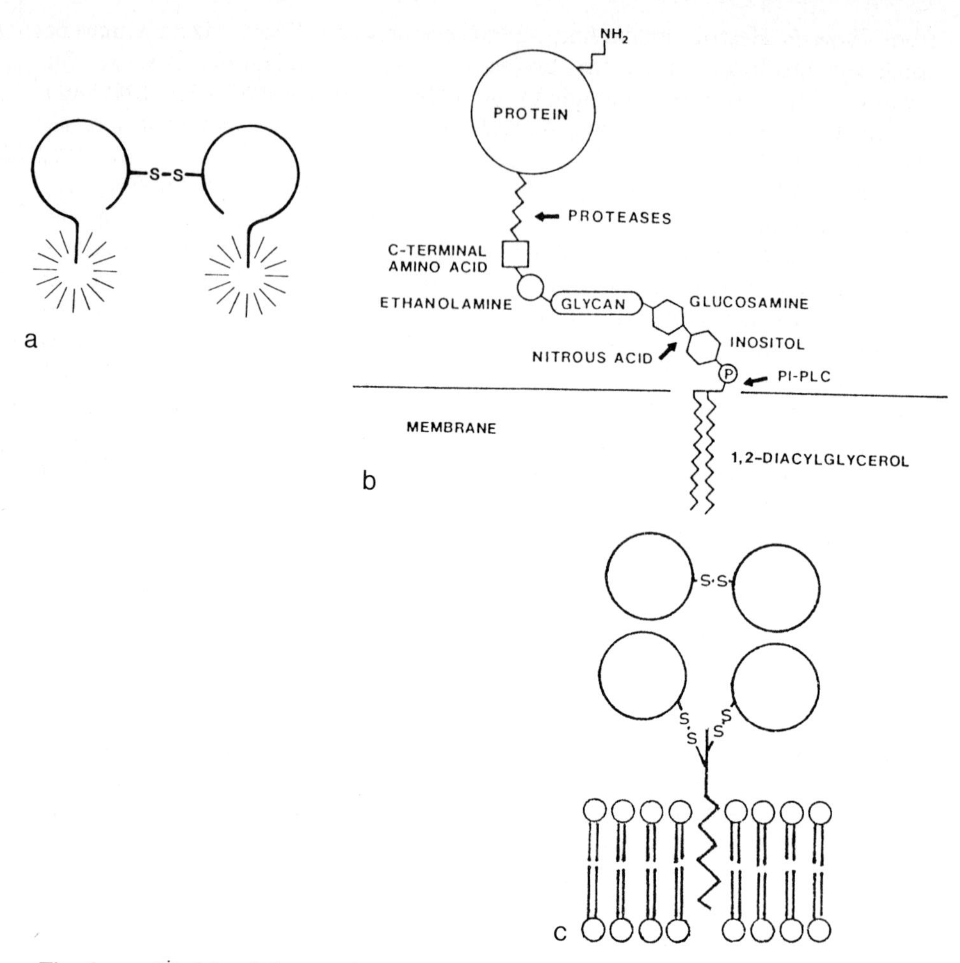

Fig. 6a–c. Models of the membrane attachment of amphiphilic AChE. **a** Diagrammatic representation of the interaction of G_2 human erythrocyte AChE with Triton X-100 micelles (after ROSENBERRY and SCOGGIN 1984). The enzyme is shown as an amphiphilic form in which the C-terminal domain of each subunit is inserted into the detergent micelle. This model is probably valid for other amphiphilic form, such as G_2 AChE from *Torpedo* electric organs (STIEGER and BRODBECK 1985; FUTERMAN et al. 1985c). **b** Model of the glycolipid anchor of amphiphilic AChE (after LOW et al. 1986). The actual composition and arrangement of the sugars is still hypothetical. *Arrows* indicate sites of action of various agents. **c** Schematic representation of the membrane-bound G_4 form of AChE from ox caudate nucleus (after INESTROSA et al. 1987). The tetramer is composed of one light dimer of disulphide-linked catalytic subunits (70 kDa×2) and one heavy dimer (160 kDa) in which two catalytic subunits are disulphide-linked to a non-catalytic component (20 kDa). The hydrophobic domain of this element mediates the attachment to the membrane

catalytic subunit (ROSENBERRY and SCOGGIN 1984; see Fig. 6). These complexes sediment at about 6 *S* and their Stokes radius is close to 8 nm (BON and MASSOULIÉ 1980; WEITZ et al. 1984; STIEGER and BRODBECK 1985). The sedimentation coefficient of the complexes depends on the nature of the detergent: 7.4 *S* in deoxycholate, 5.4 *S* in Brij 96 in the case of the *Torpedo* G_2 form (FUTERMAN et al. 1983, 1984).

In the absence of detergent, depending on its degree of purity and on its concentration, the enzyme sediments either as apparently isolated dimers (7 *S*; S. BON, unpublished results), heavier associations with other proteins (8–9 *S*; BON and MASSOULIÉ 1980), multiple polymeric forms (OTT and BRODBECK 1978) or polydisperse aggregates (BON and MASSOULIÉ 1980).

Amphiphilic G_4 AChE exists in *Torpedo* nerves and spinal cord (J.-P. TOUTANT and S. BON, unpublished results) and is a major component of AChE in mammalian brain and muscle. This enzyme may be partially solubilized in the absence of detergent, forms soluble complexes with detergent micelles and forms polymers or polydisperse aggregates with other proteins upon removal of the detergent, as also shown in the case of AChE from ox brain (GRASSI et al. 1982).

In contrast with those of asymmetric forms, the aggregating properties of globular forms of AChE are not significantly influenced by ionic strength and are not sensitive to collagenase treatment (BON and MASSOULIÉ 1980; LEE et al. 1982b). Using a crossed immunoelectrophoresis method (charge-shift electrophoresis) for characterizing the mobility of AChE, BRODBECK and his colleagues showed marked differences in the presence of anionic (deoxycholate), neutral (Triton X-100) and cationic (cetyltrimethylammonium) detergents (WEITZ et al. 1984; STIEGER and BRODBECK 1985). They also showed that the enzyme was immobilized in a phenyl-Sepharose hydrophobic matrix in crossed hydrophobic immunoelectrophoresis experiments (STIEGER and BRODBECK 1985).

Upon removal of the detergent these amphiphilic molecules readily associate with phospholipid vesicles (RÖMER-LÜTHI et al. 1980; KIM and ROSENBERRY 1985; STIEGER and BRODBECK 1985).

VI. Dimeric Forms Containing a Hydrophobic Glycolipid Anchor

1. Susceptibility of the Hydrophobic Domain to Proteases and to PIPLC

In amphiphilic G_2 dimers the subunits are linked by interchain disulphide bonds, as shown for the enzymes of *Torpedo* electric organs (BON and MASSOULIÉ 1980; LEE et al. 1982b; LI and BON 1983) and human erythrocytes (RÖMER-LÜTHI et al. 1979; OTT et al. 1982; ROSENBERRY and SCOGGIN 1984). Active amphiphilic monomers could be produced by reduction and alkylation in low yield with extensive denaturation in the case of *Torpedo* (BON and MASSOULIÉ 1980) but more efficiently in the case of erythrocytes (OTT et al. 1983).

The *Torpedo* and erythrocyte G_2 forms lose their hydrophobic properties after a limited digestion by pronase (BON and MASSOULIÉ 1980), proteinase K (WEITZ et al. 1984; STIEGER et al. 1984) or papain (KIM and ROSENBERRY 1985).

The resulting enzymes are fully active, sediment at about 7 *S*, with a Stokes radius of 5.9 nm, irrespective of the presence or absence of detergent (BON and MASSOULIÉ 1980; WEITZ et al. 1984) and cannot be reassociated with liposomes (FUTERMAN et al. 1984; KIM and ROSENBERRY 1985; STIEGER and BRODBECK 1985). Analyses of the intact and proteolysed enzymes by SDS-PAGE showed that the decrease in molecular mass of the modified catalytic subunits was very small (FUTERMAN et al. 1984). The apparent decreases were from 69 kDa to 68 kDa (MASSOULIÉ et al. 1984) and from 67 kDa to 66 kDa (STIEGER et al. 1984) in the case of *Torpedo* and from 75 kDa to 73 kDa (DUTTA-CHOUDHURY and ROSENBERRY 1984; KIM and ROSENBERRY 1985) in the case of the human erythrocytes. AChE therefore consists of a large globular hydrophilic domain and a small hydrophobic one which can be removed by proteolytic cleavage, as schematically illustrated in Fig. 6. Because of this disposition, the native enzyme should be called amphiphilic or amphipathic, rather than hydrophobic (ROSENBERRY and SCOGGIN 1984).

Because of the suspectibility of its hydrophobic domain to proteolytic cleavage, active AChE may be released from membranes by proteases as a hydrophilic form (LI and BON 1983). It can also be solubilized by PIPLC from erythrocyte membranes (LOW and FINEAN 1977a; IKEZAWA et al. 1983; TAGUCHI et al. 1984) and from sheep basal ganglia (MAJUMBAR and BALASUBRAMANIAN 1982). This effect was extensively studied by FUTERMAN et al. (1983, 1984, 1985a, b, c). These authors showed that the PIPLC-solubilized *Torpedo* G_2 form does not interact with non-denaturing detergents or with liposomes, exactly as the proteolysed hydrophilic form (FUTERMAN et al. 1983, 1984, 1985a). Other phospholipases (A_2, C or D) were found to be ineffective. PIPLC efficiently released AChE G_2 form from *Torpedo* synaptosomal membranes (FUTERMAN et al. 1985c), from rat erythrocytes (G_2 form) (FUTERMAN et al. 1984, 1985c) as well as from the erythrocytes and from the platelets of ox, pig, rabbit, sheep and horse (LOW and FINEAN 1977a; TAGUCHI et al. 1984; SHUKLA 1986) but solubilized only a small but significant fraction of the enzyme from human (LOW and FINEAN 1977a; FUTERMAN et al. 1984, 1985c; ROBERTS and ROSENBERRY 1986) and mouse (FUTERMAN et al. 1985c) erythrocytes. It did not significantly solubilize AChE from various regions of rat, mouse, ox and human brains (FUTERMAN et al. 1984, 1985c).

2. Glycolipid Nature of the Hydrophobic Domain

DUTTA-CHOUDHURY and ROSENBERRY (1984) showed that when the N-terminal residue of human erythrocyte AChE was radiolabelled by reductive methylation (HAAS and ROSENBERRY 1985), the label was not removed by papain digestion, indicating that the hydrophobic domain corresponds to the C-terminal of the molecule.

The small hydrophobic domain can be specifically labelled with the photoactivatable hydrophobic reagent 3-trifluoromethyl-3-(*m*-[^{125}I] iodophenyl) diazirine ([^{125}I] TID). The labelled fragment was identified after proteolysis of *Torpedo* G_2 AChE (STIEGER et al. 1984) and human erythrocyte G_2 AChE (ROBERTS and ROSENBERRY 1986), separated from detergent and analysed by gel filtration in

Sephadex LH60 in an ethanol-formic acid solvent. Its molecular mass, according to peptide standards, was about 3 kDa in both cases. An analysis of the fragment obtained from human erythrocyte AChE indicated the presence of only one mol each of histidine and glycine per mol (ROSENBERRY et al. 1985; ROBERTS and ROSENBERRY 1986). Further analyses showed that the structure of the hydrophobic fragment is his-gly-ethanolamine-*Z* where ethanolamine is linked to the C-terminal Gly by an amide linkage. *Z* includes glucosamine and a glycolipid (ROSENBERRY et al. 1985; HAAS et al. 1986; ROSENBERRY et al. 1986). Methanolysis releases one mol of saturated and one mol of unsaturated fatty acid per mol (ROBERTS and ROSENBERRY 1985). The non-proteic nature of this domain explains that some membrane-bound AChE forms may be solubilized by PIPLC (LOW and FINEAN 1977a; MAJUMBAR and BALASUBRAMANIAN 1982; SHUKLA 1982).

The presence of a phosphatidylinositol (PI) linkage in the glycolipidic domain has recently been confirmed by FUTERMAN et al. (1985b) in the case of the *Torpedo* amphiphilic G_2 form. They showed that most of the covalently linked inositol is removed upon cleavage of the hydrophobic domain by proteinase K but not after solubilization by PIPLC.

The striking difference observed in the activity of PIPLC on *Torpedo* and human erythrocyte AChE, in spite of the considerable similarity of the hydrophobic properties of the two enzymes, indicates that the structure of the non-proteic fragment is variable (ROBERTS et al. 1987).

In any case, the presence of a glycolipid makes it possible to anchor into membranes a protein that would otherwise be soluble. This structure raises two questions: (a) could the hydrolysis of a phosphodiester bond by an activated phospholipase operate to release AChE in certain physiological conditions, and in general, what is the relationship between secreted and amphiphilic enzyme; (b) do some membrane-bound or precursor amphiphilic AChE molecules possess a peptidic C-terminal hydrophobic domain?

3. Other Examples of Glycolipid-Conjugated Membrane-Bound Proteins

Like the G_2 forms of AChE from *Torpedo* and from mammalian erythrocytes, a dimeric AChE form from *Drosophila* is anchored by a C-terminal glycolipid (GNAGEY et al. 1987; FOURNIER et al. 1988). This mode of attachment to membranes is not unique to AChE (for reviews, see LOW et al. 1986; CROSS 1987). Previous studies have shown that other membrane-bound enzymes, alkaline phosphatase and 5′-nucleotidase, may be released by PIPLC (LOW and FINEAN 1977b). In addition, anchoring through the same type of C-terminal glycolipid has been described for a number of surface proteins (for a review, see SILMAN and FUTERMAN 1987), including the Thy-1 protein (CAMPBELL et al. 1981; TSE et al. 1985) which is also released by PIPLC (LOW and KINCADE 1985), one form of N-CAM (HE et al. 1986), trehalase (TAKESUE et al. 1986) and the decay acceleration factor (DAF) involved in the protection of cells from complement lysis (DAVITZ et al. 1986; MEDOF et al. 1986). It is interesting that in paroxysmal nocturnal haemoglobinuria, erythrocyte membranes are deficient both in DAF (NICHOLSON-WELLER et al. 1983; PANGBURN et al. 1983) and in AChE

(AUDITORE and HARTMANN 1959; METZ et al. 1960). There is no functional relationship between the two proteins (SUGARMAN et al. 1986), but there is possibly a defect in their common processing mechanism.

One of the best-documented cases of glycolipid-anchored proteins is that of the variant surface glycoproteins (VSG) of the blood stream form of trypanosomes (HOLDER 1983; BALTZ et al. 1983; FERGUSON et al. 1985). The parasites possess a large repertory of VSG genes which may be rearranged on the chromosome, so that only one is expressed at a given time. Among a population of trypanosomes individuals may switch to a new VSG and thus evade the host's immune defence. The C-terminal amino acid of these proteins is linked through an ethanolamine to the glycolipid, which contains galactose, mannose, glucosamine, phosphate, glycerol and fatty acids. The parasites possess a PIPLC and release the protein in soluble form, but it is, in fact, not known whether the switch is accompanied by a rapid clearing of the pre-existing coat protein, following activation of PIPLC.

It is interesting that the trypanosomid *Leishmania* also possesses a glycolipid-anchored surface protein; in this case it is invariant and presents a protease activity which blocks attack by the host's complement. This protein is much less susceptible to the action of phospholipase C and is not known to be released by the living parasite (ETGES et al. 1986).

The variable surface antigen of the free-living protozoan *Paramecium* and another, constant protein are both anchored by PI (CAPDEVILLE et al. 1986; C. DEREGNAUCOURT, Y. CAPDEVILLE and A.-M. KELLER, personal communication). In this species antigenic transition may be triggered by environmental conditions, and it will be interesting to examine whether this is linked to the activation of a PIPLC.

The biological significance of this type of anchoring raises a number of important questions. It is not known if glycolipid-anchored proteins are directed to their cellular destination in the same way as other membrane proteins, or if they are internalized by a specific process. This type of membrane insertion may be thought to possess distinct advantages over a hydrophobic peptide (ETGES et al. 1986):

- it probably makes it possible to accommodate a high density of surface proteins without disturbing the properties of the membrane bilayer, such as fluidity, by densely packed transmembrane helices;
- it may allow a rapid release of the protein upon activation of a phospholipase C;
- diacylglycerol, produced by hydrolysis of the anchor, may act as an intracellular signal, for example by activating protein kinase C.

4. Biosynthesis of the Glycolipid-Anchored Forms

It is interesting that the sequence of the Thy-1 protein, as deduced from the corresponding cDNA, would contain a hydrophobic peptide extending beyond the observed glycolipid conjugated C-terminal residue (SEKI et al. 1985), and that the trypanosomal VSG protein is first translated with a C-terminal hydrophobic extension (BORST and CROSS 1982). Intermediates in the biosynthesis of these molecules may therefore be anchored in cellular membranes by hydrophobic peptides which are exchanged for a glycolipid during their processing (HAAS et al. 1986). This suggests that some variants of such proteins may retain their hydrophobic peptide anchor; this is perhaps the case for a fraction of membrane-bound AChE that cannot be solubilized by PIPLC.

Although there appeared to be a short lag in the incorporation of ethanolamine into *Trypanosoma* VSG (RIFKIN and FAIRLAMB 1985), the incorporation of myristic acid was found to occur very rapidly, within 1 min after completion of the polypeptide (FERGUSON et al. 1986), and a similar observation was made for ethanolamine in the case of the DAF protein (MEDOF et al. 1986). These experiments suggest that the glycolipid anchor is added en bloc immediately after translation.

VII. Structure of G_4 Forms Containing a Distinct Hydrophobic Subunit

The amphiphilic AChE G_4 forms from rat SCG (VIGNY et al. 1979) and from human brain (GENNARI and BRODBECK 1985) were found to consist of disulphide-linked dimers, but LANDAUER et al. (1984) reported a stoichiometry of only two catalytic sites, as evaluated by DFP labelling, in the G_4 form of AChE from ox brain.

INESTROSA et al. (1987) showed that it was possible to label the hydrophobic domain of G_4 AChE from mammalian brain with radioactive TID, as in the case of *Torpedo* or erythrocyte G_2 forms. In this case, however, electrophoretic analyses of the enzyme in denaturing and reducing conditions showed that the reagent was not incorporated in the 70-kDa catalytic subunits, but in a distinct 20-kDa subunit.

Two components of 140 kDa and 160 kDa were observed in the absence of reduction, and only the latter was labelled, suggesting that it consisted of a disulphide-linked assembly of two catalytic subunits and the hydrophobic subunit, whereas the lighter one represented a disulphide-linked dimer. The proposed structure of this amphiphilic G_4 form is schematically represented in Fig. 6.

The formal similarity between this structure and the arrangement of the catalytic subunits and tail peptides in the asymmetric forms raises interesting questions regarding the structure, the biosynthesis of the different molecular forms and their relationships:

- Do these structures involve homologous cysteine residues of the catalytic subunits, and are the catalytic and structural subunits assembled in the same subcellular compartment by the same cellular machinery?
- Is there a common precursor for the catalytic subunits of amphiphilic G_2 and G_4, when they coexist, or for the globular G_4 and asymmetric forms, or are there distinct precursors for these various forms? Are the external subunits identical to those that are linked to structural subunits, and if not, do they derive from the same precursor?

Recent evidence shows that in muscle cells G_4 and asymmetric forms incorporate active monomers (G_1) (BROCKMAN et al. 1986). This form is known to be amphiphilic, but the nature of its hydrophobic domain is not known, and, on the other hand, intracellular G_1 has been shown to exist in multiple compartments which may correspond to distinct molecular entities engaged towards different assembly lines (see Chap. 8b).

VIII. Solubilization Artefacts, Modifications of the Solubilized Forms

In general the activity of AChE and BuChE was found to be extremely stable, both in tissues and in extracts. The characteristics and proportions of molecular forms appear to be more stable in intact tissues than in extracts. It has been shown for example that a post-mortem delay of about 36 h does not lead to significant changes in the molecular forms of ChEs in brain (ATACK et al. 1986). In fact, brain enzymes may be less susceptible to endogenous proteases since antiproteolytic agents had little effect on the distribution of molecular forms (SUNG and RUFF 1983; ATACK et al. 1986).

The activity and composition of a LSS fraction from ox brain depends on the nature of the buffer (GRASSI et al. 1982) but not on the number of passes in a Potter homogenizer (ATACK et al. 1986). In agreement with previous reports by HOLLUNGER and NIKLASSON (1973) and by ADAMSON (1977), we found that repeated extractions increased the overall yield of soluble G_1, apparently at the expense of the bulk of detergent-soluble G_4 (GRASSI et al. 1982). As discussed previously, the G_4 form consists of two disulphide-linked dimers, and by analogy with other enzymes we may assume that the interchain disulphide bonds may be easily reduced. The mechanism by which this structure may be disrupted during extraction is, however, not understood.

The molecular forms undergo various modifications after their solubilization. Endogenous proteolysis may of course dissociate the asymmetric forms and cleave the hydrophilic domain of amphiphilic globular forms, producing lytic hydrophilic globular molecules. Although polymeric globular forms are not readily dissociated by mild proteolysis without loss of activity (VIGNY et al. 1979; LANDAUER et al. 1984), the G_2 form of chicken BuChE may be quantitatively converted into G_1 by trypsin (ALLEMAND et al. 1981). The globular forms may also be dissociated by reduction, e. g. human erythrocyte G_2 into G_1 (OTT et al. 1983).

Light forms may also generate heavier molecules, either during storage or by aggregation upon removal of detergent. In a LSS fraction from ox caudate nucleus for example, the G_1 form was found to be converted spontaneously into a 6.5-*S* form (GRASSI et al. 1982). Similarly, the hydrophobic G_1 form of BuChE was found to give rise to a hydrophilic form, apparently G_4, in extracts from *Torpedo* heart (TOUTANT et al. 1985). LANDAUER et al. (1984) reported that upon removal of Triton X-100, a detergent-solubilized amphiphilic G_4 form purified from ox brain, produced well-defined 16-*S* and 20-*S* aggregates which could not be dissociated by non-denaturing detergents, reduction or mild proteolysis. GENNARI and BRODBECK (1985) observed similar aggregates in the case of a human brain enzyme but they were mostly dissociated when Triton X-100 was added. The polymeric nature of the modified molecules is not clear, since they may combine catalytic and non-catalytic subunits (GRASSI et al. 1982; LANDAUER et al. 1984).

IX. Catalytic Activity and Hydrophobic Interactions

BRODBECK and his colleagues have found that human erythrocyte G_2 AChE requires an amphiphilic environment for the maintenance of its catalytic activity (WIEDMER et al. 1979, DI FRANCESCO et al. 1981). This environment may be afforded by detergents, by proteins such as bovine serum albumin or by AChE itself in hydrophobic aggregates. These results were recently extended to a monomeric G_1 species obtained by reduction of the native G_2 form (OTT and BRODBECK 1984), and similar observations have been made with G_4 AChE from rat SCG (KLINAR et al. 1985). The activity of the hydrophilic G_2 form that was obtained by treating the native G_2 enzyme by papain was no longer dependent on the presence of amphiphiles (WEITZ et al. 1984). These observations suggest that dilution of detergent solutions of amphiphilic enzyme below the critical micelle concentration results in the loss of activity through surface denaturation, or that the hydrophobic domain may interfere with the active site.

Detergents may also inhibit AChE activity, under conditions which we do not believe reflect hydrophobic interactions, as discussed in the preceding section. BIRMAN (1985) showed that deoxycholate, and to a lesser degree cholate, inactivate both amphiphilic and hydrophilic *Torpedo* AChE but not *Electrophorus* AChE. The latter was found to be reversibly inhibited by Triton X-100 at high concentration, an effect apparently due to the binding of two detergent micelles per catalytic subunit; the maximal hydrolysis rate was decreased and the excess substrate inhibition was increased (MILLAR et al. 1979).

Temperature-dependence studies of membrane-bound AChE yield non-linear Arrhenius plots; the break points have often been thought to reflect phase transitions in the lipid environment of the enzyme. Recent studies have shown however that similar results were obtained with membrane-bound *Torpedo* and human AChE and with the non-hydrophobic lytic forms that were solubilized by proteinase or PIPLC (BARTON et al. 1985). This result supports the idea that AChE is not deeply embedded in lipid bilayers.

X. Thermal Stability

The thermal stability of all asymmetric and lytic globular forms from *Electrophorus* have been studied in detail (MASSOULIÉ et al. 1975). Although the denaturation did not follow first order kinetics, ΔH values could be obtained: the values were higher for asymmetric than for globular molecules.

A comparison of intact human G_4 BuChE with modified molecules in which interchain disulphide bonds had been reduced showed that the latter was less stable (LOCKRIDGE et al. 1979). Recently RAKONCZAY and BRIMIJOIN (1985) compared the thermal stability of soluble and detergent-soluble G_1 and G_4 forms from rabbit brain. G_1 was found to be less stable than G_4, and soluble forms were more resistant than the detergent-soluble ones. These results were considered to reflect structural differences between the two enzyme fractions (see below).

Active site inhibitors of AChE may protect the enzyme from denaturation under several conditions, including exposure to urea, dithiothreitol, proteases (DUTTA-CHOUDHURY and ROSENBERRY 1984) and deoxycholate (BIRMAN 1985).

G. Variability of the Catalytic Subunits

I. Catalytic Subunits of AChE and BuChE in Vertebrates: Analysis by SDS-PAGE

The hydrodynamic properties of homologous forms of AChE and BuChE are similar in fish, amphibians, reptiles and mammals, indicating that the catalytic subunits of these enzymes are of comparable molecular mass. By contrast, the sedimentation characteristics of AChE and BuChE in fowl (SILMAN et al. 1979; ALLEMAND et al. 1981) and of AChE in quail (ROTUNDO 1984a) and pigeon (CHATONNET and BACOU 1983) indicate that the molecular mass of the avian enzymes is considerably greater.

Apparent molecular masses of the catalytic subunits of AChE and BuChE have been determined in a few species by SDS-PAGE (see Tables 2 and 3). The values obtained must be considered with caution since these enzymes are glycoproteins and thus contain a significant proportion of their molecular mass as carbohydrates: 15% for AChE from *Electrophorus* (BON et al. 1976) and human erythrocytes (NIDAY et al. 1977) and 25% for BuChE from human serum (MUENSCH et al. 1976). The molecular masses obtained for AChE of *Electrophorus* (DUDAI et al. 1972; ROSENBERRY and RICHARDSON 1977; ANGLISTER and SILMAN 1978), rat (VIGNY et al. 1979), rabbit (MINTZ and BRIMIJOIN 1985a), ox (LANDAUER et al. 1984; INESTROSA et al. 1987), and man (OTT et al. 1975; RÖMER-LÜTHI et al. 1979; DUTTA-CHOUDHURY and ROSENBERRY 1984; ROSENBERRY and SCOGGIN 1984) all fall in the range of 70–80 kDa (see Tables 2 and 3). As discussed below, the electric rays present several distinct types of catalytic subunits of AChE, also in the same range (65–80 kDa).

ROTUNDO (1984b) found that fowl brain AChE presents two distinct catalytic polypeptides of molecular mass 100–105 kDa and 98–108 kDa, respectively. The difference between these polypeptides was maintained after treatment with endoglycosidase F, which removed carbohydrates and reduced the molecular mass of each subunit by about 10 kDa (ROTUNDO 1984b). RANDALL et al. (1987a) found that individual fowl may possess either or both of the two polypeptides, which suggests that they represent allelic variants of AChE. A similar result has been obtained in quail by R. ROTUNDO (personal communication).

It is remarkable that a limited tryptic digestion readily converts fowl AChE into molecules which sediment like their mammalian homologues (ALLEMAND et al. 1981). This digestion reduces the mass of the protein by about 30%, with no apparent modification of its catalytic properties or molecular interactions. The sedimentation characteristics of fowl BuChE molecular forms are also modified by trypsin, although to a lesser degree than AChE (ALLEMAND et al. 1981).

It is therefore likely that the catalytic subunits of avian cholinesterases are homologous to those of other vertebrates, but possess an additional, dispensable

Table 2. Molecular masses of AChE catalytic subunits in electric organs. These values were obtained by SDS-PAGE in reducing conditions and are given in kDa. In the case of *Electrophorus* AChE they correspond to A forms only. The electric organs of rays contain two or three types of catalytic subunits. Depending on the experimental procedure used by different authors, the LSS (low salt soluble), DS (detergent-soluble) and HSS (high salt soluble) fractions present variable compositions of molecular forms

	Electric eel *Electrophorus electricus*	Electric rays							
		Torpedo californica		*Torpedo marmorata*			*Narke japonica*		
		LSS + DS	HSS	LSS	DS	HSS	LSS	DS	HSS
Molecular forms	A_{12}, A_8, A_4	G_2	A_{12}, A_8	G_2	G_2	A_{12}, A_8	G_2, A_{12}, A_8	G_2	A_{12}, A_8
References									
DUDAI et al. (1972)	80								
DUDAI and SILMAN (1974)	82								
ROSENBERRY and RICHARDSON (1977)	75								
ANGLISTER and SILMAN (1978)	80								
VIRATELLE and BERNHARD (1980)		80	80						
LEE et al. (1982a)		66							
LEE et al. (1982b)			68						
FUTERMAN et al. (1983)		70							
LI and BON (1983)					70				
WITZEMANN and BOUSTEAD (1983)				75 – 80	69	72			
ABE et al. (1983)							66 + 70	66	70
MASSOULIÉ et al. (1984)				68 + 76	69	72			
STIEGER et al. (1984)					67				
STIEGER and BRODBECK (1985)					66	68			
SAKAI et al. (1985a, b)		66	70						
LAPPIN et al. (1987)		66	68						
BON et al. (1986)				67	67	72			

Table 3. Molecular weights of cholinesterase catalytic subunits in higher vertebrates

	Nature and origin of the enzyme								
	Human AChE		Human plasma BuChE	Rat brain AChE	Rabbit brain AChE	Rabbit plasma BuChE	Ox caudate nucleus AChE	Quail muscle cultures AChE	Fowl brain AChE
	Red blood cells	Brain							
Molecular forms	G_2	G_4+G_1	G_4	G_4+G_1	G_4		G_4+G_1	A_{12}, G_4, G_2	G_4
OTT et al. (1975)	80								
RÖMER-LÜTHI et al. (1979)	80								
ROSENBERRY and SCOGGIN (1984)	75								
DUTTA-CHOUDHURY and ROSENBERRY (1984)	75								
SØRENSEN et al. (1982)		78							
GENNARI and BRODBECK (1985)		66							
LOCKRIDGE et al. (1979)			90						
MASSON (1979)			85						
VIGNY et al. (1979)				75					
MINTZ and BRIMIJOIN (1985a)					71				
MAIN et al. (1977)						83			
LANDAUER et al. (1984)							72 – 75/66 – 68		
INESTROSA et al. (1987)							70		
ROTUNDO (1984a)								98 + 108	
ROTUNDO (1984b)									100 + 105
RANDALL et al. (1985)									110

Values are in kDa

protein domain. It is possible that the existence of allelic size variants of the subunits reflect variations in the extent of this domain. This extension probably corresponds to an N-terminal exon, since the C-terminal extremity of some subunits is involved in the linkage of the glycolipid anchor.

II. Specificity of Inhibitory Antibodies and Peptide Toxins

Special mention should be made of inhibitory antibodies. An anti-rat AChE antiserum was found to contain inhibitory antibodies; this inhibition was somewhat species-specific, being more efficient against the rat than against the human enzyme (MARSH et al. 1984).

AChE inhibition was observed with one of five monoclonal antibodies raised against human erythrocyte AChE (FAMBROUGH et al. 1982). This antibody, AE-2, was found partially to inhibit the human and ox but not the *Torpedo* or *Electrophorus* enzymes, although it showed a similar apparent titre for ox and *Torpedo* AChEs.

An inhibitory monoclonal antibody obtained against rabbit AChE (F4 43)was studied by BRIMIJOIN and his colleagues (MINTZ and BRIMIJOIN 1985b; BRIMIJOIN et al. 1985). It is somewhat surprising that this antibody shows a narrow range of specificity, since it reacts with guinea-pig AChE but not with human, rat or cat enzymes, although it is thought to bind between the catalytic site and the peripheral anionic site, that is, to probably highly conserved regions of the protein. This observation implies that structures of the enzyme that are in the direct vicinity of the active site are not strictly conserved even in relatively close species. It is interesting in this respect that fasciculin, a polypeptidic toxin from the venom of the mamba snake *Dendroaspis angusticeps*, inhibits rat AChE with a very high affinity (10 n*M*) but is totally inactive even at a high concentration on fowl AChE (RODRIGUEZ-ITHURRALDE et al. 1983; KARLSSON et al. 1984).

III. Comparison Between Catalytic Subunits of Different AChE Fractions Within a Given Species

1. *Electrophorus* and *Torpedo* Electric Organs

All molecular forms of AChE from *Electrophorus* electric organs appear to contain a single type of catalytic subunit. In particular, no differences have yet been found between the subunits that are linked by disulphide bonds to the tail peptides and the external subunits that are linked as dimers (ROSENBERRY 1985).

The catalytic subunits of asymmetric and globular forms of AChE from torpedinid electric organs are clearly different in their primary structure, as well as having a C-terminal glycolipid anchor in the amphiphilic G_2 form; as shown in Table 2 the apparent mass of the subunit is about 2–3 kDa larger in asymmetric than in globular forms in *T. californica* (LEE et al. 1982a; SAKAI et al. 1985a), *T. marmorata* (WITZEMANN and BOUSTEAD 1983; MASSOULIÉ et al. 1984; STIEGER and BRODBECK 1985) and *N. japonica* (ABE et al. 1983; SAKAI et al. 1985a). This difference is not due to N-linked carbohydrates since it is maintained after deglycosylation by endoglycosidase F (BON et al. 1986). In *T. califor-*

nica the two types of subunits showed a quasi-identity, with only a few differences in their α-chymotryptic peptides and tryptic peptides in gel electrophoresis (LEE et al. 1982a) and high performance liquid chromatography (DOCTOR et al. 1983).

Circular dichroism spectra indicated the presence of about 33% α-helical and 23% β-sheet (14% antiparallel, 9% parallel) structures in the lytic G_4 form and similar results in the amphiphilic G_2 form of *Torpedo* enzyme, whether detergent is present or not. Addition of the active-site ligand, edrophonium, and of the peripheral site ligand, propidium, induced weak changes in opposite directions in the G_4 form but no significant modifications in the G_2 form (MANAVALAN et al. 1985).

WITZEMANN and BOUSTEAD (1983) have reported that the LSS fraction from *T. marmorata* electric organs contains a third type of catalytic subunit which is heavier than those of the A and G forms and is not disulphide-linked. Recent studies showed that this 76-kDa subunit corresponds to a 5-S inactive component, which reacts with an anti-AChE polyclonal antiserum and can be labelled by DFP (STIEGER et al. 1987). This component does not bind the catalytic site ligand edrophonium, as evinced by the lack of inhibition of its reaction with DFP, and the lack of binding to edrophonium-Sepharose. It has been suggested that it might represent a precursor of the active enzyme. Its mass, however, seems too high to correspond to the precursor of the glycolipid-anchored form, which contains a C-terminal extension of 29 amino-acids, as indicated by the cDNA sequence (J. L. SIKORAV, E. KREJCI, A. ANSELMET, N. DUVAL and J. MASSOULIÉ, in preparation). The relationship of the inactive 76 kDa subunit to the active enzyme is therefore still problematic.

2. Catalytic Subunits of AChE in Other Species

RANDALL et al. (1985, 1987a) found that in fowl the apparent molecular mass of the catalytic subunit of muscle AChE in SDS-PAGE is slightly greater than that of brain AChE. This difference was maintained after deglycosylation by endoglycosidase F, which removes all N-linked polysaccharides. Since muscle contains mostly asymmetric forms and brain amphiphilic globular forms (ROTUNDO 1984b), it is tempting to assume that this difference corresponds, by analogy with that observed in *Torpedo*, to distinct catalytic subunits present in the A and G forms in this species.

Several authors have reported that G_4 AChE from rat, ox and human brain contains both light and heavy subunits (RAKONCZAY et al. 1981a, b; LANDAUER et al. 1984; GENNARI and BRODBECK 1985). Such differences have not been consistently observed (VIGNY et al. 1979; MARSH and MASSOULIÉ 1985), and it has not been excluded that they might arise from secondary proteolytic cleavage. In fact, a rapidly purified G_4 AChE from rabbit brain was found to contain a single type of 71-kDa subunit (MINTZ and BRIMIJOIN 1985a). As discussed above, the 70-kDa catalytic subunits may be engaged in both light and heavy dimers, the latter including an additional hydrophobic subunit (INESTROSA et al. 1987).

IV. Heterogeneity of AChE Globular Forms in Non-denaturing Electrophoresis

Charge-shift electrophoresis in non-denaturing conditions is useful for demonstrating the binding of detergents to amphiphilic molecules (HELENIUS and SIMONS 1977). This method is generally applicable to globular forms of ChE.

By comparing the electrophoretic migration of amphiphilic molecules in the presence of anionic (sodium deoxycholate), cationic (cetyltrimethylammonium bromide) and neutral (Triton X-100) detergents it was possible to define an operational index of hydrophobicity that indicates the binding of detergent micelles to the enzyme and reveals a striking heterogeneity of the G_4 form of fowl embryo muscle extracts (TOUTANT 1986).

Non-denaturing electrophoresis also reveals differences between molecular forms of identical hydrodynamic properties in *Torpedo.* Amphipathic G_2 AChE forms obtained from electric organ, skeletal muscle, electric nerve, spinal cord, erythrocytes and plasma of *T. marmorata*, all sedimented at 6 *S* in the presence of Triton X-100. These enzymes however were all different in their electrophoretic migrations in the presence of Triton X-100 and in some cases, e. g. in electric organs, presented two components. The lytic forms obtained after pronase digestion migrated faster than the native amphiphilic enzymes but showed the same heterogeneity. In contrast with those of the native forms their migration rates were not influenced by cationic or anionic detergents (BON et al. 1988b).

These experiments indicate differences in electric charges that possibly result from differences in glycosylation. It should be noted that this complexity does not coincide with the possible difference between the soluble and detergent-soluble molecules, because LSS and DS fractions from a given tissue presented the same migration pattern. The difference observed is physiologically significant since the two electrophoretic variants do not appear at the same stage of embryonic development of the electric organ (BON et al. 1987).

Non-denaturing electrophoresis also revealed the heterogeneity of G_1 and G_4 AChE forms in the rat superior cervical ganglion (GISIGER et al. 1978) within G_2 and G_4 forms of dystrophic mice muscles (POIANA et al. 1985) and of the G_2 form of fowl embryo muscle (CISSON et al. 1981; TOUTANT 1986). In the sciatic nerve of rat G_4 molecules subject to axonal transport were found to differ from stationary molecules in their electrophoretic migration (KÁSA and RAKONCZAY 1982).

V. Binding of Lectins and Heterogeneity of Glycosylation of AChE Molecules

By comparing the binding of AChE from ox erythrocytes, lymphocytes, brain and muscle with various lectins, we showed that otherwise identical molecular forms may differ in their glycosylation (MÉFLAH et al. 1984). For example, erythrocyte and lymphocyte membranes both contain a G_2 form, but only the former reacted with concanavalin A and *Ulex europeus* agglutinin. In muscle the G_4 form, but not the G_1 form, was found to react with wheat-germ agglutinin. This suggests that G_1 carries only high-mannose sugars and that G_4, which is preferentially exposed at the external membrane surface, contains more elaborate complex sugars.

SANES and his colleagues showed that *Dolichos biflorus* agglutinin, a lectin that recognizes terminal *N*-acetyl galactosamine, specifically binds to neuromuscular junctions in rat skeletal muscle (SANES and CHENEY 1982). Interestingly, this agglutinin binds asymmetric forms but not globular forms of rat muscle AChE (SCOTT and SANES 1984).

H. Immunochemical Studies

I. Immunochemical Cross-Reactivity Between AChE and BuChE

In spite of the high degree of homology suggested by their parallel structural polymorphism and by available peptide sequences (see Sect. I), immunochemical cross-reactivity between AChE and BuChE has not been observed with polyclonal antisera in *Torpedo* (TOUTANT et al. 1985), rat (VIGNY et al. 1978), fowl (ROTUNDO 1984b) and man (GENNARI and BRODBECK 1985; SØRENSEN et al. 1986) or with monoclonal antibodies in *Torpedo* (MUSSET et al. 1987), rat (RAKONCZAY and BRIMIJOIN 1986), man (BRIMIJOIN et al. 1983), rabbit (MINTZ and BRIMIJOIN 1985b) and fowl (ROTUNDO 1984b).

However, DOCTOR and his colleagues obtained several monoclonal antibodies raised against foetal calf serum AChE which also reacted with the AChE of *Torpedo* and *Electrophorus* electric organs and the BuChE of human serum. The latter reaction was in some cases directed against carbohydrates and did not occur after deglycosylation. They also found that AE-2, a monoclonal antibody prepared against human erythrocyte AChE (FAMBROUGH et al. 1982; see Sect. II. 2) reacted weakly with human BuChE but only if this enzyme was first deglycosylated (B. P. DOCTOR, personal communication).

II. Interspecific Cross-Reactivity

1. Polyclonal Antisera

Cross-reactivity between the AChEs of different species has been found to depend markedly upon their evolutionary relationship. For example, rabbit antisera raised against *Electrophorus* (WILLIAMS 1969, GRASSI et al. 1983) or *Torpedo* (SIKORAV et al. 1984) did not show any cross-reaction between the two species and did not recognize the enzymes of higher vertebrates. Conversely, an antiserum raised against ox caudate nucleus AChE did not react with *Electrophorus* AChE (GREENBERG et al. 1977). Similarly, a rabbit antiserum prepared against rat AChE did not react with the *Electrophorus*, *Torpedo*, fowl or rabbit enzymes but did recognize AChE from various mammalian species in the following order: rat = mouse > ox > human (MARSH et al. 1984). In a similar study SØRENSEN et al. (1985) examined the cross-reactivity of various enzymes with anti-human and anti-bovine AChE antisera. The anti-bovine AChE antiserum reacted in the following order: ox > horse > swine > dog > human.

Cross-reactivity between brain AChE fractions from man, dog, ox, swine and horse was also demonstrated, with variable binding efficiency, with antisera raised against human brain, human erythrocyte and ox brain AChE (SØRENSEN et al. 1985). The human and dog enzymes appeared to be the most closely related in these experiments.

2. Monoclonal Antibodies

Monoclonal antibodies raised against mammalian AChE (FAMBROUGH et al. 1982) or human BuChE (BRIMIJOIN et al. 1983) generally do not react with non-mammalian (e.g. fowl) enzymes. Among different mammalian enzymes the degree of interspecific cross-reactivity is very variable. For example, FAMBROUGH et al. (1982) described five monoclonal antibodies directed against human erythrocyte AChE, including the inhibitory antibody AE-2. The antibodies stained the neuromuscular junctions of monkey, but not those of rat, fowl or frog. They showed diverse specificity patterns with rabbit, dog, swine, and calf.

BRIMIJOIN and his colleagues studied the specificity of a wide variety of monoclonal antibodies raised against rabbit (MINTZ and BRIMIJOIN 1985b), rat (RACKONCZAY and BRIMIJOIN 1986) and human AChE, and against human BuChE (BRIMIJOIN et al. 1983). It is interesting that RACKONCZAY and BRIMIJOIN (1986) succeeded in raising monoclonal antibodies by injecting mice with partially heat-inactivated rat brain AChE. Some of the antibodies obtained cross-reacted with mouse AChE and showed no consistent difference in their reactivity towards enzymes from different inbred strains of mice.

Some monoclonal antibodies raised against human BuChE also reacted against monkey, dog, cat, horse and rabbit enzymes, but one monoclonal did not do so (BRIMIJOIN et al. 1983). An antibody of the first type showed identical reactivity with the enzymes of man and hominid primates, but the latter showed a decrease in affinity in the following order: man > chimpanzee > gorilla > orang-utan > gibbon (MINTZ et al. 1984). The 'usual' and 'atypical' allelic variants of human BuChE were not distinguished by polyclonal (ECKERSON et al. 1983) or monoclonal antibodies (BRIMIJOIN et al. 1983).

III. Immunochemical Differences Between AChE Fractions Within a Given Organism

In general all AChE molecular forms, in soluble as well as membrane-bound fractions, appear to share a majority of common epitopes and are not distinguished by most antisera, but monoclonal antibodies may present a more restricted specificity.

The tail peptides which differentiate the asymmetric and globular forms appear to be far less antigenic than the catalytic subunits themselves. In the case of *Electrophorus* for example, it has not been possible to obtain antisera against the tail (MASSOULIÉ and BON 1982; ROSENBERRY 1985). In addition, the collagen-like tail of *Electrophorus* and ox AChE asymmetric forms was not recognized by antisera against various types of mammalian collagen (types I, II, III, IV and V), and it did not bind laminin or fibronectin (GRASSI et al. 1983).

The asymmetric forms and the amphiphilic dimers of *Torpedo* electric organs have been extensively studied. The asymmetric and globular *Torpedo* enzymes have not been distinguished by polyclonal antisera (J. P. TOUTANT, unpublished results) and by most monoclonal antibodies (DOCTOR et al. 1983; MUSSET et al. 1987). A similar result was reported for the LSS, DS and HSS fractions from *Narke japonica* (ABE et al. 1983); however, DOCTOR et al. (1983) described one monoclonal antibody that presented a higher affinity for the G_2 form than for a preparation ($A_{12}+A_8$).

RUBIN and his colleagues have recently obtained a rabbit antiserum that recognizes the three types of subunits of *Torpedo* asymmetric AChE (55, 68 and 100 kDa). They also obtained monoclonal antibodies against this enzyme, most of which recognize the catalytic subunits (in some cases with a greater affinity for the A subunit than for the G subunit; see below) while others recognize the tail subunit (LAPPIN and RUBIN 1984; LAPPIN et al. 1987). CHANG et al. (1986) also obtained monoclonal antibodies that appear to be specific for the tail peptides. SAKAI et al. (1986) described two monoclonal antibodies produced against *N. japonica* AChE that recognize the tail subunits of both *Narke* and *T. californica* enzymes. Such antibodies will be extremely useful for studying the biosynthesis of the complex forms of AChE. It is interesting that an anti-tail and several anti-catalytic subunit antibodies bind to frog neuromuscular junctions (LAPPIN and RUBIN 1984; LAPPIN et al. 1987).

The question of an immunological difference between soluble and membrane-bound forms of mammalian AChE has been raised repeatedly by a number of authors. Some reports claimed that polyclonal antisera raised against a purified detergent-solubilized G_4 form did not recognize soluble AChE. Subsequent studies showed that there was, in fact, no difference in the reactivity of the two enzyme fractions with such sera in the case of rat, ox (MARSH et al. 1984) and human (GENNARI and BRODBECK 1985) brain. Indeed, the two types of AChE fractions from ox caudate nucleus showed very similar peptide patterns with several proteases, in agreement with the fact that their antigenic sites are largely or totally identical (MARSH and MASSOULIÉ 1985). Note that the presence or absence of a hydrophobic fragment which may correspond to distinct antigenic sites is not a systematic difference between these enzyme fractions, since most of the 'soluble' enzyme is amphiphilic.

Monoclonal antibodies raised against rabbit brain AChE were found to bind the soluble less efficiently than the detergent-soluble enzyme (MINTZ and BRIMIJOIN 1985b). One monoclonal antibody (F_443) recognized only about one-half of the soluble enzyme fraction. It is to be noted that this antibody was rather selective in its species recognition since it reacted with AChE from rabbit and guinea-pig brains but not with that from rat, cat or human brains. It bound G_4 more efficiently than G_1, and, in fact, a fraction of G_1 from the soluble fraction of rabbit brain was apparently not recognized at all, suggesting the existence of different pools of this enzyme, as discussed further below (RAKONCZAY and BRIMIJOIN 1985, 1986).

The tissue specificity of monoclonal antibodies is also of considerable interest. In general, antibodies raised against either brain or erythrocyte AChE cross-react with AChE from the other tissue in the case of human (FAMBROUGH et al. 1982), rat and rabbit (RAKONCZAY and BRIMIJOIN 1985, 1986). One monoclonal antibody (ZR 1) was however shown to bind both brain and serum AChE from rat, rabbit and ox but not erythrocyte AChE from the same species (RACKONCZAY and BRIMIJOIN 1985, 1986). The interpretation of this result must await the identification of the corresponding epitope. The authors suggest that it may be either a carbohydrate or a portion of the molecule that is masked by carbohydrate in other tissues.

We have recently characterized a monoclonal IgM antibody (Elec-39) that recognizes a carbohydrate structure present on some AChE molecules from *Electrophorus* and *Torpedo* electric organs but not on the enzyme from nervous tissue or muscle from electric fish, frog, fowl or rat (BON et al. 1987). In *Electrophorus* Elec-39 reacts with A forms, but in *Torpedo* it binds only to a subset of the G_2 form, and not at all to the A forms. The Elec-39 epitope appears progressively on AChE during embryogenesis of the *Torpedo* electric organ. Elec-39 reacts with non-AChE glycoproteins in electric organ, and also in nervous tissue, including in species where AChE is not recognized.

We found that several anti-carbohydrate IgMs (HNK-1, NC-1, NSP-4), originally raised against various antigens, reacted with the same population of *Torpedo* AChE molecules as Elec-39. They formed similar but significantly different complexes, as indicated by sedimentation analyses. The prototype of this family, HNK-1, was produced against human natural killer lymphocytes (ABO and BALCH 1981). The monoclonal antibodies L2 (KRUSE et al. 1984), NC-1 (VINCENT et al. 1983) and NSP-4 (ROUGON et al. 1983) as well as IgM antibodies that frequently occur in human neuropathies and react with the myelin-associated glycoprotein (LATOV et al. 1980; MURRAY and STECK 1984) possess a similar specificity. They appear to recognize sulphated glucuronic acid-containing glycolipids and cell surface glycoproteins (CHOU et al. 1985). The epitope is present on a subset of cell adhesion molecules such as N-CAM, L1 and J1 (homologous to avian Ng-CAM and cytotactin) (KRUSE et al. 1984, 1985; GRUMET et al. 1985) and might be directly involved in cellular interactions (KEILHAUER et al. 1985). The presence of this type of epitope on AChE in electric organs illustrates a tissue-specific differentiation of the enzyme and suggests that it might be involved in synaptic supra-molecular interactions.

The glycolipid moiety of the C-terminal hydrophobic anchors may well have been conserved with little change during evolution: antibodies directed against this 'cross-reacting determinant' from VSGs of *Trypanosoma brucei* (CARDOSO DE ALMEIDA and TURNER 1983) were recently shown to recognize an immunological determinant revealed by PIPLC in amphiphilic G_2 forms of AChE from *Torpedo* electric organs and human erythrocytes (STIEGER et al. 1986).

I. Molecular-Biological Studies

I. Characterization of AChE mRNAs and Their Translation Products

Translation products of AChE mRNA, obtained in a cell-free rabbit reticulocyte system, have been characterized by specific immunoprecipitation with polyclonal antisera. In the case of *E. electricus* SIKORAV et al. (1984) reported the presence of a single 65-kDa polypeptide, most probably a precursor of the catalytic subunit of collagen-tailed AChE. Using mRNA from the electric organ of *Discopyge tschudii*, MÉNDEZ et al. (1984) observed a major polypeptide of 62 kDa, which they assume to correspond to asymmetrical AChE; their data however suggest the presence of other polypeptides, and this may be related to the presence of both globular and asymmetrical forms in this tissue. In the case of *T. marmorata* at least two polypeptides of 61 kDa and 65 kDa were identified (SIKORAV et al. 1984). This multiplicity of translation products very likely corresponds to different types of catalytic subunit of AChE in the electric organ. It demonstrates the multiplicity of mRNA species coding for AChE catalytic subunits.

AChE mRNA may be translated into active enzyme by micro-injection into *Xenopus* oöcytes and may thus be detected specifically and sensitively (SOREQ et al. 1982; MEEDEL and WHITTAKER 1983). This made it possible to evaluate the AChE mRNA content of rat brain and *Torpedo* electric organ (SOREQ et al. 1983). SOREQ et al. (1984) demonstrated in this way that in human brain multiple mRNAs of different sizes induced the synthesis of the enzyme.

II. Characterization of AChE cDNA Clones

In *Drosophila* the AChE gene has been localized in a small region of the third chromosome, 87E1-5, using the method of segmental aneuploidy (HALL and KANKEL 1976). The subsequent characterization of non-conditional and conditional mutants allowed the definition of the *Ace* locus, the proposed structural gene for AChE (HALL et al. 1980; BENDER et al. 1983). A DNA fragment which was thought to belong to the *Ace* region was found to hybrid-select *Drosophila* and human brain mRNAs that induced an increase of AChE activity in *Xenopus* oöcytes (SOREQ et al. 1985). The *Drosophila* DNA fragment was used as a probe to isolate a human genomic fragment, which, in turn, was reported to hybridize to a 7-kb mRNA in human brain and to hybrid-select human mRNAs capable of inducing the synthesis of AChE activity in *Xenopus* oöcytes. A polypeptide of 85 kDa was immunoprecipitated from the synthesized products by an anti-human AChE monoclonal antibody (ZEVIN-SONKIN et al. 1985). Several cDNA clones were found to hybridize with the DNA fragment but were not conclusively identified as AChE coding sequences. The original *Drosophila* fragment was, in fact, shown to be close to, but not included in the *Ace* locus (SOREQ et al. 1985), and this was definitively established by the subsequent cloning of the *Drosophila* AChE structural gene (HALL and SPIERER 1986). The reason for the induction of AChE synthesis in oöcytes by hybrid-selected mRNAs remains unclear.

Polynucleotide probes were designed from the amino acid sequence of human BuChE (LOCKRIDGE et al. 1987) and used to screen human cDNA libraries. The clones obtained in several laboratories all appear to correspond to BuChE (PRODY et al. 1987; LA DU and LOCKRIDGE 1986; ROSENBERRY, personal communication).

TAYLOR and his colleagues prepared polynucleotide probes corresponding to peptidic sequences of purified lytic G_4 AChE, derived from asymmetric forms of *T. californica* electric organs (MACPHEE-QUIGLEY et al. 1985). Using these probes they identified several positive clones from a library of cDNA from *Torpedo* electric organ in RNA (SCHUMACHER et al. 1986a). One clone appeared to cover the coding sequence of the mature protein.

We recently identified a small cDNA clone corresponding to a fragment of *Torpedo* AChE catalytic subunit by immunological characterization of the translation products obtained in a λgt11 expression vector (SIKORAV et al. 1985). The recombinant protein was found to compete very efficiently with the native enzyme in an immunoassay, although the cDNA fragment codes only for amino acids 245–298 of the mature subunit (cf. Fig. 7); the immunogenicity of this region is probably due to the fact that it is particularly hydrophilic.

Using this DNA fragment as a probe, we isolated cDNA clones which allowed us to establish an entire coding sequence (SIKORAV et al. 1987; see Fig. 7), which is obviously homologous to that described by SCHUMACHER et al. (1986a, b).

Southern blot analyses with genomic restriction fragments suggested the existence of a single AChE gene in *Torpedo* (SIKORAV et al. 1987). Considering the identity of N-terminal peptide sequences (see Sect. I. III.), the difference between the A and G subunits probably lies in the C-terminal domain of the protein and may be generated by differential splicing of a primary transcript, as shown in the case of the DAF protein (CARAS et al. 1987).

In Northern blot analysis of mRNA from *Torpedo* electric organ, we found three distinct transcripts of 14500, 10500 and 5500 bp (SIKORAV et al. 1985), and mutiple mRNAs were also demonstrated by SCHUMACHER et al. (1986a). It is at present too early to ascribe these transcripts to different types of catalytic subunits. We assume however that they are fully mature mRNAs and not nuclear precursors, because such precursors have not been observed in the case of the different subunits of *Torpedo* nicotinic acetylcholine receptor (nAChR) in identical experimental conditions (NODA et al. 1983).

We have shown the existence of multiple 5′-untranslated sequences by analysing several cDNA clones, by S1 mapping and by primer extension experiments (SIKORAV et al. 1987). This multiplicity of 5′-untranslated regions was not expected; it is not *a priori* directly related to the polymorphism of AChE, but might be involved in the regulation of AChE expression (LEFF et al. 1986).

III. Primary Structures of ChEs

1. Homology Among ChEs

It has long been known that a 6-amino acid sequence that contains the active-site serine residue, Phe-Gly-Glu-Ser-Ala-Gly, is highly conserved in cholinesterases (see, for example, JANSZ et al. 1959). Interestingly, a 30-amino acid segment which includes this peptide was found to be unchanged in allelic variants of human BuChE in which the catalytic activity is modified or suppressed (LOCKRIDGE and LA DU 1986).

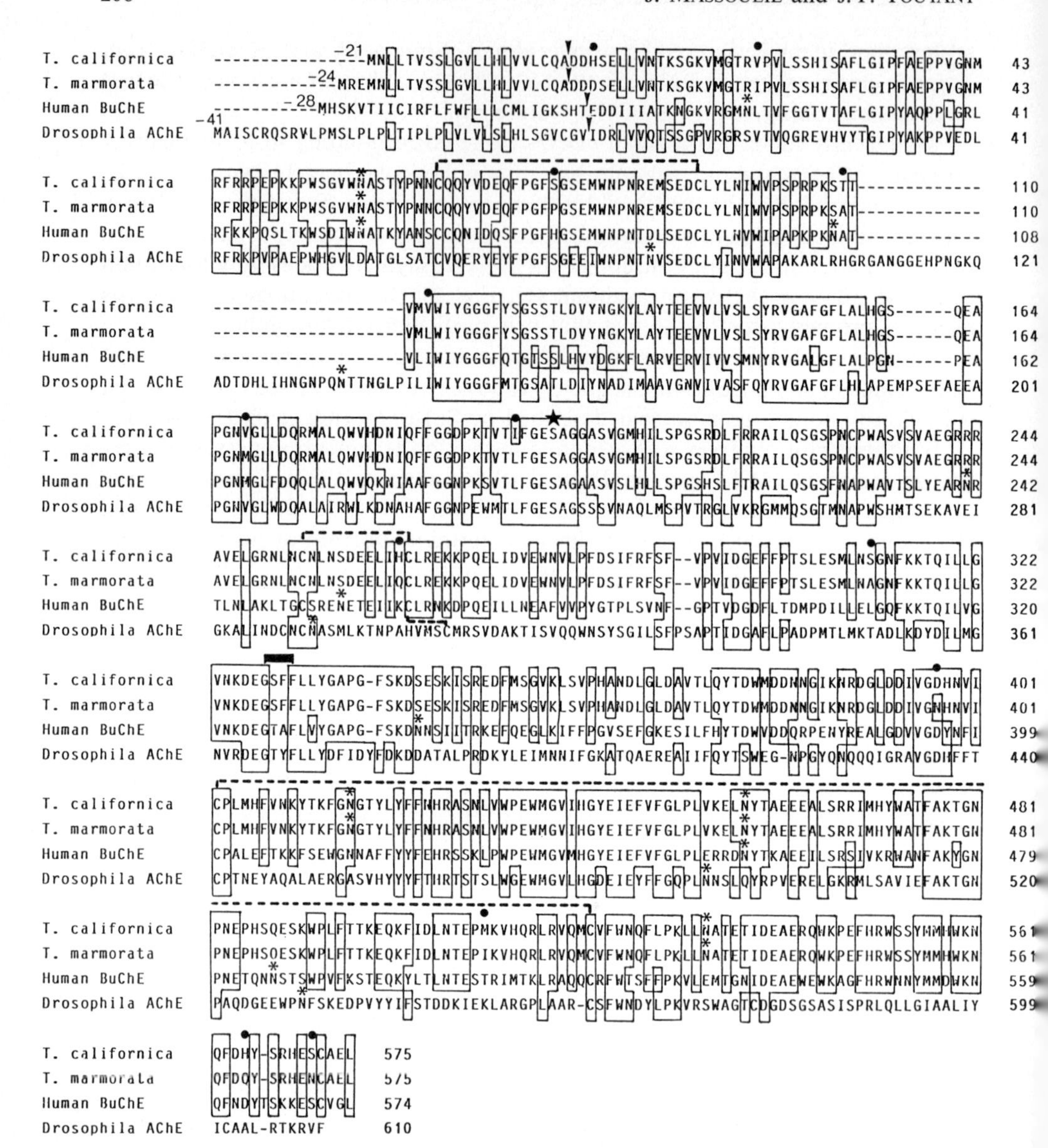

Fig. 7. Primary structure of cholinesterases. The amino acid sequences of AChE from *Torpedo californica* (SCHUMACHER et al. 1986a; P. TAYLOR, personal communication), *T. marmorata* (SIKORAV et al. 1987) were deduced from cDNA sequences. They probably correspond to the catalytic subunit of the asymmetric forms of AChE. N-terminal sequences were determined in the case of *T. californica* (MACPHEE-QUIGLEY et al. 1986) and *T. marmorata* (BON et al. 1986) and found to be identical for all forms of AChE. The primary structure of *Drosophila* AChE was deduced from the cDNA sequence (HALL and SPIERER 1986), and its N-terminal sequence was determined by ROSENBERRY and his colleagues (T.L. ROSENBERRY, personal communication). The structure of human serum BuChE was determined chemically (LOCKRIDGE et al. 1987) and from the sequence of a cDNA (PRODY et al. 1987). *Arrows* indicate the N-terminal amino acid of the mature protein; *frames*, identical residues; *star*, serine residue of active site (position 200 in *Torpedo*); *thick line*, (328 – 331 in *Torpedo*) a peptide which appears to be involved in the anionic subsite, be-

Labelling experiments using a cationic photo-activatable reagent have identified a tetrapeptide from *Electrophorus* AChE, Gly-Ser-*X*-Phe, where *X* is the undetermined residue labelled by the reagent. This peptide is probably involved in the anionic subsite responsible for the binding of the quaternary ammonium group of ACh (KIEFFER et al. 1986).

N-terminal sequences were determined for AChE from *Torpedo* electric organ (MACPHEE-QUIGLEY et al. 1985; BON et al. 1986), ox brain (BON et al. 1986) and human erythrocytes (HAAS and ROSENBERRY 1985). The complete primary structure of human serum BuChE was also determined by direct amino acid analysis (LOCKRIDGE 1984; LOCKRIDGE et al. 1987). All N-terminal sequences show a high degree of homology. It is to be noted that the monoclonal antibody AE-2 (see Section H. II. 2) which partially inhibits the activity of human and ox AChE is directed against a conserved peptide sequence close to the N-terminal extremity (B.P. DOCTOR, personal communication).

For an understanding of the origin of the diversity of ChE molecular forms it is important to note that all forms of *Torpedo* AChE, including the LSS and DS amphiphilic G_2 form and the asymmetric forms, share the same N-terminal sequence (MACPHEE-QUIGLEY et al. 1985; BON et al. 1986).

Recently analyses of cDNA sequences have considerably enriched our knowledge of AChE primary structure. The N-terminal residue of the mature catalytic subunit of *Torpedo* AChE is preceded by a hydrophobic sequence, which presumably corresponds to a signal peptide removed during the processing of the enzyme (P. TAYLOR, personal communication; SIKORAV et al. 1987). The C-terminal sequence deduced from the cDNA is identical to that of the catalytic subunit of the asymmetric forms (SCHUMACHER et al. 1986a, b), which strongly suggests that this cDNA corresponds to this type of molecule.

The primary sequence of this enzyme is strongly homologous to that of human BuChE, determined by direct protein sequencing (LOCKRIDGE 1984; LOCKRIDGE et al. 1987) and that of *Drosophila* AChE, determined from the cDNA sequence (HALL and SPIERER 1986) (Fig. 7). These enzymes obviously possess the same peptide sequence around the active-site serine (Phe-Gly-Glu-Asp-Arg-Gly), and it is also possible to identify a peptide which may be associated with the anionic subsite Gly-Ser-Phe-Phe in *Torpedo* (328–331), according to labelling experiments of *Electrophorus* AChE (KIEFFER et al. 1986).

Interestingly, the primary structures of ChEs shown in Fig. 7 are likely to correspond to three types of enzyme forms. The *Torpedo* sequences probably repre-

Fig. 7 (continued) cause of its homology to an *Electrophorus* AChE labelled peptide (KIEFFER et al. 1986); *dots*, divergences between the two *Torpedo* species; *asterisks*, potential *N*-glycosylation sites in the case of *Torpedo* and *Drosophila* AChE and observed glycosylated asparagines in the case of human BuChE; *dashed lines*, disulphide loops, as determined for *T. californica* AChE (MACPHEE-QUIGLEY et al. 1986); A (C^{67-94}), B ($C^{254-265}$), and C ($C^{402-521}$). Cysteine 231 was found to be free and cysteine 572 was found to be engaged in the formation of dimers. Note that only the Cys residues engaged in intrachain loops are conserved, at homologous or very close positions, in the four ChEs

sent the catalytic subunit of the asymmetric forms, as mentioned above. The human BuChE sequence represents a soluble G_4 enzyme, in which the subunits are arranged as two dimers, linked by one disulphide bond (LOCKRIDGE et al. 1979). *Drosophila* AChE has been characterized as a dimeric, glycolipid-anchored enzyme (GNAGEY et al. 1987; FOURNIER et al. 1988).

TAYLOR and his colleagues determined the position of intra- and intersubunit disulphide bridges in the catalytic subunit of the asymmetric forms (MACPHEE-QUIGLEY et al. 1986). They showed the existence of three intrasubunit loops (in *Torpedo*, 67–94, 254–265, 402–521). The corresponding cysteine residues are conserved in the known cholinesterase sequences and also in thyroglobulin. Two other cysteine residues are present in the *Torpedo* sequence and are not conserved: one, which is close to the C-terminal (572), has been shown to be involved in a dimeric intersubunit linkage, while the other (231) is considered to be involved in the binding of the tail peptide. This interpretation is not consistent however with the results of ROSENBERRY, who found that in *Electrophorus* the same cysteine residue is probably involved in the formation of dimers and in the binding of catalytic and tail subunits (see Sect. E. I).

Figure 7 shows that the vertebrate ChEs are more closely related to each other than to *Drosophila* AChE, which for example contains two insertions of 33 and 6 amino acids. Assuming that these are not secondary insertions but are evolutionarily significant, this suggests that the divergence between vertebrate AChE and BuChE occurred in the deuterostomian lineage, perhaps simultaneously with the emergence of vertebrates. Indeed, it is extremely likely that the two enzymes derive from an ancestor ChE which also possessed asymmetric forms, since these structures are known in vertebrates only.

2. Homology Between ChEs and Thyroglobulin

Unexpectedly, the sequence of the C-terminal domain of mammalian thyroglobulin presents a clear homology with AChE (SCHUMACHER et al. 1986a; SWILLENS et al. 1986) (Fig. 8). Thyroglobulin is a protein of 2750 amino acids, containing several approximately repeated domains and a large, unique C-terminal domain. The homology between the C-terminal region of thyroglobulin and ChEs conserves the hydropathy profile, as well as the cysteine residues involved in intrasubunit bonds and thus appears to maintain the tertiary folding of the protein, but it does not conserve either the catalytic site of the enzyme or the hormonogenic sites, three of which are close to the C-terminal (MERCKEN et al. 1985).

Ox thyroglobulin and *Torpedo* AChE share 155 identical amino acids (allowing for a few insertions or deletions). With the same alignment the number of common amino acids is essentially the same with human BuChE (147) but significantly smaller (117) with *Drosophila* AChE. In addition, thyroglobulins do not possess the amino-acid sequences that exist in *Drosophila* AChE but not in vertebrate enzymes. This suggests that the C-terminal domain of thyroglobulin evolved from a ChE or ChE-like protein after the divergence between insects and vertebrates. The remarkable maintenance of a considerable homology, and partic-

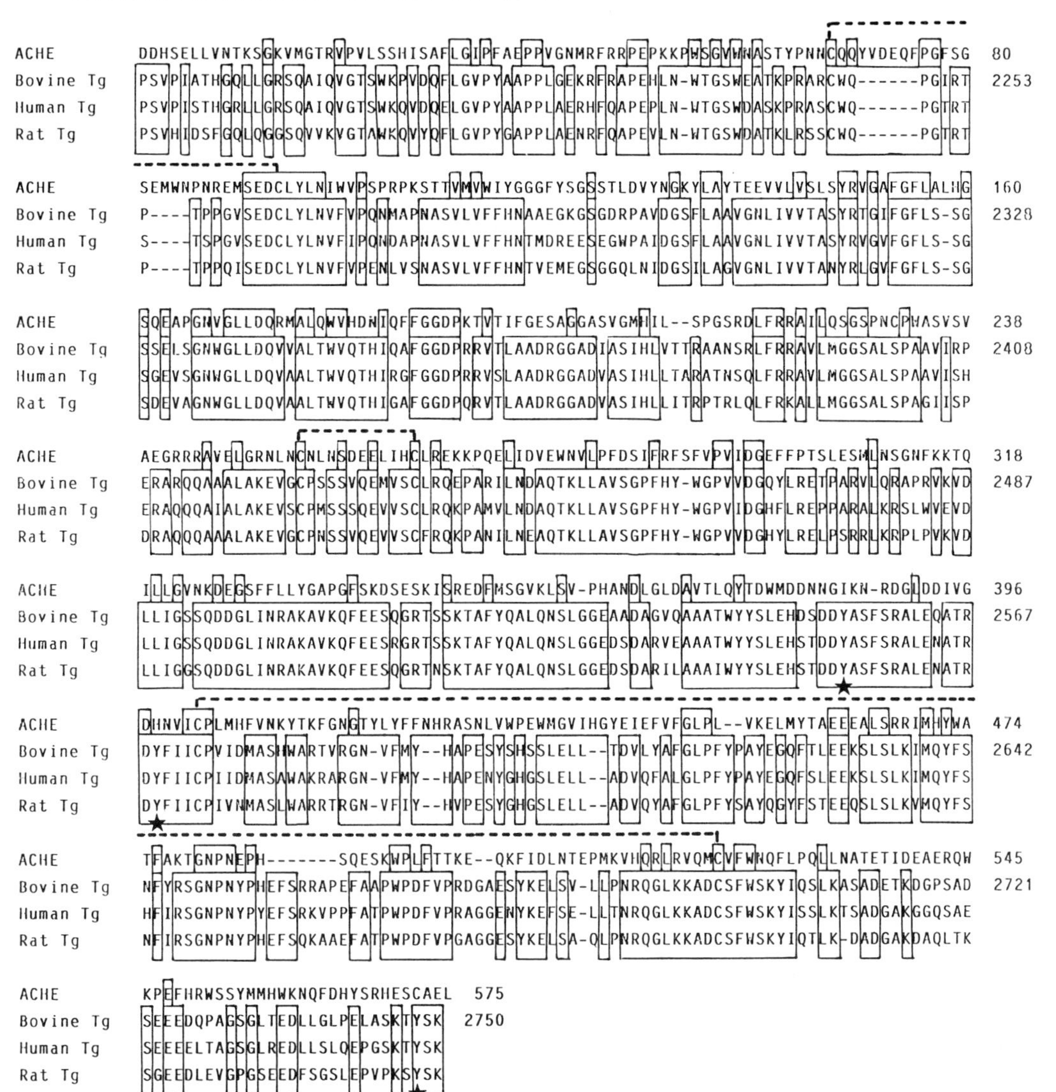

Fig. 8. Homology between *Torpedo* AChE and mammalian thyroglobulin. The sequence of mature AChE from *Torpedo californica* (SCHUMACHER et al. 1986a) is aligned with the sequence of the C-terminal region of ox (MERCKEN et al. 1985), rat (DI LAURO et al. 1985) and human (MALTHIÉRY and LISSITZKY 1987) thyroglobulins. Residues of AChE which coincide with conserved amino acids in the three thyroglobulins in *frames*; hormonogenic tyrosines indicated by *stars*. The cysteine residues of AChE involved in intrachain loops (cf. Fig. 7) are also present in the thyroglobulins

ularly in the tertiary structure of the protein, raises the question of its possible functional significance. It could have served to conserve a catalytic activity (perhaps peptidasic) or, more likely, it might be involved in specific protein-protein interactions, possibly with membrane-bound components (SIKORAV et al. 1987). It has already been proposed that the two proteins share conformations because they interact with membranes in a similar way (SWILLENS et al. 1986). A transitory attachment through a glycolipid anchor appears unlikely however since thyroglobulin does not seem to possess the appropriate C-terminal sequence.

3. Lack of Homology with Other Proteins

The sequence of *Torpedo* AChE catalytic subunit does not appear to present any homology with the subunits of the *Torpedo* ACh receptor (NODA et al. 1983).

It is surprising that ChEs present no homology with serine proteases apart from a short peptide sequence that contains the active-site serine. Particularly the imidazole and aspartic acid residues that form the charge relay system of these enzymes are not found at or near the expected homologous locations in the AChE sequences. It is likely that such a charge relay system also exists in ChEs; however the kinetics of AChE differentiate it from trypsin-like serine hydrolases: acetyl-AChE is hydrolysed about 10^6 times faster than acetyl-chymotrypsin, and deuterium oxide effects suggest the existence of a rate-limiting conformational change (induced fit) (ROSENBERRY 1975a, b).

Because the efficiency of cross-specific hybridization of AChE and BuChE genetic sequences is probably very high, additional peptide sequence data for these enzymes will certainly become available quite rapidly, allowing fruitful comparisons concerning the evolution of these enzymes and the structural basis of their catalytic specificity.

J. Conclusions

All vertebrates appear to possess two distinct ChEs, AChE and BuChE, which present similar catalytic properties and homologous sets of molecular forms. This suggests a functional similarity or equivalence of the two enzymes, but the significance of their duality remains enigmatic. The catalytic subunits of ChEs are similar in size throughout the range of vertebrates except in domestic fowl and probably other birds, in which both enzymes present an additional trypsin-sensitive domain that is not involved in the catalytic activity or in quaternary associations. AChE and BuChE do not cross-react immunologically, and interspecific cross-reactivity for each enzyme depends on the degree of phylogenetic relationship.

In a given organism AChE and BuChE present a variety of molecules. For each enzyme the different molecular forms possess the same catalytic activity but differ in a number of ways: primary structure of the catalytic subunits, degree of polymerization, presence of non-catalytic subunits, glycosylation or addition of a hydrophobic glycolipid. In order to explore this complexity, it is useful to begin by classifying the enzymes according to their general quaternary structure. One must clearly distinguish solubility characteristics and intrinsic molecular properties. The molecular forms corresponding to similar quaternary associations may be classified according to their hydrodynamic properties. These characteristics differentiate asymmetric, collagen-tailed from globular forms. The asymmetric forms (A_4, A_8, A_{12}) possess a characteristic sensitivity to collagenase. They may be inserted in extracellular structures (basal lamina) through ionic interactions, probably involving polyanions (heparan sulphate proteoglycan) and divalent ions. The globular forms are classified under the general categories of monomers (G_1), dimers (G_2), and tetramers (G_4). They may be hydrophilic or amphiphilic: amphiphilic molecules possess a small hydrophobic domain that consists of a phosphatidylinositol glycolipid, covalently linked at the C-terminal extremity or possess a distinct hydrophobic structural subunit. Amphiphilic molecules occur in soluble as well as in detergent-soluble fractions.

Detailed biochemical analyses of asymmetric and globular amphiphilic forms of *Torpedo* showed that the catalytic subunits present significant differences in their size and in their peptide maps. The binding of lectins and of monoclonal antibodies, as well as non-denaturing electrophoresis, reveal a further diversity within the catalytic subunits of ChEs in various organisms.

Asymmetric and globular forms are probably encoded by different mRNAs. The distinct ionic and hydrophobic interactions of the asymmetric and globular forms, which allow these molecules to be directed to various subcellular locations, are however conferred upon otherwise soluble hydrophilic catalytic subunits by post-translational addition of the tail peptides or of a glycolipidic hydrophobic anchor.

We now possess a fairly complete description of the molecular diversity of ChEs and the tools to apply the powerful methods of molecular biology to explore the roles of genomic diversity, mRNA splicing and post-translational processing in the generation of their multiple variants. The stage is therefore set for rapid advance in the understanding of the complexity of these important and fascinating enzymes.

Acknowledgements. We wish to thank all our colleagues for the communication of their recent work, Drs. Suzanne Bon, Terrone L. Rosenberry and Jean-Louis Sikorav for critical reading of the manuscript, and Mrs. Jacqueline Pons for her expert secretarial assistance.

References

Abe T, Sakai M, Saisu H (1983) A monoclonal antibody against catalytic subunits of acetylcholinesterase in the electric organ of an electric ray *Narke japonica*. Neurosci Lett 38:61–66

Abo T, Balch CM (1981) A differentiation antigen of human NK and K cells identified by a monoclonal antibody (HNK-1). J Immunol 127:1024–1029

Adams DH, Whittaker VP (1949) The cholinesterase of human blood. I. The specificity of the plasma enzyme and its relation to the erythrocyte cholinesterase. Biochim Biophys Acta 3:358–366

Adamson ED (1977) Acetylcholinesterase in mouse brain erythrocytes and muscle. J Neurochem 28:605–615

Allemand P, Bon S, Massoulié J, Vigny M (1981) The quaternary structure of chicken acetylcholinesterase and butyryl-cholinesterase: effect of collagenase and trypsin. J Neurochem 36:860–867

Anglister L, McMahan UJ (1983) Acetylcholinesterase in the synaptic basal lamina of damaged muscle fibers (Abstr). 2nd International Meeting on Cholinesterases, Bled, p 58

Anglister L, Silman I (1978) Molecular structure of elongated forms of electric eel acetylcholinesterase. J Mol Biol 125:293–311

Atack JR, Perry EK, Bonham JR, Candy JM, Perry RH (1986) Molecular forms of acetylcholinesterase and butyrylcholinesterase in the elderly human central nervous system. J Neurochem 47:263–277

Auditore JV, Hartmann RC (1959) Paroxysmal nocturnal hemoglobinuria. II. Erythrocyte acetylcholinesterase defect. Am J Med 27:401

Baltz T, Duvillier G, Giroud C, Richet C, Baltz D, Degaud P (1983) The variant surface glycoprotein of *Trypanosoma equiperdum*: identification of a phosphorylated glycopeptide as the cross-reacting antigenic determinant. FEBS Lett 158:174–178

Barrat A, Gómez-Barriocanal J, Ramírez G (1984) Two classes of collagen-tailed, asymmetric molecular forms of acetylcholinesterase in skeletal muscle: differential effects of denervation. Neurochem Int 6:403–412

Barat A, Escudero E, Ramírez G (1986) Heparin and the solubilization of asymmetric acetylcholinesterase. FEBS Lett 195:209–214

Barnett P, Rosenberry TL (1978) A residual subunit fragment in the conversion of 18 S to 11 S acetylcholinesterase. Fed Proc 36:485

Barton PL, Futerman AH, Silman I (1985) Arrhenius plots of acetylcholinesterase activity in mammalian erythrocytes and in *Torpedo* electric organ: effect of solubilization by proteinases and by a phosphatidylinositol-specific phospholipase C. Biochem J 231:237–240

Bender W, Spierer P, Hogness DS (1983) Chromosomal walking and jumping to isolate DNA from the *Ace* and *rosy* loci and the bithorax complex in *Drosophila melanogaster.* J Mol Biol 168:17–33

Berman HA, Yguerabide J, Taylor P (1980) Fluorescence energy transfer on acetylcholinesterase: spatial relationship between peripheral site and active center. Biochemistry 19:2226–2235

Birman S (1985) Determination of acetylcholinesterase activity by a new chemiluminescence assay with the natural substrate. Biochem J 225:825–828

Bon S (1982) Molecular forms of acetylcholinesterase in developing *Torpedo* embryos. Neurochem Int 4:577–585

Bon S, Massoulié J (1976) Molecular forms of *Electrophorus* acetylcholinesterase; the catalytic subunits: fragmentation, intra- and inter-subunit disulfide bonds. FEBS Lett 71:273–278

Bon S, Massoulié J (1978) Collagenase sensitivity and aggregation properties of *Electrophorus* acetylcholinesterase. Eur J Biochem 89:89–94

Bon S, Massoulié J (1980) Collagen-tailed and hydrophobic component of acetylcholinesterase in *Torpedo marmorata* electric organ. Proc Natl Acad Sci USA 77:4464–4468

Bon S, Huet M, Lemonnier M, Rieger F, Massoulié J (1976) Molecular forms of *Electrophorus* acetylcholinesterase: molecular weight and composition. Eur J Biochem 68:523–530

Bon S, Cartaud J, Massoulié J (1978) The dependence of acetylcholinesterase aggregation at low ionic strength upon a polyanionic component. Eur J Biochem 85:1–14

Bon S, Vigny M, Massoulié J (1979) Asymmetric and globular forms of acetylcholinesterase in mammals and birds. Proc Natl Acad Sci USA 76:2546–2550

Bon S, Chang JY, Strosberg AD (1986) Identical N-terminal peptide sequences of asymmetric forms and of amphiphilic low-salt-soluble and detergent-soluble dimers of *Torpedo* acetylcholinesterase: comparison with bovine acetylcholinesterase. FEBS Lett 209:206–212

Bon S, Méflah K, Musset F, Grassi J, Massoulié J (1987) An immunoglobulin M monoclonal antibody, recognizing a subset of acetylcholinesterase molecules from electric organs of *Electrophorus* and *Torpedo*, belongs to the HNK-1 anti-carbohydrate family. J Neurochem 49:1720–1731

Bon S, Toutant JP, Méflah K, Massoulié J (1988a) Amphiphilic and non-amphiphilic forms of *Torpedo* cholinesterases. I: Solubility and aggregation properties. J Neurochem (in press)

Bon S, Toutant JP, Méflah K, Massoulié J (1988b) Amphiphilic and non-amphiphilic forms of *Torpedo* cholinesterases. II: Electrophoretic variants, phosphatidylinositol-phospholipase C sensitive and insensitive forms. J Neurochem (in press)

Borst P, Cross GAM (1982) Molecular basis for trypanosome antigenic variation. Cell 29:291–303

Brandan E, Inestrosa NC (1984) Binding of asymmetric forms of acetylcholinesterase to heparin. Biochem J 221:415–422

Brandan E, Inestrosa NC (1986) The synaptic form of acetylcholinesterase binds to cell-surface heparan sulfate proteoglycans. J Neurosci Res 15:185–196

Brandan E, Inestrosa NC (1987) Co-solubilization of asymmetric acetylcholinesterase and dermatan sulfate proteoglycan from the extracellular matrix of the rat skeletal muscle. FEBS Lett 213:159–163

Brandan E, Maldonado M, Garrido J, Inestrosa NC (1985) Anchorage of collagen-tailed acetylcholinesterase to the extracellular matrix is mediated by heparan sulfate proteoglycans. J Cell Biol 101:985–992

Brimijoin S (1983) Molecular forms of acetylcholinesterase in brain, nerve and muscle: nature, localization and dynamics. Prog Neurobiol 21:291–322

Brimijoin S, Rakonczay Z (1986) Immunology and molecular biology of the cholinesterases: current results and prospects. Int Rev Neurobiol 28:363–410

Brimijoin S, Mintz KP, Alley MC (1983) Production and characterization of separate monoclonal antibodies to human acetylcholinesterase and butyrylcholinesterase. Mol Pharmacol 24:513–520

Brimijoin S, Mintz KP, Prendergast FG (1985) An inhibitory monoclonal antibody to rabbit brain acetylcholinesterase: evidence for binding near the catalytic site. Mol Pharmacol 28:539–545

Brockman SK, Usiak MF, Younkin SG (1986) Assembly of monomeric acetylcholinesterase in cultured rat myotubes. J Neurosci 4:131–140

Brodbeck U (1986) Amphiphilic acetylcholinesterase: properties and interactions with lipids and detergents. In: Watts A, de Pont J (eds) Progress in protein-lipid interactions 2. Elsevier, Amsterdam, p 303

Brzin M, Sketelj J, Klinar B (1983) Cholinesterases. In: Lathja A (ed) Handbook of neurochemistry. Plenum, New York, pp 251–292

Brzin M, Barnard EA, Sket D (eds) (1984) Cholinesterases: fundamental and applied aspects. de Gruyter, Berlin

Campbell DG, Gagnon J, Reid KBM, Williams AF (1981) Rat brain Thy-1 glycoprotein: the amino acid sequence, disulphide bonds and unusual hydrophobic region. Biochem J 195:15–30

Capdeville Y, Baltz T, Deregnaucourt C, Keller A-M (1986) Immunological evidence of a common structure between *Paramecium* surface antigens and *Trypanosoma* variant surface glycoproteins. Exp Cell Res 167:75–86

Caras IW, Davitz MA, Rhee L, Weddell G, Martin DW Jr, Nussenzweig V (1987) Cloning of decay-accelerating factor suggests novel use of splicing to generate two proteins. Nature 325:545–549

Cardoso de Almeida L, Turner MJ (1983) The membrane form of variant surface glycoproteins of *Trypanosoma brucei*. Nature 302:349–352

Cartaud J, Rieger F, Bon S, Massoulié J (1975) Fine structure of electric eel acetylcholinesterase. Brain Res 88:127–130

Cartaud J, Bon S, Massoulié J (1978) *Electrophorus* acetylcholinesterase: biochemical and electron microscope characterization of low ionic strength aggregates. J Cell Biol 77:315–322

Chang HW, Niebroj-Dobosz I, Bock E, Miranda AF, Bonilla E (1986) Monoclonal antibodies against two molecular species from *Torpedo californica* and their cross-reactivities with human acetylcholinesterase. Muscle Nerve 9:143

Chatonnet A, Bacou F (1983) Acetylcholinesterase molecular forms in the fast or slow muscles of the chicken and the pigeon. FEBS Lett 161:122–126

Chemnitius JM, Haselmeyer KH, Zech R (1983) Brain cholinesterases: differentiation of target enzymes for toxic organophosphorus compounds. Biochem Pharmacol 32:1693–1699

Chou KH, Ilyas AA, Evans JE, Quarles RH, Jungalwala FB (1985) Structure of a glycolipid reacting with monoclonal IgM in neuropathy and with HNK-1. Biochem Biophys Res Commun 128:383–388
Cisson CM, McQuarrie CH, Sketelj J, McNamee MG, Wilson BW (1981) Molecular forms of acetylcholinesterase in chick embryonic fast muscle: developmental changes and effects of DFP. Dev Neurosci 4:157–164
Cohen R, Barenholz Y (1984) Characterization of the association of *Electrophorus electricus* acetylcholinesterase with sphingomyelin liposomes. Biochim Biophys Acta 778:94–104
Cross GAM (1987) Eukaryotic protein modification and membrane attachment via phosphatidylinositol. Cell 48:179–181
Davitz MA, Low MG, Nussenzweig V (1986) Release of decay-accelerating factor (DAF) from the cell membrane by phosphatidylinositol-specific phospholipase C (PIPLC). J Exp Med 163:1150–1161
Di Francesco C, Brodbeck U (1981) Interaction of human red cell membrane acetylcholinesterase with phospholipids. Biochim Biophys Acta 640:359–364
Di Giamberardino L, Couraud JY (1978) Rapid accumulation of high molecular weight acetylcholinesterase in transected sciatic nerve. Nature 271:170–172
Di Lauro R, Obici S, Condliffe D, Ursini VM, Musti A, Moscatelli C, Avvedimento VE (1985) The sequence of 967 amino acids at the carboxyl-end of rat thyroglobulin: location and surroundings of two thyroxine-forming sites. Eur J Biochem 148:7–11
Doctor BP, Camp S, Gentry MK, Taylor SS, Taylor P (1983) Antigenic and structural differences in the catalytic subunits of the molecular forms of acetylcholinesterase. Proc Natl Acad Sci USA 80:5767–5771
Dreyfus PA, Rieger F, Pinçon-Raymond M (1983) Acetylcholinesterase of mammalian neuromuscular junctions: presence of tailed asymmetric acetylcholinesterase in synaptic basal lamina and sarcolemma. Proc Natl Acad Sci USA 80:6698–6702
Dreyfus PA, Friboulet A, Tran LH, Rieger F (1984) Polymorphism of acetylcholinesterase and identification of new molecular forms after sedimentation analysis. Biol Cell 51:35–41
Dudai Y, Silman I (1973) The effect of Ca^{2+} on interaction of acetylcholinesterase with subcellular fractions of electric organ tissue from the electric eel. FEBS Lett 30:49–52
Dudai Y, Silman I (1974) The effects of solubilization procedures on the release and molecular state of acetylcholinesterase from electric organ tissue. J Neurochem 23:1177–1187
Dudai Y, Silman I, Shinitzky M, Blumberg S (1972) Purification by affinity chromatography of the molecular forms of acetylcholinesterase present in fresh electric organ tissue of electric eel. Proc Natl Acad Sci USA 69:2400–2403
Dutta-Choudhury TA, Rosenberry TL (1984) Human erythrocyte acetylcholinesterase is an amphipathic protein whose short membrane binding domain is removed by papain digestion. J Biol Chem 259:5653–5660
Eckerson HW, Oseroff A, Lockridge O, La Du BN (1983) Immunological comparison of the usual and atypical human serum cholinesterase phenotypes. Biochem Genet 21:93–108
Edwards JA, Brimijoin S (1983) Thermal inactivation of the molecular forms of acetylcholinesterase and butyrylcholinesterase. Biochim Biophys Acta 742:509–516
Etges R, Bouvier J, Bordier C (1986) The major surface protein of *Leishmania* promastigotes is anchored in the membrane by a myristic acid-labeled phospholipid. EMBO J 5:597–601
Fambrough DM, Engel AG, Rosenberry TL (1982) Acetylcholinesterase of human erythrocytes and neuromuscular junctions: homologies revealed by monoclonal antibodies. Proc Natl Acad Sci USA 79:1078–1082
Ferguson MAJ, Haldar K, Cross GAM (1985) *Trypanosoma brucei* variant surface glycoprotein has a sn-1,2-dimyristyl glycerol membrane anchor at its COOH terminus. J Biol Chem 260:4963–4968
Ferguson MAJ, Duszenko M, Lamont GS, Overath P, Cross GAM (1986) Biosynthesis of *Trypanosoma brucei* variant surface glycoproteins: *N*-glycosylation and addition of a phosphatidylinositol membrane anchor. J Biol Chem 261:356–362

Fernandez HL, Duell AJ, Festoff BW (1979) Cellular distribution of 16 S acetylcholinesterases. J Neurochem 32:581–585

Fournier D, Bergé J, Cardoso de Almeida ML, Bordier C (1988) Acetylcholinesterase from *Musca domestica* and *Drosophila melanogaster* brain are linked to membranes by a glycophospholipid anchor sensitive to an endogenous phospholipase. J Neurochem (in press)

Fox GQ, Richardson GP (1978) The developmental morphology of *Torpedo marmorata*: electric organ-myogenic phase. J Comp Neurol 179:677–697

Fox GQ, Richardson GP (1979) The developmental morphology of *Torpedo marmorata*: electric organ-electrogenic phase. J Comp Neurol 185:293–314

Futerman AH, Low MG, Silman I (1983) A hydrophobic dimer of acetylcholinesterase from *Torpedo californica* electric organ is solubilized by phosphatidylinositol-specific phospholipase C. Neurosci Lett 40:85–89

Futerman AH, Fiorini RM, Roth E, Michaelson DM, Low MG, Silman I (1984) Solubilization of membrane-bound acetylcholinesterase by a phosphatidylinositol-specific phospholipase C: enzymatic and physicochemical studies. In: Brzin M, Barnard EA, Sket D (eds) Cholinesterases: fundamental and applied aspects. de Gruyter, Berlin, p 99

Futerman AH, Fiorini RM, Roth E, Low MG, Silman I (1985a) Physicochemical behaviour and structural characteristics of membrane-bound acetylcholinesterase from *Torpedo*: effect of phosphatidylinositol-specific phospholipase C. Biochem J 226:369–377

Futerman AH, Low MG, Ackermann KE, Sherman WR, Silman I (1985b) Identification of covalently bound inositol in the hydrophobic membrane-anchoring domain of *Torpedo* acetylcholinesterase. Biochem Biophys Res Commun 129:312–317

Futerman AH, Low MG, Michaelson DM, Silman I (1985c) Solubilization of membrane-bound acetylcholinesterase by a phosphatidylinositol-specific phospholipase C. J Neurochem 45:1487–1494

Gennari K, Brodbeck U (1985) Molecular forms of acetylcholinesterase from human caudate nucleus: comparison of salt soluble and detergent soluble tetrameric enzyme species. J Neurochem 44:697–704

Gisiger V, Vigny M, Gautron J, Rieger F (1978) Acetylcholinesterase of rat sympathetic ganglion: molecular forms, localization and effects of denervation. J Neurochem 30:501–516

Gnagey AL, Forte M, Rosenberry TL (1987) Isolation and characterization of acetylcholinesterase from *Drosophila*. J Biol Chem 262:13290–13298

Gómez-Barriocanal J, Barat A, Escudero E, Rodriguez-Borrajo C, Ramírez G (1981) Solubilization of collagen-tailed acetylcholinesterase from chick retina: effect of different extraction procedures. J Neurochem 37:1239–1249

Grassi J, Vigny M, Massoulié J (1982) Molecular forms of acetylcholinesterase in bovine caudate nucleus and superior cervical ganglion: solubility properties and hydrophobic character. J Neurochem 38:457–469

Grassi J, Massoulié J, Timpl R (1983) Relationship of collagen-tailed acetylcholinesterase with basal lamina components: absence of binding with laminin fibronectin and collagen types IV and V and lack of reactivity with different anticollagen sera. Eur J Biochem 133:31–38

Greenberg AJ, Parker KK, Trevor AJ (1977) Immunochemical studies of mammalian brain acetylcholinesterase. J Neurochem 29:911–917

Grumet M, Hoffman S, Crossin KL, Edelman G (1985) Cytotactin, an extracellular matrix protein of neural and non-neural tissues that mediates glia-neuron interaction. Proc Natl Acad Sci USA 82:8075–8079

Haas R, Rosenberry TL (1985) Quantitative identification of N-terminal amino acids in proteins by radiolabeled reductive methylation and amino acid analysis: application to human erythrocyte acetylcholinesterase. Anal Biochem 148:154–162

Haas R, Brandt P, Knight J, Rosenberry TL (1986) Identification of non amino acid components in the hydrophobic membrane-binding domain at the C-terminus of human erythrocyte acetylcholinesterase. Biochemistry 25:3098–3105

Hall JC, Kankel DR (1976) Genetics of acetylcholinesterase in *Drosophila melanogaster*. Genetics 83:517–533

Hall JC, Alahiotis SN, Strumpf DA, White K (1980) Behavioral and biochemical defects in temperature sensitive acetylcholinesterase mutants of *Drosophila melanogaster*. Genetics 96:939–965

Hall LMC, Spierer P (1986) Ace locus of *Drosophila melanogaster*: structural gene for acetylcholinesterase with an unusual 5′ leader. EMBO J 5:2949–2954

Hall ZW (1973) Multiple forms of acetylcholinesterase and their distribution in endplate and non-endplate regions of rat diaphragm muscle. J Neurol 4:343–361

Hall ZW, Kelly RB (1971) Enzymatic detachment of endplate acetylcholinesterase from muscle. Nature [New Biol] 232:62–64

Hasan FB, Cohen SG, Cohen JB (1980) Hydrolysis by acetylcholinesterase: apparent molar volumes and trimethyl and methyl subsites. J Biol Chem 255:3898–3904

Hasan FB, Elkind JL, Cohen SG, Cohen JB (1981) Cationic and uncharged substrates and reversible inhibitors in hydrolysis by acetylcholinesterase (EC 3.1.1.7); the trimethyl subsite. J Biol Chem 256:7781–7785

He HT, Barbet J, Chaix JC, Goridis C (1986) Phosphatidylinositol is involved in the membrane attachment of N-CAM 120, the smallest component of the neural cell adhesion molecule. EMBO J 5:2489–2494

Helenius A, Simons K (1977) Charge shift electrophoresis: simple method for distinguishing between amphiphilic and hydrophilic proteins in detergent solution. Proc Natl Acad Sci USA 74:529–532

Hodgson AJ, Chubb IW (1983) Isolation of the secretory form of soluble acetylcholinesterase by using affinity chromatography on edrophonium-sepharose. J Neurochem 41:654–662

Holder AA (1983) Carbohydrate is linked through ethanolamine to the C-terminal aminoacid of *Trypanosoma brucei* variant surface glycoprotein. Biochem J 209:261–262

Hollunger EG, Niklasson BH (1973) The release and molecular state of mammalian brain acetylcholinesterase. J Neurochem 20:821–826

Ikezawa H, Nakabayashi T, Suzuki K, Nakajima M, Taguchi T, Taguchi R (1983) Complete purification of phosphatidylinositol specific phospholipase C from a strain of *Bacillus thuringiensis*. J Biochem (Tokyo) 93:1717–1719

Inestrosa NC, Ruiz G (1985) Membrane-bound form of acetylcholinesterase activated during postnatal development of the rat somatosensory cortex. Dev Neurosci 7:120–132

Inestrosa NC, Roberts WL, Marshall T, Rosenberry TL (1987) Acetylcholinesterase from bovine caudate nucleus is attached to membranes by a novel subunit distinct from those of acetylcholinesterase in other tissues. J Biol Chem 262:4441–4444

Jansz HJ, Brons D, Warringa MGPJ (1959) Chemical nature of DFP-binding site of pseudocholinesterase. Biochim Biophys Acta 34:573–575

Karlsson E, Mbugua PM, Rodriguez-Ithurralde D (1984) Fasciculins, anticholinesterase toxins from the venom of the green mamba *Dendroaspis angusticeps*. J Physiol (Paris) 79:232–240

Kása P, Rakonczay Z (1982) Histochemical and biochemical demonstration of the molecular forms of acetylcholinesterase in peripheral nerve of rat. Acta Histochem (Jena) 70:244–257

Kaufmann K, Silman I (1980) Interaction of electric eel acetylcholinesterase with natural and synthetic lipids. Neurochem Int 2:205–207

Keilhauer G, Faissner A, Schachner M (1985) Differential inhibition of neurone-neurone, neurone-astrocyte and astrocyte-astrocyte adhesion by L1, L2 and N-CAM antibodies. Nature 316:728–730

Kieffer B, Goeldner M, Hirth C, Aebersold R, Chang JY (1986) Sequence determination of a peptide fragment from electric eel acetylcholinesterase, involved in the binding of quaternary ammonium. FEBS Lett 202:91–96

Kim BH, Rosenberry TL (1985) A small hydrophobic domain that localizes human erythrocyte acetylcholinesterase in liposomal membranes in cleaved by papain digestion. Biochemistry 24:3586–3592

Klinar B, Kamaric L, Sketelj J, Brzin M (1985) Properties of acetylcholinesterase and nonspecific cholinesterase in rat superior cervical ganglion and plasma. Neurochem Res 10:797–808

Koelle GB (ed) (1963) Cholinesterases and anticholinesterase agents. Springer, Berlin Göttingen Heidelberg (Handbch exp Pharmakol Ergw 15)
Krenz WD, Tashiro T, Wächtler K, Whittaker VP, Witzemann V (1980) Aspects of the chemical embryology of the electromotor system of *Torpedo marmorata* with special reference to synaptogenesis. Neuroscience 5:617–624
Kruse J, Mailhammer R, Wernecke H, Faissner A, Sommer I, Goridis C, Schachner M (1984) Neural cell adhesion molecules and myelin-associated glycoprotein share a common carbohydrate moiety recognized by monoclonal antibodies L2 and HNK-1. Nature 311:153–155
Kruse J, Keilhauer G, Faissner A, Timpl R, Schachner M (1985) The J1 glycoprotein – a novel nervous system cell adhesion molecule of the L2/HNK-1 family. Nature 316:146–148
La Du BN, Lockridge O (1986) Molecular biology of human serum cholinesterase. Fed Proc 45:2965–2969
Lai J, Jedrzejczyk J, Pizzey JA, Green D, Barnard EA (1986) Neural control of the forms of acetylcholinesterase in slow mammalian muscles. Nature 321:72–74
Landauer P, Ruess KP, Liefländer M (1984) Bovine nucleus caudatus acetylcholinesterase: active site determination and investigation of a dimeric form obtained by selective proteolysis. J Neurochem 43:799–805
Lappin RI, Rubin LL (1984) Characterization of monoclonal antibodies to *Torpedo* acetylcholinersterase. J Cell Biol 99:22a
Lappin RL, Rubin LL, Lieberburg IM (1987) Generation of subunit specific antibody probes for *Torpedo* acetylcholinesterase: cross-species reactivity and use in cell-free translation. J Neurobiol 18:75–100
Latov N, Sherman WH, Nemmi R, Galassi G, Shyong JS, Penn AS, Chess L, Olarte MR, Rowland LP, Osserman EF (1980) Plasma cell dyscrasia and peripheral neuropathy with a monoclonal antibody to peripheral nerve myelin. N Engl J Med 303:618
Lazar M, Vigny M (1980) Modulation of the distribution of acetylcholinesterase molecular forms in a murine neuroblastoma × sympathetic ganglion cell hybrid cell line. J Neurochem 35:1067–1079
Lazar M, Salmeron E, Vigny M, Massoulié J (1984) Heavy isotope labeling study of the metabolism of monomeric and tetrameric acetylcholinesterase forms in the murine neuronal-like T28 hybrid cell line. J Biol Chem 259:3703–3713
Lee SL, Taylor P (1982) Structural characterization of the asymmetric (17+13)S species of acetylcholinesterase from *Torpedo.* II. Component peptides obtained by selective proteolysis and disulfide bond reduction. J Biol Chem 257:12292–12301
Lee SL, Heinemann S, Taylor P (1982a) Structural characterization of the asymmetric (17+13)S forms of acetylcholinesterase from *Torpedo.* I. Analysis of subunit composition. J Biol Chem 257:12283–12291
Lee SL, Camp SJ, Taylor P (1982b) Characterization of a hydrophobic, dimeric form of acetylcholinesterase from *Torpedo.* J Biol Chem 257:12302–12309
Leff SE, Rosenfeld MG, Evans RM (1986) Complex transcriptional units: diversity in gene expression by alternative RNA processing. Annu Rev Biochem 55:1091–1117
Li ZY, Bon C (1983) Presence of a membrane-bound acetylcholinesterase form in a preparation of nerve endings from *Torpedo marmorata* electric organ. J Neurochem 40:338–349
Lockridge O (1984) Amino acid composition and sequence of human serum cholinesterase: a progress report. In: Brzin M, Barnard EA, Sket D (eds) Cholinesterases: fundamental and applied aspects. de Gruyter, Berlin, p 5
Lockridge O, La Du B (1986) Amino acid sequence of the active site of human serum cholinesterase from usual, atypical and atypical-silent genotypes. Biochem Genet 24:485–498
Lockridge O, Eckerson HW, La Du BN (1979) Interchain disulfide bonds and subunit organization in human serum cholinesterase. J Biol Chem 254:8324–8330
Lockridge O, Bartels CF, Vaughan TA, Wong CK, Norton SE, Johnson LL (1987) Complete amino-acid sequence of human serum cholinesterase. J Biol Chem 262:549–557
Low MG, Finean JB (1977a) Non-lytic release of acetylcholinesterase from erythrocytes by a phosphatidylinositol-specific phospholipase C. FEBS Lett 82:143–146

Low MG, Finean JB (1977b) Release of alkaline phosphatase from membranes by a phosphatidylinositol-specific phospholipase C. Biochem J 167:281–284
Low MG, Kincade PW (1985) Phosphatidylinositol is the membrane domain of the Thy-1 glycoprotein. Nature 318:62–64
Low MG, Ferguson MA, Futerman AH, Silman I (1986) Covalently attached phosphatidylinositol as a hydrophobic anchor for membrane proteins. Trends Biochem Sci 11:212–215
Lyles JM, Silman I, di Giamberardino L, Couraud JY, Barnard EA (1982) Comparison of the molecular forms of the cholinesterases in tissues of normal and dystrophic chickens. J Neurochem 38:1007–1021
MacPhee-Quigley K, Taylor P, Taylor S (1985) Primary structures of the catalytic subunits from two molecular forms of acetylcholinesterase: a comparison of NH2-terminal and active center sequences. J Biol Chem 260:12185–12189
MacPhee-Quigley K, Vedvick TS, Taylor P, Taylor S (1986) Profile of the disulfide bonds in acetylcholinesterase. J Biol Chem 261:13565–13570
Main AR, McKelly SC, Burgess-Miller SK (1977) A subunit-sized butyrylcholinesterase present in high concentrations in pooled rabbit serum. Biochem J 167:367–376
Majumbar R, Balasubramanian AS (1982) Essential and non essential phosphatidylinositol residues in acetylcholinesterase and arylacylamidase of sheep basal ganglia. FEBS Lett 146:335–338
Malthiéry Y, Lissitzky S (1987) Primary structure of human thyroglobulin deduced from the sequence of its 8448 base complementary DNA. Eur J Biochem 165:491–498
Manavalan P, Taylor P, Johnson WC Jr (1985) Circular dichroism studies of acetylcholinesterase conformation: comparison of the 11 S and 5.6 S species and the differences induced by inhibitory ligands. Biochim Biophys Acta 829:365–370
Marnay A, Nachmansohn D (1938) Cholinesterase in voluntary muscle. J Physiol (Lond) 92:37–47
Marsh D, Massoulié J (1985) Proteolytic digestion patterns of soluble and detergent soluble bovine caudate nucleus acetylcholinesterases. J Neurochem 44:1602–1604
Marsh D, Grassi J, Vigny M, Massoulié J (1984) An immunological study of rat acetylcholinesterase: comparison with acetylcholinesterase from other vertebrates. J Neurochem 43:204–213
Marshall LM, Sanes JR, McMahan UJ (1977) Reinnervation of original synaptic sites on muscle fiber basement membrane after disruption of the muscle cells. Proc Natl Acad Sci USA 74:3073–3077
Masson P (1979) Formes moléculaires multiples de la butyrylcholinestérase du plasma humain. I. Paramètres moléculaires apparents et ébauche de la structure quaternaire. Biochim Biophys Acta 578:493–504
Massoulié J, Bon S (1982) The molecular forms of cholinesterase in vertebrates. Annu Rev Neurosci 5:57–106
Massoulié J, Bon S, Rieger F, Vigny M (1975) Molecular forms of acetylcholinesterase. Croat Chem Acta 47:163–179
Massoulié J, Bon S, Lazar M, Grassi J, Marsh D, Méflah K, Toutant JP, Vallette F, Vigny M (1984) The polymorphism of cholinesterases: classification of molecular forms; interactions and solubilization characteristics; metabolic relationships and regulations. In: Brzin M, Barnard EA, Sket D (eds) Cholinesterases: fundamental and applied aspects. de Gruyter, Berlin, p 73
Massoulié J, Vigny M, Lazar M (1985) Expression of acetylcholinesterase in murine neural cells *in vivo* and in culture. In: Changeux JP, Hucho F, Maelicke A, Neumann E (eds) Molecular basis of nerve activity. de Gruyter, Berlin, p 619
Mays C, Rosenberry TL (1981) Characterization of pepsin-resistant collagen-like tail subunit fragments of 18 S and 14 S acetylcholinesterase from *Electrophorus electricus*. Biochemistry 20:2810–2817
McMahan UJ, Sanes JR, Marshall LM (1978) Cholinesterase is associated with the basal lamina at the neuromuscular junction. Nature 271:172–174
Medof ME, Walter EI, Roberts WL, Haas R, Rosenberry TL (1986) Decay accelerating factor of complement is anchored to cells by a C-terminal glycolipid. Biochemistry 25:6740–6747

Meedel TH, Whittaker JR (1983) Development of translationally active mRNA from larval muscle acetylcholinesterase during ascidian embryogenesis. Proc Natl Acad Sci USA 80:4761–4765

Méflah K, Bernard S, Massoulié J (1984) Interactions with lectins indicate differences in the carbohydrate composition of the membrane-bound enzymes acetylcholinesterase and 5′-nucleotidase in different cell types. Biochimie 66:59–69

Mellinger J, Belbenoît P, Ravaille M, Szabo T (1978) Electric organ development in *Torpedo marmorata* (chondrichtyes). Dev Biol 67:167–188

Mendel B, Mundell DB, Rudney H (1943) Studies on cholinesterase. III. Specific tests for true cholinesterase and pseudo-cholinesterase. Biochem J 37:473–476

Méndez B, Garrido J, Maldonado M, Jaksic FM, Inestrosa NC (1984) The electric organ of *Discopyge tschudii*: its innervated face and the biology of acetylcholinesterase. Cell Mol Neurobiol 4:125–142

Mercken L, Simons MJ, Swillens S, Massaer M, Vassart G (1985) Primary structure of bovine thyroglobulin deduced from the sequence of its 8,431-base complementary DNA. Nature 316:647–651

Metz J, Bradlow BA, Lewis SM, Dacie JV (1960) The acetylcholinesterase activity of the erythrocytes in paroxysmal nocturnal hemoglobinuria: relation to the severity of the disease. Br J Haematol 6:372

Millar DB, Christopher JP, Bishop WH (1979) Inhibition of acetylcholinesterase by TX-100. Biophys Chem 10:147–151

Mintz KP, Brimijoin S (1985a) Two-step immunoaffinity purification of acetylcholinesterase from rabbit brain. J Neurochem 44:225–232

Mintz KP, Brimijoin S (1985b) Monoclonal antibodies to rabbit brain acetylcholinesterase: selective enzyme inhibition, differential affinity for enzyme forms, and cross-reactivity with other mammalian cholinesterases. J Neurochem 45:284–292

Mintz KP, Weinshilboum RM, Brimijoin WS (1984) Evolution of butyrylcholinesterase in higher primates: an immunochemical study. Comp Biochem Physiol 79C:35–37

Mooser G, Sigman DS (1974) Ligand binding properties of acetylcholinesterase determined with fluorescent probes. Biochemistry 13:2299–2307

Muensch H, Goedde HW, Yoshida A (1976) Human serum cholinesterase subunits and number of active sites of the major component. Eur J Biochem 70:217–223

Muller F, Dumez Y, Massoulié J (1985) Molecular forms and solubility of acetylcholinesterase during the embryonic development of rat and human brain. Brain Res 331:295–302

Murray N, Steck AJ (1984) Indication of a possible role in a demyelinating neuropathy for an antigen shared between myelin and NK cells. Lancet 1 (8379):711–713

Musset F, Frobert Y, Grassi J, Vigny M, Boulla G, Bon S, Massoulié J (1987) Monoclonal antibodies against acetylcholinesterase from electric organs of *Electrophorus* and *Torpedo*. Biochimie 69:147–156

Nicholson-Weller A, March JP, Rosenfeld SI, Austen KF (1983) Affected erythrocytes of patients with paroxysmal nocturnal hemoglobinuria are deficient in the complement regulatory protein, decay acceleration factor. Proc Natl Acad Sci USA 80:5066–5070

Nicolet M, Rieger F, Dreyfus P, Pinçon-Raymond M, Tran L (1984) Tailed, asymmetric acetylcholinesterase and skeletal muscle basal lamina in Vertebrates: subcellular location: hydrophilic and hydrophobic variants. In: Brzin M, Barnard EA, Sket D (eds) Cholinesterases: fundamental and applied aspects. de Gruyter, Berlin, p 129

Nicolet M, Pinçon-Raymond M, Rieger F (1986) Globular and asymmetric acetylcholinesterase in frog muscle basal lamina sheath. J Cell Biol 102:762–768

Niday E, Wang CS, Alaupovic P (1977) Studies on the characterization of human erythrocyte acetylcholinesterase and its interaction with antibodies. Biochim Biophys Acta 459:180–193

Noda M, Takahashi H, Tanabe T, Toyosato M, Kikyotani S, Furutani Y, Hirose T, Takashima H, Inayama S, Miyata T, Numa S (1983) Structural homology of *Torpedo californica* acetylcholine receptor subunits. Nature 302:528–553

Ochoa ELM (1978) Redistribution of acetylcholinesterase activity from electroplax membrane fragments into phosphatidylcholine vesicles. FEBS Lett 89:317–320

Ott P (1985) Membrane acetylcholinesterases: purification, molecular properties and interactions with amphiphilic environments. Biochim Biophys Acta 822:375–392
Ott P, Brodbeck U (1978) Multiple molecular forms of acetylcholinesterase from human erythrocyte membranes: interconversion and subunit composition of forms separated by density gradient centrifugation in a zonal rotor. Eur J Biochem 88:119–125
Ott P, Brodbeck U (1984) Amphiphile dependency of the monomeric and dimeric forms of acetylcholinesterase from human erythrocyte membrane. Biochim Biophys Acta 775:71–76
Ott P, Jenny B, Brodbeck U (1975) Multiple molecular forms of purified human erythrocyte acetylcholinesterase. Eur J Biochem 57:469–480
Ott P, Lustig A, Brodbeck U, Rosenbusch JP (1982) Acetylcholinesterase from human erythrocyte membranes: dimers as functional units. FEBS Lett 138:187–189
Ott P, Ariano BH, Binggeli Y, Brodbeck U (1983) A monomeric form of human erythrocyte membrane acetylcholinesterase. Biochim Biophys Acta 729:193–199
Pangburn MK, Schreiber RD, Müller-Eberhard JH (1983) Deficiency of an erythrocyte membrane protein with complement regulatory activity in paroxysmal nocturnal hemoglobinuria. Proc Natl Acad Sci USA 80:5430–5434
Pezzementi L, Reinheimer EJ, Pezzementi ML (1987) Acetylcholinesterase from the skeletal muscle of the lamprey *Petromyzon marinus* exists in globular and asymmetric forms. J Neurochem 48:1753–1760
Poiana G, Scarcella G, Biagioni S, Senni MI, Cossu G (1985) Membrane acetylcholinesterase in murine muscular dystrophy *in vivo* and in cultured myotubes. Int J Dev Neurosci 3:331–340
Prody CA, Zevin-Sonkin D, Gnatt A, Goldberg O, Soreq H (1987) Isolation and characterization of full-length cDNA clones for cholinesterase from fetal human tissues. Proc Natl Acad Sci USA 84:3555–3559
Rakonczay Z (1986) Mammalian brain acetylcholinesterase. In: Boulton AA, Baker GB, Yu PH (eds) Neurotransmitter enzymes. Humana, Clifton, p 319 (Neuromethods, vol 5)
Rakonczay Z, Brimijoin S (1985) Immunochemical differences among molecular forms of acetylcholinesterase in brain and blood. Biochim Biophys Acta 832:127–134
Rakonczay Z, Brimijoin S (1986) Monoclonal antibodies to rat brain acetylcholinesterase: comparative affinity for soluble and membrane-associated enzyme and for enzyme from different vertebrate species. J Neurochem 46:280–287
Rakonczay Z, Mallol J, Schenk H, Vincendon G, Zanetta JP (1981a) Purification and properties of the membrane-bound acetylcholinesterase from adult rat brain. Biochim Biophys Acta 657:243–256
Rakonczay Z, Vincendon G, Zanetta JP (1981b) Heterogeneity of rat brain acetylcholinesterase: a study by gel filtration and gradient centrifugation. J Neurochem 37:662–669
Ralston JS, Rush RS, Doctor BP, Wolfe AD (1985) Acetylcholinesterase from fetal bovine serum. Purification and characterization of soluble G4 enzyme. J Biol Chem 260:4312–4318
Ramírez G, Gómez-Barriocanal J, Barat A, Rodriguez-Borrajo C (1984) Two classes of collagen-tailed molecular forms of acetylcholinesterase. In: Brzin M, Barnard EA, Sket D (eds) Cholinesterases: fundamental and applied aspects. de Gruyter, Berlin, p 115
Randall WR, Lai J, Barnard EA (1985) Acetylcholinesterase of muscle and nerve. In: Changeux JP, Hucho F, Maelicke A, Neumann E (eds) Molecular basis of nerve activity. de Gruyter, Berlin, p 595
Randall WR, Tsim KWK, Lai J, Barnard EA (1987) Monoclonal antibodies against chicken brain acetylcholinesterase: their use in immunopurification and immunohistochemistry to demonstrate allelic variants of the enzyme. Eur J Biochem (in press)
Razon N, Soreq H, Roth E, Bartal A, Silman I (1984) Characterization of activities and forms of cholinesterases in human primary brain tumors. Exp Neurol 84:681–695
Richardson GP, Witzemann V (1986) *Torpedo* electromotor system development: biochemical differentiation of *Torpedo* electrocytes *in vitro*. Neuroscience 17:1287–1296
Rieger F, Faivre-Bauman A, Benda P, Vigny M (1976) Molecular forms of acetylcholinesterase: their *de novo* synthesis in mouse neuroblastoma cells. J Neurochem 27:1059–1063

Rifkin MA, Fairlamb AH (1985) Transport of ethanolamine and its incorporation into the variant surface glycoprotein of bloodstream forms of *Trypanosoma brucei*. Mol Biochem Parasitol 15:245–256

Roberts WL, Rosenberry TL (1985) Identification of covalently attached fatty acids in the hydrophobic membrane-binding domain of human erythrocyte acetylcholinesterase. Biochem Biophys Res Commun 133:621–627

Roberts WL, Rosenberry TL (1986) Selective radiolabelling and isolation of the hydrophobic membrane-binding domain of human erythrocyte acetylcholinesterase. Biochemistry 25:3091–3098

Roberts WL, Kim BH, Rosenberry TL (1987) Differences in the glycolipid membrane anchors of bovine and human erythrocyte acetylcholinesterases. Proc Natl Acad Sci USA 84:7817–7821

Rodriguez-Ithurralde D, Silveira R, Barbeito L, Dajas F (1983) Fasciculin, powerful anticholinesterase polypeptide from *Dendroaspis angusticeps* venom. Neurochem Int 5:267–274

Römer-Lüthi CR, Hajdu J, Brodbeck U (1979) Molecular forms of purified human erythrocyte membrane acetylcholinesterase investigated by cross linking with diimidates. Hoppe Seylers Z Physiol Chem 360:929–934

Römer-Lüthi CR, Ott P, Brodbeck U (1980) Reconstitution of human erythrocyte membrane acetylcholinesterase in phospholipid vesicles: analysis of the molecular forms by cross-linking studies. Biochim Biophys Acta 601:123–133

Rosenberry TL (1975a) Acetylcholinesterase. Adv Enzymol 43:103–218

Rosenberry TL (1975b) Catalysis by acetylcholinesterase: evidence that the rate-limiting step for acylation with certain substrates precedes general acid-base catalysis. Proc Natl Acad Sci USA 72:3834–3838

Rosenberry TL (1985) Structural distinctions among acetylcholinesterase forms. In: Martonosi AN (ed) The enzymes of biological membranes. Plenum, New York, p 403

Rosenberry TL, Richardson JM (1977) Structure of 18 S and 14 S acetylcholinesterase: identification of collagen-like subunits that are linked by disulfide bonds to catalytic subunits. Biochemistry 16:3550–3558

Rosenberry TL, Scoggin DM (1984) Structure of human erythrocyte acetylcholinesterase: characterization of intersubunit disulfide bonding and detergent interaction. J Biol Chem 259:5643–5652

Rosenberry TL, Barnett P, Mays C (1980) The collagen-like subunits of acetylcholinesterase from the eel *Electrophorus electricus*. Neurochem Int 2:135–147

Rosenberry TL, Barnett P, Mays C (1982) Acetylcholinesterase. Methods Enzymol 82:325–339

Rosenberry TL, Haas R, Roberts WL, Kim BH (1985) The hydrophilic domain of human acetylcholinesterase contains non amino acid components. In: Changeux JP, Hucho F, Maelicke A, Neumann E (eds) Molecular basis of nerve activity. de Gruyter, Berlin, p 651

Rosenberry TL, Roberts WL, Hass R (1986) Glycolipid membrane binding-domain of human erythrocyte acetylcholinesterase. Fed Proc 45:2970–2975

Rotundo RL (1984a) Asymmetric acetylcholinesterase is assembled in the Golgi apparatus. Proc Natl Acad Sci USA 81:479–483

Rotundo RL (1984b) Purification and properties of the membrane-bound form of acetylcholinesterase from chicken brain: evidence for two distinct polypeptide chains. J Biol Chem 259:13186–13194

Rougon G, Hirsch MR, Hirn M, Guénet J-L, Goridis C (1983) Monoclonal antibody to neural cell surface protein: identification of a glycoprotein family of restricted cellular localization. Neuroscience 10:511–520

Sakai M, Saisu H, Koshigoe N, Abe T (1985a) Detergent-soluble form of acetylcholinesterase in the electric organ of electric rays: its isolation, characterization and monoclonal antibodies. Eur J Biochem 148:197–206

Sakai M, Saisu H, Abe T (1985b) Comparison of asymmetric forms of acetylcholinesterase from the electric organ of *Narke japonica* and *Torpedo californica*. Eur J Biochem 153:497–502

Sakai M, Koshigoe N, Saisu H, Abe T (1986) Monoclonal antibodies specific to asymmetric forms of acetylcholinesterase from the electric organ of electric rays. Biomed Res 7:1–5
Sanes JR, Cheney JM (1982) Laminin, fibronectin and collagen in synaptic and extrasynaptic portions of muscle fiber basement membrane. J Cell Biol 93:442–451
Schumacher M, Camp S, Maulet Y, Newton M, MacPhee-Quigley K, Taylor SS, Friedman T, Taylor P (1986a) Primary structure of *Torpedo californica* acetylcholinesterase deduced from cDNA sequence. Nature 319:407–409
Schumacher M, Camp S, Maulet Y, Newton M, MacPhee-Quigley K, Taylor SS, Friedman T, Taylor P (1986b) Primary structure of acetylcholinesterase: implications for regulation and function. Fed Proc 45:2976–2981
Scott LJ, Sanes JR (1984) A lectin that selectively stains neuromuscular junctions binds to collagen-tailed acetylcholinesterase. Neurosci Abstr 10:546
Seki T, Chang HC, Moriuchi T, Denome R, Ploegh H, Silver J (1985) A hydrophobic transmembrane segment at the carboxyl terminus of Thy-1. Science 227:649–651
Shukla SD (1982) Phosphatidylinositol-specific phospholipase C. Life Sci 30:1323–1335
Shukla SD (1986) Action of phosphatidylinositol specific phospholipase C on platelets: non-lytic release of acetylcholinesterase, effect on thrombin and PAF-induced aggregation. Life Sci 38:751–755
Sikorav JL, Grassi J, Bon S (1984) Synthesis *in vitro* of precursors of the catalytic subunits of acetylcholinesterase from *Torpedo marmorata* and *Electrophorus electricus*. Eur J Biochem 145:519–524
Sikorav JL, Vallette F, Grassi J, Massoulié J (1985) Isolation of a cDNA clone for a catalytic subunit of *Torpedo marmorata* acetylcholinesterase. FEBS Lett 193:159–163
Sikorav JL, Krejci E, Massoulié J (1987) cDNA sequences of *Torpedo marmorata* acetylcholinesterase: primary structure of the precursor of a catalytic subunit; existence of multiple 5′-untranslated regions. EMBO J 6:1865–1873
Silman I, Futerman AH (1987) Posttranslational modification as a means of anchoring acetylcholinesterase to the cell surface. Biopolymers 26:S241
Silman I, Lyles JM, Barnard E (1978) Instrinsic forms of acetylcholinesterase in skeletal muscle. FEBS Lett 94:166–170
Silman I, di Giamberardino L, Lyles J, Couraud JY, Barnard EA (1979) Parallel regulation of acetylcholinesterase and pseudocholinesterase in normal denervated and dystrophic chicken skeletal muscle. Nature 280:160–162
Silver A (1974) The Biology of cholinesterases. North-Holland, Amsterdam
Sine JP, Colas B (1987) Soluble form of acetylcholinesterase from rabbit enterocytes: comparison of its molecular properties with those of the plasma membrane species. Biochimie 69:75–80
Sketelj J, Brzin M (1985) Asymmetric molecular forms of acetylcholinesterase in mammalian skeletal musceles. J Neurochem Res 14:95–103
Sørensen K, Gentinetta R, Brodbeck U (1982) An amphiphile-dependent form of human brain caudate nucleus acetylcholinesterase: purification and properties. J Neurochem 39:1050–1060
Sørensen K, Gennari K, Brodbeck U, Landauer P, Liefländer M (1985) Polymorphism and immunochemical cross-reactivity of acetylcholinesterases from the brains of human, dog, hog, bovine and horse. Comp Biochem Physiol 80C:263–268
Sørensen K, Brodbeck U, Rasmussen AG, Norgaard-Pedersen B (1986) Normal human serum contains two forms of acetylcholinesterase. Clin Chim Acta 158:1–6
Soreq H, Parvari R, Silman I (1982) Biosynthesis and secretion of catalytically active acetylcholinesterase in *Xenopus* oocytes microinjected with mRNA from rat brain and from *Torpedo* electric organ. Proc Natl Acad Sci USA 79:830–834
Soreq H, Parvari R, Silman I (1983) Biosynthesis of acetylcholinesterase in rat brain and *Torpedo* electric organ is directed by scarce mRNA species. Prog Brain Res 58:107–115
Soreq H, Zevin-Sonkin D, Razon N (1984) Expression of cholinesterase gene(s) in human brain tissues: translational evidence for multiple mRNA species. EMBO J 3:1371–1375

Soreq H, Zevin-Sonkin D, Avni A, Hall LMC, Spierer P (1985) A human acetylcholinesterase gene identified by homology to the *Ace* region of *Drosophila*. Proc Natl Acad Sci USA 2:1827–1831

Steck A, Murray N, Meier C, Page N, Perruisseau G (1983) Demyelinating neuropathy and monoclonal IgM antibody to myelin-associated glycoprotein. Neurology (NY) 33:19–23

Stieger A, Cardoso de Almeida ML, Blatter MC, Brodbeck U, Bordier C (1986) The membrane-anchoring systems of vertebrate acetylcholinesterase and variant surface glycoproteins of African trypanosomes share a common antigenic determinant. FEBS Lett 199:182–186

Stieger S, Brodbeck U (1985) Amphiphilic detergent-soluble acetylcholinesterase from *Torpedo marmorata*: characterization and conversion by proteolysis to a hydrophilic form. J Neurochem 44:48–56

Stieger S, Brodbeck U, Reber B, Brunner J (1984) Hydrophobic labeling of the membrane binding domain of acetylcholinesterase from *Torpedo marmorata*. FEBS Lett 168:231–234

Stieger S, Brodbeck U, Witzemann V (1987) Inactive monomeric acetylcholinesterase in the low-salt-soluble extract of the elctric organ from *Torpedo marmorata*. J Neurochem 49:460–467

Sugarman J, Devine DV, Rosse WF (1986) Structural and functional difference between decay-acceleration factor and red cell acetylcholinesterase. Blood 68:680–684

Sung SC, Ruff BA (1983) Molecular forms of sucrose extractable and particulate acetylcholinesterase in the developing and adult rat brain. Neurochem Res 8:303–311

Swillens S, Ludgate M, Mercken L, Dumont JE, Vassart G (1986) Analysis of sequence and structure homologies between thyroglobulin and acetylcholinesterase: possible functional and clinical significance. Biochem Biophys Res Commun 137:142–148

Taguchi R, Suzuki K, Nakabayashi T, Ikezawa H (1984) Acetylcholinesterase release from mammalian erythrocytes by phosphatidylinositol-specific phospholipase C of *Bacillus thuringiensis* and characterization of the released enzyme. J Biochem 96:437–446

Takesue Y, Yokota K, Nishi Y, Taguchi R, Ikezawa H (1986) Solubilization of trehalase from rabbit renal and intestinal brush-border membranes by a phosphatidylinositol-specific phospholipase C. FEBS Lett 201:5–8

Taylor P, Lappi S (1975) Interaction of fluorescence probes with acetylcholinesterase. The site and specificity of propidium binding. Biochemistry 14:1989–1997

Taylor P, Schumacher M, Maulet Y, Newton M (1986) A molecular perspective on the polymorphism of acetylcholinesterase. Trends Pharmacol Sci 7:321–323

Torres JC, Inestrosa NC (1983) Heparin solubilizes asymmetric acetylcholinesterase from rat neuromuscular junction. FEBS Lett 154:265–268

Torres JC, Behrens MI, Inestrosa NC (1983) Neural 16 S acetylcholinesterase is solubilized by heparin. Biochem J 215:201–204

Toutant JP (1986) An evaluation of the hydrophobic interactions of chick muscle acetylcholinesterase by charge shift electrophoresis and gradient centrifugation. Neurochem Int 9:111–119

Toutant JP, Massoulié J (1987) Acetylcholinesterase. In: Turner AJ, Kenny AJ (eds) Mammalian ectoenzymes. Elsevier/North-Holland, Amsterdam pp 289–328

Toutant JP, Massoulié J, Bon S (1985) Polymorphism of pseudocholinesterase in *Torpedo marmorata* tissues: comparative study of the catalytic and molecular properties of this enzyme with acetylcholinesterase. J Neurochem 44:580–592

Tse AGD, Barclay AN, Watts A, Williams AF (1985) A glycophospholipid tail at the carboxyl terminus of the Thy-1 glycoprotein of neurons and thymocytes. Science 230:1003–1008

Tsim KWK, Randall WR, Barnard EA (1988) An asymmetric form of muscle acetylcholinesterase contains three subunit types and two enzymic activities in one molecule. Proc Natl Acad Sci USA (in press)

Verdenhalven J, Hucho F (1985) Inhibition and photoaffinity labeling of acetylcholinesterase by phencyclidine and triphenylmethylphosphonium. In: Changeux JP, Hucho F, Maelicke A, Neumann E (eds) Molecular basis of nerve activity. de Gruyter, Berlin, p 741

Vigny M, Gisiger V, Massoulié J (1978) 'Non specific' cholinesterase and acetylcholinesterase in rat tissues: molecular forms, structural and catalytic properties and significance of the two enzyme systems. Proc Natl Acad Sci USA 75:2588–2592

Vigny M, Bon S, Massoulié J, Gisiger V (1979) The subunit structure of mammalian acetylcholinesterase: catalytic subunits, dissociating effect of proteolysis and disulfide reduction of the polymeric forms. J Neurochem 33:559–565

Vigny M, Martin GR, Grotendorst GR (1983) Interactions of asymmetric forms of acetylcholinesterase with basement membrane components. J Biol Chem 258:8794–8798

Vincent L, Duband JL, Thiéry JP (1983) A cell surface determinant expressed early on migrating avian neural crest cells. Dev Brain Res 9:235–238

Viratelle OM, Bernhard SA (1980) Major component of acetylcholinesterase in *Torpedo* electroplax is not basal lamina associated. Biochemistry 19:4999–5007

Watkins MS, Hitt AS, Bulger JE (1977) The binding of 18 S acetylcholinesterase to sphingomyelin and the role of the collagen-like tail. Biochem Biophys Res Commun 79:640–647

Weitz M, Bjerrum OJ, Brodbeck U (1984) Characterization of an active hydrophilic erythrocyte membrane acetylcholinesterase obtained by limited proteolysis of the purified enzyme. Biochim Biophys Acta 776:65–74

Wiedmer T, di Francesco C, Brodbeck U (1979) Effects of amphiphiles on structure and activity of human erythrocyte acetylcholinesterase. Eur J Biochem 102:59–64

Williams M (1969) Antibodies to acetylcholinesterase. Proc Natl Acad Sci USA 62:1175–1180

Witzemann V, Boustead C (1982) Changes in acetylcholinesterase molecular forms during the embryonic development of *Torpedo marmorata*. J Neurochem 39:747–755

Witzemann V, Boustead C (1983) Structural differences in the catalytic subunits of acetylcholinesterase forms from the electric organ of *Torpedo marmorata*. EMBO J 2:873–878

Younkin SG, Rosenstein C, Collins PL, Rosenberry TL (1982) Cellular localization of the molecular forms of acetylcholinesterase in rat diaphragm. J Biol Chem 257:13630–13637

Zevin-Sonkin D, Avni A, Zisling R, Koch R, Soreq H (1985) Expression of acetylcholinesterase gene(s) in the human brain molecular cloning evidence for cross-homologous sequences. J Physiol (Paris) 80:221–228

CHAPTER 8b

Cholinesterases: Tissue and Cellular Distribution of Molecular Forms and Their Physiological Regulation

J.-P. TOUTANT and J. MASSOULIÉ

A. Introduction

In Chapter 8a we have described the molecular structure and the interactions of the multiple molecular forms of acetylcholinesterase (AChE, EC 3.1.1.7) and butyrylcholinesterase (BuChE, EC 3.1.1.8) in vertebrates. We show there that the two enzymes present globular forms (G_1, G_2, G_4) which exist as non-hydrophobic as well as amphiphilic molecules, and asymmetric, collagen-tailed molecules (A_4, A_8, A_{12}). These molecular forms may be further subdivided according to the binding of lectins or according to their electrophoretic migrations in non-denaturing electrophoresis.

In the present chapter we describe the tissue and cellular distributions of the various molecular forms in the adult and during embryonic development. We examine some aspects of the regulation of their biosynthesis with particular interest for the expression of asymmetric forms at neuromuscular junctions (NMJs). We also discuss the involvement of cholinesterases (ChEs) in non-cholinergic functions, e.g. in the elaboration or degradation of peptidic neurotransmitters, and we briefly review the anomalies of AChE in pathological states. Additional references may be found in previous reviews on these subjects (MASSOULIÉ and BON 1982; BRIMIJOIN 1983; BRZIN et al. 1983; RAKONCZAY 1986; BRIMIJOIN and RAKONCZAY 1986).

B. Tissue Distribution of ChE Molecular Forms

I. General Considerations

In each species each organ possesses a definite pattern of molecular forms of both AChE and BuChE (see reviews in MASSOULIÉ and BON 1982; BRIMIJOIN 1983). This has been described for example in the case of the domestic fowl (LYLES et al. 1982) and *Torpedo marmorata* (TOUTANT et al. 1985).

Globular forms (G forms) of AChE and BuChE are ubiquitous; they occur in cholinergic but also in non-cholinergic tissues where they might have other roles than the hydrolysis of choline esters (discussed below). For example erythrocytes contain membrane-bound dimers (*Torpedo*, fowl, rat, ox, human) of AChE. Globular forms of BuChE (essentially G_4) are present in the plasma of all vertebrates. In rabbit, the plasma also contains a high level of G_4 AChE. *Torpedo* plasma contains both G_2 AChE and G_4 BuChE (MASSOULIÉ et al. 1984).

In contrast with globular forms, asymmetric forms (A forms) appear to exist only in the context of cholinergic transmission: they occur in muscles, central nervous system (CNS) and peripheral nerves. In some tissues, e.g. in salivary glands, A forms of AChE are associated with the cholinergic nerve endings (KÁSA and RAKONCZAY 1982a). Ventral motor roots supplying the rat sciatic nerve contain both G and A forms of AChE, while the dorsal sensory roots are devoid of asymmetric molecules (FERNANDEZ et al. 1979).

The proportion of collagen-tailed AChE in the CNS and in nerves of mammals and birds is usually very low, but it may be underestimated when the extraction is performed with a medium lacking calcium-chelating agents, as shown by RAMÍREZ and colleagues in the fowl visual system and in the brains of various vertebrate species (GÓMEZ-BARRIOCANAL et al. 1981; RODRÍGUEZ-BORRAJO et al. 1982; BARAT et al. 1983). The proportion of asymmetric forms seems to be much higher in some lower vertebrates, e.g. in amphibians (reviewed in MASSOULIÉ and BON 1982; RODRÍGUEZ-BORRAJO et al. 1982). It is, however, only of the order of 1% in the CNS of the lamprey *Petromyzon marinus* (L. PEZZEMENTI, personal communication). In the frog *Rana temporaria* high levels of A forms are found in various parts of the CNS and in peripheral nerves (NICOLET and RIEGER 1982). In addition, in the frog sciatic nerve the A_{12} and A_8 forms of AChE are partly transported via the rapid axonal flow and are partly stationary (COURAUD et al. 1985), whereas in birds and mammals the totality of the A_{12} form is rapidly transported (DI GIAMBERARDINO and COURAUD 1978). In contrast, BuChE A forms are stationary and might be inserted in the axolemma or in Schwann cells of the frog sciatic nerve (COURAUD et al. 1985).

In skeletal muscles of birds the relative proportions of AChE molecular forms depend on the physiological properties of the fibres: A forms are predominant in fast-twitch muscles whereas G forms are the major components in slow-twitch and slow-tonic muscles (fowl, BARNARD et al. 1982a; fowl and pigeon, CHATONNET and BACOU 1983). A difference between the proportions of A and G forms is also observed, although to a lesser degree, between fast-twitch and slow-twitch muscles of the rabbit (BACOU et al. 1982). In mouse and rat muscles the differences between fast- and slow-twitch muscles concern both the proportion of the G_4 form (GISIGER and STEPHENS 1983b; GROSWALD and DETTBARN 1983) and the relative proportions of A_{12} and A_8 (GROSWALD and DETTBARN 1983; LØMO et al. 1985). These characteristic features appear to be controlled by the type of activity imposed by the nerve (LØMO et al. 1985; DETTBARN et al. 1985).

There is a remarkable similarity between the proportions of AChE forms in the muscles and in the corresponding nerves (GISIGER and STEPHENS 1983a; BACOU et al. 1985); it is tempting to suggest that this parallelism corresponds to a functional adaptation.

II. Distribution of AChE Molecular Forms at Neuromuscular Junctions

The AChE of the motor endplates of skeletal muscle, but not that of sympathetic ganglia, was shown to resist solubilization by Triton X-100 (KOELLE et al. 1970).

On the other hand, HALL and KELLY (1971) first observed that AChE could be solubilized by collagenase from endplates, but not from other sites of frog and rat muscles. This was later explained by the presence of A forms at the NMJ (HALL 1973). AChE activity is present in the synaptic basal lamina (MCMAHAN et al. 1978), and it is now assumed that the synaptic A forms are anchored by ionic interactions with or covalently cross-linked to components of the extracellular matrix (see Chap. 8a).

The A forms are probably not the only molecular forms of AChE at the synapses; SKETELJ and BRZIN (1979) observed that 35% of AChE activity was not removed by collagenase treatment from NMJ in mouse diaphragm and MILEDI et al. (1984) found that 20% of the AChE activity was collagenase-resistant at the frog sartorius endplates. This may be explained by the presence of G forms associated with the synaptic membrane. It has also been reported that after degeneration of nerve and muscle, the synaptic basal lamina contains G forms of AChE (ANGLISTER and MCMAHAN 1983; NICOLET et al. 1986).

In fact, a direct analysis of AChE in the synaptic regions of several fowl muscles led BARNARD and his colleagues to the conclusion that whereas the A_{12} form was the predominant form at the NMJ of the fast-twitch posterior latissimus dorsi muscle, globular forms were predominant at the endplates of the tonic anterior latissimus dorsi muscle (RANDALL et al. 1985). In the case of the latter muscle LYLES and BARNARD (1980) and JEDRZEJCZYK et al. (1984) showed that the A forms completely disappeared from the synaptic areas in aged animals (>600 days old); thus, the globular forms that remained were responsible for the physiological hydrolysis of ACh and the histochemical staining of these zones. Whether these synaptic globular forms are inserted into pre- and/or post-synaptic membranes or entrapped in the basal lamina remains to be elucidated.

In conclusion, it appears that the A forms are generally the predominant synaptic forms in fast-twitch muscles, and that they coexist with variable amounts of synaptic G forms at the NMJ of a number of muscle fibres including slow-twitch mammalian muscles (RANDALL et al. 1985; LAI et al. 1986). Both A and G forms are thus probably involved in the hydrolysis of the transmitter.

MILEDI et al. (1984) suggested that in physiological conditions only some synaptic AChE molecules participate in the hydrolysis of acetylcholine (ACh), depending on their location at the endplate. Studying the frog sartorius muscle, where the postsynaptic membrane presents an extensive folding, they proposed that ACh does not reach the inner part of the folds.

In the electric organ of *Torpedo*, A forms of AChE are localized in the basal lamina of the synaptic cleft, but, in addition, a significant amount of the total AChE activity corresponds to presynaptic membrane-bound hydrophobic dimers. The active sites of these molecules are exposed in the synaptic cleft (MOREL and DREYFUS 1982; LI and BON 1983; FUTERMAN et al. 1985). In this particular case it is tempting to correlate the presence of presynaptic membrane-bound AChE with that of presynaptic muscarinic receptors (mAChR) of M2 type which are possibly involved in the regulation of ACh release (for reviews, see KOELLE 1961; MICHAELSON et al. 1979; RUSSELL et al. 1985; ERULKAR 1983; KILBINGER 1984; MARQUIS and FISHMAN 1985; Chap. 10, this volume).

In addition to AChE, vertebrate NMJs also contain BuChE, as demonstrated by HESS (1961), SILVER (1963), DAVIS and KOELLE (1967) and VIGNY et al. (1978), and this enzyme presents both globular and asymmetric forms (JEDRZEJCZYK et al. 1981; BARNARD et al. 1982b). It is tempting to imagine that BuChE is particularly involved in hydrolysing the transmitter at the NMJ if its local concentration reaches values high enough to cause excess substrate inhibition of AChE (SILVER 1963; VIGNY et al. 1978). However, the possibility that the local concentration of ACh produces substrate inhibition in the cleft is not supported by quantitative simulations of endplate currents (ROSENBERRY 1979).

III. Ganglionic Synapses of the Autonomic Nervous System

The physiological role of ChEs at the autonomic cholinergic ganglionic synapses is far less clear than at the NMJ. Anti-ChE drugs do not alter cholinergic transmission as markedly as at the NMJ (VOLLE 1980; BRZIN et al. 1983). In these autonomic ganglionic synapses the physical diffusion of ACh from the synaptic cleft may play an equivalent, or possibly more important, role than enzymatic hydrolysis of ACh.

AChE is concentrated at the postsynaptic membrane (DAVIS and KOELLE 1978) but also in regions devoid of synaptic contacts. For example, concentrations of AChE activity are detected in the membrane of presynaptic endings in regions covered by Schwann cells. In postsynaptic cells similar accumulations are observed in the nuclear envelope and in the rough-surfaced endoplasmic reticulum (RER), as well as in non-synaptic areas of the membrane of perikaryon and dendrites (DAVIS and KOELLE 1978; for a review, see BRZIN et al. 1983). The subcellular distribution of ChE activity is similar in the superior cervical ganglion (SCG) (orthosympathetic) and in the ciliary ganglion (parasympathetic) of the cat (DAVIS et al. 1984).

G_1 and G_4, together with lower proportions of G_2 and A_{12}, are the major AChE forms in rat SCG and other ganglia (stellate, nodose, dorsal root ganglia) (GISIGER et al. 1978), as well as in ox SCG (GRASSI et al. 1982). There is no evidence for a preferential localization of the A_{12} form at synaptic sites. In nerve cells the asymmetric forms of AChE may be associated to the glycocalyx or to other constituents of the cellular membrane since these cells lack a basement membrane (see VERDIÈRE et al. 1982 and KLINAR et al. 1983).

BuChE is also present in ganglionic cells at the same membranous sites as AChE but is absent from the preganglionic fibres and their terminals (DAVIS and KOELLE 1978; DAVIS et al. 1984). BuChE is synthesized in the RER of the ganglionic cells and then rapidly transported to the neuropil (UCHIDA and KOELLE 1983).

C. Subcellular Distribution of AChE Molecular Forms in Muscle and Nerve Cells

I. Determination of Endo- and Ectocellular Enzyme Activities

The distinction between external and internal cellular AChE and its physiological significance was established several years ago (KOELLE 1957; KOELLE and

KOELLE 1959; MCISAAC and KOELLE 1959). It is possible to evaluate the activity of externally accessible ChE activity (ectocellular compartment) from the hydrolysis of a substrate to which the plasma membrane is impermeable, e.g. acetylthiocholine, by intact cells in culture and to determine the total activity after lysing the cells by addition of Triton X-100 in the culture dish. This method has been successfully applied to fowl and mouse muscle cells in culture (ROTUNDO and FAMBROUGH 1980a; RUBIN et al. 1985; VALLETTE et al. 1986) and to T28 neuroblastoma cells (LAZAR and VIGNY 1980). The ectocellular compartment corresponds to 40%–60% of the AChE activity, depending on the age of the cultures and state of differentiation.

II. Distribution of Molecular Forms in Ecto- and Endocellular Compartments

1. Reagents

Several methods allow the analysis of molecular forms of external and internal AChE. Echothiopate, or phospholine (O,O-diethyl-*S*-(2-trimethylammoniummethyl)phosphorothiolate iodide), a non-permeant, irreversible inhibitor of AChE, acts essentially on external enzyme and this allows the analysis of the internal pool (LAZAR and VIGNY 1980; BROCKMAN et al. 1982; VALLETTE et al. 1986).

In order to analyse the external AChE compartment muscle cells are incubated successively with BW284C51 (bis(4-allyldimethylammoniumphenyl)pentan-3-one dibromide), a reversible and impermeant inhibitor which protects the external enzyme, then with diisopropylphosphorofluoridate (DFP) which irreversibly inactivates the intracellular compartment. Reactivation of the external enzyme is achieved by washing (MCISAAC and KOELLE 1959; RIEGER et al. 1976; ROTUNDO and FAMBROUGH 1980a; TAYLOR et al. 1981; VALLETTE et al. 1986). A variant of the latter method that gives similar results consists in inhibiting the external pool of AChE by echothiopate and inactivating irreversibly the intracellular enzyme with methanesulfonyl fluoride in the presence of decamethonium. The extracellular enzyme is reactivated by treatment with 2-PAM (1-methyl-2-hydroxyiminomethylpyridinium) (BROCKMAN et al. 1982). The diethylphosphorylated AChE (external) is accurately estimated as 2-PAM reactivatable activity (BROCKMAN et al. 1982; NEWMAN et al. 1984).

After treatment the remaining activity may be biochemically analysed, or cytochemically stained and observed at the electron microscope level (DONOSO and FERNANDEZ 1985).

2. Muscle Cells *In Vitro*

All forms occur in both compartments. However the A_{12} and G_4 forms are essentially ectocellular while the A_8, G_1 and G_2 forms are predominantly intracellular, as shown in primary cultures of rat (BROCKMAN et al. 1982, 1984), quail (ROTUNDO 1984a), mouse (RUBIN et al. 1985), fowl (VALLETTE et al. 1986), and the mouse C2 line (INESTROSA et al. 1982, 1985). These observations show that the A_{12} form is assembled intracellularly and subsequently externalized (ROTUNDO 1984a).

3. Muscle Cells *In Vivo*

A similar study of the subcellular localization of the molecular forms of AChE has been performed on muscle cells *in vivo* on innervated or denervated adult rat diaphragm (YOUNKIN et al. 1982; COLLINS and YOUNKIN 1982) and in the synaptic and extrasynaptic regions of the rat gracilis muscle, which consists of a very few layers of fibres (FERNANDEZ et al. 1984), after denervation (FERNANDEZ and STILES 1984a) and during development (FERNANDEZ and SEITER 1984). ZISKIND-CONHAIM et al. (1984) also studied the localization of the A_{12} form during development of the rat intercostal muscle *in utero*.

In adult innervated rat muscles the G_1 and G_2 forms are essentially intracellular, the G_4 form is recovered both in the ecto- and endocellular compartments and the A_{12} form is predominantly ectocellular with a small percentage inside the cell.

After denervation of rat muscle the amount of intracellular G_1 form decreases rapidly and the external G_4 form more slowly (COLLINS and YOUNKIN 1982). In contrast to the observations of INESTROSA et al. (1977) and CARTER and BRIMIJOIN (1981), this suggests that denervation results primarily in a decrease of the synthesis of AChE rather than an increase of its secretion. Denervation also results in an immediate decrease of the intracellular pool of asymmetric forms, followed after a 12- to 18-h lag by a parallel decrease of the extracellular predominant pool (FERNANDEZ and STILES 1984a). The A_8 form decreases more slowly than the A_{12} form, and the A_4 form actually presents a transient increase (COLLINS and YOUNKIN 1982). This sequence of events suggests a proteolytic breakdown of the A_{12} form into A_8 and A_4 forms, and this is confirmed by the fact that an antiproteolytic agent, leupeptin, retards the effect of denervation (FERNANDEZ and DUELL 1980). Denervation does not modify the activity or the molecular forms of BuChE (BERMAN et al. 1987).

During development of the gracilis muscle the A_8 and A_{12} forms are first distributed throughout the fibre in the intracellular pool (7-day-old rat) and then progressively concentrate in the synaptic zones where they are externalized. This process was found to be essentially complete in 28-day-old rats. A forms do not appear to be externalized in non-end-plate zones (FERNANDEZ and SEITER 1984). ZISKIND-CONHAIM et al. (1984) demonstrated that during embryonic development of rat intercostal muscle, most of the A_{12} form of AChE is present at the muscle surface prior to their focal accumulation which takes place at embryonic day (ED) 16–17 (see Sect. F. II.1).

4. Nerve Cells *In Vitro*

Cultured fowl sympathetic neurons and mouse T28 cells present 80%–90% of the G_4 form extracellularly and 90% of G_2 (fowl) and G_1 (mouse) intracellularly (TAYLOR et al. 1981; LAZAR and VIGNY 1980). In rat SCG neurons in culture the G_4 form is essentially external, the G_1 form is predominantly intracellular and only 60% of A_{12} form is external (FERRAND et al. 1986; see also GOUDOU et al. 1985). In contrast, 90% of A_{12} form is extracellular in PC12 cells (INESTROSA et al. 1981). In the mutant cell line F3 which lacks heparan sulphate proteoglycan, the percentage of extracellular A_{12} is reduced to only 40% (INESTROSA et al. 1985) (see Sect. F. IV.).

5. Nerve Cells *In Vivo*

In neonatal and adult rat SCG the G_4 form is predominantly extracellular, and the A_{12} and G_1 forms are equally distributed inside and outside the cell (GOUDOU et al. 1985).

In the electric lobes of *T. marmorata* A forms are mostly intracellular (BON 1982; WITZEMANN and BOUSTEAD 1982), and this is also the case in motor nerves of rat, fowl and frog (references in COURAUD et al. 1985).

In cultures of the mouse neural cells, T28, the G_1 form is mainly intracellular and consists of distinct pools with different metabolic fates and half-lives (LAZAR et al. 1984). The G_4 form also consists of at least two distinct metabolic pools, as described in the next section.

D. Biosynthesis and Renewal of AChE Molecular Forms

I. *In Vitro* Studies

In cultured muscle cells ROTUNDO and FAMBROUGH (1980a) estimated the half-lives of the membrane-bound, externally exposed AChE and of the intracellular enzyme in chick myotubes differentiated *in vitro*, finding, respectively, 50 h and 2–3 h. In rat myotubes *in vitro* the half-life of AChE globular forms was estimated to be about 85 min (BROCKMAN et al. 1984). The turnover rate of asymmetric AChE could not be determined accurately in these conditions but was much slower than that of the globular forms.

ROTUNDO (1984b) studied the metabolism of radioactively labelled AChE in fowl muscle cultures by immunoprecipitation and sodium dodecylsulphate (SDS) electrophoresis. These experiments showed that 80% of the newly synthesized subunits do not acquire catalytic activity and are rapidly degraded within 1 h (Fig. 1). The possible significance of this massive and wasteful degradation is not apparent.

The method of heavy-isotope labelling was used to analyse the metabolism of AChE in stationary cultures of T28 cells which contain mostly intracellular G_1 and membrane-bound G_4 forms (LAZAR and VIGNY 1980). The replacement of pre-existing light molecules by newly synthesized heavy molecules was followed by a computer-aided, quantitative analysis of sedimentation profiles (LAZAR et al. 1984; MASSOULIÉ et al. 1985). From these studies it was possible to draw a number of conclusions regarding the metabolic fate of the catalytically active AChE molecules. First, the delay in the appearance of labelled enzyme indicated the existence of inactive precursors (with a half-life of about 30 min). Secondly, G_1 was found to be renewed much faster than G_4, with mean half renewal times of 5 and 40 h, respectively, and both forms presented multiple metabolic pools. Thirdly, heavy molecules (G_4) were secreted into the culture medium after a delay of about 3 h, in agreement with previous findings of ROTUNDO and FAMBROUGH (1980a).

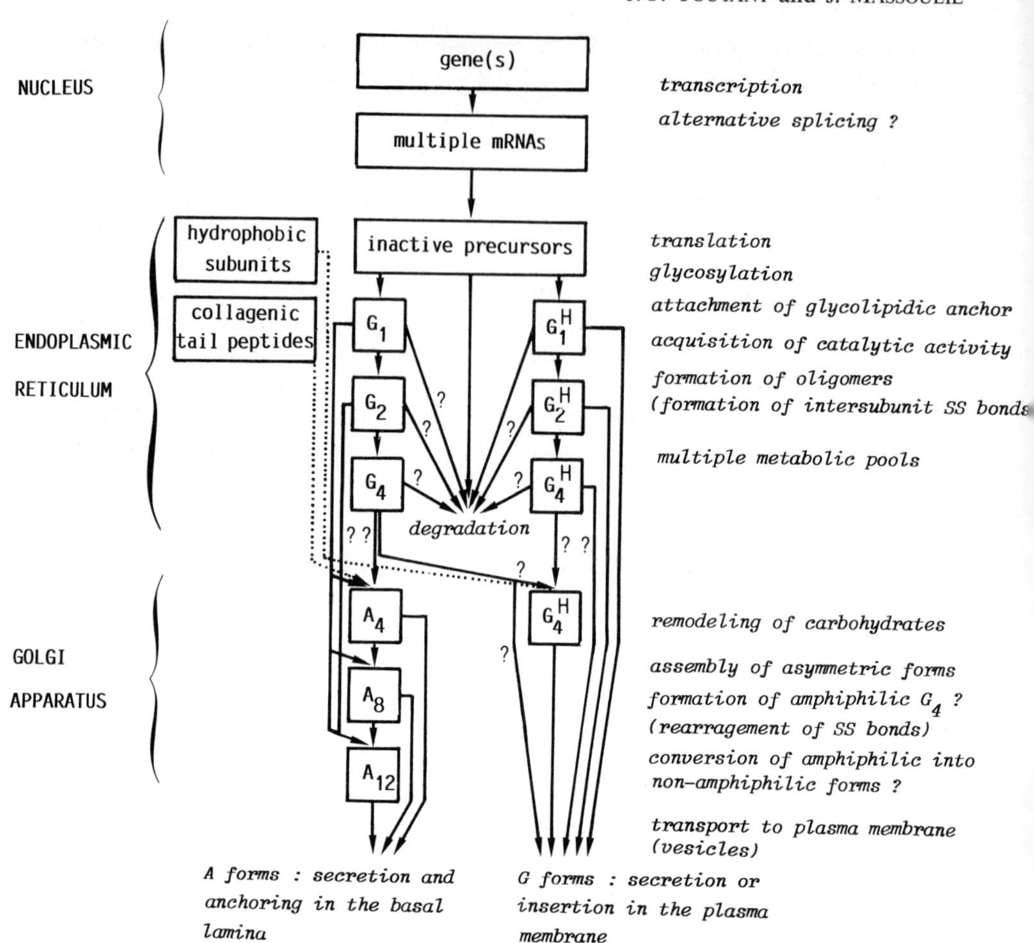

Fig. 1. Hypothetical biosynthetic relationship between AChE molecular forms incorporating features derived from the studies of ROTUNDO and FAMBROUGH (1982) and ROTUNDO (1984a, b) on primary cultures of avian embryonic muscle cells; of LAZAR et al. (1984) on a mouse neural cell line; and of BROCKMAN et al. (1986) on primary cultures of rat muscle cells. A large part of the synthesized enzyme is assumed to be degraded either as an inactive precursor or as an intracellular monomer. Monomers (G_1) and probably also dimers (G_2) and tetramers (G_4) comprise several metabolic pools which may correspond to distinct destinations. For example, monomers may be polymerized into G_2, G_4 and asymmetric forms. The enzyme may be degraded, exposed on the outer surface of the cell membrane, secreted and, in the case of A forms, anchored in a basal lamina. Amphiphilic, detergent-binding globular forms are noted G^H, where the superscript denotes hydrophobic interaction. The assembly of G_2 and G_4 forms takes place in the RER; they may be either associated with the membrane or secreted. It has been suggested that sarcolemmal globular forms may be internalized and then rapidly degraded to amino acids (NEWMAN et al. 1984). The glycolipid, for those molecules which possess this type of hydrophobic anchor at their C-terminal end, is probably added en bloc in the RER, as demonstrated for trypanosomal variant surface glycoproteins and DAF of complement (see text). The provenance of the amphiphilic G_4 form (INESTROSA et al. 1987) and the site of its assembly are not known. We tentatively consider that its biosynthetic pathway may be similar to that of the A forms because of its structural resemblance to them. A forms are assembled intracellularly in muscle cells (ROTUNDO 1984a; FERNANDEZ and SEITER

The observed kinetics of heavy-isotope labelling were fitted by different theoretical models with G_1 pools arranged in cascade or in parallel. The two models share the following common features: (a) G_4 is derived from a specific, rapidly renewed pool of precursors, probably G_1; (b) G_1 presents at least three pools with half-lives of 1, 5 and 10 h; (c) about 80% of the newly synthesized G_1 is rapidly degraded and never reaches the cellular surface either as a secretion product or as membrane-bound G_4 molecules; and (d) both G_1 and a fraction of G_4 appear to be secreted from a cellular G_1 pool.

Thus, like the fowl muscle cells studied by ROTUNDO (1984b), the T28 cells produce a large proportion of AChE molecules that are not assembled into tetramers and are rapidly degraded intracellularly. In T28 cells the rapidly turning-over pool was characterized as active enzyme, but it may also contain inactive molecules that could not be detected. There is thus a very striking similarity between these findings and those obtained by ROTUNDO (1984b) with fowl muscle cells.

In a recent study BROCKMAN et al. (1986) established the metabolic relationships between G_1, G_4, and A_{12} forms of AChE in cultures of rat myotubes. They observed that newly synthesized active enzyme was largely degraded or secreted and that only a small fraction of the G_1 monomers was assembled into G_4 and A forms.

II. Nervous Tissue and Muscle *In Vivo*

From the time course of disappearance of labelled enzyme (WENTHOLD et al. 1974; GOOSSENS et al. 1984) the half-life of total AChE in mature rat brain *in vivo* was estimated as 1.3–3.0 days. The recovery of G_1 and G_4 forms in rat brain apparently occurred at similar rates (GOOSSENS et al. 1984). Using cycloheximide as an inhibitor of protein synthesis, BRIMIJOIN and CARTER (1982) studied the turnover of AChE in the rat diaphragm. They found that the decrease of G_4 and A_{12} forms was negligible during the course of their experiments (8 h) and estimated the half-life of G_1 as 6.2 h. They concluded from the accumulation of AChE at ligatures of the phrenic nerve that axonal transport does not significantly contribute to this renewal of the synaptic enzyme.

The turnover rate of AChE in rat diaphragm was studied by following the rate of replacement of phospholine-labelled enzyme (NEWMAN et al. 1984). In normal muscle the half-life of non-end-plate AChE was 11.5 h while that of endplate AChE was too long to be determined. Denervation induced two main effects: in

1984) at the level f the Golgi apparatus (ROTUNDO 1984a). We suggest here that the catalytic subunits of A forms derive from a specific metabolic pool of G_1 molecules, different from the precursors of polymeric globular forms (see text). We have indicated that the assembly of the A forms may involve either the sequential addition of G_1 molecules to the tail subunits and the association of external dimers through quaternary interactions or the addition of tetramers G_4, with a rearrangement of the intersubunit SS bonds. The predominance of the A_8 form as well as the existence of a low proportion of A_4 form of AChE in the electric organs of early *Torpedo* embryos suggest that they represent transient intermediates in the assembly of A_{12} forms (BON 1982; WITZEMANN and BOUSTEAD 1982). They may, however, also arise as degradation products

non-end-plate regions, the rate of synthesis of G_1 was rapidly reduced, leading to a subsequent decrease of the other forms (COLLINS and YOUNKIN 1982), and in endplate regions the extracellular A forms were degraded, with a transient accumulation of A_4. The rate of degradation of non-synaptic enzyme was unaffected by the operation.

KARPRZAK and SALPETER (1985) reported that the synaptic enzyme in mouse sternocleidomastoid muscle has a half-life of around 20 days. It is possible either that all AChE molecular forms are stabilized at the synaptic zone (as already demonstrated for the ACh receptor), or that the half-life of A forms, which are concentrated at the synaptic site in this muscle, is longer than that of the globular forms. Estimations of the renewal rates of individual synaptic AChE molecular forms are still lacking.

FERNANDEZ and STILES (1984b) determined the rate at which the A_{12} form was newly synthesized after total inhibition by DFP: in the case of the end-plate regions of the rat gracilis muscle they found a value of 1.32% h^{-1}, with a total recovery in about 75 h. They also demonstrated that A forms first reappear inside the cells before being externalized.

In summary, the studies of NEWMAN et al. (1984), ROTUNDO (1984b), LAZAR et al. (1984) and BROCKMAN et al. (1986) suggest that the metabolism of AChE begins with the synthesis of distinct pools of G_1 molecules. A major part of the synthesized enzyme is rapidly degraded; a minor fraction is used as precursor in the intracellular assembly of the G_2 and G_4 forms or of the A forms. All forms may be transported to the membrane, externalized and secreted. We examine the intracellular transport and the mechanism of the secretion of AChE in the next section.

E. Intracellular Transport and Secretion of AChE Molecular Forms

I. Intracellular Transport

Studies of the recovery of AChE after DFP treatment of fowl muscle cultures showed that the newly synthesized enzyme reaches the cellular surface where it is associated with the plasma membrane or secreted after about 3 h (ROTUNDO and FAMBROUGH 1980a).

Recently ROTUNDO (1984a, b) monitored the different stages of intracellular transport of AChE molecules in quail muscle cultures by the binding of lectins to their asparagine-*N*-linked carbohydrates (about 20 kDa per subunit). Concanavalin A (ConA) recognizes mannose residues, i.e. it binds the unprocessed glycans as they are cotranslationally linked to the nascent protein; wheat germ agglutinin (WGA) binds terminal *N*-acetylglucosamine, which is added in the proximal Golgi compartment, and *Ricinus communis* agglutinin (RCA) binds galactose, which is added in the distal Golgi compartment. Quail muscle cultures contain extracellular and intracellular G_2, G_4 and A_{12} forms; all extracellular molecules bind the three lectins, indicating that they contain fully mature complex sugar moieties. When quail muscle cultures were treated with DFP, the totality of the *de novo* synthesized enzyme was bound by ConA, but WGA binding developed gradually for a fraction of the enzyme, and RCA binding appeared only after a delay of 1 h. Resistance to endoglycosidase H, which reflects the removal of two terminal mannose residues from the carbohydrate core, also appeared after 1 h. Sedimentation analyses showed that the globular forms were synthesized immediately, and that asymmetric forms first appeared after a delay of 1 h. After 2 h, G_1, G_2 and G_4 were partially bound by WGA, indicating that the polymerization of the subunits takes place before reaching the Golgi apparatus (Fig. 1). By contrast, asymmetric molecules were totally recognized by WGA but were only partially bound by RCA. This clearly showed that their site of assembly is located in the distal elements of the Golgi apparatus (ROTUNDO 1984a, b, and Fig. 1).

In the case of fowl muscle cultures ROTUNDO (1984b) found that the rapidly degraded fraction of enzyme contains monomers and dimers which never show catalytic activity and possibly also some catalytically active molecules. These molecules acquire the capacity to bind WGA, indicating that the sorting process between degradation and stabilization takes place in the Golgi apparatus (R. ROTUNDO, personal communication, and Fig. 1).

II. Secretion

AChE is secreted by nerve and muscle cells *in vitro* as well as *in situ*. This secretion is stimulated by spontaneous or evoked cellular activity, as shown for example by GREENFIELD and her colleagues in the substantia nigra and caudate nucleus of cat, rabbit and rat (GREENFIELD et al. 1980, 1983; GREENFIELD and SHAW 1982; reviewed in GREENFIELD 1985) and by BROCKMAN and YOUNKIN (1986) in cultured rat myotubes. AChE secretion by primary cultures of ox adrenal medullary cells (MIZOBE and LIVETT 1980) is enhanced by the agonist nicotine (MIZOBE and LIVETT 1983); nerve growth factor (NGF) stimulates the secretion by PC12 cells (LUCAS and KREUTZBERG 1985). In most situations the G_4 form is that predominantly secreted, sometimes with minor amounts of A_{12} form (LUCAS and KREUTZBERG 1985). The secretion of G_1 has also been reported in T28 cells (LAZAR et al. 1984) and of G_2 in fowl muscle cells together with the G_4 form (VALLETTE et al. 1986). In culture, rat SCG neurons secrete about 50% G_4, together with approximately equal proportions of G_1 and A_{12} (FERRAND et al. 1986). RUBIN et al. (1985) also reported that mouse muscle cells in culture secrete asymmetric as well as globular forms of AChE (i.e. A_{12}, A_8, G_4 and G_1). When compared to the proportions of molecular forms in the cellular pool of enzyme, secretion products contain, however, a relatively higher amount of G_4 form and a smaller contribution of A forms. A forms may be secreted as soluble aggregates that are dissociated during analysis by centrifugation in high-salt sucrose gradients (RUBIN et al. 1985).

The secretion of AChE probably involves an exocytosis of intracellular vesicles, as illustrated in the case of the sciatic nerve of the rat by KÁSA and RAKONCZAY (1982b). ROTUNDO and FAMBROUGH (1980b) and LUCAS and KREUTZBERG (1985) showed that tunicamycin, an inhibitor of glycosylation, decreases the secretion of AChE and also reduces the intracellular AChE activity, suggesting that one or several steps of the intracellular transport preceding the secretion require the addition of specific sugars to the molecule. The non-glycosylated molecules are rapidly degraded (R. ROTUNDO, personal communication). Colchicine also blocks secretion and induces an increase in the intracellular AChE content, probably by blocking the export of the synthesized enzyme, whereas cytochalasin B has no effect on secretion (LUCAS and KREUTZBERG 1985). These observations suggest that tubulin, but not actin, is involved in the process of secretion. It is likely that microtubules regulate the motility of exocytotic vesicles (ROTUNDO and FAMBROUGH 1982; LUCAS and KREUTZBERG 1985). BENSON et al. (1985) recently demonstrated the presence of AChE, probably as a soluble form, in coated vesicles in cultured muscle cells.

AChE thus appears to follow the classical route: RER to cis-Golgi apparatus to trans-Golgi apparatus to Golgi vesicles to exocytosis (Fig. 1) (see review on general aspects of intracellular protein transport in DAUTRY-VARSAT and LODISH 1983). Some aspects of the processing of membrane-bound and secreted forms are still obscure. It is interesting that cells contain very little non-hydrophobic globular forms and that certain secreted globular forms of AChE interact with detergents (LAZAR et al. 1984). In addition, the mode of attachment of a glycolipidic hydrophobic anchor at the C-terminal end of otherwise soluble subunits remains to be understood; as discussed in the preceding chapter, it might replace an initial hydrophobic terminal peptide. In addition, it should be considered that AChE may in some cases be released from plasma membranes by splitting this glycolipid anchor. As demonstrated by FUTERMAN et al. (1984, 1985) in a variety of tissues, AChE molecules treated with phosphatidylinositol-specific phospholipase C (PIPLC) are soluble and do not interact with non-denaturing detergent micelles. It is tempting to imagine that the release of some forms of AChE might be triggered *in situ* by the activation of an endogenous phospholipase C. This enzyme would probably be quite distinct from the phospholipase C which generates diacylglycerol and phosphoinositides as intracellular second messengers (BERRIDGE and IRVINE 1984).

It should be remembered that most of the secreted AChE usually corresponds to a G_4 form, e.g. in muscle and brain. In the latter tissue the membrane-bound G_4 form contains a disulphide-linked subunit that does not possess a glycolipid anchor similar to that of the erythrocyte G_2 enzyme (INESTROSA et al. 1987).

F. Regulation of the Synthesis of AChE by Muscle and Nerve Cells During Development *In Vivo* and *In Vitro*

I. Evolution of AChE Activity and Molecular Forms in Muscle and Nerve Cells During Embryogenesis

1. Muscle Cells

AChE appears in pre-muscle mononucleated cells in the myotome of 2- or 3-day-old fowl embryos while the other mesodermal cells are AChE-negative (MIKI and MIZOGUTI 1982). A number of observations suggested that the synthesis of A forms during muscle differentiation was subordinate to cell fusion. For example, INESTROSA et al. (1983) found that inhibition of fusion in a C2 line cell culture resulted in an inhibition of the synthesis of the A forms of AChE. We also found that cell fusion appears to condition the assembly of A forms in quail myoblasts, using cultures that were infected by a thermosensitive variant of Rous sarcoma virus. The cells were able to fuse at 42° but not at 35 °C and produced A_{12} AChE only under the former conditions (TOUTANT et al. 1983).

Fusion of muscle cells, however, is not an absolute prerequisite for the synthesis of A forms, as shown both *in vivo* and *in vitro*. In dermomyotomes from early quail embryos (3 days), mononucleated cells contain AChE with significant levels of asymmetric forms (VALLETTE et al. 1987). In the case of cultures of the myogenic C2 cell line, we observed an unambiguous synthesis of A forms in culture conditions in which the cells were withdrawn from the mitotic cycle but not allowed to fuse (C. PINSET and J.-P. TOUTANT, unpublished results).

During further muscle development the molecular forms of AChE are synthesized in an ordered sequence, characteristic of the animal and of the physiological fibre type.

2. Nerve Cells

Neuronal cell precursors removed from the neural plate or neural folds of *Pleurodeles* or *Ambystoma* embryos synthesize AChE *in vitro* (DUPRAT et al. 1985), including A forms (A.-M. DUPRAT, unpublished results).

In fowl embryos the first AChE-positive cells are observed at the cervical level in the mantle layer of the neural tube as early as stage 13 (2-day-old fowl embryo). The intensely AChE-positive cells are no longer capable of proliferation. In contrast, some weakly positive cells are still in the mitotic cycle (MIKI and MIZOGUTI 1982). At this time the neural crest cells which begin to migrate are AChE-positive (MIKI et al. 1983; COCHARD and COLTEY 1983); they contain mostly a G_2 AChE form (COCHARD and COLTEY 1983).

During further development each region of the nervous system displays a characteristic and specific evolution of the different molecular forms of AChE, as demonstrated for example in the case of the electric lobes of *T. marmorata* (BON 1982; WITZEMANN and BOUSTEAD 1982). Other examples are reviewed in MASSOULIÉ and BON (1982). In human brain the developmental changes which affect the molecular forms of AChE consist of an absolute increase of total activity and a relative increase of G_4 over G_1 form (MULLER et al. 1985; ZAKUT et al. 1985). The adult normal brain contains mostly G_4 with a minor component of G_1 (RAZON et al. 1984). AChE activity is highest in the telencephalic subcortical structures of the human brain (caudate nucleus and nucleus of MEYNERT) and varies dramatically from one region to another (ATACK et al. 1986; FISHMAN et al. 1986). BuChE activity is more evenly distributed and does not show a similar increase in activity during development (MULLER et al. 1985). In all regions both enzymes are present under a major G_4 form and a minor G_1 form (ATACK et al. 1986).

II. Relationship Between Synthesis of A Forms and Innervation of Muscle and Nerve Cells *In Vivo*

1. Muscle Cells

In addition to the myotendinous ends of the developing fibre (MUMENTHALER and ENGEL 1961) histochemically detectable concentrations of AChE appear in muscle cells at the sites of nerve contacts when the first neuromuscular interactions are established during embryogenesis (KUPFER and KOELLE 1951; FILOGAMO and GABELLA 1967; ATSUMI 1971). The appearance of A forms of AChE apparently coincides with the innervation of muscle cells (see review in MASSOULIÉ and BON 1982). For example, during the development of the fowl embryo A forms appear in the leg muscles (TOUTANT et al. 1983) at the time of establishment of the first neuromuscular contacts in this region (LANDMESSER and MORRIS 1975).

Thus, it is tempting to consider the A forms as 'endplate-specific' forms which could be used as topographical markers of the NMJ. A forms also exist however in non-synaptic regions of adult muscle fibres, as demonstrated in man (CARSON et al. 1979; SKETELJ and BRZIN 1985), rat diaphragm (YOUNKIN et al. 1982), rat soleus and extensor digitorum longus (LØMO et al. 1985), and slow-twitch muscles of rabbit, rat, guinea-pig (LAI et al. 1986) and fowl (VIGNY et al. 1976; JEDRZEJCZYK et al. 1981; BACOU 1982). A forms are also observed in non-synaptic areas of postnatal rat and mouse sternocleidomastoid muscles (RIEGER et al. 1984d) as well as in embryonic muscles of fowl and rat (KOENIG and VIGNY 1978b; SKETELJ and BRZIN 1980). The myotendinous regions of adult or embryonic muscles display a focalized high AChE activity including A forms, at least in adult amphibian muscles (RATHBONE et al. 1979; S. BON, unpublished results) although these regions are devoid of cholinergic innervation. Moreover, A forms may reappear at denervated synaptic sites after ectopic re-innervation (WEINBERG and HALL 1979). Thus the muscular A forms are not synaptic markers although they are concentrated at the NMJ.

The presence of A forms in aneural muscles *in vivo* has been described in a number of situations, including the following:

- HARRIS (1981) showed that in rat embryos treated *in utero* by *β*-bungarotoxin, a treatment which destroys the motoneurons, non-innervated muscles still synthesize the A forms of AChE.
- Similarly, SOHAL and WRENN (1984) observed that extraocular muscles which develop in the absence of innervation still produce asymmetric forms. We also observed that after the surgical deletion of the lumbar zone of the neural tube performed in the 2-day-old fowl embryo the innervation of leg muscles was prevented but A forms were still detectable in these muscles at 11 days *in ovo* (TOUTANT et al. 1983).
- At 3 days *in ovo* quail dermomyotomes are not innervated and consist of mononucleated cells, some of which are the precursors of muscle cells. They produce the A_{12} form of AChE even in animals deprived of neural tube before the neural crest cells started to migrate (VALLETTE et al. 1987).
- In general, denervation of adult muscles leads to the disappearance of the A forms, as originally shown in rat and fowl muscles. However, this is not always

so; thus, the levels of A forms rise after denervation in the rabbit semi-proprius muscle (BACOU et al. 1982) and in the guinea-pig soleus and semi-proprius muscles (LAI et al. 1986). These A forms occur mostly in an intracellular compartment in non-junctional regions (LAI et al. 1986).
- Regenerating rat myotubes developing in the absence of motor nerves *in vivo* are able to synthesize and accumulate the A_{12} form of AChE (SKETELJ et al. 1985).
- During the development of the electric organ in *T. marmorata* the A forms are synthesized by the electroplaques before the establishment of innervation and the onset of the electric discharge (BON 1982; WITZEMANN and BOUSTEAD 1982).

Thus the expression of A forms by muscle cells *in vivo* does not always require innervation. Ample additional evidence is given by cell cultures, as described below. During normal development the onset of the synthesis of A forms is concomitant with innervation, but it is not clear whether the two events are causally related. For example, innervation of the developing muscle could improve the selection of a particular class of myoblasts (BONNER 1978, 1980) possessing the intrinsic ability to synthesize the A forms of AChE.

2. Nerve Cells

In the SCG of the rat innervation of ganglionic cells by the preganglionic fibres takes place after birth. This is accompanied by a reduction of the A forms of AChE (VERDIÈRE et al. 1982). After decentralization of the rat SCG at birth both the amount of AChE and the relative proportion of asymmetric forms are maintained (KLINAR et al. 1983); they are even transiently increased during 2 weeks after denervation in the adult (GISIGER et al. 1978). In this case innervation appears to repress the synthesis of A forms.

III. Regulation of the Synthesis of A Forms by Muscle and Nerve Cells *In Vitro*

1. Role of Innervation of Muscle Cells

In culture, muscle cells obtained from the limbs of rat embryos synthesize A forms in the absence of nerve cells if they originate from innervated muscles (KOENIG and VIGNY 1978a). Fowl muscle cells also synthesize the A forms in aneural cultures (KATO et al. 1980; BULGER et al. 1982; TOUTANT et al. 1983; POPIELA et al. 1983; VALLETTE et al. 1986; PATTERSON and WILSON 1986). The negative results obtained with aneural fowl muscle cultures by some authors (ROTUNDO and FAMBROUGH 1979; RUBIN et al. 1980; CISSON et al. 1981) were probably due to the use of horse serum to supplement the culture medium (BULGER et al. 1982; TOUTANT et al. 1983). This serum has an inhibitory effect on the synthesis of the A forms (PATTERSON and WILSON 1986) and also contains immunoglobulins that bind fowl AChE and thus interfere with sedimentation analyses (VALLETTE et al. 1986).

Rat myoblasts originating from non-innervated muscles (i.e. before 15 days *in utero*) do not synthesize the A forms in the absence of nerve cells. They acquire this capacity when they are cocultured with nerve cells (KOENIG and VIGNY 1978a).

We found that fowl embryo myoblasts removed from non-innervated limb buds (before 5 days *in ovo*) developed in culture and fused into thin myotubes. These cells were unable to synthesize the A forms either in the absence or in the presence of nerve cells; the presence of nerve cells increased the level of AChE in muscle cells but did not induce the synthesis of A forms (TOUTANT et al. 1983). On the contrary, myoblasts removed at 7, 9 and 11 days *in ovo* synthesized the A forms of AChE even in aneural cultures. Cocultures of these myoblasts with ciliary ganglion or spinal cord cells did not increase the level of A_{12} (KATO et al. 1980; TOUTANT et al.1983).

Nerve cells therefore do not directly control the expression of asymmetric forms by muscle cells, at least in fowl muscle cultures. In fact, as already mentioned, it is likely that the capacity to synthesize the collagen-tailed forms of AChE is characteristic of some of the distinct classes of myoblasts that sequentially appear during fowl leg muscle differentiation (WHITE et al. 1975). According to this view, the earlier myoblasts are unable to produce the asymmetric molecules of AChE, either in the absence or in the presence of nerve cells (TOUTANT et al. 1983). This is consistent with the fact that these distinct classes of myoblasts, differentiated *in vitro*, also produced distinct types of myosin light chains (FISZMAN et al. 1983; TOUTANT et al. 1984).

It is possible to study the differentiation of *Torpedo* electric organ tissue explants in aneural cultures *in vitro* (RICHARDSON et al. 1981). Explants taken before the onset of synaptogenesis *in vivo* (38–47 mm stages) produced large amounts of asymmetric forms of AChE (RICHARDSON and WITZEMANN 1986). This observation confirms that the expression of the A_{12} form does not depend on direct neural influence (induction) but rather constitutes an intrinsic property of the cell.

2. Role of Spontaneous Contractile Activity of Myotubes

In rat muscle cultures the first synthesis of A forms apparently coincides with the onset of contractions. Conversely, the paralysing drug tetrodotoxin (TTX) produces a selective reduction of these forms (RIEGER et al. 1980a; BROCKMAN et al. 1984). A similar effect of TTX is observed in mouse muscle cells differentiated *in vitro* (RUBIN et al. 1985).

A series of observations indicates however that the production of A forms is not always correlated with a contractile activity of muscle fibres. The A forms are synthesized in fowl muscle cells which display very few or no contractions, and curarization or treatment by TTX does not alter this production (POPIELA et al. 1983, 1984; VIGNY and VALLETTE 1985). In addition, TTX does not specifically interfere with the synthesis of A_{12} AChE in C2 cell cultures (INESTROSA et al. 1983). In amphibians (*Xenopus*) the contractile activity of the muscle cells is not required for the *in vivo* development of physiological levels of muscle ChE (COHEN et al. 1984) and of A forms (LAPPIN and RUBIN 1985); no difference was observed between controls and tricaine-treated animals. In vitro *Xenopus* muscle cells synthesize the A_{12}, A_8 and the three globular forms of AChE. The production of A forms is not influenced by TTX, veratridine or aconitine (LAPPIN and RUBIN 1985).

3. Influence of Ionic Permeability

Various drugs that act at the level of ionic channels in the cellular membrane have been reported to modify the synthesis of AChE and particularly the A forms. Whereas paralysis by TTX inhibits the synthesis of A forms in rat muscle cultures, an opposite result was reported with veratridine, a paralyzing drug that maintains the sodium channels in the open state; both the level of AChE and the proportion of A_{12} form were dramatically increased (DE LA PORTE et al. 1984). This effect was suppressed by the divalent ions Cd^{2+} and Co^{2+} and was tentatively attributed to an increase of intracellular Ca^{2+}, indirectly triggered by the Na^+ influx. It was reproduced however neither by KCl depolarization nor by the calcium ionophore, A23187 (DE LA PORTE et al. 1984). In other recent experiments veratridine was found to increase the level of AChE A forms in cultures of fowl (VIGNY and VALLETTE 1985) but not in mouse or rat cultured muscle cells (RUBIN et al. 1985; RUBIN 1985).

Some of the differences between the preceding results may be due to the considerable variability in membrane properties, particularly between developing rat and fowl *in vivo* as well as *in vitro* (RITCHIE and FAMBROUGH 1975; SAMPSON et al. 1983). In addition, POPIELA et al. (1983) suggested that paralysing drugs, such as curare or veratridine, prevent the detachment of the more mature contractile myotubes from the dish: the cells that contain the higher proportion of A_{12} form would be therefore included in the analysis of paralysed but not of control cultures. This could account for an increase of both total AChE activity and of the proportion of A forms in quiescent cultures (POPIELA et al. 1983).

The involvement of Ca^{2+} ions in regulation of the synthesis of A forms has recently been demonstrated by RUBIN (1985) who showed that the inhibitory effect of TTX was reversed by the calcium ionophore A23187 in TTX-paralysed cultures of rat muscle cells. This regulation may involve specific calcium channels, for HAYNES recently found that cinnarizine, a calcium-channel and calcium-sequestration blocker, does not affect AChE synthesis in rat muscle cells (L. HAYNES, personal communication). By using ouabain RUBIN obtained evidence that the influx of Ca^{2+} into the cells might result from an exchange between Na^+ and Ca^{2+}, rather than from an activation of voltage sensitive Ca^{2+} channels or a release of Ca^{2+} from the sacroplasmic reticulum (L. RUBIN, unpublished results).

4. Nerve Cells

Some nerve cells, such as rat SCG neurons (VERDIÈRE et al. 1982; SWERTS et al. 1983) and a pheochromocytoma-derived cell line (PC12), synthesize A forms of AChE *in vitro* in the presence of NGF (RIEGER et al. 1980b; INESTROSA et al. 1981; LUCAS and KREUTZBERG 1985). In SCG neuronal cell cultures and in mouse spinal cell cultures the depolarization induced by an increase of extracellular KCl leads to a decrease of the AChE activity (ISHIDA and DEGUCHI 1983; VERDIÈRE et al. 1984). In SCG cultures depolarization affects the level of the A_{12} form more than that of globular forms (VERDIÈRE et al. 1984; RAYNAUD et al. 1987). It therefore appears that with respect to the level of A forms rat neurons and muscle cells respond differently to manipulations of ion permeability.

WEBER and colleagues reported that the cholinergic differentiation of rat SCG neurons was induced *in vitro* by a culture medium conditioned by rat skeletal muscle cells (SWERTS et al. 1983), by a macromolecular factor partially purified from this conditioned medium (WEBER 1981) or by sodium butyrate (SWERTS et al. 1984). Under these culture conditions the activity of choline acetyltransferase (ChAT) increases dramatically in parallel with the number of cholinergic synapses (SWERTS and WEBER 1984; SWERTS et al. 1984). Neuronal depolarization by elevated K^+, on the contrary, favours the noradrenergic differentiation (RAYNAUD et al. 1987). Although these conditions have opposite effects on the cholinergic or noradrenergic phenotype of the neurons, they both reduce the total AChE activity and, more strikingly, the proportion of the A_{12} form. The expression of A_{12} AChE cannot therefore be correlated with a particular neurotransmitter phenotype.

The level of A_{12} AChE in sympathetic neurons is also sensitive to NGF. This form is not produced by PC 12 cells in the absence of NGF and may be induced to about 5% of the total activity when they are cultured for several days in the presence of NGF (RIEGER et al. 1980b; INESTROSA et al. 1985). Similarly, rat SCG neurons grown with a suboptimal concentration of NGF contained mostly the G_4 form. Higher concentrations of NGF induced an increase of AChE activity by a factor of 2–3, in a dose-dependent manner. This effect was mainly due to an increase of the G_1, G_2 and A_{12} forms, the latter reaching a proportion of 30%–40% of the total activity (M. WEBER, personal communication).

IV. Localization of AChE on Muscle Membranes

In vivo the appearance of restricted zones ('patches') of high AChE activity is generally a sign of the establishment of neuromuscular contacts (review in FILOGAMO and GABELLA 1967). However, patches of AChE are also observed in the absence of the motor nerve: at the myotendinous junction (COUTEAUX 1953), in rat muscles after denervation at birth (LUBINSKA and ZELENA 1966; GAUTRON et al. 1983), in denervated rabbit muscles (TENNYSON et al. 1977) and in embryonic extraocular duck muscle developing in the absence of nerve (SOHAL and WRENN 1984).

In vitro the focalization of AChE is increased in the presence of neurons from embryonic spinal cord; patches of AChE activity occur at sites of contact between myotubes and neuronal processes. AChE patches always coincide with AChR clusters, but the reverse is not true (KOBAYASHI and ASKANAS 1985; DE LA PORTE et al. 1986). This has been observed in homologous co-cultures of rat and fowl nerve and muscle cells (KOENIG 1979, RUBIN et al. 1979), in heterologous co-cultures of rat muscle and fowl nerve cells (DE LA PORTE et al. 1986) and of human myotubes innervated by rat neurons (KOBAYASHI and ASKANAS 1985). The presence of nerve endings was demonstrated by immunohistochemical labelling with tetanus toxin, indicating that the focalization of AChE takes place at the site of neuromuscular contact and is probably induced by the nerve. In addition, in this type of experiment the use of polyclonal anti-rat or anti-fowl AChE antibodies led to the conclusion that the patches consisted essentially of muscle AChE (DE LA PORTE et al. 1986). An evaluation of the proportion of enzyme that is included in the patches showed that it represents less than 1% of the total activity of the A forms, so that any changes in the patched enzyme would not be reflected by significant, biochemically detectable variations in the activity of the total enzyme or of the A forms (F. VALLETTE and M. VIGNY, personal communication).

Patches of AChE activity have also been observed in aneural cultures of *Xenopus* muscle cells (MOODY-CORBETT and COHEN 1981; WELDON et al. 1981) as well as in aneural cultures of the myogenic C2 line (INESTROSA et al. 1982) and during the differentiation of mouse mdg mutant muscle cells *in vitro* (POWELL et al. 1979). Non-synaptic concentrations of AChE were also observed in the basal lamina of regenerating frog muscle fibres (ANGLISTER and MCMAHAN 1985). Although the staining intensity of these patches is usually lower than those observed in cocultures, their existence shows that the focalization of AChE cannot be considered as an exclusive marker of nerve-muscle contacts; in aneural cultures of muscle cells it is related to the presence of patches of extracellular matrix material (SILBERSTEIN et al. 1982). These are controlled by the contractile activity of the myotubes (SANES and LAWRENCE 1983) and/or by factors extracted from brain (SANES et al. 1984) or the basal lamina of *Torpedo* electric organs (WALLACE et al. 1985; WALLACE 1986).

In addition to collagen, two components of the extracellular matrix are most probably involved in the process of AChE focalization. (a) Heparan sulphate proteoglycan (HSPG) appears to control the subcellular distribution of the A forms by anchoring these molecules *in situ* via ionic interactions between its negative charges and the collagenous tail. INESTROSA et al. (1985) showed that the A forms are mostly externalized in PC12 cells but to a lesser degree in a mutant line that lacks HSPG or in cultures which were treated with β-D-xylosides, which interfere with proteoglycan assembly. These authors also obtained evidence that the A forms interact *in situ* with HSPG in the basal lamina of the electric organ of *Discopyge tschudii* (BRANDAN et al. 1985) and bind *in vitro* to intact cells rich in cell-surface HSPG (BRANDAN and INESTROSA 1986). (b) Agrin, an extracellular synaptic organizing molecule, has been shown to influence the *in situ* focalization of both receptor and AChE (MCMAHAN 1985).

G. Non-cholinergic Functions of ChEs

I. Adult Tissues: New Substrates for ChEs

The presence of ChEs in non-cholinergic tissues has led to the suggestion that they may be involved in functions other than cholinergic transmission (see reviews by SILVER 1974; RAMA-SASTRY and SADAVONGVIVAD 1979; GREENFIELD 1984; CHUBB 1984). In non-nervous tissues, for example, it has been suggested that AChE may be involved in the regulation of membrane permeability in erythrocytes and the blood-brain barrier and motility in sperm. Serum BuChE may be involved in the control of the lipoprotein metabolism of adult mammals (see review in KUTTY 1980), and in particular in the formation of low-density lipoproteins from very low density lipoproteins in human serum (KUTTY et al. 1977).

In nervous tissues the distribution of ChEs is not restricted to cholinergic domains (SILVER 1974; GRAYBIEL and RAGSDALE 1982), as became evident when discrepancies between the regional distribution of ChAT and AChE were discovered (KIMURA et al. 1981; SATOH et al. 1983; ECKENSTEIN and SOFRONIEW 1983; MESULAM et al. 1984). AChE is detected intracellularly in noradrenergic neurons of the substantia nigra (ECKENSTEIN and SOFRONIEW 1983; HENDERSON and GREENFIELD 1984) and locus ceruleus where it coexists with BuChE (SKET and PAVLIN 1985) and in serotoninergic neurons of the raphe (MIZUKAWA et al. 1986). GREENFIELD et al. (1980, 1983) showed that substantia nigra, a typical dopaminergic domain of the brain, actively secretes AChE and dopamine upon depolarization or electrical stimulation. The enzyme is also released from the dendrites of nigrostriatal cells and may either act as a neurotransmitter (GREENFIELD et al. 1981) or adjust the sensitivity of those cells to neurotransmitters other than ACh, for example by hydrolysing released neuroactive peptides (see review in GREENFIELD 1985, and discussion in CHUBB and BORSTEIN 1985).

It must therefore be considered that ChEs may act on other substrates besides ACh. It is unlikely that such substrates are higher homologues of ACh, except perhaps in the very few tissues (bovine spleen, hypobranchial glands of certain molluscs) where these occur (WHITTAKER 1963).

An amidase activity of AChE was first reported by MOORE and HESS (1975). GEORGE and BALASUBRAMANIAN (1980, 1981) showed that human erythrocyte AChE and also plasma BuChE were able to cleave *o*-nitroacetanilide into *o*-nitroaniline and acetate. This amidase activity is specifically inhibited by serotonin but activated by tyramine. The catalytic site that is responsible appears partially to overlap the cholinesteratic site (BALASUBRAMANIAN 1984).

Extensively purified, apparently homogenous preparations of both AChE and BuChE hydrolyse substance P (SP) (CHUBB et al. 1980; LOCKRIDGE 1982). Interestingly, the peptide bonds that are hydrolysed by the two enzymes differ (LOCKRIDGE 1982). In addition, AChE also hydrolyses enkephalins (CHUBB et al. 1983). The esteratic site was shown for both AChE (CHUBB et al. 1980) and BuChE (CHATONNET and MASSON 1984, 1985) to be distinct from the peptidasic site on the basis of their different sensitivities to DFP. In human plasma the number of peptidasic and esteratic sites are not equivalent, suggesting that they are carried by distinct molecules. In fact, NAUSCH and HEYMANN (1985) as well as KAEMMER et al. (1986) recently claimed that the peptidasic activity is due to a contamination by dipeptidylpeptidase IV. CHATONNET amd MASSON (1986) however observed an increase of SP hydrolysis in the plasma of human 'atypic silent' mutants of BuChE in which there was no alteration of dipeptyldipeptidase. This suggests that the silent enzyme allele, which has no esteratic activity, may possess a peptidasic activity.

In any case, it must be borne in mind that the physiological substrate(s) of the ChE-associated peptidase activity, if it exists, has not yet been identified. The presence of ChEs in non-cholinergic regions of the CNS is clearly understandable if these enzymes participate in the inactivation of neurotransmitter peptides such as SP or enkephalins. From analyses of synaptic AChE activity in relation to the morphology of presynaptic boutons on α-motoneurons in fowl ventral horn, SAKAMOTO et al. (1985) suggested that AChE is probably located at synapses where presynaptic boutons contain SP, enkephalins, and/or ACh (ATSUMI et al. 1985). ChEs might also participate in the intracellular maturation of neurotransmitters. CHUBB and MILLAR (1984) and MILLAR and CHUBB (1984) suggested, for example, that AChE is involved in the proteolytic maturation of both enkephalins and SP in retinal amacrine cells. Similarly, it has been reported that the degradation of chromogranins from adrenal chromaffin granules by purified AChE (SMALL et al. 1986) generates enkephalin-like immunoreactivity (ISMAËL et al. 1986), suggesting that AChE may act either on the production or the breakdown of enkephalins (ISMAËL et al. 1986).

II. 'Embryonic ChE': A Cholinergic Function in the Absence of Innervation?

AChE and BuChE activities are present in a variety of cell types at the beginning of embryonic development. AChE was detected in the hypoblast cells of the early gastrulating fowl embryo under G_1, G_2 and G_4 forms (LAASBERG and NEUMAN 1985).

During early leg differentiation in the fowl the ectodermal cells of the limb bud apical ridge are ChE-positive, as well as the underlying mesenchyme (DREWS 1975). The enzymatic activity is localized intracellularly and also appears as membrane-bound forms (DREWS et al. 1986). Both AChE and BuChE are also present in the basal lamina between the two tissues (FALUGI and RAINERI 1985) and in the chondrogenic core of the limb bud (SCHRÖDER 1980). This type of ChE activity ('embryonic ChE'; DREWS 1975) disappears from the cells with further differentiation and is independent of innervation.

In the developing nervous system AChE and BuChE are widely distributed and numerous cells which are devoid of activity in the adult contain the enzyme at some stage of their development (SILVER 1971). The formation of sensory thalamo-cortical synapses, particularly in the visual cortex, coincides spatially and temporally with the transient presence of AChE activity in the afferent axon terminals (ROBERTSON et al. 1985; ROBERTSON 1987). LAYER (1983) considers that the expression of AChE is an early differentiation event in the entire brain. During the morphogenesis of the eye in fowl embryos LAYER (1983) observed a double inverse gradient of AChE and BuChE in the retina. He suggested that such a distribution of ChEs might contain the information needed for the organization of developing retinal cells. He recently demonstrated a direct stimulatory action of purified embryonic BuChE (but not AChE) on the proliferation of brain and retinal cells *in vitro* (LAYER et al. 1985). Therefore early ChE, or ChE-positive, cells may play a transient role in morphogenesis, even in non-cholinergic tissues.

DREWS and his colleagues suggest that 'embryonic ChE' activity is part of a primitive muscarinic system since mAChRs (SCHMIDT 1981) and ChAT (REICH and DREWS 1983) are also expressed in the developing fowl limb buds. On stimulation with ACh or muscarinic ligands (SCHMIDT et al. 1984) the embryonic cells respond with intracellular Ca^{2+} mobilization which might, in turn, regulate cell movements and possibly cellular migrations.

Embryonic ChE might therefore possess a transient 'cholinergic' function in the regulation of morphogenetic events, independently of any cholinergic innervation.

H. Effects of Peptides on ChEs

I. Effect on the Activity of AChE

As discussed in the preceding section, peptides may be substrates of ChEs. On the other hand, peptides may exert a direct influence on both the synthesis and the activity of these enzymes.

Substance P has been reported to exert a stimulating effect on AChE in rat brain *in vivo* and on AChE from *Electrophorus in vitro* (CATALÁN et al. 1984).

On the other hand, detailed studies by HAYNES and colleagues have demonstrated a specific inhibition of some AChE forms by endorphins (HAYNES and SMITH 1982; HAYNES et al. 1984a, b, 1985). β-Endorphin exists within the cholinergic motor nerve terminals of immature rat skeletal muscles. It is released together with ACh when these nerves are stimulated (HAYNES et al. 1985). β-Endorphin was reported to inhibit the A forms of AChE under conditions where no inhibitory effect was seen either on the native globular forms or on the lytic tetramers (G_4) obtained by their collagenase treatment (HAYNES and SMITH 1982; HAYNES et al. 1984a, b). This inhibition of asymmetric AChE by β-endorphin appeared to be highly species-specific (HAYNES et al. 1984b). This interesting observation raises the possibility that β-endorphin may regulate the sensitivity of the developing muscle fibres to ACh, and this may explain why the response of denervated muscles to ACh is increased in the presence of β-endorphin (HAYNES et al. 1983). No other examples of the selective inhibition of AChE forms have been reported, and the specific inhibition of the A forms by β-endorphin deserves further experimental investigation.

II. Effect on the Synthesis of ChE

Certain proteinaceous substances extracted from nerves stimulate the synthesis of AChE in muscle cells *in vitro*. This stimulating effect is also observed *in vivo*: the neurotrophic substances are transported along the axons and released from the nerve terminals (for a review, see MASSOULIÉ and BON 1982). One of the candidates for this neurotrophic effect was first called sciatin and later identified as transferrin (OH and MARKELONIS 1982; BEACH et al. 1983).

KOELLE and his colleagues demonstrated that brain, spinal cord and sciatic nerve extracts contain a neurotrophic factor that is responsible for the maintenance of both AChE and BuChE activities in the preganglionically denervated cat SCG (KOELLE and RUCH 1983; KOELLE et al. 1983). This substance appeared to be a peptide of less than 1000 Da and its effect on the SCG was partially reproduced by cAMP (KOELLE et al. 1984). At that time HAYNES et al. (1984b) reported that β-endorphin exerted a positive neurotrophic effect on the total AChE activity of rat muscle cells in culture but a negative effect on the A_{12} form. HAYNES and SMITH (1985) further showed that the dipeptide glycyl-1-glutamine (Gly-Gln) which is a cleavage product of the C-terminus of β-endorphin induced an increase of the A_{12} form. KOELLE et al. (1985) observed that this dipeptide also reproduced the neurotrophic effect previously described on the denervated cat SCG. Glycyl-1-glutamine may be the precursor of another active substance, glycyl-1-glutamic acid (KOELLE et al. 1985).

I. ChEs in Pathological States and Their Evolution with Aging

I. Muscular Dystrophy

This topic has been reviewed by BRIMIJOIN (1983) and RAKONCZAY (1986) and is also considered in Chaps. 24, 25 and 26. However, for the sake of completeness a few examples of the effect of pathological states and aging on ChEs will be discussed here.

1. Avian Dystrophy

Detailed studies (LYLES et al. 1979, 1982) have shown that both AChE and BuChE are affected primarily in the fast-twitch muscles of dystrophic fowl.

2. Mouse Dystrophies

129/Rej. In this mutant mouse line fast-twitch but not slow-twitch muscles are affected and present a reduction of the G_4 form of AChE (for reviews, see GISIGER and STEPHENS 1983b; BRIMIJOIN 1983). A similar reduction of the G_4 form is also observed in the corresponding nerves (BRIMIJOIN and SCHREIBER 1982; GISIGER and STEPHENS 1984).

C57BL/6J dy/dy and dy^{2J}/dy^{2J}. In these mutants the fast-twitch muscles also present a reduction of the G_4 form (LINDENBAUM and LIVETT 1983). Sequential extractions with low salt, detergent and high salt revealed that the alteration affected the membrane-bound G_4 (POIANA et al. 1985). Since the amounts of G_2 and G_1 forms were normal in dystrophic muscles, it was suggested that the defect could be due to an anomaly in the assembly of the G_4 form (POIANA et al. 1985). Alternatively, an increase of G_4 secretion could also account for the lowered level of the muscle G_4 form.

Motor Endplate Disease (med and med^J*).* These mutations induce a progressive rigid paralysis of the hind limbs. In parallel a characteristic paranodal dysmyelination of the axons was described (RIEGER et al. 1984b), and a selective reduction of the G_4 form occurred in the gastrocnemius muscle of *med/med* and med^J/med^J mice (RIEGER et al. 1983). Since a similar reduction of the G_4 form of AChE is also observed after short-term denervation or tenotomy, it was assumed that the defect might be correlated with a partial loss of muscle activity.

Muscular Dysgenesis (mdg). The *mdg/mdg* embryos do not present any contractile activity and die at birth. An extensive neurite outgrowth is observed in these mutants, for example, in the diaphragm. The A_{12} form is reduced but still detectable in these inactive muscles; *mdg* myotubes apparently fail to elaborate a normal basal lamina. This is probably correlated with the low level of A_{12} AChE and may even be the direct cause of this defect (RIEGER et al. 1984a). The myotubes of these mutants acquire contractile activity *in vitro* when they are cocultured with wild type (+/+) neurons but not with *mdg/mdg* neurons: this suggests that the primary target of the mutation might be the nerve rather than the muscle cells (KOENIG et al. 1982; POWELL et al. 1984; RIEGER et al. 1984c).

Trembler. Characteristic of Trembler mice is an abnormal motor activity. The mutation results in hypomyelination and segmental demyelination of peripheral axons. The motor innervation is aberrant, with multi-innervated muscle fibres and abnormally large, complex neuronal arborizations, particularly in the soleus muscle. The levels of ACh receptors and AChE are not modified in this muscle, however, in spite of the morphological anomaly in their distribution (DO THI et al. 1986). The proportions of molecular forms of AChE are not affected in the soleus muscle but the proportion of the collagen-tailed A_{12} and A_8 forms is increased in the gastrocnemius muscle (LEFAIX et al. 1982).

II. Human Pathology

1. Senile Dementia of Alzheimer's Type

Several authors have observed a decrease in the membrane-bound G_4 form, without change in the G_1 form, in the cortex of patients with senile dementia of the Alzheimer type (SDAT) (ATACK et al. 1983; YOUNKIN et al. 1986; FISHMAN et al. 1986). In addition YOUNKIN et al. (1986) found that the level of A_{12} and A_8 forms was increased several-fold. The loss of G_4 might be a reflection of a selective destruction of cholinergic axonal processes (ATACK et al. 1983) or presynaptic elements (MARQUIS and FISHMAN 1985; FISHMAN et al. 1986). The decrease of G_4 form may also be due to an increase in its secretion, since the AChE activity in the plasma of SDAT patients is significantly higher than in controls while both the BuChE of the plasma and the AChE of erythrocytes are unchanged (ATACK et al. 1985). This observation is challenged however by the reports of MARQUIS et al. (1985) and ST CLAIR et al. (1986) who did not notice any significant difference between SDAT and control plasma AChE. In addition no increase in AChE is found in the cerebrospinal fluid of SDAT patients; low BuChE activity was found to be correlated with the severity of dementia but AChE activity did not differentiate patients with SDAT from control cases (HUFF et al. 1986).

2. Amyotrophic Lateral Sclerosis

In patients suffering from amyotrophic lateral sclerosis (ALS) the level of AChE is increased about two-fold in the plasma but remains unchanged in the erythrocytes (FESTOFF and FERNANDEZ 1981). The increased AChE of the plasma is probably related to marked modifications of the enzyme in skeletal muscle; the total activity is reduced to one-fourth its normal value, and there is a complete disappearance of both A_4 and G_4 forms of AChE. In addition, the G_1 form was reported to shift from 3.8 to 4.8 *S* (FERNANDEZ et al. 1986). These changes do not resemble those following denervation (disappearance of A forms with a transient increase of A_8 and A_4 forms). They could correspond to a defect in the assembly of the G_4 form or to an increase of the secretion of this form, which would explain the increase of AChE activity in the plasma. The tissue origin of the circulating AChE, however, remains to be elucidated (RASOOL et al. 1983). It will probably be possible to establish this point by using specific monoclonal antibodies (BRIMIJOIN and RAKONCZAY 1986).

3. Hirschsprung's Disease

The terminal colon of these patients is characterized by an absence of ganglion cells and a proliferation of unmyelinated nerve fibres. The AChE activity of biopsied colon samples is correspondingly increased (six-fold) with an elevated G_4/G_1 ratio (RAKONCZAY and NEMETH 1984; BONHAM et al. 1985). This suggests that the G_4 form could be associated preferentially with the axons (KÁSA and RAKONCZAY 1982b; COURAUD et al. 1983; MARQUIS and FISHMAN 1985).

4. Paroxysmal Nocturnal Haemoglobinuria

The erythrocytes of patients with paroxysmal nocturnal haemoglobinuria are abnormally sensitive to haemolysis. Their membranes have long been known to be deficient in AChE to a degree which is correlated with the severity of the disease (METZ et al. 1960). The sensitivity of the erythrocytes to lysis by complement however is due to the lack of another membrane protein, the decay acceleration factor (DAF) (NICHOLSON-WELLER et al. 1983; PANGBURN et al. 1983), which has no relationship to AChE (SUGARMAN et al. 1986). As discussed in Chap. 8a, the primary defect might reside in the mechanism of coupling of these proteins with their glycolipid anchor.

5. Other Pathological States

SOREQ examined the AChE content of the blood of a patient who showed considerable muscular weakness with fasciculation and muscarinic-type symptoms (diarrhoea, insomnia). The patient collapsed upon administration of anti-ChEs. The erythrocytes and serum of this patient presented an abnormally low level of AChE. His serum contained auto-antibodies against human AChE which showed a high affinity for AChE extracted from muscle by salt and detergent and appeared responsible for the functional defects (LIVNEH et al. 1988).

ENGEL et al. (1977) found that A forms were missing in muscles of patients suffering from a rare neuromuscular disease characterized by extreme fatiguability. In these muscles failure to hydrolyse ACh rapidly enough resulted in a pronounced morphological adaptation, with a diminished quantal content, despite the presence of G forms. POIANA et al. (1986) observed a complete lack of both A_{12} and G_4 forms in muscles of a majority of myopathic non-dystrophic children.

III. AChE and the Prenatal Diagnosis of Neural Tube Defects

The AChE activity of amniotic fluid has been used, together with α-foetoprotein and echography, in the prenatal diagnosis of neural tube defects (spina bifida, anencephaly) since the activity of the enzyme is generally markedly increased in such cases (SMITH et al. 1979). As with α-foetoprotein, however, the dispersion of the observed values is very large so that there is a considerable overlap between controls and pathological levels.

Sedimentation studies showed that normal amniotic fluid contains mostly G_4 BuChE together with G_1 and G_2 AChE, and that in neural tube defects there is an increase of G_4 AChE (BONHAM and ATACK 1983) and BuChE and sometimes also of G_1 and G_2 AChE (F. MULLER, unpublished results). Electrophoretic analyses of AChE proved to be useful in the diagnosis of neural tube defects. In these cases AChE of amniotic fluid presents an additional, faster migrating band (SELLER et al. 1980; AITKEN et al. 1984) which also exists in nervous tissue extracts (MULLER et al. 1985) and probably corresponds to a secreted G_4 form. Recent results (MULLER et al. 1986) have shown that all cases of neural tube defects were detected by the presence of the electrophoretically fast AChE variant, even when they were not diagnosed by echographic recordings. A similar electrophoretic pattern was also observed in other cases of foetal anomalies, i.e. omphalocele, X monosomy, and sacrococcygial teratoma, all of which can be recognized by echography. Apart from these cases false-positives were never seen if the amniotic fluids were obtained before 25 weeks of pregnancy (MULLER et al. 1986). In experimentally induced neural tube defects in chick embryo, PILOWSKI et al. (1982) did not find any anomaly in the electrophoretic mobility of AChE.

IV. Aging

It has been reported that the A_{12} form of AChE disappears in the anterior latissimus dorsi muscle of aged fowls (LYLES and BARNARD 1980). In the rat diaphragm the relative proportions of A_{12} were found to be identical in 3-month-old and 24-month-old rats (SKETELJ and BRZIN 1985), but a decrease was observed in the rat gracilis muscle around the 9th month (H. FERNANDEZ, personal communication). The evolution of the A_{12} form may therefore be controlled by the type of muscular activity and may be an interesting correlate of the physiological state of the muscle.

J. Conclusion

Cholinesterases may be either secreted, anchored in plasma membranes through their hydrophobic domain (G forms), or attached to extracellular matrix elements (A forms). The globular forms occur in non-cholinergic as well as in cholinergic tissues; in contrast, the asymmetric forms seem to exist only in muscular and nervous tissues. This is apparently true for BuChE as well as for AChE. Asymmetric forms of BuChE have so far been characterized only in mammalian and avian muscles and nerves and in *Torpedo* heart and nervous tissue.

It is tempting to assume that these different molecules are designed in such a way that they may be positioned at various subcellular levels, e.g. at strategic sites in cholinergic synapses. In particular, the evolution of collagen-tailed molecules may be functionally correlated to the cholinergic control of skeletal muscles in vertebrates, since these forms exist even in the most primitive vertebrates (Agnathas, hagfish and lampreys), but have not been observed in invertebrates (see discussion in ARPAGAUS and TOUTANT 1985). They appear to be exclusively synthesized in the context of cholinergic transmission and are subject to specific physiological regulation.

It is thus necessary do distinguish the regulation of total activity for AChE and BuChE from the regulation of expression of particular forms of these enzymes. Although this aspect has not been specifically discussed here, it must be noted that the expression of ChEs, and particularly of asymmetric forms, is not systematically correlated with that of other cholinergic proteins such as the ACh receptor, e.g. upon denervation of muscle.

The appearance of asymmetric AChE forms in embryonic muscle and their maintenance in adult muscle have long been thought to be conditioned by neuromuscular interactions. The fact that myotubes in cultures or some types of denervated muscle *in vivo* produce these molecules might be the result of a permanent nervous induction. Asymmetric forms have, however, been found at a very early stage of development in morphologically undifferentiated myoblasts in the absence of any nerve contact, e.g. *in ovo* in the dermomyotomes of 3-day-old quail embryos.

It thus appears that the capacity to assemble collagen-tailed ChE corresponds to a particular state of differentiation of the myoblasts or to particular categories of myoblasts. Although generally correlated in cultures with the fusion of myoblasts into myotubes or with the onset of contractile activity, these factors are not required as such. Recent experimental evidence suggests that the control of the biosynthesis of asymmetric forms involves the level of Ca^{2+} in intracellular compartments.

ChEs may also be involved in non-cholinergic functions. Both AChE and BuChE have been reported to cleave peptides such as substance P and enkephalins with distinct specificities. It would be possible to interpret the differential localization of AChE and BuChE in the central nervous system if they contribute with different specificities to the metabolism of neuropeptides.

An analysis of the variety of ChE molecules in a particular tissue provides a wealth of information and may differentiate normal from pathological states. It may therefore contribute useful clues in the investigation of certain diseases. Further characterization of the AChE and BuChE molecular species will certainly constitute an extremely valuable source of information on both normal cellular processes and their anomalies.

Acknowledgements. We wish to thank all our colleagues for the communication of their recent results, Dr. George B. Koelle for critical reading of the manuscript, and Jacqueline Pons for expert secretarial assistance.

References

Adamo S, Zani BM, Nervi C, Senni MI, Molinaro M, Ensebi F (1985) Acetylcholine stimulates phosphatidylinositol turnover at nicotinic receptors of cultured myotubes. FEBS Lett 190:161–164

Aitken DA, Morrison NM, Ferguson-Smith (1984) Predictive value of amniotic acetylcholinesterase analysis in the diagnosis of fetal abnormality in 3700 pregnancies. Prenat Diagn 4:329–340

Anglister L, McMahan UJ (1983) Acetylcholinesterase in the synaptic basal lamina of damaged muscle fibers (Abstr). 2nd International Meeting on Cholinesterases, Bled, p 58

Anglister L, McMahan UJ (1985) Basal lamina directs acetylcholinesterase accumulation at synaptic sites in regenerating muscle. J Cell Biol 101:735–743

Arpagaus M, Toutant JP (1985) Polymorphism of acetylcholinesterase in adult *Pieris brassicae* heads. Evidence for detergent-insensitive and Triton X-100-interacting forms. Neurochem Int 7:793–804

Atack JR, Perry EK, Bonham JR, Perry RH, Tomlinson BE, Blessed G, Fairbairn A (1983) Molecular forms of acetylcholinesterase in senile dementia of Alzheimer type: selective loss of the intermediate (10 S) form. Neurosci Lett 40:199–204

Atack JR, Perry EK, Perry RH, Wilson ID, Bober MJ, Blessed G, Tomlinson BE (1985) Blood acetyl and butyrylcholinesterases in senile dementia of Alzheimer type. J Neurol Sci 70:1–12

Atack JR, Perry EK, Bonham JR, Candy JM, Perry RH (1986) Molecular forms of acetylcholinesterase and butyrylcholinesterase in the elderly human central nervous system. J Neurochem 47:263–277

Atsumi S (1971) The histogenesis of motor neurons with special reference to the correlation of their endplate formation. I. The development of endplates in the intercostal muscle in the chick embryo. Acta Anat (Basel) 80:161–182

Atsumi S, Sakamoto H, Yokota S, Fujiwara T (1985) Substance P and 5-hydroxytryptamine immunoreactive presynaptic boutons on presumed α-motoneurons in the chicken ventral horn. Arch Histol Jpn 48:159–172

Bacou F (1982) The polymorphism of acetylcholinesterase in different animal species and motor innervation. Reprod Nutri Dev 22:227–233

Bacou F, Vigneron P, Massoulié J (1982) Acetylcholinesterase forms in fast and slow rabbit muscle. Nature 296:661–664

Bacou F, Vigneron P, Couraud JY (1985) Retrograde effect of muscle on forms of AChE in peripheral nerves. J Neurochem 45:1178–1185

Balasubramanian AS (1984) Have cholinesterases more than one function? Trends Biochem Sci 9:467–468

Barat A, Escudero E, Rodríguez-Borrajo C, Ramírez G (1983) Collagen-tailed and globular forms of acetylcholinesterase in the developing chick visual system. Neurochem Int 5:95–99

Barnard EA, Lyles JM, Pizzey JA (1982a) Fibre types in chicken skeletal muscles and their changes in muscular dystrophy. J Physiol (Lond) 331:333–354

Barnard EA, Lyles JM, Silman I, Jedrzejczyk J, Barnard PJ (1982b) Molecular forms of acetylcholinesterase and pseudocholinesterase in chicken skeletal muscles: their distribution and change with muscular dystrophy. Reprod Nutr Dev 22:261–273

Beach RL, Popiela H, Festoff BW (1983) The identification of neurotrophic factor as a transferrin. FEBS Lett 156:151–156

Benson RJJ, Porter-Jordan K, Buoniconti P, Fine RE (1985) Biochemical and cytochemical evidence indicates that coated vesicles in chick myotubes contain newly synthesized acetylcholinesterase. J Cell Biol 101:1930–1940

Berman HA, Decker MM, Sangmee J (1987) Reciprocal regulation of butyrylcholinesterase and acetylcholinesterase in mammalian skeletal muscle. Dev Biol 120:154–161

Berridge MJ, Irvine RF (1984) Inositol triphosphate, a novel second messenger in cellular signal transduction. Nature 312:315–321

Bon S (1982) Molecular forms of acetylcholinesterase in developing *Torpedo* embryos. Neurochem Int 4:577–585

Bonham JR, Atack JR (1983) A neural tube defect specific form of acetylcholinesterase in amniotic fluid. Clin Chim Acta 135:233–237

Bonham JR, Dale G, Scott D, Wagget J (1985) Molecular forms of acetylcholinesterase in Hirschsprung's disease. Clin Chim Acta 145:297–305

Bonner PH (1978) Nerve-dependent changes in clonable myoblast populations. Dev Biol 66:207–219

Bonner PH (1980) Differentiation of chick embryo myoblasts is transiently sensitive to functional denervation. Dev Biol 76:79–86

Brandan E, Inestrosa NC (1986) The synaptic form of acetylcholinesterase binds to cell-surface heparan sulfate proteoglycans. J Neurosci Res 15:185–196

Brandan E, Maldonado M, Garrido J, Inestrosa NC (1985) Anchorage of collagen-tailed acetylcholinesterase to the extracellular matrix is mediated by heparan sulfate proteoglycan. J Cell Biol 101:985–992

Brimijoin S (1983) Molecular forms of acetylcholinesterase in brain, nerve and muscle: nature, localization and dynamics. Prog Neurobiol 21:291–322

Brimijoin S, Carter J (1982) Turnover of the molecular forms of acetylcholinesterase in the rat diaphragm. J Neurochem 38:588–590

Brimijoin S, Rakonczay Z (1986) Immunology and molecular biology of the cholinesterases: current results and prospects. Int Rev Neurobiol 28:363–410

Brimijoin S, Schreiber PA (1982) Reduced axonal transport of 10 S acetylcholinesterase in dystrophic mice. Muscle Nerve 5:405–410

Brockman SK, Younkin SG (1986) Effect of fibrillation on the secretion of acetylcholinesterase from cultured embryonic rat myotubes. Brain Res 376:409–411

Brockman SK, Przybylski RJ, Younkin SG (1982) Cellular localization of the molecular forms of acetylcholinesterase in cultured embryonic rat myotubes. J Neurosci 2:1775–1785

Brockman SK, Younkin LH, Younkin SG (1984) The effect of spontaneous electromechanical activity on the metabolism of acetylcholinesterase in cultured embryonic rat myotubes. J Neurosci 4:131–140

Brockman SK, Usiak MF, Younkin SG (1986) Assembly of monomeric acetylcholinesterase into tetrameric and asymmetric forms. J Biol Chem 261:1201–1207

Brzin M, Sketelj J, Klinar B (1983) Cholinesterases. In: Lathja E (ed) Handbook of neurochemistry. Plenum, New York, p 251

Bulger JE, Randall WR, Nieberg PS, Patterson GT, McNamee MG, Wilson BW (1982) Regulation of acetylcholinesterase forms in quail and chicken muscle cultures. Dev Neurosci 5:474–483

Carson S, Bon S, Vigny M, Massoulié J, Fardeau M (1979) Distribution of acetylcholinesterase molecular forms in neural and non-neural sections of human muscle. FEBS Lett 96:348–352

Carter JL, Brimijoin S (1981) Effects of acute and chronic denervation on release of acetylcholinesterase and its molecular forms in rat diaphragms. J Neurochem 36:1018–1025

Catalán RE, Martinez AM, Aragonés MD, Miguel BG, Robles A, Godoy JE (1984) Effects of substance P on acetylcholinesterase activity. Biochem Int 8:203–208

Chatonnet A, Bacou F (1983) Acetylcholinesterase molecular forms in the fast or slow muscles of the chicken and the pigeon. FEBS Lett 161:122–126

Chatonnet A, Masson P (1984) Le site peptidasique de la butyrylcholinestérase est distinct du site estérasique. C R Séances Acad Sci [III C] 229:529–534
Chatonnet A, Masson P (1985) Study of the peptidasic site of cholinesterase: preliminary results. FEBS Lett 182:493–498
Chatonnet A, Masson P (1986) Is the peptidase activity of highly purified human plasma cholinesterase due to a specific cholinesterase isoenzyme or a contaminating dipeptylaminopeptidase? Biochimie 68:657–667
Chubb IW (1984) Acetylcholinesterase: multiple functions? In: Brzin M, Barnard EA, Sket D (eds) Cholinesterases: fundamental and applied aspects. de Gruyter, Berlin, pp 345–359
Chubb IW, Borstein JC (1985) Dopamine and acetylcholinesterase released in the substantia nigra: cooperative or coincidental? Neurochem Int 7:905–912
Chubb IW, Millar TJ (1984) Is intracellular acetylcholinesterase involved in the processing of peptide neurotransmitters? Clin Exp Hypertens [A] 6:79–89
Chubb IW, Hodgson AJ, White GH (1980) Acetylcholinesterase hydrolyzes substance P. Neuroscience 5:2065–2072
Chubb IW, Ranieri E, White GH, Hodgson AJ (1983) The enkephalins are amongst the peptides hydrolyzed by purified acetylcholinesterase. Neuroscience 10:1369–1378
Cisson CM, McQuarrie CH, Sketelj J, McNamee MG, Wilson BW (1981) Molecular forms of acetylcholinesterase in chick embryonic fast muscle: developmental changes and effects of DFP treatment. Dev Neurosci 4:157–164
Cochard P, Coltey P (1983) Cholinergic traits in the neural crest: acetylcholinesterase in crest cells of the chick embryo. Dev Biol 98:221–238
Cohen MW, Greschner M, Tucci M (1984) In vivo development of cholinesterase at a neuromuscular junction in the absence of motor activity in *Xenopus laevis*. J Physiol (Lond) 348:57–66
Collins PL, Younkin SG (1982) Effect of denervation on the molecular forms of acetylcholinesterase in rat diaphragm. J Biol Chem 257:13638–13644
Couraud JY, di Giamberardino L, Hässig R, Mira JC (1983) Axonal transport of the molecular forms of acetylcholinesterase in developing and regenerating peripheral nerve. Exp Neurol 80:94–110
Couraud JY, Nicolet M, Hässig R (1985) Rapid axonal transport of three molecular forms of acetylcholinesterase in the frog sciatic nerve. Neuroscience 14:1141–1147
Couteaux R (1953) Particularités histochimiques des zones d'insertion du muscle strié. C R Soc Biol (Paris) 147:1974–1976
Dautry-Varsat A, Lodish HF (1983) The Golgi complex and the sorting of membrane and secreted proteins. Trends Neurosci 6:484–490
Davey B, Younkin SG (1978) Effect of nerve stump length on cholinesterase in denervated rat diaphragm. Exp Neurol 59:168–175
Davis R, Koelle GB (1967) Electron microscopic localization of acetylcholinesterase and non-specific cholinesterase at the neuro-muscular junction by gold-thioacetic acid methods. J Cell Biol 34:157–171
Davis R, Koelle GB (1978) Electron microscope localization of acetylcholinesterase and butyrylcholinesterase in the superior cervical ganglion of the cat. I. Normal ganglion. J Cell Biol 78:785–809
Davis R, Koelle GB, Sanville UJ (1984) Electron microscope localization of acetylcholinesterase and butyrylcholinesterase in the ciliary ganglion of the cat. J Histochem Cytochem 32:849–861
De la Porte S, Vigny M, Massoulié J, Koenig J (1984) Action of veratridine on acetylcholinesterase in cultures of rat muscle cells. Dev Biol 106:450–456
De la Porte S, Vallette FM, Grassi J, Vigny M, Koenig J (1986) Pre-synaptic or post-synaptic origin of acetylcholinesterase at neuromuscular junctions? An immunological study in heterologous nerve-muscle cultures. Dev Biol 116:69–77
Dettbarn WD, Groswald D, Gupta RC, Misulis KE (1985) Use and disuse and the control of acetylcholinesterase activity in fast and slow twitch muscle of rat. In: Changeux JP, Hucho F, Maelicke A, Neumann E (eds) Molecular basis of nerve activity. de Gruyter, Berlin, p 567

Di Giamberardino L, Couraud JY (1978) Rapid accumulation of high molecular weight acetylcholinesterase in transected sciatic nerve. Nature 271:170–172

Donoso JA, Fernandez HL (1985) Cellular localization of cytochemically stained acetylcholinesterase activity in adult rat skeletal muscle. J Neurocytol 14:795–808

Do Thi NA, Bon C, Koenig HL, Bourre JM (1986) Acetylcholine receptors and acetylcholinesterase activity in soleus muscle of Trembler dysmyelinating mutant: a cytochemical and biochemical analysis. Neurosci Lett 65:72–78

Drews U (1975) Cholinesterase in embryonic development. Prog Histochem Cytochem 7:3

Drews U, Schmidt H, Oettling G, Vanittanakom P (1986) Embryonic cholinesterase in the chick limb bud. Acta Histochem (Jena) 32S:133–137

Duprat AM, Kan P, Foulquier F, Weber M (1985) In vitro differentiation of neuronal precursor cells from amphibian late gastrulae: morphological, immunocytochemical studies, biosynthesis, accumulation and uptake of neurotransmitters. J Embryol Exp Morphol 86:71–87

Eckenstein F, Sofroniew MV (1983) Identification of central cholinergic neurons containing both choline acetyltransferase and acetylcholinesterase and of central neurons containing only acetylcholinesterase. J Neurosci 3:2286–2291

Engel A, Lambert EH, Gomez MR (1977) New myasthenic syndrome with endplate acetylcholinesterase deficiency, small nerve terminals, and reduced acetylcholine release. Ann Neurol 1:315–330

Erulkar SD (1983) The modulation of neurotransmitter release at synaptic junctions. Rev Physiol Biochem Pharmacol 98:63–175

Falugi C, Raineri M (1985) Acetylcholinesterase (AChE) and pseudocholinesterase (BuChE) activity distribution pattern in early developing chick limbs. J Embryol Exp Morphol 86:89–108

Fernandez HL, Duell MJ (1980) Protease inhibitors reduce effects of denervation on muscle endplate acetylcholinesterase. J Neurochem 35:1166–1171

Fernandez HL, Seiter TC (1984) Subcellular distribution of acetylcholinesterase asymmetric forms during postnatal development of mammalian skeletal muscle. FEBS Lett 170:147–151

Fernandez HL, Stiles JR (1984a) Denervation induced changes in subcellular pools of 16 S acetylcholinesterase activity from adult mammalian skeletal muscle. Neurosci Lett 44:187–192

Fernandez HL, Stiles JR (1984b) Intra- versus extracellular recovery of 16 S acetylcholinesterase following organophosphate inactivation in the rat. Neurosci Lett 49:117–122

Fernandez HL, Duell MJ, Festoff BW (1979) Cellular distribution of 16 S acetylcholinesterase. J Neurochem 32:581–585

Fernandez HL, Inestrosa NC, Stiles JR (1984) Subcellular localization of acetylcholinesterase molecular forms in endplate regions of adult mammalian skeletal muscle. Neurochem Res 9:1211–1230

Fernandez HL, Stiles JR, Donoso JA (1986) Skeletal muscle acetylcholinesterase in amyotrophic lateral sclerosis. Muscle Nerve 9:399–406

Ferrand C, Clarous D, Delteil C, Weber M (1986) Cellular localization of the molecular forms of acetylcholinesterase in primary cultures of rat sympathetic neurons and analysis of the secreted enzyme. J Neurochem 46:349–358

Festoff BW, Fernandez HL (1981) Plasma and red blood cell acetylcholinesterase in amyotrophic lateral sclerosis. Muscle Nerve 4:41–47

Filogamo G, Gabella G (1967) The development of neuromuscular correlations in vertebrates. Arch Biol (Liège) 78:9–60

Fishman EB, Siek GC, MacCallum RD, Bird ED, Volicer L, Marquis JK (1986) Distribution of the molecular forms of acetylcholinesterase in human brain: alterations in dementia of the Alzheimer type. Ann Neurol 19:246–252

Fiszman M, Toutant M, Montarras D (1983) Biochemical evidence for two types of myoblasts during avian embryonic development. In: Limb development and regeneration, part 3. Liss, New York, p 401

Futerman AH, Fiorini RM, Roth E, Michaelson DM, Low MG, Silman I (1984) Solubilization of membrane bound acetylcholinesterase by a phosphatidyl inositol-specific phospholipase C: enzymatic and physicochemical studies. In: Brzin M, Barnard EA, Sket D (eds) Cholinesterases: fundamental and applied aspects. de Gruyter, Berlin, p 99

Futerman AH, Low MG, Michaelson DM, Silman I (1985) Solubilization of membrane-bound acetylcholinesterase by a phosphatidyl inositol-specific phospholipase C. J Neurochem 45:1487–1494

Gautron J, Rieger F, Blondet B, Pinçon-Raymond M (1983) Extrasynaptic accumulations of acetylcholinesterase in the rat sternocleidomastoid muscle after neonatal denervation: light and electron microscopic localization and molecular forms. Biol Cell 49:55–68

George ST, Balasubramanian AS (1980) The identity of serotonin-sensitive arylacylamidase with acetylcholinesterase from human erythrocytes, sheep basal ganglia and electric eel. Eur J Biochem 111:511–524

George ST, Balasubramanian AS (1981) The arylacylamidases and their relationship to cholinesterases in human serum, erythrocyte and liver. Eur J Biochem 121:177–186

Gisiger V, Stephens HR (1983a) Correlation between the acetylcholinesterase content in motor nerves and their muscles. J Physiol (Paris) 78:720–728

Gisiger V, Stephens HR (1983b) Asymmetric and globular forms of AChE in slow and fast muscles of 129/ReJ normal and dystrophic mice. J Neurochem 41:919–929

Gisiger V, Stephens HR (1984) Decreased G4 (10 S) acetylcholinesterase content in motor nerves to fast muscles of dystrophic 129/ReJ mice: lack of a specific compartment of nerve acetylcholinesterase? J Neurochem 43:174–183

Gisiger V, Vigny M, Gautron J, Rieger F (1978) Acetylcholinesterase of rat sympathetic ganglion: molecular forms, localization and effects of denervation. J Neurochem 30:501–516

Gómez-Barriocanal J, Barat A, Escudero E, Rodríguez-Borrajo C, Ramírez G (1981) Solubilization of collagen-tailed acetylcholinesterase from chick retina: effect of different extraction procedures. J Neurochem 37:1239–1249

Goossens P, Viret J, Leterrier F (1984) Rat brain acetylcholinesterase turnover *in vivo*: use of a radioactive methylphosphonothiate irreversible inhibitor. Biochem Biophys Res Commun 123:71–77

Goudou D, Verdière-Sahuqué M, Rieger F (1985) External and internal AChE in rat sympathetic neurons *in vivo* and *in vitro*. FEBS Lett 186:54–58

Graybiel AM, Ragsdale CW Jr (1982) Pseudocholinesterase staining in the primary visual pathway of the macaque monkey. Nature 299:439–442

Greenfield SA (1984) Acetylcholinesterase may have novel functions in the brain. Trends Neurosci 7:364–368

Greenfield SA (1985) The significance of dendritic release of transmitter and protein in the substantia nigra. Neurochem Int 7:887–901

Greenfield SA, Shaw SG (1982) Amphetamine-evoked release of acetylcholinesterase and amino-peptidase *in vivo*. Neuroscience 7:2883–2893

Greenfield SA, Chéramy A, Leviel V, Glowinski J (1980) In vivo release of acetylcholinesterase in cat substantia nigra and caudate nucleus. Nature 284:355–357

Greenfield SA, Stein JF, Hodgson AJ, Chubb IW (1981) Depression of nigral pars compacta cell discharge by exogenous acetylcholinesterase. Neuroscience 6:2287–2295

Greenfield SA, Chéramy A, Glowinski J (1983) Evoked relesed of proteins from central neurons *in vivo*. J Neurochem 40:1048–1057

Groswald DE, Dettbarn WD (1983) Characterization of acetylcholinesterase molecular forms in slow and fast muscle of rat. Neurochem Res 8:983–995

Hall ZW (1973) Multiple forms of acetylcholinesterase and their distribution in endplate and non-endplate regions of rat diaphragm muscle. J Neurol 4:343–361

Hall ZW, Kelly RB (1971) Enzymatic detachment of endplate acetylcholinesterase from muscle. Nature [New Biol] 232:62–64

Harris AJ (1981) Embryonic growth and innervation of rat skeletal muscles. II. Neural regulation of muscle cholinesterase. Philos Trans R Soc Lond [Biol] 293:280–286

Haynes LW, Smith ME (1982) Selective inhibition of 'motor endplate specific' acetylcholinesterase by beta-endorphin and related peptides. Neuroscience 7:1007–1013

Haynes LW, Smith ME (1985) Induction of endplate-specific acetylcholinesterase by beta-endorphin C-terminal dipeptide in rat and chick muscle cells. Biochem Soc Trans 13:174–175

Haynes LW, Harborne AJ, Smith ME (1983) Augmentation of acetylcholine response in denervated skeletal muscle by endorphins and spinal cord-conditioned culture media. Eur J Pharmacol 86:415–425

Haynes LW, Smith ME, Li CH (1984a) Structural requirements and species specificity of the inhibition by beta-endorphin of heavy acetylcholinesterase from vertebrate skeletal muscle. Mol Pharmacol 26:45–50

Haynes LW, Smith ME, Smyth DG (1984b) Evidence for the neurotrophic regulation of collagen-tailed acetylcholinesterase in immature skeletal muscle by beta-endorphin. J Neurochem 42:1542–1551

Haynes LW, Smith ME, Li CH (1985) The regulation by beta-endorphin and related peptides of collagen-tailed acetylcholinesterase forms in the skeletal muscles of vertebrates. In: Comparative aspects of opioid and related neuropeptides mechanisms. CRC, Boca Raton (in press)

Henderson Z, Greenfield SA (1984) Ultrastructural localization of acetylcholinesterase in substantia nigra: a comparison between rat and guinea-pig. J Comp Neurol 230:278–286

Hess A (1961) Structural differences of fast and slow extrafusal fibers and their nerve endings in chickens. J Physiol (Lond) 157:221–231

Huff FJ, Maire J-C, Growdon JH, Corkin S, Wurtman RJ (1986) Cholinesterases in cerebrospinal fluid. Correlations with clinical measures in Alzheimer's disease. J Neurol Sci 72:121–129

Inestrosa NC, Ramírez BU, Fernandez HL (1977) Effect of denervation and of axoplasmic transport blockage on the *in vitro* release of muscle endplate acetylcholinesterase. J Neurochem 28:941–945

Inestrosa NC, Reiness CG, Reichardt LF, Hall ZW (1981) Cellular localization of the molecular forms of acetylcholinesterase in rat pheochromocytoma PC12 cells treated with nerve growth factor. J Neurosci 1:1260–1267

Inestrosa NC, Silberstein L, Hall ZW (1982) Association of the synaptic form of acetylcholinesterase with extracellular matrix in cultured mouse muscle cells. Cell 29:71–79

Inestrosa NC, Miller JB, Silberstein L, Ziskind-Conhaim L, Hall ZW (1983) Development and regulation of 16S acetylcholinesterase and acetylcholine receptors in a mouse muscle cell line. Exp Cell Res 147:393–406

Inestrosa NC, Matthew WD, Reiness CG, Hall ZW, Reichardt LF (1985) Atypical distribution of asymmetric acetylcholinesterase in mutant PC12 pheochromocytoma cells lacking a cell surface heparan sulfate-proteoglycan. J Neurochem 45:86–94

Inestrosa NC, Roberts WL, Marshall T, Rosenberry TL (1987) Acetylcholinesterase from bovine caudate nucleus is attached to membranes by a novel subunit distinct from those of acetylcholinesterase in other tissues. J Biol Chem 262:4441–4444

Ishida I, Deguchi T (1983) Effect of depolarizing agents on choline acetyltransferase and acetylcholinesterase activities in primary cell cultures of spinal cord. J Neurosci 9:1818–1823

Ismaël Z, Millar TJ, Small DH, Chubb IW (1986) Acetylcholinesterase generates enkephalin-like immunoreactivity when it degrades the soluble proteins (chromogranins) from adrenal chromaffin granules. Brain Res 376:230–238

Jedrzejczyk J, Silman I, Lyles JM, Barnard EA (1981) Molecular forms of the cholinesterases inside and outside muscle endplates. Biosci Rep 1:45–51

Jedrzejczyk J, Silman I, Lai J, Barnard EA (1984) Molecular forms of acetylcholinesterase in synaptic and extrasynaptic region of avian tonic muscle. Neurosci Lett 46:283–289

Kaemmer D, Neubert K, Demuth HU, Barth A (1986) Contamination of highly purified human serum cholinesterase by dipeptidylpeptidase IV causing hydrolysis of substance P. Pharmazie 41:494–497

Karprzak H, Salpeter M (1985) Recovery of acetylcholinesterase at intact neuromuscular junctions after *in vivo* inactivation with diisopropylfluorophosphate. J Neurosci 5:951–955

Kása P, Rakonczay Z (1982a) Biochemical and histochemical evidence of 16 S acetylcholinesterase in salivary glands. J Neurochem 38:278–280
Kása P, Rakonczay Z (1982b) Histochemical and biochemical demonstration of the molecular forms of acetylcholinesterase in peripheral nerve of rat. Acta Histochem (Jena) 70:244–257
Kato AC, Vrachliotis A, Fulpius B, Dunant Y (1980) Molecular forms of acetylcholinesterase in chick muscle and ciliary ganglion: embryonic tissues and cultured cells. Dev Biol 76:222–228
Kilbinger H (1984) Presynaptic muscarinic receptors modulating acetylcholine release. In: Receptors again. Elsevier, Amsterdam, p 174
Kimura H, McGeer PL, Peng JH, McGeer EG (1981) The central cholinergic system studied by choline acetyltransferase immunohistochemistry in the cat. J Comp Neurol 200:151–201
Klinar B, Sketelj J, Sket D, Brzin M (1983) Presynaptic modulation of activity and molecular forms of acetylcholinesterase in the rat superior cervical ganglion during early postnatal development. J Neurosci Res 9:437–444
Kobayashi T, Askanas V (1985) Acetylcholine receptors and acetylcholinesterase accumulate at the nerve-muscle contacts of *de novo* grown human monolayer muscle cocultured with fetal rat spinal cord. Exp Neurol 88:327–335
Koelle GB (1957) Histochemical demonstration of reversible anticholinesterase action at selective cellular sites *in vivo*. J Pharmacol Exp Ther 120:488–503
Koelle GB (1961) A proposed dual neurohumoral role of acetylcholine: its functions at the pre- and post-synaptic sites. Nature 190:208–211
Koelle GB, Ruch GA (1983) Demonstration of a neurotrophic factor for the maintenance of acetylcholinesterase and butyrylcholinesterase in the preganglionically denervated superior cervical ganglion of the cat. Proc Natl Acad Sci USA 80:3106–3110
Koelle GB, Ruch GA, Uchida E (1983) Effects of sodium pentobarbital anesthesia and neurotrophic factor on the maintenance of acetylcholinesterase and butyrylcholinesterase in the preganglionically denervated superior cervical ganglion of the cat. Proc Natl Acad Sci USA 80:6122–6125
Koelle GB, Sanville UJ, Richard KK, Williams JE (1984) Partial characterization of the neurotrophic factor for maintenance of acetylcholinesterase and butyrylcholinesterase in the preganglionically denervated superior cervical ganglion of the cat *in vivo*. Proc Natl Acad Sci USA 81:6539–6542
Koelle GB, Sanville UJ, Wall SJ (1985) Glycyl-1-glutamine: a precursor, and glycyl-1-glutamic acid: a neurotrophic factor for maintenance of acetylcholinesterase and butyrylcholinesterase in the preganglionically denervated superior cervical ganglion of the cat *in vivo*. Proc Natl Acad Sci USA 82:5213–5217
Koelle WA, Koelle GB (1959) The localization of external or functional acetylcholinesterase at the synapses of autonomic ganglia. J Pharmacol Exp Ther 126:1–8
Koenig J (1979) Formation and maturation of nerve-muscle contacts in cultured rat embryo cells. Biol Cell 35:147–152
Koenig J, Vigny M (1978a) Neural induction of the 16 S acetylcholinesterase in muscle cell cultures. Nature 271:75–77
Koenig J, Vigny M (1978b) Formes moléculaires d'acétylcholinestérase dans le muscle lent et le muscle rapide du poulet. C R Soc Biol (Paris) 172:1069–1074
Koenig J, Bournaud R, Powell JA, Rieger F (1982) Appearance of contractile activity in muscular dysgenesis (mdg/mdg) mouse myotubes during coculture with normal spinal cord cells. Dev Biol 92:188–196
Kupfer C, Koelle GB (1951) The embryological development of cholinesterase activity at the motor endplate of skeletal muscle in the rat. J Exp Zool 116:399–414
Kutty KM (1980) Biological function of cholinesterase. Clin Biochem 13:239–243
Kutty KM, Redheeran R, Murphy D (1977) Serum cholinesterase: function in lipoprotein metabolism. Experientia 43:420–422
Laasberg T, Neuman T (1985) Changes in the acetylcholinesterase and choline acetyltransferase activities in the early development of chick embryo. Rouxs Arch Dev Biol 194:306–310

Lai J, Jedrzecjczyk J, Pizzey JA, Green D, Barnard EA (1986) Neural control of the forms of acetylcholinesterase in slow mammalian muscles. Nature 321:72–74
Landmesser L, Morris DG (1975) The development of functional innervation in the hind limb of the chick embryo. J Physiol (Lond) 249:301–326
Lappin RI, Rubin LL (1985) Molecular forms of acetylcholinesterase in *Xenopus* muscle. Dev Biol 110:269–274
Layer PG (1983) Comparative localization of acetylcholinesterase and pseudocholinesterase during morphogenesis of the chicken brain. Proc Natl Acad Sci USA 80:6413–6417
Layer PG, Girgert R, Rommel S, Sporns O (1985) Development of embryonic cholinesterases and cell proliferation in chick brain and retina. J Neurochem 44:S129
Lazar M, Vigny M (1980) Modulation of the distribution of acetylcholinesterase molecular forms in a murine neuroblastoma X sympathetic ganglion cell hybrid cell line. J Neurochem 35:1067–1079
Lazar M, Salmeron E, Vigny M, Massoulié J (1984) Heavy isotope labeling study of the metabolism of monomeric and tetrameric acetylcholinesterase forms in the murine neuronal-like T28 hybrid cell line. J Biol Chem 259:3703–3713
Lefaix JC, Koenig HL, Vigny M, Bourre JM (1982) Innervation motrice de la souris Trembler. Reprod Nutr Dev 22:275–282
Li ZY, Bon C (1983) Presence of a membrane-bound acetylcholinesterase form in a preparation of nerve endings from *Torpedo marmorata* electric organ. J Neurochem 40:338–349
Lindenbaum MH, Livett BG (1983) Acetylcholinesterase molecular forms in C57BL/6J dystrophic mice. Muscle Nerve 6:638–645
Livneh A, Sarova I, Michaeli D, Pras M, Wagner K, Zakut H, Soreq H (1988) Antibodies against acetylcholinesterase and low levels of cholinesterases in a patient with an atypical neuromuscular disorder. Clin Immunol Immunopathol (in press)
Lockridge O (1982) Substance P hydrolysis by human serum cholinesterase. J Neurochem 39:106–110
Lømo T, Massoulié J, Vigny M (1985) Stimulation of denervated rat soleus muscle with fast and slow activity patterns induces different expression of acetylcholinesterase molecular forms. J Neurosci 5:1180–1187
Lubinska L, Zelena J (1966) Formation of new sites of acetylcholinesterase activity in denervated muscles of young rats. Nature 210:39–41
Lucas CA, Kreutzberg GW (1985) Regulation of acetylcholinesterase secretion from neuronal cell cultures. I. Actions of nerve growth factor, cytoskeletal inhibitors and tunicamycin. Neuroscience 14:349–360
Lyles JM, Barnard EA (1980) Disappearance of the ‘endplate’ form of acetylcholinesterase from a slow tonic muscle. FEBS Lett 109:9–12
Lyles JM, Silman I, Barnard EA (1979) Developmental changes in levels and forms of cholinesterases in muscles of normal and dystrophic chickens. J Neurochem 33:727–738
Lyles JM, Silman I, di Giamberardino L, Couraud JY, Barnard EA (1982) Comparison of the molecular forms of the cholinesterases in tissues of normal and dystrophic chickens. J Neurochem 38:1007–1021
Marquis JK, Fishman EB (1985) Presynaptic acetylcholinesterase. Trends Pharmacol Sci 6:387–388
Marquis JK, Volicer L, Mark KA, Direnfeld LK, Freedman M (1985) Cholinesterase activity in plasma, erythrocytes and cerebrospinal fluid of patients with dementia of the Alzheimer type. Biol Psychiatry 20:605–610
Massoulié J, Bon S (1982) The molecular forms of cholinesterase in vertebrates. Annu Rev Neurosci 5:57–106
Massoulié J, Bon S, Lazar M, Grassi J, Marsh D, Méflah K, Toutant JP, Vallette F, Vigny M (1984) The polymorphism of cholinesterases: classification of molecular forms; interactions and solubilization characteristics; metabolic relationships and regulations. In: Brzin M, Barnard EA, Sket D (eds) Cholinesterases: fundamental and applied aspects. de Gruyter, Berlin, p 73

Massoulié J, Vigny M, Lazar M (1985) Expression of acetylcholinesterase in murine neural cells *in vivo* and in culture. In: Changeux JP, Hucho F, Maelicke A, Neumann E (eds) Molecular basis of nerve activity. de Gruyter, Berlin, p 619

McIsaac RS, Koelle GB (1959) Comparison of the effects of inhibition of external, internal and total acetylcholinesterase upon ganglionic transmission. J Pharmacol Exp Ther 126:9–20

McMahan UJ (1985) ESOMs: extracellular synaptic organizing molecules (abstr). International Workshop on Mechanisms of Secretion and Action of Neurotransmitters and Neuromodulators in Central and Peripheral Synapses, Jerusalem

McMahan UJ, Sanes JR, Marshall LM (1978) Cholinesterase is associated with the basal lamina at the neuromuscular junction. Nature 271:172–174

Mesulam MM, Mufson EJ, Levey AI, Wainer BH (1984) Atlas of cholinergic neurons in the forebrain and upper brainstem of the macaque based on monoclonal choline acetyltransferase immunochemistry and acetylcholinesterase histochemistry. Neuroscience 12:669–686

Metz J, Bradlow BA, Lewis SM, Dacie JV (1960) The acetylcholinesterase activity of the erythrocytes in paroxysmal nocturnal hemoglobinuria: relation to the severity of the disease. Br J Haematol 6:372

Michaelson DM, Avissar S, Kloog Y, Sokolovski M (1979) Mechanism of acetylcholine release: possible involvement of presynaptic muscarinic receptors in regulation of acetylcholine release and protein phosphorylation. Proc Natl Acad Sci USA 76:6336–6340

Miki A, Mizoguti H (1982) Acetylcholinesterase activity in the myotome of the early chick embryo. Cell Tissue Res 227:23–40

Miki A, Fujimoto E, Mizoguti H (1983) Acetylcholinesterase activity in neural crest cells of the early chick embryo. Histochemistry 78:81–93

Miledi R, Molenaar PC, Polak RL (1984) Acetylcholinesterase activity in intact and homogenized skeletal muscle of the frog. J Physiol (Lond) 349:663–686

Millar TJ, Chubb IW (1984) Sections of chick retinae treated with acetylcholinesterase show enhanced enkephalin and substance P immunoreactivity. Neuroscience 12:441–451

Mizobe F, Livett BG (1980) Production and release of acetylcholinesterase by a primary cell culture of bovine adrenal medullary chromaffin cells. J Neurochem 35:1469–1472

Mizobe F, Livett BG (1983) Nicotine stimulates secretion of both catecholamines and acetylcholinesterase from cultured adrenal chromaffin cells. J Neurosci 3:871–876

Mizukawa K, McGeer PL, Tago H, Peng JH, McGeer EG, Kimura H (1986) The cholinergic system of the human hindbrain studied by choline acetyltransferase immunochemistry and acetylcholinesterase histochemistry. Brain Res 379:39–55

Moody-Corbett F, Cohen MW (1981) Localization of cholinesterase at sites of high acetylcholine receptor density on embryonic amphibian muscle cells cultures without nerve. J Neurosci 1:596–605

Moore DE, Hess GP (1975) Acetylcholinesterase-catalyzed hydrolysis of an amide. Biochemistry 14:2386–2389

Morel N, Dreyfus P (1982) Association of acetylcholinesterase with the external surface of the presynaptic plasma membrane in *Torpedo* electric organ. Neurochem Int 4:283–288

Muller F, Dumez Y, Massoulié J (1985) Molecular forms and solubility of acetylcholinesterase during the embryonic development of rat and human brain. Brain Res 331:295–302

Muller F, Cédard L, Boué J, Giraudet P, Massoulié J, Boué A (1986) Diagnostic prénatal des défauts de fermeture du tube neural. Intérêt de l'électrophorèse des cholinestérases. Presse Med 15:783–786

Mumenthaler M, Engel WK (1961) Cytological localization of cholinesterase in developing chick embryo skeletal muscle. Acta Anat (Basel) 47:274–299

Nausch I, Heymann E (1985) Substance P in human plasma is degraded by dipeptylpeptidase IV, not by cholinesterase. J Neurochem 44:1354–1357

Newman JR, Virgin JB, Younkin LH, Younkin SG (1984) Turnover of acetylcholinesterase in innervated and denervated rat diaphragm. J Physiol (Lond) 352:305–318

Nicholson-Weller A, March JP, Rosenfeld SI, Austen KF (1983) Affected erythrocytes of patients with paroxysmal nocturnal hemoglobinuria are deficient in the complement regulatory protein, decay acceleration factor. Proc Natl Acad Sci USA 80:5066–5070

Nicolet M, Rieger F (1982) Ubiquitous presence of the tailed, asymmetric forms of acetylcholinesterase in the peripheral and central nervous systems of the frog (*Rana temporaria*). Neurosci Lett 28:67–73

Nicolet M, Pinçon-Raymond M, Rieger F (1986) Globular and asymmetric acetylcholinesterase in frog muscle basal lamina sheaths. J Cell Biol 102:762–768

Oh TH, Markelonis GJ (1982) Chicken serum transferrin duplicates the myotrophic effects of sciatin on cultured cells. J Neurosci Res 8:535–545

Pangburn MK, Schreiber RD, Müller-Eberhard HJ (1983) Deficiency of an erythrocyte membrane protein with complement regulatory activity in paroxysmal nocturnal hemoglobinuria. Proc Natl Acad Sci USA 80:5430–5434

Patterson GT, Wilson BW (1986) Serum regulation of acetylcholinesterase in cultured myotubes. Exp Neurol 91:308–318

Pezzementi L, Reinheimer EJ, Pezzementi ML (1987) Acetylcholinesterase from the skeletal muscle of the lamprey *Petromyzon marinus* exists in globular and asymmetric forms. J Neurochem 48:1753–1760

Pilowsky PM, Hodgson AJ, Chubb IW (1982) Acetylcholinesterase in neural tube defects: a model using embryo amniotic fluid. Neuroscience 7:1203–1214

Poiana G, Scarsella G, Biagioni S, Senni MI, Cossu G (1985) Membrane acetylcholinesterase in murine muscular dystrophy *in vivo* and in cultured myotubes. Int J Dev Neurosci 3:331–340

Poiana G, Leone F, Longstaff A, Scarsella G, Biagioni S (1986) Muscle acetylcholinesterase in childhood myopathies. Neurochem Int 9:239–245

Popiela H, Beach RL, Festoff BW (1983) Appearance of acetylcholinesterase molecular forms in noninnervated cultured primary chick muscle cells. Cell Mol Neurobiol 3:263–277

Popiela H, Beach RL, Festoff BW (1984) Developmental appearance of acetylcholinesterase molecular forms in cultured primary chick muscle cells. In: Serratrice G (ed) Neuromuscular diseases. Raven, New York, p 447

Powell JA, Friedman B, Cossi A (1979) Tissue culture study of murine muscular dysgenesis: role of spontaneous action potential generation in the regulation of muscle maturation. Ann NY Acad Sci 317:550–570

Powell JA, Peterson AC, Paul C (1984) Neurons induce contractions in myotubes containing only muscular dysgenic nuclei. Muscle Nerve 7:904–910

Rakonczay Z (1986) Mammalian brain acetylcholinesterase. In: Boulton AA, Baker GB, Yu PH (eds) Neurotransmitter enzymes. Humana, Clifton, p 319 (Neuromethods, vol 5)

Rakonczay Z, Nemeth P (1984) Change in the distribution of acetylcholinesterase molecular forms in Hirschsprung's disease. J Neurochem 43:1194–1196

Rama-Sastry BV, Sadavongvivad C (1979) Cholinergic system in non-nervous tissue. Pharmacol Rev 30:65–132

Randall WR, Lai J, Barnard EA (1985) Acetylcholinesterase of muscle and nerve. In: Changeux JP, Hucho F, Maelicke A, Neumann E (eds) Molecular basis of nerve activity. de Gruyter, Berlin, p 595

Rasool CG, Chad D, Bradley WG, Conolly B, Barnah JK (1983) Acetylcholinesterases and ATPases in motor neuron degenerative diseases. Muscle Nerve 6:430–435

Rathbone MP, Vickers JD, Ganagarajah M, Brown JA, Logan MD (1979) Neural regulation of cholinesterases in newt skeletal muscle. I. Distribution and characterization of the enzyme species. J Exp Zool 210:435–450

Raynaud B, Clarous D, Vidal S, Ferrand C, Weber MJ (1987) Comparison of the effects of elevated K^+ ions and muscle-conditioned medium on the neurotransmitter phenotype of cultured sympathetic neurons. Dev Biol (in press)

Razon N, Soreq H, Roth E, Bartal A, Silman I (1984) Characterization of activities and forms of cholinesterases in human primary brain tumors. Exp Neurol 84:681–695

Reich A, Drews U (1983) Choline acetyltransferase in the chick limb bud. Histochemistry 78:383–389

Richardson GP, Witzemann V (1986) *Torpedo* electromotor system development: biochemical differentiation of *Torpedo* electrocytes *in vitro*. Neuroscience 17:1287–1296

Richardson GP, Krenz WD, Kirk C, Fox GQ (1981) Organotypic culture of embryonic electromotor system tissues from *Torpedo marmorata*. Neuroscience 6:1181–1200

Rieger F, Faivre-Bauman A, Benda P, Vigny M (1976) Molecular forms of acetylcholinesterase: their *de novo* synthesis in mouse neuroblastoma cells. J Neurochem 27:1059–1063

Rieger F, Koenig J, Vigny M (1980a) Spontaneous contractile activity and the presence of the 16 S form of acetylcholinesterase in rat muscle cells in culture. Reversible suppressive action of tetrodotoxin. Dev Biol 76:358–365

Rieger F, Shelanski ML, Greene LA (1980b) The effects of nerve growth factor on acetylcholinesterase and its molecular forms in cultures of rat PC12 pheochromocytoma cells: increased total specific activity and appearance of the 16 S molecular form. Dev Biol 76:238–248

Rieger F, Shelanski ML, Sidman RL (1983) The multiple molecular forms of acetylcholinesterase in 'motor endplate disease' in the mouse (med^J and med allelic forms): sensitivity of the 10 S form to partial or total loss of muscle activity. Exp Neurol 79:299–315

Rieger F, Powell JA, Pinçon-Raymond M (1984a) Extensive nerve overgrowth and paucity of the tailed asymmetric form (16 S) of acetylcholinesterase in the developing skeletal neuromuscular system of the dysgenic (mdg/mdg) mouse. Dev Biol 101:181–191

Rieger F, Pinçon-Raymond M, Lombet A, Ponzio G, Lazdunski M, Sidman RL (1984b) Paranodal dysmelination and increase in tetrodotoxin binding sites in the sciatic nerve of the motor end-plate disease (med/med) mouse during postnatal development. Dev Biol 101:401–409

Rieger F, Cross D, Peterson A, Pinçon-Raymond M, Tretjakoff I (1984c) Disease expression in +/+ mdg/mdg mouse chimeras: evidence for an extramuscular component in the pathogenesis of both dysgenic abnormal diaphragm innervation and skeletal muscle 16 S acetylcholinesterase deficiency. Dev Biol 106:296–306

Rieger F, Goudou D, Tran LH (1984d) Increase of junctional and background 16 S (tailed, asymmetric) acetylcholinesterase during postnatal maturation of rat and mouse sternocleidomastoid muscle. J Neurochem 42:601–606

Ritchie AK, Fambrough DM (1975) Electrophysiological properties of the membrane and acetylcholine receptor in developing rat and chick myotubes. J Gen Physiol 66:4790–4799

Robertson RT (1987) A morphogenic role for transiently expressed acetylcholinesterase in thalamocortical development? Neurosci Lett 75:259–264

Robertson RT, Tijerina AA, Gallivan ME (1985) Transient patterns of acetylcholinesterase activity in visual cortex of the rat: normal development and the effects of neonatal monocular enucleation. Dev Brain Res 21:203–214

Rodriguez-Borrajo C, Barat A, Ramírez C (1982) Solubilization of collagen-tailed molecular forms of acetylcholinesterase from several brain areas in different vertebrate species. Neurochem Int 4:563–568

Rosenberry TL (1979) Quantitative simulation of endplate currents at neuromuscular junctions based on the reaction of acetylcholine receptor and acetylcholinesterase. Biophys J 26:263–290

Rotundo RL (1984a) Asymmetric acetylcholinesterase is assembled in the Golgi apparatus. Proc Natl Acad Sci USA 81:479–483

Rotundo RL (1984b) Synthesis, assembly, and processing of AChE in tissue cultured muscle. In: Brzin M, Barnard EA, Sket D (eds) Cholinesterases: fundamental and applied aspects. de Gruyter, Berlin, p 203

Rotundo RL, Fambrough DM (1979) Molecular forms of chicken embryo acetylcholinesterase *in vitro* and *in vivo*, isolation and characterization. J Biol Chem 254:4790–4799

Rotundo RL, Fambrough DM (1980a) Synthesis, transport and fate of acetylcholinesterase in cultured chick embryo muscle cells. Cell 22:583–594

Rotundo RL, Fambrough DM (1980b) Secretion of acetylcholinesterase: relation to acetylcholine receptor metabolism. Cell 22:595–602

Rotundo RL, Fambrough DM (1982) Synthesis, transport and fate of acetylcholinesterase and acetylcholine receptors in cultured muscle. In: Membranes in growth and development. Liss, New York, p 259

Rubin LL (1985) Increases in muscle Ca^{2+} mediate changes in acetylcholinesterase and acetylcholine receptors caused by muscle contraction. Proc Natl Acad Sci USA 82:7121–7125

Rubin LL, Schuetze SM, Fischbach GD (1979) Accumulation of acetylcholinesterase at newly formed nerve-muscle synapses. Dev Biol 69:46–58

Rubin LL, Schuetze SM, Weill CL, Fischbach GD (1980) Regulation of acetylcholinesterase appearance at neuromuscular junctions *in vitro*. Nature 283:264–267

Rubin LL, Chalfin NA, Adamo A, Klymkowsky M (1985) Cellular and secreted forms of acetylcholinesterase in mouse muscle cultures. J Neurochem 45:1932–1940

Russell RW, Booth RA, Jenden DJ, Roch M, Rice KM (1985) Changes in presynaptic release of acetylcholine during development of tolerance to the anticholinesterase, DFP. J Neurochem 45:293–299

Sakamoto H, Ohsato K, Atsumi S (1985) Acetylcholinesterase activity at the presynaptic boutons with presumed α-motoneurons in chicken ventral horn. Light- and electron-microscopic studies. Histochemistry 83:291–298

Salpeter M (1967) Electron microscope radioautography as a quantitative tool in enzyme cytochemistry. I. The distribution of acetylcholinesterase at motor endplates of a vertebrate twitch muscle. J Cell Biol 32:379–389

Sampson SR, Babila T, Disatnik MH, Shainberg A, Yales E (1983) Evidence for a functional role of acetylcholinesterase in cultured chick myotubes. Brain Res 289:338–341

Sanes JR, Lawrence JC Jr (1983) Activity-dependent accumulation of basal lamina by cultured rat myotubes. Dev Biol 97:123–136

Sanes JR, Feldman DH, Cheney JM, Lawrence JC Jr (1984) Brain extract induces synaptic characteristics in the basal lamina of cultured myotubes. J Neurosci 4:464–473

Satoh K, Armstrong DM, Fibiger HC (1983) A comparison of the distribution of central cholinergic neurons as demonstrated by acetylcholinesterase pharmacohistochemistry and choline acetyltransferase immunohistochemistry. Brain Res Bull 11:693–720

Schmidt H (1981) Muscarinic acetylcholine receptor in chick limb bud during morphogenesis. Histochemistry 71:89–98

Schmidt H, Oettling G, Kaufenstein T, Hartung G, Drews U (1984) Intracellular calcium mobilization on stimulation of the muscarinic cholinergic receptor in chick limb bud cells. Rouxs Arch Dev Biol 194:44–49

Schröder C (1980) Characterization of embryonic cholinesterase in chick limb bud by colorimetry and disk electrophoresis. Histochemistry 69:243–253

Seller M, Cole KJ, Fenson AH, Polani PE (1980) Amniotic fluid acetylcholinesterase and prenatal diagnosis. Br J Obstet Gynaecol 87:501–505

Silberstein L, Inestrosa NC, Hall ZW (1982) Aneural muscle cell cultures make synaptic basal lamina components. Nature 295:143–145

Silver A (1963) A histochemical investigation of cholinesterases at neuromuscular junctions in mammalian and avian muscle. J Physiol (Lond) 169:386–393

Silver A (1971) The significance of cholinesterase in the developing nervous system. Prog Brain Res 34:345–355

Silver A (1974) The biology of cholinesterases. North-Holland, Amsterdam

Sket D, Pavlin R (1985) Cholinesterases in single nerve cells isolated from locus ceruleus and from nucleus of the facial nerve of the rat: a microgasometric study. J Neurochem 45:319–323

Sketelj J, Brzin M (1979) Attachment of acetylcholinesterase to structures of the motor endplate. Histochemistry 61:239–248

Sketelj J, Brzin M (1980) 16 S acetylcholinesterase in endplate-free regions of developing rat diaphragm. Neurochem Res 5:653–658

Sketelj J, Brzin M (1985) Asymmetric molecular forms of acetylcholinesterase in mammalian skeletal muscles. J Neurochem Res 14:95–103

Sketelj J, Blinc A, Brzin M (1985) Acetylcholinesterase in regenerating skeletal muscles. In: Changeux JP, Hucho F, Maelicke A, Neumann E (eds) Molecular basis of nerve activity. de Gruyter, Berlin, p 709

Small DH, Ismaël Z, Chubb IW (1986) Acetylcholinesterase hydrolyses chromogranin A to yield low molecular weight peptides. Neuroscience 19:289–295

Smith AD, Wald NJ, Cuckle HS, Stirrat GM, Bobrow M, Lagercrantz H (1979) Amniotic fluid acetylcholinesterase as a possible test for neural tube defects in early pregnancy. Lancet 1:685–688

Sohal GS, Wrenn RW (1984) Appearance of high-molecular weight acetylcholinesterase in aneural muscle developing *in vivo*. Dev Biol 101:229–234

St Clair DM, Brock DJH, Barron L (1986) A monoclonal antibody assay technique for plasma and red cell acetylcholinesterase activity in Alzheimer's disease. J Neurol Sci 73:169–176

Sugarman J, Devine DV, Rosse WF (1986) Structural and functional difference between decay-acceleration factor and red cell acetylcholinesterase. Blood 68:680–684

Swerts JP, Weber MJ (1984) Regulation of enzyme responsible for neurotransmitter synthesis and degradation in cultured rat sympathetic neurons. III. Effects of sodium butyrate. Dev Biol 106:282–288

Swerts JP, Le Van Thai A, Vigny A, Weber MJ (1983) Regulation of enzymes responsible for the neurotransmitter synthesis and degradation in cultured rat sympathetic neurons. I. Effects of muscle conditioned medium. Dev Biol 100:1–11

Swerts JP, Le Van Thai A, Weber MJ (1984) Regulation of enzymes responsible for neurotransmitter synthesis and degradation in cultured rat sympathetic neurons. II. Regulation of 16 S acetylcholinesterase by conditioned medium. Dev Biol 103:230–234

Taylor PB, Rieger F, Shelanski ML, Greene LA (1981) Cellular localization of the multiple molecular forms of acetylcholinesterase in cultured neuronal cells. J Biol Chem 256:3827–3830

Tennyson VM, Kremzner LT, Brzin M (1977) Electron microscopic cytochemical and histochemical studies of acetylcholinesterase in denervated muscle of rabbits. J Neuropathol Exp Neurol 36:245–275

Toutant JP, Massoulié J (1987) Acetylcholinesterase. In: Turner AJ, Kelly AJ (eds) Mammalian ectoenzymes. Elsevier/North-Holland, Amsterdam (in press)

Toutant JP, Toutant M, Fiszman M, Massoulié J (1983) Expression of the A12 form of acetylcholinesterase by developing avian leg muscle cells *in vivo* and during differentiation in primary cell cultures. Neurochem Int 5:751–762

Toutant JP, Massoulié J, Bon S (1985) Polymorphism of pseudocholinesterase in *Torpedo marmorata* tissues: comparative study of the catalytic and molecular properties of this enzyme with acetylcholinesterase. J Neurochem 44:580–592

Toutant M, Montarras D, Fiszman M (1984) Biochemical evidence for two classes of myoblasts during chick embryonic muscle development. Exp Biol Med 9:10–15

Uchida E, Koelle GB (1983) Identification of the probable site of synthesis of butyrylcholinesterase in the superior cervical and ciliary ganglia of the cat. Proc Natl Acad Sci USA 80:6723–6727

Vallette FM, Vigny M, Massoulié J (1986) Muscular differentiation of chicken myotubes in a simple defined synthetic culture medium and in serum supplemented media: expression of the molecular forms of acetylcholinesterase. Neurochem Int 8:121–133

Vallette FM, Fauquet M, Teillet MA (1987) Difference in the expression of asymmetric acetylcholinesterase molecular forms during myogenesis in early avian dermomyotomes and limb buds *in ovo* and *in vitro*. Dev Biol 120:77–84

Verdière M, Dérer M, Rieger F (1982) Multiple molecular forms of rat superior cervical ganglion acetylcholinesterase: developmental aspects in primary cell culture and during postnatal maturation *in ovo*. Dev Biol 89:509–515

Verdière M, Dérer M, Poullet M (1984) Decrease of tailed, asymmetric 16 S acetylcholinesterase in rat superior cervical ganglion neurons *in vitro* after potassium depolarization: partial antagonist action of a calcium-channel blocker. Neurosci Lett 52:135–140

Vigny M, Vallette F (1985) Distribution and regulation of acetylcholinesterase forms in chick myotubes: comparison with rat myotubes. In: Changeux JP, Hucho F, Maelicke A, Neumann E (eds) Molecular basis of nerve activity. de Gruyter, Berlin, p 719

Vigny M, di Giamberardino L, Couraud JY, Rieger F, Koenig J (1976) Molecular forms of chicken acetylcholinesterase: effect of denervation. FEBS Lett 69:277–280

Vigny M, Gisiger V, Massoulié J (1978) 'Non specific' cholinesterase and acetylcholinesterase in rat tissues: molecular forms, structural and catalytic properties and significance of the two enzyme systems. Proc Natl Acad Sci USA 75:2588–2592

Volle RL (1980) Ganglionic actions of anticholinesterase agents, catecholamines, neuromuscular blocking agents and local anaesthetics. In: Kharkevich DA (ed) Pharmacology of ganglionic transmission. Springer, Berlin Heidelberg New York, p 385

Wallace BG (1986) Aggregating factor from *Torpedo* electric organ induces patches containing acetylcholine receptors, acetylcholinesterase and butyrylcholinesterase on cultured myotubes. J Cell Biol 102:783–794

Wallace BG, Nitkin RM, Reist NE, Fallon JR, Moayeri NN, McMahan UJ (1985) Aggregates of acetylcholinesterase induced by acetylcholine receptor-aggregating factor. Nature 315:574–577

Weber M (1981) A diffusible factor responsible for the determination of cholinergic functions in cultured sympathetic neurons. Partial purification and characterization. J Biol Chem 256:3447–3453

Weinberg CG, Hall ZW (1979) Junctional form of acetylcholinesterase restored at nerve free endplate. Dev Biol 68:631–635

Weldon PR, Moody-Corbett F, Cohen MW (1981) Ultrastructure of sites of cholinesterase activity on amphibian embryonic muscle cells cultured without nerve. Dev Biol 84:341–350

Wenthold RJ, Mahler HR, Moore WJ (1974) The half-life of acetylcholinesterase in mature rat brain. J Neurochem 22:941–943

White NK, Bonner PH, Nelson DR, Hauschka SD (1975) Clonal analysis of vertebrate myogenesis. IV. Medium-dependent classification of colony forming cells. Dev Biol 44:346–361

Whittaker VP (1963) Identification of acetylcholine and related esters of biological origin. In: Koelle GB (ed) Cholinesterase and anticholinesterase agents. Springer, Berlin Göttingen Heidelberg, pp 1–39 (Handbuch der experimentellen Pharmakologie, vol 15 [Suppl])

Witzemann V, Boustead C (1982) Changes in acetylcholinesterase molecular forms during the embryonic development of *Torpedo marmorata.* J Neurochem 39:747–755

Younkin SG, Rosenstein C, Collins PL, Rosenberry TL (1982) Cellular localization of the molecular forms of AChE in rat diaphragm. J Biol Chem 257:13630–13637

Younkin SG, Goodridge B, Katz J, Lockett G, Nafziger D, Usiak MF, Younkin LH (1986) Molecular forms of acetylcholinesterases in Alzheimer's disease. Fed Proc 45:2982–2988

Zakut H, Matzkel A, Schejter E, Avni A, Soreq H (1985) Polymorphism of acetylcholinesterase in discrete regions of the developing human fetal brain. J Neurochem 45:382–389

Ziskind-Conhaim L, Inestrosa NC, Hall ZW (1984) Acetylcholinesterase is functional in embryonic rat muscle before its accumulation at the sites of nerve-muscle contact. Dev Biol 103:369–377

CHAPTER 9

Structure and Function of the Nicotinic Acetylcholine Receptor

A. MAELICKE

A. Introduction and Scope

The nicotinic acetylcholine receptor (nAChR) is the prototype of an integral signal transducer: it contains in its protein moiety the binding sites for acetylcholine (ACh) and its agonists and antagonists (receptor function), the ligand-gated cation channel (response function) and several types of modulator sites (modulation function); nAChRs are components of all vertebrate neuromuscular junctions (NMJs) and of many nerve-nerve junctions. The nAChR was the first neuroreceptor to be isolated, to be biochemically and biophysically characterized, to be reconstituted into artificial bilayers and to be cloned and sequenced. Its electrical properties can be studied on the single-channel level. Ligand binding, agonist-induced ion fluxes and the modulation of these by local anaesthetics can be studied by time-resolved methods and has led to the formulation of models for the interaction of ligand binding sites and for channel activation and closing. Taken together, this remarkable progress may well make the nAChR the first signal transducer of the nervous system to be thoroughly understood on the molecular level.

In this chapter an attempt is made to provide a concise overview of recent developments in the molecular analysis of the structure and function of the nAChR. In the past 4 years more than 500 refereed papers on the nAChR have been published. Thus, to remain within the allotted space citations have had to be selective (and thus subjective). Apologies are tendered to those whose work has been inadequately referenced. For accounts of earlier work and of more specialized topics, the reader is referred to the following reviews of recent years: general aspects of structure and function (MAELICKE 1984; POPOT and CHANGEUX 1984; CHANGEUX et al. 1984; CHANGEUX and HEIDMANN 1987; ANHOLT et al. 1985; MCCARTHY et al. 1986; KARLIN et al. 1986), molecular genetics (HEINEMANN et al. 1986), kinetics of binding and response (MAELICKE and PRINZ 1983; HESS et al. 1987), electrophysiology (PEPER et al. 1982; SAKMANN and NEHER 1984; COLQUHOUN and SAKMANN 1985; COLQUHOUN 1986), cell biology (MERLIE and SMITH 1986), immunology and myasthenia gravis (LINDSTROM et al. 1983; VINCENT 1983; ENGEL 1984; HEILBRONN 1985; LINDSTROM 1985), learning and memory (CHANGEUX et al. 1987). A collection of research reports on the subject of this review is provided by the proceedings of the 1986 Santorini workshop (editor, MAELICKE 1986). If not indicated otherwise, the reviewed data relate primarily to *Torpedo* nAChR.

B. Structure of the nAChR

I. Primary Structure

The full sequences of all four subunits of the nAChR from the electric ray, *Torpedo californica* and ox and of selected subunits of other nAChRs have been determined by cDNA sequencing (Table 1 and Fig. 1). The N-termini of the translation products were identified by means of the partial protein sequences determined previously (DEVILLERS-THIERY et al. 1979; RAFTERY et al. 1980). These data showed that each subunit has a leader peptide which is removed during post-translational processing.

Following the complete sequencing of the four chains of the *Torpedo* nAChR (NODA et al. 1983c), their close relatedness became apparent. It was suggested that the different chains evolved by gene duplication from a primordial subunit, the stoichiometry of $\alpha_2\beta\gamma\delta$ having been established some 550–690 million years ago (KUBO et al. 1985). Furthermore, the different degrees of interspecies homology between subunit sequences led to the suggestion that the α-subunit has evolved more slowly than the others.

Other structural predictions based on the available primary structures concerned the number and location of asparagine-linked glycosylation and of phosphorylation sites, the main immunogenic region, and the location of the

Table 1. List of sequenced full-length cDNA of nAChR subunits

Species	Subunit	Reference
Torpedo californica	α	NODA et al. (1982)
	β	NODA et al. (1983b)
		BALLIVET et al. (1982)
	γ	NODA et al. (1983c)
		CLAUDIO et al. (1983)
	δ	NODA et al. (1983b)
Torpedo marmorata	α	DEVILLERS-THIERY et al. (1983)
Mouse	α	BOULTER et al. (1985)
		ISENBERG et al. (1986)
	γ	BOULTER et al. (1986b)
	δ	LA POLLA et al. (1984)
BC_3H-1	α	MERLIE et al. (1983)
Ox	α	NODA et al. (1983a)
	β	TANABE et al. (1984)
	γ	TAKAI et al. (1984)
	δ	KUBO et al. (1985)
	ε	TAKAI et al. (1985)
Human	α	NODA et al. (1983a)
Fowl	γ	NEF et al. (1984)
	δ	NEF et al. (1984)
PC12	α	BOULTER et al. (1986a)
Drosophila	ARD	HERMANS-BORGMEYER et al. (1986)

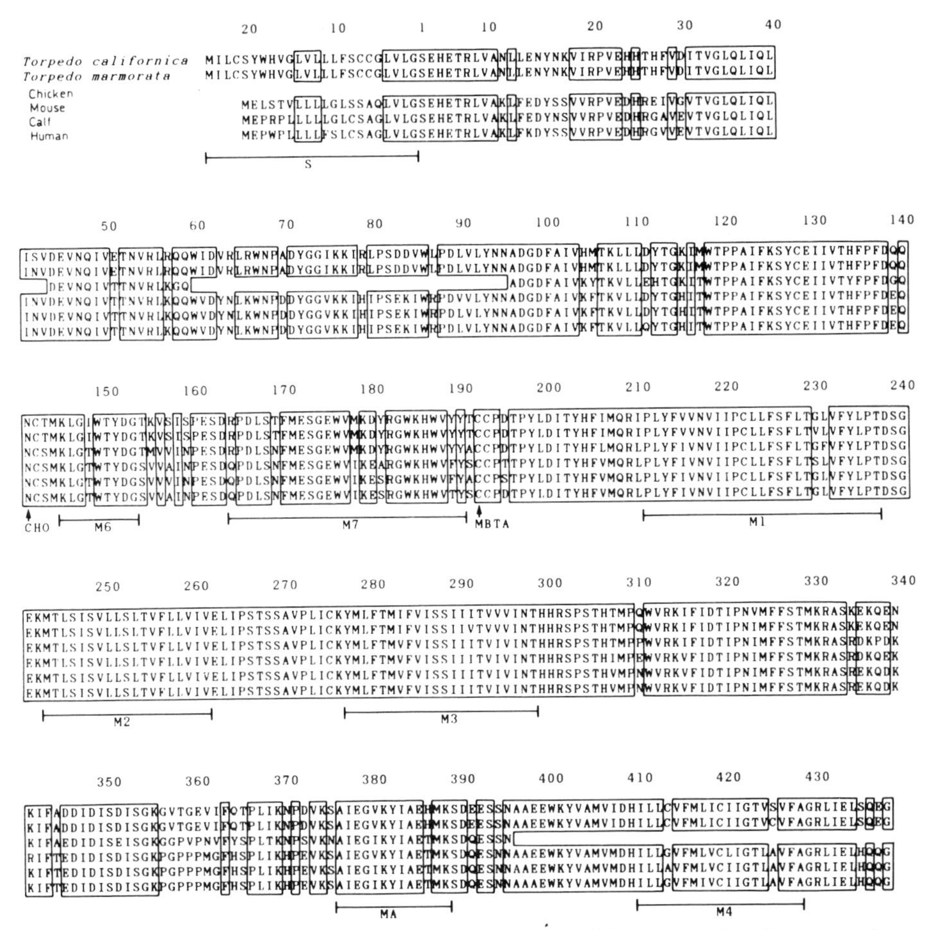

Fig. 1. Primary structures of nAChR α-subunits from different species. Sequences have been aligned so as to show maximal homology. *S*, signal sequence; *M1–M7* and *MA*, proposed transmembrane regions (see text for details); *MBTA*, site of affinity labelling; *CHO*, site of *N*-glycosylation. Although extensive sequence homology is observed, the cross-species reactivity of monoclonal antibodies against α-subunits is rather limited (LINDSTROM et al. 1981, 1983; WATTERS and MAELICKE 1983; FELS et al. 1983). Thus, minor sequence alterations in the extracellular and intracellular regions produce significant changes in tertiary structure

cysteine-cysteine disulphide bond at or near the ACh binding site (NODA et al. 1982; CLAUDIO et al. 1983; DEVILLERS-THIERY et al. 1983; SMART et al. 1984). Of these predictions, those for the main immunogenic region and the transmitter binding site(s) do not agree with recent data.

The available sequences permit the calculation of molecular masses for the protein moieties of the subunits and a comparison of these with the apparent molecular masses (M_r) obtained from migration on sodium dodecylsulphate polyacrylamide gel electrophoresis (SDS-PAGE). The considerable discrepancies observed are not due to post-translational proteolytic cleavage (NODA et al. 1982;

BARKAS et al. 1984), as was shown by protein sequencing of the mature subunit (CONTI-TRONCONI et al. 1984). Furthermore, they cannot be accounted for by the loss of the signal peptide during translocation or the contributions to the molecular masses by the carbohydrate moieties (ANDERSON and BLOBEL 1981). Other structural differences such as incomplete formation of SDS micelles due to tightly bound endogenous lipid (OLSON et al. 1984b) may be responsible.

II. Secondary Structure and Topography

Considering the high degree of sequence homology among nAChRs their secondary structures are expected to be very similar. The existence of α-helical segments in the structure of the nAChR was deduced from small angle x-ray diffraction studies (ROSS et al. 1977). Circular dichroism and resonance Raman studies of nAChR from several *Torpedo* species indicated the following structural contents: 20%–35% α-helix, 30%–50% β-structure and 30%–40% random coil (MOORE et al. 1974; MIELKE et al. 1984; ASLANIAN et al. 1983).

The hydrophobicity profiles of subunit sequences show four regions of sufficient length and hydrophobicity to span the lipid bilayer (CLAUDIO et al. 1983; DEVILLERS-THIERY et al. 1983; NODA et al. 1983a, b, c). Searching for sequences which, if α-helical, would represent one highly charged surface and one very hydrophobic surface led to the prediction of additional membrane-spanning regions (KOSOWER 1983; FINER-MOORE and STROUD 1984; GUY 1984; CRIADO et al. 1985).

Biochemical and immunological studies have provided further details of the transmembrane topography of the nAChR. The N-termini of the subunits are probably on the extracellular side of the nAChR (ANDERSON et al. 1982); but in the native receptor are inaccessible to antibodies (NEUMANN et al. 1984; RATNAM and LINDSTROM 1984). From limited proteolysis (WENNOGLE and CHANGEUX 1980; ANDERSON and BLOBEL 1981; ANDERSON et al. 1983) it was deduced that all subunits contain a large extracellular domain at their N-termini. This domain includes the likely *N*-glycosylation site at Asn-141. From antibody mapping (LINDSTROM et al. 1984; YOUNG et al. 1985; RATNAM et al. 1986a, b) it has been suggested that the carboxy termini are cytoplasmic. While these data exclude all structural models assuming an even number of membrane crosses, the correct number of transmembrane regions is still being debated.

Until recently it was commonly assumed that the N-terminal extracellular domain of the nAChR (*Torpedo* α-subunit) consist of residues 1–210 (CLAUDIO et al. 1983). Since a monoclonal antibody directed against the peptide α152–159 did not bind to native nAChR (CRIADO et al. 1985a) but competed with an antibody to the cytoplasmic site for binding to nAChR attached to plastic wells, two additional transmembrane domains were proposed for the N-terminal region (CRIADO et al. 1985). The inaccessibility for antibody of parts of one of the proposed transmembrane domains is supported by the failure of an antiserum against peptide α161–166 to bind to membrane-bound nAChR (A. MAELICKE et al., unpublished). This serum and an anti-α151–169 antibody (JUILLERAT et al. 1984) only bind to denatured nAChR. The antibody competition data, however, cannot be counted as conclusive evidence for a cytoplasmic location of the

α152–159 region as antibodies have been shown to be capable of allosteric competition (FELS et al. 1986). Thus, the existence of a hydrophilic (M6) and an amphipathic α-helical (M7) transmembrane domain remains an open question.

Based on the accessibility of cysteines α192, 193, to alkylating affinity reagents, this region must be extracellular. Antibodies raised against the peptides α235–242 and α339–346 bound to the cytoplasmic side (CRIADO et al. 1985) providing evidence for the existence of the proposed transmembrane domain M1 and circumstantial evidence for M2 and M3. The proposed amphipathic transmembrane domain M5 (GUY 1984; FINER-MOORE and STROUD 1984) is unlikely to exist as antibodies recognizing the peptides α353–359, α360–370, α371–378 and α389–396 all bind to the cytoplasmic side of the nAChR (RATNAM et al. 1986a, b). Further evidence against an amphipathic domain in this region has been provided by KORDOSSI and TZARTOS (1987) and MAELICKE et al. (1988). With the C-termini also located on the cytoplasmic side (see above), the proposed hydrophobic domain M4 cannot also be membrane spanning. From studies with photoreactive phospholipids, however, this region must be in close contact or partially embedded in the lipid phase (GIRAUDAT et al. 1985).

Site-directed mutagenesis of α-subunits co-expressed with the other subunits in *Xenopus* oöcytes (MISHINA et al. 1985) showed that the hydrophobic domains M1–M4 must all be present to form an active channel. In contrast, deletion of the whole amphipathic domain M5 had no effect on channel function. These results suggest that all four hydrophobic domains are essential for nAChR assembly (and function?) whereas the amphipathic domain M5 is not. Further data pertinent to the transmembrane orientation of the nAChR are discussed in Sect. B.VI.

Taken together, our present knowledge of the transmembrane orientation of the nAChR depends heavily on immunological data. These are not without ambiguity, as antibodies may in some cases (a) also bind to epitopes other than those defined by the linear sequence of the synthetic peptide used as antigen (PLÜMER et al. 1984; RATNAM et al. 1986a, b), and (b) shift the equilibrium mixture of receptor conformations to one which normally does not significantly contribute to the equilibrium state (MAELICKE et al. 1986b). A model for the secondary structure of the nAChR incorporating theoretical considerations and experimental data from several laboratories is presented in Fig. 2.

III. Tertiary Structure

Despite strenuous efforts crystallization and x-ray analysis of the nAChR has not yet been successful. This may in part be due to the flexibility of the molecule (see Sect. B.VI) and to the protrusion of hydrophobic segments out of the membrane (RATNAM et al. 1986b). At the present time, all information on the three-dimensional structure of the nAChR stems from electron microscopy and the statistical analysis of ultrastructural images.

From hydrodynamic (MEUNIER et al. 1972) and ultracentrifuge sedimentation studies (RÜCHEL et al. 1981) of detergent-solubilized *Torpedo* nAChR the receptor has been shown to be elongated. Neutron scattering studies yielded a radius of gyration of 4.6 nm (WISE et al. 1979). X-ray diffraction measurements of

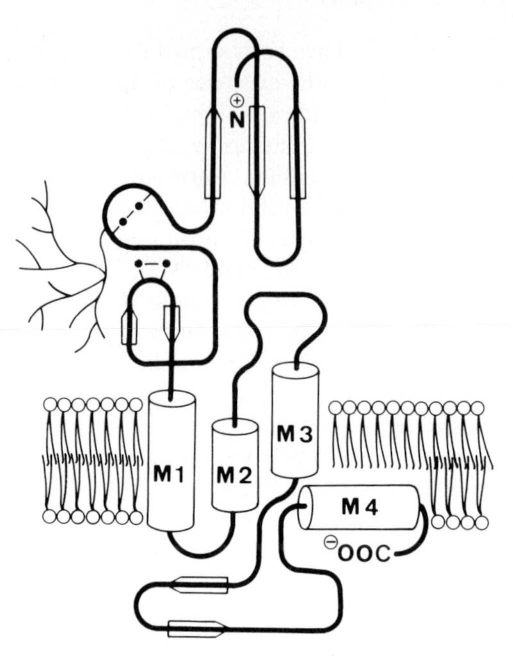

Fig. 2. Model of the transmembrane topography of the α-subunit from *Torpedo* nAChR. The model shows a large extracellular domain, the N-terminus of which in the native nAChR is inaccessible for antibody binding. The domain contains two disulphide bridges formed by cysteines 128 and 142 and by cysteines 192 and 193, respectively. Essential elements of transmitter and α-toxin binding sites have been localized close to the latter disulphide bridge. *M1–M4*, putative α-helical domains of sufficient length to span the lipid bilayer. As the N-terminus is at the extracellular surface and the C-terminus at the cytoplasmic surface of the nAChR, the number of membrane crossings must be uneven. In the absence of compelling evidence for additional α-helical (or amphipathic) segments of sufficient length, three membrane-spanning and one membrane-associated α-helical domain are proposed. Domain *M2* has been shown to contain essential elements of the nAChR- integral cation channel. The extracellular or cytoplasmic location of loop regions shown in the scheme has been determined by ultrastructural studies employing the colloidal gold double-labelling technique. For further details see text and original publications

'two-dimensional crystals', and image analysis of electron micrographs show the same gross features: the membrane-bound nAChR appears to have an overall length of 11–14 nm, extending 5.5 nm beyond the extracellular surface and 1.5–3.5 nm into the cytoplasm (ROSS et al. 1977; CARTAUD et al. 1978; KLYMKOWSKY and STROUD 1979; ZINGSHEIM et al. 1980; BRISSON and UNWIN 1984, 1985). Because some of these preparations still contained the nAChR-associated 43-kDa peripheral protein (SOBEL et al. 1977; CARTAUD et al. 1981; FROEHNER et al. 1981), a protrusion into the cytoplasmic space of more than 2 nm may be an overestimation. In terms of protein mass, 50%–60% extends into the extracellular space, 35%–45% is embedded in the membrane and only 5%–15% extend into the cytoplasm.

Viewed from above the extracellular surface of negatively stained nAChR appears as a doughnut or rosette 6.5–8.0 nm in diameter containing a central pit of 2.5–3.0 nm in diameter (NICKEL and POTTER 1973; CARTAUD et al. 1973;

ROSS et al. 1977; ZINGSHEIM et al. 1982b; KISTLER et al. 1982; BRISSON and UNWIN 1984; GIERSIG et al. 1986). As the central pit can be filled with stain almost throughout the whole length of the molecule, the pit has been proposed as the location of the ion channel (KISTLER et al. 1982). The predominantly 5-fold symmetry of the rosette (ZINGSHEIM et al. 1980; BON et al. 1984; BRISSON and UNWIN 1984; GIERSIG et al. 1986) suggests a regular arrangement of the five receptor subunits around the pit (channel). This is further supported by the even electron-density distributions determined at various cross-sections through the molecule (BRISSON 1986).

Most of the above data have been obtained under staining conditions which render the receptor biologically inactive (REINHARDT et al. 1984). When repeated in the absence of stain (MILLIGAN et al. 1984) or with stains that do not interfere with receptor activity (REINHARDT et al. 1984), the same gross features of nAChR structure were observed (BRISSON 1986; GIERSIG et al. 1986). Thus, the structure of stain-inactivated nAChR does not differ from that of active nAChR down to a resolution of approximately 0.3 nm. The position of the toxin-binding subunits (ZINGSHEIM et al. 1980, 1982a), however, certainly cannot be determined in the presence of denaturing stains.

IV. Quaternary Structure

Torpedo nAChR is a heterologous pentamer (HUCHO and CHANGEUX 1973) composed of four polypeptide chains with a stoichiometry of $\alpha_2\beta\gamma\delta$ (HUCHO and CHANGEUX 1973; REYNOLDS and KARLIN 1978; LINDSTROM et al. 1979; RAFTERY et al. 1980). This stoichiometry appears to be evolutionarily conserved up to mammalian muscle nAChR (RAFTERY et al. 1983; KUBO et al. 1985; LA POLLA et al. 1984; YU et al. 1986; HEINEMANN et al. 1986). An unequivocal demonstration of the subunit stoichiometry of mammalian receptors is, however, still missing. Adult human muscle nAChR (MOMOI and LENNON 1982; TURNBULL et al. 1985) and rat muscle nAChR (FROEHNER et al. 1977b; NATHANSON and HALL 1979) showed five-subunit patterns in SDS gels, and only three subunits have so far been observed for rabbit muscle nAChR (GOTTI et al. 1983). Furthermore, affinity labelling (see Sect. B.V.1) identified two different α-subunits in rat muscle nAChR both in innervated and denervated muscle (FROEHNER et al. 1977a; NATHANSON and HALL 1979). These uncertainties may be due to the very small amounts of these receptors available. Alternatively, the five-subunit patterns may be caused by the existence of iso-forms (FROEHNER et al. 1977b) which may be developmentally or functionally regulated (e.g. innervated junctional or denervated extrajunctional) but may coexist (MISHINA et al. 1986). Recently chicken muscle has been shown to contain a single type of nAChR with a subunit stoichiometry of 2:1:1:1, independent of whether it is innervated, denervated or embryonic (BARNARD et al. 1986). Quaternary structures differing from the *Torpedo* stoichiometry have been implicated in ganglionic and brain receptors of insects (BREER et al. 1985, 1986; GUNDELFINGER et al. 1986) and brain receptors from birds and mammals (HEINEMANN et al. 1986; WHITING and LINDSTROM 1986a, b, 1987a, b). These data will be discussed in Sect. B.VII. In summary, the quaternary structure of the nAChR has so far only been established for

Torpedo receptor. Formation of the channel by participation of all subunits (ANHOLT et al. 1980; BOHEIM et al. 1981) argues only for similarity in size and shape of nAChRs and not for an identical subunit stoichiometry.

The topography of subunits in the nAChR molecule is still being debated even for the *Torpedo* receptor. The order $\alpha\beta\alpha\delta\gamma$ was suggested on the basis of (a) nearest-neighbour analysis with cleavable cross-linkers (SCHIEBLER et al. 1980); (b) ultrastructural studies employing subunit-specific monoclonal antibodies (KISTLER et al. 1982) or α-toxins in conjunction with the known δ-δ linkage in dimers (ZINGSHEIM et al. 1982b); (c) α-neurotoxin cross-linking (HAMILTON et al. 1985); and (d) electron microscopy of two-dimensional crystalline arrays of nAChR labelled with subunit-specific Fab fragments and lectins (KUBALEK et al. 1987). A different order, $\alpha\gamma\alpha\beta\delta$, was determined from angle determinations between β and δ subunits in artificial trimers (WISE et al. 1981) in conjunction with angle determinations between nAChR-bound avidin-linked biotinylated α-toxins (HOLTZMAN et al. 1982; KARLIN et al. 1983b). As the ultrastructural studies were all performed with denaturing negative stains, their results are questionable. As a minimal conclusion of these studies, however, the two α-subunits appear to be separated from each other by the other subunits, and they therefore would reside in different environments.

From the sequences of the subunit precursors and their stoichiometry, the molecular mass of *Torpedo* nAChR is calculated to be 268 kDa. As the mature receptor is glycosylated (VANDLEN et al. 1979; ANDERSON and BLOBEL 1981) and phosphorylated (VANDLEN et al. 1979; HUGANIR et al. 1983, 1984, 1986; NOMOTO et al. 1986) on all chains and contains covalently bound lipids in the α- and β-chains (OLSON et al. 1984b), the final molecular mass should be around 290 kDa. This value agrees closely with the results from laser-light scattering (DOSTER et al. 1980). In contrast, as already mentioned, SDS-PAGE of cross-linked nAChR subunits and ultracentrifugation also underestimate the molecular mass (WEILL et al. 1974; HUCHO et al. 1978; REYNOLDS and KARLIN 1978).

Native nAChR from *Torpedo* can form dimers which are stabilized by disulphide bonds between δ-subunits. The cysteines involved are probably located at the carboxy-terminal end of each δ-chain (WENNOGLE et al. 1981) which by binding of antibody (see Sect. B.II) has been shown to be cytoplasmic. Dimers can be dissociated by disulphide-reducing agents or detergent-forced disulphide rearrangement (RÜCHEL et al. 1981). Receptor monomers have a strong tendency to noncovalent association (RÜCHEL et al. 1981). In subsynaptic areas (FELS et al. 1985) and when tightly packed (HEUSER and SALPETER 1979), nAChR dimers can form parallel (KISTLER and STROUD 1981; BON et al. 1984) and antiparallel (FAIRCLOUGH et al. 1983; BRISSON and UNWIN 1984) rows. These symmetrical forms are not due to intrinsic properties of the dimer since many more forms were observed in diluted membranes (BON et al. 1984).

So far nAChR dimers have been observed in *Torpedo, Narcine* and *Electrophorus* but not in vertebrate preparations. Their significance is unknown. They have been reported to be functionally identical with monomers in transmitter binding (FELS et al. 1982) and single-channel properties (ANHOLT et al. 1980; BOHEIM et al. 1981). Other reports suggest different binding (HAMILTON et al. 1979; CHANG et al. 1984) and channel properties (SCHINDLER et al. 1984) of dimers as compared to monomers. The problem is complicated by the tendency of monomers to associate spontaneously under a variety of conditions (RÜCHEL et al. 1981; SCHINDLER et al. 1984).

V. Functional Domains

This section deals with the location of functional domains. Their properties in the context of the receptor's mechanism of function are discussed in Sect. D.

1. Competitive Ligand-Binding Sites

ACh and its agonists and antagonists bind to the extracellular side of the nAChR. Binding of curaremimetic neurotoxins to isolated α-subunits (RÜCHEL et al. 1981; HAGGERTY and FROEHNER 1981; TZARTOS and CHANGEUX 1983) and affinity alkylation with bromoacetylcholine and 4-(*N*-maleimido)benzyltrimethylammonium (MBTA) (KARLIN 1969; DAMLE and KARLIN 1978; DAMLE et al. 1978; WOLOSIN et al. 1980) of α-subunits show that these subunits contain all or major parts of the competitive ligand-binding sites. Recently the main target cysteine for affinity alkylation has been identified as Cys-192 (KAO et al. 1984). Soon thereafter it was shown that Cys-192 forms a disulphide bridge with Cys-193 in native *Torpedo* receptor (KAO and KARLIN 1986; NEUMANN et al. 1986). Disulphide bonds between adjacent cysteines are not common but have been observed in peptides (CAPASSO et al. 1977) and proteins and are energetically permitted (MITRA and CHANDRASEKARAM 1984). The peptide bonds between disulphide-linked adjacent cysteine residues are in a non-planar *cis* rather than in the usual *trans* configuration.

The attachment site at the nAChR of the trimethylammonium head group of ACh is generally assumed to be the side-chain carboxylate of a glutamic or aspartic acid residue (SMART et al. 1984). However, labelling experiments with appropriate affinity probes have not so far been successful indicating that groups other than carboxylate groups may form the negatively charged primary attachment site for the transmitter. Photolabile diazonium affinity labels (KOTZYBA-HIBERT et al. 1985; DENNIS et al. 1986; GIRAUDAT et al. 1986a) may help to localize the anionic subsite of the transmitter binding site.

An alternative approach to localize the ligand-binding site at the α-subunit is to employ antibodies against short stretches of the nAChR in competition binding studies with ligands (PLÜMER et al. 1984; CRIADO et al. 1986; MAELICKE et al. 1986a, b) or in ligand overlays of protein blots (GERSHONI et al. 1983; BARKAS et al. 1986). By means of the latter it was demonstrated that a cholinergic binding site lies within a 15 k-Da fragment beginning behind Glu-161 of the α-subunit (NEUMANN et al. 1986). Neither Cys-128 nor Cys-142 of the intact nAChR are linked to this fragment (NEUMANN et al. 1985, 1986), indicating that the specific disulphide in the vicinity of the ligand-binding site is formed by Cys-192 and Cys-193.

α-Bungarotoxin (α-BTX) binding studies with a synthetic peptide matching in sequence residues α173–204 of *Torpedo* had previously shown that major determinants of the toxin-binding site reside in this region (WILSON et al. 1984). In another approach, making use of expression cloning in *E. coli*, proteins corresponding to several cDNA fragments of the mouse α-subunit were synthesized and tested for α-BTX binding. These studies mapped the toxin-binding site to residues α158–216 (BARKAS et al. 1986, 1987). Binding studies with the dodecapeptide α185–196 further reduced the region under question (NEUMANN et al. 1986; FUCHS et al. 1986b): the dodecapeptide bound α-BTX with an equilibrium dissociation constant of 30 μM, and this binding was competitively inhibited by cholinergic agonists and antagonists. Furthermore, carboxymethylation after reduction of the disulphide bond in the dodecapeptide or replacement of the cysteines by alanine residues both significantly reduced the binding affinity for neurotoxin (FUCHS et al. 1986b). Similar results have been obtained by RALSTON et al. (1987) with the synthetic peptide α185–199. Binding studies with synthetic peptides are usually limited to high-affinity ligands such as neurotoxins; they must be performed with utmost care in order to be conclusive (MCCORMICK and ATASSI 1984). Since there exist multiple attachment sites for neurotoxins at the nAChR (MARTIN et al. 1983; CHIBBER et al. 1983), neurotoxins do not necessarily bind only to peptides which contain elements of the transmitter binding site (OBLAS et al. 1986).

Employing oligonucleotide-directed mutagenesis, each of the five Cys residues of *Torpedo* α-subunit was converted into a serine residue. The resulting nAChR mutants were expressed in *Xenopus* oöcytes, and their binding of α-BTX and response to ACh were determined (MISHINA et al. 1985). Mutants with α192 S or α193 S had nearly normal α-BTX binding activity but were not responsive to ACh. This agrees with the chemical identification of these cysteines as close to the ACh binding site (KAO and KARLIN 1986). In addition, the result points to at least partially separate sites for ACh and α-BTX (MAELICKE et al. 1977b; KANG and MAELICKE 1980; MAELICKE 1987a). Further evidence for separate binding sites for groups of competitive ligands is discussed in Sect. D.I.

2. Noncompetitive Ligand-Binding Sites

In addition to the sites discussed above, other regulatory sites exist on the nAChR. A major group of ligands which bind noncompetitively with agonists and antagonists to the nAChR are organic amines such as the common local anaesthetics. As most competitive ligands are organic amines by structure, they can also bind to and can act as noncompetitive inhibitors (NCIs) via these additional regulatory sites. The affinity of binding to NCI sites, however, is usually lower than that for binding to the competitive ligand-binding sites.

Suitable affinity ligands to map NCI sites should preferentially bind to the open channel in a voltage-dependent manner and should promote desensitization of the nAChR (see Sect. D.III). These properties are met by quinacrine (ADAMS and FELTZ 1980) and in part by chlorpromazine (CARP et al. 1983). When rapidly mixed with agonist and *Torpedo* membranes, chlorpromazine labelled all four types of nAChR subunits (HEIDMANN and CHANGEUX 1984). Employing equilibrium labelling conditions, the site of labelling in the β- and δ-subunits was identified as Ser-262 (GIRAUDAT et al. 1986a, b, 1987). This residue lies within the putative transmembrane region M2 (Fig. 2).

The lipophilic cation triphenylmethylphosphonium (TPMP) photolabelled the α-, β- and δ-chains of desensitized *Torpedo* nAChR (MUHN and HUCHO 1983; MUHN et al. 1984; FAHR et al. 1985). Microsequencing of the major TPMP-labelled fragments identified Ser-262 of the δ-subunit and the homologous positions of the α- and β-chains as the sites of labelling (HUCHO 1986; HUCHO et al. 1986a, b; OBERTHÜR et al. 1986). These are the same serine residues that are labelled by chlorpromazine (GIRAUDAT et al. 1986), again indicating that TPMP preferentially acts as a noncompetitive inhibitor.

Time-resolved quinacrine azide labelling of the open-channel state of the nAChR targeted the α- and, to a much lesser extent, the β-subunit (COX et al. 1985). Labelling of the α-subunit has been reported to be confined to a cyanogen bromide fragment containing the segment M1 (KARLIN et al. 1987). In view of the multiplicity of NCI binding sites at the AChR (OSWALD et al. 1983) and their complex action (ADAMS 1981), heterogeneity in labelling may indeed exist (KOSOWER 1986).

3. Ion Channel

Single-channel recordings from planar lipid bilayers composed of a synthetic phospholipid and affinity-purified nAChR provided the first unequivocal proof that the transmitter-controlled cation channel is part of the nAChR molecule (BOHEIM et al. 1981; LABARCA et al. 1984a; SCHINDLER et al. 1984). Recently it was shown that co-expression of all four subunit genes is needed to produce a fully functional *Torpedo* nAChR channel in the *Xenopus* oöcyte membrane (MISHINA et al. 1984, 1985; TAKAHASHI et al. 1985).

Affinity labelling of the ion channel of the nAChR is conceptually ambiguous: NCIs may label sites other than channel sites (see Sect. B.V.1), organic ions usually possess a complicated pharmacology and the most specific probes of the channel, e.g. histrionicotoxin, are not yet available as photoreactive probes.

The most direct approach in identifying the receptor channel is site-directed mutagenesis in conjunction with oöcyte electrophysiology (MISHINA et al. 1984, 1985; SAKMANN et al. 1985; IMOTO et al. 1986). After voltage-clamp studies had established that injection of mRNA specific for the four *Torpedo* subunits leads to the appearance of fully functional nAChR channels in the oöcyte membrane, the contribution of each subunit to the channel properties was studied by interspecies exchange of subunits (see Sect. D.II). Studying the single-channel properties of hybrid *Torpedo*-ox δ-subunits in oöcyte-expressed *Torpedo* nAChR, IMOTO et al. (1986) found that part of the hydrophobic region M2 and a short sequence between M2 and M3 determine the electrical characteristics of the channel. Together with the chemical labelling experiments of GIRAUDAT et al. (1986) and HUCHO et al. (1986a, b), the hydrophobic region M2 thus appears to be (a) membrane spanning and (b) an essential component of the nAChR cation channel. As previously suggested on the basis of electrophysiological data (NEHER and STEINBACH 1978; ALBUQUERQUE et al. 1980; ADAMS et al. 1981). NCIs indeed seem to exert their channel-blocking effects by binding to parts of the channel itself rather than allosterically by binding to more distant sites (KOBLIN and LESTER 1979; COHEN et al. 1980; HEIDMANN et al. 1980b).

4. Other Functional Domains

Phosphorylation of *Torpedo* nAChR (GORDON et al. 1977a, b; TEICHBERG et al. 1977; VANDLEN et al. 1979) has been shown to be catalysed by at least four different protein kinases (SMILOWITZ et al. 1981; HUGANIR and GREENGARD 1983; PALFREY et al. 1983; ZAVOIKO et al. 1984; ERIKSSON et al. 1986, 1987). Three of these phosphorylate-specific subunits of the nAChR. These kinases and their proposed phosphorylation sites are tyrosin-specific protein kinase (βTyr-355, γTyr-364, δTyr-372), cAMP-dependent protein kinase (γSer-354, δSer-361) and protein kinase C (αSer-333, δSer-377) (HUGANIR et al. 1984). Employing the synthetic peptide antibody technique, FUCHS et al. (1986b) and SAFRAN et al. (1986) recently reported that δSer-361 is, indeed, the correct phosphorylation site for cAMP-dependent protein kinase. *In vivo* phosphorylation of the nAChR (VANDLEN et al. 1979) and the existence of endogenous protein kinases (HUGANIR et al. 1984; ERIKSSON et al. 1986) and phosphatases in *Torpedo* electric tissue has been established. The physiological significance of nAChR phosphorylation, however, is still being debated. There exists circumstantial evidence that phosphorylation affects one or more of the following receptor properties: mobility, clustering and turnover, channel gating, channel coupling in receptor clusters and desensitization (SAITOH and CHANGEUX 1981; GORDON et al. 1977a, b; HEILBRONN et al. 1985, 1986; BROWNING et al. 1985; MIDDLETON et al. 1986; ALBUQUERQUE et al. 1986; HUGANIR et al. 1986; HÄGGBLAD et al. 1987).

Glycosylation is required to render the nAChR fully functional (MEUNIER et al. 1974; MERLIE et al. 1982; MISHINA et al. 1985). So far, only *N*-glycosylation sites have been localized. The single *N*-glycosylation site of the α-chain is at Asn-141.

The nAChR of the BC_3H-1 muscle cell line has been shown to contain covalently bound lipid at its α- and β-chains (OLSON et al. 1984b). The linkage is alkali resistant and thus is probably via an amide bond. It was suggested (OLSON et al. 1984b) that the lipid modification is required for receptor assembly.

Although not a functional domain in the true sense, the main immunogenic region is an important structural feature of the nAChR. The majority of antibodies raised against native nAChR, both in the human muscle disease myasthenia gravis and in its experimental models, are directed against this region (TZARTOS and LINDSTROM 1980). It is located at the extracellular site (TZARTOS and LINDSTROM 1980) and close to the N-terminus (BARKAS et al. 1986) of the α-subunit. Employing proteolytic peptides (RATNAM et al. 1986b; BARKAS et al. 1986) and fusion proteins (BARKAS et al. 1987), essential elements of the main immunogenic region have been located within the fragments α46–127 (*Torpedo californica* nAChR) and α6–85 (mouse muscle nAChR). The sequence α63–80 appears to be the immunodominant epitope of human muscle nAChR (TZARTOS et al. 1988).

VI. Conformations

Electrophysiological and biochemical evidence suggests that the nAChR can exist in a multiplicity of conformational states. These states develop and decay in the millisecond time range and are closely related to receptor function.

The vertebrate skeletal muscle nAChR channel shows considerably greater noise than is predicted to arise from the random passage of ions through the open channel (SIGWORTH 1985). The observed subconductance states are probably caused by fluctuations of the channel protein conformation (SIGWORTH 1986). This view is supported by theoretical considerations (LÄUGER 1985a, b; BRANDL and DEBER 1986).

Based on their conductances and lifetimes a considerable number of channel states in the presence of agonists have been identified (HAMILL and SAKMANN 1981; AUERBACH and SACHS 1983; SINE and STEINBACH 1986; COLQUHOUN and OGDEN 1986). Furthermore, the nAChR channel exhibits pharmacological specificity, i.e channel open times depend on the activating ligand employed (COLQUHOUN 1981). The latter property suggests that agonist-AChR complexes assume ligand-specific conformations (MAELICKE 1984).

Ligand-specific conformational states have been demonstrated by fluorescence equilibrium (COVARRUBIAS et al. 1986) and kinetic studies (HEIDMANN and CHANGEUX 1979, 1980; PRINZ et al. 1988). Fluorescence-kinetic studies of agonist binding to the nAChR have indicated the existence of several conformational states in the course of receptor saturation (PRINZ and MAELICKE 1983a, b; PRINZ et al. 1988).

Ligand binding to the nAChR is an entropy-driven process (MAELICKE et al. 1977b) and thus appears to be coupled to conformational transitions. The cooperativity of agonist binding to the membrane-bound nAChR (NEUBIG and COHEN 1979; FELS et al. 1982) and the allosteric competition of some antibodies and ligands for receptor binding (FELS et al. 1986) are evidence for long-range allosteric coupling between functional domains.

In summary, an understanding of the static structure of the nAChR will not suffice to unravel its molecular mechanism of function in full. Unfortunately the repertoire of methods for studying protein dynamics in the subsecond time range is still very limited.

VII. Classification of ACh Receptors According to Structure

Recently the cDNA sequences of a muscarinic acetylcholine receptor (mAChR) (KUBO et al. 1986) and of a *Torpedo* acetylcholinesterase (AChE) (SCHUMACHER et al. 1986) have been reported. Neither of these ACh-binding proteins showed any significant sequence homology with the nAChR in regions believed to contain the ACh binding site. Similarly, no cross-reactive antibodies have been found for these three proteins (FELS et al. 1983). In contrast, there exists strong homology between nAChRs (Fig. 1) which apparently extends to functional homology. As an example, fully functional hybrid receptors are expressed in the toad oöcyte after co-injection of mRNA coding for the α-, β- and δ-chains from *Torpedo* and of either the γ or the (embryonic AChR) ε chain from calf (TAKAI et al. 1985). Functional interchangeability is observed although the interchanged subunits (γ or ε from calf for γ from *Torpedo*) have only about 50% sequence homology. In a similar study the δ-chain from mouse was interchanged with the same chain in *Torpedo* (WHITE et al. 1985), yielding an expressed channel resembling the mouse nAChR in its properties. These results suggest that nAChRs from muscle and electroplaque cells are structurally and functionally homologous, and that their functional differences result from local variations in the sequence of specific domains.

Synaptic and extrasynaptic vertebrate muscle receptors differ in their biochemical and antigenic properties (FROEHNER et al. 1977b; NATHANSON and HALL 1979). Their differences are apparently not due to different genes for their α-chains (GOLDMAN et al. 1985). The quaternary structure of muscle nAChRs has been discussed in Sect. B.IV.

Other than in muscle tissue nAChRs are found in sympathetic ganglion cells, adrenal medullary cells and the central nervous system (CNS) of vertebrates. These receptors differ in their pharmacological, electrophysiological, antigenic and biochemical properties both from muscle nAChR and from each other (KRNJEVIĆ 1975; FREEMAN 1977; BROWN 1978; MARSHALL 1981; OSWALD and FREEMAN 1981; BARNARD and DOLLY 1982; LUKAS 1984; MEHRABAN et al. 1984; FREEMAN and NORDEN 1984; YAMADA et al. 1985; CONTI-TRONCONI et al. 1985; MEEKER et al. 1986; CLARKE 1986; WHITING and LINDSTROM 1986a, b). The following are so far the best studied: the adrenal tumour cell line PC12 contains an α-BTX binding protein and an ACh-activated ion channel which are separate molecular entities (PATRICK and STALLCUP 1977; CARBONETTO et al. 1978; JACOB and BERG 1983; LEPRINCE 1983). Three α-BTX binding proteins (from rat and mouse brain and from domestic fowl optic lobe) have been purified and were found to be related in structure to muscle receptor (MORLEY and KEMP 1981; SCHNEIDER et al. 1985; CONTI-TRONCONI et al. 1985; BARNARD et al. 1986). In addition to the α-toxin, *Bungarus multicinctus* venom contains a minor toxin specific for neuronal acetylcholine receptors (RAVDIN and BERG 1979; LORING et al. 1984; GRANT and CHIAPINELLI 1985).

Originating from immunological approaches to examine the nature of neuronal toxin-binding sites and acetylcholine receptors WHITING and LINDSTROM have purified putative nAChRs from fowl (WHITING and LINDSTROM 1986a) and rat (WHITING and LINDSTROM 1987a) brains. Both receptors consist of two types of subunits, α and β, of which the α-subunits share antigenic determinants with α-subunits from nAChRs of electric tissue and muscle. Affinity labelling indicates, however, that the transmitter binding site and the adjacent disulphide bridge are located on the β-subunit of neuronal nAChRs (WHITING and LINDSTROM 1987b).

Several cDNA clones coding for putative vertebrate brain nAChRs have been isolated (GOLDMAN et al. 1986; BOULTER et al. 1986a; BARNARD et al. 1986; NEF et al. 1986). The results support the notion of multiple forms of nAChRs in the brain. One of the cloned putative brain receptors (BARNARD et al. 1986) appears to be a hitherto unknown type in that it shows little sequence homology in the region of domains M1 and M5 (Fig. 1) with the nAChRs from PC12 cells and muscle.

AChRs of nicotinic pharmacology also exist in invertebrate brain. In particular the insect nervous system is very rich in nAChR (FLOREY 1963; SATELLE 1980; BREER 1981 a, b). Insect neuronal nAChR has been isolated from fly and cockroach heads and from locust ganglia (BREER et al. 1984, 1985; FILBIN et al. 1983; SATELLE and BREER 1985). While the quaternary structure of these nAChRs is still debated (LUNT 1986; MAELICKE 1987c), there is some suggestion that they may form a separate class of their own: the nAChRs from *Locusta migratoria* and *Periplaneta americana* appear to consist of just a single type of subunit comigrating with the δ-chain of *Torpedo* nAChR on SDS-PAGE (BREER et al. 1985; SATTELLE and BREER 1985; HANKE and BREER 1986; BREER et al. 1986). A cDNA clone coding for a putative *Drosophila* brain nAChR has been isolated (HERMANS-BORGMEYER et al. 1986; GUNDELFINGER et al. 1986). The clone exhibits a high degree of sequence homology with the rat neural nAChR α-subunit clone isolated by BOULTER et al. (1986a). The *Drosophila* genome contains more than a single gene coding for the nAChR (A. BOSSY and M. BALLIVET, personal communication; E.D. GUNDELFINGER, personal communication). It remains to be established whether these genes code for other subunits of the same nAChR of for other nAChRs. These other nAChRs, if they exist, should be similar to brain receptors since neuromuscular transmission in insects is mediated by amino acids (GRATION 1981). Taken together, there might exist in insects an ancestral nAChR composed of only a single subunit or of fewer subunits than vertebrate muscle nAChR. In vertebrates central and peripheral receptors probably diverged very early in evolution (CONTI-TRONCONI et al. 1985; HEINEMANN et al. 1986). As the vertebrate neural α-subunit has approximately the same size as the larger muscle nAChR subunits (HEINEMANN et al. 1986), an order in evolutionary age might exist beginning with the invertebrate brain nAChR followed by the vertebrate brain nAChR and finally the vertebrate muscle nAChR. Recombinant DNA technology will probably soon provide conclusive evidence concerning the multiplicity of nAChRs in different species and tissues and their structural relatedness.

C. Molecular Genetics

I. Structure of nAChR Genes

Molecular genetics has achieved a central role within the repertoire of techniques at present available for unravelling the molecular basis of cholinergic transmission. The structural information obtained by this approach has been reviewed in the preceding section. Here we shall deal only with aspects of gene structure and gene expression.

The subunits of the nAChR have been shown to be products of discrete mRNAs rather than to be cleaved from a large precursor protein (ANDERSON and BLOBEL 1981). Full-length genomic DNA fragments encoding the human (NODA et al. 1983a) and the fowl muscle (NEF et al. 1986) nAChR α-subunit precursor have been isolated and sequenced. In both cases the protein coding sequence is divided into nine exons (Fig. 3). Most of the prepeptide is encoded by exon P1,

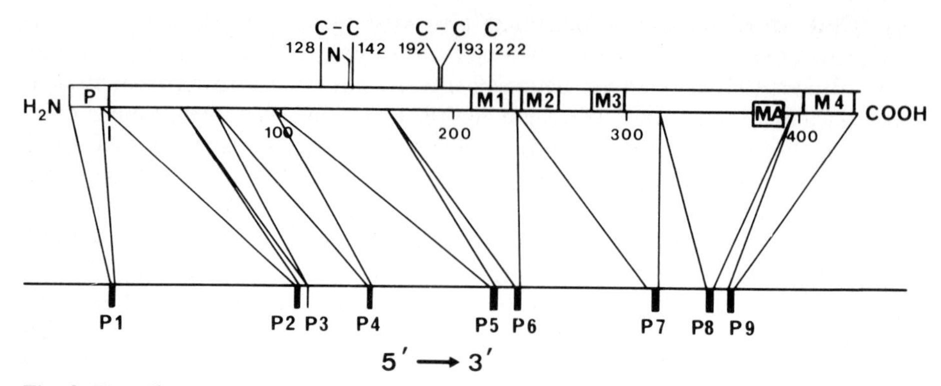

Fig. 3. Exon/intron distribution in the human α-subunit gene in comparison with the domain structure of the related prepeptide; for the original data see NODA et al. (1983a). The organization of the structural domains and disulphide bridges are according to recent data (see Sects. B.II and B.V)

the ACh binding region (Sect. B.V.1) and the hydrophobic segment M1 by P6, M2 and M3 by P7, M4 by P9, and the large cytoplasmic region by P8 (see also Sect. B.II.). The highest levels of sequence homology between nAChR α-subunits are observed in the regions of exons P3, P5, P6, P7 and P9 (FENG et al. 1985), i.e. in the regions of the ACh binding site, the hydrophobic segments, and an area of the large N-terminal region not yet identified as a particular functional site.

The two genes encoding the δ- and γ-chains of fowl muscle nAChR are encoded by the same DNA strand and are very closely linked (NEF et al. 1984). This close linkage is probably the result of gene duplication of an ancestral gene and may have been maintained for the purpose of co-regulated transcription (NEF et al. 1986). The protein coding sequences of the two genes are both divided into 12 exons. Exons of like rank are nearly identical in size and have extensive sequence homology. In addition, the splice sites between homologous exons are highly conserved. As observed for the α-genes from human and fowl, gene structure and the functional organization of the gene products are correlated in that the four hydrophobic regions are each encoded by a separate exon. Recently, several genomic and brain cDNA clones encoding α-like subunits of fowl and rat nAChRs have been isolated, sequenced and shown to be expressed in the nervous system but not in muscle (GOLDMAN et al. 1986; BOULTER et al. 1986a; BARNARD et al. 1986; NEF et al. 1986). These data show that there exist gene families of receptor subunits, some of which are expressed at the NMJ and the remainder in the nervous system. In addition to the multiplicity of specific subunit genes the receptors of the central and the peripheral nervous system probably differ in their genomic organization. For instance, genomic clones encoding nAChR subunits from fowl optic lobe and brain show considerable differences in gene organization (BARNARD et al. 1986) suggesting that their genes have long diverged.

II. nAChR Gene Expression and Processing

nAChR cDNA probes have been employed to assay the locations and levels of nAChR mRNA in various tissues. The large increase in the concentration of nAChR observed upon denervation of muscle fibres is preceded by a comparable increase in the level of α- and δ-chain mRNA (MERLIE et al. 1984a, b; KLARSFELD and CHANGEUX 1985). This confirms previous findings (reviewed in FAMBROUGH 1979) that the additional nAChRs are newly synthesized. Similarly the increase in nAChR mRNA levels in the course of muscle cell development in tissue culture is correlated with increased rates of transcription of α- and δ-chain genes (BUONANNO and MERLIE 1986). For intact muscle cells it was shown that nuclei positioned directly below an NMJ are more actively involved in nAChR synthesis than nuclei located extrasynaptically (MERLIE and SANES 1985). Recently the promotor of a fowl nAChR α-subunit gene has been isolated and sequenced (KLARSFELD et al. 1987).

Northern blot analysis employing a neural α -cDNA probe showed that homologous mRNA is present in the hypothalamus, hippocampus, cerebellum and adrenal gland but not skeletal muscle (BOULTER et al. 1986a, b). Expression in the adrenal gland was specific for the medulla, agreeing with the notion that the employed cDNA probe encodes an adrenal medullary cell nAChR. *In situ* hybridization with the same cDNA probe (BOULTER et al. 1986a; GOLDMAN et al. 1986) labelled brain areas known to contain cholinergic synapses but few or no α-BTX binding sites (CLARKE et al. 1985).

Biogenesis of the nAChR is a highly complex process leading from translation of the respective subunit mRNAs to assembly of the mature subunit complex and its insertion into the plasma membrane. Translation occurs on membrane-bound polysomes (MERLIE et al. 1981). Membrane-polysome assembly requires the interaction of a signal recognition particle (SRP), its specific membrane receptor of the rough endoplasmic reticulum (SRP receptor) and the nAChR-synthesizing polysome (WALTER and BLOBEL 1982; GILMORE et al. 1982). Membrane insertion is followed by cleavage of the signal peptide and core glycosylation (ANDERSON et al. 1983).

Newly synthesized α-subunits differ from mature ones in that they do not bind α-BTX or antibodies directed against the main immunogenic region (MERLIE and SEBBANE 1981) but soon acquire these properties (MERLIE and LINDSTROM 1983). It has been suggested that maturation of subunits may involve a conformational change induced by a covalent modification, possibly disulphide-bond formation (M.M. SMITH et al., personal communication). As a final step in subunit biosynthesis, their oligosaccharide structure is modified (LINDSTROM et al. 1979).

Assembly of subunits has been suggested to occur in the endoplasmic reticulum (CARLIN and MERLIE 1986) followed by transport to the Golgi apparatus (FAMBROUGH and DEVREOTES 1978). Transport to the cell surface has been shown to involve coated vesicles (BURSZTAJN and FISCHBACH 1984). Posttranslational processing is a regulated series of processes: refeeding of BC_3-H-1 cells with fetal calf serum leads to an increase in α-subunit synthesis but inhibition of assembly and transport (OLSON et al. 1983). Spontaneous contractions reduced the expression of nAChR on the surface of embryonic fowl myotubes. The effect was reversed by incubation with the sodium channel blocker tetrodotoxin (TTX) (BIRNBAUM et al. 1980); this is mainly due to transcriptional control (KLARSFELD and CHANGEUX 1985; KLARSFELD et al. 1986). To provide a time scale for synthesis and assembly of nAChR in muscle cell lines, synthesis (including cleavage of signal sequence, co-translational *N*-glycosylation and termination of translation) takes about 1 min; maturation is completed approximately 30 min after synthesis, assembly after 90 min; transport and cell surface expression follows about 90–150 min after synthesis (MERLIE and LINDSTROM 1983; CARLIN et al. 1986).

To study selected properties of cloned cDNA products several expression systems have been developed. mRNA-encoding nAChR subunits have been microinjected into *Xenopus* oöcytes resulting in successful translation, processing, assembly, insertion into the plasma membrane and function (BARNARD et al. 1982; MISHINA et al. 1985; WHITE et al. 1985; SAKMANN et al. 1985; CLAUDIO 1987; CLAUDIO et al. 1988). Expression of foreign mRNA in the oöcyte system occurs within 2 days after injection. The main disadvantage of this system however is that expression is transient, lasting only for a few days. Recently, *Torpedo* α-subunits have been expressed in the plasma membrane of yeast (FUJITA et al. 1986), however in an incorrect orientation. Should further experiments establish that the nAChR also functionally assembles in this system, it would be a useful addition to currently available expression systems. Stable expression in bacteria of fusion proteins containing fragments from single subunits of the nAChR has been reported by BARKAS et al. (1987).

Stable gene expression in tissue culture cell lines is a very attractive goal. It would permit large quantities of the gene product to be obtained, expression would be continuous and the cell lines could be selected for advantageous properties in single-channel recordings. Thus far, each *Torpedo* cDNA clone has been introduced individually into the chromosomes of mammalian fibroblast and/or muscle cell lines by direct transfection or virus injection (CLAUDIO 1986a, b). In addition, stable expression of functional cell surface *Torpedo* nAChR has recently been established in one fibroblast cell line (T. CLAUDIO, personal communication).

D. Function of the nAChR

I. Ligand Binding

1. Classes and Stoichiometries of Sites

Most studies indicate the presence of two agonist binding sites per nAChR monomer, independently of whether the receptor is membrane-bound or detergent-solubilized (reviewed in MAELICKE 1984; POPOT and CHANGEUX 1984). At equilibrium the two agonist binding sites on the membrane-bound *Torpedo* nAChR interact weakly, exhibiting positive cooperativity (FELS et al. 1982; COVARRUBIAS et al. 1986). In contrast, the agonist sites at the solubilized

Electrophorus nAChR are independent (or interact with negative cooperativity) (MAELICKE et al. 1977b; PRINZ and MAELICKE 1983a). Low molecular mass competitive antagonists bind to the same number of sites as agonists (CHANGEUX et al. 1975; MAELICKE et al. 1977a, b; NEUBIG and COHEN 1979; CONTI-TRONCONI and RAFTERY 1982; PRINZ and MAELICKE 1983a; COVARRUBIAS et al. 1986); the binding of agonists and antagonists to the nAChR is mutually exclusive. Depending on the nAChR and specific neurotoxin applied, the receptor displays the same (WEILAND et al. 1976; DAMLE and KARLIN 1978; BLANCHARD et al. 1979; FELS et al. 1982) or twice the number of binding sites as for agonists (MAELICKE et al. 1977b; CONTI-TRONCONI and RAFTERY 1986). Independently of the relative stoichiometry of neurotoxin sites, the binding of neurotoxins and of the other competitive ligands are mutually exclusive.

In addition to the high-affinity sites considered above, other low-affinity binding sites for agonists have been proposed (DUNN and RAFTERY 1982a, b; DUNN et al. 1983; PASQUALE et al. 1983; TAKEYASU et al. 1983, 1986). The authors suggested that the high-affinity agonist sites are only involved with desensitization while the additional low-affinity sites control channel activation. Direct evidence for low-affinity agonist binding sites with such properties is still lacking. Competition binding and ion-flux studies in the presence of an antibody which blocks one of the two high-affinity agonist sites at the nAChR argue against the existence of the proposed sites (FELS et al. 1986).

Additional low affinity sites for agonists with other than the properties just mentioned have however been shown to exist at the nAChR: (a) ligand-induced accelerated dissociation of neurotoxin-nAChR complexes (MAELICKE and REICH 1975; MAELICKE et al. 1977b; KANG and MAELICKE 1980; MAELICKE et al. 1986a, b) and of antibody-AChR complexes (MAELICKE et al. 1987a; FELS and MAELICKE 1987) require the presence of additional low-affinity ligand-binding sites to be assumed; (b) there exists electrophysiological and biochemical evidence that agonists and some antagonists can also bind to the local anaesthetic (NCI) binding sites (ADAMS and SAKMANN 1978; SINE and TAYLOR 1982; NEHER 1983; GAGE et al. 1983; COVARRUBIAS et al. 1984; SINE and STEINBACH 1984b; OGDEN and COLQUHOUN 1985; COVARRUBIAS et al. 1986). Models for the existence of multiple ligand-binding sites on the nAChR have been proposed (KANG and MAELICKE 1980; CONTI-TRONCONI et al. 1982; SMART et al. 1984; CONTI-TRONCONI and RAFTERY 1986).

Many organic amines act as noncompetitive inhibitors at the nAChR (for reviews, see CHANGEUX 1981; MAELICKE 1984). The mechanism of action of most of these is complex (NEHER 1983; GAGE and WACHTEL 1984; PAPKE and OSWALD 1986). Binding and kinetic studies suggest three types of binding sites for NCIs at the nAChR; one or few sites of high affinity at the nAChR protein (K_D in the μM range), a larger number of low affinity sites probably situated at the nAChR-lipid boundary (OSWALD et al. 1983; HEIDMANN et al. 1983b; M. COVARRUBIAS, H. PRINZ and A. MAELICKE, unpublished) and the competitive ligand-binding sites (KRODEL et al. 1979; M. COVARRUBIAS, H. PRINZ and A. MAELICKE, unpublished).

A number of natural and synthetic compounds previously not thought to be ligands of the nAChR have been shown to bind and act at the receptor. Anatoxin-*a* is a potent cholinergic agonist (KOSKINEN and RAPOPORT 1985; SWANSON et al. 1986), the reversible cholinesterase (ChE) inhibitors eserine and pyridostigmine act as agonists and channel blockers (SHAW et al. 1985; BRADLEY et al. 1986). Histamine acts as a competitive antagonist at the frog endplate (ARIYOSHI et al. 1985), chloramphenicol and forskolin have direct channel-blocking effects (HENDERSON et al. 1986; MCHUGH and MCGEE 1986; HÄGGBLAD et al. 1987) and some opiates modulate receptor function by binding directly to the nAChR (KARPEN et al. 1982; COSTA et al. 1983; OSWALD et al. 1986). It is unknown whether these ligands bind to established sites on the nAChR, or whether they exert their effects via as yet undefined binding sites.

2. Mode and Kinetics of Binding

Equilibrium agonist binding to the membrane-bound nAChR is characterized by convex Scatchard plots (NEUBIG and COHEN 1979; SINE and TAYLOR 1980; FELS et al. 1982). Quantitative analysis (FELS et al. 1982; PRINZ and MAELICKE 1986) suggests that two binding sites per nAChR monomer interact with weak positive cooperativity. The K_D values of the two sites are in the 10-nM range. Since electrophysiological results yield concentrations for half-maximal response in the order of $1-100$ µM, the equilibrium data are thought to relate to the desensitized state of the nAChR.

These findings together with the sigmoid shape of dose-response curves have led to the analysis of nAChR binding data in terms of an allosteric mechanism of the Monod type (MONOD et al. 1965). The classical allosterical model, however, is a random model and thus not really consistent with the fact that the two agonist binding sites on the nAChR are not symmetrical in their properties. The two sites differ in their interactions with affinity labels (DAMLE and KARLIN 1978), α-neurotoxins (WEBER and CHANGEUX 1974; MAELICKE et al. 1977b), tubocurare (NEUBIG and COHEN 1979; SINE and TAYLOR 1981) and lophotoxin (CULVER et al. 1984). Furthermore, kinetic studies with the fluorescent agonist *N*-7-(4-nitrobenzo-2-oxa-1,3-diazole)-5-aminopentanoic acid β-(*N*-trimethylammonium)ethyl ester (NBD-5-acylcholine) have shown that agonist interaction with the nAChR is an ordered process (PRINZ and MAELICKE 1983b).

This lack of equivalence in agonist sites and the ordered mode of agonist association and dissociation argues against the application of random models in the analysis of nAChR binding data (Sect. D.IV). In particular, multiexponential analysis of nAChR binding kinetics does not suffice to obtain quantitative data on *intermediate* reaction steps in the course of receptor saturation (MAELICKE and PRINZ 1983; PRINZ and MAELICKE 1985; PRINZ 1986). Independently of the method of data analysis applied there is agreement that the initial association of agonist and nAChR is very rapid (k in the order of $10^8\,\mathrm{l\,mol^{-1}\,s^{-1}}$), and that the late reaction step(s) with rate constant(s) (k) in the order of $50\,\mathrm{s^{-1}}$ to $0.01\,\mathrm{s^{-1}}$ (JÜRSS et al. 1979; HEIDMANN and CHANGEUX 1979, 1980a; BOYD and COHEN 1980a; DUNN et al. 1980; PRINZ and MAELICKE 1983b) are rather slow. An additional bimolecular reaction step (k in the order of $10^5\,\mathrm{l\,mol^{-1}\,s^{-1}}$) was resolved in the studies with 1-(5-dithylamino)naphthalene(sulphonamido)-hexanoic acid β-(*N*-trimethylammonium)ethyl ester (dansyl-6-choline) (HEIDMANN and

CHANGEUX 1979) and NBD-5-acylcholine (PRINZ and MAELICKE 1983b). All other results from kinetic studies appear closely linked to the particular models applied. Thus, authors favouring the concerted model of allosteric interactions (MONOD et al. 1965) analysed their data on the basis of a four-state model (one resting, one active, and two desensitized states) (HEIDMANN and CHANGEUX 1980; NEUBIG and COHEN 1980; BOYD and COHEN 1980b) while others suggested that conformational isomerizations of ligand-nAChR complexes occur in the course of agonist binding (PRINZ and MAELICKE 1983b; MAELICKE and PRINZ 1983; MAELICKE 1987a, b). Central differences between the two approaches are: (a) the allosteric model suggests that the proposed states pre-exist albeit in different relative concentrations in the absence of ligands; in contrast, the model of ordered interaction (PRINZ and MAELICKE 1983b) suggests a single state in the absence of ligand and the formation of other states (conformers) in the course of ligand interaction; (b) the allosteric model is a random model requiring the agonist-binding sites to have symmetrical properties; the model of ordered interaction, in contrast, is based on non-identical agonist binding sites. Recent models combining central features of the specific allosteric model and an ordered mechanism of binding may resolve the differences between the two approaches (COLQUHOUN 1986; PRINZ 1988).

Recent binding studies involving monoclonal antibodies (FELS et al. 1986; MAELICKE et al. 1986b; FELS and MAELICKE 1988) suggest that competitive ligands do not necessarily bind to identical sites, but that competition may in some cases be the result of allosteric coupling between separate binding sites. In the cases reported antibodies were shown to bind in a mutually exclusive fashion with agonists to the nAChR while no competition with antagonists was observed. It was suggested that agonist (or antibody) binding to the nAChR induces a conformational transition which causes the specific three-dimensional organization of the antibody (or agonist) binding site to disintegrate. This type of allosteric competition is probably also responsible for the ligand-induced accelerated dissociation of some neurotoxin-nAChR and antibody-nAChR complexes (see Sect. D.I.1). Should antibodies generally induce conformational changes, their application as structural probes (Sect. B.II) is ambiguous.

II. Channel Gating

The nAChR is a cation-selective ion channel (TAKEUCHI and TAKEUCHI 1960; LEWIS 1979; ADAMS et al. 1980). Permeability studies have shown that alkali ions pass by free diffusion and divalent cations by limited diffusion through the channel (ADAMS et al. 1980; LEWIS and STEVENS 1983). From the cut-off size of permeant organic cations the minimal diameter of the channel has been calculated to be 0.65 nm (HUANG et al. 1978; DWYER et al. 1980; SANCHEZ et al. 1986). From single-channel studies the conductance of the channel appears to be independent of the activating ligand and only weakly dependent of the membrane potential (reviewed in SAKMANN and NEHER 1984; GARDNER et al. 1984). In contrast, the mean open time of the channel strongly depends on the structure of the activating ligand applied.

It has been suggested that the mean open time of the channel is determined by the dissociation rate constant of the bound agonist (TRAUTMANN and FELTZ 1980). Recent findings argue against a simple mechanism: (a) brief partial and complete closings within single-channel events have been observed with embryonic tissue (COLQUHOUN and SAKMANN 1981) but only rarely with adult tissue (COLQUHOUN and SAKMANN 1985); (b) covalent attachment of agonist to the nAChR results in lengthened but not persistent openings on the channel (CHABALA and LESTER 1986; SIEMEN et al. 1986); (c) fluorescence-kinetic studies suggest the channel lifetime to be coupled to a transient conformational change of the nAChR (PRINZ and MAELICKE 1983b; PRINZ et al. 1988).

The association rate constant of agonist and nAChR has been determined indirectly from high-resolution single-channel studies. In agreement with the biochemical data, it was calculated to be of the order of $10^8 \, \mathrm{l \, mol^{-1} \, s^{-1}}$ (COLQUHOUN and SAKMANN 1985; SINE and STEINBACH 1986).

Dose-response studies with voltage-clamped muscle cells yielded Hill coefficients between 1.5 and 2.7 (ADAMS 1975; PEPER et al. 1975; SHERIDAN and LESTER 1977; DIONNE et al. 1978; DREYER et al. 1978). Similar results were obtained from ion-flux experiments (SINE and TAYLOR 1979; NEUBIG and COHEN 1980; CASH et al. 1981; NEUBIG et al. 1982; WALKER et al. 1982), all of which suggested that two molecules of agonist are required to activate the channel. Assuming a random process of channel activation, NEUBIG et al. (1982) estimated the probability of channel opening in the absence of agonist as $1{:}10^7$ and by occupation of a single agonist site as $1{:}10^4$. Further evidence that the simultaneous occupancy of both agonist sites is required for channel opening is provided by studies correlating ACh-mediated cation permeability with fractional occupancy of agonist sites by α-cobra toxin (SINE and TAYLOR 1980, 1981; ANHOLT et al. 1982).

Several recent reports suggest other stoichiometries for channel activation: (a) 'spontaneous' channel openings have been observed in single channel studies of cultured muscle and brain cells in the absence of added agonist (JACKSON 1984, 1986; STEINBACH et al. 1986), however, in view of the sensitivity level of single-channel recordings, the possible presence in these experiments of endogenous transmitter or other activating ligands cannot be excluded with certainty; (b) channel activation by a single agonist molecule is suggested by electrophysiological studies with agonists covalently attached to the nAChR (LESTER et al. 1980a, b; CHABALA and LESTER 1986; SIEMEN et al. 1986). As an alternative interpretation, however, affinity labelling of the gating site may have required the *simultaneous* occupancy (by a reversible agonist) of also the other agonist site. In other words, occupation of the triggering (second) site may require a conformational change induced by saturation of the first site, as is suggested by the model of ordered interaction (PRINZ and MAELICKE 1983b; MAELICKE 1984). Recent electrophysiological and biochemical models (COLQUHOUN and SAKMANN 1985; JACKSON 1986; PRINZ 1988) consider channel gating as probability events: compared to the frequency of channel opening in the presence of two bound agonist molecules the frequency of channel gating of the monoliganded nAChR was calculated to be approximately 100 times lower and that of the non-liganded receptor 10^7 times lower.

The decrease in endplate conductivity observed during prolonged exposure to transmitter has been termed 'desensitization' (KATZ and THESLEFF 1957). The rate at which this takes place depends on the nature and concentration of agonist applied (FELTZ and TRAUTMANN 1982) and is enhanced by membrane hyperpolarization, divalent cations and temperature (MAGAZANIK and VYSKOČIL 1970). On the single-channel level desensitization is manifested by burst behaviour (SAKMANN et al. 1980). From the distribution of bursts two processes of 'desensitization' were resolved (SAKMANN et al. 1980) in agreement with other electrophysiological data (MAGLEBY and PALOTTA 1981; FELTZ and TRAUTMANN 1982; AOSHIMA 1984). Recovery from the desensitized state also appears to be a two-step process in which fast onset and recovery are coupled (FELTZ and TRAUTMANN 1982). One of the two desensitization processes may require occupation of additional low-affinity agonist sites (FELTZ and TRAUTMANN 1982).

In agreement with the original prediction by KATZ and THESLEFF (1957), biochemical and ion-flux studies show an increase in agonist binding affinity correlated in time with desensitization (HEIDMANN and CHANGEUX 1979; SINE and TAYLOR 1979, HESS et al. 1979; BOYD and COHEN 1980a, b; WALKER et al. 1981b; DUNN and RAFTERY 1982b; NEUBIG et al. 1982; PASQUALE et al. 1983; HEIDMANN et al. 1983a; COVARRUBIAS et al. 1984). These are the slow reaction steps observed in ligand binding (Sect. D.I.) and ion-flux kinetics (NEUBIG and COHEN 1980; MOORE and RAFTERY 1980; WALKER et al. 1981b, 1982; KARPEN et al. 1983; KARPEN and HESS 1986; TAKEYASU et al. 1986). Assuming that the free receptor exists as an equilibrium of the same conformational states as in the presence of ligands, it has been calculated that prior to ligand binding 10%–20% of the receptor in *Torpedo* electric organ (WEILAND and TAYLOR 1979; HEIDMANN and CHANGEUX 1979; BOYD and COHEN 1980a) but only 0.01% of that in BC_3H-1 muscle cells (SINE and TAYLOR 1982) exists in the desensitized state. As a further difference from *Torpedo* nAChR 25% of the total receptor concentration in BC_3H-1 cells does not desensitize; thus a limiting level of steady-state ion flux remains even at high agonist concentrations (SINE and TAYLOR 1979).

As a further complication of the gating properties of the nAChR a multiplicity of main conductance and subconductance states has been observed (HAMILL and SAKMANN 1981; AUERBACH and SACHS 1983; LABARCA et al. 1984b; SIGWORTH 1985; SINE and STEINBACH 1986). These probably reflect the existence of corresponding conformational states of the agonist-occupied nAChR (MAELICKE 1984).

III. Modulation of Binding and Response

As the modes of modulation of nAChR properties are manifold, only the modulating effects of (a) local anaesthetics and (b) the membrane environment will be considered here.

Of the many NCIs of the nAChR the organic amines with local anaesthetic action are the ones best understood. Their action depends on the presence of agonist and is voltage dependent (KOBLIN and LESTER 1979; ALBUQUERQUE et al. 1980; LAMBERT et al. 1980). They induce burst behaviour and shorten lifetimes of the agonist-induced single-channel events (NEHER and STEINBACH

1978). These effects have led to the suggestion that local anaesthetics directly bind to and thereby occlude the channel (RUFF 1977, 1982; NEHER and STEINBACH 1978). NCIs of such properties have been shown to interact with the nAChR from the extracellular side (ARACAVA and ALBUQUERQUE 1984) and were employed for the localization of their sites (and possibly the channel) in the protein moiety of the nAChR (Sect. B.V). They also show high affinity binding to reconstituted nAChR and block the agonist-induced ion flux (EARNEST et al. 1984; BLICKENSTAFF and WANG 1985). It should be noted, however, that only a few local anaesthetics behave thus; others act in a voltage-independent manner (KOBLIN and LESTER 1979; OGDEN et al. 1981) and may exert their effects by disturbing the interaction of the nAChR with the surrounding lipid matrix (KOBLIN and LESTER 1979; ANDREASEN and MCNAMEE 1980).

Local anaesthetics, including the channel blocker histrionicotoxin, enhance the affinity of agonists for the nAChR (KATO and CHANGEUX 1976; WEILAND et al. 1977; BURGERMEISTER et al. 1977; HEIDMANN and CHANGEUX 1979; KRODEL et al. 1979; DUNN et al. 1981; OSWALD and CHANGEUX 1981 a; SINE and TAYLOR 1982; COVARRUBIAS 1984; CHEMOUILLI et al. 1985). They have therefore been described as enhancing the rate of desensitization and the concentration of desensitized conformations of the nAChR (SINE and TAYLOR 1982; COVARRUBIAS et al. 1984; OSWALD et al. 1984; BOYD and COHEN 1984).

As has recently been established by studies of ion fluxes (FONG and MCNAMEE 1986) and direct binding kinetics (PRINZ et al. 1988), the functional properties of the nAChR are sensitive to the composition and state of the surrounding lipid matrix. Channel gating requires an appropriate membrane fluidity and the presence of specific phospholipids (MCNAMEE et al. 1986). Compounds with anaesthetic action, including alcohols (GAGE et al. 1975) and detergents (BRISSON et al. 1975; BRILEY and CHANGEUX 1978), dramatically change the rate constants and mechanism of ligand binding (PRINZ et al. 1988).

A physiological role for the modulating effects of local anaesthetics and other NCIs has not yet been established. As structural homologues of ACh, they may mimic or intensify a modulating action of the transmitter; alternatively, they may mimic the modulating action of endogenous CNS neuromodulators.

IV. Molecular Mechanism of Function

1. Model of Ligand-Specific States and Ordered Ligand Interaction of the nAChR

As pointed out in Sect. D.II.2, the two competitive ligand binding sites at the nAChR differ in their properties and, in the case of the *Electrophorus* receptor, are filled and emptied in a strictly ordered fashion. These properties are accommodated in the model of ligand-specific states and ordered ligand interaction of the nAChR (PRINZ and MAELICKE 1983b; MAELICKE and PRINZ 1983; MAELICKE 1984). It therefore seems appropriate to begin this concluding section by restating this model.

A scheme of the model is shown in Fig. 4. Primarily, this was the simplest reaction scheme capable of accommodating our kinetic studies with the fluores-

$$
\begin{array}{ccccc}
R_{RC} & \overset{①}{\rightleftharpoons} & AR_{LC} & \overset{②}{\rightleftharpoons} & AR\text{-}A_{LO} \\
 & & ③\updownarrow & & ④\updownarrow \\
 & & AR_{LD} & \overset{⑤}{\rightleftharpoons} & AR\text{-}A_{LD}
\end{array}
$$

Fig. 4. Model of ligand-specific conformational states and ordered ligand interaction of the nAChR. *Large letters* refer to levels of receptor occupancy (*R*, unoccupied receptor; *AR*, monoliganded AChR; *AR-A*, diliganded AChR; *A*, agonist); *small letters* refer to the activity state of the ion channel (*RC*, resting closed; *LO*, liganded open; *LD*, liganded desensitized). From the rate constants for ACh as ligand: (a) formation of AR ($k > 10^8$ $l\ mol^{-1}\ s^{-1}$) and AR-A (k approximately 10^7 $l\ mol^{-1}\ s^{-1}$) is very rapid; (b) AR and AR-A have rather short half-lives as they rapidly decay by ligand dissociation (k approximately $10^3\ s^{-1}$) and by isomerization (k approximately $10^2\ s^{-1}$ and $10\ s^{-1}$); (c) AR-A_{LD} is several thousand-fold more stable than AR-A_{LO} (PRINZ and MAELICKE 1983b)

cent agonist NBD-5-acylcholine and the purified nAChR from *E. electricus* (PRINZ and MAELICKE 1983b). It was applied because our data contradicted all random reaction schemes, including those based on the concerted model of allosteric interactions (CHANGEUX et al. 1984). The key result requiring a new reaction scheme was our finding that agonists associate and dissociate from their two binding sites on the purified *Electrophorus* nAChR in a strictly ordered fashion, i.e. the agonist molecule which bound first dissociated last and *vice versa.* We further incorporated into the scheme the finding that although there exist only *two* competive agonist binding sites on the nAChR, at least *four* agonist-AChR complexes are formed in the course of receptor saturation. From the concentration dependence of the individual reaction steps the two additional complexes were shown to result from isomerization of initially formed monoliganded and diliganded nAChR (Fig. 4). Finally, when we correlated the properties (rate constants and affinities) of the observed agonist-AChR complexes with the electrophysiologically established 'states' of the nAChR, we observed a close correlation of one of the diliganded nAChR conformers (AR-A_{LO}) with the 'active state', and of the other one (AR-A_{LD}) with the 'desensitized state' of the nAChR. By interpreting our minimal reaction scheme on the basis of the channel states postulated to exist in the presence of low concentrations of agonists (COLQUHOUN and OGDEN 1986; STEINBACH et al. 1986) the model depicted in Fig. 4 was completed. This resembles in its basic features classical electrophysiological models. In contrast to the latter, it does not contain the obligatory isomerization step from AR-A to the open channel. In the time resolution of our experiments (milliseconds), AR-A_{LO} *is* the active conformation of the nAChR exhibiting a half-life similar to the mean open time of the ACh-activated channel. A central feature of the scheme of Fig. 4 (and the related electrophysiological schemes) is that it implies a strictly *ordered* mode of agonist association and dissociation. This is often not properly appreciated and has led to the inadmissible mingling of this scheme with the *random* allosteric model (BOYD and COHEN 1980a; HESS et al. 1983; CHANGEUX et al. 1984). As a further central difference from the allosteric model the model of ligand-specific states correlates the physiological 'states' of the nAChR with specific conformers of the *liganded* (not the unoccupied) receptor. In particular, the active (open channel) state is correlated with a conformation of the diliganded receptor ($AR\text{-}A_{LO}$) which within the saturation kinetics of the nAChR is characterized as a reaction intermediate. Thus, activation and channel closing and desensitization are identified as specific reaction steps along a pathway of conformational changes of the nAChR molecule induced (or increased in their probability) by agonist binding. Since the conformations of agonist-AChR complexes are ligand specific (COVARRUBIAS et al. 1986), individual reaction pathways and 'states' result. As an example, the stability constant of the complex $AR\text{-}A_{LO}$ is ligand dependent and thus results in agonist-specific mean channel open times.

2. Is a Comprehensive Receptor Model Possible?

The central purpose of all structural and functional studies of the nAChR is to understand its mechanism of function at the molecular level. In spite of the immense speed and success with which new facts on the nAChR have been assembled, this remains an exacting task: the nAChR is an integral and modulated signal transducer in which signal transduction and amplification is achieved by chemo-electric coupling within the same molecule. By its complex yet compact molecular organization the nAChR is also optimized for speed of signal transduction. In view of our still incomplete knowledge of its properties, all theoretical 'models' developed so far should be viewed merely as working hypotheses which will certainly require extensions and adjustments. The models proposed so far still depend heavily on the sources of the information being summarized, whether pharmacological/biochemical (NACHMANSOHN 1955; CHANGEUX and THIERY 1967; KARLIN 1967; NACHMANSOHN 1972; MAELICKE and PRINZ 1983; CHANGEUX et al. 1984), electrophysiological (KATZ 1966; MAGLEBY and STEVENS 1972; COLQUHOUN and SAKMANN 1981; SINE and STEINBACH 1984a, b, c, 1986; LÄUGER 1985a; FREDKIN et al. 1985) or molecular biological (KISTLER et al. 1982; KOSOWER 1983, 1984, 1986; FINER-MOORE and STROUD 1984; GUY 1984; CRIADO et al. 1985; RATNAM et al. 1986a). Even these confined models are still under discussion in their particular disciplines. A comprehensive model of the receptor's structure-function relationship which includes all these aspects is still some distance away. As a first step in this direction, however, it appears timely to adjust the biochemical models of nAChR function to the following properties of this integral signal transducer: (a) the nAChR can exist in a *large* (not only three or four) number of distinct active and inactive states; (b) many of these states are correlated with specific conformations of ligand nAChR complexes; (c) agonist association and dissociation is an *ordered* (not random) process in at least one system; (d) the active (open channel) state is a reaction intermediate within a chain of conformational transitions induced upon ligand occupancy. It is also clear however that statements concerning a particular mechanism of function are linked to probabilities; thus, occupancy by two agonist molecules in an ordered fashion may not be a necessary requirement of channel gating but may increase, rather, the probability of channel opening by orders of magnitude. In view of these complications comprehensive models should cover only those features of the nAChR that are associated with its most probable structural and functional states.

Acknowledgements. I thank Marc Ballivet (Geneva), Tom Barkas (Lausanne), Jean-Pierre Changeux (Paris), Toni Claudio (New Haven), David Colquhoun (London), Edith Heilbronn (Stockholm), George Hess (Ithaca, NY), Ed Kosower (Tel-Aviv), Jon Lindstrom (San Diego), Heino Prinz (Dortmund), Joe-Henry Steinbach (St. Louis), and Socrates Tzartos (Athens) for their critical reading of the manuscript and many helpful suggestions. Work from my laboratory is supported by grants from the Deutsche Forschungsgemeinschaft and the Fonds der Chemischen Industrie.

References

Adams DJ, Dwyer TM, Hille B (1980) The permeability of endplate channels to monovalent and divalent metal cations. J Gen Physiol 75:493–510

Adams PR (1975) An analysis of the dose-response curve at voltage-clamped frog endplates. Pflugers Arch 360:145–153

Adams PR (1981) Acetylcholine receptor kinetics. J Membr Biol 58:161–174

Adams PR, Feltz A (1980) Quinacrine (Mepacrine) action at frog end-plate. J Physiol (Lond) 306:261–281

Adams PR, Sakmann B (1978) Decamethonium both opens and blocks endplate channels. Proc Natl Acad Sci USA 75:2994–2998

Albuquerque EX, Tsai MC, Aronstam RS, Witkop B, Eldefrawi AT, Eldefrawi ME (1980) Phencyclidine interactions with the ionic channel of the acetylcholine receptor and electrogenic membrane. Proc Natl Acad Sci USA 77:1224–1228

Albuquerque EX, Deshpande SS, Aracava Y, Alkondon M, Daly JW (1986) A possible involvement of cyclic AMP in the expression of desensitization of the nicotinic acetylcholine receptor. A study with forskolin and its analogs. FEBS Lett. 199:113–120

Anderson DJ, Blobel G (1981) In vitro synthesis, glycosylation and membrane insertion of the four subunits of *Torpedo* acetylcholine receptor. Proc Natl Acad Sci USA 78:5598–5602

Anderson DJ, Walter P, Blobel G (1982) Signal recognition protein is required for the integration of acetylcholine receptor alpha subunit, a transmembrane glycoprotein, into the endoplasmic reticulum membrane. J Cell Biol 93:501–506

Anderson DJ, Blobel G, Tzartos S, Gullick W, Lindstrom J (1983) Transmembrane orientation of an early biosynthetic form of acetylcholine receptor subunit determined by proteolytic dissection in conjunction with monoclonal antibodies. J Neurosci 3:1773–1784

Andreasen TJ, McNamee MG (1980) Inhibition of ion permeability control properties of acetylcholine receptor from *Torpedo californica* by long-chain fatty acids. Biochemistry 19:4719–4726

Anholt R, Lindstrom J, Montal M (1980) Function equivalence of monomeric and dimeric forms of purified acetylcholine receptors from *Torpedo californica* in reconstituted lipid vesicles. Eur J Biochem 109:481–487

Anholt R, Fredkin DR, Deerinck T, Ellisman M, Montal M, Lindstrom J (1982) Incorporation of acetylcholine receptors into liposomes: vesicle structure and acetylcholine receptor function. J Biol Chem 257:7122–7134

Anholt R, Lindstrom J, Montal M (1985) The molecular basis of neurotransmission: structure and function of the nicotinic acetylcholine receptor. In: Martonosi AN (ed) The enzymes of biological membranes, vol 3. Plenum, New York, pp 335–401

Aoshima H (1984) A second, slower inactivation process in acetylcholine receptor-rich membrane vesicles prepared from *Electrophorus electricus*. Arch Biochem Biophys 235:312–318

Aracava Y, Albuquerque EX (1984) Meproadifen enhances activation and desensitization of the acetylcholine receptor-ionic channel complex (AChR): single channel studies. FEBS Lett 174:267–274

Ariyoshi M, Hasuo H, Koketsu K, Ohta Y, Tokimasa T (1985) Histamine is an antagonist of the acetylcholine receptor at the frog endplate. Br J Pharmacol 85:65–73
Aslanian D, Heidmann T, Nigrerie M, Changeux J-P (1983) Raman spectroscopy of acetylcholine receptor-rich membranes from *Torpedo mamorata* and of their isolated components. FEBS Lett 164:393–400
Auerbach A, Sachs F (1983) Flickering of a nicotinic ion channel to a subconductance state. Biophys J 42:1–10
Ballivet M, Patrick J, Lee J, Heinemann S (1982) Molecular cloning of cDNA coding for gamma subunit of *Torpedo* acetylcholine receptor. Proc Natl Acad Sci USA 79:4466–4470
Ballivet M, Nef P, Stadler R, Fulpius B (1983) Genomic sequences encoding the α-subunit of acetylcholine receptor are conserved in evolution. Cold Spring Harbor Symp Quant Biol 48:83–87
Barkas T, Juillerat M, Kistler J, Schwendimann B, Moody J (1984) Antibodies to synthetic peptides as probes of acetylcholine receptor structure. Eur J Biochem 143:309–314
Barkas T, Gabriel J-M, Juillerat M, Kokla A, Tzartos SJ (1986) Localisation of the main immunogenic region of the nicotinic acetylcholine receptor. FEBS Lett 196:237–241
Barkas T, Mauron A, Roth B, Alliod C, Tzartos SJ, Ballivet M (1987) Mapping the main immunogenic region and toxin-binding site of the nicotinic acetylcholine receptor. Science 235:77–80
Barnard EA, Dolly JO (1982) Peripheral and central nicotinic ACh receptors – how similar are they? Trends Neurosci 5:325–327
Barnard EA, Miledi R, Sumikawa K (1982) Translation of exogenous messenger RNA coding for nicotinic acetylcholine receptors produces functional receptors in *Xenopus* oocytes. Proc R Soc Lond (Biol) 215:241–246
Barnard EA, Beeson DMW, Cockcroft VB, Darlison MG, Hicks AA, Moss SG, Squire MD (1986) Molecular biology of nicotinic acetylcholine receptors from chicken muscle and brain. In: Maelicke A (ed) Structure and function of the nicotinic acetylcholine receptor. (NATO IASI series, series H, vol 3) Springer, Berlin Heidelberg New York, pp 389–415
Birnbaum M, Reis MA, Shainberg A (1980) Role of calcium in the regulation of acetylcholine receptor synthesis in cultured muscle cells. Pflugers Arch 385:37–43
Blanchard SG, Quast U, Reed K, Lee T, Schimerlik MI, Vandlen R, Claudio T, Strader CD, Moore HP, Raftery MA (1979) Interaction of [^{125}I]-alpha-bungarotoxin with acetylcholine receptor from *Torpedo californica*. Biochemistry 18:1875–1883
Blickenstaff GD, Wang HH (1985) The effect of spin-labeled local anesthetics on acetylcholine receptor-mediated ion flux. Biophys J 47:261 a
Boheim G, Hanke W, Barrantes FJ, Eibl H, Sakmann B, Fels G, Maelicke A (1981) Agonist-activated ionic channels in acetylcholine receptor reconstituted into planar lipid bilayers. Proc Natl Acad Sci USA 78:3586–3590
Bon F, Lebrun E, Gomel J, van Rapenbusch R, Cartaud J, Popot JL, Changeux J-P (1984) Image analysis of the heavy form of the acetylcholine receptor from *Torpedo marmorata*. J Mol Biol 176:205–237
Boulter J, Luyten W, Evans K, Mason P, Ballivet M, Goldman D, Stengelin S, Martin G, Heinemann S, Patrick J (1985) Isolation of a clone coding for the α-subunit of a mouse acetylcholine receptor. J Neurosci 5:2545–2552
Boulter J, Evans K, Goldman D, Martin G, Treco D, Heinemann S, Patrick J (1986a) Isolation of a cDNA clone coding for a possible neural nicotinic acetylcholine receptor alpha-subunit. Nature 319:368–374
Boulter J, Evans K, Martin G, Mason P, Stengelin S, Goldman D, Heinemann S, Patrick J (1986b) Isolation of cDNA clones coding for the precursor of the gamma subunit of mouse muscle nicotinic acetylcholine receptor. J Neurosci Res 16:37–49
Boyd ND, Cohen JB (1980a) Kinetics of binding of [^{3}H]acetylcholine to *Torpedo* postsynaptic membranes: association and dissociation rate constants by rapid mixing and ultrafiltration. Biochemistry 19:5353–5358
Boyd ND, Cohen JB (1980b) Kinetics of binding of [^{3}H]acetylcholine and [^{3}H]carbamoylcholine to *Torpedo* postsynaptic membranes: slow conformational transitions of the cholinergic receptor. Biochemistry 19:5344–5353

Boyd ND, Cohen JB (1984) Desensitization of membrane-bound *Torpedo* acetylcholine receptor by amine noncompetitive antagonists and aliphatic alcohols: Studies of [^{3}H]acetylcholine binding and $^{22}Na^+$-ion fluxes. Biochemistry 23:4023–4033
Bradley RJ, Sterz R, Peper K (1986) Agonist and inhibitory effects of pyridostigmine at the neuromuscular junction. Brain Res 376:199–203
Brandl CJ, Deber CM (1986) Hypothesis about the function of membrane-buried proline residues in transport proteins. Proc Natl Acad Sci USA 83:917–921
Breer H (1981a) Comparative studies on cholinergic activities in the central nervous system of *Locusta migratoria* J Comp Physiol 141:271–275
Breer H (1981b) Properties of putative nicotinic and muscarinic cholinergic receptors in the central nervous system of *Locusta migratoria.* Neurochem Int 3:43–52
Breer H, Kleene R, Behnke D (1984) Isolation of a putative nicotinic acetylcholine receptor from the central nervous system of *Locusta migratoria.* Neurosci Lett 46:323–328
Breer H, Kleene R, Hinz G (1985) Molecular forms and subunit structure of the acetylcholine receptor in the nervous system of insects. J Neurosci 5:3386–3392
Breer H, Hinz G, Mädler U, Hanke W (1986) Identification and reconstitution of a neuronal acetylcholine receptor from insects. In: Maelicke A (ed) Structure and function of the nicotinic acetylcholine receptor. Springer, Berlin Heidelberg New York, pp 319–332 (NATO ASI series, series H, vol 3)
Briley MS, Changeux J-P (1978) Recovery of some functional properties of the detergent-extracted cholinergic receptor protein from *Torpedo marmorata* after reintegration into a membrane environment. Eur J Biochem 84 (2):429–439
Brisson A (1986) Three-dimensional structure of the acetylcholine receptor. In: Maelicke A (ed) Structure and function of the nicotinic acetylcholine receptor. Springer, Berlin Heidelberg New York, pp 1–6 (NATO ASI series, series H, vol 3)
Brisson A, Unwin PN (1984) Tubular crystals of acetylcholine receptor. J Cell Biol 99:1202–1211
Brisson A, Unwin PN (1985) Quaternary structure of the acetylcholine receptor. Nature 315:474–477
Brisson A, Devaux PF, Changeux J-P (1975) Effect anesthésique local de plusieurs composés liposolubles sur la réponse de l' électroplaque de Gymmote à la carbamylcholine et sur la liaison de l' acétylcholine au récepteur cholinergique de Torpille. C R Seances Acad Sci (Paris) 280D:2153–2156
Browning MD, Huganir R, Greengard P (1985) Protein phosphorylation and neuronal function. J Neurochem 45:11–23
Buonanno A, Merlie JP (1986) Transcriptional regulation of nicotinic acetylcholine receptor genes during muscle development. J Biol Chem 261:11452–11455
Burgermeister W, Catterall W, Witkop B (1977) Histrionicotoxin enhances agonist-induced desensitization of acetylcholine receptor. Proc Natl Acad Sci USA 74:5754–5758
Bursztajn S, Fischbach GD (1984) Evidence that coated vesicles transport acetylcholine receptors to the surface membrane of chick myotubes. J Cell Biol 98:498–506
Capasso S, Mattia C, Mazzarella L, Puliti R (1977) Structure of a cis-peptide unit: molecular conformation of the cyclic disulphide l-cysteinyl-l-cysteine. Acta Crystallogr B33:2080–2083
Carbonetto ST, Fambrough DM, Muller KJ (1978) Nonequivalence of alpha-bungarotoxin receptors and acetylcholine receptors in chick sympathetic neurons. Proc Natl Acad Sci USA 75:1016–1020
Carlin BE, Merlie JP (1986) In: Strauss D, Kreil G, Boime I (eds) Protein compartmentalization. Springer, Berlin Heidelberg New York Tokyo
Carlin BE, Lawrence JC, Lindstrom JM, Merlie JP (1986) An acetylcholine receptor precursor *a* subunit that binds *a*-bungarotoxin but not *d*-tubocurare. Proc Natl Acad Sci USA 83:498–502
Carp JS, Aronstam RS, Witkop B, Albuquerque EX (1983) Electrophysiological and biochemical studies on enhancement of desensitization by phenothiazine neuroleptics. Proc Natl Acad Sci USA 80:310–314
Cartaud J, Sobel A, Rousselet A, Devaux PF, Changeux J-P (1981) Consequences of alkaline treatment for the ultrastructure of the acetylcholine-receptor-rich membranes from *Torpedo marmorata* electric organ. J Cell Biol 90:418–426

Cash DJ, Aoshima H, Hess GP (1981) Acetylcholine-induced cation translocation across cell membranes and inactivation of the acetylcholine receptor: chemical kinetic measurements in the millisecond time region. Proc Natl Acad Sci USA 78:3318–3322

Chabala LD, Lester HA (1986) Activation of acetylcholine receptor channels by covalently bound agonists in cultured rat myoballs. J Physiol (Lond) 379:83–108

Chang HW, Bock E, Neumann E (1984) Long-lived metastable states and hysteresis in the binding of acetylcholine to *Torpedo californica* acetylcholine receptor. Biochemistry 23:4546–4556

Changeux J-P (1981) The acetylcholine receptor: an 'allosteric' membrane protein. Harvey Lect 75:85–254

Changeux J-P, Heidmann T (1987) Allosteric receptors and molecular models of learning. In: Edelman G, Gall WE, Cowan WM (eds) New insights into synaptic function. Wiley and Sons, Chichester, pp 549–601

Changeux J-P, Thiery J (1967) On the excitability and cooperativity of the electroplax membrane. In: Jarnefelt J (ed) Regulatory functions of biological membranes. Elsevier, Amsterdam, pp 116–138 (BBA library, vol II)

Changeux J-P, Benedetti EL, Bourgeois J-P, Brisson A, Cartaud J, Devaux P, Grünhagen H, Moreau M, Popot J-P, Sobel A, Weber M (1975) Some structural properties of the cholinergic receptor protein in its membrane environment relevant to its function as a pharmacological receptor. Cold Spring Harbor Symp Quant Biol 40:211–230

Changeux J-P, Devillers-Thiery A, Chemouilli P (1984) Acetylcholine receptor: an allosteric protein. Science 225:1335–1345

Changeux J-P, Klarsfeld A, Heidmann T (1987) The acetylcholine receptor and molecular models for short- and long-term learning. In: Changeux J-P, Konishi M (eds) The neural and molecular basis of learning. Wiley and Sons, Chichester, pp 31–84

Chemouilli P, Heidmann T, Changeux J-P, Bachy A, Morre M (1985) Allosteric effects of diprobutine on acetylcholine receptors. Eur J Pharmacol 117:205–214

Chibber BA, Martin BM, Walkinshaw MD, Saenger W, Maelicke A (1983) The sites of neurotoxicity in alpha-cobratoxin. In: Toxins as tools in neurochemistry. pp 141–150

Clarke PBS (1986) Radioligand labelling of nicotinic receptors in mammalian brain. In: Maelicke A (ed) Structure and function of the nicotinic acetylcholine receptor. Springer, Berlin Heidelberg New York, pp 345–357 (NATO ASI series, series H, vol 3)

Clarke PBS, Schwartz RD, Paul SM, Pert CB, Pert A (1985) Nicotinic binding in rat brain: autoradiographic comparison of ^{3}H-acetylcholine, ^{3}H-nicotine and ^{125}I-alpha-bungarotoxin. J Neurosci 5:1307–1315

Claudio T (1986a) Recombinant DNA technology in the study of ion channels. Trends Pharmacol Sci 8:308–312

Claudio T (1986b) Establishing a system for the stable expression of *Torpedo* acetylcholine receptors. In: Maelicke A (ed) Structure and function of the nicotinic acetylcholine receptor. Springer, Berlin Heidelberg New York, pp 431–436 (NATO ASI series, series H, vol 3)

Claudio T (1987) Stable expression of *Torpedo* acetylcholine α-subunits in mouse fibroblast L cells by transfection. Proc Natl Acad Sci USA 84:5967–5971

Claudio T, Ballivet M, Patrick J, Heinemann S (1983) Nucleotide and deduced amino acid sequences of *Torpedo californica* acetylcholine receptor gamma subunit. Proc Natl Acad Sci USA 80:1111–1115

Claudio T, Paulson HL, Hartman D, Sine S, Sigworth FJ (1988) Establishing a stable expression system for studies of acetylcholine receptors. In: Molecular biology of channels. (Current topics in membranes and transport, vol 29, in press)

Cohen JB, Boyd ND, Shera NS (1980) Interaction of anesthetics with nicotinic postsynaptic membranes isolated from *Torpedo* electric tissue. In: Molecular mechanisms of anesthesia, vol 2. Raven, New York, pp 165–174

Colquhoun D (1979) The link between drug binding and response: theories and observations. In: O'Brien RD (ed) The receptors: a comprehensive treatise. Plenum, New York, pp 93–142

Colquhoun D (1981) The kinetics of conductance changes at nicotinic receptors of the muscle end-plate and of ganglia. In: Birdsall NJM (ed) Drug receptors and their effectors. MacMillan, London, pp 107–127

Colquhoun D (1986) On the principles of postsynaptic action of neuromuscular blocking agents. In: Kharkevich DA (ed) Handbook of experimental pharmacology. Springer, Berlin Heidelberg New York, pp 59–113

Colquhoun D, Ogden DC (1986) States of the nicotinic acetylcholine receptor: enumeration, characteristics and structure. In: Maelicke A (ed) Structure and function of the nicotinic acetylcholine receptor, vol 79. Springer, Berlin Heidelberg New York, pp 197–218 (NATO ASI series, series H, vol 3)

Colquhoun D, Sakmann B (1981) Fluctuations in the microsecond time range of the current through single acetylcholine receptor ion channels. Nature 294:464–466

Colquhoun D, Sakmann B (1985) Fast events in single-channel currents activated by acetylcholine and its analogues at the frog muscle end-plate. J Physiol (Lond) 369:501–557

Conti-Tronconi BM, Raftery MA (1982) The nicotinic cholinergic receptor: correlation of molecular structure with functional properties. Annu Rev Biochem 51:491–530

Conti-Tronconi BM, Raftery MA (1986) Nicotinic acetylcholine receptor contains multiple binding sites: evidence from binding of alpha-dendrotoxin. Proc Natl Acad Sci USA 83:6646–6650

Conti-Tronconi BM, Dunn SM, Raftery MA (1982) Independent sites of low and high affinity for agonists on *Torpedo californica* acetylcholine receptor. Biochem Biophys Res Commun 107:123–129

Conti-Tronconi BM, Hunkapiller MW, Raftery MA (1984) Molecular weight and structural nonequivalence of the mature alpha subunits of *Torpedo californica* acetylcholine receptor. Proc Natl Acad Sci USA 81:2631–2634

Conti-Tronconi BM, Dunn SM, Barnard EA, Dolly JO, Lai FA, Ray N, Raftery MA (1985) Brain and muscle nicotinic acetylcholine receptors are different but homologous proteins. Proc Natl Acad Sci USA 82:5208–5212

Costa E, Guidotti A, Hanbauer I, Saiani L (1983) Modulation of nicotinic receptor function by opiate recognition sites highly selective for Met5-enkephalin[Arg6Phe7]. Fed Proc 42:2946–2952

Covarrubias M, Prinz H, Maelicke A (1984) Ligand-specific state transitions of the membrane-bound acetylcholine receptor. FEBS Lett 169:229–233

Covarrubias M, Prinz H, Meyers H-W, Maelicke A (1986) Equilibrium binding of cholinergic ligands to the membrane-bound acetylcholine receptor. J Biol Chem 261:14955–14961

Covarrubias M, Prinz H, Maelicke A (1988) Interaction of local anaesthetics with the membrane-bound acetylcholine receptor from *Torpedo marmorata*. Mol Pharmacol (in press)

Cox RN, Kaldany RR, Dipaola M, Karlin A (1985) Time-resolved photolabeling by quinacrine azide of a noncompetitive inhibitor site of the nicotinic acetylcholine receptor in a transient, agonist-induced state. J Biol Chem 260:7186–7193

Criado M, Hochschwender S, Sarin V, Fox JL, Lindstrom J (1985) Evidence for unpredicted transmembrane domains in acetylcholine receptor subunits. Proc Natl Acad Sci USA 82:2004–2008

Criado M, Sarin V, Fox JL, Lindstrom J (1986) Evidence that the acetylcholine binding site is not formed by the sequence alpha 127–143 of the acetylcholine receptor. Biochemistry 25:2839–2846

Culver P, Fenical W, Taylor P (1984) Lophotoxin irreversibly inactivates the nicotinic acetylcholine receptor by preferential association at one of the two primary agonist sites. J Biol Chem 259:3763–3770

Damle VN, Karlin A (1978) Affinity labeling of one of two alpha-neurotoxin binding sites in acetylcholine receptor from *Torpedo californica*. Biochemistry 17:2039–2045

Damle VN, McLaughlin M, Karlin A (1978) Bromoacetylcholine as an affinity label of the acetylcholine receptor from *Torpedo californica*. Biochem Biophys Res Commun 84:845–851

Dennis M, Giraudat J, Kotzyba-Hibert F, Goeldner M, Hirth C, Chang JY, Changeux J-P (1986) A photoaffinity ligand of the acetylcholine binding site predominantly labels the region 179–207 of the alpha subunit on native acetylcholine receptor from *Torpedo marmorata*. FEBS Lett 207:243–247

Devillers-Thiery A, Changeux J-P, Parotaud P, Strosberg AD (1979) The amino terminal sequence of the 40 k subunit of the acetylcholine receptor protein from *Torpedo marmorata*. FEBS Lett 104:99–105

Devillers-Thiery A, Giraudat J, Bentaboulet M, Changeux J-P (1983) Complete mRNA coding sequence of the acetylcholine binding alpha-subunit of *Torpedo marmorata* acetylcholine receptor: a model for the transmembrane organization of the polypeptide chain. Proc Natl Acad Sci USA 80:2067–2071

Dionne VE, Steinbach JH, Stevens CF (1978) An analysis of the dose-response relationship of voltage-clamped frog neuromuscular junctions. J Physiol (Lond) 281:421–444

Doster W, Hess B, Watters D, Maelicke A (1980) Translational diffusion coefficient and molecular weight of the acetylcholine receptor from *Torpedo marmorata*. FEBS Lett 113:312–314

Dreyer F, Peper K, Sterz R (1978) Determination of dose-response curves by quantitative ionophoresis at the frog neuromuscular junction. J Physiol (Lond) 281:395–419

Dunn SM, Raftery MA (1982a) Multiple binding sites for agonists on *Torpedo californica* acetylcholine receptor. Biochemistry 21:6264–6272

Dunn SM, Raftery MA (1982b) Activation and desensitization of *Torpedo* acetylcholine receptor: evidence for separate binding sites. Proc Natl Acad Sci USA 79:6757–6761

Dunn SM, Blanchard SG, Raftery MA (1980) Kinetics of carbamoylcholine binding to membrane-bound acetylcholine receptor monitored by fluorescence changes of a covalently bound probe. Biochemistry 19:5645–5652

Dunn SM, Blanchard SG, Raftery MA (1981) Effects of local anesthetics and histrionicotoxin on the binding of carbamoylcholine to membrane-bound acetylcholine receptor. Biochemistry 20:5617–5624

Dunn SM, Conti-Tronconi BM, Raftery MA (1983) Separate sites of low and high affinity for agonists on *Torpedo californica* acetylcholine receptor. Biochemistry 22:2512–2518

Dwyer TM, Adams D, Hille B (1980) The permeability of the endplate-channel to organic cations in frog muscle. J Gen Physiol 75:469–492

Earnest J-P, Wang HH, McNamee MG (1984) Multiple binding sites for local anesthetics on reconstituted acetylcholine receptor membranes. Biochem Biophys Res Commun 123:862–868

Engel AG (1984) Myasthenia gravis and myasthenic syndromes. Ann Neurol 16:519–534

Eriksson H, Salmonsson R, Liljeqvist G, Heilbronn E (1986) Studies on a cAMP-dependent protein kinase obtained from nicotinic receptor-bearing microsacs. In: Hamin I (ed) Dynamics of cholinergic function. Plenum, New York, pp 933–937

Eriksson H, Salmonsson R, Liljeqvist G, Heilbronn E (1987) Pitfalls in the assay of cyclic AMP-dependent protein kinase activity in microsacs from *Torpedo marmorata*. J Neurochem 47:1127–1131

Fahr A, Lauffer L, Schmidt D, Heyn MP, Hucho F (1985) Covalent labeling of functional states of the acetylcholine receptor. Effects of antagonists on the receptor conformation. Eur J Biochem 147:483–487

Fairclough RH, Finer-Moore J, Love RA, Kristofferson D, Desmeules PJ, Stroud RM (1983) Subunit organization and structure of an acetylcholine receptor. Cold Spring Harbor Symp Quant Biol 48:9–20

Fambrough DM (1979) Control of acetylcholine receptors in skeletal muscle. Physiol Rev 59:165–227

Fambrough DM, Devreotes PN (1978) Newly synthesized acetylcholine receptors are located in the golgi apparatus. J Cell Biol 76:237–244

Fels G, Maelicke A (1988) Allosteric competition of cholinergic agonists and monoclonal antibodies for binding to the acetylcholine receptor from *Torpedo* electric organ. Biochemistry (in press)

Fels G, Wolff EK, Maelicke A (1982) Equilibrium binding of acetylcholine to the membrane-bound acetylcholine receptor. Eur J Biochem 127:31–38

Fels G, Breer H, Maelicke A (1983) Are there nicotinic acetylcholine receptors in invertebrate ganglionic tissue? In: Hucho F, Ovchinnikov YuA (eds) Toxins as tools in neurochemistry. Walter de Gruyter, Berlin, pp 127–140

Fels G, Maelicke A, Schäfer DS, Walrond J-P, Reese TS (1985) A monoclonal antibody specific for one of the two agonist binding sites at the acetylcholine receptor. In:

Changeux J-P, Hucho F, Maelicke A, Neumann E (eds) Molecular basis of nerve activity. de Gruyter, Berlin, pp 197–208
Fels G, Plümer-Wilk R, Schreiber M, Maelicke A (1986) A monoclonal antibody interfering with binding and response of the acetylcholine receptor. J Biol Chem 261:15746–15754
Feltz A, Trautmann A (1982) Desensitization at the frog neuromuscular junction: a biphasic process. J Physiol (Lond) 322:257–272
Feng DF, Johnson MS, Doolittle RF (1985) Aligning amino acid sequences: comparison of commonly used methods. J Mol Evol 21:112–125
Filbin MT, Lunt GG, Donnellan JF (1983) Partial purification and characterization of an acetylcholine receptor with nicotinic properties from the supra-oenophageal ganglion of the locust (*Schistocerca gregaria*). Eur J Biochem 132:151–156
Finer-Moore J, Stroud RM (1984) Amphipathic analysis and possible formation of the ion channel in an acetylcholine receptor. Proc Natl Acad Sci USA 81:155–159
Florey E (1963) Acetylcholine in invertebrate nervous systems. Can J Biochem Physiol 41:2619–2626
Fong TM, McNamee MG (1986) Correlation between acetylcholine receptor function and structural properties of membranes. Biochemistry 25:830–840
Fredkin DR, Montal M, Rice JA (1985) Identification of aggregated markovian models: application to the nicotinic acetylcholine receptor. In: Le Cam LM, Olshen RA (eds) Proceedings of the Berkeley conferences, vol 1. Wadsworth, Belmont, pp 269–289
Freeman JA (1977) Possible regulatory function of acetylcholine receptor in maintenance of retinotectal synapses. Nature 269:218–222
Freeman JA, Norden JJ (1984) In: Vanegas H (ed) Comparative neurology of the optic tectum. Plenum, New York, pp 469–546
Froehner SC, Karlin A, Hall ZW (1977a) Affinity alkylation labels two subunits of the reduced acetylcholine receptor from mammalian muscle. Proc Natl Acad Sci USA 74:4685–4688
Froehner SC, Reiness CG, Hall ZW (1977b) Subunit structure of the acetylcholine receptor from denervated rat skeletal muscle. J Biol Chem 252:8589–8596
Fuchs S, Neumann D, Safran A (1986) What can we learn about the acetylcholine receptor from synthetic peptides? In: Maelicke A (ed) Structure and function of the nicotinic acetylcholine receptor. Springer, Berlin Heidelberg New York, pp 49–59 (NATO ASI series, series H, vol 3)
Fujita N, Nelson N, Fox TD, Claudio T, Lindstrom J, Riezman H, Hess GP (1986) Biosynthesis of the *Torpedo californica* acetylcholine receptor alpha subunit in yeast. Science 231:1284–1287
Gage PW, Wachtel R (1984) Some effects of procaine at the toad endplate are not consistent with a simple channel blocking model. J Physiol (Lond) 346:331–339
Gage PW, McBurney RN, Schneider GT (1975) Effects of some aliphatic alcohols on the conductance change caused by a quantum of acetylcholine at the toad end-plate. J Physiol (Lond) 244:409–429
Gage PW, Hamill OP, Wachtel RE (1983) Sites of action of procaine at the motor endplate. J Physiol (Lond) 335:123–137
Gardner P, Ogden DC, Colquhoun D (1984) Conductances of single ion channels opened by nicotinic agonists are indistinguishable. Nature 309:160–162
Gershoni JM, Hawrot E, Lentz TL (1983) Binding of alpha-bungarotoxin to isolated alpha subunit of the acetylcholine receptor of *Torpedo californica*: quantitative analysis with protein blots. Proc Natl Acad Sci USA 80:4973–4977
Giersig M, Kunath W, Dack-Kongehl H, Hucho F (1986) Ultrastructural analysis of the native acetylcholine receptor. In: Maelicke A (ed) Structure and function of the nicotinic acetylcholine receptor. Springer, Berlin Heidelberg New York, pp 7–17 (NATO ASI series, series H, vol 3)
Gilmore R, Walter P, Blobel G (1982) Protein translocation across the endoplasmic reticulum. II. Isolation and characterization of the signal recognition particle receptor. J Biol Chem 95:470–477
Giraudat J, Montecucco C, Bisson R, Changeux J-P (1985) Transmembrane topology of acetylcholine receptor subunits probed with photoreactive phospholipids. Biochemistry 24:3121–3127

Giraudat J, Dennis M, Heidmann T, Changeux J-P, Bisson R, Montecucco C, Kotzyba-Hibert F, Goeldner M, Hirth C, Chang J-Y (1986a) Tertiary structure of the nicotinic acetylcholine receptor probed by photolabeling and protein chemistry. In: Maelicke A (ed) Structure and function of the nicotinic acetylcholine receptor. Springer, Berlin Heidelberg New York, pp 103–114 (NATO ASI series, series H, vol 3)

Giraudat J, Dennis M, Heidmann T, Chang J-Y, Changeux J-P (1986b) Structure of the high-affinity binding site for noncompetitive blockers of the acetylcholine receptor: serine-262 of the δ subunit is labeled by [^{3}H]chlorpromazine. Proc Natl Acad Sci USA 83:2719–2723

Giraudat J, Dennis M, Heidmann T, Haumont P-Y, Lederer F, Changeux J-P (1987) Structure of the high-affinity binding site for noncompetitive blockers of the acetylcholine receptor: [^{3}H]chlorpromazine labels homologous residues in the β and δ chains. Biochemistry 26:2410–2418

Goldman D, Boulter J, Heinemann S, Patrick J (1985) Muscle denervation increases the levels of two mRNAs coding for the acetylcholine receptor alpha-subunit. J Neurosci 5:2553–2558

Goldman D, Simmons D, Swanson LW, Patrick J, Heinemann S (1986) Mapping of brain areas expressing RNA homologous to two different acetylcholine receptor alpha-subunit cDNAs. Proc Natl Acad Sci USA 83:4076–4080

Gordon AS, Davis CG, Milfay D, Diamond I (1977a) Phosphorylation of acetylcholine receptor by endogenous membrane protein kinase in receptor-enriched membranes of *Torpedo californica*. Nature 267:539–540

Gordon AS, Davis CG, Diamond I (1977b) Phosphorylation of membrane proteins at a cholinergic synapse. Proc Natl Acad Sci USA 74:263–267

Gotti C, Casadei G, Clementi F (1983) Study of the subunit structure of rabbit nicotinic acetylcholine receptor. Neurosci Lett 35(2):143–148

Grant GA, Chiappinelli VA (1985) Kappa-bungarotoxin: complete amino acid sequence of a neuronal nicotinic receptor probe. Biochemistry 24:1532–1537

Gration KAF, Lambert JJ, Ramsey R, Usherwood PNR (1981) Non-random openings and concentration-dependent lifetimes of glutamate-gated channels in muscle membrane. Nature 291:423–425

Gundelfinger ED, Hermans-Borgmeyer I, Zopf D, Sawruk E, Betz H (1986) Characterization of the mRNA and the gene of a putative neuronal nicotinic acetylcholine receptor protein from Drosophila. In: Maelicke A (ed) Structure and function of the nicotinic acetylcholine receptor. Springer, Berlin Heidelberg New York, pp 437–446 (NATO ASI series, series H, vol 3)

Guy HR (1984) A structural model of the acetylcholine receptor channel based on partition energy and helix packing calculations. Biophys J 45:249–261

Häggblad J, Eriksson H, Hedlund B, Heilbronn E (1987) Forskolin blocks carbachol-mediated ion-permeability of chick myotube nicotine receptors and inhibits of ^{3}H-phencyclidine to *Torpedo* microsac nicotinic receptors. Naunyn-Schmiedebergs. Arch Pharmacol (in press)

Haggerty JG, Froehner SC (1981) Restoration of ^{125}I-alpha-bungarotoxin binding activity to the alpha subunit of *Torpedo* acetylcholine receptor isolated by gel electrophoresis in sodium dodecyl sulfate. J Biol Chem 256:8294–8297

Hamill OP, Sakmann B (1981) Multiple conductance states of single acetylcholine receptor channels in embryonic muscle cells. Nature 294:462–464

Hamilton SL, McLaughlin M, Karlin A (1979) Formation of disulfide-linked oligomers of acetylcholine receptor in membrane from *Torpedo* electric tissue. Biochemistry 18:155–163

Hamilton SL, Pratt DR, Eaton DC (1985) Arrangement of the subunits of the nicotinic acetylcholine receptor of *Torpedo californica* as determined by alpha-neurotoxin crosslinking. Biochemistry 24:2210–2219

Hanke W, Breer H (1986) Channel properties of an insect neuronal acetylcholine receptor protein reconstituted in planar lipid bilayers. Nature 321:171–174

Heidmann T, Changeux J-P (1979) Fast kinetic studies on the interaction of a fluorescent agonist with the membrane-bound acetylcholine receptor from *Torpedo marmorata*. Eur J Biochem 94(1):255–279

Heidmann T, Changeux J-P (1980) Interaction of a fluorescent agonist with the membrane-bound acetylcholine receptor from *Torpedo marmorata* in the millisecond time range: resolution of an 'intermediate' conformational transition. Biochem Biophys Res Commun 97:889–896

Heidmann T, Changeux J-P (1984) Time-resolved photolabeling by the noncompetitive blocker chlorpromazine of the acetylcholine receptor in its transiently open and closed ion channel conformations. Proc Natl Acad Sci USA 81:1897–1901

Heidmann T, Sobel A, Popot J-L, Changeux J-P (1980) Reconstitution of a functional acetylcholine receptor: conservation of the conformational and allosteric transitions and recovery of the permeability response; role of lipids. Eur J Biochem 110:35–55

Heidmann T, Bernhardt J, Neumann E, Changeux J-P (1983a) Rapid kinetics of agonist binding and permeability response analyzed in parallel on acetylcholine receptor rich membranes from *Torpedo marmorata*. Biochemistry 22:5452–5459

Heidmann T, Oswald RE, Changeux J-P (1983b) Multiple sites of action for noncompetitive blockers on acetylcholine receptor rich membrane fragments from *Torpedo marmorata*. Biochemistry 22:3112–3127

Heilbronn E (1985) Diseases and induced lesions of the neuromuscular endplate. In: Lajtha A (ed) Handbook of neurochemistry, vol 10. Plenum, New York, pp 241–284

Heilbronn E, Eriksson H, Salmonsson R (1985) Regulation of the nicotinic acetylcholine receptor by phosphorylation. In: Changeux J-P, Hucho F, Maelicke A, Neumann E (eds) Molecular basis of nerve activity. de Gruyter, Berlin, pp 237–250

Heilbronn E, Eriksson H, Häggblad J (1986) cAMP-dependent phosphorylation of the nicotinic acetylcholine receptor: characterization of the protein kinase in *Torpedo* electric organ, lack of correlation in myotubes between increased intracellular levels of cAMP and influx of monovalent ions, block of influx by forskolin independently of cAMP levels. In: Maelicke A (ed) Structure and function of the nicotinic acetylcholine receptor. Springer, Berlin Heidelberg New York, pp 291–303 (NATO ASI series, series H, vol 3)

Heinemann S, Boulter J, Connolly J, Goldman D, Evans K, Treco D, Ballivet M, Patrick J (1986) Molecular biology of muscle and neural acetylcholine receptors. In: Maelicke A (ed) Structure and function of the nicotinic acetylcholine receptor. Springer, Berlin Heidelberg New York, pp 359–387 (NATO ASI series, series H, vol 3)

Henderson F, Prior C, Dempster J, Marshall IG (1986) The effects of chloramphenicol isomers on the motor end-plate nicotinic receptor-ion channel complex. Mol Pharmacol 29:52–64

Hermans-Borgmeyer I, Zopf D, Ryseck R-P, Hovemann B, Betz H, Gundelfinger ED (1986) Primary structure of a developmentally regulated nicotinic acetylcholine receptor protein from *Drosophila*. EMBO J 5:1503–1508

Hess GP, Cash DJ, Aoshima H (1979) Acetylcholine receptor-controlled ion fluxes in membrane vesicles investigated by fast reaction techniques. Nature 282:329–331

Hess GP, Cash DJ, Aoshima H (1983) Acetylcholine receptor-controlled ion translocation: chemical kinetic investigations of the mechanism. Annu Rev Biophys Bioeng 12:443–473

Hess GP, Udgaonkar JB, Ulbricht WL (1987) Chemical kinetic measurements of transmembrane processes using rapid reaction techniques – acetylcholine receptor. Annu Rev Biophys Chem 16:507–534

Heuser JE, Salpeter SR (1979) Organization of acetylcholine receptors in quick-frozen deep-etched and rotary replicated *Torpedo* postsynaptic membrane. J Cell Biol 82:150–173

Holtzman E, Wise D, Wall J, Karlin A (1982) Electron microscopy of complexes of isolated acetylcholine receptor, biotinyl-toxin, and avidin. Proc Natl Acad Sci USA 79:310–314

Huang LM, Catterall WA, Ehrenstein G (1978) Selectivity of cations and non-electrolytes for acetylcholine-activated channels in cultured muscle cells. J Gen Physiol 71:397–410

Hucho F (1986) The nicotinic acetylcholine receptor and its ion channel. Eur J Biochem 158:211–226

Hucho F, Changeux J-P (1973) Molecular weight and quaternary structure of the cholinergic receptor protein extracted by detergents from *Electrophorus electricus* electric tissue. FEBS Lett 38(1):11–15
Hucho F, Bandini G, Suárez-Isla A (1978) The acetylcholine receptor as part of a protein complex in receptor-enriched membrane fragments from *Torpedo californica* electric tissue. Eur J Biochem 83:335–340
Hucho F, Oberthür W, Lottspeich F (1986a) The ion channel of the nicotinic acetylcholine receptor is formed by the homologous helices m ii of the receptor subunits. FEBS Lett 205:137–142
Hucho F, Oberthür W, Lottspeich F, Wittmann-Liebold B (1986b) A structural model of the ion channel of the nicotinic acetylcholine receptor. In: Maelicke A (ed) Structure and function of the nicotinic acetylcholine receptor. Springer, Berlin Heidelberg New York, pp 115–127 (NATO ASI series, series H, vol 3)
Huganir RL, Greengard P (1983) cAMP-dependent protein kinase phosphorylates nicotinic acetylcholine receptor. Proc Natl Acad Sci USA 80:1130–1134
Huganir RL, Miles K, Greengard P (1984) Phosphorylation of the nicotinic acetylcholine receptor by an endogenous tyrosine-specific protein kinase. Proc Natl Acad Sci USA 81:6968–6972
Huganir RL, Delcour AH, Greengard P, Hess GP (1986) Phosphorylation of the nicotinic acetylcholine receptor regulates its rate of desensitization. Nature 321:774–776
Imoto K, Methfessel C, Sakmann B, Mishina M, Mori Y, Konno T, Fukuda K, Kurasaki M, Bujo H, Fujita Y, Numa S (1986) Location of a delta-subunit region determining ion transport through the acetylcholine receptor channel. Nature 324:670–674
Isenberg KE, Mudd K, Shak V, Merlie JB (1986) Nucleotide sequence of the mouse muscle nicotinic acetylcholine receptor alpha subunit. Nucleic Acids Res 14:5111–5117
Jackson MB (1984) Spontaneous openings of the acetylcholine receptor channel. Proc Natl Acad Sci USA 81:3901–3904
Jackson MB (1986) Kinetics of unliganded acetylcholine receptor channel gating. Biophys J 49:663–672
Jacob MH, Berg DK (1983) The ultrastructural localization of alpha-bungarotoxin binding sites in relation to synapses on chick ciliary ganglion neurons. J Neurosci 3:260–271
Jürss R, Prinz H, Maelicke A (1979) NBD-5-acylcholine: fluorescent analog of acetylcholine and agonist at the neuromuscular junction. Proc Natl Acad Sci USA 76:1064–1068
Juillerat MA, Barkas T, Tzartos SJ (1984) Antigenic sites of the nicotinic acetylcholine receptor cannot be predicted from the hydrophilicity profile. FEBS Lett 168:143–148
Kang S, Maelicke A (1980) Fluorescein isothiocyanate-labeled alpha-cobratoxin. Biochemical characterization and interaction with acetylcholine receptor from *Electrophorus electricus*. J Biol Chem 255:7326–7332
Kao PN, Karlin A (1986) Acetylcholine receptor binding site contains a disulfide crosslink between adjacent half-cystinyl residues. J Biol Chem 261:8085–8088
Kao PN, Dwork AJ, Kaldany RR, Silver ML, Wideman J, Stein S, Karlin A (1984) Identification of the alpha subunit half-cystine specifically labeled by an affinity reagent for the acetylcholine receptor binding site. J Biol Chem 259:11662–11665
Karlin A (1967) On the application of a 'plausible model' of allosteric proteins to the receptor for acetylcholine. J Theor Biol 16:306–318
Karlin A (1969) Chemical modification of the active site of the acetylcholine receptor. J Gen Physiol 54:245s–265s
Karlin A, Holtzman E, Yodh N, Lobel P, Wall J, Hainfeld J (1983) The arrangement of the subunits of the acetylcholine receptor of *Torpedo californica*. J Biol Chem 258:6678–6681
Karlin A, DiPaola M, Kao PN, Lobel P (1987) Functional sites and transient states of the nicotinic acetylcholine receptor. In: Hille B, Fambrough DM (eds) Proteins of excitable membranes. Wiley and Sons, Chichester (in press)
Karpen JW, Hess GP (1986) Acetylcholine receptor inhibition by d-tubocorarine involves both a competitive and a noncompetitive binding site as determined by stopped-flow measurements of receptor-controlled ion flux in membrane vesicles. Biochemistry 25:1786–1792

Karpen JW, Aoshima H, Abood LG, Hess GP (1982) Cocaine and phencyclidine inhibition of the acetylcholine receptor: analysis of the mechanisms of action based on measurements of ion flux in the millisecond-to-minute time region. Proc Natl Acad Sci USA 79:2509–2513

Karpen JW, Sachs AB, Cash DJ, Pasquale EB, Hess GP (1983) Direct spectrophotometric detection of cation flux in membrane vesicles: stopped-flow measurements of acetylcholine-receptor-mediated ion flux. Anal Biochem 135:83–94

Kato G, Changeux J-P (1976) Studies on the effect of histrionicotoxin on the monocellular electroplax from *Electrophorus electricus* and on the binding of (^{3}H)acetylcholine to membrane fragments from *Torpedo marmorata*. Mol Pharmacol 12(1):92–100

Katz B (1966) Nerve, muscle and synapse. McGraw-Hill, New York

Katz B, Thesleff S (1957) A study of the desensitization produced by acetylcholine at the motor end-plate. J Physiol (Lond) 138:63–80

Kistler J, Stroud RM (1981) Crystalline arrays of membrane-bound acetylcholine receptor. Proc Natl Acad Sci USA 78:3678–3682

Kistler J, Stroud RM, Klymkowsky MW, Lalancette RA, Fairclough RH (1982) Structure and function of an acetylcholine receptor. Biophys J 37:371–383

Klarsfeld A, Changeux J-P (1985) Activity regulates the levels of acetylcholine receptor alpha-subunit mRNA in cultured chicken myotubes. Proc Natl Acad Sci USA 82:4558–4562

Klarsfeld A, Daubas P, Bourrachot B, Changeux J-P (1987) A 5′-flanking region of the chicken acetylcholine receptor α-subunit gene confers tissue-specificity and developmental control of expression transfected cells. Mol Cell Biol 7:951–955

Klymkowsky MW, Stroud RM (1979) Immunospecific identification three-dimensional structure of a membrane-bound acetylcholine receptor *Torpedo californica*. J Mol Biol 128:319–334

Koblin DD, Lester HA (1979) Voltage-dependent and voltage-independent blockade of acetylcholine receptors by local anesthetics in *Electrophorus* electroplaques. Mol Pharmacol 15:559–580

Kordossi AA, Tzartos SJ (1987) Conformation of cytoplasmic segments of acetylcholine receptor α- and β-subunits probed by monoclonal antibodies. EMBO J 6:1605–1610

Koskinen AM, Rapoport H (1985) Synthetic and conformational studies on anatoxin-a: a potent acetylcholine agonist. J Med Chem 28:1301–1309

Kosower EM (1983) A molecular model for the exobilayer portion of the alpha-subunit of the acetylcholine receptor with binding sites for acetylcholine and non-competitive antagonists. Biochem Biophys Res Commun 116:17–22

Kosower EM (1986) A structural and dynamic model for the nicotinic acetylcholine receptor. In: Maelicke A (ed) Structure and function of the nicotinic acetylcholine receptor. Springer, Berlin Heidelberg New York, pp 465–483 (NATO ASI series, series H, vol 3)

Kotzyba-Hibert F, Lagenbuch-Cachat J, Jaganathen J, Goeldner M, Hirth C (1985) Aryldiazonium salts as photoaffinity labels of the nicotinic acetylcholine receptor PCP binding site. FEBS Lett 182:297–301

Krnjević K (1975) Acetylcholine receptors in vertebrate central nervous system. In: Iversen LL, Iversen SD, Snyder SH (eds) Handbook of psychopharmacology, vol 6. Plenum, New York, pp 97–126

Krodel EK, Beckman RA, Cohen JB (1979) Identification of a local anesthetic binding site in nicotinic post-synaptic membranes isolated from *Torpedo marmorata* electric tissue. Mol Pharmacol 15:294–312

Kubalek E, Ralston S, Lindstrom J, Unwin N (1987) Location of subunits within the acetylcholine receptor by electron image analysis of tubular crystals from *Torpedo marmorata*. J Cell Biol 105:9–18

Kubo T, Noda M, Takai T, Tanabe T, Kayano T, Shimizu S, Tanaka K, Takahashi H, Hirose T, Inayama S, Kikuno R, Miyata T, Numa S (1985) Primary structure of delta subunit precursor of calf muscle acetylcholine receptor deduced from cDNA sequence. Eur J Biochem 149:5–13

Kubo T, Fukuda K, Mikami A, Maeda A, Takahashi H, Mishina M, Haga T, Haga K, Ichiyama A, Kangawa K, Kojima M, Matsuo H, Hirose T, Numa S (1986) Cloning,

sequencing and expression of complementary DNA encoding the muscarinic acetylcholine receptor. Nature 323:411–416
Labarca P, Lindstrom J, Montal M (1984a) Acetylcholine receptor in planar lipid bilayers: characterization of the channel properties of the purified nicotinic acetylcholine receptor from *Torpedo californica* reconstituted in planar lipid. J Gen Physiol 83:473–496
Labarca P, Lindstrom J, Montal M (1984b) The acetylcholine receptor channel from *Torpedo californica* has two open states. J Neurosci 4:502–507
Läuger P (1985a) Ionic channels with conformational substates. Biophys J 47:581–591
Läuger P (1985b) Structural fluctuations and current noise of ionic channels. Biophys J 48:369–373
Lambert JJ, Volle RL, Henderson EG (1980) An attempt to distinguish between the actions of neuromuscular blocking drugs on the acetylcholine receptor and on its associated ionic channel. Proc Natl Acad Sci USA 77:5003–5007
Lapolla RJ, Mayne KM, Davidson N (1984) Isolation and characterization of a cdna clone for the complete protein coding region of the delta subunit of the mouse acetylcholine receptor. Proc Natl Acad Sci USA 81:7970–7974
Leprince P (1983) Chemical modification of the nicotinic cholinergic receptor of PC-12 nerve cell. Biochemistry 22:5551–5556
Lester HA, Krouse ME, Nass MM, Wassermann NH, Erlanger BF (1980a) A covalently bound photoisomerizable agonist: comparison with reversibly bound agonists at *Electrophorus* electroplaques. J Gen Physiol 75:207–232
Lester HA, Nass MM, Krouse ME, Nerbonne JM, Wassermann NH, Erlanger BF (1980b) Electrophysiological experiments with photoisomerizable cholinergic compounds: review and progress report. Ann NY Acad Sci 346:475–490
Lewis CA (1979) Ion-concentration dependence of the reversal potential and single channel conductance of ion channels at the frog neuromuscular junctions. J Physiol (Lond) 286:417–445
Lewis CA, Stevens CF (1983) Acetylcholine receptor channel ionic selectivity: Ions experience an aqueous environment. Proc Natl Acad Sci USA 80:6110–6113
Lindstrom J (1984) Nicotinic acetylcholine receptors: Use of monoclonal antibodies to study synthesis, structure, function, and autoimmune response. In: Monoclonal and anti-idiotypic antibodies: probes for receptor structure and function. Liss, New York, pp 21–57
Lindstrom J (1985) Immunobiology of myasthenia gravis experimental autoimmune myasthenia gravis, and Lambert-Eaton syndrome. Annu Rev Immunol 3:109–131
Lindstrom J, Cooper J, Tzartos S (1979) Acetylcholine receptor from *Torpedo* and *Electrophorus* have similar subunit structures. Biochemistry 19:1454–1458
Lindstrom J, Einarson B, Tzartos S (1981) Production and assay of antibodies to acetylcholine receptors. Methods Enzymol 74:432–460
Lindstrom J, Tzartos S, Gullick W, Hochschwender S, Swanson L, Sargent P, Jacob M, Montal M (1983) Use of monoclonal antibodies to study acetylcholine receptors from electric organs, muscle, and brain and the autoimmune response to receptor in myasthenia gravis. Cold Springer Harbor Symp Quant Biol 48:89–99
Lindstrom J, Criado M, Hochschwender S, Fox JL, Sarin V (1984) Immunochemical tests of acetylcholine receptor subunit models. Nature 311:573–575
Loring RH, Chiappinelli VA, Zigmond RE, Cohen JB (1984) Characterization of a snake venom neurotoxin which blocks nicotinic transmission in the avian ciliary ganglion. Neuroscience 11:989–999
Lukas RJ (1984) Properties of curaremimetic neurotoxin binding sites in the rat central nervous system. Biochemistry 23:1152–1160
Lunt GG (1986) Is the insect neuronal nAChR the ancestral ACh receptor protein? Trends Neurosci 9:341–342
Maelicke A (1984) Biochemical aspects of cholinergic excitation. Angew Chem [Int Edn] 23:195–221
Maelicke A (ed) (1986) Nicotinic acetylcholine receptor: structure and function. Springer, Berlin Heidelberg New York (NATO ASI series, series H, vol 3)
Maelicke A (1987a) Allosteric interactions between distinct structural domains at the nicotinic acetylcholine receptor. Biochem Soc Trans 15:108–112

Maelicke A (1987b) The nicotinic acetylcholine receptor is an allosteric protein: the specific allosteric model does not apply to the receptor's molecular mechanism of action. In: Hucho F, Ovchinnikov YuA (eds) Receptors and channels. Walter de Gruyter, Berlin, pp 51–61

Maelicke A (1987c) The insect neuronal nAChR, a homo-oligomer or not? Trends Neurosci 10:107

Maelicke A, Prinz H (1983) The acetylcholine receptor-ion channel complex: linkage between binding and response. In: Satir BH (ed) Modern cell biology, vol 1. Liss, New York, pp 171–197

Maelicke A, Reich E (1975) On the interaction between cobra alpha-neurotoxin and the acetylcholine receptor. Cold Spring Harb Symp Quant Biol 40:231–235

Maelicke A, Fulpius BW, Reich E (1977a) Biochemical aspects of neurotransmitter receptors. In: Kandel ER (ed) Cellular biology of neurons, part 1. American Physiological Society, Bethesda, pp 493–519 (Handbook of physiology, sect 1, vol 1)

Maelicke A, Fulpius BW, Klett RP, Reich E (1977b) Acetylcholine receptor responses to drug binding. J Biol Chem 252(14):4811–4830

Maelicke A, Fels G, Plümer-Wilk R, Wolff EK, Covarrubias M, Methfessel C (1986a) Specific blockade of acetylcholine receptor function by monoclonal antibodies. In: Ritchie JM, Keynes RD, Bolis L (eds) Ion channels in neural membranes, vol 20. Liss, New York, pp 275–282

Maelicke A, Watters D, Fels G (1986b) Monoclonal antibodies as probes of acetylcholine receptor function. In: Maelicke A (ed) Structure and function of the nicotinic acetylcholine receptor. Springer, Berlin Heidelberg New York, pp 83–91 (NATO ASI series, series H, vol 3)

Maelicke A, Plümer-Wilk R, Fels G, Veltel D, Conti-Tronconi BM (1988) Antibodies raised against synthetic peptides to study the structure of the nicotinic acetylcholine receptor. Biochemistry (in press)

Magazanik LG, Vyskočil F (1970) Dependence of acetylcholine desensitization on the membrane potential of frog muscle fibre and on the ionic changes in the medium. J Physiol (Lond) 210:507–518

Magleby KL, Pallotta BS (1981) A study of desensitization of acetylcholine receptors using nerve-released transmitter in the frog. J Physiol (Lond) 316:225–250

Magleby KL, Stevens CF (1972) A quantitative description of end-plate currents. J Physiol (Lond) 223:151–169

Marshall LM (1981) Synaptic localization of α-bungarotoxin binding which blocks nicotinic transmission at frog sympathetic neurons. Proc Natl Acad Sci USA 78:1948–1952

Martin BM, Chibber BA, Maelicke A (1983) The sites of neurotoxicity in alpha-cobratoxin. J Biol Chem 258:8714–8722

McCarthy MP, Earnest JP, Young EF, Choe S, Stroud RM (1986) The molecular neurobiology of the acetylcholine receptor. Annu Rev Neurosci 9:383–413

McCormick DJ, Atassi MZ (1984) Localization and synthesis of the acetylcholine-binding site in the alpha-chain of the *Torpedo californica* acetylcholine receptor. Biochem J 224:995–1000

McHugh EM, McGee R (1986) Direct anesthetic-like effects of forskolin on the nicotinic acetylcholine receptors of PC12 cells. J Biol Chem 261:3103–3106

McNamee MG, Fong TM, Jones OT, Earnest JP (1986) Lipid-protein interactions and acetylcholine receptor function in reconstituted membranes. In: Maelicke A (ed) Structure and function of the nicotinic acetylcholine receptor. Springer, Berlin Heidelberg New York, pp 147–157 (NATO ASI series, series H, vol 3)

Meeker RB, Michels KM, Libber MT, Hayward JN (1986) Characteristics and distribution of high- and low-affinity alpha-bungarotoxin binding sites in the rat hypothalamus. J Neurosci 6:1866–1875

Mehraban F, Kemshead JT, Dolly JO (1984) Properties of monoclonal antibodies to nicotinic acetylcholine receptor from chick muscle. Eur J Biochem 138:53–61

Merlie JP, Sanes JR (1985) Concentration of acetylcholine receptor mRNA in synaptic regions of adult muscle fibres. Nature 317:66–68

Merlie JP, Sebbane R (1981) Acetylcholine receptor subunits transit a precursor pool before acquiring alpha-bungarotoxin binding activity. J Biol Chem 256:3605–3608
Merlie JP, Smith MM (1986) Synthesis and assembly of acetylcholine receptor, a multisubunit membrane glycoprotein. J Membr Biol 91:1–10
Merlie JP, Hofler JG, Sebbane R (1981) Acetylcholine receptor synthesis from membrane polysomes. J Biol Chem 256:6995–6999
Merlie JP, Sebbane R, Tzartos S, Lindstrom J (1982) Inhibition of glycosylation with tunicamycin blocks assembly of newly synthesized acetylcholine receptor subunits in muscle cells. J Biol Chem 257:2694–2701
Merlie JP, Sebbane R, Gardner S, Lindstrom J (1983) cDNA clone for the alpha subunit of the acetylcholine receptor from the mouse muscle cell line BC_3H-1. Proc Natl Acad Sci USA 80:3845–3849
Merlie JP, Isenberg KE, Russell SD, Sanes JR (1984a) Denervation supersensitivity in skeletal muscle: analysis with a cloned cDNA probe. J Cell Biol 99:332–335
Merlie JP, Sebbane R, Gardner S, Olson E, Lindstrom J (1984b) The regulation of acetylcholine receptor expression in mammalian muscle. Cold Spring Harbor Symp Quant Biol 48:135–146
Meunier JC, Olsen RW, Changeux J-P (1972) Studies on the cholinergic receptor protein from *Electrophorus electricus*. III. Effect of detergent on some hydrodynamic properties of the receptor protein in solution. FEBS Lett 24:63–68
Meunier JC, Olsen R, Sealock R, Changeux JP (1974) Purification and properties of the cholinergic receptor protein from *Electrophorus electricus* electric tissue. Eur J Biochem 45:371–378
Middleton P, Jaramillo F, Schuetze SM (1986) Forskolin increases the rate of acetylcholine receptor desensitization at rat soleus endplates. Proc Natl Acad Sci USA 83:4967–4971
Mielke DL, Kaldany R-R, Karlin A, Wallace BA (1984) Effector-induced changes in the secondary structure of the nicotinic acetylcholine receptor. Biophys J 45:205a
Milligan RA, Brisson A, Unwin PNT (1984) Molecular structure determination of crystalline specimens in frozen aqueous solutions. Ultramicroscopy 13:1–10
Mishina M, Kurosaki T, Tobimatsu T, Morimoto Y, Noda M, Yamamoto T, Terao M, Lindstrom J, Takahashi T, Kuno M, Numa S (1984) Expression of functional acetylcholine receptor from cloned cDNAs. Nature 307:604–608
Mishina M, Tobimatsu T, Imoto K, Tanaka K, Fujita Y, Fukuda K, Kurasaki M, Takahashi H, Morimoto Y, Hirose T, Inayama S, Takahasi T, Kuno M, Numa S (1985) Location of functional regions of acetylcholine receptor alpha-subunit by site-directed mutagenesis. Nature 313:364–369
Mishina M, Takai T, Imoto K, Noda M, Takahashi T, Numa S, Methfessel C, Sakmann B (1986) Molecular distinction between fetal and adult forms of muscle acetylcholine receptor. Nature 321:406–411
Miskin P, Easton TG, Maelicke A, Reich E (1978) Metabolism of acetylcholine receptor in chick embryo muscle cells: effects of RSV and PMA. Cell 15:1287–1300
Mitra, Chandrasekaram R (1984) Conformational flexibilities in malformin a. Biopolymers 23:2513–2524
Momoi MY, Lennon VA (1982) Purification and biochemical characterization of nicotinic acetylcholine receptors of human muscle. J Biol Chem 257:12757–12764
Monod J, Wyman J, Changeux J-P (1965) On the nature of allosteric transitions: a plausible model. J Mol Biol 12:88–118
Moore HP, Raftery MA (1980) Direct spectroscopic studies of cation translocation by *Torpedo* acetylcholine receptor on a time scale of physiological relevance. Proc Natl Acad Sci USA 77:4509–4513
Moore WM, Holliday LA, Puett D, Brady RN (1974) On the conformation of the acetylcholine receptor protein from *Torpedo nobiliana*. FEBS Lett 45:145–149
Morley BJ, Kemp GE (1981) Characterization of a putative nicotinic acetylcholine receptor in mammalian brain. Brain Res Reviews 3:81–104
Muhn R, Hucho F (1983) Covalent labeling of the acetylcholine receptor from *Torpedo* electric tissue with the channel blocker [^{3}H]triphenylmethylphosphonium by ultraviolet irradiation. Biochemistry 22:421–425

Muhn P, Fahr A, Hucho F (1984) Photoaffinity labeling of acetylcholine receptor in millisecond time scale. FEBS Lett 166:146–150
Nachmansohn D (1955) Metabolism and function of the nerve cell. Harvey Lect 49:57–99
Nachmansohn D (1972) Biochemistry as part of my life. Annu Rev Biochem 41:1–28
Nathanson NM, Hall ZW (1979) Subunit structure and peptide mapping of junctional and extrajunctional acetylcholine receptors from rat muscle. Biochemistry 18:3392–3401
Nef P, Mauron A, Stalder R, Alliod C, Ballivet M (1984) Structure linkage, and sequence of the two genes encoding the delta and gamma subunits of the nicotinic acetylcholine receptor. Proc Natl Acad Sci USA 81:7975–7979
Nef P, Oneyser C, Barkas T, Ballivet M (1986) Acetylcholine receptor related genes expressed in the nervous system. In: Maelicke A (ed) Structure and function of the nicotinic acetylcholine receptor. Springer, Berlin Heidelberg New York, pp 417–422 (NATO ASI series, series H, vol 3)
Neher E (1983) The charge carried by single-channel currents of rat cultured muscle cells in the presence of local anaesthetics. J Physiol (Lond) 339:663–678
Neher E, Steinbach H (1978) Local anaesthetics transiently block currents through single acetylcholine-receptor channels. J Physiol (Lond) 277:153–176
Neubig RR, Cohen JB (1979) Equilibrium binding of [^{3}H]tubocurarine and [^{3}H]acetylcholine by *Torpedo* postsynaptic membranes: stoichiometry and ligand interactions. Biochemistry 18:5464–5475
Neubig RR, Cohen JB (1980) Permeability control by cholinergic receptors in *Torpedo* postsynaptic membranes: agonist dose-response relations measured at second and millisecond times. Biochemistry 19:2770–2779
Neubig RR, Boyd ND, Cohen JB (1982) Conformations of *Torpedo* acetylcholine receptor associated with ion transport and desensitization. Biochemistry 21:3460–3467
Neumann D, Fridkin M, Fuchs S (1984) Anti-acetylcholine receptor response achieved by immunization with a synthetic peptide from the receptor sequence. Biochem Biophys Res Commun 121:673–679
Neumann D, Gershoni JM, Fridkin M, Fuchs S (1985) Antibodies to synthetic peptides as probes for the binding site on the alpha-subunit of the acetylcholine receptor. Proc Natl Acad Sci USA 82:3490–3493
Neumann D, Barchan D, Safran A, Gershoni JM, Fuchs S (1986) Mapping of the alpha-bungarotoxin binding site within the alpha-subunit of the acetylcholine receptor. Proc Natl Acad Sci USA 83:3008–3011
Nickel E, Potter LT (1973) Ultrastructure of isolated membranes of *Torpedo* electric tissue. Brain Res 57:508–517
Noda M, Takahashi H, Tanabe T, Toyosato M, Furutani Y, Hirose T, Asai M, Inayama S, Miyata T, Numa S (1982) Primary structure of alpha-subunit precursor of *Torpedo californica* acetylcholine receptor deduced from cDNA sequence. Nature 299:793–797
Noda M, Furutani Y, Takahashi H, Toyosato M, Tanabe T, Shimizu S, Kikyotani S, Kayano T, Hirose T, Inayama S, Numa S (1983a) Cloning and sequence analysis of calf cDNA and human genomic DNA encoding alpha-subunit precursor of muscle acetylcholine receptor. Nature 305:818–823
Noda M, Takahashi H, Tanabe T, Toyosato M, Kikyotani S, Hirose T, Asai M, Takashima H, Inayama S, Miyata T, Numa S (1983b) Primary structures of beta- and delta-subunit precursors of *Torpedo californica* acetylcholine receptor deduced from cDNA sequences. Nature 301:251–255
Noda M, Takahashi H, Tanabe T, Toyosato M, Kikyotani S, Furutani Y, Hirose T, Takashima H, Inayama S, Miyata T, Numa S (1983c) Structural homology of *Torpedo californica* acetylcholine receptor subunits. Nature 302:528–532
Nomoto H, Takahashi N, Nagaki Y, Endo S, Arata Y, Hayashi K (1986) Carbohydrate structures of acetylcholine receptor from *Torpedo californica* and distribution of oligosaccharides among the subunits. Eur J Biochem 157:233–242
Oberthür W, Muhn P, Baumann H, Lottspeich F, Wittmann-Liebold B, Hucho F (1986) The reaction site of a non-competitive antagonist in the δ-subunit of the nicotinic acetylcholine receptor. EMBO J 5(8):1815–1819

Oblas B, Singer RH, Boyd ND (1986) Location of a polypeptide sequence within the alpha-subunit of the acetylcholine receptor containing the cholinergic binding site. Mol Pharmacol 29:649–656

Ogden DC, Colquhoun D (1985) Ion channel block by acetylcholine, carbachol and suberyldicholine at the frog neuromuscular junction. Proc R Soc Lond [Biol] 225(1240):329–355

Ogden DC, Siegelbaum SA, Colquhoun D (1981) Block of acetylcholine-activated ion channels by an uncharged local anaesthetic. Nature 289:596–598

Olson EN, Glaser L, Merlie JP, Sebanne R, Lindstrom J (1983) Regulation of surface expression of acetylcholine receptors in response to serum and cell growth in the BC_3H-1 muscle cell line. J Biol Chem 258:13946–13953

Olson EN, Glaser L, Merlie JP, Lindstrom J (1984a) Expression of acetylcholine receptor alpha-subunit mRNA during differentiation of the BC_3H-1 muscle cell line. J Biol Chem 259:3330–3336

Olson EN, Glaser L, Merlie JP (1984b) Alpha and beta subunits of the nicotinic acetylcholine receptor contain covalently bound lipid. J Biol Chem 259:5364–5367

Oswald RE, Changeux J-P (1981a) Selective labeling of the delta subunit of the acetylcholine receptor by a covalent local anesthetic. Biochemistry 20:7166–7174

Oswald RE, Changeux J-P (1981b) Ultraviolet light-induced labeling by noncompetitive blockers of the acetylcholine receptor from *Torpedo marmorata.* Proc Natl Acad Sci USA 78:3925–3929

Oswald RE, Freeman JA (1981) Alpha-bungarotoxin binding and central nervous system nicotinic acetylcholine receptors. Neuroscience 6:1–14

Oswald RE, Heidmann T, Changeux J-P (1983) Multiple affinity states for noncompetitive blockers revealed by [^{3}H]phencyclidine binding to acetylcholine receptor rich membrane fragments from *Torpedo marmorata.* Biochemistry 22:3128–3136

Oswald RE, Bamberger MJ, McLaughlin JT (1984) Mechanism of phencyclidine binding to the acetylcholine receptor from *Torpedo* electroplaque. Mol Pharmacol 25:360–368

Oswald RE, Michel L, Bigelow J (1986) Mechanism of binding of a benzomorphan opiate to the acetylcholine receptor from *Torpedo* electroplaque. Mol Pharmacol 29:179–187

Palfrey HC, Rothlein J, Greengard P (1983) Calmodulin-dependent protein kinase and associated substrates in *Torpedo* electric organ. J Biol Chem 258:9496–9503

Papke RL, Oswald RE (1986) Effects of allosteric ligands on the gating of single channel currents in BC_3H-1 cells. In: Maelicke A (ed) Structure and function of the nicotinic acetylcholine receptor. Springer, Berlin Heidelberg New York, pp 243–257 (NATO ASI series, series H, vol 3)

Pasquale EB, Takeyasu K, Udgaonkar JB, Cash DJ, Severski MC, Hess GP (1983) Acetylcholine receptor: evidence for a regulatory binding site in investigations of suberyldicholine-induced transmembrane ion flux in *Electrophorus electricus* membrane vesicles. Biochemistry 22:5967–5973

Patrick J, Stallcup B (1977) Alpha-bungarotoxin binding and cholinergic receptor function on a rat sympathetic nerve line. J Biol Chem 252:8629–8633

Peper K, Dreyer F, Müller K-D (1975) Analysis of cooperativity of drug-receptor interaction by quantitative iontophoresis at frog motor end plates. Cold Spring Harbor Symp Quant Biol 40:187–192

Peper K, Bradley RJ, Dreyer F (1982) The acetylcholine receptor at the neuromuscular junction. Physiol Rev 62:1271–1340

Plümer R, Fels G, Maelicke A (1984) Antibodies against preselected peptides to map functional sites on the acetylcholine receptor. FEBS Lett 178:204–208

Popot JL, Changeux J-P (1984) Nicotinic receptor of acetylcholine: structure of an oligomeric integral membrane protein. Physiol Rev 64:1162–1239

Prinz H (1986) A general treatment of ligand binding to the acetylcholine receptor. In: Maelicke A (ed) Structure and function of the nicotinic acetylcholine receptor. Springer, Berlin Heidelberg New York, pp 129–146 (NATO ASI series, series H, vol 3)

Prinz H (1988) Agonist binding to the nicotinic acetylcholine receptor and probability of channel opening. Neurochem Int (in press)

Prinz H, Maelicke A (1983a) Interaction of cholinergic ligands with the purified acetylcholine receptor protein. I. Equilibrium binding studie. J Biol Chem 258:10263–10271

Prinz H, Maelicke A (1983b) Interaction of cholinergic ligands with the purified acetylcholine receptor protein. II. Kinetic studies. J Biol Chem 258:10273–10282
Prinz H, Maelicke A (1985) Do sigmoid dose-response curves require the assumption of allostericity? In: Changeux J-P, Hucho F, Maelicke A, Neumann E (eds) Molecular basis of nerve activity. de Gruyter, Berlin, pp 437–444
Prinz H, Maelicke A (1986) Quantitative analysis of receptor-ligand interactions at equilibrium. J Biol Chem 261:14962–14964
Prinz H, Covarrubias M, Maelicke A (1988) Slow conformational transitions in the course of ligand interaction of the nicotinic acetylcholine receptor from *Torpedo marmorata.* J Biol Chem (in press)
Raftery MA, Hunkapiller MW, Strader CD, Hood LE (1980) Acetylcholine receptor: complex of homologous subunits. Science 208:1454–1457
Raftery MA, Dunn SMJ, Conti-Tronconi BM, Middlemas DS, Crawford RD (1983) The nicotinic acetylcholine receptor: Subunit structure, functional binding sites, and ion transport properties. Cold Spring Harbor Symp Quant Biol 48:21–23
Ralston S, Sarin V, Thanh HL, Rivier J, Fox JL, Lindstrom J (1987) Synthetic peptides used to locate the α-bungarotoxin binding site and immunogenic regions on α subunits of the nicotinic acetylcholine receptor. Biochemistry 26:3261–3266
Ratnam M, Lindstrom J (1984) Structural features of the nicotinic acetylcholine receptor revealed by antibodies to synthetic peptides. Biochem Biophys Res Commun 122:1225–1233
Ratnam M, Le Nguyen DL, Sargent PB, Lindstrom J (1986a) Transmembrane topology of nicotinic acetylcholine receptor: immunochemical tests contradict theoretical predictions based on hydrophobicity profiles. Biochemistry 25:2633–2643
Ratnam M, Sargent PB, Sarin V, Fox JL, Le Nguyen DL, Rivier J, Criado M, Lindstrom J (1986b) Location of antigenic determinants on primary sequences of subunits of nicotinic acetylcholine receptor by peptide mapping. Biochemistry 25:2621–2632
Ravdin PN, Berg DK (1979) Inhibition of neuronal acetylcholine sensitivity by α-toxins from *Bungarus multicinctus* venom. Proc Natl Acad Sci USA 76:2072–2076
Reinhardt S, Schmiady H, Tesche B, Hucho F (1984) Functional and structural analysis of acetylcholine receptor-rich membranes after negative staining. FEBS Lett 173:217–221
Reynolds JA, Karlin A (1978) Molecular weight in detergent solution of acetylcholine receptor from *Torpedo californica.* Biochemistry 17:2035–2038
Ross MJ, Klymkowsky MW, Agard DA, Stroud RM (1977) Structural studies of a membrane-bound acetylcholine receptor from *Torpedo californica.* J Mol Biol 116:635–659
Rüchel R, Watters D, Maelicke A (1981) Molecular forms and hydrodynamic properties of acetylcholine receptor from electric tissue. Eur J Biochem 119:215–223
Ruff RL (1977) A quantitative analysis of local anaesthetic alteration of miniature end-plate currents and end-plate current fluctuations. J Physiol (Lond) 264: 89–124
Ruff RL (1982) The kinetics of local anesthetic blockade of end-plate channels. Biophys J 37:625–631
Safran A, Neumann D, Fuchs S (1986) Analysis of acetylcholine receptor phosphorylation sites using antibodies to synthetic peptides and monoclonal antibodies. EMBO J 5:3175–3178
Saitoh T, Changeux J-P (1981) Change in state of phosphorylation of acetylcholine receptor during maturation of the electromotor synapse in *Torpedo marmorata* electric organ. Proc Natl Acad Sci USA 78:4430–4434
Sakmann B, Neher E (1984) Patch-clamp techniques for studying ionic channels in excitable membranes. Annu Rev Physiol 46:455–472
Sakmann B, Patlak J, Neher E (1980) Single acetylcholine-activated channels show burst-kinetics in presence of desensitizing concentrations of agonist. Nature 286:71–73
Sakmann B, Methfessel C, Mishina M, Takahashi T, Takai T, Kurasaki M, Fukuda K, Numa S (1985) Role of acetylcholine receptor subunits in gating of the channel. Nature 318:538–543
Sanchez JA, Bani JA, Siemen D, Hille B (1986) Slow permeation of organic cations in acetylcholine receptor channels. J Gen Physiol 87:985–1001

Sattelle DB (1980) Acetylcholine receptors of insects. Adv Insect Physiol 15:215–316

Sattelle DB, Breer H (1985) Purification by affinity-chromatography of a nicotinic acetylcholine receptor from the CNS of the cockroach periplaneta americana. Comp Biochem Physiol 82:349–352

Schiebler W, Bandini G, Hucho F (1980) Quaternary structure and reconstitution of acetylcholine receptor from *Torpedo californica*. Neurochem Int 2:281–290

Schindler H, Spillecke F, Neumann E (1984) Different channel properties of *Torpedo* acetylcholine receptor monomers and dimers reconstituted in planar membranes. Proc Natl Acad Sci USA 81:6222–6226

Schneider M, Adee C, Betz H, Schmidt J (1985) Biochemical characterization of two nicotinic receptors from the optic lobe of the chick. J Biol Chem 260:14505–14512

Schuhmacher M, Camp S, Manlet Y, Newton M, MacPhee-Quigley K, Taylor SS, Friedmann T, Taylor P (1986) Primary structure of *Torpedo californica* acetylcholinesterase deduced from its cDNA sequence. Nature 319:407–409

Shaw KP, Aracava Y, Akaike A, Daly JW, Rickett DL, Albuquerque EX (1985) The reversible cholinesterase inhibitor physostigmine has channel-blocking and agonist effects on the acetylcholine receptor-ion channel complex. Mol Pharmacol 28:527–538

Sheridan RE, Lester HA (1977) Rates and equilibria at the acetylcholine receptor of Electrophorus electroplaques: a study of neurally evoked postsynaptic currents and of voltage-jump relaxations. J Gen Physiol 70:187–219

Shibahara S, Kubo T, Perski HJ, Takahashi H, Noda M, Numa S (1985) Cloning and sequence analysis of human genomic DNA encoding gamma subunit precursor of muscle acetylcholine receptor. Eur J Biochem 146:15–22

Siemen D, Hellmann S, Maelicke A (1986) Single channel studies of acetylcholine receptors covalently alkylated with acetylcholine. In: Maelicke A (ed) Structure and function of the nicotinic acetylcholine receptor. Springer, Berlin Heidelberg New York, pp 233–241 (NATO ASI series, series H, vol 3)

Sigworth FJ (1985) Open channel noise. I. Noise in acetylcholine receptor currents suggests conformational fluctuations. Biophys J 47:709–720

Sigworth FJ (1986) Open channel noise. II. A test for coupling between current fluctuations and conformational transitions in the acetylcholine receptor. Biophys J 49:1041–1046

Sine SM, Steinbach JH (1984a) Activation of a nicotinic acetylcholine receptor. Biophys J 45:175–185

Sine SM, Steinbach JH (1984b) Agonists block currents through acetylcholine receptor channels. Biophys J 46:277–283

Sine SM, Steinbach JH (1986) Acetylcholine receptor activation by a site-selective ligand: nature of brief open and closed states in BC_3H-1 cells. J Physiol (Lond) 370:357–379

Sine SM, Taylor P (1979) Functional consequences of agonist-mediated state transitions in the cholinergic receptor. J Biol Chem 254:3315–3325

Sine SM, Taylor P (1980) The relationship between agonist occupation and the permeability response of the cholinergic receptor revealed by bound cobra alpha-toxin. J Biol Chem 255:10144–10156

Sine SM, Taylor P (1981) Relationship between reversible antagonist occupancy and the functional capacity of the acetylcholine receptor. J Biol Chem 256:6692–6699

Sine SM, Taylor P (1982) Local anesthetics and histrionicotoxin are allosteric inhibitors of the acetylcholine receptor: studies of clonal muscle cells. J Biol Chem 257:8106–8114

Smart L, Meyers H-W, Hilgenfeld R, Saenger W, Maelicke A (1984) A structural model for the ligand-binding sites at the nicotinic acetylcholine-receptor. FEBS Lett 178:64–68

Smilowitz H, Hadjian RA, Dwyer J, Feinstein MB (1981) Regulation of acetylcholine receptor phosphorylation by calcium and calmodulin. Proc Natl Acad Sci USA 78:4708–4712

Sobel A, Weber M, Changeux J-P (1977) Large scale purification of the acetylcholine receptor protein in its membrane-bound and detergent extracted forms from *Torpedo marmorata* electric organ. Eur J Biochem 80:215–224

Steinbach JH, Covarrubias M, Sine SM, Steele J (1986) Function of mammalian nicotinic acetylcholine receptors. In: Maelicke A (ed) Structure and function of the nicotinic acetylcholine receptor. Springer, Berlin Heidelberg New York, pp 219–232 (NATO ASI series, series H, vol 3)

Swanson KL, Allen CN, Aronstam RS, Rapoport H, Albuquerque EX (1986) Molecular mechanisms of the potent and stereospecific nicotinic receptor agonist (+)-anatoxin-a. Mol Pharmacol 29:250–257

Takahashi T, Kuno M, Mishina M, Numa S (1985) A physiological study on acetylcholine receptor expressed in *Xenopus* oocytes from cloned cDNAs. J Physiol (Paris) 80:229–232

Takai T, Noda M, Furutani Y, Takahashi H, Notake M, Shimizu S, Kayano T, Tanabe T, Tanaka K, Hirose T, Inayama S, Numa S (1984) Primary structure of subunit precursor of calf-muscle acetylcholine receptor deduced from the cDNA sequence. Eur J Biochem 143:109–115

Takai T, Noda M, Mishina M, Shimizu S, Furutani Y, Kayano T, Ikeda T, Kubo T, Takahashi H, Takahashi T, Kuno M, Numa S (1985) Cloning, sequencing and expression of cDNA for a novel subunit of acetylcholine receptor from calf muscle. Nature 315:761–764

Takeuchi A, Takeuchi N (1960) On the permeability of end-plate membrane during the action of transmitter. J Physiol (Lond) 154:52–67

Takeyasu K, Udgaonkar JB, Hess GP (1983) Acetylcholine receptor: evidence for a voltage-dependent regulatory site for acetylcholine; chemical kinetic measurements in membrane vesicles using a voltage clamp. Biochemistry 22:5973–5978

Takeyasu K, Shiono S, Udgaonkar JB, Fujita N, Hess GP (1986) Acetylcholine receptor: characterization of the voltage-dependent regulatory (inhibitory) site for acetylcholine in membrane vesicles from *Torpedo californica* electroplax. Biochemistry 25:1770–1776

Tanabe T, Noda M, Furutani Y, Takai T, Takahashi H, Tanaka K, Hirose T, Inayama S, Numa S (1984) Primary structure of β subunit precursor of calf-muscle acetylcholine receptor deduced from the cDNA sequence. Eur J Biochem 144:11–17

Teichberg VI, Sobel A, Changeux J-P (1977) In vitro phosphorylation of the acetylcholine receptor. Nature 267:540–542

Trautmann A, Feltz A (1980) Open time of channels activated by binding of two distinct agonists. Nature 286:291–293

Turnbull GM, Harrison R, Lunt GG (1985) Nicotinic acetylcholine receptor from foetal human skeletal muscle. Int J Dev Neurosci 3:123–134

Tzartos SJ, Changeux J-P (1983) High affinity binding of alpha-bungarotoxin to the purified alpha-subunit and to its 27000-dalton proteolytic peptide from *Torpedo marmorata* acetylcholine receptor requirement for sodium dodecyl. EMBO J 2:381–387

Tzartos SJ, Lindstrom J (1980) Monoclonal antibodies used to probe acetylcholine receptor structure: localization of the main immunogenic region and detection of similarities between subunits. Proc Natl Acad Sci USA 77:755–759

Tzartos SJ, Kokla A, Walgrave S, Conti-Tronconi B (1988) The main immunogenic region of human muscle acetylcholine receptor is located within residues 63–80 of the α-subunit. Biochemistry (in press)

Vandlen RL, Wu WC, Eisenach JC, Raftery MA (1979) Studies of the composition of purified *Torpedo californica* acetylcholine receptor and of its subunits. Biochemistry 18:1845–1854

Vincent A (1983) Acetylcholine receptors and myasthenia gravis. Clin Endocrinol Metab 12:57–78

Walker JW, Lukas RJ, McNamee MG (1981a) Effects of thio-group modifications on the ion permeability control and ligand binding properties of *Torpedo californica* acetylcholine receptor. Biochemistry 20:2192–2199

Walker JW, McNamee MG, Pasquale E, Cash DJ, Hess GP (1981b) Acetylcholine receptor inactivation in *Torpedo californica* electroplax membrane vesicles: detection of two processes in the millisecond and second time regions. Biochem Biophys Res Commun 100:86–90

Walker JW, Takeyasu K, McNamee MG (1982) Activation and inactivation kinetics of *Torpedo californica* acetylcholine receptor in reconstituted membranes. Biochemistry 21:5384–5389

Walkinshaw MD, Saenger W, Maelicke A (1980) Three-dimensional structure of the 'long' neurotoxin from cobra venom. Proc Natl Acad Sci USA 77:2400–2404

Walter P, Blobel G (1982) Signal recognition particle contains a 7s RNA essential for protein translocation across the endoplasmic reticulum. Nature 299:691–698

Watters D, Maelicke A (1983) Organization of ligand binding sites at the acetylcholine receptor: a study with monoclonal antibodies. Biochemistry 22:1811–1819

Weber M, Changeux J-P (1974) Binding of *Naja nigricollis* [^{3}H]alpha-toxin to membrane fragments from *Electrophorus* and *Torpedo* electric organs. III. Effects of local anaesthetics on the binding of the tritiated alpha-neurotoxin. Mol Pharmacol 10:35–40

Weiland G, Taylor P (1979) Ligand specificity of state transitions in the cholinergic receptor: behavior of agonists and antagonists. Mol Pharmacol 15:197–212

Weiland G, Georgia B, Wee VT, Chignell CF, Taylor P (1976) Ligand interactions with cholinergic receptor-enriched membranes from *Torpedo*: influence of agonist exposure on receptor properties. Mol Pharmacol 12:1091–1105

Weiland G, Georgia B, Lappi S, Chignell CF, Taylor P (1977) Kinetics of agonist-mediated transitions in state of the cholinergic receptor. J Biol Chem 252:7648–7656

Weill CL, McNamee MG, Karlin A (1974) Affinity labeling of purified acetylcholine receptor from *Torpedo californica*. Biochem Biophys Res Commun 61:997–1003

Wennogle LP, Changeux J-P (1980) Transmembrane orientation of proteins present in acetylcholine receptor-rich membranes from *Torpedo marmorata* studied by selective proteolysis. Eur J Biochem 106:381–393

Wennogle LP, Oswald R, Saitoh T, Changeux J-P (1981) Dissection of the 66000-dalton subunit of the acetylcholine receptor. Biochemistry 20:2492–2497

White MM, Mayne KM, Lester HA, Davidson N (1985) Mouse-*Torpedo* hybrid acetylcholine receptors: functional homology does not equal sequence homology. Proc Natl Acad Sci USA 82:4852–4856

Whiting PJ, Lindstrom JM (1986a) Purification and characterization of a nicotinic acetylcholine receptor from chick brain. Biochemistry 25:2082–2093

Whiting PJ, Lindstrom JM (1986b) Pharmacological properties of immuno-isolated neuronal nicotinic receptors. J Neurosci 6:3061–3069

Whiting P, Lindstrom J (1987a) Purification and characterization of a nicotinic acetylcholine receptor from rat brain. Proc Natl Acad Sci USA 84:595–599

Whiting P, Lindstrom J (1987b) Affinity labelling of neuronal acetylcholine receptors localizes acetylcholine binding sites to their β-subunits. FEBS Lett 213:55–60

Wilson PT, Gershoni JM, Hawrot E, Lentz TL (1984) Binding of alpha-bungarotoxin to proteolytic fragments of the alpha subunit of *Torpedo* acetylcholine receptor analyzed by protein transfer on positively charged membrane filters. Proc Natl Acad Sci USA 81:2553–2557

Wise DS, Karlin A, Schoenborn BP (1979) An analysis by low-angle neutron scattering of the structure of the acetylcholine receptor from *Torpedo californica* in detergent solution. Biophys J 28:473–496

Wise DS, Wall J, Karlin A (1981) Relative locations of the beta and delta chains of the acetylcholine receptor determined by electron microscopy of isolated receptor trimer. J Biol Chem 256:12624–12627

Wolosin JM, Lyddiatt A, Dolly JO, Barnard EA (1980) Stoichiometry of the ligand-binding sites in the acetylcholine-receptor oligomer from muscle and from electric organ: measurement by affinity alkylation with bromoacetylcholine. Eur J Biochem 109:495–505

Yamada S, Isogai M, Kagawa Y, Takayanagi N, Hayashi E, Tsuji K, Kosuge T (1985) Brain nicotinic acetylcholine receptors: biochemical characterization by neosurugatoxin. Mol Pharmacol 28:120–127

Young EF, Ralston E, Blake J, Ramachandran J, Hall ZW, Stroud RM (1985) Topological mapping of acetylcholine receptor: evidence for a model with five transmembrane segments and a cytoplasmic cooh-terminal peptide. Proc Natl Acad Sci USA 82:626–630

Yu L, Lapolla RJ, Davidson N (1986) Mouse muscle nicotinic acetylcholine receptor gamma subunit: cDNA sequence and gene expression. Nucleic Acids Res 14:3539–3555

Zavoico GB, Comerci C, Subers E, Egan JJ, Huang C-K, Feinstein MB, Smilowitz H (1984) cAMP, not Ca^{2+}/calmodulin, regulates the phosphorylation of acetylcholine receptor in *Torpedo californica* electroplax. Biochem Biophys Acta 770:225–229

Zingsheim HP, Neugebauer D-C, Barrantes FJ, Frank J (1980) Structural details of membrane-bound acetylcholine receptor from *Torpedo marmorata*. Proc Natl Acad Sci USA 77:952–956

Zingsheim HP, Barrantes FJ, Frank J, Haenicke W, Neugebauer D-C (1982a) Direct structural localization of two toxin recognition sites on an acetylcholine receptor protein. Nature 299:81–84

Zingsheim HP, Neugebauer D-C, Frank J, Haenicke W, Barrantes FJ (1982b) Dimeric arrangement and structure of the membrane-bound acetylcholine receptor studied by electron microscopy. EMBO J 1:541–547

CHAPTER 10

Muscarinic Acetylcholine Receptors

J. JÄRV * and T. BARTFAI

A. Introduction

I. Two Main Types of ACh Receptor

The central and peripheral actions of the endogenous neurotransmitter acetylcholine (ACh) are exerted at two main types of receptor, present in both the central and peripheral nervous system of all mammals. Their definition and pharmacological classification dates back to the work of Sir Henry DALE (1914), who demonstrated that certain actions of ACh on smooth muscle can be mimicked by muscarine while other actions of ACh on, for example, striated muscle can be mimicked by nicotine. The acetylcholine receptors at which muscarine is an agonist are termed muscarinic acetylcholine receptors (mAChRs). DALE also pointed out that the muscarinic actions of ACh can be blocked by atropine. Up to now the use of muscarine as agonist and of atropine as antagonist have provided the best definition of a muscarinic action in all contexts, physiological, pharmacological and biochemical.

This chapter will describe the various compounds that can bind to muscarinic receptors. It will also provide details of a commonly used *in vitro* method of determining the equilibrium binding of labelled ligands and of some of the *in vivo* techniques. Several other methods for measuring muscarinic effects will only be listed due to limitations of space.

II. Biochemistry of mAChRs

The mAChR is an integral membrane protein which occurs in the plasma membranes of smooth muscle, nerve and secretory cells. The receptor is accessible to muscarinic ligands from the extracellular space. Studies of the chemical structure of its ligands and the effect of different protein-modifying agents on the ligand-binding properties of mAChR showed the presence in the ligand-binding site of at least one disulphide bridge, several thiol groups (some of which may be vicinal, since Cd^{2+} ions bind avidly), some strong nucleophilic groups and one or more tyrosine residues (for reviews, see BIRDSALL and HULME 1983; SOKOLOVSKY et al. 1983; SOKOLOVSKY 1984). Studies on antagonist binding showed a strong correlation between the affinity of the ligands for the mAChR and their hydro-

* From Department of Biochemistry, Arrhenius Laboratory, University of Stockholm (current address see list of contributors).

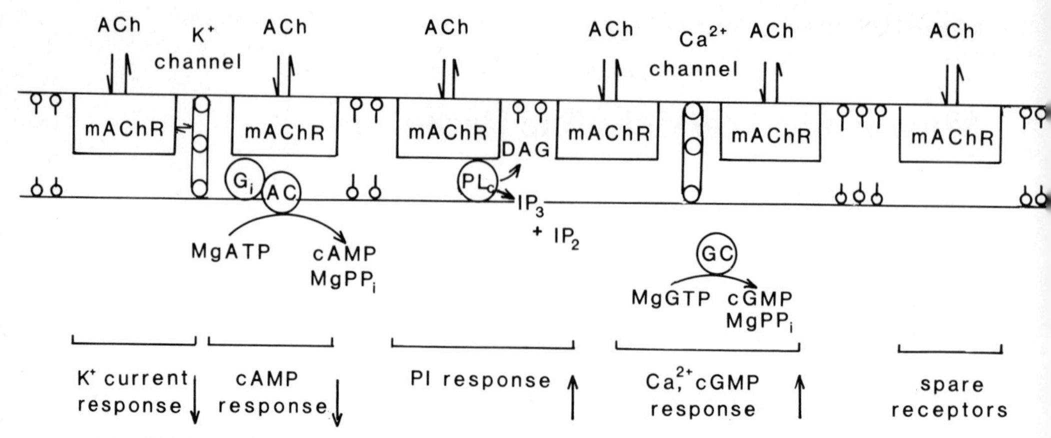

Fig. 1. Summary of biochemical events accompanying ligand binding to the mAChR. Not all of these responses are measurable in each cell type possessing mAChRs, although more than one response is usually present. (a) A K^+ *channel* is shut off producing the M current. (b) *cAMP levels* decrease due to the inhibitory muscarinic input. (c) PI turnover increases, producing diacylglycerol (*DAG*) which activates the protein kinase C, thus increasing the concentration of soluble inositol di-(IP_2) and triphosphates (IP_3) which in turn mobilize Ca^{2+}. (d) *cGMP* levels increase due to muscarinic stimulation. (e) Occupancy of spare receptors produces no response by definition

phobicity, indicating that binding of the bulky antagonists involves hydrophobic interactions with the receptor (JÄRV and BARTFAI 1982).

Purification of mAChR to homogeneity has only recently been achieved by HAGA and HAGA (1983, 1985) who utilized a tropa-acid ester as the ligand for affinity chromatography of the digitonin-solubilized mAChR from ox brain. HAGA et al. (1985) found that the purified mAChR, which is a polypeptide of molecular mass 70 kDa as determined by sodium dodecylsulphate gel electrophoresis, retained at least two of the features of mAChR *in situ*: (a) the recognition site for agonists and antagonists and (b) the site(s) coupling it to another membrane protein, the guanine nucleotide binding protein (G_i or N_i) (RODBELL 1980) (Fig. 1). Binding of the agonist-occupied mAChR to G_i is an obligatory step in the mAChR-mediated inhibition of adenylate cyclase activity. This topic is considered in more detail in Sect. B.II.2.

The primary structure of the swine cerebral mAChR has been deduced from its cDNA sequence by KUBO et al. (1986). This sequence of 460 amino acid residues suggests that the N-terminal portion of the molecule faces the extracellular space and is followed by seven hydrophobic, transmembrane segments while the C-terminal is thought to be exposed to the cytosol. The C-terminal segment carries several potential protein phosphorylation sites (Ser, Thr) while the N-terminal portion carries at least two possible glycosylation sites. This transmembrane organization of the seven hydrophobic segments closely parallels the structure of the β-adrenergic receptor (DIXON et al. 1986) and of rodopsin (OVCHINNIKOV 1982). The very strongly charged, long intracellular segment may play a role in coupling the receptor to other proteins.

Table 1. Distribution of muscarinic receptors and their affinity for [^{3}H]3-QNB

Tissue	Receptor concentration[a] (pmol mg^{-1} protein)	K_D for L-[^{3}H]3-QNB (n*M*)	Reference[d]
Rat brain			
Cerebral cortex	1.10	0.16	2
Hippocampus	0.92	0.19	1
Striatum	1.80	0.18	2
Hypothalamus	1.00	0.19	2
Cerebellum	0.20	0.17	2
Medulla Pons	0.30	0.20	2
Rat heart	0.25	0.55[b]	3
Rat ileum	0.64	0.23[b]	4
Human brain[c]			
Parietal cortex	0.21	0.05	5
Frontal cortex	0.26	0.05	5
Hippocampus	0.12	0.05	5
Caudate nucleus	0.44	0.05	5
Human lymphocytes	0.15	20.0[b]	6
Human erythrocytes	0.02	1.3[b]	7
NG 108-15 Neuroblastoma--glioma hybrid cells	0.04	0.06[b]	8
N1E-115 Neuroblastoma cells	0.01	0.06[b]	9
Cat submandibular salivary gland	0.30	0.20[b]	10

[a] All data (excluding those on cells) from measurements on membrane (P2) preparations.
[b] DL-[^{3}H]3-QNB
[c] Frozen material used
[d] 1, WESTLIND et al. (1981); 2, YAMAMURA and SNYDER (1974a); 3, ROBBERECHT et al. (1981); 4, JÄRV et al. (1979); 5, DANIELSSON et al. (1988); 6, LOPKER et al. (1980); 7, ARONSTAM et al. 1977; 8, NATHANSON et al. 1978; 9, KLEIN et al. (1979); 10, LUNDBERG et al. (1982)

III. Localization of mAChRs

mAChRs are abundant throughout the central nervous system, heart, gastrointestinal tract, smooth muscle and secretory organs; they also occur on circulating cells such as erythrocytes and T-lymphocytes (Table 1).

The localization of mAChRs within the synapse has also been examined. Lesion studies suggest that most (80%–90%) of the receptors are postsynaptic, but that 10%–20% are present as presynaptic autoreceptors (WESTLIND et al. 1981) involved as part of a feedback mechanism regulating the release of ACh (cf. SZERB 1964; POLAK and MEEUWS 1966; KILBINGER and WAGNER 1979; NORDSTRÖM and BARTFAI 1980). The presynaptic muscarinic autoreceptors play an important role in determining ACh levels and turnover (see Chaps. 6, 7).

Table 2. Ligands used in studies of membrane-bound muscarinic receptors

Ligand	Tissue	$-\log K_D$	Notes	Reference
Agonists				
Acetylcholine	Rat Brain	7.70	0 °C	Uchida et al. (1979)
		7.47	25 °C, 25% of total binding sites	Gurwitz et al. (1984a)
	Rat cerebral cortex	7.12	25 °C	Gurwitz et al. (1984c)
cis-Methyldioxolane	Rat forebrain	8.7; 6.9	K_H; K_L	Ehlert et al. (1980a)
Oxotremorine-M	Human cerebral cortex	8.12	–	Wood et al. (1983)
	Rat heart	8.68	25 °C	Chatelain et al. (1983)
Pilocarpine	Rat cerebral cortex	8.3; 6.7	25 °C	Hedlund and Bartfai (1981)
Antagonists				
Atropine	Rat brain	8.7	25 °C	Bartfai et al. (1974)
	Ox trachea	8.58	20 °C	Beld and Ariens (1974)
	Mouse brain	8.88	25 °C	Kloog and Sokolovsky (1978)
	Rat cerebral cortex	9.20	30 °C; (−)-ligand	Hulme et al. (1978)
	Rat striatum	8.30	20 °C	Kato et al. (1978)
Dexetimide	Ox trachea	9.38	–	Laduron et al. (1979)
	Rat forebrain	8.96	37 °C	Vauquelin et al. (1982)
N-Methylatropine	Rat cerebral cortex	9.45	30 °C	Birdsall et al. (1983a)
		9.71	30 °C; (−)-ligand	Birdsall et al. (1983b)
		9.76	30 °C; (−)-ligand	Hulme et al. (1978)
N-Methyl-4-piperidinyl benzilate	Human striatum			Danielsson et al. (1985)
	Human cerebral cortex			Danielsson et al. (1986)
	Human brain	8.92	25 °C	Gurwitz et al. (1984b)
	Mouse brain	9.41; 9.34;	10 °C; 25 °C	Kloog and Sokolovsky (1978)
		9.38	37 °C	Kloog and Sokolovsky (1978)
	Rat adenohypophysis	8.61	25 °C	Avissar et al. (1982)
N-Methylscopolamine	Calf heart	9.23	30 °C	Hammer and Giachetti (1982)
	Cell culture, chick embryo heart	9.70	37 °C	Nathanson (1983)
	Neuroblastoma, N1E-115	10.20	25 °C	Milligan and Strange (1984)
	Human hippocampus	9.05	25 °C	Lang and Henke (1983)
	Human heart	8.97	25 °C	Repke (1982)

	Rat cerebral cortex	9.52	30 °C	Birdsall et al. (1983a)
		9.61	30 °C; (−)-ligand	Birdsall et al. (1983b)
	Rat submandibular gland	9.60	30 °C	Hammer et al. (1980)
	Neuroblastoma-glioma, NG 108-15	8.39	25 °C	Akiyama et al. (1984)
Pirenzepine	Human stellate ganglia	7.85		Watsonet al. (1984)
	Rat brain	7.79	32 °C	Luthin and Wolfe (1984)
		7.74	Room temperature	Wamsley et al. (1984)
	Rat cerebellum	8.43	–	Watson et al. (1984)
	Guinea-pig small intestine; rat cerebral cortex	8.3	30 °C	Burgen et al. (1974a, b)
N-Propylbenzilyl-choline mustard	Rat brain	7.94	30 °C	Birdsall et al. (1980b)
	Rat cerebral cortex	7.92	30 °C	Hulme et al. (1978)
	Rat medulla	8.08	30 °C	Kuhar et al. (1980)
3-Quinuclidinyl benzilate	Ox brain	9.74	25 °C	Hurko (1978)
	Cell culture, human astro-cytome	10.75	37 °C	Meeker and Harden (1982)
	Neuroblastoma, N1E-115	10.40	30 °C	Strange et al. (1978)
	Drosophila melangaster, head	9.15	–	Dudai (1980)
		8.69	25 °C	Dudai and Ben-Barak (1977)
	Guinea-pig ileum	9.51	25 °C	Nilvebrant and Sparf (1982)
		9.52	35 °C	Yamamura and Snyder (1974b)
	Human cerebral cortex	10.42	37 °C	Nordström et al. (1982)
		10.19	37 °C	Wastek and Yamamura (1978)
	Human lymphocytes	7.17	22 °C	Zalcman et al. (1981)
	Human lung	10.20	37 °C	Raaijmakers et al. (1983)
	Rat brain	9.77	20 °C	Aronstam and Eldefrawi (1979)
				Yamamura and Snyder (1979a)
	Salamander olfactory epithelium	9.10	–	Hedlund and Shepherd (1983)
	Torpedo electric organ	9.52	25 °C	Kloog et al. (1978)
Scopolamine	Mouse brain	9.30	IC_{50}; 25 °C	Kloog and Sokolovsky (1978)
	Rat cerebral cortex	9.21	25 °C	Kloog et al. (1979)
[^{125}I]3-Quinuclidinyl iodobenzilate	Dog striatum	10.74	25 °C	Gibson et al. (1984)
	Dog neutricular muscle	9.90	25 °C	Gibson et al. (1984)

B. Assays of Muscarinic Receptors and Muscarinic Ligands

I. Binding Studies

1. Labelled Ligands

Equilibrium binding studies designed to characterize and quantify mAChRs can be carried out by using a series of radioactive muscarinic antagonists labelled to high specific activity (Table 2). The ^{3}H-labelled agonists were used to characterize the high-affinity agonist binding sites. Not listed in Table 2 are the monoclonal antibodies to mAChR produced by STROSSBERG's and VENTER's groups (VENTER et al. 1984) which can also be used to label the receptors.

The binding of labelled muscarinic ligands to specific binding sites not only yields information about their affinity to the receptors; if can be used to predict their interaction with muscarinic cholinergic functions *in vivo*. This aspect will be discussed in detail below. The simplicity of the equilibrium binding method has led to its general use for estimating receptor concentration. There are some pitfalls however that need to be considered. Thus, the number of receptors present per milligram of tissue, determined by antagonist (e.g. [^{3}H]3-quinuclidinylbenzylate, [^{3}H]3-QNB) binding includes *all* receptor sites which can recognize this muscarinic ligand and includes the so-called spare receptors which are not coupled to any biochemical or tissue responses.

2. Binding of Muscarinic Antagonists

The equilibrium binding of muscarinic antagonists of the benzilic acid and tropaacid ester classes (cf. Table 2) obeys a single Langmuir isotherm, indicating the presence of a single class of binding sites, present in finite numbers, with equal affinity for the receptor. The amount of bound ligand b is given by

$$b = \frac{B_{max}\,[L]}{K_D + [L]} \tag{1}$$

where B_{max} is the maximum number of binding sites, K_D is the equilibrium dissociation constant of the [L·mAChR] complex and [L] is the free ligand concentration. The B_{max} and K_D values for the binding of [^{3}H]3-QNB for a number of human and rat tissues are listed in Table 1.

In contrast to benzilates, the recently introduced muscarinic antagonist pirenzepine (HAMMER et al. 1980) showed binding to two classes of mAChRs in the heart and in the salivary and other glands. The dissociation constant (affinity) of the high-affinity binding site, K_D^H, for pirenzepine is around 20 nM in most tissues while that of the low-affinity site K_D^L is over 200 nM (cf. Table 2). [^{3}H]pirenzepine has become a popular ligand in experiments aimed at defining subclasses of muscarinic antagonist binding sites (GOYAL and RATTAN 1978; HAMMER and GIACHETTI 1982; WATSON et al. 1986; BERRIE et al. 1986; MCCORMICK and PRINCE 1986). Other antagonists such as bretylium tosylate can also distinguish between high- and low-affinity antagonist binding sites, which are indistinguishable for the traditional muscarinic antagonists like atropine, scopolamine and 3-QNB. It has been suggested that muscarinic receptors with a high affinity for pirenzepine be designated M_1-muscarinic receptors while those with lower affinity should be called M_2-receptors (HAMMER and GIACHETTI 1982; LEVINE et al. 1986).

3. Binding of Muscarinic Agonists

Muscarinic agonist binding studies using a constant concentration of a labelled antagonist and variable concentrations of the unlabelled agonist reveal low- and high-affinity components. The percentage of displaced antagonist b conforms to the equation:

$$b = \frac{a \cdot [A]}{K_D^H + [A]} + \frac{(1-a)\ [A]}{K_D^L + [A]} \tag{2}$$

where a is the fraction of high-affinity agonist binding sites with dissociation constant K_D^H; K_D^L is the low-affinity dissociation constant, and [A] is the free agonist concentration. K_D^H values are in the range of $0.01-1.0\ \mu M$, those of K_D^L are $1\ \mu M$ to $1.0\ mM$.

It should be noted that even super-high affinity binding sites for muscarinic agonists with nanomolar equilibrium binding constants have been reported (BIRDSALL et al. 1980b). In most tissues and under most experimental conditions this phenomenon is not observed and only the high- and low-affinity sites are apparent.

With labelled agonists (Table 2) such as $[^3H]$ACh (GURWITZ et al. 1984c), $[^3H]$*N*-methyloxotremorine (BIRDSALL et al. 1980b), $[^3H]$pilocarpine (HEDLUND and BARTFAI 1981) or $[^3H]$*cis*-methyldioxalane (EHLERT et al. 1980a) only the high-affinity agonist binding site can be studied, because at concentrations required for saturation of the low-affinity site the ratio of specific to nonspecific binding becomes very unfavourable, resulting in large experimental errors.

It was shown in some tissues (e.g. heart, cerebellum) that the high-affinity agonist binding conformation can be shifted into a low-affinity conformation in the presence of GTP or its analogues (BERRIE et al. 1979; ROSENBERGER et al. 1979; SOKOLOVSKY et al. 1980). In these tissues the muscarinic receptor-agonist complex activates the G_i units in the membranes with resulting inhibition of the adenylate cyclase and activation of the GTPase activity of the G_i protein (Fig. 1). Pertussis toxin-mediated ribosylation of the G_i protein prevents the inhibitory coupling of the mAChR-agonist complex to the adenylate cyclase (KUROSE and UI 1985). Even the transformation of low-affinity agonist binding sites into high-affinity sites can be induced by appropriate means, e.g. Mn^{2+} ions (GURWITZ and SOKOLOVSKY 1980).

4. Assessment of the Binding of any Substance to mAChRs

Binding of a given substance to mAChR may be assessed by studying the competition between the substance and any of the labelled, high-affinity muscarinic antagonists (or agonists) such as $[^3H]$3-QNB, $[^3H]$4-*N*-methylpiperidinylbenzilate, $[^3H]$*N*-methylscopolamine, $[^3H]$pirenzepine, $[^3H]$*N*-methyloxotremorine or

[^{3}H]ACh. Data from this type of experiments on a variety of substances are listed in the Appendix.

Briefly, the assay is carried out *in vitro* using membranes from the tissue of interest. These are prepared under hypo-osmotic conditions by homogenization of the tissue in 10 vol 50 m*M* sodium phosphate buffer, pH 7.4. The homogenates are centrifuged at 100000 g for 1 h and the pellet is resuspended in the same buffer to yield 0.1 – 1.0 mg ml^{-1} protein. These membranes can be stored below –20 °C (preferably –70 °C) for years without decline of muscarinic binding capacity. The binding assay can most conveniently be carried out by incubating in a final volume of 500 µl 0.1 mg^{-1} protein, 0.1 n*M* or 0.2 n*M* [^{3}H]3-QNB and various concentrations of the substance tested (or 1 µ*M* atropine to define specific binding of [^{3}H]3-QNB).

The incubation should be carried out for at least 60 min at room temperature, stopped by the addition of 5 ml of ice-cold buffer to the incubation mixture and filtering onto a Whatman GF/B glass fibre filter under vacuum; the filter is washed with two 5-ml portions of buffer, dried and counted in a scintillation spectrometer.

The concentration which displaces half of the specifically bound [^{3}H]3-QNB corresponds to the IC'_{50} value for the substance. Specifically bound [^{3}H]3-QNB is defined as the difference between [^{3}H]3-QNB bound in the presence and absence of atropine (1 µ*M*). If the concentration of [^{3}H]3-QNB is negligible as compared to its own K_D, the IC_{50} value closely approaches the dissociation constant (K_D^A) of the substance (A) with mAChR, i.e.:

$$\frac{[\mathrm{A}]\,[\mathrm{mAChR}]}{[\mathrm{A}\cdot\mathrm{mAChR}]} = K_D^A \simeq IC_{50} \ . \tag{3}$$

To take into account the effect of [^{3}H]3-QNB concentration on the values of IC_{50} the exact value for the latter can be extrapolated using the formula of CHENG and PRUSOFF (1973):

$$IC_{50} = IC'_{50}\left(1 + \frac{[^{3}\mathrm{H}]3\text{-QNB}}{K_D\ [^{3}\mathrm{H}]3\text{-QNB}}\right) \tag{4}$$

where IC_{50} is the concentration of the drug which causes 50% inhibition of the specific binding of [^{3}H]3-QNB at zero [^{3}H]3-QNB concentration; K_D^A is the equilibrium dissociation constant of the [mAChR·A] complex; and [A], [mAChR] and [^{3}H]3-QNB are drug, receptor and antagonist concentrations, respectively. IC'_{50} is the apparent value of IC_{50} obtained at a finite concentration of [^{3}H]3-QNB. K_D [^{3}H]3-QNB is the dissociation constant of the mAChR·[^{3}H]3-QNB complex. The values of this constant are listed for several tissues in Table 1.

5. Drugs that Bind to mAChRs

The procedure described above is most often used to assess the anticholinergic-antimuscarinic properties of drugs. The Appendix lists in tabular form a number of commonly used pharmaceutical substances which have intermediate to high afinities for mAChR (i.e. K_D or IC_{50} in the range of 10 μM to 1 nM); thus these compounds qualify as mAChR ligands, some of which are specific for mAChR (i.e. they bind with 10- to 50-fold lower affinity to other receptors). Other compounds listed in the Appendix are commonly used pharmaceutical agents which, besides their other effects, also have considerable affinity for the mAChR. These drugs most often act as antagonists and produce symptoms such as dryness of mouth, blurred vision and confusion (see Sect. C. III). The equilibrium binding experiment described above provides a tissue-specific, rapid and cheap test of the affinity of any drug for mAChR or its antagonist binding site and therefore is useful in drug development (see SNYDER et al. 1974). Recent study (DANIELSSON et al. 1985) shows that using membranes from different regions of the rat brain (SNYDER et al. 1974) is a valid model for determining the relative affinities of drugs for mAChRs in the corresponding region of the human brain, i.e. affinity of these drugs shows little species-specific differences.

6. Changes in mAChR Concentration

Parasympathetic denervation or chronic blockade of mAChRs by antagonists can cause increases in the concentration of mAChR. It is of great pharmacological importance that several of the most active antidepressant agents of the tricyclic type (see Appendix) and several of the most commonly used antipsychotic agents of the phenothiazine and butyrophenon type have high affinities for mAChR (SNYDER et al. 1974; REHAVI et al. 1977c; GOLDS et al. 1980); when applied chronically these agents cause increases in the concentration of the receptor in both the central nervous system (CNS) and the periphery (REHAVI et al. 1980). This increased receptor concentration may also account for the appearance of super-sensitive muscarinic responses (MILLER and HILEY 1975). Development of tardive dyskinesia upon treatment with antipsychotic drugs is one of the most important unsolved problems in the drug therapy of schizophrenia (cf. BARCHAS et al. 1977). Chronic treatment with antagonists caused increases in the concentration of mAChR in the CNS of 15%–20% whereas the same drug regimen may cause 50%–80% increase in the periphery (e.g. heart).

A decrease in the concentration of mAChR can be caused by chronic elevation of muscarinic agonist concentrations in the synaptic cleft via inhibition of AChE or by administration of exogenous muscarinic agonists. Inhibition of AChE by the reversibly acting inhibitor eserine does not, however, result in a decrease in mAChR concentration because of the very short half-life of the inhibition (0.5–1.0 h). By contrast, inhibition of AChE by the irreversibly acting organophosphorus compounds, such as are commonly used as pesticides, can cause a decrease in the concentration of mAChR (UCHIDA et al. 1979) and also corresponding decreases in functional muscarinic responses (subsensitivity) (MAAYANI et al. 1977).

7. Development of Tolerance

Several studies show that treatment with muscarinic agonists or with AChE-inhibitors can cause development of tolerance with respect to all muscarinic responses. Cross-tolerance has been noted in mice treated with either eserine or oxotremorine (MAAYANI et al. 1977). The time course of development of tolerance is rapid, within 7–10 days. One correlate of the attenuated response is a decrease in the number of muscarinic binding sites (WISE et al. 1980). These decreases alone (10%–30%) cannot, however, explain the degree of tolerance observed: responses may be reduced by as much as 60%–80%. It is assumed that post-receptor changes are also involved, e.g. in G_i proteins (Fig. 1) (TJÖRNHAMMAR and BARTFAI 1984).

It should also be noted that the number of available, functional mAChRs can also be reduced as a consequence of poisoning with Mn^{2+}, Cd^{2+} or Pb^{2+}, which can bind to the essential thiol groups of the mAChRs (HEDLUND et al. 1979).

II. Assay of Muscarinic Agonist and Antagonist Actions

1. Biochemical Responses

The different biochemical events elicited by binding of agonists to mAChR (Fig. 1) all permit determination of the agonist or partial agonist character of a substance which in binding studies is found to act as a ligand to the mAChR. Muscarinic agonists elicit simultaneously one or more of the following responses: inhibition of adenylate cyclase, stimulation of cGMP synthesis and phosphatidylinositol (PI) turnover, and changes in K^+ or (Ca^{2+}, Na^+) conductance. However not all such responses are easy to detect in every tissue, thus when mentioning a response the tissue and species in which it has been observed should always be given.

The biochemical (and electrophysiological) responses to muscarinic stimulation are as follows:

1. *Muscarinic inhibition of the adenylate cyclase activation of GTPase.* Muscarinic agonist-mediated inhibition of adenylate cyclase and activation of GTPase in membrane and whole cells or tissue slices (rabbit heart, EHLERT 1985; rat striatum, ONALI et al. 1984).
2. *Muscarinic stimulation of cGMP synthesis.* Muscarinic agonist-mediated increases in cGMP levels in excitable cells (heart, cerebral cortex slices, NIE 115 neuroblastoma and T-lymphocytes or in intact animals) (GEORGE et al. 1970; GOLDBERG and HADDOX 1977; BARTFAI et al. 1977; RICHELSON and DIVINETZ-ROMERO 1977; BARTFAI et al. 1978).
3. *Muscarinic stimulation of phosphatidylinositol turnover.* Muscarinic agonist-mediated increase in PI turnover in whole cells (pancreas) or tissue slices (rat cerebral cortex, heart) (HOKIN and HOKIN 1952; MICHELL 1975; BROWN and BROWN 1984).
4. *Muscarinic regulation of ion fluxes.* Muscarinic agonist-mediated changes in cation fluxes (Rb^+, K^+, Ca^{2+}, Na^+) measured by electrodes or isotopes in intact cells (BURGEN and SPERO 1968; PUTNEY and PAROD 1978; BROWN and ADAMS 1982; NORTH 1986).
5. *Regulation of neurotransmitter release via presynaptic muscarinic receptors.* Muscarinic agonist/antagonist effect on the release of neurotransmitters exerted at presynaptic muscarinic receptors in synaptosomes or tissue slices, or

in intact animals (in rat cerebral cortex, hippocampus, striatum, heart and guinea-pig ileum) (cf. Chap. 20, this volume).

2. Bioassay

Several important organ bath assays have also been used in the past 70 years to study muscarinic receptor-mediated processes. These are:

1. Atropine-sensitive contraction of the guinea-pig, rat or rabbit ileum and (RINGDAHL 1984) urinary bladder.
2. Atropine-sensitive effects on the isolated rat or rabbit heart (EHLERT 1985).
3. Muscarinic agonist-stimulated and atropine-sensitive secretion of salivary and gastric juices quantified, for example, by measurement of amylase activity (cf. BABKIN 1950; EMMELIN 1967; LESLIE et al. 1976).

An excellent detailed description of the two first methods is given by BURN (1952) and two recent detailed papers are also quoted. The sensitivity of these assays offers several advantages over biochemical methods.

In the past two years the importance of mAChR-coupled K^+ conductances in regulating the excitability of ganglion cells and of neurons in the CNS has been recognized (BROWN and ADAMS 1982; HARTZELL 1981). This applies especially to the muscarinic agonist-mediated inhibition of the K^+ current: this 'M-current' has been described in detail and studied with respect to agonist and antagonist specificity (NORTH 1986).

III. Muscarinic Agonist Effects in Intact Animals

Intraperitoneal or subcutaneous administration of muscarinic agonists may induce pupillary contraction, increase salivary secretion, tremor, hypothermia, analgesia in the tail flick assay, and changes in performance in avoidance and make learning, and may influence both retention and recall.

The *in vivo* muscarinic agonist property of a substance is most conveniently determined by examining whether pretreatment with atropine ($0.1-1.0\ mg\ kg^{-1}$, intraperitoneally) blocks its effects on salivation, tremor and hypothermia; the results can be assessed as shown in Fig. 2.

Conversely, the muscarinic-antagonist properties of substances are inferred from their ability to block the effects of $20-500\ \mu g\ kg^{-1}$ oxotremorine, a muscarinic agonist, and/or $20-200\ \mu g\ kg^{-1}$ eserine, separately or in combination. The drugs may be administrated subcutaneously into the neck of rats or mice or intraperitoneally. The reversible inhibition of AChE by eserine produces cholinergic (both nicotinic and muscarinic) stimulation by causing the accumulation of endogenous ACh and thus provides a more realistic model for an assessment of the drug's antimuscarinic effects than does the administration of the exogenous agonist oxotremorine.

Separate application of oxotremorine and eserine also permits one to decide whether the muscarinic or antimuscarinic effects of a substance *in vivo* are indirect and result from an inhibition of oxotremorine breakdown by the compound under study or are due to a true interaction of the drug with mAChRs. In the lat-

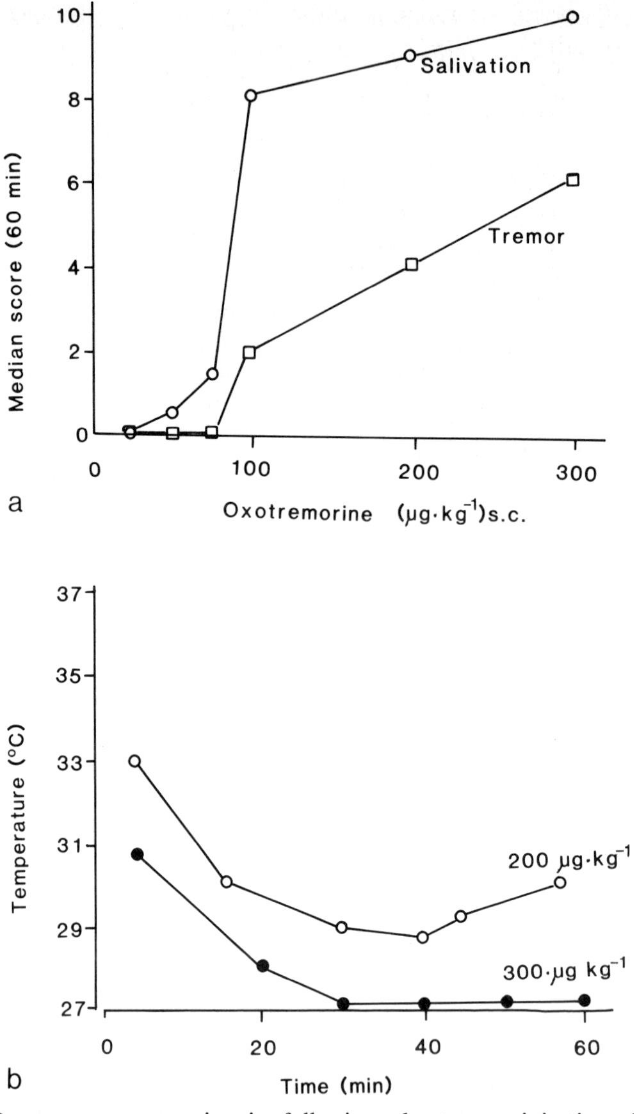

Fig. 2 a, b. Dose-response curves in mice following subcutaneous injection of oxotremorine into the neck. **a** Salivation and tremor; **b** hypothermia. (Modified from ÖGREN et al. 1985; ENGSTRÖM et al. 1986). Each dose of the drug was tested on groups of 3–5 mice. Measurements were conducted every 10 min after injection of oxotremorine or eserine subcutaneously into the neck. Tremor was scored as described by JENDEN (1965) and ÖGREN et al. (1985): *0*, no apparent tremor; *1*, moderate or discontinuous tremor of the head and forelimbs; *2*, strong tremor, intense continuous tremor involving the whole body. Salivation was scored according to the following scale: *0*, no salivation; *1*, moderate salivation, visible salivation but no dripping of saliva; *2*, strong salivation, saliva dripping. Scoring was conducted for 60 min. Hypothermia was evaluted by recording rectal temperature with an AMR-Therm 2210-3 NiCr-Ni digital thermometer (Holzkirchen, FRG), with a rubber stopper placed 10 mm from the tip of the termistor to enable reproducible penetration into the anus. The temperature reading, recorded digitally, stabilized within 10 s after insertion and gave a reading with a precision of ±0.1 °C

ter case the drug will have similar effects on eserine- and oxotremorine-evoked muscarinic responses.

The easiest to perform among the *in vivo* tests of muscarnic and antimuscarinic action are the measurements of hypothermia, tremor, salivation and analgesia in the tail flick test. The sites at which these effects are primarily generated are known: hypothermia is induced in the preoptic area of the hypothalamus, tremor in the striatum and the salivary response peripherally in the salivary glands. Thus these measurements provide information about both central and peripheral muscarinic actions of a given substance.

C. Clinical Application of Muscarinic Drugs

I. Therapeutic Uses of Muscarinic Drugs

The muscarinic agonist pilocarpine is used as a locally applied agent in treatment of glaucoma. Several mild AChE inhibitors are also utilized to evoke muscarinic actions at the muscles of the pupil. Efforts are now being made to utilize centrally active muscarinic agonists to improve cognitive function in Alzheimer's disease (BARTUS et al. 1982; WURTMAN et al. 1983).

Muscarinic antagonists (especially atropine) are utilized accutely in conjunction with major surgery to prevent profuse salivation and gastric secretion and to relax smooth muscle; they are used chronically as the first choice for controlling tremor and rigidity in Parkinson's disease before introducing L-dopa therapy (TAYLOR 1980). Some reports describe the use of muscarinic antagonists for the treatment of torticollis.

II. Acute Poisoning with Muscarinic Agonists

Poisoning (BECKER et al. 1979) with the fugus *Amanita muscaria* (from which muscarine was originally isolated) or with species of *Inocybe* presents muscarinic agonist peripheral effects such as salivation, hypotension, diarrhoea and central effects such as restlessness, mania and hallucination. Treatment is usually atropine (2 mg, parenterally) (TAYLOR 1980).

III. Acute Poisoning with Muscarinic Antagonists

Overdose of the muscarinic antagonist atropine (used to reduce salivary and bronchial secretion during surgery) seldom occurs. The most common cause of cholinergic (i.e. antimuscarinic) intoxication or of the 'Cholinergic Syndrome' is an overdose of antipsychotic agents of the phenothiazine class which have high affinity ($K_d = 100-1$ nM, see Table 3) for the peripheral and central muscarinic receptors (YAMAMURA and SNYDER 1974a, b; DANIELSSON et al. 1985). The cholinergic syndrome presents symptoms such as confusion, delirium and coma (TINKLENBERG and BERGER 1977). Its immediate treatment is by intravenous administration of eserine (HEISER and GILLIN 1971; RUMACK 1973; AQUILONIUS 1978).

Appendix. Binding of drugs to membrane-bound muscarinic receptor from brain tissues (only selected examples are given, data bank is available)

Ligand	Tissue	$-\log K_D$	Notes	References
Aceptomazine; 1[10-[3-(dimethylamino)propyl]-10H-phenothiazin-2-yl]-ethanone, Merck Index, 26	Rat brain	6.46	IC_{50}; 30 °C	Fjalland et al. (1977)
Acetylcholine; 2-(acetyloxy)-*N*,*N*,*N*-trimethyl-ethanaminium, Merck Index, 80, 81	Rat brain	5.35	IC_{50}; 25 °C	Nukina (1983)
		6.29	K_D; 37 °C	Speth and Yamamura (1979)
		7.05; 4.46	K_H; K_L; 37 °C	Vauquelin et al. (1982)
		5.70	K_D; 23 °C	Ward and Trevor (1981)
		5.52	IC_{50}; 25 °C	Yamamura and Snyder (1974a)
Adiphenine; *α*-phenylbenzene-acetic acid 2-(diethylamino)ethyl ester, Merck Index, 150	Rat cerebral cortex	5.92	K_D	Aronstam et al. (1980)
Adamantadine; tri-[3,3,1,1]decan-1-amine, Merck Index, 373	Mouse brain	3.52	K_D; 25 °C	Gabrielewitz et al. (1980)
	Rat cerebral cortex	3.57	K_D	Aronstam et al. (1980)
Amiodarone; (2-butyl-3-benzofuranyl)[4-[2-(di-methylamino)ethoxy]-3,5-diiodophenyl]-methanone, Merck Index, 491	Rat cerebral cortex	4.35	IC_{50}; 25 °C	Cohen-Armon et al. (1985)
Amitrityline; 3-(10,11-dihydro-5H-dibenzo[a,d]-cyclohepten-5-ylidene)-*N*,*N*-dimethyl-1-propanamine, Merck Index, 496	Mouse brain	7.45	IC_{50}; 37 °C	Bohman et al. (1981)
		7.48	K_D; 25 °C	Rehavi et al. (1977a)
	Rat brain	7.33	IC_{50}; 30 °C	Fjalland et al. (1977)
		7.46	K_D; 30 °C	Golds et al. (1980)
		6.82	IC_{50}; 25 °C	Hyslop and Taylor (1980)
		8.10	IC_{50}; 25 °C	Nukina (1983)
		8.15	K_D; 29 °C	Smith and Huger (1983)
Amoxapine; 2-chloro-11-(1-piperazinyl)-diben [b, f] [1,4]-oxazepine, Merck Index, 598	Human caudate nucleus	6.00	K_D; 37 °C	El-Fakahany and Richelson (1983)
Amphetamine; *α*-methylbenzenethanamine, Merck Index, 606	Rat brain	4.78	IC_{50}	Goldman and Erickson (1983)
Apomorphine; 5,6,6a,7-tetrahydro-6-methyl-4H-dibenzo[de,g]quinoline-10,11-diol, Merck Index, 769	Sheep caudate nucleus	9.52	IC_{50}; 23 °C	Bylund (1981)

Arecoline; 1,2,5,6-tetrahydro-1-methyl-3-pyridine-carboxylic acid methyl ester, Merck Index, 795	Rat brain	5.26	IC_{50}; 20 °C	SMITH and HUGER (1983)
		5.70	IC_{50}; 25 °C	YAMAMURA and SNYDER (1974a)
	Rat cerebral cortex	6.92; 4.98	K_H; K_L; 30 °C	BIRDSALL et al. (1983a)
		6.26	K_D; 25 °C	GURWITZ et al. (1984a)
Atropine; *α*-(hydroxymethyl)benzeneacetic acid 8-methyl-8-azabicyclo-[3,2,1]-oct-3-yl ester, Merck Index, 878	Human brain	9.52	IC_{50}	ENNA et al. (1976)
	Rat brain	8.8	K_D; 25 °C	BARTFAI et al. (1974)
		9.15	IC_{50}	ENNA et al. (1976)
		9.38	K_D; 20 °C	SMITH and HUGER (1983)
		8.96	IC_{50}; 37 °C	SPETH and YAMAMURA (1979)
		8.57	K_D; 37 °C	VAUQUELIN et al. (1982)
		8.43	IC_{50}; 25 °C	VINCENT et al. (1978)
		8.82	IC_{50}; 25 °C	YAMAMURA and SNYDER (1974a)
Azaperone; 1-(4-fluorophenyl)-4-[4-(2-pyridinyl)-1-piperazinyl]-1-butanone, Merck Index, 10, 903	Rat striatum	4.27	IC_{50}; 37 °C	LOPKER et al. (1980)
Benactyzine; *α*-hydroxy-*α*-phenylbenzeneacetic acid 2-(diethylamine)ethyl ester, Merck Index, 1028	Rat brain	8.63	IC_{50}; 25 °C	SNYDER and YAMAMURA (1977)
Benzechol; benzoylcholine	Rat brain	5.87	K_D; 32 °C	LUTHIN and WOLFE (1984)
	Rat cerebral cortex	8.26	K_D; 30 °C	HULME et al. (1976)
Benzotropine; 3-(diphenylmethoxy)-8-methyl-8-arabicyclo[3,2,1]octane, Merck Index, 1127	Rat brain	8.63	IC_{50}; 30 °C	FJALLAND et al. (1977)
		7.69	IC_{50}; 30 °C	GOSSUIN et al. (1984)
		9.26	IC_{50}; 20 °C	SMITH and HUGER (1983)
		8.82	IC_{50}; 25 °C	SNYDER and YAMAMURA (1977)
Bethanechol; 2-[(aminocarbonyl)-oxy]-*N*,*N*,*N*-trimethyl-1-propanaminium, Merck Index, 1200	Rat brain	5.42; 3.74	K_H; K_L; 30 °C	BIRDSALL et al. (1983a)
		4.28	IC_{50}; 32 °C	LUTHIN and WOLFE (1984)
		5.35; 3.39	K_H;K_L; 37 °C	VAUQUELIN et al. (1982)
Biperiden; 2-bicyclo[2,2,1]-hept-5-en-2-yl-d-phenyl-1-piperidinepropanol, Merck Index, 1231	Rat brain	8.52	IC_{50}; 25 °C	NUKINA (1983)
Bretylium tosylate; 2-bromo-*N*-ethyl-*N*,*N*-dimethylbenzene methanaminium-4-methylbenzene sulphonate, Merck Index, 1348	Rat cerebral	4.86	K_D; 25 °C	SCHREIBER and SOKOLOVSKY (1985)
		4.74	K_D	SCHREIBER et al. (1984)
Brompheniramine; 1-(4-bromophenyl)-*N*,*N*-dimethyl-2-pyridinepropanamine, Merck Index, 1417	Rat brain	4.60; 4.35	IC_{50}; *d* and *l* isomers	YAMAMURA and SNYDER (1974a)

Appendix (continued)

Ligand	Tissue	$-\log K_D$	Notes	References
Bupropion; 1-(3-chlorophenyl)[(1,1-dimethyl)-amino]-1-propanone, Merck Index, 1465	Human caudate nucleus	4.32	IC_{50}; 37 °C	EL-FAKAHANY and RICHELSON (1983)
		4.27	IC_{50}; 37 °C	TOLLEFSON and SENOGLES (1983)
Butriptyline; 10,11-dihydro-*N*,*N*-*β*-trimethyl-5H-dibenzo[a,d]cycloheptene-5-propanamine, Merck Index, 1506	Human caudate nucleus	7.46	IC_{50}; 37 °C	EL-FAKAHANY and RICHELSON (1983)
Caramiphen; 2[(amino-carbonyl)oxy]-*N*,*N*,*N*-trimethylethanaminium chloride, Merck Index, 1754	Rat brain	6.59; 4.40	K_H; K_L	FARRAR and HOSS (1984)
		4.22	IC_{50}; 37 °C	GRAMMAS et al. (1983)
		3.49	IC_{50}	GORISSEN et al. (1978)
		5.00	IC_{50}; 25 °C	NUKINA (1983)
		6.92; 3.60	K_H, K_L; 37 °C	VAUQUELIN et al. (1982)
		4.60	IC_{50}; 25 °C	YAMAMURA and SNYDER (1974a)
Chlorpheniramine; *γ*-[4-chlorophenyl]-*N*,*N*-dimethyl-2-pyridinepropanamine, Merck Index, 2157	Rat brain	4.60; 4.35	IC_{50}; *d* and *l* isomers	YAMAMURA and SNYDER (1974a)
Chlorpromazine; 2-chloro-*N*,*N*-dimethyl-10H-phenothiazine-10-propanamine, Merck Index, 2163	Rat brain	6.00	IC_{50}; 30 °C	FJALLAND et al. (1977)
		6.18	IC_{50}; 37 °C	LADURON and LEYSEN (1978)
		5.70	IC_{50}; 25 °C	NUKINA (1983)
		6.00	IC_{50}; 25 °C	SNYDER and YAMAMURA (1977)
Chlorprothixene; 3-(2-chloro-9H-thioxanthen-9-ylidene)-*N*,*N*-dimethyl-1-propanamine, Merck Index, 2166	Rat brain	7.25; 6.96	IC_{50}; *trans* and *cis* isomers 30 °C	FJALLAND et al. (1977)
Chlomiphene; 2[*p*-(2-chloro-1,2-diphenylvinyl)-phenoxy]-triethylamine, Merck Index, 2349	Rat hypothalamus	6.68; 4.15	K_1, K_2; 25 °C	BEN-BARUCH et al. (1982)
Choline; 2-hydroxy-*N*,*N*,*N*-trimethylethan-aminium, Merck Index, 2166	Rat brain	3.04	K_D; 30 °C	BIRDSALL et al. (1983)
		4.00; 3.40; 3.00	K_H; K_H; K_L; 30 °C	BIRDSALL et al. (1980)
		2.69	IC_{50}; 25 °C	PALACIOS and KUHAR (1979)
		2.87	IC_{50}; 37 °C	SPETH and YAMAMURA (1979)

Clomipramine; 3-chloro-10,11-dihydro-*N*,*N*-dimethyl-5H-dibenz[b,f]azepine-5-propanamine, Merck Index, 2350	Mouse brain	6.91	IC_{50}; 37 °C	Bohman et al. (1981)
	Rat brain	7.30	K_D	Aronstam and Norayanan (1981)
		6.51	IC_{50}; 30 °C	Fjalland et al. (1977)
		6.84	K_D; 25 °C	Hulme et al. (1978)
Clopenthixol; 4[3-(2-chloro-9H-thioxanthen-9-ylidene)-propyl]-1-piperazine-ethanol, Merck Index, 2357	Rat brain	5.22	IC_{50}; 30 °C	Fjalland et al. (1977)
Clozapine; 8-chloro-11-(5-methyl-1-piperazinyl)-5H-dibenzo-[b,e][1,4]-diazepine, Merck Index, 2378	Rat brain	7.31	IC_{50}; 30 °C	Fjalland et al. (1977)
		6.92	IC_{50}; 37 °C	Laduron and Leysen (1978)
		8.21	K_D; 20 °C	Smith and Huger (1983)
		7.59	IC_{50}; 25 °C	Snyder and Yamamura (1977)
Cyproheptadine; 4-(5H-dibenzo[a,d]-cyclohepten-5-ylidene)-1-methylpiperidine, Merck Index, 2766	Rat brain	7.24	IC_{50}; 30 °C	Fjalland et al. (1977)
Deanol; 2(dimethylamino)ethanol, Merck Index, 2825	Rat brain	2.68	K_I; 37 °C	Speth and Yamamura (1979)
Decamethonium; decamethylene bis[trimethylammoniumbromide], Merck Index, 2831	Rat brain	4.30	K_I; 25 °C	Dudai and Yavin (1978)
Desipramine; 10,11-dihydro-5-[3-(methylamino)-propyl]-5H-dibenz[b,f]azepine, Merck Index, 2886	Mouse brain	6.42	IC_{50}; 37 °C	Bohman et al. (1981)
	Rat brain	6.25	IC_{50}	Goldman and Erickson (1983)
		6.42	K_I; 30 °C	Golds et al. (1980)
		6.34	K_I; 25 °C	Hulme et al. (1978)
		6.77	IC_{50}; 25 °C	Snyder and Yamamura (1977)
Dexetimide; 3-phenyl-1′-(phenylmethyl)-[3,4′-bipipezidine]-2,6-dione, Merck Index, 2909	Human frontal cortex	8.70	IC_{50}; 37 °C	Wastek and Yamamura (1978)
	Mouse brain	8.72	IC_{50}; 25 °C	Mantione and Hanin (1980)
	Rat brain	10.00	K_D; 25 °C	Dudai and Yavin (1978)
		10.00	IC_{50}; 37 °C	Ehlert et al. (1980a)
		8.95	IC_{50}; 4 °C	Watson et al. (1982)
Dibenzepin; 10-[2-(dimethylamino)-ethyl]-5,10-dihydro-5-methyl-11H-dibenzo-[b,c][1,4]diazepin-11-one, Merck Index, 2979	Mouse brain	5.74	IC_{50}	Rehavi et al. (1977a)
Dibucaine; 2-butoxy-*N*-(2-diethylaminoethyl)-cinchoniamide, Merck Index, 3008	Rat cerebral cortex	4.51	K_D	Aronstam et al. (1980)
Dimethisoquin; 3-butyl-1-[2(dimethylamino)-ethoxy]-isoquinoline, Merck Index, 3208	Rat cerebral cortex	5.70	K_D	Aronstam et al. (1980)

Appendix (continued)

Ligand	Tissue	$-\log K_D$	Notes	References
2-Dimethylaminoethylbenzilate, Merck Index, 3225	Rat cerebral cortex	7.92	K_D; 30 °C	BIRDSALL et al. (1983a)
Disopyramide; [2-(di*iso*propylamino)ethyl]-2-phenyl-2-pyridineacetamide, Merck Index, 3378	Rat cerebral cortex	5.80	K_D; 25 °C	SCHREIBER et al. (1985)
Doxepin; 3-dibenz-[b,c]oxepin-11-(6H)-ylidene-*N*,*N*-dimethyl-1-propanamine, Merck Index, 3434	Mouse brain	6.97	IC_{50}; 37 °C	BOHMAN et al. (1981)
	Rat brain	6.12	IC_{50}; 30 °C	FJALLAND et al. (1977)
		6.74	K_D; 30 °C	GOLDS et al. (1980)
		7.36	IC_{50}; 25 °C	SNYDER and YAMAMURA (1977)
Flupentixol; 4-[3-[2-(trifluoromethyl)-3H-thioxanthen-9-ylidene]propyl]-1-piperazineethanol, Merck Index, 4092	Rat brain	5.41	IC_{50}; 30 °C	FJALLAND et al. (1977)
		5.32	K_D; 30 °C	GOLDS et al. (1980)
Fluphenazine; 4-[3-[2-(trifluoromethyl)-10H-phenothiazin-10-yl]-propyl]-1-piperazineethanol, Merck Index, 4094	Rat brain	5.10	IC_{50}; 30 °C	FJALLAND et al. (1977)
Furmethide; *N*,*N*,*N*-trimethyl-2-furanmethanaminium, Merck Index, 4190	Rat cerebral cortex	6.09; 4.32	K_H; K_L; 30 °C	BIRDSALL et al. (1983a)
		6.79; 5.70 4.40	K_{IH}; K_H; K_L; 30 °C	BIRDSALL et al. (1980)
		4.89	IC_{50}	BIRDSALL et al. (1978)
		5.40	K_D; 30 °C	BURGEN et al. (1974)
Haloperidol; 4-[4-(4-chloro-phenyl)-4-hydroxy-1-piperidinyl]-1-(4-fluorophenyl)-1-butanone, Merck Index, 4480	Rat brain	5.06	IC_{50}; 30 °C	FJALLAND et al. (1977)
		4.08	IC_{50}; 25 °C	NUKINA (1983)
Imipramine; 10,11-dihydro-*N*,*N*-dimethyl-5H-dibenz[b,f]-azepine-*S*-propanamine, Merck Index, 4817	Mouse brain	6.70	K_D	REHAVI et al. (1977b)
		6.70	IC_{50}; 37 °C	ROTTER et al. (1979)
	Rat brain	6.36	IC_{50}; 39 °C	FJALLAND et al. (1977)
		6.42	IC_{50}	GOLDMAN and ERICKSSON (1983)
		6.74	IC_{50}; 30 °C	GOLDS et al. (1980)
		7.03	K_D; 25 °C	HULME et al. (1978)
		7.29	K_D; 20 °C	SMITH and HUGER (1983)
		7.11	IC_{50}; 25 °C	SNYDER and YAMAMURA (1977)

Iprindole; 5-[3-(dimethylamono)propyl]-6,7,8,9,10,11-hexahydro-5H-cyclooct[6]-indole, Merck Index, 4930	Rat brain	5.19	IC_{50}	GOLDMAN and ERICKSSON (1983)
		5.26	IC_{50}; 30 °C	GOLDS et al. (1980)
Isopropamide; *γ*-(aminocarbonyl)-*N*-methyl-*N*,*N*-bis-(1-methylethyl)-*γ*-phenyl-benzene-propanaminium, Merck Index, 5051	Rat brain	8.52	IC_{50}; 37 °C	GOSSUIN et al. (1984)
		8.82	IC_{50}; 25 °C	YAMAMURA and SNYDER (1974a)
Lachesine; ethyl-(2-hydroxyethyl)dimethyl-ammonium benzilate, Merck Index, 5168	Rat brain	8.84	K_D; 30 °C	BIRDSALL et al. (1983a)
		8.28	IC_{50}	BIRDSALL et al. (1978)
		8.05	K_D; 30 °C	HULME et al. (1978)
		8.78	K_D; 30 °C	HULME et al. (1976)
Levetimide; (R)(−)-isomer of dexetimide	Mouse brain	4.79	IC_{50}; 25 °C	MANTIONE and HANIN (1980)
	Rat brain	6.00	IC_{50}; 37 °C	EHLERT et al (1980a)
		5.09	IC_{50}; 37 °C	GOSSUIN et al. (1984)
Lidocaine; 2-diethylamino-2′,6′-acetoxylidine, Merck Index, 5310	Rat brain	3.64	K_D	ARONSTAM et al. (1980)
		3.10	IC_{50}	COHEN-ARMON et al. (1985)
		4.09	K_D; 25 °C	FAIRHURST et al. (1980)
Loxapine; 2-chloro-11-(4-methyl-1-piperazinyl)-dibenz[b,f][1,4]oxyzepine, Merck Index, 5404	Rat brain	6.35	IC_{50}; 30 °C	FJALLAND et al. (1977)
Maprotiline; *N*-methyl-1,10-ethanoanthracene-9-(MOH)-propanamine, Merck Index, 5573	Rat brain	6.20	K_D; 30 °C	GOLDS et al. (1980)
		5.62	IC_{50}; 25 °C	NUKINA (1983)
Methacholine; 2-(acetoxy)-*N*,*N*,*N*-trimethyl-1-propanaminium, Merck Index, 5793, 5794 or 5795	Rat brain	6.45; 4.54	K_H; K_L (+)-isomer	BIRDSALL et al. (1983a)
		3.32; 3.32	K_H; K_L (−)-isomer	BIRDSALL et al. (1983a)
		6.93; 4.24	IC_{50}; *l*- and *d*-isomers, 37 °C	EHLERT et al. (1980a)
		4.67	IC_{50}; 20 °C	SMITH and HUGER (1983)
		4.58	IC_{50}; 4 °C	WATSON et al. (1982)
Methixene; 1-methyl-3-(9H-thioxanthen-9-yl-methyl)piperidine, Merck Index, 5855	Rat brain	7.31	IC_{50}; 30 °C	FJALLAND et al. (1977)
		8.52	IC_{50}; 25 °C	NUKINA (1983
Methotrimeprazine; 2-methoxy-*N*,*N*-trimethyl-10H-phenothiazine-10-propanamine, Merck Index, 5862	Rat brain	5.49	IC_{50}; 25 °C	NUKINA (1983)
Methscopolamine; 7-(3-hydroxy-1-oxo-2-phenyl-propoxy)-9,9-dimethyl-3-oxo-9-azonitricyclo-[3,3,1,0,2,4]-nonane bromide, Merck Index, 5881	Frog brain	6.90	K_D	BIRDSALL et al. (1980)
	Rat brain	9.46	K_D; 30 °C	HULME et al. (1976)

Appendix (continued)

Ligand	Tissue	$-\log K_D$	Notes	References
***N*-Methylatropine**; Merck Index, 878	Rat brain	9.15	IC_{50}	BIRDSALL et al. (1978)
		9.49	K_D; 30 °C	HULME et al. (1978)
***N*-Methyl-4-piperidinyl benzilate**	Rat cerebral cortex	9.26	IC_{50}; 25 °C	GURWITZ et al. (1984c)
Mianserin; 1,2,3,4,10,14b-hexahydro-2-methyl-dibenzo[c,f]pyrazino[1,2]azepine, Merck Index, 6050	Human caudate nucleus	7.02	K_D; 37 °C	TOLLEFSON et al. (1982)
	Rat brain	6.33	K_D; 30 °C	GOLDS et al. (1980)
Muscarine; tetrahydro-4-hydroxyl-*N*,*N*,*N*,5-tetramethyl-2-furanmethanaminium, Merck Index, 6156	Rat brain	4.70	K_D; D- and L-isomers	DUDAI 1980
		5.15	K_D; D- und L-isomers, 25 °C	DUDAI and YAVIN (1978)
Nortriptyline; 3-[10,11-dihydro-5H-dibenzo[a,d]-cyclohepten-5-ylidene]-*N*-methyl-1-propanamine, Merck Index 6558	Mouse brain	6.87	IC_{50}; 37 °C	BOHMAN et al. (1981)
		6.66	K_D; 25 °C	REHAVI et al. (1977b)
	Rat brain			
Oxapropanium iodide; *N*,*N*,*N*-trimethyl-1,3-dioxolane 4-methanaminium iodide, Merck Index, 6796	Rat cerebral cortex	5.34; 3.82	K_H; K_L	BIRDSALL et al. (1978)
Oxotremorine; 1-[4-(1-pyrrolidinyl)-2-butynyl]-2-pyrrolidione, Merck Index, 6819	Frog brain	6.44	IC_{50}	BIRDSALL et al. (1980)
	Human frontal cortex	6.10	IC_{50}; 37 °C	WASTEK and YAMAMURA (1978)
	Mouse brain	7.19; 6.01	K_H; K_L; 27 °C	COSTA et al. (1982)
		7.70; 5.79	K_D; K_L; 37 °C	SOKOLOVSKY et al. (1980)
	Rabbit brain	5.96	IC_{50}; 25 °C	YAVIN and HAREL (1979)
	Rat brain	6.30	K_D; 25 °C	DUDAI and YAVIN (1978)
		5.19	IC_{50}	GORISSEN et al. (1978)
		6.18	IC_{50}; 37 °C	GOSSUIN et al. (1984)
		6.00	K_D; 37 °C	GRAMMAS et al. (1983)
		5.18	IC_{50}; 37 °C	LADURON (1980)
		6.79	K_D; 32 °C	LUTHIN and WOLFE (1984)
		5.52	IC_{50}; 25 °C	NUKINA (1983)
		6.54	K_D; 20 °C	SMITH and HUGER (1983)
		6.21	K_D; 37 °C	SPETH and YAMAMURA (1979)
		7.46; 5.55	K_D; 37 °C	VAUQUELIN et al. (1982)

Oxotremorine M	Rat brain	5.32	IC_{50}; 4 °C	HULME et al. (1983)
	Rat cerebral cortex	7.44; 4.88	K_H; K_L; 30 °C	BIRDSALL et al. (1983b)
		8.40; 7.30; 5.11	K_H; K_H; K_L; 39 °C	BIRDSALL et al. (1980b)
		5.09	IC_{50}; 37 °C	JACOBSON et al. (1985)
Penfluridol; 1-[4,4-bis(4-fluorophenyl)butyl]-4-[4-chloro-3-(trifluoromethyl)phenyl]-4-piperidinol, Merck Index, 6939	Rat brain	5.05	IC_{50}; 30 °C	FJALLAND et al. (1977)
Perlapine; 6-(4-methyl-1-piperazinyl)merphantrhidine, Merck Index, 7040	Rat striatum	6.44	IC_{50}; 37 °C	LADURON and LEYSEN (1978)
Perphenazine; 4-[3-(2-chlorophenolthiazin-10-yl)-propyl]-1-piperazinethanol, Merck Index, 7044	Rat brain	4.96	IC_{50}; 30 °C	FJALLAND et al. (1977)
		5.00	IC_{50}; 25 °C	SNYDER and YAMAMURA (1977)
Phencyclidine; 1-(1-phenylcyclohexyl)piperidine, Merck Index, 7087	Mouse brain	5.04	K_D; 25 °C	GABRIELEWITZ et al. (1980)
	Rat brain	5.82	K_D	ARONSTAM et al. (1980)
		4.52	IC_{50}; 25 °C	VINCENT et al. (1978)
		5.79	K_D; 23 °C	WARD and TREVOR (1981)
		4.96	K_D; 25 °C	WEINSTEIN et al. (1983)
Physostigmine; 1,2,3,3a,8,8a-hexahydro-1,3,8-trimethylpyrrolo[2,3,6]indol-5-d-methylcarbamate, Merck Index, 7267	Rat brain	5.74	K_D	BIRDSALL et al. (1978)
Pilocarpine; 3-ethyldihydro-4-[(1-methyl-1H-imidazol-5-yl)methyl]-2-(3H)-furanone, Merck Index, 7301	Human frontal cortex	5.30	IC_{50}; 37 °C	WASTEK and YAMAMURA (1978)
	Rat brain	6.74; 5.80	K_H; K_L; 22 °C	ELLIS and HOSS (1980)
		5.15	K_D; 37 °C	GRAMMAS et al. (1983)
		5.10	IC_{50}; 25 °C	NUKINA (1983)
		5.24	K_D; 20 °C	SMITH and HUGER (1983)
		4.22	IC_{50}; 25 °C	YAMAMURA and SNYDER (1974a)
Pimozide; 1-[1-[4,4-bis(*p*-fluorophenyl)butyl)]-4-piperidyl]-2-benzimidazolinone, Merck Index, 7310	Rat brain	5.59	IC_{50}; 30 °C	FJALLAND et al. (1977)
	Rat striatum	5.35	IC_{50}; 37 °C	LADURON and LEYSEN (1978)
Pirenzepine; 5,11-dihydro-11-[(4-methyl-1-piperazinyl)-methyl]-6H-pyrido[2,36][1,4]benzodiazepin-6-one, Merck Index, 7365	Rat brain	7.72; 6.60	K_H; K_L; 30 °C	BIRDSALL et al. (1983)
		7.72; 6.40	K_H; K_L; 30 °C	BIRDSALL et al. (1983)
		7.46	K_D; 32 °C	LUTHIN and WOLFE (1984)
		7.93	IC_{50}; 0 °C	WATSON et al. (1982)

Appendix (continued)

Ligand	Tissue	$-\log K_D$	Notes	References
Procaine; 4-amino-benzoic acid 2-(diethylamino)-ethyl ester, Merck Index, 7659	Rat cerebral cortex	4.85	K_D	ARONSTAM et al. (1980)
Procyclidine; α-cyclohexyl-α-phenyl-1-pyrrolidine-propanol, Merck Index, 7667	Rat cerebral cortex	8.77	K_D; 30 °C	HULME et al. (1978)
Promazine; *N*,*N*-dimethyl-10H-phenothiazine-10-propanamine, Merck Index, 7688	Rat striatum	6.11	IC_{50}; 37 °C	LADURON and LEYSEN (1978)
Propiomazine; 1-[10-(2-dimethylaminopropyl)-phenothiazin-2-yl]-1-propanone, Merck Index, 7723	Rat cerebral cortex	7.96	K_D; 30 °C	HULME et al. (1978)
Protriptyline; *N*-methyl-5H-dibenzo[a,d]-cycloheptene-5-propylamine, Merck Index, 7804	Mouse brain	7.29	IC_{50}; 37 °C	BOHMAN et al. (1981)
Quinacrine; 6-chloro-9-[[4-(diethylamino)-1-methylbutyl]amino]-2-methoxyacridine, Merck Index, 7946	Rat cerebral cortex	6.02	K_D	ARONSTAM et al. (1980)
Quinidine; 6-methoxycinchonan-9-ol, Merck Index, 7965	Rat brain cortex	4.40	IC_{50}	COHEN-ARMON et al. (1985)
		4.85	IC_{50}; 25 °C	SCHREIBER et al. (1985)
3-Quinuclidinylacetate; Merck Index, 8005	Rat brain	5.75; 4.80	K_H; K_L; 30 °C	BIRDSALL et al. (1983b)
		6.47; 5.18	K_H; K_L; (+)-isomer 37 °C	RINGDAHL et al. (1982)
		4.92	K_D; (−)-isomer, 37 °C	RINGDAHL et al. (1982)
3-Quinuclidinylbenzilate; Merck Index, 8005	Human brain	9.70	IC_{50}	ENNA et al. (1976)
	Mouse brain	9.74	K_D; (−)-isomer	REHAVI et al. (1977b)
	Rat brain	9.70	IC_{50}	ENNA et al. (1976)
		10.15	K_D; 32 °C	LUTHIN and WOLFE (1984)
		9.52	IC_{50}; 25 °C	SNYDER and YAMAMURA (1977)
	Rat cerebral cortex	9.43; 8.12	K_D; (−)- and (+)-isomers, 25 °C	KLOOG et al. (1978)

Scopolamine; 6β,7β-epoxy-1αH, 5αH-tropan-3α-ol (−)-tropate, Merck Index, 8254	Human frontal cortex	9.09	IC_{50}; 37 °C	WASTEK and YAMAMURA (1978)
	Mouse brain	9.40	K_D	REHAVI et al. (1977a)
		9.30	K_D	REHAVI et al. (1977b)
	Rabbit brain	10.52	K_D; 25 °C	YAVIN and HAREL (1979)
	Rat brain	8.52	K_D	BAUMGOLD et al. (1977)
		8.00	IC_{50}; 37 °C	CHAKRABARTI et al. (1982)
		8.44	IC_{50}; 30 °C	FJALLAND et al. (1977)
		8.39	IC_{50}	GOLDMAN and ERICKSON (1983)
		8.70	IC_{50}; 25 °C	NUKINA (1983)
		9.47	K_D; 20 °C	SMITH and HUGER (1983)
		9.52	IC_{50}; 25 °C	SNYDER and YAMAMURA (1977)
Spiperone; 8-[4-(4-fluorophenyl)-4-oxobutyl]-1-phenyl-1,3,8-triazaspiro[4,5]-decan-4-one, Merck Index, 8596	Rat striatum	4.82	IC_{50}; 37 °C	LADURON and LEYSEN (1978)
Tetracaine; 4-(butylamino)benzoic acid 2-(dimethylamino) ethyl ester, Merck Index, 9014	Rat cerebral cortex	4.70	IC_{50}	ARONSTAM et al. (1980)
Tetraethylammonium bromide, Merck Index, 9024	Rat cerebral cortex	3.59	K_D	ARONSTAM et al. (1980)
Tetramethylammonium iodide, Merck Index, 9048	Rat cerebral cortex	4.28; 2.82	K_H; K_L; 30 °C	BIRDSALL et al. (1983b)
		4.15; 3.08	K_H; K_L; 30 °C	BIRDSALL et al. (1978)
Thiothixene; *N*,*N*-dimethyl-9-[3-(4-methyl-1-piperazinyl)propylidene] thioxanthene-2-sulphonamide, Merck Index, 9209	Rat brain	5.08	IC_{50}	FJALLAND et al. (1977)
Trihexylphenidylhydrochloride; α-cyclohexyl-α-phenyl-1-piperidinepropanol hydrochloride, Merck Index, 9501	Rat cerebral cortex	8.14	K_D; 30 °C	HULME et al. (1978)
Trimipramine; 10,11-dihydro-*N*,*N*-trimethyl-5H-dibenz[b,f]azepine-5-propanamine, Merck Index, 9528	Rat brain	6.98	IC_{50}; 25 °C	NUKINA (1983)
Verapamil; 5-[(3,4-dimethoxyphenethyl) methylamino]-2-(3,4-dimethoxyphenyl)-2-isopropylvaleronitride, Merck Index, 9747	Rat brain	5.05	K_D; 25 °C	FAIRHURST et al. (1980)

Merck Index, 10th edn. used throughout

Acknowledgements. The authors' research was supported by grants from the Swedish Medical Research Council and the National Institute of Mental Health, Bethesda, Maryland. The authors are indebted to Drs Öie Nordström and Björn Ringdahl for helpful discussions.

References

Akiyama K, Watson M, Roeske WR, Yamamura HI (1984) High-affinity [^{3}H]pirenzepine binding to putative M_1 muscarinic sites in the neuroblastoma x glioma hybrid cell line (NG 108-15). Biochem Biophys Res Commun 119:289–297

Aquilonius SM (1978) Physostigmine in the treatment of drug overdose. In: Jenden DJ (ed) Cholinergic mechanisms and psychopharmacology. Plenum, New York, pp 817–825 (Advances in behavioral biology, vol 24)

Aronstam RS, Eldefrawi ME (1979) Transition and heavy metal inhibition of ligand binding to muscarinic acetylcholine receptors from rat brain. Toxicol Appl Pharmacol 48:489–496

Aronstam RS, Narayanan L (1981) 3-Chlorodibenzazepine binding to cholinergic receptors and ion channels. Res Commun Chem Pathol Pharmacol 34:551–554

Aronstam RS, Abood LG, MacNeil MK (1977) Muscarinic cholinergic binding in human erythrocyte membranes. Life Sci 20:1175–1180

Aronstam RS, Eldefrawi AT, Eldefrawi ME (1980) Similarities in the binding of muscarinic receptor and the ionic channel of the nicotinic receptor. Biochem Pharmacol 29:1311–1314

Avissar S, Moscona-Amir E, Sokolovsky M (1982) Photoaffinity labeling reveals two muscarinic receptor macromolecules associated with the presence of calcium in rat adenohypophysis. FEBS Lett 150:343–346

Babkin BP (1950) Secretary mechanisms of the digestive glands, 2nd edn. Hoeber, New York

Barchas JD, Berger PA, Ciaranello RD, Elliott GR (1977) Physopharmacology. Oxford University Press, New York

Bartfai R, Anner J, Schultzberg M, Montelius J (1974) Partial purification and characterization on the muscarinic acetylcholine receptor from rat cerebral cortex. Biophys Biochem Res Commun 59:725–733

Bartfai R, Study RF, Greengard P (1977) Muscarinic stimulation and cGMP synthesis. In: Jenden DE (ed) Cholinergic mechanisms and psychopharmacology. Plenum, New York, pp 285–295

Bartfai T, Breakefield XO, Greengard P (1978) Regulation of synthesis of guanosine 3′5′-cyclic monophosphate in neuroblastoma cells. Biochem J 176:119–127

Bartus RT, Dean RL III, Beer B, Lippa AS (1982) The cholinergic hypothesis of geriatric memory dysfunction. Science 217:408–417

Baumgold J, Abood LG, Aronstam R (1977) Studies on the relationship of binding affinity to psychoactive and anticholinergic potency of a group of psychotomimetic glycolates. Brain Res 124:331–340

Becker CE, Tong TG, Boerner U, Roe RL, Scott RAT, MacQuarrie MB, Bartter F (1979) Diagnosis and treatment of Amanita phalloides type mushroom poisoning. West J Med 125:100–109

Beld AJ, Ariens EJ (1974) Stereospecific binding as a tool in attempts to localize and isolate muscarinic receptors. II. Binding of (+)benzetimide, (−)benzetimide and atropine to a fraction from bovine tracheae, smooth muscle and to bovine caudate nucleus. Eur J Pharmacol 25:203–209

Ben-Baruch G, Schreiber G, Sokolovsky M (1982) Cooperativity pattern in the interaction of the antiestrogen drug clomiphene with the muscarinic receptors. Mol Pharmacol 21:287–293

Berrie CP, Birdsall NJM, Burgen ASV, Hulme EC (1979) Guanine nucleotides modulate muscarinic receptor binding in the heart. Biochem Biophys Res Commun 87: 1000–1005

Berrie CP, Birdsall NJM, Hulme EC, Keen M, Stockton JM, Wheatley M (1986) Characteristics of two biochemical responses to stimulation of muscarinic cholinergic receptors. Trends Pharmacol Sci [Suppl]78:8–13

Birdsall NJM, Hulme EC (1983) Muscarinic receptor subclasses. Trends Pharmacol Sci 4:459–463

Birdsall NJM, Burgen ASV, Hulme EC (1978) The binding of agonists to brain muscarinic receptors. Mol Pharmacol 14:723–736

Birdsall NJM, Burgen ASV, Hulme EC (1980a) The binding properties of muscarinic receptors in the brain of the frog (*R. temporaria*). Brain Res 184:385–393

Birdsall NJM, Hulme EC, Burgen A (1980b) The character of muscarinic receptors in different regions of the rat brain. Proc R Soc Lond [Biol] 207 (1166):1–12

Birdsall NJM, Burgen ASV, Hulme EC, Wong EHF (1983a) The effects of p-chloromercuribenzoate on muscarinic receptors in the cerebral cortex. Br J Pharmacol 80:187–196

Birdsall NJM, Burgen ASV, Hulme EC, Wong EHF (1983b) The effect of p-chloromercuribenzoate on structure-binding relationships of muscarinic receptors in the rat cortex. Br J Pharmacol 80:197–204

Bohman B, Halaris A, Karbowski M (1981) Effects of tricyclic antidepressants on muscarinic cholinergic receptor binding in mouse brain. Life Sci 29:833–841

Brown DA, Adams PR (1982) Muscarinic suppression of a novel voltage-sensitive K^+-current in a vertebrate neuron. Nature 283:673–676

Brown JH, Brown SL (1984) Agonists differentiate muscarinic receptors that inhibit cyclic AMP formation from those that stimulate phosphoinositide metabolism. J Biol Chem 259:3777–3781

Burgen ASV, Spero L (1968) The action of acetylcholine and other drugs on the efflux of potassium and rubidium from smooth muscle of guinea pig intestine. Br J Pharmacol 34:98–115

Burgen ASV, Hiley CR, Young JM (1974a) The properties of muscarinic receptors in mammalian cerebral cortex. Br J Pharmacol 51:279–285

Burgen ASV, Hiley CR, Young JM (1974b) Binding of ^{3}H propyl benzilyl choline mustard by longitudinal muscle strips from guinea pig small intestine. Br J Pharmacol 50:145–151

Burn JH (1952) Practical pharmacology. Blackwell, Oxford

Bylund DB (1981) Interactions of neuroleptic metabolites with dopaminergic, alpha-adrenergic and muscarinic cholinergic receptors. J Pharmacol Exp Ther 217:81–86

Chakrabarti JK, Hotten TM, Morgan SE, Pullar IA, Rackham DM, Risius FC, Wedley S, Chaney MO, Jones ND (1982) Effects of conformationally restricted 4-piperazinyl-10H thienobenzodiazepine neuroleptics on central dopaminergic and cholinergic systems. J Med Chem 25:1133–1140

Chatelain P, Waelbroeck M, De Neef P, Robberecht P, Christophe J (1983) Increased number of high-affinity muscarinic receptors in rat heart after phenylhydrazine treatment. Eur J Pharmacol 91:279–282

Cheng Y-C, Prusoff WH (1973) Relationship between the inhibition constant (K_i) and the concentration of inhibitor which causes 50 per cent inhibition (IC_{50}) of an enzymatic reaction. Biochem Pharmacol 22:3099–3108

Cohen-Armon M, Henis YI, Kloog Y, Sokolovsky M (1985) Interactions of quinidine and lidocaine with rat brain and heart muscarinic receptors. Biochem Biophys Res Commun 127:326–332

Costa LG, Schwab BW, Murphy SD (1982) Differential alterations of cholinergic muscarinic receptors during chronic and acute tolerance to organophosphorus insecticides. Biochem Pharmacol 32:3407–3413

Dale HH (1914) The action of certain esters and ethers of choline and their relation to muscarine. J Pharmacol 6:147–190

Danielsson E, Peterson L-L, Grundin R, Ögren S-O, Bartfai T (1985) Anticholinergic potency of psychoactive drugs in human and rat cerebral cortex and striatum. Life Sci 36:1451–1457

Danielsson E, Unden A, Nordström Ö, Ögren S-O, Bartfai T (1986) Interactions of alaproclate, a selective 5-HT-uptake blocker, with muscarinic receptors: *in vivo* and *in vitro* studies. In: Hanin J (ed) Dynamics and function of the cholinergic systems. Plenum, New York, pp 415–422

Danielsson E, Eckernäs S-Å, Westlind-Danielsson A, Nordström Ö, Bartfai T, Wallin A, Gottfries C-G (1988) VIP-sensitive adenylate cyclase, guanylate cyclase, muscarinic receptors, choline acetyltransferase and acetylcholinesterase, in brain tissue afflicted by Alzheimer's disease/senile dementia of the Alzheimer type. Neurobiol Aging 9:153–162

Dixon RAF, Kobilka BK, Strader DJ, Benovik JL, Dohlman HG, Frielle R, Bolanowski MA, Bennett CD, Rands E, Diehl RE, Lefkowitz RJ, Strader CD (1986) Cloning of the gene and cDNA for mammalian β-receptors and homology with rhodopsin. Nature 321:75–79

Dudai Y (1980) Cholinergic receptors of Drosophila. In: Satelle DB (ed) Receptors, neurotransmitters, hormones, and pheromones of insects. Proc workshop, Cambridge, 1979. Elsevier, Amsterdam, pp 93–110

Dudai Y, Ben-Barak J (1977) Muscarinic receptor in Drosophila melanogaster demonstrated by binding of [^{3}H]quinuclidinyl benzilate. FEBS Lett 81:134–136

Dudai Y, Yavin E (1978) Ontogenesis of muscarinic receptors and acetylcholinesterase in differentiating rat cerebral cells in culture. Brain Res 155:368–373

Ehlert FJ (1985) The relationship between muscarinic receptor occupancy and adenylate cyclase inhibition in the rabbit myocardium. Mol Pharmacol 28:410–421

Ehlert FJ, Dumont Y, Roeske WR, Yamamura HI (1980a) Muscarinic receptor binding in rat brain using the agonist [^{3}H]cis-methyldioxolane. Life Sci 26:961–967

Ehlert FJ, Roeske WR, Yamamura HI (1980b) Regulation of muscarinic receptor binding by guanine nucleotides and N-ethylmaleimide. J Supramol Struct 14:149–162

El-Fakahany E, Richelson E (1983) Antagonism by antidepressants of mAChR of human brain. Br J Pharmacol 78:97–102

Ellis J, Hoss W (1980) Analysis of regional variations in the variations in the affinities of muscarinic agonists in the rat brain. Brain Res 193:189–198

Emmelin N (1967) Nervous control of salivary glands. In: Lode CF (ed) The handbook of physiology, sect 6, vol 2. American Physiological Society, Washington, pp 595–635

Engström C, Unden A, Ladinsky H, Consolo S, Bartfai T (1986) BM-5, a centrally active partial muscarinic agonist with low tremorogenic activity . In vivo and *in vitro* studies. Psychopharmacology (in press)

Enna SJ, Bird ED, Bennett JP, Bylund DB, Yamamura HI, Iversen LL, Snyder SH (1976) Huntington's chorea. Changes in neurotransmitter receptors in the brain. N Engl J Med 294:1305–1309

Fairhurst SD, Whittaker ML, Ehlert FJ (1980) Interactions of D600 (methoxyverapamil) and located anesthetics with rat brain -adrenergic and muscarinic receptors. Biochem Pharmacol 29:155–162

Farrar J, Hoss W (1984) Effects of copper on the binding of agonists and antagonists to muscarinic receptors in rat brain. Biochem Pharmacol 33:2849–2856

Fjalland B, Christensen AV, Hyttel J (1977) Peripheral and cental muscarinic receptor affinity of psychotropic drugs. Naunyn Schmiedebergs Arch Pharmacol 301:5–9

Gabrielewitz A, Kloog Y, Kalir A, Balderman D, Sokolovsky M (1980) Interaction of phecyclidine and its new adamantyl derivatives with muscarinic receptors. Life Sci 26:89–95

George W, Polson JB, O'Toole AG, Goldberg ND (1970) Elevation of guanosine 3'5'-cyclic phosphate in rat heart after perfusion with acetylcholine. Proc Natl Acad Sci USA 66:398–403

Gibson RE, Rzeszotarski WJ, Jagoda EM, Francis BE, Reba RC, Eckelman WC (1984) [^{125}I]3-quinuclidinyl-4-iodobenzilate: a high affinity high specific activity radioligand for the M_1 and M_2 acetylcholine receptors. Life Sci 34:2287–2296

Goldberg ND, Haddox MK (1977) Cyclic GMP metabolism and involvement in biological regulation. Annu Rev Biochem 46:823–896

Goldman ME, Erickson CK (1983) Effects of acute and chronic administration of antidepressant drugs on the central cholinergic nervous system. Comparison with anticholinergic drugs. Neuropharmacology 22:1215–1222

Golds PR, Przyslo FR, Strange PG (1980) The binding of some antidepressant drugs to brain muscarinic acetylcholine receptors. Br J Pharmacol 68:541–549
Gorissen H, Aerts G, Laduron P (1978) Characterization of digitonin-solubilized muscarinic receptor from rat brain. FEBS Lett 96:64–68
Gossuin A, Maloteaux JM, Trouet A, Laduron P (1984) Differentiation between ligand trapping into intact cells and binding on muscarinic receptors. Biochem Biophys Acta 804:100–106
Goyal RK, Rattan S (1978) Neurohormonal, hormonal and drug receptors for the lower esophagal spincter. Gastroenterology 74:598–619
Grammas P, Diglio AA, Marks BH, Giacomelli F, Wiener J (1983) Identification of muscarinic receptor in rat cerebral cortial microvessels. J Neurochem 40:645–651
Gurwitz D, Sokolovsky M (1980) Agonist-specific regulation of muscarinic receptors by transition metal ions and guanine nucleotides. Biochem Biophys Res Commun 94:1296–1304
Gurwitz D, Baron B, Sokolovsky M (1984a) Copper ions and diamide induce a high affinity guanine-nucleotide-insensitive state for muscarinic agonists. Biochem Biophys Res Commun 120:271–277
Gurwitz D, Razon N, Sokolovsky M, Soreq H (1984b) Expression of muscarinic binding sites in primary human brain tumors. Dev Brain Res 14:61–70
Gurwitz D, Kloog Y, Sokolovsky M (1984c) Recognition of the muscarinic receptor by its endogenous neurotransmitter: binding of [^{3}H]acetylcholine and its modulation by transition metal ions and guanine nucleotides. Proc Natl Acad Sci USA 81:3650–3654
Haga K, Haga T (1983) Affinity chromatography of the muscarinic acetylcholine receptor. J Biol Chem 258:13575–13578
Haga K, Haga T (1985) Purification of the muscarinic acetylcholine receptor from porcine brain. J Biol Chem 260:7927–7935
Haga K, Haga T, Ichiyama A, Katada T, Kurose H, Ui M (1985) Functional reconstitution of purified muscarinic receptors and inhibitory guanine nucleotide regulatory protein. Nature 316:731–733
Hammer R, Giachetti A (1982) Muscarinic receptor subtype: M_1 and M_2; biochemical and functional characterization. Life Sci 31:2991–2998
Hammer R, Berrie CP, Birdsall NJM, Burgen ASV, Hulme EC (1980) Pirenzepine distinguishes between different subclasses of muscarinic receptors. Nature 283:90–92
Hartzell HC (1981) Mechanisms of slow postsynaptic potentials. Nature 291:539–544
Hedlund B, Bartfai R (1981) Binding of ^{3}H-pilocarpine to membranes from rat cerebral cortex. Naunyn Schmiedebergs Arch Pharmacol 317:126–130
Hedlund B, Shepherd GM (1983) Biochemical studies on muscarinic receptors in the salamander olfactory epithelium. FEBS Lett 162:428–431
Hedlund B, Gamarra M, Bartfai T (1979) Inhibition of striatal muscarinic receptors *in vivo* by cadmium. Brain Res 168:216–218
Heiser JF, Gillin JC (1971) The reversal of anticholinergic drug induced delirium and coma with physostigmine. Am J Psychiatry 127:1050–1052
Hokin MR, Hokin LE (1952) Enzyme secretion and the incorporation of P^{32} into phospholipides of pancreas slices. J Biol Chem 203:967–977
Hulme EC, Burgen ASV, Birdsall NJM (1976) Interactions of agonist and antagonist with the muscarinic receptor. In: Smooth muscle pharmacology and physiology. INSERM 50:49–70
Hulme EC, Birdsall NJM, Burgen ASV, Mehta P (1978) The binding of antagonists to brain muscarinic receptors. Mol Pharmacol 14:737–750
Hulme EC, Berrie CP, Haga T, Birdsall NJM, Burgen ASV, Stockton J (1983) Solubilization and molecular characterization of muscarinic acetylcholine receptors. J Recept Res 3:301–311
Hurko O (1978) Specific [^{3}H]-quinuclidinyl benzilate binding activity in digitonin-solubilized preparations from bovine brain. Arch Biochem Biophys 190:434–445
Hyslop DK, Taylor DP (1980) The interaction of trazodone with rat brain muscarinic cholinoreceptors. Br J Pharmacol 71:359–361
Jacobson MD, Wusterman M, Downes CP (1985) Muscarinic receptors and hydrolysis of inositol phospholipids in rat cerebral cortex and parotial gland. J Neurochem 44:465–472

Järv J, Bartfai T (1982) Importance of hydrophobic interactions in the antagonist binding to the muscarinic acetylcholine receptor. Acta Chem Scand [B] 36:487–490

Järv J, Hedlund B, Bartfai T (1979) Kinetic studies on muscarinic antagonist-agonist competition. J Biol Chem 255:2649–2651

Jenden DJ (1965) Studies on tremorine antagonism by structurally related compounds. In: Costa E, Cote LJ, Yaln MD (eds) Biochemistry and pharmacology of the basal ganglia. Raven, New York, pp 159–170

Kato G, Carson S, Kemel ML, Glowinski J, Giorguieff MF (1978) Changes in striatal specific 3-H-atropine binding after unilateral 6-hydroxydopamine lesions of nigrostriatal dopaminergic neurones. Life Sci 22:1607–1614

Kilbinger H, Wagner B (1979) The role of presynaptic receptors in regulating acetylcholine release from peripheral cholinergic neurons. In: Langer SH, Starke K, Dubocowich ML (eds) Presynaptic receptors. Pergamon, Oxford, pp 347–351

Klein WL, Nathanson N, Nirenberg M (1979) Muscarinic acetylcholine receptor regulation by accelerated rate of receptor loss. Biochim Biophys Res Commun 90:506–512

Kloog Y, Sokolovsky M (1978) Studies on muscarinic AChR from mouse brain: characterization of the interaction with antagonists. Brain Res 144:31–48

Kloog Y, Michaelson DM, Sokolovsky M (1978) Identification of muscarinic receptors in the *Torpedo* electric organ. Evidence for their presynaptic localization. FEBS Lett 95:331–334

Kloog Y, Egozi Y, Sokolovsky M (1979) Characterization of muscarinic acetylcholine receptors from mouse brain: evidence for regional heterogenity and isomerization. Mol Pharmacol 15:545–558

Kubo T, Fukuda K, Mikami A, Maeda A, Takahashi H, Mishina M, Haga T, Haga K, Ichiyama A, Kangawa K, Kojima M, Matsuo H, Hirose T, Numa S (1986) Cloning, sequencing and expression of complementary DNA encoding the muscarinic acetylcholine receptor. Nature 323:411–416

Kuhar MJ, Birdsall NJM, Burgen ASV, Hulme EC (1980) Ontogeny of muscarinic receptors in rat brain. Brain Res 184:375–383

Kurose H, Ui M (1985) Dual pathways of receptor-mediated cyclic GMP generation in NG 108-15 cells as differentiated by susceptibility of islet-activating protein, pertussis toxin. Arch Biochem Biophys 238:424–434

Laduron P (1980) Dopamine and muscarinic receptors *in vivo* identification, axonal transport and solubilization. In: Yamamura HI, Olsen U (eds) Psychopharmacology and biochemistry of neurotransmitter receptors. Elsevier North-Holland, Amsterdam, pp 115–132

Laduron PM, Leysen JE (1978) Is the low incidence of extrapyramidal side-effects of antipsychotics associated with antimuscarinic properties? J Pharm Pharmacol 30:120–122

Laduron PM, Verwimp M, Leysen JE (1979) Stereospecific *in vitro* binding of [^{3}H]dexetimide to brain muscarinic receptors. J Neurochem 32:421–427

Lang W, Henke H (1983) Cholinergic receptor binding and autoradiography in brains of non-neurological and senile dementia of Alzheimer-type patients. Brain Res 267:271–280

Leslie EBA, Putney JW Jr, Sherman JM (1976) Adrenergic, β-adrenergic and cholinergic mechanisms for amylase secretion by rat parotid gland *in vitro*. J Physiol (Lond) 260:351–370

Levine RR, Birdsall NJM, Giachetti A, Hammer R, Iversen LL, Jenden DJ, North RA (eds) (1986) Subtypes of muscarinic receptors II. Trends Pharmacol Sci [Suppl] 78

Lopker A, Abood LG, Hoss W, Lionetti FJ (1980) Stereoselective muscarinic acetylcholine and opiate receptors in human phagocytic leucocytes. Biochem Pharmacol 29:1361–1365

Lundberg JM, Hedlund B, Bartfai T (1982) Vasoactive intestinal polypeptide enhances muscarinic ligand binding in cat submandibular salivary gland. Nature 295:147–149

Luthin GR, Wolfe BB (1984) Comparison of [^{3}H]pirenzepine and [^{3}H]quinuclidinylbenzilate binding to muscarinic cholinergic receptors in rat brain. J Pharmacol Exp Ther 228:648–655

Maayani S, Egazi Y, Pinchasi J, Sokolovsky M (1977) On the interactions of drugs with the cholinergic nervous system – IV. Biochem Pharmacol 26:1681–1687

Mantione CR, Hanin I (1980) Further characterization of human red blood cell membrane cholinergic receptors. Mol Pharmacol 18:28–32
McCormick DA, Prince DA (1986) Pirenzepine discriminates among ionic responses to acetylcholine in guinea-pig cerebral cortex and reticular nucleus of thalamus. Trends Pharmacol Sci [Suppl] 7:72–77
Meeker RB, Hardent TK (1982) Muscarinic cholinergic receptor-mediated activation of phosphodiesterase. Mol Pharmacol 22:310–319
Michell RH (1975) Inositol phospholipids and cell surface receptor function. Biochim Biophys Acta 415:81–147
Miller R, Hiley R (1975) Antimuscarinic actions of neuroleptic drugs. Adv Neurol 9:141–154
Milligan G, Strange PG (1984) Muscarinic acetylcholine receptors in neuroblastoma cells: lack of effect of veratrum alkaloids on receptor number. J Neurochem 43:33–41
Nathanson NM (1983) Binding of agonists and antagonists to muscarinic acetylcholine receptor on intact cultured heart cells. J Neurochem 41:1545–1549
Nathanson NM, Klein WL, Nirenberg M (1978) Regulation of adenylate cyclase activity mediated by muscarinic acetylcholine receptors. Proc Natl Acad Sci USA 75:1788–1791
Nilvebrant L, Sparf B (1982) Muscarinic receptor binding in the parotid gland. Different affinities of some anticholinergic drugs between the parotid pland and ileum. Scand J Gastroenterol [Suppl 72] 17:69–76
Nordström Ö, Bartfai T (1980) Muscarinic autoreceptor regulates acetylcholine release in rat hippocampus: *in vitro* evidence. Acta Physiol Scand 108:347–354
Nordström Ö, Westlind A, Unden A, Meyerson B, Sachs C, Bartfai T (1982) Pre- and postsynaptic muscarinic receptors in surgical samples from human cerebral cortex. Brain Res 234:287–297
North RA (1986) Muscarinic receptors and membrane ion conductances. Trends Pharmacol Sci [Suppl] 7:19–22
Nukina I (1983) Characteristics of muscarinic acetylcholine receptors in rat brain. Acta Med Okayama 37:179–191
Ögren S-O, Carlsson S, Bartfai T (1985) Serotonergic potentiation of muscarinic agonist evoked tremor and salivation in rat and mouse. Psychopharmacology 86:258–264
Onali P, Olianas MC, Schwatz JP, Costa P (1984) Molecular mechanisms involved in the inhibitory coupling of striatal muscarinic receptors to adenylate cyclase. Adv Biosci 48:163–174
Ovchinnikov Y (1982) Rhodopsin and bacriorhodopsins structure, function, relationships. FEBS Lett 148:179–191
Palacios JM, Kuhar MJ (1979) Choline: binding studies provide some evidence for a weak, direct agonist action in brain. Mol Pharmacol 16:1084–1088
Polak RL, Meeuws MM (1966) The influence of atropine on the release and uptake of acetylcholine by the isolated cerebral cortex of the rat. Biochem Pharmacol 15:989–992
Putney JW, Parod RJ (1978) Calcium-mediated effects of carbachol on cation pumping and sodium uptake in rat parotid gland. J Pharmacol Exp Ther 205:449–458
Raaijmakers JAM, Terpstra GK, van Rozen AJ, Witter A, Kreukniet J (1983) Muscarinic cholinergic receptors in peripheral lung tissue of normal subjects and of patients with chronic obstructive lung disease. Clin Sci 66:585–590
Rehavi M, Maayani S, Goldstein L, Assael M, Sokolovsky M (1977a) Antimuscarinic properties of antidepressants: dibenzepin (Noveril). Psychopharmacology 54:35–38
Rehavi M, Maayani S, Sokolovsky M (1977b) Tricyclic antidepressants as anti-muscarinic drugs: In vivo and *in vitro* studies. Biochem Pharmacol 26:1559–1567
Rehavi M, Maayani S, Sokolovsky M (1977c) Enzymatic resolution and cholinergic properties of ±3-quinuclidinol derivates. Life Sci 21:1293–1302
Rehavi M, Ramat O, Yavetz B, Sokolovsky M (1980) Amitryptiline: long-term treatment elevates alpha-adrenergic and muscarinic receptor binding in mouse brain. Brain Res 194:443–453

Repke H (1982) Regulatory process at the muscarinic acetylcholine receptor – an approach to a receptor model. Wiss Z Humboldt Univ Berlin Math Naturwiss Reihe 31:525–527
Richelson E, Divinetz-Romero S (1977) Blockade by psychotropic drugs of the muscarinic acetylcholine receptor in cultured nerve cells. Biol Psych 12:771–785
Ringdahl B (1984) Structural requirements for muscarinic receptor occupation and receptor activation by oxotremorine and analogs in the guinea pig ileum. Pharmacol Exp Ther 232:67–73
Ringdahl B, Ehler FJ, Jenden DJ (1982) Muscarinic activity and receptor binding of the enantiomers of acelidine and its methiodide. Mol Pharmacol 21:594–599
Robberecht P, Winaud J, Chatelain P, Poloczek P, Camus J-C, DeNeef P, Christophe JP (1981) Comparison of beta-adrenergic receptors and adenylate cyclase system with muscarinic receptors and guanylate cyclase activities in heart of spontaneously hypertensive rats. Biochem Pharmacol 30:385–387
Rodbell M (1980) The role of hormone receptors and GTP-regulatory proteins in membrane transduction. Nature 284:17–22
Rosenberger LB, Roeske WR, Yamamura HI (1979) The regulation of muscarinic receptor binding. J Biol Chem 255:820–823
Rotter A, Birdsall NJM, Burgen ASV, Field PM, Hulme EC, Raisman G (1979) Muscarinic receptors in the central nervous system of the rat. I. Technique for autoradiographic localization of the binding of [^{3}H]-propylbenzilylcholine mustard and its distribution in the forebrain. Brain Res 1:141–165
Rumack BH (1973) Anticholinergic poisoning: treatment with physostigmine. Pediatrics 52:449–451
Schreiber G, Sokolovsky M (1985) Muscarinic receptor heterogeneity revealed by interaction with bretylium tosylate. Different ligand-receptor conformations versus different receptor subclasses. Mol Pharmacol 27:27–31
Schreiber G, Friedeman M, Sokolovsky M (1984) Interactions of bretylium tosylate with rat cardiac muscarinic receptors: possible pharmacological relevance to antiarrhythmic action. Circ Res 55:653–659
Schreiber G, Barak A, Sokolovsky M (1985) Disopyramide and quinidine bind with inverse selectivity to muscarinic receptors in cardiac and extracardiac rat tissues. J Cardiovasc Pharmacol 7:390–393
Smith CP, Huger FP (1983) Effect of zinc on [^{3}H]QNB displacement by cholinergic agonists and antagonists. Biochem Pharmacol 32:377–380
Synder S, Greenberg D, Yamamura HJ (1974) Antischizophrenic drugs and brain cholinergic receptors: affinity for muscarinic sites predicts extrapyramidal effects. Arch Gen Psychiatry 31:58–61
Snyder SH, Yamamura HJ (1977) Antidepressants and the muscarinic acetylcholine receptor. Arch Gen Psychiatry 34:236–239
Sokolovsky M (1984) Muscarinic receptors. Int Rev Neurobiol 25:139–183
Sokolovsky M, Gurwitz D, Galron R (1980) Muscarinic receptor binding in mouse brain: regulation by guanine nucleotides. Biochem Biophys Res Commun 94:487–492
Sokolovsky M, Gurwitz D, Kloog Y (1983) Biochemical characterization of the muscarinic receptors. Adv Enzymol 137–196
Speth RC, Yamamura HI (1979) On the ability of choline and its analogues to interact with muscarinic cholinergic receptors in the rat brain. Eur J Pharmacol 58:197–201
Strange PG, Birdsall NJM, Burgen ASV (1978) Ligand-binding properties of the muscarinic acetylcholine receptor in mouse neuroblastoma cells. Biochem J 72:495–501
Szerb JC (1964) The effect of tertiary and quaternary atropine on cortical acetylcholine output and on the electroencephalogram in cats. Can J Physiol Pharmacol 42:303–314
Taylor P (1980) Cholinergic agonists. In: Goodman, Gilman (eds) The pharmacological basis of therapeutics. Macmillan, New York, pp 91–99
Tinkelberg JR, Berger PA (1977) Treatment of abusers of nonadditive drugs in psychopharmacology. In: Barchas JD et al. (eds) Oxford University Press, New York, pp 386–402
Tjörnhammar M-L, Bartfai T (1984) Interaction of drugs with the cyclic nucleotide generating enzymes in the central nervous system. Med Res Rev 3:513–534

Tollefson GD, Senogles SE (1983) A comparison of first and second generation antidepressants at the human muscarinic cholinergic receptor. J Clin Psychopharmacol 3:231–234
Tollefson GD, Senogles SE, Frey WH II, Tuason VB, Nicol SE (1982) A comparison of peripheral and central muscarinic cholinergic receptor affinities for psychotropic drugs. Biol Psychiatry 17:555–567
Uchida S, Takeyasu K, Matsuda T, Yoshida H (1979) Changes in muscarinic acetylcholine receptors of mice by chronic administrations of diisopropylfluorophosphate and papaverine. Life Sci 24:1805–1812
Vauquelin G, Andre C, De Backer J-P, Laduron P, Strosberg AD (1982) Agonist-mediated conformational changes of muscarinic receptors in rat brain. Eur J Biochem 125:117–124
Venter CJ, Fraser CM, Lindström J (1984) Monoclonal and anti-idiotypic antibodies. Probes for receptor structure and function. In: Venter CJ, Harrison L (eds) Receptor biochemistry and methodology, vol 4. Liss, New York, pp 117–140
Vincent JP, Cavey D, Kamenka JM, Geneste P, Lazdunski M (1978) Interaction of phencyclidines with the muscarinic and opiate receptors in the central nervous system. Brain Res 152:176–182
Wamsley JK, Gehlert DR, Roeske WR, Yamamura HI (1984) Muscarinic antagonist binding site heterogeneity as evidenced by autoradiography after direct labeling with [^{3}H]-QNB and [^{3}H]-pirenzepine. Life Sci 34:1395–1402
Ward D, Trevor A (1981) Phenylcyclidine-induced alteration in rat muscarinic cholinergic receptor regulation. Eur J Pharmacol 74:189–193
Wastek GJ, Yamamura HI (1978) Biochemical characterization of the muscarinic cholinergic receptor in human brain: alterations in Huntington's disease. Mol Pharmacol 14:768–780
Watson M, Roeske WR, Yamamura HI (1982) [^{3}H]pirenzepine selectively identifies a high affinity population of muscarinic cholinergic receptors in the rat cerebral cortex. Life Sci 31:2019–2023
Watson M, Roeske W, Johnson PC, Yamamura HI (1984) [^{3}H]pirenzepine identifies putative M_1 muscarinic receptors in human stellate ganglia. Brain Res 290:179–182
Watson M, Roeske WR, Vickroy TW, Smith TL, Akiyama K, Gulya K, Duckles SP, Serra M, Adem A, Nordberg A, Gehlert DR, Wamsley JK, Yamamura HI (1986) Mapping of subtypes of muscarinic receptors in the human brain with receptor autoradiographic techniques. Trends Pharmacol Sci [Suppl] 7:46–55
Weinstein H, Maayani S, Pazhenchevsky B (1983) Multiple actions of phencyclidine: discriminant structure-activity relationship from molecular conformations and assay conditions. Fed Proc 42:2574–2578
Westlind A, Grynfarb M, Hedlund B, Bartfai T, Fuxe K (1981) Muscarinic supersensitivity induced by septal lesion or chronic atropine treatment. Brain Res 225:131–141
Wise BC, Shoji M, Kuo J, Fazi T (1980) Decrease or increase in cardiac muscarinic cholinergic receptor number in rats treated with methacholine or atropine. Biochem Biophys Res Commun 92:1136–1142
Wood PL, Etienne P, Lal S, Nair NPV, Finlayson MH, Gauthier S, Palo J, Haltia M, Paetau A, Bird ED (1983) A post-mortem comparison of the cortical cholinergic system in Alzheimer's disease and Pick's disease. J Neurol Sci 62:211–217
Wurtman RJ, Corkin SH, Crowclen (eds) (1983) Alzheimer's disease: advances in basic research and therapeutics. Proceedings of the 3rd Meeting of the International Study Group on the Treatment of Memory Disorders with Ageing, Zürich
Yamamura HI, Snyder SH (1974a) Muscarinic cholinergic binding in rat brain. Proc Natl Acad Sci USA 71:1725–1729
Yamamura HI, Snyder SH (1974b) Muscarinic cholinergic receptor binding in the longitudinal muscle of the guinea pig ileum with [^{3}H]quinuclidinyl benzilate. Mol Pharmacol 10:861–867
Yavin E, Harel S (1979) Muscarinic binding sites in the developing rabbit brain: regional distribution and ontogenesis in the prenatal and early neonatal cerebellum. FEBS Lett 97:151–154
Zalcman SJ, Neckers LM, Kaayalp O, Wyatt RJ (1981) Muscarinic cholinergic binding sites on intact human lymphocytes. Life Sci 29:69–73

Part IV Cellular Organization of the Cholinergic System

CHAPTER 11

Cholinergic Synaptic Vesicles

H. ZIMMERMANN

A. Introduction

Axon terminals of cholinergic neurons are typically filled with electron-lucent synaptic vesicles. In the central nervous system, the ganglionic synapses and the neuromuscular junction (NMJ) the diameter of these vesicles is in the range of 40–50 nm. In the cholinergic nerve terminals of the electric organs of electric fishes, important as model cholinergic synapses (see Chap. 2), the diameter of electron-lucent vesicles may be greater (58 nm, electric eel; 90 nm, electric rays; for references, see VOLKNANDT and ZIMMERMANN 1986). It is this type of electron-lucent synaptic vesicle which is normally referred to as the cholinergic synaptic vesicle. It should be kept in mind however that cholinergic nerve terminals also typically contain a small number of dense-cored synaptic vesicles of a diameter of 80–150 nm. Since very little is known about the function and fate of these dense-cored vesicles and since the electric organ of electric rays as the source par excellence of cholinergic vesicles does not contain any, they have been neglected in the study of the biochemistry of cholinergic synaptic transmission.

The emphasis of the present review is on the biochemical, biophysical and functional properties of the cholinergic vesicles. In view of the limited space available special emphasis is placed upon recent biochemical developments. Various aspects of the cell biology and biochemistry of cholinergic synaptic vesicles have previously been reviewed (MACINTOSH and COLLIER 1976; BOYNE 1978; MELDOLESI et al. 1978; ZIMMERMANN 1979, 1981, 1982a, b, c; ISRAËL et al. 1979; KELLY et al. 1979; MORRIS 1980; MACINTOSH 1980; WHITTAKER and STADLER 1980; ZIMMERMANN et al. 1981; KELLY and HOOPER 1982; WHITTAKER 1984a, b).

B. Isolation of Cholinergic Synaptic Vesicles

Cholinergic synaptic vesicles may be isolated from a variety of vertebrate tissues innervated by cholinergic neurons. These include brain (WHITTAKER et al. 1963, 1964; DEROBERTIS et al. 1963), ox superior cervical ganglion (SCG) (WILSON et al. 1973), guinea-pig small intestine myenteric plexus (DOWE et al. 1980), rat diaphragm (VOLKNANDT and ZIMMERMANN 1986) and the electric organs of the electric catfish (*Malapterurus electricus*), electric eel (*Electrophorus electricus*) (VOLKNANDT and ZIMMERMANN 1986) and a variety of species of electric rays (Torpedinidae) (SHERIDAN et al. 1966; for further references, see ZIMMERMANN 1982a).

Whereas synaptic vesicles from brain and sympathetic ganglia are isolated by hypo-osmotic shock and subsequent density-gradient centrifugation of a synaptosome fraction previously prepared from these tissues, vesicles are liberated from the other tissues by breaking open nerve terminals directly on homogenization.

Since synaptic vesicles are osmotically labile structures (BREER et al. 1978) the various isolation steps are performed under iso-osmotic conditions. The introduction of zonal centrifugation (WHITTAKER et al. 1972) and column chromatography of synaptic vesicles prepurified by density-gradient centrifugation using either porous glass beads (MORRIS 1973) or Sephacryl-1000 (STADLER and TSUKITA 1984) were significant steps forward in obtaining a homogenous population of isolated particles. The principal experimental procedures for isolating cholinergic synaptic vesicles have been extensively reviewed (ZIMMERMANN 1982a). Although brain tissue was the first source from which synaptic vesicles were isolated, and acetylcholine (ACh) was the first neurotransmitter shown to be contained in brain synaptic vesicles, progress in understanding the molecular structure and function of cholinergic synaptic vesicles was only made when the electric organ of electric rays was introduced for vesicle isolation (*Torpedo*, SHERIDAN et al. 1966; ISRAËL et al. 1970). In contrast to brain tissue, with nerve terminals belonging to many different types of neurotransmitters and yielding mixed populations of synaptic vesicles, electric organs carry a purely cholinergic innervation. Furthermore, in electric rays the density of innervation is very high (1000 nmol ACh g^{-1} wet weight) and a large amount of tissue is available from one specimen (500–1000 g). Thus the report on properties of cholinergic vesicles will largely (and unless otherwise specified) review results obtained from electric ray electric organs.

C. General Properties

A summary of synaptic vesicle properties is presented in Table 1. Synaptic vesicles have a diameter of about 80 nm, with the membrane contributing about 5–7 nm. Intact vesicles from either *Torpedo* or *Narcine* electric organs sediment on top of a solution of 0.4 M sucrose and have a density of about 1.05 g ml^{-1}. This is considerably less than for adrenergic vesicles and reflects both the enrichment of the vesicle membrane in lipids and the absence of a proteinaceous core. The value for the membrane density (about 1.1 g ml^{-1}) is consistent with a low lipid/protein ratio. The vesicle membrane is the major vesicle constituent by mass.

From density-gradient measurements of intact and osmotically shocked synaptic vesicles (ghosts) in media exchanging different volume fractions of vesicle water (glycerol, dimethylsulphoxide, deuterium oxide) the volume fractions of the various vesicle compartments could be determined. The volume fraction of the hydrated membrane amounts to about 20% and that of the vesicle core to 80%. Osmotic lysis increases the fraction of osmotically active water from 65% to 76%, indicating a loss of some highly hydrated core material. This might reflect the water of hydration of ACh, ATP or metal ions or of the proteoglycan in the vesicle core (GIOMPRES et al. 1981a).

Synaptic vesicles from the *Narcine* electric organ are negatively charged at a pH of 6.4 and have an electrophoretic mobility of $3.8 \times 10^{-4}\ cm^2\ V^{-1} s^{-1}$.

Table 1. Structure and composition of electric ray synaptic vesicles

Physical parameters	Units	*Torpedo*		*Narcine*	
Diameter	(nm)				
Calculated from physical data		≈100[a]		81[b]	
Measured from electron micrographs		84–88[a]		91[c]	
Calculated membrane thickness	(nm)	4–5[a]		7[b]	
Particle density	(g ml^{-1})	1.057[a]		1.048[b]	
Density of membrane		1.132[a]		1.090[b]	
Osmotically active water	(glycerol space, % of volume)	65[a]		74[a]	
Internal pH		5.2–5.5[d]		–	
Membrane potential	(mV, negative inside)	60–80[e]		–	
Electrophoretic mobility	(cm^2 V $s^{-1} \times 10^{-4}$)	–		3.8[b]	
			calculated conc. (m*M*)		calculated conc. (m*M*)
Chemical composition					
Lipid	(10^{-18} g)	135[a]		55[b]	
Protein	(10^{-18} g)	80[a]		8.5[b]	
Proteoglycan	(10^{-18} g)	10[f]		2[g]	
ACh	10^4 molecules	19.9[a]	880[a]	5.1[b]	520[b]
ATP	10^4 molecules	2.7[a]	120[a]	1.8[b]	184[b]
GTP	10^4 molecules	–	–	0.3[b]	31[b]
Ca^{2+}	10^4 molecules	4.1[h] 2.7[i]	120[h]	–	–
Mg^{2+}	10^4 molecules	1.5[h] 1.4[i]	62[h]	–	–
Na^{2+}	10^4 molecules	7.9[h] 2.6[i]	115[h]	–	–
K^+	10^4 molecules	3.0[h] 0.6[i]	27[h]	–	–

Values are for individual particles. [a]From Table I of MORRIS (1980) and derived from the work of BREER et al. (1978), TASHIRO and STADLER (1978), OHSAWA et al. (1979). [b]WAGNER et al. (1978); [c]determined by MORRIS (1980); [d]MICHAELSON and ANGEL (1980), FÜLDNER and STADLER (1982); [e]LUQMANI (1981), ANGEL and MICHAELSON (1981); [f]STADLER and DOWE (1982); [g]CARLSON and KELLY (1983); [h]SCHMIDT et al. (1980), calculated from values obtained with vesicles harvested from a NaCl/sucrose gradient or [i]vesicles subsequently subjected to Millipore filtration; based on the ratio mol mol^{-1} ACh and assuming 19.9×10^4 molecules of ACh per vesicle

D. Vesicle Core

I. Constituents of the Core

Synaptic vesicles contain in their core high concentrations of ACh, ATP, metal ions such as Ca^{2+}, Mg^{2+}, Na^{2+} and K^{+} and also a proteoglycan with glycosaminoglycan side chains of the heparan sulphate type. Normal electron-lucent vesicles may also contain peptides, although peptide storage (possibly together with ACh) is usually regarded as typical for dense-cored synaptic vesicles.

II. ACh and ATP

Molar ratios of ACh to ATP for electric ray synaptic vesicles are about 5:1 (DOWDALL et al. 1974) (Table 1), whereas ratios for vesicles isolated either from the electric organs of electric catfish and electric eel or the diaphragm of the rat are around 10:1 (VOLKNANDT and ZIMMERMANN 1986). Both ACh and ATP are stable for many hours in isolated vesicles. Vesicles from electric rays also contain small amounts of ADP, AMP and GTP (ZIMMERMANN 1978; CARLSON et al. 1978). Whereas vesicular Ach can be regarded as the true signal-transmitting substance, the molecular function of ATP in vesicular storage, release or signal transmission is not yet established. ATP is released from the activated nerve terminal together with ACh (MOREL and MEUNIER 1981; SCHWEITZER 1987), and there is evidence that ATP may have a direct effect on the postsynaptic current influx (for references, see GRONDAL and ZIMMERMANN 1986). Recently it has been demonstrated that externally applied ATP causes inositol triphosphate accumulation in cultured myotubes, suggesting that ATP co-released with ACh could be involved in mobilizing Ca^{2+} in the postsynaptic muscle cell (HÄGGBLAD and HEILBRONN 1987).

Highly purified fractions of synaptic vesicles from electric ray electric organs may contain 7–8 μmol ACh mg^{-1} protein or 1–2 μmol ACh mg^{-1} phospholipid (TASHIRO and STADLER 1978, *Torpedo,* step gradient plus zonal gradient; WAGNER et al. 1978, *Narcine,* step gradient plus column chromatography). This has been calculated to correspond to 200000 (*Torpedo,* OHSAWA et al. 1979) or 47000 (*Narcine,* WAGNER et al. 1978) ACh molecules per vesicle or to ACh concentrations of 880 m*M* and 520 m*M*, respectively. Both values are in the order of magnitude of the osmolarity of the extracellular body fluid in electric rays (800–900 mOsm) and in principle would be consistent with a storage model in which vesicular ACh and ATP are in solution in the core, particularly since the solutions at this high ionic strength are non-ideal.

From evidence derived from proton-NMR and ^{31}P-NMR spectra of isolated *Torpedo* vesicles it has been concluded that ACh and ATP are dissolved in an essentially fluid phase within the vesicle core (STADLER and FÜLDNER 1980; FÜLDNER and STADLER 1982). However weak interactions with a solid matrix cannot be excluded. This is in accordance with results obtained with ACh-selective electrodes (KOBOS and RECHNITZ 1976). They indicate only minor molecular interactions of ACh and ATP under the conditions prevailing in the vesicle core, insufficient to reduce chemical or osmotic gradients significantly across the vesicle membrane.

Specific contents of ACh (μmol mg^{-1} protein) are lower for vesicles isolated from other sources (e.g. 13.9, guinea-pig cerebral cortex, NAGY et al. 1976; 4.4, ox SCG, WILSON et al. 1973; 31, guinea-pig small intestine myenteric plexus, DOWE et al. 1980; 36, electric eel electric organ, VOLKNANDT and ZIMMERMANN 1986). This does not necessarily imply that the osmotic load is reduced in these types of vesicles. Their radius is only about one-half that of electric ray vesicles and their tissue fluids have one-third the ray's osmolality.

Taking into account appropriate corrections, 1600 and 2000 ACh molecules, respectively, were assigned to synaptic vesicles isolated from guinea-pig cerebral cortex and calf SCG (WHITTAKER and SHERIDAN 1965; WILSON et al. 1973). These latter values would be well in agreement with electrophysiological estimates of the size of quantal packages released at the neuromuscular junction (1000–6000; for references, see ZIMMERMANN 1979). Unfortunately, corresponding electrophysiological estimates are missing for the electric ray electric organs. It may, however, be possible that larger transmitter packages need to be released at ray electroplaque cells in order to create a signal of identical amplitude as in the muscle subsynaptic membrane.

III. Metal Ions

The exact determination of vesicular metal ion contents (Table 1) is limited by the partial permeability of the vesicle membrane to these ions, in particular Na^+ and K^+ (SCHMIDT et al. 1980). Subcellular fractionation may cause redistribution of ions and result in an artifactual final vesicle composition in metal ions.

The concentration in Ca^{2+} ions is of particular interest since cholinergic vesicles contain an ATP-dependent Ca^{2+} uptake system (see below) which may be involved in the Ca^{2+} homeostasis of the nerve terminal cytoplasm. A Ca^{2+} level of between 60 and 183 nmol mg^{-1} protein has been estimated for *Torpedo* electric organ vesicles (ISRAËL et al. 1980; MICHAELSON et al. 1980; SCHMIDT et al. 1980). If a molar ratio for ACh to ATP of 5 is assumed, that of ATP to Ca^{2+} would be about 1 : 1 and the calculated intravesicular Ca^{2+} concentration in the range of 150 m*M*. This is unlikely to be the free intravesicular Ca^{2+} concentration as it would precipitate vesicular ATP. The vesicle core contains potential Ca^{2+} binding sites (such as the intravesicular proteoglycan and possibly gangliosides, see below) which are revealed as an electron-dense spot in vesicle profiles of intact tissue or after vesicle isolation when Ca^{2+}-containing fixatives are applied (BOYNE et al. 1974). Presumably the free intravesicular Ca^{2+} concentration is in the low micromolar range as in chromaffin granules (BULENDA and GRATZL 1985). On osmotic shock 75% of vesicular Ca^{2+} content is lost whereas Mg^{2+} (molar ratio ATP to Mg^{2+}, 0.35) remains tightly bound to vesicles. Stimulation of the nerve increases vesicular Ca^{2+} content by a factor of 2.5 (SCHMIDT et al. 1980).

IV. pH and Membrane Potential

The intravesicular pH was determined by following the self-quenching of fluorescence of a solution of 9-aminoacridine on addition of synaptic vesicles

(MICHAELSON and ANGEL 1980) or from the ^{31}P-NMR spectrum of model solutions containing ATP and Mg^{2+} (FÜLDNER and STADLER 1982). At an external pH of 7.4 vesicular pH values of 5.2 and 5.5 were calculated, respectively. The pH gradient is likely to be established by a Mg-ATP-dependent proton pump situated in the synaptic vesicle membrane. Both uptake and storage of ACh are affected by the vesicular proton electrochemical gradient (see below). The acidic intravesicular pH will also contribute toward stabilizing stores of vesicular ACh. The stability of ACh has an optimum at a pH of 4 and considerably decreases towards more alkaline pH values. Furthermore the release of H^+ ions into the synaptic cleft accompanying vesicle exocytosis could result in a transient inhibition of the pH-sensitive acetylcholinesterase (AChE).

As determined by analysis of the uptake of [^{3}H]triphenylmethylphosphonium or [^{3}H]tetraphenylphosphonium ions *Torpedo* vesicles have a resting membrane potential of 60–80 mV and are negative inside (LUQMANI 1981; ANGEL and MICHAELSON 1981).

V. Proteoglycan

Synaptic vesicles isolated from *Torpedo, Narcine* and *Discopyge* electric organs contain in their core a highly negatively charged proteoglycan (STADLER and DOWE 1982; CARLSON and KELLY 1983). The proteoglycan is located inside the vesicles since antibodies directed against it form immunoprecipitates only after lysis by detergent. The proteoglycan is lost from vesicles on prolonged dialysis. In *Torpedo* the glycosaminoglycan moiety has molar ratios for glucosamine: uronic acid: sulphate close to 1:1:1 which is characteristic of heparan sulphate. The greater part of the proteoglycan has a molecular weight of 350000 and higher. It amounts to about 20%–25% of the vesicular protein content and (in *Narcine*) this would correspond to about 10–20 proteoglycan molecules per vesicle.

The proteoglycan is highly antigenic, and antisera (WALKER et al. 1982, 1983) or monoclonal antibodies (tor 70, CARLSON and KELLY 1983) have been produced against it. The antibodies have been successfully employed to identify the proteoglycan immunohistochemically in a variety of cholinergic tissues. The antisera not only stained nerve terminals in the *Torpede* electric organ but also motor nerve terminals in the rat diaphragm (WALKER et al. 1982) and the electric organs of the electric eel and electric catfish (VOLKNANDT and ZIMMERMANN 1986) and presumptive cholinergic nerve endings in the inner and outer plexiform layers of the retinas of rat, ox and frog (OSBORNE et al. 1982). Using anti-proteoglycan antisera the presence of proteoglycan was also demonstrated in synaptic vesicles isolated from the electric organs of electric eel and electric catfish as well as from the diaphragm of the rat (VOLKNANDT and ZIMMERMANN 1986).

This interesting vesicle constituent whose antigenic sites are conserved from rays to mammals could serve a number of functions. Due to its high negative charge it could sequester intravesicular Ca^{2+} or ACh and contribute both to the osmotic balance and to the balance of charge inside the particle. Further possible functions would be in stabilizing the patch of vesicular membrane after exocytosis or in its release into the synaptic cleft and subsequent incorporation into the basal lamina (BUCKLEY et al. 1983; STADLER et al. 1985).

VI. Peptides

The presence of biologically active peptides in association with classical low molecular weight transmitters has led to the concept of transmitter co-existence and co-transmission at the nerve terminal. Typically peptides are packed in dense-cored synaptic vesicles rather than in electron-lucent synaptic vesicles. The occurrence of neuropeptides in cholinergic nerve terminals is the subject of a separate chapter (Chap. 17). On subcellular fractionation of cat submandibular glands using sucrose density gradients, vasoactive intestinal polypeptide (VIP) was found to co-sediment with a peak of ACh at 0.8 – 1.0 *M* sucrose (dense-cored vesicles) but not with a lighter peak of ACh at 0.3 – 0.4 *M* sucrose (electron-lucent vesicles) (LUNDBERG et al. 1981). The molar ratio of ACh to VIP in the dense fraction was 70 : 1. Similarly on subcellular fractionation of guinea-pig ileum a denser peak of ACh and VIP could be separated from a lighter ACh peak (ÁGOSTON et al. 1986). Although these studies are far from conclusive, they point to the possibility that in cholinergic nerve terminals the dense-cored vesicles may contain both peptide and neurotransmitter whereas the electron-lucent vesicles contain only ACh.

Torpedo nerve terminals and their electron-lucent synaptic vesicles appear to contain one or more peptides. According to ÁGOSTON and CONLON (1986) perikarya, axons and nerve terminals of the *Torpedo* electromotor system contain VIP-like immunoreactivity. MICHAELSON et al. (1984) describe an enkephalin-like substance in *Torpedo* synaptosomes, and according to STADLER et al. (1985) isolated vesicles contain a 7-kDa peptide which can be released by osmotic shock under acidic conditions.

E. Vesicle Membrane

I. Composition

Besides lipids which do not reveal any unusual composition the cholinergic synaptic vesicles from electric rays contain a set of proteins whose polypeptide pattern has been analysed. Generally the vesicle membrane has a rather low protein content. A number of these proteins either integrated or associated with the membrane have now been characterized with regard to molecular properties and/or function.

II. Membrane Lipids

The lipid composition has been determined for both *Torpedo* and *Narcine* synaptic vesicles (Table 2). The membrane is rich in lipids with a lipid to protein ratio (weight : weight) of between 3.5 : 1 and 5 : 1. Cholesterol is present in synaptic vesicle membranes in relatively high concentrations (cholesterol to phospholipid molar ratio, 0.4 to 0.5). Although earlier studies had reported the virtual absence

Table 2. Lipid composition of electric ray synaptic vesicles

	Torpedo	*Narcine*	Whole electric organ
General			
Lipid/protein (wt:wt)	0.7[a], 3,5[b], 2.1[e]	5.3[d]	
Cholesterol/phospholipid (molar ratio)	0.42[a]	0.39[d], 0.5[c]	
silaic acid/protein (μg mg^{-1})	2.3[g], 58 – 106[f]	–	
Phospholipids			
Phosphatidylcholine	46.6[b]	48.2[c]	38.5[c]
Phosphatidylethanolamine	29.5	30.0	24.5
Phosphatidylserine	12.6	9.2	–
Sphingomyelin	5.1	4.5	12.7
Phosphatidylinositol	5.1	2.6	–
Lysophosphatidylcholine	0.4	–	–
Lysophosphatidylethanolamine	–	3.0	–
Fatty acids			
16:0	–	31.2	23.9
18:0	–	12.7	12.7
16:1	–	5.6	6.8
18:1	–	10.8	13.9
20:4, 20:5, 22:5, 24:1, 22:4	–	7.7	22.2
22:6	–	29.1	15.4

[a]NAGY et al. (1976); [b]TASHIRO and STADLER (1978); [c]DEUTSCH and KELLY (1981); [d]WAGNER et al. (1978); [e]MICHAELSON et al. (1980); [f]LEDEEN et al. (1987); [g]calculated from NAGY et al. (1976)

of gangliosides from cholinergic vesicles (NAGY et al. 1976) a recent study (LEDEEN et al. 1987) suggests that gangliosides are a significant constituent of synaptic vesicles with polysialo species predominating. This finding could be of significance in relation to the conservation of the identity of vesicle lipids during recycling.

Synaptic vesicles contain all the usual phospholipids, with phosphatidylcholine (PC), phosphatidylethanolamine (PE), phosphatidylserine (PS), sphingomyelin and phosphatidylinositol (PI) (in this order) as the major constituents. Of particular interest is the virtual absence of lysolipids; their concentration would be too low to act as a fusogen for synaptic vesicle exocytosis. An analysis of fatty acids of *Narcine* synaptic vesicles revealed a high degree of unsaturation. Judged by their accessibility to membrane-impermeable reagents aminolipids such as PE and PS are present in significant amounts at the vesicle surface (60% – 85% and 40% – 100% respectively, DEUTSCH and KELLY 1981). Both would facilitate vesicle fusion and may be of importance in a mechanism of exocytosis which is dominated by lipid interaction. The overall lipid composition of synaptic vesicles is significantly different from those of the bulk of electric organ membranes, suggesting segregation of vesicle lipids through cycles of exo- and endocytosis (DEUTSCH and KELLY 1981).

III. Polypeptide Composition

The observed protein composition of synaptic vesicles is likely to be more sensitive to the degree of purity of the vesicle fraction than the lipid composition, which appears to be more uniform between the various types of cellular membranes. Conditions of tissue homogenization and media used throughout the purification procedure, e.g. addition of high concentrations of EGTA, can result in the loss of surface components associated with synaptic vesicles *in situ*. Even vesicles purified by both density-gradient centrifugation and column chromatography are likely to be no more than 80%–90% pure (WAGNER et al. 1978).

Analysis of the vesicular protein composition by sodium dodecylsulphate-polyacrylamide gel electrophoresis (SDS-PAGE) revealed only a limited number of protein species (Table 3) with about 300 copies per vesicle (WHITTAKER and STADLER 1980). The two most extensive studies (on *Torpedo,* TASHIRO and STADLER 1978; STADLER and TASHIRO 1979; and on *Narcine,* WAGNER et al. 1978; WAGNER and KELLY 1979) are only partly reconcilable. In *Torpedo* among the 13 components identified in the vesicle fraction, five (with molecular masses between 160 kDa and 25 kDa) appear to be particularly prominent and reproducible. Interestingly, synaptic plasma membranes are different in all major components (except actin) and appear to share only some minor components with synaptic vesicles. The estimated contents in the plasma membrane of less than 10% of vesicle proteins may indicate the presence of remnants of vesicular fusion in localized patches of the presynaptic plasma membrane, as has also become apparent using immunochemical techniques (BUCKLEY et al. 1983). In *Narcine* 8 major and 15 minor protein components were identified, of which six (75.5–19.5 kDa were enriched in fractions of chromatographically purified vesicles. Of these proteins five are exposed on the outer surface of the vesicle. It is not clear why high molecular weight components are absent from *Narcine* vesicles.

IV. Vesicular Proteins

1. Overview

Some of the protein molecular species which are an integral part of or associated with cholinergic synaptic vesicles have now been characterized (Table 4). They need not be identical with the major polypeptides identified by SDS-PAGE. All vesicular proteins hitherto identified have a membrane location, and even the vesicular proteoglycan with its luminal orientation is firmly anchored in the membrane via its protein moiety.

2. ATPase

An ATPase activity insensitive to ouabain and oligomycin (inhibitors of the Na^+, K^+-ATPase and the mitochondrial proton pump, respectively) is associated with synaptic vesicles. It is activated by Mg^{2+} and to a somewhat lesser degree by Ca^{2+} in the millimolar range (BREER et al. 1977; ROTHLEIN and PARSONS 1979; MICHAELSON and OPHIR 1980). The enzyme is modestly stimulated by

Table 3. Polypeptides present in cholinergic vesicles and plasma membranes isolated from the electric organ of *Torpedo marmorata*

Synaptic vesicles		Estimated no. of copies per vesicle	Synaptosomal plasma membrane		Preliminary identification
M_r ($\times 10^{-3}$)	Intensity (%)		M_r ($\times 10^{-3}$)	Intensity (%)	
220	12		220	5	
200	14		200	9	
160/147	100	13	160/147	16	Calmodulin binding, ATPase? Ca^{2+} transporter?
115	9		115	12	
100	<1		100	204	Na^+, K^+-ATPase
90	4		85	8	
76	16		76	112	Acetylcholinesterase
52	11		52	32	Subunit of ATPase?
42	101	102	42	72	Actin
37	19				
34	100	126	35 – 33	100	Atractyloside-binding site (nucleotide transporter)
32	2				
25	45	85	25	23	

Values represent densitometric tracings of protein patterns after gel electrophoresis. Peak hights of bands at molecular mass 34 kDa and 35 – 33 kDa, respectively, were taken as 100%. Major components are underlined. From STADLER and TASHIRO (1979), WHITTAKER and STADLER (1980) and references of Table 4

Table 4. Proteins of cholinergic vesicles

Type of protein	Presumed function	References
Ca^{2+}, Mg^{2+}-ATPase	Proton pumping	BREER et al. (1977), ROTHLEIN and PARSONS (1979), MICHAELSON and OPHIR (1980), ANDERSON et al. (1982), HARLOS et al. (1984), DIEBLER and LAZAREG (1985)
Atractyloside-binding protein	Nucleotide carrier	LUQMANI et al. (1981), STADLER and FENWICK (1983)
Protein kinase	Vesicle phosphorylation and Ca^{2+} uptake	REPHAELI and PARSONS (1982)
Actin	Connection to cytonet? endocytosis?	TASHIRO and STADLER (1978), WAGNER and KELLY (1979), ZECHEL and STADLER (1982)
Calmodulin	Activation of proteinkinase, exocytosis?	HOOPER and KELLY (1984a, b), WALKER et al. (1984)
Synapsin I	Exocytosis?	VOLKNANDT et al. (1987)
Proteoglycan	Storage of ACh, Ca^{2+}? synaptic modulation?	STADLER and DOWE (1982), CARLSON and KELLY (1983), WALKER et al. (1982, 1983b), JONES et al. (1982), BUCKLEY et al. (1983), VOLKNANDT and ZIMMERMANN (1986)

ACh (BREER et al. 1977) and also by bicarbonate ion (ROTHLEIN and PARSONS 1980). Bicarbonate stimulation (20%–50%) is observed in the presence of either Ca^{2+} or Mg^{2+} ions, Ca^{2+} being about half as effective as Mg^{2+}. It does not require Na, K or Cl. Bicarbonate is thought to act at an allosteric site outside the synaptic vesicle. Both high-affinity and low-affinity K_m values for the ATPase have been reported (Table 5). Bicarbonate converts the enzyme to a single high-affinity-high-velocity state (ROTHLEIN and PARSONS 1982).

The enzyme activity comigrates with the vesicle marker ATP under various experimental conditions such as equilibrium density and velocity centrifugation in glycerol gradients (ROTHLEIN and PARSONS 1979) or after valinomycin-induced ACh depletion which causes a massive shift in vesicle density (DIEBLER and LAZAREG 1985). The reduction of enzyme activity by tryptic digestion (MICHAELSON and OPHIR 1980) and the stability of ATP stored inside isolated vesicles both suggest an outward orientation of the ATPase. Inhibition experiments imply that the vesicle fraction may contain more than one type of ATPase. It is not clear at present whether this reflects an endogenous vesicular property or a contamination. Part of the ATPase activity detected in the vesicle fraction by the analysis of phosphate may be due to the presence of protein kinase and phosphatase in the fraction (REPHAELI and PARSONS 1982). The vesicular ATPase has been partially purified and assigned an apparent molecular mass of 250 kDa by gel filtration (HARLOS et al. 1984). Analysis of solubilized ATPase reveals a single 50-kDa band which can be photoaffinity-labelled with azido-ATP.

Table 5. Properties of ATPase associated with synaptic vesicles isolated from *Torpedo* electric organ

Species	K_m ATP (mM)	Activity (Mg^{2+}) (nmol Pi min^{-1} mg^{-1} protein)	Stimulation	Inhibition	References
Torpedo marmorata	–	100	Ca^{2+}, Mg^{2+}, ACh	–	BREER et al. (1977)
	0.01	–	Mg^{2+}	tris-*n*-butyltin, NBD-Cl, SITS, vanadate	DIEBLER and LAZAREG (1985)
Torpedo ocellata	1.5	235	Ca^{2+}, Mg^{2+}	DCCD, mersalyl, quercitin	MICHAELSON and OPHIR (1980), MICHAELSON et al. (1980)
Torpedo californica	0.01, 1.2 0.05[a], –[a]	52, 89, 232[a], –[a]	Ca^{2+}, Mg^{2+}, HCO_3^-, (K_m, 5 mM), FCCP, valinomycin, gramicidin	NBD-Cl, DCCD, SITS, DIDS, HCO_3^--stimulation blocked by ouabain and oligomycin	ROTHLEIN and PARSONS (1979, 1980, 1982), ANDERSON et al. (1982)

[a] Bicarbonate (20 mM) converts the enzyme into a single-high-affinity, high-velocity state; DCCD, *N,N'*-dicyclohexylcarbodiimide; NBD-Cl, 4-chloro-7-nitrobenzooxidazole; SITS, 4-acetamido-4'-isothiocyanostilbene-2,2'-disulphonic acid; DIDS, 4,4'-diisothiocyanostilbene-2,2'-disulphonic acid

With regard to subunit size and its solubility in dichloromethane it resembles the mitochondrial F_1-ATPase (HARLOS et al. 1984). Binding of the vesicle ATPase to a variety of lectins suggests that it is a glycoprotein, or that it interacts with glycolipids (BATTEIGER and PARSONS 1986).

The function of the vesicular ATPase has been implicated in proton pumping (ANDERSON et al. 1982; HARLOS et al. 1984) and also in vesicular Ca^{2+} transport (MICHAELSON et al. 1980). In contrast to previous reports that proton and monovalent ionophores have no significant effect on ATPase activity (MICHAELSON et al. 1980) the observation by ANDERSON et al. (1982) that carbonyl cyanide *p*-(trifluoromethoxy)phenylhydrazone (FCCP), valinomycin or gramicidin stimulate ATPase activity (up to 40%) implies that the ATPase is at least partially inhibited by an opposing electrochemical gradient of protons.

3. Atractyloside-Binding Protein

Atractyloside binds to a synaptic vesicle component of a molecular mass of 34 kDa and a basic isoelectric point (STADLER and FENWICK 1983). Atractyloside is known to bind to the ADP/ATP translocase of the inner mitochondrial membrane, a complex formed by two basic subunits of relative molecular mass of 30 kDa. Since atractyloside also inhibits vesicular uptake of ATP, it has been suggested that the 34 kDa polypeptide constitutes the vesicular nucleotide carrier, and that the mitochondrial and vesicular nucleotide translocase have a common evolutionary origin.

4. Protein Kinase

Synaptic vesicles isolated from the *Torpedo* electric organ in the presence of millimolar concentrations of EGTA and EDTA become phosphorylated on addition of Ca^{2+} and calmodulin (four polypeptides between 41 kDa and 64 kDa) (REPHAELI and PARSONS 1982). This suggests that a protein kinase is either closely associated with or endogenous to synaptic vesicles. Phosphorylation is not enhanced by cyclic AMP. Phosphorylation of synaptic vesicles results in an increase in vesicular Ca^{2+} uptake.

5. Actin

Actin is associated with synaptic vesicles from both *Torpedo* (TASHIRO and STADLER 1978) and *Narcine* (WAGNER and KELLY 1979). Based on two-dimensional gel electrophoresis and comparison of tryptic peptide maps it was concluded that the vesicle actin is an isoform different from the main component of the electric organ actin (ZECHEL and STADLER 1982).

6. Calmodulin

The calcium-binding protein calmodulin is associated with synaptic vesicles from *Discopyge* and *Narcine* as an integral membrane component independent of

Ca^{2+}. It can be solubilized only with detergents (HOOPER and KELLY 1984a). Two to four calmodulin molecules were calculated per vesicle. Using sheep anti-calmodulin antibodies synaptic vesicles can be precipitated, suggesting a surface location of the calmodulin. Furthermore, binding sites for calmodulin could be detected in isolated vesicles using a gel overlay technique and ^{125}I-labelled calmodulin (*Torpedo*, WALKER et al. 1984; *Narcine*, HOOPER and KELLY 1984b). Amongst the three to six molecular mass species which bound calmodulin only a 150/160-kDa band was common between the two studies. The 150-kDa component has been suggested to be a subunit of the vesicular ATPase (WALKER et al. 1984). Calmodulin binding is Ca^{2+}-dependent with an apparent dissociation constant of 10 n*M* and a maximal binding capacity of 80 pmol mg^{-1} protein (HOOPER and KELLY 1984b). Since intact synaptic vesicles can be bound to immobilized calmodulin a cytoplasmically oriented binding site is deduced. Calmodulin stimulates vesicular protein phosphorylation and Ca^{2+} transport.

7. Synapsin I

As revealed by immunocytochemistry, binding activity to an affinity-purified antibody directed against ox brain synapsin I (protein I) is present selectively in cholinergic electromotor nerve terminals of the electric organs of *Torpedo, Electrophorus* and *Malapterurus* as well as in the motor nerve terminals of the rat diaphragm. Analysis by immunoblots reveals synapsin I bands in these tissues, and synaptic vesicles isolated at low ionic strength from the *Torpedo* electric organ retain bound synapsin I (VOLKNANDT et al. 1987). This suggests that the phosphoprotein synapsin I is located at the vesicle surface also of cholinergic vesicles where it may be involved in the control of transmitter release, as has been implied for other synapses.

8. Calelectrin

Calelectrin, a calcium-binding protein of an apparent molecular mass of 34 kDa was found to bind to *Torpedo* synaptic vesicles and to aggregate them when applied in millimolar concentrations (WALKER et al. 1983a). It is not yet clear to what extent calelectrin interacts with synaptic vesicles *in situ*.

9. Proteins Common to Cholinergic and Non-cholinergic Vesicles

The availability of antisera specific to cholinergic vesicles and of monoclonal antibodies specific to determinants of cholinergic and non-cholinergic vesicles has allowed the comparison of antibody binding sites in a variety of tissues and vesicle preparations (see also Chap. 15). From these studies it can be concluded that a number of constituents are conserved in cholinergic vesicles from electric ray to mammals (e.g. proteoglycan and synapsin I). Furthermore, the concept is evolving that synaptic and a variety of other secretory vesicles may (despite distinct differences in molecular structure) carry common molecular constituents. An antiserum to *Narcine* synaptic vesicles, made vesicle specific by adsorption to non-vesicle membrane, was used to scan binding activity to a variety of tissues

(HOOPER et al. 1980). Immunofluorescence is found not only in the electric organ but also in the mammalian NMJ, sympathetic ganglionic and parasympathetic ganglionic cells and areas of the hippocampus and cerebellum that stain with AChE. They are also found in some non-cholinergic regions, including adrenergic sympathetic postganglionic terminals, peptidergic terminals in the neuroendocrine posterior pituitary and adrenal chromaffin cells. Interestingly, the endocrine anterior lobe of the pituitary is not stained.

For comparison a monoclonal antibody direct againt a *Discopyge* synaptic vesicle transmembrane glycoprotein (100 kDa) was found to be a general marker for neurons and endocrine cells including the anterior lobe of the pituitary and the islet cells in the pancreas. Exocrine tissues such as the pancreas acinar cells, the rat submaxillary gland and liver and the steroid-producing cells of the adrenal cortex, were not stained (BUCKLEY and KELLY 1985).

Similarly, an antiserum directed against *Torpedo* synaptic vesicles recognized a 86-kDa protein not only in *Torpedo* tissue but also in rat brain, but not in liver or adrenal glands. The 86-kDa antigen was greatly enriched in a rat brain synaptic vesicle fraction (WALKER et al. 1986). Binding activity for a monoclonal antibody (ASV 48) directed against a 65-kDa rat synaptic vesicle protein is common to both neurons and endocrine cells but not to exocrine cells (MATTHEW et al. 1981). This monoclonal antibody also binds to cholinergic synaptic vesicles isolated from rat diaphragm but not to those derived from electric organ, suggesting that the molecular species is either not present in electric organ vesicles or that is not recognized by the antibody (VOLKNANDT and ZIMMERMANN 1986).

F. Vesicular Transport Systems

I. Types of System Present

Cholinergic synaptic vesicles possess transport systems for protons, Ca^{2+} ions, ATP and ACh and possibly also for opioid peptides (Table 6). These processes are only incompletely analysed at the moment, and the carrier molecules involved are not characterized.

II. Vesicular Proton Pump

Since cholinergic synaptic vesicles have an internal acidic pH, it is likely that they possess a Mg-ATP-dependent proton pump as described for a variety of vesicle types, including chromaffin granules, coated vesicles and lysosomes. Vesicle ghosts obtained after lysis and resuspended in a medium iso-osmotic to the lysis medium (but not to intact vesicles) show uptake of [^{14}C]methylamine in the presence of Mg-ATP (HARLOS et al. 1984), suggesting that the vesicular ATPase is capable of pumping protons to acidify the vesicle interior. This notion is further supported by the finding that a number of mitochondrial uncouplers can stimulate vesicular ATPase activity (ANDERSON et al. 1982).

Table 6. Vesicular transport systems

Type of transport system	Kinetic properties					Stimulation	Inhibition	References
	K_T (mM)				V_{max} (nmol min^{-1} mg^{-1} protein)			
	Ca^{2+}	ATP	ACh	E[b]				
Proton pump	–	–	–	–		Mitochondrial uncouplers	–	ANDERSON et al. (1982), HARLOS et al. (1984)
Ca^{2+} transport	0.050	0.005	–	–	3	–	DCCD, mersalyl, quercitin	MICHAELSON et al. (1980)
	0.005	0.030	–	–	1.4	oxalate	–	ISRAËL et al. (1980)
	0.052	–	–	–	3.4	calmodulin	–	REPHAELI and PARSONS (1982)
	0.012[a]	–	–	–	5.2[a]			
ATP transport	–	1.15	–	–	1.2	–	atractyloside	LUQMANI (1981)
ACh transport	–	–	3.1	–	0.5	–	–	GIOMPRESS and LUQMANI (1980)
	–	–	–	–	3.9	–	–	DIEBLER and MROT-GAUDRY (1981)
	–	–	0.3, 10	–	4.7, 138	HCO_3^-	–	MICHAELSON and ANGEL (1981)
	–	–	0.3	–	1.6	HCO_3^-	AH 5183 (vesamicol) K_I, 40 nM	PARSONS and KOENIGSBERGER (1980), BAHR and PARSONS (1986a, b)
Enkephalin transport	–	–	–	0.012	0.00085	Mg^{2+}-ATP above 10 mM	–	DAY et al. (1985), MICHAELSON and WIEN-NAOR (1987)

[a] In the presence of 20 µg ml^{-1} calmodulin; abbreviations as in Table 5
[b] [D-Ala^2,D-LEU^5]-enkephalin

III. Vesicular Uptake of Ca^{2+}

Cholinergic synaptic vesicles display uptake of Ca^{2+}; this is stimulated by Mg-ATP (MICHAELSON et al. 1980; ISRAËL et al. 1980; REPHAELI and PARSONS 1982). Michaelis constants for ATP-dependent Ca^{2+} uptake between 5 and 50 μM were reported. Uptake is stimulated by oxalate and the Ca^{2+} taken up is released by Ca^{2+} ionophore.

The molecular mechanism underlying Ca^{2+} uptake is not clear. Since Ca^{2+} is expected to be present in the nerve terminal cytoplasm at concentrations of 1 μM or lower, a vesicular Mg-ATPase driving Ca^{2+} transport should be activated by low (micromolar) Ca^{2+} concentrations. This however is not so (REPHAELI and PARSONS 1982; DIEBLER and LAZAREG 1985). Interestingly, vesicular ATP-dependent Ca^{2+} uptake is stimulated by exogenous calmodulin (REPHAELI and PARSONS 1982). Calmodulin lowers the Michaelis constant for Ca^{2+} uptake by a factor of 5 (from 52 to 12 μM) with a nearly 2-fold increase in uptake rate. Four vesicle polypeptides are phosphorylated under these conditions. Most importantly, vesicles which have previously been phosphorylated and washed still exhibit an enhanced uptake of Ca^{2+} independent of ATP. However, vesicular ATPase activity is not stimulated by calmodulin. These results suggest that the function of ATP in vesicular Ca^{2+} uptake may be in the activation (phosphorylation) of the porter rather than in directly providing the energy for the Ca^{2+} transport.

IV. Vesicular Uptake of Opioid Peptides

Cholinergic neurons in the *Torpedo* electric organ have been reported to contain an enkephalin-like peptide, regulatory presynaptic opiate receptors and aminopeptidase and enkephalinase activities. Recently the uptake of tritiated [D-Ala^{2},D-Leu^{5}]-enkephalin as a non-hydrolysable analogue of Leu-enkephalin into both synaptosomes and synaptic vesicles from the *Torpedo* electric organ has been described (DAY et al. 1985; MICHAELSON and WIEN-NAOR 1987). The vesicular uptake system has a low Michaelis constant K_T, 12 μM) and is saturable and temperature-dependent. It is not affected by morphine and NaCl, which enhance synaptosomal enkephalin uptake. The maximal uptake rate is very much smaller than that of the other vesicle components (Table 6). The study implies that *Torpedo* nerve terminals can take up enkephalin-like peptides and translocate them from the cytosol into synaptic vesicles.

V. Vesicular Uptake of ATP

Loading of synaptic vesicles with ATP takes place in the intact nerve terminal. Vesicular ATP can be labelled by extracellular application of [^{3}H]adenosine (ZIMMERMANN 1978), and vesicular refilling with ATP in previously stimulated nerve terminals is independent of reloading with ACh (ZIMMERMANN and BOKOR 1979). Isolated and intact synaptic vesicles take up ATP (K_T, 1.15 mM, LUQMANI 1981); uptake is strongly temperature dependent and is greatest when the external pH is acidic in the range between 6.0 and 6.3. It is not entirely specific for ATP, but ATP is taken up in preference to GTP, ADP and AMP. Nucleosides show no uptake. The uptake of radiolabelled ATP was estimated to be equivalent to 20%–50% of the initial endogenous ATP content of vesicles. Atractyloside blocks vesicular uptake of ATP (STADLER and FENWICK 1983), and an antiserum directed against the 34-kDa atractyloside-binding protein of synaptic vesicles stimulates ATP-uptake severalfold.

VI. Vesicular Uptake of ACh

1. Types of *In Vitro* Uptake Observed

In intact nerve terminals of resting or stimulated *Torpedo* electric organ, synaptic vesicles can be loaded with ACh derived from extracellularly applied radiolabelled choline or acetate (see Chap. 7 and TUČEK 1978). The demonstration of an ACh-uptake process under *in vitro* conditions has been beset by great difficulties, presumably originating from the lability of the synaptic vesicle structure and the lack of control of the effect on the vesicular ion and/or protein composition by the various isolation procedures. Washing of isolated vesicles, for example, leads to a substantial loss in transport activity. Vesicular ACh uptake may depend on the concentration of vesicles in the incubation medium and vesicle fractions can vary in their characteristics from one preparation to another (BAHR and PARSONS 1986a).

Studies originally demonstrated passive uptake of ACh into either intact or lysed synaptic vesicles (e.g. CARPENTER and PARSONS 1978; CARPENTER et al. 1980; GIOMPRES and LUQMANI 1980; MICHAELSON and ANGEL 1981; DIEBLER and MOROT-GAUDRY 1981). This uptake was not substantially concentrative and a significant feature observed in the various studies was the lack of specificity for uptake of ACh over uptake of choline. Using the voltage-sensitive dye 3,3′-dipropylthiadicarbocyanine [diS-C_3-(5)], vesicles isolated in high NaCl concentrations and resuspended in a low NaCl medium could be shown to become hyperpolarized by gramicidin. When ACh was present in the external solution as well, spontaneous depolarization due to ACh uptake occurred (CARPENTER et al. 1980). This was the first evidence that ACh (but also other monovalent cations) are taken up via an electrogenic process which can be driven by the efflux of cations from vesicles.

2. Concentrative Uptake

In the presence of bicarbonate ion and either Mg-ATP or Ca-ATP (but not in the absence of bicarbonate) cholinergic vesicles take up ACh by a temperature-dependent and saturable process resulting in concentrative ACh accumulation (2- to 4-fold). The uptake is independent of external Na^+, K^+ or Cl^-. The presence of bicarbonate and Mg-ATP selectively stimulates uptake of ACh whereas (passive) uptake of choline remains low (KOENIGSBERGER and PARSONS 1980; PARSONS and KOENIGSBERGER 1980; PARSONS et al. 1982). The dissociation constant of the bicarbonate-induced uptake process is 0.3 mM with a maximum velocity of 1.6 nmol ACh min^{-1} mg^{-1} protein (ANDERSON et al. 1983a, b; BAHR and PARSONS 1986a, b). Up to 320 molecules may be actively transported per synaptic vesicle (ANDERSON et al. 1986). This is approximately 0.2% of the total vesicular ACh content.

Further experiments demonstrate a close linkage between the vesicular ATPase (as a possible proton-motive force) and vesicular ACh uptake. Whereas mitochondrial uncouplers like FCCP, nigericin or valinomycin all stimulate vesicular ATPase by uncoupling it from an opposing electrochemical proton gradient, they inhibit ACh uptake. Accordingly, the uptake of ACh is increased when the external pH is raised from 6.6 to 7.6, achieving a 12-fold selectivity in uptake of ACh over choline. In addition exogenous ammonium ion which dissipates any vesicular proton gradient inhibits the active uptake of ACh. Finally, ADP and adenosine-5′-(β,γ-methylenetriphosphate) as a non-hydrolysable ATP analogue do not support uptake of ACh (ANDERSON et al. 1982). The question is still unresolved as to why concentrative uptake of ACh is supported only in the presence of both Mg/Ca-ATP and bicarbonate although vesicular ATPase is activated by Mg/Ca-ATP and only moderately stimulated by bicarbonate. Similarly studies on intact PC-12 cells which store ACh in granules imply that vesicular ATPase and a proton gradient are involved in vesicular uptake and storage (TOLL and HOWARD 1980; MELEGA and HOWARD 1981). This is supplemented by the observation (MICHAELSON and ANGEL 1980; ANGEL and MICHAELSON 1981) that vesicular storage of ACh depends on an intact proton electrochemical gradient.

The studies suggest that under passive conditions (vesicles hyperpolarized, internally negative) ACh spontaneously enters vesicles with little selectivity over other monovalent cations incuding choline. Vesicles ‘energized’ by the ATPase would internally be acid and positively charged. This would increase the proton motive force in the direction of proton efflux, and this could be used to drive a selective and active uptake of ACh via antiport through a separate ACh porter. Possibly the structure of the porter depends on the vesicle membrane potential, and in the more ‘energized’ vesicle the conformation of the ACh porter changes so as to become more specific for ACh (ANDERSON et al. 1982). In this respect ACh uptake and storage would be linked to proton electrochemical gradients similar to catecholamines. However in contrast to catecholamines, ACh cannot be protonated on uptake, suggesting differences in uptake and/or storage in the two cases.

3. Specific Inhibition

On screening a large number of potential inhibitors of ACh uptake, DL-trans-2-(4-phenylpiperidino)cyclohexanol (AH 5183, vesamicol) which has previously been shown to block neuromuscular transmission was found to be very potent (ANDERSON et al. 1983a, b; MARSHALL and PARSONS 1987). It is a noncompetitive and reversible inhibitor with respect to ACh, with an inhibition constant of 41 n*M*, and has no effect on the vesicular ATPase. It is not clear at the moment whether the substance binds directly to the presumed ACh transporter, since neither choline nor Ach compete effectively with the radiolabelled drug for vesicle binding. About four enantioselective binding sites for vesamicol have been calculated per vesicle (BAHR and PARSONS 1986a, b). Vesamicol has the property of selectively releasing new vesicular ACh from previously loaded vesicles (ANDERSON et al. 1986), indicating some degree of heterogeneity in the intravesicular storage of ACh (for references, see ZIMMERMANN 1982b).

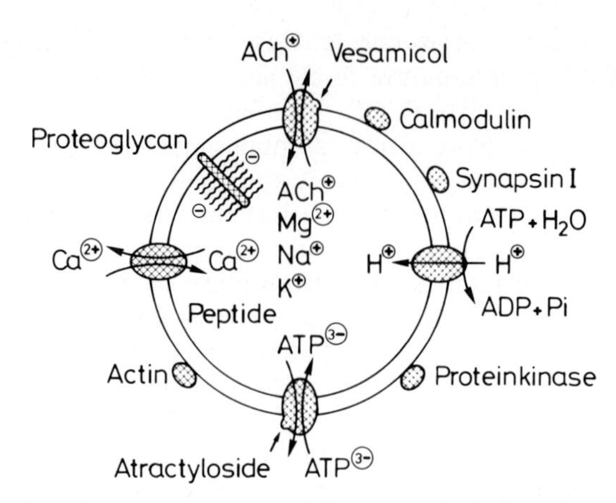

Fig. 1. Aspects of molecular structure and function of cholinergic synaptic vesicles

Figure 1 summarizes some of the basic molecular properties of cholinergic synaptic vesicles.

G. Dynamic Properties of Synaptic Vesicles

I. Morphological Alterations on Nerve Stimulation

1. Changes in Vesicle Number and Uptake of Marker

Synaptic vesicles have long been implicated in exocytotic transmitter release, and the vesicle hypothesis of neurotransmitter release is largely based on work on cholinergic synapses. This aspect has been intensively reviewed (ZIMMERMANN 1979, 1982c; CECCARELLI and HURLBUT 1980), and only recent developments are discussed here.

The fine-structure of the cholinergic nerve terminal changes on stimulation and the degree of alteration depends on the stimulation conditions used (for references, see MELDOLESI et al. 1978; ZIMMERMANN 1979).

High-frequency stimulation causes a significant fall in the number of synaptic vesicles per unit area of presynaptic profile at the NMJ and at electric organ synapses. This is accompanied by an increase in the area of the nerve terminal plasma membrane and is indicative of an exocytotic incorporation of the synaptic vesicle membrane into the plasma membrane. Moderate stimulation does not result in depletion of synaptic vesicle numbers. The uptake of extracellular markers such as dextran, ferritin or horseradish peroxidase into the lumen of a large proportion of synaptic vesicles demonstrates, however, that exocytotic fusion and retrieval of synaptic vesicles is also taking place under these experimental conditions. Clathrin coats are likely to be involved in synaptic vesicle retrieval, but the demonstration of coated vesicles in electron micrographs of stimulated tissue appears to vary with the fixation methods used (HEUSER and REESE 1973; CECCARELLI et al. 1973). Experiments using black widow spider venom for inducing transmitter release suggest that extracellular Ca^{2+} is essential for synaptic vesicle retrieval (CECCARELLI et al. 1979a).

2. Quick-Freezing Evidence for Vesicle Exocytosis

Additional unequivocal evidence for a fusion of cholinergic synaptic vesicles with the presynaptic plasma membranes comes from freeze-fracture experiments at the NMJ (e.g. CECCARELLI et al. 1979b; HEUSER et al. 1979). Activation of the nerve terminals causes dimples (P-face) and corresponding protuberances (E-face) at a double row of particles in the presynaptic plasma membrane corresponding to the active zone of the synapse. Further, the appearance of rows of omega-shaped fusion profiles of synaptic vesicles at the active zone has also been shown in normal thin sections of stimulated and quick-frozen nerve endings (TORRI-TARELLI et al. 1985). Interestingly, fusion profiles of dense-cored vesicles could not be observed in this study. Possibly exocytosis of dense-cored vesicles takes place outside the active zone of nerve terminals (ZHU et al. 1986).

3. Immunocytochemical Demonstration of Vesicle Fusion and Retrieval

The incorporation of synaptic vesicle constituents into the presynaptic plasma membrane on induced transmitter release can be monitored by immunocytochemical techniques (see also Chap. 5). On application of anti-proteoglycan antibodies to intact nerve terminals in the *Torpedo* electric organ (JONES et al. 1982) or of *Narcine* vesicle-specific antisera to the frog NMJ (VON WEDEL et al. 1981) stimulation causes a strong increase in immunofluorescence. Since in the case of the proteoglycan the antibody binding sites were inside synaptic vesicles, this demonstrates vesicle fusion and the exposure of the vesicle's inner membrane face to the synaptic cleft. During a period of rest following stimulation, immunofluorescence fades (JONES et al. 1982), a sequel consistent with a retrieval of synaptic vesicles, recovery of synaptic vesicle membranes and reestablishment of the original synaptic fine-structure as previously observed with the electron microscope.

All these results suggest that activation of the synapse at physiological levels induces vesicle exo- and endocytosis. In the light of the biochemical findings that the protein composition of synaptic vesicles and synaptic plasma membranes is quite different, the results furthermore imply that vesicle membrane constituents are not dispersed after fusion (e.g. by lateral diffusion), but that the vesicle membrane retains its identity and is retrieved from the plasma membrane by a specific process.

4. Changes in Vesicle Diameter

Another alteration in vesicle structure which has been observed both at the NMJ and the electric ray electric organ is a decrease in diameter on stimulation. This change is particularly prominent in the *Torpedo* electric organ where on continued low-frequency stimulation a population of vesicles with reduced diameter (by 25%) makes its appearance (ZIMMERMANN and DENSTON 1977a). The smaller vesicles are of endocytotic origin since they can be shown to take up extracellular dextran. During a period of rest following stimulation the vesicles regain the original diameter (GIOMPRES et al. 1981 c).

II. Biochemical Alterations on Nerve Stimulation

1. Intracellular Transmitter Pools

The vesicle hypothesis postulates that evoked neurotransmitter release is by exocytosis of synaptic vesicle contents. To gain evidence for this various lines of experimental approach have been adopted. These include for example: analysis of net changes in nerve terminal neurotransmitter pools; labelling of nerve terminal and especially of vesicular ACh with radioactive precursors and comparison of the specific radioactivities (dpm of radiolabelled ACh per mole ACh) of isolated synaptic vesicles and other subcellular fractions with that of the released neurotransmitter; and application of precursors of false transmitters or of drugs which interfere with vesicular storage of ACh. Results obtained by the various groups were not always congruent and this has resulted in discussions as to whether synaptic vesicles were at all involved in evoked transmitter release (for references, see ZIMMERMANN 1979, 1982c; ISRAËL et al. 1979). A major reason for some of the problems that arose is the poor understanding of the redistribution of the cellular pools of ACh existing *in situ* when the various subcellular fractionation procedures are applied to the tissue. Neurotransmitter pools obtained by subcellular fractionation referred to as 'free' or 'cytoplasmic' and 'bound' or 'vesicular' may not correspond exactly to the previous *in situ* cellular compartments. Estimates for the size of the non-vesicular, cytoplasmic pool of ACh vary between 50% (DUNANT et al. 1972) and 20% (SUSZKIW 1980; WEILER et al. 1982). (For an extensive review of the methodological aspects see ZIMMERMANN 1982a, c.)

2. Net Changes in Transmitter Pool

Nerve stimulation causes a net loss of ACh from the vesicular pool from various tissues, including brain, NMJ and electric organ (e.g. ZIMMERMANN and WHITTAKER 1974a; SALEMOGHADDAN and COLLIER 1976; CARROL and NELSON 1978; MILEDI et al. 1982). However for the electric organ preferential loss of ACh from a free (cytoplasmic) rather than a vesicular pool has also been reported (DUNANT et al. 1972; CORTHAY et al. 1982). The loss of vesicular ACh is paralleled by a decrease in synaptic vesicle numbers. On subsequent recovery vesicular ACh is restored (ZIMMERMANN and WHITTAKER 1974b). Interestingly synaptic vesicle numbers recovered first and only subsequently were the vesicles refilled with ACh and ATP (SUSZKIW 1980).

3. Synaptic Vesicles and Preferential Release of Newly Synthesized ACh

Newly synthesized ACh is released from nerve endings in preference to previously stored ACh. This implies a mechanism of metabolic channelling of ACh inside the nerve ending. The initial approach to identifying the cellular compartment from which ACh was released was by analysis of the specific radioactivities of various transmitter pools obtained by subcellular fractionation.

Comparison of the specific radioactivities of radiolabelled ACh in subcellular fractions with that of released ACh suffers a 2-fold handicap. First, the redistribution of ACh pools on subcellular fractionation is uncontrolled, and, secondly, fractions of synaptic vesicles sedimenting at a similar density on density-gradient centrifugation may not be homogenous metabolically. The measurement of specific radioactivity in a fraction would then yield only the average value over all types or metabolic states of vesicles present. Furthermore, vesicular ACh may become labelled more intensively on tissue stimulation or in an awake animal than under conditions of rest and in an anaesthetized animal (e.g. MOLENAAR and POLAK 1973; MOLENAAR et al. 1973). Thus, the ample literature quotes results where the specific radioactivity of ACh released on synapse activation was either higher or the same as in the isolated vesicle fraction and either the vesicles or a free, cytoplasmic pool of ACh have been implicated in ACh release (for references, see ZIMMERMANN 1979).

4. Metabolic Heterogeneity of Synaptic Vesicles

Evidence for vesicular metabolic heterogeneity came first from studies on mammalian brain (BARKER et al. 1972) and has been extensively demonstrated for the *Torpedo* electric organ (ZIMMERMANN and DENSTON 1977a, b; ZIMMERMANN 1978; SUSZKIW et al. 1978; GIOMPRES et al. 1981b, c). In *Torpedo* a population of smaller vesicles formed on induced transmitter release (fraction VP_2) can be physically separated from a population of larger vesicles (fraction VP_1) which have not yet been involved in exo- and endocytosis. The recycled vesicles are not only smaller but also denser and therefore can be separated either by density-gradient centrifugation or column chromatography. The density of the recycled vesicles (VP_2) increases from 1.056 to 1.067 g ml^{-1} and this is accompanied by a decrease in water space from 65% to 42%. They retain less ACh and ATP but have a high capacity to take up both newly synthesized ACh and ATP. Subsequent recovery of the stimulated tissue results in reuptake of ACh and ATP, an increase in vesicle diameter and a corresponding decrease in vesicular density.

The decrease in vesicle size and the corresponding increase in density reflect the loss in the soluble core constituents ACh and ATP, resulting in a lower content of osmotically active solutes. The observations that the vesicle radius is increased on vesicle refilling, and that recycled VP_2-type vesicles are more resistant to hypo-osmotic lysis than fully charged vesicles (VP_1-type) (GIOMPRES and WHITTAKER 1984), are consistent with this view.

After labelling of perfused blocks of electric tissue the recycled VP_2-type vesicles have a specific radioactivity of ACh 10–15 times higher than vesicles sedimenting of lower density (VP_1). A similar result is obtained on labelling of vesicular ATP (ZIMMERMANN and DENSTON 1977b; ZIMMERMANN 1978). The use of columns of porous glass beads for separating vesicle populations from larger membrane fragments has excluded the possibility that the VP_2-type vesicle fraction could contain synaptosome-like contaminants retaining part of their cytoplasmic, non-vesicular ACh (GIOMPRES et al. 1981b). Restimulation of previously labelled tissue causes release of highly labelled ACh with a specific radioactivity corresponding to that of the VP_2-type vesicles and significantly greater than that of the VP_1-type vesicles or total tissue (SUSZKIW et al. 1978). This would be in accordance with the assumption that the recycled synaptic vesicles are a major source of released ACh.

Interestingly a difference in density has also been observed in cholinergic vesicle populations isolated from the field-stimulated myenteric plexus of the guinea-pig ileum (ÁGOSTON et al. 1985). Application of labelled precursor resulted in a preferential labelling of the denser and smaller (VP_2-like) subpopulation as in *Torpedo.*

Thus, vesicles can be present in the axon terminal in different metabolic states. Recycled vesicles are recharged with newly synthesized ACh and ATP. The high specific radioactivity of released ACh may be accounted for by a preferential recycling of those synaptic vesicles (VP_2) which have previously released their contents (ZIMMERMANN 1982c).

5. False Transmitters

Studies using false transmitters have consolidated the view that synaptic vesicles are the source of ACh released on synapse activation; these are reviewed in Chap. 16. A first example for the storage of a cholinergic false transmitter inside synaptic vesicles was acetylpyrrolcholine (ZIMMERMANN and DOWDALL 1977). In the *Torpedo* electric organ false transmitters such as acetylhomocholine are preferentially incorporated into recycled vesicles (VP_2), corresponding to results with the real transmitter. In the case of homocholine both the precursor and the acetylated derivative are taken up into vesicles *in situ*, and the isotopic ratio of homocholine to acetylhomocholine is identical in synaptic vesicles and in the perfusate of stimulated blocks of electric tissue (LUMANI et al. 1980). Using mouse brain tissue and homocholine as a precursor for a false transmitter the molar ratio of acetylhomocholine to ACh in the isolated fraction of synaptic vesicles corresponds to that of material released on Ca^{2+}-dependent stimulation. The ratio in the 'labile bound' (cytoplasmic?) fractionation compartment corresponds to the ratio in the medium during resting release (CARROL and ASPRY 1980).

6. Drugs

That the drug vesamicol (AH5183) selectively blocks uptake of ACh into synaptic vesicles opens up the possibility of interfering with vesicular reloading in experiments with induced transmitter release. Experiments with a variety of tissues including perfused tissue blocks or synaptosomes from *Torpedo* electric organ, rat cortical synaptosomes, rat brain slices, mouse forebrain minces, guinea-pig myenteric plexus, SCG and cultured PC12 cells all yield similar results (for references, see MARSHALL and PARSONS 1987; SUSZKIW and MANALIS 1987; MICHAELSON et al. 1987). The drug neither effects the uptake of precursors of ACh or the influx of Ca^{2+} into the nerve terminals nor does it impair intraterminal synthesis of radiolabelled ACh. It inhibits the uptake *in situ* of radiolabelled ACh into synaptic vesicles but has little effect on previously loaded ACh. When the preparation is stimulated after application of vesamicol, the release of newly synthesized (radiolabelled) ACh but not that of preexisting ACh stores is markedly reduced. When the drug is added after accumulation of radiolabelled ACh, it does not affect the release of labelled ACh. This is all strong support for the view that newly synthesized ACh needs first to be taken up into synaptic vesicles before it can be released on nerve stimulation.

Interestingly, vesamicol also reduces spontaneous (non-evoked) release (leakage) of ACh from the NMJ (EDWARDS et al. 1985) and from nerve terminals of the *Torpedo* electric organ (MICHAELSON et al. 1987). This implies the presence of additional binding sites in the plasma membrane. Possibly these binding sites are the result of spontaneous vesicle fusion with the presynaptic plasma membrane. In this case the transmitter may leak through the vesicular ACh porter incorporated into the plasma membrane.

III. Mechanism of Exocytosis

The evidence quoted above strongly supports a mechanism for the evoked release of ACh in which synaptic vesicles fuse with the presynaptic plasma membrane to release their soluble contents. These would include ACh, ATP but also metal ions like Ca^{2+} and peptides. Whereas there is evidence for a co-release of ACh and ATP (see above), release of metal ions or peptides has not yet been demonstrated. Similarly, convincing evidence is still lacking for a release of vesicular proteoglycan. The relative decrease of proteoglycan in the fraction of recycled vesicles (VP_2) as well as the presence of binding sites against anti-proteoglycan antibodies in fractions of basal lamina isolated from the electric organ (KIENE and STADLER 1987; STADLER and KIENE 1987) would be supportive of an (at least partial) loss of this compound into the synaptic cleft. On the other hand, a recent report by CARLSON et al. (1986) suggests that the vesicular proteoglycan and a glycoprotein anchored in both the presynaptic plasma membrane and the basal lamina share a common epitope and are thus recognized by the same monoclonal antibody but are different molecular species.

Since biochemical analysis normally operates at a slow time scale the demonstration of vesicle fusion at the neuromuscular junction within the synaptic delay time (2.5 ms, TORRI-TARELLI et al. 1985) added a final piece of evidence for the vesicular mechanism of ACh release, and contradicts reports on the *Torpedo* electric organ (GARCIA-SEGURA et al. 1986) that synaptic vesicle fusion may occur only when the neurotransmitter has already been released from the synapse.

The molecular mechanism of vesicle fusion with the presynaptic plasma membrane is still unresolved. Since it must occur within a millisecond time scale, it is likely to involve only a very limited number of biochemical or biophysical reactions. Exocytosis is known to be Ca^{2+}-dependent. When *Torpedo* synaptosomes are permeabilized by high-voltage discharge, ACh is released on elevating the Ca^{2+} concentrations in the medium to the micromolar range. This corresponds to the Ca^{2+}-dependent release also observed in a number of other systems like chromaffin cells or blood platelets (KNIGHT and SCRUTTON 1986). The specific requirements for a protein or lipid interaction in the fusion of cholinergic synaptic vesicles need to be elucidated (for references, see REICHARDT and KELLY 1983).

IV. Gated Release of ACh

Previous experiments which suggested a preferential release of ACh from the nerve terminal cytoplasm rather than from synaptic vesicles (for references, see ISRAËL et al. 1979) have led to a search for a presynaptic membrane gate mediating release of ACh. In reconstitution experiments presynaptic membrane constituents were incorporated into lipid vesicles. When preloaded with ACh the proteoliposomes displayed Ca^{2+}-dependent release of ACh (addition of Ca^{2+} ionophore A23187) in preference to choline or ATP. The Ca^{2+} dependency of ACh release from proteoliposomes was similar (K_m for Ca^{2+}, 0.2–2.0 mM) to that of intact synaptosomes, although one might expect that the proteoliposomes as 'empty' sacs without a Ca^{2+}-sequestering system should have responded to Ca^{2+} in the micromolar range as do permeabilized synaptosomes. A molecule mediating the release (mediatophore) has been isolated. It has an apparent molecular mass of 200–600 kDa and is likely to be an aggregate of several subunits (ISRAËL et al. 1984; ISRAËL and MANARANCHE 1985; BIRMAN et al. 1986). A final evaluation of the data must await an analysis of the purity of the membrane fraction involved. It is known that plasma membranes of synaptosomes isolated from electric organ contain constituents of synaptic vesicles (BUCKLEY et al. 1983).

V. Life-Cycle of the Synaptic Vesicle

Little is known about the cellular origin and fate of synaptic vesicles. On labelling of both ACh and phospholipid by intraventricular injection of [^{3}H]choline into guinea-pig brains the turnover of vesicular ACh was found to be several orders of magnitude faster than that of synaptic vesicle membrane lipid (DOWDALL et al. 1972). This suggests that the synaptic vesicle is a rather conservative structure of the nerve terminal and may be reused repeatedly for transmitter release. Using *Torpedo* vesicle-specific antiserum axonal transport of vesicular membrane constituents was observed by immunofluorescence techniques in the rat sciatic nerve (DAHLSTRÖM et al. 1985) and also in the nerve innervating the electric organ (JONES et al. 1982). Vesicular proteoglycan can be labelled by injection of [^{35}S]sulphate in the electric lobe of electric rays. Outflow of radioactively labelled proteoglycan into the nerve was by fast axonal transport, suggesting its vesicular transport. Radioactively labelled vesicles isolated from the electric nerves elute on zonal centrifugation at a significantly smaller density than the VP_1-type of vesicles. Similarly, synaptic vesicles isolated from the electric organ after injection of labelled sulphate into the lobe contain labelled proteoglycan. They are also less dense than normal (VP_1-type) vesicles and contain little ACh and ATP. There is a time-dependent increase in the density of labelled vesicles suggesting that the immature vesicles derived from the axon become charged with ACh and ATP only in the nerve terminal (STADLER and DOWE 1982; STADLER and KIENE 1987). These authors suggest that vesicles are transported as preformed particles in the axon, and that the newly arrived vesicles constitute an imma-

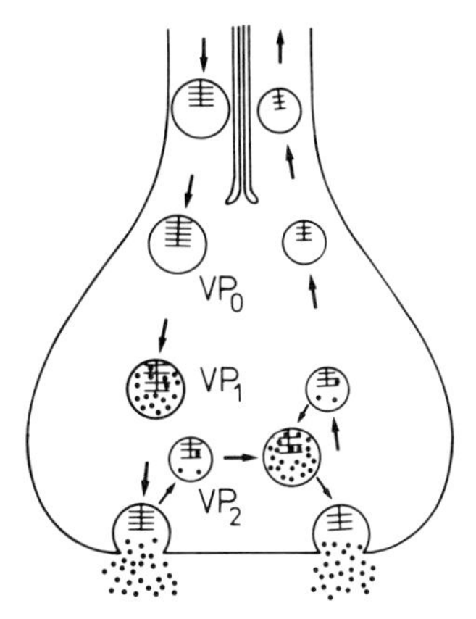

Fig. 2. Dynamics of cholinergic synaptic vesicles. VP_0, metabolic state of axonally transported vesicles containing proteoglycan but not ACh and ATP; VP_1, metabolic state of mature vesicles filled with ACh and ATP; VP_2, metabolic state of vesicles recycled after exocytosis. Modified from STADLER and KIENE (1987)

ture population of vesicles (VP_0-type) which lacks ACh and ATP. On charging with ACh and ATP they would be transformed into VP_1-type vesicles which after induced transmitter release would form the VP_2-type. On refilling these could again become the VP_1-type. This would extend the notion of vesicular metabolic heterogeneity as developed previously and calls for further experiments to define the biochemical and functional properties of vesicles in the various functional states (Fig. 2).

Vesicular constituents are expected eventually to return to the lysosomal compartments of the cell body by retrograde axonal transport. However this aspect has not yet been documented.

References

Ágoston DV, Conlon JM (1986) Presence of vasoactive intestinal polypeptide-like immunoreactivity in the electromotor system of *Torpedo marmorata.* J Neurochem 47:446–453

Ágoston DV, Kosh, JW, Lisziewicz J, Whittaker VP (1985) Separation of recycling and reserve synaptic vesicles from cholinergic nerve terminals of the myenteric plexus of guinea-pig ileum. J Neurochem 44:299–305

Ágoston DV, Ballmann M, Conlon JM, Dowe GHC, Whittaker VP (1986) Isolation of neuropeptide-containing vesicles from the guinea-pig ileum. J Neurochem 45:398–406

Anderson DC, King SC, Parsons SM (1982) Proton gradient linkage to active uptake of [^{3}H]acetylcholine by *Torpedo* electric organ synaptic vesicles. Biochemistry 21: 3037–3043

Anderson DC, King SC, Parsons SM (1983a) Pharmacological characterization of acetylcholine transport system in purified *Torpedo* electric organ synaptic vesicles. Mol Pharmacol 24:48–54

Anderson DC, King SC, Parsons SM (1983b) Inhibition of [^{3}H]acetylcholine active transport by tetraphenylborate and other anions. Mol Pharmacol 24:55–59

Anderson DC, Bahr BA, Parsons SM (1986) Stoichiometrics of acetylcholine uptake, release, and drug inhibition in *Torpedo* synaptic vesicles; heterogeneity in acetylcholine transport and storage. J Neurochem 46:1207–1213

Angel I, Michaelson DM (1981) Determination of $\Delta\psi$, ΔpH and the proton electrochemical gradient in isolated cholinergic synaptic vesicles. Life Sci 29:411–416

Bahr BA, Parsons SM (1986a) Acetylcholine transport and drug inhibition in *Torpedo* synaptic vesicles. J Neurochem 46:1214–1218

Bahr BA, Parons SM (1986b) Demonstration of a receptor in *Torpedo* synaptic vesicles for the acetylcholine storage blocker L-trans-2-(4-phenyl[3,4-^{3}H]-piperidino)cyclohexanol). Proc Natl Acad Sci USA 83:2267–2270

Barker LA, Dowdall MJ, Whittaker VP (1972) Choline metabolism in the cerebral cortex of guinea pig. Biochem J 130:1063–1080

Batteiger DL, Parsons SM (1986) The ATPase of cholinergic synaptic vesicles is associated with sugars. Neurochem Int 8:249–253

Birman S, Israël M, Lesbats B, Morel N (1986) Solubilization and partial purification of a presynaptic membrane protein ensuring calcium-dependent acetylcholine release from proteoliposomes. J Neurochem 47:433–444

Boyne AF (1978) Neurosecretion: integration of recent findings into the vesicle hypothesis. Life Sci 22:2057–2066

Boyne AF, Bohan TP, Williams TH (1974) Effect of calcium-containing fixation solutions on cholinergic synaptic vesicles. J Cell Biol 63:780–785

Breer H, Morris SJ, Whittaker VP (1977) Adenosine triphosphatase activity associated with purified cholinergic synaptic vesicles of *Torpedo marmorata.* Eur J Biochem 80:313–318

Breer H, Morris SJ, Whittaker VP (1978) A structural model of cholinergic synaptic vesicles from the electric organ of *Torpedo marmorata* deduced from density measurements at different osmotic pressures. Eur J Biochem 87:453–458

Buckley K, Kelly RB (1985) Identification of a transmembrane glycoprotein specific for secretory vesicles of neural and endocrine cells. J Cell Biol 100:1284–1294

Buckley KM, Schweitzer ES, Miljanich GP, Clift-O'Grady L, Kustner PP, Reichardt LF, Kelly RB (1983) A synaptic vesicle antigen is restricted to the junctional region of the presynaptic plasma membrane. Proc Natl Acad Sci USA 80:7342–7346

Bulenda D, Gratzl M (1985) Matrix-free Ca^{2+} in isolated chromaffin vesicles. Biochemistry 24:7760–7765

Carlson SS, Kelly RB (1983) A highly antigenic proteoglycan-like component of cholinergic synaptic vesicles. J Biol Chem 258:11082–11091

Carlson SS, Wagner JA, Kelly JS (1978) Purification of synaptic vesicles from elasmobranch electric organ and the use of biophysical criteria to demonstrate purity. Biochemistry 17:1188–1199

Carlson SS, Caroni P, Kelly RB (1986) A nerve terminal anchorage protein from electric organ. J Cell Biol 103:509–520

Carpenter RS, Parsons SM (1978) Electrogenic behavior of synaptic vesicles from *Torpedo californica.* J Biol Chem 253:326–329

Carpenter RS, Koenigsberger R, Parsons SM (1980) Passive uptake of acetylcholine and other organic cations by synaptic vesicles from *Torpedo* electric organ. Biochemistry 19:4373–4379

Carrol PT, Aspry JM (1980) Subcellular origin of cholinergic transmitter release from mouse brain. Science 210:641–642

Carrol PT, Nelson SH (1978) Cholinergic vesicles: ability to empty and refill independently of cytoplasmic acetylcholine. Nature 199:85–86

Ceccarelli B, Hurlbut WP (1980) Vesicle hypothesis of the release of quanta of acetylcholine. Physiol Rev 6:396–441

Ceccarelli B, Hurlbut WP, Mauro A (1973) Turnover of transmitter and synaptic vesicles at the frog neuromuscular junction. J Cell Biol 57:499–524

Ceccarelli B, Grohovaz F, Hurlbut WP, Jezzi N (1979a) Freeze-fracture studies of frog neuromuscular junctions during intense release of neurotransmitter. I. Effects of black widow spider venom and Ca^{2+}-free solutions on the structure of the active zone. J Cell Biol 81:163–177

Ceccarelli B, Grohovaz F, Hurlbut WP, Jezzi N (1979b) Freeze-fracture studies of frog neuromuscular junctions during intense release of neurotransmitter – II. Effects of electrical stimulation and high potassium. J Cell Biol 81:178–192

Corthay J, Dunant Y, Loctin F (1982) Acetylcholine changes underlying transmission of a single nerve impulse in the presence of 4-aminopyridine in *Torpedo.* J Physiol (Lond) 325:461–479

Dahlström A, Larsson P-A, Carlson SS, Bööj S (1985) Localization and axonal transport of immunoreactive cholinergic organelles in rat motor neurons – an immunofluorescent study. Neuroscience 14:607–625

Day NC, Wien D,, Michaelson DM (1985) Saturable [D-Ala^2, D-Leu^5]-enkephalin transport into cholinergic synaptics vesicles. FEBS Lett 183:25–28

De Robertis E, Rodriguez de Lores Arnaiz G, Salganicoff L, Pellegrino de Iraldi A, Zieher LM (1963) Isolation of synaptic vesicles and structural organization of the acetylcholine system within brain nerve endings. J Neurochem 10:225–235

Deutsch JW, Kelly RB (1981) The lipids of synaptic vesicles: relevance to the mechanisms of membrane fusion. Biochemistry 20:378–385

Diebler, MF, Lazereg S (1985) Mg-ATPase and *Torpedo* cholinergic synaptic vesicles. J Neurochem 44:1633–1641

Diebler MF, Morot-Gaudry (1981) Acetylcholine incorporation by cholinergic synaptic vesicles from *Torpedo marmorata.* J Neurochem 37:467–475

Dowdall MJ, Barker LA, Whittaker VP (1972) Choline metabolism in the cerebral cortex of guinea pigs. Phosphorylcholine and lipid choline. Biochem J 130:1081–1094

Dowdall MJ, Boyne AF, Whittaker VP (1974) Adenosine triphosphate, a constituent of cholinergic synaptic vesicles. Biochem J 140:1–12

Dowe GHC, Kilbinger H, Whittaker VP (1980) Isolation of cholinergic synaptic vesicles from the myenteric plexus of guinea-pig small intestine. J Neurochem 35:993–1003

Dunant Y, Gautron J, Israël M, Lesbats B, Manaranche R (1972) Les compartiments d'acétylcholine de l'organe électrique de la *Torpille* et leurs modifications par la stimulation. J Neurochem 19:1987–2002

Dunant Y, Jones GJ, Loctin F (1982) Acetylcholine measured at short time intervals during transmission of nerve impulses in the electric organ of *Torpedo.* J Physiol (Lond) 325:441–460

Edwards C, Doležal V, Tuček S, Zemková H, Yyskočil F (1985) Is an acetylcholine transport system responsible for nonquantal release of acetylcholine at the rodent myoneural junction? Proc Natl Acad Sci USA 82:3514–3518

Füldner HH, Stadler H (1982) ^{31}P-NMR analysis of synaptic vesicles. Status of ATP and internal pH. Eur J Biochem 121:519–524

Garcia-Segura LM, Muller D, Dunant Y (1986) Increase in the number of presynaptic large intramembrane particles during synaptic transmission at the *Torpedo* nerve-electroplaque junction. Neuroscience 19:63–79

Giompres PE, Luqmani YA (1980) Cholinergic synaptic vesicles isolated from *Torpedo marmorata:* demonstration of acetylcholine and choline uptake in an *in vitro* system. Neuroscience 5:1041–1052

Giompres PE, Whittaker VP (1984) Differences in the osmotic fragility of recycling and reserve synaptic vesicles from the cholinergic electromotor nerve terminals of *Torpedo* and their possible significance for vesicle recycling. Biochem Biophys Acta 770: 166–170

Giompres PE, Morris SJ, Whittaker VP (1981 a) The water spaces in cholinergic synaptic vesicles from *Torpedo* measured by changes in density induced by permeant substances. Neuroscience 6:757–763

Giompres PE, Zimmermann H, Whittaker VP (1981 b) Purification of small dense vesicles from stimulated *Torpedo* electric tissue by glass bead chromatography. Neuroscience 6:765–774

Giompres PE, Zimmermann H, Whittaker VP (1981 c) Changes in the biochemical and biophysical parameters of cholinergic synaptic vesicles on transmitter release and during a subsequent period of rest. Neuroscience 6:775–785

Grondal EJM, Zimmermann H (1986) Ectonucleotidase activities associated with cholinergic synaptosomes isolated from *Torpedo* electric organ. J Neurochem 47:871–881

Häggblad J, Heilbronn E (1987) Externally applied ATP causes inosital triphosphate accumulation in cultured myotubes. Neurosci Lett 74:199–204

Harlos P, Lee DA, Stadler H (1984) Characterization of a Mg^{2+}-ATPase and a proton pump in cholinergic synaptic vesicles from the electric organ of *Torpedo marmorata*. Eur J Biochem 144:441–446

Heuser JE, Reese TS (1973) Evidence for recycling of synaptic vesicle membrane during transmitter release at the frog junction. J Cell Biol 57:315–344

Heuser JE, Reese TS, Dennis MJ, Jan Y, Jan L, Evans L (1979) Synaptic vesicle exocytosis captured by quick freezing and correlated with quantal transmitter release. J Cell Biol 81:275–300

Hooper JE, Kelly RB (1984a) Calcium-dependent calmodulin binding to cholinergic synaptic vesicles. J Biol Chem 259:141–147

Hooper JE, Kelly RB (1984b) Calmodulin is tightly associated with synaptic vesicles independent of calcium. J Biol Chem 259:148–153

Hooper JE, Carlson SS, Kelly RB (1980) Antibodies to synaptic vesicles purified from *Narcine* electric organ bind a sub-class of mammalian nerve terminals. J Cell Biol 87:104–113

Israël M, Manaranche R (1985) The release of acetylcholine: from a cellular towards a molecular mechanism. Biol Cell 55:1–14

Israël M, Gautron J, Lesbats B (1970) Fractionnement de l'organe éléctrique de la Torpille: localisation subcellulaire de l'acetylcholine. J Neurochem 17:1441–1450

Israël M, Dunant Y, Manaranche R (1979) The present status of the vesicular hypothesis. Prog Neurobiol 13:237–275

Israël M, Manaranche R, Marsal J, Meunier FM, Morel N, Frachon P, Lesbats B (1980) ATP-dependent calcium uptake by cholinergic synaptic vesicles isolated from *Torpedo* electric organ. J Membr Biol 54:115–126

Israël M, Lesbats B, Morel N, Manaranche R, Gulik-Krzywicki T, Dedieu JC (1984) Reconstitution of a functional synaptosomal membrane possessing the protein constituents involved in acetylcholine translocation. Proc Natl Acad Sci USA 81:277–281

Jones RT, Walker JH, Stadler H, Whittaker VP (1982) Immunohistochemical localization of a synaptic vesicle antigen in a cholinergic neuron under conditions of stimulation and rest. Cell Tissue Res 223:117–126

Kelly RB, Hooper JE (1982) Cholinergic vesicles. In: Poisner AM, Trifaró JM (eds) The secretory granule. Elsevier, Amsterdam, pp 81–119

Kelly RB, Deutsch JW, Carlson SS, Wagner JA (1979) Biochemistry of neurotransmitter release. Annu Rev Neurosci 2:399–446

Kiene M-L, Stadler H (1987) Synaptic vesicles in electromotoneurones. I. Axonal transport, site of transmitter uptake and processing of a core proteoglycan during maturation. EMBO J 6:2209–2215

Knight DE, Scrutton MC (1986) Gaining access to the cytosol: the technique and some applications of electropermeabilization. Biochem J 234:497–506

Kobos RK, Rechnitz GA (1976) Acetylcholine-ATP binding by direct membrane electrode measurement. Biochem Biophys Res Commun 71:762–767

Koenigsberger R, Parsons SM (1980) Bicarbonate and magnesium ion-ATP dependent stimulation of acetylcholine uptake by *Torpedo* electric organ synaptic vesicles. Biochem Biophys Res Commun 94:305–312

Ledeen RW, Sbaschnig-Agler M, Aquino DA, Diebler MF, Lazereg S, Parsons SM (1987) Ganglioside transport and function in the nervous system. Evidence for gangliosides in synaptics vesicles. In: Tuček S (ed) Metabolism and development of the nervous system. Academia, Praha, pp 106–112

Lundberg JM, Fried G, Fahrenkrugh J, Holmstedt B, Hökfelt T, Lagercrantz H, Lundgren G, Ånggård A (1981) Subcellular fractionation of cat submandibular gland: comparative studies on the distribution of acetylcholine and vasoactive intestinal polypeptide (VIP). Neuroscience 6:1001–1010

Luqmani YA (1981) Nucleotide uptake by isolated cholinergic synaptic vesicles: evidence for an ATP carrier. Neuroscience 6:1011–1021

Luqmani YA, Sudlow G, Whittaker VP (1980) Homocholine and acetylcholine: false transmitters in the cholinergic electromotor system of *Torpedo*. Neuroscience 5:153–160

MacIntosh FC (1980) The role of vesicles in cholinergic systems. In Brzin M, Sket D, Bachelard H (eds) Synaptic constituents in health and disease. Mladinska, Lubljana; Pergamon, Oxford, pp 11–50
MacIntosh FC, Collier B (1976) Neurochemistry of cholinergic nerve terminals. In: Zamis E (ed) Neuromuscular junction. Springer, Berlin Heidelberg New York, pp 99–228 (Handbook of experimental pharmacology, vol 42)
Marshall IG, Parsons SM (1987) The vesicular acetylcholine transport system. Trends Neurosci 10:174–177
Matthew WD, Tsavaler L, Reichardt LF (1981) Identification of a synaptic vesicle-specific membrane protein with a wide distribution in neuronal and neurosecretory tissue. J Cell Biol 91:257–269
Meldolesi J, Borgese N, de Camilli P, Ceccarelli B (1978) Cytoplasmic membranes and the secretory process. In: Poste G, Nicolson GL (eds) Membrane fusion. Elsevier/North-Holland, Amsterdam, pp 509–627 (Cell surface reviews, vol. 5)
Melega WP, Howard BD (1981) Choline and acetylcholine metabolism in PC 12 secretory cells. Biochemistry 20:4477–4483
Michaelson DM, Angel I (1980) Determination of Δ pH in cholinergic synaptic vesicles: its effect on storage and release of acetylcholine. Life Sci 27:39–44
Michaelson DM, Angel I (1981) Saturable acetylcholine transport into purified cholinergic vesicles. Proc Natl Acad Sci USA 78:2048–2052
Michaelson DM, Ophir I (1980) Sideness of (calcium, magnesium) adenosine triphosphatase of purified *Torpedo* synaptic vesicles. J Neurochem 34:1483–1490
Michaelson DM, Wien-Naor D (1987) Enkephalin uptake into cholinergic synaptic vesicles and nerve terminals. Ann NY Acad Sci 493:234–251
Michaelson DM, Ophir I, Angel I (1980) ATP-stimulated Ca^{2+} transport into cholinergic *Torpedo* synaptic vesicles. J Neurochem 35:116–124
Michaelson DM, McDowall C, Sarne Y (1984) The *Torpedo* electric organ is a model for opiate regulation of acetylcholine release. Brain Res 305:173–176
Michaelson DM, Licht R, Burstein M (1987) The effects of the vesicular uptake blocker AH5183 on evoked and spontaneous release of acetylcholine from cholinergic nerve terminals. In: Dowdall MJ, Hawthorne JN (eds) Cellular and molecular basis of cholinergic function. Horwood, Chichester, pp 316–322
Miledi R, Molenaar PC, Polak RL (1982) Free and bound acetylcholine in frog muscle. J Physiol (Lond) 333:189–199
Molenaar PC, Polak RL (1973) Newly formed acetylcholine in synaptic vesicles in brain tissue. Brain Res 62:537–542
Molenaar PC, Nickolson VJ, Polak RL (1973) Preferential release of newly synthetized [^{3}H]acetylcholine from rat cerebral cortex slices *in vitro*. Br J Pharmacol 47:97–108
Morel N, Meunier FR (1981) Simultaneous release of acetylcholine and ATP from stimulated cholinergic synaptosomes. J Neurochem 36:1766–1773
Morris SJ (1973) Removal of residual amounts of acetylcholinesterase and membrane contamination from synaptic vesicles isolated from the electric organ of *Torpedo*. J Neurochem 21:713–715
Morris SJ (1980) The structure and stoichiometry of electric ray synaptic vesicles. Neuroscience 5:1509–1516
Nagy A, Baker RR, Morris SJ, Whittaker VP (1976) The preparation and characterization of synaptic vesicles of high purity. Brain Res 109:285–309
Ohsawa K, Dowe GHC, Morris SJ, Whittaker VP (1979) The lipid and protein content of cholinergic synaptic vesicles from the electric organ of *Torpedo marmorata* purified to constant composition: implications for vesicle structure. Brain Res 161:447–457
Osborne NN, Beale R, Nicholas D, Stadler H, Walker JH, Jones RT, Whittaker VP (1982) An antiserum to cholinergic synaptic vesicles from *Torpedo* recognizes nerve terminals in retinas from a variety of species. Cell Mol Neurobiol 2:157–163
Parsons SM, Koenigsberger R (1980) Specific stimulated uptake of acetylcholine by *Torpedo* electric organ synaptic vesicles. Proc Natl Acad Sci USA 77:6234–6238
Parsons SM, Carpenter RS, Koenigsberger R, Rothlein JE (1982) Transport in the cholinergic synaptic vesicle. Fed Proc 41:2765–2768

Reichardt LF, Kelly RB (1983) A molecular description of nerve terminal function. Annu Rev Biochem 52:871–926

Rephaeli A, Parsons SM (1982) Calmodulin stimulation of $^{45}Ca^{2+}$ transport and protein phosphorylation in cholinergic synaptic vesicles. Proc Natl Acad Sci USA 79:5783–5787

Rothlein JE, Parsons SM (1979) Specificity of association of a Ca^{2+}/Mg^{2+} ATPase with cholinergic synaptic vesicles from *Torpedo* electric organ. Biochem Biophys Res Commun 88:1069–1079

Rothlein JE, Parsons SM (1980) Bicarbonate stimulation of the Ca^{2+}/Mg^{2+} ATPase of *Torpedo* electric organ synaptic vesicles. Biochem Biophys Res Commun 95: 1869–1874

Rothlein JE, Parsons SM (1982) Origin of the bicarbonate stimulation of *Torpedo* electric organ synaptic vesicle ATPase. J Neurochem 39:1660–1668

Salehomoghaddam SH, Collier B (1976) The relationship between acetylcholine release from brain slices and the acetylcholine content of subcellular fractions prepared from brain. J Neurochem 27:71–76

Schmidt R, Zimmermann H, Whittaker VP (1980) Metal ion content of cholinergic vesicles isolated from the electric organ of *Torpedo:* effect of stimulation induced transmitter release. Neuroscience 5:625–638

Schweitzer E (1987) Coordinated release of ATP and ACh from cholinergic synaptosomes and its inhibition by calmodulin antagonists. J Neurosci 7:2948–2956

Sheridan MN, Whittaker VP, Israël M (1966) The subcellular fractionation of the electric organ of *Torpedo.* Z Zellforsch 74:291–307

Stadler H, Dowe GHC (1982) Identification of a heparan sulfate-containing proteoglycan as a specific core component of cholinergic synaptic vesicles from *Torpedo marmorata.* EMBO J 1:1381–1384

Stadler H, Fenwick EM (1983) Cholinergic synaptic vesicles from *Torpedo marmorata* contain an atractyloside-binding protein related to the mitochondrial ADP/ATP carrier. Eur J Biochem 136:377–382

Stadler H, Füldner HH (1980) Proton NMR detection of acetylcholine status in synaptic vesicles. Nature 286:293–294

Stadler H, Kiene M-L (1987) Synaptic vesicles in electromotoneurones. II. Heterogeneity of populations is expressed in uptake properties, exocytosis and insertion of a core proteoglycan into the extracellular matrix. EMBO J 6:2217–2221

Stadler H, Tashiro T (1979) Isolation of synaptosomal plasma membranes from cholinergic nerve terminals and a comparison of their proteins with those of synaptic vesicles. Eur J Biochem 101:171–178

Stadler H, Tsukita K (1984) Synaptic vesicles contain an ATP-dependent proton pump and show 'knob-like' protrusions on their surface. EMBO J 3:3333–3337

Suszkiw JB (1980) Kinetics of acetylcholine recovery in *Torpedo* electromotor synapses depleted of synaptic vesicles. Neuroscience 5:1341–1349

Suszkiw JB, Manalis RS (1987) Acetylcholine mobilization: effects of vesicular ACh uptake blocker, AH5183, on ACh release in rat brain synaptosomes, *Torpedo* electroplax and frog neuromuscular junction. In: Dowdall MJ, Hawthorne JN (eds) Cellular and molecular basis of cholinergic function. Horwood, Chichester, pp 323–332

Suszkiw JB, Zimmermann H, Whittaker VP (1978) Vesicular storage and release of acetylcholine in *Torpedo* elecroplaque synapses. J Neurochem 30:1269–1280

Tashiro T, Stadler H (1978) Chemical composition of cholinergic synaptic vesicles from *Torpedo marmorata* based on improved purification. Eur J Biochem 90:479–487

Toll T, Howard BD (1980) Evidence that an ATPase and a protonmotive force function in the transport of acetylcholine into storage vesicles. J Biol Chem 255:1787–1789

Torri-Tarelli F, Grohovaz F, Fesce R, Ceccarelli B (1985) Temporal coincidence between synaptic vesicle fusion and quantal secretion of acetylcholine. J Cell Biol 101: 1386–1399

Tuček S (1978) Acetylcholine synthesis in neurones. Chapman and Hall, London

Volknandt W, Zimmermann H (1986) Acetylcholine, ATP, and proteoglycan are common to synaptic vesicles isolated from the electric organs of electric eel and electric catfish as well as from rat diaphragm. J Neurochem 47:1449–1462

Volknandt W, Naito S, Ueda T, Zimmermann H (1987) Synapsin I is associated with cholinergic nerve terminals in the electric organs of *Torpedo, Electrophorus* and *Malapterurus* and copurifies with *Torpedo* synaptic vesicles. J Neurochem 49:342–347

Von Wedel R, Carlson SS, Kelly RB (1981) Transfer of synaptic vesicle antigens to the presynaptic plasma membrane during exocytosis. Proc Natl Acad Sci USA 78:1014–1018

Wagner JA, Kelly RB (1979) Topological organization of proteins in an intracellular secretory organelle: the synaptic vesicle. Proc Natl Acad Sci USA 76:4126–4130

Wagner JA, Carlson SS, Kelly RB (1978) Chemical and physical characterization of cholinergic synaptic vesicles. Biochemistry 17:1199–1206

Walker JH, Jones RT, Obrocki J, Richardson GP, Stadler H (1982) Presynaptic plasma membranes and synaptic vesicles of cholinergic nerve endlings demonstrated by means of specific antisera. Cell Tissue Res 223:101–116

Walker JH, Obrocki J, Südhoff TC (1983a) Calelectrin, a calcium-dependent membrane-binding protein associated with secretory granules in *Torpedo* cholinergic electromotor nerve endings and rat adrenal medulla. J Neurochem 41:139–145

Walker JH, Obrocki J, Zimmermann CW (1983b) Identification of a proteoglycan antigen characteristic of cholinergic synaptic vesicles. J Neurochem 41:209–216

Walker JH, Stadler H, Witzemann VL (1984) Calmodulin binding proteins of the cholinergic electromotor synapse: synaptosomes, synaptic vesicles, receptor-enriched membranes, and cytoskeleton. J Neurochem 42:314–320

Walker JH, Kristjansson GI, Stadler H (1986) Identification of a synaptic vesicle antigen (M_r 86000) conserved between *Torpedo* and rat. J Neurochem 46:875–881

Weiler M, Roed IS, Whittaker VP (1982) The kinetics of acetylcholine turnover in a resting cholinergic nerve terminal and the magnitude of the cytoplasmic compartment. J Neurochem 38:1187–1191

Whittaker VP (1984a) The structure and function of cholinergic synaptic vesicles. Biochem Soc Transact 12:561–575

Whittaker VP (1984b) The synaptic vesicle. In: Lajtha A (ed) Structural elements of the nervous system. Plenum, New York, pp 41–69 (Handbook of Neurochemistry, 2nd edn, vol 7)

Whittaker VP, Sheridan MN (1965) The morphology and acetylcholine content of cerebral cortical synaptic vesicles. J Neurochem 12:363–372

Whittaker VP, Stadler G (1980) The structure and function of cholinergic synaptic vesicles. In: Bradshaw RA, Schneider DM (eds) Proteins of the nervous system, 2nd edn. Raven, New York, pp 231–255

Whittaker VP, Michaelson IA, Kirkland RJA (1963) The separation of synaptic vesicles from disrupted nerve ending particles. Biochem Pharmacol 12:300–302

Whittaker VP, Michaelson IA, Kirkland RJA (1964) The separation of synaptic vesicles from nerve-ending particles ('synaptosomes'). Biochem J 90:293–303

Whittaker VP, Essman WB, Dowe GHC (1972) The isolation of pure cholinergic synaptic vesicles from the electric organs of elasmobranch fish of the family *Torpedinidae.* Biochem J 128:833–846

Wilson WS, Schulz RA, Cooper JT (1973) The isolation of cholinergic synaptic vesicles from bovine superior cervical ganglion and estimation of their acetylcholine content. J Neurochem 20:659–667

Zechel K, Stadler H (1982) Identification of actin in highly purified synaptic vesicles from the electric organ of *Torpedo marmorata.* J Neurochem 39:788–795

Zhu PC, Thureson-Klein Å, Klein RL (1986) Exocytosis from large dense cored vesicles outside the active synaptic zones of terminals within the trigeminal subnucleus caudalis: a possible mechanism for neuropeptide release. Neuroscience 19:43–54

Zimmermann H (1978) Turnover of adenine nucleotides in cholinergic synaptic vesicles of the *Torpedo* electric organ. Neuroscience 3:827–836

Zimmermann H (1979) Vesicle recycling and transmitter release. Neuroscience 4:1773–1804

Zimmermann H (1981) Mechanism(s) of secretion of chemical neurotransmitters: chairman's overview: In: Stjärne L, Hedquist P, Lagercrantz H, Wennmalm Å (eds) Chemical neurotransmission 75 years. Academic, New York, pp 179–185

Zimmermann H (1982a) Isolation of cholinergic nerve vesicles. In: Klein RL, Lagercrantz H, Zimmermann H (eds) Neurotransmitter vesicles. Academic, New York, pp 241–269
Zimmermann H (1982b) Biochemistry of the isolated cholinergic vesicles. In: Klein RL, Lagercrantz H, Zimmermann H (eds) Neurotransmitter vesicles. Academic, New York, pp 271–304
Zimmermann H (1982c) Insights into the functional role of cholinergic vesicles. In: Klein RL, Lagercrantz H, Zimmermann H (eds) Neurotransmitter vesicles. Academic, New York, pp 305–359
Zimmermann H, Bokor JT (1979) ATP recycles independently of ACh in cholinergic synaptic vesicles. Neurosci Lett 13:319–324
Zimmermann H, Denston CR (1977a) Recycling of synaptic vesicles in the cholinergic synapses of the *Torpedo* electric organ during induced transmitter release. Neuroscience 2:695–714
Zimmermann H, Denston CR (1977b) Separation of synaptic vesicles of different functional states from the cholinergic synapses of the *Torpedo* electric organ. Neuroscience 2:715–730
Zimmermann H, Dowdall MJ (1977) Vesicular storage and release of a cholinergic false transmitter (acetylpyrrolcholine) in the *Torpedo* electric organ. Neuroscience 2:731–739
Zimmermann H, Whittaker VP (1974a) Effect of electrical stimulation on the yield and composition of synaptic vesicles from the cholinergic synapses of the electric organ of *Torpedo:* a combined biochemical, electrophysiological and morphological study. J Neurochem 22:435–450
Zimmermann H, Whittaker VP (1974b) Different recovery rates of the electrophysiological, biochemical and morphological parameters in the cholinergic synapses of the *Torpedo* electric organ after stimulation. J Neurochem 22:1109–1114
Zimmermann H, Stadler H, Whittaker VP (1981) Structure and function of cholinergic synaptic vesicles. In: Stjärne L, Hedquist P, Lagercrantz H, Wennmalm A (eds) Chemical neurotransmission 75 years. Academic, New York, pp 91–111

CHAPTER 12

Isolation of Cholinergic Nerve Terminals

P. J. RICHARDSON

A. Introduction

Nerve terminal preparations derived from the central nervous system (CNS) have proved to be ideal starting points for many studies of cholinergic function. This is largely because they contain all the required components of nerve endings, and exhibit many of the properties expected of such nerve terminals *in vivo*.

This chapter describes the development of subcellular fractionation techniques to the point at which purely cholinergic nerve terminals can be prepared from both mammalian and non-mammalian tissues, thus making possible a more detailed study of the structure and function of these components.

B. Historical Aspects

In the early 1950s when subcellular fractionation techniques were applied to the study of nervous tissue (e.g. BRODY and BAIN 1951; ABOOD et al. 1951) it was not realized that nerve terminals could be sheared from their axons and isolated as intact entities. The discovery that this was so arose largely as a consequence of investigations into the nature of bound acetylcholine, i.e. the pool of latent acetylcholine (ACh) which was not liberated upon mild homogenization of nervous tissue in isotonic media. At this time, there was already some evidence that choline acetyltransferase (ChAT) also existed in a latent, or membrane bound, state (FELDBERG 1945). In the latter part of the decade procedures of mild homogenization and differential centrifugation were applied specifically to the study of cholinergic biochemistry, and resulted in the demonstration that both ChAT (HEBB and SMALLMAN 1956) and ACh (HEBB and WHITTAKER 1958) sedimented in the mitochondrial fraction. Further analysis of this mitochondrial fraction on sucrose density gradients showed that both the bound ACh and ChAT were not localized in mitochondria, but rather in particles of a lower density (HEBB and WHITTAKER 1958). These particles were intact nerve terminals, although in the absence of suitable morphological techniques, this was not immediately recognized. For a detailed historical review, see WHITTAKER (1965).

C. Heterogeneous Mammalian Nerve-Terminal Preparations

After the identification of the nerve terminals (GRAY and WHITTAKER 1960; DE ROBERTIS et al. 1960), a wide variety of homogenization and centrifugation conditions were investigated (reviewed in JONES 1975) in order to optimize both the yield and metabolic activity of these terminals (named 'synaptosomes', WHITTAKER et al. 1964). There were two basic protocols: those of GRAY and WHITTAKER (1962) and DE ROBERTIS et al. (1962) which were based on that of HEBB and WHITTAKER (1958): both utilize differential osmotic dehydration in a hyperosmotic step gradient to sharpen the separation of synaptosomes from other particles (WHITTAKER 1984). Normal morphology and metabolism, including the re-establishment of normal concentration gradients, is achieved by brief incubation in Krebs-Ringer solution (HAJÓS et al. 1979). High-resolution density gradients were also used, both in swinging bucket (FONNUM 1968a) and zonal (BETZ et al. 1974) rotors. One of the main aims of later modifications has been to find a method which avoided subjecting the terminals to the hyperosmotic stress induced by high-sucrose concentrations. A number of alternative density gradients have been developed, e. g. Ficoll, Ficoll/sucrose mixtures (KUROKAWA et al. 1965; BOOTH and CLARK 1978), and Percoll (NAGY and DELGADO-ESCUETA 1984).

The preparation of intact nerve terminals opened numerous lines of investigation (reviewed in WHITTAKER 1984) including the preparation of subsynaptic organelles, i.e. synaptic vesicles (DE ROBERTIS et al. 1963; WHITTAKER et al. 1964), plasma membranes (COTMAN and MATTHEWS 1971), intraterminal mitochondria (BOOTH and CLARK 1979), and junctional complexes (DE ROBERTIS et al. 1967; COHEN et al. 1977). Studies of neurotoxin action (DOWDALL et al. 1979) and many investigations into transmitter metabolism and pharmacology have also been reported. One great advantage inherent in the synaptosome preparation is the ability of the constituent particles to function as 'mini-cells', i.e. they are able to oxidise glucose (BRADFORD 1969), maintain transmembrane ion gradients (CAMPBELL 1976), and, when stimulated, to release transmitters in a Ca^{2+}-dependent manner (DE BELLEROCHE and BRADFORD 1972). However, since *de novo* protein synthesis appears to be restricted to neuronal cell bodies (DROZ and BARONDES 1969) the terminals' metabolic life *in vitro* is limited (WHITTAKER 1973).

The nerve terminals prepared from mammalian brain are derived from many different cell types and so contain a large number of transmitters. This results in a considerable metabolic diversity in the synaptosome preparation which limits the nature of the biochemical investigations possible. Studies of mammalian cholinergic nerve terminal biochemistry have therefore been limited to those aspects which are purely cholinergic, for example the subcellular localization of ACh and its specific enzymes: ChAT and acetylcholinesterase (AChE), on the uptake of choline from the extraterminal medium, and on the stimulation-induced release of ACh. For instance, hypo-osmotic lysis of synaptosomes resulted in the release of approximately 50% of the ACh (presumably of cytoplasmic origin) while the remainder was occluded in the synaptic vesicles (WHITTAKER 1971, 1984). In addition ChAT has been shown to be a soluble enzyme in the cytoplasm of cholinergic terminals (FONNUM 1967, 1968b), while AChE is predominantly membrane bound, being found both at cholinergic synapses and elsewhere (SILVER 1974; Chaps. 8a, 8b, this volume). The high-affinity choline uptake (HACU) system (Chap. 14), whose distribution in the brain follows that of cholinergic innervation

(KUHAR et al. 1973) appears to be localized in cholinergic terminals. This carrier-mediated system has a K_m of approximately 5 μM, and is absolutely dependent upon the presence of a transmembrane ion gradient (see JOPE 1979; Chap. 14, this volume). Investigations have also been made into the effect of exogenous agents – usually other transmitters and/or neuropharmacologically active drugs – on the release of ACh and its rate of synthesis (e.g. DE BELLEROCHE et al. 1982). The heterogeneity of the synaptosome preparation has however severely hindered studies in a number of areas, e.g. mechanisms controlling ACh synthesis and release. Thus, biochemical studies of ACh synthesis have been limited by the nature of the reactants involved, i.e. choline and acetylcoenzyme A (the precursors of ACh) which are present in all neuronal cells (see Chap. 7). The heterogeneity of the nervous system has also hindered other investigations into the molecular basis of cholinergic function. For instance the inability to prepare transmitter-specific synaptic vesicles and synaptic membranes from mammalian brain has prevented studies into mechanisms of mammalian vesicle filling and release, as well as the nature of the surface molecules responsible for recognition between pre- and postsynapse. Attempts to purify cholinergic nerve terminals by density gradient centrifugation have been hindered by the similarities in density and size of most terminals, although populations partially enriched in cholinergic synaptosomes have been detected (DE ROBERTIS et al. 1962; BRETZ et al. 1974). However, prior digestion of superior cervical ganglia with collagenase enabled WILSON and COOPER (1972) to produce a nerve terminal population greatly enriched in cholinergic endings.

D. Heterogeneous Non-mammalian Nerve Terminals

Nerve terminals have been isolated from a wide variety of species, some of whose nervous systems contain regions highly enriched in cholinergic nerves. For instance, the optic lobes of cephalopods have 50–100 times as much ACh as the mammalian central nervous system (BACQ 1935), which presumably reflects an increased proportion of cholinergic nerve terminals (see Chap. 1). However, the percentage yield of synaptosomal ACh from such tissues as squid brain (DOWDALL and WHITTAKER 1973) is less than that from the mammalian CNS. This implies that squid terminals are more fragile, although if the isolation media used are nearly isotonic with sea water, then good morphological preservation is observed. Indeed, squid brain synaptosomes exhibit a HACU system, which does however show greater susceptibility to metabolic poisons than that described in mammalian brain (DOWDALL and SIMON 1973). The variation in properties observed in the choline uptake systems of squid and mammals, together with the fragile nature of the squid terminals and the lack of availability of squid in inland laboratories, has restricted attempts to use the cephalopod preparation as a model cholinergic system (Chap. 1). It should also be mentioned, that although cephalopod terminals (from e.g. squid and octopus, JONES 1970) are highly enriched in cholinergic content, other putative transmitters are also present, e.g. dopamine, noradrenaline and serotonin (JUORIO 1971).

E. Purely Cholinergic Nerve-Terminal Preparations from Electric Organs

I. Types of Preparation Available

The heterogeneity inherent in both mammalian and non-mammalian CNS led to the search for alternative sources of cholinergic nerve terminals. One of the richest sources of such terminals is the electric organ of torpedine fish. High levels of ACh and AChE and a high density of synaptic contacts have been known to exist in the electric organ since the work of MARNAY (1937), NACHMANSOHN and LEDERER (1939) and FELDBERG and FESSARD (1942). However these advantages are counterbalanced by the extreme toughness of the tissue which makes mild homogenization techniques very difficult to use. Indeed the original homogenization conditions applied to the tissue resulted in rupture of the presynaptic plasma membrane (SHERIDAN et al. 1966) and yields of particles identified as synaptosomes were low. However techniques were developed which permitted the preparation of purely cholinergic synaptic vesicles (SHERIDAN et al. 1966; ISRAËL et al. 1968; WHITTAKER et al. 1972) from this organ. This preparation has been widely used in the study of the structure and metabolism of synaptic vesicles; thus experiments involving the use of radiolabelled substrates have been used to show that the vesicular pool of ACh is heterogeneous, and to study the mechanisms of vesicle loading (see ZIMMERMANN 1979; Chap. 11, this volume).

Two independent approaches were developed to overcome the toughness of this organ. The one used by DOWDALL and co-workers (DOWDALL et al. 1975; DOWDALL and ZIMMERMANN 1976, 1977) was to use the much softer juvenile tissue. After chopping the electric organ the subsequent homogenization and fractionation, carried out in 0.8 *M* glycine to maintain to correct osmotic pressure, were modifications of the standard scheme of GRAY and WHITTAKER (1962). The same scheme was also adapted for use on adult tissue. The resulting so-called 'nerve terminal sacs' or 'T-sacs' showed many to the properties of cholinergic endings, i.e. they contain ChAT and ACh, are able to concentrate choline via a HACU system (DOWDALL and ZIMMERMANN 1977), and release ACh upon depolarization (MICHAELSON and SOKOLOVSKY 1976).

A slightly different approach was used by ISRAËL and co-workers who had also noticed that the electric organ could be chopped into very small segments without rupturing the presynaptic terminals. The finely chopped tissue was forced through a series of grids of progressively finer mesh and the terminals isolated by density-gradient centrifugation (ISRAËL et al. 1976). The resulting synaptosomes were larger (3 µm in diameter) than those obtained from mammalian brain (0.6 µm, GRAY and WHITTAKER 1962; JONES 1975) and from juvenile *Torpedo* (1 µm, DOWDALL and ZIMMERMANN 1977). In common with the T-sac preparation, but unlike mammalian nerve terminals, the large synaptosomes do not have fragments of postsynaptic membrane adhering to them, but otherwise retain all the metabolic and physiological properties expected of cholinergic endings (MOREL et al. 1977, 1979). Although electron micrographs of the two types of synaptosomes show smaller profiles and fewer vesicles within the T-sacs than

Table 1. Properties of *Torpedo* electromotor synaptosomes

ACh (nmol)	ChAT (nmol h^{-1})	HACU (pmol min^{-1})	References
Small (T-sac)			
18	360	7–140	DOWDALL and ZIMMERMANN (1977)
29–40	–	70–390	DOWDALL (1977)
74	1100	–	MICHAELSON and SOKOLOVSKY (1976)
Large			
–	–	23	MOREL et al. (1977)
130	450	–	MOREL et al. (1979)
260	680	19–90	ZIMMERMANN et al. (1979)
–	–	74	LUQMANI et al. (1980)
180	560	55	RICHARDSON and WHITTAKER (1981)

Values are given per mg protein and rounded to two significant figures

within the larger synaptosomes, biochemical comparisons (Table 1) show that there is little difference between the two preparations.

These two cholinergic nerve terminal preparations have been widely used in the study of cholinergic biochemistry. For instance the demonstration of high levels of ATP in cholinergic vesicles (DOWDALL et al. 1974) led to the demonstration of ATP release from cholinergic nerve terminals (MOREL and MEUNIER 1981) and to the discovery of a purine salvage pathway at the cholinergic ending (ZIMMERMANN et al. 1979; Chap. 11, this volume). Components of the nerve terminal which may be involved in the coupling between depolarization and ACh release have also been isolated. For instance the Ca^{2+}-dependent membrane-binding protein calelectrin which was first found in these nerve terminals (WALKER 1982) has been shown to aggregate membranes in a Ca^{2+}-dependent manner (SÜDHOF et al. 1982), and a variety of calmodulin-binding proteins have also been identified (WALKER et al. 1984). In addition these cholinergic nerve terminals have been used to study the HACU system (DOWDALL 1977) and also the possible role and mechanism of opioid peptides (MICHAELSON et al. 1984) and muscarinic receptors (MICHAELSON et al. 1979) in the control of ACh release.

II. Purity and Functional State of Electromotor Synaptosome Preparations

The limiting concentration of electromotor vesicular ACh is 6–7 μmol mg^{-1} protein (TASHIRO and STADLER 1978; OHSAWA et al. 1979). Synaptic vesicles have been estimated to occupy 4% of mammalian synaptosomal volume and terminal mitochondria 24% (for references, see WHITTAKER 1984). Since mitochondria are rarely seen within electromotor synaptosomal profiles, vesicles may occupy more space in these synaptosomes – say 28%. If the hydrated lipid and protein content of vesicles and synaptsomes is comparable, the synaptosomes would have an ACh content of, say, 6500×0.28 or 1800 nmol mg^{-1}. Since the tissue content is about 60 nmol mg^{-1}, this would imply that just over 3% of the

Table 2. Estimated number of intact synaptosomes in a large synaptosome preparation

Preparation	ACh (nmol mg^{-1})	ChAT (nmol h^{-1} mg^{-1})	HACU (pmol min^{-1} mg^{-1})
Whole tissue	58	450	21
Synaptosomes			
expected[a]	1800	13500	630
found[b]	260	680	74
% Intact	15	5	9

[a] Assuming that terminals occupy 3% of the tissue protein
[b] Highest values for large synaptosomes in Table 1

tissue protein is present in the terminals. This is reasonably consistent with the volume occupied by the terminals as deduced by electron microscopy.

The highest ACh content reported for electromotor synaptosomes is 260 nmol mg^{-1}, suggesting that the number of well-sealed, ACh-rich synaptosomes, even in the best preparations, is quite small – about 15%. Calculations based on ChAT or HACU give even lower proportions (Table 2). This may account for the poor ion-pumping performance of electromotor synaptosomes when measured by bulk analysis (RICHARDSON and WHITTAKER 1981). It is not inconsistent with the results obtained with fluorescent probes by MEUNIER (1984), which recorded normal membrane potentials, since the probes would only report from the well-sealed subpopulation.

III. Presynaptic Plasma Membranes

Extremely pure presynaptic plasma membranes have been isolated from these preparations and their protein constituents analysed (STADLER and TASHIRO 1979). This analysis suggested that synaptic vesicle proteins are different from, and remain segregated from, the plasma membrane proteins. Such an interpretation has been confirmed using immunological techniques in a mammalian system by VON WEDEL et al. (1981). However, as with the mammalian preparation, there has been some debate as to the purity of *Torpedo* synaptosomes. Mammalian nerve terminal preparations are contaminated with both axonal (LEMKEY-JOHNSTON and DEKIRMENJIAN 1970) and glial membranes (COTMAN et al. 1971). Similarly the *Torpedo* preparation also appears to contain membrane fragments of unknown origin – perhaps derived from both glia and electrocytes (FRETTEN 1982). Owing to such impurities and the problems involved in extrapolating metabolic data from elasmobranchs to mammals, it may be that the greatest contribution of the *Torpedo* preparations will have been in the development of purely cholinergic probes. Antibodies raised to *Torpedo* fractions have been used to probe and analyse cholinergic function in more complex systems (Chap. 15). For example an antiserum raised to *Narcine* synaptic vesicles cross-reacts with a subset of mammalian nerve terminals (HOOPER et al. 1980). Similarly an antiserum recognizing a conjugated glycosaminoglycan constituent of *Torpedo* vesicles recognizes an antigen in mammalian motor end-plates (WALKER et al. 1982).

Additionally an antiserum raised to the *Torpedo* electromotor presynaptic plasma membrane recognizes an antigen present on cholinergic nerve terminals of a wide variety of species (JONES et al. 1981; RICHARDSON et al. 1982; WALKER et al. 1982). The antigen recognized is known as Chol-1 (for cholinergic antigen 1) and appears to be present specifically on mammalian cholinergic nerve terminals (RICHARDSON 1981, 1983). The discovery of this specificity suggested that it would be possible to purify cholinergic nerve terminals from mammalian brain using immunoaffinity techniques. Such a purification would then circumvent the problems inherent in both heterogeneous tissue preparations, and in the use of model cholinergic systems (e.g. *Torpedo* preparations) which are widely separated phylogenetically, morphologically and functionally from mammalian terminals.

F. Affinity Purification of Mammalian Cholinergic Nerve Terminals

I. Separation Procedure

Immunoaffinity techniques are extremely selective purification methods. However, when applied to cells and subcellular organelles, non-specific binding to the immunomatrix can cancel out the specificity conferred by the use of antibodies (AU and VARON 1979). In order to overcome this problem a high-capacity matrix was designed for use in the purification of cholinergic nerve terminals. This consisted of a mouse monoclonal antibody covalently coupled to cellulose, capable of binding 290 μg sheep IgG mg^{-1} cellulose, which was well above the capacity of previously used matrices (RICHARDSON 1985; RICHARDSON and LUZIO 1986). Briefly, the protocol involved incubation of a crude mitochondrial fraction (P_2, GRAY and WHITTAKER 1962) with sheep anti-(Chol-1) serum, followed by isolation of those terminals bearing the sheep IgG on the anti-(sheep IgG) matrix (see RICHARDSON et al. 1984). The purification achieved using a variety of enzyme markers is shown in Table 3, under conditions optimized to give a non-specific binding of less than 0.5% for most of the enzyme markers. The purification of the rat ChAT achieved is equivalent to a 26-fold purification over homogenate values compared to the 3- to 4-fold obtained by density-gradient centrifugation. The low amounts of glutamate decarboxylase (a marker for terminals using γ-aminobutyrate as transmitter) and dihydroxyphenylalanine decarboxylase (a catecholaminergic marker) (RICHARDSON et al. 1984) recorded in the purified terminals implies that cholinergic terminals have been purified away from the others. Indeed, the ratio of the ChAT to lactate dehydrogenase activities in the purified terminals suggests that they have been purified between 3- and 13-fold depending on the amount of cholinergic innervation in the brain region used. Since this is similar to the ratios of the activities released upon complement-mediated lysis with anti-(Chol-1) serum, it suggests that this purification is the best available with this serum. However, if expressed in terms of milligrams of protein, and assuming that about 30% of the total protein in a P_2 fraction is synaptosomal, the isolation of pure cholinergic synaptosomes from a cortex-derived P_2 fraction (in which cholinergic synaptosomes comprise about 6% of the total) should be

Table 3. Affinity purification of cholinergic nerve terminals from rat and human striatum

Parameter	Units	No. of expts.	Rat			Human		
			Total P_2 content	Purification	Yield (%)	Total P_2 content	Purification	Yield (%)
ChAT	nmol h^{-1}	21	506 ± 67	13.0	14.6	291 ± 15	30.1	14.8
AChE	μmol h^{-1}	8	56 ± 12	6.8	7.6	119 ± 10	16.4	8.1
Cytochrome oxidase	μmol h^{-1}	6	14 ± 5	6.2	7.0	17 ± 0.8	4.4	8.8
Lactate dehydrogenase	μmol h^{-1}	17	145 ± 21	4.2	4.8	185 ± 16	14.7	7.1
Glutamate decarboxylase	nmol h^{-1}	6	94 ± 4	0.5	0.5	–	–	–
2′3′-Cyclic nucleotide phosphodiesterase	μmol h^{-1}	8	912 ± 85	0.6	0.6	5760 ± 400	2.1	1.0
Protein	μg	6	573 ± 73	–	1.1	850 ± 35	–	0.5

Purification is expressed as the ratio of the specific activity in the purified terminals to that in the P_2 fraction. Results (per 100 mg tissue) are expressed as means ± SEM for the numbers of experiments in col. 3 (rat), or means ± range of two experiments (human)

accompanied by an approximately 57-fold increase in the concentration of cholinergic marker with respect to protein. However, the purification obtained in the affinity procedure is approximately 18-fold (RICHARDSON et al. 1984). One reason for this relatively low purification is contamination by the anti-(Chol-1) immunoglobulin. If allowance is made for this (approximately 3.0 μg per preparation from 100 mg striatum, unpublished data), then the purification obtained rises to 36-fold. Alternatively, the purity can be assessed by estimating the likely ACh content of pure cholinergic nerve terminals. Using the estimate that 4% of the terminal is occupied by vesicles, and a vesicular ACh limiting concentration of 6.5 $\mu mol\ mg^{-1}$ protein, then pure terminals should have an ACh content of 260 $nmol\ mg^{-1}$. In fact, the concentration found is $248 \pm 37\ nmol\ mg^{-1}$ (RICHARDSON and BROWN 1987) which rises to approximately 330 $nmol\ mg^{-1}$ after allowance for the anti-(Chol-1) immunoglobulin. This high estimate may be a consequence of a substantial pool of cytoplasmic ACh (WHITTAKER 1984). The difference in these estimates of purity probably reflects the problems associated with measuring small amounts of protein bound to a matrix already bearing very large amounts of protein. The lower purification of AChE reflects this enzyme's presence at non-cholinergic synapses (SILVER 1974), as does the low purification of the more generally distributed enzymes lactate dehydrogenase and cytochrome oxidase. The greater purification of ChAT obtained with human material reflects the contamination of the original tissue with white matter as indicated by the high activity of the myelin enzyme 2′,3′-cyclic nucleotide phosphodiesterase. Cholinergic nerve terminals purified in this fashion show characteristic synaptosomal profiles with mitochondria and clear synaptic vesicles enclosed by a plasma membrane bearing postsynaptic adhesions (RICHARDSON et al. 1984).

II. Characteristics of the Preparation

The available evidence for the specificity of the anti-(Chol-1) serum based on complement-mediated lysis tests (RICHARDSON 1981; RICHARDSON 1983) and immunohistochemical studies (JONES et al. 1981; WALKER et al. 1982), together with the degree of purification obtained, demonstrates that cholinergic terminals can be purified to homogeneity from mammalian brain.

However, before it is possible to use terminals isolated in this manner for the study of the fine details of cholinergic biochemistry, it is necessary to show that they are metabolically competent and show all the known characteristics of cholinergic nerve endings. Thus, for example, Fig. 1 is a double reciprocal plot illustrating the ability of the terminals to oxidise [^{14}C]glucose to $^{14}CO_2$ in a concentration-dependent manner. The apparent K_m for glucose was 0.25 ± 0.03 (2) mM (variances here and subsequently are SEMs or ranges for two observations), somewhat higher than that observed for a mixed population of terminals (0.08 ± 0.02 (2) mM). The HACU system is also present in the purified terminals, where it shows a dependence on a transmembrane Na^+ gradient and is inhibited by hemicholinium-3. The apparent K_m for choline was $2.7 \pm 0.5(6)$ μM with a V_{max} of $794 \pm 150(6)\ pmol\ min^{-1}\ mg^{-1}$ protein (RICHARDSON 1986). There was however evidence for a low-affinity choline transport system in the purified terminals with a K_m of $58.2 \pm 7.5(4)$ μM (see Fig. 2). The purified ter-

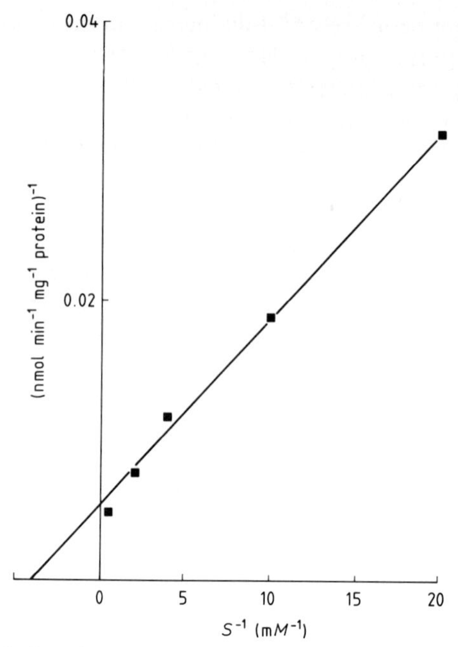

Fig. 1. Glucose oxidation in affinity-purified cholinergic nerve terminals. Purified cholinergic nerve terminals were incubated for 30 min at 37 °C with varying glucose concentrations containing 5 μCi [U-^{14}C] glucose. The evolved CO_2 was absorbed in 0.2 ml of 1 *M* methylbenzethonium hydroxide. Apparent $K_m = 0.25 \pm 0.03$ m*M*, apparent $V_{max} = 180 \pm 16$ nmol CO_2 min^{-1} mg^{-1} protein. Values are means ± SEM of 3 determinations

minals are able to synthesize ACh from glucose and choline – the proportion of the choline acetylated being dependent on the metabolic state of the terminals (RICHARDSON 1986). These terminals also release ACh upon depolarization by both elevated extracellular K^+ levels (25 m*M*) (RICHARDSON 1985) as well as by the alkaloid veratridine. In both cases the release of ACh is inhibited in the absence of extracellular Ca^{2+} (Fig. 3). It is apparent therefore that the affinity-purified nerve terminals are able to respire, take up choline, and to synthesize and release ACh upon stimulation. This preparation should therefore be suitable for detailed studies of the regulatory mechanisms operating at the cholinergic nerve terminal. It is also hoped that, by increasing the yield of the process, the molecular constituents of the nerve terminal can be characterized.

G. Summary

It has now become possible to study acetylcholine metabolism in isolated cholinergic terminals, without contamination by other non-cholinergic systems, i.e. other terminals, glia etc. The development of the immunoaffinity-purification technique made possible by the initial isolation of *Torpedo* terminals, may prove to be one example of an approach of widespread applicability in neuronal sub-

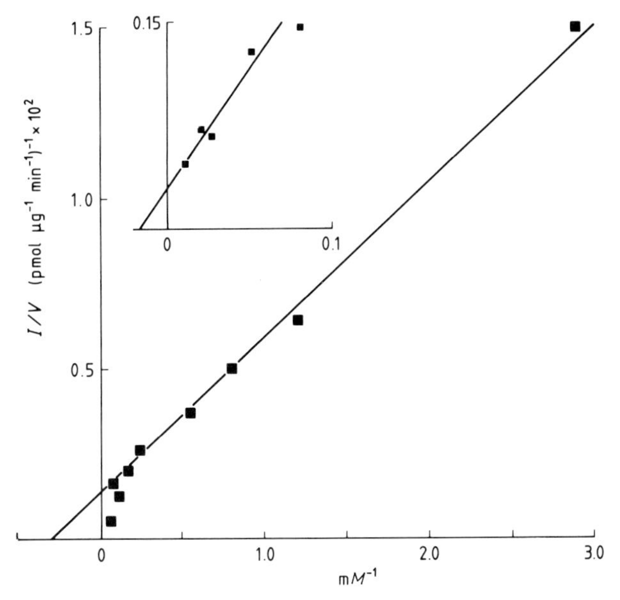

Fig. 2. Uptake of choline by affinity-purified cholinergic nerve terminals. Purified cholinergic nerve terminals were incubated for 4 min at 37 °C with choline concentrations between 0.4 and 40 μM containing 2 μCi [^{3}H]choline. The uptake was stopped by dilution in ice-cold saline and filtration. After washing, the radioactivity on the filter was determined. *Insert*, choline uptake plotted on an expanded scale to show uptake at high extracellular choline concentrations

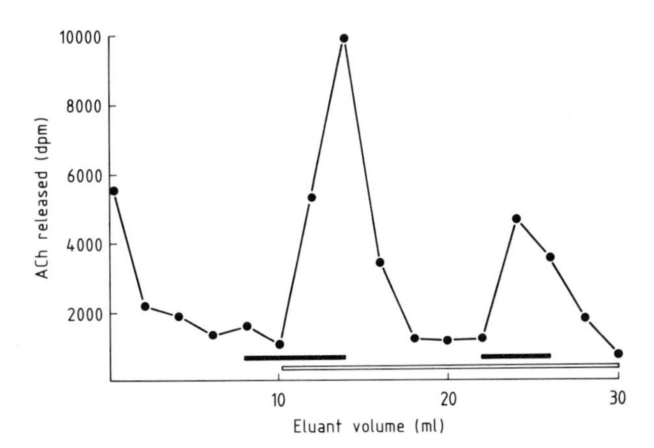

Fig. 3. Effect of veratridine and Ca^{2+} on the release of acetylcholine from affinity-purified nerve terminals. Purified nerve terminals were preincubated with 5 μCi [^{3}H]choline for 15 min at 37 °C, washed and placed in a low-volume superfusion chamber. Terminals were superfused (1 ml min^{-1}) with a Krebs-Ringer-Hepes solution (RICHARDSON 1986) with and without 50 μ*M* veratridine and 5 m*M* EGTA. The eluant was collected and radioactivity determined. *Black bars* represent the presence of 50 μ*M* veratridine; *white bar* the presence of 1.3 m*M* Ca^{2+}

fractionation. For example, it may prove possible to purify cholinergic terminals from other heterogeneous tissues, e.g. muscle using a similar procedure. Additionally a judicious search for transmitter-specific surface antigens throughout the phyla may permit the isolation of many different populations of terminals and neurons from mammalian nervous systems. This modification of standard tissue fractionation techniques would then lay the groundwork for the study of many aspects of neurobiology, e.g. control mechanisms of synaptogenesis and transmitter metabolism, the identification of those components responsible for intercellular recognition and potentially even the mechanisms involved in the control of transmitter-specific gene expression.

References

Abood L, Cavanaugh JB, Tschirgi RD, Gerard RW (1951) Metabolic and pharmacological studies in glia and nerve cells and components. Fed Proc 10:3
Au AMJ, Varon S (1979) Neural cell sequestration on immunoaffinity columns. Exp Cell Res 120:269–276
Bacq ZM (1935) Recherches sur la physiologie et la pharmacologie du système nerveux autonome: XVII. Les esters de la choline dans les extraits de tissus des invertébrés. Arch Int Physiol Biochim 42:24–42
Betz V, Baggiolini M, Hauser R, Hodel C (1974) Resolution of three distinct populations of nerve endings from rat brain homogenates by zonal isopycnic centrifugation. J Cell Biol 61:466–480
Booth RFG, Clark JB (1978) A rapid method for the preparation of relatively pure metabolically competent synaptosomes from rat brain. Biochem J 176:365–370
Booth RFG, Clark JB (1979) A method for the rapid separation of soluble and particulate components of rat brain synaptosomes. FEBS Lett 107:387–392
Bradford HF (1969) Respiration *in vitro* of synaptsomes from mammalian cerebral cortex. J Neurochem 16:675–684
Brody TM, Bain JA (1951) Effect of barbiturates on oxidative phosphorylation. Proc Soc Exp Biol Med 77:50–53
Campbell CWB (1976) The Na^+, K^+, Cl^- contents and derived membrane potentials of presynaptic nerve endings *in vitro*. Brain Res 101:594–599
Cohen RS, Blomberg F, Berzins K, Siekevitz P (1977) The structure of postsynaptic densities isolated from dog cerebral cortex: I. Overall morphology and protein composition. J Cell Biol 74:181–203
Cotman CW, Matthews DA (1971) Synaptic plasma membranes from rat brain synaptosomes: isolation and partial characterization. Biochim Biophys Acta 249:380–394
Cotman C, Herschman H, Taylor D (1971) Subcellular fractionation of cultured glial cells. J Neurobiol 2:169–180
De Belleroche JS, Bradford HF (1972) Metabolism of beds of mammalian cortical synaptosomes: response to depolarizating influences. J Neurochem 19:585–602
De Belleroche J, Coutinho-Netto J, Bradford HJ (1982) Dopamine inhibition of the release of endogenous acetylcholine from corpus striatum and cerebral cortex in tissue slices and synaptosomes: a presynaptic response? J Neurochem 39:217–222
De Robertis E, Pellegrino de Iraldi A, Rodrigues G, Gomez CJ (1960) Sesiones de la Sociedad Argentina de Biologia, Mendoza, Argentina
De Robertis E, Pellegrino de Iraldi A, Rodriguez De Lores Arnaiz G, Salganicoff L (1962) Cholinergic and non-cholinergic nerve endings in rat brain: I. J Neurochem 9:23–35
De Robertis E, Rodriguez D Lores Arnaiz G, Salganicoff L, Pellegrino de Iraldi A, Zieher LM (1963) Isolation of synaptic vesicles and structural organization of the acetylcholine system within brain nerve endings. J Neurochem 10:225–235
De Robertis E, Azcurra JM, Fiszers S (1967) Ultrastructure and cholinergic binding capacity of junctional complexes isolated from rat brain. Brain Res 5:45–56

Dowdall MJ (1977) The biochemistry of *Torpedo* cholinergic neurons. In: Osborne NN (ed) Biochemistry of characterised neurons. Pergamon, Oxford, pp 177–216
Dowdall MJ, Simon EJ (1973) Comparative studies on synaptosomes: uptake of [*N*- *Me*-^{3}H]choline by synaptosomes from squid optic lobes. J Neurochem 21:969–982
Dowdall MJ, Whittaker VP (1973) Comparative studies in synaptosome formation: the preparation of synaptosomes from the head ganglion of the squid, *Loligo pealii*. J Neurochem 20:921–935
Dowdall MJ, Zimmermann H (1976) The preparation and properties of 'nerve terminal sacs' from the electric organ of *Torpedo*. Exp Brain Res 24:8–9
Dowdall MJ, Zimmermann H (1977) The isolation of pure cholinergic nerve terminal sacs (T-sacs) from the electric organ of juvenile *Torpedo*. Neuroscience 2:405–421
Dowdall MJ, Boyne AF, Whittaker VP (1974) Adenosine triphosphate, a constituent of cholinergic synaptic vesicles. Biochem J 140:1–12
Dowdall MJ, Wächtler K, Henderson GM (1975) Comparative studies on the high affinity choline uptake system. Vth Meet Int Soc Neurochem, Abstr, Barcelona, p 122
Dowdall MJ, Fohlman JP, Watts A (1979) Presynaptic action of snake venom neurotoxins on cholinergic systems. Adv Cytopharmacol 3:63–76
Droz B, Barondes SH (1969) Nerve endings: rapid appearance of labelled proteins shown by electron microscope radioautography. Science 165:1131–1133
Feldberg W (1945) Present views on the mode of action of acetylcholine in the central nervous system. Physiol Rev 25:596–642
Feldberg W, Fessard A (1942) The cholinergic nature of the nerves of the electric organ of the *Torpedo (Torpedo marmorata)*. J Physiol (Lond) 101:200–216
Fonnum F (1967) The compartmentation of choline acetyltransferase. Biochem J 103:262–270
Fonnum F (1968a) The distribution of glutamate decarboxylase and aspartate transaminase in subcellular fractions of rat and guinea-pig brain. Biochem J 106:401–412
Fonnum F (1968b) Choline acetyltransferase binding to and release from membranes. Biochem J 109:389–398
Fretten P (1982) The effects of presynaptic neurotoxins on *Torpedo* synaptosomes. PhD thesis, University of Nottingham
Gray EG, Whittaker VP (1960) The isolation of synaptic vesicles from the central nervous system. J Physiol (Lond) 153:35–37*P*
Gray EG, Whittaker VP (1962) The isolation of nerve endings from brain: an electron-microscopic study of cell fragments derived by homogenization and centrifugation. J Anat 96:79–87
Hajós F, Csillag A, Kálmán M (1979) The morphology of microtubules in incubated synaptosomes: effect of low temperature and vinblastine. Exp Brain Res 35:387–393
Hebb CO, Smallman BN (1956) Intracellular distribution of choline acetylase. J Physiol (Lond) 134:385–392
Hebb CO, Whittaker VP (1958) Intracellular distributions of acetylcholine and choline acetylase. J Physiol (Lond) 142:187–196
Hooper JE, Carlson SS, Kelly RB (1980) Antibodies to synaptic vesicles purified from *Narcine* electric organ binds a subclass of mammalian nerve terminals. J Cell Biol 81:275–300
Israël M, Gautron J, Lesbats B (1968) Isolement des vésicules synaptiques de l'organe électrique de la torpille et localisation de l'acétylcholine a leur niveau. C R Acad Sci III (Paris) 266:273–275
Israël M, Manaranche R, Mastour-Frachon P, Morel N (1976) Isolation of pure cholinergic nerve endings from the electric organ of *Torpedo marmorata*. Biochem J 160:113–115
Jones DG (1970) A study of the presynaptic network of *Octopus* synaptosomes. Brain Res 20:145–158
Jones DG (1975) Synapses and synaptosomes. Chapman and Hall, London, pp 44–97
Jones RT, Walker JH, Richardson PJ, Fox GQ, Whittaker VP (1981) Immunohistochemical localization of cholinergic nerve terminals. Cell Tissue Res 218:355–373
Jope RS (1979) High affinity choline transport and acetyl CoA production in brain and their roles in the regulation of acetylcholine synthesis. Brain Res Rev 1:313–344

Juorio AV (1971) Catecholamines and 5-hydroxytryptamine in nervous tissue of cephalopods. J Physiol (Lond) 216:213–226
Kuhar MJ, Sethy VH, Roth RH, Aghajanian GK (1973) Choline selective accumulation by central cholinergic neurons. J Neurochem 20:581–593
Kurokawa M, Sakamoto T, Kato M (1965) A rapid isolation of nerve ending particles from brain. Biochim Biophys Acta 94:307–309
Lemkey-Johnston N, Dekirmenjian H (1970) The identification of fractions enriched in nonmyelinated axons from rat whole brain. Exp Brain Res 11:392–410
Luqmani YA, Sudlow G, Whittaker VP (1980) Homocholine and acetylhomocholine: false transmitters in the cholinergic electromotor system of *Torpedo*. Neuroscience 5:153–160
Marnay A (1937) Cholinesterase dans l'organe électrique de la torpille. C R Soc Biol (Paris) 126:573
Meunier FM (1984) Relationship between presynaptic membrane potential and acetylcholine release in synaptosomes from *Torpedo* electric organ. J Physiol (Lond) 354:121–137
Michaelson DM Sokolovsky M (1976) Neurotransmitter release from viable purely cholinergic *Torpedo* synaptosomes. Biochem Biophys Res Commun 73:25–31
Michaelson DM, Avissar S, Kloog Y, Sokolovsky M (1979) Mechanism of acetylcholine release: possible involvement of presynaptic muscarinic receptors in the regulation of acetylcholine release and protein phosphorylation. Proc Natl Acad Sci USA 76:6336–6340
Michaelson DM, McDowall G, Sarne Y (1984) Opiates inhibit acetylcholine release from *Torpedo* nerve terminals by blocking Ca^{2+} influx. J Neurochem 43:614–618
Morel N, Meunier FM (1981) Simultaneous release of acetylcholine and ATP from stimulated cholinergic synaptosomes. J Neurochem 36:1766–1773
Morel N, Israël M, Manaranche R, Mastour-Frachon P (1977) Isolation of pure cholinergic nerve endings from *Torpedo* electric organ. J Cell Biol 75:43–55
Morel N, Israël M, Manaranche R, Lesbats B (1979) Stimulation of cholinergic synaptosomes isolated from *Torpedo* electric organ. Prog Brain Res 49:191–202
Nachmansohn D, Lederer E (1939) Sur la biochimie de la cholinesterase: 1. Préparation de l'enzyme, rôle des groupements-SH. Bull Soc Chim Biol (Paris) 21:797–808
Nagy A, Delgado-Escueta AV (1984) Rapid preparation of synaptosomes from mammalian brain using nontoxic iso-osmotic gradient material (Percoll). J Neurochem 43:1114–1123
Ohsawa K, Dowe GHC, Morris SJ, Whittaker VP (1979) The lipid and protein content of cholinergic synaptic vesicles from the electric organ of *Torpedo marmorata* purified to constant composition: implications for vesicle structure. Brain Res 161:447–457
Richardson PJ (1981) Quantitation of cholinergic synaptsomes from guinea-pig brain. J Neurochem 37:258–260
Richardson PJ (1983) Presynaptic distribution of the cholinergic specific antigen Chol-1 and 5′-nucleotidase in rat brain, as determined by complement-mediated release of neurotransmitters. J Neurochem 41:640–648
Richardson PJ (1985) Use of antibodies to purify cholinergic nerve terminals. In: Reid E, Cook GMW, Morre DJ (eds) Investigation and exploitation of antibody combining sites. Plenum, New York, pp 207–214 (Methodological Surveys in Biochemistry, vol 9)
Richardson PJ (1986) Choline uptake and metabolism in affinity purified cholinergic nerve terminals from rat brain. J Neurochem 46:1251–1255
Richardson PJ, Brown SJ (1987) ATP release from affinity purified rat cholinergic nerve terminals. J Neurochem (in press)
Richardson PJ, Luzio JP (1986) Immunoaffinity purification of subcellular particles and organelles. Appl Biochem Biotechnol 13:133–145
Richardson PJ, Whittaker VP (1981) The Na^+ and K^+ content of isolated *Torpedo* synaptosomes and its effect on choline uptake. J Neurochem 36:1536–1542
Richardson PJ, Walker JH, Jones RT, Whittaker VP (1982) Identification of a cholinergic specific antigen Chol-1 as a ganglioside. J Neurochem 38:1605–1614
Richardson PJ, Siddle K, Luzio JP (1984) Immunoaffinity purification of intact, metabolically active, cholinergic nerve terminals from mammalian brain. Biochem J 219:647–654

Sheridan MN, Whittaker VP, Israël M (1966) The subcellular fractionation of the electric organ of *Torpedo*. Z Zellforsch 74:291–307
Silver A (1974) The biology of cholinesterases. North Holland, Amsterdam
Stadler H, Tashiro T (1979) Isolation of synaptosomal plasma membranes from cholinergic nerve terminals and a comparison of their proteins with those of synaptic vesicles. Eur J Biochem 101:171–178
Südhof TC, Walker JH, Obrocki J (1982) Calelectrin self-aggregates and promotes membrane aggregation in the presence of Ca^{2+}. EMBO J 1:1167–1170
Tashiro T, Stadler H (1978) Chemical composition of cholinergic synaptic vesicles from *Torpedo marmorata* based on improved purification. Eur J Biochem 90:479–487
Wedel RJ von, Carlson SS, Kelly RB (1981) Transfer of synaptic vesicle antigens to the presynaptic plasma membrane during exocytosis. Proc Natl Acad Sci USA 78:1014–1018
Walker JH (1982) The isolation from cholinergic synapses of a protein that binds to membranes in a Ca^{2+} dependent manner. J Neurochem 39:815–823
Walker JH, Jones RT, Obrocki J, Richardson GP, Stadler H (1982) Presynaptic plasma membranes and synaptic vesicles of cholinergic nerve endings demonstrated by means of specific antisera. Cell Tissue Res 223:101–116
Walker JH, Stadler H, Witzemann V (1984) Calmodulin-binding proteins of the cholinergic electromotor synapse: synaptosomes, synaptic vesicles, receptor enriched membranes and cytoskeleton, J Neurochem 42:314–320
Whittaker VP (1965) The application of subcellular fractionation techniques to the study of brain function. Prog Biophys Mol Biol 15:39–96
Whittaker VP (1971) Subcellular localization of neurotransmitters. Adv Cytopharmacol 1:319–330
Whittaker VP (1973) The biochemistry of synaptic transmission. Naturwissenschaften 60:281–289
Whittaker VP (1984) The synaptosome. In: Lajtha A (ed) Handbook of neurochemistry, vol 7. Plenum, New York, pp 1–30
Whittaker VP, Michaelson IA, Kirkland RJA (1964) The separation of synaptic vesicles from nerve-ending particles ('synaptosomes'). Biochem J 90:293–303
Whittaker VP, Essman WE, Dowe GHC (1972) The isolation of pure synaptic vesicles from the electric organs of elasmobranch fish of the family Torpedinidae. Biochem J 128:833–846
Wilson WS, Cooper JR (1972) The preparation of cholinergic synaptosomes from bovine superior cervical ganglia. J Neurochem 19:2779–2790
Zimmermann H (1979) Vesicle recycling and transmitter release. Neuroscience 4:1773–1804
Zimmermann H, Dowdall MJ, Lane DA (1979) Purine salvage at the cholinergic nerve endings of the *Torpedo* electric organ: the central role of adenosine. Neuroscience 4:979–993

CHAPTER 13

Axonal Transport in Cholinergic Neurons

M. TYTELL * and H. STADLER

A. Introduction

Since 1980 there have been six major reviews of various aspects of axonal transport, each of which has included significant portions of the literature concerning the transport of acetylcholine (ACh) and its associated enzymes (GRAFSTEIN and FORMAN 1980; THOENEN and KREUTZBERG 1981; KARLSSON 1982; WEISS 1982; IQBAL 1986; SMITH and BISBY 1987). Based on those reviews and the current literature, it seems that transport in cholinergic axons received the most attention during the 1970s and that the rate at which additional information has been uncovered slowed since 1980. Most of the work published within the last few years consists of studies that use the transport of cholinergic constituents as indicators of the state of neuron function following challenges such as exposure to neurotoxins or inducement of disease-related pathological changes (see for example, BRIMIJOIN and SCHREIBER 1982; VITADELLO et al. 1983; MAYER and TOMLINSON 1983; OIKARINEN and KALIMO 1984). These latter studies have not added much to our understanding of how cholinergic components are delivered to the synaptic terminal. Thus, to avoid simply repeating the information summarized in the earlier reviews, this report will cover selected works with the goals of focusing attention on the gaps and key problems in the existing literature and identifying the types of experiments that have the potential to increase our understanding of the form and manner of the transport of cholinergic constituents in axons.

B. General Protein Transport in Cholinergic Nerves

Investigations of axonal transport generally fall into one of two categories: (a) those that provide information about the transport of an identified substance or (b) those that present an overview of the transport of a large number of polypeptides which are uncharacterized except for their relative molecular masses (M_r) and sometimes their apparent isoelectric points. Usually there is little information presented that bridges the gap between these two types of studies and relates the transport of an identified substance to the general picture of protein transport in axons. Research on axonal transport in cholinergic neurons exhibits this same dichotomy of effort. Several groups of investigators have employed as their model systems nerves that are at least partly cholinergic, such as the vagus (BLACK and LASEK 1978; TASHIRO and KUROKAWA 1982; MCLEAN et al. 1983), the hypo-

* From Abt. Neurochemie, Max-Planck-Institut Göttingen (current address see list of contributors).

glossal (BLACK and LASEK 1978), and the motor portion of the sciatic (HOFFMAN and LASEK 1975; STONE and WILSON 1979; BISBY 1981). Comparisons of the classes of transported proteins revealed in these different studies, as was done to a limited extent in a report by LASEK and his colleagues on fast axonal transport in a purely cholinergic nerve of the toadfish (BARKER et al. 1975), support the idea that anterograde protein transport in nerves is generally similar, regardless of the nerve type or the species. In fact, if one discounts the difference in transport rates between the peripheral and central nervous systems, one can appreciate the overall similarity between cholinergic nerves and the optic nerve (compare for example, BLACK and LASEK 1980; BAITINGER et al. 1982; TYTELL et al. 1981; TYTELL et al. 1985). Furthermore, HOOPER and co-workers have shown that antibodies raised against purified cholinergic vesicles from the electric fish, *Narcine*, react with material in mammalian adrenergic synapses (HOOPER et al. 1980). The significance of these observations may be that the materials which the neurons use to produce their distinctive neurotransmitters probably represent a relatively minor proportion of the total set of proteins that all nerve cells use to package those neurotransmitters and construct the cytoplasmic ground substance. The apparent similarity means that one should be able to apply what is known about the general characteristics of axonal protein transport from a more thoroughly studied model like the visual system to the planning and interpretation of studies of the transport of ACh and its related enzymes.

In the discussions of the transport of molecules specific to cholinergic neurons that follow, reports will be cited that bring out the existing areas of disagreement and some experiments will be suggested which could resolve those controversies. In the last section some of the relevant general principles of axonal transport will be described to illustrate the kinds of questions that need to be answered to integrate the data on transport of cholinergic constituents with the general aspects of axonal protein transport.

C. Transport of Distinctive Components of the Cholinergic Neuron

I. Acetylcholine

A variety of approaches has been employed in widely different organisms to examine the anterograde transport of ACh itself. These have included the microinjection of both [^{3}H]choline and [^{3}H]ACh directly into the large neurons of the invertebrate, *Aplysia* (KOIKE et al. 1972; KOIKE and NAGATA 1979), the injection of [^{3}H]choline in the vicinity of cholinergic neuronal cell bodies (WOODWARD and LINDSTRÖM 1977; WALDENLIND and ANDERSON 1978), and the direct measurement of endogenous ACh as it accumulates adjacent to a ligature in the nerve (ZIMMERMANN and WHITTAKER 1973; HEILBRONN and PETTERSSON 1973; WHITTAKER et al. 1975; WIDLUND et al. 1975; HEIWALL et al. 1976; DAVIES 1978; DAHLSTRÖM et al. 1982a, b). All these studies support the idea that neurons are capable of the rapid anterograde transport of some proportion of their ACh in a vesicle-like compartment that includes ATP (DOWDALL et al. 1974), but the relative amount of ACh that is actually supplied to the synaptic terminal in this manner remains a matter of controversy (compare DAHLSTRÖM et al. 1974; WALDENLIND and ANDERSON 1978). Attempts to resolve this question have been confounded by two factors: (a) ACh can be synthesized locally and (b) the membranous organelle in which ACh is transported seems to be extremely labile (DAHLSTRÖM et al. 1982b). The former makes it difficult in accumulation studies to distinguish between what is transported and what is locally synthesized, whereas the latter raises doubts about the various estimates of the proportions of soluble versus vesicular ACh.

In a recent publication by KIENE and STADLER (1987) the fast axonal transport of cholinergic synaptic vesicles from *Torpedo* electromotoneurons is described. Vesicles move anterogradely as fast components but as 'empty organelles' and accumulate ACh only after arrival in the nerve terminal.

There is a suggestion that ACh is also transported retrogradely, based upon ligation studies which showed the accumulation of both ACh itself and cholinergic vesicle antigens (DAHLSTRÖM et al. 1981, 1985). However, such a conclusion must be regarded as questionable until the contribution of local synthesis of ACh at the ligation site is clarified.

The observations on ACh transport to date leave unresolved the issue of the form or forms in which ACh exists within the living axon. A knowledge of the magnitude of any soluble and/or stationary pool is essential in order to relate ACh transport to the general scheme of axonal transport. One way to obtain this information would be to devise a histochemical method for detecting ACh *in situ*; another would be to apply to purely axonal material the non-invasive technique used by WEILER et al. (1982) for determining the size of the cytoplasmic pool of ACh in the terminal regions of electromotor axons in electric organ. Both approaches avoid the pitfalls of subcellular fractionation.

Related to the question of the form of transported ACh is an even more fundamental consideration. Since ACh can be readily synthesized at the synaptic terminal, is ACh transport a necessary and functionally significant process? This issue has also been raised by DAHLSTRÖM et al. (1985). The exploration of this point will require the discovery of a means selectively of depleting ACh from the axons of a cholinergic nerve while leaving the terminal unaffected and then testing for a decrement in synaptic ACh content. If synaptic ACh remained unchanged in such an experiment, this would suggest that local ACh synthesis and not transport is the primary source of ACh in the terminal.

II. Acetylcholinesterase

By far the most frequently measured of the cholinergic neuron-specific substances is acetylcholinesterase (AChE), probably because its endogenous activity can be quantified using a relatively simple assay. Thus, the typical approach to the study of AChE transport has been to interrupt the nerve by a crush, cut, or ligation and then calculate the transport rate of the enzyme on the basis of its rate of accumulation at the interruption with time. Because of the ease of this method the other often used means for measuring transported materials, the incorporation of a radioactive precursor, has not been applied. Therefore, almost all the available data on the quantity and rate of AChE transport are derived from accumulation studies. These data, obtained from a variety of nerves and organisms, all indicate that a portion of the AChE in a nerve, about 20%, is rapidly transported anterogradely and that a fraction of the anterogradely transported material is returned to the cell bodies via retrograde transport (FONNUM et al. 1973; BRIMIJOIN and WIERMAA 1978; DROZ et al. 1979; DAHLSTRÖM et al. 1982b; COURAUD et al. 1983). Like ACh, the bulk of the material is not rapidly transported and, within the time scale of most experiments, does not move appreciably, so it is considered to be stationary. However, the situation for AChE is more complex than for ACh because there are at least four different forms of the enzyme found within a nerve (BON et al. 1979; Chap. 8a), this volume) and these forms are differentially distributed between the moving and the stationary fractions (COURAUD and DI GIAMBERARDINO 1982). The forms are distinguished by their sedimentation coefficients and their behaviour upon molecular sieve chromatography. Briefly, there are three globular forms, G1, G2 and G4, comprised of one, two and four monomers, respectively, and three asymmetric forms consisting of one (A4), two (A8) or three (A12) tetramers connected to a collagen-like tail (BON et al. 1979; MASSOULIÉ et al. 1980; ALLEMAND et al. 1981). In the fowl sciatic nerve most (80%) of the AChE is in the G4 form, A12 makes up 3% and the remainder (17%) consists of the G1 and G2 forms (VIGNY et al. 1976).

Most of the studies of AChE transport go into some detail concerning the differences in the rate of accumulation of the various forms of the enzyme on either side of a nerve ligature. An extensive study using the fowl sciatic nerve by COURAUD and DI GIAMBERARDINO (1982) revealed the following specific points. G4 and A12 are the main transported forms of AChE. A little more than one-tenth of the total nerve G4 and about two-thirds of the total nerve A12 are rapidly transported anterogradely, whereas slightly less than one-tenth of the former and about one-third of the latter are transported retrogradely. G1 and G2, on the other hand, constitute the stationary or slowly moving forms.

The implications of the above observations are not clear because little is known about the functional significance of the different forms of AChE, though it has been suggested that their structures are related to their locations within the tissue (BON et al. 1979). A further complication arises from the fact that the functions may vary in different species. The relative amounts of each form of AChE are apparently different in the frog sciatic nerve from those in the fowl (COURAUD et al. 1985). However, it does appear that G4 and A12 are membrane-associated, and that G1 and G2 are relatively soluble (VIGNY et al. 1976). Additionally, the A12 form within the nerve is distinctive in that most of it appears to be transported to the terminals (COURAUD and DI GIAMBERARDINO 1982).

An important consideration in any interpretation of these observations was raised by a careful set of experiments performed by BRIMIJOIN et al. (1978). They showed that about 80% of the AChE content of the rat peroneal nerve is located outside the axon, most probably bound in some way to the axolemma. This fraction of the enzyme accounts for the large amount of apparently stationary AChE. Their data also indicated that intra-axonally transported forms of AChE continually enter the external compartment, where they have a half-life of about 5 days. Thus, studies of AChE transport must be interpreted in the context of the presence of this large, external pool of the enzyme.

III. Choline Acetyltransferase

Like AChE, choline acetyltransferase (ChAT) transport has been studied only by the technique of measuring accumulation at a nerve crush. However, it has not been the subject of nearly so many investigations as AChE. Most of the observations are concisely reviewed in DAHLSTRÖM et al. (1982b). The data obtained from the sciatic nerve favour the hypothesis that ChAT is contained mainly in a compartment within the axon different from that of ACh and AChE because it is largely a soluble enzyme which is slowly transported at about 2 mm day^{-1} toward the axon terminal (SAUNDERS et al. 1973; TUČEK 1975; DAVIES et al. 1977; O'BRIEN 1978). However, in the cervical vagus and hypoglossal nerves, there is some evidence that ChAT may be partly associated with membranes and thereby rapidly transported (FONNUM and MALTHE-SØRENSSSEN 1973). Furthermore, in rats rendered diabetic for weeks, ChAT transport was reduced in the sciatic nerve, but not in the vagus (MAYER and TOMLINSON 1983). This variation in the transport rate of ChAT in different nerves under normal and pathological conditions may be a consequence of the presence of multiple forms of the enzyme with differing membrane affinities (FONNUM and MALTHE-SØRENSSEN 1973).

IV. Cholinergic Synaptic Vesicle-Associated Proteins

Currently it is not known in what form the cholinergic vesicle exists while being transported to the synaptic terminal. Elucidation of this point will aid in understanding better the present observations concerning the transport of other cholinergic-specific substances and will very likely contribute to a clearer view of neurotransmitter transport in general. A pioneering attempt to detect transport of vesicles in *Torpedo* cholinergic electromotor axons was made by ULMAR and WHITTAKER (1974), who showed the apparent accumulation of vesicle marker proteins using an immunochemical detection method. More recent work from DAHLSTRÖM and colleagues has also been directed at this question of vesicle transport. Using a polyclonal anti-cholinergic vesicle antibody and a monoclonal antibody against a synaptic vesicle glycoprotein, they showed that immunoreactive material was distributed in rat sciatic nerve in a manner consistent with the rapid transport of the vesicles in both directions; it was present only weakly in the neuronal somata, was undetectable in intact axons, and accumulated rapidly on both sides of a ligature (BÖÖJ et al. 1985; DAHLSTRÖM et al. 1985). Interestingly, the material which accumulated proximally was of higher density than typical synaptic vesicles (BÖÖJ et al. 1985), which suggests that the transported form of vesicles is different from that present in the terminal.

D. Interpretation of the Transport of Cholinergic-Specific Substances in the Context of General Axonal Protein Transport

The general scheme of axonal protein transport as it is understood today includes the existence of five distinct rate components of anterogradely, and one component of retrogradely transported material (reviewed in GRAFSTEIN and FORMAN 1980); WEISS 1982). Of the four components of anterograde transport the best characterized are the fast component (FC; rate of 100–400 mm day^{-1}) and the two slow components with rates of 1–4 mm day^{-1} (SCb of LASEK and co-workers or Group IV of WILLARD and co-workers) and 0.1–1.0 mm day^{-1} (SCa of LASEK and co-workers or Group V of WILLARD and co-workers) (reviewed in WEISS 1982). It is generally accepted that the constituents of the FC are primarily membrane proteins or membrane-associated proteins (reviewed in GRAFSTEIN and FORMAN 1980), those of SCb are actin and easily solubilized cytoplasmic proteins like the enzymes of intermediary metabolism (LASEK et al. 1984), and those of SCa are primarily the cytoskeletal proteins, consisting of the tubulins and the subunits of neurofilaments (HOFFMAN and LASEK 1975; BLACK and LASEK 1980). All the constituents of each of these groups are transported within the axon in the form of macromolecular complexes of functionally related proteins, according to the structural hypothesis of axonal transport (TYTELL et al. 1981; LASEK and BRADY 1982).

In attempting to integrate into this general scheme of axonal transport the observations on the transport of cholinergic-specific molecules, one primary problem becomes apparent. Most of the observations on the transport of ACh, AChE and ChAT utilize the techniques of accumulation at a ligature, whereas the general work on axonal protein transport is based on the analysis of the movement of materials labelled with radioactive precursors. Since it is recognized that a ligature may have untoward affects on the process of axonal transport (BISBY 1981, 1982), including premature initiation of retrograde transport and altered turnover rate, and since no direct comparisons have been made between the two approaches, there is insufficient information available to resolve the points where the techniques yield different conclusions. One of the more puzzling contradictions that arises from a comparison of the results of the two techniques is that the accumulation method often supports the existence of a major stationary component of the transported substance, whereas such material appears minor or non-existent when the movement of radioactively labelled proteins is followed.

To begin to investigate the basis for the conflicting observations, we have initiated a study of the transport of [^{35}S]methionine-labelled proteins in the electric nerves of *T. marmorata* (TYTELL et al. 1985). Preliminary results have confirmed the technical feasibility of this procedure and the presence of both rapidly and slowly transported labelled polypeptides with relative molecular weights that are similar to FC and SC proteins characterized in the more commonly used mammalian neural pathways. Thus, it should be possible in future work to bridge the gap in the current literature by determining which of the labelled proteins are contained in a typical cholinergic vesicle fraction derived from the nerve terminals in the electric organ and to investigate how those proteins behave in a ligated electric nerve.

Acknowledgements. The authors are indebted to Prof. V.P. WHITTAKER for supporting their collaboration and for his constructive comments during the preparation of this review. In addition, the support of M. TYTELL as Visiting Scientist during the summer of 1984 in part by a stipend from the Max Planck Gesellschaft and by a Foreign Travel Fellowship from the Bowman Gray School of Medicine is gratefully acknowledged.

References

Allemand P, Bon S, Massoulié J, Vigny M (1981) The quaternary structure of chicken acetylcholinesterase and butyrylcholinesterase: effect of collagenase and trypsin. J Neurochem 36:860–867

Baitinger C, Levine J, Lorenz T, Simon C, Skene P, Willard M (1982) Characteristics of axonally transported proteins. In: Weiss DG (ed) Axoplasmic transport. Academic, New York, pp 110–120

Barker JL, Hoffman PN, Gainer H, Lasek RJ (1975) Rapid transport of proteins in the sonic motor system of the toadfish. Brain Res 97:291–301

Bisby M (1981) Reversal of axonal transport: similarity of proteins transported in anterograde and retrograde directions. J Neurochem 36:741–745

Bisby M (1982) Ligature techniques. In: Weiss DG (ed) Axoplasmic transport. Springer, Berlin Heidelberg New York, pp 437–441

Black MM, Lasek RJ (1978) A difference between the proteins conveyed in the fast component of axonal transport in guinea pig hypoglossal and vagus motor neurons. J Neurobiol 9:433–443

Black MM, Lasek RJ (1980) Slow components of axonal transport: two cytoskeletal networks. J Cell Biol 86:616–623

Bon S, Vigny M, Massoulié J (1979) Asymmetric and globular forms of acetylcholinesterase in mammals and birds. Proc Natl Acad Sci USA 76:2546–2550

Bööj S, Carlson SS, Dahlström A, Fried G, Larsson PA (1985) Distribution and axonal transport of cholinergic synaptic vesicle-like organelles in rat sciatic nerve. J Neurochem 44 (Suppl):S27

Brimijoin S, Schreiber PA (1982) Reduced axonal transport of 10S acetylcholinesterase in dystrophic mice. Muscle Nerve 5:405–410

Brimijoin S, Weirmaa MJ (1978) Rapid orthograde and retrograde axonal transport of acetylcholinesterase as characterized by the stop-flow technique. J Physiol (Lond) 285:129–142

Brimijoin S, Skau K, Wiermaa MJ (1978) On the origin and fate of external acetylcholinesterase in peripheral nerve. J Physiol (Lond) 285:143–158

Couraud JY, Di Giamberardino L (1982) Axonal transport of the molecular forms of acetylcholinesterase: its reversal at a nerve transection. In: Weiss DG (ed) Axoplasmic transport. Springer, Berlin Heidelberg New York, pp 144–152

Couraud JY, Di Giamberardino L, Hassig R, Mira JC (1983) Axonal transport of the molecular forms of actylcholinesterase in developing and regenerating peripheral nerve. Exp Neurol 80:94–110

Couraud JY, Nicolet M, Hässig R (1985) Rapid axonal transport of three molecular forms of acetylcholinesterase in the frog sciatic nerve. Neuroscience 14:1141–1147

Dahlström A, Häggendal CJ, Heilbronn E, Heiwall P-O, Saunders N (1974) Proximo-distal transport of acetylcholine in peripheral cholinergic neurons. In: Fuxe K, Olson L, Zotterman Y (eds) Dynamics of degeneration and growth in neurons. Pergamon, Oxford, pp 275–289

Dahlström A, Bööj S, Carlson SS, Larsson PA (1981) Rapid accumulation and axonal transport of cholinergic vesicles in rat sciatic nerve, studied by immunohistochemistry. Acta Physiol Scand 111:217–219

Dahlström A, Bööj S, Larsson PA, McLean WG (1982a) The turnover of acetylcholine in ligated sciatic nerves of the rat. Acta Physiol Scand 115:493–498

Dahlström A, Bööj S, Larsson PA (1982b) Axonal transport of enzymes and transmitter organelle in cholinergic neurons. In: Weiss DG (ed) Axoplasmic transport. Springer, Berlin Heidelberg New York, pp 131–138

Dahlström A, Larsson PA, Carlson SS, Bööj S (1985) Localization and axonal transport of immunoreactive cholinergic organelles in rat motor neurons: An immunofluorescent study. Neuroscience 14:607–625

Davies LP (1978) ATP in cholinergic nerves: evidence for the axonal transport of a stable pool. Exp Brain Res 33:149–157

Davies LP, Whittaker VP, Zimmerman H (1977) Axonal transport in the electromotor nerves of *Torpedo marmorata*. Exp Brain Res 30:493–510
Dowdall MJ, Boyne AF, Whittaker VP (1974) Adenosine triphosphate, a constituent of cholinergic synaptic vesicles. Biochem J 140:1–12
Droz B, Koenig L, Di Giamberardino L, Couraud JY, Chretien M, Souyri F (1979) The importance of axonal transport and endoplasmic reticulum in the function of cholinergic synapse in normal and pathological conditions. In: Tuček S (ed) The cholinergic synapse. Elsevier, Oxford, pp 23–44 (Progress in brain research, vol 49)
Fonnum F, Malthe-Sørenssen D (1973) Membrane affinities and subcellular distribution of the different molecular forms of choline acetyltransferase from rat. J Neurochem 20:1351–1359
Fonnum F, Frizell M, Sjöstrand J (1973) Transport, turnover and distribution of choline acetyltransferase and acetylcholinesterase in the vagus and hypoglossal nerves of the rabbit. J Neurochem 21:1109–1120
Grafstein B, Forman DS (1980) Intracellular transport in neurons. Physiol Rev 60: 1167–1283
Heilbronn E, Pettersson H (1973) Acetylcholine and related enzymes in normal and ligated cholinergic nerves from *Torpedo marmorata*. Acta Physiol Scand 88:590–592
Heiwall PO, Saunders NR, Dahlström A, Häggendal J (1976) The effect of local application of vinblastine or colchicine on acetylcholine accumulation in rat sciatic nerve. Acta Physiol Scand 96:478–485
Hoffman PN, Lasek RJ (1975) The slow component of axonal transport: identification of major structural polypeptides of the axon and their generality among mammalian neurons. J Cell Biol 66:351–366
Hooper JE, Carlson SS, Kelly RB (1980) Antibodies to synaptic vesicles purified from *Narcine* electric organ bind a subclass of mammalian nerve terminals. J Cell Biol 87:104–113
Iqbal Z (ed) (1986) Axoplasmic transport. CRC Press, Boca Raton, FL
Karlsson JO (1982) Axonal transport of macromolecules. In: Brown IR (ed) Molecular approaches to neurobiology. Academic, New York, pp 131–157
Kiene M-L, Stadler H (1987) Synaptic vesicles in electromotoneurones: Axonal transport, site of transmitter uptake and processing of a core proteoglycan during maturation. EMBO J 6:2217–2221
Koike H, Nagata Y (1979) Intra-axonal diffusion of [^{3}H]acetylcholine and [^{3}H]-aminobutyric acid in a neurone of *Aplysia*. J Physiol (Lond) 295:397–417
Koike H, Eisenstadt M, Schwartz JH (1972) Axonal transport of newly synthesized acetylcholine in an identified neuron of *Aplysia*. Brain Res 37:152–159
Lasek RJ, Brady ST (1982) The axon: a prototype for studying expressional cytoplasm. Cold Spring Harbor Symp Quant Biol 46:113–124
Lasek RJ, Garner JA, Brady ST (1984) Axonal transport of the cytoplasmic matrix. J Cell Biol 99:212s–221s
Massoulié J, Bon S, Vigny M (1980) The polymorphism of cholinesterase in vertebrates. Neurochem Int 2:161–184
Mayer JH, Tomlinson DR (1983) Axonal transport of cholinergic transmitter enzymes in vagus and sciatic nerves of rats with acute experimental diabetes mellitus: correlation with motor nerve conduction velocity and effects of insulin. Neuroscience 9:951–957
McLean WG, McKay AL, Sjöstrand J (1983) Electrophoretic analysis of axonally transported proteins in rabbit vagus nerve. J Neurobiol 14:227–236
O'Brien RAD (1978) Axonal transport of acetylcholine, choline acetyltransferase and cholinesterase in regenerating peripheral nerve. J Physiol (Lond) 282:91–103
Oikarinen R, Kalimo H (1984) Acetylcholinesterase activity and its fast axonal transport in rabbit sciatic nerves during the recovery phase of experimental allergic neuritis. Neuropathol Appl Neurobiol 10:163–171
Saunders NR, Dziegielewska K, Häggendal CJ, Dahlström AB (1973) Slow accumulation of choline acetyltransferase in crushed sciatic nerves of the rat. J Neurobiol 4:95–103
Smith RS, Bisby MA (eds) (1987) Axonal transport. Liss, New York
Stone GC, Wilson DL (1979) Qualitative analysis of proteins rapidly transported in ventral horn motoneurons and bidirectionally from dorsal root ganglia. J Neurobiol 10:1–12

Tashiro T, Kurokawa M (1982) Rapid transport of a calmodulin-related polypeptide in the vagal nerve. In: Weiss DG (ed) Axoplasmic transport. Academic, New York, pp 153–160

Thoenen H, Kreutzberg GW (1981) The role of fast transport in the nervous system. Neurosci Res 20:1–138

Tuček S (1975) Transport of choline acetyltransferase and acetylcholinesterase in the central stump and isolated segments of a peripheral nerve. Brain Res 86:259–270

Tytell M, Black MM, Garner JA, Lasek RJ (1981) Axonal transport: each major rate component reflects the movement of distinct macromolecular complexes. Science 214:179–181

Tytell M, Kiene M-L, Stadler H (1985) Axonal protein transport in electric nerves of *Torpedo*. J Neurochem 44:S122

Ulmar G, Whittaker VP (1974) Immunohistochemical localization and immunoelectrophoresis of cholinergic synaptic vesicle protein constituents from the *Torpedo*. Brain Res 71:155–159

Vigny M, Di Giamberardino L, Couraud JY, Reiger F, Koenig J (1976) Molecular forms of chicken acetylcholinesterase: effect of denervation. FEBS Lett 69:277–280

Vitadello M, Couraud JY, Hässig R, Gorio A, Di Giamberardino L (1983) Axonal transport of acetylcholinesterase in the diabetic mutant mouse. Exp Neurol 82:143–147

Waldenlind L, Anderson P (1978) Subcellular distribution and transport of acetylcholine in cat ventral roots and sciatic nerve. J Neurochem 30:1067–1076

Weiler M, Roed I, Whittaker VP (1982) The kinetics of acetylcholine turnover in a resting cholinergic nerve terminal and the magnitude of the cytoplasmic compartment. J Neurochem 38:1187–1191

Weiss DG (ed) (1982) Axoplasmic transport. Springer, Berlin Heidelberg New York

Whittaker VP, Zimmermann H, Dowdall MJ (1975) The biochemistry of cholinergic synapses as exemplified by the electric organ of *Torpedo*. J Neural Transm [Suppl] 12:39–60

Widlund L, Pettersson H, Heilbronn E (1975) Origin of presynaptic acetylcholine studied on cholinergic neurons from *Torpedo marmorata*. In: Waser PG (ed) Cholinergic mechanisms. Raven, New York, pp 187–198

Woodward WR, Lindström SH (1977) A potential screening technique for neurotransmitters in the CNS: model studies in the cat spinal cord. Brain Res 137:37–52

Zimmermann H, Whittaker VP (1973) Evidence for axonal flow of acetylcholine (ACh) in cholinergic nerves. 4th Int Meet Int Soc Neurochem, Abstr Tokyo, p 245

CHAPTER 14

The High-Affinity Choline Uptake System

I. DUCIS

A. Introduction

BROWN and FELDBERG (1936) first observed the importance of choline as a precursor of the neurotransmitter acetylcholine (ACh) during their studies of the latter's metabolism in the stimulated sympathetic ganglion of the cat. The suggestion that the choline supplied to the ganglion during perfusion was transported by a specific mechanism located in the cholinergic nerve terminals came, however, much later (BIRKS and MACINTOSH 1961) as did the first indication of the sodium-dependent nature of this uptake (BIRKS 1963). In spite of these earlier observations a number of subsequent studies on nervous tissue (for references, see reviews in KUHAR and MURRIN 1978; SPETH and YAMAMURA 1979) reported that choline uptake exhibited saturation kinetics but that it was not particularly sodium-dependent nor did it appear that there was a specific choline uptake mechanism unique to cholinergic nerve terminals since only a small portion of the transported choline was converted to ACh. HAGA (1971) observed however that in the presence of increasing concentrations of sodium, the proportion of the choline transported into synaptosomes and converted to ACh increased as choline concentrations were decreased. He therefore proposed that only a small fraction of the heterogeneous pool of brain synaptosomes may exhibit a sodium-dependent mechanism for choline uptake and have the capacity to synthesize ACh, and again suggested, as BIRKS and MACINTOSH had years earlier, that a specific uptake system for choline existed in cholinergic nerve terminals. Subsequently, both high- ($K_T < 10\ \mu M$) and low- ($K_T > 20\ \mu M$) affinity neuronal uptake systems were directly demonstrated in a number of laboratories (WHITTAKER et al. 1972; YAMAMURA and SNYDER 1972, 1973; HAGA and NODA 1973; DOWDALL and SIMON 1973). In mammalian synaptosomes high-affinity choline uptake (HACU) was characterized as being temperature-, energy- and sodium-dependent, associated with effective ACh synthesis, and inhibited by low concentrations of hemicholinium-3 (HC-3). In contrast, low-affinity choline uptake (LACU), which has been observed in many types of tissues and preparations, exhibited less energy and sodium dependence, diminished sensitivity to HC-3 and was not associated with a significant synthesis of ACh.

Although the presence of a sodium-dependent HACU system has now been accepted as an intrinsic constituent of cholinergic nerve terminals, little is known about its molecular properties and behaviour. These will undoubtedly be areas of active investigation in the near future. In particular, recent reports have shown that resealed presynaptic plasma membrane fragments (membrane sacs) (BREER

1983; BREER and LUEKEN 1983; DUCIS and WHITTAKER 1985) or reconstituted systems (MEYER and COOPER 1982; VYAS and O'REGAN 1985; DUCIS and WHITTAKER 1985) are useful for studying the kinetics of choline uptake. Such preparations should also be useful in identifying and isolating the HACU system and in gaining insight into some of its molecular properties such as the identification of functional ligands and components, its configuration in the membrane, and the conformational changes associated with HACU.

In this chapter information obtained from various preparations, including brain, ganglia, electric organ and the vertebrate myoneural junction, will be discussed. Although each tissue provides specific advantages over others for certain types of experiments, the information obtained from each is complementary and provides a more complete picture of presynaptic cholinergic processes.

B. Characteristics of HACU

I. Tissue and Site Specificity

Much evidence indicates that the sodium-dependent HACU system is primarily localized in cholinergic nerve terminals. The types of studies which support the close association between sodium-dependent HACU and cholinergic neurons are the following: the observation, usually using synaptosomal preparations, that the majority of choline transported by the high-affinity system is synthesized into ACh (YAMAMURA and SNYDER 1972, 1973; HAGA and NODA 1973; JOPE and JENDEN 1977) and that this ACh is releasable (MURRIN et al. 1977); that the regional distribution of HACU in brain corresponds with choline acetyltransferase (ChAT) activity and ACh concentration (YAMAMURA and SNYDER 1973; KUHAR et al. 1973); and, finally, denervation of specific cholinergically innervated tissues, whether centrally or peripherally, results in decreased sodium-dependent HACU in these preparations (KUHAR et al. 1973, 1975; PERT and SNYDER 1974; SUSZKIW and PILAR 1976; SUSZKIW et al. 1976) which can be attributed directly to the loss of cholinergic nerve terminals and not to the loss of the cell soma (EISENSTADT et al. 1975; SUSZKIW and PILAR 1976). Contrary to the latter observations, however, denervation of superior cervical (COLLIER and KATZ 1974; BOWERY and NEAL 1975) and ciliary (SUSZKIW et al. 1976) ganglia, resulting in a presumed loss of cholinergic nerve terminals, does not result in any noticeable decrease in choline uptake. This could be due to difficulties in observing sodium-dependent HACU in these types of preparations since only a fraction of the total choline uptake in ganglia may be directed to cholinergic nerve terminals (SUSZKIW et al. 1976). During stimulation of the preganglionic nerve, however, a sodium-dependent (O'REGAN et al. 1982) and HC-3 sensitive (COLLIER and ILSON 1977) activation of choline transport in sympathetic ganglia can be readily observed.

Although it appears that HACU plays an integral role in presynaptic cholinergic processes, it may not be a property unique to cholinergic nerve terminals *per se* since HACU systems have been found in a number of non-cholinergic tissues, such as undifferentiated glial cells (MASSARELLI et al. 1974a), neuroblastoma clones (RICHELSON and THOMPSON 1973; LANKS et al. 1974; MASSARELLI et al. 1974b), human erythrocytes (SHERMAN et al. 1984), photoreceptor cells of the rabbit retina (MASLAND and MILLS 1980), human retinoblastoma cells (HYMAN and SPECTOR 1982), and the cardiac ganglion of the horseshoe crab (IVY et al. 1985). In most of these preparations choline uptake has been reported to exhibit two-compartment kinetics, that is, high- and low-affinity uptake mechanisms. However, K_T and V_{max} values may differ with the different cell types from those found in cholinergic neurons. The dependence of HACU on sodium, its sensitivity and reaction (e.g. stimulation and not inhibition) to HC-3, and whether it is present during all phases of cell growth or only during particular periods may also differ. Since an association with ChAT, and hence with ACh synthesis, is not observed in these tissues, it has been suggested that the function of the HACU system in some of these preparations may be to ensure a supply of substrate for the synthesis of choline-containing membrane phospholipids (IVY et al. 1985), particularly in actively dividing or cultured cells where HACU may be related to the metabolic state of a cell (SUSZKIW and PILAR 1976; OIDE 1982). It has recently been directly demonstrated that the HACU system in rabbit retinal photoreceptor cells subserves the synthesis of membrane phospholipids (PU and MASLAND 1984). Of particular interest, however, is the cardiac ganglion preparation of IVY et al. (1985). Although HACU was reported to be associated with phosphorylcholine and not ACh formation in this preparation, the uptake system exhibited the following characteristics observed in cholinergic neurons: sodium dependency, inhibition by HC-3 at low concentrations, and stimulation of choline uptake after pre-exposure of the ganglion to elevated levels of potassium. This indicates, unequivocally, that activation of HACU can be dissociated from ACh release and decreased intraterminal levels of ACh. Since the sodium dependency and HC-3 sensitivity of the uptake system is similar to that found in cholinergic nerve terminals, the cardiac ganglion of the horseshoe crab could prove to be an interesting and promising preparation in which to investigate the dynamics of choline uptake further, under conditions where synthesis and release of ACh do not occur.

Although it cannot be concluded that a specific sodium-dependent HACU system is a characteristic of cholinergic nerve terminals universally, it can be stated that mammalian central and peripheral cholinergic neurons do appear to have a unique transport system for choline which is present in the nerve terminal and is highly sodium-dependent and HC-3 sensitive. It has also been reported that the rate of this transport system (YAMAMURA et al. 1974) and its ontogenetic development (SORIMACHI et al. 1974; COYLE and YAMAMURA 1976) are correlated with ChAT activity.

II. Physiological Significance

1. Supply of Choline for Neurotransmitter Synthesis

High-affinity, sodium-dependent choline transport is an integral element of cholinergic nerve endings. Most workers support the idea that the functional significance of the HACU system is to provide substrate for ACh synthesis during neuronal activity since the latter, unlike most neurotransmitters, is itself not recycled.

A number of studies have demonstrated that the rate of synthesis of ACh is directly proportional to changes in its rate of release. This observation has led to the idea that ACh synthesis is somehow regulated and that the HACU system may be one of the rate-limiting steps.

Although there is evidence that choline can be synthesized *de novo* in nerve terminals via the step-wise methylation of phosphatidylethanolamine (BLUSZTAJN and WURTMAN 1981), it is not apparent to what extent, if any, this choline contributes to the intraterminal pool of free choline and the subsequent synthesis of ACh. It has also been reported that choline released during phospholipid turnover in nerve terminals may be utilized for ACh synthesis (FRIESEN et al. 1967; COLLIER and LANG 1969). Nevertheless, evidence to date supports the view that the main source of choline for the synthesis of releasable ACh is the HACU system (WEILER et al. 1981). In particular, the principal source of substrate for this system is the immediate recapture of choline produced by the hydrolysis of released ACh by acetylcholinesterase (AChE) (COLLIER and MACINTOSH 1969; BENNETT and MCLACHLAN 1972; COLLIER and KATZ 1974). Another significant source of choline for the HACU system is the circulating free choline in plasma which can readily enter the brain (SCHUBERTH et al. 1969; DIAMOND 1971; ROSS and JENDEN 1973; SAELENS et al. 1973; JENDEN et al. 1974) or cerebrospinal fluid (GARDINER and DOMER 1968) and can be modified by dietary intake (COHEN and WURTMAN 1975, 1976; HIRSCH and WURTMAN 1978).

2. Acetylcholine Production at the Nerve Terminal

Most, but not all, workers support the idea that the functional significance of the HACU system is as the principal supplier of choline for ACh synthesis during neuronal activity. It has been reported that choline transported by the high-affinity system supplies most of the choline needed for ACh synthesis (GUYENET et al. 1973a; POLAK et al. 1977). The ACh synthesized *in vitro* from choline taken up by the high-affinity system is also releasable (HAGA 1971; MULDER et al. 1974; MURRIN et al. 1977) under conditions similar to those associated with endogenous ACh release (HUBBARD 1970; RUBIN 1970).

However, while it is generally believed that the HACU system supplies adequate amounts of choline for ACh synthesis under resting conditions, it has been questioned whether this transport system can supply enough choline to the nerve terminal to keep up with the higher metabolic demands under stimulated conditions, particularly during continuous depolarization (CARROLL and GOLDBERG 1975; GOLDBERG et al. 1982). After nerve stimulation the concentration of choline in the synaptic cleft may be between 0.1 and 1.0 m*M* (SIMON et al. 1976). Therefore, during increased neuronal activity the capacity of the high-affinity system may not be sufficient to transport enough choline to form the needed ACh. It has been observed that the rate of choline transported into depolarized tissue, conditions under which the capacity of the HACU system is reduced, exceeds that into non-depolarized tissue. In addition, there is evidence that choline can be acetylated after transport by the low-affinity system in ciliary ganglia soma

(SUSZKIW et al. 1976) and striatal slices (BURGESS et al. 1978). It has therefore been proposed (CARROLL and GOLDBERG 1975; GOLDBERG et al. 1982) that the LACU system is able to supply choline for ACh synthesis at the nerve terminal, and that this contribution may be significant during periods of increased metabolic demand.

3. Coupling of the Capacity of the HACU System to Neuronal Activity

The HACU system has been shown to be coupled to cholinergic neuronal activity both centrally and peripherally. The velocity of the HACU system measured *in vitro* has been shown to be correlated with ACh turnover in different parts of the brain. The rate of choline uptake can be altered by drug treatments (SIMON and KUHAR 1975; ATWEH et al. 1975; SIMON et al. 1976; JENDEN et al. 1976; VALLANO et al. 1982; SHIH and PUGSLEY 1985) and by lesions or direct electrical stimulation of cholinergic nerves (SIMON and KUHAR 1975; SIMON et al. 1976; VACA and PILAR 1979; O'REGAN et al. 1982; VELDSEMA-CURRIE et al. 1984) that alter ACh turnover *in vivo*. A number of *in vivo* drug treatments have been tested and have been observed to alter the *in vitro* rate of HACU (for a summary, see the review by JOPE 1979). Generally drugs that increase ACh turnover in the brain, such as convulsants, cause increased uptake of choline *in vitro* whereas drug treatments that decrease ACh turnover in the brain, such as hypnotics, result in a decreased capacity of choline uptake by the HACU system. These changes in the capacity of the HACU system have been correlated to changes in the V_{max}. When an increase, or conversely, a decrease in HACU is observed, there is also a concomitant increase or decrease, respectively, in the V_{max} (SIMON and KUHAR 1975; SIMON et al. 1976). Under conditions where changes in the HACU system are described, no changes in either ChAT activity or the LACU system have been observed (SIMON et al. 1976). This suggests that the number of high-affinity choline transport sites *per se* changes or that the activity of the existing ones changes in parallel with neuronal activity (BARKER 1976).

In vitro experiments have shown that pre-exposure of synaptosomes to high potassium, with the concurrent presence of calcium or other divalent cation, before measuring HACU in normal potassium media, results in stimulation of choline uptake (MURRIN and KUHAR 1976; MURRIN et al. 1977; WEILER et al. 1978; however, see BARKER 1976). Stimulation of HACU after depolarization of synaptosomes by other means also exhibits calcium dependency (MURRIN and KUHAR 1976). In addition, ACh release appears not to be necessary for the activation of the HACU system (BARKER 1976; COLLIER and ILSON 1977; MURRIN et al. 1977). These observations have led to the suggestion that the stimulation of the high-affinity transport system depends on some ionic flux of divalent cations which occurs during depolarization (MURRIN et al. 1977) rather than on reduction of intraterminal ACh levels.

In mammalian cholinergic nerve endings the synthesis and release of ACh appear to be closely coupled. During neuronal inactivity ACh synthesis is slowed, but it is increased during nerve stimulation in order to meet the extra metabolic needs of the tissue (BIRKS and MACINTOSH 1961). Since tissue levels of ACh remain fairly constant, even during periods of relatively prolonged nerve stimulation, this indicates that ACh synthesis is under some type of metabolic regulation. The demonstration of a high-affinity uptake mechanism for choline in cholinergic nerve endings has led to the suggestion by various investigators that the transport of choline is a rate-limiting factor in the synthesis of ACh (Yamamura and Snyder 1972; Barker and Mittag 1975; Whittaker and Dowdall 1975). It has been reported that most of the choline transported by this high-affinity system is used for the synthesis of ACh (Yamamura and Snyder 1973) and that the amount of ACh synthesized is directly proportional to the amount of choline transported (Guyenet et al. 1973a). The observation that the HACU system changes in parallel with neuronal activity gives further support to the idea that choline uptake is one of the rate-limiting and regulatory steps in neurotransmitter synthesis.

Further evidence to suggest that the HACU system is a rate-limiting step comes from inhibition studies. The inhibition of HACU by HC-3, a specific inhibitor of HACU at low concentrations, and by other pharmacological agents, causes a parallel decrease in the rate of ACh synthesis and release (Guyenet et al. 1973a, b). Also, studies with choline analogues (Barker et al. 1975; Barker et al. 1977) indicate that HACU is one of the rate-limiting steps in ACh synthesis.

Although the HACU system meets the requirements of a regulatory mechanism in ACh synthesis, it is not possible to rule out completely that other such mechanisms do not exist. In particular, intraterminal availability of acetyl coenzyme A (AcCoA) may also play a role in regulating ACh synthesis. Jope and Jenden (1977) have shown that inhibition of AcCoA production in synaptosomes results in a marked inhibition of ACh synthesis. However, before it can be determined whether this is a regulatory step in ACh synthesis, it should be determined whether such rate-limiting concentrations of co-factor are physiologically significant.

ChAT has also been considered as a possible site of ACh synthesis regulation (Kaita and Goldberg 1969; Glover and Potter 1971; Sharkawi 1972). However, product-feedback inhibition of ChAT by ACh seems unlikely since the high cytoplasmic concentrations of ACh that would cause a significant inhibition of ChAT (Kaita and Goldberg 1969; Glover and Potter 1971) are not likely to be attained intraterminally. Coenzyme A, however, is a more effective inhibitor of ChAT than is ACh (Glover and Potter 1971; White and Wu 1973) and its dissociation from the enzyme may well play a role in controlling ACh synthesis (Barker et al. 1977; Hersh et al. 1978).

A further relevant issue to ACh synthesis control is the determination of the precise nature of the association between HACU and ChAT. Although the exact nature of this coupling is unknown, there appears to be a close association between HACU and ChAT since the majority of the choline transported by the HACU system is converted to ACh (Yamamura and Snyder 1972, 1973). Both this close association with ChAT and the very efficient conversion of choline or its analogues to acetylated derivatives after their transport by the high-affinity system has also led to the suggestion that choline availability is rate-limiting in the synthesis of ACh (Yamamura and Snyder 1972; Barker and Mittag 1975; Barker et al. 1975). Although it has been suggested that the two processes may be physically coupled *in vivo* (Barker and Mittag 1975), with choline never existing in a free state intraterminally, most findings have shown that choline transport and acetylation can be dissociated (BARKER et al. 1975; SUSZKIW and PILAR 1976; BARKER et al. 1977; COLLIER et al. 1977; JOPE and JENDEN 1977; DOWDALL 1977; JOPE et al. 1978; SHELTON et al. 1979; JOPE and JENDEN 1980). In particular, studies with proteoliposomes show directly that transport of choline can occur without any detectable concomitant acetylation (KING and MARCHBANKS 1982; MEYER and COOPER 1982; VYAS and O'REGAN 1985). Even studies with synaptosomal preparations have demonstrated that there is a pool of free choline within nerve terminals (JOPE and JENDEN 1980).

Although at present a kinetic rather than a direct coupling of HACU and ChAT is favoured by most workers, the exact nature of this association is not known. Model systems such as proteoliposomes or membrane sacs could, however, be useful in elucidating this relationship. Recently a synaptosomal plasma membrane fragment reconstitution carried out in the presence of an ACh-generating system demonstrated the synthesis of [^{3}H]ACh in proteoliposomes after [^{3}H] choline uptake (MEYER and COOPER 1983). It has also been reported that membrane sacs derived from osmotically lysed insect synaptosomes are capable of synthesizing ACh when loaded with AcCoA (BREER and LUEKEN 1983).

III. Structure-Activity Relationship Studies

A number of studies have been carried out with different inhibitory compounds which have provided insight into the specific structural requirements for the choline-binding site of the transporter. Although the structural requirements are strict (FISHER and HANIN 1980), many compounds have been found to inhibit the HACU system. The most potent known inhibitors of choline uptake are presynaptic snake venom toxins, with notexin ($K_i = 40$ p*M*) as the most potent inhibitor (DOWDALL et al. 1977). However, these toxins act not on the choline transport system *per se* but on the structure of the presynaptic membrane (DOWDALL and FRETTEN 1981).

To date, the most potent synthetic inhibitors of choline uptake have been shown to be certain quaternary ammonium compounds which are structurally related to troxonium. These *N*-pentyl derivatives in the acetylenic piperidinium and pyrrolidinium series are even more potent inhibitors (K_i approximately 5 n*M*) of the HACU system than the extensively studied bis-quaternary base HC-3 (LINDBORG et al. 1984). However, although they are competitive inhibitors of choline uptake into mouse brain synaptosomes, it was not reported whether the compounds themselves are transported. Most studies of transportable choline analogues, with at least one exception (BARKER 1978), suggest that there is a strict structural specificity for a quaternary nitrogen and a free hydroxyl group by the HACU mechanism (SIMON et al. 1975; COLLIER et al. 1979; COLLIER 1981). This explains why ACh is only a weak inhibitor of the HACU system (GUYENET et al. 1973a; DOWDALL et al. 1976; BATZOLD et al. 1980). In fact, ACh accumulation in brain tissue is not specifically associated with cholinergic neurons (KATZ et al. 1973; KUHAR and SIMON 1974). Moreover, without drastically decreasing the affinity for the HACU system, it appears that the methyl groups on the nitrogen of choline can be extended by only one carbon atom (BATZOLD et al. 1980) and that the molecule itself can be lengthened by only one methylene group (COLLIER 1981), giving homocholine. Homocholine is readily transported across the synaptic plasma membrane, as are some other analogues of choline (COLLIER et al. 1979), and serves as a false transmitter (COLLIER et al. 1977). The forementioned restrictions, however, do not always apply when only high-affinity binding of ligands to the choline transporter are being measured, as may be the case with the *N*-pentyl derivatives in the acetylenic piperidinium and pyrrolidinium series of the troxonium-like inhibitors. These compounds have an electron-rich acetylene bond that replaces the oxygen atom of choline and have substituents five carbons long attached to the nitrogen. That the acetylene group may substitute for the oxygen atom of choline is supported by the demonstration that *N,N,N*-trimethyl-*N*-(2-propynyl)ammonium is transported by the HACU mechanism (BARKER 1978).

From studies with substrates and inhibitory compounds, much insight into the architecture of the active centre of the choline transport system has been gained. It appears to be composed of an anionic site to which the quaternary nitrogen binds and an electron-poor site where the hydroxyl group of choline or another electron-rich moiety such as an acetylenic bond interacts. The configuration of the substrate during interaction with the active centre must be such that the centre of the electron-rich moiety is 0.29–0.33 nm (COLLIER 1981; BATZOLD et al. 1980; LINDBORG et al. 1984) from the quaternary nitrogen, suggesting that a cisoid staggered conformation of choline is preferred by the transporter (LINDBORG et al. 1984). It has been proposed that next to the anionic and electron-poor sites of the active centre, lipophilic sites may be present (LINDBORG et al. 1984). The active centre also exhibits stereoselectivity which was directly demonstrated by using 3-quinuclidinol derivatives (RINGDAHL et al. 1984).

IV. Bioenergetics of Uptake

Studies using synaptosomal preparations have reported that the HACU system has specific ionic requirements (DOWDALL and SIMON 1973; YAMAMURA and SNYDER 1973; SIMON and KUHAR 1976; KUHAR and ZARBIN 1978; BREER 1982). For example, the system is highly and specifically sodium-dependent since other monovalent cations cannot substitute for sodium and it also exhibits sensitivity to chloride and potassium ions (SIMON and KUHAR 1976; KUHAR and ZARBIN 1978). In addition, HACU in innervated longitudinal muscle minces of guinea-pig small intestine has been reported to be calcium dependent (PERT and SNYDER 1974). Although omission of calcium from the incubation medium does not reduce choline uptake in rat brain synaptosomes under normal conditions, depolarization-induced activation of HACU is calcium dependent in the brain (MURRIN and KUHAR 1976; MURRIN et al. 1977) as well as the rat phrenic nerve-hemidiaphragm preparation (VELDSEMA-CURRIE et al. 1984), the superior cervical ganglion (SCG) of the cat (O'REGAN et al. 1982) and the domestic fowl ciliary nerve-iris preparation (VACA and PILAR 1979).

HACU in synaptosomes appears to be active since it is also temperature and energy dependent (DOWDALL and SIMON 1973; YAMAMURA and SNYDER 1973;

SIMON and KUHAR 1976). Incubation of synaptosomes with metabolic inhibitors such as cyanide and 2,4-dinitrophenol results in decreased choline uptake (DOWDALL and SIMON 1973; YAMAMURA and SNYDER 1973; SIMON and KUHAR 1976). Ouabain, which interferes with the utilization of ATP to generate ion gradients (CHRISTENSEN 1975), has also been shown to inhibit choline uptake in synaptosomes (DOWDALL and SIMON 1973; YAMAMURA and SNYDER 1973; SIMON and KUHAR 1976; BREER 1982). These observations suggest that a sodium ion gradient, generated by a Na^+, K^+-ATPase (SCHULTZ and CURRAN 1970), may provide the driving force for HACU into cholinergic nerve terminals.

Recently, in order to study choline uptake under more defined conditions, there has been a move away from the relatively complex synaptosome to resealed presynaptic plasma membrane fragments (membrane sacs) obtained after osmotic lysis of synaptosomes. However, it has not been possible to use mammalian brain preparations successfully to isolate membrane sacs since HACU appears to be lost upon osmotic lysis of synaptosomes (MEYER and COOPER 1982; I. DUCIS, unpublished observations). To date, membrane sacs with efficient HACU activity have been exclusively derived from synaptosomes isolated from non-mammalian preparations such as insect head ganglia (BREER 1983; BREER and LUEKEN 1983) and the electromotor system of *Torpedo marmorata* (DUCIS and WHITTAKER 1985). The lack of success with mammalian brain preparations could be due to the low yield of synaptosomes derived from cholinergic nerve terminals (RICHARDSON 1981), whereas insect head ganglia have a high degree of cholinergic innervation (BREER 1981) and the innervation of the electric organ of elasmobranch fish of the family Torpedinidae is purely cholinergic (FELDBERG and FESSARD 1942; Chap. 2, this volume).

The vesiculated membrane fragments derived from synaptosomes from such preparations have been useful for studying the HACU system since there is no interference from subcellular storage and metabolism, and ion gradients and energy sources can be readily defined. Since the composition of both the external and internal milieu of membrane sacs can be manipulated relatively easily, it is possible to define not only the dependence of the transport system on specific ions but to also test directly whether choline accumulation into nerve terminals can be achieved solely by cotransport with ions which move down their electrochemical potential gradients (MITCHELL 1963; CRANE 1965).

In addition to the use of membrane sacs for studying HACU, reconstitution of the choline transporter into proteoliposomes has been achieved using both presynaptic plasma membrane fragments (MEYER and COOPER 1982; VYAS and O'REGAN 1985) and the solubilized transporter (DUCIS and WHITTAKER 1985). Solubilization of the transporter with sodium cholate-EDTA and its subsequent incorporation into liposomes during dialysis results in proteoliposomes which are capable of displaying choline uptake with similar characteristics to those in membrane sacs (DUCIS and WHITTAKER 1985). It has been possible to drive HACU solely with artifically imposed transmembranous ion gradients in membrane sacs (BREER and LUEKEN 1983; DUCIS and WHITTAKER 1985) and proteoliposomes (MEYER and COOPER 1982; VYAS and O'REGAN 1985; DUCIS and WHITTAKER 1985). This suggests that HACU at the nerve terminal is a secondary, not a primary, active transport system (CHRISTENSEN 1975). Although choline accumulation in model systems is driven primarily by an inwardly directed transmembranous sodium gradient, HACU in membrane sacs is not inhibited by ouabain (BREER and LUEKEN 1983) as it is in synaptosomes (BREER 1982) where a transmembranous sodium gradient is, presumably, maintained by the ouabain-sensitive Na^+, K^+-ATPase. In membrane sacs HACU is also not influenced by the addition of ATP to the loading medium (BREER and LUEKEN 1983), whereas in proteoliposomes it was observed to inhibit choline uptake (VYAS and O'REGAN 1985).

Ionophores like gramicidin and monensin (MEYER and COOPER 1982; BREER and LUEKEN 1983; VYAS and O'REGAN 1985) or alkaloid neurotoxins such as aconitine and veratridine (BREER and LUEKEN 1983; DUCIS and WHITTAKER 1985) which dissipate the sodium ion gradient more rapidly across the membrane, inhibit HACU. The inhibition observed with veratridine, which selectively increases the permeability of the nerve membrane to sodium (OHTA et al. 1973), is almost completely reversed by the addition of tetrodotoxin (TTX) (BREER and LUEKEN 1983). This indicates that the preparation of BREER and LUEKEN (1983) is of neuronal origin, and that their membrane sacs are predominantly right-side-out since TTX blockade of sodium channels and, thus, reversal of veratridine inhibition can only be effective when the former (which is not lipid soluble) is applied on the outer surface of the neuron (NARAHASHI et al. 1966). The preparations of DUCIS and WHITTAKER (1985) appear also to be predominantly right-side-out, as suggested by certain observations during measurements of the plasma membrane marker enzymes ouabain-sensitive Na^+, K^+-ATPase and 5'-nucleotidase (adenosine 5'-α,β-methylene]diphosphate-sensitive phosphatase) whose substrate sites face the cytoplasm and external milieu, respectively. Na^+, K^+-ATPase activity was not detectable in various membrane fractions without the addition of saponin, while addition of detergent to membrane samples during measurements of 5'-nucleotidase activity did not result in increased orthophosphate production (I. DUCIS, unpublished observations). This indicates that after osmotic lysis of synaptosomes the resulting membrane fragments seal right-side-out and the substrate site of the ATPase is not accessible to ATP in sealed membrane sacs while, in the case of 5'-nucleotidase, the substrate site faces outwards and is freely accessible to AMP.

Valinomycin, an ionophore that specifically increases the permeability of membranes to potassium, creates or enhances an interior negative membrane potential in potassium-loaded membrane sacs (BREER 1983; BREER and LUEKEN 1983; DUCIS and WHITTAKER 1985) and proteoliposomes (VYAS and O'REGAN 1985; DUCIS and WHITTAKER 1985) when there is an outwardly directed potassium chemical gradient. In these systems, both the initial rate as well as the extent of choline transport is stimulated in a sodium-containing medium. This indicates that the membrane potential contributes to the driving force for HACU, and that HACU is an electrogenic process. These observations fit with previous studies using synaptosomes where HACU was observed to be diminished or lost in depolarized preparations (CARROLL and GOLDBERG 1975; MURRIN and KUHAR 1976). The proton ionophore CCCP, which abolishes the negative membrane potential created by an outwardly directed potassium chemical gradient, inhibits HACU in membrane sacs (BREER 1983; BREER and LUEKEN 1983). However, it has not been tested whether CCCP stimulates HACU when an outwardly directed proton chemical gradient is applied.

It appears that both the sodium concentration gradient and the membrane potential act as driving forces for the HACU system. Direct evidence that HACU is sodium-coupled and depends on its electrochemical gradient comes from model systems such as right-side-out membrane sacs of neuronal origin. In these preparations, HACU exhibits an absolute dependence on a sodium gradient; other monovalent cations such as lithium, potassium, rubidium, caesium or Tris are completely ineffective as substitutes (BREER and LUEKEN 1983; DUCIS and WHITTAKER 1985). In addition, HACU in the presence of an inwardly directed sodium gradient and outwardly directed potassium gradient exhibits an overshoot phenomenon (DUCIS and WHITTAKER 1985). This overshoot is characteristic of an uptake system driven by an imposed ion gradient. However, sodium uptake has not yet been quantitated in membrane sacs or proteoliposomes that exhibit HACU as has been done with other sodium cotransport systems (KANNER 1980; DUCIS and KOEPSELL 1983).

Studies with model systems leave little doubt that the choline transporter is capable of catalysing net influx. However, it is not known whether membrane sacs or proteoliposomes also catalyse dilution-induced net efflux of choline as has been found for L-glutamate (KANNER and MARVA 1982; KANNER and BEN-DAHAN 1982) and γ-aminobutyric acid (GABA) (KANNER and KIFER 1981). In addition, it is not known whether the mechanism of homoexchange (CHRISTENSEN 1975) is operative in these more simplified systems which unequivocally demonstrate sodium-dependent HACU but which are free of subcellular storage and metabolism. Although homoexchange has been observed in other high-affinity uptake systems (LEVI and RAITERI 1974; SIMON et al. 1974), no evidence for this mechanism has been found for choline in synaptosomal preparations (SIMON et al. 1976).

Studies with model systems have revealed that in addition to its sodium-gradient dependency, the HACU system is dependent on the simultaneous presence of external chloride, although unlike sodium, where no replacements are permitted, bromide can partially substitute for chloride (BREER and LUEKEN 1983; DUCIS and WHITTAKER 1985). In membrane sacs maximal accumulation of choline is achieved when both ionic gradients are present (BREER and LUEKEN 1983; DUCIS and WHITTAKER 1985). However, DUCIS and WHITTAKER (1985) reported only a partial dependence on the chloride gradient in membrane sacs, while BREER and LUEKEN (1983) observed an almost complete loss of choline uptake at chloride equilibrium. It has also been reported that choline uptake in proteoliposomes, in the presence of a chloride gradient, is not enhanced over that observed at chloride equilibrium (VYAS and O'REGAN 1985). These variations in results may represent species differences or leakiness of the preparations to chloride.

Since HACU is an electrogenic process, the requirement for chloride, which is a fairly permeant anion in nerve cells, could be simply to create an internal negative membrane potential which is needed as a driving force for the transport system and to compensate for the positive charge (or charges) moving in. Such a role for chloride was suggested previously from work with synaptosomes

(KUHAR and ZARBIN 1978). However, from observations made with model systems, this explanation appears unlikely since more permeant anions like thiocyanate and nitrate are not able to substitute for chloride (DUCIS and WHITTAKER 1985). Furthermore, the chloride requirement persists in the presence of potassium-loaded membrane sacs and proteoliposomes (BREER and LUEKEN 1983; VYAS and O'REGAN 1985; DUCIS and WHITTAKER 1985). Potassium, which is also a permeant ion in nervous tissue (BAKER et al. 1962) would be expected to move out to compensate for the positive charge (or charges) moving in at least as effectively as inwardly moving chloride.

Since chloride lowers the K_T for transport in synaptosomes (KUHAR and ZARBIN 1978), its role in HACU, then, appears to be due to a specific interaction with this system. The transporter may actually cotransport chloride along with sodium and choline or, alternately, since chloride appears to activate the transporter somehow, it may only interact with the system at the external surface. In order to distinguish between these two possibilities and, assuming an asymmetry of the transport system in the membrane, the effect of internal ions on the efflux of choline should be investigated in model systems. If cotransport with chloride occurs, then efflux of internal choline via the HACU system should require not only sodium but also internal chloride, whereas, if chloride only activates at some external site of the membrane so that sodium and choline can bind effectively to the transporter, this would not be expected to be the case. Consistent with the view that chloride may be cotransported along with sodium and choline is the observation that the transmembranous chloride gradient influences the maximal level of choline accumulation in membranes. For example, reversing the chloride gradient, $[Cl^-]_i \gg [Cl^-]_O$ results in an inhibition of HACU (DUCIS and WHITTAKER 1985). Furthermore, an inwardly directed chloride gradient serves as a driving force for choline uptake (BREER and LUEKEN 1983). However, as was previously stated, this was not observed in all preparations (VYAS and O'REGAN 1985). To provide a final test of the idea that chloride, sodium and choline are cotransported, studies should be carried out in a relatively non-leaky system, such as, perhaps, purified transporter incorporated into liposomes, in order to observe whether sodium- and choline-dependent chloride and also chloride- and choline-dependent sodium uptake occur. If net choline transport is chloride-gradient dependent, as well as sodium-dependent, it would not be unique to this neuronal membrane transport mechanism since a number of reuptake systems appear to exhibit a similar dependency (KANNER 1978; KANNER and KIFER 1981; MAYOR et al. 1981; NELSON and RUDNICK 1982). Since in nervous tissue chloride ions are thought to be close to electrochemical equilibrium, it is not presently known why it is advantageous to have an inwardly directed chloride gradient serving as a driving force for HACU. Because little is known about the nature of the microenvironment surrounding neuronal transport systems, it is not yet possible to answer this question; however, it is plausible that chloride may not actually be at its electrochemical equilibrium at the specific membrane sites where transport takes place.

Influx of choline is stimulated by, but not absolutely dependent upon, internal potassium ions. Univalent cations can replace potassium to varying extents as the loading ion in the presence of an inwardly directed sodium gradient. Ions are effective in the following order: potassium > rubidium ≫ ammonium > caesium > lithium, with the last named being completely ineffective (DUCIS and WHITTAKER 1985). Since HACU appears to be electrogenic (BREER 1983; BREER and LUEKEN 1983; VYAS and O'REGAN 1985; DUCIS and WHITTAKER 1985), it is possible that the observed stimulation by potassium is due only to the ability of the outwardly directed potassium gradient to generate an interior negative membrane potential which serves as a driving force for choline uptake. Although sufficient information is not yet available for the HACU system since efflux studies have not been carried out, there are indications that there may be a specific interaction of potassium with the transporter. Evidence for this comes from studies with membrane sacs which show that in the absence of internal potassium, inwardly directed gradients of sodium and chloride do not give rise to any choline transport (BREER 1983; BREER and LUEKEN 1983). Since chloride is a permeant anion in nervous tissue, its gradient should generate a sufficient internal negative membrane potential for HACU to take place even in the absence of potassium.

External potassium can inhibit choline influx into proteoliposomes (MEYER and COOPER 1982; VYAS and O'REGAN 1985) and synaptosomes (SIMON and KUHAR 1976). Consequently, VYAS and O'REGAN (1985) have suggested that internal potassium may interact with the transporter directly and block choline efflux. However, it was not tested in their study whether external potassium enhances efflux of choline as it does with serotonin and L-glutamate (NELSON and RUDNICK 1979; KANNER and BENDAHAN 1982; KANNER and MARVA 1982). Efflux studies with model systems could provide direct evidence that there is a specific interaction between potassium and the transporter, and that stimulation of HACU by potassium is not just due to its ability to generate an internal negative membrane potential. These studies would also indicate whether potassium is translocated by the neuronal HACU system, as has been previously shown for the choline uptake mechanism in erythrocytes (MARTIN 1972).

It is plausible that the HACU system in nerve terminals cotransports sodium, chloride and choline inwardly while potassium is transported outwardly. The HACU system may be analogous to the serotonin uptake mechanism (KEYES and RUDNICK 1982) where the reorientation of the empty transporter, facilitated by potassium, is thought to be the rate-limiting step in the transport cycle. KEYES and RUDNICK (1982) have reported that protons can effectively substitute for potassium ions in the absence of the latter but that there is competition for binding sites when they are both present. For example, the stimulation of serotonin uptake by internal potassium is diminished considerably at lower internal pH values, and, conversely, the stimulation of serotonin uptake, observed in the presence of internal protons, is reduced when potassium is present. If such a mechanism is operative in the HACU system, it is expected that potassium ions (or protons) would inhibit choline translocation when they are on the same side of the membrane (which they do) but that they would stimulate choline movement across the membrane when they are on the opposite side.

The electrogenicity and the ionic requirements of HACU place limitations on the possible stoichiometry of the process. Studies with synaptosomal preparations (SIMON and KUHAR 1976; WHEELER 1979a) and membrane sacs (BREER and LUEKEN 1983) have shown that the sodium-dependent pathway of choline uptake probably involves more than one sodium ion since a sigmoid-shaped curve is obtained when the initial rate of HACU is plotted against the external sodium ion concentration. When the membrane potential and the chloride ion gradient are kept constant, a log – log plot of the choline gradient versus the sodium concentration (imposed as a gradient across the membrane) gives a straight line with a slope equal to 2.0 (BREER and LUEKEN 1983), indicating that two sodium ions are translocated per choline molecule.

By a similar approach the stoichiometry for chloride can also be determined. When BREER and LUEKEN (1983) plotted the initial rate of choline uptake against the chloride ion concentration, a hyperbola was obtained, suggesting that only one type of chloride-binding site is involved. When the values for the initial rate of choline uptake in the presence of chloride ions and the corresponding concentrations of the latter (up to 80 mM) are estimated from Fig. 1 of their report, and their determined V_{max} value of 62.5 pmol min^{-1} mg^{-1} protein is used for calculations, a log – log plot of the choline gradient $[v/(V_{max} - v)]$ versus the chloride ion concentration results in a straight line ($r^2 > 0.997$) with a slope of 1.0, suggesting a stoichiometry of one chloride ion per molecule of choline.

Like the serotonin transport system, the HACU mechanism appears to be even more complex than those of GABA (KANNER 1978) and L-glutamate (KANNER and SHARON 1978a, b) since it requires both external sodium and chloride as well as internal potassium. Furthermore, it appears that these ions are very likely participating directly in the translocation cycle. Assuming that choline is transported in its positively charged form, the predominant species at physiological pH, the above observations designate that the stoichiometry $l Na^+ : m K^+ : n Cl^-$: choline is restricted and should be such as to fulfil the criterion of electrogenicity. Since, as indicated previously, mere charge compensation fails to account for the role of chloride and potassium in HACU, it may very likely be that the HACU process is $(Na^+ + K^+ + Cl^-)$-coupled. Although the simplest stoichiometry consistent with such a model would be $2 Na^+ : K^+ : Cl^-$: choline, more complex stoichiometries cannot at present be excluded. One possibility for the HACU system is that two sodium ions are moving in with one chloride ion and one choline molecule while one potassium ion is countertransported per translocation cycle. Further studies, however, would be required to decide this issue unequivocally.

It should be possible to gain insight into the binding order of substrates for the HACU process via exchange and kinetic studies as has been done previously in other systems (WHEELER 1979b; NELSON and RUDNICK 1979; WHEELER 1980; KANNER and BENDAHAN 1982). Kinetic experiments with synaptosomes have evaluated the role of sodium and choline only in HACU and have suggested that two sodium ions translocate with one choline molecule per cycle (WHEELER 1979a). It was also proposed that the transporter can first combine with either sodium or choline. In addition, the model allowed for the binding of choline to the transporter and its translocation in the absence of sodium.

Further studies, however, need to be carried out on the ionic dependency of net fluxes and exchange of choline in model systems. Questions such as the following should be addressed: Does efflux of choline require internal sodium and chloride and external potassium? Does external choline enhance efflux and, if so, does it require external sodium and chloride? And, finally, what are the ionic requirements, both internal and external, for choline exchange? Such studies with native membranes should also help determine if the binding order to the transporter is different on the inside than on the outside, thus demonstrating asymmetry of the HACU system.

V. Possible Role of Sialocompounds, Lipids and Steroids

1. Sialocompounds

A possible involvement of membrane sialocompounds in HACU has been demonstrated (STEFANOVIĆ et al. 1975; MASSARELLI et al. 1982a, b; VYAS and O'REGAN 1985). Since neuraminidase treatment of cultured nerve cells and synaptosomes has been shown to abolish HACU completely (MASSARELLI et al. 1982a), both sialoglycoproteins and gangliosides have been implicated. Possibly, negatively charged sialic acid residues serve as recognition- or binding-sites for the positively charged choline molecule. However, MASSARELLI et al. (1982a) have demonstrated that the tetanus toxin binding sites (GD 1 b and GT 1 b gangliosides) are not involved in choline uptake. Although the results of MASSARELLI et al. (1982a, b) further suggest that GM 1 ganglioside does not participate in HACU, PEDATA et al. (1984) have reported that GM 1 directly facilitates the recovery of HACU activity in rat cerebral cortex in the lesioned brain.

Although it has been demonstrated that sialocompounds can serve as specific and functional cellular receptors (LEEDEN and MELLANBY 1977; STAERK et al. 1974; HOLMGREN et al. 1974; MELDOLESI et al. 1977), they may not necessarily be specific recognition sites for choline but may serve only to increase the negative surface membrane potential. Surface potential effects may, therefore, influence the kinetics of choline uptake. If so, the kinetic parameters of the HACU system cannot be accurately determined by unmodified Michaelis-Menten equations (THEUVENET and BORST-PAUWELS 1976). It is possible that the translocation of choline across membranes exhibits an apparent diminishing affinity, or negative cooperativity, with increasing concentrations due to a concomitant decrease in the surface potential of the membrane (THEUVENET and BORST-PAUWELS 1976; MCLAUGHLIN 1977). This may be why MASSARELLI et al. (1982a) have reported a loss of HACU and an increase in the K_T after neuraminidase treatment, which would result in the loss of negatively charged surface sialic acid residues. It is possible that the true affinity of the high-affinity transporter for choline may be much less than has been reported to date.

2. Lipids

An effect of the membrane fatty acid composition on HACU has been demonstrated in cultured human retinoblastoma cells (HYMAN and SPECTOR 1982). Enrichment of the membrane in polyunsaturated fatty acids such as arachidonic, linolenic or docosahexaenoic acids resulted in reductions in both the K_T and V_{max}. A 2-fold increase in the affinity of choline was observed while the V_{max} was reduced by 22% – 39%. HACU in this cell line, however, is not sodium-dependent, associated with ACh formation or inhibited by low concentrations of HC-3 (HYMAN and SPECTOR 1982). A possible effect of the lipid environment on the sodium-dependent HACU system derived from electric organ, which is associated with ACh synthesis and inhibited by low concentrations of HC-3, was observed after reconstitution of the transporter into proteoliposomes (DUCIS and WHITTAKER 1985). Although K_T and V_{max} were not determined in the reconstituted system, the initial (1 min) rate of choline uptake per milligram protein was increased approximately 10-fold over that observed in native membranes. This increase in transport activity could be due to stimulation of choline transport molecules, a purification of the transport protein(s), a loss of some inhibitory factor during the reconstitution procedure, or to a combination of these factors. When Arrhenius plots of the initial transport rates measured at 0°–25 °C in both membranes and proteoliposomes (DUCIS and WHITTAKER 1985, Figs. 5 and 16, respectively) are graphed (I. DUCIS, unpublished), straight lines are obtained. Apparent activation energies, derived from the slope of the lines, are 22.7 and 10.6 kcal mol^{-1} for membranes and proteoliposomes, respectively. Because enough data points were not available, no distinct break in the Arrhenius plots could be determined to indicate a change in the membrane lipid configuration. Nevertheless, if transport molecules are being activated after reconstitution into liposomes, this would suggest that the HACU system in electric organ may be sensitive to its lipid environment and warrants a systematic study under well-defined conditions. Furthermore, ACh release and, therefore, choline uptake may be sensitive to phosphatidylserine (PS). ACh release has been reported to be increased from rat cerebral cortex after intravenous injection of PS, whereas phosphatidylethanolamine (PE) and phosphatidylcholine (PC) had no effect (AMADUCCI et al. 1976).

3. Steroids

Stimulation of HACU by corticosteroids has been observed in the rat SCG (SZE et al. 1983) and in the rat hemidiaphragm preparation (LEEUWIN et al. 1981; VELDSEMA-CURRIE et al. 1985). Low concentrations of glucocorticoids added *in vitro* stimulate uptake of choline and its incorporation into ACh (SZE et al. 1983; VELDSEMA-CURRIE et al. 1985) and antagonize the inhibitory effects of HC-3 (VELDSEMA-CURRIE et al. 1976). These actions are not only concentration-dependent but also selective since all glucocorticoids do not show an effect on HACU (SZE et al. 1983). The effects of corticosteroids on these systems may be directly on the HACU system and not mediated by a cytoplasmic glucocorticoid receptor (TOWLE and SZE 1982; SZE et al. 1983). Evidence for this comes from a study by TOWLE and SZE (1983) who have identified and characterized specific synaptic plasma membrane binding sites for a variety of steroid hormones and have shown that the binding properties of these sites are different from those of the cytoplasmic receptor for a particular steroid.

In vivo pretreatment with corticosteroids results in an increase in choline uptake in mammalian brain synaptosomes (RIKER et al. 1979) as well as other transport systems (NECKERS and SZE 1975; MILLER et al. 1978). In contrast to tryptophan uptake, however, no increase in choline uptake was observed when brain synaptosomes were incubated directly with glucocorticoids (RIKER et al. 1979; KING and DAWSON 1985). So far, the direct effects of corticosteroids on the HACU system have only been demonstrated peripherally. The reason for this discrepancy is not known; however one possibility is that the HACU mechanism is different at these steroid-sensitive sites than at the central nerve terminal.

C. Identification of the HACU System

I. Criteria for Identification

To date no conclusive evidence has been presented as to the identity of the choline transporter. As far as is known, only one report has attempted to identify the transporter (DUCIS and WHITTAKER 1985). This is not surprising since mammalian brain synaptosomal preparations are heterogeneous and only a small fraction are cholinergic (RICHARDSON 1981). Even with a purely cholinergic system (DUCIS and WHITTAKER 1985) tissue yields after fractionation, membrane isolation and solubilization are very low. For example, it has been estimated that only about 0.02% of the total protein in electric organ homogenates is recovered in proteoliposomes (I. DUCIS, unpublished observations). Isolation of the choline transporter may, therefore, be a difficult task. However, since there is evidence to indicate that proteinaceous species are involved in HACU (VYAS and O'REGAN 1985), the criteria on which a polypeptide (or polypeptides) can be identified as participating in HACU, and the objectives that must be achieved are the following: (a) there should be a direct relationship between the transport properties of a preparation and the extent of modification or loss of specific membrane polypeptides; (b) a specific binding should be demonstrated between a particular polypeptide and choline or ligands, such as HC-3, which alter HACU activity; and finally (c) it should be possible to purify and isolate the transporter after solubilization and fractionation of membrane components. Fractions enriched in the HACU system should be incorporated into liposomes and should exhibit initial uptake activity proportional to the amount of transporter present. The transport properties of the reconstituted system should reflect those observed in native membranes.

So far none of these objectives has been fully or satisfactorily met for the identification of the choline transport system. Recently, however, advances have begun to be made in a number of laboratories so that it may be possible to meet at least some of these goals in the near future. Progress in identifying the molecular nature of the choline transporter is being made by the use of the following techniques: enzymatic modification of membrane proteins, characterization of the binding properties of HC-3 and certain covalent probes at brain HACU sites, and finally, the development of reconstituted systems which demonstrate highly efficient HACU activity with characteristics similar to those of the native membrane.

II. Attempted Identification of Presynaptic Plasma Membrane Components and Their Modification

The polypeptide composition of the presynaptic plasma membrane (PSPM), derived from synaptosomes which have been isolated from electric organ, has been investigated using both one-dimensional (MOREL et al. 1982) and two-dimensional (STADLER and TASHIRO 1979; MILJANICH et al. 1982) gel electrophoresis. The lack of a specific biochemical or morphological membrane marker for the PSPM makes evaluation of its purity difficult. However, synaptosomes which have been isolated from electric organ, a tissue which is embryologically derived from muscle (KEYNES and MARTINS-FERREIRA 1953), have been assessed to be 20%–40% pure using the criterion of enzyme-marker measurements (KELLY et al. 1979). PSPM derived from synaptosomes after osmotic lysis of the latter have been shown to be contaminated by postsynaptic elements (MOREL et al. 1982) and synaptic vesicle components (MILJANICH et al. 1982; DUCIS and WHITTAKER 1985). Nevertheless, two-dimensional gel electrophoresis has shown that the polypeptide composition of the synaptic vesicle and PSPM are very different (STADLER and TASHIRO 1979; MILJANICH et al. 1982). If the polypeptide composition of the PSPM is considered only insofar as it relates to polypeptides which may participate in HACU, an unequivocal identification is not yet possible. Only one report (DUCIS and WHITTAKER 1985) has made a tentative identification of the transporter. The authors, assuming the transporter is a protein of relative molecular mass (M_r) of less than 200000, as most identified transport polypeptides are, determined that the best correlation of transport activity observed in proteoliposomes and incorporation of a specific polypeptide was exhibited by a 54-kDa component.

When enzymatic modification of membrane components was carried out using either pronase (VYAS and O'REGAN 1985) or neuraminidase (STEFANOVIĆ et al. 1975; MASSARELLI et al. 1982a, b; VYAS and O'REGAN 1985), HACU activity was subsequently diminished. This implies that the transporter is a proteinaceous substance and, perhaps, a glycoprotein. No attempt, however, was made in these studies to identify which specific membrane proteins were modified by exposure to the enzymes.

III. Binding of Hemicholinium-3

One of the most potent competitive inhibitors of the HACU system (HOLDEN et al. 1975) is the respiratory paralysant (SCHUELER 1955) HC-3. This alkyl bisquaternary ammonium compound has a greater affinity for the HACU system than does choline itself (HOLDEN et al. 1975) and is characterized (SCHUELER 1955) by the presence of a choline moiety cyclized through hemiacetal formation. Although it has been reported that HC-3 can accumulate in nervous tissue, this is not via the HACU system (COLLIER 1973; SLATER and STONIER 1973a, b). In addition, LACU appears to be relatively insensitive to HC-3 (YAMAMURA and SNYDER 1973).

Recently, the high-affinity binding properties of [^{3}H]HC-3 have been investigated in membranes (VICKROY et al. 1984a, b; SANDBERG and COYLE 1985) and frozen sections (RAINBOW et al. 1984; MANAKER et al. 1986) derived from rat brain. Its binding was observed to be saturable, reversible, specific and ion dependent (RAINBOW et al. 1984; VICKROY et al. 1984b; SANDBERG and COYLE 1985; MANAKER et al. 1986). Optimal high-affinity binding of [^{3}H]HC-3 occurred in the presence of sodium and chloride (VICKROY et al. 1984b; SANDBERG and COYLE 1985). When sodium was substituted by potassium, lithium, rubidium, calcium or caesium, [^{3}H]HC-3 binding was significantly reduced (VICKROY et al. 1984b; SANDBERG and COYLE 1985). Although fluoride, iodide, sulphate and phosphate were poor substitutes, bromide could partially substitute for chloride (SANDBERG and COYLE 1985). These ionic dependencies for [^{3}H]HC-3 binding are similar to those previously described for HACU (SIMON and KUHAR 1976; MEYER and COOPER 1982; BREER and LUEKEN 1983; DUCIS and WHITTAKER 1985). Furthermore, increasing the concentration of NaCl in the medium increased the affinity of [^{3}H]HC-3 with no significant effect on the maximal number of specific binding sites (B_{max}) (SANDBERG and COYLE 1985). This observed decrease in K_D in the presence of increasing concentrations of NaCl (up to 200 m*M*) was due solely to a selective slowing of the rate of dissociation of [^{3}H]HC-3 from its membrane binding sites (SANDBERG and COYLE 1985). The binding of HC-3 was also reported to be relatively insensitive to metabolic inhibitors (VICKROY et al. 1984b) but was temperature and pH sensitive (SANDBERG and COYLE 1985) and inhibited by choline (RAINBOW et al. 1984; VICKROY et al. 1984b; SANDBERG and COYLE 1985; MANAKER et al. 1986) and compounds known to block HACU (SANDBERG and COYLE 1985). There was a highly significant correlation between the efficacy of the compounds to block [^{3}H]HC-3 binding and their potency in inhibiting HACU in synaptosomes (SANDBERG and COYLE 1985). Direct analysis of [^{3}H]HC-3 binding and indirect competition studies suggest the existence of a single class of independent high-affinity binding sites for HC-3 with an apparent K_D of 2–65 n*M* (RAINBOW et al. 1984; VICKROY et al. 1984b; SANDBERG and COYLE 1985; MANAKER et al. 1986).

A direct correlation has also been made between the activity of ChAT, a presynaptic marker for cholinergic nerve terminals, and the number of [^{3}H]HC-3 binding sites in rat neocortex (VICKROY et al. 1984a). It was reported that after chemical destruction of cell bodies in the nucleus basalis magnocellularis, a parallel reduction in the density of specific binding sites for [^{3}H]HC-3 (41%–48%) and ChAT activity (39%–42%) occurred in the anterior cerebral cortex (VICKROY et al. 1984a). It was previously observed that a comparable reduction in the magnitude of HACU (40%–50%) and ChAT activity (40%) occurred in rat cerebral cortex following destruction of this nucleus (PEDATA et al. 1982).

The localization of [^{3}H]HC-3 receptor sites and the pharmacology and ionic dependence of high-affinity [^{3}H]HC-3 binding, suggest that the specific binding of this ligand is to the presynaptic membrane and that it labels some moiety of the HACU system. Thus, the use of [^{3}H]HC-3 to quantify HACU sites *in vitro* provides an approach for the identification and, perhaps, purification of the HACU system. If it is assumed that the density of [^{3}H]HC-3 recognition sites accurately reflects the quantity of HACU sites, the choline transporter appears, then, to represent only a small fraction of the total mammalian presynaptic plasma membrane protein since the density of specific membrane binding sites for [^{3}H]HC-3 was estimated to be only in the femtomolar range. For example, the B_{max} for [^{3}H]HC-3 was estimated to be approximately 184 (Vickroy et al. 1984b) and 312 (Sandberg and Coyle 1985) fmol mg^{-1} membrane protein in rat striatum where the highest binding of [^{3}H]HC-3 was observed (Vickroy et al. 1984b; Sandberg and Coyle 1985).

IV. Labelling with Covalent Probes

The development of a site-directed covalent probe for the HACU system would be a significant advance towards quantification, positive identification and, perhaps, isolation of the transporter. A number of covalently binding nitrogen mustard derivatives such as analogues of hexamethonium (Barker et al. 1977), HC-3 (Smart 1981) and choline (Rylett and Colhoun 1984; Pedder and Prince 1985) have been developed and tested for their selectivity for the HACU system. Differential labelling (Phillips 1977) experiments have also been carried out with *N*-ethylmaleimide (NEM), a covalently binding sulphydryl reagent (Sukumar and Townsel 1983). Selective protection of HACU sites during the initial labelling period with non-radioactive NEM was achieved by incubating synaptosomes in the presence of low concentrations of HC-3 (Sukumar and Townsel 1983). To date, however, affinity (Wold 1977) or differential labelling (Phillips 1977) procedures have not resulted in an attempted identification of the HACU system, and preferential binding of these labels to specific membrane components has not yet been demonstrated.

Up to the present, most characterization work with irreversible inhibitors has been done with nitrogen mustard analogues of choline and their effects on HACU in synaptosomes. Choline mustard aziridinium ion (CMA) has been investigated extensively by Rylett and Colhoun (1979, 1980, 1982, 1984), and the characteristics of the monoethylcholine mustard aziridinium ion (ECMA) on HACU have been described by several laboratories (Rylett and Colhoun 1980, 1982; Curti and Marchbanks 1984; Pedder and Prince 1984, 1985). These mustards are potent and irreversible inhibitors of choline transport in rat brain synaptosomes (Rylett and Colhoun 1980, 1984), CMA being the most effective (Rylett and Colhoun 1982, 1984). Differential inhibition of sodium-dependent and independent choline uptake has been observed with CMA (Rylett and Colhoun 1984) and ECMA (Pedder and Prince 1985) and irreversible inhibition of choline uptake by these mustard compounds can be prevented by choline and HC-3 (Curti and Marchbanks 1984; Rylett and Colhoun 1984).

Although it has been shown that the major site of action for CMA is at the nerve terminal (Rylett and Colhoun 1984), these nitrogen mustard analogues of choline are not completely selective for the HACU system (Clement and Colhoun 1975; Rylett and Colhoun 1979; Barlow and Marchbanks 1984; Pedder and Prince 1984, 1985).

Ideally, in order to identify or purify a transporter, it would be desirable that a site-directed covalent probe exhibit complete selectivity for the system. Degree of transport inhibition could then be unambiguously related to the amount of analogue bound to a particular membrane component. However, since other nerve terminal components besides the HACU system are known to interact with the nitrogen mustard analogues of choline, membrane receptor labelling techniques (ZISAPEL and SOKOLOVSKY 1977) that control for non-specific binding of affinity labels could prove to be potentially useful for the HACU system.

V. Reconstitution of Choline Uptake Activity

The reconstitution of the choline transport system into liposomes is a direct approach to the identification of the HACU system. Reconstitution of membrane components serves as an assay system after specific purification procedures have been carried out, provided that the reconstitution of different polypeptide fractions and the transport assay are performed under identical experimental conditions. Ideally, purification of presynaptic membrane components should lead to isolation of only those elements that are involved in choline uptake.

Since the permeability of bilayers to charged compounds is very low, liposomes are potentially very sensitive assay systems for the HACU process. Several liposomal systems have been shown to have low permeability to choline but, after incorporation of presynaptic membrane components, the proteoliposomes showed significant and specific changes in permeability to choline (VYAS and O'REGAN 1985; DUCIS and WHITTAKER 1985). Although a number of reconstitution experiments with liposomes have been carried out with the choline transporter (KING and MARCHBANKS 1980, 1982; MEYER and COOPER 1982; VYAS and O'REGAN 1985; DUCIS and WHITTAKER 1985), only two laboratories have reported the reconstitution of solubilized systems (KING and MARCHBANKS 1980, 1982; DUCIS and WHITTAKER 1985). DUCIS and WHITTAKER (1985) have demonstrated highly efficient choline transport rates after incorporation of solubilized membrane components into liposomes. Their reconstituted system exhibits similar characteristics of HACU to those reported in native membranes (BREER 1983; BREER and LUEKEN 1983; DUCIS and WHITTAKER 1985) and synaptosomes (YAMAMURA and SNYDER 1973; SIMON and KUHAR 1976). The results of KING and MARCHBANKS (1980, 1982) are more ambiguous, however. The ionic specificity of choline uptake, both cationic and anionic, was not thoroughly investigated in their studies and only partial inhibition of choline uptake in the presence of high concentrations (100 μM) of HC-3 was observed. In addition, the authors generally reported very low rates of uptake in their preparations with considerable variability of the experimental data within one experiment. The variability observed may reflect, at least in part, leakiness in their system.

Judging from initial rates, the most active reconstituted system to date is the preparation of DUCIS and WHITTAKER (1985). Their system exhibited initial choline uptake rates of 183–761 pmol min^{-1} mg^{-1} protein. Preparations from other laboratories exhibited considerably lower uptake rates of substrate, ranging from 0.16–0.32 (VYAS and O'REGAN 1985) to 61–71 (KING and MARCHBANKS 1982) pmol min^{-1} mg^{-1} protein. Furthermore, choline uptake in their system exhibited sodium and chloride dependence; an overshoot, indicating Na/choline cotransport and lack of leakiness of the proteoliposomes to sodium; high-affinity inhibition by HC-3; and stimulation by valinomycin and temperature sensitivity. Their study also led to a tentative identification of the choline transporter. Ability of proteoliposomes to take up choline in various preparations was correlated to three polypeptides of M_r ($\times 10^{-3}$) of 34, 54 and > 200 from Coomassie blue stained gels. Of these, only the 54- and > 200-kDa polypeptides copurified with choline binding sites in membranes. Since the authors identified a > 200-kDa component immunochemically as the synaptic vesicle-specific proteoglycan, they tentatively identified the 54-kDa polypeptide as the choline transporter. These results, however, are far from conclusive. Future studies should involve purifying membrane components further before their incorporation into liposomes as well as using covalent probes to identify membrane moieties involved in choline uptake.

To date, of the identification techniques discussed, the most promising results have been with reconstituted systems. The unequivocal identification of the transporter, however, rests on the isolation of the system in a pure and functional state. The criterion of reasonable purity is necessary before definite conclusions about the identity of the choline transporter can be made. The transport system may well involve more than one type of polypeptide, or it may form functional aggregates in the membrane to give dimers or larger complexes. Clearly the continued production of increasingly purified polypeptides and their reproducible incorporation into liposomes should be helpful in determining the molecular components of the HACU system.

Interestingly, the reconstitution of choline transport activity across a carbon tetrachloride phase has been recently reported by KING and SHARP (1985). Although no movement of choline across the organic phase was observed without the addition of brain extracts, it was not determined whether this was a nonspecific effect. Before this technique can be considered useful for monitoring choline uptake activity, it should first be demonstrated that this effect depends on selective incorporation of tissue extracts or proteins since it was reported that very high concentrations of HC-3 (1.0 m*M*) are needed to block [^{3}H]choline (50 n*M*) movement. Caution should, furthermore, be exercised with this technique since choline uptake and diffusion were previously reported in liposomes and across a chloroform layer, respectively, in the absence of incorporated membrane extracts (KING 1984). In both cases, the phenomenon was saturable, could be inhibited by high concentrations of HC-3 and was increased by unlabelled choline on the opposite side of the membrane or chloroform phase. Significantly, uptake of 5 μM [^{3}H]choline into liposomes was inhibited by high concentrations of HC-3 (100 μM) and unlabelled choline (10 m*M*) and exhibited *trans* stimulation, characteristics of choline uptake that were previously used as evidence for reconstitution of the choline transporter (KING and MARCHBANKS 1980, 1982). The results of KING (1984) suggest that care must be taken when interpreting transport data when very low rates of choline uptake are observed and uptake cannot be measured over a long enough time period against diffusion and binding blanks under appropriate experimental conditions.

D. Areas for Future Investigation

I. Functional Ligands and Components

Future research will undoubtedly be focused on the molecular aspects of the choline-transport mechanism since it has now been demonstrated that the system can be successfully extracted from membranes and reconstituted into artificial systems. In particular, specific chemical probes would be invaluable to further explore the molecular aspects of the transport system. By utilizing such probes selectively to modify and label cell surface components and by determining how, or if, they affect choline uptake, it should be possible to characterize some of the molecular properties of the HACU system, including the identification of critical protein ligands in the active centre, the identification, isolation and characterization of functional components, the arrangement of the system in the membrane, the role, if any, of phospholipids and cholesterol, and conformational changes associated with choline transport.

The identification of critical protein ligands which are essential for HACU function can be carried out by the use of group-directed covalent probes. Furthermore, when used in their radiolabelled form, these compounds could potentially help to identify the functional components of the HACU system, particularly when used in conjunction with techniques such as differential labelling (SUKUMAR and TOWNSEL 1983). Reagents that are specific for sulphydryl groups such as organomercurials, iodoalkyl derivatives and SS-SH exchangers such as dithiothreitol and 5,5′-dithiobis (2-nitrobenzoic acid) (DTNB) should be tested for their effect on HACU. NEM, which is a somewhat less specific sulphydryl reagent, has been shown to inhibit choline uptake in *Limulus* central nervous system synaptosomes (SUKUMAR and TOWNSEL 1983). In order to determine whether carboxyl or amino groups are involved in HACU, carbodiimides and imidoesters, respectively, could be tested. Measuring HACU activity as a function of pH may also indicate whether hydrogen ion titration of a ligand in the active site affects choline uptake. However, for most group-directed modifiers, the possibility that inhibition occurs indirectly via their interaction with other membrane components, and not the critical ligand associated with HACU, must not be disregarded until conclusive evidence is available to the contrary.

Ideally, in order to identify functional components of the HACU system, site-directed radiolabelled or fluorescent covalent probes, specific for the transport system, would be extremely useful and could serve as specific markers for these active site structures during purification and characterization, assuming, as with group-directed probes, that the covalent products are stable. Potentially, compounds such as the covalently binding nitrogen mustard analogues of choline (RYLETT and COLHOUN 1984; PEDDER and PRINCE 1985), although not completely specific for the HACU system, could be useful. These compounds contain a very reactive three-membered aziridinium ring and can cause alkylation of amino, sulphydryl and carboxyl groups as well as phosphate-containing compounds (FRUTON et al. 1946). In order to gain insight into the specific protein ligands at the active site of the HACU system, pretreatment with sulphydryl, amino or carboxyl specific reagents could be carried out to see if any of these group-directed compounds prevent irreversible blocking of HACU by the nitrogen mustard analogues of choline. Since these mustard compounds can be transported across the nerve terminal membrane (PEDDER and PRINCE 1984, 1985), they could react with the HACU system (or other membrane components) on the inner surface. This could potentially complicate the interpretation of labelling data and the conclusions drawn about external active site components of the HACU system. If possible, incubation conditions should always be chosen so that the nitrogen mustard probe reacts only at the outer surface. It is also not presently clear whether other proteins and transport systems, unrelated to choline metabolism and uptake, or if any lipid moieties are affected by these probes. Theoretically, the specificity of the nitrogen mustard analogues of choline could be enhanced by reacting them with negatively purified synaptosomes or right-side-out membrane sacs. Negative purification could entail mild proteolysis of externally exposed membrane proteins. It is possible that proteolytic treatments which do not interfere with choline uptake would remove many non-specific components of labelling at the outer membrane surface. Reacting the covalent probes with proteoliposomes, after solubilized nerve terminal membrane components have been incorporated into liposomes, may also be effective. However, a certain fraction of transport polypeptides would be expected to be inside-out in this system. Finally, it should be possible to carry out differential labelling experiments in the near future with nitrogen mustard analogues of choline using low concentrations of HC-3 to protect the HACU system selectively during labelling of non-specific protein sites.

II. Arrangement of Transport System in the Membrane

Although studies on nerve cells and synaptosomes with, presumably, impermeable agents such as neuraminidase (MASSARELLI et al. 1982a) suggest that the HACU system can be directly modified at the external membrane surface, it should also be determined whether the transporter is susceptible to alterations on the inner surface. Experiments need to be carried out with model systems to determine the effects of both chemical and biochemical modification of membrane lipids, carbohydrates and proteins on the transport system as well as the effects of reversible inhibitors. Ideally, studies should be carried out at both faces of the membrane. In general, however, to obtain unequivocal results, compounds being tested for their inhibitory effect on HACU, whether from the outside or the inside, should be impermeable or only slowly penetrating. Decreased membrane permeability to certain agents can perhaps be achieved by manipulating environmental factors such as temperature. Some compounds may have different inhibitory effects on the HACU system depending on the surface of application. This would then suggest a structural asymmetry of the transporter.

Arrangement of the HACU system in the membrane can be potentially assessed by the use of proteolytic enzymes and chemical probes. Information that would be pertinent to determining the arrangement of the transport polypeptide (or polypeptides) in the membrane would be the following: the hydrophobicity of different segments of the transporter; whether the polypeptide crosses the membrane and is exposed to the aqueous environment on both the external and cytoplasmic sides and is therefore accessible from either side to hydrophilic probes; and, thirdly, whether it is asymmetric. Attempts can be made to determine the arrangement of the choline transporter in the membrane by segmenting it with proteolytic enzymes at both the inside and outside of the presynaptic membrane. The membrane fragments can then be identified by use of endogenous markers or by chemical labels. Segmenting of the choline transport protein or proteins can be carried out in conjunction with prior negative purification of the membrane by extracting agents or proteolytic enzymes to minimize the number of contaminating peptides associated with the choline transporter. Ultimately, segmentation combined with peptide mapping and amino acid sequence studies could probably define the precise arrangement of the choline transporter in the membrane. Some possibilities for endogenous markers would be: sites of phosphorylation by protein kinases or S-S cross-linking sites, if any, and presence of sialic acid residues. Chemical labels could include binding of HC-3, NEM, CMA and ECMA.

The hydrophobicity of different segments of the choline transporter can be determined by the necessity to extract them with detergents after treatment of the membrane with proteolytic enzymes. Asymmetry of the transport system can be verified if, after exposure of the outside or only the inside face of the membrane to impermeable probes, different peptides are labelled. The effects on HACU during segmentation should be monitored to see if any hydrolysis products remaining in the membrane are able to operate at full transport capacity and have the ability to bind inhibitors specific for the HACU system.

III. Isolation of Transport System

As was previously stated, direct assignment of choline-transport properties to specific membrane polypeptides depends on their isolation in a pure and reasonably functional form. This can be accomplished either by isolating the transport system in its natural lipid environment after negative purification procedures or by reconstituting the extracted and purified polypeptides into artificial systems. Since reconstitution of choline uptake activity into liposomes has been successfully carried out, the goal of future research should be to utilize the potential of these systems, particularly as they relate to exploring the detailed mechanisms of choline transport.

Negative purification entails leaving behind membrane fragments containing the HACU system in a reasonably pure form together with membrane lipids. In principle, purification of the HACU mechanism could be carried out with various solvents, detergents, salts or by selective removal of membrane polypeptides by proteolytic cleavage. So far, negative purification procedures have not yet been attempted with the HACU system. With this approach, it is possible that higher protein yields may be achieved than via reconstitution.

IV. Possible Role of Lipids

There is evidence that lipids can play a dynamic role in the function of membrane-transport processes (GRISHAM and BARNETT 1973; BARNETT and PALAZZOTTO 1975; SHINITZKY 1984). It is, therefore, possible that membrane lipids may also play a role in the HACU mechanism at the nerve terminal. The role of lipids in choline uptake could be of the following nature: (a) lipid *per se* or in combination with membrane protein serves as the recognition site for choline; (b) the boundary layer of lipids that surrounds the transport molecules, the lipid annulus, modulates the three-dimensional structure of the transport system and regulates the substrate affinity (K_T) for the transport system; and (c) membrane-fluidity properties determine the accessibility and mobility of the transport molecules in the membrane and, hence, regulate the capacity (V_{max}) of the HACU system.

To determine conclusively whether a specific lipid (e.g. ganglioside) is the cellular recognition site for choline, the substrate must bind to the purified lipid with high affinity and the latter must compete with the biological membrane containing the receptor site for choline. Furthermore, the relative affinities of various ligands for the lipid should correspond to their affinities as substrates for the HACU system. Addition and incorporation of the exogenous lipid into nerve terminal membranes would be expected to increase choline uptake and the cellular response to choline (ACh synthesis), whereas removal or absence of the lipid from presynaptic membranes should have the opposite effect. Finally, a positive correlation should be observed between the amount of [^{3}H]HC-3 bound or choline transported in different brain regions and the density of the specific membrane lipid at these locations. Similar experiments have been previously carried out with cholera toxin and the membrane ganglioside GM1 (Cuatrecasas 1973; Holmgren et al. 1975; Gill 1977).

Experiments that can be carried out to determine the participation, if any, of phospholipids as the membrane recognition sites for choline involve the hydrolysis of synaptic membrane phospholipids by phospholipases and the subsequent effect of these manipulations on HACU. Such experiments have been carried out in frog and human erythrocyte membranes to show the involvement of phospholipids in the recognition of adrenergic ligands (Limbird and Lefkowitz 1976) and in D-glucose uptake (Kahlenberg and Banjo 1972), respectively.

Evidence suggesting a direct substrate-phospholipid interaction should result in a decrease in the number of binding sites and not in a change in the affinity of the receptor site for the ligand. However, if changes in affinity are observed after phospholipase treatment, this would suggest that the phospholipids surrounding the transport proteins in the membrane may regulate transporter-substrate interactions. For example, it is possible that the polar head groups of the phospholipids may interact with substrates of the HACU system. If so, choline uptake may be enhanced by phospholipase C (from *Clostridium perfringens*) treatment which results in hydrolysis of neutral lipids such as PE, PC and sphingomyelin. It is possible that choline, as well as ethanolamine (Zelinski and Choy 1984) polar head groups would complete with substrates of the HACU system for binding sites on the transport proteins. Therefore, besides affecting the fluidity or lipid order of the membrane, phospholipids may also interact competitively with a ligand for its binding site. Perhaps this could be tested in the case of choline by exposing synaptosomes or membrane sacs to PS decarboxylase which converts PS to PE.

Exposing synaptic membranes to PS decarboxylase was found to inhibit opiate binding; however, binding activity could be restored by the addition of PS

(ABOOD et al. 1978). Choline uptake, like opiate binding, may be enhanced after exposure of nerve terminal membranes to PS or, perhaps, other types of acidic lipids. Further evidence for choline (or other ligand of the HACU system) interaction with particular membrane lipids can be obtained by showing that it modulates the partitioning of a lipid between an aqueous and an organic phase, indicating that there is an interaction between the two and subsequent alteration of the physical properties of the lipid. Such experiments have been previously carried out with opiates and cerebroside sulphate (CHO et al. 1976).

The importance of the lipid environment on choline uptake can be tested by altering the lipid composition of the native membrane or incorporation of the transporter into liposomes of defined constitution. The HACU system may be specific for certain lipids since the use of a binary lipid system, PS/cholesterol (DUCIS and KOEPSELL 1983), for its reconstitution resulted in no observable choline uptake (DUCIS and WHITTAKER 1985). Modification of the types and ratios of phospholipids, with defined fatty acid chains, in proteoliposomes would determine whether HACU activity is sensitive to the presence of particular membrane lipids or fatty acids. These membrane lipid components may affect the properties of the HACU system directly or via membrane fluidity changes which may play a vital role in regulating nerve terminal choline transport activity. The use of fluorescent probes such as perylene and 1,6-diphenyl-1,3,5-hexatriene (SHINITZKY and BARENHOLZ 1974) would show, respectively, a direct correlation between choline uptake and changes in the order of the lipid matrix and microviscosity of the membrane. Besides a dependency on temperature and the lipid and fatty acid composition of the membrane, membrane fluidity can also be modified with benzyl alcohol (DIPPLE and HOUSLAY 1978), divalent cations (BASHFORD et al. 1975), and changes in pH or membrane potential (SHINITZKY 1984).

The possible effects of the cholesterol content of membranes on HACU function have also not been investigated. Since cholesterol influences the fluidity properties of membranes and can eliminate their phase transition, it is conceivable that it may also participate in the regulation of the HACU system. Furthermore, cholesterol effects on membrane proteins may also be expressed by modification of their vertical displacement or degree of exposure (BOROCHOV and SHINITZKY 1976; SHINITZKY and RIVNAY 1977). An increase in membrane microviscosity, caused by increased membrane cholesterol content, could result in the exposure of masked choline transport sites to the outside of the cell, thus enhancing substrate binding-site accessibility and, perhaps, HACU activity if mobility of transport molecules is, concomitantly, not too severely restricted. Optimal activity for the HACU system may thus be at a defined lipid microviscosity where V_{max} reaches a peak value, as has been found for other membrane transport processes (SHINITZKY 1984). The effects of cholesterol on HACU activity could be studied by modulating the membrane cholesterol content of nerve cells or synaptosomes by incubating them in the presence of PC or PC/cholesterol vesicles or by reconstituting the solubilized choline transporter into phospholipid vesicles containing increasingly greater concentrations of cholesterol. Such techniques have been successfully applied in other systems (GRUNZE and DEUTICKE 1974; BOROCHOV and SHINITZKY 1976; DUCIS and KOEPSELL 1983).

Since V_{max}, but not K_T, changes in choline uptake appear to be coupled to neuronal activity (SIMON and KUHAR 1975; SIMON et al. 1976), HACU may be regulated by changes in membrane lipid fluidity. Changes in membrane microviscosity can lead to the vertical or lateral displacement of membrane proteins (SHINITZKY 1984). These effects may have direct bearing on HACU activation after nerve stimulation and the influx of calcium into the nerve terminal since divalent cations have been shown to modulate membrane lipid fluidity properties (BASHFORD et al. 1975; SHINITZKY 1984). In most studies *in vitro*, HACU activation is dependent on the presence of calcium (MURRIN and KUHAR 1976; MURRIN et al. 1977; WEILER et al. 1978) and not on the release of ACh (BARKER 1976; COLLIER and ILSON 1977; MURRIN et al. 1977).

Membrane sacs, derived from cholinergic neurons (BREER 1983; BREER and LUEKEN 1983; DUCIS and WHITTAKER 1985), would be particularly useful in elucidating the role, if any, that calcium and cholesterol play in modulating HACU. These resealed presynaptic plasma membrane fragments would be expected to retain a lipid and protein composition similar to that of the intact nerve terminal and factors such as internal compartmentation of substances, presence of intracellular ACh or choline and undefined ion fluxes would not interfere with the interpretation of results.

V. Conformational Changes of the Transporter Associated with Choline Uptake

Future studies should also address the question of whether the transporter is fixed in the plane of the membrane or has translocational freedom, as well as whether its inside-outside orientation is fixed. It is theoretically possible to demonstrate transmembrane mobility of the transporter by showing a depletion of choline transport sites on the outer face of the membrane. If a slowly penetrating substrate, with high affinity for the choline transport site, and which binds at the inner surface, is present only on the inside of the membrane, it could cause mobile binding sites to sequester at the cytoplasmic surface with a consequent depletion of sites on the outer surface. The transport sites available on the membrane surface should subsequently be titrated with irreversible and non-penetrating probes. This approach, however, requires that the irreversible probe does not perturb the distribution of sites to any great extent during the time required for the titration, conditions which may not easily be met.

Physical techniques to find evidence of conformational changes associated with transport can also be employed. Studies with membrane-bound paramagnetic or fluorescent probes could potentially show the interaction of the HACU system with substrates by quantitative changes of the electron spin resonance spectrum or by alterations in fluorescence, respectively. Such measurements should show directly that changes in the state of the HACU system proteins are associated with binding or transport of choline.

References

Abood LG, Salem N, MacNiel M, Butler M (1978) Phospholipid changes in synaptic membranes by lipolytic enzymes and subsequent restoration of opiate binding with phosphatidylserine. Biochim Biophys Acta 530:35–46

Amaducci L, Mantovani P, Pepeu G (1976) Effects of some phospholipids on acetylcholine output from the cerebral cortex in the rat. Br J Pharmacol 56:379P–380P

Atweh S, Simon JR, Kuhar MJ (1975) Utilization of sodium-dependent high affinity choline uptake *in vitro* as a measure of the activity of cholinergic neurons *in vivo*. Life Sci 17:1535–1544

Baker PF, Hodgkin AL, Shaw TI (1962) The effects of changes in internal ionic concentrations on the electrical properties of perfused giant axons. J Physiol (Lond) 164:355–374

Barker LA (1976) Modulation of synaptosomal high affinity choline transport. Life Sci 18:725–731

Barker LA (1978) Synaptosomal uptake of (^{3}H-N-Me)-N,N,N-trimethyl-N-propynyl-ammonium (TMPYA). Fed Proc 37:649 (Abstract 2293)

Barker LA, Mittag TW (1975) Comparative studies of substrates and inhibitors of choline transport and choline acetyltransferase. J Pharmacol Exp Ther 192:86–94

Barker LA, Dowdall MJ, Mittag TW (1975) Comparative studies on synaptosomes: high-affinity uptake and acetylation of N-[Me-^{3}H]choline and N-[Me-^{3}H] N-hydroxyethyl-pyrrolidinium. Brain Res 86:343–348

Barker LA, Mittag TW, Krespan B (1977) Studies on substrates, inhibitors and modifiers on the high affinity choline transport-acetylation system present in rat brain synaptosomes. In: Jenden DJ (ed) Cholinergic mechanisms and psychopharmacology. Plenum, New York, pp 465–480

Barlow P, Marchbanks RM (1984) Effect of ethylcholine mustard on choline dehydrogenase and other enzymes of choline metabolism. J Neurochem 43:1568–1573

Barnett RE, Pallazzotto J (1975) Mechanism of the effects of lipid phase transitions on the Na^+, K^+-ATPase, and the role of protein conformational changes. Ann NY Acad Sci 242:69–76

Bashford CL, Harrison SJ, Radda GK, Mehdi Q (1975) The relation between lipid mobility and the specific hormone binding of thyroid membrane. Biochem J 146:473–479

Batzold F, De Haven R, Kuhar MJ, Birdsall, N (1980) Inhibition of high affinity choline uptake, structure activity studies. Biochem Pharmacol 29:2413–2416

Bennett MR, McLachlan EM (1972) An electrophysiological analysis of the synthesis of acetylcholine in preganglionic nerve terminals. J Physiol (Lond) 221:669–682

Birks RI (1963) The role of sodium ions in the metabolism of acetylcholine. Can J Biochem Physiol 41:2573–2597

Birks R, MacIntosh FC (1961) Acetylcholine metabolism of a sympathetic ganglion. Can J Biochem Physiol 39:787–827

Blusztajn JK, Wurtman RJ (1981) Choline biosynthesis by a preparation enriched in synaptosomes from rat brain. Nature 290:417–418

Borochov H, Shinitzky M (1976) Vertical displacement of membrane proteins mediated by changes in microviscosity. Proc Natl Acad Sci USA 73:4526–4530

Bowery NG, Neal MJ (1975) Failure of denervation to influence the high affinity uptake of choline by sympathetic ganglia. Br J Pharmacol 55:278P

Breer H (1981) Comparative studies on cholinergic activities in the central nervous system of *Locusta migratoria*. J Comp Physiol 141:271–275

Breer H (1982) Uptake of [N-Me-^{3}H]-choline by synaptosomes from the central nervous system of *Locusta migratoria*. J Neurobiology 13:107–117

Breer H (1983) Choline transport by synaptosomal membrane vesicles isolated from insect nervous tissue. FEBS Lett 153:345–348

Breer H, Lueken W (1983) Transport of choline by membrane vesicles prepared from synaptosomes of insect nervous tissue. Neurochem Int 5:713–720

Brown GL, Feldberg W (1936) The acetylcholine metabolism of a sympathetic ganglion. J Physiol (Lond) 88:265–283

Burgess EJ, Atterwill CK, Prince AK (1978) Choline acetyltransferase and the high affinity uptake of choline in corpus striatum of reserpinised rats. J Neurochem 31:1027–1033
Carroll PT, Goldberg AM (1975) Relative importance of choline transport to spontaneous and potassium depolarized release of ACh. J Neurochem 25:523–527
Cho TM, Cho JS, Loh HH (1976) ^{3}H-cerebroside sulfate re-distribution induced by cation, opiate or phosphatidylserine. Life Sci 19:117–123
Christensen H (1975) Biological transport. Reading, Benjamin
Clement JG, Colhoun EH (1975) Inhibition of choline transport into human erythrocytes by choline mustard aziridinium ion. Can J Physiol Pharmacol 53:1089–1093
Cohen EL, Wurtman RJ (1975) Brain acetylcholine: increase after systemic choline administration. Life Sci 16:1095–1102
Cohen EL, Wurtman RJ (1976) Brain acetylcholine: control by dietary choline. Science 191:561–562
Collier B (1973) The accumulation of hemicholinium by tissues that transport choline. Can J Physiol Pharmacol 51:491–495
Collier B (1981) The structural specificity of choline transport into cholinergic nerve terminals. J Neurochem 36:1292–1294
Collier B, Ilson D (1977) The effect of preganglionic nerve stimulation on the accumulation of certain analogues of choline by a sympathetic ganglion. J Physiol (Lond) 264:489–509
Collier B, Katz HS (1974) Acetylcholine synthesis from recaptured choline by a sympathetic ganglion. J Physiol (Lond) 238:639–655
Collier B, Lang C (1969) The metabolism of choline by a sympathetic ganglion. Can J Physiol Pharmacol 47:119–126
Collier B, MacIntosh FC (1969) The source of choline for acetylcholine synthesis in a sympathetic ganglion. Can J Physiol Pharmacol 47:127–135
Collier B, Lovat S, Ilson D, Barker LA, Mittag TW (1977) The uptake, metabolism and release of homocholine: studies with rat brain synaptosomes and cat superior cervical ganglion. J Neurochem 28:331–339
Collier B, Boksa P, Lovat S (1979) Cholinergic false transmitters. Prog Brain Res 49:107–121
Coyle JT, Yamamura HI (1976) Neurochemical aspects of the ontogenesis of cholinergic neurons in the rat brain. Brain Res 118:429–440
Crane RK (1965) Na^{+}-dependent transport in the intestine and other animal tissues. Fed Proc 24:1000–1006
Cuatrecasas P (1973) Interaction of *Vibrio cholerae* enterotoxin with cell membranes. Biochemistry 12:3547–3558
Curti D, Marchbanks RM (1984) Kinetics of irreversible inhibition of choline transport in synaptosomes by ethylcholine mustard aziridinium. J Membr Biol 82:259–268
Diamond I (1971) Choline metabolism in brain: the role of choline transport and the effects of phenobarbital. Arch Neurol 24:333–339
Dipple I, Houslay MD (1978) The activity of glucagon-stimulated adenylate cyclase from rat liver plasma membranes is modulated by the fluidity of its lipid environment. Biochem J 174:179–190
Dowdall MJ (1977) Nerve terminal sacs from *Torpedo* electric organ: a new preparation for the study of presynaptic cholinergic mechanisms at the molecular level. In: Jenden DJ (ed) Cholinergic mechanisms and psychopharmacology. Plenum, New York, pp 359–375
Dowdall MJ, Fretten P (1981) Inhibition of membrane transport systems in synaptosomes from *Torpedo* electric organ by snake neurotoxins. In: Pepeu G, Ladinsky H (eds) Cholinergic mechanisms: phylogenetic aspects, central and peripheral synapses, and clinical significance. Plenum, New York, pp 249–259
Dowdall MJ, Simon EJ (1973) Comparative studies on synaptosomes: uptake of [N-Me-^{3}H]-choline by synaptosomes from squid optic lobes. J Neurochem 21:969–982
Dowdall MJ, Barrantes FJ, Stender W, Jovin TM (1976) Inhibitory action of 1-pyrene butyrylcholine and related compounds on choline uptake by cholinergic nerve endings. J Neurochem 27:1253–1255

Dowdall MJ, Fohlman JP, Eaker D (1977) Inhibition of high-affinity choline transport in peripheral cholinergic endings by presynaptic snake venom neurotoxins. Nature 269:700–702

Ducis I, Koepsell H (1983) A simple liposomal system to reconstitute and assay highly efficient Na^+/D-glucose cotransport from kidney brush-border membranes. Biochim Biophys Acta 730:119–129

Ducis I, Whittaker VP (1985) High-affinity, sodium-gradient-dependent transport of choline into vesiculated presynaptic plasma membrane fragments from the electric organ of *Torpedo marmorata* and reconstitution of the solubilized transporter into liposomes. Biochim Biophys Acta 815:109–127

Eisenstadt ML, Treistman SN, Schwartz JH (1975) Metabolism of acetylcholine in the nervous system of *Aplysia californica*. II. Regional localization and characterization of choline uptake. J Gen Physiol 65:275–291

Feldberg W, Fessard A (1942) The cholinergic nature of the nerves to the electric organ of the *Torpedo* (*Torpedo marmorata*). J Physiol (Lond) 101:200–216

Fisher A, Hanin I (1980) Choline analogs as potential tools in developing selective animal models of central cholinergic hypofunction. Life Sci 27:1615–1634

Friesen AJD, Ling GM, Nagai M (1967) Choline and phospholipid-choline in a sympathetic ganglion and their relationship to acetylcholine synthesis. Nature 214: 722–724

Fruton JS, Stein WH, Stahmann MA, Golumbic C (1946) Chemical reactions of the nitrogen mustard gases. VI The reactions of the nitrogen mustard gases with chemical compounds of biological interest. J Org Chem 11:571–580

Gardiner JE, Domer FR (1968) Movement of choline between the blood and cerebrospinal fluid in the cat. Arch Int Pharmacodyn Ther 175:482–496

Gill DM (1977) Mechanism of action of cholera toxin. Adv Cyclic Nucleotide Res 8:85–118

Glover VAS, Potter LT (1971) Purification and properties of choline acetyltransferase from ox brain striate nuclei. J Neurochem 18:571–580

Goldberg AM, Wecker L, Bierkamper G, Millington W, Kselzak-Redding H (1982) Acetylcholine release – interaction of choline and neuronal activity. European Symposium on Cholinergic Transmission Presynaptic Aspects, Strasbourg, p 51

Grisham CM, Barnett RE (1973) The role of lipid-phase transitions in the regulation of the (sodium+potassium) adenosine triphosphatase. Biochemistry 12:2635–2637

Grunze M, Deuticke B (1974) Changes of membrane permeability due to extensive cholesterol depletion in mammalian erythrocytes. Biochim Biophys Acta 356:125–130

Guyenet P, Lefresne P, Rossier J, Beaujouan JC, Glowinski J (1973a) Inhibition by hemicholinium-3 of [^{14}C]acetylcholine synthesis and [^{3}H]choline high-affinity uptake in rat striatal synaptosomes. Mol Pharmacol 9:630–639

Guyenet P, Lefresne P, Rossier J, Beaujouan JC, Glowinski J (1973b) Effect of sodium, hemicholinium-3 and antiparkinson drugs on [^{14}C]acetylcholine synthesis and [^{3}H]choline uptake in rat striatal synaptosomes. Brain Res 62:523–529

Haga T (1971) Synthesis and release of [^{14}C]-acetylcholine in synaptosomes. J Neurochem 18:781–798

Haga T, Noda H (1973) Choline uptake systems of rat brain synaptosomes. Biochim Biophys Acta 291:564–575

Hersh LB, Barker LA, Rush B (1978) Effect of sodium chloride on changing the rate-limiting step in the human placental choline acetyltransferase reaction. J Biol Chem 253:4966–4970

Hirsch MJ, Wurtman RJ (1978) Lecithin consumption increases acetylcholine concentrations in rat brain and adrenal gland. Science 202:223–225

Holden JT, Rossier J, Beaujouan JC, Guyenet P, Glowinski J (1975) Inhibition of high-affinity choline transport in rat striatal synaptosomes by alkyl bisquaternary ammonium compounds. Mol Pharmacol 11:19–27

Holmgren J, Månsson J-E, Svennerholm L (1974) Tissue receptor for cholera exotoxin: structural requirements of G_{M1} ganglioside in toxin binding and inactivation. Med Biol 52:229–233

Holmgren J, Lönnroth I, Månsson J-E, Svennerholm L (1975) Interaction of cholera toxin and membrane G_{M1} ganglioside of small intestine. Proc Natl Acad Sci USA 72:2520–2524
Hubbard JI (1970) Mechanism of transmitter release. Prog Biophys Mol Biol 21:33–124
Hyman BT, Spector AA (1982) Choline uptake in cultured human Y79 retinoblastoma cells: effect of polyunsaturated fatty acid compositional modifications. J Neurochem 38:650–656
Ivy MT, Sukumar R, Townsel JG (1985) The characterization of a sodium-dependent high affinity choline uptake system unassociated with acetylcholine biosynthesis. Comp Biochem Physiol 81C:351–357
Jenden DJ, Choi L, Silverman RW, Steinborn JA, Roch M, Booth RA (1974) Acetylcholine turnover estimation in brain by gas chromatography/mass spectrometry. Life Sci 14:55–63
Jenden DJ, Jope RS, Weiler MH (1976) Regulation of acetylcholine synthesis: does cytoplasmic acetylcholine control high affinity choline uptake? Science 194:635–637
Jope RS (1979) High affinity choline transport and acetyl CoA production in brain and their roles in the regulation of acetylcholine synthesis. Brain Res Rev 1:313–344
Jope RS, Jenden DJ (1977) Synaptosomal transport and acetylation of choline. Life Sci 20:1389–1392
Jope RS, Jenden DJ (1980) The utilization of choline and acetyl coenzyme A for the synthesis of acetylcholine. J Neurochem 35:318–325
Jope RS, Weiler MH, Jenden DJ (1978) Regulation of acetylcholine synthesis: control of choline transport and acetylation in synaptosomes. J Neurochem 30:949–954
Kahlenberg A, Banjo B (1972) Involvement of phospholipids in the D-glucose uptake activity of isolated human erythrocyte membranes. J Biol Chem 247:1156–1160
Kaita AA, Goldberg AM (1969) Control of acetylcholine synthesis – the inhibition of choline acetyltransferase by acetylcholine. J Neurochem 16:1185–1191
Kanner BI (1978) Active transport of γ-aminobutyric acid by membrane vesicles isolated from rat brain. Biochemistry 17:1207–1211
Kanner BI (1980) Modulation of neurotransmitter transport by the activity of the action potential sodium ion channel in membrane vesicles from rat brain. Biochemistry 19:692–697
Kanner BI, Bendahan A (1982) Binding order of substrates to the sodium and potassium ion coupled L-glutamic acid transporter from rat brain. Biochemistry 21:6327–6330
Kanner BI, Kifer L (1981) Efflux of γ-aminobutyric acid by synaptic plasma membrane vesicles isolated from rat brain. Biochemistry 20:3354–3358
Kanner BI, Marva E (1982) Efflux of L-glutamate by synaptic plasma membrane vesicles isolated from rat brain. Biochemistry 21:3143–3147
Kanner BI, Sharon I (1978a) Active transport of L-glutamate by membrane vesicles isolated from rat brain. Biochemistry 17:3949–3953
Kanner BI, Sharon I (1978b) Solubilization and reconstitution of the L-glutamic acid transporter from rat brain. FEBS Lett 94:245–248
Katz HS, Salehmoghaddam S, Collier B (1973) The accumulation of radioactive acetylcholine by a sympathetic ganglion and by brain: failure to label endogenous stores. J Neurochem 20:569–579
Kelly RB, Deutsch JW, Carlson SS, Wagner JA (1979) Biochemistry of neurotransmitter release. Annu Rev Neurosci 2:399–446
Keyes SR, Rudnick G (1982) Coupling of transmembrane proton gradients to platelet serotonin transport. J Biol Chem 257:1172–1176
Keynes RD, Martins-Ferreira H (1953) Membrane potentials in the electroplates of the electric eel. J Physiol (Lond) 119:315–351
King RG (1984) Choline uptake by lipid liposomes and its diffusion across chloroform are saturable and inhibited by hemicholinium-3. Gen Pharmacol 15:419–421
King RG, Dawson S (1985) Effects of dexamethasone on [^{3}H]choline uptake by rat forebrain synaptosomes. Naunyn-Schmiedebergs Arch Pharmacol 331:117–118
King RG, Marchbanks RM (1980) Solubilization of the choline transport system and reincorporation into artificial membranes. Nature 287:64–65

King RG, Marchbanks RM (1982) The incorporation of solubilized choline-transport activity into liposomes. Biochem J 204:565–576

King RG, Sharp JA (1985) Choline transport across a carbon tetrachloride phase containing a chloroform-methanol extract of brain. Biochim Biophys Acta 815:505–509

Kuhar MJ, Murrin LC (1978) Sodium-dependent, high affinity choline uptake. J Neurochem 30:15–21

Kuhar MJ, Simon JR (1974) Acetylcholine uptake: lack of association with cholinergic neurons. J Neurochem 22:1135–1137

Kuhar MJ, Zarbin MA (1978) Synaptosomal transport: a chloride dependence for choline, GABA, glycine and several other compounds. J Neurochem 31:251–256

Kuhar MJ, Sethy VH, Roth RH, Aghajanian GK (1973) Choline: selective accumulation by central cholinergic neurons. J Neurochem 20:581–593

Kuhar MJ, De Haven RN, Yamamura HI, Rommelspacher H, Simon JR (1975) Further evidence for cholinergic habenulo-interpeduncular neurons: pharmacologic and functional characteristics. Brain Res 97:265–275

Lanks K, Somers L, Papirmeister B, Yamamura H (1974) Choline transport by neuroblastoma cells in tissue culture. Nature 252:476–478

Leeden RW, Mellanby J (1977) Gangliosides as receptors for bacterial toxins. In: Bernheimer AW (ed) Perspectives in toxinology. Wiley, New York, pp 16–42

Leeuwin RS, Veldsema-Currie RD, van Wilgenburg H, Ottenhof M (1981) Possible direct effects of corticosteroids in myasthenia gravis. In: Grob D (ed) Myasthenia gravis: pathophysiology and management. Ann NY Acad Sci 377:840–841

Levi G, Raiteri M (1974) Exchange of neurotransmitter amino acid at nerve endings can stimulate high affinity uptake. Nature 250:735–737

Limbird LE, Lefkowitz RJ (1976) Adenylate cyclase-coupled beta-adrenergic receptors: effect of membrane lipid-perturbing agents on receptor binding and enzyme stimulation by catecholamines. Mol Pharmacol 12:559–567

Lindborg B, Crona K, Dahlbom R (1984) Troxonium-like inhibitors of the high affinity uptake of choline in mouse brain synaptosomes *in vitro*. Acta Pharm Suec 21:271–294

Manaker S, Wieczorek CM, Rainbow TC (1986) Identification of sodium-dependent, high-affinity choline uptake sites in rat brain with [^{3}H]hemicholinium-3. J Neurochem 46:483–488

Martin K (1972) Extracellular cations and the movement of choline across the erythrocyte membrane. J Physiol (Lond) 224:207–230

Masland RH, Mills JW (1980) Choline accumulation by photoreceptor cells of the rabbit retina. Proc Natl Acad Sci USA 77:1671–1675

Massarelli R, Ciesielski-Treska J, Ebel A, Mandel P (1974a) Choline uptake in glial cell cultures. Brain Res 81:361–363

Massarelli R, Ciesielski-Treska J, Ebel A, Mandel P (1974b) Kinetics of choline uptake in neuroblastoma clones. Biochem Pharmacol 23:2857–2865

Massarelli R, Wong TY, Harth S, Louis JC, Freysz L, Dreyfus H (1982a) Possible role of sialocompounds in the uptake of choline into synaptosomes and nerve cell cultures. Neurochem Res 7:301–316

Massarelli R, Gorio A, Dreyfus H (1982b) Influx, metabolism and efflux of choline in nerve cells and synaptosomes; role of sialocompounds and glycoconjugates. J Physiol (Paris) 78:392–398

Mayor F Jr, Marvizón JG, Aragón MC, Gimenez C, Valdivieso F (1981) Glycine transport into plasma-membrane vesicles derived from rat brain synaptosomes. Biochem J 198:535–541

McLaughlin S (1977) Electrostatic potentials at membrane-solution interfaces. Curr Top Membr Transp 9:71–144

Meldolesi MF, Fishman PH, Aloj SM, Ledley FD, Lee G, Bradley RM, Brady RO, Kohn LD (1977) Separation of the glycoprotein and ganglioside components of thyrotropin receptor activity in plasma membranes. Biochem Biophys Res Commun 75:581–588

Meyer EM, Cooper JR (1982) High-affinity choline transport in proteoliposomes derived from rat cortical synaptosomes. Science 217:843–845

Meyer EM, Cooper JR (1983) High affinity choline uptake and calcium-dependent acetylcholine release in proteoliposomes derived from rat cortical synaptosomes. J Neurosci 3:987–994
Miljanich GP, Brasier AR, Kelly RB (1982) Partial purification of presynaptic plasma membrane by immunoadsorption. J Cell Biol 94:88–96
Miller AL, Chaptal C, McEwen BS, Peck EJ Jr (1978) Modulation of high affinity GABA uptake into hippocampal synaptosomes by glucocorticoids. Psychoneuroendocrinology 3:155–164
Mitchell P (1963) Molecule, group and electron translocation through natural membranes. Biochem Soc Symp 22:142–168
Morel N, Manaranche R, Israël M, Gulik-Krzywicki T (1982) Isolation of a presynaptic plasma membrane fraction from *Torpedo* cholinergic synaptosomes: evidence for a specific protein. J Cell Biol 93:349–356
Mulder AH, Yamamura HI, Kuhar MJ, Snyder SH (1974) Release of acetylcholine from hippocampal slices by potassium depolarization: dependence on high affinity choline uptake. Brain Res 70:372–376
Murrin LC, Kuhar MJ (1976) Activation of high-affinity choline uptake *in vitro* by depolarizing agents. Mol Pharmacol 12:1082–1090
Murrin LC, De Haven RN, Kuhar MJ (1977) On the relationship between [^{3}H]choline uptake activation and [^{3}H]acetylcholine release. J Neurochem 29:681–687
Narahashi T, Anderson NC, Moore JW (1966) Tetrodotoxin does not block excitation from inside the nerve membrane. Science 153:765–767
Neckers L, Sze PY (1975) Regulation of 5-hydroxytryptamine metabolism in mouse brain by adrenal glucocorticoids. Brain Res 93:123–132
Nelson PJ, Rudnick G (1979) Coupling between platelet 5-hydroxytryptamine and potassium transport. J Biol Chem 254:10084–10089
Nelson PJ, Rudnick G (1982) The role of chloride ion in platelet serotonin transport. J Biol Chem 257:6151–6155
Ohta M, Narahashi T, Keeler RF (1973) Effects of veratrum alkaloids on membrane potential and conductance of squid and crayfish giant axons. J Pharmacol Exp Ther 184:143–154
Oide M (1982) Uptake and metabolism of choline by the embryonic heart of the chick *in vitro*. Comp Biochem Physiol 72B:493–499
O'Regan S, Collier B, Israël M (1982) Studies on presynaptic cholinergic mechanisms using analogues of choline and acetate. J Physiol (Paris) 78:454–460
Pedata F, Lo Conte G, Sorbi S, Marconcini-Pepeu I, Pepeu G (1982) Changes in high affinity choline uptake in rat cortex following lesions of the magnocellular forebrain nuclei. Brain Res 233:359–367
Pedata F, Giovannelli L, Pepeu G (1984) GM1 ganglioside facilitates the recovery of high-affinity choline uptake in the cerebral cortex of rats with a lesion of the nucleus basalis magnocellularis. J Neurosci Res 12:421–427
Pedder EK, Prince AK (1984) The reaction of rat brain choline acetyltransferase (ChAT) with ethylcholine mustard aziridinium (ECMA) and phenoxybenzamine (PB). Br J Pharmacol [Suppl] 81:134P
Pedder EK, Prince AK (1985) Ethylcholine mustard aziridinium: inhibition of rat brain synaptosome choline uptake and choline acetyltransferase. Br J Pharmacol [Suppl] 84, 114P
Pert CB, Snyder SH (1974) High affinity transport of choline into the myenteric plexus of guinea-pig intestine. J Pharmacol Exp Ther 191:102–108
Phillips AT (1977) Differential labeling: a general technique for selective modification of binding sites. Methods Enzymol 46:59–69
Polak RL, Molenaar PC, van Gelder M (1977) Acetylcholine metabolism and choline uptake in cortical slices. J Neurochem 29:477–485
Pu GA, Masland RH (1984) Biochemical interruption of membrane phospholipid renewal in retinal photoreceptor cells. J Neurosci 4:1559–1576
Rainbow TC, Parsons B, Wieczorek CM (1984) Quantitative autoradiography of [^{3}H]hemicholinium-3 binding sites in rat brain. Eur J Pharmacol 102:195–196

Richardson PJ (1981) Quantitation of cholinergic synaptosomes from guinea pig brain. J Neurochem 37:258–260

Richelson E, Thompson EJ (1973) Transport of neurotransmitter precursors into cultured cells. Nature [New Biol] 241:201–204

Riker DK, Sastre A, Baker T, Roth RH, Riker WF Jr (1979) Regional high-affinity [^{3}H]choline accumulation in cat forebrain: selective increase in the caudate-putamen after corticosteroid pretreatment. Mol Pharmacol 16:886–899

Ringdahl B, Jope RS, Jenden DJ (1984) Inhibition of high affinity choline transport by stereoisomers of some 3-quinuclidinol derivatives. Biochem Pharmacol 33:2819–2822

Ross SB, Jenden DJ (1973) Failure of hemicholinium-3 to inhibit the uptake of ^{3}H-choline in mouse brain *in vivo.* Experientia 29:689–690

Rubin RP (1970) The role of calcium in the release of neurotransmitter substances and hormones. Pharmacol Rev 22:389–428

Rylett BJ, Colhoun EH (1979) The interactions of choline mustard aziridinium ion with choline acetyltransferase. J Neurochem 32:553–558

Rylett BJ, Colhoun EH (1980) Kinetic data on the inhibition of high-affinity choline transport into rat forebrain synaptosomes by choline-like compounds and nitrogen mustard analogues. J Neurochem 34:713–719

Rylett RJ, Colhoun EH (1982) Monoethylcholine mustard aziridinium ion, a cholinergic neurochemical probe: comparison with choline mustard aziridinium ion. Soc Neurosci 8:773 (Abstract 221.7)

Rylett RJ, Colhoun EH (1984) An evaluation of irreversible inhibition of synaptosomal high-affinity choline transport by choline mustard aziridinium ion. J Neurochem 43:787–794

Saelens JK, Simke JP, Allen MP, Conroy CA (1973) Some of the dynamics of choline and acetylcholine metabolism in rat brain. Arch Int Pharmacodyn Ther 203:305–312

Sandberg K, Coyle JT (1985) Characterization of [^{3}H]hemicholinium-3 binding associated with neuronal choline uptake sites in rat brain membranes. Brain Res 348:321–330

Schuberth J, Sparf B, Sundwall A (1969) A technique for the study of acetylcholine turnover in mouse brain *in vivo.* J Neurochem 16:695–700

Schueler FW (1955) A new group of respiratory paralyzants. I. The "hemicholiniums." J Pharmacol Exp Ther 115:127–143

Schultz SG, Curran PF (1970) Coupled transport of sodium and organic solutes. Physiol Rev 50:637–718

Sharkawi M (1972) Effects of some centrally acting drugs on acetylcholine synthesis by rat cerebral cortex slices. Br J Pharmacol 46:473–479

Shelton DL, Nadler JV, Cotman CW (1979) Development of high affinity choline uptake and associated acetylcholine synthesis in the rat fascia dentata. Brain Res 163:263–275

Sherman KA, Gibson G, Blass J (1984) Human red blood cell choline uptake with age and Alzheimer's disease. Age (Omaha) 7:145–146 (Abstract 47)

Shih YH, Pugsley TA (1985) The effects of various cognition-enhancing drugs on *in vitro* rat hippocampal synaptosomal sodium dependent high affinity choline uptake. Life Sci 36:2145–2152

Shinitzky M (1984) Membrane fluidity and cellular functions. In: Shinitzky M (ed) Physiology of membrane fluidity, vol 1. CRC, Boca Raton, pp 2–39

Shinitzky M, Barenholz Y (1974) Dynamics of the hydrocarbon layer in liposomes of lecithin and sphingomyelin containing dicetylphosphate. J Biol Chem 249:2652–2657

Shinitzky M, Rivnay B (1977) Degree of exposure of membrane proteins determined by fluorescence quenching. Biochemistry 16:982–986

Simon JR, Kuhar MJ (1975) Impulse-flow regulation of high affinity choline uptake in brain cholinergic nerve terminals. Nature 255:162–163

Simon JR, Kuhar MJ (1976) High affinity choline uptake: ionic and energy requirements. J Neurochem 27:93–99

Simon JR, Martin DL, Kroll M (1974) Sodium-dependent efflux and exchange of GABA in synaptosomes. J Neurochem 23:981–991

Simon JR, Mittag TW, Kuhar MJ (1975) Inhibition of synaptosomal uptake of choline by various choline analogs. Biochem Pharmacol 24:1139–1142

Simon JR, Atweh S, Kuhar MJ (1976) Sodium-dependent high affinity choline uptake: a regulatory step in the synthesis of acetylcholine. J Neurochem 26:909–922
Slater P, Stonier PD (1973a) The uptake of hemicholinium-3 by rat brain cortex slices. J Neurochem 20:637–639
Slater P, Stonier PD (1973b) The uptake of hemicholinium-3 by rat diaphragm and isolated perfused heart. Arch Int Pharmacodyn 204:407–414
Smart L (1981) Hemicholinium 3-bromo mustard: a new high affinity inhibitor of sodium-dependent high affinity choline uptake. Neuroscience 6:1765–1770
Sorimachi M, Miyamoto K, Kataoka K (1974) Postnatal development of choline uptake by cholinergic terminals in rat brain. Brain Res 79:343–346
Speth RC, Yamamura HI (1979) Sodium-dependent high-affinity neuronal choline uptake. In: Barbeau A, Growdon JH, Wurtman RJ (eds) Nutrition and the brain, vol 5. Raven, New York, pp 129–139
Stadler H, Tashiro T (1979) Isolation of synaptosomal plasma membranes from cholinergic nerve terminals and a comparison of their proteins with those of synaptic vesicles. Eur J Biochem 101:171–178
Staerk J, Ronneberger HJ, Wiegandt H, Ziegler W (1974) Interaction of ganglioside $G_{Gtet}1$ and its derivatives with choleragen. Eur J Biochem 48:103–110
Stefanović V, Massarelli R, Mandel P, Rosenberg A (1975) Effect of cellular desialylation on choline high affinity uptake and ecto-acetylcholinesterase activity of cholinergic neuroblasts. Biochem Pharmacol 24:1923–1928
Sukumar R, Townsel JG (1983) Selective protection of high affinity [^{3}H]choline uptake sites by hemicholinium-3 against N-ethylmaleimide binding in *Limulus* synaptosomes. Comp Biochem Physiol 75C:317–320
Suszkiw JB, Pilar G (1976) Selective localization of a high affinity choline uptake system and its role in ACh formation in cholinergic nerve terminals. J Neurochem 26:1133–1138
Suszkiw JB, Beach RL, Pilar GR (1976) Choline uptake by cholinergic neuron cell somas. J Neurochem 26:1123–1131
Sze PY, Marchi M, Towle AC, Giacobini E (1983) Increased uptake of [^{3}H]choline by rat superior cervical ganglion: an effect of dexamethasone. Neuropharmacology 22:711–716
Theuvenet APR, Borst-Pauwels GWFH (1976) The influence of surface charge on the kinetics of ion-translocation across biological membranes. J Theor Biol 57:313–329
Towle AC, Sze PY (1982) ^{3}H-corticosterone binding in rat superior cervical ganglion. Brain Res 253:221–229
Towle AC, Sze PY (1983) Steroid binding to synaptic plasma membrane: differential binding of glucocorticoids and gonadal steroids. J Steroid Biochem 18:135–143
Vaca K, Pilar G (1979) Mechanisms controlling choline transport and acetylcholine synthesis in motor nerve terminals during electrical stimulation. J Gen Physiol 73:605–628
Vallano ML, Spoerlein MT, Vanderwende C (1982) Morphine enhances high-affinity choline uptake in mouse striatum. Pharmacol Biochem Behav 17:419–423
Veldsema-Currie RD, Wolters EChMJ, Leeuwin RS (1976) The effect of corticosteroids and hemicholinium-3 on choline uptake and incorporation into acetylcholine in rat diaphragm. Eur J Pharmacol 35:399–402
Veldsema-Currie RD, van Marle J, Langemeijer MWE, Lind A, van Weeren-Kramer J (1984) Radioactive choline uptake in the isolated rat phrenic nerve-hemidiaphragm preparation: a biochemical and autoradiographic study. J Neurochem 43:1032–1038
Veldsema-Currie RD, van Marle J, Langemeijer MWE, Lind A, van Weeren-Kramer J (1985) Dexamethasone affects radioactive choline uptake in rat diaphragm nerve endings and not in muscle fibres. Brain Res 327:340–343
Vickroy TW, Fibiger HC, Roeske WR, Yamamura HI (1984a) Reduced density of sodium-dependent [^{3}H]hemicholinium-3 binding sites in the anterior cerebral cortex of rats following chemical destruction of the nucleus basalis magnocellularis. Eur J Pharmacol 102:369–370

Vickroy TW, Roeske WR, Yamamura HI (1984b) Sodium-dependent high-affinity binding of [^{3}H]hemicholinium-3 in the rat brain: a potentially selective marker for presynaptic cholinergic sites. Life Sci 35:2335–2343
Vyas S, O'Regan S (1985) Reconstitution of carrier-mediated choline transport in proteoliposomes prepared from presynaptic membranes of *Torpedo* electric organ, and its internal and external ionic requirements. J Memb Biol 85:111–119
Weiler MH, Jope RS, Jenden DJ (1978) Effect of pretreatment under various cationic conditions on acetylcholine content and choline transport in rat whole brain synaptosomes. J Neurochem 31:789–796
Weiler MH, Gundersen CB, Jenden DJ (1981) Choline uptake and acetylcholine synthesis in synaptosomes: investigations using two different labeled variants of choline. J Neurochem 36:1802–1812
Wheeler DD (1979a) A model of high affinity choline transport in rat cortical synaptosomes. J Neurochem 32:1197–1213
Wheeler DD (1979b) A model of high affinity glutamic acid transport by rat cortical synaptosomes – a refinement of the originally proposed mode. J Neurochem 33:883–894
Wheeler DD (1980) A model for GABA and glutamic acid transport by cortical synaptosomes. Pharmacology 21:141–152
White HL, Wu JC (1973) Kinetics of choline acetyltransferases (E.C. 2.3.1.6) from human and other mammalian central and peripheral nervous tissues. J Neurochem 20:297–307
Whittaker VP, Dowdall MJ (1975) Current state of research on cholinergic synapses. In: Waser PG (ed) Cholinergic mechanisms. Raven, New York, pp 23–42
Whittaker VP, Dowdall MJ, Boyne AF (1972) The storage and release of acetylcholine by cholinergic nerve terminals: recent results with nonmammalian preparations. Biochem Soc Symp 36:49–68
Wold F (1977) Affinity labeling – an overview. Methods Enzymol 46:3–14
Yamamura HI, Snyder SH (1972) Choline: high affinity uptake by rat brain synaptosomes. Science 178:626–628
Yamamura HI, Snyder SH (1973) High affinity transport of choline into synaptosomes of rat brain. J Neurochem 21:1355–1374
Yamamura HI, Kuhar MJ, Greenberg D, Snyder SH (1974) Muscarinic cholinergic receptor binding: regional distribution in monkey brain. Brain Res 66:541–546
Zelinski TA, Choy PC (1984) Ethanolamine inhibits choline uptake in the isolated hamster heart. Biochim Biophys Acta 794:326–332
Zisapel N, Sokolovsky M (1977) Affinity labeling of receptors. Methods Enzymol 46:572–582

CHAPTER 15

Cholinergic-Specific Antigens

V. P. WHITTAKER and E. BORRONI

A. Introduction

I. General Strategy

In recent years immunochemical techniques have begun to play a much larger role in neurobiology than formerly. They have been applied in various ways and for different purposes. Antibodies raised against components specific for a particular type of nerve cell may be used to trace its location and connections within the nervous system. Antibodies raised against known transmitter-specific components may be used to trace pathways utilizing neurons of this particular transmitter type. In the cholinergic system, choline acetyltransferase (ChAT) is the obvious component to select, and both polyclonal and monoclonal antibodies which are specific for ChAT are now available. In Chap. 22 the results of such studies are reviewed. Such methods have tended to replace histochemical methods which rely on a chemical reation; thus methods based on ChAT immunohistochemistry have tended to displace older methods which detected acetylcholinesterase (AChE), in itself not always a reliable cholinergic marker, by precipitating the thioacetate generated by the enzymic hydrolysis of acetylthiocholine as the copper salt.

Antibodies may also be raised to membrane- or organelle-specific components. Such antibodies may then be used not only to provide new empirical immunohistochemical methods based on such organelles which, in favourable circumstances, may also be transmitter-specific; they may also be used in functional studies to investigate such problems as the genesis, recycling and retrieval of the organelle or membrane and, by applying immunoadsorption techniques, to develop new methods of subcellular fractionation.

This chapter will summarize recent work on these lines in the cholinergic system. The strategies used are summarized in Fig. 1. In the cholinergic system, we are particularly fortunate in having, in the electromotor system of *Torpedo,* an abundant source of purely cholinergic neurons. Methods have been developed (Fig. 1, box a; and Chap. 2) for isolating a variety of membrane preparations and organelles from various parts of the electromotor system (cell bodies from the electric lobe, axons, terminals in the electric organ) which can be used as a source of potentially cholinergic-specific antigens. Specific antigens may be isolated and antisera or monoclonal antibodies may be raised to whole membranes or organelles, assessed with respect to their specificity for the membrane or organelle in question and the antigen(s) subsequently identified.

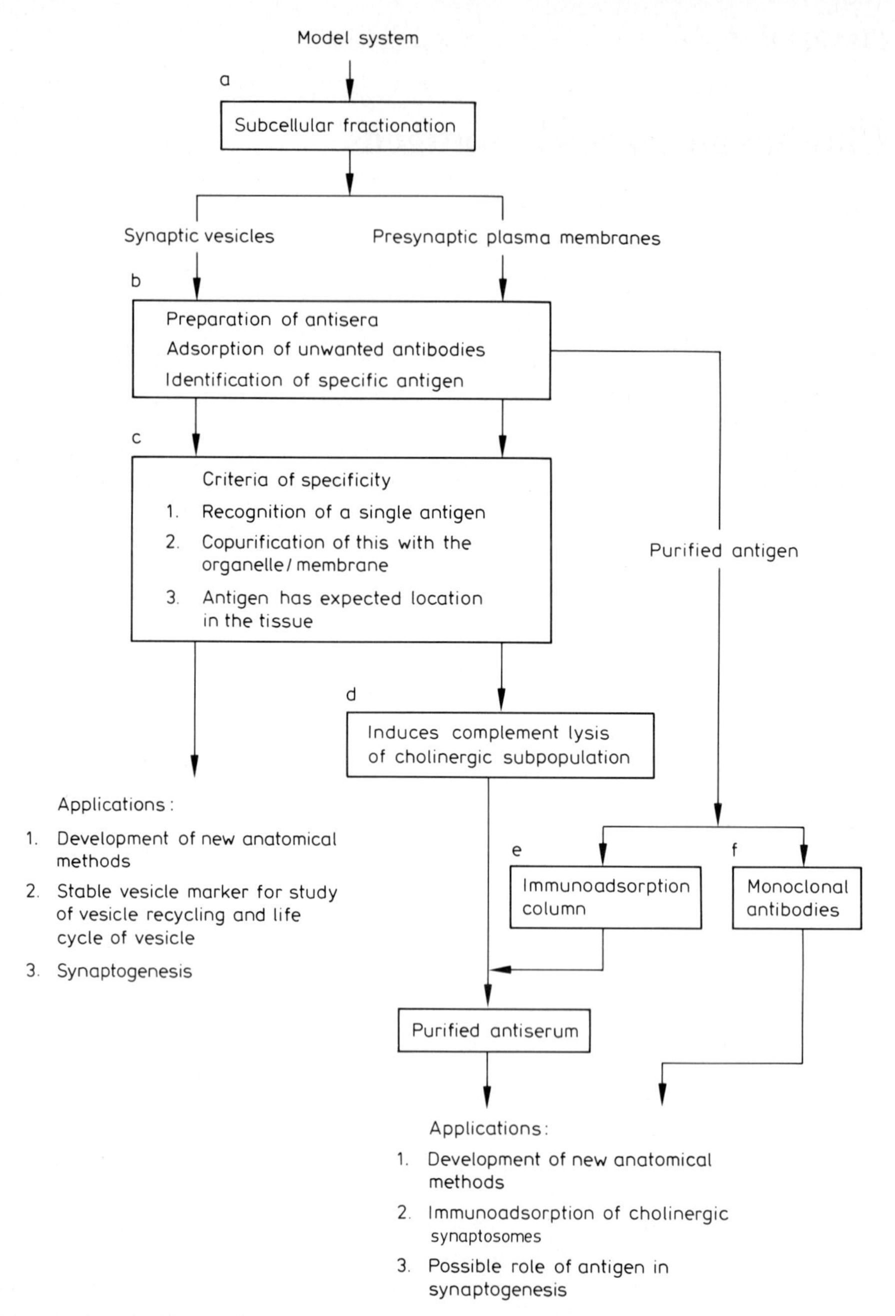

Fig. 1. Diagram illustrating the strategy of the research of the Göttingen group described in this chapter. (From WHITTAKER et al. 1985)

II. The Problem of Unwanted Antibodies

In such work a potent source of difficulty may be the presence, in biochemically speaking relatively insignificant amounts, of highly antigenic contaminants. The non-specific antibodies generated by such contaminants may sometimes be removed from polyclonal antibody preparations by adsorption (Fig. 1, box b). However, stringent tests must be applied to establish that the antibody or antiserum is specific for the membrane or organelle against which it has been raised. One important criterion (Fig. 1, box c) is copurification of the antigen with the organelle or membrane in question, another that the antigen has the expected location in the tissue. Here, the highly polarized structure of the electric organ is a distinct advantage: a putative nerve terminal-specific antiserum should react only with the densely innervated ventral surface of the electrocyte and not with the non-innervated dorsal surface.

When the antigens are known and are available in pure or enriched form or have been subsequently identified and purified, they may be immobilized and used for the separation of immunoglobulins enriched in antibodies to them by the technique of immunoaffinity chromatography (Fig. 1, box e). This is particularly useful for raising the titre of antibodies raised to weak antigens. Alternatively, the purified antigen may be used to raise monospecific polyclonal or monoclonal antibodies (Fig. 1, box f).

III. Complement Lysis Test

A useful technique for detecting transmitter-related antigens present on the surface of nerve terminals is that of selective complement-induced lysis of synaptosomes (Fig. 1, box d; Fig. 2). Synaptosomes are osmotically sensitive, sealed structures, comparable to miniature non-nucleated cells. In the presence of an antibody recognizing a surface antigen, components of the complement system are attached and lysis ensues. If the antigen is common to all terminal plasma membranes, lysis will be complete, but if it is restricted to a subpopulation of the total synaptosome population, partial lysis will occur. This will be detectable by partial release of a common cytoplasmic marker. This may be an endogenous marker

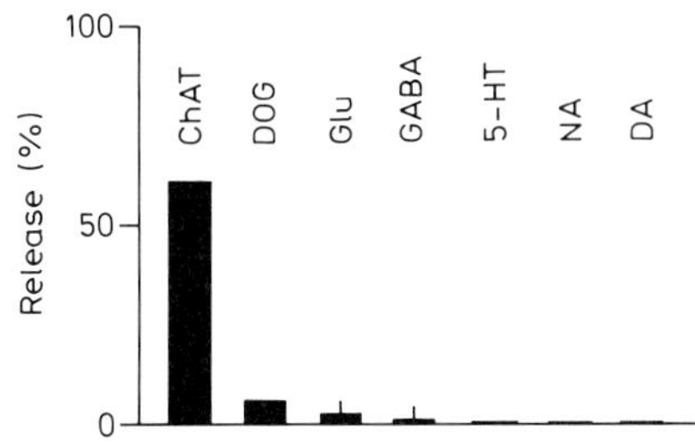

Fig. 2. Selective complement lysis of the cholinergic subpopulation of guinea-pig cortical synaptosomes. Synaptosomes were loaded beforehand with the general cytoplasmic marker deoxyglucose (*DOG*) or with glutamata (*Glu*), γ-aminobutyrate (*GABA*), 5-hydroxytryptamine (*5-HT*), noradrenaline (*NA*) and dopamine (*DA*). Only the cholinergic marker ChAT is released as a result of anti-Chol-1 induced complement lysis. (Results of RICHARDSON 1983)

such as lactate dehydrogenase or an exogenous one, such as labelled deoxyglucose (DOG). Synaptosomes may be preloaded with DOG (RICHARDSON 1983) and with transmitter-specific markers such as the labelled transmitters themselves or their precursors, which are in general taken up by high-affinity transporters confined to the subpopulation utilizing the transmitter in question. Addition of antigen to the antiserum reduces its effectiveness in stimulating complement-mediated lysis. The shift in the lysis curve that ensues may be made the basis of an assay for the antigen which is useful in monitoring purification.

B. Surface Antigens Specific for Cholinergic Nerve Terminals

I. Identification of Cholinergic-Specific Gangliosides

The technique just described has been successfully used to detect cholinergic-specific antigens with epitopes common to *Torpedo* and mammals.

Electromotor presynaptic plasma membranes were isolated from *Torpedo* electric organs and injected into sheep in the presence of Freund's adjuvant (WALKER et al. 1982). The resulting antisera recognized many components of these plasma membranes, among them cholinesterase (ChE). Immunohistochemical studies on electric organ showed that the antisera contained antibodies which recognized membrane components not confined to presynaptic plasma membranes. Adsorption with electric organ homogenates removed these non-specific antibodies as well as ChE and several other presynaptic plasma membrane components. The specificity of the antisera for mammalian cholinergic-specific surface antigens was shown by their ability both to induce the selective complement lysis of the cholinergic subpopulation of mammalian synaptosomes (JONES et al. 1981; RICHARDSON 1981; RICHARDSON et al. 1982, and Fig. 2) and selectively to stain both peripheral and putative central cholinergic nerve terminals in immunofluorescence histochemical studies (JONES et al. 1981). When the fimbria were cut, there was a decline in the number of fluorescent nerve terminals in the hippocampus on the operated side as the cholinergic input degenerated, which was well correlated with a concomitant fall in ChAT (Fig. 3).

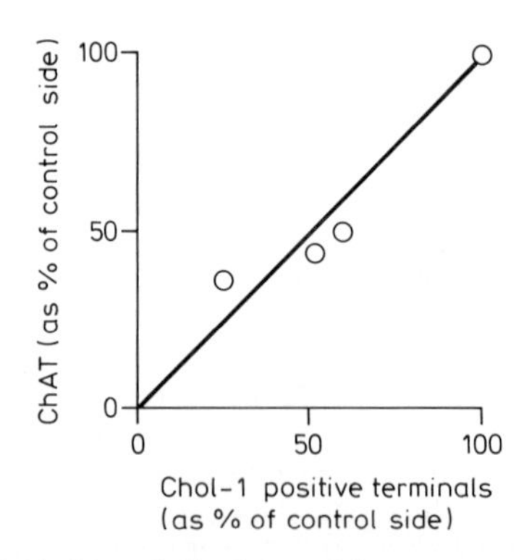

Fig. 3. Concomitant loss of ChAT and Chol-1-positive nerve terminals in the hippocampus of the rat after sectioning the fimbria. (From JONES et al. 1981)

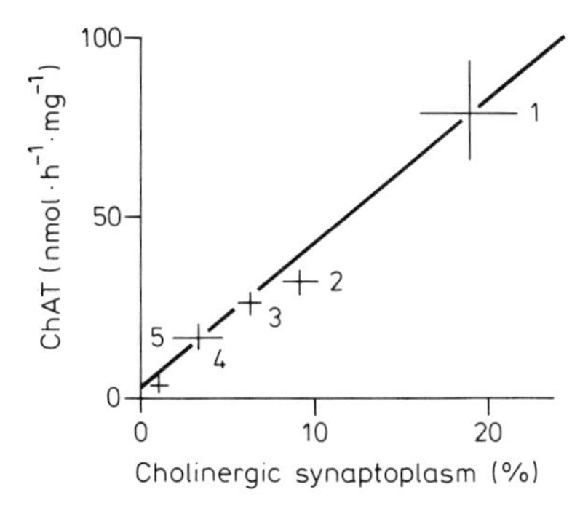

Fig. 4. ChAT content and proportion of synaptic cytoplasm attributable to cholinergic nerve terminals in different regions of the guinea-pig brain (plotted from RICHARDSON 1981). *1,* caudate nucleus; *2,* hippocampus; *3,* cortex; *4,* medulla and pons; *5,* cerebellum

The proportion of a general cytoplasmic marker, lactate dehydrogenase, released in the complement lysis procedure, is a measure of the proportion of synaptosomes derived from cholinergic nerve terminals in the total synaptosome population. This varies considerably from one brain region to another but is well correlated with the ChAT content of the various regions (RICHARDSON 1981, and Fig. 4). There is no release from appropriately labelled non-cholinergic synaptosomes (RICHARDSON 1983; and Fig. 2).

Further work showed that the antiserum does not precipitate protein from solubilized mammalian cerebral cortical synaptosomes and that the antigen is gangliosidic in nature (RICHARDSON et al. 1982, *Torpedo;* FERRETTI and BORRONI 1986, guinea pig). In *Torpedo* electric organ several gangliosides are inhibitory in the complement lysis test and are immunoreactive, but in mammals immunoreactivity is mainly confined to two minor gangliosides now designated Chol-1 α and β. In thin-layer chromatography (TLC) Chol-1 α runs behind the tetrasialoganglioside fraction (GQ) and Chol-1 β behind the trisialoganglioside (GT 1b) fractions (Fig. 5).

The recognition of these gangliosides by immunostaining of TLC plates and by immunocytochemical staining of tissue sections has been facilitated by the purification and concentration of the anti-Chol-1 antibodies from the crude antiserum by passing it through a column of *Torpedo* electric organ gangliosides bound to alkylamine-glass beads. The anti-*Torpedo* electric organ ganglioside antibodies are selectively adsorbed and after washing the beads free of unwanted serum components the adsorbed antibodies are eluted with KCNS.

II. Applications of Anti-Chol-1 Antisera

1. Immunocytochemical Identification of Cholinergic Neurons

The purified antibodies now stain not only nerve terminals but also the axons and cell bodies of mammalian central cholinergic neurons (OBROCKI 1987; OBROCKI and BORRONI 1988). In sections of rat spinal cord (Fig. 6a–c), three densely

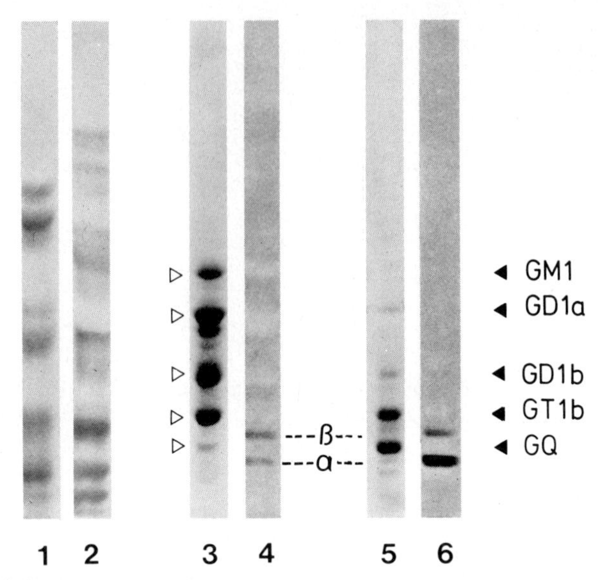

Fig. 5. Distribution of (*lanes 1, 3, 5*) Ehrlich-positive gangliosides of (*lane 1*) *Torpedo* electric organ (*lanes 3, 5*) guinea-pig brain and (*lanes 2, 4, 6*) immunoblots showing distribution of Chol-1-positive gangliosides. *Lanes 5* and *6* are from a guinea-pig brain preparation enriched in polysialogangliosides. Note (a) the enrichment in Chol-1 *α* along with other polysialogangliosides in *lane 6;* (b) the numerous Chol-1 components in *Torpedo;* (c) the non-identity between the various Chol-1 components and the main Ehrlich-positive ganglioside fractions, indicated *on the right* and by *arrows*

stained areas are seen. One of these, the area of the ventral horns, contains numerous motor neuron somata; these are densely stained, as are their dendrites and axons, cross-sections of which are numerous in the sections. The terminals of recurrent collaterals may also be seen clustered on small unstained interneurons, presumably Renshaw cells. Another area owes its dense staining to the presence of heavily stained cell bodies of cholinergic preganglionic autonomic neurons. The third area is rich in cholinergic spinal afferents which are also densely stained. Densely Chol-1 positive cells are found in other areas rich in cholinergic neurons, such as the diagonal band of Broca and the median forebrain bundle (Fig. 6d, e).

2. Quantitative Assay of Cholinergic Representation in Different Brain Areas

When the amount of selective complement-mediated lysis of cholinergic synaptosomes is measured for different brain areas, marked differences are found. The extent of the release of the cytoplasmic marker is a measure of the proportion of total terminal cytoplasm accounted for by cholinergic nerve terminals. This is well correlated with the proportion of cholinergic innervation as shown by the level of ChAT in the area (RICHARDSON 1981; and Fig. 4).

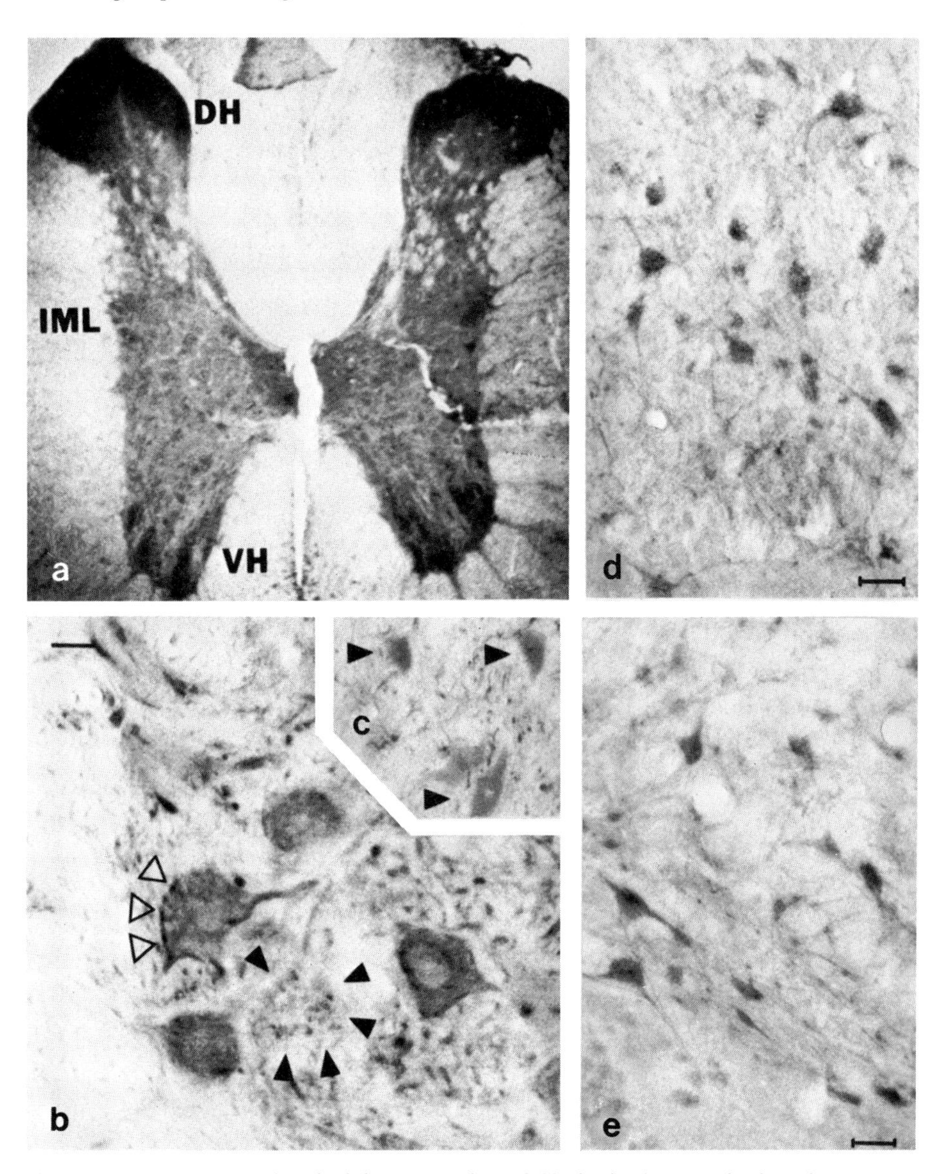

Fig. 6a–e. Immunocytochemical demonstration of Chol-1 in the rat spinal cord. **a** Intense staining of ventral horn cells (*VH*), preganglionic autonomic neuron somata in the intermediate lateral nuclei (*IML*), spinal afferents in dorsal horns (*DH*). **b** Somata of α-motoneurons with nerve terminals of recurrent collaterals (*outline arrows*), unstained, non-cholinergic cell bodies with cholinergic terminals, identified as Renshaw cells (*filled arrows*). **c** Smaller somata of γ-motoneurons without nerve terminals of recurrent collaterals. **d** Putative cholinergic neuron somata in the diagonal band of Broca. **e** Similar cells in the median forebrain bundle

3. Separation of Cholinergic Synaptosomes by Immunoadsorption

When anti-Chol-1 antiserum is added to a brain synaptosome preparation, the antiserum binds selectively to cholinergic synaptosomes and causes these to adhere to immobilized antiserum directed against the IgG of the anti-Chol-1 antiserum. The availability of a high-titre anti-sheep IgG monoclonal antibody immobilized on cellulose enabled cholinergic synaptosomes to be separated from non-cholinergic (RICHARDSON et al. 1984). Such synaptosomes are enriched in the high-affinity choline uptake system (Chap. 14, this volume; RICHARDSON 1986) and have been used to study of the role of ATP in cholinergic synaptic activity (RICHARDSON and BROWN 1987). The topic is fully dealt with in Chap. 12.

4. A New Test for Co-localization?

There is much evidence that at least a proportion of cholinergic nerve terminals in the mammalian nervous system contain other neuroactive substances, specifically neuropeptides (Chap. 17). Affinity purification of cholinergic synaptosomes and their selective complement lysis induced by anti-Chol-1 antisera should permit co-localization to be tested.

These tests have been tried with respect to vasoactive intestinal polypeptide (VIP) (AGOSTON et al. 1988). VIP copurified with ChAT when cholinergic synaptosomes were purified by immunoadsorption as described above. It was estimated that 75% of brain VIP was present in cholinergic endings and that 50% of such endings contained VIP. Complement-induced lysis mediated by anti-Chol-1 antiserum was not as effective in releasing VIP as ChAT or ACh but that would be accounted for if VIP were packaged in a sub-set of vesicles which were too large to escape from the complement-permeabilized synaptosomes and unlike the ACh-containing vesicles were stable under the lysis conditions.

5. Are There Gangliosides or Other Surface Markers Specific for Other Transmitters?

The discovery of Chol-1 has depended on (a) the availability of a rich source of purely cholinergic nerve terminals and (b) the conservation of the epitope between a lower vertebrate and mammals. These conditions may not be easy to fulfil in the case of other transmitters, though the electric lobe of *Torpedo* might furnish a pure source of glutamergic terminals (Chap. 2). However, the monoclonal antibody technique might provide a means of obtaining other transmitter-specific antibodies. Monoclonal antibodies should be raised to synaptosomal plasma membranes and screened for their ability to induce selective complement lysis of other subpopulations of synaptosomes. Some indication that this approach might be successful is provided by the work of DODD and JESSELL (1985) which demonstrates that dorsal root ganglion cells exhibit transmitter-specific carbohydrates that can be recognized by monoclonal antibodies.

III. Possible Functional Significance of Chol-1

1. Expression During Synaptogenesis and Effect of Denervation

The appearance of Chol-1 during synaptogenesis has been studied in the electromotor synapse of *Torpedo* (FIEDLER et al. 1986; and Fig. 7). Staining first becomes apparent at 93-mm embryo length, thence increases steadily in intensity as neurite coverage of the ventral (innervated) surface of the electrocyte increases. The electromotor axons begin to invade the inter-electrocyte spaces at 55-mm embryo length. The formation of the highly distinctive mature presynaptic structure does not occur until later, at the 90 mm stage. Electric organ discharges in response to presynaptic stimulation begin to be detected at 80–90 mm (KRENZ et al. 1980; and Chaps. 2, 19 this volume). In the rat, Chol-1 is detected in intercostal muscles and diaphragm at 3 days postnatal, again at a time when mature synapses are forming (E. DERRINGTON and E. BORRONI, personal communication). These authors have also shown that when reactive synaptogenesis is induced in the cholinergic input to the dentate gyrus of the hippocampus by interrupting

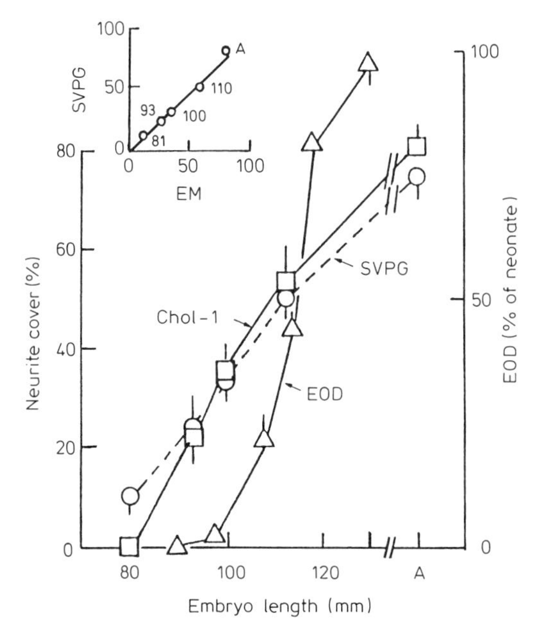

Fig. 7. Neurite coverage of the ventral surfaces of *Torpedo* electrocytes obtained by linear analysis following staining with anti Chol-1 antibodies (*squares* and *continuous lines*) and anti-vesicle proteoglycan (*SVPG*) antiserum (*circles* and *dashed lines*). *Triangles* show magnitude of electric organ discharge (*EOD*), a measure of functional synaptogenesis. Although Chol-1 was detected later than vesicular proteoglycan, the closeness of the two curves suggests that both begin to be expressed just before functional synaptogenesis occurs and are necessary concomitants of the formation of mature functional synapses. *Insert,* regression plot of neurite coverage as measured by linear analysis of electron micrographs (FOX and KÖTTING 1984) (*EM, abscissae*) and anti-vesicular proteoglycan staining (*SVPG, ordinates*) (Data of FIEDLER et al. 1986). Small figures are embryo lengths in mm. Correlation coefficient is 0.99

the non-cholinergic entorhinal input the Chol-1 content of the dentate gyrus significantly increases but not that of the rest of the hippocampus. The expression of Chol-1 seems therefore to be associated with the maturation of synapses.

Another clue to the possible function of Chol-1 is provided by the results of denervation experiments (FERRETTI and BORRONI 1984). Here it was found that while the Chol-1 epitope did not disappear with the degeneration of the presynaptic nerve terminals, it was no longer associated primarily with the more highly sialidated ganglioside fractions nor with the ventral, innervated surface of the electrocyte. It is known from *in vitro* studies (RICHARDSON et al. 1982) that the ability of Chol-1 to inhibit the complement lysis of mammalian synaptosomes is not impaired by desialidation with neuraminidase. It is also known that gangliosides have a high affinity for fibronectin, a component of the basal lamina. It has thus been suggested (FERRETTI and BORRONI 1984) that Chol-1 is effectively a bivalent component of the presynaptic plasma membrane; it is anchored to the membrane by its lipid portions and to the basal lamina by its putative interaction with fibronectin. On disappearance of the plasma membrane, Chol-1 remains anchored to the basal lamina, but is now free to spread over the entire surface of the electrocyte; at the same time it becomes vulnerable to neuraminidase action.

2. Possible Role of Chol-1 in Synaptogenesis

The foregoing suggests that the function of Chol-1 is to consolidate cholinergic synapse formation by interacting with receptor molecules on the target cell. If these were Chol-1-specific, exploring non-cholinergic terminals would be unable to develop synapses on non-appropriate cholinoceptive target areas. There is however also the possibility that Chol-1 is itself a receptor, possibly for a cholinotrophic factor. Such a factor has been identified in embryonic electric organ beginning at 80 mm (RICHARDSON et al. 1985; and Chaps. 2, 5, this volume).

3. Possible Role in Neuropathology

Antibodies to Chol-1 recognize an immunoreactively identical substance in the cholinergic neurons of human spinal cord; this immunoreactivity is decreased in amyotrophic lateral sclerosis (ALS) and is accompanied by severe degeneration of the cell bodies of cholinergic motoneurons in the ventral horn (J. OBROCKI and E. BORRONI, personal communication). The possibility must now be considered that a derangement in the expression of Chol-1 may be involved in some way in the aetiology of a group of neurological disorders (e.g. ALS, Alzheimer's disease) in which the primary target apprears to be central cholinergic neurons. Clearly, much more work will be required to establish the role of Chol-1, if any, in the aetiology of such diseases.

IV. Complement-Mediated Lysis of Synaptosomes Induced by Anti-ChAT Antisera

BRADFORD and coworkers (DOCHERTY et al. 1985a, b) have obtained selective complement lysis of subpopulations of cholinergic, GABAergic, serotonergic and dopaminergic synaptosomes by the addition of antisera to the soluble cytoplasmic enzymes ChAT, glutamate decarboxylase, tryptophan hydroxylase and tyrosine hydroxylase, respectively. This is at first sight a surprising result, since there is no reason to believe that the epitopes of cytoplasmic enzymes concerned with transmitter synthesis would be expressed on the outside of the synaptosomal plasma membrane – a sine qua non for the initiation of complement-induced lysis. It is, however, possible that the antisera used are not completely specific for the enzymes against which they have been raised but recognize an epitope or epitopes which are common to other functional proteins involved in the metabolism or storage of the transmitter concerned and which *are* expressed on the surface of the synaptosomal plasma membrane. Such components could have escaped detection when the antisera were evaluated for their specificity by immunoblotting and immunocytochemical methods. In the case of cholinergic synaptosomes a possible candidate for such a functional protein would be the high-affinity choline uptake (HACU) system which resembles ChAT in having a binding site for choline.

It is therefore of considerable interest that in a recent publication DOCHERTY and BRADFORD (1986) have shown that two anti-ChAT antisera which induce complement lysis of cholinergic synaptosomes also inhibit the HACU system. It is also of interest that the anti-ChAT antisera induce the release of about 10% of the general cytoplasmic marker, lactate dehydrogenase, about the proportion attributable to cholinergic synaptosomes in the total synaptosome population, but release only 27% of ChAT; comparable figures for anti-Chol-1 antisera are 6% and 85% or more. DOCHERTY and BRADFORD (1986) explain this by postulating that the released ChAT reacts with antiserum and is sedimented with it when the lysis reaction is terminated by centrifuging, thus reducing the amount of ChAT recorded as having been released.

C. Synaptic Vesicle Proteoglycans

I. From *Torpedo marmorata*

Antibodies have been raised to highly purified synaptic vesicles prepared from the cholinergic electromotor nerve terminals of *Torpedo marmorata* (WALKER et al. 1982, 1983; JONES et al. 1982a). The most antigenic component is a proteoglycan consisting of a protein backbone conjugated to a glycosaminoglycan of the heparan sulphate type (STADLER and DOWE 1982; and Chap. 11, this volume). The proteoglycan is rather heterogeneous with respect to molecular mass as measured by its migration in sodium dodecylsulphate polyacrylamide gel electrophoresis (SDS-PAGE): part of it enters the gel producing a rather broad band of $M_r \sim 200000$; the rest is retained in the stacking gel ($M_r > 300000$). Recent work (KIENE 1986) suggests that the higher molecular mass material ($M_r > 300000$) is preferentially localized in immature vesicles (the VP_0 fraction) that have recently arrived in the terminal by fast axonal transport. Recruitment of these vesicles into the VP_1 and VP_2 pools may involve the release of some glucosaminoglycan. The staining pattern obtained in immune blots of SDS-PAGE gels and the distribution pattern of the antigen in zonal-density gradients of isolated synaptic vesicles tends to vary, according to which forms of proteoglycan are preferentially recognized by the antiserum employed.

The epitope of the vesicular proteoglycan is directed to the core of the vesicle, because anti-proteoglycan antibodies do not precipitate vesicles unless they are lysed (STADLER and DOWE 1982). Copurification with vesicles (Fig. 1, box c) has been demonstrated (WALKER et al. 1983; ÁGOSTON et al. 1986; KIENE 1986; KIENE and STADLER 1987).

Antibodies to vesicular proteoglycan have proved useful in immunofluorescence cytochemical studies of vesicular genesis, axonal transport and vesicle recycling at the terminal. The crude anti-vesicle antiserum was adsorbed with electric organ supernatant and liver membranes to remove antibodies to non-vesicular components such as AChE (cf. Fig. 1, box b). To allow the purified antiserum access to the antigen, it was necessary to treat the tissue with detergent. Under these circumstances the vesicular proteoglycan was detected in the cell bodies of the electromotor neurons in the electric lobe (Chap. 2) as punctate spots in the cytoplasm, with accumulations in the axon hillocks, in the electromotor axons and in the nerve terminals (JONES et al. 1982a; AGOSTON and CONLON 1986). Ligating the axons caused an accumulation of immunoreactive material above the ligature (JONES et al. 1982a; AGOSTON and CONLON 1986) and a reduction of staining intensity in the electromotor terminals (AGOSTON and CONLON 1986). The fast transport of the vesicular proteoglycan was demonstrated by ^{35}S-labelling (KIENE 1986; KIENE and STADLER 1987).

In the absence of detergent treatment the response of the electromotor terminals to the adsorbed antiserum is little above background. A strong reaction is however obtained with tissue stimulated under conditions known to cause extensive vesicle exocytosis and vesicle depletion (JONES et al. 1982a). At rest, for time intervals known to allow recovery in vesicle numbers, the antigen is reinternalized but is re-externalized and again reinternalized during a second cycle of stimulation and recovery (JONES et al. 1982b).

In developmental studies of the formation of electromotor synapses (FIEDLER et al. 1986), vesicular proteoglycan was detected immunocytochemically before Chol-1 at a time (80 mm embryo length) 1–2 weeks earlier than the formation of mature synapses but when synaptic vesicles are beginning to appear in considerable numbers in the electromotor terminals. Thereafter the time course of neurite coverage measured either by anti-vesicle proteoglycan or anti-Chol-1 antisera was quite similar.

Other types of cholinergic vesicle cross-react with the *Torpedo* electric organ vesicular proteoglycan antibodies: motoneuron terminals in *Torpedo* back muscle and rat diaphragm (WALKER et al. 1982) and isolated synaptic vesicles from the electromotor nerve terminals of the electric eel *Electrophorus electricus,* electric catfish *Malapterurus* and rat diaphragm (VOLKNANDT and ZIMMERMANN 1986). However, with the possible exception of retina (OSBORNE et al. 1982) the antibodies have not proved to be directed exclusively against cholinergic-specific components in the mammalian CNS (C.W. ZIMMERMANN, personal communication). They react with an electrocyte basal lamina component which may be vesicle-derived (STADLER and KIENE 1987) or simply expressing a common epitope (cf. the *Discopyge* electric organ vesicular glycoprotein recognized by the SV4 monoclonal antibody discussed in the next section). The vesicular proteoglycan is also a constituent of a larger, denser vesicle present in *Torpedo* electromotor neurons which stores a VIP-like neuropeptide (D.V. ÁGOSTON, personal communication). Nevertheless, when one considers the ubiquitous nature of proteoglycans, the relatively restricted distribution of the epitopes of the *Torpedo* electromotor vesicle proteoglycan and its high concentration in other cholinergic vesicles is impressive.

II. From Other Torpedine Fish

In work which to some extent parallels that of the Göttingen group in its approach. CARLSON, KELLY and associates have raised an antiserum against synaptic vesicle from *Narcine brasiliensis* (CARLSON and KELLY 1980) and several monoclonal antibodies against vesicles from *Torpedo californica* (CARLSON and KELLY 1983) and *Discopyge ommata* (CARONI et al. 1985; and Table 1) in efforts to identify cholinergic-specific vesicle components. The antigen recognized by the anti-vesicle antiserum is a proteoglycan which copurifies with synaptic vesicles and whose epitope is directed to the vesicle core. The antiserum also recognizes a similar or identical immunoreactive substance in terminals at the amphibian neuromuscular junction (SANES 1979; HOOPER et al. 1980; VON WEDEL 1981) and in the superior cervical ganglion, in those of parasympathetic postganglionic neurons and in AChE-positive regions of the hippocampus and cerebellum; it is, however, absent from the molecular layer of the cerebellum and the laminae of the

Table 1. Anti-synaptic vesicle antibodies from *Torpedo californica, Narcine* and *Discopyge* electric organ synaptic vesicles

Species of torpedine fish	Monoclonal antibody or antiserum	Code	Antigen	$M_r \times 10^{-3}$	Reference[a]
Narcine	AS	–	PG	–	1–5
T. californica	MAB	tor 70	PG	100–200	6, 7
Discopyge	MAB	SV 1	PG	–	8
	MAB	SV 2	GP	100	8, 9
	MAB	SV 3	–	–	8
	MAB	SV 4	GP	400 2000	8, 10

Abbreviations: PG, proteoglycan; GP, glycoprotein; SV, synaptic vesicle; MAB, monoclonal antibody; AS, antiserum

[a] 1 VON WEDEL et al. (1978)
2 CARLSON and KELLY (1980)
3 HOOPER et al. (1980)
4 DAHLSTRÖM et al. (1985)
5 SANES et al. (1979)
6 CARLSON and KELLY (1983)
7 BUCKLEY et al. (1983)
8 CARONI et al. (1985)
9 BUCKLEY and KELLY (1985)
10 CARLSON et al. (1986)

dentate gyrus that receive hippocampal and commissural input (HOOPER et al. 1980). Adrenal medullary cells also show a positive response. The antiserum is thus clearly not fully cholinergic-specific; on the other hand not all types of terminal react with it.

The antiserum has been used to demonstrate La^{3+}-induced vesicle exocytosis in frog muscle (von WEDEL et al. 1981) but did not detect stimulus-induced exocytosis and reinternalization of the antigen as was possible with the *T. marmorata* antiserum (JONES et al. 1982a, b). It has also been used to show antero- and retrograde axonal transport of the epitope in ligated sympathectomized rat sciatic nerves (DAHLSTRÖM et al. 1985).

Of the antigens recognized by the five monoclonal antibodies reported on by this group, the vesicular proteoglycan recognized by tor 70 seems to correspond most closely to that described by WALKER et al. (1983). Its copurification with synaptic vesicles, the core location of its epitope, its M_r of 100000–200000 and the heparan sulphate-like nature of its constituent glycosaminoglycan (CARLSON and KELLY 1983; BUCKLEY et al. 1983) are all points of resemblance. However, the epitope is also expressed on the outer surface of the presynaptic plasma membrane; in motor nerve terminals and adrenal medullary cells, stimulation increases the amount of epitope associated with the plasma membrane (BUCKLEY et al. 1983) but as reported, this phenomenon does not seem to be comparable to the transient stimulus-induced incorporation of the *T. marmorata* vesicle-specific proteoglycan into the plasma membrane reported by JONES et al. (1982a, b). Indeed BUCKLEY et al. (1983) postulate that the function of the antigen in the plasma membrane is to interact with the basal lamina or postsynaptic membrane in such a way as to keep the presynaptic release sites in register with receptor clusters on the postsynaptic side.

The epitopes recognized by four monoclonal antibodies raised against *Discopyge* electric organ synaptic vesicles are divergent in nature and location (CARONI et al. 1985; BUCKLEY and KELLY 1985; CARLSON et al. 1986). Those recognized by antibodies SV1, SV3 and SV4 are located in the vesicle core and that by SV2 is on the surface. The most fully studied is the SV4 epitope. About 90% of this is not associated with the synaptic vesicle at all but with a large glycoprotein of the extracellular matrix. It has been suggested that exocytosing synaptic vesicles supply this material, designated a 'nerve terminal anchorage protein', to the basal lamina, but it is far from established that such an exchange actually takes place since the molecular mass of the synaptic vesicle material is estimated to be only one-fifth that of the extracellular material; possibly the two antigens simply share a common epitope (CARLSON and WIGHT 1987).

The SV2 epitope has also been studied (BUCKLEY and KELLY 1985). It has been classified as a glycoprotein and is not cholinergic-specific; it has a wide distribution and recognizes all synaptic vesicles and endocrine secretory granules tested, though it is absent from exocrine cells. The antigen copurifies with *Discopyge* electric organ synaptic vesicles and has a molecular mass of about 65 kDa.

The above work well illustrates the complexity, power and limitations of the monoclonal antibody approach to cell structure and function. Large numbers of antibodies can be raised against a subcellular organelle, but it is often difficult to know if the epitopes they recognize are specific for the organelle used to generate them or not. When the epitope is present in other more or less functionally related structures, it is difficult to tell by immunochemical or immunocytochemical methods alone whether there is a precursor-product or other functional relationship between the related structures or not. There seems little doubt that proteoglycans are conveyed, in cholinergic neurons, from cell body to terminal in synaptic vesicles and possibly also in synaptic vesicle-like structures that are not classical synaptic vesicles, i.e. are not transmitter-storing organelles. There is also good evidence that proteoglycans are exteriorized onto the presynaptic plasma membrane and that at least part of this process is stimulus-dependent and reversed during recovery. The ultimate fate of the vesicular proteoglycan is less certain. Part may be transferred to the extracellular matrix, but the evidence for this is not at present very convincing. It also seems that the specificity of vesicular proteoglycan for cholinergic vesicles is only relative and that all available mono- and polyclonal antibodies derived from electromotor vesicles are rather unselective in the mammalian nervous system. Nevertheless the results do point to some interesting classifications and family relationships between diverse storage organelles and to some possibly functionally significant interactions between synaptic vesicles, plasma membranes and extracellular matrix. Much more work and a broader spectrum of techniques, including pulse labelling with radioactive precursors, will be required to clarify these.

References

Agoston DV, Conlon MJ (1986) Presence of VIP-like immunoreactivity in the cholinergic electromotor system of *Torpedo marmorata.* J Neurochem 47:445–453

Agoston DV, Dowe GHC, Fiedler W, Giompres PE, Roed IS, Walker JH, Whittaker VP, Yamaguchi T (1986) The use of synaptic vesicle proteoglycan as a stable marker in kinetic studies of vesicle recycling. J Neurochem 47:1584–1592

Agoston DV, Borroni E, Richardson PJ (1988) Cholinergic surface antigen Chol-1 in present in a subclass of VIP-containing rat cortical synaptosomes. J Neurochem (in press)

Buckley K, Kelly RB (1985) Identification of a transmembrane glycoprotein specific for secretory vesicles of neural and endocrine cells. J Cell Biol 100:1284–1294

Buckley KM, Schweitzer ES, Miljanich GP, Clift-O'Grady L, Kushner PD, Reichardt LF, Kelly RB (1983) A synaptic vesicle antigen is restricted to the junctional region of the presynaptic plasma membrane. Proc Natl Acad Sci USA 80:7342–7346

Carlson SS, Kelly RB (1980) An antiserum specific for cholinergic synaptic vesicles from electric organ. J Cell Biol 87:98–103

Carlson SS, Kelly RB (1983) A highly antigenic proteoglycan-like component of cholinergic synaptic vesicles. J Biol Chem 258:11082–11091

Carlson SS, Wight TN (1987) Nerve terminal anchorage protein 1 (TAP-1) is a chondroitin sulfate proteoglycan: biochemical and electron microscopic characterization. J Cell Biol 105:3075–3086

Carlson SS, Caroni P, Kelly RB (1986) A nerve terminal anchorage protein from electric organ. J Cell Biol 103:509–520

Caroni P, Carlson SS, Schweitzer E, Kelly RB (1985) Presynaptic neurones may contribute a unique glycoprotein to the extracellular matrix at the synapse. Nature 314:441–443

Dahlström A, Larsson PA, Carlson SS, Bööj S (1985) Localization and axonal transport of immunoreactive cholinergic organelles in rat motor neurons – an immunofluorescent study. Neuroscience 14:607–625

Docherty M, Bradford HF (1986) A cell-surface antigen of cholinergic nerve terminals recognized by antisera to choline acetyltransferase. Neurosci Lett 70:234–238

Docherty M, Bradford HF, Cash CD, Maitre M (1985a) Specific immunolysis of serotonic nerve terminals using an antiserum against tryptophan hydroxylase. FEBS Lett 182:489–492

Docherty M, Bradford HF, Wu JY, Joh TH, Reis DJ (1985b) Evidence for specific immunolysis of nerve terminals using antisera against choline acetyltransferase, glutamate decarboxylase and tyrosine hydroxylase. Brain Res 339:105–113

Dodd J, Jessell TM (1985) Lactoseries carbohydrates specify subsets of dorsal root ganglion neurons projecting to the superficial dorsal horn of rat spinal cord. J Neurosci 5:3278–3294

Ferretti P, Borroni E (1984) Effect of denervation on a cholinergic-specific ganglioside antigen (Chol-1) present in *Torpedo* electromotor presynaptic plasma membrane. J Neurochem 42:1085–1093

Ferretti P, Borroni E (1986) Putative cholinergic-specific gangliosides in guinea-pig forebrain. J Neurochem 46:1888–1894

Fiedler W, Borroni E, Ferretti F (1986) An immunohistochemical study of synaptogenesis in the electric organ of *Torpedo* using antisera to vesicular and presynaptic plasma membrane components. Cell Tissue Res 246:439–446

Fox GQ, Kötting D (1984) *Torpedo* electromotor system development: a quantitative analysis of synaptogenesis. J Comp Neurol 224:337–343

Hooper JE, Carlson SS, Kelly RB (1980) Antibodies to synaptic vesicles purified from *Narcine* electric organ bind a sub-class of mammalian nerve terminals. J Cell Biol 87:104–113

Jones RT, Walker JH, Richardson PJ, Fox GQ, Whittaker VP (1981) Immunohistochemical localization of cholinergic nerve terminals. Cell Tissue Res 218:355–373

Jones RT, Walker JH, Stadler H, Whittaker VP (1982a) Immunohistochemical localization of a synaptic vesicle antigen in a cholinergic neuron under conditions of stimulation and rest. Cell Tissue Res 223:117–126

Jones RT, Walker JH, Stadler H, Whittaker VP (1982b) Further evidence that glycosaminoglycan specific to cholinergic vesicles recycles during electrical stimulation of the electric organ of *Torpedo marmorata*. Cell Tissue Res 224:685–688

Kiene ML (1986) Charakterisierung eines Proteoglykans acetylcholinspeichernder synaptischer Vesikel und Beiträge zur Dynamik der Vesikel in der Nervenendigung. Dissertation, Göttingen

Kiene ML, Stadler H (1987) Synaptic vesicles in electromotoneurones: axonal transport, site of transmitter uptake and processing of a core proteoglycan during maturation. EMBO J 6:2209–2215

Krenz WD, Tashiro T, Wächtler K, Whittaker VP (1980) Aspects of the chemical embryology of the electromotor system of *Torpedo marmorata* with special reference to synaptogenesis. Neuroscience 5:617–624

Obrocki J (1987) Localization of Chol-1 in the central nervous system of the rat. In: Dowdall MJ, Hawthorne (eds) Cellular and molecular basis of cholinergic function. Horwood, Chichester, pp 431–435

Obrocki J, Borroni E (1988) Immunocytochemical evaluation of a putative cholinergic-specific ganglioside antigen (Chol-1) in the central nervous system of the rat. Exp Brain Res (in press)

Osborne NN, Beale R, Nicholas D, Stadler H, Walker JH, Jones RT, Whittaker VP (1982) An antiserum to cholinergic vesicles from *Torpedo* recognizes nerve terminals in retinas for a variety of species. Cell Mol Neurobiol 2:157–163

Richardson PJ (1981) Quantitation of cholinergic synaptosomes from guinea pig brain. J Neurochem 37:258–260

Richardson PJ (1983) Presynaptic distribution of the cholinergic-specific antigen Chol-1 and 5′-nucleotidase in rat brain, as determined by complement-mediated release of neurotransmitters. J Neurochem 41:640–648

Richardson PJ (1986) Choline uptake and metabolism in affinity purified cholinergic nerve terminals from rat brain. J Neurochem 46:1251–1255

Richardson PJ, Brown SJ (1987) ATP release from affinity-purified rat cholinergic nerve terminals. J Neurochem 48:622–630
Richardson PJ, Walker JH, Jones RT, Whittaker VP (1982) Identification of a cholinergic-specific antigen Chol-1 as a ganglioside. J Neurochem 38:1605–1614
Richardson PJ, Siddle H, Luzio JP (1984) Immunoaffinity purification of intact, metabolically active cholinergic nerve terminals from mammalian brain. Biochem J 219:647–654
Richardson GP, Rinschen B, Fox GQ (1985) *Torpedo* electromotor system development: developmentally regulated neuronotrophic activities of electric organ tissue. J Comp Neurol 231:339–352
Sanes JR, Carlson SS, von Wedel RJ, Kelly RB (1979) Antiserum specific for motor nerve terminals in skeletal muscle. Nature 280:403–404
Stadler H, Dowe GHC (1982) Identification of a heparan sulfate-containing proteoglycan as a specific core component of cholinergic synaptic vesicles from *Torpedo marmorata.* EMBO J 1:1381–1384
Stadler H, Kiene ML (1987) Synaptic vesicles in electromotoneurones: heterogeneity of populations is expressed in uptake properties, exocytosis and insertion of a core proteoglycan into the extracellular matrix. EMBO J 6:2217–2221
Volknandt W, Zimmermann H (1986) Acetylcholine, ATP and proteoglycan are common to synaptic vesicles isolated from the electric organs of electric eel and electric catfish as well as from rat diaphragm. J Neurochem 47:1449–1462
Von Weddel R, Carlson SS, Kelly RB (1981) Transfer of synaptic vesicle antigens to the presynaptic plasma membrane during exocytosis. Proc Natl Acad Sci USA 78:1014–1018
Walker JH, Jones RT, Obrocki J, Richardson GP, Stadler H (1982) Presynaptic plasma membranes and synaptic vesicles of cholinergic nerve endings demonstrated by means of specific antisera. Cell Tissue Res 223:101–116
Walker JH, Obrocki J, Zimmermann CW (1983) Identification of a proteoglycan antigen characteristic of cholinergic synaptic vesicles. J Neurochem 41:209–216
Whittaker VP, Borroni E, Ferretti P, Richardson GP (1985) Cholinergic-specific nerve terminal antigens. In: Reid E (ed) Investigation and exploitation of antibody combining sites. Plenum, New York, pp 189–206 (Methodological surveys in biochemistry and analysis, vol 15)

CHAPTER 16

Cholinergic False Transmitters

V. P. WHITTAKER

A. Definition and Structural Requirements

I. Definition

A false transmitter may be defined as an analogue of an endogenous chemical neurotransmitter which is incorporated into the releasable pool of transmitters and released on stimulation in the same Ca^{2+}-dependent manner as the endogenous transmitter (KOPIN 1968). Since in general the agonist action of the false transmitter will differ from that of the endogenous transmitter – normally its pharmacological potency will be lower – the presence of the false transmitter may have significant pharmacological consequences: partial blockade of transmission may ensue. This may result from the displacement of the endogenous transmitter from its storage sites by a less potent false transmitter, but competition for postsynaptic receptors may also be important.

II. Adrenergic System

False transmitters were first described for the adrenergic system (KOPIN 1968) and proved to be extremely useful for identifying the transmitter pool from which transmitter is released on stimulation. A summary of this work has been given by SMITH (1972).

A large number of analogues of noradrenaline are known which are taken up by adrenergic nerve terminals and enter the cytoplasmic pool of transmitter. From there they may enter the vesicular pool. The vesicular amine transporter has a different, effectively more restricted specificity than the plasma membrane transporter so that several amines are known that label the cytoplasmic compartment without entering the vesicles. On stimulation, only those false transmitters are released that enter the vesicular compartment, thus identifying the vesicular transmitter pool as the source of transmitter for stimulus-evoked transmitter release. Further evidence that phenylethylamines can only be released on stimulation through the vesicular compartment was the observation (for references, see SMITH 1972) that noradrenaline itself is not released from the nerve terminals of a reserpinized animal treated with a monoamine inhibitor, even though, after infusion with noradrenaline, tissue levels of the transmitter are normal. Reserpine blocks uptake into vesicles and the monoamine oxidase inhibitor stabilizes cytoplasmic noradrenaline, allowing its concentration to rise in response to infusion while the vesicles remain depleted. Under such conditions noradrenaline itself behaves like any other phenylethylamine that cannot gain access to the vesicular compartment and like them is not released on stimulation. That reserpine *per se* interferes with the release process was excluded by experiments with meteraminol, a phenylethylamine derivative which is taken up into vesicles by a reserpine-resistent process; reserpinization had little effect on the stimulus-evoked release of this substance.

Such experiments provide convincing evidence, in the adrenergic system at least, for the release of transmitters directly from the vesicular compartment.

III. Cholinergic System

1. Choline Analogues

In the cholinergic system matters are complicated by the fact that esters of choline are very poor substrates for the high-affinity choline uptake (HACU) system and, in the absence of a cholinesterase (ChE) inhibitor, are in general rapidly hydrolysed by ChEs active in the extracellular space round the terminal. Thus one is limited in practice to precursors of potential false transmitters. A number of near analogues of choline (Table 1) are fairly rapidly taken up by the HACU system (see Chap. 14) and enter the cytoplasm. At least two such choline analogues, homocholine (HC) (LUQMANI et al. 1980) and diethylHC (WELNER and COLLIER 1985), are substrates – albeit poor ones – for the vesicular uptake system, but most analogues studied are not effective substrates for vesicular uptake unless they are first acetylated. This is effected by the enzyme choline acetyltransferase (ChAT) (Chap. 7). This enzyme has a fairly broad specificity for near-choline analogues, which, however, differs from that of the HACU system (Table 2). Several of the acetylcholine analogues that result are substrates for the vesicular uptake system. Thus an acetate ester in the cholinergic system can only gain access to the vesicle compartment if its precursor satisfies the specificity requirements of both (a) the HACU system and (b) ChAT, and if (c) it satisfies the specificity of the vesicular uptake system.

An unexpected finding is that the kinetics of the acetylation of HC and diethylHC by ChAT *in situ* differ from those of the purified enzyme. Both bases are acetylated *in situ* but are acetylated more slowly (*Torpedo*, HC, LUQMANI and RICHARDSON 1982) or not at all (brain, HC, COLLIER et al. 1977; BOKSA and COLLIER 1980a; diethyl HC, WELNER and COLLIER 1984) by solubilized, partially purified ChAT.

2. Acyl-Group Variants

Many esters of choline other than acetylcholine (ACh) are pharmacologically active (for review, see WHITTAKER 1963); some of them, interestingly enough, occur in nature (though not in nervous tissue) (for review, see WHITTAKER 1963, 1973), and tissues possess enzymes which synthesize them, including some esters whose natural occurrence has not been demonstrated (BERRY and WHITTAKER 1959).

Some of these compounds may be potential false transmitters. The limiting factors will, once again, be the capacity of the cholinergic nerve terminal to take

Table 1. Some cholinergic false transmitter precursors

Structure			Trivial name (code)	Preparations tested	References[a]
R_1	R_2	R_3			
Choline analogues ($R_1R_2R_3N^+CH_2CH_2OH$)					
Et	Me	Me	Monoethylcholine (MEC)	ED SCG NMJ B	1 2, 3 4 5, 6
Et	Et	Me	Diethylcholine (DEC)	EO SCG B	1 3, 5 5, 6, 7
Et	Et	Et	Triethylcholine (TEC)	EO SCG B	1 3, 8, 9 5, 6, 9
CH_2- | CH_2- (R₁R₂)		Me	Pyrrolidinecholine (pyrrolcholine; PyC)	EO SCG MP B	10 2, 3 11 12, 13
NH_2	Me	Me	*N*-aminodeanol	B	14
Homocholine and analogues ($R_1R_2R_3N^+CH_2CH_2CH_2OH$)					
Me	Me	Me	Homocholine (HC)	EO SCG B	3, 15 2, 16 6, 13, 16, 17
Et	Me	Me	Monoethylhomocholine	SCG B	18 19
Et	Et	Me	Diethylhomocholine	SCG B	18 6
Et	Et	Et	Triethylhomocholine	SCG B	18 19

Abbreviations: SCG, superior cervical ganglion; EO, electric organ; NMJ, neuromuscular junction; B, brain; MP, myenteric plexus

[a] 1 WHITTAKER and LUQMANI (1980); ROED (1986)
2 COLLIER et al. (1976)
3 COLLIER (1979)
4 COLQUHOUN et al. (1977); LARGE and RANG (1978a, b)
5 BOKSA and COLLIER (1980b)
6 COLLIER and WELNER (1986)
7 CHIOU (1974)
8 ILSON and COLLIER (1975)
9 ILSON et al. (1977)
10 ZIMMERMANN and DOWDALL (1977)
11 KILBINGER (1977)
12 VON SCHWARZENFELD (1978)
13 VON SCHWARZENFELD et al. (1979); VON SCHWARZENFELD (1979)
14 NEWTON and JENDEN (1986)
15 LUQMANI et al. (1980)
16 COLLIER et al. (1977)
17 BOKSA and COLLIER (1980a)
18 WELNER and COLLIER (1985)
19 WELNER and COLLIER (1984)

up the requisite precursor organic acid and of a choline acyltransferase to acylate choline with it. Since all the known choline acyltransferases accept the acyl moiety only in the form of its coenzyme A derivative, an added complication is the ability of the terminal to generate the requisite acylcoenzyme A. This is extremely restricted.

The only acylcholine other than ACh that has been proposed as a false transmitter is propionylcholine. This is pharmacologically quite active and rapidly hydrolysed by ChEs. Propionate is taken up, rather inefficiently, by electric organ synaptosomes (O'REGAN 1983, 1984). O'REGAN (1982) claimed that electric organ contains endogenous propionylcholine and that the concentration can be increased by the administration of propionate. This is surprising: propionylcholine was not detected in an earlier, careful characterization of the endogenous transmitter by SHERIDAN et al. (1966) and non-ruminants lack the means of converting propionate to propionylcoenzyme A. A careful reappraisal of O'REGAN's findings using gas chromatography-mass spectrometry has failed to confirm them (KOSH and WHITTAKER 1985).

B. Differential Labelling of Transmitter Pools by False Transmitters

I. Electric Organ

1. False Transmitters Are Taken up by Recycling Vesicles but Not by Reserve Vesicles

In Chaps. 2 and 11 it has been pointed out that under appropriate circumstances, vesicles caused to recycle by stimulation (referred to in kinetic studies as the loading stimulus) may be separated from the reserve vesicles by making use of physical methods of separation based on differences in density (isopycnic density-gradient centrifugation; ZIMMERMANN and WHITTAKER 1977; ZIMMERMANN and DENSTON 1977a, b) or size (chromatography through columns of porous glass beads; GIOMPRES et al. 1981a).

The recycling vesicles may be additionally characterized by their ability – not possessed by the reserve vesicles – to take up ACh newly synthesized in the cytoplasm by cytoplasmic ChAT. The newly synthesized ACh is identified by the use of radiolabelled (ZIMMERMANN and DENSTON 1977b; GIOMPRES et al. 1981a, b) or deuterated (KOSH and WHITTAKER 1985; ÁGOSTON et al. 1986) precursors. These, if perfused through a block of tissue, are rapidly incorporated into the continuously recycling cytoplasmic compartment (Chap. 1, Fig. 2), but the newly synthesized, labelled ACh does not appreciably exchange with the ACh of the reserve vesicles. Only when vesicles are caused to recycle by a loading stimulus can the labelled ACh enter them, and it is confined to this subpopulation.

During a subsequent period of rest, the recycled vesicles gradually reacquire the biophysical characteristics of the reserve pool (GIOMPRES et al. 1981b). If, before this happens, the tissue is restimulated (referred to in kinetic studies as the release stimulus), the newly synthesized ACh is preferentially released, and its specific radioactivity (isotopic ratio) matches that of the recycling vesicle population (SUSZKIW et al. 1978). This is a confirmation of an important prediction of the 'vesicle theory'; however, in itself, it does not exclude the possibility that the evoked release could have been directly from the cytoplasmic compartment, since the ACh in this compartment is in approximate isotopic equilibrium with that of

Table 2. Comparison of the specificity of HACU and ChAT

Compound[b]	Preparation[c]	HACU[a]			ChAT[a]		
		$1/K_T$	V_{max}	Reference[d]	$1/K_m$	V_{max}	Reference[d]
MEC	T	27 ± 1 (3)	52 ± 1 (3)	1	–	–	
	B	37	–	2	14 to 50	67 to 100	2
DEC	T	17 ± 1 (7)	24 ± 1 (7)	1	–	–	
	B	40	–	2	2	70	2
TEC	T	6 ± 1 (4)	10 ± 1 (4)	1	–	–	
	B	20	–	2	0.6	70	2
PyC	T	32	67	3	–	–	
	B	18	–	2	2	53	2
HC	T	57	62	4	–	–	
	B	33	–	2	–	0	2
Monoethyl HC	B	95	30	5	–	–	
Diethyl HC	B	64	23	5	–	–	
Triethyl HC	B	–	0	5	–	–	
N-aminodeanol	B	48	33	6	–	–	

[a] Values are either single values, ranges or means ± SD (no. of experiments in parentheses) and are expressed as percentages of the values for choline. Those for ChAT are for the enzyme of mammalian brain
[b] Abbreviations: MEC, monoethylcholine; DEC, diethylcholine; TEC, triethylcholine; PyC, pyrrolidinecholine; HC, homocholine
[c] Preparations are synaptosomes prepared from *Torpedo* electric organ (T) or mammalian brain (B)
[d] References: 1 ROED (1980, 1986); 2 data collected by COLLIER et al. (1979); 3 ZIMMERMANN and DOWDALL (1977); 4 LUQMANI et al. (1980); 5 WELNER and COLLIER (1984); 6 NEWTON and JENDEN (1986)

the recycling vesicles and presumably has the same or nearly the same specific radioactivity.

The use of false transmitters provides a way out of this ambiguity. Acetate esters of choline analogues, being derived from an exogenous precursor, are of necessity 'recently synthesized', and they too are preferentially taken up by the subpopulation of recycling vesicles (Fig. 1; and LUQMANI et al. 1980). To distinguish between the cytoplasmic and recycling vesicular pools it is simply necessary to find a false transmitter which has a distribution between cytoplasm and recycling vesicles different from that of ACh. In practice, this requires one that is a poorer substrate for vesicular uptake than ACh and accumulates in the cytoplasm, since no false transmitter so far known is a better substrate than ACh for vesicular uptake. The ensuing large difference in the ratios in which the false transmitter and ACh are present in the cytoplasmic and recycling vesicular pools facilitates a clear-cut answer when these ratios are matched with that in which the transmitters are released. By using labelled ACh, the necessity for separating the recycling and reserve pools is avoided since the ratio of labelled ACh to false transmitter determined on the total vesicle fraction will be that of the recycling vesicles. It is also possible to utilize two false transmitters, both of which accumulate in the cytoplasm, but one of which is a much better substrate for vesicular uptake than

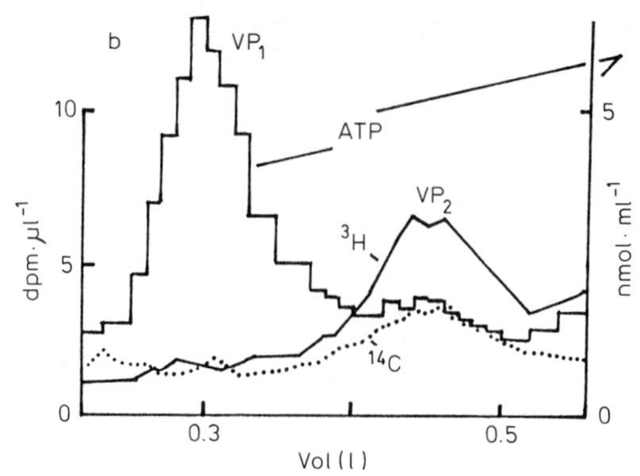

Fig. 1. Incorporation of a false transmitter into recycling vesicles. Blocks of electric organ tissue were stimulated at 0.1 Hz and perfused with [^{14}C]choline and [^{3}C]HC. After a 5-h recovery period, synaptic vesicles were isolated from a cytoplasmic extract of the crushed and frozen tissue and separated on a density gradient in a zonal rotor. The labelled bases were recovered in the fraction of recycling vesicles (VP_2), largely as their acetate esters. ATP was used as a vesicle marker and detected both reserve (VP_1) and recycling (VP_2) vesicles

the other. This also eliminates the necessity for separating the reserve and recycling vesicular pools.

2. Identification of the Source of Transmitter Released by Stimulation

The first false transmitter precursor to be studied for differential vesicular uptake was pyrrolidinecholine ('pyrrolcholine'; PyC) (Table 1) (ZIMMERMANN and DOWDALL 1977), a compound introduced by BARKER and MITTAG (1975). When blocks of electric organ were perfused with [^{3}H]PyC and [^{14}C]choline and submitted to a low-frequency loading stimulus which did not deplete vesicle numbers, the acetylated products of both bases were detected in the tissue and in the vesicles. On stimulation the ratio in which the two labelled transmitters were released agreed with that of the vesicles; however there was insufficient difference between this ratio and that of whole tissue to distinguish clearly between the cytosolic and vesicular pools. Thus the experiments provided little advance on those with labelled choline alone.

Further work showed that triethylcholine (TEC) and HC were more suitable for this kind of study (WHITTAKER and LUQMANI 1980; LUQMANI et al. 1980; LUQMANI and WHITTAKER 1981). Experiments with TEC were performed with essentially the same technique as had been used with PyC except that [^{3}H]acetate was used to label both acetylTEC and newly synthesized ACh. The esters were separated from excess [^{3}H]acetate by the FONNUM (1969) liquid ion-exchange technique and from each other by paper chromatography. Incorporation of the radioactive esters into the recycling vesicles was again promoted by a low-frequency (0.1 Hz) loading stimulus. The ratio (Fig. 2, left) in which the two labelled transmitters ACh and acetylTEC were released from the resting block after the loading stimulus was close to that of whole tissue. On applying a 10-Hz release stimulus the output of both radiolabelled bases greatly increased but that of

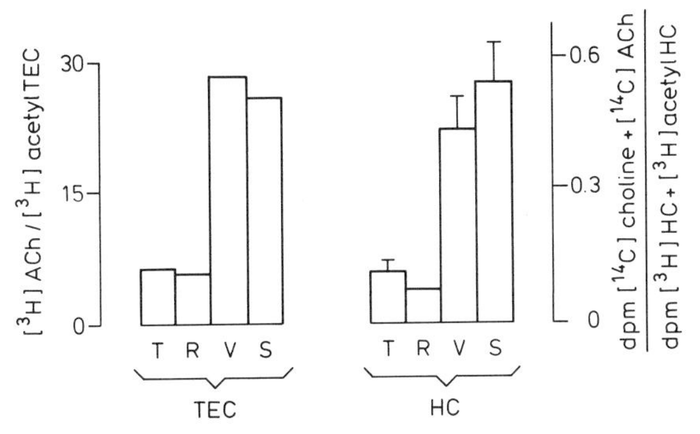

Fig. 2. Incorporation of false transmitters into blocks of perfused electric organ tissue subjected to a loading stimulus and their release in response to a release stimulus. *Blocks* show the ratio in which the labelled transmitters are present in whole tissue (*T*), perfusate before applying the release stimulus (*R*) synaptic vesicles isolated immediately after application of a release stimulus (*V*) and perfusate collected during the release stimulus (*S*). The change of ratio as a result of the evoked release and the closeness of *V* and *S* demonstrate that release was from the recycling vesicular and not the cytoplasmic pool. In the case of HC, ratios refer to total (free and acetylated) bases, but in V and S much of the [^{3}H]HC and almost all of the [^{14}C]choline was acetylated. Plotted from data of WHITTAKER and LUQMANI (1980) and LUQMANI et al. (1980)

[^{3}H]ACh increased much more that that of [^{3}H]acetylTEC, so that the ratio of [^{3}H]ACh to [^{3}H]acetylTEC in the perfusate rose nearly 4-fold. This ratio now greatly exceeded that of whole tissue but was not significantly different from that of isolated vesicles. Since the distribution of ACh between vesicles and cytoplasm is known, it was possible to calculate the proportion of acetylTEC accumulating in the cytoplasm; this was about 80%.

The experiments with HC (LUQMANI et al. 1980) supported the findings with TEC. As shown in Fig. 2 (right), the release stimulus again provoked a large (over 5-fold) change in the ratio in which [^{14}C]choline and [^{3}H]HC (free and acetylated) appeared in the perfusate of a labelled tissue block at rest and after a release stimulus. The direction of change again indicated that the normal transmitter was preferentially released. A comparison with the ratios in vesicles and whole tissue once more showed that it is the ratio in vesicles and not that in whole tissue (which includes the cytoplasmic fraction) which determines the ratio in which the transmitters are released by stimulation.

Separation of the labelled bases by paper chromatography on high-voltage paper electrophoresis revealed that isolated vesicles contained appreciable amounts of unacetylated HC (though the proportion of total HC in this form was much less than in whole tissue), and that HC was released on stimulation along with acetyl HC in the same ratio as in vesicles. At rest, the composition of the perfusate with respect to HC and acetylHC reflected more closely that of whole tissue. This showed that with HC acetylation was not essential for vesicular uptake or for acting as a false transmitter, though acetylation greatly increases its effectiveness in both respects.

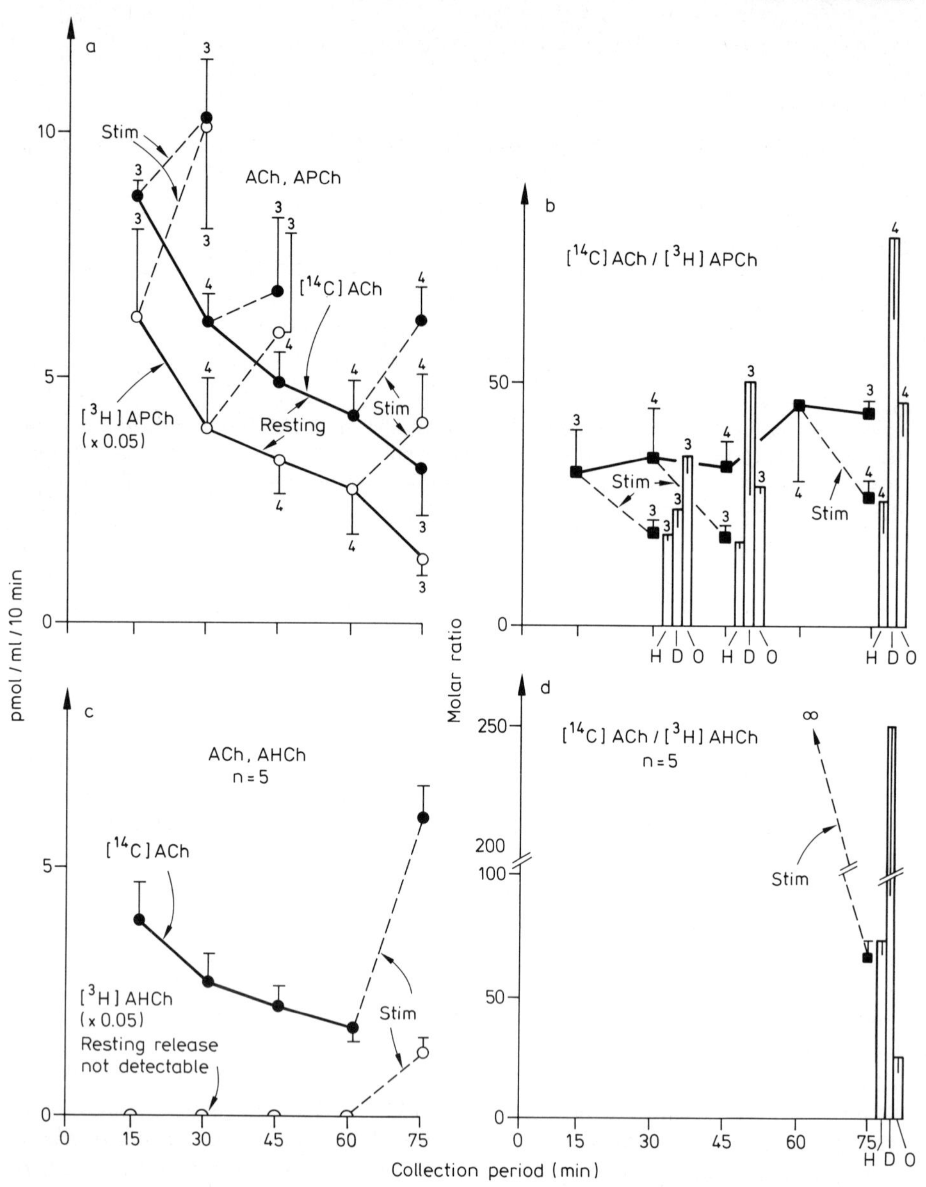

Fig. 3a–d. Release of false transmitters from the surface of guinea-pig cortex (from VON SCHWARZENFELD et al. 1979).

a, c Resting (*continuous lines*) or stimulated (*broken lines*) release of [^{14}C]acetylcholine (ACh) (*filled circles*) and either **(a)** [^{3}H]acetylPyC (APCh) or **(b)** [^{3}H]acetylHC (AHCh) (*open circles*) into cups containing paraoxonized (100 μM) Locke's solution applied to the surface of the cortex. The precursors [^{14}C]choline and [^{3}H]PyC or [^{3}H]HC had been previously applied by injection into the cortical tissue and their incorporation enhanced by electrical stimulation at 0.1 Hz for 40 min. After rinsing the cups, a series of 15-min collection periods (plotted as *abscissae*) was begun and a release stimulus (30 Hz) applied during the second, third or fifth collection period.

b, d Ratio of acetyl esters in the cortical superfusate (*squares*) at rest (*continuous lines*) or after stimulation (*broken lines*). *Blocks* give ratios in fractions containing terminal

II. Mammalian Preparations

1. Brain Cortex

The Oborin cup technique (Chap. 1) was used in combination with subcellular fractionation by VON SCHWARZENFELD (1979) to study the behaviour of PyC and HC as potential false transmitters in mammalian central cholinergic synapses. In two series of experiments [^{3}H]PyC or [^{3}H]HC was injected along with [^{3}H]choline into the tissue underlying the cup, and stimulation (0.1 Hz for 40 min) was applied to promote incorporation. Thereafter, the output of labelled acetyl esters into the cup fluid was monitored over successive 10-min periods before a final release stimulus was applied. At this point, the underlying cortex was fractionated to give samples of nerve-terminal cytoplasm, monodisperse (reserve) vesicles and membrane-bound (recycling) vesicles. The results are shown in Fig. 3, from which it will be seen that the ratio in which the native and false transmitters are released was always indistinguishable from that in which they were packaged in the recycling vesicles (fraction H in the fractionation scheme of WHITTAKER et al. 1964) but significantly different from that in which they were present in reserve vesicles (fraction D) or cytoplasm (fraction O). AcetylHC discriminated between the fractions relative to ACh to a greater extent than acetylPyC. An analysis of the ratios indicates that both false transmitters accumulate in the cytoplasm and gain little access to the reserve vesicles but are effectively taken up by recycling vesicles and effectively function as false transmitters. In other experiments (VON SCHWARZENFELD 1978) using intraventricular injection and separation of subcellular fractions in a zonal rotor, the time course of labelling of the various fractions and the status of fraction H as the recycling vesicle fraction were clearly demonstrated.

In contrast to VON SCHWARZENFELD, BOKSA and COLLIER (1980c) obtained equivocal results with [^{14}C]choline and [^{3}H]HC in brain. They used slices *in vitro* followed by subcellular fractionation. They did not use a loading stimulus and perhaps for this reason obtained insufficient differences in the ratio of acetyl esters in the three subcellular fractions representing terminal cytoplasm, reserve vesicles and recycling vesicles to be able to discriminate between them. The ratio of released labelled ACh to labelled acetylHC under resting conditions was much larger than that of the cytoplasmic fraction, suggesting that they were not working under the steady-state conditions essential for meaningful comparisons of ratios.

In experiments with mono-, di- and triethylHC (WELNER and COLLIER 1984) the first two of these substances were found to be taken up by rat brain synaptosomes and to be acetylated, monoethylHC less readily than diethylHC. The latter was a poor substrate for vesicular uptake and was not released along with ACh by K^+-depolarization. This supports the idea that in the cholinergic system as in

---

soluble cytoplasm (*O*), monodispersed synaptic vesicles (*D*) or synaptic vesicles associated with external presynaptic membrane fragments (*H*) prepared at the end of stimulation from the cortical tissue subjacent to the cup essentially by the method of WHITTAKER et al. (1964). All *points* and *blocks* are mean values of the number of experiments indicated; *bars* are SEMs. The ratios for stimulated release have been corrected by extrapolation for the contribution made to total release by resting release. It will be seen that the ratios after stimulation are not significantly different from those in fraction *H* but quite different from those in fraction *O*

the adrenergic system, transmitter analogues must gain entry to vesicles to be released by stimulation.

2. Superior Cervical Ganglion

DiethylHC is also taken up into the preganglionic nerve terminals of the cat superior cervical ganglion (SCG) and part of it is acetylated (WELNER and COLLIER 1985). On stimulation the unacetylated base is liberated in a Ca^{2+}-dependent manner but the acetylated base is retained. This suggests that diethylHC is a better substrate for vesicular uptake than its acetylated product and shows, as was found with HC in *Torpedo*, that a choline analogue does not necessarily have to be acetylated to be a false transmitter. Stimulation increased the uptake of diethylHC, like choline, and enhanced its acetylation, in spite of the fact that acetyldiethylHC is not a transmitter. This suggests that the effect is mediated by the influx of Ca^{2+} accompanying stimulation, and not via depletion of transmitter as has been proposed for ACh.

3. Neuromuscular Junction

In a detailed study, COLQUHOUN et al. (1977) were able to distinguish between miniature end-plate potentials (mEPPs) generated by ACh and by acetylmonoethylcholine (acetylMEC) at the neuromuscular junctions (NMJs) of the frog cutaneous pectoralis muscle and the rat diaphragm by the reduced amplitude and increased decay rate of the latter. They argued that these characteristics were due to a lower affinity of the ACh analogue for the nicotinic ACh receptor (nAChR) rather than differences in the rate of synthesis or hydrolysis of the two esters or their efficacy, and they estimated that acetylMEC had about 25% of the affinity of ACh for nAChR. MEC had no direct effect on the time course of mEPPs in the absence of stimulation. When the nerve was stimulated in the presence of MEC, there was a reduction in the amplitude of the end-plate potential (EPP) which was reversed by choline. After stimulation in the presence of MEC, there was a reduction in the amplitude and decay time of mEPPs, indicating that ACh had to be released and the releasable stores (i.e. the recycling vesicles) filled with acetylMEC before any effect on the mEPPs could be observed. With frog muscle about 3×10^5 quanta had to be released before acetylMEC exerted its maximum effect; this is within the range of $3-7 \times 10^5$ quanta released from this junction by black widow spider venom. The change in the characteristics of the mEPP is a smooth one, with no detectable bimodality (LARGE and RANG 1978a, b). This indicates that the recycling vesicles draw upon a common pool (the cytoplasmic) in which the two transmitters are free to mix and whose composition gradually changes as the false transmitter replaces the endogenous one.

The pharmacological effect of PyC on transmission at the rat gastrocnemius NMJ *in situ* was studied by VON SCHWARZENFELD and WHITTAKER (1977). A neuromuscular blocking effect was seen at relatively high concentrations of PyC, but was judged to be a direct effect of the drug and not due to uptake by the presynaptic nerve terminals, followed by conversion to a false transmitter of low potency and displacement of ACh from presynaptic stores by this weaker agonist, since hemicholinium-3, which blocks uptake, did not reduce the effect.

4. Myenteric Plexus

KILBINGER (1977) has demonstrated that acetylPyC can act as a false transmitter in the myenteric plexus of the guinea-pig ileum (MPLM). This ester comprised about 15% of the total tissue content of true and false transmitters. Stimulation in the presence of PyC increased this to 21%. PyC inhibited the synthesis of ACh from endogenous choline in resting preparations and acetylPyC replaced this extra ACh. Stimulation of the preparation released a mixture of acetylPyC and ACh; a comparison of the ratio in which they were released with that in which they were present in the tissue showed that the newly synthesized acetylPyC was preferentially released, but in these experiments, newly synthesized ACh was not distinguished from ACh in the reserve vesicle store.

Since ÁGOSTON et al. (1985) have demonstrated that it is possible to separate recycling from reserve cholinergic vesicles from MPLM, this preparation would make an excellent mammalian system in which to identify the transmitter pool from which ACh is released on stimulation using the same techniques as have been applied to the electromotor nerve terminals. So far this has not been done.

C. Conclusions

In general, false transmitters or their precursors have not proved to be particularly interesting from the pharmacological point of view, with the possible exception of *N*-aminodeanol (Table 1), which is readily incorporated into brain tissue and acetylated in amounts comparable to ACh and has some behavioural effects (reviewed in NEWTON and JENDEN 1986). The main application of these compounds has been to give us a better understanding of the dynamics of transmitter storage and release, and as a tool for identifying the pool from which the transmitter released on stimulation is derived. The results obtained unequivocally point to the pool of recycling vesicles and would be very difficult to account for on any non-vesicular theory without falling foul of Occam's razor: *hypotheses non sunt multiplicanda nisi necessitas.*

References

Ágoston DV, Kosh JW, Lisziewicz J, Whittaker VP (1985) Separation of recycling and reserve synaptic vesicles from the cholinergic nerve terminals of the myenteric plexus of guinea pig ileum. J Neurochem 45:398–406

Ágoston DV, Dowe GHC, Fiedler W, Giompres PR, Roed IS, Walker JH, Whittaker VP, Yamaguchi T (1986) The use of synaptic vesicle proteoglycan as a stable marker in kinetic studies of vesicle recycling. J Neurochem 47:1580–1592

Barker LA, Mittag TW (1975) Comparative studies of substrates and inhibitors of choline transport and choline acetyltransferase. J Pharmacol 192:86–94

Berry JF, Whittaker VP (1959) The acyl-group specificity of choline acetylase. Biochem J 73:447–458

Boksa P, Collier B (1980a) Acetylation of homocholine by rat brain: subcellular distribution of acetylhomocholine and studies on the ability of homocholine to serve as substrate for choline acetyltransferase *in situ* and *in vitro.* J Neurochem 34:1470–1482

Boksa P, Collier B (1980b) *N*-ethyl analogues of choline as precursors to cholinergic false transmitters. J Neurochem 34:1099–1104

Boksa P, Collier B (1980c) Spontaneous and evoked release of acetylcholine and a cholinergic false transmitter from brain slices: comparison to true and false transmitter in subcellular stores. Neuroscience 5:1517–1532

Chiou CY (1974) Studies on action mechanism of a possible false cholinergic transmitter (2-hydroxyethyl methyldiethylammonium). Life Sci 14:1721–1733

Collier B, Welner SA (1986) Synthesis, storage and release of choline analogue esters. In: Hanin I (ed) Dynamics of cholinergic function. Plenum, New York, pp 1161–1168

Collier B, Barker LA, Mittag TV (1976) The release of acetylated choline analogues by a sympathetic ganglion. Mol Pharmacol 12:340–344

Collier B, Lovat S, Ilson D, Barker LA, Mittag TW (1977) The uptake, metabolism and release of homocholine: studies with rat brain synaptosomes and cat superior cervical ganglion. J Neurochem 28:331–339

Collier B, Boksa P, Lovat S (1979) Cholinergic false transmitters. Prog Brain Res 49:107–121

Colquhoun D, Large WA, Rang HP (1977) An analysis of the action of a false transmitter at the neuromuscular junction. J Physiol (Lond) 266:361–395

Fonnum F (1969) Radiochemical microassays for the determination of choline acetyltransferase and acetylcholinesterase activities. Biochem J 115:465–472

Giompres PE, Zimmermann H, Whittaker VP (1981a) Purification of small dense vesicles from stimulated *Torpedo* electric tissue by glass bead column chromatography. Neuroscience 6:765–774

Giompres PE, Zimmermann H, Whittaker VP (1981b) Changes in the biochemical and biophysical parameters of cholinergic synaptic vesicles on transmitter release and during a subsequent period of rest. Neuroscience 6:775–785

Ilson D, Collier B (1975) Triethylcholine as a precursor to a cholinergic false transmitter. Nature 254:618–620

Ilson D, Collier B, Boksa P (1977) Acetyltriethylcholine: a cholinergic false transmitter in cat superior cervical ganglion and rat cerebral cortex. J Neurochem 28:371–381

Kilbinger H (1977) Formation and release of acetyl-pyrrolidinecholine (*N*-methyl-*N*-acetoxyethylpyrrolidinium) as a false cholinergic transmitter in the myenteric plexus of the guinea-pig small intestine. Naunyn Schmiedebergs Arch Pharmacol 296:153–158

Kopin IJ (1968) False adrenergic transmitters. Annu Rev Pharmacol Toxicol 8:377–394

Kosh JW, Whittaker VP (1985) Is propionylcholine present in or synthesized by electric organ? J Neurochem 45:1148–1153

Large WA, Rang HP (1978a) Factors affecting the rate of incorporation of a false transmitter into mammalian motor nerve terminals. J Physiol (Lond) 285:1–24

Large WA, Rang HP (1978b) Variability of transmitter quanta released during incorporation of a false transmitter into cholinergic nerve terminals. J Physiol (Lond) 285:25–34

Luqmani YA, Richardson PJ (1982) Homocholine and short-chain *N*-alkyl choline analogues as substrates for *Torpedo* choline acetyltransferase. J Neurochem 38:368–374

Luqmani YA, Whittaker VP (1981) A vesicular site of origin for the release of a false transmitter at the *Torpedo* synapse. In: Pepeu G, Ladinsky H (eds) Cholinergic mechanisms: phylogenic aspects, central and peripheral synapses and clinical significance. Plenum, New York, pp 47–58

Luqmani YA, Sudlow G, Whittaker VP (1980) Homocholine and acetylhomocholine: false transmitters in the cholinergic electromotor system of *Torpedo.* Neuroscience 5:153–160

Newton MW, Jenden DJ (1986) False transmitters as presynaptic probes for cholinergic mechanisms and function. Trends Pharmacol Sci 7:316–320

O'Regan S (1982) The synthesis, storage and release of propionylcholine by the electric organ of *Torpedo marmorata.* J Neurochem 39:764–772

O'Regan S (1983) Uptake of acetate and propionate by isolated nerve endings from the electric organ of *Torpedo marmorata* and their incorporation into choline esters. J Neurochem 41:1596–1601

O'Regan S (1984) Evaluation of acetate uptake and its relative conversion to acetylcholine in *Torpedo* electric organ synaptosomes under different ionic and metabolic conditions. Neurochem Int 6:339–346

Roed, IS (1980) Uptake of false transmitter precursors into synaptosomes derived from a purely cholinergic source. Hoppe-Seyler's Z Physiol Chem 361:1331

Roed IS (1986) Anwendung von integrierter Gaschromatographie-Massenspektrometrie zur Untersuchung der Transmitterkompartimente in rein cholinergen Synapsen von *Torpedo marmorata.* Dissertation, Göttingen

Schwarzenfeld I von (1978) The uptake of acetyl-pyrrolidinecholine – a false cholinergic transmitter – into mammalian cerebral cortical synaptic vesicles. In: Jenden DJ (ed) Cholinergic mechanisms and psychopharmacology. Plenum, New York, pp 657–672

Schwarzenfeld I von (1979) Origin of transmitters released by electrical stimulation from a small, metabolically very active vesicular pool of cholinergic synapses in guinea-pig cerebral cortex. Neuroscience 4:477–493

Schwarzenfeld I von, Whittaker VP (1977) The pharmacological properties of the cholinergic false transmitter *N*-2-acetoxyethyl-*N*-methylpyrrolidinium and its precursor, *N*-2-hydroxyethyl-*N*-methylpyrrolidinium. Br J Pharmacol 59:69–74

Schwarzenfeld I von, Sudlow G, Whittaker VP (1979) Vesicular storage and release of cholinergic false transmitters. Prog Brain Res 49:163–174

Sheridan MN, Whittaker VP, Israël M (1966) The subcellular fractionation of the electric organ of *Torpedo.* Z Zellforsch 74:291–307

Smith AD (1972) Cellular control of the uptake, storage and release of noradrenaline in sympathetic nerves. Biochem Soc Symp 36:103–131

Suszkiw JB, Zimmermann H, Whittaker VP (1978) Vesicular storage and release of acetylcholine in *Torpedo* electroplaque synapses. J Neurochem 30:1269–1280

Welner SA, Collier B (1984) Uptake, metabolism, and releasability of ethyl analogues of homocholine by rat brain. J Neurochem 43:1143–1151

Welner SA, Collier B (1985) Accumulation, acetylation and releasability of diethylhomocholine from a sympathetic ganglion. J Neurochem 45:210–218

Whittaker VP (1963) Identification of acetylcholine and related esters of biological origin. In: Koelle GB (ed) Cholinesterases and anticholinesterase agents. Springer, Berlin Göttingen Heidelberg, pp 1–39 (Handbook of experimental pharmacology, vol 15)

Whittaker VP (1973) Choline esters other than acetylcholine. In: Michaelson MJ (ed) Int enc pharmacol therap sect 85 comparative pharmacology, vol 1. Pergamon, Oxford, pp 229–240

Whittaker VP, Luqmani YA (1980) False transmitters in the cholinergic system: implications for the vesicle theory of transmitter storage and release. Gen Pharmacol 11:7–14

Whittaker VP, Michaelson IA, Kirkland RJA (1964) The separation of synaptic vesicles from nerve ending particles ('synaptosomes'). Biochem J 90:293–303

Zimmermann H, Denston CR (1977a) Recycling of synaptic vesicles in the cholinergic synapses of the *Torpedo* electric organ during induced transmitter release. Neuroscience 2:695–714

Zimmermann H, Denston CR (1977b) Separation of synaptic vesicles of different functional states from the cholinergic synapses of the *Torpedo* electric organ. Neuroscience 2:715–730

Zimmermann H, Dowdall MJ (1977) Vesicular storage and release of a false cholinergic transmitter (acetylpyrrolcholine) in the *Torpedo* electric organ. Neuroscience 2:731–739

Zimmermann H, Whittaker VP (1977) Morphological and biochemical heterogeneity of cholinergic synaptic vesicles. Nature 267:633–635

CHAPTER 17

Cholinergic Co-transmitters

D. V. AGOSTON *

A. Introduction

I. Scope of the Chapter: ATP and Neuropeptides

The concept of cholinergic neurotransmission *sensu stricto*, that cholinergic neurons transmit information solely by releasing the small molecular mass endogenous neurotransmitter acetylcholine (ACh), has been increasingly questioned during the last few years.

Our general view of neurocommunication has been changing as a result of recent findings suggesting that many neurons are able to synthesize, store, release and detect more than one 'neuroactive' or 'informational' substance (SCHMITT 1983). Growing numbers of neurons previously identified on the basis of their classical transmitter content have been found to contain neuroactive peptides. These appear to participate as endogenous ligands in the communication process. Central and peripheral cholinergic neurons have proved to be no exceptions and some at least have been found to manufacture, process and release neuroactive compounds other than ACh. Many questions are raised by these findings. Is the concentration of neuroactive compounds in cholinergic neurones constant or subject to variation with time? Does it reflect changes in the metabolic condition of the neuron, or is it a response to significant but as yet unidentified external stimuli? Is it related to the aging of the neuron, or determined by environmental factors? How are the genes coding for these co-existing peptides related to the cholinergic genome? Unfortunately, in the present state of knowledge it is not possible to answer most of them.

The aim of this chapter is to present immunocytochemical, cell biological, physiological and pharmacological evidence for the presence of neuroactive substances in certain cholinergic neurons. This description will cover a number of such compounds belonging to two chemical classes: peptides of molecular mass less than 5000 daltons and ATP, a nucleotide with ubiquitous distribution, metabolism and function in all living cells. As far as neuropeptides are concerned discussion will be limited to those known to be present in cholinergic neurons and thus presumed to be involved in cholinergic transmission. The combinations that must be considered are: (1) ACh + ATP; (2) ACh + classical transmitter (?); (3) ACh + peptide(s); (4) ACh + peptide(s) + ATP (?). For more general aspects of biologically active peptides including their function as neurotransmitters the

---

* From Abt. Neurochemie, Max-Planck-Institut Göttingen (current address see list of contributors).

reader is directed to other recent reviews and books (BROWNSTEIN 1982; COOPER and MARTIN 1982; EIPER et al. 1986; EMSON 1979; HÖKFELT et al. 1980; KRIEGER 1983; KRIEGER and MARTIN 1981; KRIEGER et al. 1983; MUTT 1985; SNYDER 1980). Similarly, the general metabolism of ATP and the concept of purinergic nerves and purinergic transmission are not covered in this chapter; the discussion will confine itself to focusing on the presence of ATP in cholinergic nerve terminals and its possible interaction with cholinergic neurotransmission.

Almost all the evidence for co-transmission in cholinergic systems is immunohistochemical; there is an unfortunate lack of functional studies with neuropeptides morphologically demonstrated to be present in cholinergic neurons.

II. Terminology

During recent years a new terminology has been created to describe the phenomena which the presence of more than one neuroactive substance within an individual neuron has generated.

1. The term 'neuroactive substance' – later 'informational substance' – was invented by SCHMITT (1983) to replace the terms 'neurotransmitter' and 'neuromodulator' in order to overcome difficulties generated by the discovery of literally dozens of new, pharmacologically active substances whose status as classical neurotransmitters is questionable. Neuroactive substances may be defined as substances which activate processes within nerve cells needed for them to communicate with each other or with their target cells. 'Classical transmitters' are neuroactive compounds discovered early in the history of chemical neurotransmission. They are small molecular mass compounds, their synthetic machinery is present in the terminal, their re-synthesis and re-loading into synaptic vesicles takes place at the terminal; thus their short-term renewal and recycling is not dependent on axonal transport. 'Co-transmitters' are neuroactive substances which are present along with a classical transmitter and operate simultaneously with it in an individual neuron. Their capacity for recycling may be more restricted.

2. The term 'co-expression' has been used to describe the phenomenon that products of different genes coding for different neuroactive substances, peptide precursors and enzymes synthesizing the classical transmitters as well as their messages are present simultaneously in an individual neuron. This implies that both the neuroactive substances and the mRNAs coding for them can be detected using appropriate methods. However, this co-expression does not imply any functional interaction of the molecules being expressed.

3. 'Co-localization' is restricted to morphological observations involving cytochemical or immunocytochemical techniques, whereas 'co-existence' is a more general term describing the simultaneous presence of multiple neuroactive substances – or their key synthesizing enzymes – in the same neuron or in the same terminal. This situation must be distinguished from cytochemical identification of multiple transmitter substances in a group of nerve cells, e.g. a nucleus. Co-

localization in this restricted meaning does not imply, in the absence of evidence to the contrary, the presence of truly functioning molecules; see also co-expression.

4. 'Co-packaging' and 'co-storage' are extended implications of co-existence at the subcellular, i.e. subterminal level. Co-packaging occurs when more than one neuroactive compound, peptide or propeptide, usually products of the same gene, using the same synthetic machinery, are being sorted into the same – Golgi – vesicle. Co-storage implies, at the subterminal level, the storage of more than one neuroactive compound of different chemical composition in one granule or vesicle; this may occur partly by loading in the terminal.

5. 'Co-release' is a cascade of physiological processes resulting in simultaneous secretion of multiple neuroactive substances from the same terminal of an individual neuron upon appropriate stimulus.

6. The term 'co-transmission' comprises all the events of simultaneous chemical signalling between neurons by multiple neuroactive substances. It implies the presence of more than one neuroactive substance in the same neuron, possibly in the same granule. Co-release can be expected to occur simultaneously in response to the appropriate stimulus, resulting in interaction between the neuroactive substances on pre- and postsynaptic receptors.

III. Criteria to Be Satisfied

1. By a Neurotransmitter

As early as 1877 Du Bois Reymond expressed the opinion that the effect of nerve on muscle was either electrical or chemical in nature. Ammonia and lactic acid were suggested as chemical mediators.

A number of criteria must be met to consider a substance as a neurotransmitter. None of these criteria is adequate by itself, but it is also unlikely that all can be clearly established for each substance. Here an integrated view is presented based on several previous formulations (Burnstock 1972; McGeer et al. 1978; Carpenter and Reese 1981; Krieger et al. 1983).

1. The neurotransmitter must be present in nerve terminals whether synthesized *in situ* – classical transmitters – or transported from the perikaryon – peptides – where it was synthesized, processed and packaged.

2. A transmitter must be released upon appropriate stimuli.

3. A neurotransmitter, when exogenously administered, must act on the receptor cell in a manner that is quantitatively and qualitatively similar to the action of the substance released by appropriate stimulation.

4. A mechanism must exist for the rapid termination of the effect of the neurotransmitter.

5. In peripheral organs, the transmitter stores should disappear subsequent to denervation by severing mechanically or chemically the appropriate axons.

2. By a Co-transmitter

1. The co-transmitter should be present simultaneously with the classical – primary – neurotransmitter in the same terminal. It may be stored in the same granule or synaptic vesicle; however co-storage is not an essential requirement.

2. A co-transmitter should satisfy the same criteria as the primary neurotransmitter in respect of its release from the nerve terminal. However, the stimulation conditions for the release of co-transmitters may differ from those of the primary transmitter; different, e.g. in respect of the frequency optimum of the release.

3. The co-transmitter should have a function in signal transmission. Possible effects would be activation or inactivation of the postsynaptic cells by a specific receptor or presynaptic modulation of the release of the classical neurotransmitter, or both.

4. There should be a mechanism for terminating the action of the co-transmitter.

5. On anatomical or chemical denervation the disappearance of the co-transmitter from the terminal should follow that of the primary transmitter; however there may be differences in the time-course of their disappearance.

B. ATP as a Neuroactive Compound

I. ATP and Its Derivatives in Neuronal Function

Much evidence has accumulated that nervous activity is associated with the release of both nucleotides and nucleosides at central and peripheral synapses (for review, see FREDHOLM and HEDQUIST 1980; STONE 1981). When ATP is liberated it is rapidly hydrolysed to adenosine, inosine and hypoxanthine in the extraterminal space (FREDHOLM and HEDQUIST 1980; STONE 1981). Electrical stimulation of peripheral nerves result in a Ca^2-dependent release of ATP (BURNSTOCK 1972). Chemical stimuli such as elevated K^+ or veratridine can also cause the release of nucleotides from a wide range of preparations, however to very different extents (STONE 1981). The site of origin of the released ATP or its derivatives has been a major subject of controversy, especially as in addition to the synaptic release a non-synaptic, axonal release was demonstrated (ABOOD et al. 1962; KUPERMAN et al. 1964). ATP or adenosine can be released from adrenergic as well as cholinergic nerves on appropriate stimulation, and ATP is known to be co-stored in both systems with the primary transmitter (DOUGLAS and POISNER 1966; DOWDALL et al. 1974). According to BURNSTOCK ATP may be a transmitter in its own right, and be released from a subset of neurons which he has called purinergic. The origin of the released ATP and/or its derivatives may also be postsynaptic and non-neuronal even under physiological conditions. Thus stimulation of the sympathetic nerves to the heart, kidney and adipose tissue is accompanied by the release of purine compounds of an almost exclusively postsynaptic origin (FREDHOLM 1976; FREDHOLM and HEDQUIST 1980). Apart from neuronal stimuli a pathological condition such as hypoxia can also induce purine release from non-neuronal tissues (FREDHOLM 1976).

A working hypothesis, compatible with the available evidence, is that membrane depolarization and/or local disparity in the balance between energy expediture and energy production is responsible for the release of adenine compounds during nerve activity (FREDHOLM and HEDQUIST 1980). Neuronal function in general can be affected in many ways by extraterminal purines, originating from whatever sources; thus ATP may alter neuronal excitability, increase the A1 receptor-mediated potassium permeability (SCHUBERT et al. 1985) or regulate receptor sensitivity for the main transmitters (for review, see STONE 1981; WITZEMANN 1987). The most frequently and clearly described action of adenosine is its ability to depress the evoked release of transmitters from many different types of neuron, central and peripherial. The exact mechanism of this inhibition is unknown. It is strongly frequency dependent; at high frequency stimulation the inhibitory effect is small or non-existent (VIZI and KNOLL 1976). Very likely the presynaptic effects of adenosine are not mediated via cAMP (STONE 1978) but mediation of the free concentration of terminal Ca^{2+} has been proposed as the key regulatory step (RIBERIO et al. 1979). Numerous experimental data are consistent with an effect via an increased K^+ permeability (STJÄRNE 1979).

II. ATP in Chromaffin Granules

It has been known for more than 20 years that ATP is stored and released together with catecholamines in the adrenal medullary cells (DOUGLAS and POISNER 1966). In the acidified interior of a chromaffin granule some 22000 molecules of catecholamines are stored together with nucleotides in a ratio of about 4.5:1. Among them are 4900 molecules of ATP (70% of the total nucleotide content) 700 molecules of ADP (10%) and some 210 molecules of AMP (2.5%) (WINKLER and WESTHEAD 1980). It is important to note here that during the biogenesis of the granules the ATP accumulates first and the catecholamines only subsequently (WINKLER 1977). The total solute concentration inside the granules, mostly catecholamine and nucleotides, is about 0.75 *M* (WINKLER and WESTHEAD 1980). Nuclear magnetic resonance (NMR) studies have explained how this high concentration of osmotically active solutes can be kept in isoosmotic relationship with the surrounding cytoplasm. The co-stored nucleotides, but primarily ATP, form a labile soluble complex with the catecholamines thereby lowering their effective concentration and averting osmotic disruption of the granule (for review, see WINKLER and WESTHEAD 1980). This model can be extended, with some reservations, to catecholamine-containing vesicles isolated from sympathetic nerves (for review, see LAGERCRANTZ 1976). ATP appears to gain entry to vesicles by a carrier-mediated mechanism: the Mg^{2+} activated ATPase associated with the vesicular membrane is responsible for the establishment of a proton gradient which acts as a driving force for the uptake process (ABERER et al. 1978). As mentioned earlier the important question is that of the site of origin of the released material, in particular whether release occurs from pre- or postsynaptic sites. In the case of the adrenal medulla the co-stored ATP is released together with the catecholamine from the chromaffin cells (DOUGLAS 1968; WINKLER 1976). The situation in neurons is far from clear due to the massive efflux of purines from effector cells following nervous stimulation which may mask the neuronal release (STJÄRNE et al. 1970; LAGERCRANTZ 1976; STJÄRNE and ASTRAND 1984).

Extensive pharmacological investigations have shown that extracellular nucleotides are potent inhibitors of catecholamine release in a frequency dependent manner (HEDQUIST and FREDHOLM 1979). The ability of nucleotides and nucleosides in blocking catecholamine release varies very much however from one tissue to another and is dependent on the density of various purine receptors on the presynaptic site (STARKE 1977).

For further details extensive reviews of the subject are available (BURNSTOCK 1982; CARMICHAEL and WINKLER 1985; FREDHOLM and HEDQUIST 1980; LAGERCRANTZ 1976; PHILLIS and WU 1981; STONE 1981; WINKLER 1976; WINKLER and WESTHEAD 1980; WINKLER et al. 1983).

III. ATP in the Cholinergic System of *Torpedo*

Early work by DOWDALL et al. (1974) demonstrated that highly purified cholinergic vesicles isolated from the electromotor terminals in the electric organ of *Torpedo marmorata* store ATP in high concentrations along with the transmitter ACh. Since a separate chapter describes in detail the structure of these vesicles only a brief description will be given here. According to the established model (OHSAWA et al. 1979) some 200000 molecules of ACh per vesicle are stored together with some 30000 molecules of ATP which gives a molar ratio of around 6. This molar ratio was found not only in synaptic vesicles but also in synaptosomes (ZIMMERMANN 1982). The combined concentration of ACh and ATP in the *Torpedo* synaptic vesicles is almost 1.0 *M*, higher than in chromaffin granules, but it must be noted that the body fluids of the fish have an average osmolarity of 800 mOsm. In the acidic core of the vesicles (pH 5.5) ACh and ATP are in an essentially fluid phase together with some 10000 molecules per vesicle of Mg^{2+} as was demonstrated in proton NMR studies (STADLER and FÜLDNER 1980; FÜLDNER and STADLER 1981).

Repetitive electrical stimulation of the electromotor nerve releases both substances from the vesicular pool; their depletion at various frequencies of stimulation is remarkably well correlated (ZIMMERMANN 1978; GIOMPRES et al. 1981; ZIMMERMANN and DENSTON 1977a, b). It can therefore be concluded that ATP as the second major vesicular solute is released upon electrical stimulation via the same exocytotic mechanism as ACh.

In isolated electromotor nerve terminals (synaptosomes) prepared from *Torpedo* electric organ the release of ACh and ATP can also be triggered by different chemical stimuli but according to MOREL and MEUNIER (1981) the ratio between the two released substances was dependent on the applied depolarizing agent.

Controversial reports have also appeared on synaptosomal ACh and ATP release in *Torpedo* electric organ. Under certain conditions ACh is apparently released without a corresponding release of detectable amounts of ATP (MICHAELSON 1978). In another series of experiments ATP release was observed without release of detectable amount of ACh (SCHMIDT 1978). However, in the latter case a systematic occurrence of a technical artifact cannot be excluded.

It was also demonstrated by analogy with the adrenergic system that effector cells – electroplaques – might play a much greater role in the release of ATP and its derivates than the nerves (MEUNIER et al. 1975). At the *Torpedo* electroplaque a massive ATP release was observed from the postsynaptic surface. This is thought to be the result of depolarization rather than interaction with receptors and it can be reduced by curare and enhanced with eserine (ISRAËL et al. 1976).

The released ATP is rapidly degraded as was demonstrated by adding labelled ATP to a suspension of sealed synaptosomes (ZIMMERMANN et al. 1979). Responsible for this rapid degradation of ATP are the different nucleotidases present in the plasma membrane that exhibit ecto-enzyme characteristics. An early report from DOWDALL (1978) demonstrated the presence of an ecto-5′-nucleotidase on the electromotor presynaptic plasma membrane. Recent studies show that these terminals possess an ATPase activity which is localized on the extracellular face of the membrane (KELLER and ZIMMERMANN 1984). Very likely the real terminating event – at least in electromotor nerve endings – is the uptake mechanism for adenosine described below. Although systematic investigations from ZIMMERMANN and coworkers have described an ectonucleotidase ‘chain’ associated with highly purified electromotor synaptosomes (KELLER and ZIMMERMANN 1984; GRONDAL and ZIMMERMANN 1986), speculations regarding the real function(s) of these ecto-ATPases are continuing. Several possibilities may be considered: thus the energy provided by the hydrolysis of the ATP may be used to activate uptake processes of ions, be coupled to purine salvage or contribute to the so-called adenosine cycle (for details, see ZIMMERMANN and GRONDAL 1985; WITZEMANN 1987).

The major physiological consequence of this nucleotide release from whatever source is an inhibition of the release of the principle transmitter ACh. Externally applied adenine derivates inhibit ACh release from electromotor nerve terminals (ISRAËL et al. 1977; ISRAËL and MEUNIER 1978). The mechanism of this feedback is unknown, but since adenosine released from the postsynaptic surface could act on the presynaptic terminal the phenomenon could be described as ‘retrograde transmission’ (ISRAËL et al. 1980).

On subsequent recovery after low frequency stimulation (0.1 Hz) replenishment of vesicular ACh and ATP takes place *pari passu* suggesting that ACh and ATP are also involved in some aspect of the reloading mechanism (GIOMPRES et al. 1981).

Cholinergic synaptosomes isolated from *Torpedo* electric organ have a high-affinity uptake mechanism for adenosine and adenine (ZIMMERMANN et al. 1979; ZIMMERMANN and BOKOR 1979). However, that for adenosine seems to resemble that for choline more closely than that for adenine. The adenosine system has a K_T of 2 μM and a V_{max} of 30 pmol min^{-1} mg^{-1} protein and can be competitively inhibited by 2-deoxyadenosine (ZIMMERMANN et al. 1979). It is distinct from the choline system since it cannot be blocked by hemicholinium-3 (HC-3) – a potent inhibitor of the latter (ZIMMERMANN and DENSTON 1977a). Adenine or adenosine, once internalized, can serve as a precursor for ATP in the cholinergic terminal cytoplasm. Adenosine is rapidly phosphorylated by adenosine kinase and forms AMP. The phosphorylation of adenine is slower so that it initially accumulates in synaptosomes but it is eventually converted to

AMP, ADP and ATP (ZIMMERMANN et al. 1979). One final destination of the newly formed ATP is the synaptic vesicle (ZIMMERMANN and DENSTON 1977b). The newly synthesized ATP, like the newly synthesized ACh, is preferentially taken up by the actively recycling (VP_2) pool of synaptic vesicles (ZIMMERMANN 1979). Isolated synaptic vesicles also have the ability to take up nucleotides, principally ATP (LUQMANI 1981). The uptake system is saturable and transports ATP with a K_T value of 1.15 m*M*. However ADP, AMP, GTP and UTP were competitive substrates for the saturable system. The carrier is not inhibited by the choline uptake inhibitor, HC-3, but atractyloside, which blocks ATP-ADP translocation in mitochondria and chromaffin granules, also acted as a specific inhibitor of the vesicular system without affecting ACh transport (LUQMANI 1981). The carrier was identified as a protein with a molecular mass of 34 kDa and a basic isoelectric point. Like the mitochondrial carrier it can also bind atractyloside (STADLER and FENWICK 1983; LEE and WITZEMANN 1983).

IV. ATP in Mammalian Cholinergic Systems

Evidence that ATP and ACh are released in parallel from the mammalian skeletal neuromuscular junction was first presented by SILINSKY (1975). In a rat phrenic nerve-hemidiaphragm preparation treated with tubocurarine to eliminate all postsynaptic electrical activity he was able to record ATP and ADP release on electrical stimulation. The release of adenine nucleotides were Ca^{2+}-dependent and using carbachol it was demonstrated that the release was not produced by a secondary action of released ACh. Treatment of the preparations with HC-3, to deplete the vesicular ACh pool, resulted in a failure of nerve stimulation to release adenine nucleotides. This strongly suggests that the released ATP is of neuronal indeed probably of vesicular origin.

Several attempts have been made to purify mammalian cholinergic vesicles from tissues with a relatively rich cholinergic innervation since there is no counterpart in the mammalian cholinergic system to the purely adrenergic adrenal medulla.

Although in the mammalian forebrain cholinergic terminals account for only some 6% of the total (RICHARDSON 1981) attempts to purify cholinergic vesicles by using differences in physico-chemical properties of synaptic vesicles storing different transmitter substances (NAGY et al. 1977) has been partially successful. Cholinergic synaptic vesicles isolated from guinea-pig cerebral cortex indeed contain ATP besided the transmitter ACh, however with a much lower molecular ratio close to 1:1 (NAGY et al. 1976, 1977) suggesting a selective lost of the vesicular ACh pool during the long preparation. The successful isolation of mammalian cholinergic vesicles from the peripheral nervous system has been reported, namely from rat diaphragm (VOLKNANDT and ZIMMERMANN 1986) and from the myenteric plexus of the guinea-pig small intestine (DOWE et al. 1980; ÁGOSTON et al. 1985b). Cholinergic synaptic vesicles isolated from rat diaphragm also store ATP; the ACh:ATP ratio was found to be around 10:1 (VOLKNANDT and ZIMMERMANN 1986). Cholinergic synaptic vesicles isolated from the guinea-pig myenteric plexus preparation likewise contain ATP. The molar ratio was found to be around 10:1; however, due to contamination derived from muscle this value may alter on further purification (ÁGOSTON et al. 1985b; D.V. AGOSTON, unpublished work).

The availability of purely cholinergic synaptosomes isolated by immunoaffinity chromatography using an antiserum to a cholinergic-specific ganglioside (Chol-1) (see Chaps. 12, 16) permitted the study of the interaction of ACh and ATP at the level of the isolated mammalian cholinergic nerve terminal (RICHARDSON et al. 1984). Such immunoaffinity-purified cholinergic synaptosomes release ATP in a Ca^{2+}-dependent manner in association with ACh upon depolarization with elevated K^+ or veratridine (RICHARDSON and BROWN 1987a). The ratio of released ACh to ATP (9:1; 25 m*M* KCl and 11:1; veratridine) was much closer to that measured in isolated synaptic vesicles than to that in whole terminals (3:1). Since the molar ratio of the released ACh to ATP from *Torpedo* synaptosomes depends on the method of depolarization (MOREL and MEUNIER 1981) ATP and ACh may be released by different mechanisms. Given that the time-courses of the degradation of the released ACh and ATP are quite different it is difficult to estimate the true molar ratios in which they are released. By contrast, experiments with cortical synaptosomes gave a negative correlation between depolarization-induced ATP and ACh release. In the presence of type A botulinum toxin the depolarization-induced release of ACh was drastically reduced without a significant reduction in the ATP release (MACDONALD and WHITE 1985; WHITE et al. 1980). However, these contradictions could be resolved by taking into account the fact that only some 6% of the cerebral nerve endings are cholinergic; the release of ATP from other terminals could mask the release of ATP from cholinergic nerve endings.

The ATP released from cholinergic synaptosomes is hydrolysed to adenosine outside the terminal; this participates in a negative feedback loop and inhibits further ACh and ATP release (RICHARDSON and BROWN 1987a). This is consistent with previous electrophysiological studies which showed that adenine nucleotides and adenosine depress neuronal firing in various neuronal systems (PHILLIPS et al. 1975; SCHUBERT and MITZDORF 1979; DUNWIDDIE and HOFFER 1980). It is also known that extraterminal adenosine inhibits the release of different transmitters in the mammalian central nervous system (CNS) (STONE 1981). The rate limiting factor in the production of adenosine is the hydrolysis of AMP. In this feed-back loop the key step is apparently the interaction between the released ATP and the enzyme 5′-nucleotidase. Inhibition of the enzyme with specific anti 5′-nucleotidase antibodies (STANLEY et al. 1983) stimulates ACh release and the external addition of immobilized enzyme causes inhibition of the ACh release (RICHARDSON and BROWN 1987a). The direct nucleoside release was demonstrated to be insufficient to activate a receptor of the A1 type. The absence of the key regulatory enzyme 5′-nucleotidase from the presynaptic plasma membrane indicates that parasynaptic elements such as glial cells play an important role in controlling the extraterminal adenosine concentration in the neighbourhood of the cholinergic terminals. Thus, the increased extraterminal adenosine concentration then regulates both ACh and ATP release via the presynaptic adenosine receptors (A1 type).

Various mechanisms have been suggested for the inhibition of ACh release by adenosine derivatives. These include: reduction of nerve terminal action potential

(AKASHU et al. 1981); blockade of voltage-dependent Ca^{2+} entry (KURODA 1978; RIBERIO et al. 1979); a decrease in the availability of Ca^{2+} (GINSBORG and HIRST 1972); stimulation of the rate of Ca^{2+} uptake into storage sites (RIBERIO and DOMINGUEZ 1978); and decrease in the affinity for Ca^{2+} of a structural component of the secretory process (SILINSKY 1981). In frog skeletal muscle preparations the activation of extracellular adenosine receptors on adenylate cyclase inhibits evoked ACh release by reducing the affinity for Ca^{2+} of an intracellular, so far unidentified, component of the secretory apparatus (SILINSKY 1984).

By analogy with the *Torpedo* electromotor system the inactivation of the ligand adenosine in the preparation of immunoaffinity-purified cholinergic synaptosomes probably occurs via an adenosine uptake process. The transporter is dipyridamol sensitive and has an apparent K_T for adenosine uptake of about 17 μM. Following its uptake adenosine is rapidly metabolized via ATP synthesizing pathways in the terminals and appears as adenine nucleotides (RICHARDSON and BROWN 1987b; STONE 1981).

V. The Concept of Independent Purinergic Terminals

HOLTON and HOLTON (1954) were the first to suggest a transmitter role for ATP but BURNSTOCK (1972) has crystallized this into the concept of 'purinergic transmission'. For more information about this controversial subject readers are referred to detailed reviews by BURNSTOCK (1972, 1975, 1978), STONE (1981) and PHILLIS and WU (1981). Here the arguments for and against the view that ATP is a primary transmitter are briefly outlined. In favour are the following observations (for more details, see BURNSOCK 1972, 1975; FYFFE and PERL 1984; JAHR and JESSEL 1983):

1. In the mammalian gut the synthesis and storage of ATP in nerve terminals is such that it is available for release upon stimulation.
2. Release of the endogenous nucleotides during nerve activity has been demonstrated.
3. Degrading enzymes – ATPase, 5′nucleotidase, adenosine deaminase – which could terminate any transmitter action by ATP are present.
4. Various drugs can block, potentiate or mimic the action of directly applied ATP.
5. Large opaque vesicles (LOVs) 80–200 nm in diameter present in some neurons, e.g. those of Auerbach's plexus in the mammalian small intestine and the area postrema in the CNS, could be the storage sites for ATP in purinergic nerves.
6. Receptors of A 1 and A 2 types which can bind nucleotide compounds are present on some autonomic nerves and in the CNS.
7. ATP can induce a response similar to that mediated by nerve stimulation in many peripheral nervous system (PNS) preparations.

The arguments against purinergic transmission are as follows (for more details, see FREDHOLM and HEDQUIST 1980; STONE 1981; PHILLIS and WU 1981):

1. Unlike neurons containing classical transmitters where synthesizing enzymes are confined to the neuron which utilizes them the transmitter ATP and its synthesizing system are ubiquitously distributed in all living cells including neurons.

2. As has already been mentioned, the released ATP and/or its derivatives could originate from neurons containing classical transmitters, e.g. ACh or catecholamines. In some situations this release of nucleotides accompanies that of the primary transmitter and must be regarded as co-release. Release of nucleotides without the secretion of the primary transmitter and *vice versa* may also occur differentially depending on the type of stimulus applied. Release of nucleotides is not restricted to neurons but is a common feature of all excitable cells as a response to various type of stimulus including noxious ones.
3. The activity of enzymes degrading ATP (see above) is not localized to special regions rich in purinergic nerves.
4. Purinergic receptors are not restricted to specialized regions of the CNS or peripheral nervous system (PNS) rich in purinergic neurons. They are also present in groups of neurons receiving cholinergic or adrenergic inputs.
5. While isolation and characterization of storage granules with different transmitter or peptide content has become a standard neurochemical procedure no report has appeared so far on the isolation of the LOV-type vesicles or the demonstration that they contain only ATP. Attempts to isolate these granules from Auerbach's plexus of the guinea-pig ileum failed (D. V. AGOSTON, unpublished work).

In summary, the balance of evidence is in favour of the view that ATP is a 'metabolic cell-cell messenger' or co-transmitter rather than a primary neurotransmitter.

VI. Must ATP Be Reckoned as a Cholinergic Co-transmitter?

It has been known for several years that with changes in the degree of activation of postsynaptic receptors, there results a compensatory adaptation of the presynaptic system so as to change the transmitter output, thus restoring the receptor activity towards normal. Nucleotides liberated into the extracellular space from whatever source would therefore make excellent candidates for a mediator role in a feedback regulation of transmitter release. As the firing frequency of a neuron increases so the extracellular nucleotide concentration would be expected to rise, and could be regarded as a means of restricting otherwise excessive primary transmitter release.

The physiological significance of ATP in cholinergic synaptic vesicles could be:

1. In the storage of ACh, as a counter ion to ACh.
2. Involvement in Ca^{2+} sequestration in the cholinergic terminal.
3. Involvement in energy-dependent exocytotic fusion and recycling of the cholinergic vesicle. It should be noted that while ATP may keep exocytotic sites in a primed state, it is not required for membrane fusion as demonstrated in *Paramecium* cells (VILMART-SEUWEN et al. 1986).

The physiological significance of ATP in cholinergic signal transduction could include (see also WITZEMANN 1987):

1. Presynaptic regulation; free adenosine may block the release of ACh through binding to adenosine autoreceptors involving adenylate cyclase (KILBINGER

and WESSLER 1980; RICHARDSON and BROWN 1987b; SAWYNOK and JAHMADAS 1976), or by reducing the affinity of Ca^{2+} for an unidentified intracellular secretory component (SILINSKY 1984).

2. Postsynaptic regulation; ATP may have a direct effect on ACh receptors of the nicotinic type (nAChR) via specific binding sites altering the sensitivity for the primary transmitter ACh (EWALD 1976; GORDON et al. 1979; AKATSU et al. 1981).
3. In embryonic muscular cells – myotubes – ATP can act as a true transmitter; it may have an ion-translocating effect via nAChRs but at a later stage in the development only a modulator role of ATP may remain (KOLB and WAKELAM 1983; HÄGGBLAD et al. 1985).

However, ATP as a co-transmitter or modulator is not restricted to cholinergic neurons; it may exert an analogous function in adrenergic transmission (FREDHOLM and HEDQUIST 1980; STONE 1981; PHILLIS and WU 1981).

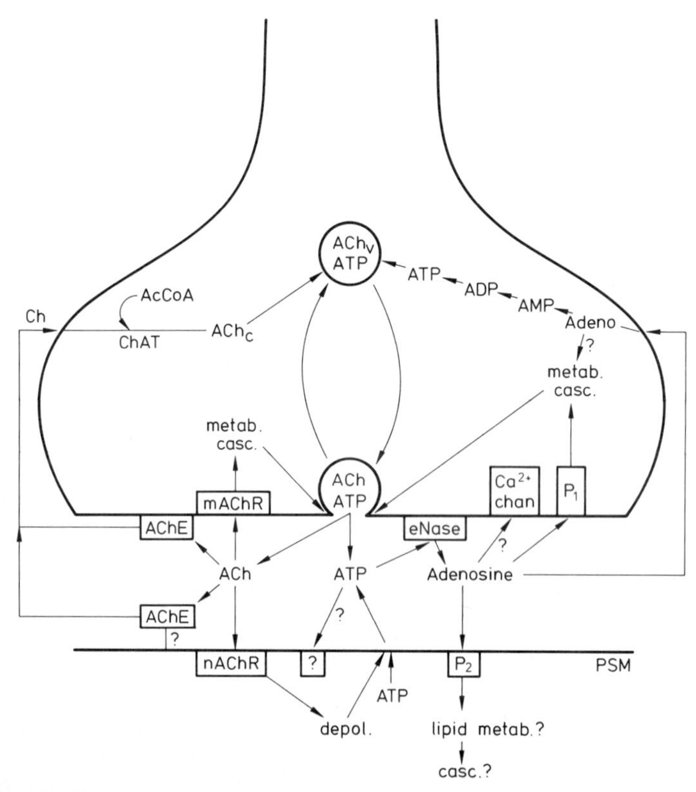

Fig. 1. Speculative model of ATP as co-transmitter in the cholinergic neuron. *mAChR,* muscarinic acetylcholine receptor; *nAChR,* nicotinic acetylcholine receptor; *depol.,* depolarization; *metab. casc.,* metabolic cascade; *PSM,* postsynaptic membrane; *P1* and *P2,* pre- and postsynaptic adenosine receptors; *lipid metab.,* lipid metabolism; *casc.,* cascade; *eNase,* ectonuleotidase; *Adeno.,* adenosine; *Ca chan.,* calcium channel; *Ch,* choline; *ACh_c,* cytoplasmic acetylcholine; *ACh_v,* vesicular acetylcholine; *?,* components and routes for which the evidence could be better

It remains uncertain whether the nucleotides of presynaptic origin or those released from effector cells in much higher amounts are primarily responsible for the pharmacological actions known to be triggered by extracellular nucleotides (FORRESTER and WILLIAMS 1977; ISRAËL and MEUNIER 1978; RIBERIO 1979). However, the balance of evidence is in favour of assigning a role to ATP as a co-transmitter in cholinergic neurons (see Fig. 1). In the electromotor system, there is nerve terminal synthesis, vesicular uptake and storage, very likely vesicular release, extracellular hydrolysis, re-uptake of hydrolysis products followed by nerve terminal resynthesis and vesicular reloading. However, the established role of ATP as a 'metabolic cell-cell messenger' which is apparently a feature of all excitable cells may certainly mask its cholinergic co-transmitter function. These two related functions of ATP in the cholinergic system are therefore from a pharmacological point of view indistinguishable.

C. Evidence for Co-localization of Certain Neuropeptides and ACh

I. Discovery of Neuropeptides

Neurosecretion, the striking ability of neurons to secrete hormone-like substances, was discovered by SPEIDEL in 1919 when he described giant neurons in the posterior spinal cord of elasmobranch fish with anatomical features associated with secretory activity. The accumulated evidence that nerves can secrete hormones led to the concept of the 'glandular neuron' formulated by the SCHARRERS (1940). The secretion products of neurons, consisting of short chains of amino acids, have been – *nomen est omen* – called neuropeptides. Dozens of biologically active peptides with hormone-like action were then isolated and characterized from the hypothalamo-hypophyseal axis (GUILLEMIN 1978). The discovery of substance P (SP) – the doyenne of the 'gut peptides' – had widened the concept of neurosecretion. In contrast to the hypothalamo-hypophyseal peptides, SP was first isolated from the gastrointestinal tract and subsequently described as being present also in brain (VON EULER and GADDUM 1931). There it was found to be highly localized in synaptosomes (CLEUGH et al. 1964) and synaptic vesicles (CUELLO et al. 1977; FLOOR and LEEMAN 1981) indicating a transmitter or co-transmitter function. The discovery of SP was followed by the isolation and characterization of a large number of biologically active peptides. They comprise the ever growing class of gut-brain peptides. Alternatively, they are also called regulatory peptides on the basis of their metabolic role in the gastrointestinal tract.

New strategies such as a search for peptides with a C-terminal amide which is indicative of biological activity were developed (MUTT 1983, 1985). An alternative and very powerful approach is to use the techniques of molecular genetics to characterize peptide precursor mRNAs or the genes that encode them (AMARA et al. 1982; ROSENFELD et al. 1983; EVANS et al. 1983; KOLLER and BROWNSTEIN 1987). The discovery of the many biologically active peptides and their availability in pure form, often as synthetic products available in large quantities, have coincided with new and sensitive immunochemical techniques (for review see KRIEGER et al. 1983; BJÖRKLUND and HÖKFELT 1983).

Different aspects of neuropetide research have been extensively discussed in numerous publications. For detailed information readers are referred to recent reviews and books, e.g. BARKER and SMITH (1984); BROWNSTEIN et al. (1982); DOUGLASS et al. (1984); EIPER et al. (1986); HÖKFELT et al. (1986b); IVERSEN et al. (1983); KRIEGER et al. (1983); LOH et al. (1984); LYNCH and SNYDER (1986); PALKOVITS (1984); PICKERING (1978); SCHARRER (1987); TURNER (1986); WHITE et al. (1985).

II. Discovery of Co-localization

1. Immunocytochemistry

Much of the evidence for the simultaneous presence of more than one neuroactive compound in a neuron is histochemical or immunohistochemical (HÖKFELT et al. 1982, 1986a). However, the term 'coexistence' was first used by BROWNSTEIN et al. (1974) when they described the existence of several putative neurotransmitters in a single neuron. Neurons – with the exception of the giant neurons of *Aplysia* – are much too small to allow the simultaneous measurement of different components with present techniques at the level of a single cell, nor it is possible to dissect out a neuron from its environment with all of its processes intact but uncontaminated by neighbouring cells. Hence immunohistochemistry represents a promising alternative in terms of resolving power, specificity and localization for studying co-localization, since the technique permits the simultaneous visualization of several antigens in one cell. The methodology has recently been reviewed (PRIESTLY and CUELLO 1982; BJÖRKLUND and HÖKFELT 1983; LUNDBERG and HÖKFELT 1983, 1986; HÖKFELT et al. 1986a).

2. Methodology

Generally speaking, the presence of more than one antigen in a single neuron can be demonstrated by three techniques based on the classical method developed by COONS (1958):

1. Under favourable conditions, portions of the same neuron will be present in several adjacent sections and can be identified as such. The consecutive sections can then be incubated with different antisera.
2. Secondly, the 'elution-restaining' method can be used. After staining for the first antigen and photographing the staining pattern the antiserum can be eluted from the section with acidic potassium permanganate and the section reincubated with a new antiserum.
3. Thirdly, the 'direct double-staining' method can be used if antibodies against the different antigens have been raised in different, immunologically unrelated species. Then by application of differentially labelled (e.g. by fluorescine and rhodamine) second antibodies the distribution of more than one antigen can be visualized by changing the wavelength of the exciting light (for more technical details see BJÖRKLUND and HÖKFELT 1983; HÖKFELT et al. 1982).

In spite of the numerous advantages of these methods and their virtually unlimited potential, problems of specificity and sensitivity do arise. Since structurally similar peptides (so-called peptide families) exist, an antiserum raised to a given peptide also can react with others having similar amino acid sequences. Therefore such terms as 'X-like-immunoreactivity', 'X-immunoreactive' or simply 'X-positive' where X is the peptide against which the serum has been raised are used and are generally accepted. Immunopositivity uncontrolled by a demonstration that the original antiserum after adsorption with the original antigen gives no reaction cannot be accepted.

Sensitivity is an equally important problem. Since the sections undergo considerable chemical transformation as a result of fixation, uncontrollable immunological modifications could result. Application of colchicine by blocking the axonal drainage (DAHLSTRÖM 1968) can improve the staining of cell bodies, but then negative result should be interpreted with caution.

In Table 1 a number of selected cases are listed where peptides have been demonstrated to be present in cholinergic neurons. A distinction has been made between direct and indirect evidence. Evidence was accepted as direct when the demonstration of the peptide immunoreactivity of a given neuron was accompanied by the simultaneous demonstration of the presence of one or more histological cholinergic markers (see below) in the same neuron or in the same terminal. When no such simultaneous demonstration was presented but the neuron or fibres are known to be cholinergic on the basis of previous histological or other evidence, the evidence for co-localization was accepted as indirect.

3. Histochemical and Immunocytochemical Definition of Cholinergic Neurons

With the exception of catecholamines (FALCK et al. 1962) where the direct demonstration of the transmitter is possible, the histochemical identification of neurons containing classical transmitters is based on visualization of enzymes specifically related to the metabolism of the transmitter in question. In cholinergic histochemistry two such enzymes have been used to mark cholinergic cells, the enzyme which hydrolyses ACh, acetylcholinesterase (AChE) (see Chap. 8) and the key synthesizing enzyme, choline acetyltransferase (ChAT) (see Chap. 7).

The first specific and reliable method for the localization of AChE-containing structures was described by KOELLE and FRIEDENWALD (1949). It became evident however after extensive use as a marker for cholinergic structures, that this enzyme may also occur in several non-cholinergic and non-cholinoceptive cells, while ChAT is demonstrated to be more closely correlated to the distribution of the transmitter ACh (HEBB 1956). The histochemical detection of the enzyme by KÁSA (1970) and KÁSA et al. (1970) promised a more specific method for identifying cholinergic neurons. However, the results were impaired by poor localization and lability of enzyme activity. The method of choice today is immunohistochemistry using poly- or monoclonal antibodies raised against ChAT (KÁSA 1986; see also Chap. 22). Reports concerning discrepancies in the co-localization of AChE and ChAT have appeared (LEVEY et al. 1983; MESULAM et al. 1983). Whatever immunochemical method for ChAT is used and however specific and sensitive this may be difficulties regarding localization and detection always occur. The use of

Table 1. Immunocytochemical evidence for coexistence of ACh and peptides (selected cases)

Tissue/region/ species	Cholinergic		Neuroactive compound(s)/ peptide(s)		Evidence for co-localization		Notes	References
	Marker or evidence (Reference)	Function	Name	Demonstration of co-localization	Direct[a]	Indirect[b]		
Nerves in exocrine glands of rat	AChE EHC	Secretovasomotor	VIP	ICC of the peptide followed by AChE EHC on the same sections	Yes	–	Only AChE EHC but no ChAT ICC	1
Nerves in ganglia of rat				IEM demonstration of VIP-immunoreactive LDC vesicles in established cholinergic terminals	–	Yes	Lack of simultaneous cholinergic marker	2
Sympathetic nerves in sweat gland of rat	Ref. 4	Secretovasomotor	CGRP and VIP	ICC on consecutive sections in an established cholinergic system	–	Yes	Lack of simultaneous cholinergic marker	3
Neurons in myenteric and submucosus plexus of guinea-pig ileum	ChAT ICC	Motor	CGRP NPY CCK SOM	Simultaneous multiple double labelling ICC on whole mount preparations	Yes	–		5
Neurons in submucosus ganglia of guinea-pig ileum	ChAT ICC	Secretomotor	CCK SOM NPY	Simultaneous multiple double labelling ICC on whole mount preparations	Yes	–	Peptide immunoreactivity is lacking only in one subpopulation among the three different ChAT-positive neuronal groups investigated	6
			SP		Yes	–		

Sympathetic ganglia in bull-frog	AChE EHC	?	LHRH	?	Yes?	–	Only AChE EHC but no ChAT ICC was performed	7
Preganglionic nerves in rat	AChE EHC	?	Neuro-tensin	?	Yes?	–	Only AChE EHC but no ChAT ICC was performed	8
Motor end-plates in skeletal muscle in mammals and neurons in hypoglossal nuclei and spinal cord of mammals	ChAT ICC, AChE ICC and AChE EHC	Motor	CSADCase (taurine); tyrosine hydroxylase (dopamine); GADase (GABA)	ICC on consecutive sections combined with elution followed by restaining	Yes	–	No presence of the neuroactive compounds was demonstrated; only that of the key synthesizing enzyme	9, 10
Neurons in hypoglossal-facial and ambiguous nuclei of rat	ChAT ICC	Motor	CGRP	ICC on consecutive sections	Yes	–		11
Neurons in nervus terminalis in foetal and embryonic rat	ChAT ICC	Secretovaso-motor?	VIP	ICC on separated but not consecutive sections	–	Yes	Similar distribution of the two immunoreactivities	12
Neurons in medial septal nucleus and diagonal band nuclei of rat	ChAT ICC	Higher brain functions (memory and learning?)	GAL	ICC on consecutive sections	Yes	–	Colchicine pretreatment	13
Fibres in hippocampal afferents of rat	AChE EHC	Higher brain functions	GAL	ICC and AChE EHC on consecutive sections	Yes	–	The study was combined with anatomical and chemical lesions in order to identify the origin of the peptide reactivity	14

Table 1 (continued)

Tissue/region/species	Cholinergic		Neuroactive compound(s)/peptide(s)		Evidence for co-localization		Notes	References
	Marker or evidence (Reference)	Function	Name	Demonstration of co-localization	Direct [a]	Indirect [b]		
Neurons in the laterodorsal segmental nucleus of rat	ChAT ICC	Ascending cholinergic reticular system to the cortex	SP	ICC on consecutive sections and double labelling	Yes	–	Colchicine pretreatment	15
Neurons and fibres in the superior olivary complex of guinea-pig	AChE EHC	Sensory	Enkephalin-like	ICC followed by AChE EHC	Yes	–	No ChAT ICC was performed	16
Interneurons in the cerebral cortex of rat	ChAT ICC	Vasomotor?	VIP	ICC on consecutive sections	Yes	–	Close association of ChAT-VIP-positive neurons with blood vessels was observed	17
Neurons and fibres of the cochlea in guinea-pig and rat	Ref. 18	Auditory feed-back loop	Enkephalin-like	ICC in an established cholinergic system	–	Yes	The system, as previously demonstrated by AChE EHC, is known to be cholinergic	18
Nerves and fibres of cerebral arteries of the rat	AChE EHC	Vasomotor	VIP	ICC and AChE EHC on similar but not on the same whole mount preparations	–	Yes	Similar distribution of the immunoreactivities was proved	19

Neurons and fibres in parasympathetic ganglion of cat	AChE ECH	Motor	VIP	ICC followed by AChE EHC on the same sections	Yes	–	A combined morphological and electrophysiological study	20
Preganglionic terminals in avian ciliary ganglion	–	Motor	SP enkephalin	ICC in an established cholinergic system	–	Yes		21
Neurons in heart of toad	–	Motor?	SOM	ICC in an established cholinergic system	–	Yes	A combined morphological and electrophysiological study	22
Electromotor neurons, axons and terminals in *Torpedo marmorata*	Ref. 23	Modified motor	VIP-LI	ICC in an established cholinergic system	Yes	–	Homogeneous cholinergic model system	23

ICC, immunocytochemistry; EHC, enzyme histochemistry; IEM, immunoelectron microscopy
[a] Direct evidence: simultaneous localization by any technique (double labelling, consecutive sections, etc.) in the presence of a cholinergic marker (ChAT or AChE) and the peptide in the same neuron or its processes
[b] Indirect evidence: the detection of the presence of a neuropeptide in a previously but simultaneously identified cholinergic neuron or its processes

References:
1 Lundberg (1981)
2 Johansson and Lundberg (1981)
3 Landis and Frediew (1986)
4 LeBlanc et al. (1986)
5 Furness et al. (1985)
6 Furness et al. (1984)
7 Jan and Jan (1983a)
8 Gleser and Basbaum (1980)
9 Chan-Palay et al. (1982a)
10 Chan-Palay et al. (1982b)
11 Takami et al. (1985)
12 Schwanzel-Fukuda et al. (1986)
13 Melander et al. (1985)
14 Melander et al. (1986)
15 Vincent et al. (1983)
16 Altschucher et al. (1983)
17 Eckenstein and Baugham (1984)
18 Fex and Altschucher (1981)
19 Kobayashi et al. (1983)
20 Kawatani et al. (1985)
21 Erichsen et al. (1982)
22 Campbell et al. (1982)
23 Agoston and Conlon (1986b)

monoclonal antibodies raised against ChAT has undoubtedly increased the specificity of the method, but unfortunately this carries with it the risk of false-negative reactions due to alteration of the epitope during fixation and tissue preparation. The controversy over ChAT positivity has increased. Recent reviews dealing with the critical assessment of the central cholinergic pathways with special reference to controversial ChAT immunohistochemistry are available (KÁSA 1986; and Chap. 22, this volume).

Promising alternatives for visualizing cholinergic structures are the use of markers other than ChAT such as Chol-1 (see Chap. 18) or antibodies directed against the transmitter molecule itself (GEFFARD et al. 1985a, b).

III. Histochemical and Cytochemical Evidence for Co-localization

1. Mammals

a) Peripheral Autonomic Nervous System

Systematic immunohistochemical mapping of the cat autonomic nervous system for vasoactive intestinal polypeptide (VIP)-like immunoreactivity demonstrated that a large proportion of cells in the ganglia of the sympathetic chain are immunoreactive. The most strongly immunopositive are the seventh lumbar and stellate ganglia (LUNDBERG et al. 1979). The relative proportions of the VIP-like immunoreactive neurons in these ganglia have been well correlated with much earlier figures for AChE-positivity (SJÖQUIST 1963) and are consistent with the results obtained from microdissection followed by biochemical analysis for ChAT activity (BUCKLEY et al. 1967). A systematic histological analysis then showed that VIP-like immunoreactive neurons correspond to AChE positive cells suggesting that a VIP-like peptide is present in cholinergic neurons (LUNDBERG et al. 1979). In a parallel analysis of the parasympathetic division of the autonomic system it was shown that in submandibular ganglia a high proportion of neurons are VIP-positive, and this immunoreactivity is well correlated with AChE-positive neurons (LUNDBERG 1981). However, AChE staining is not considered to be absolutely specific for cholinergic neurons (KÁSA 1986; Chap. 22, this volume). A final and very convincing piece of evidence for the coexistence of VIP and ACh in the same neuron was the positive VIP-immunoreactivity of neurons of the sphenopalatine ganglion, all cells of which are considered to be cholinergic (LUNDBERG et al. 1981). VIP-like immunoreactivity is also consistently found in the cholinergic innervation of exocrine glands and indeed represents a common feature of these tissues (BRYANT et al. 1976; LARSON et al. 1978; LUNDBERG et al. 1979a, 1980b, 1981a, 1981; UDDMAN 1980). Extirpation of the sphenopalatine ganglion caused parallel losses in AChE-positive and VIP-immunoreactive fibres as well as decreased ChAT and VIP levels in the nasal mucosa which is innervated by this ganglion (LUNDBERG et al. 1981).

Immunocytochemistry at the electron microscope level has showed that VIP-positivity is present in nerve terminals and that the immunoreactivity is localized mainly if not entirely in large dense-cored vesicles (LDCs) of 100–120 nm diameter and not in the typical small electrolucent vesicles (SELs) of diameter 50–60 nm, characteristic of cholinergic nerve endings and also present in large numbers in the terminals (JOHANSSON and LUNDBERG 1981).

In the gastrointestinal tract multiple overlapping between immunostainings has been reported using the whole mount technique (COSTA et al. 1980). In the submucosus plexus of the guinea-pig ileum any combination of antisera against ChAT, cholecystokinin (CCK), calcitonin gene-related peptide (CGRP), neuropeptide Y (NPY) and growth hormone release-inhibitory hormone or somatostatin (SOM) shows complete overlap in nerve cell bodies (FURNESS et al. 1984, 1985). In the submucosus ganglia of the same preparation some 55% of the cell bodies show ChAT-positive immunoreactivity. Only 16% of these ChAT-positive neurons fail to show a positive immunoreactivity to antisera to other peptides. Of the ChAT-positive neurons 11% showed a positive response when probed for SP and 28% for NPY; the latter were also CCK-, CGRP- and SOM-positive (COSTA et al. 1986a, b).

It must be emphasized that the co-occurrence of these immunoreactivities is restricted to the cell bodies and in most cases only after treatment with colchicine which by arresting axonal transport causes a damming-up of immunopositive material in the cell bodies to detectable levels.

In contrast, due to sensitivity and other technical problems, the demonstration of the co-occurrence of different immunoreactivities in nerve fibres and terminals remains difficult (for review, see COSTA et al. 1986a, b).

The submucosus ganglia of the guinea-pig ileum was investigated for ChAT and some peptide (CCK, NPY, SOM, SP and VIP) immunoreactivities by FURNESS et al. (1984). Some 45% of the cell bodies showed VIP-like immunoreactivity; a separate group of nerve cells comprising some 54% of the total population contained ChAT-immunoreactivity with no overlapping between the two groups of neurons. However, some 30% of the ChAT-positive cells also showed immunoreactivity for CCK, SOM and NPY. In some 11% of neurons there was positive staining for both ChAT and SP. Cell bodies, morphologically classified as Dogiel type III in the myenteric and submucosus plexus of guinea-pig small intestine, are found to exhibit simultaneous immunoreactivities to five different neuroactive substances of ChAT, CGRP, CCK, NPY and SOM (FURNESS et al. 1984).

By contrast, dense VIP-, SOM- and CCK-immunoreactive groups of neurons were identified in the submucosus plexus of guinea-pig ileum using transverse sections (HÖKFELT et al. 1984), but their relationship to cholinergic neurons was not established with certainty due to difficulties with the histochemical identification of cholinergic fibres and neurons (SCHULTZBERG 1984).

In the myenteric ganglia some 5% of the cell bodies show immunoreactivity for ChAT, CCK, CGRP, NPY, SOM simultaneously (FURNESS et al. 1985; COSTA et al. 1986a, b).

Co-localization of CGRP and VIP in the functionally cholinergic rat sweat gland was reported by LANDIS and FREDIEU (1986). The use of adjacent serial sections showed identical distribution and density for CGRP- and VIP-like immunoreactivity in the plexus of the gland. It should be noted, however, that no cholinergic marker for identifying the cells or fibres was used.

b) Neuromuscular Junction

Some neuromuscular junctions (NMJs) in rat tongue show CGRP-like immunoreactivity as well as positive staining for AChE. However, the overlap between

AChE positivity and CGRP-like immunoreactivity was not complete (TAKAMI et al. 1985).

Motor end-plates in NMJs of five primate species including man exhibited immunoreactivity not only to ChAT but to three other synthesizing enzymes, tyrosine β-hydroxylase, cysteine sulphinic acid decarboxylase and glutamate decarboxylase, enzymes that respectively catalyse the rate-limiting steps in the formation of catecholamines, taurine and γ-aminobutyric acid (GABA) (CHAN-PALAY et al. 1982a, b; CHAN-PALAY and PALAY 1984). The study was performed using poly- and monoclonal antibodies in light- and electron-microscopic immunocytochemistry. Its most striking aspect is that despite the positive immunostaining of synthesizing enzymes the presence of the three non-cholinergic transmitter substances at the NMJ has not been proved biochemically. Such results indicate that sensitivity immunohistochemical techniques may detect a range of different substances in neurons but this does not necessarily imply any functional activity. Some of the immunoreactivities may represent 'sleeping' substances, in a functionally inactive or partially repressed state.

All neurons contain the information, encoded in the genome, required for the expression of any particular transmitter function. Normally, all but one of these functions is repressed. This repression may not always be as complete as has hitherto been assumed and some component of a repressed function, but in amounts or in a form insufficient to assert it may escape repression and be expressed in low concentration and in a functionless state.

Careful assessment is needed of a number of reports in which peptide immunoreactivities were demonstrated in putative cholinergic systems without simultaneous demonstration of the cholinergic nature of the peptide-immunopositive neurons (GLAZER and BASBAUM 1980; LANDIS and FREDIEU 1986; ERICHSEN et al. 1982; TAKAMI et al. 1985; see also Table 1).

c) Central Nervous System

VIP-like immunoreactivity and ChAT-immunopositivity was simultaneously demonstrated in bipolar interneurons of the rat cerebral cortex (ECKENSTEIN and BAUGHMAN 1984). The coexistence of VIP and ChAT was remarkably high in neurons which are intimately associated with blood-vessels. Close association of VIP and ChAT positive nerves and fibres has been demonstrated in whole mount preparations of cerebral arteries (KOBAYASHI et al. 1983). However, due to methodological problems direct proof of this coexistence is lacking.

VIP-like immunoreactivity was also found to be distributed similarly to ChAT immunoreactivity in neurons of the nervus terminalis of foetal and embryonic rats (SCHWANZEL-FUKUDA et al. 1986). Enkephalin-like immunoreactivities have been found in the olivocochlear neurons and fibres of the cochlea in guinea-pig and cat (FEX and ALTSCHULER 1981) and in AChE-positive neurons and fibres of the superior olivary complex of the guinea-pig (ALTSCHULER et al. 1983).

The light- and electron-microscopic evidence for the coexistence of synthesizing enzymes for four neuroactive substances present in the primate NMJ was extended to central motoneurons. Positive immunostaining was found in neurons of the hypoglossal nucleus and the spinal cord for ChAT, cysteine sulphinic acid decarboxylase, glutamate decarboxylase and tyrosine-β-hydroxylase (CHAN-PALAY et al. 1982b; CHAN-PALAY and PALAY 1984). As with the NMJ no evidence is forthcoming for the actual presence of the transmitters synthesized by these enzymes.

The alternative RNA processing of the calcitonin gene in neuronal tissues generates the novel 37 amino acid long peptide CGRP (AMARA et al. 1982; ROSENFELD et al. 1983; EVANS et al. 1983). A systematic search for CGRP-like immunoreactivity in selected central cholinergic nuclei (hypoglossal and facial nuclei and nucleus ambiguus) using ChAT immunocytochemistry as the cholinergic marker has demonstrated that CGRP-like and ChAT-like immunoreactivities coexist in the motoneuron cell bodies of these nuclei (TAKAMI et al. 1985). CGRP-like immunoreactivity was also found in some motoneurons innervating the striated muscle of the tongue (ROSENFELD et al. 1983).

Galanin (GAL), a novel biologically active peptide isolated from swine intestine and named after its N-terminal glycine and C-terminal alanine residues, causes smooth-muscle contraction (TATEMOTO et al. 1983) and is present in central and peripheral neurons (RÖKAEUS et al. 1984). Neurons in the medial septal nucleus and the nuclei of the diagonal band as well as fibres in hippocampal afferents exhibit immunoreactivity for GAL which is accompanied by positive immunoreactivity for ChAT and AChE (MELANDER et al. 1985, 1986).

2. Non-mammalian Systems

a) Amphibia

The presence of SOM has been demonstrated in cholinergic fibres of the cardiac vagus and intracardiac neurons in the heart of a toad, *Bufo marinus* (CAMPBELL et al. 1982). It should be noted that the cholinergic nature of these neurons was predicted from previous pharmacological studies and not by histochemical colocalization. The role of SOM in vagal transmission was also demonstrated (see below) (CAMPBELL et al. 1982).

Sympathetic ganglia in bullfrog have been shown to contain a luteinizing hormone releasing hormone (LHRH)-like immunoreactivity (JAN and JAN 1983a). Essentially all terminals of preganglionic C fibres show LHRH-like immunoreactivity. Physiological experiments indicate (see below) that most of the fibres also contain (and release) ACh. However, the cholinergic nature of the peptide-positive neurons has not been proved by histochemical means (BRANTON et al. 1986).

b) Birds

Preganglionic terminals in avian ciliary ganglion which are known to be purely cholinergic have been demonstrated to contain overlapping immunoreactivities of both SP and enkephalin (ERICHSEN et al. 1982).

c) *Torpedo* Electromotor System

VIP-like immunoreactivity has been detected (AGOSTON and CONLON 1986, 1987a) in the electromotor neurons of the electric ray *Torpedo marmorata,* widely used (Chap. 2) as a model cholinergic system. A VIP-like antigen is present in the electromotor cell bodies in the electric lobes, their axons and nerve terminals. The distribution of this VIP-like immunoreactivity closely resembles that of the *Torpedo* cholinergic synaptic vesicle marker proteoglycan (ÁGOSTON et al. 1986a). In the electric organ the immunoreactivity is localized on the ventral, innervated surface of the electrocytes; in the electric lobes the perikarya of the electromotor neurons are positively stained. The nuclear staining also observed may be artefactual. The electromotor axons showed weak staining without nerve ligation, but insertion of a ligature caused a marked proximal accumulation of the VIP-like immunoreactivity.

Simultaneously, a considerable reduction of the immunostaining was observed in the terminals covering the ventral surface of the electrocytes. In the electric lobe, the loss of the axonal drainage caused a marked accumulation of the immunoreactive material in the perikaryal cytoplasm and at the axon hillock. The redistribution of the VIP-like immunostaining following the ligation of the nerves was quite comparable to the redistribution of the cholinergic vesicle marker proteoglycan and is evidence that both synaptic vesicles and granules storing the VIP-like peptide are elaborated in the cell body and transported via the axons to the terminal. Preliminary subcellular fractionation studies (AGOSTON 1987) indicate that the two storage organelles are indeed distinct (see following section).

IV. Cell-Biological and Biochemical Evidence

1. Methodology

Attempts have also been made to analyse the subcellular distribution and storage of multiple neuroactive compounds by means of fractionation studies. To discover whether more than one neuroactive substance is stored in the same (co-storage) or separate storage organelles, subcellular fractionation combined with biochemical, immunochemical and electron-microscopic analysis has been used. Table 2 summarizes selected examples with special reference to histochemically demonstrated coexistence.

2. Mammals

Subcellular fractionation of cat submandibular gland, as already mentioned, where the coexistence of VIP with ACh had previously been demonstrated immunohistochemically, revealed the existence of distinct storage particles for the two neuroactive compounds (LUNDBERG et al. 1981). While VIP was exclusively found in the heavy fractions of the sucrose density gradient (0.8–0.97 *M* sucrose) ACh showed a bimodal distribution. The majority of the ACh-containing vesicles was found in the lighter part of the gradient (0.28–0.44 *M* sucrose) but also a relative enrichment was found in a denser fraction (0.87 *M* sucrose). The electron-microscopic analysis showed SELs in the light ACh-containing peak, whereas the fractions containing VIP consisted of large vesicles, sometimes LDCs. A small proportion of the particle-bound ACh was apparently localized in the LDCs. However, it was not clear from the study whether separate populations of large vesicles exist, one storing VIP and the other ACh, or whether true co-storage is occurring.

Table 2. Cell biological and biochemical evidence for coexistence of ACh and peptides (selected cases)

Tissue/region/ species	Cholinergic marker or evidence (reference)	Neuroactive peptide(s)		Evidence for cotransmission, co-storage or specific interaction	Notes	References
		Name	Method of identification			
Submandular gland of the cat	ACh	VIP	HPLC, RIA	LDC storage granules were isolated containing VIP and ACh	No direct proof whether VIP and ACh are stored in the same granule or in two similar but distinct types of vesicles	1
Myenteric plexus of guinea-pig small intestine	ACh	VIP	HPLC, RIA	LDC vesicles were isolated containing VIP and ACh	Osmotic fragility measurements reveal same storage mechanism for ACh and VIP in the LDC vesicles	2, 3
Anterior hypothalamus and hippocampus of the rat	ChAT	VIP 10^{-7} *M*	External addition of synthetic peptide	Activation of ChAT observed in homogenates from different brain areas	V_{max} of choline for ChAT is increased and K_m is decreased	4
Electric organ of *Torpedo*	Ref. 4	Leu-enkephalin	[^{3}H] D-Ala^2, D-Leu^5 enkephalin (the stable analogue)	Isolated nerve terminals and synaptic vesicles exhibit the presence of an enkephalin uptake system	$K_T = 12\ \mu M$; $V_{max} = 0.85$ pmol (mg^{-1} protein min^{-1}) D-Ala^2, D-Leu^5 enkephalin stimulated by high concentration of Mg^{2+} ATP (10 m*M*)	5, 6

Table 2 (continued)

Tissue/region/species	Cholinergic marker or evidence (reference)	Neuroactive peptide(s)		Evidence for cotransmission, co-storage or specific interaction	Notes	References
		Name	Method of identification			
Electromotor system of *Torpedo*	ACh	VIP-LI	HPLC, RIA	Large vesicles, partly dense-cored, were prepared from the electric organ, electromotor nerves. Electric lobes of *Torpedo* containing only peptide but no ACh	No co-storage was observed	7
Electric organ of *Torpedo*	Ref. 7	?	Membrane-bound peptide-hydrolysing activity was observed	Captopril-sensitive metallopeptidase with resemblance to mammalian angiotensin-converting enzyme was purified and characterized	The specific endogenous substrate	8

References:
1 LUNDBERG (1981)
2 ÁGOSTON et al. (1985a)
3 AGOSTON (1987b)
4 LEVINE et al. (1984)
5 DAY et al. (1985)
6 MICHAELSON and WIEN-NAOR (1987)
7 AGOSTON (1987c)
8 TURNER et al. (1986)

A similar investigation has been made with the myenteric plexus-longitudinal muscle preparation of the guinea-pig ileum using a high-resolution zonal rotor (ÁGOSTON et al. 1985a). Under certain conditions bimodal distribution of the particulate ACh was demonstrated. The lighter fraction of the particulate ACh equilibrated at a density of 1.066 g ml^{-1}. Electron microscopy showed that the particles in this fraction had the typical appearance of the classical SELs with a mean profile diameter of 48 nm (ÁGOSTON et al. 1985b). The second ACh peak which almost coincident with the VIP peak equilibrated at a density of 1.144 g ml^{-1}. Electron microscopy showed large vesicles; many of them were of the LDC type with a profile diameter of 110 nm. Homogenization at a high rate of shear reduced the size of the VIP peak and caused the disappearance of the second ACh peak, indicating, however, indirectly the possible co-storage of the two substances.

Further evidence was obtained from osmotic fragility studies. Fractions from the peak of SELs containing ACh only and others from that of LDCs apparently co-storing ACh and VIP were separately pooled and submitted to varying degrees of hypo-osmotic stress. ACh and VIP were released *pari passu* from the LDC fraction indicating a common storage mechanism in the same granule, and the osmotic dependency of the release clearly differed from that of ACh in the SELs. These were less sensitive to hypo-osmotic conditions than the LDCs (WHITTAKER and AGOSTON 1986; AGOSTON 1988b; AGOSTON and COLON 1987b).

The synaptic vesicle marker synaptophysin, also known as p38, is an integral synaptic vesicle membrane protein (WIEDENMANN and FRANKE 1985; JAN et al. 1985). The distribution of this antigen was studied in the zonal gradient by means of semi-quantitative dot-blot analysis. A bimodal distribution of synaptophysin was observed in experiments where ACh was also bimodally distributed. Disappearance of the second ACh peak in parallel with the reduction of the corresponding VIP peak was accompanied by loss of synaptophysin from the remaining vesicles (mostly LDC type) further indicating the existence of a distinct subpopulation of LDCs co-storing ACh and VIP (AGOSTON 1988a). Whether this co-storage is a phenomenon related to mechanisms present in the nerve terminals or in the perykaria remains to be answered.

Addition of VIP to homogenates obtained from dorsal and ventral hippocampus and anterior hypothalamus, but not from the parietal cortex and posterior hypothalamus increased the activity of ChAT by 27%–42%. The presence of VIP in a concentration of 0.1 μM increased the V_{max} of the enzyme and the K_m for choline but that for acetylcoenzyme A was decreased (LUINE et al. 1984). No strict correlation of the effect with the distribution of VIP-binding sites in different brain areas could be demonstrated. However, muscarinic ACh receptor (mAChR) involvement at the presynaptic site via increased cAMP levels cannot be excluded.

3. *Torpedo* Electromotor System

The VIP-like immunoreactivity localized by immunohistochemistry in the cholinergic electromotor system of *T. marmorata* was quantified and characterized by means of specific radioimmune assay (RIA) and reverse-phase high-performance liquid chromatography (HPLC) (AGOSTON and COLON 1986a, b). The peptide as revealed by HPLC and dilution experiments followed by RIA is similar but not identical to that of mammalian VIP. Also molecular diversity was found between the VIP-like immunoreactivities extracted from the lobe and the electric organ respectively, indicating a post-translational processing of the peptide during its passage down the axon.

The storage organelles containing the VIP-like peptide were isolated from the *Torpedo* electric organ, electromotor nerves and the electric lobe using high-resolution zonal sucrose-density centrifugation (AGOSTON 1987b). They form a distinct population of storage organelles as compared to the classical synaptic vesicles containing ACh. The peptide-containing granules equilibrated at a sucrose density of $1.140\,g\,ml^{-1}$. Big vesicles, partly of the LDC type with a diameter of 120 nm were present in this part of the gradient. The stable cholinergic synaptic vesicle marker vesicular proteoglycan (AGOSTON et al. 1986b) was associated with these vesicles too. No ACh and ATP were found in this region of the gradient, indicating that in the *Torpedo* electromotor system ACh is not even partially co-stored with the peptide.

An enkephalin-like peptide has been reported to occur in extracts of the electric organ of *T. ocellata*. The peptide co-eluted with tritiated leu-enkephalin on Sephadex G10 chromatography (MICHAELSON et al. 1984). The same immunoreactivity was found in P2 fractions prepared from minced electric organ and some enrichment was observed in the synaptosomal fraction when the P2 fraction was further fractionated. The uptake of a stable enkephalin analogue into isolated synaptic vesicles and nerve terminals was also observed (MICHAELSON and WIEN-NAOR 1987; see also Chap. 5, this volume).

The physiological relevance of one or more peptides in the electromotor system of *T. marmorata* have been further strengthened by the isolation from the electric organ of a membrane-bound peptidase (TURNER and DOWDALL 1984; TURNER et al. 1986). The enzyme is a captopril-sensitive metallopeptidase and acts as a peptidyl-dipeptidase in cleaving dipeptides from the C-terminus of a variety of peptide substrates including angiotensin I, bradykinin and (D-Leu^5)-enkephalin. The enzyme was recognized by a polyclonal antibody to pig kidney angiotensin-converting enzyme. However, the specific, endogenous substrate of the enzyme has not yet been identified.

V. Physiological and Pharmacological Evidence

1. Methodology

Interaction between ACh and peptides can occur when one or more peptides are present in and released from cholinergic neurons as confirmed by other techniques (e.g. immunocytochemistry). This situation allows the terminal to respond to different stimuli with the preferential release of one or other of the neuroactive substances and autoregulation can also take place. This will be called the intrinsic mode of interaction. By contrast, where coexistence has not been established, the neuroactive compounds may be released from different neurons but interact on the same target cell. This could involve competition for a single binding site or interaction between binding sites. This will be called the extrinsic mode of interaction. Interactions can also be classified as short- or long-term on the basis of their duration; these might represent different cellular mechanisms (see Table 3).

2. Intrinsic Short-Term Interactions of ACh and Peptides *In Vivo*

a) PNS of Mammals

Parasympathetic nerve stimulation (PNS) of the cat submandibular gland releases VIP followed by salivary secretion and vasodilatation (LUNDBERG 1981). The VIP overflow is higher at 15 Hz than at 5 Hz. Muscarinic blockade, e.g. by atropine, abolishes the salivary secretion caused by parasympathetic stimulation. The effect of atropine on vasodilatation is frequency-dependent. At 2- and 6-Hz stimulation the vasodilatation is reduced while at 15 Hz the blood flow response is increased and lasts longer. This can be explained by the fact that atropine increases the VIP output at high frequency stimulation about eight- to ten-fold. This potentiated VIP output also occurs at 6-Hz stimulation, however to a lesser extent. This potentiating effect could be the result of the blockade of muscarinic autoreceptors. These observations suggest that both ACh and VIP are involved in vasodilatation (LUNDBERG et al. 1981 b). The presence of anti-VIP antiserum in the perfusate reduces vasodilatation and secretion caused by parasympathetic nerve stimulation, indicating the involvement of VIP in both responses. However a similar reduction can be obtained in both secretion and vasodilatation if avian pancreatic polypeptide (APP) is included in the perfusate. This could be the result of a blockade of 'VIPergic' mechanisms on the effector cells (LUNDBERG et al. 1981 a). Both vasodilatation and secretion can be blocked by hexamethonium, a nicotinic agonist, indicating that the preganglionic nerves are cholinergic. Perfusion with exogenous VIP causes an atropine-resistant vasodilatation but not salivary secretion, while ACh perfusion induces both vasodilatation and secretion. VIP is demonstrated to be 100 times more potent as a vasodilator agent than ACh. The secretory effect of the infused ACh can be enhanced by addition of VIP. In addition, ACh can modulate the release of VIP (LUNDBERG et al. 1982). In cat submandibulary gland ACh and VIP seem to be involved in both secretory and vasodilatatory responses. Thus, for secretory responses ACh acts as a classical neurotransmitter, while VIP serves as a neuromodulator. In vasodilatation both ACh and VIP can act as neurotransmitters.

In vesical ganglia of the cat's urinary bladder, where over 90% of the VIP-positive neurons also contains AChE, muscarinic firing is selectively facilitated by VIP (KAWATANI et al. 1985; DEGROAT et al. 1986). The VIP immunoreactivity found in the ganglia originates from intrinsic neurons as indicated by denervation studies. Intra-arterial administration of VIP ($1-50\ \mu g\ kg^{-1}$) enhances the postganglionic discharge elicited by the muscarinic agonist acetyl-β-methylcholine but has no effect on nicotinic transmission. Also VIP suppresses the muscarinic inhibition of ganglionic transmission produced by acetyl-β-methylcholine without altering the response to other inhibitory agents (norepinephrine; leu-enkephalin and GABA). VIP also has a direct inhibitory effect on bladder smooth muscle. The VIP-induced facilitation of muscarinic postganglionic firing in the bladder

Table 3. Physiological and pharmacological evidence for coexistence of ACh and peptides (selected cases)

Tissue/region/ species	Cholinergic ligand	Neuroactive peptide(s)			Coexistence confirmed (techniques)	Mode of interaction		Notes	References
		Name (conc.)	Effect	Mode of action, receptors involved		In-trinsic[a]	Ex-trinsic[b]		
Submandibular ganglion of the cat	ACh	VIP (intra-arterial infusion)	Atropine-resistant vasodilatation; potentiated salivary secretion induced by ACh	Presynaptic control to the release of ACh and VIP via muscarinic autoreceptors	ICC and subcellular fractionation (see also Table 1 and Table 2)	+ +			1
	Atropine	VIP (release)	10× increased VIP release at high-frequency stimulation			+			
	[^{3}H]-4-NMPB (muscarinic (agonist)	VIP (0.1 pmol to 0.01 nmol)	Enhanced binding of the muscarinic agonist	Shift from low- to high-affinity receptor conformation (?); VIP and muscarinic receptor interaction (?)		+			2

Superior cervical ganglia of the cat	ACh	VIP (0.03 – 0.12 nmol)	Dose-dependent prolonged depolarization	Selective muscarinic excitatoric	No	?	+		3
		VIP (1.8 – 10 nmol)	Enhanced muscarinic transmission						
Parasympathetic ganglia of the cat	Acetyl-β-methyl-choline (muscarinic agonist)	VIP	Facilitatory effect on muscarinic firing	VIP facilitates muscarinic responses via M1 muscarinic receptors	ICC and EHC	+	(?)	No alteration in the nicotinic transmission	4
Intracardiac neurons in toad	ACh and hyoscyan (muscarinic antagonist)	SOM	Hyoscyan-resistant release of SOM on high frequency vagal stimulation	Cardioinhibitory action; common binding sites?	ICC	+ (?)	(?)	No double labelling ICC	5
Sympathetic ganglia of the frog	ACh	LHRH	Preganglionic fibres contain and release ACh and LHRH	Fast EPSP is mediated by ACh; slow EPSP is mediated by LHRH	ICC	+ (?)	?	No double labelling ICC was performed	6

Table 3 (continued)

Tissue/region/ species	Cholinergic ligand	Neuroactive peptide(s)			Coexistence confirmed (techniques)	Mode of interaction		Notes	References
		Name (conc.)	Effect	Mode of action, receptors involved		Intrinsic[a]	Extrinsic[b]		
Salivary gland of the rat	Atropine and NMPB	VIP	Depleted VIP pool; large increase in the numbers of VIP and muscarinic receptors	VIP release is regulated by negative feedback via muscarinic receptors	ICC	?	+	In vivo chronic atropine treatment	7
Cerebral cortex of the rat	Atropine (1 μ*M*)	^{125}I-VIP	Reduced ^{125}I-VIP binding in the presence of atropine	Occupied muscarinic autoreceptors by atropine; no free binding site for VIP	ICC	+	?		8
Electric organ of *Torpedo*	ACh	Opiates	Inhibition of ACh release (*in vitro*)	Ca^{2+}-channel blockade; opiate receptor (?)	Enkephalin-like substance was found in tissue extracts by RIA	?	+	The existence of opiate receptors is not confirmed	9

Cerebral cortex of the rat	Atropine	VIP (receptors)	Tissue VIP level lowered by 20%	75% increase in the number of VIP receptors	ICC	?	+	Chronic *in vitro* atropine treatment was performed	10
Cerebral cortex of the rat (corticospinal neurons)	ACh	VIP (1 μ*M*)	Long-lasting excitation of ACh receptive neurons	Common binding site for ACh and VIP(?); (muscarinic receptor involvement)	ICC	?	+	Peptide iontophoretically applied	11
Cerebral cortex of the rat	ACh	VIP SP CCK	ACh receptive neurons also excited by these peptides	Common binding sites (?)	ICC only for VIP	?	+		12
Neuromuscular junction of the frog	ACh	VIP (10 – 100 n*M*)	Dose-dependent increase in quantal content of the EPPs	Mediation via voltage sensitive Ca^{2+} channels?	No	?	?	No evidence for VIP to be present in frog neuromuscular junction	13,14

Table 3 (continued)

Tissue/region/ species	Cholinergic ligand	Neuroactive peptide(s)			Coexistence confirmed (techniques)	Mode of interaction		Notes	References
		Name (conc.)	Effect	Mode of action, receptors involved		In-trinsic[a]	Ex-trinsic[b]		
Parasympathetic ganglia of the cat	ACh	Enkephalin-like	Depressed ganglionic transmission at high-frequency stimulation	Autoinhibition via delta opiate receptors	ICC combined with anatomical lesion	?	+		15
Small intestine of the rabbit and guinea-pig	Atropine	VIP (1 – 100 nM)	VIP-induced muscle contractions antagonized by atropine; enhanced ACh release; on field stimulation; increased cAMP and cGMP levels	VIP interacts with presynaptic muscarinic receptors; VIP receptor involvement	No	?	+		16
Myenteric plexus of the guinea-pig ileum	ACh	VIP (100 pM to 1 μM)	Muscle contraction; scopolamine and hexamethonium insensitive release of ACh	VIP receptors on postganglionic cholinergic neurons	Subcellular fractionation Ref. 18	?	+		17

MPLM preparation of guinea-pig small intestine	ACh	atropine	VIP, VIP-antiserum	Reduced release of ACh in the presence of anti-VIP anti-sera; increased VIP release in the presence of atropine at high frequency stimulation	Muscarinic control of VIP release; (voltage-sensitive Ca-channel involvement?)	Subcellular fractionation Ref. 18	+(?)	+(?)	Frequency-dependent release of ACh and VIP	19

ICC, immunocytochemistry; EHC, enzyme histochemistry; NMPB, LHRH, luteinizing hormone-releasing hormone; EPP, end-plate potential; CCK, cholecystokinin; MPLM, myenteric plexus-longitudinal muscle

[a] Intrinsic interaction: ACh and the peptide are released from the same neuron (coexistence is confirmed by other techniques) as the prerequisite of autoregulation

[b] Extrinsic interaction: no confirmed coexistence; the substances are released from different neurons, but interact on common or closely associated binding sites

References:
1 LUNDBERG (1981)
2 LUNDBERG et al. (1981)
3 KAWATANI et al. (1985)
4 KAWATANI et al. (1986)
5 CAMPBELL et al. (1982)
6 JAN and JAN (1983)
7 HEDLUND et al. (1983)
8 HEDLUND et al. (1986)
9 MICHAELSON et al. (1984)
10 ABENS et al. (1984)
11 PHILLIS et al. (1978)
12 LAMOUR et al. (1983)
13 GOLD (1982)
14 GOLD (1984)
15 DeGROAT et al. (1986)
16 COHEN and LANDRY (1980)
17 KUSUNOKI et al. (1986)
18 AGOSTON et al. (1985)
19 AGOSTON et al. (1987)

ganglia resembles the VIP effect on muscarinic transmission in the salivary glands just mentioned.

In parasympathetic ganglia (e.g. urinary bladder of cat) VIP abolishes the effect of the irreversible anticholinesterase drug 217 AO on the muscarinic transmission (KAWATANI et al. 1986). The VIP-induced discharges are not blocked by hexamethonium but are by atropine or a selective muscarinic antagonist pirenzepine. This facilitatory effect of VIP lasts for 5–10 min. Nicotinic transmission remains unaltered during VIP administration. It is supposed that VIP facilitates muscarinic responses via a specific interaction with muscarinic receptors of the M1 type.

b) PNS of Lower Vertebrates

Intracardiac postganglionic cholinergic neurons in the heart of the toad *Bufo marinus* contain SOM which is released on vagal stimulation (CAMPBELL et al. 1982). The application of synthetic SOM inhibits the rate and force of beat of atrial preparations. After muscarinic blockade with hyoscine, vagal stimulation at 3 Hz or above still caused inhibition of the pacemaker and the atrium. This effect of vagal stimulation is hyoscine resistant and can be diminished by about 60% after induction of tachyphylaxis to SOM. Intermittent stimulation of the vagus nerves for 1 h at 10 Hz reduced the effect of SOM. The stimulation of the vagus at 3 Hz or more released sufficient SOM to inhibit the pacemaker and atrial muscle, so SOM may be the compound which controls the heart rate in the toad.

In sympathetic ganglia of the bullfrog both ACh and LHRH serve as transmitters (JAN and JAN 1983a, b). Preganglionic C fibres appear to supply ACh and LHRH because thresholds for the cholinergic fast excitatory postsynaptic potentials (EPSP) correlate well with the thresholds for the peptidergic late show EPSP recorded in the same C cell. Immunocytochemistry revealed the presence of LHRH in preganglionic nerve terminals on C cells (JAN et al. 1983a). The LHRH-like peptide is released together with ACh from the preganglionic C fibres upon stimulation causing the late slow EPSP in C cells. The LHRH-like immunoreactivity is similar but not identical to the mammalian hormone (JAN and JAN 1982). Unlike ACh the LHRH-like immunoreactivity diffuses tens of micrometres after being released before activating sympathetic neurons (JAN et al. 1983b).

3. Intrinsic Short-Term Interactions of ACh and Peptides *In Vitro*

a) The Mammalian PNS

In membrane preparations of cat submandibular gland VIP induces cAMP formation in a dose-dependent manner (FREDHOLM and LUNDBERG 1982). The addition of carbachol (1–10 μ*M*) further enhances cAMP formation. The two drugs potentiate each other and the mechanisms are apparently identical to that involved in the induction of salivary secretion in the presence of ACh and VIP. The potentiating effect of VIP on salivary secretion (LUNDBERG et al. 1982a) induced by ACh infusions is very likely to be due to the enhancement of muscarinic ligand binding (LUNDBERG et al. 1982b). The association rate of the muscarinic antagonist [^{3}H]4-*N*-methylpipendinyl benzilate ([^{3}H]4-NMPB) is increased in the presence of low (5-nM) concentrations of VIP; the number of receptors is unchanged, only their affinity as demonstrated by equilibrium binding experiments. These suggest the existence of two or three interconvertible receptor populations having high and low affinities. Since VIP is rapidly degraded ($t_{1/2} = 1 - 1.5$ min) after exposure to plasma or tissue extracts, and a stable VIP analogue has not yet been synthesized the desensitization and equilibrium studies needed are not yet possible. Similar effects have not been found in experiments using membranes prepared from rat salivary gland (HEDLUND et al. 1983, 1986). The VIP-activated adenylate cyclase in membranes from the cat submandibulary gland can be further activated by the presence of muscarinic agonists, while no effect has been found in similar experiments using membranes from rat salivary gland (HEDLUND et al. 1986).

Much less work has been done on the mammalian CNS than on the PNS; however HEDLUND et al. (1986) found that the binding of VIP to membranes of rat cerebral cortex appears to be inhibited by atropine (1 μM).

b) Electromotor System of *Torpedo*

ACh release from isolated nerve terminals of *T. marmorata* induced by high K^+ concentrations is found to be inhibited by morphine in a dose-dependent, naloxone-reversible fashion (MICHAELSON et al. 1984a). Further experiments revealed that morphine inhibits $^{45}Ca^{2+}$ influx into K^+-depolarized *Torpedo* synaptosomes, and that this effect can be blocked by naloxone (MICHAELSON et al. 1984b). This mechanism presupposes the existence in the terminals of presynaptic enkephalin receptors controlling the voltage-sensitive Ca^{2+} channels. This is supported by the finding that morphine does not inhibit ACh release when the Ca^{2+} channels are bypassed by Ca^{2+} ionophores which introduce Ca^{2+} directly into the nerve terminals. The endogenous ligand of this regulation could be that of the enkephalin-like immunoreactivity found in extracts of the electric organ (MICHAELSON et al. 1984b).

4. Intrinsic Long-Term Interactions of ACh and VIP *In Vitro*

Chronic atropine treatment (14 days, 20 mg kg^{-1}) results in a 75% increase in the number of VIP receptors in the rat cerebral cortex (ABENS et al. 1984). The affinity of these receptors for VIP remains unaltered, but the level of VIP in the tissue is reduced by 26%. Also, the number of mAChRs is simultaneously increased by 26%.

5. Extrinsic Short-Term Interactions of ACh and Peptides *In Vivo*

a) PNS of Mammals

In the superior cervical ganglion (SCG) of the cat VIP induces depolarization and muscarinic excitation (KAWATANI et al. 1985a). At doses of 0.03–0.12 nmol per ganglia, VIP produces a dose-dependent, prolonged (3- to 15-min) depolarization of the ganglion and enhances the ganglionic depolarization elicited by a muscarinic agonist (acetyl-β-methylcholine). In the 1.8- to 10-nmol per ganglia range VIP prolonged the postganglionic discharge elicited by the muscarinic agonist, increased muscarinic transmission in anti-ChE-treated ganglia and also potentiated the late muscarinic discharge elicited by ACh. No nicotinic components of the transmission were shown to be involved. The mode of action is thought to be an interaction of VIP with M-channels to decrease membrane conductance to potassium ions, thereby producing depolarization.

In sympathetic and parasympathetic ganglia of the cat the selective potentiation of muscarinic excitation by VIP has also been demonstrated (KAWATANI et al. 1986).

b) CNS of Mammals

Iontophoretically applied VIP excites deep, spontaneously active cortical neurons, including identified corticospinal neurons in rat sensory motor cerebral cortex (PHILLIS et al. 1978). This excitation has a latency of onset varying from several seconds to over 1 min and often lasts for a minute or longer after cessation of the application. Neurons which can be excited by VIP are also shown to be excited by ACh. However, atropine which can antagonize the excitant action of ACh on the identified neurons does not affect the excitation elicited by VIP on the same cells.

Rat cortical neurons in infragranular layers are excited by different peptides as well as ACh applied microiontophoretically (LAMOUR et al. 1983). Among peptides SP and VIP excite neurons in the largest number. When the neuronal excitation is induced by ACh the simultaneous application of peptides (SP, CCK and VIP) depresses this excitation. The possible involvement of presynaptic sites of action is ruled out by the observation that the effects of SP and VIP are not antagonized by atropine.

c) Non-mammalian Tissues

VIP present in the organ bath (10 – 100 n*M*) or applied iontophoretically facilitates neuromuscular transmission at the frog NMJ (GOLD 1982, 1984). Bath application of VIP produces a dose-dependent increase in the quantal content of the end-plate potential, but has no effect on the quantal size or on the resting membrane potential of the muscle fibres. VIP also increases miniature end-plate potential frequency when the nerve terminal is depolarized with K^+. This effect can be abolished by removing Ca^{2+} from the bath. When VIP is iontophoretically applied to the end-plate the quantal content also increased, but no effect was seen on quantal size, input resistance or resting membrane potential. An enhancement by VIP of voltage-sensitive calcium influx or of some later step in the stimulus-secretion coupling has been postulated. Since VIP, like many peptides, lacks competitive antagonists, the specificity of the effect cannot be tested. However, neither SP nor leu-enkephalin when tested under the same conditions produce any of the responses observed with VIP. Whether VIP is present as an endogenous ligand at the frog NMJ remains to be established.

6. Extrinsic Short-Term Interactions of ACh and Peptides *In Vitro*

Parasympathetic ganglia of the cat's urinary bladder have been shown to contain dense networks of leu-enkephalin-positive neurons (DEGROAT et al. 1986). Indirect evidence (retrograde dye tracing and denervation) reveals that ACh and leu-enkephalin may coexist in the same neuron and are released at the same terminals. Intraarterial administration of leu-enkephalin to bladder ganglia *in situ* depresses the postganglionic action potentials elicited by electric stimulation. This depression appears to be dose-dependent, has a duration of 5 – 10 min and can be antagonized by naloxone administration. Also, the magnitude of enkephalinergic inhibition is found to be frequency-dependent. The inhibition is maximal at low frequencies (0.25 – 0.5 Hz) and negligible at frequencies of 5 – 7 Hz even with large doses of leu-enkephalin. It has been supposed that *in situ* preganglionic stimulation at high frequencies can release sufficient concentrations of leu-enkephalin at the synapse to activate δ opiate receptors to suppress the release of ACh.

VIP contracts the guinea-pig ileum and rabbit jejunum in a concentration-dependent manner in concentrations ranging from 1 – 100 n*M* (COHEN and LANDRY 1980). This effect can be partly antagonized by the presence of atropine indicating that one component of the contractile response is due to the indirect release of ACh. VIP elevates cAMP levels in both rabbit jejunum and ileal smooth muscle. VIP can also provoke ACh release from the myenteric plexus of the guinea-pig ileum (KUSUNOKI et al. 1986). This indicates that VIP induces contractions of the longitudinal muscle both directly and indirectly by the stimulation of both cholinergic neurons and non-cholinergic excitatory neurons.

Frequency-dependent release of ACh and VIP from myenteric plexus longitudinal muscle strips has been demonstrated (AGOSTON et al. 1988). The release of ACh is maximal at low frequency ($<$ 1.0 Hz) electrical field stimulation, while VIP is preferentially released at high frequency ($>$ 25 Hz) stimulation. Maximal release of VIP is found at 50 Hz while maximal ACh release appears at 0.1 Hz field stimulation. The presence of anti-VIP antisera in the perfusate reduces the release of ACh at an intermediate (5 Hz) frequency. The addition of atropine to the bath increases the amount of the VIP released at a high (50 Hz) frequency.

7. Extrinsic Long-Term Interactions of ACh and Peptides *In Vitro*

In the rat submandibulary gland long-term atropine treatment causes a big increase in the number of muscarinic and VIP receptors and depletion of VIP from tissue stores (HEDLUND et al. 1983). The depletion of VIP from the tissue following chronic atropine treatment indicates that VIP release is subject to a so far unidentified negative feed-back involving muscarinic binding sites.

VI. Consequences of Coexistence of ACh and Neuropeptides

1. Ionotropic and Metabotropic Transmission

The coexistence of ACh with neuroactive peptides implies the simultaneous operation in one and the same neuron of distinct metabolic processes normally

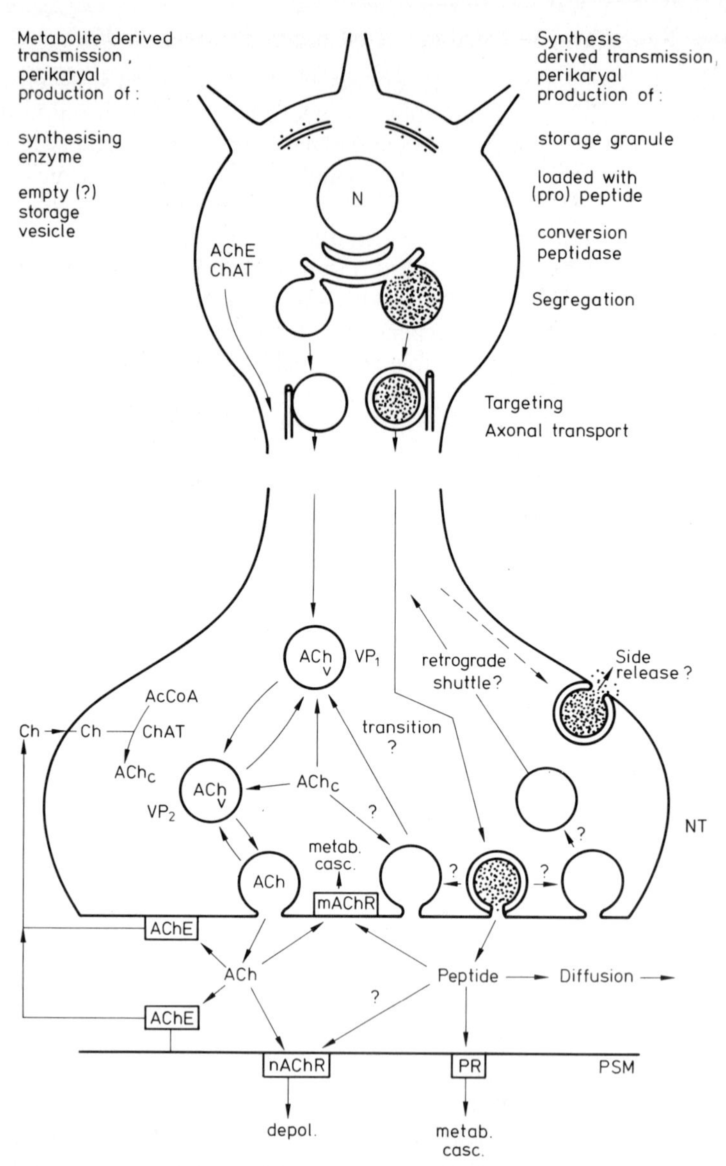

Fig. 2. Integrative view of the cholinergic multitransmitter neuron. *mAChR,* muscarinic acetylcholine receptor; *nAChR,* nicotinic acetylcholine receptor; ACh_c, cytoplasmic ACh; ACh_v, vesicular ACh; *depol.,* depolarization; *NT,* nerve terminal; *PSM,* postsynaptic plasma membrane; *metab. casc.,* metabolic cascade; *PR,* peptide receptor; VP_2, recycling synaptic vesicle; VP_1, reserve synaptic vesicle; *N,* nucleus; *?,* components and routes for which the evidence could be better or whose existence is at present only speculative

confined to cholinergic or peptidergic neurons. This subtype of cholinergic neuron (see Fig. 2) should perhaps be called a cholinergic multitransmitter neuron to distinguish it from cholinergic neurons containing only ACh which could be called cholinergic unitransmitter neurons.

According to the classification of McGEER et al. (1976) and ECCLES and McGEER (1979) ACh is a classical ionotropic transmitter. This type acts specifically to open ionic gates in the postsynaptic membrane. The latency is very short; the duration of the potential is only slightly longer than the transmitter action.

By contrast, peptidergic transmission may represent the metabotropic (more correctly metabolotropic; WHITTAKER 1979) type of transmission acting primarily on the metabolism of the target cell. The effect is long lasting and may constitute a type of chemical signalling operating on processes critical for the maintenance of the neuron itself.

In contrast to the sharply focused action of ACh, that of peptides is more general resembling more that of hormones circulating in the blood stream. This mode of action of the neuropeptides can also be termed (SCHMITT 1984) parasynaptic. Due to their special properties peptides are prime candidates for the so-called trophic actions of neurotransmitters.

2. Metabolism of Peptides and ACh

In order to maintain neurotransmitter levels in the nerve terminal neurons have evolved several specific mechanisms. Those regulating the levels of neuropeptides, which are derived from proteins, are fundamentally different from those regulating the synthesis of ACh which is derived from metabolites transported into the cell (REICHARDT and KELLY 1983). The unusual geometry of the neuronal cell restricts protein biosynthesis to the perikaryon, so peptide transmitters have to be transported from there to the nerve terminal. The neuropeptides (which might be called synthesis-derived neurotransmitters) are derived from larger precursors; the post-translational processing of these propeptides (or protransmitters) very likely occurs in the secretory granules as with the peptide hormones (BROWNSTEIN et al. 1982; GAINER and BROWNSTEIN 1981; LOH and GAINER 1983; LOH et al. 1984).

Intracellular compartmentation following synthesis of the protein on the rough endoplasmic reticulum (RER) is determined, in part, by the translational process itself. The initial N-terminal amino acid sequence of the protein serves as a signal for the protein to traverse the RER membrane and enter the cisternae, where the signal sequence is immediately cleaved off. The propeptide plus its transient N-terminal sequence, which usually contains 20–30 amino acids, is referred to as the prepropeptide (BROWNSTEIN et al. 1982; GAINER and BROWNSTEIN 1981; LOH and GAINER 1983). By contrast, the regulation of the metabolite-derived transmitters such as ACh occurs in the nerve terminals (see Chap. 7). The transporters and enzymes required for the synthesis of ACh are present in the terminal; thus tight coupling between transmitter release and replacement is possible. ACh synthesis is regulated by the changes in activity of the choline transporter which can limit precursor availability (Chap. 7). The ability of the cholinergic terminal to maintain *in situ* synthesis of ACh on demand by reutilizing the split products on the released transmitter makes the cholinergic terminal relatively independent of the perikaryon.

3. Release: Translation of an Electrical Code into a Chemical One

As already mentioned, evidence has accumulated that the release of neuropeptides from putative cholinergic multitransmitter neurons often has a different stimulus frequency dependence from that of ACh (LUNDBERG 1981; AGOSTON and LISZIEWICZ 1988).

Changes in the frequency of stimulation of a cholinergic unitransmitter neuron will change the amount of ACh released but the nature of the chemical message is not changed. By contrast, changes in firing frequency of a cholinergic multitransmitter neuron may alter which of the co-transmitters is released or in what proportion they are released. In this case the chemical nature of the signal could be altered (BARTFAI et al. 1986). The multitransmitter cholinergic neuron permits the direct translation of the frequency code into a chemical code at the level of the individual nerve terminal (AGOSTON et al. 1988). Cholinergic multitransmitter neurons may therefore release different transmitters or different combinations of transmitters dependent on the frequency of stimulus (LUNDBERG and HÖKFELT 1986; AGOSTON et al. 1988). This indicates that the concept of the neuron as binary unit – when it fires the signal value is one, when it is mute the signal value is nought – should be changed.

The preferential secretion of peptides at a high frequency of electrical stimulation, at which ACh release is negligible and vice versa (LUNDBERG 1981; AGOSTON et al. 1988; AGOSTON and LISZIEWICZ 1988) implies the existence of separate release mechanisms.

According to the widely accepted concept of stimulus-secretion coupling the arrival of an action potential in the region of the nerve terminals gives rise to the depolarization of the plasma membrane followed by the entry of Ca^{2+} into the cytoplasm (DOUGLAS 1968). It is assumed that the resulting increase in the free cytoplasmic Ca^{2+} concentration ($[Ca^{2+}]_i$) triggers neurohormone and transmitter release.

Recent investigations have demonstrated that neurons have multiple voltage-sensitive Ca channels (VSCCs); each type has unique properties and pharmacology (for reviews, see MILLER 1987). The dihydropyridine (DHP) insensitive N-type VSCC is intimately associated with the active zones in the nerve terminal. The DHP-sensitive L-type of VSCC differs electrophysiologically from the N-type and predominantly localized outside the active zone.

The release of SP (a model peptide transmitter) from rat cultured sensory neurons is extremely sensitive to modulation by DHP; by contrast the release of noradrenaline (a classical low molecular mass transmitter) from sympathetic neurons is completely DHP-resistant. This indicates that activation of different VSCCs with different channel properties localized at different places on the nerve terminal could result in a transient localized increase of Ca^{2+} followed by the exocytosis of storage granules of different types at different loci (AGOSTON and LISZIEWICZ 1988).

Recent observation (ZHU et al. 1986) that the exocytosis of LDCs containing SP occurs predominantly outside the active zone is further evidence for the existence of separate secretory mechanisms localized at spacially separated sites on the terminal membrane. Provided that the different Ca channels are indeed present in cholinergic multitransmitter neurons, they could play a key role in translating the electrical code into chemical coding.

4. Membrane Utilization of Different Types of Synaptic Vesicle

The vesicle membrane of cholinergic SELs is retrieved after exocytosis as a functional organelle and takes up newly synthesized ACh (for review, see WHITTAKER 1984a, b). When peptide-containing vesicles release their content, they presumably cannot be refilled with the peptide in the terminal. The fate of the vesicle membrane is unknown but might be (a) converted after retrieval into a classical synaptic vesicle and be loaded with ACh; (b) left unretrieved in the plasma membrane, thereby supplying the latter with components transported from the perikarya to maintain the molecular composition of the nerve terminal; or (c) retrieved and returned to the cell body by retrograde axonal transport for reprocessing as peptide storage vesicles.

5. Receptors and Inactivation of ACh and Neuropeptides

The peptide receptors so far studied appear to have high affinities for peptides, the dissociation constants being in the range 0.1 – 1.0 nM, which is four orders of magnitude less than that of ACh (JAN and JAN 1983b). This is also reflected in the concentrations of neuropeptides in the CNS which appear to be several orders of magnitude lower (1 pmol to 1 fmol mg^{-1} protein) than the classical neurotransmitters such as ACh (TURNER 1986).

An important regulatory mechanism particularly for cholinergic multitransmitter neurons is the regulation of transmitter release via autoreceptors (BARTFAI et al. 1986). The release of ACh from nerves containing ACh and VIP is feedback-regulated through muscarinic autoreceptors (LUNDBERG 1981). However, the molecular mechanism by which chronic treatment with atropine, a classical muscarinic antagonist, which depletes VIP stores and elevates the number of VIP binding sites, is not known.

ACh is inactivated immediately after being released by the highly specific degrading enzyme AChE (Chap. 8) to the pharmacologically inactive metabolites choline and acetate which however can be re-utilized. By contrast, once released by exocytosis, neuropeptides are degraded, not recycled, and inactivated at a much slower rate than ACh (TURNER 1986). This extended biologically active life-time allows peptides to diffuse in the extracellular space and act on receptors many micrometres away from the site of the secretion (JAN and JAN 1983). The observation that no overlap was found between the distribution of peptidergic neurons and peptide receptors raised the problem of receptor mismatch (KUHAR 1985; SCHULTZBERG and HÖKFELT 1986; HÖKFELT and TERENIUS 1987). In marked contrast to the classical transmitters, the long-acting peptides can thus have a variety of different actions on receptors far from the place of release.

6. Peptides as Factors in Non-neuronal Tissues

Recent evidence suggests that neuropeptides may function in a communication network that links the nervous, endocrine and immune systems. Each peptide may function as a neurotransmitter, peptide hormone or lymphokine depending on the site of release of the peptide and the target tissue with which it interacts. The two peptides that have been extensively studied in this connexion, and for which there is increasing evidence for diverse immunoregulating properties are SP and VIP (O'DORISIO 1987; McGILLIS et al. 1987). These peptides have been demonstrated to modulate lymphocyte migration, mast cell mediator response and natural killer cell activity.

Bombesin and bombesin-like peptides (BLPs) such as gastrin-releasing peptide can function as autocrine growth factors in human small-cell lung cancer (CUTTITTA et al. 1985; CARNEY et al. 1987). Cell lines of this type of cancer secrete BLPs and can express a single class of high-affinity receptors for BLPs. Exogenously added BLPs can also stimulate the clonal growth and DNA synthesis of the tumour cells, while monoclonal antibodies against BLPs inhibit tumour growth *in vitro* and *in vivo.*

The intracellular mechanism responsible for the effect of bombesin as a tumour growth factor may be the stimulation of the production of inositol phosphate through a guanine nucleotide regulatory protein (G-protein) encoded by the N-*ras* oncogene (WAKELAM et al. 1986).

7. Developmental Aspects

During development one of the most important decisions that a developing neuron makes is which transmitter to use. Since each neuron is supplied with a complete set of genes, it possesses the potential to synthesize the machinery for all transmitters. The particular transmitter or combination of transmitters that appears at the end of differentiation is determined by trophic factors released from other neurons or other cells in a particular environment, which control the expression of the appropriate genetic program (CHUN et al. 1987; GERSHON et al. 1984; LANDIS 1984; PATTERSON 1978; POTTER et al. 1986). Coexistence of neuropeptides with cholinergic neurons could therefore be one of the end-results of such an environmental determination. The lack of any consistent coexistence pattern (e.g. VIP and ACh) for cholinergic neurons such as was believed in the early days of the coexistence phenomenon to exist, supports the idea that environmental factors determine not only the expression of the synthetic machinery for ACh but also the type of coexistence. However, it cannot be excluded that coexistence of multiple transmitters is a paraphenomenon, which could represent an evolutionary vestige. Neuropeptides may have been important messengers early in evolution and/or during ontogenesis but have been replaced later on by more efficient, small molecule transmitters like ACh; the peptides would then have survived as 'silent passengers' (HÖKFELT et al. 1986b). However, due to their wide biological activity they are prime candidates as mediators in other long-term functions of nerve activity such as trophic action.

8. Conclusions and Future Perspective

Multiple transmitters can enormously increase the capacity for information transfer in the nervous system. However, the discrepancy between the extensive immunocytochemical and the meagre physiological evidence calls attention to a need for the critical evaluation of the coexistence phenomenon and for extensive functional studies.

The number of peptides two to ten amino acids long that might theoretically exist is astronomical ($>10^{13}$). There are 10^{10} neurons in the human brain, so it is easy to imagine – though this is unlikely – that each neuron makes its own unique peptide.

Application of DNA technology is already beginning to reveal the existence of new peptides and will undoubtedly continue to do so. The use of this technology also permits the study of the co-expression of peptides in a cholinergic multitransmitter neuron.

Studies on the plasticity of transmitter phenotype will provide the clue as to what factors or environmental conditions are responsible for the expression of multitransmitter systems in the cholinergic neuron. It could also eventually answer the question as to whether coexistence is a transient phenomenon or one of long duration, whether it is related to the normal differentiation of the nervous system or reflects pathological conditions.

Acknowledgement. The author is supported by a grant from the Deutsche Forschungsgemeinschaft (Co 130/1-1 and Wh 1/4-2).

References

Abens J, Westlind A, Bartfai T (1984) Chronic atropine treatment causes increase in VIP receptors in rat cerebral cortex. Peptides 5:375–377

Aberer W, Kostron H, Huber E, Winkler H (1978) A characterization of the nucleotide uptake by chromaffin granules of bovine adrenal medulla. Biochem J 172:353–360

Abood LG, Koketsu K, Miyomoto S (1962) Outflux of various phosphates during membrane depolarisation of excitable tissues. Am J Physiol 202:469–474

Agoston DV (1988a) Biophysical and biochemical characterization of synaptic vesicles storing neuropeptides and acetylcholine isolated from myenteric neurons. J Neurochem (in press)

Agoston DV (1988b) Isolation and characterization of secretory granules storing a VIP-like peptide in *Torpedo* electromotor nerve terminals. J Neurochem (in press)

Agoston DV, Conlon JM (1986) Presence of vasoactive intestinal polypeptide-like immunoreactivity in the cholinergic electromotor system of *Torpedo marmorata.* J Neurochem 47:445–453

Agoston DV, Conlon JM (1987a) Presence of a neuropeptide in a model cholinergic system. Ann NY Acad Sci 493:135–138

Agoston DV, Conlon JM (1987b) Association of neuropeptides with cholinergic neurons. In: Dowdall MJ, Hawthorne JM (eds) Molecular and cellular mechanisms of cholinergic function. Horwood, Chichester, pp 460–471

Agoston DV, Lisziewicz J (1987) The translation of electrical coding into chemical coding leads to the secretion of different types of transmitter in myenteric neurons. J Neurochem (in press)

Ágoston DV, Ballmann M, Conlon JM, Dowe GHC, Whittaker VP (1985a) Isolation of neuropeptide-containing vesicles from the guinea pig ileum. J Neurochem 45:398–406

Ágoston DV, Kosh JW, Lisziewicz J, Whittaker VP (1985b) Separation of recycling and reserve synaptic vesicles from cholinergic nerve terminals of the myenteric plexus of guinea pig ileum. J Neurochem 44:299–305
Ágoston DV, Dowe GHC, Fiedler W, Giompres PE, Roed IS, Walker JH, Yamaguchi T, Whittaker VP (1986) The use of synaptic vesicle proteoglycan as a stable marker in kinetic studies of vesicle recycling. J Neurochem 47:1584–1592
Agoston DV, Borroni E, Richardson PJ (1987) Cholinergic suface antigen Chol-1 is present in a subclass of VIP-containing rat cortical synaptosomes. J Neurochem 50:1659–1662
Agoston DV, Conlon JM, Whittaker VP (1988) Selective depletion of the acetylcholine and VIP of the guinea-pig myenteric plexus by differential mobilization of distinct transmitter pools. Exp Brain Res (in press)
Akasu T, Hirai K, Koketsu K (1981) Increase of acetylcholine-receptor sensitivity by adenosine triphosphate: a novel action of ATP on ACh-sensitivity. Br J Pharmacol 74:505–507
Altschuler RA, Parakkal MH, Fex J (1983) Localization of enkephalin-like immunoreactivity in acetylcholinesterase positive cells in the guinea-pig lateral superior olivary complex that project to the cochlea. Neuroscience 9:621–630
Amara SG, Jonas V, Rosenfeld MG, Ong ES, Evans RM (1982) Alternative RNA processing in calcitonin gene expression generates mRNAs encoding different polypeptide products. Nature 298:240–245
Barker JL, Smith TG Jr (eds) (1983) The role of peptides in neuronal function. Dekker, New York
Bartfai T, Iverfeldt K, Brodin E, Ogren S-O (1986) Functional consequences of coexistence of classical and peptide neurotransmitters. Prog Brain Res 68:321–330
Björklund A, Hökfelt T (eds) (1983) Methods in chemical neuroanatomy. Elsevier, Amsterdam
Branton WD, Phillips HS, Jan YN (1986) The LHRH family of peptide messengers in the frog nervous system. Prog Brain Res 68:205–215
Brownstein MJ (1982) Post-translational processing of neuropeptide precursors. TINS 5:318–320
Brownstein MJ, Saavedra JM, Axelrod J, Zeman GH, Carpenter DO (1974) Coexistence of several putative neurotransmitters in single identified neurons of *Aplysia.* Proc Natl Acad Sci USA 7:4662–4665
Brownstein MJ, Russel JT, Gainer H (1982) Biosynthesis of posterior pituitary hormones. In: Ganong WF, Martini L (eds) Frontiers in endocrinology, vol 7. Raven, New York, pp 31–43
Bryant MG, Polak JM, Modlin J, Bloom SR, Albuquerque RJ, Pearse AGE (1976) Possible dual role for vasoactive intestinale peptide as gastrointestinal hormone and neurotransmitter substance. Lancet 1:991–993
Buckley G, Consolo S, Giacobini E, Sjöqvist F (1967) Choline acetylase in innervated and denervated sympathetic ganglia and ganglion cells of the cat. Acta Physiol Scand 71:348–356
Burnstock G (1972) Purinergic nerves. Pharmacol Rev 24:509–581
Burnstock G (1975) Purinergic transmission. In: Iversen LL, Iversen SD, Snyder SH (eds) Handbook of psychopharmacology, vol 5. Plenum, New York, pp 131–194
Burnstock G (1978) A basis for distinguishing two types of purinergic receptor. In: Straub RW, Bolis L (eds) Cell membrane receptors for drugs and hormones. Raven, New York, pp 107–118
Burnstock G (1982) The co-transmitter hypothesis, with special reference to the storage and release of ATP with noradrenaline and acetylcholine. In: Cuello AA (ed) Co-transmission. Macmillan, London, pp 151–163
Campbell G, Gibbins IL, Morris JL, Furness JB, Costa M, Oliver JR, Beardsley AM, Murphy R (1982) Somatostatin is contained and released from cholinergic nerves in the heart of the toad *Bufo marinus.* Neuroscience 7:2013–2023
Carmichael SW, Winkler H (1985) The adrenal chromaffin cell. Sci Am 253:30–39
Carney DN, Cuttitta F, Moody TW, Minna JD (1987) Selective stimulation of small cell lung cancer clonal growth by bombesin and gastrin-releasing peptide. Cancer Res 47:821–825

Chan-Palay V, Palay SL (1984a) Coexistence in human and primate neuromuscular junctions of enzymes synthesizing four neuroactive substances. In: Chan-Palay V, Palay SL (eds) Coexistence of neuroactive substances in neurons. Wiley and Sons, New York, pp 141–155

Chan-Palay V, Palay SL (eds) (1984b) Coexistence of neuroactive substances in neurons. Wiley and Sons, New York

Chan-Palay V, Engel AG, Palay SL, Wu J-Y (1982a) Synthesizing enzymes for four neuroactive substances in motor neurons and neuromuscular junctions: light and electronmicroscopic immunocytochemistry. Proc Natl Acad Sci USA 79:6717–6721

Chan-Palay V, Engel AG, Wu J-Y, Palay SL (1982b) Coexistence in human and primate neuromuscular junctions of enzymes synthesizing acetylcholine, catecholamine, taurine and γ-aminobutyric acid. Proc Natl Acad Sci USA 79:7027–7030

Chun JJ, Nakamura MJ, Shatz CJ (1987) Transient cells of the developing mammalian telencephalon are peptide-immunoreactive neurons. Nature 325:617–620

Cleugh J, Gaddum JH, Mitchell AA, Smith MW, Whittaker VP (1984) Substance P in brain extracts. J Physiol (Lond) 170:69–85

Cohen ML, Landry AS (1980) Vasoactive intestinal polypeptide: increased tone, enhancement of acetylcholine release, and stimulation of adenylate cyclase in intestinal smooth muscle. Life Sci 26:811–822

Coons AH (1958) Fluorescent antibody methods. In: Danielli JF (ed) General cytochemical methods. Academic, New York, pp 399–422

Cooper PE, Martin JB (1982) Neuroendocrinology and brain peptides. TINS 5:186–189

Costa M, Buffa R, Furness JB, Solcia EL (1980) Immunohistochemical localization of polypeptides in peripheral autonomic nerves using whole mount preparation. Histochemistry 65:157–165

Costa M, Furness JB, Gibbins IL (1986a) Chemical coding of enteric neurons. Prog Brain Res 68:217–241

Costa E, Furness JB, Llewellyn-Smith I (1986b) Histochemistry of the enteric nervous system. In: Johnson LR (ed) Physiology of the gastrointestinal tract, 2nd edn. Raven, New York

Cuello AC (ed) (1981) Co-transmission. Macmillan, London

Cuello AC, Jessel TM, Kanazawa I, Iversen LL (1977) Substance P: localization in synaptic vesicles in the rat central nervous system. J Neurochem 29:747–751

Cuttitta F, Carney DN , Mulshine J, Moody TW, Fedorko J, Fischer A, Minna JD (1985) Bombesin-like peptide can function as autocrine growth factors in human small-cell lung cancer. Nature 316:823–826

Dahlström A (1968) Effect of colchicine on amine storage granules in sympathetic nerves of rat. Eur J Pharmacol 5:111–113

Day NC, Wien D, Michaelson DM (1985) Saturable [D-Ala2,D-Leu5]-enkephalin transport into cholinergic synaptic vesicles. FEBS Lett 183:25–28

deGroat WC, Kawatani M, Booth AM (1986) Enkephalinergic modulation of cholinergic transmission in parasympathetic ganglia of the cat urinary bladder. In: Hanin I (ed) Dynamics of cholinergic function. Plenum, New York, pp 1007–1017

Douglas WW (1968) Stimulus-secretion coupling: the concept and clues from chromaffin and other cells. Br J Pharmacol Chemother 34:451–474

Douglas WW, Poisner AM (1966) On the relation between ATP splitting and secretion in the adrenal chromaffin cell: extrusion of ATP – unhydrolysed – during release of catecholamines. J Physiol (Lond) 183:249–256

Douglass J, Civelli O, Herbert E (1984) Polyprotein gene expression: generation of diversity of neuroendocrine peptides. Annu Rev Biochem 53:15

Dowdall MJ (1978) Adenine nucleotides in cholinergic transmission: presynaptic aspects. J Physiol (Paris) 74:497–501

Dowdall MJ, Boyne AF, Whittaker VP (1974) Adenosine triphosphate, a constituent of cholinergic synaptic vesicles. Biochem J 140:1–12

Dowe GHC, Kilbinger H, Whittaker VP (1980) Isolation of cholinergic synaptic vesicles from the myenteric plexus of guinea-pig small intestine. J Neurochem 35:993–1003

Dunwiddie TV, Hoffer BJ (1980) Adenine nucleotides and synaptic transmission in the *in vitro* rat hippocampus. Br J Pharmacol 69:59–68
Eccles JC, McGeer PL (1979) Ionotropic and metabotropic neurotransmission. TINS 2:39–40
Eckenstein F, Baughman RW (1984) Two types of cholinergic innervation in cortex, one co-localized with vasoactive intestinal polypeptide. Nature 309:153–155
Eiper BA, Mains RE, Herbert E (1986) Peptides in the nervous system. TINS 9:463–468
Emson PC (1979) Peptides as neurotransmitter candidates in the mammalian CNS. Prog Neurobiol 13:61–116
Erichsen JT, Karten HJ, Eldred WD, Brecha NC (1982) Localization of substance P-like and enkephalin-like immunoreactivity within preganglionic terminals of the avian ciliary ganglion: light and electron microscopy. J Neurosci 2:994–1003
Evans RM, Amara S, Rosenfeld MG (1983) Molecular events in developmental regulation of neuroendocrine genes: characterization of the novel neuropeptide CGRP. Cold Spring Harbor Symp Quant Biol 48:413–417
Ewald DA (1976) Potentiation of postjunctional cholinergic sensitivity of rat diaphragm muscle by high-energy phosphate adenine nucleotides. J Membr Biol 29:47–65
Falck B, Hillarp NA, Thieme G, Torp A (1962) Fluorescence of catecholamines and related compounds condensed with formaldehyde. J Histochem Cytochem 10:348–354
Fex J, Altschuler RA (1981) Enkephalin-like immunoreactivity of olivocochlear nerve fibers in cochlea of guinea pig and cat. Proc Natl Acad Sci USA 78:1255–1259
Floor E, Leeman SE (1982) Synaptic vesicles containing Substance P purified by chromatography on controlled pore glass. Neuroscience 7:1647–1655
Forrester T, Williams CA (1977) Release of adenosine triphosphate from adult heart cells in response to hypoxia. J Physiol (Lond) 268:371–390
Fredholm BB (1976) Release of adenosine-like material from isolated perfused dog adipose tissue following sympathetic nerve stimulation and its inhibition by adrenergic A-receptor blockade. Acta Physiol Scand 96:422–430
Fredholm BB, Hedquist P (1980) Modulation of neurotransmission by purine nucleotides and nucleosides. Biochem Pharmacol 29:1635–1643
Fredholm BB, Lundberg JM (1982) VIP-induced cyclic AMP formation in the cat submandibular gland. Potentiation by carbacholine. Acta Physiol Scand 114:157–159
Füldner H-H, Stadler H (1981) The storage of acetylcholine and ATP in synaptic vesicles. Hoppe Seylers Z Physiol Chem 362:198–205
Furness JB, Costa M, Keast JR (1984) Choline acetyltransferase and peptide immunoreactivity of submucosus neurons in the small intestine of the guinea-pig. Cell Tissue Res 237:329–336
Furness JB, Costa M, Gibbins IL, Llewellyn-Smith IJ, Oliver JR (1985) Neurochemically similar myenteric and submucosus neurons directly traced to the mucosa of the small intestine. Cell Tissue Res 241:155–163
Fyffe REW, Perl ER (1984) Is ATP a central synaptic mediator for certain primary afferent fibers from mammalian skins? Proc Natl Acad Sci USA 81:6890–6893
Gainer H, Brownstein MJ (1981) Neuropeptides. In: Siegel GJ, Albers RW, Agranoff BW, Katzman R (eds) Basic neurochemistry, 3rd edn. Little, Brown, Boston, pp 269–296
Geffard M, Vieillemaringe J, Heinrich-Roch A-M, Duris P (1985a) Anti-acetylcholine antibodies and first immunocytochemical application in insect brain. Neurosci Lett 57:1–6
Geffard M, McRae-Degueurce A, Souan ML (1985b) Immunocytochemical detection of acetylcholine in the rat central nervous system. Science 229:77–79
Gershon MD, Payette R, Teitelman G, Rothman TP (1984) Neuronal commitment and phenotypic expression by developing enteric neurones. In: Chan-Palay V, Palay S (eds) Coexistence of neuroactive substances in neurons. Wiley and Sons, New York, pp 181–204
Ginsborg BL, Hirst GDS (1972) The effect of adenosine on the release of the transmitter from the phrenic nerve of the cat. J Physiol (Lond) 224:629–645
Giompres PE, Zimmermann H, Whittaker VP (1981) Changes in the biochemical and biophysical parameters of cholinergic synaptic vesicles on transmitter release and during a subsequent period of rest. Neuroscience 6:775–785

Glazer EJ, Basbaum AI (1980) Leucine enkephalin: Localization in and axoplasmic transport by sacral parasympathetic preganglionic neurons. Science 208:1479–1481

Gold MR (1982) The effect of vasoactive intestinal polypeptide on neuromuscular transmission in the frog. J Physiol (Lond) 327:325–335

Gold MR (1984) The action of vasoactive intestinale polypeptide at the frog's neuromuscular junction. In: Chan-Palay V, Palay S (eds) Coexistence of neuroactive substances in neurons. Wiley and Sons, New York, pp 161–170

Gordon AS, Guillory RJ, Diamond I, Hucho F (1979) ATP-binding proteins in acetylcholine receptor-enriched membranes. FEBS Lett 108:37–39

Grondal EJM, Zimmermann H (1986) Ectonucleotidase activities associated with cholinergic synaptosomes isolated from *Torpedo* electric organ. J Neurochem 47:871–881

Guillemin R (1978) Peptides in the brain: the new endocrinology of the neuron. Science 202:390–402

Häggblad J, Eriksson H, Heilbronn E (1985) Effects of extracellular ATP on ^{86}Rb influx in chick myotubes; indications of a cotransmitter role in neuromuscular transmission. In: Changeux J-P, Hucho F, Maelicke A, Neumann A (eds) Molecular basis of nerve activity. de Gruyter, Berlin, pp 185–193

Hebb CO (1956) Choline acetylase in the developing nervous system of the rabbit and guinea-pig. J Physiol (Lond) 133:566–570

Hedlund B, Abens J, Bartfai T (1983) Vasoactive intestinal polypeptide and muscarinic receptors: supersensitivity induced by long-term atropine treatment. Science 220:519–521

Hedlund B, Abens J, Westlind A, Bartfai T (1986) Vasoactive intestinal polypeptide-muscarinic cholinergic interactions. In: Hanin I (ed) Dynamics of cholinergic function. Plenum, New York, pp 1019–1025

Hedquist P, Fredholm BB (1979) Inhibitory effect of adenosine on adrenergic neuroeffector transmission in the rabbit heart. Acta Physiol Scand 105:120–122

Hökfelt T, Terenius L (1987) More on receptor mismatch. TINS 10:22–23

Hökfelt T, Johansson O, Ljungdahl A, Lundberg JM, Schultzberg M (1980) Peptinergic neurones. Nature 284:515–521

Hökfelt T, Lundberg JM, Skirboll L, Johansson O, Schultzberg M, Vincent SR (1982) Coexistence of classical transmitters and peptides in neurones. In: Cuello AC (ed) Coexistence. Macmillan, London, pp 77–125

Hökfelt T, Fuxe K, Pernow B (eds) (1986a) Coexistence of neuronal messengers: a new principle in chemical transmission. Prog Brain Res 68

Hökfelt T, Holets VR, Staines W, Meister B, Melander T, Schalling M, Schultzberg M, Freedman J, Björklund H, Olson L, Lindh B, Elfvin L-G, Lundberg JM, Lindgren JA, Samuelsson B, Pernow B, Terenius L, Post C, Everitt B, Goldstein M (1986b) Coexistence of neuronal messengers – an overview. Prog Brain Res 68:33–70

Holton FA, Holton P (1954) The capillary dilatator substances in dry powders of spinal roots: a possible role of ATP in chemical transmission from nerve endings. J Physiol (Lond) 126:124–140

Israël M, Meunier FM (1978) The release of ATP triggered by action and its possible physiological significance: retrograde transmission. J Physiol (Paris) 74:485–490

Israël M, Lesbats B, Meunier FM, Stinnakre J (1976) Postsynaptic release of adenosine triphosphate induced by single impulse transmitter action. Proc R Soc Lond [Biol] 193:461–468

Israël M, Lesbats B, Manaranche R, Marsal J, Mastour-Franchon P, Meunier FM (1977) Related changes in amounts of ACh and ATP in resting and active *Torpedo* electroplaque synapses. J Neurochem 28:1259–1267

Israël M, Lesbats B, Manaranche R, Meunier FM, Frachon P (1980) Retrograde inhibition of transmitter release by ATP. J Neurochem 34:923–932

Iversen LL, Iversen SD, Snyder SH (eds) (1983) Neuropeptides. Plenum, New York (Handbook of psychopharmacology)

Jahn R, Schiebler W, Ouimet C, Greengard P (1985) A 38000-dalton membrane protein (p 38) present in synaptic vesicles. Proc Natl Acad Sci USA 82:4137–4141

Jahr CE, Jessel TM (1983) ATP excites as subpopulation of rat dorsal horn neurones. Nature 304:730–733

Jan YN, Jan LY (1983a) Coexistence and co-release of acetylcholine and the LHRH-like peptide from the same preganglionic fibers in frog sympathetic ganglia. Fed Proc 42:2929–2938
Jan YN, Jan LY (1983b) Some features of peptidergic transmission. Prog Brain Res 58:49–59
Jan YN, Bowers CW, Branton D, Evans L, Jan LY (1983) Peptides in neuronal function: Studies using frog autonomic ganglia. Cold Spring Harbour Symp Quant Biol 48:363–374
Johansson O, Lundberg JM (1981) Ultrastructural localization of VIP-like immunoreactivity in large dense-core vesicles of 'cholinergic-type' nerve terminals in cat exocrine glands. Neuroscience 6:847–862
Kása P (1970) Identification of cholinergic neurons in the spinal cord: an electron histochemical study of choline acetyltransferase. J Physiol (Lond) 210:89–90
Kása P (1986) The cholinergic systems in brain and spinal cord. Prog Neurobiol 26:211–272
Kása P, Mann SP, Hebb CO (1970) Localization of choline acetyltransferase. Histochemistry at the light microscope level. Nature 226:814–816
Kawatani M, Rutigliano M, deGroat WC (1985a) Depolarization and muscarinic excitation induced in a sympathetic ganglion by vasoactive intestinale polypeptide. Science 229:879–881
Kawatani M, Rutigliano M, deGroat WC (1985b) Selective facilitatory effect of vasoactive intestinal polypeptide on muscarinic firing in vesical ganglia of the cat. Brain Res 336:223–234
Kawatani M, Rutiglian M, de Groat WC (1986) Selective facilitatory effects of vasoactive intestinal polypeptide on muscarinic mechanisms in sympathetic and parasympathetic ganglia of the cat. In: Hanin I (ed) Dynamics of cholinergic function. Plenum, New York, pp 1057–1066
Keller F, Zimmermann H (1984) EctoATPase activity at the cholinergic nerve endings of the *Torpedo* electric organ. Life Sci 33:2635–2641
Kilbinger H, Wessler I (1980) Inhibition by acetylcholine of the stimulation-evoked release of ^{3}H-acetylcholine from the guinea pig myenteric plexus. Neuroscience 5:1331–1340
Kobayashi S, Kyoshima K, Olschowska JA, Jakobowitz DM (1983) Vasoactive intestinal polypeptide immunoreactive and cholinergic nerves in the whole mount preparation of the major cerebral arteries of the rat. Histochemistry 79:377–381
Koelle GB, Friedenwald JS (1949) A histochemical method for localizing cholinesterase activity. Proc Soc Exp Biol Med 70:612–622
Kolb H-A, Wakelam MJO (1983) Transmitter-like action of ATP on patched membranes of cultured myoblasts and myotubes. Nature 303:621–623
Koller KJ, Brownstein MJ (1987) Use of a cDNA to identify a supposed precursor protein containing valosin. Nature 325:542–545
Kondo H, Kuramoto H, Wainer BH, Yanaihara N (1985) Evidence for the coexistence of acetylcholine and enkephalin in the sympathetic preganglionic neurons of rats. Brain Res 335:309–314
Krieger DT (1983) Brain peptides: what, where, why? Science 222:975–985
Krieger DT, Martin JB (1981) Brain peptides. N Engl J Med 304:876–951
Krieger DT, Brownstein MJ, Martin JB (1983) Introduction. In: Krieger DT, Brownstein MJ, Martin JB (eds) Brain peptides. Wiley and Sons, New York, pp 1–15
Kuhar MJ (1985) The mismatch problem in receptor mapping studies. TINS 8:190–191
Kuperman AS, Volpert VA, Okamoto M (1964) Release of adenine nucleotide from nerve axons. Nature 204:1000–1001
Kuroda Y (1978) Physiological role of adenosine derivatives which are released during neurotransmission in mammalian brain. J Physiol (Paris) 74:463–470
Kusunoki M, Tsai LH, Taniyama K, Tanaka C (1986) Vasoactive intestinal polypeptide provokes acetylcholine release from the myenteric plexus. Am J Physiol 251:G51–G55
Lagercrantz H (1976) On the composition and function of large dense cored vesicles in sympathetic nerves. Neuroscience 1:81–92

Lamour Y, Dutar P, Jobert A (1983) Effects of neuropeptides on rat cortical neurons: laminar distribution and interaction with the effect of acetylcholine. Neuroscience 10:107–117

Landis ST (1984) Neurotransmitter plasticity and coexistence during the development of cholinergic-sympathetic neurons in culture and *in vivo*. In: Chan-Palay V, Palay SL (eds) Coexistence of neuroactive substances in neurons. Wiley and Sons, New York, pp 205–216

Landis SC, Fredieu JR (1986) Coexistence of calcitonin gene-related peptide and vasoactive intestinal peptide in cholinergic sympathetic innervation of rat sweat glands. Brain Res 377:177–181

Larsson L-I, Fahrenkrug J, Holst JJ, Schaffalitzky de Muckadell OB (1978) Innervation of the pancreas by vasoactive intestincal polypeptide (VIP) immunoreactive nerves. Life Sci 22:773–780

Leblanc GG, Landis SC (1986) Development of choline acetyltransferase activity in the cholinergic sympathetic innervation of sweat glands. J Neurosci 6:220–226

Lee DA, Witzemann V (1983) Photoaffinity labeling of a synaptic vesicle specific nucleotide transport system from *Torpedo marmorata*. Biochemistry 22:6123–6130

Levey AI, Wainer BH, Mufson EJ, Mesulam M-M (1983) Co-localization of acetylcholinesterase and choline acetyltransferase in the rat cerebrum. Neuroscience 9:9–22

Loh YP, Gainer H (1983) Biosynthesis and processing of neuropeptides. In: Krieger DT, Brownstein MJ, Martin JB (eds) Brain peptides. Wiley and Sons, New York, pp 80–116

Loh YP, Brownstein MJ, Gainer H (1984) Proteolysis in neuropeptide processing and other neural function. Annu Rev Neurosci 7:189–222

Luine VN, Rostene W, Rhodes J, McEwen BS (1984) Activation of choline acetyltransferase by vasoactive intestinal peptide. J Neurochem 42:1131–1134

Lundberg JM (1981) Evidence for coexistence of vasoactive intestinal polypeptide and acetylcholine in neurons of cat exocrine glands. Morphological, biochemical and functional studies. Acta Physiol Scand [Suppl] 496:1–57

Lundberg JM, Hökfelt T (1983) Coexistence of peptides and classical neurotransmitters. TINS 4:325–333

Lundberg JM, Hökfelt T (1986) Multiple co-existence of peptides and classical transmitters in peripherial autonomic and sensory neurons – functional and pharmacological implications. Prog Brain Res 68:241–262

Lundberg JM, Hökfelt T, Schultzberg M, Uvnäs-Wallensten K, Kohler C, Said S (1979) Occurrence of vasoactive intestinal polypeptide (VIP)-like immunoreactivity in certain cholinergic neurons of the cat: evidence from combined immunohistochemistry and acetylcholinesterase staining. Neuroscience 4:1539–1559

Lundberg JM, Fahrenkrug J, Brimijoin S (1981a) Characteristics of the axonal transport of vasoactive intestinal polypeptide – VIP – in nerve of the cat. Acta Physiol Scand 112:427–436

Lundberg JM, Fried G, Fahrenkrug J, Holmstedt B, Hökfelt T, Lagercrantz H, Lundgren G, Anggard A (1981b) Subcellular fractionation of cat submandibular gland: comparative studies on the distribution of acetylcholine and vasoactive intestinal polypeptide. Neuroscience 6:1001–1010

Lundberg JM, Hedlund B, Bartfai T (1982) Vasoactive intestinal polypeptide (VIP) enhances muscarinic ligand binding in the cat submandibular salivary gland. Nature 295:147–149

Luqmani YA (1981) Nucleotide uptake by isolated cholinergic synaptic vesicles: evidence for a carrier of adenosine 5′-triphosphate. Neuroscience 6:1011–1021

Lynch DR, Snyder SH (1986) Neuropeptides: multiple molecular forms, metabolic pathways and receptors. Annu Rev Biochem 55:773–799

McDonald WF, White DT (1985) Nature of extrasynaptosomal accumulation of endogenous adenosine evoked by K^+ and veratridine. J Neurochem 45:791–797

McGeer PL, Eccles JC, McGeer EG (1978) Molecular neurobiology of the mammalian brain. Plenum, New York

McGillis JP, Organist ML, Payan DG (1987) Substance P and immunoregulation. Fed Proc 46:196–199
Melander T, Stainers WA, Hökfelt T, Rokaeus A, Eckenstein F, Salvaterra PM, Wainer BH (1985) Galanin-like immunoreactivity in cholinergic neurons of the septum-basal forebrain complex projecting to the hippocampus of the rat. Brain Res 360:130–138
Melander T, Staines WM, Rokaeus A (1986) Galanin-like immunoreactivity in hippocampal afferents in the rat, with special reference to cholinergic and noradrenergic inputs. Neuroscience 19:223–240
Mesulam MM, Mufson EJ, Wainer BH, Levey AI (1983) Central cholinergic pathways in the rat: an overview based on an alternative nomenclature (Ch1–Ch6). Neuroscience 10:1185–1201
Meunier FM, Israël M, Lesbats B (1975) Release of ATP from stimulated nerve electroplaque junctions. Nature 257:407–408
Michaelson DM (1978) Is presynaptic acetylcholine release accompanied by secretion of the synaptic vesicle contents? FEBS Lett 89:51–53
Michaelson DM, Wien-Naor D (1987) Enkephalin uptake into synaptic cholinergic vesicles and nerve terminals. Ann NY Acad Sci (in press)
Michaelson DM, McDowall G, Sarne Y (1984a) The *Torpedo* electric organ is a model for opiate regulation of acetylcholine release. Brain Res 305:173–176
Michaelson DM, McDowall G, Sarne Y (1984b) Opiates inhibit acetylcholine release from *Torpedo* nerve terminals by blocking Ca^{2+} influx. J Neurochem 43:614–618
Miller RJ (1987) Multiple calcium channels and neuronal function. Science 235:46–52
Morel N, Meunier FM (1981) Simultaneous release of acetylcholine and ATP from stimulated cholinergic synaptosomes. J Neurochem 36:1766–1773
Mutt V (1982) Gastrointestinal hormones: a field of increasing complexity. The Brohee Lecture 1982. Scand J Gastroenterol [Suppl] 77:133–152
Mutt V (1983) New approaches to the identification and isolation of hormonal polypeptides. TINS 6:357–360
Nagy A, Baker RR, Morris SJ, Whittaker VP (1976) The preparation and characterization of synaptic vesicles of high purity. Brain Res 109:285–309
Nagy A, Varady G, Joó F, Rakonczay Z, Pilz A (1977) Separation of acetylcholine and catecholamine containing synaptic vesicles from brain cortex. J Neurochem 29:449–459
O'Dorisio SM (1987) Biochemical characteristics of receptors for vasoactive intestinal polypeptide in nervous, endocrine and immune systems. Fed Proc 46:192–195
Ohsawa K, Dowe GHC, Morris SJ, Whittaker VP (1979) The lipid and protein content of cholinergic synaptic vesicles from the electric organ of *Torpedo marmorata* purified to constant composition: implications for vesicle structure. Brain Res 161:447–457
Palkovits M (1984) Distribution of neuropeptides in the central nervous system: a review of biochemical mapping studies. Prog Neurobiol 23:151–189
Patterson PH (1978) Environmental determination of autonomic neurotransmitter functions. Annu Rev Neurosci 1:1–17
Phillis JW, Wu PH (1981) The role of adenosine and its nucleotides in central synaptic transmission. Prog Neurobiol 16:187–239
Phillis JW, Kostopoulos GK, Limacher JJ (1975) A potent depressant action of adenine derivatives on cerebral neurons. Eur J Pharmacol 52:1226–1229
Phillis JW, Kirpatrick JR, Said SI (1978) Vasoactive intestinal polypeptide excitation of central neurons. Can J Physiol Pharmacol 56:337–340
Pickering BT (1978) The neurosecretory neurone: a model system for the study of secretion. In: Campbell PN, Aldridge WN (eds) Essays in biochemistry, vol 14. Academic, London, pp 45–81
Potter DD, Matsumoto SG, Landis SC, Sah DWY, Furshpan EJ (1986) Transmitter status in cultured sympathetic principal neurons: plasticity, graded expression and diversity. Prog Brain Res 68:103–121
Priestly JV, Cuello AC (1982) Coexistence of neuroactive substances as revealed by immunohistochemistry with monoclonal antibodies. In: Cuello AC (ed) Co-transmission. Macmillan, London, pp 165–189
Reichardt LF, Kelly RB (1983) A molecular description of nerve terminal function. Annu Rev Biochem 52:871–926

Riberio JA (1979) Purinergic modulation of transmitter release. J Theor Biol 80:259–270

Riberio JA, Dominguez ML (1978) Mechanism of depression of neuromuscular transmission by ATP and adenosine. J Physiol (Paris) 74:491–496

Riberio JA, Sa-Almeida AM, Namorado JM (1979) Adenosine and adenosine triphosphate decrease $^{45}Ca^{2+}$ uptake by synaptosomes stimulated by potassium. Biochem Pharmacol 28:1297–1300

Richardson PJ (1981) Quantitation of cholinergic synaptosomes from guinea pig brain. J Neurochem 37:258–260

Richardson PJ, Brown SJ (1987a) Adenosine and ATP at the cholinergic nerve terminal. In: Dowdall MJ, Hawthorne JH (eds) Cellular and molecular biology of cholinergic function. Horwood, Chichester, pp 436–443

Richardson PJ, Brown SJ (1987b) ATP release from affinity purified rat cholinergic nerve terminals. J Neurochem 48:622–630

Richardson PJ, Siddle K, Luzio JP (1984) Immunoaffinity purification of intact, metabolically active, cholinergic nerve terminals from mammalian brain. Biochem J 219:647–654

Rokaeus A, Melander T, Hökfelt T, Lundberg JM, Tatemoto K, Carlquist M, Mutt V (1984) A galanine-like peptide in the central nervous system and intestine of the rat. Neurosci Lett 47:161–166

Rosenfeld MG, Mermod JJ, Amara SG, Swanson LW, Sawchenko PE, Rivier J, Vale WW, Evans RE (1983) Production of a novel neuropeptide encoded by the calcitonin gene via tissue-specific RNA processing. Nature 304:129–135

Sawynok J, Jahmadas KH (1976) Inhibition of acetylcholine release from cholinergic nerves by adenosine, adenine nucleotides and morphine: antagonism by theophylline. J Pharmacol Exp Ther 197:379–390

Scharrer B (1987) Neurosecretion: beginnings and new directions in neuropeptide research. Annu Rev Neurosci 10:1–17

Scharrer E, Scharrer B (1940) Secretory cells within the hypothalamus. Res Publ Assoc Res Nerv Ment Dis 20:170–194

Schmidt R (1978) Sequential release of ATP and ACh from cholinergic synaptic vesicles modulated by Ca^{2+} and synaptic membranes *in vitro*. Hoppe Seylers Z Physiol Chem 359:316

Schmitt FO (1983) Molecular regulators of brain function: a new view. Neuroscience 13:991–1001

Schubert P, Mitzdorf U (1979) Analysis and quantitative evaluation of the depressive effect of adenosine on evoked potentials in hippocampal slices. Brain Res 172:186–190

Schubert P, Lee KS, Tetzlaff W, Kreutzberg GW (1985) Postsynaptic modulation of neuronal firing pattern by adenosine. In: Changeux J-P, Hucho F, Maelicke A, Neumann E (eds) Molecular basis of nerve activity. de Gruyter, Berlin, pp 283–292

Schultzberg M (1984) Overviews of colocalization of peptides in the peripheral nervous system. In: Chan-Palay V, Palay SL (eds) Coexistence of neuroactive substances in neurons. Wiley and Sons, New York, pp 225–244

Schultzberg M, Hökfelt T (1986) The mismatch problem in receptor autoradiography and the coexistence of multiply messengers. TINS 9:109–110

Schwanzel-Fukuda M, Morrel JI, Pfaff DW (1986) Localization of choline acetyltransferase and vasoactive intestinal polypeptide-like immunoreactivity in the nervus terminalis of the fetal and neonatal rat. Peptides 7:899–906

Silinsky EM (1975) On the association between transmitter secretion and the release of adenine nucleotides from mammalian motor nerve terminals. J Physiol (Lond) 247:145–162

Silinsky EM (1984) On the mechanism by which adenosine receptor activation inhibits the release of acetylcholine from motor nerve endings. J Physiol (Lond) 346:243–256

Sjöquist F (1963) Pharmacological analysis of acetylcholinesterase-rich ganglion cells in the lumbosacral sympathetic system of the cat. Acta Physiol Scand 57:352–362

Snyder SH (1980) Brain peptides as neurotransmitters. Science 209:976–983

Speidel CC (1919) Gland-cells of internal secretion in the spinal cord of the skates. Carnegie Inst Washington Publ 13:1–31

Stadler H, Fenwick E (1983) Cholinergic synaptic vesicles from *Torpedo marmorata* contain an atractyloside binding protein related to the mitochondrial ADP ATP carrier. Eur J Biochem 136:377–382
Stadler H, Füldner HH (1980) Proton NMR detection of acetylcholine status in synaptic vesicles. Nature 286:293–294
Stanley KK, Burke P, Pitt P, Siddle K, Luzio JP (1983) Localization of 5′nucleotidase in a rat liver cell line using a monoclonal antibody and indirect immunofluorescent labelling. Exp Cell Res 144:39–46
Starke K (1977) Regulation of noradrenaline release by presynaptic receptor systems. Rev Physiol Biochem Pharmacol 77:1–124
Stjärne L, Astrand P (1984) Discrete events measure single quanta of 5′-triphosphatase secreted from sympathetic nerves of guinea pig and mouse vas deferens. Neuroscience 13:21–28
Stjärne L, Hedquist R, Labercrantz H (1970) Catecholamines and adenine nucleotide material in effluent from stimulated adrenal medulla and spleen: a study of the exocytosis hypothesis for hormone secretion and neurotransmitter release. Biochem Pharmacol 19:1147–1158
Stone TW (1981) Physiological role for adenosine and adenosine 5′-triphosphate in the nervous system. Neuroscience 6:523–555
Takami K, Kawai Y, Shiosaka S, Lee Y, Girgis S, Hillyard CJ, Macintyre I, Emson PC, Tohyama M (1985) Immunohistochemical evidence for the coexistence of calcitonin gene-related peptide and choline acetyltransferase-like immunoreactivity in neurons of the rat hypoglossal, facial and ambiguus nuclei. Brain Res 328:386–389
Tatemoko K, Rokaeus A, Jornvall H, McDonald TJ, Mutt V (1983) Galanin – a novel biologically active peptide from porcine intestine. FEBS Lett 164:124–128
Turner AJ (1986) Processing and metabolism of neuropeptides. Essays Biochem 22:69–119
Turner AJ, Dowdall MJ (1984) The metabolism of neuropeptides. Both phosphoramidon-sensitive and Captopril-sensitive metallopeptidases are present in the electric organ of *Torpedo marmorata.* Biochem J 222:255–259
Turner AJ, Hryszko J, Hooper NM, Dowdall MJ (1987) Purification and characterization of peptidyl dipeptidase resembling angiotensin converting enzyme from electric organ of *Torpedo marmorata.* J Neurochem 48:910–916
Uddman R (1980) Vasoactive intestinal polypeptide: distribution and possible role in the upper respiratory and digestive region. MD thesis, University of Lund
Vilmart-Seuwen J, Kersken H, Strutzl R, Plattner H (1986) ATP keeps exocytotic sites in a primed state but is not required for membrane fusion: an analysis with *Paramecium* cells *in vivo* and *in vitro.* J Cell Biol 103:1279–1288
Vincent SR, Satoh K, Armstrong DM, Fibiger HC (1983) Substance P in the ascending cholinergic reticular system. Nature 306:688–691
Vizi ES, Knoll J (1976) The inhibitory effect of adenosine and related nucleotides on the release of acetylcholine. Neuroscience 1:391–398
Volknandt W, Zimmermann H (1986) Acetylcholine, ATP, and proteoglycan are common to synaptic vesicles isolated from the electric organs of electric eel and electric catfish as well as from rat diaphragm. J Neurochem 47:1449–1462
Von Euler US, Gaddum JH (1931) An unidentified depressor substance in certain tissue extracts. J Physiol (Lond) 72:74–87
Wakelam MJO, Davies SA, Houslay MD, McKay I, Marshall CJ, Hall A (1986) Normal p21 N-*ras* couples bombesin and other growth factor receptors to inositol phosphate production. Nature 323:173–176
White JD, Stewart KM, Krause JE, McKelvy JF (1985) Biochemistry of peptide-secreting neurons. Physiol Rev 65:553–606
White T, Potter P, Wonnacott S (1980) Depolarization-induced release of ATP from cortical synaptosomes is not associated with acetylcholine release. J Neurochem 34:1109–1112
Whittaker VP (1979) Re: metabotropic neurotransmission. Trends Neurosci 2:VII

Whittaker VP (1984a) The structure and function of cholinergic synaptic vesicles. The 3rd Thiudicum lecture. Biochem Soc Trans 12:561–576
Whittaker VP (1984b) The synaptic vesicle. In: Lajtha A (ed) Handbook of neurochemistry, vol 7. Plenum, New York, pp 41–69
Whittaker VP, Agoston DV (1986) Association of neuropeptides and acetylcholine in guinea pig myenteric plexus and electromotor neurons of *Torpedo* investigated biochemically and histologically. In: Scharrer B, Korf HW, Hartwig HG (eds) Functional morphology of neuroendocrine systems: evolutionary and environmental aspects, pp 9–21
Wiedemann B, Franke WW (1985) Identification and localization of an integral membrane glycoprotein of M_r 38000 (synaptophysin) characteristic of presynaptic vesicles. Cell 41:1017–1028
Winkler H (1976) The composition of adrenal chromaffin granules: an assessment of controversial results. Neuroscience 1:65–80
Winkler H (1977) The biogenesis of adrenal chromaffin granules. Neuroscience 2:657–683
Winkler H, Westhead E (1980) The molecular organization of adrenal chromaffin granules. Neuroscience 5:1803–1827
Winkler H, Schmidt W, Fischer-Colbrie R, Weber A (1983) Molecular mechanisms of neurotransmitter storage and release: a comparison of the adrenergic and cholinergic systems. Prog Brain Res 58:11–21
Witzemann V (1987) Photoaffinity labelling of the adenosine nucleotide transporter of cholinergic vesicles. J Pharmacol Exp Ther 33:287–302
Zhu PC, Thureson-Klein A, Klein RL (1986) Exocytosis from large dense cored vesicles outside the active synaptic zones of terminals within the trigeminal subnucleus caudalis: a possible mechanism for neuropeptide release. Neuroscience 19:43–54
Zimmermann H (1978) Turnover of adenine nucleotides in cholinergic synaptic vesicles of the *Torpedo* electric organ. Neuroscience 3:827–836
Zimmermann H (1979) Vesicular heterogeneity and turnover of acetylcholine and ATP in cholinergic synaptic vesicles. Prog Brain Res 49:141–151
Zimmermann H (1982) Coexistence of adenosine 5′-triphosphate and acetylcholine in the electromotor synapse. In: Cuello AA (ed) Co-transmission. Macmillan, London, pp 243–259
Zimmermann H, Bokor JT (1979) ATP recycles independently of ACh in cholinergic synaptic vesicles. Neurosci Lett 13:319–324
Zimmermann H, Denston CR (1977a) Recycling of synaptic vesicles in the cholinergic synapses of the *Torpedo* electric organ during induced transmitter release. Neuroscience 2:695–714
Zimmermann H, Denston CR (1977b) Separation of synaptic vesicles of different functional states from the cholinergic synapses of the *Torpedo* electric organ. Neuroscience 2:715–730
Zimmermann H, Grondall EJM (1985) Adenosine-5′-triphosphate at the cholinergic synapse: a co-transmitter? In: Changeux JP, Hucho F, Maelicke A, Neumann E (eds) Molecular basis of nerve activity. de Gruyter, Berlin, pp 91–105
Zimmermann H, Dowdall MJ, Lane DA (1979) Purine salvage at the cholinergic nerve endings of the *Torpedo* electric organ: the central role of adenosine. Neuroscience 4:979–993

Part V Peripheral Cholinergic Synapses

CHAPTER 18

The Neuromuscular Junction

M. E. KRIEBEL

A. Introduction

Physiological studies on the neuromuscular junction (NMJ) are usually concerned with either postsynaptic receptor and channel mechanisms (Chap. 9) or with presynaptic characteristics of transmitter release (Chap. 1). The present chapter will discuss the quantal basis of transmitter release at the NMJ, emphasizing the multiplicity of classes of quanta observed and their pharmacological differences. Quanta are detected as miniature end-plate potentials (mEPPs) which may be divided into at least three main classes: skew-mEPPs, bell-mEPPs, and giant-mEPPs.

The mEPP was first analysed by FATT and KATZ (1952) who showed that it represents the smallest event (quantum) that is evoked with nerve stimulation. They noted that mEPPs form a bell-shaped distribution in an amplitude-frequency plot (Fig. 1). These have been named bell-mEPPs to distinguish them from the smaller skew-mEPPs (a designation which includes sub-mEPPs) that are also found in normal, adult frog and mouse preparations albeit usually at low frequencies (Figs. 2, 3) (KRIEBEL and GROSS 1974; KRIEBEL et al. 1976).

The skew class was first recognized by DENNIS and MILEDI (1971, 1974a) at the reinnervating amphibian NMJ and by HARRIS and MILEDI (1972) in the botulinum toxin poisoned preparation. MUNIAK et al. (1982) using the mouse junction showed that the skew-mEPP class was replaced by the bell-mEPP class during morphogenesis and reinnervation. By contrast, during botulinum toxin poisoning the bell-mEPP class is blocked leaving the skew-mEPP class (KRIEBEL et al. 1976; cf. SPITZER 1972).

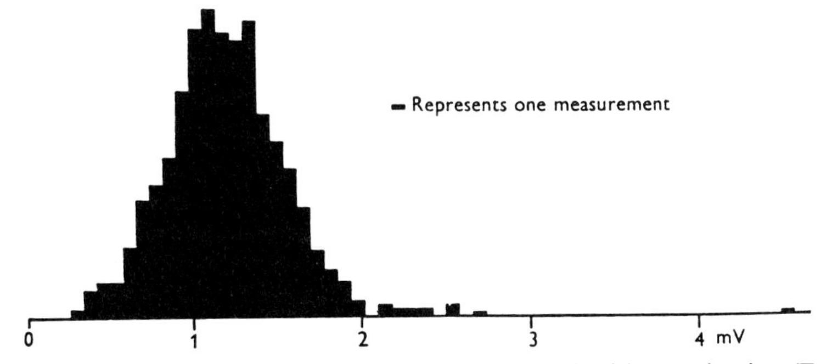

Fig. 1. Distribution of mEPP amplitudes; muscle treated with prostigmine (FATT and KATZ 1952)

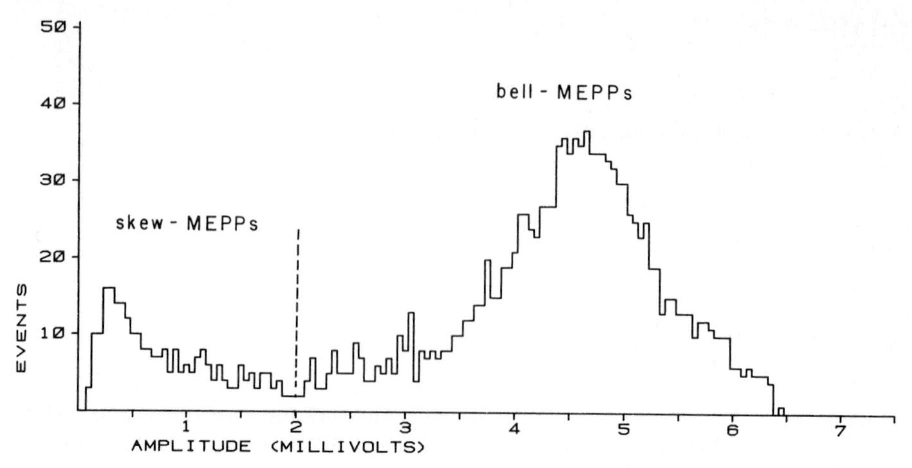

Fig. 2. Typical amplitude histogram from 21-day-old mouse preparation maintained at 21 °C. Note the two classes of mEPP amplitudes; the class to the *right* of the *broken line* is bell-shaped (bell-mEPPs) whereas the one to the *left* of the *broken line* is skewed towards the smaller mEPPs (skew-mEPPs). The peak of the bell-mEPPs (at 4.7 mV) is about 15 times that of the skew-mEPPs (at 0.3 mV). Resting potential was −80 mV at beginning and −75 mV at the end of the 22-min period of recording. 55 mEPPs min^{-1}. The mode of skew-MEPPs is the sub-MEPP value, cf. first peak in Fig. 8a–e (MUNIAK et al. 1982)

The giant-mEPP class was first described by LILEY (1956a, 1957) who demonstrated that they were composed of two or more mEPPs (bell-mEPPs). KRIEBEL and coworkers have followed changes in mEPP amplitudes and frequencies under various treatments at a given electron-microscopically identified junction and have discussed these in terms of the changes in the ratio of skew- to bell-mEPPs. Consequently what appears to be a wide range of pharmacological effects on mEPP and end-plate potential (EPP) amplitude distributions and mEPP frequencies resulting from the application of various poisons and drugs can be readily explained by changes in the relative frequency of the three classes of normally occurring mEPPs and there is no need to postulate the induction of quantal classes with various treatments.

A brief description of Schwann cell mEPPs found at the Schwann-cell-muscle junction (after denervation) will be included because of the similarity in size and time characteristics to sub-mEPPs. However, there are major pharmacological differences between the Schwann-cell mEPPS and the sub-mEPP class of quanta and it seems reasonable to conclude that Schwann-cell quanta are not released at the normal NMJ (KRIEBEL and GROSS 1974; BEVAN et al. 1976; KRIEBEL et al. 1980).

The subunit hypothesis is an extension of the quantal hypothesis as developed by KATZ (1962, 1977, 1978) in that the mEPP and the unitary EPP are composed of ten subunits the size of sub-mEPPs. The sub-unit hypothesis explains many pharmacological treatments and disease states that alter mEPP histogram amplitude profiles and/or change in mEPP size. It is important to stress that studies concerned with correlating vesicle numbers with fatigue at the NMJ show many different results. For example, a reduced EPP has been interpreted to indicate two

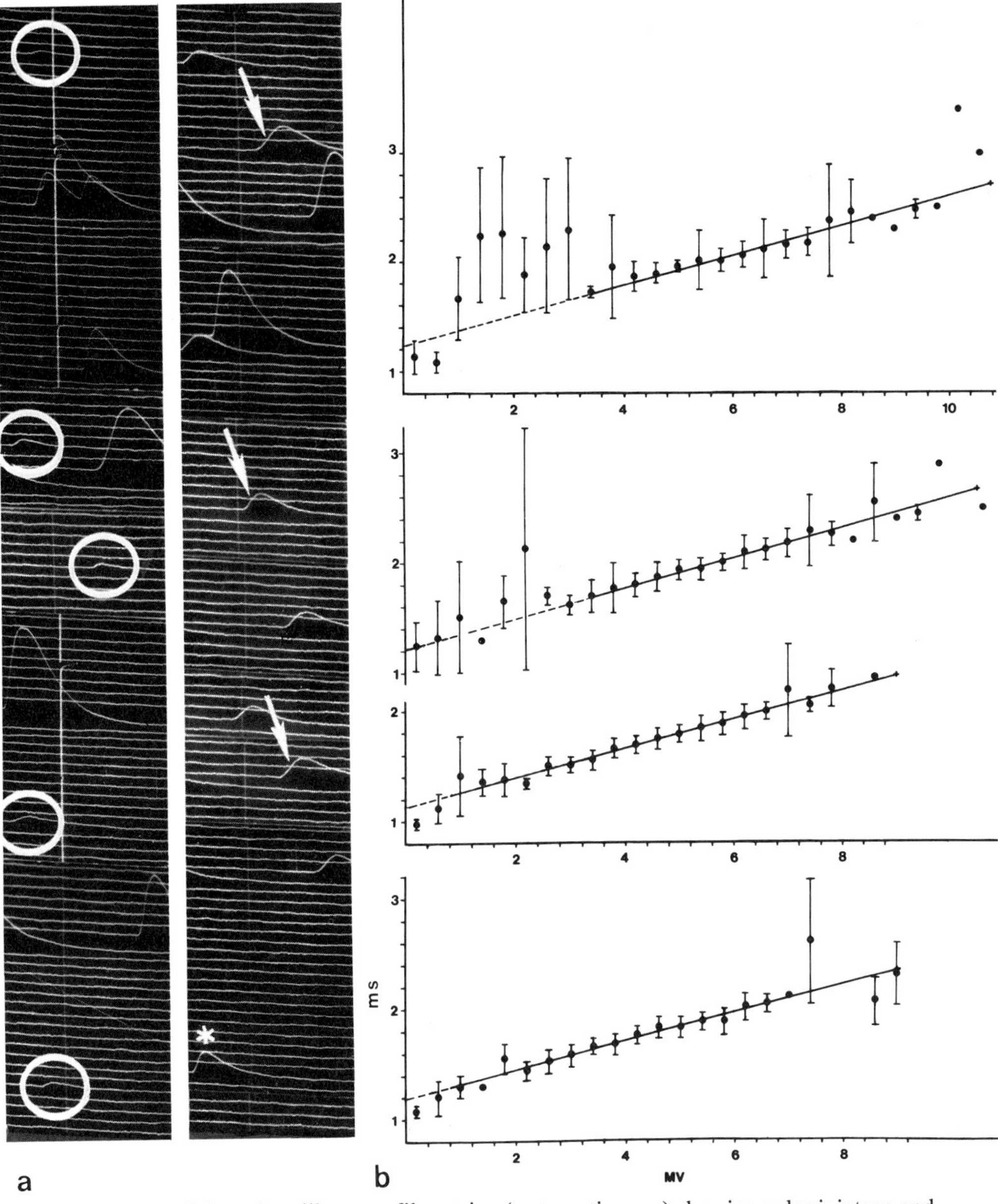

Fig. 3a, b. a Selected oscilloscope film strips (not continuous) showing subminiature end-plate potentials (sub-mEPPs, *circled*) and slow time-to-peak intermediate-sized mEPPs. Offsets on rising phase of slow time-to-peak mEPPs, suggesting two events, are indicated by *arrows*. Intermediate-sized mEPP with normal time-to-peak is indicated by an *asterisk*. Calibration: 1.4 mV between traces; each trace is 20 ms long. **b** mEPP amplitude versus time-to-peak plots. The mEPPs used for these curves are from one preparation. Regression lines were calculated from the range of mEPPs indicated by the *continuous part of the line*. Many mEPPs of intermediate amplitudes at the start of the experiment (*top two graphs*) had a long time-to-peak and were probably composed of two or more subunits. Many of these appeared to have offsets on the rising phase (indicated by *arrows* on the oscilloscope records). Each *point* on the regression line indicates mean rise time of all mEPPs in a 400-μV interval. *Bars* indicate mean ±1 SD of observation. Sub-mEPPs are represented by the *first two points* on the plots (KRIEBEL et al. 1976)

very different vesicular conditions. HEUSER and REESE (1973) thought that a fatigued junction with a small EPP should result in a depletion of synaptic vesicles and they found examples of depleted junctions. On the other hand, ROSE et al. (1978) found that identified junctions that were fatigued to yield small EPPs were full of vesicles; they concluded that the fatigue was in the release mechanism, and found vesicular depletion only when the EPP was reduced to about half. Vesicles were well described by DÜRING (1967) and the vesicle hypothesis has recently been reviewed (CECCARELLI and HURLBUT 1980; MONOLOV and OVTSCHAROFF 1982; ZIMMERMANN 1982; WHITTAKER 1984). Nonvesicular release and arguments against the vesicle hypothesis have been discussed by TAUC (1983). TREMBLAY et al. (1983) have reviewed the subunit hypothesis with regard to vesicles. It should be mentioned that KRIEBEL and FLOREY (1982) and KRIEBEL et al. (1986) have shown that there is only one class of vesicles based on diameter regardless of the percentage of skew-mEPPs.

B. Quantal Basis of Transmitter Release

FATT and KATZ (1952) discovered the occurrence of mEPPs in isolated frog muscle and noted that their time characteristics were similar to EPPs. They were found only at the end-plate and were either potentiated with eserine or blocked by curare. These observations demonstrated that mEPPs were generated at the postsynaptic muscle fibre membrane by acetylcholine (ACh). Moreover, they found that mEPPs were not generated by the addition of acetylcholine to the bath at a concentration sufficient to depolarize the muscle fibre smoothly to a value several times that of the mEPP amplitude. Thus ACh must be released as a packet (quantum) if it is to generate a mEPP with a fast rise time and an amplitude greater than background noise. Later, KATZ and MILEDI (1977) demonstrated the effect of molecular ACh leakage with the observation that the muscle fibre hyperpolarized with the addition of curare to the junctional region. Normal mEPP frequencies contribute little to the background release of ACh from the nerve muscle preparation. Molecular ACh leakage is not related to the skew- (or sub-)mEPP class because skew-mEPPs form a mode in a frequency-amplitude histogram (KRIEBEL and GROSS 1974; KRIEBEL et al. 1976, 1982). Therefore, the frequency and distribution of the skew-mEPP class can be described. Recently BLIGHT and PRECHT (1982) demonstrated in the intact frog that bell-mEPPs are rare events so there is no doubt that mEPP frequency depends on the trauma of dissection, stretch of the preparation and the position of the recording micropipette. Nevertheless, it is appreciated that a bell-mEPP represents a quantum of ACh which is available for evoked release. The frequency of mEPPs (bell- and/or skew-class) is readily altered by various treatments such as changes in temperature, ions, tonicity, drugs and poisons but caution must be exercised in attempting to quantify these changes because the mEPP frequency and the ratio of skew- to bell-mEPPs can change 'spontaneously' (KRIEBEL et al. 1976). This can be readily appreciated by penetrating the muscle fibre near a nerve terminal; this produces a cascade of mEPPs (FATT and KATZ 1952). It is difficult to achieve stable mEPP frequencies (FATT and KATZ 1952; KRIEBEL et al. 1976; KRIEBEL and STOLPER

1975; KRIEBEL 1978), and it is disconcerting to notice that this fact is usually ignored.

FATT and KATZ (1952) also noted that the amplitude-frequency distribution of mEPPs was bell-shaped with no secondary peaks or shoulders. This permitted them to conclude that mEPPs represent a single class of events (here termed bell-mEPPs) which could thus be described with a mean and variance. Moreover, FATT and KATZ noted that bell-mEPPs usually occurred independently of one another but could do so in bursts indicating periods of cooperativity. DEL CASTILLO and KATZ (1954) reduced the amplitude of the EPP with a low Ca^{2+}, high Mg^{2+} saline and found that the smallest nerve-evoked response (unitary evoked responses) equalled mEPPs in size and shape (Fig. 4). They proposed that

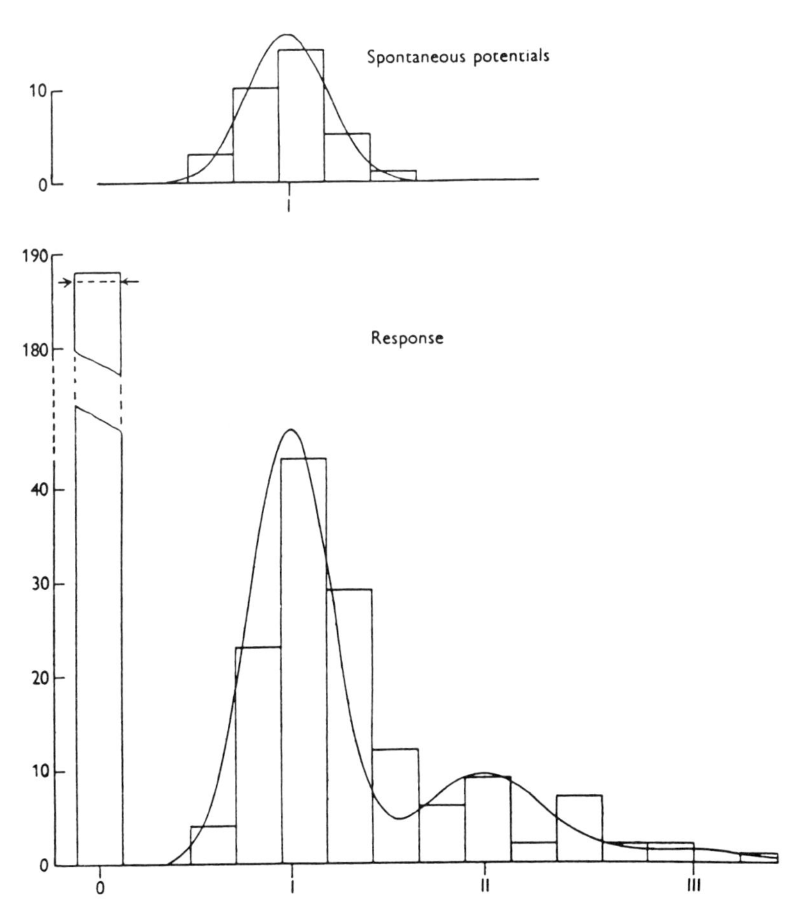

Fig. 4. Distribution of amplitudes of spontaneous miniature potentials and end-plate responses at a Ca^{2+}-deficient junction (experiment of FATT and KATZ 1952a, pp. 119–120). In the *lower part,* the *continuous curve* has been calculated on the hypothesis that the responses are built up statistically of units whose mean size and amplitude distribution are identical with those of the spontaneous potentials. Expected number of failures is shown by *arrows. Abscissae:* scale units = mean amplitude of spontaneous potentials (0.875 mV) (DEL CASTILLO and KATZ 1954)

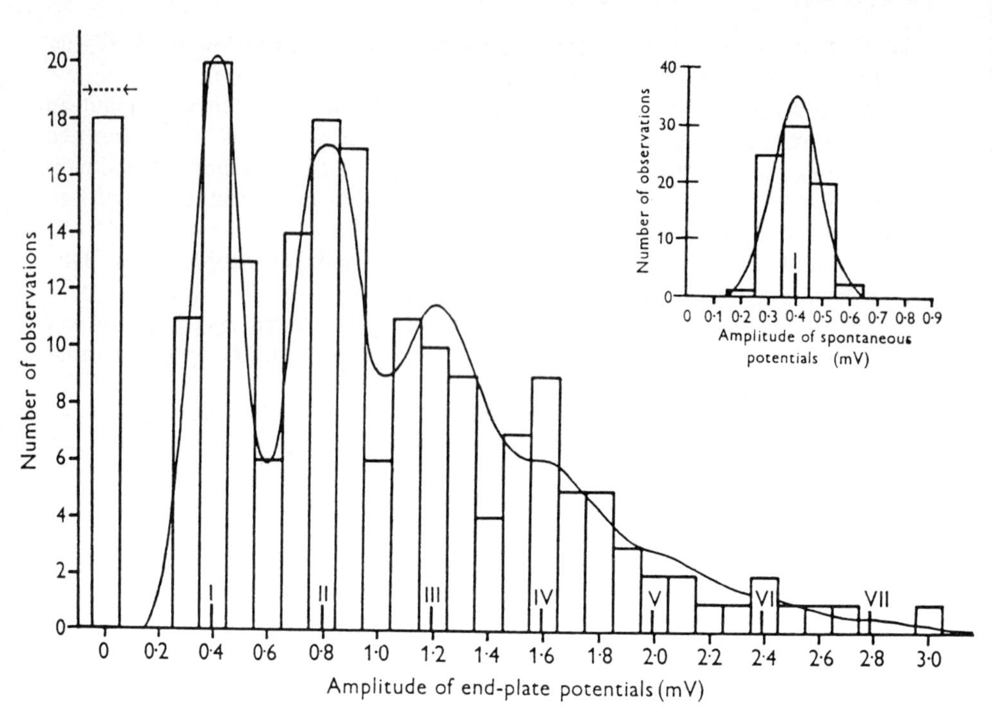

Fig. 5. EPP and spontaneous potential amplitude distribution in a fibre in which neuromuscular transmission was blocked by increasing the magnesium concentration of Krebs's solution to 12.5 m*M*. Peaks in EPP amplitude distribution occur at 1, 2, 3 and 4 times the mean amplitude of the spontaneous miniature potentials. Gaussian curve is fitted to spontaneous potential distribution and used to calculate theoretical distribution of EPP amplitude (*continuous curve*). *Arrows* indicate expected number of failures of response to nerve stimuli (BOYD and MARTIN 1956)

normal EPPs were composed of summed quanta the size of mEPPs (bell-mEPPs). This is the quantal hypothesis of transmitter release which was proved by BOYD and MARTIN (1956) and LILEY (1956b). These authors published EPP amplitude-frequency histograms of reduced EPP responses showing four peaks (Fig. 5). The first peak is equal in amplitude and distribution to that of mEPPs (bell-mEPPs) and the subsequent peaks are integral multiples of the first peak. In addition, they determined that the number of quanta composing EPPs, as well as the number of failures in release, fitted Poisson statistics. This demonstrates that each quantum is released independently, and that during low quantal contents the probability of release is low and the number available for release is high. The normal EPP is composed of several hundred quanta so the EPP is normally large enough to pass threshold for spike initiation. Larger fibres have larger junctional areas and thus larger quantal contents (KUNO et al. 1971). Under normal conditions the numbers of quanta generating the EPP fit binomial statistics (DEL CASTILLO and KATZ 1954; MARTIN 1966; MIYAMOTO 1975).

Early studies concerned with analyses of quantal content (number of quanta in EPP) and quantal size (number of ACh molecules in a quantum) were under experimental conditions which reduced the EPP size to the point at which the numbers of quanta released during an experiment were relatively small. Under such experimental conditions the quantal size was found to remain remarkedly constant throughout an experiment (KATZ 1969; see DOHERTY et al. 1984). Nevertheless, LONGENECKER et al. (1970) pointed out that at high frequencies induced with black widow spider venom (BWSV) mEPP amplitudes can decline by 50%. mEPP size does vary with the input resistance of the muscle fibre (KATZ and THESLEFF 1957): smaller fibres have a higher input resistance than larger ones and thus larger mEPPs. Voltage-clamp studies have shown that the size and variance of miniature endplate currents (mEPCs) are the same regardless of muscle fibre size which demonstrates the uniform character of the presynaptic quantum (bell-mEPP). The skew-mEPP currents are also uniform across fibres. Since the ratio of skew- to bell-mEPPs is readily changed and since the size of the bell-mEPP can change, it is imperative to define these two classes of normal quanta.

C. Classes of Quanta

I. Bell-mEPPs and Skew- (sub-)mEPPs

DENNIS and MILEDI (1971, 1974a) and BENNETT and PETTIGREW (1975) reported the appearance of small mEPPs at the reinnervating or developing NMJ which form a skew distribution into the noise (hence skew-mEPPs). With very good signal-to-noise ratios, KRIEBEL and GROSS (1974) found the skew-mEPP mode (mode is the sub-mEPP size) of the frog NMJ to be about one-seventh that of the larger bell-mEPP class (see also KULLBERG et al. 1977). During synaptogenesis KRIEBEL and GROSS (1974) demonstrated that the percentage of bell-mEPPs is initially very low: their proportion then increases until the profile of the amplitude-frequency distribution exhibits a discontinuity between the smaller skew-mEPPs and the larger bell-mEPPs; and finally the skew-mEPPs almost disappear. MUNIAK et al. (1982) found a similar change at the NMJ of the mouse diaphragm in neonates and during reinnervation (Figs. 6, 7). Skew-mEPPs have been reported by MILEDI and SLATER (1970), KOENIG and PÉCOT-DECHAVASSINE (1971), MCARDLE and ALBUQUERQUE (1973), TONGE (1974) and BENNETT and PETTIGREW (1975). Since the receptor density is high in the developing mouse junction, changes in skew- to bell-mEPP ratios cannot be attributed to a receptor deficiency (MATTHEWS-BELLINGER and SALPETER 1983). DENNIS and MILEDI (1971, 1974a) and BENNETT and PETTIGREW (1975) with the frog preparation, MUNIAK et al. (1982) with the mouse preparation and COLMÉUS et al. (1982) with the rat preparation found that only quanta of the bell-mEPP class are evoked with nerve stimulation. The bell- and skew-classes of mEPPs may correspond to the two classes found in *Xenopus* nerve-muscle cultures (KIDOKORO 1984). SUN and POO (1985) also report small mEPPs in culture which may be similar to sub-mEPPs (see YOUNG and POO 1983; ZISKIND-CONHAIN and DENNIS 1981).

GROSS and KRIEBEL (1973) and KRIEBEL and GROSS (1974) defined the skew-mEPPs (sub-mEPP is mode) in frog preparations because they were able to estab-

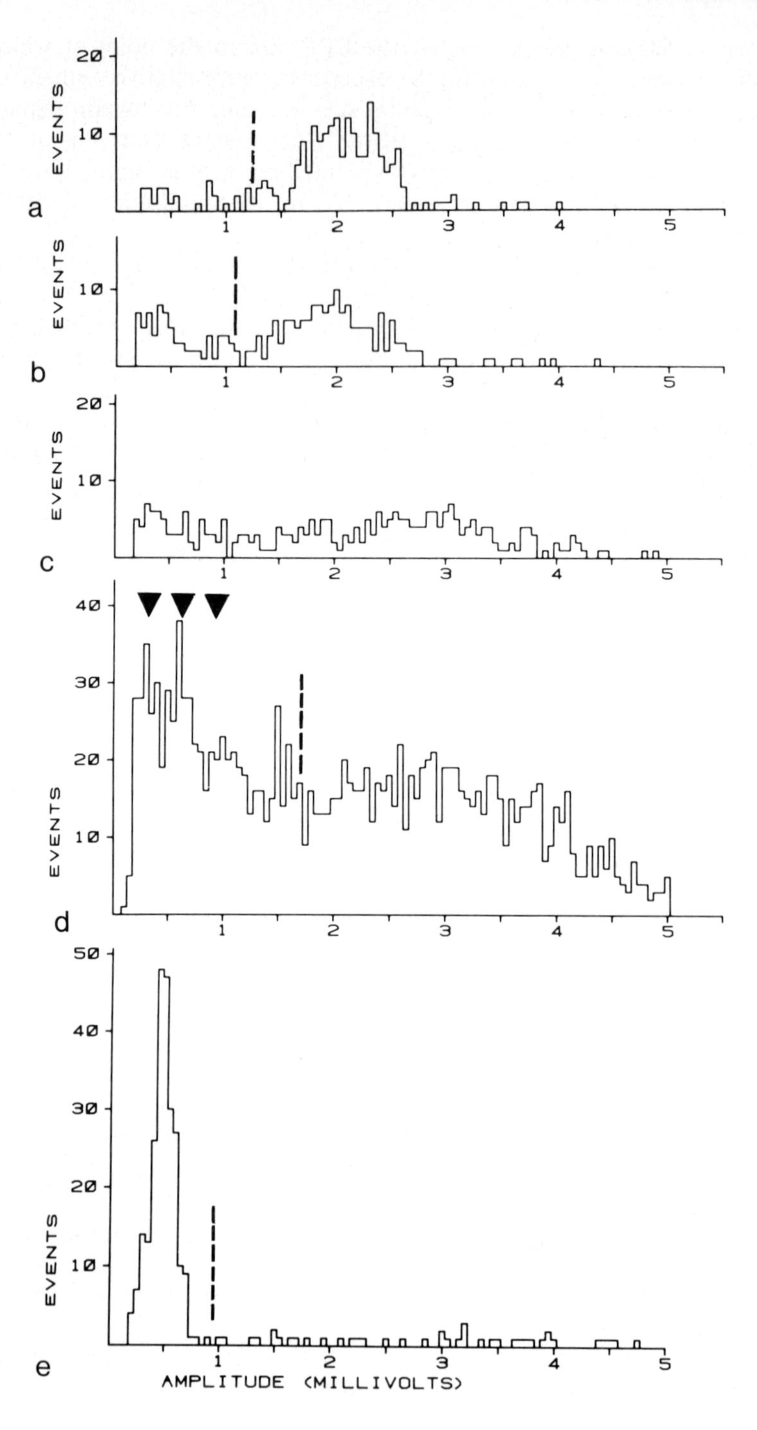
a
b
c
d
e
EVENTS
AMPLITUDE (MILLIVOLTS)

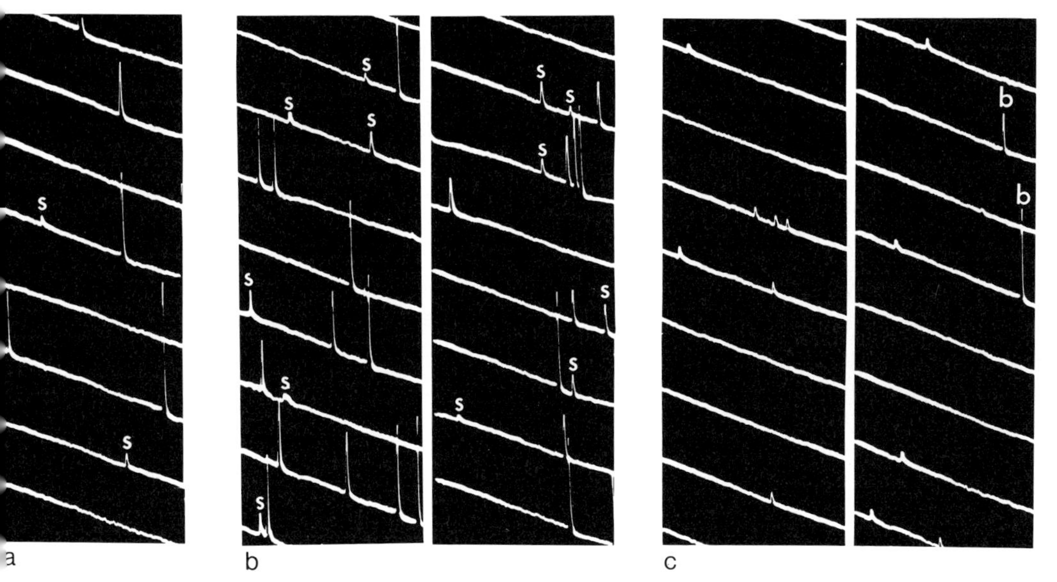

Fig. 7a–c. Oscilloscope records of some mEPPs used to construct histograms in Fig. 6a, d, e. Oscilloscope traces are 0.5 s long and 2 mV apart. Noise levels are about 75 μV peak-to-peak. **a** mEPPs for Fig. 6a. **b** mEPPs for Fig. 6d. **c** mEPPs for Fig. 6e. *b*, bell-mEPPs; *s*, skew-mEPPs (MUNIAK et al. 1982)

lish a mode and thus ascertain the frequency of the skew class. Moreover, they found a discontinuity between the skew-mEPP class and the bell-mEPP class in the amplitude-frequency histograms. And, because of high signal-to-noise ratios they determined that the time characteristics of both bell- and skew-(sub-)mEPPs were the same (see ERXLEBEN and KRIEBEL 1983, 1988a). Since the frequency of the sub-mEPPs in the normal adult and unstressed frog preparation is only 1/50

Fig. 6a–e. mEPP amplitude distributions from reinnervated junctions 5 days after the phrenic nerve was crushed. This series of histograms was chosen to demonstrate the increase in the relative numbers of bell-mEPPs that occurred during reinnervation. Fibre a was closest to the point of nerve entry and fibre **e** was farthest away (4 m*M* Co^{2+}, 22 °C). *Broken lines* divide skew- and bell-mEPPs. **a** The ratio of the number of bell-mEPPs to skew-mEPPs at this junction is in the control range. Nerve stimulation evoked EPPs of the bell-mEPP size. Resting potential was −62 mV; 60 mEPPs min^{-1}, 5 min of recording. **b** This junction shows more skew-mEPPs than control. Nerve stimulation evoked EPPs. Resting potential was −60 mV throughout 7 min of recording; 45 mEPPs min^{-1}. **c** Uniform mEPP amplitude distribution of skew- and bell-mEPPs. Nerve stimulation evoked EPPs. Resting potential was −80 mV; 30 mEPPs min^{-1}, 11 min of recording. **d** Overall mEPP distribution is skewed since the bell-mEPPs are relatively few, and there is an equal number of skew and bell-mEPPs although the exact demarcation between the two classes is not clear. Note that the skew-mEPPs show three integral peaks (*arrowheads*). Resting potential was −78 mV; 30 min of recording, 60 mEPPs min^{-1}. **e** The skew-mEPP amplitudes form a bell-shaped distribution and the bell-mEPP amplitudes form a flat distribution. Over 80% of the mEPPs are skew-mEPPs. Nerve stimulation did not evoke mEPPs. Resting potential was −60 mV throughout 14 min of recording; 20 mEPPs min^{-1} (MUNIAK et al. 1982)

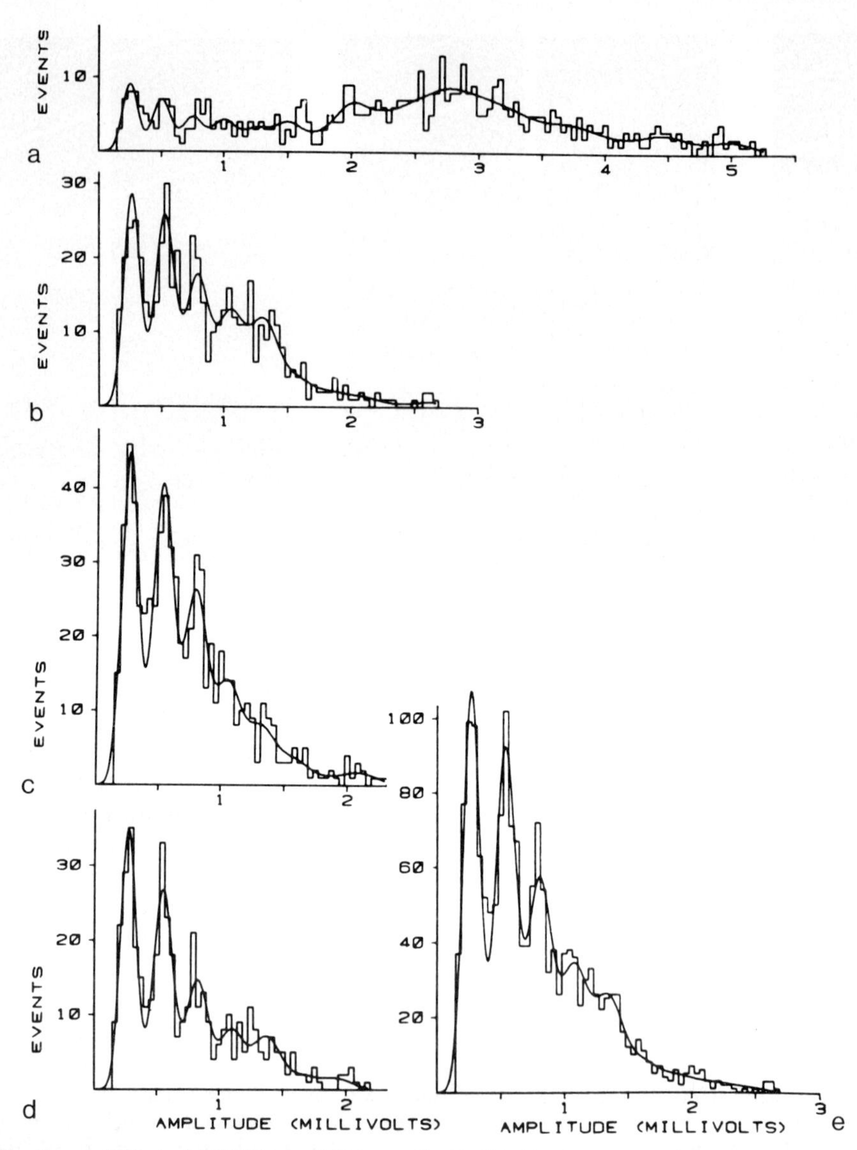

Fig. 8a–e. Effect of β-BTX on mEPP amplitude distribution. **a** Control histogram, 20 min of recording. After the addition of 20 μg ml^{-1} β-BTX there was a period of depression lasting approximately 5 min, followed by a 4-fold increase in frequency that reached its peak 40 min after the drug was added. **b–d** Successive 12-min periods 1 h after the addition of β-BTX. **e** represents **b–d** combined. Temperature was kept at 26 °C during the experiment. Resting potential was 70 mV at the start and 66 mV at the end; 21-day-old mouse. The curves are predicted from a model derived under the following assumptions: (a) sub-mEPPs result from the release of a subunit of transmitter. The subunit amplitude is assumed to be normally distributed with mean (μ) and variance ($\sigma^2+\sigma_m^2$), where (σ^2) is the actual variance of the subunit and (σ_m^2) represents measurement error (with noise). (b) Larger amplitude mEPPs result from the synchronous release of two or more subunits. Subunits sum linearly; therefore, the release of j subunits would produce a normally distributed population of mEPP amplitudes with mean (j^{μ}) and variance ($j\sigma^2+\sigma_m^2$).

that of the bell-mEPP class, and since the sub-mEPP amplitudes are only one-seventh to one-tenth that of bell-mEPPs, this class was simply lost to the background noise in earlier studies (KRIEBEL and GROSS 1974; KRIEBEL and MOTELICA-HEINO 1987). It is probable that synaptogenesis (synapse remodelling) continues in the adult so the skew-class may originate from newly formed terminal (BARKER and IP 1966; BIESER et al. 1984; ROTSCHENKER and TAL 1985). Thus the ratio of skew- to bell-mEPPs in both neonates and adults may correlate to the amount of newly formed nerve terminal.

COOKE and QUASTEL (1973) report that the skew-class in the normal adult rat diaphragm preparation has the same time characteristics as the larger quanta. Since COOKE and QUASTEL (1973) were able to increase the frequency of the skew-mEPPs along with that of the bell-mEPP class by means of focal depolarization they nicely demonstrated that both classes of mEPPs were released from the same presynaptic site and that the same postsynaptic membrane generated both classes. AUGUSTINE and LEVITAN (1983) were also able to rule out postsynaptic differences as a basis for the appearance of the skew-mEPP class: with 10 μM erythrosin they were able to increase both skew-mEPPs and giant-mEPPs leaving the bell class unchanged. CARLSON et al. (1982) demonstrated that both classes of quanta result from different release mechanisms with different frequency temperature characteristics. KRIEBEL et al. (1976, 1982) from the unstressed young and adult mouse diaphragm preparation demonstrated both mEPP classes with distinct modes. The boundary between the two classes must be established with an amplitude histogram because different treatments alter the ratio of skew- to bell-mEPPs. Among these are heat, lanthanum ions, BWSV, β-bungarotoxin (βBTX) (Fig. 8), botulinum toxin (Fig. 9), tetanus toxin or an increase in osmotic pressure (DUCHEN and TONGE 1973; CULL-CANDY et al. 1976; KRIEBEL et al. 1976; MACDERMOT et al. 1978). Moreover, the skew- to bell-mEPP ratio may 'spontaneously' change (KRIEBEL et al. 1976). In the normal mouse diaphragm preparation many junctions show a relatively low mEPP frequency and thus a high proportion of skew-mEPPs (KRIEBEL et al. 1982). In these junctions the unitary evoked quantum is much larger than the average mEPP because of the high proportion of skew-mEPPs which are not evoked (Fig. 10) and in extreme cases the bell-mEPP class is not spontaneously represented but is evoked with nerve stimulation (Fig. 11; KRIEBEL et al. 1982). Normally unitary evoked quanta of the size of skew-mEPPs (sub-mEPPs) are not evoked (BEVAN 1976; KRIEBEL 1978) but they can be evoked in the fatigued preparation

◄

(c) The overall amplitude distribution is, therefore, composed of the sum of these subpopulations. Each subpopulation must be multiplied by a weighting factor (W_j), which represents the conditional probability that a mEPP belongs to the jth population. The probability density function for k subpopulations is, therefore, as follows: $f(x) = W_1 \cdot N_1(x) + W_2 \cdot N_2(x) + \ldots + W_k \cdot N_k(x)$, where $f(x)$ represents the probability of observing a mEPP of amplitude x, and $N_j(x)$ is the normal probability density function with mean (j^{μ}) and variance ($j\sigma^2 + \sigma_m^2$). For any observed amplitude histogram, estimates of the parameters in this function (W_j, μ_x and σ) are obtained as maximum likelihood estimates. The probability density function was used to produce a fit to the observed histograms (*smooth lines*). The data were also fitted to a bimodal model, and the fit for the skew-mEPPs was not as significant as that of the summed release model (see MATTENSON et al. 1981 for details; LLADOS et al. 1980)

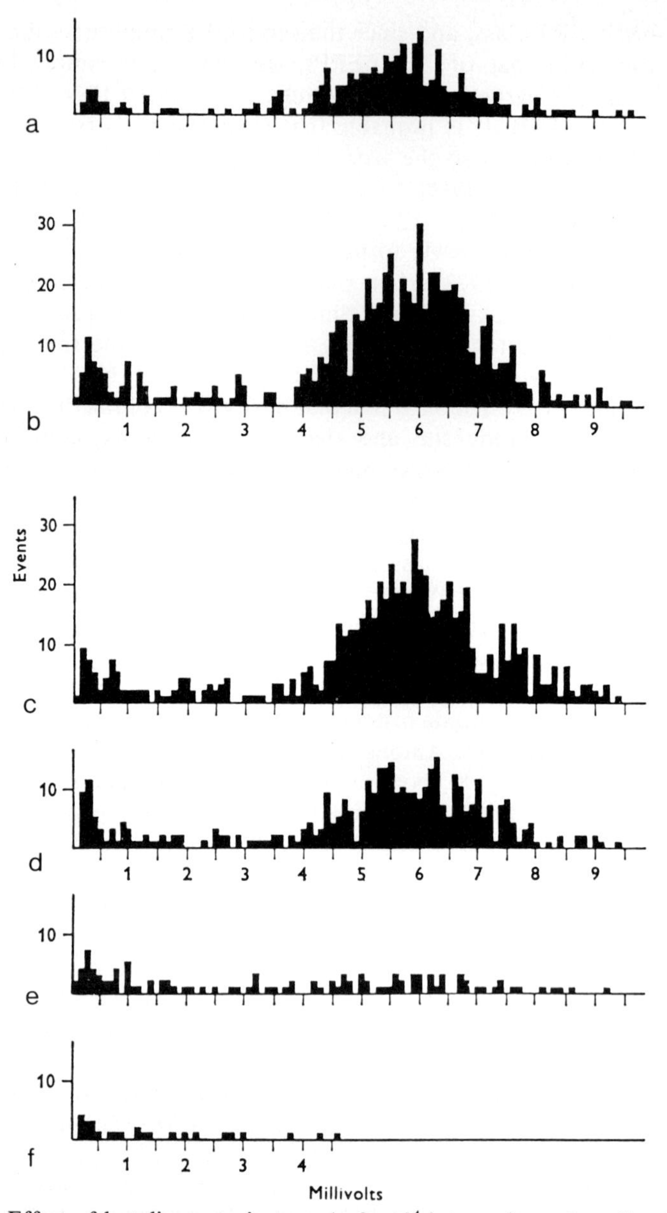

Fig. 9a–f. Effect of botulinum toxin type A. 3×10^4 intraperitoneal median lethal doses per milliliter, on mEPP amplitude distributions. These are successive histograms showing all recorded mEPPs during 96 min (4 min of control) of continuous recording. Temperature, 33 °C; 23-day-old mouse; neostigmine, 0.1 μg ml^{-1}; normal Ca^{2+} concentration in saline. Resting potential at start was −64 mV, at end −60 mV. **a** Control histogram. The mean mEPP amplitude and frequency were stationary; 266 mEPPs in 4 min, 66.5 mEPPs min^{-1}. **b** Botulinum toxin added to bath; 666 mEPPs in 12 min. **c** 647 mEPPs in 12 min, 54 mEPPs m^{-1}. **d** 378 mEPPs in 12 min, 31.5 mEPPs min^{-1}. Note that the means of major mode mEPPs and sub-mEPPs are the same as in the control, **a.** **e** 112 mEPPs in 24 min, 4.6 mEPPs min^{-1}. **f** 27 mEPPs in 32 min, 0.9 mEPPs min^{-1}. The frequency was dropping through the sequence of histograms **b–f** (KRIEBEL et al. 1976)

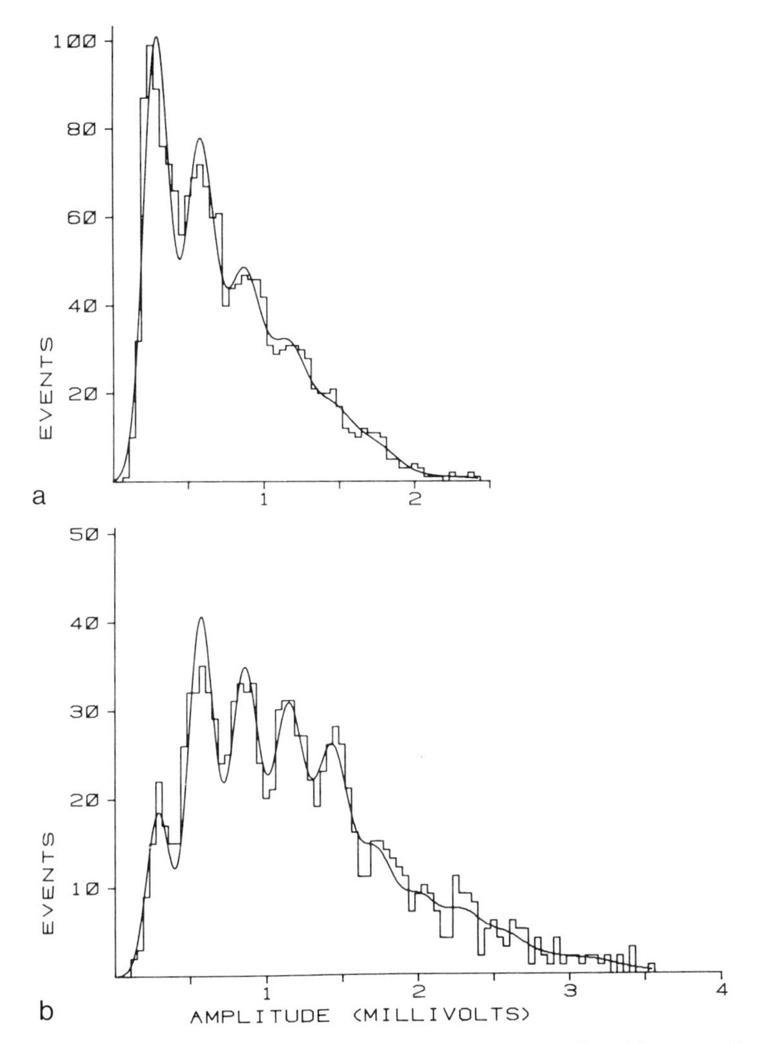

Fig. 10 a, b. mEPP and EPP amplitude distributions from a 15-day-old mouse diaphragm at 35 °C; an extreme example where the mEPPs are mainly of the skew-class. Neostigmine, 1.0 μg ml^{-1}. The phrenic nerve was stimulated at 2 Hz, and all mEPPs and EPPs generated during the 25 min of continuous recording are shown. *Continuous lines* show the predicted distributions based on the subunit hypothesis. The standard deviation of the noise and measurement (σ_m^2) is 0.067 mV. Membrane potential was −62 mV throughout. Co^{2+} was approximately 4 m*M*. **a** mEPPs. Bell-mEPPs do not form a class distinct from skew-mEPPs because their frequency was too low. The first four peaks were stationary throughout the recording period. The peak intervals are 7.0 histobars. There are 1.8×10^3 mEPPs in this histogram (72 mEPPs min^{-1}); μ was 0.29 mV and σ was 0.053 mV. $\chi^2_{LRT} = 177.7$ with 4 df; $P < 0.001$. $\chi^2_{GTF} = 26.9$ with 26 df; $P > 0.3$. **b** EPPs. The high frequency of failure of release (80%) indicates that most EPPs are classical unitary evoked potentials. Note that the overall distribution is greatly different from that of the mEPPs in that the overall EPP distribution is bell-shaped. However, there are peaks that closely match the peak intervals in the mEPP distribution (μ was 0.29 mV and σ was 0.040 mV). Small EPPs the size of sub-mEPPs or skew-mEPPs were evoked, but the numbers more or less fit those expected from an overall bell-shaped distribution. There are 1.2×10^3 EPPs in this distribution. The few larger EPPs forming a right-hand skew would result from two or three classical unitary evoked potentials. However, there were not enough larger potentials to show peaks comparable to those reported by BOYD and MARTIN (1956) and LILEY (1956). $\chi^2_{LRT} = 248.1$ with 8 df; $P < 0.001$. $\chi^2_{GFT} = 19.2$ with 37 df; $P > 0.98$ (KRIEBEL et al. 1982). Abbreviation: *df*, degrees of freedom

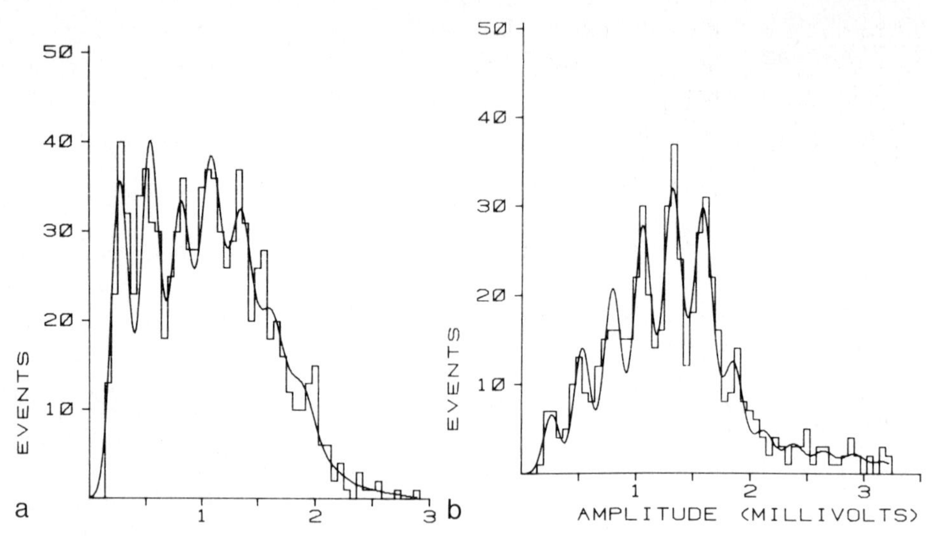

Fig. 11 a, b. mEPP and EPP amplitude distributions from a 13-day-old mouse. Temperature, 35 °C; neostigmine, 1.0 μg ml^{-1}. *Continuous lines* show the predicted distribution based on the subunit hypothesis. The standard deviation of the noise and measurement error (σ_m^2) is 0.067 mV. The resting potential dropped from −67 to −65 mV for the 35 min of recording time. Co^{2+} ca. 4 m*M*. **a** mEPPs. This mEPP distribution was chosen because the ratio of skew- to bell-mEPPs was approximately unity. The peak intervals are 4.75 histobars; 9.0×10^2 mEPPs are in this histogram (24 mEPPs min^{-1}). μ was 0.27 mV and σ was 0.037 mV. $\chi^2_{LRT} = 40.9$ with 6 df; $P < 0.001$. $\chi^2_{GFT} = 25.2$ with 22 df; $P > 0.2$. **b** EPPs 80% evoked failure rate (EPP reduced with Co^{2+}) indicates that most EPPs were the classical unitary evoked potentials. Peak intervals are the same as in the mEPP histogram ($\mu = 0.26$ mV, $\sigma = 0.021$ mV). Note that the overall shape is more or less bell-shaped and the number of evoked potentials the size of sub-mEPPs fits a bell-shaped envelope. 5.0×10^2 EPPs are in this histogram. $\chi^2_{LRT} = 103.7$ with 8 df; $P < 0.001$. $\chi^2_{GFT} = 8.87$ with 24 df; $P > 0.9$ (KRIEBEL et al. 1982)

(KRIEBEL 1978). KELLY and ROBBINS (1984) report small mEPPs that are evoked with nerve stimulation but these appear to be larger than the normal skew-class.

II. Schwann-Cell mEPPs

It is conceivable that sub-mEPPs arise from Schwann-cell protrusions between the nerve filament and the muscle fibre because it is known from the work of Birks et al. (1960) that small mEPPs appear in denervated muscle at the same time as the Schwann-cell muscle junction develops. However, the Schwann-cell mEPPs differ greatly in their pharmacological properties from the skew-mEPP class (BEVAN et al. 1976). KRIEBEL and GROSS (1974) confirmed the presence of a 'quiescent period' between nerve quanta and those appearing with the formation of the Schwann-cell muscle junction. These studies were extended to newly formed Schwann-cell-muscle fibre junctions that were either generating mEPPs or were quiescent and were subsequently identified in the electron microscope, and both active and quiescent Schwann-cell junctions appeared similar and essentially devoid of vesicles (KRIEBEL et al. 1982). KRIEBEL et al. (1982) used very

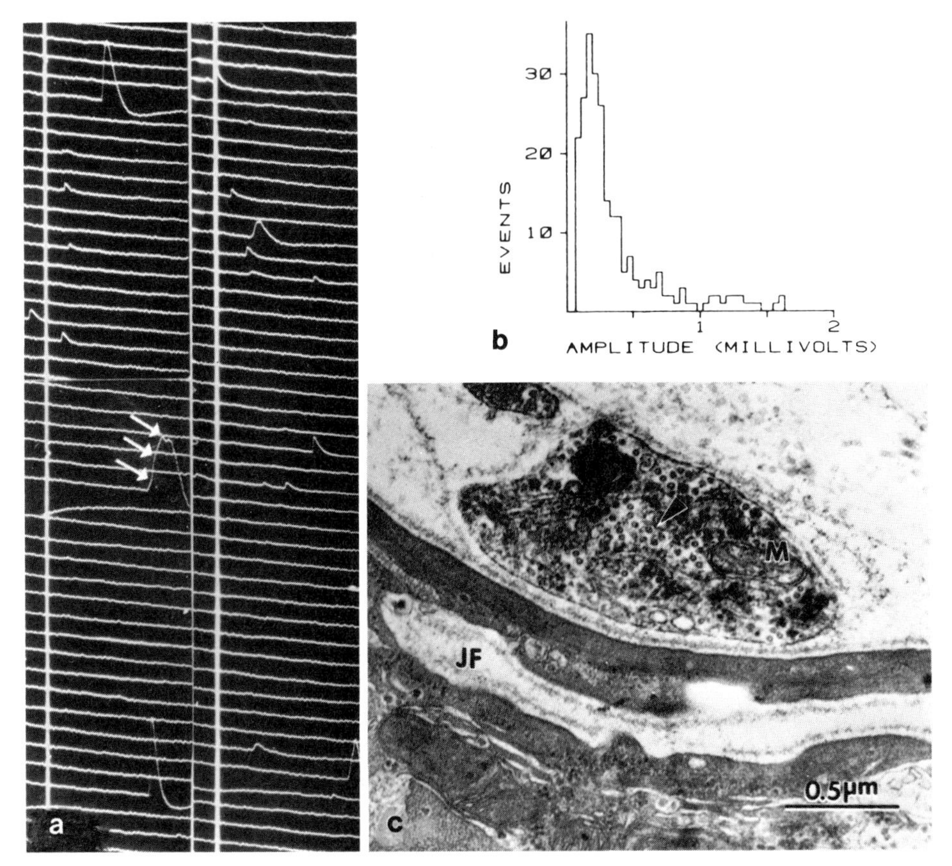

Fig. 12a–c. Schwann-cell mEPPs from a 10-μm-diameter muscle cell. Six-day-postoperative preparation. **a** Schwann-cell mEPPs (resting potential, −83 mV). Note three breaks on the rising phase of a large mEPP (*three arrows*) indicating 'coupled' mEPPs. Calibration: each trace 500 ms long, 0.25 mV between traces (note the low noise level). **b** Schwann-cell mEPP amplitude histogram showing a peak. This histogram indicates that few mEPPs were lost to the noise. Frequency, 11 s^{-1}. **c** The Schwann-cell-muscle junction from which **a** was recorded. The Schwann cell contains numerous microtubules (*arrowhead*) and a mitochondrion (*M*). *JF*, junctional fold (KRIEBEL et al. 1980)

small muscle fibres so that the mean Schwann-cell mEPP amplitude and frequency could be determined. Their frequency is surprisingly higher (up to 100 min^{-1}) than that of mEPPs from an innervated junction (Fig. 12). Since no mEPPs were recorded in some junctions (later identified by electron microscopy) for 30 min, it is concluded that the Schwann cell is induced to release quanta only after nerve degeneration and is probably quiescent in the presence of a nerve filament (CULL-CANDY et al. 1976). This notion is supported by the work of BEVAN et al. (1976) who found that actinomycin D prevented Schwann-cell mEPPs. The Schwann-cell mEPP of denervated junctions is smaller (0.2 mV) than the sub-mEPP (0.5 mV) of the innervated junction of a muscle fibre of similar diameter (ca. 6–10 μm diameter; there is little change in input resistance early in denervation; NICHOLLS 1956). The Schwann-cell mEPP appears slower than the bell-mEPP, but this results from the small size of the Schwann-cell mEPP in relation to noise, which has the effect of slowing small mEPPs (ERXLEBEN and KRIEBEL 1988a). DENNIS and MILEDI (1974) depolarized the Schwann cell in the denervated preparation and induced the release of ACh as evinced by a depolarization of the muscle fibre; however, these induced responses are not related to the Schwann-cell mEPP since they were usually larger and appear to represent ACh release through holes punched through the Schwann-cell membrane by the depolarizing electrolytic current. The pharmacological properties of Schwann-cell mEPPs are different from bell-mEPPs in that La^{3+}, Ca^{2+}, BWSV, ethyl alcohol, diamide, ouabain, theophylline and acid Ringer's solution fail to increase the Schwann-cell mEPP frequency (BEVAN et al. 1976). Many Schwann-cell mEPPs have notches or breaks on their rising phases indicating interaction between release sites (KRIEBEL et al. 1982). The Schwann-cell release mechanism is difficult to characterize because the influence of temperature and tonicity is variable (KRIEBEL et al. 1982). Nevertheless ITO and MILEDI (1977) found that a Ca^{2+} ionophore decreased the Schwann-cell mEPP frequency whereas Ca^{2+} or Ca^{2+} ionophore greatly increased the skew-mEPP frequency (LLADOS et al. 1980).

III. Giant mEPPs and Doublet mEPPs

LILEY (1956a, 1957) reported large mEPPs from the rat diaphragm that are multiples of the bell-mEPP. Giant mEPPs can certainly occur by chance coincidence of two or more bell-mEPPs (FATT and KATZ 1952). However, in very low frequency fibres, two bell-mEPPs often occur too close together for chance, and these have been termed 'doublet-mEPPs' (KRIEBEL and STOLPER 1975). Giant and doublet mEPPs show a continuum in that the second mEPP may occur on the falling phase, on the peak of the first mEPP or along the rising phase of the first mEPP (producing a notch). And, if no break is observed on the rising phase, the doublet is termed a giant (HEINONEN et al. 1982). Thus doublet mEPPs simply grade into the giant class. The presence of giant mEPPs and doublet mEPPs in normal, unstressed preparations indicates an interaction between release mechanisms (LILEY 1956a, 1957; MARTIN and PILAR 1964). However, preparations with relatively high mEPP frequencies show no more doublets or giants than predicted by chance, which demonstrates that the interactive mechanism is weak and easily suppressed in the normal preparation at usual mEPP frequencies. HUBBARD and JONES (1973) found no 'drag' effect and proposed that the giant mEPPs arise from large vesicles (cf. the La^{3+} effect by HEUSER and MILEDI 1971). Moreover, a 'drag' mechanism is not activated by the nerve impulse because the number of events composing the EPP fits Poisson statistics during low quantal contents (DEL CASTILLO and KATZ 1954; BOYD and MARTIN 1956; LILEY 1956b) and nerve stimulation does not release giant mEPPs (MOLGÓ and THESLEFF 1982). Vinblastine (PÉCOT-DECHAVASSINE 1976), La^{3+} (HEUSER and MILEDI 1971), 3,4-diaminopyridine (DURANT and MARSHALL 1980), crotoxin (BRAZIL et al. 1971) and chronic botulinum toxin poisoning (KIM et al. 1984) increase the number of giant and doublet mEPPs. MOLGÓ and

THESLEFF (1982) have carefully studied the potentiation of giant-mEPPs with 4-aminoquinoline and found a positive correlation between rise time and amplitude. All points on the rise time-amplitude graph fall on a straight line, which indicates a single class of events (cf. NEGRETE et al. 1972; KRIEBEL et al. 1976). I suggest that agents such as vinblastine, botulinum toxin, 4-aminoquinoline and La^{3+} potentiate the drag process that is normally present, and that this process may occur at a secondary release site as suggested by THESLEFF and MOLGÓ (1983). DENNIS and MILEDI (1974b) recorded both skew- and bell-mEPPs from the same extracellular focal position at the regenerating frog NMJ but suggested that the skew-mEPPs could originate far from the receptors. However, ERXLEBEN and KRIEBEL (1988a) found that both skew- and bell-mEPCs have the same time characteristics proving that both quantal classes are released near the receptor membrane (cf. BENNETT et al. 1974).

IV. Slow Skew-mEPPs

With favourable signal-to-noise ratios, KRIEBEL et al. (1976) reported that many mEPPs of the skew class appear as doublets and have notches on their rising phases. Two sub-mEPPs may occur at the same time which would yield a fast skew-mEPP so there would be a continuum between doublet skew-mEPPs and larger skew-mEPPs with a fast, smooth rising phase (Fig. 3a). Preparations with excellent signal-to-noise ratios yield the greatest number of doublet skew-mEPPs whereas relatively poor signal-to-noise ratios yield slow skew-mEPPs. Small signals just above background noise levels appear slower than larger signals because of the noise contribution to the small signal (ERXLEBEN and KRIEBEL 1988a). BENNETT, FLORIN and WOOG (1974) also noted that slow mEPPs did not originate at a distant point. However, caution is needed in interpreting mEPP classes because BENNETT and PETTIGREW (1974a) show different classes with multiple innervations. BWSV (KRIEBEL and STOLPER 1975), La^{3+} (KRIEBEL and FLOREY 1983), short temperature challenges (CARLSON et al. 1982) and probably any treatment that increases the number of skew-mEPPs also increases the number of skew-mEPPs that have breaks on their rising phases and produces clusters of sub-mEPPs. This drag effect becomes stronger with time and is probably not related to the mechanism which produces giant and doublet mEPPs. Skew-mEPPs (slow or fast) are not evoked with nerve stimulation (KRIEBEL et al. 1976) and are probably the same class as those described by GAGE and MCBURNEY (1975) as 'tent' mEPPs. The percentage of slow skew mEPPs is variable in the developing synapse (MUNIAK et al. 1982) which has broad mEPP amplitude distributions because both skew- and bell-mEPPs are present in about equal numbers (MUNIAK et al. 1982; COLMÉUS et al. 1982). However, MUNIAK et al. (1982) found no discontinuities in the skew-mEPP amplitude histogram in regard to time-to-peak analyses. Thus, the slow skew- and skew-mEPPs with breaks are all of the same class. Since skew-mEPP amplitude-frequency histograms show multiple peaks (Figs. 8, 10a, 11a) KRIEBEL et al. (1976) and MATTESON et al. (1980) proposed that the larger skew-mEPPs are composed of subunits the same size as the bell-mEPPs (see subunit hypothesis). When subunits are not released synchronously, either a slow skew- or a doublet skew-mEPP is generated. This in-

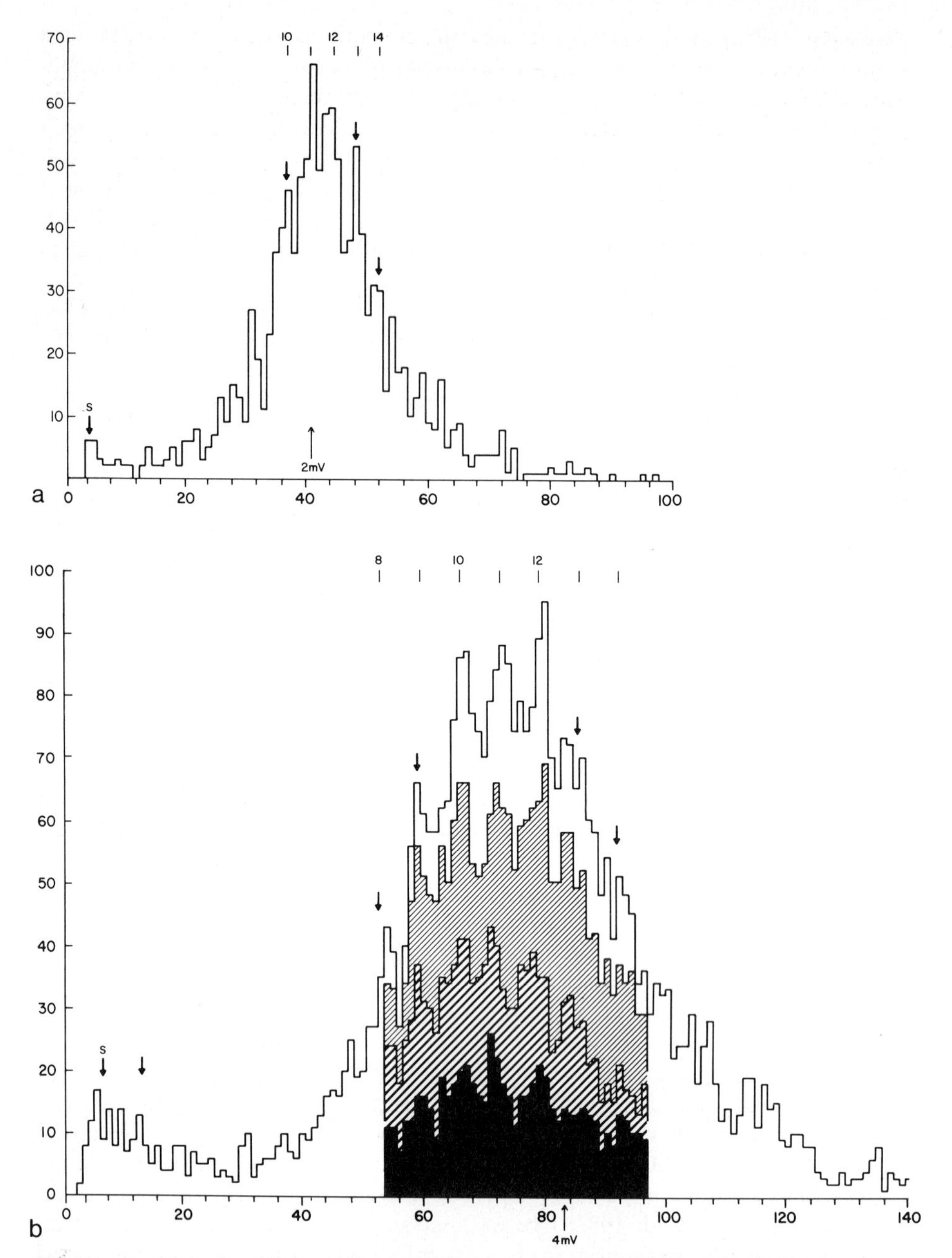

Fig. 13a, b. Control mEPP amplitude histogram. 1.2×10^3 mEPPs; $4.6\,s^{-1}$. Temperature, 36 °C; approximately 22-day-old mouse. Peak intervals are 3.7 bins. *Arrow* (*s*) indicates the subunit size equal to the peak intervals. The first blank bins represent noise. Bell-mEPP is 2.06 mV. **b** mEPP amplitude histogram after neostigmine. 3.7×10^3 mEPPs. Note that central peaks are noticeable with the first 950 mEPPs (*filled histogram*), and that the peaks become more pronounced with larger sample size. The average bell-mEPP (3.55 mV) increased by a factor of 1.70 times the control value, and the peak interval increased by a factor of 1.73 times the control value (CARLSON and KRIEBEL 1985)

terpretation can explain the larger slow mEPPs and small fast mEPPs found by BENNETT et al. (1973) and BENNETT and PETTIGREW (1974a, b) at the reinnervating mammal junction (see MUNIAK et al. 1982). Botulinum toxin intoxication simply promotes the skew-mEPPs along with giant mEPPs which were found to persist after the bell-mEPP class was blocked (KRIEBEL et al. 1976). This proposal fits the data of KIM et al. (1984) and COLMÉUS et al. (1982) because their figures show a wide range of time-to-peak values with no discontinuities that indicate different classes. However, it is also possible that botulinum toxin intoxication could induce a slower class of receptors (SELLIN and THESLEFF 1981) thus yielding a slow skew-mEPP.

D. Subunit Hypothesis of the Quantum of Transmitter Release

With favourable signal-to-noise ratios GROSS and KRIEBEL (1973) and KRIEBEL and GROSS (1974) found that profiles of mEPP amplitude-frequency histograms from frog preparations showed integral peaks that were multiples of the sub-mEPP mode (see WERNING and STIRNER 1977; VAUTRIN and MAMBRINI 1982). In addition the overall bell-mEPP profile and/or the number of peaks were changed after a short heat challenge or elevated Ca^{2+}, both of which also increased the mEPP frequency. KRIEBEL et al. (1976) extended these studies on the mouse hemidiaphragm preparation to demonstrate that amplitude distributions had peaks in the same position with successive samples, and they become clearer with increasing sample size (Fig. 13b) (not found by MILLER et al. 1978). WERNIG and STIRNER (1977) found that peak intervals of the frog preparation were increased in size with an anticholinesterase (anti-ChE) agent whereas the number of peaks remained constant. CARLSON and KRIEBEL (1985) have confirmed this on the mouse diaphragm preparation (Fig. 13). In addition, WERNIG and MOTELICA-HEINO (1978) were able to 'sharpen' the peaks of the mEPP amplitude-frequency histogram by selecting only those mEPPs generated near the recording electrode. Since the frog junction has relatively long nerve filaments in regard to the muscle fibre space constant, there is appreciable signal attenuation of mEPPs generated at a site distant from a recording electrode and this property may mask integral peaks (TREMBLY et al. 1983). Another problem encountered with the frog preparation is that quantal size may be 10% – 60% smaller at the distal ends of the terminal (BIESER et al. 1984; but see D'ALONZO and GRINNEL 1985). A space constant problem does not exist at the mouse neuromuscular endplate since the endplate size does not exceed the muscle fibre diameter. KRIEBEL and coworkers found that histogram profiles of the skew-mEPP class of the mouse preparation showed integral peaks that had the same interval as the peaks in the bell-mEPP distribution (KRIEBEL et al. 1976; MATTESON et al. 1979, 1981). In addition, the ratio of skew-mEPPs to bell-mEPPs was readily altered, the shape of the bell-mEPP class and the number of peaks on the bell-mEPP distribution were changed with various experimental conditions that altered mEPP frequency. KRIEBEL et al. (1982) published amplitude histograms of mEPPs and unitary EPPs (EPPs reduced with 4 m*M* Co^{2+}) that have the same number of peaks and similar profiles except that quanta of the skew class of mEPPs are not

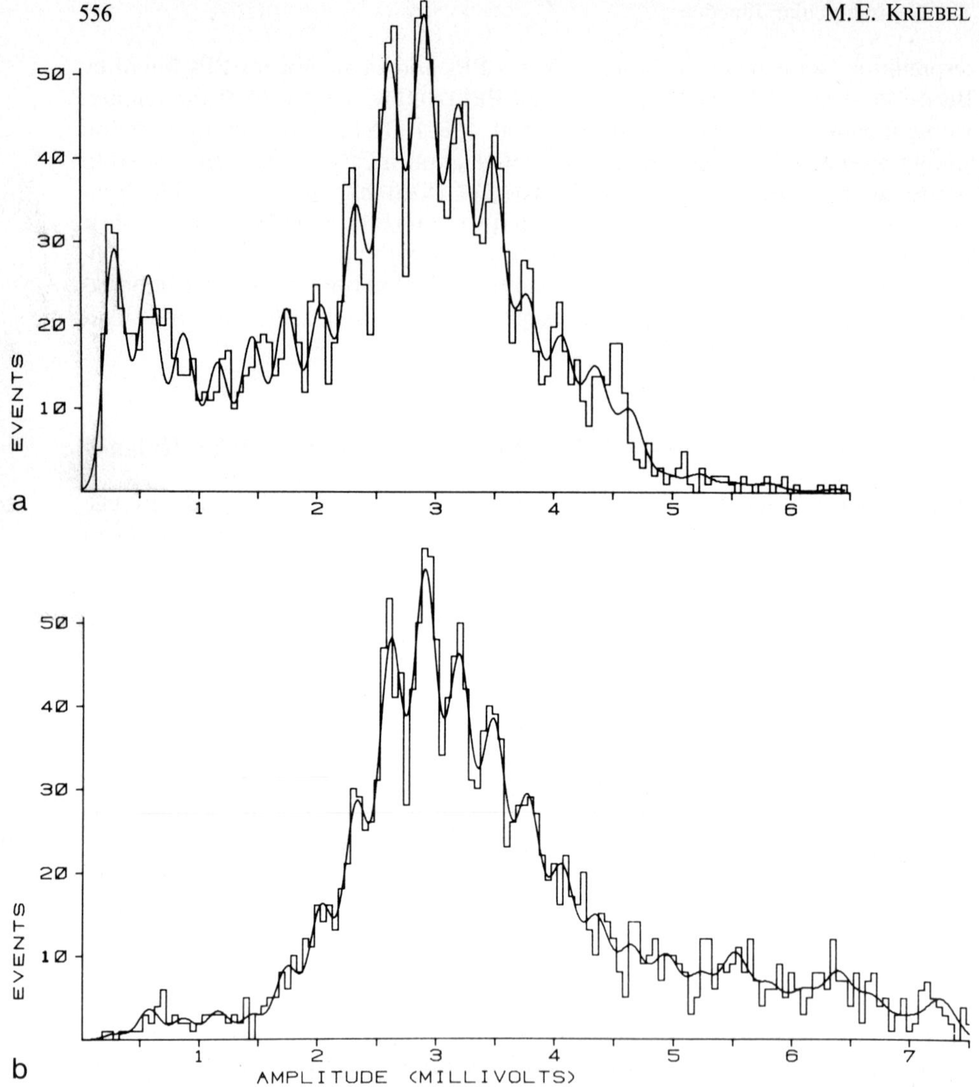

Fig. 14a, b. mEPP and EPP amplitude distribution from a 14-day-old mouse diaphragm junction at room temperature, 25 °C. No neostigmine. The phrenic nerve was stimulated at 2 Hz, and all mEPPs and EPPs were measured during the 70 min of continuous recording. *Continuous lines* show the predicted distribution based on the subunit hypothesis. The noise and measurement standard deviation (σ_m^2) was 0.09 mV. Membrane potential was −64 mV throughout; Co^{2+} was approximately 4 m*M*. **a** mEPPs. Note that there are two general classes of mEPPs. The smaller ones from a skew distribution and, in this case, show two distinct peaks. 2.4×10^3 mEPPs (34 min^{-1}) compose this histogram, of which 20% are skew-mEPPs and 5% are sub-mEPPs. The estimated value of subunit size (μ) was 0.29 mV and of subunit standard deviation (σ) was 0.016 mV. $\chi^2_{LRT} = 155.7$ with 18 df; $P < 0.001$. $\chi^2_{GFT} = 59.74$ with 54 df; $P > 0.2$. **b** EPPs. Since most stimuli resulted in failures (75%), most EPPs correspond to the classical unitary evoked potential. The few larger EPPs (6 mV) represent two classical unitary evoked potentials. Note that few EPPs the size of sub-mEPPs or skew-mEPPs are present. The EPP histogram is not smooth but shows multiple peaks which are in register with those in the mEPP amplitude distribution. There are 2.5×10^3 EPPs in this histogram. $\mu = 0.29$ mV and $\sigma = 0.018$ mV. $\chi^2_{LRT} = 558.8$ with 21 df; $P < 0.001$. $\chi^2_{GFT} = 46.7$ with 71 df; $P > 0.98$ (KRIEBEL et al. 1982)

evoked (Fig. 14). Since the peak intervals of the spontaneous bell-mEPPs (random events) and those of the unitary EPPs (time responses) matched, the underlying subunit of the quantum of transmitter release is the same for spontaneous mEPPs and evoked unitary quanta. It is true that histograms constructed from data randomly drawn from a smooth bell-shaped distribution can show peaks due to random variations in sample size (MAGLEBY and MILLER 1981; CARLSON and KRIEBEL 1985; F. LLADOS and M. E. KRIEBEL, unpublished work); however the chance that the same peaks would be present on both mEPP and unitary EPP amplitude profiles would be vanishingly small. With a sample size of 2×10^3 events drawn from a Gaussian distribution, peaks may occur in about 10% of the traces by chance (F. LLADOS and M. E. KRIEBEL, unpublished work). Therefore the chance that both mEPP and unitary EPP histograms could show the same number of peaks (modal number 10) would be less than one in a thousand. We obtained about eight cases and published these (Figs. 10, 11, 13) (KRIEBEL et al. 1982). VAUTRIN and MAMBRINI (1981) also report seven to eight integral peaks and they found that the different classes of unitary evoked responses also had different latencies.

The subunit hypothesis has been criticized because the integral peaks of the raw data appear too sharp to be composed of subunits with an amplitude distribution and variance of the sub-mEPP. The sub-mEPP coefficient of variance of the measured data is indeed high at about 50%, and we found that a model based on this subunit variance did not yield an amplitude distribution with multiple peaks (MATTERSON et al. 1980). However, much of the variance of the sub-mEPP distribution results from measurement error and background noise. MAGLEBY and MILLER (1981) proposed that variance in the postsynaptic membrane response could also contribute significantly to the overall variance. However PENNEFATHER and QUASTEL (1981) found that over 80% of the postsynaptic receptors had to be blocked with α-bungarotoxin in order to reduce the mEPC amplitude by half. This observation implies that almost all ACh would become bound. The dense packing of receptor molecules ensures that ACh is bound over a small postsynaptic area which would give a high binding efficiency; further, the forward rate of ACh binding to Ach-receptor is much faster than the back reaction such that little ACh will remain unbound or escape the cleft (LAND et al. 1980). In addition, a small postsynaptic influence on the contribution of the variance of the response is to be expected from anti-ChE experiments because the mEPP height is increased by only 20% indicating that receptors are not normally saturated (HARTZELL et al. 1975). KUFFLER and YOSHIKAMI (1973) found that the variability in the response induced by pulsed ACh was only a few percent.

MATTESON et al. (1981) tested a subunit model which fits the raw data histograms by considering that the distribution of a given peak has variance due to the noise-measurement errors plus the subunit variance multiplied by the number of subunits composing the peak. The noise-measurement values were determined by reading calibration signals. Curve fitting and maximum likelihood estimates were used to calculate the subunit size and variance which best fit all peaks (Figs. 8, 10, 11, 14). The multimodal model fit the raw data whereas a bimodal model based on two populations of mEPPs (sub-mEPPs and bell-mEPPs) does not fit the data (Fig. 15). The subunit variance was 12% – 16%, which is similar to the

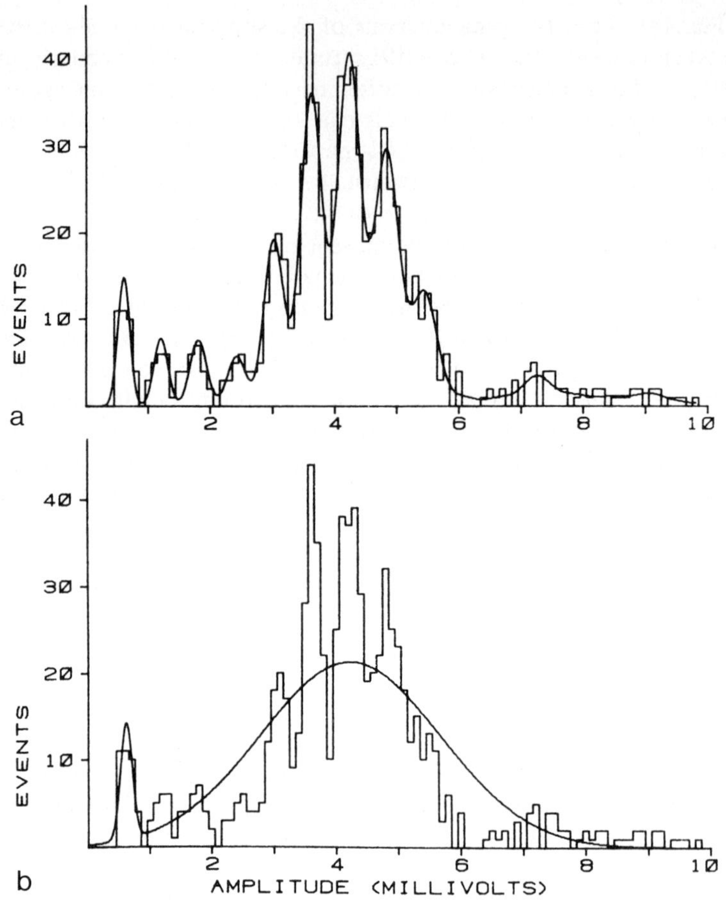

Fig. 15. a The fit produced by the multimodal model. The number of subpopulations in this case was 16. For each histogram this number was simply determined by the amplitude range of the histogram, under the assumption that the weighting factors of any larger amplitude subpopulations are zero. The maximum likelihood estimates of the parameters are as follows: $W_1 = 4.52 \times 10^{-2}$; $W_2 = 2.85 \times 10^{-2}$; $W_3 = 3.19 \times 10^{-2}$; $W_4 = 2.67 \times 10^{-2}$; $W_5 = 9.78 \times 10^{-2}$; $W_6 = 1.97 \times 10^{-1}$; $W_7 = 2.36 \times 10^{-1}$; $W_8 = 1.81 \times 10^{-1}$; $W_9 = 8.49 \times 10^{-2}$; $W_{10} = 7.13 \times 10^{-3}$; $W_{11} = 6.83 \times 10^{-3}$; $W_{12} = 2.54 \times 10^{-2}$; $W_{13} = 9.57 \times 10^{-3}$; $W_{14} = 8.36 \times 10^{-3}$; $W_{15} = 1.15 \times 10^{-2}$; $W_{16} = 3.12 \times 10^{-3}$. $\mu = 0.60$ mV, $\sigma = 0.065$ mV, $\sigma_m^2 = 0.071$ mV. **b** The fit produced by the bimodal model. Maximum likelihood estimates of the parameters are as follows: $W = 3.96 \times 10^{-2}$, $\mu = 0.61$ mV, $\sigma_1 = 0.094$, $\mu_2 = 4.2$ mV, $\sigma_2 = 1.4$ mV. Using the likelihood ratio test criterion the fit in **a** is significantly better ($P < 0.001$) (MATTESON et al. 1981)

value calculated by WERNIG and STIRNER (1978). CARLSON and KRIEBEL (1985) and MONTELICO-HEINO and KRIEBEL (1987) calculated the same subunit variance directly from the measured sub-mEPP distribution by subtracting the noise and measurement variance determined from calibration signals (cf. KATZ 1977). Therefore the contribution of noise-measurement errors is constant for each peak and contributes substantially to the variance of the smaller peaks but relatively little to the larger peaks. This was first appreciated by BOYD and MARTIN (1956) who noted that their histogram of evoked EPPs had sharper peaks than the theoretical histogram which did not take into account the noise-measurement error.

In order to obtain profiles of mEPP amplitude-frequency distributions that show integral peaks several experimental criteria must be satisfied. Firstly, the signal-to-noise ratio must be better than 50:1. Experimentally this means that bell-mEPPs should be 2–3 mV, and that the noise level must be less than 100 μV peak-to-peak. In addition, the sample size must be over 10^3 events and preferably $3-4 \times 10^3$ mEPPs, otherwise variations in sample size may either obscure or generate peaks (MAGLEBY and MILLER 1981). Sample sizes of several thousand events permit a check on the invariance of peak positions. Of course, the muscle fibre membrane potential must not vary over 2%–3%, or peaks in the central region of the histogram will shift position and will fill previous existing troughs (KRIEBEL et al. 1982). BEVAN (1976) found sub-mEPPs but did not resolve the sub-mEPP mode, and with sample sizes under 10^3 mEPPs used to construct histograms showing 10–14 bins to the mEPP mode it is useless to comment on his lack of peaks of the size reported by KRIEBEL and GROSS (1974). MAGLEBY and MILLER (1981) also looked for integral peaks but their sample sizes were either relatively small and/or the noise levels were large (2%–6%). They concluded accordingly that their peaks resulted from random variations in sample size. Recently, ERXLEBEN and KRIEBEL (1988b) and ERXLEBEN et al. (1983) have voltage-clamped the mouse diaphragm junction and have analysed the mEPCs by means of a computer program. Bell-mEPCs and skew-mEPCs from different preparations are the same size, and both classes have the same time characteristics. There is a gradual decrease in the time to peak of both classes with synapse development, but no abrupt changes in time characteristics were noted (see also SAKMANN and BRENNER 1978). VICINI and SCHUETZE (1985) report that fast and slow channels are mixed together at developing rat end-plates. ERXLEBEN and KRIEBEL (1988b) found that amplitude histograms of mEPCs from most experiments show integral peaks similar to those of mEPP amplitude histograms. The subunit size was found to be 0.44 nA (30 °C, –140 mV holding potential, 25 m*M* Ca^{2+} saline). The excess Ca^{2+} reduced the mEPPs by about 30% but was needed to increase penetration stability (MAGLEBY and WEINSTOCK 1980; MCLARNON and QUASTEL 1983). ERXLEBEN and KRIEBEL (1988b) found that the peak intervals were increased by adding an anti-ChE agent, increasing the holding potential to –170 mV or increasing the bath temperature to 38 °C and by increasing both the bath temperature and the holding potential. The peak intervals were reduced by lowering the bath temperature to either 25 °C or to 20 °C or by reducing the holding potential (KRIEBEL and ERXLEBEN 1984).

In summary, the profiles of mEPP or mEPC amplitude histogram profiles analysed either by eye and hand or with a computer program show integral peaks that are multiples of the sub-mEPP size. The coefficient of variance of the sub-mEPP and subunit is about 16%. The subunit hypothesis implies a mechanism capable of synchronously releasing and/or packaging on the average ten subunits to generate the quantum. The release mechanism of skew-mEPPs and bell-mEPPs has different properties but both utilize the same subunit of release. Only the bell-mEPP class is available for nerve-evoked release. The subunit hypothesis of the quantum of release can explain the changes and differences in mEPP and unitary EPP amplitude profiles and the changes in quantal size induced by many experimental treatments.

E. Morphological Correlates of Bell-mEPPs and Skew-mEPPs

The observation that skew mEPPs are not evoked and the pharmacological differences between the skew- and bell-mEPP classes indicate two different release mechanisms. Nevertheless, both bell-mEPPs and larger skew-mEPPs are composed of the same subunit because the profiles of frequency-amplitude distributions show the same integral values between peaks in the skew- and bell-mEPP classes; and, the variances of both skew- and bell-mEPP peaks are predicted with the same subunit variance (coefficient of variance 12%–18%) (WERNIG and MOTELICA-HEINO 1978; MATTESON et al. 1981; KRIEBEL and CARLSON 1985; KRIEBEL and MOTELICA-HEINO 1987). The subunit hypothesis readily explains changes in mean bell-mEPP amplitude and distribution. Extended treatments reduce mEPPs to the sub-mEPP class (heat challenge, KRIEBEL and GROSS 1974; CARLSON et al. 1982; nerve stimulation, KRIEBEL 1978; botulinum toxin, KRIEBEL 1976; β-bungarotoxin, LLADOS et al. 1980; La^{3+}, KRIEBEL and FLOREY 1983; BWSV, KRIEBEL and STOLPER 1975; Ca^{2+} ionophore, KRIEBEL et al. 1980) and the sub-mEPP modes remain within 20%. The small variance and stable size of the sub-mEPP implies a stable mechanism in the production of the subunit. By contrast, the mechanism(s) of combining subunits to form the bell-mEPP class is labile. It is conceivable that there are 10–15 vesicle classes based on volume to correspond to the peaks on the mEPP amplitude histograms. This notion would not explain the discontinuity between skew- and bell-mEPPs. A morphological basis for the large percentage of sub-mEPPs during synaptogenesis would predict a class of vesicles with a diameter of 25 nm and these were not found (MUNIAK et al. 1982; KRIEBEL et al. 1986). In addition vesicles remain at the same diameter with treatments that reduce almost all mEPPs to the sub-mEPP class (nerve stimulation, Rose et al. 1978; BWSV, KRIEBEL and STOLPER 1975; La^{3+}, KRIEBEL and FLOREY 1983). Thus there is one vesicle class based on diameter regardless of the percentage of skew-mEPPs and/or the number of mEPPs generated.

WERNIG (1975) found that the number of available quanta is proportional to the number of active release sites (ridges) and he proposed that the ridge functions as a unit to generate the bell-mEPP with the synchronous release of ten vesicles (WERNIG and STIRNER 1977). However, HEUSER et al. (1979) found no interaction between fused and/or collapsed vesicles during release; and KATZ and MILEDI (1979) found that over 10^4 quanta are released by one nerve impulse from a diaminopyridine-treated junction. This is larger than the number of 'touching' vesicles (vesicles in contact with the plasma membrane). HEUSER et al. (1979) report that the quantal content is equal to the number of fused and 'collapsed' vesicles with potentiated release. These observations support the original proposal that one vesicle generates one mEPP and they may indicate that subunits are contained in two different classes of vesicles. One class is subject to evoked nerve release and generates spontaneous bell-mEPPs whereas the second class is available only for spontaneous release and generates the skew-class of mEPPs.

F. Concluding Remarks

The profiles of mEPP amplitude-frequency histograms of normal adult mouse and frog preparations show a discontinuity demonstrating two classes of normal mEPPs. The profile of the smaller class skews (skew- or sub-mEPPs) towards the noise in such a way that the mode is one-tenth that of the larger class which forms a bell-shaped distribution (bell-mEPPs). mEPPs of both classes usually have the same time characteristics but sometimes the skew-class shows many slower mEPPs. Various treatments (e.g. stimulation of the nerve to fatigue, heat challenges, La^{3+}, Ca^{2+} ionophores and Ca^{2+} ions, botulinum toxin, β-bungarotoxin, BWSV) increase the percentage of skew-mEPPs. Giant mEPPs form a right-hand skew on the profile of the bell-mEPPs and may represent the simultaneous occurrence of two or more bell-mEPPs; they grade into doublet-mEPPs. Giant and doublet-mEPPs indicate interactions in the release mechanism ('drag effect' of MARTIN and PILAR 1964). The drag effect is not strong and is lost during high mEPP frequencies, but it can be potentiated by botulinum intoxication (KIM et al. 1984), 4-diaminoquinoline (MOLGÓ and THESLEFF 1982) or vinblastine (PÉCOT-DECHAVASSINE 1976).

During synaptogenesis the skew-mEPP class appears first, and the bell class gradually increases to the normal adult skew-mEPP:bell-mEPP ratio of 1 : 50. The distribution of the unitary EPPs matches that of the bell-mEPP class (BOYD and MARTIN 1956; KRIEBEL et al. 1982) and quanta the size of skew-mEPPs not are evoked (DENNIS and MILEDI 1974; BEVAN 1976; KRIEBEL et al. 1982; MOLGÓ and THESLEFF 1982; COLMÉUS et al. 1982).

Both the larger skew-mEPPs and the bell-mEPPs are composed of subunits the size of sub-mEPPs because the mEPP and unitary EPP amplitude histograms show the same integral peaks. The mechanisms of release for the skew-mEPP, bell-mEPP and giant class have different properties.

Morphological or molecular correlates of the subunit have not been discovered. There is only one class of vesicle in the frog or mouse regardless of the percentage of sub-mEPPs generated (KRIEBEL and FLOREY 1983; MUNIAK et al. 1982). Thus, a vesicle may contain one subunit of Ach which would generate a sub-mEPP or it may contain multiples of the subunit (9 – 11) which would produce the physiological quantum of release.

Acknowledgements. I acknowledge the support afforded by a Senior Fellowship from the Alexander von Humboldt Stiftung and National Sciences Foundation grant no. 19694.

References

Augustine GJ, Levitan H (1983) Neurotransmitter release and nerve terminal morphology at the frog neuromuscular junction affected by the dye erythrosin B. J Physiol (Lond) 334:47–63

Barker D, Ip MC (1966) Sprouting and degeneration of mammalian motor axons in normal and de-afferentated skeletal muscle. Proc R Soc Lond [Biol] 163:538–554

Bennett MR, Pettigrew AG (1974a) The formation of synapses in striated muscle during development. J Physiol (Lond) 241:515–545

Bennett MR, Pettigrew AG (1974b) The formation of synapses in reinnervated and cross-reinnervated striated muscle during development. J Physiol (Lond) 241:547–573

Bennett MR, Pettigrew AG (1975) The formation of synapses in amphibian striated muscle during development. J Physiol (Lond) 252:203–239

Bennett MR, McLachlan EM, Taylor RS (1973) The formation of synapses in reinnervated mammalian striated muscle. J Physiol (Lond) 233:481–500

Bennett MR, Florin T, Woog R (1974) The formation of synapses in regenerating mammalian striated muscle. J Physiol (Lond) 238:79–92

Bevan S (1976) Sub-miniature end-plate potentials at untreated frog neuromuscular junctions. J Physiol (Lond) 258:145–155

Bevan S, Grampp W, Miledi R (1973) Further observation on Schwann cell min.e.p.p.s. J Physiol (Lond) 232:88*P*–89*P*

Bevan S, Grampp W, Miledi R (1976) Properties of spontaneous potentials at denervated motor end-plates of the frog. Proc R Soc Lond [Biol] 194:195–210

Bieser A, Wernig A, Zucker H (1984) Different quantal responses within single frog neuromuscular junctions. J Physiol (Lond) 350:401–412

Birks R, Katz B, Miledi R (1960) Physiological and structural changes at the amphibian myoneural junction, in the course of nerve degeneration. J Physiol (Lond) 150:145–168

Blight AR, Precht W (1982) Miniature endplate potentials related to neuronal injury. Brain Res 238:233–238

Boyd IA, Martin AR (1956) The end-plate potential in mammalian muscle. J Physiol (Lond) 132:74–91

Brazil OV, Excell BJ (1971) Action of crotoxin and crotactin from the venom of *Crotalus durissus terrificus* (South American rattlesnake) on the frog neuromuscular junction. J Physiol (Lond) 212:34*P*–35*P*

Carlson CG, Kriebel ME (1985) Neostigmine increases the size of subunits composing the quantum of neurotransmitter release at mouse neuromuscular junction. J Physiol (Lond) 367:489–502

Carlson CG, Kriebel ME, Muniak CG (1982) The effect of temperature on MEPP amplitude distributions in the mouse diaphragm. Neuroscience 7:2537–2549

Colméus C, Gomez S, Molgó J, Thesleff S (1982) Discrepancies between spontaneous and evoked synaptic potentials at normal, regenerating and botulinum toxin poisoned mammalian neuromuscular junctions. Proc R Soc Lond [Biol] 215:63–74

Cull-Candy SG, Lundh H, Thesleff S (1976) Effects of botulinum toxin on neuromuscular transmission in the rat. J Physiol (Lond) 260:177–203

D'Alonzo A, Grinnell AD (1985) Profiles of evoked release along the length of frog motor nerve terminals. J Physiol (Lond) 359:235–258

Del Castillo J, Katz B (1954) Quantal components of the end-plate potential. J Physiol (Lond) 124:560–573

Dennis MJ, Miledi R (1971) Lack of correspondence between the amplitudes of spontaneous potentials and unit potentials evoked by nerve impulses at regenerating neuromuscular junctions. Nature [New Biol] 232:126–128

Dennis MJ, Miledi R (1974a) Non-transmitting neuromuscular junctions during an early stage of end-plate reinnervation. J Physiol (Lond) 239:553–570

Dennis MJ, Miledi R (1974b) Characteristics of transmitter release at regenerating frog neuromuscular junctions. J Physiol (Lond) 239:571–594

Dennis MJ, Miledi R (1974c) Electrically induced release of acetylcholine from denervated Schwann cells. J Physiol (Lond) 237:431–452
Doherty P, Hawgood BJ, Smith ICH (1984) Changes in miniature end-plate potentials after brief nervous stimulation at the frog neuromuscular junction. J Physiol (Lond) 356:349–358
Duchen LW, Tonge DA (1973) The effects of tetanus toxin on neuromuscular transmission and on the morphology of motor end-plates in slow and fast skeletal muscle of the mouse. J Physiol (Lond) 228:157–172
Duering MV (1967) Über die Feinstruktur der motorischen Endplatte von höheren Wirbeltieren. Z Zellforsch 81:74–90
Durant NN, Marshall JD (1980) The effect of 3,4-diaminopyridine on acetylcholine release at the frog neuromuscular junction. Eur J Pharmacol 67:201–208
Erxleben C, Kriebel ME (1983) Characteristics of miniature and sub-miniature endplate currents at the mouse diaphragm endplate. Naunyn Schmiedebergs Arch Pharmacol [Suppl] 322: R62
Erxleben C, Kriebel ME (1988a) Characteristics of spontaneous miniature and sub-miniature end-plate currents at the neonate and adult mouse neuromuscular junction. J Physiol (Lond) (in press)
Erxleben C, Kriebel ME (1988b) Subunit composition of the spontaneous miniature end-plate currents at the mouse neuromuscular junction. J Physiol (Lond) (in press)
Erxleben C, Carlson G, Kriebel ME (1983) Studies of miniature endplate currents show that the quantum of release is composed of subunits. Soc Neurosci Abstr 9:88
Fatt P, Katz B (1952) Spontaneous subthreshold activity at motor nerve endings. J Physiol (Lond) 117:109–128
Gage PW, McBurney RN (1975) Effects of membrane potential, temperature and neostigmine on the conductance change caused by a quantum of acetylcholine at the toad neuromuscular junction. J Physiol (Lond) 244:385–407
Gross CE, Kriebel ME (1973) Multimodal distribution of mepp amplitudes: the changing distribution with denervation, nerve stimulation and high frequencies of spontaneous release. J Gen Physiol 62:658–659a
Harris AJ, Miledi R (1971) The effect of type D botulinum toxin on frog neuromuscular junctions. J Physiol (Lond) 217:497–515
Hartzell HC, Kuffler SW, Yoshikami D (1975) Post-synaptic potentiation: Interaction between quanta of acetylcholine at the skeletal neuromuscular synapse. J Physiol (Lond) 251:427–463
Heinonen E, Jansson SE, Tolppanen EM (1982) Independent release of supranormal acetylcholine quanta at the rat neuromuscular junction. Neuroscience 7:21–24
Heuser J, Miledi R (1971) Effect of lanthanum ions on function and structure of frog neuromuscular junctions. Proc R Soc Lond [Biol] 179:247–260
Heuser JE, Reese TS (1973) Evidence for recycling of synaptic vesicle membrane during transmitter release at the frog neuromuscular junction. J Cell Biol 57:315–344
Heuser JE, Reese TS, Dennis MJ, Jan Y, Jan L, Evans L (1979) Synaptic vesicle exocytosis captured by quick freezing and correlated with quantal transmitter release. J Cell Biol 81:275–300
Hubbard JI, Jones SF (1973) Spontaneous quantal transmitter release: A statistical analysis and some implications. J Physiol (Lond) 232:1–21
Ito Y, Miledi R (1977) The effect of calcium ionophores on acetylcholine release from Schwann cells. Proc R Soc Lond [Biol] 196:51–58
Katz B (1962) The transmission of impulses from nerve to muscle, and the subcellular unit of synaptic action. Proc R Soc Lond [Biol] 155:455–477
Katz B (1977) Prologue. In: Cottrell GA, Usherwood PNR (eds) Synapses. Academic, New York, pp 1–5
Katz B (1978) The release of the neuromuscular transmitter and the present state of the hypothesis. In: Porter R (ed) Studies in neurophysiology. Cambridge University Press, Cambridge, pp 1–21
Katz B, Thesleff S (1957) On the factors which determine the amplitude of the 'miniature end-plate potential'. J Physiol (Lond) 137:267–278

Katz B, Miledi R (1977) Transmitter leakage from motor nerve endings. Proc R Soc Lond [Biol] 196:59–72
Katz B, Miledi R (1979) Estimates of quantal content during ‘chemical potentiation’ of transmitter release. Proc R Soc Lond [Biol] 205:369–378
Kelly SS, Robbins N (1984) Bimodal miniature and evoked end-plate potentials in adult mouse neuromuscular junctions. J Physiol (Lond) 346:353–363
Kidokoro Y (1984) Two types of miniature endplate potentials in *Xenopus* nerve-muscle cultures. Neurosci Res 1:157–170
Kim YI, Lømo T, Lupa MT, Thesleff S (1984) Miniature end-plate potentials in rat skeletal muscle poisoned with botulinum toxin. J Physiol (Lond) 356:587–599
Koenig J, Pécot-Dechavassine M (1971) Relations entre l’apparition des potentiels miniatures spontanés et l’ultrastructure des plaques motrices en voie de réinnervation et de néoformation chez le rat. Brain Res 27:43–57
Kriebel ME (1978) Small mode miniature endplate potentials are increased and evoked in fatigued preparations and in high Mg^{2+} saline. Brain Res 148:381–388
Kriebel ME, Erxleben C (1986) Sub-mepps, skew-mepps and the subunit hypothesis of quantal transmitter release at the neuromuscular junction. In: Rahamimoff R, Katz B (eds) Calcium, neuronal function and transmitter release. Nijhoff, Dordrecht, pp 299–329
Kriebel ME, Florey E (1983) Effect of lanthanum ions on the amplitude distributions of miniature endplate potentials and on synaptic vesicles in frog neuromuscular junctions. Neuroscience 9:535–547
Kriebel ME, Gross CE (1974) Multimodal distribution of frog miniature endplate potentials in adult, denervated, and tadpole leg muscle. J Gen Physiol 64:85–103
Kriebel ME, Motelica-Heino I (1987) Description of the sub-mepp distribution, determination of subunit size and number of subunits in the adult frog neuromuscular bell-mepp. Neuroscience 23:757–766
Kriebel ME, Llados F, Matteson DR (1976) Spontaneous subminiature end-plate potentials in mouse diaphragm muscle: evidence for synchronous release. J Physiol (Lond) 262:553–581
Kriebel ME, Hanna RB, Pappas GD (1980) Spontaneous potentials and fine structure of identified frog denervated neuromuscular junction. Neuroscience 5:97–108
Kriebel ME, Llados F, Carlson CG (1980a) Effect of the Ca^{2+} ionophore X-537A and a heat challenge on the distribution of mouse mepp amplitude histograms. J Physiol (Paris) 76:435–441
Kriebel ME, Llados F, Matteson DR (1982) Histograms of the unitary evoked potentials of the mouse diaphragm show multiple peaks. J Physiol (Lond) 322:211–222
Kriebel ME, Hanna R, Muniak C (1986) Synaptic vesicle diameters and synaptic cleft widths at the mouse diaphragm in neonates and adults. Dev Brain Res 27:19–29
Kuffler SW, Yoshikami D (1975) The number of transmitter molecules in a quantum: an estimate from iontophoretic application of acetylcholine at the neuromuscular synapse. J Physiol (Lond) 251:465–482
Kullberg RW, Lentz TL, Cohen MW (1977) Development of the myotomal neuromuscular junction in *Xenopus laevis:* an electrophysiological and fine-structural study. Dev Biol 60:101–129
Land BR, Salpeter EE, Salpeter MM (1980) Acetylcholine receptor site density affects the rising phase of miniature endplate currents. Proc Natl Acad Sci USA 77:3736–3740
Liley AW (1956a) An investigation of spontaneous activity at the neuromuscular junction of the rat. J Physiol (Lond) 132:650–666
Liley AW (1956b) The quantal components of the mammalian end-plate potential. J Physiol (Lond) 133:571–587
Liley AW (1957) Spontaneous release of transmitter substance in multiquantal units. J Physiol (Lond) 136:595–605
Llados F, Kriebel ME, Matteson DR (1980) β-Bungarotoxin preferentially blocks one class of miniature endplate potentials. Brain Res 192:598–602
Longenecker HE, Hurlbut WP, Mauro A, Clark AW (1970) Effects of a black widow spider venom on the frog neuromuscular junction: effects on endplate potential, miniature endplate potential and nerve terminal spike. Nature 225:701–703

MacDermot J, Westgaard RH, Thompson EJ (1978) Separation of two discrete proteins with different synaptic actions. Biochem J 175:271–279
Magleby KL, Miller DC (1981) Is the quantum of transmitter release composed of subunits? A critical analysis of the mouse and frog. J Physiol (Lond) 311:267–287
Magleby KL, Weinstock MM (1980) Nickel and calcium ions modify the characteristics of the acetylcholine receptor-channel complex at the frog neuromuscular junction. J Physiol (Lond) 299:203–218
Martin AR (1966) Quantal nature of synaptic transmission. Physiol Rev 46:51–66
Martin AR, Pilar G (1964) Quantal components of the synaptic potential in the ciliary ganglion of the chick. J Physiol (Lond) 175:1–16
Matteson DR, Kriebel ME, Llados F (1981) A statistical model indicates that miniature end-plate potentials and unitary evoked end-plate potentials are composed of subunits. J Theor Biol 90:337–363
Matthews-Bellinger JA, Salpeter MM (1983) Fine structural distribution of acetylcholine receptors at developing mouse neuromuscular junctions. J Neurosci 3:644–657
McArdle JJ, Albuquerque EX (1973) A study of the reinnervation of fast and slow mammalian muscles. J Gen Physiol 61:1–23
McLarnon JG, Quastel DMJ (1983) Postsynaptic effects of magnesium and calcium at the mouse neuromuscular junction. J Neurosci 3:1626–1633
Miledi R, Slater CR (1970) On the degeneration of rat neuromuscular junctions after nerve section. J Physiol (Lond) 207:507–528
Miller DC, Weinstock MM, Magleby KL (1978) Is the quantum of transmitter release composed of subunits? Nature 274:388–390
Miyamoto MD (1975) Binomial analysis of quantal transmitter release at glycerol treated frog neurotransmitter junctions. J Physiol (Lond) 250:121–142
Molgó J, Thesleff S (1982) 4-aminoquinoline-induced 'giant' miniature endplate potentials at mammalian neuromuscular junctions. Proc R Soc Lond [Biol] 214:229–247
Monolov S, Ovtscharoff W (1982) Structure and cytochemistry of the chemical synapses. Int Rev Cytol 77:243–284
Muniak CG, Kriebel ME, Carlson CG (1982) Changes in mepp and epp amplitude distributions in the mouse diaphragm during synapse formation and degeneration. Dev Brain Res 5:123–138
Nicholls JG (1956) The electrical properties of denervated skeletal muscle. J Physiol (Lond) 131:1–12
Pécot-Dechavassine M (1976) Action of vinblastine on the spontaneous release of acetylcholine at the frog neuromuscular junction. J Physiol (Lond) 261:31–48
Pennefather P, Quastel DMJ (1981) Relation between subsynaptic receptor blockade and response to quantal transmitter at the mouse neuromuscular junction. J Gen Physiol 78:313–344
Rose SJ, Pappas GD, Kriebel ME (1978) The fine structure of identified frog neuromuscular junctions in relation to synaptic activity. Brain Res 144:213–239
Rotshenker S, Tal M (1985) The transneuronal induction of sprouting and synapse formation in intact mouse muscles. J Physiol (Lond) 360:387–396
Sakmann B, Brenner HR (1978) Changes in synaptic channel gating during neuromuscular development. Nature 276:401–402
Sellin LC, Thesleff S (1981) Pre- and post-synaptic actions of botulinum toxin at the rat neuromuscular junction. J Physiol (Lond) 317:487–495
Spitzer N (1972) Miniature end-plate potentials at mammalian neuromuscular junctions poisoned by botulinum toxin. Nature [New Biol] 237:26–27
Sun Y-A, Poo M-M (1985) Non-quantal release of acetylcholine at developing neuromuscular synapse in culture. J Neurosci 5:634–642
Tonge DA (1974) Physiological characteristics of re-innervation of skeletal muscle in the mouse. J Physiol (Lond) 241:141–153
Vautrin J, Mambrini J (1981) Caractéristiques du potentiel unitaire de plaque motrice de la grenouille. J Physiol (Paris) 77:999–1010
Vicini S, Schuetze SM (1985) Gating properties of acetylcholine receptors at developing rat endplates. J Neurosci 5:2212–2224

Wernig A (1975) Estimates of statistical release parameters from crayfish and frog neuromuscular junctions. J Physiol (Lond) 244:207–221
Wernig A, Motelica-Heino I (1978) On the presynaptic nature of the quantal subunit. Neurosci Lett 8:231–234
Wernig A, Stirner H (1977) Quantum amplitude distributions point to functional unity of the synaptic 'active zone'. Nature 269:820–822
Whittaker VP (1984) the synaptosome. In: Lajtha A (ed) Handbook of neurochemistry, vol 7. Plenum, New York, pp 1–39
Young SH, Poo M (1983) Spontaneous release of transmitter from growth cones of embryonic neurones. Nature 305:634–637
Zimmermann H (1982) Insights into the functional role of cholinergic vesicles. In: Klein RL, Lagercrantz H, Zimmermann H (eds) Neurotransmitter vesicles. Academic, London, pp 305–359
Ziskind-Conhaim L, Dennis MJ (1981) Development of rat neuromuscular junctions in organ culture. Dev Biol 85:243–251

CHAPTER 19

The Electromotor Synapse

W.-D. KRENZ

A. Introduction

Evolution has provided neurobiologists with a variety of neuronal model systems which have greatly facilitated the progress the neurosciences have made during the last 40 years. In particular our understanding of the microphysiology of chemical synaptic transmission might have been delayed considerably if for example the neuromuscular junctions (NMJs) of frog and crayfish and the giant synapse of the squid stellar ganglion would not have been available and introduced (for review, see ECCLES 1964; KATZ 1966). As this volume demonstrates, the electromotor junction (EMJ) of the Torpedinidae, long known through its literally powerful external manifestation (for recent reviews, see BRAZIER 1984; WU 1984; ZIMMERMANN 1985), has also contributed to our knowledge on cholinergic synaptic transmission (for earlier reviews, see WHITTAKER and ZIMMERMANN 1976; WHITTAKER 1977). Results on the neurochemistry of the highly concentrated pre- and postsynaptic components of the EMJ have important implications for cholinergic transmission in general. Consequently it is of interest to know, to what extent EMJ physiology is comparable to that of other central and peripheral cholinergic synapses which may be better studied physiologically but are less accessible for neurochemical investigations.

B. Gross Physiology

BERNSTEIN (1912) gave the first satisfactory explanation for the mechanism enabling the high voltages of the electric organ discharges (EOD) of strongly electric fishes which in principle is valid today. From his work and that done by numerous scientists during the nineteenth and the first half of this century (for reviews, see GRUNDFEST 1957; BRAZIER 1984; WU 1984; ZIMMERMANN 1985; Chap. 1, this volume) the following functional picture emerged.

Within each of the approximately 500 parallel columns of a *Torpedo* electric organ about 400 disk-shaped electrocytes are stacked serially in a morphologically and physiologically polarized manner. Only their entire ventral surfaces are densely innervated and activated on nervous stimulation. When synchronous activation occurs, the voltage discharges of the individual electrocytes of a column are summated serially (BERNSTEIN 1912).

Torpedo electrocytes are electrically inexcitable. A fatigued or curarized electric organ of the electric eel *Electrophorus* that no longer responds to stimulation via its afferent nerves, can still be activated by direct electrical stimulation of the electrocytes. This direct stimulation of electrocytes is not possible in the electric organs of Torpedinidae. Furthermore, as was noticed early (see FESSARD 1947), in *Torpedo* stimulation evokes an EOD only after a latency which cannot be reduced to less than about 1 ms. This again is in contrast to the electric organ of *Electrophorus.*

Activation of the afferent nerves triggers the presynaptic release of acetylcholine (ACh) (FELDBERG and FESSARD 1942) which at the subsynaptic membrane of the electrocyte binds to nicotinic acetylcholine receptors (nAChR) which in turn opens ion channels enabling excitatory current to flow. All this has been summarized in considerable detail several times in the past by various authors (GRUNDFEST 1957, 1967; GRUNDFEST and BENNETT 1961; BENNETT 1970, 1971) and is also considered in Chap. 9.

The most important consequence resulting from the unique characteristics of *Torpedo* electric organ is that EODs from electric organs or from fragments of electrocyte columns, which can be studied with simple extracellular recording techniques, represent pure postsynaptic potentials without interference with action potentials. Thus under favourable conditions, i.e. synchronous activation of all electrocytes in a column, study of the EOD may give detailed information about the electrophysiology of the EMJ. This has enabled in a convenient way the parallel investigation of physiological and biochemical parameters of cholinergic synaptic function (see ZIMMERMANN 1975; WHITTAKER and ZIMMERMANN 1976).

C. Physiology of Single Electrocytes

Single electrocytes from electric rays are difficult objects for electrophysiological analysis with microelectrodes. They are flat and disk-shaped with a diameter of 3–4 mm but only about 15 μm thickness. This makes intracellular recording difficult. FESSARD and TAUC (1951) were the first to record intracellularly from single electrocytes from *T. marmorata.* They measured resting membrane potentials (RMPs) of between −50 and −70 mV. So far the most thorough intracellular microelectrode studies on single electrocytes from different Torpedinidae were carried out by BENNETT et al. (1961) in *T. nobiliana* and by BENNETT and GRUNDFEST (1961) in *Narcine brasiliensis* main and accessory organs. Later studies on single electrocytes used either extracellular recording techniques (MILEDI et al. 1971; DUNANT et al. 1972; WALTHER 1974) or investigated specific aspects of electromotor transmission (MOREAU and CHANGEUX 1976; SORIA 1983; ERDÉLYI and KRENZ 1984; DUNANT and MULLER 1986). Table 1 summarizes the basic results from those electrophysiological experiments in which microelectrodes were utilized either intra- or extracellularly to study single electrocytes from different Torpedinidae.

Due to the technical difficulties outlined above it remains doubtful whether the low RMPs from the various studies summarized in Table 1 represent true RMPs or whether they are reduced values resulting from faulty penetrations or microelectrodes not clearly located intracellularly. With the exception of the *Narcine* electrocytes the values from the different studies lie within the range between

Table 1. Summary of electrophysiological data on single electrocytes from different Torpedinidae

Reference	Species[a]	Recording technique	E_m (mV)	EpP (mV)	mEpP (mV)	t_{mEpC} (ms)	τ_{mEpC} (ms)
FESSARD and TAUC (1951)	*T. marmorata*	Intracellular	– 50 to – 70				
BENNETT et al. (1961)	*T. nobiliana*	Intracellular	– 50 to – 70	80			
BENNETT and GRUNDFEST (1961)	*Narcine brasiliensis*[b]	Intracellular	– 20 to – 60	>60			
MILEDI et al. (1971)	*T. marmorata* (23 °C)	Extracellular			≈0.3 – 1.5[c]		
DUNANT et al. (1972)	*T. marmorata* (17 ° – 20 °C)	Extracellular		≈4[c]	≈0.3 – 2[c]		
WALTHER (1974)	*T. marmorata* (RT)	Extracellular			≈0.2 – 1[c]		
MOREAU and CHANGEUX (1976)	*T. marmorata* (14 °C)	Intracellular	– 40 to – 70 – 52 ± 7.2				
SORIA (1983)	*T. marmorata* (20 °C)	Intracellular	51.3 ± 12.7				
		Extracellular			0.79 ± 0.68[d]	0.65 ± 0.27	0.92 ± 0.67
	T. californica (20 °C)	Extracellular			0.3 ± 0.49		
ERDÉLYI and KRENZ (1984)	*T. marmorata* (18 ° – 20 °C)	Intracellular	– 48 to – 75 – 57 ± 10.5	60 – 85 75 ± 8.4	0.296 ± 0.08		
		Extracellular				0.43 ± 0.14	0.75 ± 0.16

[a] Genus *Torpedo* (*T.*) unless otherwise stated; temperature of recording, when given, in parentheses. RT, room temperature
[b] Data from main electric organ
[c] Data extracted from text figures
[d] Data of giant mEpCs not included: t, rise time; τ, time constant of decay. E_m, resting membrane potential; EpP, evoked electroplaque potential; mEpP, miniature electroplaque potential; mEpC, miniature electroplaque current

about −50 and −75 mV. In this context it is a peculiar detail in all studies so far, that the stimulus-evoked intracellularly recorded electroplaque potential (EpP) always clearly overshoots the zero value by up to 30 mV. This finding is unusual insofar as electrocytes from Torpedinidae do not generate overshooting action potentials but produce exclusively chemically mediated EpPs. Although theoretically one could speculate about an ionic mechanism responsible for an overshooting postsynaptic potential, this possibility has never been investigated further and with present knowledge it seems more likely that the overshooting EpP is an artefact resulting from unreliable RMP measurements.

BENNETT et al. (1961) were also the first to recognize the low resistance of the dorsal electrocyte membrane. In experiments using small pieces of electrocyte columns arranged with their ventral innervated surface up and with nerve fibres to the uppermost electrocyte stimulated electrically they obtained the following results which are also illustrated in Fig. 1. As long as the microelectrode rested extracellularly on top of the preparation no potential was recorded. As soon as the electrode tip penetrated the innervated electrocyte membrane an RMP was registered and on nerve stimulation a full amplitude EpP was recorded. When the electrode was further advanced the resting potential would disappear indicating

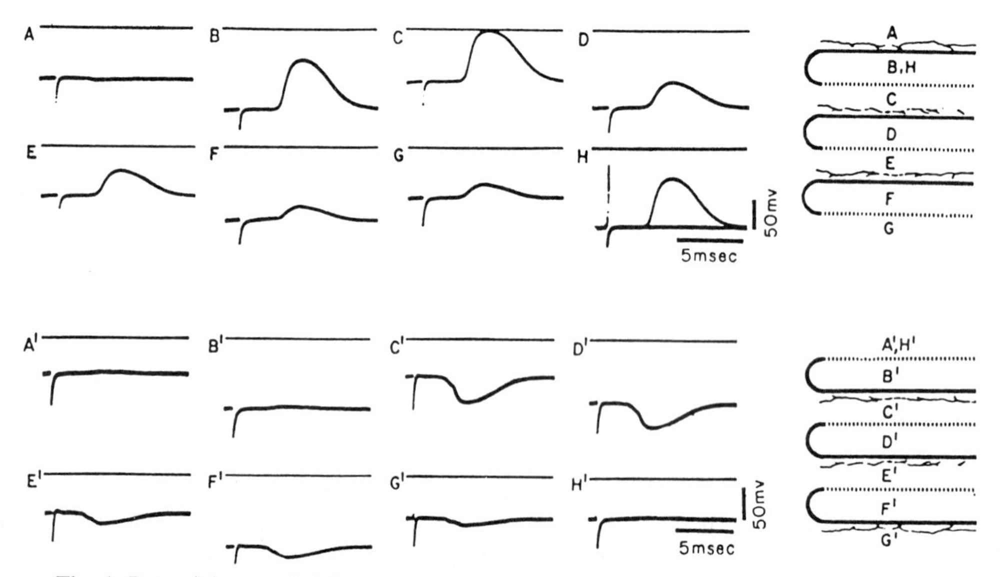

Fig. 1. Potentials recorded from individual electrocytes during activity of the superficial cell. Two experiments are shown. In *A–H* the preparation, a portion excised from the electric organ, was mounted with the ventral, innervated surfaces uppermost (diagram to the *right*). In *A′–H′*, another preparation was mounted in its normal orientation, with the non-innervated, dorsal surfaces of the electroplaques uppermost. The *letters* of the records correspond to the recording sites as indicated in the diagram to the *right* as a microelectrode was pushed from the surface into the column of electrocytes. *H* represents the return of the electrode to the interior of the uppermost cell, and *H′*, the return to the dorsal surface. The *upper trace* in each record is a reference line; the *lower trace*, the potential detected by the microelectrode. Zero potential is indicated by the distance of the potential trace from the reference line in *A* and *A′*. (From BENNETT et al. 1961)

that the microelectrode had left the electrocyte whilst the EpP could still be recorded with unchanged full amplitude. This demonstrates first that the noninnervated membrane remained passive on nerve stimulation, and secondly that it possesses a much lower resistance than the innervated one. When the microelectrode was again advanced the underlying electrocyte was penetrated, resulting again in the recording of an RMP. However on stimulation only a reduced amplitude EpP was now recorded, originating from the overlying activated electrocyte. Together these results provide a cellular basis for BERNSTEIN's (1912) model of series summation of voltages of simultaneously discharging electrocytes during EODs and they confirm the view of the electric organ of the Torpedinidae as a system designed to generate current.

Other important results from the work of BENNETT et al. (1961) are the confirmation on the single electrocyte level that the EpP can be influenced in amplitude by application of transmembrane current (see also ALBE-FESSARD 1961). Further, BENNETT and his colleagues were able to show that although the conductance of the active membrane increases during an EpP, no rectification can be observed over a wide range of applied polarizing current flow. Absence of rectification has also been confirmed in electrically inexcitable embryonic electrocytes (W. D. KRENZ, unpublished) that had differentiated *in vitro* from previously excitable myotubes in an explant culture system (RICHARDSON et al. 1981).

D. Microphysiology of Transmission at the EMJ

I. Extracellular Electroplaque Currents

With growing interest in the electric organs of Torpedinidae as a model cholinergic system several neuroscientists have attempted to measure and analyse spontaneous miniature electroplaque potentials (mEpPs) and miniature electroplaque currents (mEpCs). MILEDI et al. (1971) were the first to publish recordings of extracellular mEpCs followed by DUNANT et al. (1972), WALTHER (1974), DUNANT et al. (1980), SORIA (1983), and ERDÉLYI and KRENZ (1984). Up to now measurements of electroplaque currents (EpCs) under voltage clamp conditions have not been possible in *Torpedo* electrocytes due to technical and biological constraints, thus only extracellular recordings of EpCs (GAGE, 1976) are available. In a recent study DUNANT and MULLER (1986) applied the loose patch clamp technique (STÜHMER et al. 1983) to the innervated electrocyte membrane and obtained interesting new results on stimulus-evoked EpCs (see below).

The results on extracellular mEpCs from different studies are summarized in Table 1. While in the earlier studies (MILEDI et al. 1971; DUNANT et al. 1972) extracellular mEpCs have been reported in a more or less anecdotal fashion as functional indicators, WALTHER (1974) measured the drastic increase of the resting frequency in mEpCs of 0.5–50 Hz following application of high K^+-, La^{3+}- or black widow spider venom-containing saline. DUNANT et al. (1980) reported a low resting mEpC frequency of about 0.1 Hz at 20 °C which sharply increased with rising temperature. SORIA (1983), ERDÉLYI and KRENZ (1984), and recently DUNANT and MULLER (1986) have carried out more detailed investigations on extracellular mEpCs in electrocytes of *T. marmorata;* one measurement is available from SORIA (1983) for *T. californica* electrocytes. Frequencies described are low with 0.1 – 1 Hz at 20 °C (SORIA 1983) and 0.05 – 5 Hz at 18 °– 20 °C (ERDÉLYI and KRENZ 1984) and random succession (FATT and KATZ 1952) was confirmed (Fig. 2).

Although absolute amplitude values of extracellular mEpCs are meaningless, relative amplitude distributions from given recording sites can be useful. While ERDÉLYI and KRENZ (1984) recorded only unimodal amplitude distributions for spontaneous mEpCs, SORIA (1983) described a distinct second population of giant mEpCs occurring spontaneously mostly in bursts. According to SORIA these giant mEpCs were always resistant to the application of tetrodotoxin (TTX) even in extraordinarily high concentrations (300 m*M*), and their probability of occurrence increased in partially exhausted preparations, either by experimental manipulation or during long term incubation. On several occasions ERDÉLYI and KRENZ (1984) also measured bursts of apparently spontaneous mEpCs of higher amplitudes. However these could always be correlated with abnormal physiological states like pressing the surface of the terminal with the microelectrode (TURKANIS 1973) or KCl leakage from broken microelectrode tips, and they were never observed in TTX at low concentrations ($10^{-6} - 10^{-8}$ *M*).

In NMJs the time course of extracellular mEpCs closely approximates to those of real transmembrane current (see GAGE 1976). Specifically, the time constant of exponential current decay may be equal to the mean open time of the ACh receptor channel (ANDERSON and STEVENS 1973; KATZ and MILEDI 1973). In the study of ERDÉLYI and KRENZ (1984) mEpCs were always unimodally distributed with rise times between 0.24 and 0.66 ms (mean 0.43±0.14 ms, $N = 157$, 20 °C).

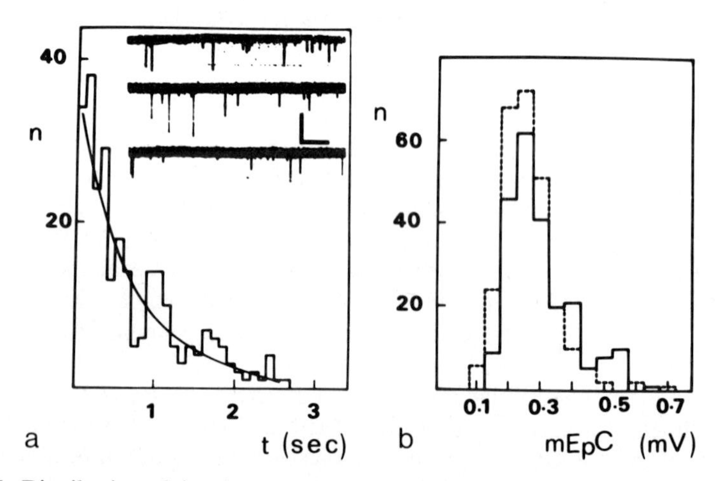

Fig. 2 a, b. Distribution of time interval *t* between spontaneous mEpCs. The observed intervals were grouped in 100-ms bins. The curve was calculated from the equation $n = N[\Delta t/T \exp(-t/T)]$ to fit the distribution for short intervals ($\Delta t = 100$ ms, $T = 700$ ms, $N = 296$). *Inset* shows spontaneous mEpCs recorded at slow sweep speed. Calibration: 0.2 mV, 1 s. **b** Unimodal amplitude-frequency histogram for mEpCs recorded during two successive measurement periods. The mean amplitudes were 0.25±0.075 mV, $N = 223$ (*solid line*) and 0.24±0.08 mV, $N = 261$ (*broken line*). (From ERDÉLYI and KRENZ 1984)

Decay time constants τ varied between 0.4 and 1.2 ms (mean 0.75 ± 0.16 ms, $N = 157$, 20 °C). Variation of time constants τ around the mean were also observed in stimulus-evoked unitary EpCs in low-Ca^{2+}, high-Mg^{2+} saline. Dependence of the time characteristics of mEpCs on membrane potential was not investigated due to the clamping problems mentioned above. Plots of rise time versus decay time constant τ showed poor correlation and those of decay time constant τ versus amplitude displayed continuous distribution of values. In contrast SORIA (1983) distinguished between two populations of small mode mEpCs besides the giant mEpCs. However in his study distributions of rise time (mean 0.65 ± 0.27 ms, $N = 338$, 20 ± 1 °C) and decay time constant τ (mean 0.92 ± 0.67 ms, $N = 319$, 20 ± 1 °C) were also unimodal. The two populations could only be distinguished in plots of frequency versus ratio of amplitude to decay time constant τ. Considering the uncertainties of extracellular recordings of mEpCs in electrocytes, voltage clamp studies will have to be carried out in order to evaluate the significance of the above observations. Recently SAKMANN et al. (1985) have measured mean open times of *Torpedo* nAChR channels expressed in *Xenopus* oöcytes after injection of specific mRNA. Their mean open time of 0.6 ± 0.2 ms at -110 to -70 mV membrane potential and 19 °–21 °C is relatively close to the mean of the time constant of decay τ (0.75 ± 0.16 ms) determined by ERDÉLYI and KRENZ (1984).

Recently MULLER and DUNANT (1985) described a population of extracellularly recorded subminiature EpCs (sub-mEpCs) in the electrocytes of *T. marmorata,* which were insensitive to *Botulinus* toxin in contrast to the normal mEpCs. Apart from the above-mentioned technical problems in analysing extracellularly recorded EpC amplitudes quantitatively, a major problem with the particular arrangement of *Torpedo* electrocytes is that at any recording site within an electrocyte column almost any kind of potential may be recorded with either polarity, depending on where the activity is originating relative to the microelectrode (see BENNETT et al. 1961). Furthermore DUNANT and MULLER (1986) did not detect sub-mEpCs in their recent loose patch clamp study. These apparently different classes of mEpCs need further investigation to learn for example whether they are comparable to the quantal subclass described by KRIEBEL and GROSS (1974), or whether they may be correlated with the presence of synaptic vesicles of different functional states (ZIMMERMANN 1979; ZIMMERMANN and WHITTAKER 1977).

II. Intracellular Electroplaque Potentials

Intracellularly recorded EpPs closely resemble the electric organ EOD however with much lower amplitude (60–85 mV) and shorter duration (BENNETT et al. 1961; ERDÉLYI and KRENZ 1984). They appear in an all-or-none fashion, indicating that once an afferent fibre has been activated its entire terminal arborization is depolarized and is releasing ACh which then activates the entire subsynaptic membrane underneath. However, new data using improved techniques (DUNANT and MULLER 1986) imply that only part of the presynaptic ending is generating action potentials.

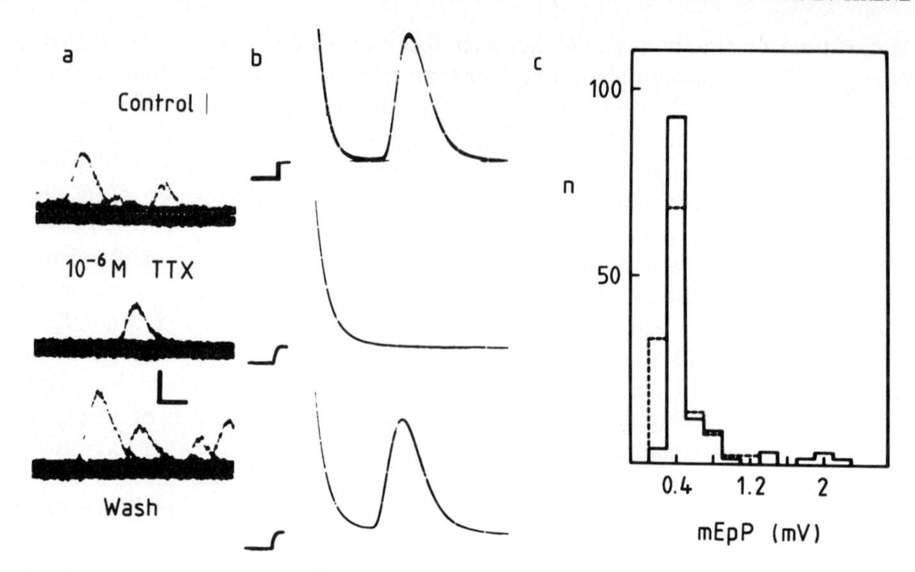

Fig. 3a–c. The effect of 10^{-6} *M* TTX on intracellularly recorded evoked EpPs (**b**) and spontaneous mEpPs (**a**). Calibration: in **a**, 0.5 mV, 1 ms; in **b**, 10 mV, 1 ms. **c** Amplitude-frequency histogram of spontaneous mEpPs in normal saline (*solid line,* mean amplitude 0.4 ± 0.13 mV, $N = 126$) and in 10^{-6} *M* TTX (*broken line,* mean amplitude 0.34 ± 0.19 mV, $N = 126$). (From ERDÉLYI and KRENZ 1984)

Frequency-dependent homosynaptic facilitation has been observed (BENNETT et al. 1961; DUNANT et al. 1980; ERDÉLYI 1982, 1984), on continued stimulation it is, however, rapidly superseded by frequency-dependent depression, the latter having also been well studied at the level of the EOD (ZIMMERMANN and WHITTAKER 1974; DUNANT et al. 1974; ZIMMERMANN 1975; KRENZ and ZIMMERMANN 1978). As would be expected for cholinergic neurotransmission and as observed in NMJs, the EpP can be specifically influenced by application of substances like 4-aminopyridine (4-AP), cholinergic drugs, and cholinesterase-blocking agents (AUGER and FESSARD 1941; BENNETT et al. 1961; MOREAU and CHANGEUX 1976; DUNANT et al. 1980; ERDÉLYI and KRENZ 1984; MULLER 1986). As Fig. 3b shows, the EpP is completely and reversibly abolished by superfusion with TTX in low concentrations (0.01 – 1 μ*M*) (ERDÉLYI and KRENZ 1984).

Up to now only ERDÉLYI and KRENZ (1984) have analysed spontaneous mEpPs intracellularly. They are regularly recorded in electrocytes with reasonable RMPs and like mEpCs appear at low frequencies of 0.05 – 5 Hz. They are almost identical in time characteristics with stimulus-evoked EpPs (Fig. 3a), however with a lower amplitude of 0.296 ± 0.08 mV (SD, 6 electrocytes). Their amplitude-frequency distribution was always unimodal (Fig. 3c). In contrast to evoked EpPs the mEpPs are insensitive to application of TTX, only potentials of amplitudes higher than 1.2 mV were reversibly abolished, which did not significantly change the mean value (Fig. 3c). This indicates that a small fraction of the apparently spontaneous mEpPs may be due to local regenerative activity in the presynaptic terminal arborization, triggering multiquantal release.

III. Quantal Analysis of Electroplaque Potentials

DUNANT et al. (1980) demonstrated that in the *Torpedo* electromotor synapse, as in other chemical synapses, decreasing the Ca^{2+} and increasing the Mg^{2+} concentration in the saline results in a decrease in the evoked release of acetylcholine (ACh) and consequently in a decrease of the EpP amplitude. Furthermore, when ERDÉLYI and KRENZ (1984) superfused electrocytes with low-Ca^{2+}, high-Mg^{2+} saline they observed that during 60–80 min the stimulus-evoked EpP can be reversibly reduced to 1%–2% of its initial amplitude (Fig. 4A, B) leaving a unitary potential with an amplitude of between 0.24 and 0.55 mV [0.32±0.13 mV (SD), 5 electrocytes]. This unitary potential is almost identical in amplitude and time characteristics with the mean of the spontaneous mEpPs [0.296±0.08 mV (SD), 6 electrocytes]. On continued stimulation at 1 Hz the amplitude of the EpP fluctuated around singles, doubles, triples, and multiples of the unitary potential, and often it failed (Fig. 4b). In contrast to stimulus-evoked EpPs, the mean amplitude of spontaneous mEpPs did not change significantly in low-Ca^{2+}, high-Mg^{2+} saline.

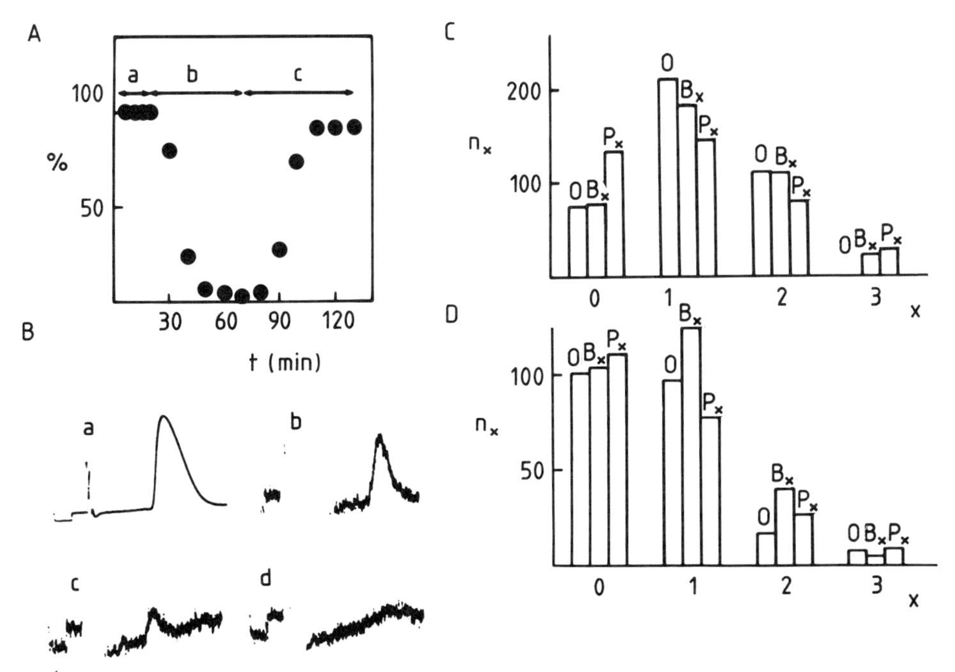

Fig. 4 A–D. Effect of low-Ca^{2+}, high-Mg^{2+} saline on stimulus-evoked EpPs. **A** Development of the EpP amplitude, before (*a*), during (*b*), and after (*c*) application of low-Ca^{2+}, high-Mg^{2+} saline, expressed as a percentage of the initial value. **B** Recordings of stimulus evoked EpPs before (*a*), and during (*b, c* and *d*) application of low-Ca^{2+}, high-Mg^{2+} saline; *b, c* and *d* represent multiple, single, and failure of quantal release. Calibrations: *a*, 5 mV, 1 ms; *b, c* and *d,* 0.2 mV, 1 ms. **C, D** Histograms of quantal distributions for two experiments. The numbers of observations and the theoretically expected values containing x quanta are plotted versus failures, singles, doubles, and triples of the unitary EpP size in low-Ca^{2+}, high-Mg^{2+} saline. *O,* observed occurrence; B_x calculated binomial; P_x, calculated Poisson distribution. (From ERDÉLYI and KRENZ 1984)

This behaviour of stimulus-evoked EpPs and spontaneous mEpPs encouraged a statistical analysis of synaptic transmission at the electromotor synapse of *Torpedo* (ERDÉLYI and KRENZ 1984). The average quantum content of EpPs ranged from 150–400 (mean 250) in normal saline and decreased to 0.65–1.36 (mean 0.95±0.29) in low-Ca^{2+}, high-Mg^{2+} saline. The fitting of the quantal fluctuations in low-Ca^{2+}, high Mg^{2+} was tested by comparing observed values with those expected for binomial and Poisson distributions. As Fig. 4C,D demonstrates for two experiments, the observed values were well represented by both binomial and Poisson statistics.

In their study using the loose patch clamp technique DUNANT and MULLER (1986) measured stimulus-evoked EpCs and spontaneous mEpCs generated under the patch pipette opening. Quantal analysis of stimulus-evoked EpCs showed that their number and distribution corresponds closely to that predicted by Poisson statistics. Since with this technique electrical stimulation can be restricted to the area under the pipette opening, it became possible to estimate the quantal output per unit area of presynaptic membrane. A constant value of about 1.3 quanta μm^{-2} was found. The value of about 7000 ACh molecules released per quantum which DUNANT and MULLER (1986) estimated from charge transfer calculations for EpCs may be larger in reality since the mean open time of 1.3 ms they considered for single nAChR channels appears to be overestimated (ERDÉLYI and KRENZ 1984; SAKMANN et al. 1985).

All this demonstrates that in spite of some functional specializations, synaptic transmission at the electromotor synapse of *Torpedo* follows the same quantal mechanism as first described by DEL CASTILLO and KATZ (1954) in the NMJ of the frog, and meanwhile confirmed for a variety of central and peripheral chemical synapses (KATZ and MILEDI 1963; KUNO 1964; MILEDI 1967; CASTELLUCCI and KANDEL 1974).

E. Conclusions

From the physiological data presented here it is evident that the electromotor synapse of *Torpedo* does not deviate in principle from our general view of the physiology of chemical synaptic transmission. A large, flattened subsynaptic membrane, most densely packed with nAChRs, is innervated by a terminal which forms an equally flattened widespread presynaptic membrane leaving only one-fifth (FOX, unpublished) or one-half (DUNANT and MULLER 1986) of the ventral electrocyte surface uninnervated. On nervous stimulation large quanta are released onto the subsynaptic membrane by means of large vesicles, highly loaded with ACh (OHSAWA et al. 1979) by a mechanism described by binomial statistics. These properties of the EMJ, together with electrocyte specializations like unexcitability, low dorsal membrane resistance and short space constant (BENNETT et al. 1961) enable the EMJ to serve its function in the electric organs of Torpedinidae – the generation of electric current.

References

Albe-Fessard D (1951) Modifications de l'activité des organs électriques par des courants d'origine extérieure. Arch Sci Physiol 5:45–73

Anderson CR, Stevens CF (1973) Voltage clamp analysis of acetylcholine produced end-plate current fluctuations at frog neuromuscular junction. J Physiol (Lond) 235:655–691

Auger D, Fessard A (1941) Actions due curare, de l'atropine, et de l'éserine sur le prisme électrogène de la Torpille. CR Soc Biol (Paris) 135:76

Bennett MVL (1970) Comparative physiology: electric organs. Annu Rev Physiol 32:471–528

Bennett MVL (1971) Electric organs. In: Hoar WS, Randall DJ (eds) Fish physiology, vol 5. Academic, New York, pp 347–491

Bennett MVL, Grundfest H (1961) The electrophysiology of electric organs of marine electric fishes. II. The electroplaques of main and accessory organs of *Narcine brasiliensis.* J Gen Physiol 44:805–818

Bennett MVL, Wurzel M, Grundfest H (1961) The electrophysiology of electric organs of marine electric fishes. I. Properties of electroplaques of *Torpedo nobiliana.* J Gen Physiol 44:757–804

Bernstein J (1912) Elektrobiologie. Vieweg, Braunschweig

Brazier MAB (1984) A history of neurophysiology in the 17th and 18th centuries. Raven, New York

Castellucci V, Kandel ER (1974) A quantal analysis of the synaptic depression underlying habituation of the gill-withdrawal reflex in *Aplysia.* Proc Natl Acad Sci USA 71:5004–5008

Del Castillo J, Katz B (1954) Quantal components of the endplate potential. J Physiol (Lond) 124:560–573

Dunant Y, Muller D (1986) Quantal release of acetylcholine evoked by focal depolarization at the *Torpedo* nerve-electroplaque junction. J Physiol (Lond) 379:461–478

Dunant Y, Gautron J, Israël M, Lesbats B, Manaranche R (1972) Les compartiments d'acetylcholine de l'organe électrique de la torpille et leurs modifications par la stimulation. J Neurochem 19:1987–2002

Dunant Y, Gautron J, Israël M, Lesbats B, Manaranche R (1974) Evolution de le décharge de l'organe électrique de la torpille et variations simultanées de l'acetycholine au cours de la stimulation. J Neurochem 23:635–643

Dunant Y, Eder L, Hirt L (1980) Acetylcholine release evoked by single or few nerve impulses in the electric organ of *Torpedo.* J Physiol (Lond) 298:185–203

Eccles JC (1964) The physiology of synapses. Springer, Berlin Göttingen Heidelberg New York

Erdélyi L (1984) Electrophysiological study of transmission at electromotor synapse of *Torpedo marmorata.* In: Vizi ES, Magyar K (eds) Regulation of transmitter function. Akadémiai Kiadó, Budapest, pp 81–82

Erdélyi L (1985) The quantal nature of transmission and spontaneous potentials at the *Torpedo* electromotor junction. Acta Physiol Hung 65:81–93

Erdélyi L, Krenz WD (1984) Electrophysiological aspects of synaptic transmission at the electromotor junction of *Torpedo marmorata.* Comp Biochem Physiol 79A:505–511

Fatt P, Katz B (1952) Spontaneous subthreshold activity at motor nerve endings. J Physiol (Lond) 117:109–128

Feldberg, W, Fessard A (1942) The cholinergic nature of the nerves to the electric organ of the *Torpedo* (*Torpedo marmorata*). J Physiol (Lond) 101:200–216

Fessard A (1947) Recherches sur le fonctionnement des organs électriques. I. Analyse des formes de décharge obtenue par divers procédés d'excitation. Arch Int Physiol 55:1–26

Fessard A, Tauc L (1951) Détermination microélectrométrique du potentiel de repos de l'élément électrogène chez *Torpedo marmorata.* CR Seances Acad Sci 233:1228–1230

Gage PW (1976) Generation of endplate potentials. Physiol Rev 56:177–247

Grundfest H (1957) The mechanism of discharge of the electric organs in relation to general and comparative electrophysiology. Prog Biophys Biophys Chem 7:1–85

Grundfest H (1967) Comparative physiology of electric organs of elasmobranch fishes. In: Gilbert PW, Mathewson RF, Rall DP (eds) Sharks, skates, and rays. Hopkins, Baltimore, pp 399–432

Grundfest H, Bennett MVL (1961) Studies on the morphology and electrophysiology of electric organs. I. Electrophysiology of marine electric fishes. In: Chagas C, Paes de Carvalho A (eds) Bioelectrogenesis. Elsevier, Amsterdam, pp 57–101

Katz B (1966) Nerve, muscle, and synapse. McGraw-Hill, New York

Katz B, Miledi R (1963) A study of spontaneous miniature potentials in spinal motoneurones. J Physiol (Lond) 168:389–422

Katz B, Miledi R (1973) The binding of acetylcholine to receptors and its removal from the synaptic cleft. J Physiol (Lond) 231:549–574

Krenz WD, Zimmermann H (1978) Long term depression of *Torpedo* electric organ response on spaced stimulation. Comp Biochem Physiol [A] 61:531–538

Kriebel ME, Gross CE (1974) Multimodal distribution of frog miniature endplate potentials in adult, denervated, and tadpole leg muscle. J Gen Physiol 64:85–103

Kuno M (1964) Quantal components of excitatory synaptic potentials in spinal motoneurones. J Physiol (Lond) 175:81–99

Miledi R (1967) Spontaneous synaptic potentials and quantal release of transmitter in the stellate ganglion of the squid. J Physiol (Lond) 192:379–406

Miledi R, Molinoff P, Potter LT (1971) Isolation of the cholinergic receptor protein of *Torpedo* electric tissue. Nature 229:554–557

Moreau M, Changeux J-P (1976) Studies on the electrogenic action of acetylcholine with *Torpedo marmorata* electric organ. I. Pharmacological properties of the electroplaque. J Mol Biol 106:457–467

Muller D (1986) Potentiation by 4-aminopyridine of quantal acetylcholine release at the *Torpedo* nerve-electroplaque junction. J Physiol (Lond) 379:479–493

Muller D, Dunant Y (1985) Subminiature electroplaque potentials are present in *Torpedo* electric organ (Abstr). 17th Annual Meeting of the Union of the Swiss Societies for Experimental Biology, Geneva, p 51

Ohsawa K, Dowe GHC, Morris SJ, Whittaker VP (1979) The lipid and protein content of cholinergic synaptic vesicles from the electric organ of *Torpedo marmorata* purified to constant composition: implications for vesicle structure. Brain Res 161:447–457

Richardson GP, Krenz WD, Kirk C, Fox GQ (1981) Organotypic culture of embryonic electromotor system tissues from *Torpedo marmorata.* Neuroscience 6:1181–1200

Sakmann B, Methfessel C, Mishina M, Takahashi T, Takai T, Kurasaki M, Fukuda K, Numa S (1985) Role of acetylcholine receptor subunits in gating of the channel. Nature 318:538–543

Soria B (1983) Properties of miniature post-synaptic currents at the *Torpedo marmorata* nerve-electroplate junction. Q J Exp Physiol 68:189–202

Stühmer B, Roberts WM, Almers W (1983) The loose patch clamp. In: Sakmann B, Neher E (eds) Single-channel recording. Plenum, New York, pp 123–132

Turkanis SA (1973) Effects of muscle stretch on transmitter release at end-plates of rat diaphragm and frog sartorius muscle. J Physiol (Lond) 230:391–403

Walther C (1974) Effects of potassium, lanthanum and black widow spider venom on miniature synaptic potentials in the *Torpedo* electroplax. J Comp Physiol 90:71–73

Whittaker VP (1977) The electromotor system of *Torpedo.* A model cholinergic system. Naturwissenschaften 64:606–611

Whittaker VP, Zimmermann H (1976) The innervation of the electric organ of Torpedinidae: a model cholinergic system. In: Malins DC, Sargent JR (eds) Biochemical and biophysical perspectives in marine biology. Academic, London, pp 67–116

Wu CH (1984) Electric fish and the discovery of animal electricity. Am Sci 72:598–607

Zimmermann H (1975) Dynamics of synaptic transmission in the cholinergic synapses of the electric organ of *Torpedo.* In: Waser PG (ed) Cholinergic mechanisms. Raven, New York, pp 169–176

Zimmermann H (1979) Vesicle recycling and transmitter release. Neuroscience 4:1773–1804

Zimmermann H (1985) Die elektrischen Fische und die Neurobiologie: Über die Bedeutung einer naturgeschichtlichen Kuriosität für die Entwicklung einer Wissenschaft. Funkt Biol Med 4:156–172

Zimmermann H, Whittaker VP (1974) Effect of electrical stimulation on the yield and composition of synaptic vesicles from the cholinergic synapses of the electric organ of *Torpedo:* a combined biochemical, electrophysiological and morphological study. J Neurochem 22:435–450

Zimmermann H, Whittaker VP (1977) Morphological and biochemical heterogeneity of cholinergic synaptic vesicles. Nature 267:633–635

CHAPTER 20

The Autonomic Cholinergic Neuroeffector Junction

H. KILBINGER

A. Introduction

In this article the term cholinergic neuroeffector *junction* is used to distinguish this site of transmission from the cholinergic *synapses* of skeletal muscle and autonomic ganglia. The characteristics of the neuroeffector junction are the wide and variable cleft (up to micrometres), the lack of discrete postjunctional specializations and the even distribution of receptors over the whole surface of the effector cell. The typical synapse, on the other hand, has a narrow cleft (20 – 50 nm) with characteristic pre- and postsynaptic structural specializations and a small area of the postsynaptic cell where the receptors are concentrated (for literature, see GABELLA 1981; VIZI 1984; BURNSTOCK 1985). The terminal portions of the autonomic cholinergic nerve fibres innervating smooth muscles are varicosities which are filled with agranular vesicles. These varicosities are the actual sites of acetylcholine (ACh) release, and constitute *en passage* junctions rather than terminal synapses (for literature, see KILBINGER 1982; LÖFFELHOLZ and PAPPANO 1985). The basic mechanisms of ACh synthesis and release are similar at cholinergic junctions and synapses. Cholinergic transmission at peripheral ganglionic synapses has been dealt with in recent reviews (COLLIER and KWOK 1982; MCAFEE 1982) and in a handbook of this series (KHARKEVICH 1980) and is therefore not considered here. This chapter reviews some of the recent discoveries about the release of ACh from peripheral neuroeffector junctions with particular emphasis on the autoreceptor mediated control of release (KILBINGER 1984), and about the actions of ACh on subtypes of muscarinic ACh receptors (mAChRs) at various peripheral junctions.

B. Release of ACh

Transmission from cholinergic nerves to autonomic neuroeffectors has been studied most extensively in the myenteric plexus-longitudinal muscle preparation of the guinea pig and, more recently, also in isolated mammalian and avian hearts. The basic mechanisms of ACh release from these tissues have been described in detail in several recent reviews (LÖFFELHOLZ 1981; KILBINGER 1982; SZERB 1982; LÖFFELHOLZ and PAPPANO 1985). Only selected aspects are considered here.

I. Resting Release

Resting release of ACh has been measured after inhibition of the tissue cholinesterase (ChE) activity. The reported values for release per minute expressed as per cent of the tissue ACh content are in fair agreement: 0.1% – 0.2% in the myenteric plexus (see KILBINGER 1982) and 0.4% in the fowl heart (DIETERICH et al. 1976). In the myenteric plexus preparation hexamethonium or tetrodotoxin (TTX) reduce the resting release by about 50%. This suggests that about half of the resting release of ACh in the presence of ChE inhibition depends on the activation of ganglionic nicotinic receptors and on the propagation of action potentials. The hexamethonium- and TTX-insensitive fraction of the resting release does not depend on the presence of Ca^{2+} in the extracellular medium and may therefore represent leakage of cytoplasmic ACh through the axonal membrane. Such cytoplasmic ACh accumulates (as surplus ACh) in the myenteric plexus preparation during incubation with a ChE inhibitor (KILBINGER and WAGNER 1975). In the heart of the domestic fowl the resting release of ACh is not affected by lowering the Ca^{2+} concentration of the medium (see LÖFFELHOLZ and PAPPANO 1985). In both fowl heart (LÖFFELHOLZ et al. 1984) und guinea-pig myenteric plexus (KILBINGER and WAGNER 1975) the Ca^{2+}-independent fraction of the resting release cannot be modulated through muscarinic autoreceptors.

II. Electrically Evoked Release

The release of ACh evoked by electrical stimulation of myenteric neurons or of cardiac vagal nerves shows the characteristics of an exocytotic release, i.e. it is dependent on the presence of extracellular Ca^{2+} and is inhibited by excess Mg^{2+} (KILBINGER 1982; LÖFFELHOLZ and PAPPANO 1985). The neural organization of the myenteric plexus is so complex that it has not been possible, so far, to stimulate postganglionic cholinergic fibres selectively. During field stimulation of the myenteric plexus the axon hillock will be brought to threshold more readily than the varicosities or the intervaricose fibres (NORTH 1982); consequently, the varicose terminal part of the neurons of the myenteric plexus is always invaded orthodromically. If nerve conduction is blocked by TTX the release evoked by field stimulation is abolished. There is comprehensive evidence that a substantial portion of the ACh collected in the medium is released from the postganglionic terminals innervating the longitudinal muscle. Ultrastructural studies (GABELLA 1972) indicate that the myenteric plexus contains many more terminals presumed to be varicosities of the neuroeffector site than terminals associated with interneuronal synapses. Hexamethonium does not reduce either the smooth muscle contractions or ACh release evoked by stimulation at 0.3 and 1 Hz (KILBINGER 1977; KILBINGER and WESSLER 1980b). Thus, nicotinic ganglionic transmission between the point of stimulation and the neuroeffector junction is not involved if ACh release evoked by field stimulation is measured.

Release of ACh from hearts has been evoked by electrical stimulation of the cervical vagus nerves or by field stimulation of the isolated heart (LÖFFELHOLZ 1981). As in the myenteric plexus the release caused by field stimulation is not affected by hexamethonium or by tubocurarine, which points to the postganglionic origin of the released transmitter. In order to measure release of endogenous ACh from myenteric neurons or from mammalian hearts it is necessary to inhibit the ChE activity. However, the use of ChE inhibitors in release experiments is disadvantageous for the following reasons. ChE inhibitors activate the negative feedback mechanism of ACh release and thus strongly reduce the evoked release (see Sect. C). Moreover, incubation of the myenteric plexus preparation (KILBINGER and WAGNER 1975) or perfusion of isolated hearts (DIETERICH et al. 1976) with a ChE inhibitor leads to the formation of a new compartment, the so-called surplus ACh (probably cytoplasmic) which is not present in the tissue under physiological conditions. Finally, ChE inhibitors cause a permanent contraction of the ileal smooth muscle so that it is not possible to measure the effect of electrical stimulation on both muscle contraction and release of endogenous ACh in one and the same preparation. In order to avoid all these complications release of ACh from mammalian hearts and myenteric nerves is best determined in the absence of ChE inhibition after loading of the tissue with radioactively labelled choline. Electrical stimulation of the cholinergic nerves causes an increased outflow of the label that originates from the release of radioactive ACh (SZERB 1976; KILBINGER and WESSLER 1980a; LINDMAR et al. 1980). The electrically evoked release of ACh from hearts of frog and birds is so large that it can easily be measured in the absence of a ChE inhibitor (LÖFFELHOLZ 1981).

III. Release of Vasoactive Intestinal Peptide as a Co-transmitter in Cholinergic Nerves

The coexistence of different neurotransmitters has been described in central and peripheral nerves (for reviews see O'DONOHUE et al. 1985; BARTFAI 1985). In the case of peripheral postganglionic cholinergic nerves the evidence is most compelling for the coexistence of ACh and vasoactive intestinal peptide (VIP). Electrical stimulation of the parasympathetic nerves supplying the cat submandibular salivary gland caused the concurrent release of ACh and VIP (LUNDBERG 1981; EDWARDS and BLOOM 1982). The ratio of ACh to VIP released from the nerve depends upon the frequency of stimulation. At low frequencies there is a preferential release of ACh while release of VIP (but not that of ACh) increases markedly at high stimulation frequencies. Atropine facilitates the release of VIP evoked by parasympathetic nerve stimulation at 6 or 15 Hz which indicates that the autoreceptor for ACh also modulates the release of the coexisting VIP. Subcellular fractionation of the cat submandibular gland (LUNDBERG 1981) or of the guinea-pig myenteric plexus preparation (ÁGOSTON et al. 1985) revealed that ACh is present not only in small clear vesicles but, in addition, in larger vesicles which appear in the same region of the density gradient as VIP-containing vesicles. This might indicate that ACh and VIP are stored within the same vesicle. It has, however, been pointed out that co-sedimentation of the two types of vesicles one containing ACh and the other VIP cannot be excluded (ÁGOSTON et al. 1985). Furthermore, the differential release of ACh and VIP at different frequencies strongly suggests that ACh and VIP are stored in the parasympathetic nerves of the submandibular gland in different vesicles whose contents are released by selective secretory mechanisms. The finding of a preferential release of either VIP or ACh by variation in stimulation frequency led EDWARDS and BLOOM (1982) to reassess the optimum stimulation patterns for the release of transmitters at autonomic neuroeffector junctions. They suggested that autonomic nerve fibres cannot any longer be characterized by low natural discharge patterns, and that intermittent stimulation in bursts may be a more physiological pattern of activity in autonomic nerves than continuous stimulation at a low frequency.

The exocytotic release of ACh from autonomic postganglionic fibres can be modulated by receptors which are sensitive to the neuron's own transmitter (autoreceptors). Muscarinic autoreceptors have been shown to be located on nerve terminals innervating the effector cells, i.e. prejunctionally, and on the soma-dendritic region of postganglionic neurons. This is the subject of the next section. Prejunctional nicotinic receptors on peripheral autonomic cholinergic nerves have not been described so far.

C. Modulation of ACh Release by Muscarinic Autoreceptors

I. Prejunctional Autoreceptors

1. Muscarinic Agonists and Antagonists, and Cholinesterase Inhibitors

a) Myenteric plexus

Muscarinic agonists including ACh inhibit the electrically evoked release of ACh. This depression is overcome by small concentrations of atropine or scopolamine. Muscarinic antagonists alone facilitate the electrically evoked release by overcoming the inhibitory action of the released endogenous ACh (KILBINGER 1977; SAWYNOK and JHAMANDAS 1977; KILBINGER and WESSLER 1980a, b; FOSBRAEY and JOHNSON 1980b; SZERB 1980; KILBINGER and KRUEL 1981; ALBERTS et al. 1982; VIZI et al. 1984). The negative feedback mechanism of ACh release is also seen if release is elicited by high K^+ or by the nicotinic agonist dimethylphenylpiperazinium (DMPP) (DZIENISZEWSKI and KILBINGER 1978).

ChE inhibitors lead to an accumulation of ACh in the biophase and thus depress its evoked release from myenteric neurons (KILBINGER and WESSLER 1980a). Muscarinic antagonists markedly increase the release of endogenous ACh by overcoming the inhibitory effect of the antiChEs, but the facilitatory effect of the antagonists is less pronounced in the absence of ChE inhibition. This indicates that the prejunctionally located ChE serves as a barrier to protect the nerve ending from the effects of the released ACh on mAChRs. A recent study with the enantiomers of bethanechol (SCHWÖRER et al. 1985) suggests that the access of choline esters to prejunctional sites is more limited by the activity of ChEs than the access to postjunctional sites. (S)-(+)-bethanechol was found to be a substrate, and (R)-(−)-bethanechol a weak inhibitor of the overall activity of the ChEs of the myenteric plexus longitudinal muscle preparation. In the presence of eserine (S)-(+)-bethanechol caused a much stronger inhibition of the evoked ACh release than in its absence, whilst the contractions of the smooth muscle elicited by (S)-(+)-bethanechol were not significantly affected by eserine.

Similarly, the (R)-enantiomer in the racemic mixture of bethanechol strongly reinforced only the prejunctional but not the postjunctional effects of (S)-(+)-bethanechol. It has been assumed that pre- and postjunctional sites may be characterized by different molecular forms of AChEs (MARQUIS and FISHMAN 1985). This raises the interesting possibility that pre- and postjunctional ChEs may be inhibited differentially by antiChEs.

There is circumstantial evidence that stimulation of neuronal mAChRs induces an inhibition of phasic intestinal activity (FOX et al. 1985). When the dog small intestine was actively contracting, intra-arterially injected ACh caused a contraction followed by prolonged inhibition of contractions. Both inhibitory and excitatory responses were reduced by atropine but not by hexamethonium or reserpine pretreatment. FOX et al. (1985) concluded that prejunctional muscarinic autoreceptors are responsible for the inhibition of intestinal motility by exogenous ACh.

b) Heart

In the isolated perfused fowl heart muscarinic agonists inhibit, and atropine facilitates the release of ACh evoked by preganglionic vagus nerve stimulation or by field stimulation in the absence of ChE inhibition (KILBINGER et al. 1981; LÖFFELHOLZ et al. 1984). Eserine reduces the electrically evoked release (LINDMAR et al. 1980). Consequently, the inhibitory effect of oxotremorine is attenuated, and the facilitatory action of atropine is enhanced if release is determined in the presence of eserine (KILBINGER et al. 1981). Feedback regulation of ACh release has also been examined in mammalian hearts. Thus, the release evoked by vagus nerve stimulation from perfused rabbit atria is inhibited by eserine (MUSCHOLL and MUTH 1982). Likewise, neostigmine reduced, and atropine enhanced the ACh release elicited by field stimulation or by high K^+ from rat atria (WETZEL and BROWN 1985).

c) Other Organs

There is some knowledge pertaining to the prejunctional actions of muscarinic agonists and antagonists at other peripheral cholinergic neuroeffector junctions. Autoreceptors occur in the *iris* since atropine and scopolamine facilitated the release evoked by electrical stimulation of ox (GUSTAFSSON et al. 1980) or rat (RICHARDSON et al. 1984) iris. On rat *urinary bladder* strips eserine inhibited the release of ACh in response to field stimulation, and this effect was antagonized by atropine. Atropine alone enhanced, whereas carbachol and muscarine decreased the depolarization-induced release (D'AGOSTINO et al. 1986). A muscarinic inhibition of ACh release was also observed on isolated rat *bronchial muscle* where oxotremorine reduced and scopolamine increased the release evoked by high K^+ (AAS and FONNUM 1986). Indirect evidence for a muscarinic inhibition of ACh release from pulmonary vagal nerves was provided in *in vivo* experiments in which the effects of gallamine and pilocarpine on the bronchoconstriction induced by electrical vagus nerve stimulation were studied (FRYER and MACLAGAN 1984; BLABER et al. 1985). Gallamine increased, and pilocarpine inhibited the vagally mediated bronchoconstrictor response in guinea pig and cat. On the other hand, gallamine did not increase the bronchoconstriction produced by injected ACh. This suggests that the receptors mediating the modulatory effect of gallamine are neuronal mAChRs in the pulmonary vagal nerves.

Taken together, the muscarinic inhibition of ACh release seems to be a general characteristic of peripheral autonomic cholinergic nerves. An exception are the cholinergic nerve terminals in sympathetic ganglia where muscarinic autoreceptors have not been detected (COLLIER and KWOK 1982).

2. Location

The muscarinic modulation of ACh release from the isolated iris (GUSTAFSSON et al. 1980; RICHARDSON et al. 1984) is obviously mediated through receptors which are located at postganglionic parasympathetic nerves since the iris sphincter is a ganglion-free tissue. Most other parasympathetically innervated tissues, however, contain both pre- and postganglionic fibres. Muscarinic agonists may then affect ACh release either by acting on the axon terminals of pre- or postganglionic neurons or by inhibiting the excitability of the neuron soma. Whilst there is, so far, no evidence for the occurrence of inhibitory mAChRs on the soma-dendritic region of peripheral cholinergic neurons the following studies suggest a prejunctional location of muscarinic autoreceptors:

1. In the myenteric plexus the release evoked by 40 mM K^+ in the presence of TTX as well as the release evoked by 80 or 108 mM K^+ (which is TTX insensitive; PATON et al. 1971) is increased by atropine and inhibited by oxotremorine (DZIENISZEWKSI and KILBINGER 1978; ALBERTS et al. 1982). These findings strongly favour a prejunctional localization of the autoreceptor since high concentrations of K^+ depolarize the nerve endings directly under these conditions. It should however be noted that the effects of atropine and oxotremorine are less marked when ACh release is elicited by high K^+ as compared to electrical field stimulation of the myenteric nerves. A similar phenomenon has been observed in experiments on isolated hearts. In the fowl heart atropine caused a 2-to-4-fold increase in ACh release evoked by vagal or field stimulation (KILBINGER et al. 1981; LÖFFELHOLZ et al. 1984), but the antagonist did not significantly enhance the release induced by either 45 or 108 mM K^+ (KILBINGER and WAGNER 1979). Likewise, oxotremorine (10 μM) inhibited the field-stimulated release of ACh from rat atria to a much greater extent (by 75%) than it inhibited the K^+ (57 mM)-stimulated release (by 25%) (WETZEL and BROWN 1985). In order to explain these differences ALBERTS et al. (1982) have proposed that autoinhibition of ACh release operates mainly through receptors which reside on the axonal varicosities and whose activation restricts impulse propagation into the distal parts of the terminal fibres. Accordingly, muscarinic agonists and antagonists modulate the release evoked by orthodromic activation of the cholinergic nerve to a greater extent than the release induced by direct depolarization of the nerve terminals. It cannot, however, be excluded that the transmitter released electrically and by high concentrations of K^+ originates from different neuronal compartments, and that drugs such as oxotremorine which are likely to reduce the Ca^{2+} entry into the nerve terminal differ in their sensitivity to both modes of release.

2. A muscarinic modulation of ACh release has been demonstrated in suspensions of synaptosomes obtained from guinea-pig myenteric plexus preparations (BRIGGS and COOPER 1982). Oxotremorine decreased the release of ACh evoked by the nicotinic agonist DMPP. Atropine antagonized the action of oxotremorine without having a significant effect of its own. The latter finding suggests that endogenous ACh does not appear to reach a concentration high enough to activate the synaptosomal mAChR.

3. Autoradiographic studies on cultured myenteric neurons revealed that mAChRs occurred on cell bodies and along the length of myenteric neurites. Both varicosities and intervaricose regions of the neurites were labelled. mAChRs were not detected on fibroblasts or on glial cells (BUCKLEY and BURNSTOCK 1984; BUCKLEY 1985).

4. Further evidence for a prejunctional location of inhibitory mAChRs comes from experiments in which the muscle contraction of the guinea-pig ileum was taken as a measure of ACh released from postganglionic fibres towards the smooth muscle. The twitch response of the longitudinal muscle to stimulation at 0.1 Hz was diminished by exogenous ACh and by oxotremorine (FOSBRAEY and JOHNSON 1980a; KILBINGER and WESSLER 1980b). This inhibition was partly overcome by scopolamine and was assumed to result from stimulation of prejunctional receptors on postganglionic neurons.

3. Mechanism of Inhibition

It has been proposed that the stimulation of prejunctional mAChRs decreases the excitability of cholinergic neurons and thus prevents the invasion of impulses to the terminal varicosities (ALBERTS et al. 1982). This hypothesis is based on the finding that the electrically evoked release of ACh which depends on the propagation of action potentials is more efficiently modulated by muscarinic agonists and antagonists than the release evoked by high K^+.

Another possibility, however, is that muscarinic agonists inhibit ACh release by reducing the Ca^{2+} entry into the nerve terminal or by reducing the intracellular utilization of Ca^{2+} for the depolarization-secretion coupling. In the myenteric plexus oxotremorine inhibits only that fraction of the resting release of ACh that is Ca^{2+}-dependent (KILBINGER and WAGNER 1975). A kinetic analysis of the electrically evoked ACh release from myenteric neurons revealed that oxotremorine or a low concentration of Ca^{2+} reduced the rate of release but not the size of the pool from which release originated (SZERB 1980). This suggests that muscarinic stimulation does not reduce the number of varicosities invaded by action potentials but decreases the fraction of ACh released by each action potential reaching the varicosities. Furthermore, muscarinic agonists and antagonists become less effective in inhibiting or in enhancing, respectively, the electrically evoked ACh release as the extracellular Ca^{2+} concentration is elevated (ALBERTS and STJÄRNE 1982; WESSLER et al. 1987). All this is consistent with a mechanism in which muscarinic agonists are assumed to decrease the availability of Ca^{2+} for electrosecretory coupling in cholinergic nerves.

4. Comparison with Postjunctional mAChRs

Compounds which behave as selective postjunctional agonists or prejunctional antagonists would be expected to enhance muscarinic cholinergic transmission and might, therefore, be of considerable therapeutic interest. However, attempts to discriminate subgroups of muscarinic drugs with selective pre- or postjunctional actions have yielded negative results so far. In studies on the myenteric plexus-longitudinal muscle preparation of the guinea pig the effects of various agonists on smooth muscle contraction and on the electrically evoked ACh release have been compared, and it was found that the potencies of the agonists in inhibiting release and in contracting smooth muscle were similar (KILBINGER and WESSLER 1980b; KILBINGER and KRUEL 1981; SCHWÖRER et al. 1985). Recently, the oxotremorine analogue BM-5 (RESUL et al. 1982) has been claimed to act as an agonist on postjunctional and as an antagonist at prejunctional mAChRs (NORDSTRÖM et al. 1983). It was found that BM-5 contracted the smooth muscle of the myenteric plexus longitudinal muscle preparation and, in the presence of 10 μM eserine, increased the electrically evoked release of ACh. In the absence of a ChE inhibitor, however, BM-5 reduced the evoked release via stimulation of prejunctional mAChRs (KILBINGER 1985). In comparison with oxotremorine BM-5 was a partial agonist on both pre- and postjunctional receptors. It is known that partial agonists may behave as antagonists in the presence of a large concentration of a full agonist. Thus, the alleged selective prejunctional antagonistic action of BM-5 is apparent only under rather unphysiological conditions, namely when the biophase concentration of ACh is increased by a ChE inhibitor.

Differences between pre- and postjunctional mAChRs have also not been detected if antagonists have been used. The affinity constants (pA_2 values) for pre- and postjunctional effects of seven antagonists in the myenteric plexus preparation did not differ significantly (KILBINGER et al. 1984; FUDER et al. 1985). mAChRs have tentatively been classified into M1 and M2 receptors according to their affinity for pirenzepine (HAMMER and GIACHETTI 1982). Pre- and postjunctional mAChRs in the guinea-pig ileum have the same low affinity to pirenzepine and can thus be regarded as M2 receptors.

In conclusion, the experiments with agonists and antagonists in the guinea-pig ileum provide no evidence for a difference in the pharmacological properties of pre- and postjunctional mAChRs. In this context it might be pertinent to mention that pre- and postjunctional mAChRs probably differ in the degree of desensitization. A diminished contractile response of the guinea-pig ileum subsequent to the initial action of a muscarinic agonist is a well-established phenomenon (for literature, see HURWITZ and MCGUFFEE 1979). In contrast, the prejunctional mAChR is not readily desensitized since the inhibitory actions of exogenous ACh or oxotremorine were similar after superfusion of the tissue for 21 min or for 30–40 min (KILBINGER and WESSLER 1980a, b).

5. Subsensitivity

Chronic treatment with ChE inhibitors results in changes in receptor concentrations at prejunctional as well as postjunctional mAChRs. In rats and mice treated for 4–10 days with a ChE inhibitor the affinity of postjunctional ileal mAChRs for oxotremorine, carbachol or atropine was reduced (FOLEY and MCPHILLIPS 1973; EHLERT et al. 1980) and, in addition, the concentration of [^{3}H]3-quinuclidinylbenzylate ([^{3}H]3-QNB) binding sites was decreased (EHLERT et al. 1980; COSTA et al. 1982). A recent study on the rat myenteric plexus-longitudinal muscle preparation has shown that the subsensitivity of postjunctional receptors develops in parallel with subsensitivity to prejunctional mAChRs (RUSSELL et al. 1985). After 4–10 days of treatment with diisopropylphosphorofluoridate (DFP) the electrically evoked release of ACh was significantly greater than the release from myenteric plexus preparations of rats that had received acute injections of DFP. AChE activity was reduced to the same degree under both DFP conditions. These results are consistent with chronic depression of ChE activity leading to down-regulation of prejunctional mAChRs which, in turn, results in an elevated release of ACh.

6. Physiological Role

The finding that muscarinic antagonists increase in the absence of ChE inhibition the evoked release of ACh from guinea-pig myenteric neurons (KILBINGER and WESSLER 1980a; ALBERTS et al. 1982), and from fowl and rat heart (KILBINGER et al. 1981; WETZEL and BROWN 1985) indicates that autoreceptors at cholinergic nerve endings are a target for the endogenously released transmitter. Stimulation of prejunctional mAChRs at myenteric neurons is partly responsible for the decrease in ACh release per impulse with increasing train length. The decline in volley output during high-frequency bursts (e.g. 3 or 6 impulses at 10 Hz) is antagonized by scopolamine which suggests that the release of ACh is controlled by the transmitter liberated by the preceding pulses (KILBINGER and WESSLER 1983).

II. Soma-Dendritic Autoreceptors

Muscarinic agonists cause a TTX-sensitive increase in spontaneous ACh release from guinea-pig myenteric neurons (KILBINGER and NAFZIGER 1985). This effect is probably due to the activation of soma-dendritic receptors. Electrophysiological experiments have shown that myenteric ganglia are endowed with mAChRs whose activation causes a slow depolarization of the neuron (MORITA et al. 1982; NORTH et al. 1985). The effects of muscarinic agonists on both spontaneous release of ACh and membrane depolarization are competitively antagonized by scopolamine or by pirenzepine. Pirenzepine is more potent as an antagonist at these soma-dendritic receptors (pA_2 8.4–8.5; KILBINGER and NAFZIGER 1985; NORTH et al. 1985) than at pre- or postjunctional receptors (pA_2 6.7–6.9; KILBINGER et al. 1984). Hence, the soma-dendritic receptor corresponds to the M1 receptor subtype. There is evidence that synaptically released ACh mediates a slow excitatory postsynaptic potential in the myenteric plexus through activation of M1 receptors (NORTH et al. 1985). A physiological role for the soma-dendritic mAChRs is indicated by the finding that muscarinic agonists cause a reduction in the membrane potassium conductance (MORITA et al. 1982), which could lead to a long-lasting increase in synaptic excitability.

Table 1. Classification of mAChRs by the affinity constants (pA_2 values) of competitive antagonists

Subtype	Tissue	Response	Scopolamine	Pirenzepine	HHSD	4-DAMP	Himbacine	AF-DX116
M1	Sympathetic ganglia	Depolarisation	–	8.4[d]	7.3[g]	8.6[d]	–	–
	Myenteric ganglia	Depolarisation	–	8.4[e]	–	–	–	–
		Increase in spontaneous ACh release	9.5[a]	8.5[a]	–	–	–	–
M2	Ileum	Contraction	9.4[b]	6.7[b]	8.2[h]	8.6[b]	7.3[j]	6.5[k]
	Ileum	Inhibition of evoked ACh release	9.1[b]	6.9[b]	8.2[h]	8.5[b]	–	6.7[k]
M2	Atrium	Negative ino-tropic or chronotropic	9.1[c]	6.7[f]	6.3[h]	7.6[i]	8.2[j]	7.3[l]
	Heart	Inhibition of evoked NA release	9.1[c]	6.6[f]	6.9[h]	–	–	–

HHSD, hexahydrosiladifenidol; 4-DAMP, 4-diphenylacetoxy-*N*-methylpiperidine; NA, noradrenaline.

[a] Kilbinger and Nafziger (1985)
[b] Kilbinger et al. (1984)
[c] Fuder et al. (1981)
[d] Brown et al. (1980)
[e] North et al. (1985)
[f] Fuder et al. (1982)
[g] Lambrecht et al. (1987)
[h] Fuder et al. (1985)
[i] Barlow et al. (1976)
[j] Gilani and Cobbin (1986)
[k] Kilbinger (1987)
[l] Giachetti et al. (1986)

D. Effects of ACh on Effector Organs: Mediation Through Subtypes of mAChRs

An important and continuing question is whether the effects of ACh in different tissues are mediated by subtypes of mAChRs. The existence of more than one class of mAChR has been proposed in the past by various workers (see review in BIRDSALL and HULME 1985) but strong evidence for such a heterogeneity has been provided only recently with the development of new antagonists. Currently, a classification is widely accepted according to which two classes of mAChRs are distinguished which exhibit high affinity (M1) and low affinity (M2) towards the antagonist pirenzepine (HAMMER and GIACHETTI 1982). Receptors with high affinity for pirenzepine have been detected not only in binding studies on central nervous system or sympathetic ganglia (see BIRDSALL and HULME 1985) but also in functional experiments on sympathetic ganglia (BROWN et al. 1980; ASHE and YAROSH 1984; NEWBERRY et al. 1985) and on myenteric ganglia (KILBINGER and NAFZIGER 1985; NORTH et al. 1985). The subdivision of mAChRs into M1 and M2 subtypes has however been criticized on the grounds that this classification is totally dependent upon one compound, pirenzepine (EGLEN and WHITING 1985). Indeed, the mAChRs on the effector cells of ileum, heart or exocrine glands, and on the terminals of peripheral autonomic cholinergic or adrenergic fibres all exhibit the same low affinity for pirenzepine and are thus not distinguishable with this drug (FUDER et al. 1982; SZELENYI 1982; KILBINGER et al. 1984). Yet, cardiac and ileal mAChRs can be discriminated by several other compounds. The competitive antagonists hexahydrosiladifenidol (MUTSCHLER and LAMBRECHT 1984) and 4-diphenylacetoxy-*N*-methylpiperidine (BARLOW et al. 1976) have a higher affinity for the receptors of the ileum than to those of the heart. On the other hand, himbacine (GILANI and COBBIN 1986) and the pirenzepine analogue AF-DX 116 (GIACHETTI et al. 1986) are more selective for cardiac than for ileal receptors (Table 1). Likewise, some neuromuscular blocking agents (notably gallamine, pancuronium and stercuronium) exhibit a higher affinity for cardiac mAChRs than for those in the ileum (for review see MITCHELSON 1984). Since these compounds did not produce a linear Schild-plot when tested on atrial muscle it was suggested that they interact allosterically rather than competitively with the binding site for muscarinic agonists.

Taken together, the weight of evidence suggests that at least three types of mAChRs can be distinguished in peripheral ganglia, ileum and heart. The properties of these types are summarized in Table 1.

References

Aas P, Fonnum F (1986) Presynaptic inhibition of acetylcholine release. Acta Physiol Scand 127:335–342

Ágoston DV, Ballmann M, Conlon JM, Dowe GHC, Whittaker VP (1985) Isolation of neuropeptide-containing vesicles from the guinea-pig ileum. J Neurochem 45:398–406

Alberts P, Stjärne L (1982) Role of calcium in muscarinic autoinhibition of [^{3}H]acetylcholine secretion in guinea-pig ileum myenteric plexus. Acta Physiol Scand 115:487–491

Alberts P, Bartfai T, Stjärne L (1982) The effects of atropine on [3H]acetylcholine secretion from myenteric plexus evoked electrically or by high potassium. J Physiol (Lond) 329:93–112

Ashe JH, Yarosh CA (1984) Differential and selective antagonism of the slow-inhibitory postsynaptic potential and slow-excitatory postsynaptic potential by gallamine and pirenzepine in the superior cervical ganglion of the rabbit. Neuropharmacology 11:1321–1329

Barlow RB, Berry KJ, Glenton PAM, Nikolaou NM, Soh KS (1976) A comparison of affinity constants for muscarine-sensitive acetylcholine receptors in guinea-pig atrial pacemaker cells at 29 °C and in ileum at 29 °C and 37 °C. Br J Pharamcol 58:613–520

Bartfai T (1985) Presynaptic aspects of the coexistence of classical neurotransmitters and peptides. Trends Pharmacol Sci 6:331–334

Birdsall NJM, Hulme EC (1985) Multiple muscarinic receptors: further problems in receptor classification. In: Kalsner S (ed) Trends in autonomic pharmacology, vol 3. Taylor and Francis, London, pp 17–34

Blaber LC, Freyer AD, Maclagan J (1985) Neuronal muscarinic receptors attenuate vagally-induced contraction of feline bronchial smooth muscle. Br J Pharmacol 86:723–728

Briggs CA, Cooper JR (1982) Cholinergic modulation of the release of [3H]acetylcholine from synaptosomes of the myenteric plexus. J Neurochem 38:501–508

Brown DA, Forward A, Marsh S (1980) Antagonist discrimination between ganglionic and ileal muscarinic receptors. Br J Pharmacol 71:362–364

Buckley NJ (1985) Autoradiographic localization of muscarinic receptors in the gut. In: Lux G, Daniel EE (eds) Muscarinic receptor subtypes in the GI tract. Springer, Berlin Heidelberg New York, pp 1–13

Buckley NJ, Burnstock G (1984) Distribution of muscarinic receptors on cultured myenteric neurons. Brain Res 310:133–137

Burnstock G (1985) Nervous control of smooth muscle by transmitters, cotransmitters and modulators. Experientia 41:869–874

Collier B, Kwok YN (1982) Superior cervical ganglion: Chemical considerations. In: Hanin I, Goldberg AM (eds) Progress in cholinergic biology: Model cholinergic synapses. Raven, New York, pp 169–190

Costa LG, Schwab BW, Murphy SD (1982) Muscarinic receptor alterations following neostigmine treatments. Eur J Pharmacol 80:275–278

D'Agostino G, Kilbinger H, Chiari MC, Grana E (1986) Presynaptic inhibitory muscarinic receptors modulating [3H]acetylcholine release in the rat urinary bladder. J Pharmacol Exp Ther 239:522–528

Dieterich HA, Kaffei H, Kilbinger H, Löffelholz K (1976) The effects of physostigmine on cholinesterase activity, storage and release of acetylcholine in the isolated chicken heart. J Pharmacol Exp Ther 199:236–246

Dzieniszewski P, Kilbinger H (1978) Muscarinic modulation of acetylcholine release evoked by dimethylphenylpiperazinium and high potassium from guinea-pig myenteric plexus. Eur J Pharmacol 50:385–391

Edwards AV, Bloom SR (1982) Recent physiological studies of the alimentary autonomic innervation. Scand J Gastroenterol [Suppl 71] 17:77–89

Eglen RM, Whiting RL (1985) Muscarinic receptors subtypes: problems of classification. Trends Pharmacol Sci 6:357–358

Ehlert FJ, Kokka N, Fairhurst AS (1980) Muscarinic receptor subsensitivity in the longitudinal muscle of the rat ileum following chronic anticholinesterase treatment with diisopropylfluorophosphate. Biochem Pharmacol 29:1391–1397

Foley DJ, McPhillips JJ (1973) Response of the rat ileum, uterus and vas deferens to carbachol and acetylcholine following repeated daily administration of a cholinesterase inhibitor. Br J Pharmacol 48:418–425

Fosbraey P, Johnson ES (1980a) Release-modulating acetylcholine receptors on cholinergic neurones of the guinea-pig ileum. Br J Pharmacol 68:289–300

Fosbraey P, Johnson ES (1980b) Modulation by acetylcholine of the electrically-evoked release of 3H-acetylcholine from the ileum of the guinea-pig. Br J Pharmacol 69:145–149

Fox JET, Daniel EE, Jury J, Robotham H (1985) Muscarinic inhibition of canine small intestinal motility *in vivo*. Am J Physiol 248:G526–G531

Fryer AD, Maclagan J (1984) Muscarinic inhibitory receptors in pulmonary parasympathetic nerves in the guinea-pig. Br J Pharmacol 83:973–978

Fuder H, Meiser C, Wormstall H, Muscholl E (1981) The effects of several muscarinic antagonists on pre- and postsynaptic receptors in the isolated heart. Naunyn Schmiedebergs Arch Pharmacol 316:31–37

Fuder H, Rink D, Muscholl E (1982) Sympathetic nerve stimulation on the perfused rat heart. Affinities of *N*-methylatropine and pirenzepine at pre- and postsynaptic muscarine receptors. Naunyn Schmiedebergs Arch Pharmacol 318:210–219

Fuder H, Kilbinger H, Müller H (1985) Organ selectivity of hexahydrosiladifenidol in blocking pre- and postjunctional muscarinic receptors studied in guinea-pig ileum and rat heart. Eur J Pharmacol 113:125–127

Gabella G (1972) Fine structure of the myenteric plexus in the guinea-pig ileum. J Anat 111:69–97

Gabella G (1981) Structure of smooth muscles. In: Bülbring E, Brading AF, Jones AW, Tomita T (eds) Smooth muscle. Arnold, London, pp 1–46

Giachetti A, Micheletti R, Montaga E (1986) Cardioselective profile of AF-DX 116, muscarine M_2 receptor antagonist. Life Sci 38:1663–1672

Gilani SAH, Cobbin LB (1986) The cardio-selectivity of himbacine: a muscarine receptor antagonist. Naunyn Schmiedebergs Arch Pharmacol 332:16–20

Gustafsson L, Hedqvist P, Lundgren G (1980) Pre- and postjunctional effects of prostaglandin E_2, prostaglandin synthetase inhibitors and atropine on cholinergic neurotransmission in guinea-pig ileum and bovine iris. Acta Physiol Scand 110:401–411

Hammer R, Giachetti A (1982) Muscarinic receptor subtypes: M1 and M2. Biochemical and functional characterization. Life Sci 31:2991–2998

Hurwitz L, McGuffee LJ (1979) Desensitization phenomena in smooth muscle. In: Kalsner S (ed) Trends in autonomic pharmacology, vol 1. Urban and Schwarzenberg, München, pp 251–264

Kharkevich DA (ed) (1980) Pharmacology of ganglionic transmission. Springer, Berlin Heidelberg New York (Handbook of experimental pharmacology, vol 53)

Kilbinger H (1977) Modulation by oxotremorine and atropine of acetylcholine release evoked by electrical stimulation of the myenteric plexus of the guinea-pig ileum. Naunyn Schmiedebergs Arch Pharmacol 300:145–151

Kilbinger H (1982) The myenteric plexus – longitudinal muscle preparation. In: Hanin I, Goldberg A (eds) Progress in cholinergic biology: Model cholinergic synapses. Raven, New York, pp 137–167

Kilbinger H (1984) Presynaptic muscarine receptors modulating acetylcholine release. Trends Pharmacol Sci 5:103–105

Kilbinger H (1985) Subtypes of muscarinic receptors modulating acetylcholine release from myenteric nerves. In: Lux G, Daniel EE (eds) Muscarinic receptor subtypes in the GI tract. Springer, Berlin Heidelberg New York, pp 37–42

Kilbinger H (1987) Control of acetylcholine release by muscarinic autoreceptors. In: Cohen S, Sokolovsky M (eds) Muscarinic cholinergic mechanisms. Freund, Tel Aviv, pp 219–228

Kilbinger H, Kruel R (1981) Choline inhibits acetylcholine release via presynaptic muscarine receptors. Naunyn Schmiedebergs Arch Pharmacol 316:131–134

Kilbinger H, Nafziger M (1985) Two types of neuronal muscarine receptors modulating acetylcholine release from guinea-pig myenteric plexus. Naunyn Schmiedebergs Arch Pharmacol 328:304–309

Kilbinger H, Wagner P (1975) Inhibition by oxotremorine of acetlycholine resting release from guinea-pig ileum longitudinal muscle strips. Naunyn Schmiedebergs Arch Pharmacol 287:47–60

Kilbinger H, Wagner B (1979) The role of presynaptic muscarine receptors in regulating acetylcholine release from peripheral cholinergic neurones. In: Langer SZ, Starke K, Dubocovich ML (eds) Presynaptic receptors. Pergamon, Oxford, pp 347–351

Kilbinger H, Wessler I (1980a) Inhibition by acetylcholine of the stimulation-evoked release of ^{3}H-acetylcholine from the guinea-pig myenteric plexus. Neuroscience 5:1331–1340
Kilbinger H, Wessler I (1980b) Pre- and postsynaptic effects of muscarinic agonists in the guinea-pig ileum. Naunyn Schmiedebergs Arch Pharmacol 314:259–266
Kilbinger H, Wessler I (1983) The variation of acetylcholine release from myenteric neurones with stimulation frequency and train length. Role of presynaptic muscarine receptors. Naunyn Schmiedebergs Arch Pharmacol 324:130–133
Kilbinger H, Krieg C, Tiemann J, Wessler I (1981) The effects of muscarinic agonists and antagonists on acetylcholine release from peripheral cholinergic nerves in the absence and presence of a cholinesterase inhibitor. In: Vizi ES (ed) Modulation of neurochemical transmission. Pergamon, Oxford, pp 281–290 (Advances in pharmacological research and practice, vol 2)
Kilbinger H, Halim S, Lambrecht G, Weiler W, Wessler I (1984) Comparison of affinities of muscarinic antagonists to pre- and postjunctional receptors in the guinea-pig ileum. Eur J Pharmacol 103:313–320
Lambrecht G, Mutschler E, Moser U, Riotte J, Wagner M, Wess J, Gmelin G, Tacke R, Zilch H (1987) Heterogeneity in muscarinic receptors: evidence from pharmacological and electrophysiological studies with selective antagonists. In: Cohen S, Sokolovsky M (eds) Muscarinic cholinergic mechanisms. Freund, Tel Aviv, pp 245–253
Lindmar R, Löffelholz K, Weide W, Witzke J (1980) Neuronal uptake of choline following release of acetylcholine in the perfused heart. J Pharmacol Exp Ther 215:710–715
Löffelholz K (1981) Release of acetylcholine in the isolated heart. Am J Physiol 240:H431–H440
Löffelholz K, Pappano AJ (1985) The parasympathetic neuroeffector junction of the heart. Pharmacol Rev 37:1–24
Löffelholz K, Brehm R, Lindmar R (1984) Hydrolysis, synthesis and release of acetylcholine in the isolated heart. Fed Proc 43:2603–2606
Lundberg JM (1981) Evidence for coexistence of vasoactive intestinal polypeptide and ACh in neurons of cat exocrine glands. Acta Physiol Scand [Suppl] 496:1–57
Marquis JK, Fishman EB (1985) Presynaptic acetylcholinesterase. Trends Pharmacol Sci 6:387–388
McAfee DA (1982) Superior cervical ganglion: physiological considerations. In: Hanin I, Goldberg AM (eds) Progress in cholinergic biology: model cholinergic synapses. Raven, New York, pp 191–211
Mitchelson F (1984) Heterogeneity in muscarinic receptors: evidence from pharmacologic studies with antagonists. Trends Pharmacol Sci [Suppl] 5:12–16
Morita K, North RA, Tokimasa T (1982) Muscarinic agonists inactivate potassium conductance of guinea-pig myenteric neurones. J Physiol (Lond) 333:125–139
Muscholl E, Muth A (1982) The effect of physostigmine on the vagally induced muscarinic inhibition of noradrenaline release from the isolated perfused rabbit atria. Naunyn Schmiedebergs Arch Pharmacol 320:160–169
Mutschler E, Lambrecht G (1984) Selective muscarinic agonists and antagonists in functional tests. Trends Pharmacol Sci [Suppl] 5:39–44
Newberry NR, Priestly T, Woodruff GN (1985) Pharmacological distinction between two muscarinic responses on the isolated superior cervical ganglion of the rat. Eur J Pharmacol 116:191–192
Nordström Ö, Alberts P, Westlind A, Unden A, Bartfai T (1983) Presynaptic antagonist – postsynaptic agonist at muscarinic cholinergic synapses. Mol Pharmacol 24:1–5
North A (1982) Electrophysiology of the enteric nervous system. Neuroscience 7:315–325
North A, Slack BE, Surprenant A (1985) Muscarinic M1 and M2 receptors mediate depolarization and presynaptic inhibition in guinea-pig enteric nervous system. J Physiol (Lond) 368:435–452
O'Donohue TL, Millington WR, Handelmann GE, Contreras PC, Chronwall BM (1985) On the 50th anniversary of Dale's law: multiple neurotransmitter neurons. Trends Pharmacol Sci 6:305–308

Paton WDM, Vizi ES, Zar MA (1971) The mechanism of acetylcholine release from parasympathetic nerves. J Physiol (Lond) 215:819–848
Resul B, Dahlbom R, Ringdahl B, Jenden DJ (1982) *N*-alkyl-*N*-(4-tert-amino-1-methyl-2-butynyl)carboxamides, a new class of potent oxotremorine antagonists. Eur J Med Chem 17:317–322
Richardson JS, Mattio TG, Giacobini E (1984) Amitryptiline and imipramine inhibit the release of acetylcholine from parasympathetic nerve terminals in the rat iris. Can J Physiol Pharmacol 62:857–859
Russell RW, Booth RA, Jenden DJ, Roch M, Rice KM (1985) Changes in presynaptic release of acetylcholine during development of tolerance to the anticholinesterase, DFP. J Neurochem 45:293–299
Sawynok J, Jhamandas K (1977) Muscarinic feedback inhibition of acetylcholine release from the myenteric plexus in the guinea-pig ileum and its status after chronic exposure to morphine. Can J Physiol Pharmacol 55:909–916
Schwörer H, Lambrecht G, Mutschler E, Kilbinger H (1985) The effects of racemic bethanechol and its (R)- and (S)-enantiomers on pre- and postjunctional muscarine receptors in the guinea-pig ileum. Naunyn Schmiedebergs Arch Pharmacol 331:307–310
Szelenyi I (1982) Does pirenzepine distinguish between "subtypes" of muscarinic receptors? Br J Pharmacol 77:567–569
Szerb J (1976) Storage and release of labelled acetylcholine in the myenteric plexus of the guinea-pig ileum. Can J Physiol Pharmacol 54:12–22
Szerb J (1980) Effect of low calcium and of oxotremorine on the kinetics of the evoked release of ^{3}H acetylcholine from the guinea-pig myenteric plexus; comparison with morphine. Naunyn Schmiedebergs Arch Pharmacol 311:119–127
Szerb JC (1982) Correlation between acetylcholine release and neuronal activity in the guinea-pig ileum myenteric plexus; effect of morphine. Neuroscience 7:327–340
Vizi ES (1984) Non-synaptic interactions between neurons: modulation of neurochemical transmission. Wiley, Chichester
Vizi ES, Ono K, Adam-Vizi V, Duncalf D, Földes FF (1984) Presynaptic inhibitory effect of met-enkephalin on [^{14}C]acetylcholine release from the myenteric plexus and its interaction with muscarinic negative feedback inhibition. J Pharmacol Exp Ther 230:493–499
Wessler I, Eschenbruch V, Halim S, Kilbinger H (1987) Presynaptic effects of scopolamine, oxotremorine, noradrenaline and morphine on [^{3}H]acetylcholine release from the myenteric plexus at different stimulation frequencies and calcium concentrations. Naunyn Schmiedebergs Arch Pharmacol 335:597–604
Wetzel GT, Brown JH (1985) Presynaptic modulation of acetylcholine release from cardiac parasympathetic neurons. Am J Physiol 248:H33–H39

Part VI Central Cholinergic Systems

CHAPTER 21

Central Cholinergic Pathways: The Biochemical Evidence

F. FONNUM

A. Introduction

The biochemical approach to identifying cholinergic pathways in the brain is based on the determination of cholinergic markers in well-defined anatomical structures. With highly sensitive biochemical methods it is possible to determine the cholinergic markers in small regions, even single cells, and thereby provide an accurate quantitative description of the cholinergic system. As will be seen in this chapter the determination of choline acetyltransferase (ChAT) in combination with chemical or surgical lesions has been of particular importance in gaining such information. The strength of the histochemical approach, particularly the immunocytochemistry of ChAT, has not been in its ability to discover new cholinergic pathways, but to identify and characterize cholinergic cells or cell groups (Chap. 22). The application of immunohistochemical methods at the electron microscopic level may in the future reveal new and exciting details of the central cholinergic nerve terminals.

B. Biochemical Markers

In principle several cholinergic markers can be selected for the identification of cholinergic pathways. ChAT is most often selected because it is specifically localized in cholinergic structures and sensitive assays are available (FONNUM 1969a, 1975; KATO 1984). Acetylcholinesterase (AChE) has also been used for this purpose for many years because of the availability of many biochemical and histochemical methods (SHUTE and LEWIS 1961). It is, however, also concentrated in noncholinergic cells such as the dopaminergic cells in the substantia nigra and the ventral tegmental area or the non-cholinergic structures in the cerebellum. ACh itself can be determined (Chap. 10) by sensitive bioassay methods, by gaschromatography-mass spectrometry (FREEMAN et al. 1975), radiochemically (GOLDBERG and MCCAMAN 1973), or enzymically, by hydrolysing the ACh by AChE and oxidising the choline so generated by choline oxidase. The hydrogen peroxide formed as a result is assayed by bioluminescence (ISRAËL and LESBATS 1981) or electrochemically. In the latter method ACh is first separated from any choline present in the sample by high performance liquid chromatography and the enzymes are immobilized on a solid phase (EVA et al. 1984; DAMSMA et al. 1985; STADLER and NESSELHUT 1986).

The tissues for ACh assays are often prepared by microwave irradiation which precludes good dissection. Sodium-dependent, high-affinity choline uptake

(HACU) (Chap. 14) is specific for cholinergic terminals and it has even been claimed to reflect ACh turnover (KUHAR and MURRIN 1978). It requires fresh tissue and more material than assays for the other markers. An alternative to choline uptake measurement is hemicholinium-binding which can also be determined autoradiographically (RAINBOW et al. 1984; VICKROY et al. 1984, 1985; MANAKER et al. 1986). HACU forms the basis of a method utilizing the uptake of choline in the terminal region and its retrograde transport to cholinergic cell bodies (BAGNOLI et al. 1981; DEMÊMES et al. 1983; VILLANI et al. 1983). Very often one sees the combination of a biochemical and histochemical approach; this is a powerful combination which makes use of both the quantitative advantages of biochemical techniques and the precise localization of the histochemical method.

C. Preparation of Tissue

I. Single Cell Preparations

In invertebrate ganglia the nerve cell bodies are large (up to 200 μm), devoid of somatic terminals and can be easily isolated. Some cells are easily recognized; the same cell can be isolated from several animals, using fine needles to tease it free. Cells can be washed free of contaminants in iso-osmotic sucrose and the ChAT or AChE activity determined by microassay. The localization of cholinergic cell bodies has been described in the nervous system of *Aplysia* (GILLER and SCHWARTZ 1971; McCAMAN and DEWHURST 1970), *Helix* (EMSON and FONNUM 1974), and locust (EMSON et al. 1974). The ChAT activity in *Helix aspersa* ranges from 20 to 100 pmol h^{-1} per cell.

By the use of an extremely sensitive method for ChAT based on coenzyme A (CoA) cycling KATO (1984) and KATO and MURASHIMA (1985) have been able to

Table 1. ChAT activity of single anterior horn cells of the spinal cord from different species. From KATO and MURASHIMA (1985)

Species	Sex	Age (years)	ChAT activity [ACh synthesized in mmol kg^{-1} dry weight h^{-1}, mean ± SD] (No. of cells)
Mouse	–	–	155 ± 86 (33)
Cat	–	–	221 ± 36 (15)
Rabbit	–	–	152 ± 19 (21)
Rat	–	–	273 ± 37 (20)
Domestic fowl	–	–	161 ± 25 (16)
Bullfrog	–	–	36 ± 5.6 (15)
Yellow tail	–	–	21 ± 4.3 (16)
Human	Male	70	48.4 ± 46.5 (15)[a]
	Female	43	36.2 ± 30.6 (11)[a]
	Female	11	28.2 ± 18.5 (13)[a]

[a] Spinal cord

determine the ChAT activity in single spinal cord cells from a variety of species (Table 1).

II. Microdissection from Freeze-Dried Sections

The use of freeze-dried sections allows an accurate dissection of well-defined anatomical structures. The technique is ideal for studying the distribution of transmitter markers in layered structures such as the hippocampus, cerebral or cerebellar cortex, olfactory bulb and retina (GOLDBERG and MCCAMAN 1967; FONNUM 1970; ROSS and MCDOUGAL 1976). This sections (20–40 µm thick) are prepared at −15 °C in a cryostat and the sections freeze-dried for 12–15 h. Details of the different brain structures can easily be recognized and samples can be dissected out with an accuracy of less than 100 µm (LOWRY and PASSONEAN 1972). Samples weighing around 1 µg dry weight can be handled and the ChAT activity determined.

III. Micropunches from Frozen Sections

This preparation does not permit the same delicate dissection as freeze-dried sections. The method has, however, been shown to be very powerful for mapping out the cholinergic markers in larger structures such as the different brain nuclei (PALKOVITS 1973; BROWNSTEIN et al. 1976a; HOOVER et al. 1979).

Brain sections are prepared at −10 °C in a cryostat. The brain sections are most often 200–400 µm thick and samples are punched out from the frozen tissue in the cold with small bore punches of inner diameters of 300–1000 µm depending on the structure studied. In this technique the slice appears opaque, but certain landmarks, such as the optic nerve, can easily be recognized. The samples are homogenized and analysed by standard methods.

IV. Micropunches from Fresh Tissue

Brain slices (300–400 µm thick) can be prepared by a Vibratome or McIlwain tissue slicer. The best-preserved slices for dissection are prepared at 3°–4 °C. Brain structures are easily recognized with this technique (JACOBOWITZ 1974). In general fresh slices are more difficult to handle than frozen ones. The samples can also be used for subcellular fractionation on a microscale. The ratio between soluble and particulate ChAT in iso-osmotic sucrose homogenates should reflect the ratio between ChAT in cell bodies and terminals. The proportion of soluble ChAT by this technique was 81% from ventral roots, 58% from the ventral horn, 43% from nucleus ruber and 10%–30% from various brain regions (FONNUM 1969b; AQUILONIUS et al. 1981).

V. Lesions

Well-defined surgical lesions can be made by the stereotactic application of an electrical or mechanical tool, but often for biochemical purposes ablations or transections of fibre bundles are more suitable. The execution of a lesion is a compromise: it should be extensive enough to provide a clear-cut result, but selective enough to avoid collateral damage. Often the results are improved by making use of the point-to-point organization which occurs in many brain projections (FONNUM et al. 1974).

The introduction of kainic acid (COYLE and SCHWARTZ 1976; McGEER and McGEER 1976), which induces selective degeneration of certain cell bodies but spares traversing fibres and terminals, has been of great importance. It has been particularly useful in differentiating between intrinsic and extrinsic cholinergic input in the striatum (COYLE and SCHWARTZ 1976), tuberculum olfactorium (GORDON and KRIEGER 1983), and nucleus accumbens (WALAAS and FONNUM 1979b).

The mapping of the monoamine system has been helped considerably through the use of aminotoxic agents such as 6-hydroxydopamine and 5,7-dihydroxyindolamine. These agents are selectively transported by the high-affinity uptake system into the dopaminergic and serotonergic structures respectively. Destruction of the neuron occurs probably by subsequent oxidation. Ethylcholine mustard aziridium ion (AF64A) has been proposed as a selective presynaptic cholinergic neurotoxin (MANTIONE et al. 1981, 1983); it acts primarily by irreversible alkylation of the HACU system (MANTIONE et al. 1983), but liver choline dehydrogenase is also inhibited at low concentrations. Intraventricular or intracerebral injections have been reported selectively to reduce cholinergic markers in hippocampus or striatum (FISHER et al. 1982). Administration of the agent to the rat interpeduncular nucleus showed an initial selective decrease in HACU, but after 5 days other transmitter markers also decreased (CONTESTABILE et al. 1986a). Ultrastructural studies confirmed that non-cholinergic structures were also destroyed. In substantia nigra, too, AF64A affects noncholinergic structures (LEVY et al. 1984). At present the interpretation of results obtained with AF64A on the localization of cholinergic structures should be treated with caution.

D. Distribution of Cholinergic Markers in the Brain

The more important references to the distribution of cholinergic markers in different brain regions are listed in Table 2. The highest level of ChAT is found in nucleus interpeduncularis, caudate nucleus, putamen, olfactory tubercle, nucleus accumbens, nucleus lateralis and nucleus basalis posterior of amygdala and the cranial motor nerve nuclei. There are also high activities in the cholinergic cell groups in the magnocellular nucleus of substantia innominata and the nucleus of the diagonal band. Low levels are found in white matter and cerebellar cortex. ACh is distributed similarly to ChAT with a correlation coefficient of 0.78 (HOOVER et al. 1978). There has been some discussion of the functional consequences of different ACh to ChAT ratios (CHENEY et al. 1975); in general ACh shows a less marked regional distribution than ChAT and tends to be comparatively higher in areas with low ChAT activity. It has been argued that in cell bodies and axons the presumed low turnover of ACh would lead to a higher level of ACh and therefore a high ACh to ChAT ratio would indicate a high proportion

Table 2. Publications giving extensive data on the distribution of cholinergic markers in the central nervous system

Markers[a]	Species	Region	References
ACh, ChAT	Rat	Forebrain	CHENEY et al. (1975)
ACh, AChE, ChAT	Rat	Forebrain	HOOVER et al. (1978)
ACh	Rat	Tel-diencephalon	KOSLOW et al. (1974)
ChAT	Cat	Forebrain	NIEOULLON and DUSTICIER (1980)
ChAT	Rat	Diencephalon	BROWNSTEIN et al. (1976a)
ChAT	Rat	Basal ganglia	FONNUM et al. (1977)
ChAT	Rat	Brain stem	KOBAYASHI et al. (1978)
ChAT	Rat	Limbic nuclei	PALKOVITZ et al. (1974)
ChAT	Mouse	Brain	GORDON and FINCH (1985)
ChAT, AChE, HACU, mAChR	Rat	Medulla oblongata	SIMON et al. (1981)
ChAT, HACU, AChE, mAChR	Monkey	Whole brain	YAMAMURA et al. (1974)
mAChR	Dog	Brain	HILEY and BURGEN (1974)
mAChR	Rat	Forebrain	ROTTER et al. (1979b)
mAChR	Rat	Mid- and hindbrain	ROTTER et al. (1979a)
Hemicholinium binding	Rat	Forebrain	RAINBOW et al. (1984)
Hemicholinium binding	Rat	Forebrain	VICKROY et al. (1985)

[a] For abbreviations see text

of cell bodies and axons relative to terminals (CHENEY et al. 1975). Such conclusions cannot, however, be extrapolated through the extensive survey by HOOVER et al. (1978).

The distribution of muscarinic receptors (mAChRs) shows a very distinct pattern which does not, however, run parallel to that of ChAT (ROTTER et al. 1979a, b). High binding activity was found in the olfactory system, hippocampus, neocortex, amygdala and the striatum. But whereas the ratio between caudate-putamen and neocortex for ChAT is 10, it is 1.2 for mAChR.

HACU is localized to cholinergic structures in the brain since it is decreased after depletion of cholinergic nerve terminals (KUHAR et al. 1973); however, this marker has not been of importance in characterizing cholinergic pathways.

E. Localization of Cholinergic Pathways in the Brain

I. Olfactory Bulb

The olfactory bulb is composed of well-defined anatomical layers which are ideal for biochemical studies of microdissected samples. ChAT and AChE activities were localized predominantly in the glomerula and the external plexiform layers. An intermediate activity was found in the mitral cell layer, with low or no activity in the granular cell layer or in the olfactory nerve (GODFREY et al. 1980; JAFFÉ and CUELLO 1980; MACRIDES et al. 1981; WENK et al. 1976, 1977). mAChRs had a widespread distribution with a peak in the external plexiform layer (ROTTER et al. 1979b). It has long been suspected that the cholinergic terminals in the olfactory bulb were extrinsic and derived from the nucleus of the diagonal band, particularly the horizontal limb (SHUTE and LEWIS 1967). In agreement with this, the ChAT and AChE activities were drastically reduced after transection of the centrifugal input (ROSS et al. 1978; GODFREY et al. 1980) or after a lesion of the medial preoptic area, including the diagonal band (YOUNGS et al. 1979). Combined biochemical and neuroanatomical studies indicated that the cholinergic input to the olfactory bulb is derived from the medial septum-nucleus of the diagonal band complex (MACRIDES et al. 1981).

II. Cerebral Cortex

Neocortex contains intermediate levels of ACh and ChAT (HOOVER et al. 1978). In the different rat cortical layers ChAT is concentrated in layers 1, 3 and 4 (BEESLEY and EMSON 1975). In aged human brain the highest concentration of ChAT was found in layers 2 and 3 (PERRY et al. 1984). mAChR had a broad distribution but peaked in layers 3 and 6 (ROTTER et al. 1979a).

It was shown many years ago by undercutting the cortex that the main cholinergic activity in cortex was of extrinsic origin (HEBB et al. 1963). It is generally agreed that a large electrolyte lesion comprising the magnocellular region of nucleus basalis will reduce ChAT in neocortex. The magnocellular nuclei in rat comprise the nucleus of the diagonal band, the nucleus basalis and large cells in the medial and lateral preoptic nuclei, in globus pallidus and entopeduncular nucleus. There appears to be a topographical relationship between the cholinergic activity in the cortical areas and the various parts of the magnocellular complex. Following lesions of the posterior part, ChAT was markedly reduced in all regions but the reduction was most pronounced in the frontal and motor-sensory cortex. Lesions in the anterior part selectively affected auditory and visual cortex (WENK et al. 1980). Lesions of the medial system and nucleus of the diagonal band area reduced ChAT in the posterior cortex, mainly the entorhinal and occipital cortex (KUHAR et al. 1973); there were only small reductions after lesions avoiding nucleus basalis but affecting globus pallidus or lentiform and entopeduncular nucleus (KELLY and MOORE 1978; LEHMANN et al. 1980).

The small remaining activity in neocortex (15%) may be due to cholinergic interneurons which have been identified by immunohistochemistry (ECKENSTEIN and THOENEN 1983). In some of these intrinsic cholinergic cells ChAT is colocalized with the vasoactive intestinal polypeptide (Chap. 17). Although a small number of cholinergic interneurones has been identified in rodent brain, it is still uncertain whether they are present in human and other primate brains.

III. Hippocampal Region

This region comprises hippocampus proper, subiculum, presubiculum and area dentata. There seems to be an excellent correlation between topographical localization of AChE and ChAT activities in this region (FONNUM 1970; STORM-MATHISEN 1970). The enzyme activities in hippocampus proper were concentrated in two narrow layers on both sides of the pyramidal cells. The activities declined from the pyramidal cells towards alveus and radiatum, respectively. In area dentata the enzyme activities were also concentrated in two narrow layers around the granular cells. High activities were also found in the outer molecular layer and the hilus (FONNUM 1970; STORM-MATHISEN 1970). In the guinea pig the AChE activities in subiculum are concentrated in two layers and in presubiculum as a wedge (GENESER-JENSEN and BLAKSTAD 1971).

Subcellular fractionation indicated that around 75% of the ChAT was recovered in synaptosomes (FONNUM 1970). After transection of the fimbria or after an extensive lesion of the septum, there was an almost total loss of ChAT and AChE in the hippocampal region (LEWIS et al. 1967; STORM-MATHISEN 1972; KUHAR et al. 1973). Detailed lesion studies have revealed that most of the extrinsic cholinergic input is derived from cells in the medial septum and particularly from the nucleus of the diagonal band (MALTHE-SØRENSSEN et al. 1980; Chap. 9, this volume). A minor intrinsic cholinergic activity localized in the molecular layer of hippocampus may be derived from cells in the subiculum (MELLGREN and SREBRO 1973).

IV. Amygdaloid Complex

The highest ChAT activity was found in the lateral and basolateral nuclei (BEN ARI et al. 1977; PALKOVITS et al. 1974). From the finding that ChAT decreases in amygdala after kainic acid injections in the lateral preoptic area but not in the diagonal band, it was concluded that the cholinergic fibres were derived from the ventrolateral part of the magnocellular nuclei (EMSON et al. 1979). The study underlines the strict topographical organization of cells in the magnocellular complex.

V. Striatum

For the purpose of this review the striatum is regarded as comprising the caudate nucleus, putamen, globus pallidus and the ventral striatum which mainly consists of nucleus accumbens and olfactory tubercle (HEIMER 1978). ACh, ChAT and AChE are very high in these regions with the exception of globus pallidus (FONNUM et al. 1977; HOOVER et al. 1978). A very high ChAT activity was also localized to the small substriatal grey region (WALAAS and FONNUM 1979b). Half of the AChE, at least in rat caudato-putamen, was derived from the dopaminergic fibres (LEHMANN and FIBIGER 1978).

It is characteristic for some afferents to caudato-putamen that they terminate in a patchy manner (GOLDMAN and NAUTA 1977). GRAYBIEL and RAGSDALE (1978) have also shown a patchy distribution of AChE staining in several species particularly during development. This has lead to the idea that the transmitters in neostriatum are concentrated in patches called striasomes. Extensive lesions of all major afferents to neostriatum did not decrease ChAT (MCGEER et al. 1971). This finding was important because in all earlier work the cholinergic input, due to the AChE staining of the dopaminergic fibres, were assumed to be derived from the ventral tegmental pathway (SHUTE and LEWIS 1961). The loss of ChAT after intrastriatal injection of kainic acid provided the final evidence that the cholinergic fibres in neostriatum were intrinsic (COYLE and SCHWARTZ 1976; MCGEER and MCGEER 1976). This has been confirmed by ChAT immunohistochemistry (Chap. 22). A possible exception to this has been the claim by SIMKE and SAELENS (1977) and SAELENS et al. (1979) that the cholinergic activity in the most rostral tip of rat caudato-putamen was derived from cells in the parafascicular nucleus. It has also been claimed that a lesion in cat centromedian nucleus would reduce ChAT in neostriatum (HASSLER et al. 1980; KIM 1978). These suggestions have not been confirmed in other laboratories. It was opposed by FIBIGER (1982), who did not find cholinergic cell bodies in these thalamic nuclei by the pharmaco-histochemical method for AChE. It was also rejected following a carefully executed study based on the distribution of ChAT in different parts of the neostriatum after a lesion in this thalamic region (BARRINGTON-WARD et al. 1984).

A very high ChAT activity was found in nucleus accumbens, particularly in the medial part (FONNUM et al. 1977). The activity was unaffected by transection of major afferent fibres, but was reduced by a local kainic acid lesion (WALAAS and FONNUM 1979b); consequently, the cholinergic fibres in nucleus accumbens were assumed to be intrinsic. AChE activity was less affected than ChAT by a kainic acid lesion and is probably in part present in the dopaminergic terminals.

The ventral olfactory tubercle contained a higher ChAT activity than the dorsal part (FONNUM et al. 1977). More detailed dissection showed that the peak activity was localized in the deep part of the pyramidal cell and granular cell layers. This peak activity was sensitive to local kainic acid administration and therefore probably derived from intrinsic neurons (GORDON and KRIEGER 1983).

Globus pallidus contained an intermediate level of ChAT, whereas the ventral pallidal structures displayed higher activities. This was particularly true for the lateral and to a lesser extent the medial substantia innominata (WALAAS and FONNUM 1979a). In the monkey a high level of ChAT activity was found in samples dissected from the magnacellular region of nucleus basalis (MCKINNEY et al. 1982).

VI. Thalamus

The different thalamic nuclei differ widely in their ChAT activities. Particularly high activity was found in the anteroventral, posterior and pretectal nuclei (HOOVER et al. 1978). According to SHUTE and LEWIS (1967) the AChE staining of the thalamic nuclei was due to the dorsal tegmental pathway arising from cells in cuneiform nucleus. This finding has been extended by HOOVER and JACOBOWITZ (1979) who showed that ChAT was lost in the ipsilateral thalamic nuclei after a lesion comprising both cuneiform nucleus and the dorsal and ventral parabrachial nuclei. A decrease was also noted in the contralateral side, particularly in the lateral geniculate body.

VII. Habenulo-Interpeduncular System

The nucleus interpeduncularis contains the highest level of ChAT in the brain. The activity is concentrated in the central core and supplied by each fasciculus retroflexus (CONTESTABILE et al. 1986b); in the habenulae the medial part dominates (CONTESTABILE and FONNUM 1983). The origin of the cholinergic input to the habenulae and nucleus interpeduncularis has been a controversial issue. Bilateral transection of stria medullaris reduced ChAT in the habenulae and nucleus interpeduncularis by more than 50% (CONTESTABILE and FONNUM 1983; GOTTESFELD and JACOBOWITZ 1978, 1979). This activity has been thought to originate in the septum either from the nucleus of the diagonal band (GOTTESFELD and JACOBOWITZ 1978, 1979) or the nucleus triangularis and fimbrialis (CONTESTABILE and FONNUM 1983; FONNUM and CONTESTABILE 1984); the remaining cholinergic activity is believed to originate from cells in the medial habenulae (CONTESTABILE et al. 1986b; CONTESTABILE and FONNUM 1983; KUHAR et al. 1975; KATAOKA et al. 1973) and not from the lateral habenulae as suggested by CUELLO et al. (1978). The interpretation of the data from nucleus interpeduncularis has been hampered by the finding that the cholinergic neurons in the habenulae were resistant to kainic acid (CONTESTABILE and FONNUM 1983) and that the main part of the AChE activity in nucleus interpeduncularis was linked to non-cholinergic structures (FLUMERFELT and CONTESTABILE 1982).

VIII. Hypothalamus

In the hypothalamic region there is a high ChAT activity in the median eminence and arcuate nucleus (WALAAS and FONNUM 1978; HOOVER et al. 1978; BROWNSTEIN et al. 1976b). Very little seems to be known at present about the origin of the cholinergic activity in the hypothalamus. Kainic acid injection into the supraoptic nucleus reduced ChAT in this nucleus by 90%. It was concluded that the cholinergic fibres originated either in the vicinity of or within the nucleus itself (MEYER and BROWNSTEIN 1980). After surgical isolation of the medial basal hypothalamus, there was a partial fall of ChAT in nucleus arcuatus and the centrobasal nucleus, but not in the median eminence (BROWNSTEIN et al. 1976b). Parenteral treatment of glutamate to neonates resulted in a substantial reduction of both ChAT activity and AChE staining in the medial basal hypothalamus and in the median eminence (CARSON et al. 1977; WALAAS and FONNUM 1978). The results indicate the existence of cholinergic elements in the tuberoinfundibular tract.

IX. Cerebellum

The cerebellum is a region with high AChE and low ChAT activity. There are large species differences with regard to both the distribution of ChAT among the cortical regions and its topographical distribution (GOLDBERG and MCCAMAN 1967; KÁSA and SILVER 1969). In most species the most intense AChE staining and most of the ChAT activity were present in certain of the mossy fibres, the glomeruli and the cerebellar nuclei (KÁSA et al. 1965; KÁSA and SILVER 1969; GOLDBERG and MCCAMAN 1967; FONNUM 1972). In agreement with this, subcellular fractions enriched in mossy fibers or glomeruli contained a relatively high level of ACh (ISRAËL and WHITTAKER 1967). The cholinergic terminals to both the cortex and the nuclei were extrinsic to the cerebellum, since ChAT fell sharply after transection of the cerebellar peduncles (FONNUM 1972) or after surgical isolation of the vermis (KÁSA and SILVER 1969).

X. Brain Stem

Some 50 areas in the brain stem have been analysed for their ChAT activity. Very high levels were found in the motor nerve nuclei, particularly the hypoglossus and facialis nuclei. In the sensory nuclei the levels were low as they were in the lateral vestibular nucleus and the acoustic nuclei (KOBAYASHI et al. 1975). In the vestibular complex, the highest activity was found in the medial nucleus (BURKE and FAHN 1985). Transection of the motor nerves was accompanied by large losses of ChAT in the motor nuclei. A decrease of 70% has been shown after transections of the facial, hypoglossus and vagus nerves (GOTTESFELD and FONNUM 1977; HODES et al. 1983; HOOVER and HANCOCK 1985).

Lesion of the cerebellar nuclei in the cat was accompanied by a decrease of ChAT in the contralateral nucleus ruber and the ventrolateral and ventro-anterior thalamic nuclei (NIEOULLON and DUSTICIER 1981). This finding correlates with a similar decrease in AChE staining after such lesions (MARSHALL et al. 1980). Consistently with this, injection of [^{3}H]choline into the ventral thalamic nucleus labelled cell bodies in the deep cerebellar nuclei (STANTON and ORR 1985). The results may indicate that the cerebello-rubral and cerebello-thalamic pathways are, at least in part, cholinergic.

XI. Spinal Cord

Detailed maps exist of the topographical distribution of ChAT in the spinal cords of cat (KANAZAWA et al. 1979), ox and man (AQUILONIUS et al. 1981). High activity was found in all these species in the lateral part of the ventral horn. This probably corresponds to the ventral root region.

In ox and human a high activity was also located in the most apical part of the dorsal horn (AQUILONIUS et al. 1981). This was also found in the cat by GWYN et al. (1972), but not by KANAZAWA et al. (1979) or AQUILONIUS et al. (1981).

XII. Retina

ROSS and MCDOUGAL (1976) showed a peak of ChAT in the inner plexiform layer of eight different species including frog, goldfish and rabbit. The activity in the photoreceptor layers and in the outer plexiform layer was very low. They concluded that most of the ChAT activity was present in the amacrine cells. LUND-KARLSEN and FONNUM (1976) showed that parenteral administration of glutamate to rat neonates almost completely abolished the ChAT activity in the retina. These results would also agree with a localization of ChAT to amacrine cells, which is strongly supported by the electrophysiological evidence (Chap. 23).

References

Aquilonius SM, Eckernäs SÅ, Gulberg P-G (1981) Topographical localization of choline acetyltransferase within the human spinal cord and with some other species. Brain Res 211:329–340

Bagnoli PA, Beaudet M, Stella M, Cuénod M (1981) Selective retrograde labeling of cholinergic neurons with [^{3}H]choline. J Neurosci 1:691–695

Barrington-Ward SJ, Kilpatric IC, Phillipson OT, Pycock CJ (1984) Evidence that thalamic efferent neurones are non-cholinergic – a study in the rat with special reference to the thalamostriatal pathway. Brain Res 299:146–151

Beesley PY, Emson PC (1975) Distribution of transmitter related enzymes in the rat sensori-motor cortex. Biochem Soc Trans 3:936–939

Ben-Ari Y, Zigmond RE, Shute CCD, Lewis PR (1977) Regional distribution of choline acetyltransferase and acetylcholinesterase within the amygdaloid cortex and stria terminalis system. Brain Res 120:435–445

Brownstein M, Kobayashi R, Palkovits M, Saavedra JM (1976a) Choline acetyltransferase levels in the diencephalic nuclei of the rat. J Neurochem 24:35–38

Brownstein M, Palkovits M, Tappaz ML, Saavedra JM, Kizer JS (1976b) Effect of surgical isolation of the hypothalamus on its neurotransmitter content. Brain Res 117:287–295

Burke RE, Fahn S (1985) Choline acetyltransferase activity of the principal vestibular nuclei of rat studied by micropunch technique. Brain Res 328:196–199

Carson KA, Nemeroff CB, Rone MS, Young-Blood WW, Prange AJ Jr, Hanker JS, Kezer JS (1977) Biochemical and histochemical evidence for the existence of a tuberonfundubular cholinergic pathway in the rat. Brain Res 129:169–173

Cheney DL, Lefevre HF, Racagni G (1975) Choline acetyltransferase activity and mass fragmentographic measurements in acetylcholine specific nuclei and tracts of rat brain. Neuropharmacology 14:801–809

Contestabile A, Fonnum F (1983) Cholinergic and GABAergic forebrain projections to the habenula and nucleus interpeduncularis: surgical and kainic acid lesions. Brain Res 275:287–297

Contestabile A, Villani L, Fonnum F (1986a) Neurochemical and ultrastructural study of the effect of the cholinergic toxin AF6417 in nucleus interpeduncularis. Brain Res 379:223–231

Contestabile A, Villani L, Fasolo A, Franzoni MF, Gribaudo L, Øktedalen O, Fonnum F (1986b) Topography of cholinergic and substance P pathways in the habenulo-interpeduncular system of the rat: an immunocytochemical and microchemical approach. Neuroscience (in press)

Coyle JT, Schwartz R (1976) Lesion of striatal neurons with kainic acid model for Huntington's chorea. Nature 271:178–180

Cuello AC, Emson PC, Paxines C, Jessel T (1978) Substance P containing and cholinergic projections from the habenulae. Brain Res 149:413–429

Damsma BH, Westerink BHC, Horn AS (1985) A simple, sensitive and economic assay for choline and acetylcholine using HPLC, an enzyme reactor, and an electrochemical detector. J Neurochem 45:1649–1653

Demêmes D, Raymond J, Sano A (1983) Selective retrograde labelling of vestibular efferent neurons with [^{3}H]choline. Neuroscience 8:290–295

Eckenstein F, Thoenen H (1983) Cholinergic neurons in the rat cerebral cortex demonstrated by immunohistochemical localization of choline acetyltransferase. Neurosci Lett 36:211–255

Emson PC, Fonnum F (1974) Choline acetyltransferase, acetylcholinesterase and aromatic L-amino acid decarboxylase in single identified nerve cells bodies from snail *Helix aspersa.* J Neurochem 22:1079–1088

Emson PC, Burrows M, Fonnum F (1974) Levels of glutamate decarboxylase, choline acetyltransferase and acetylcholinesterase in identified motorneurons of the locust. J Neurobiol 5:33–42

Emson PC, Paxinos G, Le Gal La Salle G, Ben Ari Y, Silver A (1979) Choline acetyltransferase and acetylcholinesterase containing projections from the basal forebrain to the amygdaloid complex of the rat. Brain Res 165:271–282

Eva C, Hadjiconstantinou M, Neff NM, Meek JL (1984) Acetylcholine measurement by high performance liquid chromatography using an enzyme-loaded post-column reactor. Anal Biochem 143:320–324
Fibiger HC (1982) The organization and some projections of cholinergic neurons of the mammalian forebrain. Brain Res Rev 4:324–388
Fisher A, Mantione CR, Abraham DJ, Hanin I (1982) Long term central cholinergic hypofunction induced in mice by ethylcholine aziridinium (AF64A) *in vivo*. J Pharmacol Exp Ther 222:140–145
Flumerfelt BA, Contestabile A (1982) Acetylcholinesterase histochemistry of the habenulo-interpeduncular pathway in the rat and the effect of electrolytic and kainic acid lesions. Anat Embryol (Berl) 163:435–446
Fonnum F (1969a) Radiochemical micro assays for the determination of choline acetyltransferase and acetylcholinesterase activities. Biochem J 115:465–472
Fonnum F (1969b) Subcellular localization of choline acetyltransferase in brain. In: Heilbronn E, Winter A (eds) Drugs and cholinergic metabolism in the CNS. FOA, Stockholm, pp 83–96
Fonnum F (1970) Topographical and subcellular localization of choline acetyltransferase in rat hippocampal region. J Neurochem 17:1029–1037
Fonnum F (1972) Application of microchemical analysis and subcellular fractionation technique to the study of neurotransmitters in discrete areas of mammalian brain. Adv Biochem Psychopharmacol 6:75–88
Fonnum F (1975) A rapid radiochemical method of the determination of choline acetyltransferase. J Neurochem 24:407–409
Fonnum F, Contestabile A (1984) Colchicine neurotoxicity demonstrates the cholinergic projection from the supracommissural system to the habenula and nucleus interpeduncularis in the rat. J Neurochem 42:881–884
Fonnum F, Grofová I, Rinvik E, Storm-Mathisen J, Walberg F (1974) Origin and distribution of glutamate decarboxylase in the substantia nigra of the rat. Brain Res 71:77–92
Fonnum F, Walaas I, Iversen E (1977) Localization of GABAergic, cholinergic and aminergic structures in the mesolimbic system. J Neurochem 29:221–230
Freeman JJ, Choi RL, Jenden DY (1975) Plasma choline: its turnover and exchange with brain choline. J Neurochem 24:729–734
Geneser-Jensen FA, Blackstad TW (1971) Distribution of acetylcholinesterase in the hippocampal region of the guinea-pig I. Enthorinal area, parasubiculum and presubiculum. Z Zellforsch 114:460–481
Giller E, Schwartz JH (1971) Choline acetyltransferase in identified neurons of adominal ganglion of *Aplysia california*. J Neurophysiol 34:93–107
Godfrey DA, Ross CD, Herrmann AD, Matchinsky FM (1980) Distribution and derivation of cholinergic elements in the rat olfactory bulb. Neuroscience 5:273–292
Goldberg AM, McCaman RE (1967) A quantitative microchemical study of choline acetyltransferase and acetylcholinesterase in the cerebellum of several species. Life Sci 6:1493–1500
Goldberg AM, McCaman RE (1973) The determination of picomole amounts of acetylcholine in mammalian brain. J Neurochem 20:1–8
Goldman PS, Nauta WJH (1977) An intricately patterned prefrontocaudate projection in the rhesus monkey. J Comp Neurol 171:369–386
Gordon CR, Krieger NR (1983) Localization of choline acetyltransferase in laminae of the rat olfactory tubercle. J Neurochem 40:79–83
Gordon MN, Finch CE (1985) Topochemical localization of choline acetyltransferase and acetylcholinesterase in mouse brain. Brain Res 308:364–368
Gottesfeld Z, Fonnum F (1977) Transmitter synthesizing enzymes in the hypoglossal nucleus and the cerebellum – effect of acetylpyridine and surgical lesions. J Neurochem 28:237–239
Gottesfeld Z, Jacobowitz DM (1978) Cholinergic projection of the diagonal band to the interpeduncular nucleus of the rat brain. Brain Res 156:329–332
Gottesfeld Z, Jacobowitz DM (1979) Cholinergic projections from the septal-diagonal band area to the habenular nuclei. Brain Res 176:391–394

Graybiel AM, Ragsdale CW (1978) Histochemically distinct compartments in the striatum of human, monkey and cat demonstrated by acetylcholinesterase staining. Proc Natl Acad Sci USA 75:5723–5726

Gwyn DG, Wolstencroft JH, Silver A (1972) The effect of hemisection on the distribution of acetylcholinesterase and choline acetyltransferase in the spinal cord of cat. Brain Res 47:289–301

Hassler R, Nitsch C, Lee HL (1980) The role of eight putative transmitters in the nine types of synapses in rat caudate-putamen. In: Rinne U, Klinger M, Stanson G (ed) Parkinson's disease – current progress, problems and management. Elsevier, Amsterdam, pp 61–91

Hebb CO, Krnjević K, Silver A (1963) Effect of undercutting on the acetylcholinesterase and choline acetyltransferase activity in the cat's cerebral cortex. Nature 198:692

Heimer L (1978) The olfactory cortex and the ventral striatum. In: Livingston KE, Hornykiewicz O (eds) Limbic mechanisms. Plenum, New York, pp 95–187

Hiley CR, Burgen ASV (1974) The distribution of muscarinic receptor sites in the nervous system of the dog. J Neurochem 22:159–162

Hodes HI, Rea MA, Felten DL, Aprison MH (1983) Specific binding of the muscarinic antagonist [^{3}H]quinuclidinyl benzilate is not associated with preganglionic motor neurons in the dorsal motor nucleus of the vagus. Neurochem Res 8:73–87

Hoover DB, Hancock JC (1985) Effect of facial nerve transection on acetylcholinesterase, choline acetyltransferase and [^{3}H]quinuclidinyl benzilate binding in rat facial nuclei. Neuroscience 15:481–487

Hoover DB, Jacobowitz DM (1979) Neurochemical and histochemical studies of the effect of a lesion of the nucleus cuneiformis on the cholinergic innervation of discrete areas of the rat brain. Brain Res 170:113–122

Hoover DB, Muth EA, Jacobowitz DM (1978) A mapping of the distribution of acetylcholine, choline acetyltransferase and acetylcholinesterase in discrete areas of rat brain. Brain Res 153:295–306

Israël M, Lesbats B (1981) Chemiluminescent determination of acetylcholine and continuous detection of its release from *Torpedo* electric organ synapses and synaptosomes. Neurochem Int 3:81–90

Israël M, Whittaker VP (1967) The isolation of mossy fibre endings from the granular layer of the cerebellar cortex. Experientia 21:325–326

Jaffé EH, Cuello AC (1980) The distribution of catecholamines, glutamate decarboxylase and choline acetyltransferase in layers of the rat olfactory bulb. Brain Res 186:232–237

Jacobowitz DM (1974) Removal of discrete fresh regions of the rat brain. Brain Res 80:111–115

Kanazawa I, Sutoo D, Oshima I, Saito S (1979) Effect of transection on choline acetyltransferase, thyrotropin releasing hormone and substance P in the cervical spinal cord. Neurosci Lett 13:325–330

Kása P, Silver A (1969) The correlation between choline acetyltransferase and acetylcholinesterase activity in different areas of the cerebellum of rat and guinea pig. J Neurochem 16:389–396

Kása P, Joó F, Csillik B (1965) Histochemical localization of acetylcholinesterase in the cat cerebellar cortex. J Neurochem 12:31–35

Kataoka K, Nakamura Y, Hassler R (1973) Habenulo-interpeduncular tract: a possible cholinergic neuron in rat brain. Brain Res 62:264–267

Kato T (1984) Enzymatic determination of choline acetyltransferase by coenzyme A cycling and its application to analysis of single mammalian neurons. J Neurochem 42:903–910

Kato T, Murashima YL (1985) Choline acetyltransferase activities in single motor neurons from vertebrate spinal cords. J Neurochem 44:675–679

Kelly PH, Moore KE (1978) Decrease of neocortical choline acetyltransferase after lesion of the globus pallidus in the rat. Exp Neurol 61:479–484

Kim J-S (1978) Transmitters for the afferent and efferent systems of the neostriatum and their possible interactions. Adv Biochem Psychopharmacol 19:217–233

Kobayashi RM, Brownstein M, Saavedra JM, Palkovits M (1975) Choline acetyltransferase content in discrete regions of the rat brain stem. J Neurochem 24:637–640

Kobayashi RM, Palkovits M, Hruska RE, Rothschild R, Yamamura H (1978) Regional distribution of muscarinic cholinergic receptors in rat brain. Brain Res 154:13–23
Koslow SH, Racagni C, Costa E (1974) Mass fragmentographic measurements of norephinephrine, dopamine, serotonin and acetylcholine in seven discrete nuclei of the rat tel-diencephalon. Neuropharmacology 13:1123–1130
Kuhar MJ, Murrin LC (1978) Sodium dependent high affinity choline uptake. J Neurochem 30:15–22
Kuhar MJ, Setting VH, Roth RH, Aghajanian GK (1973) Choline: selective accumulation by central cholinergic neurons. J Neurochem 20:581–593
Kuhar MJ, Dehaven RN, Yamamura HI, Rommelspacher H, Simon JR (1975) Further evidence for cholinergic habenulo-interpeduncular neurons: pharmacologic and functional characteristics. Brain Res 97:265–275
Lehmann NJ, Fibiger HC (1978) Acetylcholinesterase in the substantia nigra and caudate-putamen of the rat: properties and localization in dopaminergic neurons. J Neurochem 30:615–624
Lehmann NJ, Nagy JJ, Atmadja, Fibiger HC (1980) The nucleus basalis magnocellularis is the origin of a cholinergic projection to the neocortex of the rat. Neuroscience 5:1161–1174
Levy A, Kant GJ, Meyerhoff JL, Jarrard LE (1984) Non-cholinergic neurotoxic effects of AF64A in the substantia nigra. Brain Res 305:169–172
Lewis PR, Shute CCD, Silver A (1967) Confirmation from choline acetylase analysis of a massive cholinergic innervation of the rat hippocampus. J Physiol (Lond) 191:215–224
Lowry OH, Passoneau JV (1972) A flexible system of enzymatic analysis. Academic, New York
Lund-Karlsen R, Fonnum F (1976) The toxic effect of sodium glutamate on rat retina: changes in putative transmitters and their corresponding enzymes. J Neurochem 27:1437–1441
Macrides F, Davis BY, Youngs WM, Nach MS, Margolis FL (1981) Cholinergic and catecholaminergic afferents to the olfactory bulb in the hamster. A neuroanatomical, biochemical and histochemical investigation. J Comp Neurol 203:497–516
Malthe-Sørenssen D, Odden E, Walaas I (1980) Selective destruction by kainic acid of neurons innervated by putative glutamergic afferents in septum and nucleus of the diagonal band. Brain Res 182:461–465
Manaker S, Wieczorek CM, Rainbow TC (1986) Identification of sodium-dependent, high-affinity choline uptake sites in rat brain with [^{3}H]hemicholinum-3. J Neurochem 46:483–488
Mantione CR, Fisher A, Hanin I (1981) AF64A neurotoxicity: a potential animal model of central cholinergic hypofunction. Science 213:579–580
Mantione CR, Zigmand MY, Fisher A, Hanin I (1983) Selective presynaptic cholinergic neurotoxicity following intrahippocampal AF64A injection in rats. J Neurochem 41:251–255
Marshall RC, Flumberfelt BA, Gwyn DG (1980) Acetylcholinesterase activity and acetylcholine effects in the cerbello-rubro-thalamic pathway of the cat. Brain Res 190:493–504
McCaman RE, Dewhurst SA (1970) Choline acetyltransferase in individual neurones of *Aplysia california*. J Neurochem 17:1421–1426
McGeer EG, McGeer PL (1976) Distribution of biochemical changes of Hungtington's chorea by intrastriatal injections of glutamic and kainic acids. Nature 263:517–512
McGeer PL, McGeer EG, Fibiger HG, Wickson V (1971) Neostriatal choline acetylase and cholinesterase following selective brain lesions. Brain Res 35:308–314
McKinney M, Struble RG, Prue DL, Coyle JT (1982) Monkey nucleus basalis is enriched with choline acetyltransferase. Neuroscience 7:2363–2368
Mellgren SI, Srebro B (1973) Changes in acetylcholinesterase and distribution of degenerating fibres in the hippocampal region after septal lesions in the rat. Brain Res 52:19–36

Meyer DK, Brownstein MJ (1980) Effect of surgical deafferentation of the supraoptic nucleus on its choline acetyltransferase content. Brain Res 193:566–569

Nieoullon A, Dusticier N (1980) Choline acetyltransferase activity in discrete regions of the cat brain. Brain Res 196:139–149

Nieoullon A, Dusticier N (1981) Decrease in choline acetyltransferase activity in the red nucleus of the cat after cerebellar lesion. Neuroscience 6:1633–1641

Palkovits M (1973) Isolated removal of hypothalamic or other brain nuclei of the rat. Brain Res 59:449–450

Palkovits M, Saavedra JM, Kobayashi RM, Brownstein M (1974) Choline acetyltransferase content of limbic nuclei of the rat. Brain Res 79:443–450

Perry EL, Atack JR, Perry RH, Hardy JA, Dodd PC, Edwardson JA, Blessed G, Tomlinson BE, Fairbarn DF (1984) Intralaminar neurochemical distributions in human midtemporal cortex; comparison between Alzheimer's disease and the normal. J Neurochem 42:1402–1410

Rainbow TC, Parsons B, Wieczerok CM (1984) Quantitative autoradiography of [^{3}H]hemicholinum-3 binding sites in rat brain. Eur J Pharmacol 102:195–196

Ross CD, McDougal DM Jr (1976) The distribution of choline acetyltransferse activity in vertebrate retina. J Neurochem 26:521–526

Ross CD, Godbrey DA, Williams AD, Matschinsky FM (1978) Evidence for a central origin of cholinergic structures in the olfactory bulb. Neurosci Abstr 4:91

Rotter A, Birdsall NJM, Field PM, Raisman G (1979a) Muscarinic receptors in the central nervous system of the rat. II. Distribution of binding [^{3}H]propylbenzilylcholine mustard in the midbrain and hindbrain. Brain Res Rev 1:167–183

Rotter A, Birdsall NJM, Burgen ASV, Field PM, Hulme EC, Raisman G (1979b) Muscarinic receptors in the central nervous system of the rat. I. Technique for autoradiographic localization of the binding of [^{3}H]propylbenzilylcholine mustard and its distribution in the forebrain. Brain Res Rev 1:141–165

Saelens JK, Edwards-Neale S, Simke JP (1979) Further evidence for cholinergic thalamostriatal neurons. J Neurochem 32:1093–1094

Shute CCD, Lewis PR (1961) The use of cholinesterase techniques combined with operative procedures to follow neuron pathways in the brain. Bibl Anat 2:34–39

Shute CCD, Lewis PR (1967) The ascending cholinergic reticular system: neocortical, olfactory and subcortical projections. Brain 90:497–520

Simke JP, Saelens JK (1977) Evidence for a cholinergic fibre tract connecting the thalamus with the head of the striatum of the rat. Brain Res 126:487–495

Simon JR, Oderfeld-Nowak B, Felten DL, Aprison MH (1981) Distribution of choline acetyltransferase, acetylcholinesterase, muscarinic receptor binding and choline uptake in discrete areas of the rat medula oblongata. Neurochem Res 6:497–505

Stadler H, Nesselhut T (1986) Single and rapid measurement of acetylcholine and choline by HPLC and enzymatic-electrochemical detection. Neurochem Int 9:127–129

Stanton GB, Orr A (1985) [^{3}H]choline labeling of cerebellothalamic neurons with observations on the cerebello-thalamo-parietal pathway in cats. Brain Res 335:237–243

Storm-Mathisen J (1970) Quantitative histochemistry of acetylcholinesterase in rat hippocampal region correlated to histochemical staining. J Neurochem 17:739–750

Storm-Mathisen J (1972) Glutamate decarboxylase in the rat hippocampal region after lesions of the afferent fibre systems. Brain Res 40:215–235

Vickroy TW, Fibiger HC, Roeske WR, Yamamura HI (1984) Reduced density of sodium-dependent [^{3}H]hemicholinum-3 binding sites in the anterior cerebral cortex of rats following chemical destruction of the nucleus basalis magnocellularis. Eur J Pharmacol 102:369–370

Vickroy TW, Roeske WR, Gehlert DR, Wamsley JK, Yamamura HI (1985) Quantitative light microscope autoradiography of [^{3}H]hemicholinum-3 binding sites in the rat central nervous system: a biochemical marker for mapping the distribution of cholinergic nerve terminals. Brain Res 329:368–373

Villani L, Contestabile A, Fonnum F (1983) Autoradiographic labeling of the cholinergic habenulo-interpeduncular projection. Neurosci Lett 42:261–266

Walaas I, Fonnum F (1978) The effect of parenteral glutamate treatment on the localization of neurotransmitters in the mediobasal hypothalamus. Brain Res 153:549–562
Walaas I, Fonnum F (1979a) The distribution and origin of glutamate decarboxylase and choline acetyltransferase in ventral pallidum and other basal forebrain regions. Brain Res 177:325–336
Walaas I, Fonnum F (1979b) The effects of surgical and chemical lesions on neurotransmitter candidates in the nucleus accumbens of the rat. Neuroscience 4:209–216
Wenk H, Meyer U, Bigl V (1976) Zur Histochemie cholinerger Systeme in ZNS. II. Topochemische und quantitative Veränderungen cholinerger Transmitterenzyme (AChE, ChAc) im olfactorischen System bei Ratten nach Zwischenhirnläsion. Z Mikrosk Anat Forsch 90:940–958
Wenk H, Meyer U, Bigl V (1977) Centrifugal cholinergic connections in the olfactory system of rats. Neuroscience 2:797–800
Wenk H, Bigl V, Meyer U (1980) Cholinergic projection from magnocellular nuclei of the basal forebrain to cortical areas in rats. Brain Res Rev 2:295–316
Yamamura HI, Kuhar MY, Greenberg D, Snyder SH (1974) Muscarinic cholinergic receptor binding: regional distribution in monkey brain. Brain Res 66:541–546
Youngs WM, Nadi NS, Davis BJ, Margolis FL, Macrides F (1979) Evidence for a cholinergic projection to the olfactory bulb from the magnocellular preoptica area. Neurosci Abstr 5:135

CHAPTER 22

Central Cholinergic Pathways: The Histochemical Evidence

E. G. McGEER and P. L. McGEER

A. Introduction

Many details of peripheral cholinergic systems have long been part of classical neuroanatomy. The main challenge for some time has been to understand the distribution of cholinergic pathways in the central nervous system. Considerable progress has been made recently, partly from the development and application of the biochemical techniques discussed in the previous chapter but more from histochemical studies which have involved either localization of choline acetyltransferase (ChAT) or of acetylcholinesterase (AChE) following administration of the irreversible cholinesterase inhibitor diisopropylphosphorofluoridate (DFP). Ancillary histological techniques include immunohistochemistry for acetylcholine (ACh), autoradiographic localization of axonally transported radioactive choline, autoradiographic localization of receptors using labelled ligands presumed to be specific for ACh receptors, and double-labelling methods aimed at identifying cholinergic projections or synaptic connections between cholinergic and other neuronal systems. The immunohistochemical localization of Chol-1, a membrane ganglioside apparently specific to cholinergic nerve endings (RICHARDSON et al. 1982; Chaps. 12, 15, this volume), and autoradiography for [^{3}H]hemicholinium-3 ([^{3}H]HC-3), which is bound specifically to the high-affinity choline uptake sites characteristic of cholinergic synapses (VICKROY et al. 1985), offer promise for the future but have not yet been widely applied. Cloning of cRNA for ChAT (STRAUSS et al. 1985) may open up another avenue for investigating ACh cells.

In this chapter we will first discuss the principal and ancillary techniques and then outline the cholinergic cells and projections that have been described. Since each of the techniques has its limitations, extensive confirmation of all data is necessary for there to be firm confidence in the results.

B. Techniques

I. Immunohistochemistry for ChAT

The earliest attempts to localize cholinergic neurons by an unequivocal technique were based on histochemistry for the specific synthetic enzyme ChAT (KÁSA 1978). Results were impaired by the lability of enzymic activity to fixation; the long incubations in highly ionic solutions that were required to obtain detectable activity resulted in very poor localization. The method of choice is now immuno-

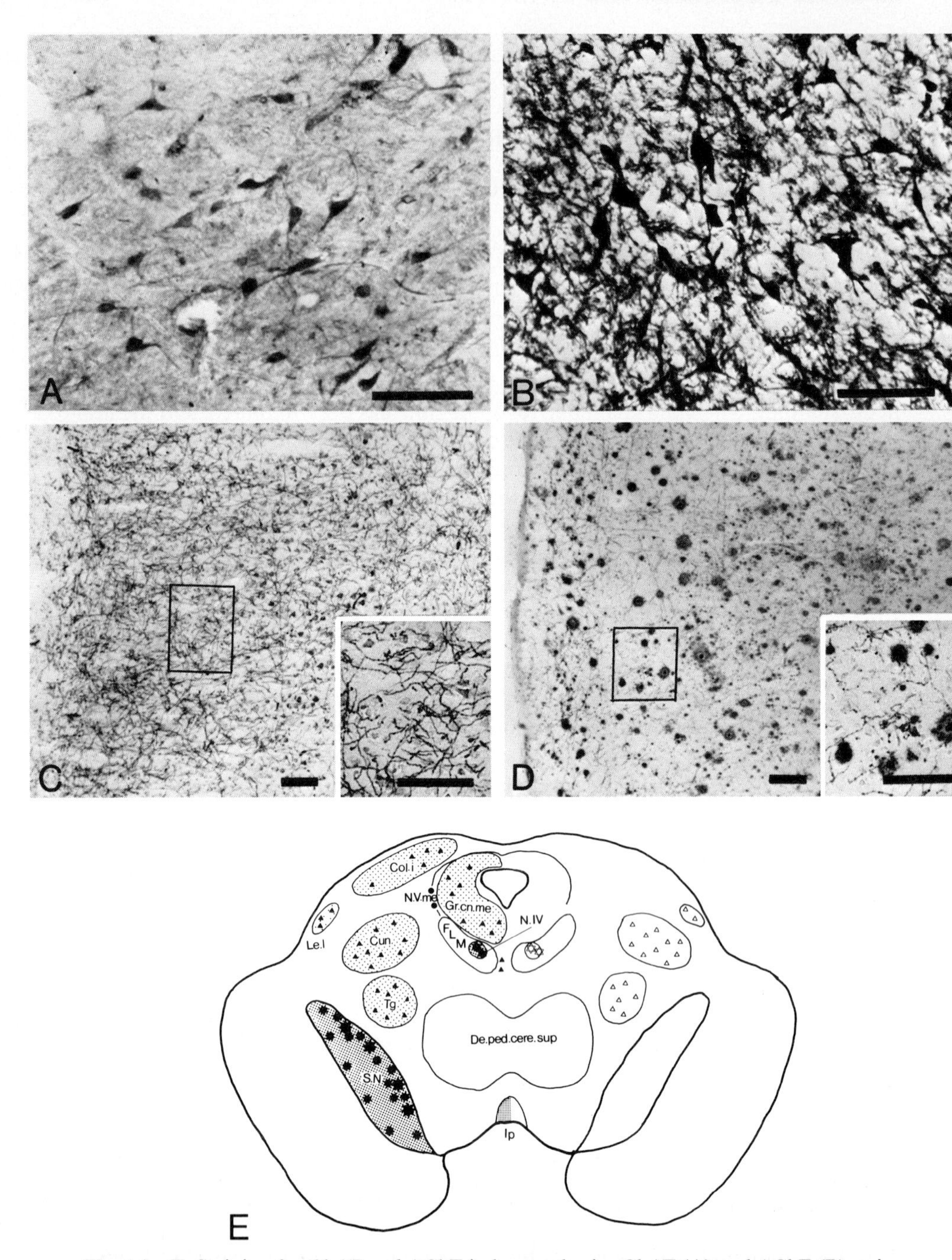

Fig. 1 A–E. Staining for ChAT and AChE in human brain. ChAT (**A**) and AChE (**B**) staining in the nucleus tegmentalis pedunculopontinus; *bars,* 50 μm. AChE staining in the cortex of a normal person (**C**) and a person with Alzheimer's disease (**D**); *bars,* 150 μm. **E** One of a series of cross-sectional diagrams of human brain mapping ChAT staining. (*right side*) and AChE staining (*left side*). *Col.i.,* inferior colliculus; *Cun,* cuneiform

histochemistry for this enzyme (Fig. 1 A). The technique has imperfections. Good immunohistochemical localization must be performed on fixed tissue. This is particularly true for ChAT because cholinergic neurons apparently make large numbers of dendrosomatic contacts and diffusion of ChAT from nerve endings in poorly fixed tissue makes it difficult to distinguish cholinoceptive from cholinergic neurons (Kimura et al. 1981). The fixation process can alter the nature of the recognition sites on the enzyme and thus reduce the intensity of the antigen-antibody reaction. If this intensity is low, it cannot overcome the background staining which is always introduced by the various proteins involved in the complicated sandwich techniques. That this is a particular problem with ChAT is evidenced by several reports from laboratories that have raised monospecific antibodies only to have weak or negative immunohistochemical results.

Access to the fixed enzyme may be another problem. Our laboratory has had greater success using the smaller Fab fragments of a polyclonal IgG serum than with other preparations (Peng et al. 1981). Penetration problems may help to explain why there have been more reports of successful staining of cell bodies than of fibres or nerve endings. Another possible complication, however, is that much of the ChAT in nerve ending fractions may be in a membrane-bound form while most antibodies are prepared against the readily solubilized form. The recent purification of membrane-bound ChAT and preparation of antibodies thereto should allow determination as to the importance of this problem for histochemical results (Peng et al. 1986).

With any immunohistochemical procedure specificity of the antibody is always a problem. The literature on ChAT immunohistochemistry suggests, however, that the greater problem may be with false-negative rather than with false-positive reactions.

Some differences between laboratories in immunohistochemical results should be anticipated since particular antibodies may recognize different sites on ChAT or because the enzyme conformation may vary according to the method of fixation.

ChAT immunohistochemistry can also give information at the ultrastructural level, as indicated by work on the striatum (Hattori et al. 1976; Phelps and Vaughn 1985; Phelps et al. 1985), amygdala (Carlsen and Heimer 1985), interpeduncular nucleus (Hattori et al. 1977), cortex (Houser et al. 1983), hippocampus (Matthews et al. 1983), diagonal band (Armstrong et al. 1983a), nucleus tractus solitarii (Armstrong et al. 1985), thalamus (Isaacson and Tanaka 1985) and motor nuclei (Ross et al. 1983).

nucleus; *De.ped.cere.sup.,* decussation of the superior cerebellar peduncles; *FLM,* fasciculus longitudinalis medialis; *Gr.cn.me.,* griseum centale mesencephali; *Le.I,* lateral lemniscus nucleus; *N.V.me,* nucleus mesencephalicus nervi trigemini; *N.IV,* nucleus nervi trochlearis; *S.N.,* substantia nigra; *Tg,* nucleus tegmentalis pedunculopontinus. Symbols: ✱ or ✲ large-sized, intensely AChE- or ChAT-positive cells with prominent processes; ▲ or △ small- or medium-sized, AChE- or ChAT-positive cells with less prominent processes; ● small- or medium-sized, AChE-positive cells with few processes; intense AChE fibre staining; moderate AChE fibre staining; light AChE fibre staining. (From Mizukawa et al. 1986)

II. AChE Histochemistry

Since the early work of SHUTE and LEWIS (1967) histochemistry of AChE has been a frequently used procedure for the localization of possible cholinergic systems. The advantage is that AChE activity in brain is far greater than ChAT and much more stable to storage and fixation. The disadvantage is that AChE is present in many non-cholinergic neurons and glia, in addition to cholinergic systems. An improvement to the original procedure was introduced by LYNCH et al. (1972), who showed that cell bodies with the capacity for rapid synthesis of AChE could be identified if the animal were treated with an irreversible AChE inhibitor and killed after a period of hours. Only cells with rapid regeneration capabilities had detectable amounts of AChE. Subsequently, many groups have used this pharmacohistochemical method to study AChE-intensive neurons in brain.

In the forebrain such AChE-intensive neurons seem to be cholinergic, as indicated by ChAT immunohistochemistry. In the hindbrain, however, many more cell populations stain intensively for AChE than for ChAT and many of the AChE-positive, ChAT-negative cell groups are known to be non-cholinergic. Well-known examples include the dopaminergic neurons of the substantia nigra, the noradrenergic neurons of the locus ceruleus and the serotonergic neurons of the raphe. LEHMANN and FIBIGER (1979) suggested that the capacity for rapid regeneration of AChE might be a necessary, but not sufficient, characteristic of cholinergic neurons, but this more limited generalization may have numerous exceptions. There are reports from a number of laboratories on ChAT-positive neurons which are not AChE-intensive in many areas of the brain (Table 1). Although AChE histochemistry is not definitive, it can be very useful in studying cholinergic systems defined by other methods. Figure 1B–D exemplifies the excellent staining of processes and cells which is possible with a highly sensitive staining technique which is a modification of the standard procedure.

AChE staining can be examined at the ultrastructural as well as the light microscopic level (SATOH et al. 1983b).

III. Ancillary Techniques

1. Immunohistochemistry for ACh

GEFFARD et al. (1985a, b) have recently reported immunohistochemical staining, comparable to that seen by ChAT immunohistochemistry, in the rat medial septal area and insect brain, using antibodies raised against ingeniously designed conjugates of choline with glutaraldehyde and a protein such as bovine serum albumin (BSA) [e.g. $(CH_3)_3N^+CH_2CH_2O\text{-}C(O)(CH_2)_3CONH\text{-}BSA$]. Preparation of antibodies to such a synthetic material avoids the laborious procedures of ChAT purification which is a necessary first step prior to immunizing for production of polyclonal anti-ChAT serum. Moreover, such antibodies should not show the species preference characteristic of some antibodies to ChAT. On the other hand, much further work is required to define the specificity and sensitivity of the method. ACh is much more labile *post mortem* than ChAT and steps must be taken to protect against neurotransmitter destruction and diffusion.

2. Autoradiography for Axonally Transported Radioactive Choline

This method is based on the general hypothesis that transmitters or transmitter-related molecules will be selectively taken up into the homotypic nerve endings and then retrogradely transported to the cell bodies. High-affinity uptake of [^{3}H]choline into cholinergic nerve endings is a well-studied process and retrograde transport of the radioactivity in some well-defined cholinergic pathways has been reported. The technique allows the demonstration of pathways as well as cell bodies but the specificity is not yet certain. CUÉNOD et al. (1982a, b) pointed out that there was also diffuse neuropil labelling in areas receiving projections from the injection site. Moreover, the technique is not good for careful morphological studies since the retrogradely labelled radioactivity can only be seen if the brains are unfixed, sectioned in a frozen state and processed according to a dry mount autoradiographic procedure like that used in binding studies.

3. Autoradiography for Muscarinic or Nicotinic Binding Sites

Autoradiography on tissue slices incubated with ^{3}H-labelled antagonists, particularly [^{3}H]3-quinuclidinylbenzylate ([^{3}H]3-QNB) or [^{3}H]pirenzepine, has been used to localize muscarinic binding sites at the light microscopic level (KUHAR and YAMAMURA 1976; ROTTER et al. 1979a, b). The regional distributions found are generally similar to those obtained in binding studies on brain homogenates; those areas rich in choline uptake or ChAT are also rich in binding sites. The results obtained depend greatly upon the exact conditions and ligand used. Similar work with muscarinic agonists such as (+)-*cis*-[^{3}H]methyldioxolane has indicated a different regional distribution from that found with antagonists (YAMAMURA et al. 1985): this presents a problem in interpreting the significance of the technique.

Some similar studies aimed at localization of nicotinic binding sites in brain have been carried out using, for example, radioactive nicotine (LONDON et al. 1985) or α-bungarotoxin (ARIMATSU et al. 1981). There is considerable controversy as to how much of the binding which is observed relates to ACh systems.

Morphological definition in binding studies is poor and there is no guarantee that all the binding sites indicated as such in *in vitro* studies are physiologically active receptors *in vivo*. Even if such could be assumed, the information would relate primarily to cholinoceptive structures rather than cholinergic ones. Binding studies are therefore unlikely to give detailed information on the location of ACh systems even though they can be applied *in situ*ations where information is difficult to obtain, such as *post-mortem* human brain. Data on regional changes in diseases such as Huntington's are of interest (PENNEY and YOUNG 1982).

4. Double-Labelling Methods

Combination of a retrogradely transported label, such as horseradish peroxidase (HRP) or wheat-germ agglutinin (WGA), with either AChE pharmacohistochemistry or, better, ChAT immunohistochemistry has been used to define projections of the basal forebrain and, to a much more limited extent, brain-stem cholinergic neurons.

Table 1. Central cholinergic cell groups reported in various species using ChAT immunohistochemistry[a]

ChAT staining (non-controversial)	ChAT-positive staining		AChE-intensive	
			All	Most
Large cells				
Medial forebrain complex[b]	H3, 4; B22; M11, 16; C2, 37; R1, 5, 6, 9, 12, 16, 20, 26, 36; G17		H21; B22; M11, 29; R12, 20, 28	M16; R16, 36
Giant cells of striatum and nucleus accumbens	H4; B22; M16; C2, 37; R5, 6, 9, 12, 16, 20, 28, 36; G17		H21; B22; M16, 29; R12, 16, 20, 30	
Motor nuclei of cranial nerves	H10; B22; C2, 37, 25; R6, 7, 9, 12, 15, 16, 26, 36; G17		H21; B22; R12, 16, 36	
Parabrachial-pedunculopontine tegmental complex	H10; B22; C2, 37; R6, 12, 15, 26, 31, 36		H21; B22; R12	R36
ChAT staining (controversial)	**ChAT-positive**	**ChAT-negative**	**AChE-intensive**	
			All	None
Large cells				
Reticular formation	H10; C2, 25; R6; Co18	B22; R12	H21; R12, 32	
Vestibular nuclei / Lateral reticular formation / Superior olivary complex	H10; C2; R6	B22; R12	H21	
Red nucleus	C2; R5, 6, 7	B22; R12	H21	
Small cells				
Striatum	R24, 26	B22; C2; R6, 12		B22; R14
Cortex[c]	R8, 9, 23, 26, 28, 32, 34, 35	B22; C2, 34; R6, 12; G34		H19; B22; M29; R12

Hippocampus	R8, 13, 33	B22; C2; R6, 12		B22; M29; R12
Medial habenula	C37; R9, 13, 24	B22; C2; R6, 12		B22; R12
Amygdala	C37; R32	C2; R6		
Hypothalamus	H38; M38; C37; R36, 38			H38; M38; R38, 38

[a] Letters before the reference numbers indicate species: H, human; B, baboon; M, monkey; C, cat; R, rat; G, guinea pig; and Co, cow. Only a small part of the AChE pharmacohistochemical literature could be included. VINCENT and REINER [37] also report some ChAT-positive cells in the superior colliculus and various other areas not included in the table because there has not yet been any confirmation.

[b] HEDREEN et al. (1984) described topography of the large neurons revealed by Nissl staining in humans and our group not only mapped the cells using ChAT immunohistochemistry [3, 4] but have shown that all large neurons in this area in the human are ChAT-positive [3].

[c] In rat and mouse some cortical VIP cells contain ChAT but this is not found in rabbit, guinea pig or cat [34].

References:
1, KIMURA et al. (1980)
2, KIMURA et al. (1981)
3, MCGEER et al. (1984b); NAGAI et al. (1983b)
4, NAGAI et al. (1983a)
5, SOFRONIEW et al. (1982)
6, KIMURA et al. (1984)
7, ROSS et al. (1983)
8, ECKENSTEIN and THOENEN (1983)
9, HOUSER et al. (1983)
10, MIZUKAWA et al. (1985)
11, MESULAM et al. (1983a)
12, SATOH et al. (1983a)
13, MATTHEWS et al. (1983)
14, SATOH et al. (1983b)
15, MUFSON et al. (1982)
16, LEVEY et al. (1982)
17, MALEY et al. (1982)
18, COZZARI and HARTMAN (1980)
19, KOSTOVIC (1982)
20, ECKENSTEIN and SOFRONIEW (1983)
21, TAGO et al. (1985)
22, SATOH and FIBIGER (1985)
23, MCGEER et al. (1974)
24, HATTORI et al. (1977)
25, SHIROMANI et al. (1985)
26, ARMSTRONG et al. (1983b)
27, CARLSEN and HEIMER (1985)
28, ARMSTRONG et al. (1982)
29, MESULAM and VAN HOESEN (1976)
30, VINCENT et al. (1983c)
31, VINCENT et al. (1983b)
32, PARNAVELAS et al. (1985)
33, WAINER et al. (1985)
34, ECKENSTEIN et al. (1985)
35, LYSAKOWSKI et al. (1985)
36, GOULD et al. (1985), BUTCHER et al. (1985)
37, VINCENT and REINER (1987)
38, TAGO et al. (1987)

Double labelling involving immunohistochemistry for substance P (SP) combined with ChAT immunohistochemistry or AChE pharmacohistochemistry has been used at the light-microscopic level to demonstrate the coexistence of SP in some of the ascending cholinergic projections from reticular neurons (Vincent et al. 1983 a) and, at the electron-microscopic level, to demonstrate SP innervation of cholinergic neurons in the striatum and medial basal forebrain (Izzo et al. 1985; Ingram et al. 1985).

Many further studies using these various double labelling techniques are required to supply information on central cholinergic systems which is missing at present.

C. Central Cholinergic Systems

I. Cholinergic Cell Groups

Five major systems of large ChAT-positive and AChE-intensive neurons have been described by several laboratories (Table 1) and hence can be considered as almost certainly cholinergic neurons. The work has been done in a number of species and the general pattern appears the same although there may be some differences in exact location and extent of cellular spread. Minor variations in the anatomical description may be attributable, not only to species, but to nomenclature differences. This applies particularly to the cell groups in the brain stem. A fairly precise visual appreciation of the neuronal locations can be obtained by studying the maps which have been produced for the cat, rat, baboon and human (Fig. 1 E).

There are also a number of minor, as yet controversial, ChAT-positive groups, some of which may not be AChE-intensive (Table 1). The major, morphologically confirmed cholinergic cell groups in brain are as follows.

Medial Forebrain Complex. This is a more or less continuous sheet of giant cells which starts anteriorly near the medial surface of the cortex and extends in a caudolateral direction, always maintaining its position close to the medial and ventral surfaces of the brain, and terminating towards the caudal aspect of the lentiform nucleus. The names usually given to the various subregions of this complex from rostral to caudal order are the following: medial septal nucleus, nucleus of the vertical limb of the diagonal band of Broca, nucleus of the horizontal limb of the diagonal band of Broca and the nucleus basalis of Meynert or substantia innominata (Fig. 2). The cholinergic cells of the olfactory tubercle may be considered part of this complex which has been shown to be ChAT-positive in a number of species using both polyclonal and monoclonal antibodies (Table 1) and has been well mapped in rats, cats, baboons and humans. In addition, this is a very prominent cell complex revealed by the AChE-DFP histochemical technique (Table 1).

Striatal and Nucleus Accumbens Interneurons. Giant cells within the striatum and nucleus accumbens have been identified as being cholinergic, both by ChAT immunohistochemistry and by AChE-DFP histochemistry (Table 1). The ChAT

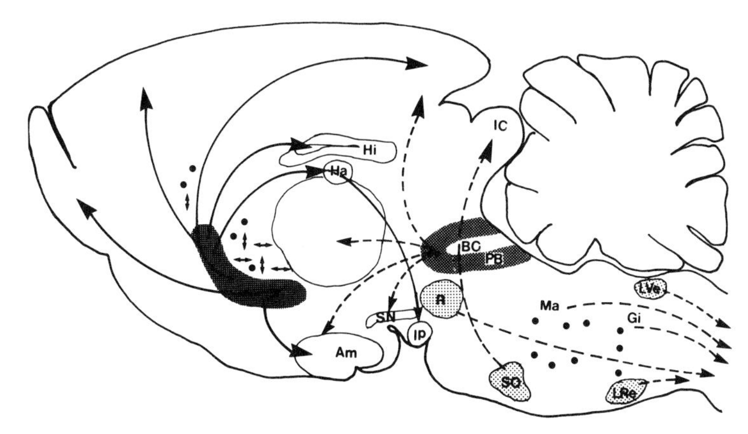

Fig. 2. Sagittal view of rat brain indicating ChAT-containing neurons and established (*solid lines*) or possible (*broken lines*) cholinergic pathways. Major cholinergic cell systems are indicated by *black dots* or *heavy stippling*, minor ones by *light stippling*. *Am*, amygdala; *BC*, brachium conjunctivum; *Gi*, gigantocellular division of the reticular formation; *Ha*, habenula; *Hi*, hippocampus; *IC*, inferior colliculus; *IP*, interpeduncular nucleus; *LRe*, lateral reticular nucleus; *LVe*, lateral vestibular nucleus; *Ma*, magnocellular division of the reticular formation; *PB*, parabrachial complex; *R*, red nucleus; *SN*, substantia nigra; *SO*, superior olive. (From McGeer et al. 1984a)

and AChE levels of the caudate, putamen and accumbens are extremely high and these giant neurons represent no more than 1%–2% of the neuronal population (Kemp and Powell 1971; Phelps et al. 1985). Smaller ChAT-positive cells have also been reported in the striatum (Table 1) and, while the giant cells are certainly the most prominent ones, they probably do not represent the entire cholinergic population.

Motor Nuclei for Peripheral Nerves. Cholinergic cells, equivalent to anterior horn cells, exist in cranial nerve nuclei III–VI and IX–XII as the source of efferent fibres to skeletal muscle and autonomic ganglia (Table 1). The counterparts in the spinal cord in the anterior and lateral horns have also been shown to stain positively in many laboratories.

Parabrachial System. This is the most intense and concentrated ChAT-positive cell group in the brain stem (Table 1). It surrounds the brachium conjunctivum commencing in the most rostral aspects of the pons and follows the direction of the brachium conjunctivum (superior cerebellar peduncles) in a caudodorsal direction (Fig. 2). Various subnuclei are separately identifiable in this particular region. The most commonly described nucleus is the pedunculopontine tegmental nucleus in the lateral aspect at the most rostral portion of the complex. Caudal to this are three adjoining nuclei, the dorsal parabrachial, the Kolliker-Fuse, and the ventral parabrachial nuclei. They surround the cerebellar peduncles in the dorsal, lateral and ventral regions, respectively. Slightly more dorsally placed are the cuneiform nucleus and the dorsolateral tegmental nucleus. Most or all of these neurons also stain intensely by the AChE-DFP method (Table 1) and have been reported to stain selectively for NADPH-diaphorase (Vincent et al. 1983b).

Reticular Formation. A scattered collection of very large ChAT- and AChE-positive cells extends throughout the gigantocellular and magnocellular tegmental fields of the reticular formation (Fig. 2). Caudally the large ChAT-containing neurons are gradually aggregated medially towards the granular layer of the raphe and ventrally to the area near the inferior olivary nucleus. Thus, these cells extend continuously from the pons into the medulla as a longitudinally oriented cluster. Only two laboratories have reported this group as staining positively for ChAT, but the cells do stain intensely by the AChE-DFP method (Table 1) and have been shown to concentrate labelled choline following spinal cord injections (Pare et al. 1982).

Minor groups which are probably cholinergic include neurons in the lateral reticular nucleus, superior olivary complex, lateral vestibular nucleus, magnocellular red nucleus and central canal region in the spinal cord (Table 1).

In addition, small bipolar interneurons staining for ChAT have been described in the cerebral cortex and hippocampus (Table 1) but others have not found either such staining or any AChE-intensive neurons in these areas. The ChAT-positive cells are a subpopulation of neurons containing vasoactive intestinal polypeptide (VIP) in the rat and mouse and may only be found in some species (Table 1).

Parnavelas et al. (1985) report ultrastructural evidence for ChAT-positive labelling of vascular endothelial cells in rat cortex. At the light-microscopic level, Kimura et al. (1981) reported extensive ChAT-positive punctate staining around small cerebral blood vessels in the cat and, at the ultrastructural level, Armstrong et al. (1985) found ChAT-positive dendrites in direct apposition to the basal lamina of blood vessels in the rat. The exact source of the ACh release which is believed to play an important role in controlling cerebral circulation remains to be determined.

Some but not all laboratories have also reported positive staining for ChAT in some ventromedial cells of the medial habenula. Once more no AChE-rich cells are reported (Table 1) and biochemical evidence is equivocal (see Chap. 21).

The recent demonstration in our laboratory of small to medium ChAT-positive cells in the hypothalamus of rat, monkey and human helps to account for previous biochemical and pharmacological evidence which strongly indicated the presence of intrinsic cholinergic neurons in the hypothalamus.

Using autoradiographic localization of [^{3}H]choline retrogradely transported from the thalamus of cats, Stanton and Orr (1985) found labelled cells in the deep cerebellar nuclei and adjacent vestibular nucleus, leading them to suggest a cholinergic cerebellothalamic path. This requires confirmation by some other technique of proven specificity, especially since the cells indicated do not appear to correspond with reported ChAT-positive populations.

II. Cholinergic Pathways

At this stage less is known about cholinergic pathways than about cholinergic cell groups. ACh and ChAT are ubiquitous in brain and every area contains some cholinergic terminals from one of the main cell groups. ChAT-positive staining of neuropil has been mapped in rat and cat (KIMURA et al. 1981, 1984), studied extensively in rabbit forebrain (CHAO et al. 1982) and described by others in selected areas such as the visual system (STICHEL and SINGER 1985), cortex, hippocampus, interpeduncular nucleus (HOUSER et al. 1983; ECKENRODE et al. 1985), olfactory tubercle (TALBOT et al. 1985), cochlea (ALTSCHULER et al. 1985) and amygdala (CARLSEN and HEIMER 1985). The ChAT-positive terminals and cholinoceptive neurons seen in various regions and lamina have been mapped in detail in cat cortex (KIMURA et al. 1983), and LYSAKOWSKI et al. (1985) report a detailed comparison of AChE and ChAT staining in rat cortex. The relation between cell groups and terminal fields is not completely defined and much of the available information has come from lesion and other biochemical methods (see Chap. 21).

The major cholinergic cell groups and their suspected or known projections are indicated in Fig. 2.

The most extensively studied projections are those of the mediobasal forebrain which go to cortical and limbic areas. Although much of the early information came from biochemical studies, many of these pathways were suggested by SHUTE and LEWIS (1967) on the basis of AChE staining. The distribution to the cortical areas has been extensively mapped using retrograde tracing methods and the cholinergic nature of many of the projections has been shown by retrograde tracing combined with ChAT immunohistochemistry and with AChE-DFP histochemistry in the monkey (MESULAM et al. 1983a) or with only AChE-DFP histochemistry in the rat (JONES et al. 1976; LEHMANN et al. 1980; MUFSON et al. 1982; MCKINNEY et al. 1982; MESULAM and van HOESEN 1976). Projections, shown by similar double labelling, also go to the amygdala (NAGAI et al. 1982) and olfactory bulb (MESULAM et al. 1983a). Another histological approach to defining these projections has been to lesion parts of the complex and examine cortical changes in AChE histochemistry (WENK et al. 1980; KRISTT et al. 1985). These ascending projections of the basal forebrain complex are more or less topographically arranged but with considerable overlap of the various cortical fields. Small projections from AChE-rich neurons of the nucleus basalis to the spinal cord (HARING and DAVIS 1982) and from ChAT-positive neurons of the medial septal area to the subfornical organ (WHEELER and SIMPSON 1985) have also been suggested on the basis of double labelling experiments.

Double labelling allows an estimate of the percentage of the ACh contribution to a given projection. WAINER et al. (1985), for example, used a combination of ChAT immunohistochemistry and retrograde tracing of HRP-conjugated WGA to estimate that at least 50% of hippocampal afferents arising in the septal/diagonal band region are non-cholinergic.

Rats with fornix lesions show changes in ChAT staining in the interpeduncular nucleus (ECKENRODE et al. 1985) which are in accord with biochemical work on the intense projection from the mediobasal forebrain complex to that nucleus.

Biochemical work in lesioned animals demonstrated the intrinsic nature of the cholinergic systems of the striatum and nucleus accumbens (see Chap. 21).

Major projection areas of brain-stem nuclei that contain ACh cells have been revealed by tracing techniques. The parabrachial-pedunculopontine system projects to medial thalamic nuclei, substantia nigra, subthalamus, cortex, amygdala and hypothalamus. There are also some descending projections (SAPER and LOEWY 1982). Magnocellular and gigantocellular cells of the reticular formation project to the spinal cord and thalamus. The lateral reticular nucleus projects to the cerebellum and the spinal cord. The superior olivary nucleus projects to the inferior colliculus (CARPENTER and SUTIN 1983). However, since populations other than the ChAT-positive ones are undoubtedly contained in these nuclei, the projection fields can only be considered as possibly cholinergic until other evidence is adduced. In a few cases radiolabelled choline transport or double labelling has provided firmer evidence. Neurons in the gigantocellular reticular formation, lateral reticular nucleus and red nucleus accumulate choline following administration via the spinal cord (PARE et al. 1982), while peduculopontine neurons accumulate it following administration via the thalamus (SUGIMOTO and HATTORI 1983). Double labelling experiments indicate cholinergic projections from the parabrachial-pedunculopontine area to the cortex (VINCENT et al. 1983a), from the parabigeminal nucleus in the brainstem to the superior colliculus (MUFSON et al. 1985), and from cells near but not in the Edinger-Westphal nucleus to the ciliary ganglion (BREZINA et al. 1985). Some of the ascending ACh neurons from the tegmental area have been reported to contain SP (VINCENT et al. 1983a).

The positively staining cells in the superior olivary complex are presumably associated with relay pathways of the acoustic system and may be involved in cochlear function. The vestibular system as well would appear to contain a cholinergic component. The projections of these and other minor cholinergic cell groups remain, however, to be explored.

D. Conclusions

ChAT immunohistochemistry has made possible the definition of five major cholinergic systems in brain. These are (a) medial basal forebrain, (b) the extrapyramidal, (3) the parabrachial, (d) the cranial motor, and (e) the reticular. There are still points of controversy and it seems probable that more cholinergic cell groups than those now generally accepted will eventually be definitively identified. The projections of many of the forebrain cholinergic cells have been studied but much remains to be done on pathways, particularly with regard to the hindbrain cells. Double-labelling experiments should be especially useful at the light-microscopic level to define pathways and at the electron-microscopic level to reveal synaptic interconnections. It can be anticipated that much further work along these lines will soon be done. Sufficient is now known about central cholinergic systems, however, to support speculations as to their roles in various brain functions (McGEER et al. 1984a).

References

Altschuler RA, Kachar B, Rubio JA, Parakkal MH, Fex J (1985) Immunocytochemical localization of choline acetyltransferase-like immunoreactivity in the guinea pig cochlea. Brain Res 338:1–11

Arimatsu Y, Seto A, Amano T (1981) An atlas of α-bungarotoxin binding sites and structures containing acetylcholinesterase in the mouse central nervous system. J Comp Neurol 198:603–631

Armstrong DM, Saper CB, Levey AI, Wainer BH, Terry RD (1982) Immunocytochemical localization of choline acetyltransferase in the rat brain. Soc Neurosci Abstr 8:662

Armstrong DM, Kress Y, Terry BD (1983a) Electron microscopic immunocytochemical characterization of cholinergic neurons in the rat brain. Soc Neurosci Abstr 9:80

Armstrong DM, Saper CB, Levey AI, Wainer BH, Terry RD (1983b) Distribution of cholinergic neurons in rat brain; demonstration by the immunocytochemical localization of choline acetyltransferase. J Comp Neurol 216:53–68

Armstrong DM, Rottler A, Pickel VM (1985) Ultrastructural characterization of choline acetyltransferase neurons in the nucleus tractus solitarius of the rat. Soc Neurosci Abstr 11:864

Brezina L, Strassman A, Mason P, Eckenstein F, Maciewicz R (1985) Ciliary ganglion afferent neurons: location and choline acetyltransferase staining in the cat. Soc Neurosci Abstr 11:1040

Butcher LL, Gould E, Woolf NJ (1985) Morphologic characteristics of central cholinergic neurons in the rat. Soc Neurosci Abstr 11:1238

Carlsen J, Heimer L (1985) Ultrastructural localization of choline acetyltransferase in rat basolateral amygdala. Soc Neurosci Abstr 11:203

Carpenter MB, Sutin J (1983) Human neuroanatomy, 8th edn. Williams and Wilkins, Baltimore, pp 291, 296, 334, 372

Chao L-P, Karen Kan K-S, Hung F-M (1982) Immunohistochemical localization of choline acetyltransferase in rabbit forebrain. Brain Res 235:65–82

Cozzari C, Hartman BK (1980) Preparation of antibodies specific to choline acetyltransferase from bovine caudate nucleus and immunocytochemical localization of the enzyme. Proc Natl Acad Sci USA 77:7453–7457

Cuénod M, Bagnoli P, Beaudet A, Rustioni A, Wiklund L, Streit P (1982a) Retrograde migration of transmitter-related molecules. In: Weiss DG, Gorio A (eds) Axoplasmic transport in physiology and pathology. Springer, Berlin Heidelberg New York, pp 160–166

Cuénod M, Bagnoli P, Beaudet A, Rustioni A, Wiklund L, Streit P (1982b) Transmitter-specific retrograde labelling of neurons. In: Chan-Palay V, Palay SL (eds) Cytochemical methods in neuroanatomy. Liss, New York, pp 17–44

Eckenrode TC, Battisti W, Murray M (1985) Distribution of choline acetyltransferase (ChAT) in the interpeduncular nucleus. Soc Neurosci Abstr 11:373

Eckenstein F, Sofroniew MV (1983) Identification of central cholinergic neurons containing both choline acetyltransferase and acetylcholinesterase and of central neurons containing only acetylcholinesterase. J Neurosci 3:2286–2291

Eckenstein F, Thoenen H (1983) Cholinergic neurons in the rat cerebral cortex demonstrated by immunohistochemical localization of choline acetyltransferase. Neurosci Lett 36:211–215

Eckenstein FE, Quinn J, Baughman RW (1985) An immunohistochemical analysis of cholinergic innervation in the cerebral cortex. Soc Neurosci Abstr 11:864

Geffard M, McRae-Degueurce A, Souan ML (1985a) Immunocytochemical detection of acetylcholine in the rat central nervous system. Science 229:77–79

Geffard M, Vieillemaringe J, Heinrich-Rock A-M, Duris P (1985b) Anti-acetylcholine antibodies and first immunocytochemical application in insect brain. Neurosci Lett 57:1–6

Gould E, Woolf NJ, Butcher LL (1985) Choline acetyltransferase neurons and their acetylcholinesterase staining patterns. Soc Neurosci Abstr 11:371

Haring JH, Davis JN (1982) Acetylcholinesterase-containing nucleus basalis neurons of the lateral hypothalamus project to the spinal cord. Soc Neurosci Abstr 8:517

Hattori T, Singh VK, McGeer EG, McGeer PL (1976) Immunohistochemical localization of choline acetyltransferase containing neostriatal neurons and their relationship with dopaminergic synapses. Brain Res 102:164–173

Hattori T, McGeer EG, Singh VK, McGeer PL (1977) Cholinergic synapse of the interpeduncular nucleus. Exp Neurol 55:102–111

Hedreen JC, Struble RG, Whitehouse PJ, Price DL (1984) Topography of the magnocellular basal forebrain system in human brain. J Neuropathol Exp Neurol 43:1–21

Houser CR, Crawford GD, Barber RP, Salvaterra PM, Vaughn JE (1983) Organization and morphological characteristics of cholinergic neurons: an immunocytochemical study with a monoclonal antibody to choline acetyltransferase. Brain Res 266:97–119

Ingram CA, Bolam JP, Wainer BH, Smith AD (1985) Cholinergic basal forebrain neurons with reference to their afferent and efferent connections: a correlated light and electron microscopic study. Neurosci Lett 22:S55

Isaacson LG, Tanaka D Jr (1985) Ultrastructural identification of cholinergic terminals onto thalamostriate projection neurons in the dog. Soc Neurosci Abstr 11:204

Izzo PN, Wainer BH, Rye DB, Levey AI, Bolam JP (1985) Substance P-immunoreactive terminals in synaptic contact with cholinergic neurons in the neostriatum. Neurosci Lett 22:S410

Jones EG, Burton H, Saper CB, Swanson LW (1976) Midbrain, diencephalic and cortical relationships of the basal nucleus of Meynert and associated structures in primates. J Comp Neurol 167:385–420

Kása P (1978) Distribution of cholinergic neurons within the nervous system revealed by histochemical methods. In: Biesold D (ed) Synaptic transmission. Karl Marx University, Leipzig, pp 45–55

Kemp JM, Powell TPS (1971) The structure of the caudate nucleus of the cat: light and electron microscopy. Philos Trans R Soc Lond [Biol] 262:383–401

Kimura H, McGeer PL, Peng JH, McGeer EG (1980) Choline acetyltransferase containing neurons in rodent brain by immunohistochemistry. Science 208:1057–1059

Kimura H, McGeer PL, Peng JH, McGeer EG (1981) The central cholinergic system studied by choline acetyltransferase immunohistochemistry in the cat. J Comp Neurol 200:151–201

Kimura H, McGeer EG, Peng F, McGeer PL (1983) Cholinergic systems in the cat cortex studied by choline acetyltransferase immunohistochemistry. In: Sano Y, Ibata Y, Zimmerman EZ (eds) Structure and function of peptidergic and aminergic neurons. Japan Scientific Societies, Tokyo, pp 263–274

Kimura H, McGeer PL, Peng JH, McGeer EG (1984) Choline acetyltransferase-containing neurons in the rat brain. In: Björklund A, Hökfelt T, Kuhar MJ (eds) Handbook of chemical neuroanatomy. Elsevier, Amsterdam, pp 51–67

Kostovic I (1982) Distribution of acetylcholinesterase-reactive cell bodies and fibers in the frontal granular cortex of the human brain. Soc Neurosci Abstr 8:933

Kristt DA, McGowan RA Jr, Martin-MacKinnon N, Solomon J (1985) Basal forebrain innervation of rodent neocortex: studies using acetylcholinesterase histochemistry, Golgi and lesions strategies. Brain Res 337:19–39

Kuhar MJ, Yamamura HI (1976) Localization of cholinergic muscarinic receptors in rat brain by light microscopic radioautography. Brain Res 110:229–243

Lehmann JC, Fibiger HC (1979) Acetylcholinesterase and the cholinergic neuron. Life Sci 25:1939–1947

Lehmann JC, Nagy JI, Atmadja S, Fibiger HC (1980) The nucleus basalis magnocellularis: the origin of a cholinergic projection to the neocortex of the rat. Neuroscience 5:1161–1174

Levey AI, Mufson EJ, Mesulam MM, Wainer BH (1982) Co-localization of choline acetyltransferase and acetylcholinesterase in mammalian forebrain. Soc Neurosci Abstr 8:134

London Ed, Waller SB, Wamsley JK (1985) Autoradiographic localization of [^{3}H]nicotine binding sites in the rat brain. Neurosci Lett 53:179–184
Lynch GS, Lucas PA, Deadwyler SA (1972) The demonstration of acetylcholinesterase containing neurons within the caudate nucleus of the rat. Brain Res 45:617–621
Lysakowski A, Wainer BH, Rye DB, Bruce G, Hersh LB (1985) Cholinergic cortical innervation: a correlative study of choline acetyltransferase-immunoreactive versus acetylcholinesterase positive fiber distribution. Soc Neurosci Abstr 11:979
Maley B, Elde R, Wainer B, Levey A (1982) The immunohistochemical localization of choline acetyltransferase-like immunoreactivity at the light and electron microscopic levels within the central nervous system of the guinea pig. Soc Neurosci Abstr 8:518
Matthews DA, Salvaterra PM, Crawford GD, Houser CR, Vaughn JE (1983) Distribution of choline acetyltransferase positive neurons and terminals in hippocampus. Soc Neurosci Abstr 9:79
McGeer PL, McGeer EG, Singh VK, Chase WH (1974) Choline acetyltransferase localization in the central nervous system by immunohistochemistry. Brain Res 81:373–379
McGeer PL, McGeer EG, Peng JH (1984a) Choline acetyltransferase: purification and immunohistochemical properties. Life Sci 34:2319–2338
McGeer PL, McGeer EG, Suzuki J, Dolman CE, Nagai T (1984b) Aging, Alzheimer's disease and the cholinergic system of the basal forebrain. Neurology (NY) 34:741–745
McKinney M, Hedreen J, Coyle JT (1982) Topographic analysis of the innervation of the rat neocortex and hippocampus by the basal forebrain cholinergic fibers. Soc Neurosci Abstr 8:518
Mesulam MM, van Hoesen GW (1976) Acetylcholinesterase-rich projections from the basal forebrain of the rhesus monkey to neocortex. Brain Res 109:152–157
Mesulam MM, Mufson EJ, Levey AI, Wainer BH (1983a) Cholinergic innervation of cortex by basal forebrain cytochemistry and cortical connections of the septal area, diagonal band nuclei, nucleus basalis (substantia innominata), and hypothalamus in the rhesus monkey. J Comp Neurol 214:170–197
Mesulam MM, Mufson EJ, Wainer BH, Levey AI (1983b) Central cholinergic pathways in the rat: an overview based on an alternative nomenclature (Ch1–Ch6). Neuroscience 10:1185–1201
Mizukawa K, McGeer PL, Tago H, Peng JH, McGeer EG, Kimura H (1986) The cholinergic system of the human hindbrain studied by choline acetyltransferase immunohistochemistry and acetylcholinesterase histochemistry. Brain Res 379:39–55
Mufson EJ, Levey A, Wainer B, Mesulam MM (1982) Cholinergic projections from the mesencephalic tegmentum to neocortex in rhesus monkey. Soc Neurosci Abstr 8:135
Mufson EJ, Martin TL, Mas DC, Wainer BH, Mesulam MM (1985) Cholinergic projections from the parabigeminal nucleus (Ch8) to the superior colliculus in the mouse. Soc Neurosci Abstr 11:1238
Nagai T, Kimura H, Maeda T, McGeer P, McGeer EG (1982) Cholinergic projections from the basal forebrain of rat to the amygdala. J Neurosci 2:513–520
Nagai T, McGeer PL, Peng JH, McGeer EG, Dolman CE (1983a) Choline acetyltransferase immunohistochemistry in brains of Alzheimer's disease patients and controls. Neurosci Lett 36:195–199
Nagai T, Pearson T, Peng JH, McGeer EG, McGeer PL (1983b) Immunohistochemical staining of the human forebrain with monoclonal antibody to human choline acetyltransferase. Brain Res 265:300–306
Pare MG, Jones BE, Beaudet A (1982) Application of a selective retrograde labelling technique to the identification of acetylcholine subcorticospinal neurons. Soc Neurosci Abstr 8:517
Parnavelas JG, Kelly W, Burnstock G (1985) Ultrastructural localization of choline acetyltransferase in vascular endothelial cells in rat brain. Nature 316:724–725
Peng JH, Kimura H, McGeer PL, McGeer EG (1981) Anti-choline acetyltransferase fragments antigen binding (Fab) for immunohistochemistry. Neurosci Lett 21:281–285
Peng JH, McGeer PL, McGeer EG (1986) Membrane-bound choline acetyltransferase from human brain; preparation and properties. Neurochem Res 11:959–971

Penney JB Jr, Young AB (1982) Quantitative autoradiography of neurotransmitter receptors in Huntington's disease. Neurology (NY) 32:1391–1395
Phelps PE, Vaughn JE (1985) Variations in choline acetyltransferase distribution within rat ventral striatum; an immunocytochemical study. Soc Neurosci Abstr 11:203
Phelps PE, Houser CR, Vaughn JE (1985) Immunocytochemical localization of choline acetyltransferase within the rat neostriatum: a correlated light and electron microscopic study of cholinergic neurons and synapses. J Comp Neurol 238:286–307
Richardson PJ, Walker JH, Jones RT, Whittaker VP (1982) Identification of a cholinergic-specific antigen Chol-1 as a ganglioside. J Neurochem 38:1605–1614
Ross ME, Park DH, Teitelman G, Pickel VM, Reis DJ, Joh TH (1983) Immunohistochemical localization of choline acetyltransferase using a monoclonal antibody: a radioautographic method. Neuroscience 10:907–922
Rotter A, Birdsall NJM, Burgen ASV, Field PM, Hulme EC, Raisman G (1979a) Muscarinic receptors in the central nervous system of the rat. I. Technique for autoradiographic localization of the binding of [^{3}H]propylbenzilylcholine mustard and its distribution in the forebrain. Brain Res Rev 1:141–165
Rotter A, Birdsall NJM, Field PM, Raisman G (1979b) Muscarinic receptors in the central nervous system of the rat. II. Distribution of binding of [^{3}H]propylbenzilylcholine mustard in the midbrain and hindbrain. Brain Res Rev 1:167–183
Saper CB, Loewy AD (1982) Reciprocal parabrachial-cortical connections in the rat. Brain Res 242:33–40
Satoh K, Fibiger HC (1985) Distribution of central cholinergic neurons in the baboon (*Papio papio*). I. General morphology. J Comp Neurol 236:197–214
Satoh K, Armstrong DM, Fibiger HC (1983a) A comparison of the distribution of central cholinergic neurons as demonstrated by acetylcholinesterase pharmacohistochemistry and choline acetyltransferase imunohistochemistry. Brain Res Bull 11:693–720
Satoh K, Staines WA, Atmadja S, Fibiger HC (1983b) Ultrastructural observations of the cholinergic neuron in the rat striatum as identified by acetylcholinesterase pharmacohistochemistry. Neuroscience 10:1121–1136
Shiromani P, Sutin E, Armstrong DA, Gillin JC (1985) Choline acetyltransferase containing cells in the brainstem: relationship to carbachol microinfusion sites. Soc Neurosci Abstr 11:1281
Shute CCD, Lewis PR (1967) The ascending cholinergic reticular system: neocortex, olfactory and subcortical projections. Brain 90:497–520
Sofroniew MV, Eckenstein F, Thoenen H, Cuello AC (1982) Topography of choline acetyltransferase-containing neurons in the forebrain of the rat. Neurosci Lett 33:7–12
Stanton GB, Orr A (1985) [^{3}H]Choline labelling of cerebellothalamic neurons with observations on the cerebello-thalamo-parietal pathway in cats. Brain Res 335:237–243
Stichel CC, Singer W (1985) Organization and morphological characteristics of choline acetyltransferase-containing fibers in the visual thalamus and striate cortex of the cat. Neurosci Lett 53:155–160
Strauss W, Hilt D, Puckett C et al. (1985) Cloning of a human choline acetyltransferase cDNA. Soc Neurosci Abstr 11:1114
Sugimoto T, Hattori T (1983) The nucleus tegmenti pedunculopontinus pars compacta in the rat: organization, efferent projections and cholinergic aspects. Soc Neurosci Abstr 9:283
Tago H, McGeer PL, Mizukawa K, McGeer EG (1985) Cholinergic systems of the human brain studied by acetylcholinesterase histochemistry. Soc Neurosci Abstr 11:863
Tago H, McGeer PL, Bruce G, Hersh LB (1987) Distribution of choline acetyltransferase-containing neurons of the hypothalamus. Brain Res 415:49–62
Talbot K, Woolf NJ, Butcher LL (1985) Basic organization and cholinergic features of the feline islands of Calleja complex. Soc Neurosci Abstr 11:1223
Vickroy TW, Roeske WR, Gehlert DR, Wamsley JK, Yamamura HI (1985) Quantitative light microscopic autoradiography of [^{3}H]hemicholinium-3 binding sites in the rat central nervous system: a novel biochemical marker for mapping the distribution of cholinergic nerve terminals. Brain Res 329:368–373

Vincent SR, Reiner PB (1987) The immunohistochemical localization of choline acetyltransferase in the cat brain. Brain Res Bull 18:371–415

Vincent SR, Satoh K, Armstrong DM, Fibiger HC (1983a) Substance P in the ascending cholinergic reticular system. Nature 306:668–671

Vincent SR, Satoh K, Armstrong DM, Fibiger HC (1983b) NADPH-diaphorase: a selective histochemical marker for the cholinergic neurons of the pontine reticular formation. Neurosci Lett 43:31–36

Vincent SR, Staines WA, Fibiger HC (1983c) Histochemical demonstration of separate populations of somatostatin and cholinergic neurons in the rat striatum. Neurosci Lett 35:111–114

Wainer BH, Levey AI, Rye DB, Mesulam MM, Mufson EJ (1985) Cholinergic and non-cholinergic septohippocampal pathways. Neurosci Lett 54:45–52

Wenk H, Bigl V, Meyer U (1980) Cholinergic projections from magnocellular nuclei of the basal forebrain to cortical areas in rats. Brain Res Rev 2:295–316

Wheeler DA, Simpson JB (1985) Cholinergic afferents to the subfornical organ in rats. Soc Neurosci Abstr 11:178

Yamamura HI, Vickroy TW, Gehlert DR, Wamsley JK, Roeske WR (1985) Autoradiographic localization of muscarinic agonist binding sites in the rat central nervous system with (+)-cis-[^{3}H]methyldioxolane. Brain Res 325:340–344

CHAPTER 23

Central Cholinergic Transmission: The Physiological Evidence

K. KRNJEVIĆ

A. Introduction

It must be made clear at the outset that this chapter deals with cholinergic transmission only in the vertebrate central nervous system (CNS) (including retina), mainly as studied by electrophysiological techniques. There is a vast literature on cholinergic synapses in invertebrate ganglia, and interested readers should consult some appropriate articles as those of GERSCHENFELD (1973), MARDER and PAUPARDIN-TRITSCH (1978) and KEHOE and MARTY (1980).

Another necessary clarification is that, there is no such thing as a single type of central cholinergic synapse. In the first place, in spite of the biochemical and cytochemical evidence presented in the previous two chapters there is still no central synapse that is incontrovertibly cholinergic – where all the essential criteria for demonstrating the nature of the transmitter have been fulfilled (adequate presynaptic release and a post-synaptic action identical with that of the natural transmitter, supported by relevant neurochemical and pharmacological evidence). Secondly, there is not just one kind of cholinergic action in the CNS. Judging by the variety of excitatory and inhibitory effects of acetylcholine (ACh), this is patently not the case. Just as in the peripheral nervous system, cholinergic synapses in the CNS will undoubtedly prove to have a variety of quite different actions, according to the type of receptor and ionic channel that they activate (or inactivate).

Two main pharmacological types are evident in the CNS (as elsewhere), the nicotinic and the muscarinic. To a perhaps surprising extent they conform closely to the characteristics originally described by DALE (1914, 1938): nicotinic actions are predominantly excitatory, with a rapid time course, whereas muscarinic excitations are typically slow in onset and prolonged. Cholinergic inhibitions are mainly muscarinic.

Although extensive surveys have demonstrated binding sites for various cholinergic ligands in all parts of the CNS (e.g. BIRDSALL and HULME 1976; OSWALD and FREEMAN 1981; BURGEN 1984), by themselves they cannot be taken as significant evidence of anything more than the *possibility* of cholinergic transmission.

B. Nicotinic Synapses

I. Characteristics of Peripheral Synapses

1. General Features

Nicotinic excitatory transmission is by far the most extensively studied mode of synaptic transmission in vertebrates, widely familiar as a result of a long history of systematic investigations of the neuromuscular junction (NMJ) (DALE 1938; DEL CASTILLO and KATZ 1956; KATZ 1966; TAKEUCHI 1977; HOYLE 1983) and sympathetic ganglia (BIRKS and MACINTOSH 1961; GINSBORG 1973; MACINTOSH and COLLIER 1976; BROWN et al. 1981). This is a characteristically quick action: end-plate potentials (EPPs) and ganglionic excitatory postsynaptic potentials (EPSPs) have a sharp onset and a duration no longer than expected from the membrane time constant. The fast time course of action is confirmed by the brief duration of synaptic currents under voltage clamp (TAKEUCHI and TAKEUCHI 1959) or in studies of single channels (KATZ and MILEDI 1972; FENWICK et al. 1982).

2. Ionic Mechanisms

FATT and KATZ (1951) first demonstrated a dramatic increase in muscle end-plate permeability which they described as a non-selective short-circuiting of the membrane. A closer examination by TAKEUCHI (1963) revealed a more selective process involving only cations: the relative increase in G_{Na}, G_K and G_{Ca} gives a reversal potential somewhat on the negative side of 0 mV. Because the normal resting potential is nearly 60–70 mV more negative than this reversal potential, the nicotinic action generates a sharp inward current and therefore has a strong depolarizing effect. Typically, the action is very rapid, and quickly reversible as ACh rapidly dissociates from the receptors and is removed by diffusion and by the action of acetylcholinesterase (AChE). In the continued presence of ACh, nicotinic ACh receptors (nAChRs) readily change to an inactive state (KATZ and THESLEFF 1957; ADAMS 1981), but there is little evidence that such desensitization can occur as a result of natural ACh release.

A similar nicotinic mechanism operates in sympathetic ganglia (BLACKMAN et al. 1963; KUBA and KOKETSU 1978) and other peripheral nicotinic synapses (e.g. on chromaffin cells, DOUGLAS et al. 1967; FENWICK et al. 1982). The ganglionic nAChRs show some distinct properties, notably in their differential sensitivity to various nicotinic antagonists: they are not readily blocked by α-bungarotoxin (α-BTX), are relatively insensitive to curare, but are susceptible to hexamethonium, tetraethylammonium and mecamylamine (BROWN 1979; FENWICK et al. 1982). Moreover, they are resistant to desensitization (HAEFELY 1974).

II. Central Nicotinic Synapses

Contrary to expectations, the best-known peripheral cholinergic mechanism has proved to be quite rare in the CNS. So far, really strong evidence of the operation of nicotinic synapses has been obtained at only two sites in mammals, on Renshaw cells in the spinal cord and ganglion cells in the retina.

1. Motor Axon Collaterals: Renshaw Cell Pathway

a) Pathway and Morphology

This recurrent pathway was identified in electrical recordings from the spinal cord by RENSHAW (1946) long before it was first suspected that the synapses in question – like those made by motor axons on skeletal muscle – might be cholinergic (ECCLES et al. 1954). Indeed, the very existence of Renshaw cells remained open to question for two decades (WILLIS 1971) before convincing morphological evidence became available (JANKOWSKA and LINDSTRÖM 1971; VAN KEULEN 1979; LAGERBÄCK and RONNEVI 1982).

The synapses made by motor and axon collaterals have been identified after injection of horseradish peroxidase (HRP) into motoneurons (LAGERBÄCK et al. 1978, 1981). According to these authors, the synaptic boutons on presumed Renshaw cells are of two kinds: 'en passant' on dendrites and terminal on cell bodies (Fig. 1a). The boutons contain spherical vesicles fully comparable to the vesicles seen at peripheral cholinergic junctions and at other synapses in the CNS (Fig. 1b). [Curiously enough, similar synapses are present on motoneurons (CULLHEIM and KELLERTH 1978), indicating a possible monosynaptic recurrent cholinergic input to motoneurons, but its physiological role is unknown.]

The inhibitory function of Renshaw cells is now well established (RENSHAW 1946; ECCLES 1957; HULTBORN et al. 1971; RYALL et al. 1971; HULTBORN and PIERROT-DESEILLIGNY 1979). There is reason to believe that it is mediated by the release of both glycine and γ-aminobutyrate (GABA) (CULLHEIM and KELLERTH 1981). Perhaps its most important role is to inhibit the Ia inhibitory interneuron (HULTBORN et al. 1971), as well as other Renshaw cells (RYALL et al. 1971), but what part it plays in locomotion remains obscure (PRATT and JORDAN 1980).

b) Synaptic Electrophysiology and Pharmacology

The cholinergic nature of synapses on Renshaw cells was first indicated by the excitatory effects of a variety of cholinergic agonists as well as the potentiation of transmission by anticholinesterases (anti-ChEs) (Fig. 2a) and its depression by classical nicotine antagonists (Fig. 2b), all administered by close arterial or systemic injection (ECCLES et al. 1954, 1956). These observations were confirmed in subsequent microiontophoretic experiments (CURTIS and ECCLES 1958; CURTIS and RYALL 1966a). As illustrated in Fig. 3, Renshaw cells are highly sensitive to brief applications of ACh and nicotinic agonists, nicotine itself being the most potent. All evoke a fast discharge, after a very short latency; the effect of ACh typically stops almost instantly at the end of the application, but the other agonists show a brief after-effect (Fig. 3a). The rapid firing that can thus be evoked does not approach, however, the very high frequency (1000 Hz or more) firing that is characteristically elicited by even moderate ventral root stimulation (Fig. 2a, b; ECCLES et al. 1954; ECCLES 1957; CURTIS and ECCLES 1958; CURTIS and RYALL 1966c).

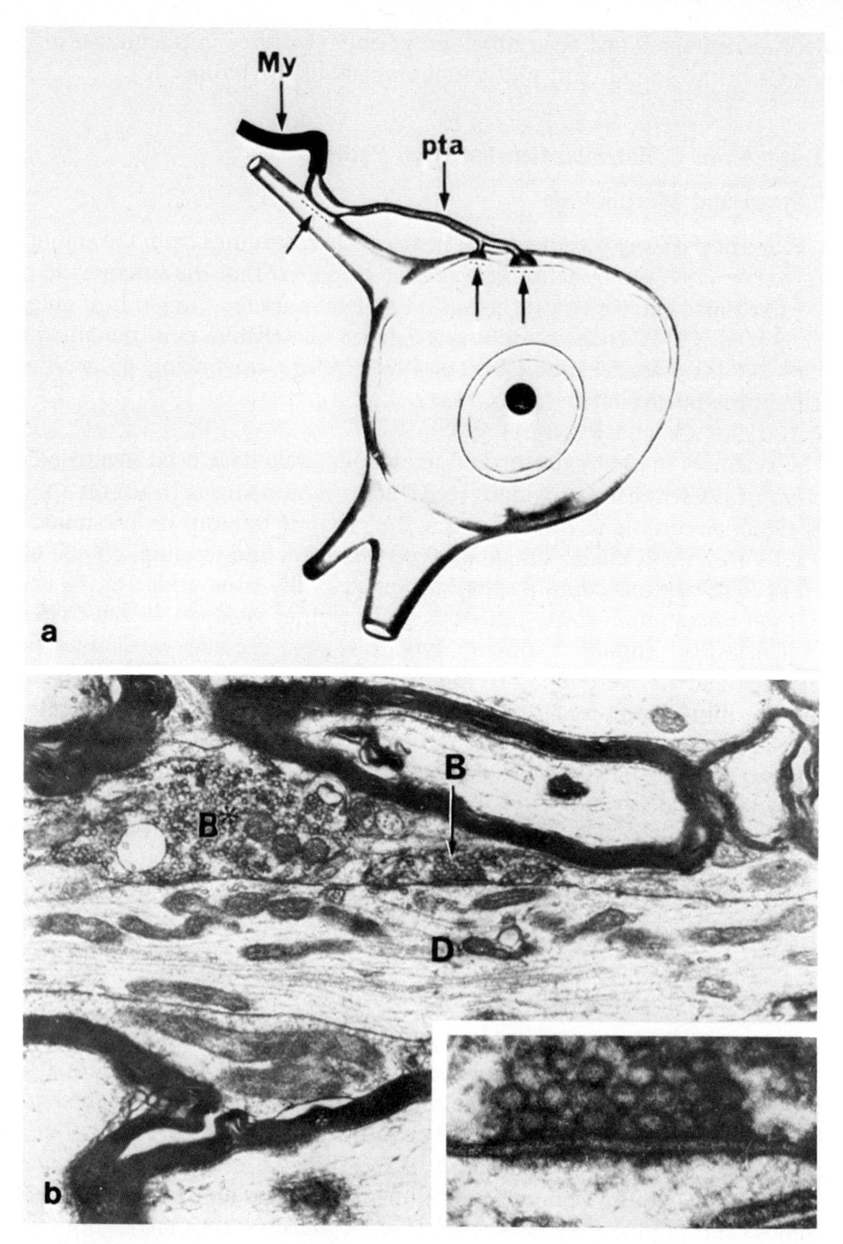

Fig. 1 a, b. Cholinergic synapses on Renshaw cells in the cat's spinal cord. **a** Reconstruction from electron micrographs of motor axon collateral and synaptic contacts (*arrows*) with dendrite and cell body of presumed Renshaw cell. *My,* myelinated part of axon collateral; *pta*, preterminal region. Magnification ×1400. **b** Electron micrograph showing end-bouton (*B*) of HRP-stained motor axon collateral making synaptic contact with dendrite (*D*) in lamina 7. *B** marks an 'en passant' bouton (×21000). *Inset* shows synaptic complex of end-bouton (×100000). (From LAGERBÄCK et al. 1981, Figs. 8, 3)

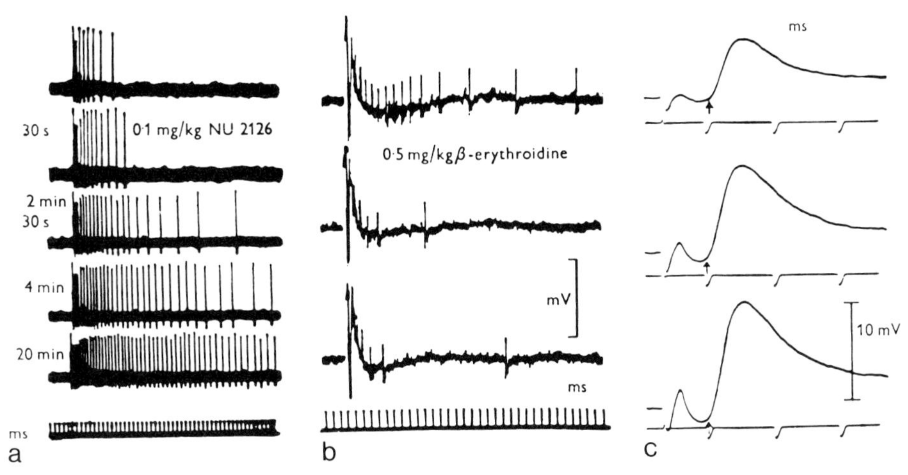

Fig. 2a–c. Cholinergic synaptic activation of Renshaw cells. **a** Extracellular records of responses of two Renshaw cells, evoked by constant antidromic stimulation of ventral root, show great enhancement after intravenous injection of the anti-ChE agent NU 2126. **b** Block by cholinergic antagonist β-erythroidine hydrochloride also administered intravenously (From ECCLES et al. 1956, Figs. 8, 11). **c** EPSPs recorded intracellularly from another Renshaw cell, evoked by increasing intensity of ventral root stimulation from *top* to *bottom*. (From ECCLES et al. 1961, Fig. 5)

The receptors on Renshaw cells are much more susceptible to blocking by mecamylamine (UEKI et al. 1961), hexamethonium and tetraethylammonium (CURTIS and RYALL 1966a) than by curare, and not at all sensitive to α-BTX (DUGGAN et al. 1976). As pointed out earlier KRNJEVIĆ 1974; BROWN 1979), they therefore conform more closely to nAChRs in ganglia than those in muscle, but they differ from ganglionic receptors in showing clear desensitization during longer (and relatively large) applications (CURTIS and RYALL 1966a).

c) Membrane Mechanisms

Occasional and only short-lasting, intracellular records have been reported from Renshaw cells. ECCLES et al. (1961) described graded EPSPs corresponding in duration to the high frequency bursts of spikes seen in extracellular recordings (Fig. 2c); and ZIEGLGÄNSBERGER and REITER (1974) observed in three Renshaw cells a conductance increase during applications of ACh. These meagre data are in keeping with a synaptic mechanism similar to that of the nicotinic EPSP in sympathetic ganglia (BLACKMAN et al. 1963; KUBA and KOKETSU 1978); but they hardly provide very compelling evidence. They do not explain why these neurons fire in the characteristic high-frequency bursts. It has been suggested that an impermeable barrier may prevent the rapid diffusion of ACh and thus prolong its action – which might also explain some apparently anomalous pharmacological properties (ECCLES et al. 1954; CURTIS and ECCLES 1958). But there is no morphological evidence of such a barrier; it is more likely that the 'anomalous', low potency of some curariform antagonists reflects the ganglion cell-like character of Renshaw cells, and that the high frequency of firing is made possible by intrin-

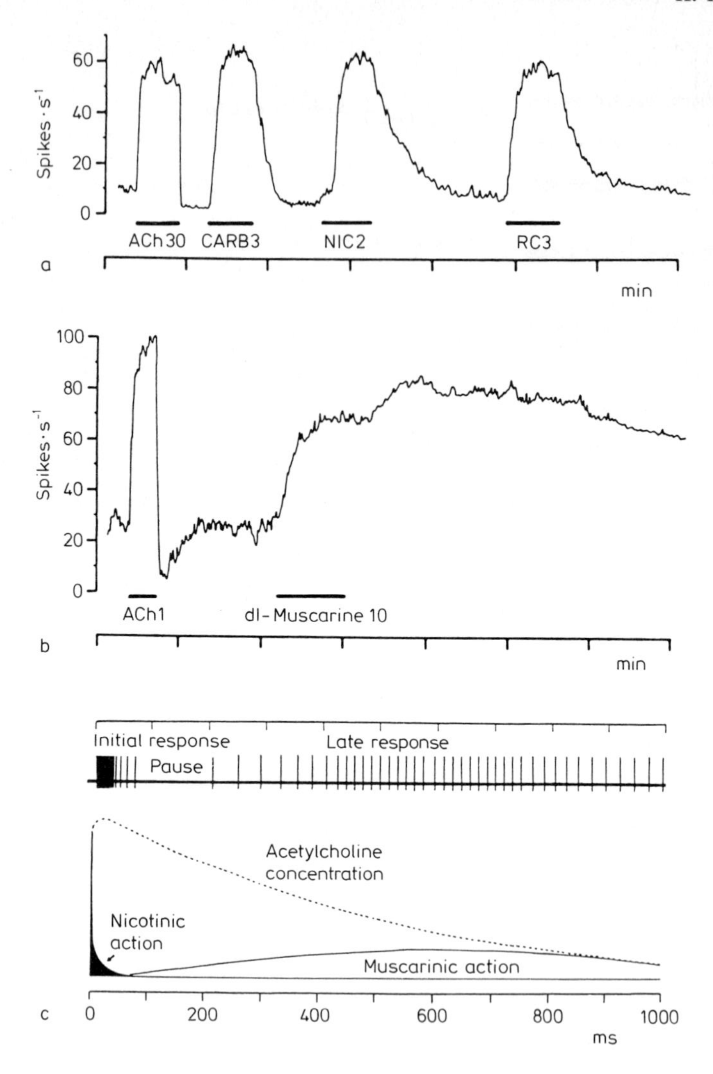

Fig. 3a–c. Renshaw cells have both nAChRs and mAChRs. **a** Ratemeter traces showing sharp excitation by ACh, carbachol (*CARB*), nicotine (*NIC*) and β-carbomethoxyethyl-trimethylammonium (*RC*). Iontophoretic currents (in nA) and times of application are indicated below ratemeter traces. Note sharp time course of action of ACh and greater potency of nicotine. **b** Muscarine also excites a Renshaw cell, but the effect is slower in onset and very much longer lasting. (From CURTIS and RYALL, Fig. 3). **c** Biphasic response of Renshaw cells to ACh. Note the initial fast nicotinic, and later, slower muscarinic responses. (From CURTIS and RYALL 1966c, Fig. 7)

sic membrane properties (e.g. the absence of A- or C-type K currents that tend to prevent repetitive firing).

d) Muscarinic Receptors on Renshaw Cells

Renshaw cells resemble sympathetic neurons in another respect: in both cases muscarinic receptors (mAChRs) are also present, which give rise to slower and much more prolonged discharges. Hence, as illustrated in Fig. 3b, Renshaw cells are also excited by muscarine and related agonists which evoke slower and much more prolonged responses (CURTIS and RYALL 1966b). In addition, atropine depresses a late component of the synaptically evoked firing without changing the early burst. CURTIS and RYALL (1966c) therefore concluded that nAChRs are responsible for the early firing elicited synaptically and muscarinic receptors for the later discharge (as illustrated diagrammatically in Fig. 3c): these would roughly correspond to the fast and slow EPSPs respectively in sympathetic ganglia. When tested by stimulating the spinal roots electrically, the mAChRs appear to be of only secondary importance; but they may play a much greater role during physiological activation (RYALL and HAAS 1975).

The mechanism of the slow muscarinic excitation has not been investigated in Renshaw cells. It is likely to be of the same type as the muscarinic excitations seen elsewhere in the CNS, which are discussed extensively below.

2. Cholinergic Amacrine Cells

a) Cholinergic Transmission in Retina

As a highly accessible offshoot of the brain – with a regular organization in layers that is readily studied *in vitro* – the retina provides some of the best conditions for systematic studies of central synaptic transmission, suitable for neurochemical, electrophysiological and pharmacological experiments. An important advantage over almost any other *in vitro* preparation of vertebrate CNS is that the retina contains its own sensory (photo-)receptors and therefore can be activated by precisely controlled physiological stimuli – in contrast to the usual, highly unphysiological electrical (or high K^+) stimulation.

Retinal transmitter mechanisms have been – and continue to be – extensively studied (cf. DAW et al. 1982; EHINGER 1982; STELL 1985). Most of the mammalian central transmitters have been identified in the retina. It has long been suspected that ACh has an important role there because of the retina's exceptionally high content of ACh and other cholinergic markers (e.g. FELDBERG 1945; HEBB 1957), making it comparable in this respect to the basal ganglia and spinal ventral roots. On the other hand, since the optic nerve is just as conspicuously *free* of cholinergic markers, which are also poorly evident in the outer retinal layers, the general consensus has placed the cholinergic elements somewhere in the inner layers. There is now strong evidence that the neurons in question are a particular type of amacrine cell. Some 22 varieties of amacrine cells have been described, possibly utilizing some eight different transmitters, including GABA, glycine, 5-hydroxytryptamine, dopamine, substance P and several other peptides (STERLING 1983; STELL 1985).

b) Target Cells

Having confirmed that in the isolated rabbit retina, ACh release is enhanced by light stimulation and that this release is Ca^{2+} dependent (MASLAND and LIVINGSTONE 1976), MASLAND and his collaborators went on to look at the effect of ACh and related agents on possible targets cells (MASLAND and AMES 1976). Many ganglion cells proved to be highly sensitive to ACh, especially those that responded to the onset of stimulation by a light spot in the centre of their receptive field ('on-centre cells') or to a spot moving in a certain preferred direction across their respective field ('directionally selective cells'). Most 'off-centre cells' – which are excited when a spot of light shining on the centre of their receptive field is switched off – were not very sensitive. ACh had a quick effect; it was not abolished by removal of Ca^{2+} and the addition of 20 mM Mg^{2+} but was very effectively blocked by tubocurarine (as illustrated in Fig. 4a), hexamethonium,

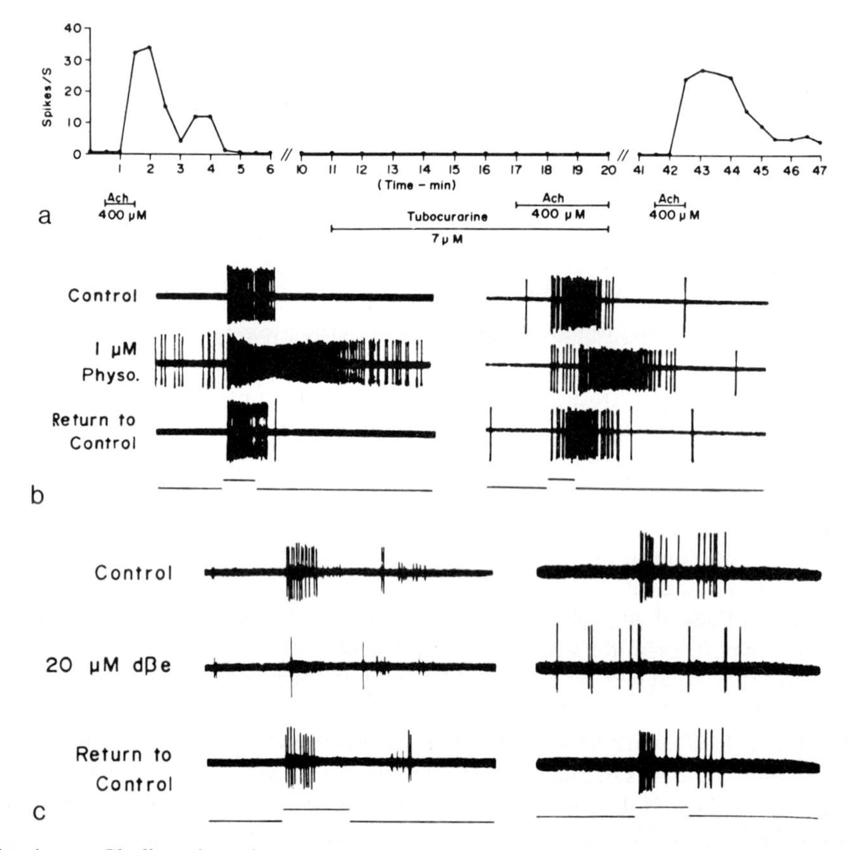

Fig. 4a–c. Cholinergic action on ganglion cells in rabbit retina *in vitro*. **a** Excitation of directionally selective ganglion cell by ACh; nicotinic character is demonstrated by blocking effect of tubocurarine. (From MASLAND and AMES 1976, Fig. 4). **b** Two on-centre cells: light-evoked responses are strongly potentiated by physostigmine (*Physo.*). (From MASLAND and AMES 1976, Fig. 3). **c** Two other on-centre cells whose responses are blocked by dihydro-*β*-erythroidine (*dβe*). (From MASLAND and AMES 1976, Fig. 5)

dihydro-β-erythroidine and mecamylamine. It therefore had all the familiar characteristics of peripheral nAChRs. Moreover, as can be seen in Fig. 4b, c, the responses evoked by light stimulation were strongly enhanced by eserine and depressed by nicotine antagonists. Clearly, many ganglion cells must receive a cholinergic innervation that is activated by appropriate light signals. Its nicotinic character was later confirm by ARIEL and DAW (1982b) when they showed that most ganglion cells are very sensitive to nicotine itself.

c) Identity of the Retinal Cholinergic Cells

The fact that cholinoceptive ganglion cells were excited especially by the onset of light made it likely that the ACh-releasing cells belong to the family of either bipolar or amacrine cells – both being depolarized in response to light, in contrast to photoreceptors and horizontal cells, which are hyperpolarized.

The identity of these cells has been revealed by a systematic series of investigations. Using ACh radioautography, MASLAND and MILLS (1979) first showed that ACh-rich cells were distributed in two groups symmetrically situated on each side of the inner plexiform layer. The outer group clearly belonged to the family of amacrine cells. The inner group, on the other hand, was in the *ganglion* cell layer, of which it made up almost a third. This was all the more surprising in view of the absence of cholinergic markers in the optic nerve (which is made up by the axons of ganglion cells). The puzzle was solved when it turned out that one-third of the neurons in the ganglion cell layer survive optic nerve lesions, and therefore cannot be true ganglion cells. Evidently, these are displaced amacrine cells, of which the majority are probably cholinergic. They appear to be identical with the striking 'starburst' amacrines – characterized by an exceptionally rich, radially oriented dendritic plexus – which are also known to be arranged symmetrically on each side of the inner plexiform layer (FAMIGLIETTI 1983a, b). This was directly demonstrated by first showing that true ganglion cells and the cholinergic amacrines can be selectively labelled when a fluorescent cell marker is applied to the optic nerve or injected directly into the eye (MASLAND et al. 1984b). Individual cells were then filled with Lucifer Yellow (TAUCHI and MASLAND 1984). Their rich, radial system of dendrites can be seen in Fig. 5.

The cholinergic amacrine cells thus bracket the inner plexiform layer. Their dendrites form a dense and overlapping plexus in two distinct sublaminae, in the inner and outer halves of the plexiform layer (Fig. 6), in keeping with the idea that 'on' and 'off' responses are mediated by separate pathways (ARIEL and DAW 1982a, b; FAMIGLIETTI 1983a, b; MASLAND et al. 1984a).

The retinal cholinergic synapse is unusual in having an exclusively nicotinic character. There is no evidence as yet concerning the mechanism of action of ACh on ganglion cells. One can only speculate that it is likely to be similar to the nicotinic actions on peripheral structures.

3. Other Central Nicotinic Synapses

The only serious evidence of a predominantly nicotinic pathway is for retino-tectal fibres in some lower vertebrates – in fish and amphibia (OSWALD et al. 1979;

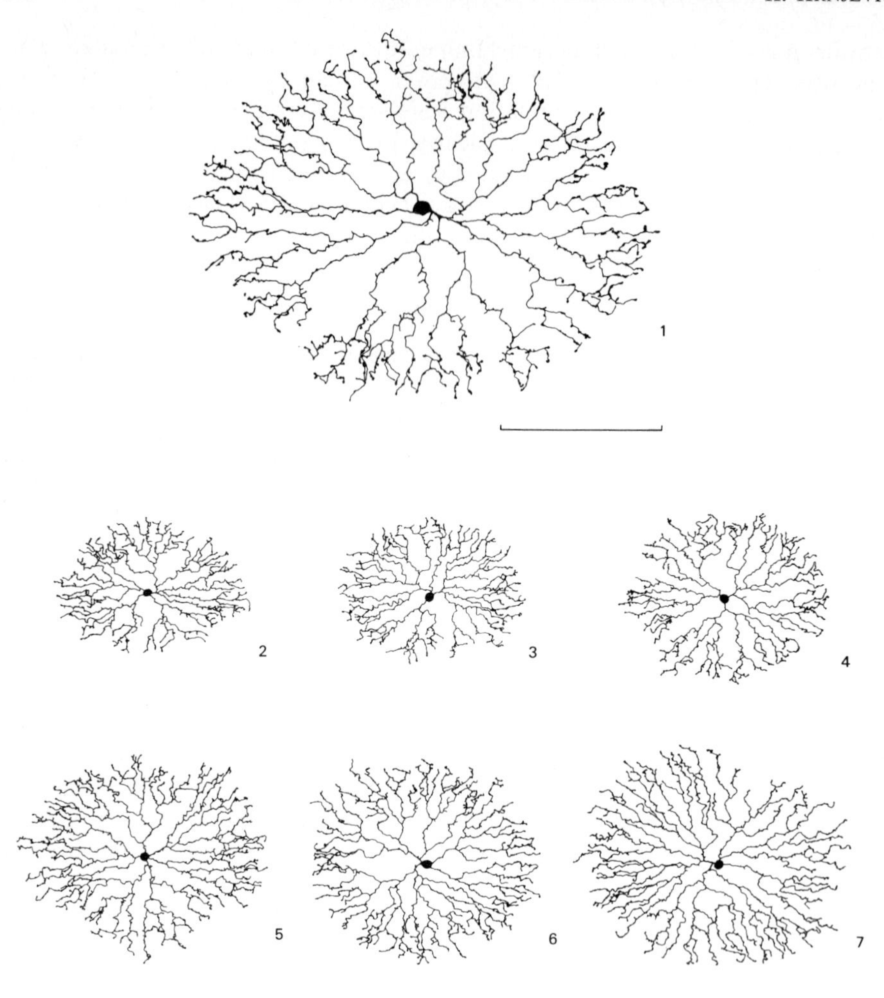

Fig. 5. Examples of cholinergic 'starburst' amacrine cells in rabbit retina. Calibrations: 100 μm. (From TAUCHI and MASLAND 1984, Fig. 7)

OSWALD and FREEMAN 1981; but compare HENLEY et al. 1986). According to these authors, transmission is blocked by α-BTX, a unique characteristic for central cholinergic pathways.

Many other central neurons are sensitive to nicotinic agonists and antagonists and therefore presumably have nAChRs. They are usually more or less mixed with mAChRs (e.g. in the thalamus, ANDERSEN and CURTIS 1964; PHILLIS 1971; the cortex, STONE 1972; and the hypothalamic supraoptic nucleus, DREIFUSS and KELLY 1972) but the cholinergic pathways involved have not been identified. nAChRs are also present on several types of presynaptic terminals, including cholinergic ones (CHESSELET 1984). They often work in opposition to presynaptic mAChRs (see below), and therefore, when activated, enhance transmitter release.

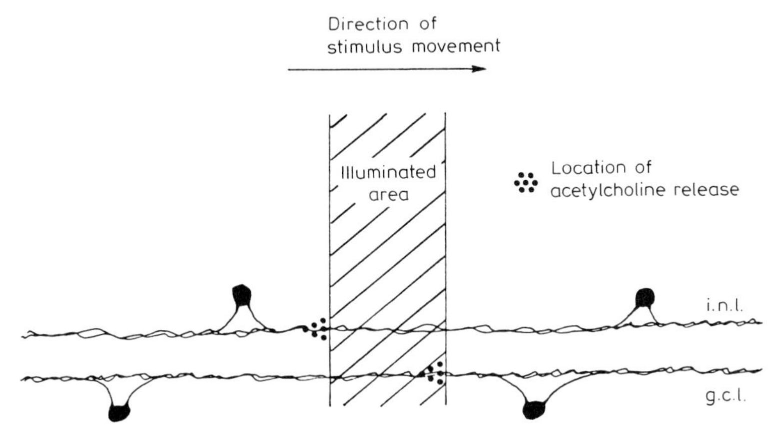

Fig. 6. Local dendritic responses generated selectively at two different levels by cholinergic amacrine cells activated by a moving light. *i.n.l.,* inner nuclear layer; *g.c.l.,* ganglion cell layer. (From MASLAND et al. 1984a, Fig. 9)

C. Muscarinic Synapses

I. General Considerations

In contrast to nicotinic actions, which have so far proved to be associated with highly specific and localized pathways, muscarinic actions in the CNS are very widespread throughout the brain and spinal cord. This was apparent even with studies of systemically administered drugs (FELDBERG 1945; STONE 1957) and became particularly evident when more precise testing of cellular response was made possible by microiontophoresis. In cat cortex (KRNJEVIĆ and PHILLIS 1963a, b) ACh receptors conform remarkably closely to the muscarinic characteristics of peripheral junctions (DALE 1938) with respect to time course of action and to the main pharmacological agonists and antagonists; comparable properties have subsequently been observed at numerous sites in the CNS (PHILLIS 1971; KRNJEVIĆ 1974; TEBECIS 1974), albeit sometimes in association with nAChRs (as in the case of Renshaw cells).

The central muscarinic synapses form a contrast to the nicotinic ones in another respect: the membrane mechanisms of the muscarinic action of ACh have been analysed and are known in substantial detail, but the relevant cholinergic pathways have so far not been amenable to precise electrophysiological investigation.

The characteristics and wide distribution of mAChRs have been extensively studied (BIRDSALL and HULME 1976; BURGEN 1984): this is the topic of Chap. 10 and therefore needs no further discussion here. Further evidence of cholinergic pathways is provided by neurochemical and histochemical data indicating the presence of ACh, choline acetyltransferase (ChAT) and AChE activity (see Chaps. 21, 22). By these criteria numerous cholinergic pathways should be present in the CNS. Perhaps the best defined is a wide system of presumed cholinergic fibres that originate from the basal forebrain and project to all parts of the cortex (including the hippocampal archicortex). This will therefore be described in particular detail.

II. Forebrain Cholinergic System

1. Origin, Distribution and Synaptic Morphology

A cholinergic innervation of the cortex had long been indicated by the presence and release of ACh both in neocortex and hippocampus (MacIntosh 1941; MacIntosh and Oborin 1953; Mitchell 1963; Dudar and Szerb 1969; Smith 1972; Dudar 1977). Histochemical techniques were needed, however, to identify the probable cholinergic cells and fibres. The pioneering observations were made using mainly AChE as a marker (Krnjević and Silver 1963, 1965, 1966; Hebb et al. 1963; Shute and Lewis 1963, 1967; Lewis et al. 1967). They demonstrated a profuse system of AChE-rich projections that was quite distinct from the principal cortical afferents in the internal capsule and appeared to have its origin in deep nuclei at the base of the forebrain. The telencephalic origin of these neurons is particularly clear during embryonic development (Krnjević and Silver 1966).

These pathways have been systematically examined by a combination of neuroanatomical (HRP) and histochemical techniques (including AChE and monoclonal antibodies for ChAT). From the work of Mesulam et al. (1984; see also Chaps. 21, 22, this volume), the following picture emerges. By far the greater part of the cholinergic innervation for cortical and limbic structures originates from four cell groups in the basal forebrain. The medial septal area and the vertical limb of the diagonal band of Broca provide most of the cholinergic innervation of the hippocampus, and the nucleus basalis-substantia innominata complex is the principal source of cholinergic fibres for the neopallium (about 90% of its large multipolar cells appear to be cholinergic, a total of 220000 per hemisphere in the human (Arendt et al. 1985)).

An important feature from an experimental point of view is that the nucleus basalis cholinergic neurons and fibres do not form compact nuclei and tracts, being very much interspersed with other cell groups and fibres in, or travelling through, the base of the forebrain. Therefore one cannot easily stimulate them selectively in isolation from other pathways. Even the medial septal/diagonal band region, which is relatively compact, contains large numbers of non-cholinergic cells that also project to the hippocampus (Baisden et al. 1984; Wainer et al. 1985).

The cholinergic projection is probably much less diffuse than was originally believed. Subgroups of basal forebrain neurons project to different cortical areas; moreover, according to Price and Stern (1983) individual neurons do not have extensive collaterals and thus probably innervate only a small region of cortex in the rat, not more than 1.0–1.5 mm wide. Another point of interest is that the putative cholinergic fibres end in classical synapses (Fibiger 1982; Wainer et al. 1984; Frotscher and Laranth 1986), having the usual attributes: synaptic vesicles, pre- and post-synaptic thickenings and narrow synaptic gap. These synapses, however, cannot be simply categorized in terms of function; according to FROTSCHER and LARANTH (1986), cholinergic synapses in the fascia dentata can be of both symmetric and asymmetric types, and the corresponding vesicles flattened or rounded. The majority are found in direct contact with the granule cell bodies and proximal dendrites, but a significant number (of both kinds) are on dendritic spines.

2. Muscarinic Excitation in Neocortex and Hippocampus

We now know that ACh has multiple actions on cortical cells; the predominant effect, however, is excitatory and muscarinic, as first became evident in the initial iontophoretic studies (KRNJEVIĆ and PHILLIS 1963a, b; SPEHLMANN 1963). In contrast to glutamate, which excited practically all detectable units, ACh had a slow and prolonged action, mainly on cells in the infragranular layers of the neocortex. It was also more labile, being markedly depressed during deep anaesthesia. Similar observations have been made by numerous authors in both neocortex or hippocampus (KRNJEVIĆ 1974; STRAUGHAN 1975; ANDERSEN and LANGMOEN 1980; LAMOUR et al. 1982).

3. Postsynaptic Actions of ACh in Cortex

a) Depression of K Outward Currents

Intracellular recordings from cortical cells (KRNJEVIĆ 1969; KRNJEVIĆ et al. 1971 b) revealed a new depolarizing mechanism, characterized by a fall rather than increase in membrane conductance; moreover it was enhanced by depolarization and reduced by hyperpolarization and so had a reversal potential close to −90 mV. As it was also insensitive to large changes in Cl^- gradient, it could be explained only by a diminished K conductance. In addition, there was evidence of a slowing of spike repolarization, suggesting some depression of an active K current. Remarkably similar effects were produced by Ba^{2+} (KRNJEVIĆ et al. 1971 a) which is well known as a blocker of K channels. They would satisfactorily account for an increased excitability and greatly prolonged after-discharges, which can enormously enhance responses evoked by either electrical (Fig. 7a) or synaptic inputs (KRNJEVIĆ 1974; DODD et al. 1981; SILLITO and KEMP 1983).

A similar depolarizing action, acompanied by a fall in conductance has been repeatedly observed in the cortex (WOODY et al. 1978), the hippocampus (DODD et al. 1981; BEN-ARI et al. 1981; BENARDO and PRINCE 1981, 1982; SEGAL 1982) and other areas (ZIEGLGÄNSBERGER and REITER 1974; CREPEL and DHANJAL 1982; NOWAK and MACDONALD 1983). According to WOODY et al. (1978) direct excitation of a cortical neuron during an application of ACh greatly prolongs the facilitating effect: this was seen as possibly of importance in the context of conditioned learning. A comparable prolongation of the action of ACh has recently been observed in extracellular recordings when cortical neurons were excited by glutamate *during* applications of ACh (METHERATE et al. 1986). Another significant observation made by BENARDO and PRINCE (1982) and later confirmed by GÄHWILER and DREIFUSS (1982) and by COLE and NICOLL (1983) is that ACh markedly depresses the after-hyperpolarization (AHP) that follows the action potential, thus clarifying the mechanism of enhancement of repetitive firing. A

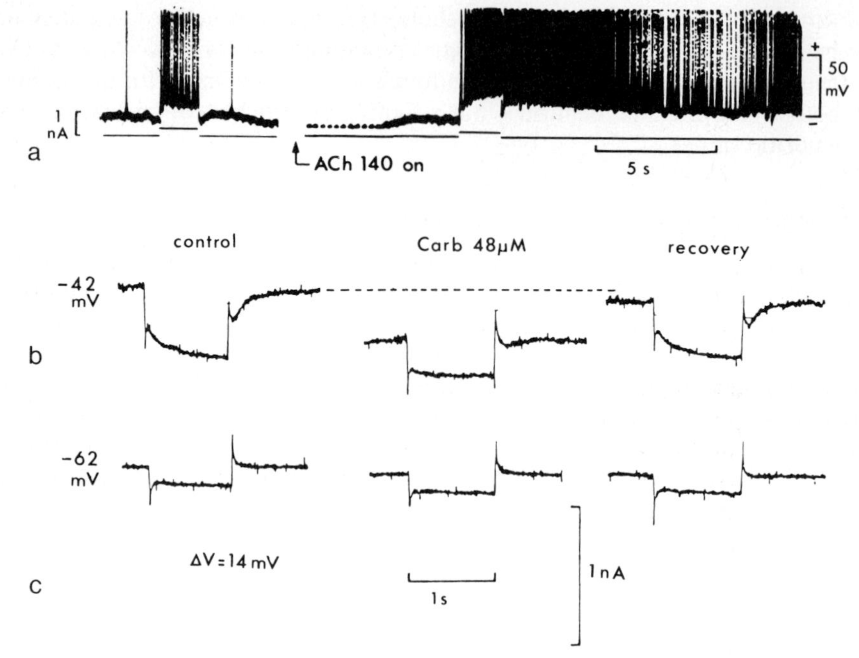

Fig. 7a–c. a Great enhancement of neuronal responsiveness in neocortex of cat by locally applied ACh. *Left,* Brief firing evoked by intracellular current pulse; *right,* application of ACh by itself causes no firing but greatly augments and especially prolongs response evoked by identical intracellular current pulse (latter is monitored in *lower trace*). (From KRNJEVIĆ et al. 1971 b, Fig. 5). **b, c** Evidence for M current and its depression by carbachol (*Carb*) in pyramidal cell in hippocampal slice. **b** When cell is voltage-clamped at relatively depolarized level of −42 mV, a 4-mV repolarizing pulse generates a slow current relaxation that reflects turning-off of voltage- and time-dependent M current; during application of carbachol, trace is lower, indicating a reduced total outward current, and slow current relaxation has disappeared. **c** Near resting potential (−62 mV); identical voltage step generates no current relaxations, and there is no effect of carbachol. (From HALLIWELL and ADAMS 1982, Fig. 1)

further point of interest was reported by GÄHWILER and DREIFUSS (1982): in hippocampal explants, the depolarizing action of ACh is abolished by Ca antagonists.

b) M Current Blockage

The next important advance came from voltage-clamp studies of hippocampal neurons by HALLIWELL and ADAMS (1982) which revealed an M current comparable to that originally discovered in frog sympathetic ganglia by BROWN and ADAMS (1980). This is a low threshold, non-inactivating K current, activated by small depolarizations from the resting level and which very effectively opposes any inward currents, thus reducing both excitability and any tendency to repetitive firing. A crucial feature of this current is that it is highly sensitive to block by muscarinic agonists. By suppressing the voltage-dependent M current (Fig. 7b, c), ACh would very effectively remove a strong brake on excitation and thus greatly facilitate any depolarizing input (of synaptic or other origin). Except for the change in spike repolarization or AHP, various features of the action of ACh on cortical and hippocampal neurons were very satisfactorily explained by the properties of the M current. In particular, the voltage-dependence of the M current accounted for the variability of the action of ACh. Little if any overt effect might be seen when ACh is applied to quiescent cells (Figs. 7a, c, 10) the M current being 'switched off' when the resting potential is high (very negative); and of course this would make it impossible to reverse the action of ACh by large hyperpolarizing currents. Nevertheless, ACh would always increase excitability and enhance repetitive firing (as in Figs. 7a, 8c) by suppressing the stabilizing effect of the M current.

c) Other Possible Mechanisms of K Current Depression

The M current may not provide the sole (or perhaps not even the principal) explanation for the muscarinic excitatory action. In the first place, there is clear evidence that muscarine can block voltage-independent K channels, for example in myenteric neurons (MORITA et al. 1982; NORTH and TOKIMASA 1982) and in spinal cord cells in culture (NOVAK and MACDONALD 1983). More information is needed regarding the voltage-dependence of muscarinic actions on central neurons.

In the second place, as already mentioned, ACh suppresses the after-hyperpolarization (AHP) of hippocampal neurons (BENARDO and PRINCE 1982; GÄHWILER and DREIFUSS 1982; GÄHWILER 1984; COLE and NICOLL 1984a, b). Somehow it must interfere with a Ca-activated K current, in keeping with observations in the myenteric plexus (MORITA et al. 1982) and even in frog sympathetic neurons (KUBA and KOKETSU 1978; PENNEFATHER et al. 1985). The block of the AHP, which normally tends to prevent repetitive firing, would also facilitate multiple spike responses (Fig. 8).

In some recent experiments on rat sympathetic ganglia, BELLUZZI et al. (1985) were unable to identify a muscarine-sensitive current having the properties of the M current [such as described by BROWN and ADAMS (1980)]. What they observed instead, was a depression of a Ca-sensitive K current, which appeared to be secondary to a block of Ca inward current. This was in agreement with some earlier findings in bullfrog sympathetic neurons by KUBA and KOKETSU (1978) and by AKASU and KOKETSU (1982). A primary depression of Ca influx is an interesting possibility, which could explain the suppression of AHPs the Ca dependence of the effect of ACh in hippocampal explants (GÄHWILER and DREIFUSS 1982), and perhaps also presynaptic actions of ACh, which appear to diminish transmitter release (see below).

d) Direct Facilitation of Inward Currents

A further complication however is that a muscarinic action may also *enhance* inward currents (KUBA and KOKETSU 1978; AKASU et al. 1984), although it is not fully clear whether this is a direct or indirect effect.

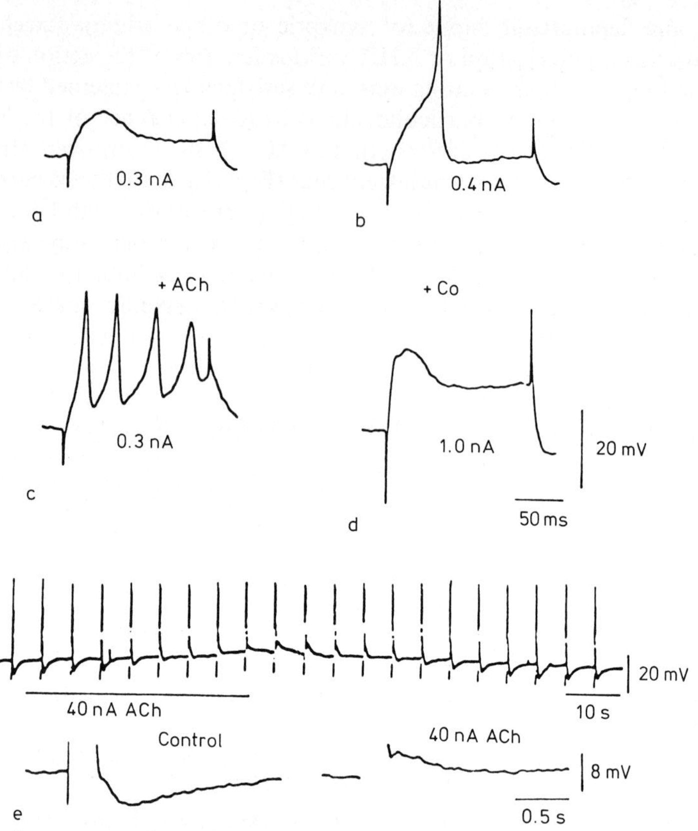

Fig. 8a–e. In presence of tetrodotoxin, ACh facilitates Ca inward current and suppresses post-spike hyperpolarization (AHP). **a, b** Thresholds for activation of Ca spike. **c** In presence of ACh (10^{-5} *M*) multiple spikes are evoked by previously subthreshold current pulse. **d** Ca spikes are blocked by cobalt (2 m*M*). **e** Iontophoretic application of ACh causes temporary disappearance and even reversal of AHP. (From GÄHWILER 1984, Figs. 1, 6)

e) Internal or Membrane Messengers for ACh

There is good evidence that mAChRs accelerate phospholipid turnover (HOKIN and HOKIN 1963; MCKINNEY and RICHELSON 1984; MASTERS et al. 1984) and may also be coupled to an inhibitory guanine nucleotide regulatory protein that controls cyclic AMP formation (HAGA et al. 1985). Alterations in Ca^{2+} influx and Ca^{2+} release from membrane (or internal) sites may therefore be triggered by changes in phosphatidylinositide or cyclic AMP metabolism. Possibly related is the evidence that cyclic GMP acts as an internal messenger for the muscarinic excitatory action on cortical neurons (WOODY et al. 1978; SWARTZ and WOODY 1984); though a comparable effect is not regularly seen in hippocampal neurons (BENARDO and PRINCE 1982; SEGAL 1982; COLE and NICOLL 1984b). Even though it is not clear that cGMP is a universal or even widespread muscarinic internal messenger, muscarinic actions may generally be mediated by changes in phospholipid and/or cyclic nucleotide metabolism, which would explain their long latency and persistence.

4. Presynaptic Muscarinic Actions

The predominant muscarinic effect is to suppress the release of various transmitters, including ACh itself (MACINTOSH and OBORIN 1953; SZERB 1978). Presumably, it involves a depression of Ca^{2+} influx into the nerve terminal and could therefore be explained by a direct interference with Ca inward current (similar to the mechanism of K current depression discussed above).

In the hippocampus this effect was first detected as a depression of EPSPs, evidently caused by an action of ACh on excitatory synaptic terminals (YAMAMOTO and KAWAI 1967; HOUNSGAARD 1978), but when ACh is applied in the pyramidal cell layer, where inhibitory terminals are concentrated, it has a marked disinhibitory action (KRNJEVIĆ et al. 1980, 1981; BEN-ARI et al. 1981; HAAS 1982; GÄHWILER and DREIFUSS 1982) which is probably caused by a re-

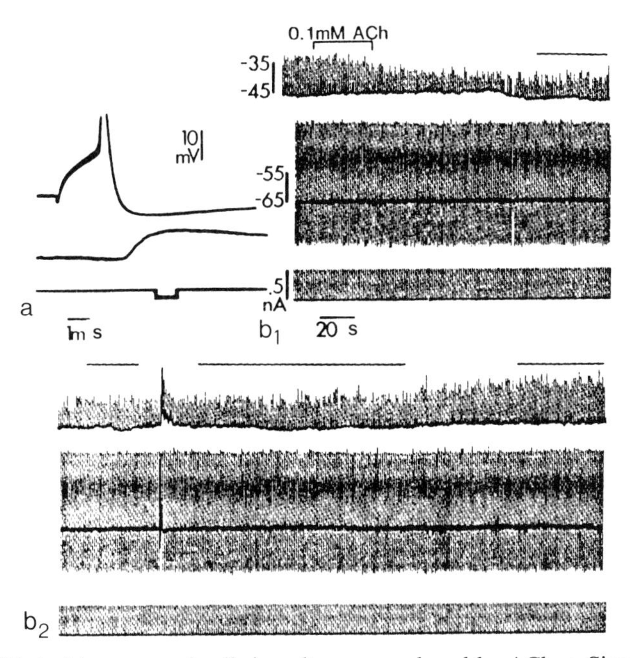

Fig. 9a, b. IPSPs in hippocampal cells in culture are reduced by ACh. **a** Simultaneous recordings from two separate neurons are illustrated by fast traces. When the first cell was directly stimulated by an intracellular depolarizing pulse (*upper trace*), an IPSP was evoked in the second 'follower' cell (*middle trace*). (IPSP was depolarizing owing to leakage of Cl^- from recording microelectrode.) **b** Slow records, illustrating amplitude of IPSPs in 'follower' cell, sampled at time indicated in *lowest trace* in **a** (*uppermost*); action potentials generated in first cell (*middle*); depolarizing pulse that triggers these action potentials (*lowermost*). When ACh was briefly infused (**b**$_1$), IPSPs were markedly diminished and then recovered slowly (**b**$_1$ and **b**$_2$ are continuous). Calculated IPSP quantal content was initially 50, but dropped to 14 in the latter part of **b**$_1$ and rose progressively to 29 and 77 in **b**$_2$ (at times indicated by *bars* above traces). (From SEGAL 1983, Fig. 10)

duced efficacy of GABA release from inhibitory terminals. According to SEGAL'S (1983) study of hippocampal cells in culture (Fig. 9), ACh reduces the quantal content of transmitter release, and therefore must act directly on the GABA-releasing nerve terminal.

5. Muscarinic Inhibition

Muscarinic inhibitions are typically mediated by an increase in K conductance, as is well known from peripheral parasympathetic junctions (BURGEN and TERROUX 1953; HARTZELL et al. 1977; GALLAGHER et al. 1982). In the CNS, in spite of many suggestive observations (RANDIĆ et al. 1964; PHILLIS 1971; STONE 1972; DODD et al. 1981; SEGAL 1982), there has been little convincing evidence of a similar inhibitory action of ACh. When tested more critically, the depressant effects of ACh have usually proved to be mediated synaptically (e.g. BENARDO and PRINCE 1982) through excitation of nearby inhibitory interneurons (MCCORMICK and PRINCE 1985). An interesting feature of interneuronal mAChRs is that the responses are quicker and shorter lasting than those of pyramidal cells and may be mediated by an increase in membrane conductance; moreover, they may belong to a different subgroup (M_2) that is relatively insensitive to the agonist pilocarpine and the antagonist pirenzipine and may have some nicotinic properties.

Although such muscarinic inhibitory receptors remain to be demonstrated in the cortex, they seem to be present in some deeper structures, notably the reticular nucleus of the thalamus (BEN-ARI et al. 1976; SILLITO et al. 1983; MCCORMICK and PRINCE 1986) and the pontine parabrachial nucleus (EGAN and NORTH 1986).

6. Muscarinic Postsynaptic Potentials

Apart from the pathway to Renshaw cells practically none of the central putative cholinergic pathways has lent itself to selective electrical stimulation. For this reason there is very little direct information about cholinergic synaptic actions in the CNS. The typically slow time course of muscarinic actions probably makes cholinergic postsynaptic potentials rather inconspicuous.

One possible approach is to search for 'slow EPSPs', as was done by COLE and NICOLL (1983, 1984a) in hippocampal slices: they stimulated regions, such as the stratum oriens, the fornix and fimbria, through which septal cholinergic (as well as many non-cholinergic) fibres run and could detect in pyramidal cells slow EPSPs that resembled the effects produced by applications of ACh: delayed slow depolarizations (lasting 10 s or more) that are enhanced by *depolarization* and reduced by hyperpolarization (Fig. 10) and are accompanied by an increase in input resistance and depression of AHPs. These slow effects were probably cholinergic, being much enhanced by eserine and blocked by atropine (Fig. 11).

Stimulations of the medial septum *in situ* can be shown to facilitate population spikes in the hippocampus, even though such stimulation by itself produces only minimal field responses (Fig. 12a) (ALVAREZ-LEEFMANS and GARDNER-MEDWIN 1975; KRNJEVIĆ and ROPERT 1982; FANTIE and GODDARD 1982).

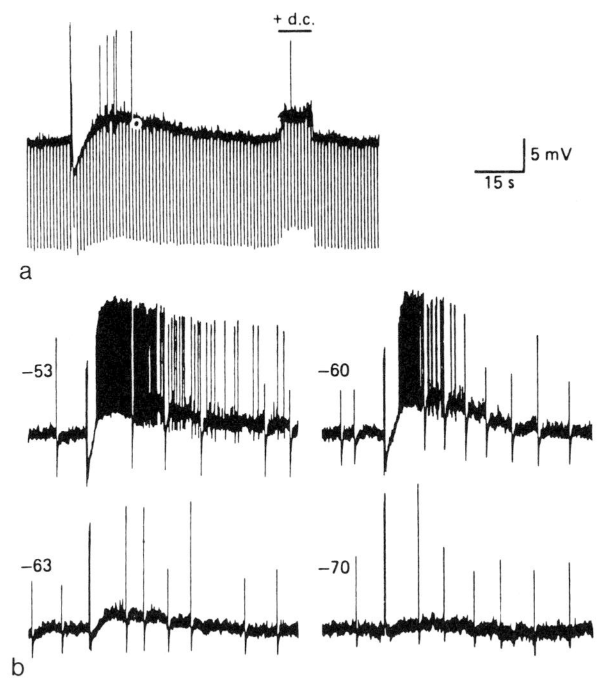

Fig. 10a, b. Properties of slow EPSPs evoked by electrical stimulation of a hippocampal slice. **a** During slow EPSP input resistance is increased (shown by larger amplitude of constant current hyperpolarizing pulses, seen as downward deflections); compare with no change in resistance during direct current (*d.c.*) depolarization. **b** Amplitude of slow EPSP and amount of firing elicited increases with depolarization and decreases with hyperpolarization. Membrane potential is indicated in mV. (From COLE and NICOLL 1984a, Fig. 8)

Similar effects are produced by local applications of ACh in the CA1/CA3 region. The facilitatory effects of septal stimulation could therefore be ascribed to a cholinergic action – at least in CA1/CA3, where the septal actions were potentiated by anticholinesterases and much reduced by muscarine antagonists or inhibitors of ACh synthesis (Fig. 12) (KRNJEVIĆ and ROPERT 1982; ROPERT and KRNJEVIĆ 1982; GLAVINOVIĆ et al. 1983) – though not in the dentate gyrus, where muscarine antagonists were not effective (FANTIE and GODDARD 1982). According to both extra- and intracellular observations (KRNJEVIĆ and ROPERT 1982; ROPERT 1985; KRNJEVIĆ et al. 1988), the predominant effect of medial septal stimulation is depression of inhibitory postsynaptic potentials (IPSPs), in accordance with the previous finding that ACh suppresses IPSPs (see above). But the frequent contamination of likely cholinergic by presumably non-cholinergic (including substantial inhibitory) inputs significantly complicates the observations.

D. Conclusions

Although half a century has passed since central cholinergic transmission was first envisaged (DALE 1938), we are far from having anything like a definitive

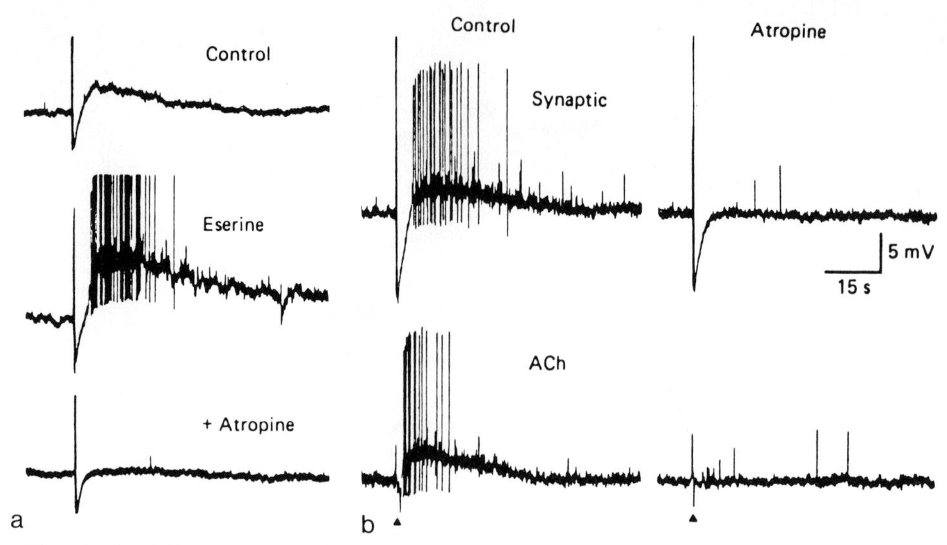

Fig. 11 a, b. Evidence that the slow EPSP evoked by direct hippocampal stimulation is cholinergic. **a** It is enhanced by an anti-ChE (eserine, 2 μ*M*) and suppressed by atropine (0.1 μ*M*). **b** A similar response is produced by an iontophoretic pulse of ACh (*arrow*). Both effects are blocked by atropine. (From COLE and NICOLL 1984a, Fig. 4)

understanding of central cholinergic synapses. Some of the best information concerns highly specific and localized nicotinic synapses in what may be viewed as outlying regions, the spinal cord and the retina. Because they involve very select populations of cells and have a brief time course, they do not loom large in the overall picture of CNS function. If, as seems probable, there are other such synapses, these may well be also detected only by very selective stimulations.

The muscarinic actions are potentially very widespread. One important possible function is to facilitate cell firing in response to other inputs by suppressing the stabilizing effect of K outward currents. This may be viewed as the paradigm of the kind of heterosynaptic modulation that *could* be the basis for long-term changes in synaptic potency and therefore for learning. The well-known loss of 'higher' cortical function in Alzheimer-type dementias, where the most prominent pathological feature is degeneration of the cholinergic forebrain system (WHITEHOUSE et al. 1982), seems in keeping with this idea. More recent evidence, however, points to a more general pathology in senile dementias, involving some other forebrain pathways as well, especially in younger patients (ROSSOR et al. 1984). A unique role of the muscarinic system is therefore by no means certain.

One major difficulty facing investigators has been the diffuse nature of many cholinergic pathways, such as those in the forebrain. At least partly as a result, there is very little convincing evidence about *synaptically* evoked muscarinic responses. Another problem is the quite unexpectedly poor correlation between, on the one hand, neurochemical markers of cholinergic pathways and, on the other, more direct evidence of cholinergic transmission. Of the regions that could

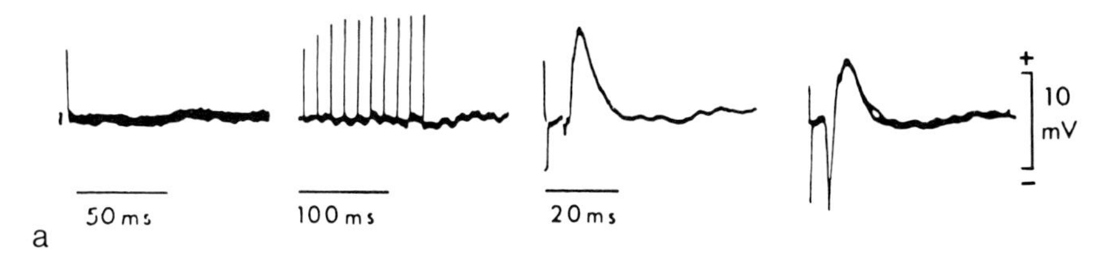

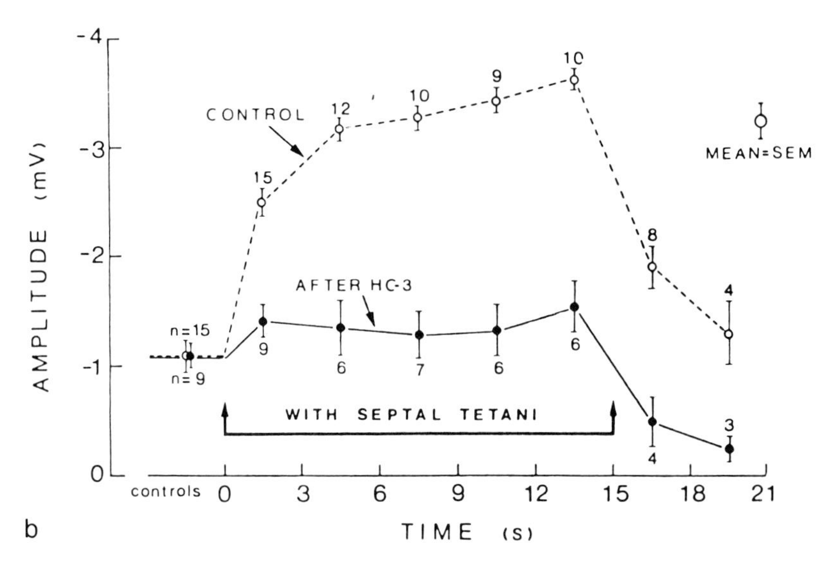

Fig. 12a, b. a Medial septal stimulation which evokes no field response in the CA1 region of rat hippocampus strongly facilitates population spike evoked by commissural stimulation. From *left* to *right; first two traces* show absence of responses to single and tetanic stimulation of medial septum; *third trace,* large, mainly positive field evoked by single commissural stimulus; *fourth trace,* marked enhancement of negative population spike when commissural stimulus was preceded by septal tetanus, as in *second trace.* (From KRNJEVIĆ and ROPERT 1982, Fig. 4). **b** Septal facilitation of population spike is practically abolished after intraventricular hemicholinium-3 (HC-3, 50 µg) injection. *Open circles,* population spikes facilitated by brief septal tetani (as in **a**); *closed circles,* corresponding data 30–45 min after HC-3 injection. (From GLAVINOVIĆ et al. 1983, Fig. 3)

be rated as primary candidates for cholinergic transmission, so far only the retina has yielded convincing electrophysiological supporting evidence. The precise role of ACh in the striatum is far from established: there is reason to believe that numerous cholinergic interneurons may act mainly presynaptically to modulate the release of other putative transmitters, such as dopamine (CHESSELET 1984). Neurochemically by far the outstanding central cholinergic pathway – judging by the exceptionally high concentration of all cholinergic markers (ACh and the related enzymes and receptors) – is the habenulo-interpeduncular system (KRNJEVIĆ and SILVER 1965; LEWIS et al. 1967; MORLEY 1986). And yet even such an expert team as that of BROWN et al. (1983) was quite unable to find any electrophysiological evidence of cholinergic transmission and had to conclude that the natural transmitter in this pathway is probably an excitatory amino acid. The only possible cholinergic action was some depression of transmitter release, via presynaptic nicotinic receptors (BROWN et al. 1984).

As discussed above, a prominent aspect of cholinergic faciliation in the hippocampus is disinhibition, apparently caused by a presynaptic action of ACh on GABAergic terminals. It is perhaps not fortuitous that in these three regions, conspicuously rich in cholinergic markers, perhaps the clearest (if not the only) targets for cholinergic transmission appear to be presynaptic terminals. It is important to bear this in mind in further attempts to unravel central cholinergic mechanisms.

Another lesson, of course, is that undue weight should not be placed on the neurochemical evidence alone. As was made strikingly clear by splendid studies on sympathetic neurons *in vitro* by FURSHPAN et al. (1982), the neurochemical properties of a given neuron only indicate the *possible* transmitters but not which transmitter is actually released.

A final point that needs re-emphasizing is that *the* central cholinergic synapse is a myth. There is wide variety of possible excitatory, inhibitory and modulatory actions, some mediated via nicotinic, some via muscarinic receptors, that may be situated post- or presynaptically. There is at present only a very tenuous factual basis for broad generalizations about central cholinergic function.

Acknowledgements. The author's research is financially supported by the Medical Research Council of Canada.

Note added in proof

To bring this chapter up to date, some significant recent developments need to be mentioned. A muscarinic sensitive (M-type) current has now been observed in human neocortical neurons (HALLIWELL 1986). But a more general block of K channels is indicated by the muscarinic suppression of a fast-inactivating (A-type) outward current in hippocampal neurons (NAKAJIMA et al. 1986), in agreement with the early observations by KRNJEVIĆ et al. 1971. It may well be that, as in sympathetic ganglion cells (CASSELL and MCLACHLAN 1987), most types of K channels are blocked by muscarinic activation, though possibly via different subtypes of receptors (e.g. MCCORMICK and PRINCE 1985; MASH et al. 1985; MÜLLER and MISGELD 1986; HASUO et al. 1988). In addition, the block of Ca current, originally reported from sympathetic ganglia by BELLUZZI et al. 1985 (see also WANKE et al. 1987), has now also been seen in striatal and hippocampal neurons (MISGELD et al. 1986; GÄHWILER and BROWN 1987). This interesting action may contribute to the well-known depression of after-hyperpolarization (BENARDO and PRINCE 1982; MCCORMICK and PRINCE 1987; SCHWINDT et al. 1988), and could explain some of the presynaptic disfacilitatory and disinhibitory actions already discussed (see also JIANG and DUN 1986; HASUO et al. 1988). The complexity of possible cholinergic actions in the thalamus is emphasized by MCCORMICK and PRINCE (1987), who found that ACh has both hyperpolarizing and slow depolarizing effects (direct and muscarinic) on medial and lateral geniculate neurons in *guinea pigs*; but in *cats* there is in addition a rapid *nicotinic* excitatory action in the same nuclei. Other sites where nicotinic actions may also be significant include the supraoptic nucleus (HATTON et al. 1983) and the cerebellum, where even an α-bungarotoxin may be an effective antagonist (DE LA GARZA et al. 1987). There is now intense interest in the possibility that at least some muscarinic actions involve a G protein and one or more internal messengers (MALENKA et al. 1986; CRAIN et al. 1987; LLANO and MARTY 1987). Finally, further information is accumulating on the long-term facilitatory (and 'plastic') function of ACh in the cortex (MURRAY and FIBIGER 1985; METHERATE et al. 1987; SATO et al. 1987): the most impressive perhaps is the demonstration of an essentially permissive action of ascending cholinergic fibres on the NMDA receptor-mediated functional reorganization of visual inputs following monocular deprivation in the visual cortex of young kittens (SINGER 1988).

References

Adams PR (1981) Acetylcholine receptor kinetics. J Membr Biol 58:161–174

Akasu T, Koketsu K (1982) Modulation of voltage-dependent currents by muscarinic receptor in sympathetic neurones of bullfrog. Neurosci Lett 29:41–45

Akasu T, Gallagher JP, Koketsu K, Shinnick-Gallagher P (1984) Slow excitatory postsynaptic currents in bull-frog sympathetic neurones. J Physiol (Lond) 351:583–593

Alvarez-Leefmans FJ, Gardner-Medwin AR (1975) Influences of the septum on the hippocampal dentate area which are unaccompanied by field potentials. J Physiol (Lond) 249:14–16 *P*

Andersen P, Curtis DR (1964) The pharmacology of the synaptic and acetylcholine-induced excitation of ventrobasal thalamic neurones. Acta Physiol Scand 61:100–120

Andersen P, Langmoen IA (1980) Intracellular studies on transmitter effects on neurons in isolated brain slices. Q Rev Biophys 13:1–18

Arendt T, Bigl V, Tennsted A, Arendt A (1985) Neuronal loss in different parts of the nucleus basalis is related to neuritic plaque formation in cortical target areas in Alzheimer's disease. Neuroscience 14:1–14

Ariel M, Daw NW (1982a) Effects of cholinergic drugs on receptive field properties of rabbit retinal ganglion cells. J Physiol (Lond) 324:135–160

Ariel M, Daw NW (1982b) Pharmacological analysis of directionally sensitive rabbit retinal ganglion cells. J Physiol (Lond) 324:161–185

Baisden RH, Woodruff ML, Hoover DB (1984) Cholinergic and non-cholinergic septo-hippocampal projections: a double-label horseradish peroxidase-acetylcholinesterase study in the rabbit. Brain Res 290:146–151

Belluzzi O, Sacchi O, Wanke E (1985) Identification of delayed potassium and calcium currents in the rat sympathetic neurone under voltage clamp. J Physiol (Lond) 358: 109–129

Benardo LS, Prince DA (1981) Acetylcholine induced modulation of hippocampal pyramidal neurons. Brain Res 211:227–234

Benardo LS, Prince DA (1982) Cholinergic excitation of mammalian hippocampal pyramidal cells. Brain Res 249:315–331

Ben-Ari Y, Dingledine R, Kanazawa I, Kelly JS (1976) Inhibitory effects of acetylcholine on neurones in the feline nucleus reticularis thalami. J Physiol (Lond) 261:647–671

Ben-Ari Y, Krnjević K, Reinhardt W, Ropert N (1981) Intracellular observations on the disinhibitory action of acetylcholine in the hippocampus. Neuroscience 6:2475–2484

Birdsall NJM, Hulme EC (1976) Biochemical studies on muscarinic acetylcholine receptors. J Neurochem 27:7–16

Birks R, MacIntosh FC (1961) Acetylcholine metabolism of a sympathetic ganglion. Can J Biochem Physiol 39:787–827

Blackman JG, Ginsborg BL, Ray C (1963) Synaptic transmission in the sympathetic ganglion of the frog. J Physiol (Lond) 167:355–373

Brown DA (1979) Neurotoxins and the ganglionic (C6) type of nicotinic receptor. Cytopharmacol 3:225–230

Brown DA, Adams PR (1980) Muscarinic suppression of novel voltage-sensitive K^+ current in a vertebrate neurone. Nature 283:673–676

Brown DA, Constanti A, Adams PR (1981) Slow cholinergic and peptidergic transmission in sympathetic ganglia. Fed Proc 40:2625–2630

Brown DA, Docherty RJ, Halliwell JV (1983) Chemical transmission in the rat interpeduncular nucleus *in vitro*. J Physiol (Lond) 341:655–670

Brown DA, Docherty RJ, Halliwell JV (1984) The action of cholinomimetic substances on impulse conduction in the habenulointerpeduncular pathway of the rat *in vitro*. J Physiol (Lond) 353:101–109

Burgen ASV (1984) Muscarinic receptors – an overview. Trends Pharmacol Sci [Suppl] 1–3

Burgen ASV, Terroux KG (1953) On the negative inotropic effect in the cat's auricle. J Physiol (Lond) 120:449–464

Cassell JF, McLachlan EM (1987) Muscarinic agonists block five different potassium conductances in guinea-pig sympathetic neurones. Br J Pharmacol 91:259–261

Chesselet M-F (1984) Presynaptic regulation of neurotransmitter release in the brain: facts and hypothesis. Neuroscience 12:347–375

Cole AE, Nicoll RA (1983) Acetylcholine mediates a slow synaptic potential in hippocampal pyramidal cells. Science 221:1299–1301

Cole AE, Nicoll RA (1984a) Characterization of a slow cholinergic postsynaptic potential recorded *in vitro* from rat hippocampal pyramidal cells. J Physiol (Lond) 352:173–188

Cole AE, Nicoll RA (1984b) The pharmacology of cholinergic excitatory responses in hippocampal pyramidal cells. Brain Res 305:283–290

Crain SM, Crain B, Makman MH (1987) Pertussis toxin blocks depressant effects of opioid, monoaminergic and muscarinic agonists on dorsal-horn network responses in spinal cord-ganglion cultures. Brain Res 400:185–190

Crepel F, Dhanjal SS (1982) Cholinergic mechanisms and neurotransmission in the cerebellum of the rat. An *in vitro* study. Brain Res 244:59–68

Cullheim S, Kellerth J-O (1978) A morphological study of the axons and recurrent axon collaterals of cat α-motoneurones supplying different hind-limb muscles. J Physiol (Lond) 281:285–299

Cullheim S, Kellerth J-O (1981) Two kinds of recurrent inhibition of cat spinal α-motoneurones as differentiated pharmacologically. J Physiol (Lond) 312:209–224

Curtis DR, Eccles RM (1958) The excitation of Renshaw cells by pharmacological agents applied electrophoretically. J Physiol (Lond) 141:435–445

Curtis DR, Ryall RW (1966a) The excitation of Renshaw cells by cholinomimetics. Exp Brain Res 2:49–65

Curtis DR, Ryall RW (1966b) The acetylcholine receptors of Renshaw cells. Exp Brain Res 2:66–80

Curtis DR, Ryall RW (1966c) The synaptic excitation of Renshaw cells. Exp Brain Res 2:81–96

Dale HH (1914) The action of certain esters and ethers of choline, and their relation to muscarine. J Pharmacol Exp Ther 6:147–190

Dale HH (1938) Acetylcholine as a chemical transmitter of the effects of nerve impulses. J Mt Sinai Hosp 4:401–429

Daw NW, Ariel M, Caldwell JH (1982) Function of neurotransmitters in the retina. Retina 2:322–331

de la Garza R, McGuire TJ, Freeman R, Hoffer BJ (1987) Selective antagonism of nicotine actions in the rat cerebellum with alpha-bungarotoxin. Neuroscience 23:887–891

Del Castillo J, Katz B (1956) Biophysical aspects of neuromuscular transmission. Prog Biophys 6:121–170

Dodd J, Dingledine R, Kelly JS (1981) The excitatory action of acetylcholine on hippocampal neurones of the guinea pig and rat maintained *in vitro*. Brain Res 207:109–127
Douglas WW, Kanno T, Sampson SR (1967) Effects of acetylcholine and other medullary secretagogues and antagonists on the membrane potential of adrenal chromaffin cells: an analysis employing techniques of tissue culture. J Physiol (Lond) 188:107–120
Dreifuss JJ, Kelly JS (1972) The activity of identified supraoptic neurones and their response to acetylcholine applied by iontophoresis. J Physiol (Lond) 220:105–118
Dudar JD (1977) The role of the septal nuclei in the release of acetylcholine from the rabbit cerebral cortex and dorsal hippocampus and the effect of atropine. Brain Res 129:237–246
Dudar JD, Szerb JC (1969) The effect of topically applied atropine on resting and evoked cortical acetylcholine release. J Physiol (Lond) 203:741–762
Duggan AW, Hall JG, Lee CY (1976) Alpha-bungarotoxin, cobra neurotoxin and excitation of Renshaw cells by acetylcholine. Brain Res 107:166–170
Eccles JC (1957) The physiology of nerve cells. Hopkins, Baltimore
Eccles JC, Fatt P, Koketsu K (1954) Cholinergic and inhibitory synapses in a pathway from motor-axon collaterals to motoneurones. J Physiol (Lond) 126:524–562
Eccles JC, Eccles RM, Fatt P (1956) Pharmacological investigations on a central synapse operated by acetylcholine. J Physiol (Lond) 131:154–169
Eccles JC, Eccles RM, Iggo A, Lundberg A (1961) Electrophysiological investigations on Renshaw cells. J Physiol (Lond) 159:461–478
Egan TM, North RA (1986) Acetylcholine hyperpolarizes central neurons by acting on an M2 muscarinic receptor. Nature 319:405–407
Ehinger B (1982) Neurotransmitter systems in the retina. Retina 2:305–321
Famiglietti EV Jr (1983a) ON and OFF pathways through amacrine cells in mammalian retina: the synaptic connections of 'starburst' amacrine cells. Vision Res 23:1265–1279
Famiglietti EV Jr (1983b) 'Starburst' amacrine cells and cholinergic neurons: mirror-symmetric ON and OFF amacrine cells of rabbit retina. Brain Res 261:138–144
Fantie BD, Goddard GV (1982) Septal modulation of the population spike in the fascia dentata produced by perforant path stimulation in the rat. Brain Res 252:227–237
Fatt P, Katz B (1951) An analysis of the end-plate potential recorded with an intra-cellular electrode. J Physiol (Lond) 115:320–370
Feldberg W (1945) Present views on the mode of action of acetylcholine in the central nervous system. Physiol Rev 25:596–642
Fenwick EM, Marty A, Neher E (1982) A patch-clamp study of bovine chromaffin cells and of their sensitivity to acetylcholine. J Physiol (Lond) 331:577–597
Fibiger HC (1982) The organization and some projections of cholinergic neurons of the mammalian forebrain. Brain Res Rev 4:327–388
Frotscher M, Leranth C (1986) The cholinergic innervation of the rat fascia dentata: identification of target structures on granule cells by combining choline acetyltransferase immunocytochemistry and Golgi impregnation. J Comp Neurol 243:58–70
Furshpan EJ, Potter D, Landis SC (1982) On the transmitter repertoire of sympathetic neurons in culture. Harvey Lect 76:149–191
Gähwiler BH (1984) Facilitation by acetylcholine of tetrodotoxin-resistant spikes in rat hippocampal pyramidal cells. Neuroscience 11:381–388
Gähwiler BH, Brown DA (1987) Muscarinic affects calcium-currents in rat hippocampal pyramidal cells in vitro. Neurosci Lett 76:301–306
Gähwiler BH, Dreifuss JJ (1982) Multiple actions of acetylcholine on hippocampal pyramidal cells in organotypic explant cultures. Neuroscience 7:1243–1256
Gallagher JP, Griffith WH, Schinnick-Gallagher P (1982) Cholinergic transmission in cat parasympathetic ganglia. J Physiol (Lond) 332:473–486
Gerschenfeld HM (1973) Chemical transmission in invertebrate central nervous system and neuromuscular junctions. Physiol Rev 53:1–119
Ginsborg BL (1973) Electrical changes in the membrane in junctional transmission. Biochim Biophys Acta 300:289–317
Glavinović M, Ropert N, Krnjević K, Collier B (1983) Hemicholinium impairs septo-hippocampal facilitatory action. Neuroscience 9:319–330

Haas HL (1982) Cholinergic disinhibition in hippocampal slices of the rat. Brain Res 233:200–204
Haefely W (1974) The effects of various 'nicotine-like' agents in the cat superior cervical ganglion *in situ*. Naunyn Schmiedebergs Arch Pharmacol 281:93–117
Haga K, Haga T, Ichiyama A, Katada T, Kurose H, Ui M (1985) Functional reconstitution of purified muscarinic receptors and inhibitory guanine nucleotide regulatory protein. Nature 316:731–733
Halliwell JV (1986) M-current in human neocortical neurones. Neurosci Lett 67:1–6
Halliwell JV, Adams PR (1982) Voltage-clamp analysis of muscarinic excitation in hippocampal neurons. Brain Res 250:71–92
Hartzell HC, Kuffler SW, Stickgold R, Yoshikami D (1977) Synaptic excitation and inhibition resulting from direct action of acetylcholine on two types of chemoreceptors on individual amphibian parasympathetic neurones. J Physiol (Lond) 271:817–846
Hasuo H, Gallagher JP, Shinnick-Gallagher P (1988) Disinhibition in the rat septum mediated by M1 muscarinic receptors. Brain Res 438:323–327
Hatton GI, Ho YW, Mason WT (1983) Synaptic activation of phasic bursting in rat supraoptic nucleus neurons recorded in hypothalamic slices. J Physiol (Lond) 345:297–317
Hebb CO (1957) Biochemical evidence for the neural function of acetylcholine. Physiol Rev 37:196–220
Hebb CO, Krnjević K, Silver A (1963) Effect of undercutting on the acetylcholinesterase and choline acetyltransferase activity in the cat's cerebral cortex. Nature 198:692
Henley JM, Lindstrom JM, Oswald RE (1986) Acetylcholine receptor synthesis in retina and transport to optic tectum in goldfish. Science 232:1627–1629
Hokin LE, Hokin MR (1963) Phosphatidic acid metabolism and active transport of sodium. Fed Proc 22:8–18
Hounsgaard J (1978) Presynaptic inhibitory action of acetylcholine in area CA1 of the hippocampus. Exp Neurol 62:787–797
Hoyle G (1983) Muscles and their neural control. Wiley and Sons, New York
Hultborn H, Pierrot-Deseilligny E (1979) Input-output relations in the pathway of recurrent inhibition to motoneurones in the cat. J Physiol (Lond) 297:267–287
Hultborn H, Jankowska E, Lindström S (1971) Recurrent inhibition of interneurones monosynaptically activated from group Ia afferents. J Physiol (Lond) 215:613–636
Jankowska E, Lindström S (1971) Morphological identification of Renshaw cells. Acta Physiol Scand 81:428–430
Jiang ZG, Dun NJ (1986) Presynaptic suppression of excitatory post-synaptic potentials in rat ventral horn neurons by muscarinic agonists. Brain Res 381:182–186
Katz B (1966) Nerve, muscle and synapse. McGraw-Hill, New York
Katz B, Miledi R (1972) The statistical nature of the acetylcholine potential and its molecular components. J Physiol (Lond) 224:665–699
Katz B, Thesleff S (1957) A study of the 'desensitization' produced by acetylcholine at the motor end-plate. J Physiol (Lond) 138:63–80
Kehoe J, Marty A (1980) Certain slow synaptic responses: their properties and possible underlying mechanisms. Annu Rev Biophys Bioeng 9:437–465
Krnjević K (1969) ACh and transmission in the CNS. In: Morphological, biochemical and physiological studies of acetylcholine release at synapses. NATO Advanced Study Institute, Varenna, pp 62–65
Krnjević K (1974) Chemical nature of synaptic transmission in vertebrates. Physiol Rev 54:418–540
Krnjević K, Phillis JW (1963a) Acetylcholine-sensitive cells in the cerebral cortex. J Physiol (Lond) 166:296–327
Krnjević K, Phillis JW (1963b) Pharmacological properties of acetylcholine-sensitive cells in the cerebral cortex. J Physiol (Lond) 166:328–350
Krnjević K, Ropert N (1982) Electrophysiological and pharmacological characteristics of facilitation of hippocampal population spikes by stimulation of the medial septum. Neuroscience 7:2165–2183

Krnjević K, Silver A (1963) Cholinesterase staining in the cerebral cortex. J Physiol (Lond) 165:3–4*P*

Krnjević K, Silver A (1965) A histochemical study of cholinergic fibres in the cerebral cortex. J Anat 99:711–759

Krnjević K, Silver A (1966) Acetylcholinesterase in the developing forebrain. J Anat 100:63–89

Krnjević K, Pumain R, Renaud L (1971a) Effects of Ba^{2+} and tetraethylammonium on cortical neurones. J Physiol (Lond) 215:223–245

Krnjević K, Pumain R, Renaud L (1971b) The mechanism of excitation by acetylcholine in the cerebral cortex. J Physiol (Lond) 215:247–268

Krnjević K, Reiffenstein RJ, Ropert N (1980) Disinhibitory action of acetylcholine in the hippocampus. J Physiol (Lond) 308:73–74*P*

Krnjević K, Reiffenstein RJ, Ropert N (1981) Disinhibitory action of acetylcholine in the rat's hippocampus: extracellular observations. Neuroscience 12:2465–2474

Krnjević K, Ropert N, Casullo J (1988) Septo-hippocampal disinhibition. Brain Res 438:182–192

Kuba K, Koketsu K (1978) Synaptic events in sympathetic ganglia. Prog Neurobiol 11:77–169

Lagerbäck P-A, Ronnevi L-O (1982) An ultrastructural study of serially sectioned Renshaw cells. I. Architecture of the cell body, axon hillock, initial axon segment and proximal dendrites. Brain Res 235:1–15

Lagerbäck P-A, Ronnevi L-O, Cullheim S, Kellerth J-O (1978) Ultrastructural characteristics of a central cholinergic synapse in the cat. Brain Res 148:197–201

Lagerbäck P-A, Ronnevi L-O, Cullheim S, Kellerth J-O (1981) An ultrastructural study of the synaptic contacts of α1-motoneuron axon collaterals. II. Contacts in lamina VII. Brain Res 222:29–41

Lamour Y, Dutar P, Jobert A (1982) Excitatory effect of acetylcholine on different types of neurons in the first somatosensory neocortex of the rat: laminar distribution and pharmacological characteristics. Neuroscience 7:1483–1494

Lewis PR, Shute CCD, Silver A (1967) Confirmation from choline acetylase analyses of a massive cholinergic innervation to the rat hippocampus. J Physiol (Lond) 191:215–224

Llano I, Marty A (1987) Protein kinase C activators inhibit the inositol trisphosphate-mediated muscarinic current responses in rat lacrimal cells. J Physiol (Lond) 394:239–248

MacIntosh FC (1941) The distribution of acetylcholine in the peripheral and the central nervous system. J Physiol (Lond) 99:436–442

MacIntosh FC, Collier B (1976) Neurochemistry of cholinergic terminals. In: Zaimis E (ed) Neuromuscular junction. Springer, Berlin Heidelberg New York, pp 99–228 (Handbook of experimental pharmacology, vol 42)

MacIntosh FC, Oborin PE (1953) Release of acetylcholine from intact cerebral cortex (Abstr) 19th International Physiological Congress, Montréal, pp 580–581

Malenka RC, Madison DV, Andrade R, Nicoll RA (1986) Phorbol esters mimic some cholinergic actions in hippocampal pyramidal neurons. J Neurosci 6:475–480

Marder E, Paupardin-Tritsch D (1978) The pharmacological properties of some crustacean neuronal acetylcholine, γ-aminobutyric acid, and L-glutamate responses. J Physiol (Lond) 280:213–236

Mash DC, Flynn DD, Potter LT (1985) Loss of M2 muscarine receptors in the cerebral cortex in Alzheimer's disease and experimental cholinergic denervation. Science 228:1115–1117

Masland RH, Ames A (1976) Responses to acetylcholine of ganglion cells in an isolated mammalian retina. J Neurophysiol 39:1220–1235

Masland RH, Livingstone CJ (1976) Effect of stimulation with light on synthesis and release of acetylcholine by an isolated mammalian retina. J Neurophysiol 39:1210–1219

Masland RH, Mills JW (1979) Autoradiographic identification of acetylcholine in the rabbit retina. J Cell Biol 83:159–178

Masland RH, Mills JW, Cassidy C (1984a) The functions of acetylcholine in the rabbit retina. Proc R Soc Lond [Biol] 223:121–139

Masland RH, Mills JW, Hayden SA (1984b) Acetylcholine-synthesizing amacrine cells: identification and selective staining by using radioautography and fluorescent markers. Proc R Soc London [Biol] 223:79–100

Masters SB, Harden TK, Brown JH (1984) Relationships between phosphoinositide and calcium responses to muscarinic agonists in astrocytoma cells. Mol Pharmacol 26:149–155

McCormick DA, Prince DA (1985) Two types of muscarinic responses to acetylcholine in mammalian cortical neurons. Proc Natl Acad Sci USA 82:6344–6348

McCormick DA, Prince DA (1986) Acetylcholine induces burst firing in thalamic reticular neurones by activating a potassium conductance. Nature 319:402–405

McCormick DA, Prince DA (1987) Actions of acetylcholine in the guinea-pig and cat medial and lateral geniculate nuclei, in vitro. J Physiol (Lond) 392:147–165

McKinney M, Richelson E (1984) The coupling of the neuronal muscarinic receptor to responses. Annu Rev Pharmacol Toxicol 24:121–146

Mesulam M-M, Mufson EJ, Levey AI, Wainer BH (1984) Atlas of cholinergic neurons in the forebrain and upper brainstem of the macaque based on monoclonal choline acetyltransferase immunochemistry and acetylcholinesterase histochemistry. Neuroscience 12:669–686

Metherate R, Tremblay N, Dykes RW (1985) Changes in neuronal function produced in cat primary somatosensory cortex by the iontophoretic application of acetylcholine. Soc Neurosci Abstr 11:753

Metherate R, Tremblay N, Dykes RW (1987) Acetylcholine permits long-term enhancement of neuronal responsiveness in cat primary somatosensory cortex. Neuroscience 22:75–81

Misgeld U, Calabresi P, Dodt HU (1986) Muscarinic modulation of calcium dependent plateau potentials in rat neostriatal neurons. Pfluegers Arch 407:482–487

Mitchell JF (1963) The spontaneous and evoked release of acetylcholine from the cerebral cortex. J Physiol (Lond) 165:98–116

Morita K, North RA, Tokimasa T (1982) Muscarinic agonists inactivate potassium conductance of guinea-pig myenteric neurones. J Physiol (Lond) 333:125–139

Morley BJ (1986) The interpeduncular nucleus. Int Rev Neurobiol 28:157–182

Müller W, Misgeld U (1986) Slow cholinergic excitation of guinea pig hippocampal neurons is mediated by two muscarinic receptor subtypes. Neurosci Lett 67:107–112

Murray CL, Fibiger HC (1985) Learning and memory deficits after lesions of the nucleus basalis magnocellularis: reversal by physostigmine. Neuroscience 14:1025–1032

Nakajima Y, Nakajima S, Leonard RJ, Yamaguchi K (1986) Acetylcholine raises excitability by inhibiting the fast transient potassium current in cultured hippocampus neurons. Proc Natl Acad Sci USA 83:3022–3026

North RA, Tokimasa T (1982) Muscarine synaptic potentials in guinea-pig myenteric plexus neurones. J Physiol (Lond) 333:151–156

Nowak LM, Macdonald RL (1983) Ionic mechanisms of muscarinic cholinergic depolarization of mouse spinal cord neurons in cell culture. J Neurophysiol 49:793–803

Oswald RE, Freeman JA (1981) Alpha-bungarotoxin binding and central nervous system nicotinic acetylcholine receptors. Neuroscience 6:1–14

Oswald RE, Schmidt DE, Freeman JA (1979) Assessment of acetylcholine as an optic nerve neurotransmitter in *Bufo marinus.* Neuroscience 4:1129–1136

Pennefather P, Lancaster B, Adams PR, Nicoll RA (1985) Two distinct Ca-dependent K currents in bullfrog sympathetic ganglion cells. Proc Natl Acad Sci USA 82:3040–3044

Phillis JW (1971) The pharmacology of thalamic and geniculate neurons. Int Rev Neurobiol 14:1–48

Pratt CA, Jordan LM (1980) Recurrent inhibition of motoneurons in decerebrate cats during controlled treadmill locomotion. J Neurophysiol 44:489–500

Price JL, Stern R (1983) Individual cells in the nucleus basalis-diagonal band complex have restricted axonal projections to the cerebral cortex in the rat. Brain Res 269:352–356

Randić M, Siminoff R, Straughan DW (1964) Acetylcholine depression of cortical neurones. Exp Neurol 9:236–242
Renshaw B (1946) Central effects of centripetal impulses in axons of spinal ventral roots. J Neurophysiol 9:191–204
Ropert N (1985) Modulation of inhibition in the hippocampus *in vivo*. Can J Physiol Pharmacol 63:838–842
Ropert N, Krnjević K (1982) Pharmacological characteristics of facilitation of hippocampal population spikes by cholinomimetics. Neuroscience 7:1963–1977
Rossor MN, Iversen LL, Reynolds GP, Mountjoy CQ, Roth M (1984) Neurochemical characteristics of early and late onset types of Alzheimer's disease. Br Med J 288:961–964
Ryall RW, Haas HL (1975) On the physiological significance of muscarinic receptors on Renshaw cells: a hypothesis. In: Waser PG (ed) Cholinergic mechanisms. Raven, New York, pp 335–341
Ryall RW, Piercey MF, Polosa C (1971) Intersegmental and intrasegmental distribution of mutual inhibition of Renshaw cells. J Neurophysiol 34:700–707
Sato H, Hata Y, Hagihara K, Tsumoto T (1987) Effects of cholinergic depletion on neuron activities in the cat visual cortex. J Neurophysiol 58:781–794
Schwindt PC, Spain WJ, Foegring RC, Chubb MC, Crill WE (1988) Slow conductances in neurons from cat sensorimotor cortex in vitro and their role in slow excitability changes. J Neurophysiol 59:450–467
Segal M (1982) Multiple actions of acetylcholine at a muscarinic receptor studied in the rat hippocampal slice. Brain Res 246:77–87
Segal M (1983) Rat hippocampal neurons in culture: responses to electrical and chemical stimuli. J Neurophysiol 50:1249–1264
Shute CCD, Lewis PR (1963) Cholinesterase-containing systems of the brain of the rat. Nature 199:1160–1164
Shute CCD, Lewis PR (1967) The ascending cholinergic reticular system: neocortical, olfactory and subcortical projections. Brain 90:497–520
Sillito AM, Kemp JA (1983) Cholinergic modulation of the functional organization of the cat visual cortex. Brain Res 289:143–155
Sillito AM, Kemp JA, Berardi N (1983) The cholinergic influence on the function of the cat dorsal lateral geniculate nucleus (dLGN). Brain Res 280:299–307
Singer W (1988) Neural mechanisms of experience-dependent self-organization of cerebral cortex. Pfluegers Arch 411:R3
Smith CM (1972) The release of acetylcholine from rabbit-hippocampus. Br J Pharmacol 45:172P
Spehlmann R (1963) Acetylcholine and prostigmine electrophoresis at visual cortical neurons. J Neurophysiol 26:127–139
Stell WK (1985) Putative peptide transmitters, amacrine cell diversity, and function in the inner plexiform layer. In: Gallego A, Gouras P (eds) Neurocircuitry of the retina: A Cajal memorial symposium. Elsevier, Amsterdam, pp 171–187
Sterling P (1983) Microcircuitry of the rat retina. Annu Rev Neurosci 6:149–185
Stone TW (1972) Cholinergic mechanisms in the rat somatosensory cerebral cortex. J Physiol (Lond) 225:485–499
Stone WE (1957) The role of acetylcholine in brain metabolism and function. Am J Phys Med 36:222–255
Straughan DW (1975) Neurotransmitters and the hippocampus. In: Isaacson RL, Pribram KH (eds) The hippocampus, vol 1. Plenum, New York, pp 239–268
Swartz BE, Woody CD (1984) Effects of intracellular antibodies to cGMP on responses of cortical neurons of awake cats to extracellular application of muscarinic agonists. Exp Neurol 86:388–404
Szerb JC (1978) Characterization of presynaptic muscarinic receptors in central cholinergic neurons. Adv Behav Biol 24:49–60
Takeuchi A (1977) Junctional transmission. I. Postsynaptic mechanisms. In: Brookhart JM, Mountcastle VB, Kandel ER, Geiger SR (eds) The nervous system, vol 1/1. American Physiological Society, Bethesda, pp 295–327 (Handbook of physiology, sect 1)

Takeuchi A, Takeuchi N (1959) Active phase of frog's end-plate potential. J Neurophysiol 22:395–411

Takeuchi N (1963) Some properties of conductance changes at the end-plate membrane during the action of acetylcholine. J Physiol (Lond) 167:128–140

Tauchi M, Masland RH (1984) The shape and arrangement of the cholinergic neurons in the rabbit retina. Proc R Soc Lond [Biol] 223:101–119

Tébecis AK (1974) Transmitters and identified neurons in the mammalian central nervous system. Scientechnica, Bristol

Ueki S, Koketsu K, Domino EF (1961) Effects of mecamylamine on the Golgi recurrent collateral-Renshaw cell synapse in the spinal cord. Exp Neurol 3:141–148

Van Keulen LCM (1979) Axon trajectories of Renshaw cells in the lumbar spinal cord of the cat, as reconstructed after intracellular staining with horseradish peroxidase. Brain Res 167:157–162

Wainer BH, Bolam JP, Freund TF, Henderson Z, Totterdell S, Smith AD (1984) Cholinergic synapses in the rat brain: a correlated light and electron microscopic immunohistochemical study employing a monoclonal antibody against choline acetyltransferase. Brain Res 308:69–76

Wainer BH, Levey AI, Rye DB, Mesulam M-M, Mufson EJ (1985) Cholinergic and non-cholinergic septohippocampal pathways. Neurosci Lett 54:45–52

Wanke E, Ferroni A, Malgaroli A, Ambrosini A, Pazzan T, Meldolesi J (1987) Activation of a muscarinic receptor selectively inhibits a rapidly inactivated Ca^{2+} current in rat sympathetic neurons. Proc Natl Acad Sci USA 84:4314–4317

Whitehouse PJ, Price DL, Struble RG, Clark AW, Coyle JT, DeLong MR (1982) Alzheimer's disease and senile dementia: loss of neurons in the basal forebrain. Science 215:1237–1239

Willis WD (1971) The case for the Renshaw cell. Brain Behav Evol 4:5–52

Woody CD, Swartz BE, Gruen E (1978) Effects of acetylcholine and cyclic GMP on input resistance of cortical neurons in awake cats. Brain Res 158:373–395

Yamamoto C, Kawai N (1967) Presynaptic action of acetylcholine in thin sections from the guinea pig dentate gyrus *in vitro*. Exp Neurol 19:176–187

Zieglgänsberger W, Reiter C (1974) A cholinergic mechanism in the spinal cord of cats. Neuropharmacology 13:519–527

Part VII Neuropathology of Cholinergic Transmission

CHAPTER 24

The Cholinergic System in Aging

E. GIACOBINI

A. The Process of Aging of Cholinergic Synapses: Hypotheses and Facts

Based on findings of age-dependent modifications of the cholinergic synapse, two major hypotheses on aging of the cholinergic system have been advanced. The first is a general hypothesis of senescence of synaptic transmission which considers as a primary effect of aging a breakdown of membrane mechanisms resulting in a reduced availability of precursors and in a defect in transmitter release (GIACOBINI 1982a, 1983). This condition in its cholinergic manifestation leads initially to a decreased capacity of neurotransmitter synthesis and then to a decreased content and release of acetylcholine (ACh). Together, these may have an effect similar to 'chemical denervation'. It is not known whether or not this hypothesis applies only to normal conditions of aging or can also be extended to pathological conditions such as Alzheimer's disease. The second is a cholinergic hypothesis of geriatric memory dysfunction which was suggested by BARTUS et al. (1982, 1985a, b) and applies to the deficits seen in Alzheimer's disease. While the first hypothesis is based mainly on data from experimental animals, the second is supported exclusively by findings on humans since there is no adequate animal model of the disease. There are, however, limitations in the use of *post-mortem* cerebral tissue which apply particularly to biochemical investigations in humans.

This chapter attempts to integrate experimental findings and human clinical data in order to present a more complete picture of our present understanding of age-related changes in the cholinergic system.

B. Aging of the Neuromuscular Junction

I. Skeletal Muscle of Rodents

Studies of motor end-plates in aging animals and humans provide models for analysing changes in cholineric mechanisms affecting neuromuscular function during old age. Inactivity and a decrease of muscle mass during senescence cannot alone explain the deficit in muscular function observed during old age. The pioneering work of GUTMANN and HANZLIKOVA (1965, 1966) and of Russian authors (FROLKIS et al. 1976) has provided the basis for a quantitative analysis of such complex phenomena as reduction of neuromuscular transmission, modified neurotrophic relationship and muscular atrophy in old age. The abun-

Table 1. Aging of the NMJ: changes in motoneurons, motor units and motor axons

Parameter	Species	Age		Variation	References
		Years	Months		
Motoneurons	Human	>60	–	Decreased	REBEIZ et al. (1972)
Motor units	Human	>60	–	Decreased	CAMPBELL et al. (1973)
Ventral root axons	Rat	–	27	10% Decrease	DUNCAN (1934)
Motor axons	Rat	–	24	Unchanged	GUTMANN and HANZLIKOVA (1966)
Motor axons	Cat	10–38	–	Unchanged	MOYER and KALISZEWSKI (1958)
Preterminal nerves	Rat	–	24	Deformed	FROLKIS et al. (1976)
Terminal innervation ratio	Human	>60	–	Unchanged	COERS et al. (1973)
Preterminal axons small terminals endplates	Rat	– –	25 25	Shrunken multiple	FUJISAWA (1976)
Terminal branches/ endplates and nodal sprouts	Cat	6	–	Increased	TUFFERY (1971)
Axons entering the junctions	Mouse	–	20	Unchanged	FAHIM et al. (1983)
End-plate size[a]	Rat	–	24	Unchanged or decreased	GUTMANN and HANZLIKOVA (1965)

[a] In extensor digitorum longus and soleus muscles

dance of information provided by the physiology and biochemistry of the cholinergic apparatus at the neuromuscular junction (NMJ) has made this structure a favourite object of detailed interdisciplinary observation. The changes in the basic components of the NMJ which are summarized in Table 1 provide a necessary first background survey. The number of motor axons both in rat and cat does not seem to change with age (GUTMANN and HANZLIKOVA 1966; MOYER and KALISZEWSKI 1958) while in humans a decrease in the number of motoneurons and motor units has been observed (REBEIZ et al. 1972; CAMPBELL et al. 1973). In rodents, preterminal axons are shrunken (FUJISAWA 1976), and terminal branches and nodal sprouts are increased (TUFFERY 1971). The size of the endplate in the 24-month-old rat is either unchanged or decreased (GUTMANN and HANZLIKOVA 1965). These findings suggest a disturbance of neuromuscular relations during old age and a diminished control of the cholinergic apparatus over the muscle.

Following these basic observations, a detailed analysis of ultrastructure has revealed a number of significant changes in the architecture of the end-plate (Table 2a, b). Three functionally different muscles, the diaphragm (D), the extensor digitorum longus (EDL) and the soleus (S), were mostly studied in rodents from young adulthood through senescence. In the EDL and S, agglutination and polymorphism of synaptic vesicles, neurotubules and neurofilaments have been observed (GUTMANN et al. 1971). In the rat D, the total number of synaptic vesicles is increased (SMITH and ROSENHEIMER 1982). However, in EDL and S,

Table 2a. Aging of the NMJ in the rat: ultrastructure and morphometry

Parameter	Muscle			References
	D	EDL	S	
Synaptic vesicles, neurotubules, neurofilaments	–	Agglutination		1
Size of dense-cored vesicles		Polymorphic		2
Average number of terminals	+	ND	ND	3
Size of end-plate	0	ND	ND	3
Sprouting	–	ND	ND	3
Synaptic vesicle number	+	ND	ND	3
Synaptic vesicle diameter	+	ND	ND	3
Schwann-cell cytoplasm	ND	ND	+	4
Liposomes, dense cored vesicles	ND	ND	+	4
Exposed folds	ND	ND	+	4
Terminal branches/end-plate	0	–	+	5
Sprouting terminal/end-plate	0	–	+	5
Degenerating terminal/end-plate	–	+	0	5
Total terminal branches/end-plate	+	0	+	5
Axon branches/end-plate	+	0	+	5
End-plate area	–	–	+	5

D, diaphragm; EDL, extensor digitorum longus muscle; S, soleus muscle; +, increase; –, decrease; 0, no change; ND, no data
References: 1, GUTMANN et al. (1971); 2, FUJISAWA (1976); 3, SMITH and ROSENHEIMER (1982); 4, CARDASIS (1983); 5, ROSENHEIMER and SMITH (1985a)

Table 2b. Aging of the NMJ in the mouse: ultrastructure and morphometry

Parameter	Muscle		References
	EDL	S	
Nerve terminal area	–	–	1
Mitochondria, synaptic vesicles	–	–	1
Smooth endoplasmic reticulum	+	+	1
Coated vesicles, cisternae	+	+	1
Neurofilaments	+	ND	2
Nerve terminal length	+	ND	2
Number of intrasynaptic branches	+	ND	2
Collateral sprouting	0	ND	2
Partial denervation	0	ND	2

EDL, extensor digitorum longus muscle; S, soleus muscle; +, increase; –, decrease; 0, no change; ND, no data
References: 1, FAHIM and ROBBINS (1982); 2, FAHIM et al. (1983)

presynaptic terminals of old mice show a decrease in nerve terminal area, mitochondria and vesicles but an increase in endoplasmic reticulum, coated vesicles, cisternae, microtubules and neurofilaments (FAHIM and ROBBINS 1982). Although the presence of occasional denervated postsynaptic regions is encountered in old NMJs, the predominant characteristics are not of denervation but rather of an adaptive process of remodelling characterized by simultaneous degeneration and regeneration. These adaptive phenomena are reflected by increases in nerve terminal length and intrasynaptic markers with no change in the number of axons entering the junction (FAHIM et al. 1983). The loss in synaptic vesicles and neurotransmitter function in aging synapses may be compensated for by simultaneous regenerative changes (GIACOBINI et al. 1984) and the diminution in nerve terminal cross-sectioned area (FAHIM and ROBBINS 1982) by an increase in total length of the nerve terminal branches (FAHIM et al. 1983), resulting in a

Table 3. Aging of the NMJ: ACh content, storage, release and leakage

Parameter	Muscle	Age (months)	Variation	References
Rat				
mEPP frequency	D	28	Doubled	1
	D, S	30	–	2
mEPP amplitude	D	28	Smaller	1
	D, S	30	+	3
Quantal release	D	28	Larger	1
Extracellular K^+ diffusion	D	28	Slower	4
Neuromuscular transmission	D	28	Increased synaptic delay	5
Number of axons	D	28	0	5
Fibre diameter	D	28	0	5
End-plate potential, frequency and amplitude	G	24	±	6
ACh sensitivity	D, S	30	0	2
ACh content				
Total	D	ND	–	1, 7
End-plate	D	ND	–	5
ACh release	D	ND	0 or –	7
ACh release				
End-plate	D	ND	0	5
Evoked	D	ND	+	1
Resting	D	ND	+	1
ACh leakage	D	ND	+	1
Quantal size	D	ND	–	7
ACh synthesis, storage	D	ND	More rapid leakage	7
Mouse				
ACh release, evoked	EDL, S	32	Decreased sensitivity to low Ca^{2+} and high Mg^{2+}	8
Quantal content	S	ND	+	9
ACh release, vesicle density	S	ND	–	9
Rabbit				
Contraction to applied ACh	G	62	+	6

D, diaphragm; EDL, extensor digitorum longus muscle; G, gastrocnemius muscle; S, soleus muscle; +, increase; –, decrease; ±, variable; 0, no change; ND, no data
References: 1, SMITH (1984); 2, GUTMANN et al. (1971); 3, VYSKOČIL and GUTMANN (1969); 4, SMITH (1982a); 5, SMITH and ROSENHEIMER (1984); 6, FROLKIS et al. (1976); 7, WEILER and SMITH (1985); 8, KELLY and ROBBINS (1983); 9, FAHIM and ROBBINS (1985)

Table 4. Aging of the NMJ in the rat: biochemical changes

Parameter	Muscle	Age (months)	Variation	Reference
AChE[a]	EDL, S	24	–	1
	D	28	+	2
	D	28	0	3
	D	28	0	4
ChAT	EDL, S	29	– (40%)	5
	D	29	0	5
	D	29	0	3
	G	28	– (40%)	6
Choline uptake	D	28	0	4
Glucose uptake	G	28	–	7
ATPase	G	26	–	8
	EDL, S	36	–	9

D, diaphragm; EDL, extensor digitorum longus; G, gastrocnemius; S, soleus; +, increase; –, decrease; 0, no change (for other abbreviations see text)
References: 1, GUTMANN and HANZLÍKOVÁ (1965); 2, SMITH (1984); 3, SMITH (1982a); 4, SMITH and ROSENHEIMER (1984); 5, TUČEK and GUTMANN (1973); 6, FROLKIS et al. (1973); 7, ROSENHEIMER and SMITH (1985b); 8, ROCKSTEIN and BRANDT (1962); 9, SYROVÝ and GUTMANN (1970)
[a] First line relates to histochemical results

preservation of the terminal volume of the NMJ. Thus, a growth of new or an increase in the old nerve terminals within the junctional region may compensate for the observed reduced transmitter release by providing more release sites (GIACOBINI et al. 1984).

The remarkable difference in the way that different muscles react to the aging process is exemplified by the recent studies of ROSENHEIMER and SMITH (1985a) as summarized in Table 2b. They indicate that while sprouting and regeneration of terminals is maintained in D through aging, both EDL and S experience a general decline in sprouting. The turnover of terminal branches in the aging NMJ seem, therefore, to be related to the particular function of that muscle (see D versus S), showing a less extensive arborization in hind-limb muscles (PESTRONK et al. 1980; TUFFERY 1971) and a more extensive one in D (ROSENHEIMER and SMITH 1985a; and Table 2b). In rat D, NMJs, ACh levels and quantal size decline but evoked release remains constant during aging (WEILER and SMITH 1985). This is associated with in increased leakage of ACh from the cytoplasmic pool in aged animals. ACh seems to be synthesized more rapidly in the older NMJ but the newly synthesized ACh is not released as readily (WEILER and SMITH 1985; and Table 3). According to these authors the reduced ACh level during aging in the D cholinergic terminals is due primarily to increased leakage of the transmitter. This is a different situation from the fowl iris NMJ, where the decrease in levels of ACh seems to be principally due to a decreased transport of precursor (MARCHI et al. 1980b; GIACOBINI 1982b).

The remodelling process of the senescent NMJ represents a balance between growth and degeneration; it controls the efficiency of synaptic transmission but is subject to age-dependent factors such as synthesis of membrane components and transport of precursors (GIACOBINI 1983).

In the phrenic-diaphragm preparation of rats (SMITH and ROSENHEIMER 1984), there are no age-related changes in the number of myelinated or non-myelinated axons, capillaries, myelin sheath invaginations, fibre diameter or myelin sheath width (Table 2a). Speed of conduction is not changed, but the average synaptic delay is increased (Table 3). The mean quantal content of ACh is also increased in the D NMJ of the aged rat while ACh content per end-plate is decreased by almost 50% (Table 3), but choline uptake and ChAT and AChE activity show no difference (SMITH and ROSENHEIMER 1984; and Table 4).

As shown in Table 3, the frequency of miniature end-plate potentials (mEPPs) is decreased in the 30-month-old rat (D, S) but the mEPP amplitude is increased (VYSKOČIL and GUTMANN 1969; GUTMANN et al. 1971). By contrast, in the diaphragm, mEPP frequency is increased 100%, amplitude is decreased, while quantal release is greater (SMITH 1984; Table 3).

Age-related changes in evoked ACh release begin in mid-life and progress more rapidly in certain muscles than other, e.g. more rapidly in EDL than in S (KELLY and ROBBINS 1983). This may be due to the difference in muscle type or to different usage. Decreased sensitivity of transmitter release to low Ca and/or high Mg is an additional early sign of altered synaptic transmission in aging mouse muscle (KELLY and ROBBINS 1983; and Table 3).

Major biochemical changes at the NMJ of various skeletal muscles of aged rats are summarized in Table 4. These include mainly reduction in ChAT and AChE activities in some muscles and increase or no changes in others reflecting functional differences in NMJs. Of interest are other changes indirectly related to ACh metabolism such as glucose uptake and ATPase activity which are all decreased (Table 4).

II. Avian Iris NMJ

The NMJ of the iris of the aging domestic fowl has been extensively studied biochemically and ultrastructurally in the author's laboratory (for review, see MARCHI and GIACOBINI 1981; GIACOBINI 1982a, b, 1983). This structure is innervated by cholinergic (ciliary ganglion), adrenergic (superior cervical), as well as peptidergic fibres (trigeminal) (Table 5) and presents a unique possibility of experimental approaches both *in vitro* and *in vivo* (MARCHI and GIACOBINI 1981; GIACOBINI 1982a, b). Biochemical and structural changes in cholinergic synapses of ciliary ganglia and irises of the fowl (Chap. 3) during aging are summarized in Table 6. The nerve terminal arborization on the iris muscle and their synaptic boutons can be stained and rendered visible in light microscopy; from such preparations quantitative results can be obtained (Fig. 1).

Based on the results summarized in Table 6 (GIACOBINI 1982a, b), a hypothesis of aging of the cholinergic synapse has been formulated (GIACOBINI 1983). This hypothesis contemplates age-related changes in carrier-mediated mechanisms of uptake and release of the neurotransmitter and its precursor (choline), leading to a chemical denervation at the cholinergic synapse (Fig. 2). In order firmly to establish this hypothesis several basic questions need to be answered.

Table 5. Neurotransmitter systems in the iris: a comparison of rodent and avian

Neurotransmitter	Species	Type of fibres	Function	Origin	References
ACh	Rat	Motor	Constrictor pupillis, Dilator pupillis	Ciliary ganglion, Superior cervical ganglion	EHINGER and FALCK (1966)
Norepinephrine	Rat	Motor	Dilator pupillis	Superior cervical ganglion	MALMFORS (1965)
Substance P	Rat	Sensory	Sensory	Trigeminal	SEIGER et al. (1985)
Leu-enkephalins	Rat	Sensory	Sensory	Trigeminal	BJÖRKLUND et al. (1984a)
Vasoactive intestinal polypeptide	Rat	Sensory	Co-transmitter	Trigeminal	SCHULTZBERG et al. (1980)
Other:					
Neurofilament protein	Rat	Sensory	Transport	Trigeminal	SEIGER et al. (1985)
Glial fibrillary acid protein	Rat	Schwann cells	Not known	Iris nerves	BJÖRKLUND et al. (1984b)
ACh	Fowl	Motor	Constrictor pupillis	Ciliary ganglion	ZENKER and KRAMMER (1967); LUCCHI et al. (1974); CHIAPPINELLI et al. (1976)
Norepinephrine	Fowl	Motor	Dilator pupillis	Superior cervical ganglion	YURKEWICZ et al. (1981); EHINGER (1967); KIRBY et al. (1978)
Substance P	Fowl	Sensory	Not known	Trigeminal	E. GIACOBINI, unpublished
Enkephalins	Fowl	Sensory	Not known	Trigeminal	E. GIACOBINI, unpublished

In the cholinergic terminals of the fowl iris signs of aging have been identified morphologically and biochemically as early as 18–24 months of age (GIACOBINI et al. 1984; MUSSINI and GIACOBINI 1985). Polymorphic signs of damage were seen in the nerve ending as well as in the axoplasm of the nerve fibre (GIACOBINI et al. 1984). Kinetic changes (V_{max}) in the uptake process of choline have been described in the same period, leading to a decreased access of the precursor (MARCHI et al. 1980b).

Qualitative ultrastructural analysis of nerve terminals and nerve fibres in iris muscles confirmed previous findings of morphological alterations common to both 3- and 5-year-old domestic fowl (GIACOBINI et al. 1984; MUSSINI and

Table 6. ACh metabolism in the peripheral autonomic nervous system of aging animals

Parameter	Species	Organ	Variation	References
ChAT, AChE, BuChE sensitivity to ACh	Rat	Autonomic ganglia	–	FROLKIS et al. (1973)
ACh and choline levels	Fowl	Ciliary ganglion, iris	–	MARCHI et al. (1979, 1981a); MARCHI and GIACOBINI (1980)
AChE and ChAT	Fowl	Ciliary ganglion, iris	–	MARCHI et al. (1980d)
Nicotinic binding	Fowl	Ciliary ganglion, iris	–	MARCHI et al. (1981b)
ACh release	Fowl	Ciliary ganglion, iris	–	GIACOBINI et al. (1984, 1986a)
Choline uptake	Fowl	Ciliary ganglion, iris	–	MARCHI et al. (1980b, 1980c)
No. of boutons, no. of vesicles, apposition of membrane, vesicle volume	Fowl	Iris	–	GIACOBINI et al. (1984, 1986a); MUSSINI and GIACOBINI (1985)
Neurofilaments	Fowl	Iris	+	MUSSINI and GIACOBINI (1985)

–, decrease; +, increase

GIACOBINI 1985). These alterations included the presence of dense granular lysosomes in the adaxonal space of a large number of nerve fibres (Fig. 3A). Lysosomes and mitochondria accumulated in the axoplasm, and occasionally, enormously swollen, degenerating axons were observed. Irregularities at the level of the nerve terminal were also seen. Polymorphic synaptic vesicles with little tendency to cluster and form active zones were observed (Fig. 3B, C). Lysosomes were also found in the terminals together with accumulations of neurofilaments. Of major interest was the observation that Schwann cell cytoplasmic processes progressively enveloped the terminal (Fig. 3B), infiltrating and protruding into the bouton (Fig. 3C). This process leads to a progressively wider synaptic cleft which can be filled by a multilayered basal lamina (Fig. 3B).

Quantitative morphometric analysis suggested a possible deficit in the ability of the 3- and 5-year-old animal actively to sustain neurotransmission. The decrease in vesicular volume and in synaptic membrane length confirmed the hypothesis that one of the consequences of aging, at the neuronal terminal, is a decreased ability for neurotransmitter release (chemical denervation). Experiments were designed to study the release of ACh from 2- to 4-month and 3-year fowl irises (Fig. 4) under more (Type II) or less (Type I) severe conditions (GIACOBINI et al. 1984).

The main effects of age showed that the 3-year tissue released significantly less [^{3}H]ACh than the 4-month tissue as determined by the area under the release curve (peak area, Fig. 5). Also, the 3-year tissue showed a lower peak release of [^{3}H]ACh than the 4-month tissue. The time needed for the 3-year tissue to reach

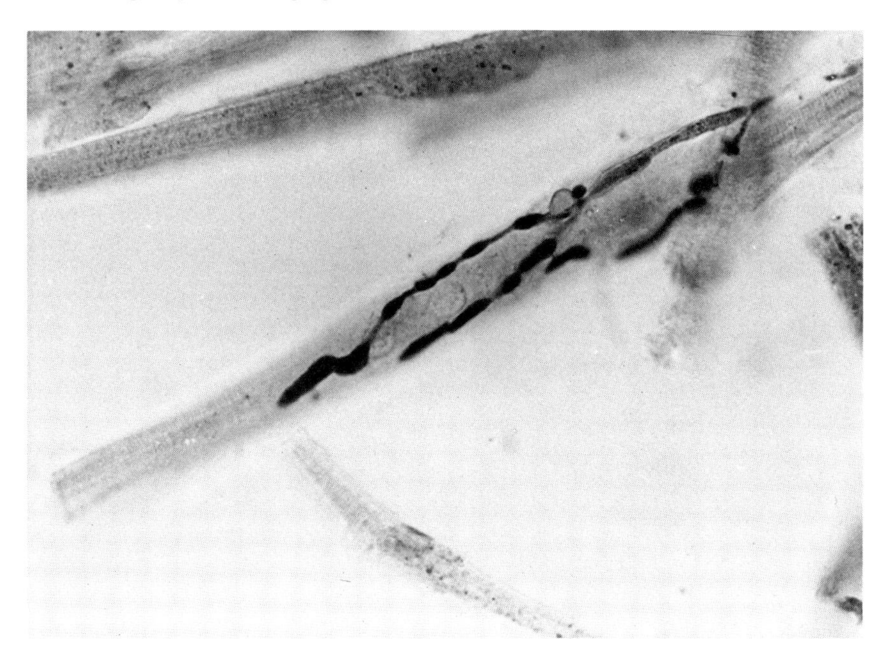

Fig. 1. Light micrograph of a nerve terminal arborization on iris muscle fibres of the 2-month-old domestic fowl, dissociated and stained with the zinc iodide-osmic acid impregnation method. Preterminal axons and their synaptic boutons are intensely *stained in black*. The average number of boutons per arborization, as well as the average length of boutons, can be determined by this method

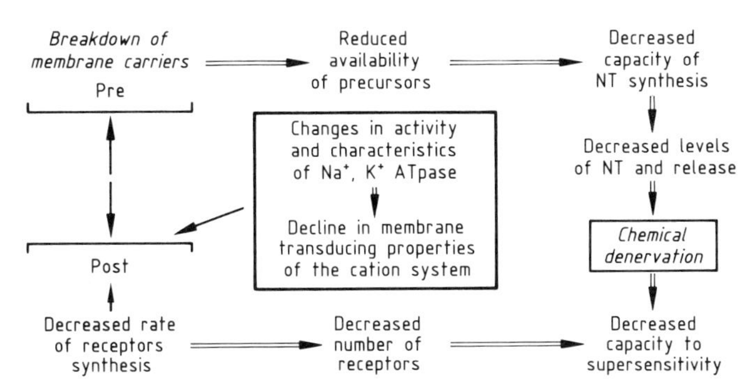

Fig. 2. A new hypothesis on aging of the cholinergic synapse. (Modified from GIACOBINI 1983)

its peak release was significantly longer than at 4 months and its rate of release was significantly slower in both types of experiments.

These neurochemical results demonstrate a decreased ability, dependent only on age and not experimental design (I or II) of the 3-year tissue to release [^{3}H]ACh when compared to a 4-month tissue (Fig. 5). This correlates well with the morphological data which demonstrate that two important features for neurotransmitter release (vesicular volume and length of the appositional membrane)

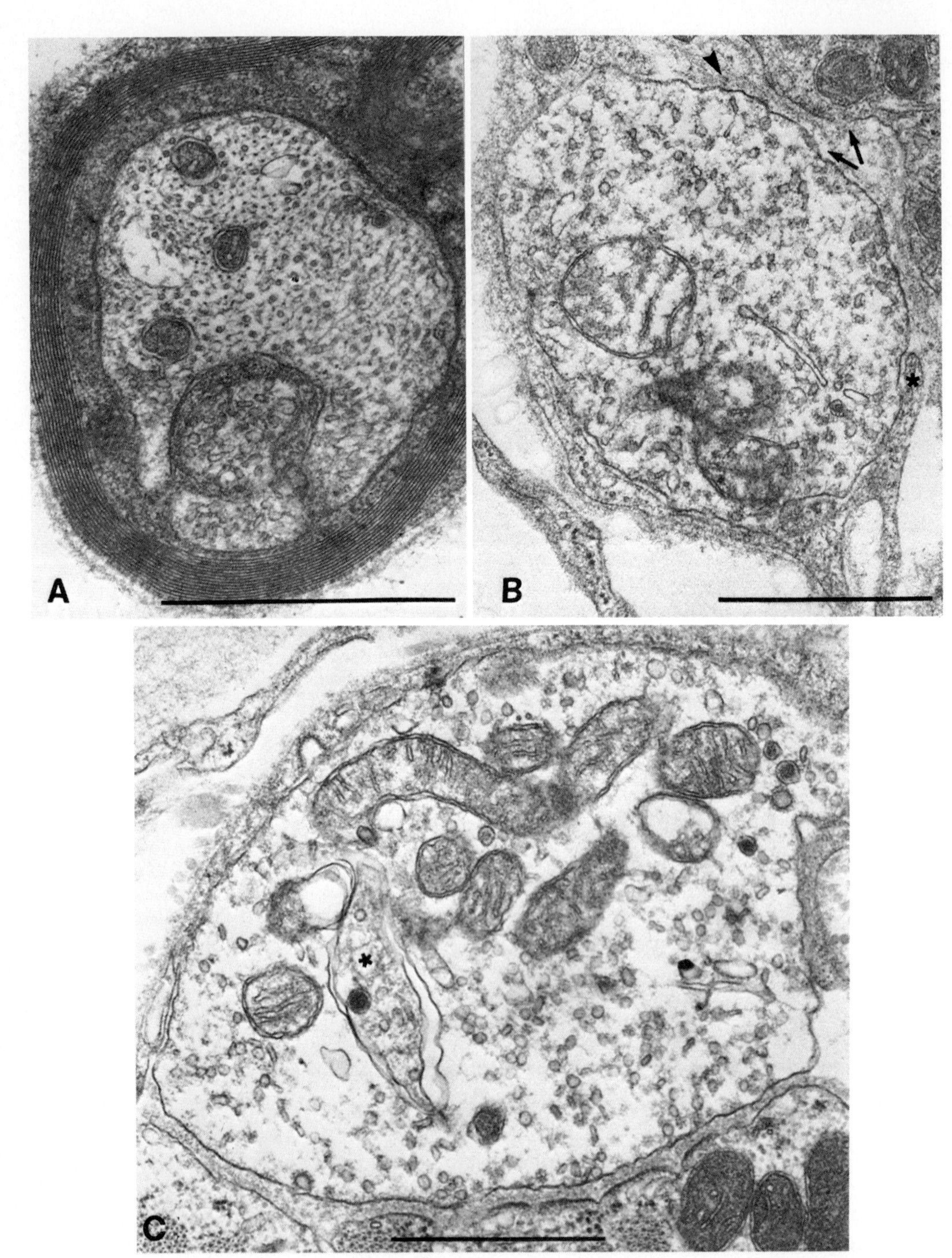

Fig. 3 A–C. Iris muscle of a 3-year-old fowl. **A** Several lysosomes have accumulated in the adaxonal space of myelinated nerve fibres. The nerve terminals (**B, C**) contain a reduced number of vesicles which are polymorphic. Schwann cell cytoplasmic processes (*asterisks*) infiltrate the synaptic cleft (**B**) and protrude into the bouton (**C**). *Arrow* points to a multilayered basal lamina. *Calibration bars* = 1 μm. (Modified from GIACOBINI et al. 1984)

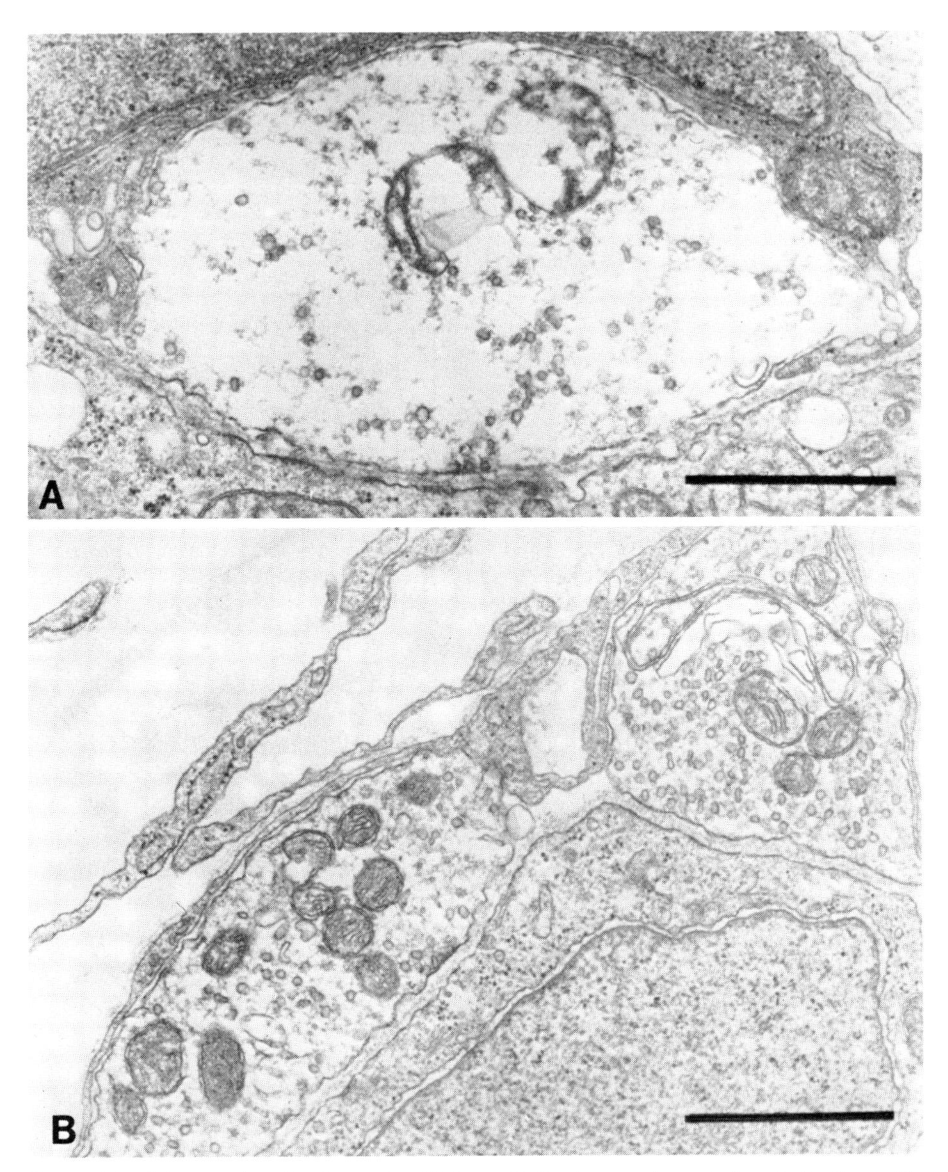

Fig. 4A, B. Iris muscle of a 2-month-old fowl. Following prolonged depolarization with high K^+ concentration. **A**, the nerve ending appears almost depleted of synaptic vesicles. **B**, After reincubation in physiological medium a nearly total retrieval of synaptic vesicles is achieved. *Calibration bars* = 1 μm. (Modified from GIACOBINI et al. 1984)

were decreased in the 3-year (or 5-year) tissue. These results support the hypothesis that cholinergic transmission in the fowl iris is age-dependent and declines with increasing age. This decline is related to modifications of presynaptic mechanisms of release and uptake of the neurotransmitter and its precursor (GIACOBINI 1983).

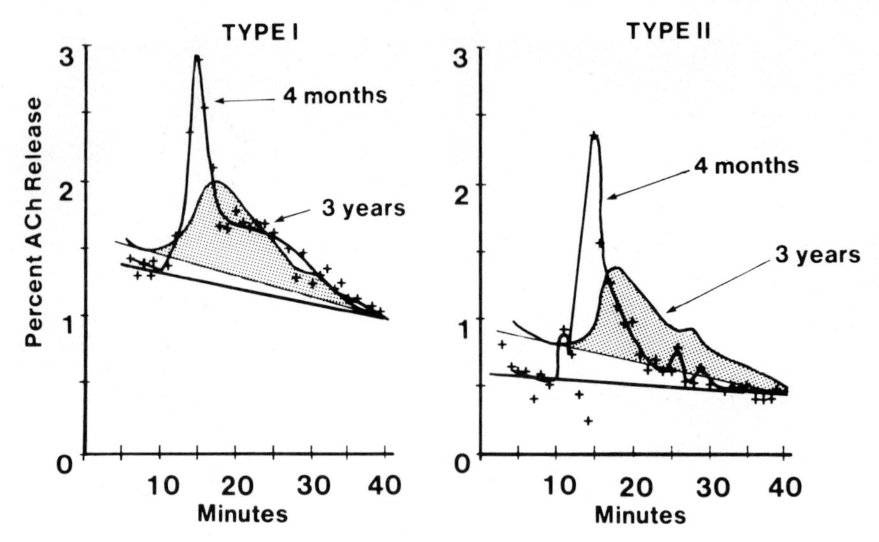

Fig. 5. Typical release curves in Type I and Type II experiments. ACh release is represented as the percent [^{3}H]ACh released of the total [^{3}H]ACh present in the tissue at the time of release. (See text; modified from GIACOBINI et al. 1984)

A diagram of the changes in ACh synthesis and release in the aging cholinergic synapse is presented in Fig. 6.

As shown in the diagram, the decrease in endogenous levels of ACh and choline in the aged iris correlates with changes in the kinetics of choline uptake (MARCHI et al. 1980b). These release experiments clearly demonstrate that only the younger tissue is able to replenish its ACh levels following periods of repeated depletion and release. Not only is this ability progressively lost with age but the release mechanism itself becomes less and less efficient with time. The diagram suggests adaptive changes in choline mobilization.

These data indicate that cholinergic endings in older tissues handle choline and ACh in a different manner than in the younger tissue. Thus, it appears that the control mechanisms which regulate and maintains ACh and choline levels in cholinergic synapses are impaired in older animals. The results of such release experiments clearly show that the capability of the older junction to sustain prolonged depolarization is decreased and the amount of ACh released declines with age. These changes are associated with severe alterations in the structure of the synapse, mainly strongly reduced number of vesicles and decreased appositional area. Together with the demonstrated reduction in neurotransmitter level and in uptake of choline (MARCHI et al. 1980b), the reduction in release forms a characteristic triad of age-related defects which may have important implications for our understanding of aging processes. Age-dependent cholinergic changes in peripheral organ innervation are not unique but are seen for other neurotransmitters as well.

It appears that in noradrenergic terminals of the peripheral nervous system (PNS), as in the central nervous system (CNS), the terminals are more affected than the cell bodies during aging since total tyrosine hydroxylase activity in the

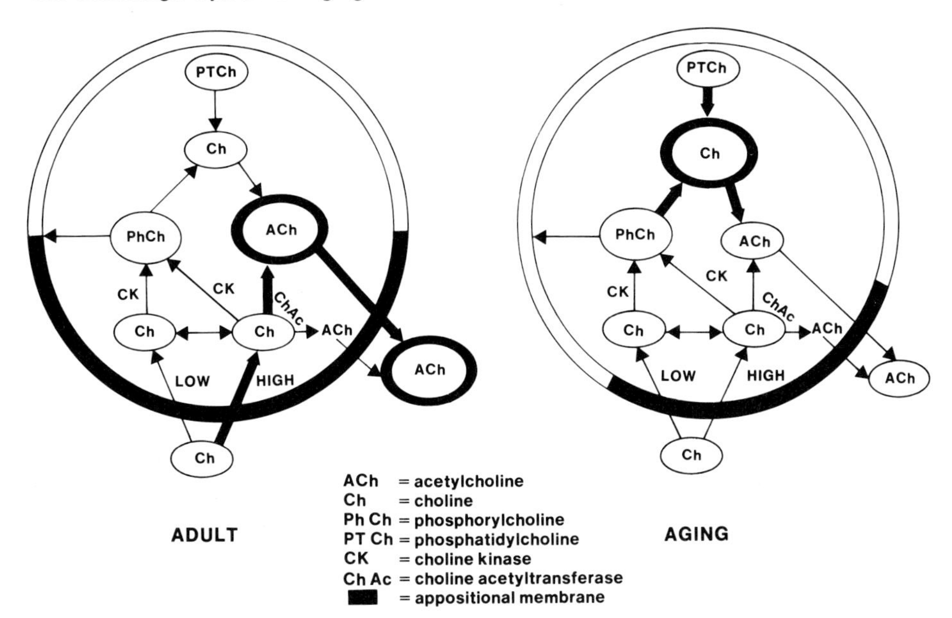

Fig. 6. Schematic diagram of an adult and an aging cholinergic ending, showing various steps involved in the regulation of ACh synthesis. The intracellular ACh is shown as deriving from (a) intrasynaptic choline formed by AChE and transported into the ending (this uptake of choline is impaired in the aging synapse); and (b) neuronal phosphorylcholine and phosphatidylcholine. In spite of an increased intracellular pool of choline and a decreased release of ACh, the levels of ACh are decreased in the aging ending

iris declines, whereas it continually increases in the sympathetic ganglia (YURKEWICZ et al. 1981). This may be explained by a decrease in enzyme synthesis or in transport to the terminals as was suggested for ChAT in the ciliary ganglion and iris during aging (MARCHI and GIACOBINI 1981). However, when comparing cell bodies and terminals, the PNS displays more marked differences in total tyrosine hydroxylase activity than does the CNS.

C. ACh Synthesis and Metabolism in Brains of Aging Animals

Several studies suggest that impaired cholinergic function may be a factor contributing to memory and cognitive deficits in old age both in animals and humans (BARTUS et al. 1982). This concept has stimulated a large number of investigations directed to detect early and irreversible changes in the function of central cholinergic synapses. Anatomical data have contributed to support this hypothesis as structural changes in specific cholinergic populations of neurons have mostly coincided with neurochemical deficits (COYLE et al. 1983).

Biochemical changes in the synthesis and metabolism of ACh in brain are summarized in Table 7a. Levels of ACh have been found to be little changed in old mice and rats; however, the rate of incorporation of [^{3}H]choline and the

Table 7a. ACh synthesis and metabolism in brain of aging animals

Parameter	Species	Region	Variation	References
ACh and choline levels, ChAT	Rat	HP, FC, CS	0	SHERMAN et al. (1981)
Turnover rate	Rat	wb	0	DOMINO et al. (1978)
Choline uptake; ACh synthesis and release	Rat	S	0	MARCHI et al. (1983)
Choline uptake	Rat	S	−	SHERMAN et al. (1981)
ChAT, AChE activity	Rat	wb	− (33%)	SASTRY et al. (1983)
ACh release		Slices	− (63%)	SASTRY et al. (1983)
Accumulation of [^{3}H]ACh	Rat	SS	−	THOMPSON et al. (1983)
Choline uptake	Rat	HP	− (22%)	SHERMAN et al. (1981)
ChAT activity	Rat	CD	−	MCGEER et al. (1971)
	Rat		0	MCGEER et al. (1971)
	Rat	CC	0	LIPPA et al. (1979)
ChAT and AChE activity	Rat	NB, MS, FC, DG, SUB	+ (28% − 46%)	SPRINGER et al. (1985)
AChE activity	Rat		0	MCGEER et al. (1971)
ACh and choline levels	Mouse	wb	0	KINDEL and KARCZMAR (1985)
ACh synthesis[a]	Mouse	wb	−	GIBSON et al. (1981)
Levels and turnover	Mouse	wb	0	KINDEL and KARCZMAR (1981, 1982)
Turnover rate	Mouse	wb	− (30%)	KINDEL and KARCZMAR (1981, 1982)
	Mouse	SC	+ (27%)	SAITO et al. (1985)
Incorporation [^{3}H]choline into ACh				
ChAT activity	Mouse	CC, CS	−	STRONG et al. (1980)
	Mouse		+	WALLER et al. (1983); WALLER and LONDON (1983)
ChAT activity	Fowl	wb	−	VERNADAKIS (1984)
ChAT and AChE activity	Fowl	wb	0 or +	VERNADAKIS (1984)
BuChE activity	Fowl	wb	+	VERNADAKIS (1973)

CC, cerebral cortex; CD, caudate nucleus; CS, corpus striatum; DG, dentate gyrus; FC, frontal cortex; HP, hippocampus; MS, medial septum; NB, nucleus basalis; S, synaptosomes; SC, cortical slices; SS, striatal slices; SUB, subiculum; wb, whole brain; +, increase; −, decrease; 0, no change
[a] [^{14}C]glucose, [$^2H_4$]choline

turnover of ACh are decreased (by 30%) in the mouse (KINDEL and KARCZMAR 1981, 1982; SAITO et al. 1985). This difference is not drastic but consistent and, therefore, may have potential significance in the control of central cholinergic function in the aged animal. ChAT activity has been found to be decreased while AChE activity is either unchanged or only slightly decreased. As shown in Table 7b, ACh release shows either no changes when measured in cortical or striated slices after K^+ stimulation or is decreased when measured in synaptosomes after electrical stimulation (Table 7a). Choline uptake is either unchanged in striatal

Table 7b. ACh release in brain of aging animals

Parameter	Species	Region	Variation	References
Release	Rat	SC	0 after K^+, veratridine	SPENCER et al. (1985)
Release	Rat	SS	0 after K^+	SIMS et al. (1982)
Release	Rat	SS	0 after K^+	THOMPSON et al. (1984)
Release, vesicle mobilization, vesicle depletion	Rat	HP	Reduced mobilization and recycling after electrical stimulation	APPLEGATE and LANDFIELD (1985)
Release	Rat	SS	–, APO modulation decrease	THOMPSON et al. (1984)
Release	Mouse	SC	–	GIBSON et al. (1981)

APO, apomorphine; SC, cortical slices; SS, striatal slices; HP, hippocampus; –, decrease; 0, no change

Table 8. Energy metabolism and ACh synthesis in brain of aging animals

Parameter	Region	Species	Method	Effect	References
ACh synthesis	Whole brain	Mouse	[U-^{14}C]glucose incorporation	–	GIBSON et al. (1981)
ACh synthesis	HP-CS-CC	Mouse	[U-^{14}C]glucose incorporation	–	PETERSON and GIBSON (1984)
ACh synthesis, oxidative metabolism, ACh release	Brain slices	Mouse	$^{14}CO_2$ production	–	GIBSON and PETERSON (1981)
ATP and Mg^2 and Na^+, K^+ATPase	Whole brain synaptosomes	Rat	Enzymatic	–	BAQUER et al. (1983)
ACh synthesis	Brain slices (CS, CC, HP)	Rat	[U-^{14}C]glucose incorporation	–	SIMS et al. (1982)

HP, hippocampus; CS, corpus striatum; CC, cerebral cortex; –, decrease

slices or decreased in synaptosomes. The molecular and membrane mechanisms for such changes are largely unknown.

Brain oxidative metabolism is decreased with senescence in brain slices, homogenates, mitochondria and *in vivo* and in aged patients decreases in cerebral oxygen consumption correlate with the degree of dementia (cf. GIBSON et al. 1981).

The synthesis of ACh measured by means of incorporation of [U^{14}]glucose into ACh was found to be reduced in two strains of senescent mice by GIBSON et al. (1981) with a decline of 60% and 73% in 10- and 30-month-old animals, respectively (Table 8).

According to GIBSON and PETERSON (1981), this decrease in oxidative metabolism contributes to a reduction in synthesis and release of ACh in the senescent brain. Energy-linked enzymes and ATP have been reported to decrease in whole brain synaptosomal preparations (BAQUER et al. 1983; Table 8).

D. Cholinergic Receptors and Their Pharmacology

Cholinergic muscarinic receptor sites which bind to 3-quinuclidine benzylate (3-QNB) are generally decreased in some regions (CC, CS, FC, HT, CB; for key to abbreviations see Tables 8, 9); however, in HP and sometimes in CC, no change in binding has been reported (Table 9). Binding to α-bungarotoxin (α-BTX) in mouse brain was found to be unchanged or increased depending on the brain region (Table 9).

Pharmacological studies (Table 10) have been performed to test the function of cholinergic receptors in whole animals, brain slices or in brain. Muscarinic receptor (mAChR) density declined with advanced age in FC, CS and HP but no change in receptor affinity was found by PEDIGO et al. (1984). Older rats were more sensitive to the pharmacological effect of cholinergic agonists such as oxotremorine used to induce hypothermia and antinociception (PEDIGO et al. 1984). However, they were less responsive to the muscarinic antagonist scopolamine used to induce locomotion and rearing (PEDIGO et al. 1984). Thus there was no apparent dissociation between the reduced levels of brain mAChRs and the functional response of senescent rats to cholinergic drugs. SEGAL (1982) found a

Table 9. Cholinergic receptors in brain of aging animals

Region (Ligand)[a]	Species	Variation	References
CC	Mouse	–	STRONG et al. (1980)
CS	Mouse	–	STRONG et al. (1980)
CS	Rabbit	0	MAKMAN et al. (1979)
CC	Rabbit	0	MAKMAN et al. (1979)
CS	Rat	–	PEDIGO et al. (1984)
FC	Rat	–	PEDIGO et al. (1984)
HP	Rat	0	PEDIGO et al. (1984)
HT	Rat	–	PEDIGO et al. (1984)
TB	Mouse	–	FREUND (1980)
CS	Rat	–	MORIN and WASTERLAIN (1980)
CB	Rat	–	MORIN and WASTERLAIN (1980)
SA	Mouse	–	DEAN et al. (1979)
SA (α-BTX)	Mouse	0	DEAN et al. (1979)
HP ([^{3}H]3-QNB)	Rat	0, +	SPRINGER et al. (1985)
HP	Rat	–	LIPPA et al. (1981)
HP	Rat	–	LIPPA et al. (1979)
CC	Rat	0	LIPPA et al. (1979)

α-BTX, α-bungarotoxin; CB, cerebellum; CC, cerebral cortex; CS, corpus striatum; FC, frontal cortex; HP, hippocampus; HT hypothalamus; SA, subcortical areas; TB, total brain; –, decrease; 0, no change; +, increase

[a] Ligand was 2-quinuclidine benzylate (3-QNB) unless otherwise stated in parentheses

Table 10. Pharmacological studies of aging in the rat

System	Drug	Test	Effect	References
mAChR	Oxotremorine	Hypothermia; antinociception	+	PEDIGO et al. (1984)
mAChR	Scopolamine	Locomotion; rearing	−	PEDIGO et al. (1984)
Hippocampal slices	ACh	Intracellular activity	−	SEGAL (1982)
Hippocampus	ACh	Extracellular activity	−	HAIGLER et al. (1985)
Hippocampus	ACh	Iontophoresis	−	LIPPA et al. (1979)
Presynaptic mAChR in cortical slices	Oxotremorine	Release of ACh	0	SPENCER et al. (1985)

−, reduced; 0, no change; +, increased

marked difference in the response to ACh from pyramidal cells of HP. This consisted of an initial hyperpolarization followed by a late, slow depolarization. In cells from old animals both hyperpolarization and depolarization were markedly reduced, yet the decreased reactivity to afferent stimulation was maintained. HAIGLER et al. (1985) found that single units response to ACh was significantly decreased in the CA-1 and CA-3-4 areas of rat HP.

LIPPA et al. (1979) found that the response to ACh was significantly reduced in dorsal hippocampal pyramidal cells. One possible explanation is that hippocampal neurons have fewer mAChRs (LIPPA et al. 1981).

These data are important as the HP is considered to play a role in recent memory (BARTUS et al. 1982) and lesions in area CA-1 of the HP and subiculum may contribute to short-term memory disorders in Alzheimer's patients. This is one of the three hippocampal areas in which a significant cell loss occurs in the course of the disease (HYMAN et al. 1984).

E. Experimental Models of Dementia and Aging Based on Chemical Lesions of Cholinergic Systems

Numerous neurochemical changes associated with Alzheimer's disease suggest a major involvement of ACh-containing synapses and neurons (HARDY et al. 1985): (a) cholinergically innervated cortical regions and the hippocampus of patients who have died with the disease show a significant reduction of ChAT (Table 11) and to some extent of AChE activity (Table 12); (b) these biochemical changes correlate with the cognitive and memory impairment shown by the patients; (c) the loss of cholinergic activity in the cortex is highest in those areas with a high density of neuritic plaques and neurofibrillary tangles; (d) *post-mortem* examination of the brain of Alzheimer's patients has demonstrated a selective loss of cholinergic neurons from the nucleus basalis of MEYNERT (Table 13; cf. GIACOBINI et al. 1986a). There is also pharmacological evidence that the cerebral cholinergic

Table 11. ChAT activity in CNS of aging humans

Region	Normal	Alzheimer	References
HP	Decrease	Decrease	McGEER (1981); NORDBERG and WINBLAD (1985)
TH	Decrease	No change	McGEER (1981); NORDBERG and WINBLAD (1985)
CC	Decrease	Decrease	BOWEN et al. (1976); ROSSOR et al. (1980); GOTTFRIES et al. (1983); VERNADAKIS (1984)
CN	Decrease	Decrease	McGEER et al. (1973); McGEER and McGEER (1975)
PT	No change	–	McGEER et al. (1973); McGEER and McGEER (1975)
GP	–	No change	ROSSOR et al. (1980)
MBR	Decrease	Decrease	PERRY et al. (1977)
MBR	Decrease	Decrease	DAVIES (1979)
MBR	Decrease	Decrease	DAVIES and MALONEY (1976)
NBM	Decrease	Decrease	WHITEHOUSE et al. (1982)

CC, cerebral cortex; CN, caudate nucleus; GP, globus pallidus; HP, hippocampus; MBR, many brain regions; NBM, nucleus basalis magnocellularis; PT, putamen; TH, thalamus

Table 12. AChE activity in CNS of aging humans

Region	Normal	Alzheimer	References
CN	Decrease	Decrease	McGEER and McGEER (1975); McGEER and McGEER (1976)
CC	Decrease	–	McGEER and McGEER (1975); McGEER and McGEER (1976)
CC	–	Decrease[a]	ATACK et al. (1983)

CC, cerebral cortex; CN, caudate nucleus
[a] G_4 molecular form

system plays a role in mechanisms of learning and memory. The AChE inhibitor eserine administered via various routes brings about a short-lasting but significant improvement in the memory of Alzheimer patients while drugs which block central mAChR produce cognitive deficits reminiscent of those seen in Alzheimer patients (cf. GIACOBINI 1986a, b).

Attempts to reproduce the disease-dependent damage to specific subcortical areas, mostly to the nucleus basalis magnocellularis (NBM) and the septo-hippocampal structures have been made (Table 14). Chemical lesions with ibotenic acid, kainic acid (an excitotoxin) AF64A (an ethylcholine mustard aziridinium ion), or aluminium have been most commonly used. Lesions through medial septal ablation, fimbrial section, electrolysis or removal of pia-arachnoid have also been utilized. Brain transplants to compensate for these lesions have been attempted (Table 15).

The behavioural effects of lesions, specific or non-specific, to the damaged area have been examined, such as acquisition of spatial memory tasks, ameliora-

Table 13. Changes in cell density and activity of cholinergic enzymes in NBM of control and SDAT brains

	Control	SDAT	Decrease (%)	References
Cell number (no. of cells per grid)	5.9±0.9[a]	1.6±0.3[a]	73	WHITEHOUSE et al. (1981, 1982)
Total no. of cells per section	345 ±65[a]	72 ±15[a]	79	WHITEHOUSE et al. (1982)
	–	–	<50	McGEER et al. (1983)
Cell density	10	15	33	CANDY et al. (1983)
Mean area per neurone ($\mu m^2 \times 10^3$)	10	30[b]	66[b]	CANDY et al. (1983)
	–	–	33	PERRY et al. (1982)
ChAT positive cells	104	35	66	NAGAI et al. (1983)
Immunohistochemistry	–	–	<50	McGEER et al. (1983)
ChAT activity[c] ($nmol\ h^{-1}\ mg^{-1}$)	–	–	89	PERRY et al. (1982)
	–	–	70	ROSSOR et al. (1982a)
	197 ±88[a]	13 ±4[a]	94	CANDY et al. (1983)
	82[d]	26[d]	64	ROSSOR et al. (1980, 1982b)
AChE activity[c] histochemistry	Intense staining	–	–	ROSSOR et al. (1982b)
$\mu mol\ min^{-1}\ mg^{-1}$	67 ±10[a]	142 ±34[a]	79	CANDY et al. (1983)

SDAT, Senile dementia of Alzheimer type; for other abbreviations see text
[a] SEM are reported ($n > 10$), [b] Parkinson patients, [c] per mg of protein, [d] Median value

tion by physostigmine or worsening by scopolamine or atropine. None of the chemicals used so far has been proved to be specific only for cholinergic neurons as the decrease on cholinergic markers (ChAT activity, ACh content or release and choline uptake) has been only partial. The behavioural effects have been specific but partially reversible (PEPEU 1984). Chemical lesions (ibotenic acid) to the NBM produced profound and selective disturbances of the memory of recent events; these persisted for several weeks but completely recovered (BARTUS et al. 1985a, b). No deficit was found after recovery for tests normally sensitive to NBM lesions (Table 14). No evidence of neuronal recovery to the cholinergic system in the cortical projection areas was found. Therefore, any singly relationship between the function of the NBM and the mediation of recent memory seens unlikely (BARTUS et al. 1985a, b). The lasting neurochemical effects of bilateral lesions found by BARTUS et al. (1985b) contrast with earlier reports of recovery of cortical ChAT activity and choline uptake following unilateral lesions (Table 14; WENK and OLTON 1984). In general, the various models appeared to exhibit only some topographical (but non-pathological) similarities with Alzheimer's disease and produced only moderate impairment in behavioural tasks. More specific cholinergic neurotoxins such as AF64A injected I.C.V. in the mouse or directly into the rat hippocampus induced a decrease in cholinergic parameters lasting up to 3 weeks (PEPEU 1984). Following this period there was a complete recovery of ChAT activity and ACh release in the cerebral cortex. Therefore, none of the

Table 14. Experimental models of dementia and aging in the rat: lesions and drugs

Site of lesion	Method	Test	Drug	Effect	References
NBM	Ibotenic acid	Memory and learning	Eserine	Reversal	MURRAY and FIBIGER (1985)
NBM	Ibotenic acid	Passive avoidance	–	Decrease	FLICKER et al. (1983)
Septum-hippocampus	Pulse-stimulation	Evoked field potential	Eserine	Increase	HETTINGER and GONZALEZ (1984)
Hippocampus	Medial-septal ablation	Glucose metabolism	Atropine	Decrease	HARREL and DAVIS (1985)
Septum-hippocampus	Microinjection	Memory	Scopolamine	Decrease	BRITO et al. (1983)
Intracerebral	Microinjection	Memory	AF64A	Decrease	WALSH et al. (1984)
Intraventricular	Microinjection	Memory	AF64A	Decrease	JARRARD et al. (1984)
Intraventricular	Injection	ChAT, choline uptake	AF64A	Decrease	FISHER et al. (1982)
Hippocampus	Injection	ChAT, choline uptake	AF64A	Decrease	MANTIONE et al. (1983)
Septo-temporal hippocampus	Fimbrial lesion	AChE, ChAT	–	Decrease	DRAVID and VANDEUSEN (1984)
Cortex, neurofibrillary tangles	Intracerebral, intracisternal, or chronic systemic	Conditioned avoidance, electrical activity	Al	Encephalopathy not related to dementia	PEPEU (1984)
NBM	Ibotenic acid	Cortical ChAT, choline uptake		Decrease and recovery	WENK and OLTON (1984)
Basal forebrain or intraventricular	Electrolysis unilaterally, neurotoxin bilaterally, kainic acid, ibotenic acid, AF64A	ChAT, ACh content and release	–	45% decrease to recovery	PEPEU (1984)
		Choline uptake, TO, exploratory activity, passive avoidance	AF64A	Decrease and partial recovery	PEPEU (1984)
NBM	Ibotenic acid bilaterally	Radial arm		Short-term memory	BARTUS et al. (1985b)
		Maze task		Decrease	
		ChAT		Decrease	
		Choline uptake		Decrease	
		QNB binding		Unchanged	
Cortical unilateral	Removal of pia-arachnoid	ChAT histochemistry, nucleus basalis	–	Degeneration	SOFRONIEW et al. (1983)
Substantia innominata	Electrolysis	Apomorphine stimulation, locomotion	Apomorphine	Decrease	SWERDLOW et al. (1983)

Table 15. Experimental models of dementia in the aging rat: brain transplants and growth factors

Site of transplant	Method	Test	Transplant	Effect	References
Hippocampus	Transection of fimbria-fornix or AF64A	Open field, passive avoidance, water maze, electrophysiology	Septal cells	Viable connections	SEGAL et al. (1985)
Occipital cortex	Implantation of tissue	Morphology, temporal cortex	Alzheimer's astrocytes spongiform encephalitis	Necrotic cells	VAN DEN BOSCH DE AGUILAR et al. (1985)
Septo-hippocampal	Intraventricular NGF or cell cultures	Morphology	–	Trophic effect	HEFTI (1985)
Hippocampus	Implantation of iris	ChAT, QNB, administration of neurotrophic factor	Iris	Stimulation of cholinergic innervation	SCHONFELD and KATZMAN (1983)
Caudate-putamen, hippocampus	Dissociated cells injection	Behavioural, motor, fluorescence	Embryonic SN, embryonic septum	New terminals, improvement of behaviour	GAGE et al. (1983)
Hippocampus	Graft	Spatial learning	Septal tissue	Improvement	GAGE et al. (1984); GAGE and BJÖRKLUND (1985)
Dentate gyrus	Graft	ChAT Immunohistochemistry	Basal forebrain tissue	Fibre outgrowth, synaptic contacts on hippocampal target neurons	CLARKE et al. (1985)
Nucleus basalis	Cell cultures	Immunohistochemistry and neurophysiology	–	Steady depolarization responses to substance P	NAKAJIMA et al. (1985)

NGF, nerve growth factor; SN, substantia nigra

experimental models of Alzheimer's disease so far developed appears to be satisfactory and to be able to reproduce even partially the characteristic pathological changes of the disease. Nevertheless, the various approaches have all confirmed the correlation between anatomical localization, cholinergic impairment and behavioural deficit making the cholinergic hypothesis worth pursuing therapeutically.

F. Brain Transplants into ACh-Deficient Animals

The procedure of cell or tissue transplantation could represent a future means of treating cholinergic deficits. This would be justified if there were evidence implicating disruption of cholinergic subcortical function. Specific memory deficit mimicking age-dependent defects can be produced in mammals by either damage to the hippocampus or to the septo-hippocampal connections. As mentioned in the previous section (cf. Table 14), transection of the fimbria-fornix, selective lesions to the medial septal area or to the NBM can be performed. Surgical or aspiration lesions, electrolysis or the use of neurotoxins such as kainic acid or AF64A have been employed (Table 14). Table 15 summarizes recent studies with grafting of tissues, injections of dissociated cells or use of nerve growth factor to stimulate trophic effects in transplants following lesions of different kinds and to different regions of the CNS. The results of these treatments has been tested by examining (a) morphological recovery of pathways or establishment of new connections, (b) biochemical recovery of specific markers, (c) reversal of behaviour (sensorimotor or cognitive function) or (d) recovery of neurophysiological function by means of electrophysiology. These studies demonstrate that dissociated cell suspensions or grafts of tissue fragments isolated from embryonic animals implanted into the depth of the brain of aged recipient animals, generally rats, can survive for a long period and reinnervate target areas. Improvement of motor coordination abilities, passive avoidance performance and spatial learning and memory has been demonstrated in such aging animals by different authors (SEGAL et al. 1985; GAGE et al. 1983, 1984). In a recent study of GAGE and BJÖRKLUND (1985), grafting of embryonic cholinergic neurons to the hippocampus of aged rats, resulted in a significant post-graft amelioration of cognitive function (spatial and non-spatial reference memory). A demonstration of atropine and physostigmine effects substantiated the specificity of the cholinergic involvement in the initial deficit. Injection of atropine reversed the positive effect of the grafts on both spatial and non-spatial reference memory. Other results showed (CLARKE et al. 1985) that grafted ChAT-immunoreactive neurons were capable of forming extensive synaptic contacts with the neuronal targets in the dentate gyrus. Staining with ChAT-antibodies revealed grafts rich in cholinergic perikarya and fibre outgrowth into the host hippocampal formation. Ultrastructurally, the graft-derived boutons were seen to have formed predominantly symmetrical synaptic contacts with neuronal elements in the dentate gyrus. These experiments suggest that the ability of cholinergic grafts to compensate for behavioural deficits in aged rats may be mediated via direct synaptic action on host target neurons.

Dissociated cell cultures of cholinergic neurons from NBM (or from other forebrain nuclei) can now be grown *in vitro* and tested both biochemically and electrophysiologically (NAKAJIMA et al. 1985). These neurons are functionally alive and can be tested also for drug effects.

G. Cholinergic Deficits in Normal Aging and in Alzheimer's Disease

The main biochemical finding with regard to Alzheimer's disease and dementia is a considerable decrease of ChAT activity localized to three brain regions, NBM, HP and temporal cortex (Tables 11, 13). The degree of the reported loss varies quantitatively from author to author but is significant in all reports. It should be noted that decreases in ChAT activity of various degrees have also been seen in several regions of normal aging subjects (Table 11). In Alzheimer's disease, however, ChAT decrease is accompanied by an related to cell loss and AChE decrease in the NBM (Table 13). Table 12 shows changes in AChE activity in normal aging subjects and Alzheimer patients. Of special interest is the loss of a particular molecular form of AChE, the 10 S intermediate form (ATACK et al. 1983). This loss is reflected in a decrease of the same form in the cerebrospinal fluid of patients showing senile dementia of Alzheimer's type (SDAT) (GIACOBINI et al. 1986b, c). Other damages to the cholinergic system are also evident. Synaptosomal choline uptake is reduced in SDAT brains (RYLETT et al. 1983). The loss is greater in HP than in neocortex. This finding supports our hypothesis of a selective cholinergic damage during aging which is mainly localized to presynaptic structures (GIACOBINI 1982a, b, 1983; Fig. 2). Reduction of ACh synthesis and release has been reported by SIMS et al. (1983). The histological correlates of the findings are constituted mainly by a sharply reduced number of cholinergic cells in the NBM (Table 13; WHITEHOUSE et al. 1982) and the presence of numerous tangles in hippocampal, cortical, brainstem and hypothalamic regions (ISHII 1966). Of particular interest also is the relationship between the degree of pathological abnormalities (cell death, tangles and plaques), loss of cholinergic markers (mainly ChAT activity) in cortex (Table 11) and intellectual impairment (high degree of dementia).

Presynaptic cholinergic damage does not appear to be mirrored by postsynaptic membrane damage in the form of loss of mAChRs or nAChRs (Table 16). Most reports have shown either normal levels (NORDBERG et al. 1980, 1982) or reductions in subpopulations of SDAT patients (WOOD et al. 1983; Table 16). NORDBERG et al. (1983) have raised the possibility that a mechanism of hypersensitivity, compensating the effect of denervation may occur in humans. This is known to take place in experimental animals following cortical cholinergic denervation. The identification of tangle-containing neurons of cholinergic origin has not been demonstrated in all brain regions, particularly not in the cerebral cortex. This missing link is important in order firmly to establish SDAT as a truly cholinergic disease. The possibility is still open that other neuronal non-cholinergic systems may be involved in subgroups of SDAT or at different phases of progress of the disease (HARDY et al. 1985).

Table 16. Cholinergic receptors in CNS of aging humans

Brain area	Ligand	Subjects		References
		Normal	Alzheimer	
	Muscarinic ligands			
HP, CC, FC	QNB	Decrease	Decrease or no change	NORDBERG and WINBLAD (1985)
TL, CN, TCo	QNB	No change	No change	NORDBERG and WINBLAD (1985)
A, ACC	QNB	–	Decrease	NORDBERG and WINBLAD (1985)
PT, GP	QNB	–	No change	NORDBERG and WINBLAD (1985)
TH	QNB	Increase	Decrease	NORDBERG and WINBLAD (1985)
CC, A	QNB or ACh +CYT	–	No change	WHITEHOUSE et al. (1985)
	Meth-scop	–	Decrease	WHITEHOUSE (1985)
	Nicotinic ligands			
HP	TC	Decrease	No change	NORDBERG and WINBLAD (1985)
HP	BTX	No change	No change	NORDBERG and WINBLAD (1985)
HP	NIC	Increase	–	NORDBERG and WINBLAD (1985)
FC	BTX	–	No change	NORDBERG and WINBLAD (1985)
TH	TC	No change	No change	NORDBERG and WINBLAD (1985)
TH	BTX	Decrease	Decrease	NORDBERG and WINBLAD (1985)
CC	ACh +ATR	–	Decrease	WHITEHOUSE (1985) WHITEHOUSE et al. (1985)

A, amygdala; ACC, accumbens; ACh, acetylcholine; ATR, atropine; BTX, α-bungarotoxin; CC, cerebral cortex; CN, caudate nucleus; CYT, cytosine; FC, frontal cortex; GP, globus pallidus; HP, hippocampus; meth-scop, methscopolamine; NIC, nicotine; PT, putamen; QNB, quinuclidine benzylate; TC, tubocurarine; TCo, temporal cortex; TH, thalamus; TL, temporal lobe

Findings of cholinergic damage have stimulated a number of therapeutical approaches attempting to compensate for the deficit by enhancement of either pre- or postsynaptic function. It is still too early to establish how successful these attempts will be. Meanwhile, a large body of new and useful information is being accumulated through these investigations on the general subject of cholinergic function in the brain making ACh one of the better-known neurotransmitters in the CNS.

Acknowledgements. The investigations carried on in the author's laboratory were supported by Public Health Service Grants, no. NS-11496, NS-11430, NS-15086 and NSF GB-41475 to E.G.; grants from the University of Connecticut Research Foundation to E.G.; National Institute of Aging grant #AG05416-01A1 to E.G.; Air Force grant #83-0051 to E.G.; grants from the Southern Illinois University School of Medicine including Nowatski Eye Foundation, Central Research Committee; and funds from the Southern Illinois University's Alzheimer and E.F. Pearson Foundations.

References

Applegate MD, Landfield PW (1985) Vesicle mobilization and depletion during frequency potentiation and depression in the hippocampus of aged and young rats. Soc Neurosci Abstr 895:261.13

Atack JR, Perry EK, Bonham JR, Perry RH, Tomlinson BE, Blessed G, Fairbairn A (1983) Molecular forms of acetylcholinesterase in senile dementia of Alzheimer type: selective loss of the intermediate (10S) form. Neurosci Lett 40:199–205

Baquer NZ, Duddridge RJ, Hothersall J (1983) The effect of aging of ATP and energy linked enzymes in rat brain. J Neurochem 41:S22

Bartus RT, Dean RL, Beer B, Lippa AS (1982) The cholinergic hypothesis of geriatric memory dysfunction. Science 217:408–417

Bartus RT, Dean RL, Pontecorvo MJ, Flicker C (1985a) The cholinergic hypothesis: a historical overview, current perspective and future directions. Ann NY Acad Sci 444:332–358

Bartus RT, Flicker C, Dean RL, Pontecorvo M, Figueiredo JC, Fisher SK (1985b) Selective memory loss following nucleus basalis lesions: long-term behavioral recovery despite persistent cholinergic deficiencies. Pharm Biochem Behav 23:125–135

Björklund H, Hoffer B, Olson L, Palmer M, Seiger A (1984a) Enkephalin immunoreactivity in iris nerves: distribution in normal and grafted irides, persistence and enhanced fluorescence after denervations. Histochemistry 80:1–7

Björklund H, Dahl D, Olson L, Seiger A (1984b) Glial fibrillary acidic protein-like immunoreactivity in the iris: development, distribution and reactive changes following transplantation. J Neurosci 4 (4):978–988

Bowen DM, Smith CB, White P, Davison AN (1976) Neurotransmitter-related enzymes and indices of hypoxia in senile dementia and other abiotrophies. Brain 99:459–496

Brito GNO, Davis BJ, Stopp LC, Stanton ME (1983) Memory and the septo-hippocampal cholinergic system in the rat. Psychopharmacology 81:315–320

Campbell MJ, McComas AJ, Petito F (1973) Physiological changes in ageing muscles. J Neurol Neurosurg Psychiatry 36:174–182

Candy JM, Perry RH, Perry EK, Irving D, Blessed G, Fairbairn AF, Tomlinson BE (1983) Pathological changes in the nucleus of Meynert in Alzheimer's and Parkinson's diseases. J Neurol Sci 59:277–289

Cardasis Ca (1983) Ultrastructural evidence of continued reorganization at the aging (11–26 months) rat soleus neuromuscular junction. Anat Rec 207:399–415

Chiappinelli V, Giacobini E, Pilar G, Uchimura H (1976) Induction of cholinergic enzymes in chick ciliary ganglion and iris muscle cells during synapse formation. J Physiol (Lond) 257:749–766

Clarke DJ, Gage FH, Björklund A (1985) Formation of cholinergic synapses in the dentate gyrus of behaviorally impaired young and aged rats by grafted basal forebrain neurons. Soc Neurosci Abstr
Coers C, Telerman-Toppet M, Gerard JM (1973) Terminal innervation ratio in neuromuscular disease. Arch Neurol 29:210–222
Coyle JT, Price DL, DeLong MR (1983) Alzheimer's disease: a disorder of cortical cholinergic innervation. Science 219:1184–1190
Davies P (1979) Neurotransmitter-related enzymes in senile dementia of the Alzheimer type. Brain Res 1971:319–327
Davies P, Maloney AFJ (1976) Selective loss of central cholinergic neurons in Alzheimer disease. Lancet 2:1403
Dean RL, Goas JA, Lippa AS, Bartus RT (1979) Changes in behaviour and receptor binding in aged vs. young mice. Soc Neurosci Abstr 9:5
Domino EG, Dren AT, Giardina WJ (1978) Biochemical and neurotransmitter changes in the aging brain. In: Lipton MA, DiMascio A, Killam KF (eds) Psychopharmacology: a generation of progress. Raven, New York, pp 1507–1515
Dravid AR, VanDeusen EB (1984) Recovery of enzyme markers for cholinergic terminals in septo-temporal regions of the hippocampus following selective fimbrial lesions in adult rats. Brain Res 324:119–128
Duncan D (1934) Determination of number of nerve fibers in eighth thoracic and largest lumbar ventral roots. J Comp Neurol 59:47–60
Ehinger B (1967) Adrenergic nerves in the avian eye and ciliary ganglion. Z Zellforsch 82:577–588
Ehinger B, Falck B (1966) Concomitant adrenergic and parasympathetic fibres in the rat iris. Acta Physiol Scand 67:201–207
Fahim MA, Robbins N (1982) Ultrastructural studies of young and old mouse neuromuscular junctions. J Neurocytol 11:641–656
Fahim MA, Robbins N (1985) Transmitter release and ultrastructure of young and old mouse neuromuscular junctions in low calcium solutions. Soc Neurosci Abstr 15:733
Fahim MA, Holley JA, Robbins N (1983) Scanning and light microscopic study of age changes at a neuromuscular junction in the mouse. J Neurocytol 12:13–15
Fisher A, Mantione CR, Abraham DJ, Hanin I (1982) Long-term central cholinergic hypofunction induced in mice by ethylcholine aziridinium ion (AF64A) *in vivo.* J Pharmacol Exp Ther 222:140–145
Flicker C, Dean RL, Watkins DL, Fisher SK, Bartus RT (1983) Behavioral and neurochemical effects following neurotoxic lesions of a major cholinergic input to the cerebral cortex in the rat. Pharmacol Biochem Behav 18:973–981
Freund G (1980) Cholinergic receptor loss in brains of aging mice. Life Sci 26:371–375
Frolkis VV (1970) Regulation, adaptation and aging (in Russian). Nauka, Leningrad
Frolkis VV, Bezrukof VV, Duplenko YK, Shghegoleva IV, Shevtchuk VG, Verkhratsky NS (1973) Acetylcholine metabolism and cholinergic regulation of functions in aging. Gerontologia 19:45
Frolkis VV, Martynenko OA, Zamostyan VP (1976) Aging of the neuromuscular apparatus. Gerontology 22:244–279
Fujisawa K (1976) Some observations on the skeletal musculature of aged rats. Exp Gerontol 11:43
Gage FH, Björklund A (1985) Specificity of intrahippocampal graft-induced improvements on cognitive performance in aged rats. Soc Neurosci Abstr
Gage FH, Dunnett SB, Stenevi U, Björklund A (1983) Aged rats: recovery of motor impairments by intrastriatal nigral grafts. Science 221:966–969
Gage FH, Björklund A, Stenevi U (1984) Intrahippocampal septal grafts ameliorate learning impairments in aged rats. Science 225:533–536
Giacobini E (1982a) Cellular and molecular mechanisms of aging of the nervous system: toward a unified theory of neuronal aging. In: Giacobini E, Filogamo G, Giacobini G, Vernadakis A (eds) The aging brain: cellular and molecular mechanisms of aging in the nervous system. Raven, New York, pp 271–284 (Aging, vol 20)

Giacobini E (1982b) Aging of autonomic synapses. In: Fedoroff S, Hertz L (eds) Advances in cellular neurobiology, vol 3. Academic, New York, pp 173–214

Giacobini E (1983) A new hypothesis of aging of the cholinergic synapse. In: Samuels D, Algeri S, Gershon S, Grimm V, Toffano G (eds) Aging of the brain. Raven, New York, pp 197–210

Giacobini E, Mussini I, Mattio T (1984) Aging of cholinergic synapses in the avian iris. In: Caciagli F, Giacobini E, Paoletti R (eds) Developmental neuroscience: physiological, pharmacological and clinical aspects. Elsevier, Amsterdam, pp 89–93

Giacobini E, Mussini I, Mattio T (1986a) Aging of cholinergic synapses: fiction or reality? In: Hanin I (ed) Dynamics of cholinergic function. Plenum, New York (in press)

Giacobini E, Becker R, Elble R, Mattio T, McIlhany M (1986b) Acetylcholine metabolism in brain: is it reflected by CSF changes? In: Fischer A, Hanin I, Lachman C (eds) Alzheimer and Parkinson diseases. Plenum, New York, pp 309–316

Giacobini E, Becker R, Elble R, Mattio T, McIlhancy M, Scarsella G (1986c) Brain acetylcholine – a view from the cerebrospinal fluid. In: Dun N (ed) Neurobiology of acetylcholine symposium. Plenum, New York (in press)

Gibson GE, Peterson C (1981) Aging decreases oxidative metabolism and the release and synthesis of acetylcholine. J Neurochem 34 (4):978–984

Gibson GE, Peterson C, Jenden DJ (1981) Brain acetycholine synthesis declines with senescence. Science 213:674–676

Gottfries CG, Adolfsson R, Aquilonius SM, Carlsson A, Eckernas SA, Nordberg A, Oreland L, Svennerholm L, Wiberg A, Winblad B (1983) Biochemical changes in dementia disorders of Alzheimer type (AC/SDAT). Neurobiol Aging 4:261–271

Gutmann E, Hanzlíková V (1965) Age changes of motor end plates in muscle fibers of the rat. Gerontology 11:12–24

Gutmann E, Hanzlíková V (1966) The motor unit in old age. Nature 209:921

Gutmann E, Hanzlíková V, Myskočil F (1971) Age changes in cross striated muscle of the rat. J Physiol (Lond) 219:331–343

Haigler HJ, Cahill L, Crager M, Charles E (1985) Acetylcholine, aging and anatomy: differential effects in the hippocampus. Brain Res 362:157–160

Hardy J, Adolfsson R, Alafuzoff I, Bucht G, Marcusson J, Nyberg P, Perdahl E, Wester P, Winblad B (1985) Transmitter deficits in Alzheimer's disease. Neurochem Int 7:545–563

Harrell LE, Davis JN (1985) Cholinergic influences on hippocampal glucose metabolism. Neuroscience 15:359–369

Hefti F (1985) Nerve growth factor (NGF): a survival factor for cholinergic neurons of the rat septum (Abstr). OHOLO Biol Conference

Hettinger MK, Gonzalez LP (1984) Effects of physostigmine on septo-hippocampal averaged evoked field potentials. Brain Res 323:148–153

Hyman BT, van Hoeven GW, Damasis AR, Branes CS (1984) Alzheimer's disease: cell specific pathology isolates the hippocampal formation. Science 235:1168–1170

Ishii T (1966) Distribution of Alzheimer's neurofibrillary changes in the brain stem and the hypothalamus of senile dementia. Acta Neuropathol (Berl) 6:181–187

Jarrard LE, Kant GJ, Meyerhoff JL, Levy A (1984) Behavioral and neurochemical effects of intraventricular AF64A administration in rats. Pharmacol Biochem Behav 21:273–280

Kelly SS, Robbins N (1983) Progression of age changes in synaptic transmission at mouse neuromuscular junctions. J Physiol (Lond) 343:375–383

Kindel G, Karczmar AG (1981) Effect of single and repeated choline administration on brain choline, acetylcholine and acetylcholine turnover of CF-1 mice. Fed Proc 40:269

Kindel G, Karczmar AG (1982) Effect of choline administration on brain choline and acetylcholine levels and acetylcholine turnover of young and old mice. Fed Proc 41:1323

Kindel G, Karczmar AG (1985) Effect of choline administration on brain choline and acetylcholine levels and acetylcholine turnover of young and old mice. Fed Proc 41:1323

Kirby ML, Diab IM, Mattio TG (1978) Development of adrenergic innervation of the iris and fluorescent ganglion cells in the choroid of the chick eye. Anat Rec 191 (3):311–319

Lippa AS, Bartus RT, Pelham RW (1979) Age-related alterations in hippocampal cholinergic functioning. Am Neurochem Soc Abstr 32

Lippa AS, Critchett DJ, Ehlert F, Yamamura HI, Enna SJ, Bartus RT (1981) Age related alterations in neurotransmitter receptors: an electrophysiological and biochemical analysis. Neurobiol Aging 2:3–8

Lucchi ML, Bortolami R, Callegari E (1974) Fine structure of intrinsic eye muscles of birds: development and postnatal changes. J Submicrosc Cytol 6:205–218

Makman MH, Ahn HS, Thal LJ, Sharpless NS, Dourkin B, Horowitz SG, Rosenfeld M (1979) Aging and monoamine receptors in brain. Fed Proc 38:1922–1926

Malmfors T (1965) Studies on adrenergic nerves: the use of rat and mouse iris for direct observations on their physiology and pharmacology at cellular and subcellular levels. Acta Physiol Scand 64 (Suppl 248)

Mantione CR, Zigmond MJ, Fisher A, Hanin I (1983) Selective presynaptic cholinergic neurotoxicity following intrahippocampal AF64A injection in rats. J Neurochem 41:251–255

Marchi M, Giacobini E (1980) Development and aging of cholinergic synapses. II. Continuous growth of acetylcholine and choline levels in autonomic ganglia and iris of the chick. Dev Neurosci 3:38–48

Marchi M, Giacobini E (1981) Aging of cholinergic synapses in the peripheral autonomic nervous system. In: Pepeu G, Ladinsky H (eds) Cholinergic mechanisms. Plenum, New York, pp 25–45

Marchi M, Giacobini E, Hruschak K (1979) Development and aging of cholinergic synapses. I. Endogenous levels of acetylcholine and choline in developing autonomic ganglia and iris of the chick. Dev Neurosci 2:201–212

Marchi M, Hoffman DW, Giacobini E (1980a) Aging of the peripheral nervous system: age dependent changes in cholinergic synapses. In: Neural regulatory mechanisms during aging. Liss, New York, pp 215–219

Marchi M, Hoffman DW, Giacobini E, Fredrickson T (1980b) Age dependent changes in choline uptake of the chick iris. Brain Res 195:423–431

Marchi M, Hoffman DW, Giacobini E (1980c) The peripheral nervous system: a model system of aging. In: Amaducci L et al. (eds) Aging of the brain and dementia. Raven, New York, pp 159–166 (Aging, vol 13)

Marchi M, Hoffman DW, Giacobini E, Fredrickson T (1980d) Development and aging of cholinergic synapses. IV. Acetylcholinesterase and choline acetyltransferase activities in autonomic ganglia and iris of the chick. Dev Neurosci 3:2335–2341

Marchi M, Hoffman DW, Giacobini E, Volle R (1981a) Development and aging of cholinergic synapses. VI. Mechanisms of acetylcholine biosynthesis in the chick iris. Dev Neurosci 4:442–450

Marchi M, Yurkewicz L, Giacobini E, Fredrickson T (1981b) Development and aging of cholinergic synapses. V. Changes in nicotinic cholinergic receptor binding in ciliary ganglia and irises of the chicken. Dev Neurosci 4:258–266

Marchi M, Paudice P, Raiteri M (1983) Choline uptake, acetylcholine synthesis and regulation of acetylcholine release through autoreceptors in the aging rat cortex. In: Samuel D (ed) Aging of the brain. Raven, New York, pp 191–196 (Aging, vol 22)

McGeer EG (1981) Neurotransmitter system in aging and senile dementia. Prog Neuropsychopharmacol 5:435–445

McGeer EG, McGeer PL (1975) Age changes in the human for some enzymes associated with metabolism of catecholamines, GABA and acetylcholine. Adv Behav Biol 16:287–327

McGeer EG, McGeer PL (1976) Neurotransmitter metabolism in the aging brain. Neurobiol Aging 3:389–403

McGeer EG, Fibiger HC, McGeer PL, Wickson V (1971) Aging and brain enzymes. Exp Gerontol 6:391–396

McGeer PL, McGeer EG, Fibiger HC (1973) Choline acetylase and glutamic acid decarboxylase in Huntington's chorea. Neurology (Minneap) 23 (1):912–917
McGeer PL, McGeer EG, Peng JH, Kimura H, Pearson T, Suzuki J, Dolman C (1983) Biochemical and immunohistochemical evaluation of ChAc in aging (Abstr). J Neurochem 41:S20D
Morin AM, Wasterlain CG (1980) Aging and rat brain muscarinic receptors as measured by quinuclidinyl benzilate binding. Neurochem Res 5 (3):301–308
Moyer EK, Kaliszewski BF (1958) The number of nerve fibers in motor spinal nerve roots of young, mature and aged rats. Anat Rec 131:681–699
Murray CL, Fibiger HC (1985) Learning and memory deficits after lesions of the nucleus basalis magnocellularis: reversal by physotstigmine. Neuroscience 14 (4):1025–1032
Mussini I, Giacobini E (1985) Changes with age in neuromuscular junction. In: Carraro U, Angelini C (eds) Proceedings of cell biology and clinical management in functional electro stimulation of neurones and muscles. Cleup, Padova, pp 1–4
Nagai T, McGeer PL, Peng JH, McGeer EG, Dolman CE (1983) Choline acetyltransferase immunohistochemistry in brains of Alzheimer's patients and controls. Neurosci Lett 36:195–199
Nakajima Y, Nakajima S, Obata K, Carlson CG, Yamaguchi K (1985) Dissociated cell culture of cholinergic neurons from nucleus basalis of Meynert and other basal forebrain nuclei. Proc Natl Acad Sci USA 82:6325–6329
Nordberg A, Winblad B (1985) Brain nicotinic and muscarinic receptors in normal aging and dementia. In: Hanin I, Fisher A (eds) Basic and therapeutic strategies in Alzheimer's and other age-related neuropsychiatric disorders. Plenum, New York, pp 95–108
Nordberg A, Adolfsson R, Aquilonius SM, Marklund S, Oreland L, Winblad B (1980) Brain enzymes and acetylcholine receptors in dementia of Alzheimer type and chronic alcohol abuse. In: Amaducci C, Davison AN, Antuono P (eds) Aging of the brain and dementia. Raven, New York, pp 169–172 (Aging, vol 13)
Nordberg A, Adolfson R, Marcusson J, Winblad B (1982) Cholinergic receptors in the hippocampus in normal aging and dementia of Alzheimer type. In: Giacobini E (ed) The aging brain: cellular and molecular mechanisms. Raven, New York, pp 231–245
Nordberg A, Larsson C, Adolfsson R, Alafuzoff I, Winblad B (1983) Muscarinic receptor compensation in hippocampus of Alzheimer patients. J Neural Transm 56:13
Pedigo NW Jr, Minor LD, Krumrei TN (1984) Cholinergic drug effects and brain muscarinic receptor binding in aged rats. Neurobiol Aging 5:227–233
Pepeu G (1984) Animal models for the study of senile dementia of Alzheimer type. In: Knook DL, Calderini G, Amaducci L (eds) Aging of the brain and senile dementia: the inventory of EEC potentialities. Eurage, pp 15–22
Perry EK, Gibson PH, Blessed G, Perry RH, Tomlinson BE (1977) Neurotransmitter enzyme abnormalities in senile dementia. J Neurol 34:247–265
Perry RH, Candy JM, Perry EK, Irving D, Blessed G, Fairbairn AF, Tomlinson BE (1982) Extensive loss of choline acetyltransferase activity is not reflected by neuronal loss in the nucleus of Meynert in Alzheimer's disease. Neurosci Lett 33:311–315
Pestronk A, Drachman DB, Griffin JW (1980) Effects of aging on nerve sprouting and regeneration. Exp Neurol 70:65–82
Peterson C, Gibson GE (1984) Regional acetylcholine metabolism in senescent mice during hypoxia. Soc Neurosci Abstr 316
Rebeiz JJ, Moore MJ, Holden EM, Adams RD (1972) Variations in muscle status with age and systemic diseases. Acta Neuropathol (Berl) 22:127–144
Rockstein M, Brandt K (1962) Muscle enzyme activity and changes in weight in ageing with rats. Nature 196:142–143
Rosenheimer JL, Smith DO (1985a) Differential changes in the end-plate architecture of functionally diverse muscles during aging. J Neurophysiol 53:1567–1581
Rosenheimer JL, Smith DO (1985b) Electrophysiological measurement of glucose uptake into motor nerve terminals of mature adult and aged rats. Brain Res 330:373–377
Rossor M, Fahrenkrug J, Emson P, Mountjoy C, Iversen L, Roth M (1980) Reduced cortical choline acetyltransferase in senile dementia of Alzheimer type is not accompanied by changes in vasoactive intestinal peptide. Brain Res 201:249–253

Rossor MN, Garrett NJ, Johnson AL, Mountjoy CQ, Roth M, Iversen LL (1982a) A post-mortem study of the cholinergic and GABA systems in senile dementia. Brain 105:313–330

Rossor MN, Svendsen C, Hunt SP, Mountjoy CQ, Roth M, Iversen LL (1982b) The substantia innominata in Alzheimer's disease: an histochemical and biochemical study of cholinergic marker enzymes. Neurosci Lett 28:217–222

Rylett RT, Ball MJ, Colhoun EH (1983) Evidence for high affinity choline transport in synaptosomes prepared from hippocampus and neocortex of patients with Alzheimer's disease. Brain Res 2389:169–175

Saito M, Kindel G, Karczmar AG, Rosenberg A (1985) Metabolism of choline in brain of the adged CBF-1 mouse. J Neurosci Res (in press)

Sastry BVR, Janson VE, Jaiswal N, Tayeb OS (1983) Changes in enzymes of the cholinergic system and acetylcholine release in the cerebra of aging male Fischer rats. Pharmacology 26:61–72

Schonfeld AR, Katzman R (1983) In vivo cholinotrophic activities. In: Katzman R (ed) Banbury report 15: biological aspects of Alzheimer's disease. Cold Spring Harbor Laboratory, Cold Spring Harbor

Schultzberg M, Hökfelt T, Olson L, Alund M, Nilsson G, Terenius L, Elde R, Goldstein M, Said S (1980) Substance P, VIP, enkephalin and somatostatin immunoreactive neurons in intestinal tissue transplanted to the anterior eye chamber. J Auton Nerv Syst 1:293–303

Segal M (1982) Changes in neurotransmitter actions in the aged rat hippocampus. Neurobiol Aging 3:121–124

Segal M, Milgram NW, Feinsod Y, Rotengerg M (1985) Brain transplantation: an alternative? (Abstr) OHOLOL Biol Conf

Seiger A, Dahl D, Ayer-LeLievre C, Björklund H (1984) Appearance and distribution of neurofilament immunoreactivity in iris nerves. J Comp Neurol 223:457–470

Seiger A, Selin U-B, Kessler J, Black I, Ayer-LeLievre C (1985) Substance P-containing sensory nerves in the rat iris: normal distribution, ontogeny and innervation of intraocular iris grafts. Neuroscience 15:519–528

Sherman KA, Kuster JE, Dean RL, Bartus RT, Friedman E (1981) Presynaptic cholinergic mechanisms in brain of aged rats with memory impairments. Neurobiol Aging 2:99–104

Sims NR, Marek KL, Bowen DM, Davison AN (1982) Production of [^{14}C]acetylcholine and [^{14}C]carbon dioxide from [U^{14}C]glucose in tissue prisms from aging rat brain. J Neurochem 38:488–492

Sims NR, Bowen DM, Allen SJ, Smith CCT, Neary D, Thomas DJ, Davison AN (1983) Presynaptic cholinergic dysfunction in patients with dementia. J Neurochem 40:503–509

Smith DO (1982a) Physiological and structural changes at the neuromuscular junction during aging. In: Giacobini E, Filogamo G, Giacobini G, Vernadakis A (eds) The aging brain: cellular and molecular mechanisms of aging in the nervous system. Raven, New York, pp 123–138 (Aging, vol 20)

Smith DO (1982b) Restricted diffusion of extracellular potassium at the neuromuscular junction of aged rats. Brain Res 239:668–673

Smith DO (1984) Acetylcholine storage, release and leakage at the neuromuscular junction of mature adult and aged rats. J Physiol (Lond) 347:161–176

Smith DO, Rosenheimer JL (1982) Decreased sprouting and degeneration of nerve terminals of active muscles in aged rats. J Neurophysiol 48(1):100–109

Smith DO, Rosenheimer JL (1984) Factors governing speed of action potential conduction and neuromuscular transmission in aged rats. Exp Neurol 83:358–366

Sofroniew MV, Pearson RCA, Eckenstein F, Cuello AC, Powell TPS (1983) Retrograde changes in cholinergic neurons in the basal forebrain of the rat following cortical damage. Brain Res 289:370–374

Spencer CJ, Schwarz RD, Pugsley TA (1985) Presynaptic control of ^{3}H-acetylcholine (^{3}H-ACh) release from cortical slices of young and aged Fisher 344 rats. Soc Neurosci Abstr 1097:(317.1)

Springer JE, Loy R, Tayrien M (1985) Age-related changes in cholinergic enzymes and muscarinic binding sites in rate brain. Soc Neurosci Abstr 895:251.14
Strong R, Hicks P, Hsu L, Bartus RT, Enna SJ (1980) Age-related alterations in the rodent brain cholinergic system and behavior. Neurobiol Aging 1:59
Swerdlow NR, Swanson LW, Koob GF (1983) Electrolytic lesions of the substantia innominata and lateral preoptic area attenuate the 'supersensitive' locomotor response to apomorphine resulting from denervation of the nucleus accumbens. Brain Res 306:141–148
Syrový I, Gutmann E (1970) Changes in speed of contraction and ATP-ase activity in striated muscle during old age. Exp Gerontol 5:31–35
Thompson JM, Makino C, Whitaker J, Joseph JA (1983) Aging and apomorphine modulation of acetylcholine release from rat striatal slices. J Neurochem 41:S21A
Thompson JM, Makino CL, Whitaker JR, Joseph JA (1984) Age-related decrease in apomorphine modulation of acetylcholine release from rat striatal slices. Brain Res 299:169–173
Tuček S, Gutmann E (1973) Choline acetyltransferase activity in muscles of old rats. Exp Neurol 38:349–360
Tuffery AR (1971) Growth and regeneration of motor end plates in normal cat and hind limb muscles. J Anat 110:221
Van den Bosch de Aguilar PH, Janssens de Varebeke PH, de Paermentier F (1985) Implantation of Alzheimer's disease brain tissue in young rat cortex: parenchymal and vascular reactions (Abstr). OHOLO Biol Conf
Vernadakis A (1973) Changes in nucleic acid content and butyrylcholinesterase activity in CNS structures durin the life-span of the chicken. J Gerontol 28:281–286
Vernadakis A (1984) The aging brain. In: Geokas MC (ed) Clinics in geriatric medicine. Saunders, New York
Vyskočil F, Gutmann E (1969) Spontaneous transmitter release from nerve endings. Experientia 25:945–946
Waller SB, London ED (1983) Age-differences in choline acetyltransferase activities and muscarinic receptor binding in brain regions of C57BL/6J mice. Exp Gerontol 18:419–425
Waller SB, Donald KI, Reynolds MA, London ED (1983) Age and strain comparisons of neurotransmitter synthetic enzyme activities in the mouse. J Neurochem 41:1421–1427
Walsh TJ, Tilson HA, DeHaven DL, Mailman RB, Fisher A, Hanin I (1984) AF64A, a cholinergic neurotoxin, selectively depletes ACh in hippocampus and cortex, and produces long-term passive avoidance and radial-arm maze deficits in the rat. Brain Res 321:91–102
Weiler MH, Smith DO (1985) Metabolism of acetylcholine at the neuromuscular junction of mature adult and aged rats. Soc Neurosci Abstr 15:732
Wenk GL, Olton DS (1984) Recovery of neocortical choline acetyltransferase activity following ibotenic acid injection into the nucleus basalis of Meynert in rats. Brain Res 293:184–186
Whitehouse PJ (1985) Alterations in cholinergic neurons and cholinergic neurotransmitter receptors in Alzheimer's disease. Biol Psych Mtg, p 115
Whitehouse PJ, Price DL, Clark AW, Coyle JT, DeLong MR (1981) Alzheimer disease: evidence for selective loss of cholinergic neurons in the nucleus basalis. Ann Neurol 10:122–126
Whitehouse PJ, Price DL, Struble RG, Clark AW, Coyle JT, DeLong MR (1982) Alzheimer's disease and senile dementia: loss of neurons in the basal forebrain. Science 215:1237–1239
Whitehouse PJ, Martino AM, Antuono PG, Coyle JT, Price DL, Kellart KJ (1985) Reduction in nicotinic cholinergic receptors measured using (^{3}H) acetylcholine in Alzheimer's disease. Soc Neurosci Abstr 11:134
Wood PL, Etienne P, Lal S, Nair NPV, Finlayson MH, Gauthier S, Palo J, Haltia M, Paetou A, Bird ED (1983) A postmortem comparison of the cortical cholinergic system in Alzheimer's disease and Pick's disease. J Neurol Sci 62:201–207
Yurkewicz L, Marchi M, Lauder JM, Giacobini E (1981) Development and aging of noradrenergic cell bodies and axon terminals in the chicken. J Neurosci Res 6:621–641
Zenker W, Krammer E (1967) Untersuchungen über Feinstruktur und Innervation der inneren Augenmuskulatur des Huhnes. Z Zellforsch Mikrosk Anat 83:147–168

CHAPTER 25

Disorders of Cholinergic Synapses in the Peripheral Nervous System

K. V. TOYKA

A. Introduction

Cholinergic pathways may be affected by a number of disorders including natural diseases of men and animals and other diseases induced experimentally. Study of the latter 'model' disorders allows better understanding of the naturally occurring ones. Our understanding of the basic pathological mechanisms ranges from a more profound knowledge of molecular defects in a few disorders to mere description of symptoms in others. This chapter will focus on the structural or molecular dysfunction present in disorders of the somatic motor pathway; other diseases are only mentioned in passing and the reader will be referred to pertinent monographs and review articles.

Table 1 lists disorders of the cholinergic neuromuscular junction (NMJ) with a brief outline of their main features. All these conditions have in common the primary involvement of the cholinergic NMJ but only a few are limited to the process of acetylcholine (ACh) release, ACh breakdown, or to the nicotinic ACh receptor (nAChR). In the group of intoxications only the snake venom neurotoxins seem to affect the NMJ predominantly; other toxins have a more widespread action on cholinergic synapses elsewhere.

In Table 2 disorders are mentioned in which the cholinergic neurons, fibres or synapses are not exclusively affected. In this group cholinergic pathology is only one aspect of the clinical disease. The molecular mechanism of these defects is not yet resolved.

In the following paragraphs the characteristic features of the disorders will be outlined and current views of the pathogenic mechanisms given.

B. Presynaptic Disorders of the Neuromuscular Junction

I. Lambert-Eaton Myasthenic Syndrome

1. Characteristics

Lambert-Eaton myasthenic syndrome (LEMS) is a rare disorder of neuromuscular transmission that was first described only in association with malignant bronchial neoplasm (reviewed by ENGEL 1984; NEWSOM-DAVIS 1985). Between 30% and 60% of patients may not develop tumours. The patients experience weakness and fatiguability of proximal limb muscles. Typically muscle strength improves on sustained maximal contraction or on repetitive contraction but on further effort

Table 1a. Disorders with predominant or exclusive involvement of the cholinergic neuromuscular synapse: presynaptic defects

Disorder	Functional defect	Physiological signs	Structural defect	Proposed mechanism	Animal model
Lambert-Eaton myasthenic syndrome (LEMS)	Muscle weakness; fatiguability improved on maximal effort and tetanic stimuli for several seconds	Low CMAP; increment on nerve stimulation at 20 Hz; reduced quantum content of EPP	Elongated motor nerve terminals; active zones disorganized	Autoantibodies (IgG) to active zones; defect in ACh release process	Passive transfer of IgG to mouse
Congenital myasthenia (some cases)	Muscle weakness; fatiguability on continuous exertion	A: repetitive CMAP to single stimulus; decremental response; small and prolonged EPP	A: small terminals; reduced postsynaptic membrane density	A: mixed genetic or dysplastic (pre- and postsynaptic); no AchE; defect in ACh release	None
		C: decrement only after exercise; reduced EPP amplitude on 10 Hz stimulation	C: none	C: autosomal recessive; (?) deficient ACh synthesis	Hemicholinium-3
Botulism	Cramps; paralysis; mixed autonomic PNS and CNS signs	Similar to LEMS; prolonged post-tetanic facilitation; mEPP frequency low	None in acute intoxication	Exogenous bacterial toxin; defect in ACh release mechanism	Botulinum intoxication in rodents
Bite of black widow spider	Hyperexcitability; twitching; pain (mixed autonomic signs)	Muscle fibre activity; high mEPP and EPP frequency until complete block	Terminals depleted of vesicles	Opening of Ca-channels or ACh recycling defect	Intoxication by bite or experimentally
Bite of snake *Bungarus multicinctus*	Paralysis[a]; mixed autonomic and general signs	β-bungarotoxin; mEPP frequency increased then decreased	Terminal membrane degeneration	Defect of ACh release process	Experimental intoxication
Aminoglycoside antibiotics	Muscle weakness in preexisting neuromuscular transmission disorders	Reduced quantum content of EPP	None	Reversible defect in ACh release	Aminoglycoside intoxication

[a] cf. Table 1b.

Table 1b. Disorders with predominant or exclusive involvement of the cholinergic neuromuscular synapse: postsynaptic defects

Disorder	Functional defect	Physiological signs	Structural defect	Proposed mechanism	Animal model
Myasthenia gravis (MG)	Muscle weakness; fatiguability	Decremental response, reduced mEPP and EPP amplitude	Reduced postsynaptic membrane density; membrane lysis	Reduced number of nAChR due to autoantibodies with mediation of complement	EAMG; passive transfer of IgG to recipient animals; natural MG in dog
Congenital myasthenia (some cases)	As in MG; progressive or non-progressive	B: repetitive CMAP to single stimulus; reduced mEPP amplitude; prolonged mEPP and EPP	B: focal degeneration of postsynaptic membrane and muscle fibres	Abnormal nAChR(?) with prolonged open time of ion channel; genetic	
		D: as in MG	D: as in MG	D: abnormal nAChR or disordered assembly	D: natural disease in dogs
		A: see above Others: as in MG	A: see above Variable	A: see above Usuall reduced nAChR; no autoantibodies	Natural disease in dogs
AChE inhibition[a] medical inhibitors organophosphates	Hyperexcitabiity (cramps, twitching); later paralysis and muscarinic signs PNS and CNS signs with ultimate blockade	Repetitive responses to single stimuli; spontaneous discharge; mEPP/EPP prolonged and increased, later reduced	None in acute toxication; reduced postsynaptic density and focal degeneration in chronic intoxication	Block of AChE direct binding to nAChR; added CNS signs (organophosphates cross blood-brain barrier)	Intoxication of experimental animals
Snake neurotoxins[b]	Muscle weakness				
Cobra toxin α-bungarotoxin	Reversible blockade Irreversible blockade of nAChR	As in MG or complete absence of synaptic activity	None in acute intoxication	Non-depolarizing block of nAChR α-subunit	Intoxication of experimental animals

[a] HOBBIGER (1976) [b] CAMPBELL (1979)

Table 2. Disorders with relevant contribution of the peripheral cholinergic system

Disorder	Functional defect	Physiological signs	Structural defect	Proposed mechanism	Animal model
Polyneuropathies and motor neuron diseases	Muscle atrophy and weakness; reduced monosynaptic reflexes	Slowed or blocked nerve conduction; low CMAP	Cell fibre degeneration (neuronopathy, axonopathy); demyelination ('Schwannopathy')	Toxic (TOCP[a], acrylamide, Vinca etc.); metabolic (diabetes, uremia, malnutrition); heredodegenerative; inflammatory-autoimmune	Experimental intoxication; mutants; streptozotocin etc; mutants, e.g. dogs, mice; experimental allergic neuritis; Coonhound paralysis etc.
Autonomic neuropathies					
Acute pandysautonomia	Autonomic dysregulation	Reduced autonomic reflexes	Cell and fibre loss of ganglia and plexus	Inflammatory-autoimmune (?); mAChR (?)	Experimental autonomic neuropathy
Chronic pandysautonomia (Riley-Day)	Progressive autonomic failure	Reduced or absent autonomic reflexes; cholinergic supersensitivity	Cell and fibre loss	Heredodegenerative, autosomal recessive	None
Systemic amyloidosis	Progressive autonomic failure and sensorimotor defect	Reduced or absent autonomic reflexes	Secondary degeneration due to intercellular storage	Heredodegenerative or acquired (monoclonal gammopathy etc.)	None
Diabetic autonomic neuropathy[b]	As in amyloidosis	As in amyloidosis	Nerve fibre degeneration	Metabolic (myoinositol, sorbitol); inflammatory-autoimmune (?)	Mutants, e.g. BB rat: Streptozotocin
Generalized smooth muscle disease (some cases)	Smooth muscle dysfunction	Defect in smooth muscle regulation	Variable	Unknown; autoimmune reaction to mAChR (?)	None

[a] Triorthocresylphosphate and certain other organophosphates are neurotoxic independently from their cholinergic blocking activity
[b] Other polyneuropathies may also include peripheral autonomic involvement

shows progressive weakness. The physiological hallmark of LEMS is a reduced compound muscle action potential amplitude (CMAP, see Glossary) on supramaximal stimulation of a motor nerve which shows a decremental response at slow rates (2–5 Hz) of stimulation but increases in amplitude by 50%–800% on repetitive stimulation at higher rates (20–50 Hz). The site of altered transmission is at the presynaptic nerve terminal. At the single end-plate level, the stimulus-induced quantal release resulting in an end-plate potential (EPP) is reduced and can be markedly augmented by nerve stimulation at high repetition rate. By contrast, spontaneous quantal ACh release (miniature end-plate potentials, mEPPs), the ACh content of the terminal, and choline acetyltransferase (ChAT) activity are normal. In high calcium solutions, the quantum content of the EPP increases in particular after repetitive stimulation (LAMBERT and ELMQUIST 1971; MOLENAAR et al. 1982). Experimental intoxication with botulinum toxin (see Sect. B II) simulates these alterations in some aspects except for the reduced size of spontaneous quantal ACh release (mEPPs) (SPITZER 1972). Some of the pathological clinical and electrophysiological signs of LEMS can be diminished by application of guanidine hydrochloride or 3,4-diaminopyridine (NEWSOM-DAVIS 1985). It has been hypothesized that guanidine serves as a monovalent cation which can reduce negative charges on the nerve terminal membrane. This may modify calcium channels in such a way that calcium entry is increased (MATTHEWS and WICKELGREN 1977). 3,4-diaminopyridine (3,4-DAMP) and 4-aminopyridine (4-AP) also act by allowing calcium entry into presynaptic terminals (MOLGO et al. 1980).

Ultrastructurally only minor changes in the anatomy of the end-plate have been noted involving increased postsynaptic membrane density (ENGEL 1984). Stereomorphometric analysis of the presynaptic terminal has disclosed a structural change which may well constitute the underlying defect: active zone particles which are usually arranged in parallel rows are disorganized and reduced in number (FUKUNAGA et al. 1982). It is possible that these particles are the calcium channels needed to initiate the ACh release process, consistent with the results of neurophysiological experiments relating impairment of ACh release to a dysfunction of calcium channels (NEWSOM-DAVIS 1985).

The association of LEMS with malignant neoplasms has stimulated the search for circulating factors secreted by the tumour. A tumour extract seemed capable of inducing a transmission defect in the recipient mice on passive transfer (ISHIKAWA et al. 1977). There is indeed now substantial evidence that the IgG of patient's sera contain a factor which can reproduce all the characteristic features of the disease after being passively transferred for days to weeks into healthy mice (LANG et al. 1983) including structural disorganization of active zones similar to that seen in the human disease (FUKUNAGA et al. 1983; for similar experiments, cf. Sect. C.I.1).

In cell culture IgG can inhibit calcium uptake in a tumour-cell line which indicates a role for the calcium channel as the putative autoantigen (ROBERTS et al. 1985).

2. Treatment

On the assumption that a pathogenic humoral factor is a causative agent, the human disease has been treated by plasma exchange with good effects in tumour associated and idiopathic cases. In addition, medical long-term immunosuppression has proved beneficial (DAU 1984). Symptomatic treatment consists of medication with guanidine hydrochloride and, more recently, 3,4-DAMP. Acetylcholinesterase (AChE) inhibitors such as pyridostigmine (see Sect. C.I.8) may be effective possibly by improving the efficiency of transmitter action. The effects of AChE-inhibitors can be augmented when applied with 3,4-DAMP (LUNDH et al. 1984).

II. Intoxication with Botulinum Toxin in Man

Intoxications with botulinum toxin are now rare in developed countries. It was believed that most occurred as a result of the ingestion of insufficiently preserved home-canned food, or, less commonly, from tissue injuries with toxin produced by the anaerobic *Clostridium botulinum* (WAPEN and GUTMANN 1974). Infantile botulism may now be the most common form of the disease. The toxin is produced by the bacteria resting in the infantile intestinal tract (PICKETT et al. 1976). The diagnosis is established by typical clinical signs including dry mouth and accommodation failure (blockade of cholinergic autonomic synapses) and hypotonia, weakness (blockade of neuromuscular synapses), and vertigo. Botulinum is one of the most toxic poisons known and as few as ten molecules may be sufficient to block a synapse (BOROFF et al. 1974).

Neurophysiological signs in patients resemble those seen in LEMS, and depending on the severity of intoxication, partial or complete muscle blockade may occur. Post-tetanic facilitation occurs but lasts longer than in LEMS and post-tetanic exhaustion may be missing (GUTMANN and PRATT 1976). At slow rates of motor nerve stimulation, decremental responses typical of LEMS and myasthenia gravis (MG) (see Sect. C.I) are not observed (GUTMANN and PRATT 1976; CORNBLATH et al. 1983). The neurophysiology of botulinum-poisoned NMJs has been extensively studied (BOROFF et al. 1974; KAO et al. 1976; CULL-CANDY et al. 1976). The results indicate that botulinum toxin interferes with the transmitter mobilizing process possibly by a direct action on calcium release or the calcium internalization mechanism. Long-term blockade produces denervation-like changes in ACh receptors and muscle fibres.

Treatment of the disease is largely symptomatic with support of vital functions, particularly respiration. Guanidine and the aminopyridines may provide some therapeutic benefit (CHERINGTON 1974). Antitoxins are available but are only effective against toxin not yet bound to its presynaptic binding site. Prophylactic vaccination with botulinum toxoid is available for research workers.

III. Other Toxins Acting on Presynaptic Terminals

Black widow spider venom and β-bungarotoxin (β-BTX) have strong blocking effects which are discussed in Chaps. 1, 9 and 18.

IV. Congenital Myasthenia

1. Types of Congenital Myasthenia

The congenital myasthenias are a heterogenous group of disorders in which there appear to be inherited abnormalities of the NMJ. They are very rare diseases with an incidence of less than one in ten million. An immune mechanism as in LEMS or in acquired MG (see Sect. C) is not likely on present evidence (VINCENT et al. 1981; ENGEL et al. 1981, 1984). Only recently have patients with congenital myasthenia been studied in sufficient detail to begin to understand underlying mechanisms. In type C (according to ENGEL et al. 1981) the congenital clinical signs of weakness and fatiguability of skeletal muscle including the extraocular muscles are likely to be caused by a presynaptic defect. In type A both pre- and postsynaptic abnormalities have been implicated. There may be a small number of less well defined cases of congenital myasthenia with an underlying presynaptic mechanism (ENGEL 1984). The remaining cases of congenital myasthenia (types B and D) are possibly postsynaptic in origin and are described in Sect. C.2.

2. Congenital Myasthenia Type C

In this familial syndrome clinical weakness is pronounced in infancy and childhood and may improve later. Inhibition of junctional AChE by neostigmine or pyridostigmine medication may reduce the exercise-induced fatiguability. Neurophysiologically, repetitive supramaximal stimulation of the motor nerve at 2–5 Hz leads to a decremental response (see Glossary) but sometimes only after previous exhaustion of transmission by sustained muscle activity. *In vitro* studies on excised intercostal nerve-muscle preparations from these patients revealed normal spontaneous and stimulus-evoked quantal ACh release with no indication of nAChR dysfunction. After continuous 10-Hz stimulation however, there was a rapid progressive decrement in mEPP amplitude and block of many fibres within the muscle bundle. Since these findings resembled the effects of *in vitro* application of hemicholinium-3, leading to formation of a 'false transmitter', it was hypothesized that ACh synthesis may be deficient in this condition (ENGEL et al. 1981).

3. Congenital Myasthenia Type A

In the only patient described with this congenital disorder, stimulation of motor nerves caused a decremental response in CMAPs at 2 Hz (see Glossary), but the most striking feature was a repetitive muscle action potential following a single stimulus. *In vitro* studies on intercostal muscle from this patient revealed spontaneous quantal ACh release at the lower margin of normal but the quantal content of stimulus-evoked EPPs was markedly reduced. Moreover, the duration of mEPPs and EPPs was prolonged. Immunocytochemical and biochemical investigation showed complete absence of synaptic (16-S) AChE. The prolonged local potentials may explain the repetitive excitation seen on clinical electrophysiological testing following a single stimulus. There was also a structural abnormality in that nerve terminals were only 25%–35% of normal size on stereomorphometric analysis (ENGEL et al. 1981). There is no effective treatment known to date and weakness was not reduced by therapeutic AChE inhibitors.

C. Postsynaptic Disorders of the Neuromuscular Junction

I. Myasthenia Gravis (Acquired Autoimmune Myasthenia)

1. Characteristics

Myasthenia gravis (MG) is the most common disorder of the NMJ. With an incidence of 2–4 in 100000, it is still relatively rare. Twenty years ago it was considered a presynaptic disorder by many experts (GROB 1963b). In the past decade is has become clear that MG is a prototype receptor disease involving the postsynaptic nAChR (VINCENT 1980; DRACHMAN 1981; ENGEL 1984).

Clinically MG causes weakness and abnormal fatiguability of skeletal muscles. Symptoms often begin in the muscles of the head including the extraocular muscles giving rise to drooping eyelids and double vision. The disease may start at any age. Muscle weakness can be temporarily improved by medication with AChE inhibitors, e.g. neostigmine or pyridostigmine bromide. In contrast to LEMS, but similar to the congenital myasthenias, muscle strength deteriorates rapidly on sustained or repeated maximum contractures.

The neurophysiological test most commonly used for diagnostic purposes is repetitive supramaximal stimulation of motor nerves; this causes a decremental response of the CMAPs (see Glossary). The initial potential is not reduced as in LEMS and decremental responses are seen with all repetition rates above 2 Hz. After a tetanus a brief phase with post-tetanic facilitation is followed by post-tetanic exhaustion. These abnormal features can be partially reversed by AChE inhibitors. Ischaemia and elevation of muscle temperature augment the abnormality (OOSTERHUIS 1984).

In vitro examination of nerve-muscle preparations reveal reduced mEPP and EPP amplitudes with virtually normal rise- and decay-time constants. The synaptic ACh sensitivity is markedly reduced (ALBUQUERQUE et al. 1976), a finding which is confirmed by single channel conductance analysis (CULL-CANDY et al. 1979). These findings were indicative of reduced number and/or impairment of nAChRs and were consistent with earlier studies in patients showing reduced ACh sensitivity after intra-arterial injection of ACh (GROB and NAMBA 1976). By contrast, ACh release is not reduced. The quantum content of the EPP and mEPP frequencies are normal (VINCENT 1980).

Structurally there are typical abnormalities in chronic cases including reduction in synaptic folds and widening of the synaptic cleft (ENGEL and SANTA 1971). The latter augments the transmission defect by imposing an added diffusion barrier. Historically our present understanding of the pathologic mechanisms underlying the transmission disorder in MG came from biochemical and immunological studies (VINCENT 1980). The snake neurotoxins α-bungarotoxin (α-BTX) and cobra toxin were found to be specific ligands for the nAChRs in electric organs of electric rays, as well as the mammalian NMJ (LEE 1972) and allowed a purification (KARLSSON et al. 1972). Using labelled α-BTX it was demonstrated that NMJs from patients with MG had markedly reduced binding, probably indicating reduced numbers of nAChR molecules per end-plate (FAMBROUGH et al. 1973). Furthermore, experimental intoxication with cobra toxin, a reversible ligand of ACh receptor, binding near to or at the ACh binding site, gave clinical and physiological signs resembling human MG (SATYAMURTI et al. 1975). Additional evidence that MG was basically a disorder of the nAChR came from the observation that immunization of rabbits with purified AChR from electric organs of electric eel or ray not only generated antibodies to nAChR but also induced a condition resembling MG. This was named experimental autoimmune MG (EAMG) (PATRICK and LINDSTROM 1973; HEILBRONN and MATTSSON 1974; VINCENT 1980).

2. Pathogenic Role of Antibodies Directed to nAChR

On the basis of this mounting evidence for an autoimmune pathogenesis, a search for antibodies to nAChR in patients was instituted using as a probe α-BTX which revealed circulating autoantibodies in the serum (ALMON et al. 1974; AHARONOV et al. 1975; LINDSTROM et al. 1976). The pathogenic role of such antibodies was shown by passive transfer of MG from patients to mice following the injection of the patient's IgG into the animals. The recipient mice when treated with doses sufficient to maintain human IgG levels in them for days to weeks developed transmission failure with reduced mEPP and EPP amplitude and reduced binding of α-BTX to NMJs, thus reproducing the typical features of MG. This effect could be augmented by activation of the complement system (TOYKA et al. 1977).

The naturally occurring counterpart of the passive transfer model is human neonatal transient myasthenia. This results from the trans-placental transfer of maternal IgG-autoantibodies to the foetus (KEESEY et al. 1977). The concept of humoral antibodies was supported by immunocytochemical studies showing IgG and activated complement bound to the postsynaptic membrane in human muscle and in mice with passively transferred myasthenia (ENGEL 1984). Moreover, elimination of IgG by therapeutic plasma exchange leads to a dramatic improvement in critically ill patients (DAU 1984).

In human MG different groups of antibodies to nAChR have been detected, most of which are directed to the main immunogenic region of the α-subunit as determined by a panel of monoclonal antibodies (TZARTOS et al. 1982). Clinically, measurement of autoantibody titres has proved useful as a diagnostic aid because they are highly specific for MG (LINDSTROM et al. 1976; VINCENT 1980) and may also be helpful to assess the action of immunosuppressive medication (HOHLFELD et al. 1985). Detergent extracts of human limb-muscle rich in nAChR provide a sensitive probe for the detection of circulating autoantibodies with elevated titres in more than 98% of patients with generalized disease (TOYKA and HEININGER 1986).

In EAMG passive transfer of hyperimmune sera or IgG-containing antibody to nAChR to healthy recipient animals leads to acute clinical and electrophysiological signs of MG and this is critically complement dependent (LENNON et al. 1978). Ultrastructurally, the chronic stage of EAMG resembles human chronic MG in all major aspects (ENGEL et al. 1976). In contrast, the man-to-mouse passive transfer model which displays no morphologic alterations (TOYKA et al. 1978) is similar to mild human MG or MG of recent onset (ENGEL et al. 1976).

3. Mechanisms of Antibody-Induced Loss of nAChR

Theoretically the antibody-induced loss of available nAChR could be produced by several possible mechanisms:

1. Immunopharmacological blockade at the ACh binding site
2. Increased rate of degradation of nAChR
3. Decreased synthesis of nAChR
4. Membrane lysis at nAChR sites possibly in conjunction with complement
5. Alteration of the ion permeability changes induced by ACh binding to the nAChR.

Present evidence suggests that all these mechanisms operate in human MG, but the relative role of each probably varies from patient to patient and even within an individual patient over time (DRACHMAN 1981; ENGEL 1984).

Membrane lysis induced by antibodies in conjunction with activation of the complement system may be the principal mechanism in long-term disease. This is supported by immunocytochemical and electron-microscopic observations in patients with myasthenia and in EAMG (ENGEL 1984) and by passive transfer experiments with polyclonal (TOYKA et al. 1977; LENNON et al. 1978) and monoclonal (RICHMAN et al. 1980; LENNON and LAMBERT 1981) antibodies to nAChR into complement-depleted recipients. Freeze-fracture studies of human NMJ revealed not only a loss of P-phase intermembraneous particles but also a new class of large-diameter E-phase particles which may represent antibody-complement complexes (RASH et al. 1981). The remaining functional nAChR may be at a steady state for several days (FUMAGALLI et al. 1982). An increased degradation rate of nAChR induced by human antibodies and referred to as antigenic modulation has been shown in cultured muscle (KAO and DRACHMAN 1977) and in intact NMJ (STANLEY and DRACHMAN 1978). In the former this seems to depend on the ability of the IgG antibody to cross-link the binding sites of two adjacent nAChR molecules (DRACHMAN et al. 1978), most likely the α-subunits. However, monovalent Fab fragments not capable of cross-linking can also reduce nAChR *in vivo* (STERZ et al. 1986). The functional implications of accelerated degradation *in vivo* may be even more complex since increased re-synthesis partly compensates for receptor loss (WILSON et al. 1983).

Some degree of immunopharmacological blockade of α-bungarotoxin binding to nAChR by human MG antibodies is present in 30%–80% of human myasthenic sera with appropriate assays (VINCENT 1980; ENGEL 1984). Incubation of nerve-muscle preparations with human myasthenic serum can reduce mEPP amplitudes, but surprisingly the antibody-containing IgG fraction does not (LERRICK et al. 1983). The clinical observation of improvement in muscle strength within hours in occasional patients after plasma exchange would be compatible with such blocking factors (DAU 1984). Nevertheless, it is not yet clear to what extent nAChR antibody produces immunopharmacological blockade at the ACh binding site.

Minor alterations of the nAChR ion channel, as assessed by the quantitative microiontophoretic application of ACh, has been shown to occur in EAMG (HOHLFELD et al. 1981) and mouse passive transfer models (DUDEL et al. 1979; PENNEFATHER and QUASTEL 1980) but not in human myasthenic muscle (CULL-CANDY et al. 1979).

4. Relation Between nAChR Antibody Titre and the Severity of Disease

As may be inferred from the complex mechanisms of the action of autoantibodies, clinical weakness is not a simple function of the plasma concentration of circulating antibodies. For a given muscle end-plate the net result of the antibody attack is influenced by many factors: (a) antibodies are directed to various parts (epitopes) of the nAChR molecule and may thus have various complementary effects; (b) antibodies may act with or without mediation of complement, the former leading to microlesions of the membrane; (c) the capability to re-synthesize nAChR or to repair membrane lesions may vary; (d) the altered end-plate geometry, with widening of the synaptic cleft and broadening of the infoldings, set in motion by sustained immunopathological attack, can further decrease the efficiency of neuromuscular transmission; (e) changes in the muscle fibres from disuse or by an alteration of still hypothetical trophic factors may influence their contractile properties; (f) the contribution of affected fibres to the motor unit pool; and (g) the fatiguability of these motor units.

A loose correlation exists between clinical severity and the ability of the antibodies (a) to degrade nAChR at an increased rate (CONTI-TRONCONI et al. 1981) and (b) to block α-BTX binding in muscle cell cultures (DRACHMAN et al. 1982). In an individual patient a much closer correlation exists between circulating antibody titres and clinical weakness, probably because some of the factors discussed above are less variable in a given patient than from patient to patient (BESINGER et al. 1983).

5. Monoclonal Antibodies to nAChR

A review of the many monoclonal antibodies raised against native or denatured nAChR from electric organs or mammalian muscle now available is beyond the scope of this chapter (TZARTOS and LINDSTROM 1980; FUCHS et al. 1981; WATTERS and MAELICKE 1983; Chap. 9, this volume). Monoclonals have been described which induce clinical myasthenia on passive transfer and thereby allow the effects of a single antibody to be dissected out. This is impossible with polyclonal antibodies which recognize many different epitopes (RICHMAN et al. 1980; LENNON and LAMBERT 1981).

A small group of monoclonal antibodies block the ACh binding site and induce complete paralysis in less than 1 h (WATTERS and MAELICKE 1983; GOMEZ and RICHMAN 1983).

6. Aetiology of the Autoimmune Response in Myasthenia Gravis

As with other naturally occurring immunopathological disorders of man, the origin of the autoimmune response remains an unsolved problem despite considerable research. A number of hypotheses are currently entertained; these are by no means exclusive; indeed in an affected individual several mechanisms may reinforce one another (WEKERLE et al. 1981; ENGEL 1984). Before these hypotheses are discussed it should be stressed that MG is a heterogenous disorder although the principle target is the nAChR. There are differences between patients of different sex, age, histocompatibility type, abnormalities of the thymus gland, and associations with other autoimmune diseases (NEWSOM-DAVIS 1982) suggesting the existence of multiple aetiologies.

7. The Thymus in Myasthenic Gravis

The thymus is hyperplastic in 75% of patients and practically in every patient below age 40. Numerous bone marrow derived (B) lymphocytes are present and are capable of producing autoantibodies *in vitro*. Other thymus cells can stimulate peripheral blood lymphocytes to produce autoantibody (NEWSOM-DAVIS 1982). Furthermore, the thymus may be the site where the critical induction of the autoimmune response takes place, possibly long before clinical signs are evident. The 'dual hypothesis' (WEKERLE and KETELSEN 1977) proposes a two-step action: first, myoid cells (the epitheloid cells of Bargmann) express nAChR on their surface (nAChR differentiation and possibly expression by these cells is under genetic control); second, lymphoid cells are made responsive to nAChR and are also 'educated' to recognize the self-antigens of the histocompatibility system (human leucocyte antigen-system, HLA in man). This step is also genetically determined. If in later life these responsive thymus-dependent (T) lymphocytes become activated by an unknown mechanism, they may stimulate B-lymphocytes to differentiate to plasma cells and produce auto-antibodies.

The possibility that a viral infection of the thymus could trigger an autoimmune response has long been discussed but evidence is still lacking. An alternative would be some alteration in the nAChR molecule on antigen-presenting cells, thus forming neo-antigens not recognized as self and therefore causing an autoimmune reaction. The currently favoured hypothesis is a defect in the immunoregulatory network, i.e. in the equilibrium between positive ('helper') and negative ('suppressor') effects of regulatory T-lymphocytes upon the antibody-producing B-lymphocytes (NEWSOM-DAVIS 1982; HOHLFELD and TOYKA 1985). Indeed, activated helper T cells specific for nAChR have recently been demonstrated to occur in human MG (HOHLFELD et al. 1984). An attractive new hypothesis for the *in vivo* activation of autoimmune helper T cells invokes as a trigger signal an atopical expression of class II histocompatibility antigens which are not normally present on the cell surface.

A different mechanism by which autoantibodies could be induced is 'idiotype mimicry'. A recent study has demonstrated this for autoimmune reactions to nAChR. The ACh-analogue *trans*-3,3′-bis[α-(trimethylammonio)methyl] azobenzene bromide (BisQ) is used to immunize an animal resulting in the production of antibodies against BisQ. With these antibodies other animals are immunized and produce antibodies, called anti-idiotype antibodies, some of which are directed to the antigen-binding region of the first antibody that binds BisQ. Therefore the antigen-binding part of this anti-idiotypic antibody, the idiotype, structurally resembles BisQ. With these anti-idiotypic antibodies not only binding to nAChR occurs but also clinical signs of MG can be induced in passive transfer experiments (CLEVELAND et al. 1983).

With the advent of genetic engineering it will soon become possible to synthesize nAChR subunits with appropriate genomic DNA probes (MISHINA et al. 1984) and to deal in more detail with the T-cell antigen receptor (ACUTO and REINHERZ 1985). These approaches will offer new insights into the interaction between nAChR and the immune system in MG.

8. Treatment of Myasthenia Gravis

Modern treatment has improved the outlook in MG remarkably. Before the advent of immunosuppressive drugs only AChE inhibitors such as neostigmine or pyridostigmine were available leaving many patients with generalized myasthenia markedly disabled (GROB 1963b). These drugs probably improve transmission efficiency by prolonging the action of ACh. This leads to a rise of small EPPs to above the threshold for an action potential and allows the muscle fibres to contract. Long-term administration of AChE inhibitors leads in itself to loss of nAChR and to alterations of the end-plate ultrastructure (HUDSON et al. 1978), but the practical implication of this in patients is not clear. Other drugs which increase ACh release have been used such as 4-AP but side effects have discouraged its use as a treatment for MG. 3,4-DAMP apparently has less severe side-effects (LUNDH et al. 1984). Ephedrine may have a role in adjuvant symptomatic treatment (GROB 1963b). Potassium salts and spironolactone may also improve myasthenic weakness but this is now rarely used (OOSTERHUIS 1984).

The therapeutic mainstay is now long-term immunosuppression with glucocorticosteroids and/or azathioprine, a precursor of 6-mercaptopurine. These drugs probably act by suppressing the autoimmune response to nAChR with a detectable fall in circulating autoantibodies (BESINGER et al. 1983) and of sensitized lymphocytes (HOHLFELD et al. 1985). The cytotoxic drugs such as cyclophosphamide and isophosphamide are only used in refractory cases. Cyclosporin A, a new immunosuppressant, is also effective in MG.

Discontinuing immunosuppressive treatment with azathioprine, after an initial remission, may leave half of the patients in sustained remission for years (HOHLFELD et al. 1985), in contrast to patients treated with glucocorticosteroids alone, where discontinuance is less often possible without relapse (OOSTERHUIS 1984). The reasons for these differences are not yet understood.

Glucocorticosteroids have direct depressive effects on neuromuscular transmission which may augment myasthenic weakness before the beneficial effects are manifested. Studies with the mouse passive transfer model indicate that myasthenic end-plates are particularly sensitive to the depressive action of steroids (DUDEL et al. 1979).

Another therapeutic procedure is surgical thymectomy which leads to moderate improvement, often of limited duration. The mode of action of this treatment is not understood (GROB 1963b; OOSTERHUIS 1984).

The most rapid way of removing pathogenic antibodies from the circulation is by therapeutic plasma exchange (DAU 1984). Because of various side effects, this procedure is reserved for patients in myasthenic crisis. If not combined with medial immunosuppression, the improvement induced by removal of pathogenic antibodies is limited to 3–4 weeks. More specific ways of selectively or preferentially removing antibodies to nAChR are under study.

Ideally the goal of therapy should be to eliminate or turn off the autoimmune lymphocytes specifically and selectively. Such procedures have been carried out in the EAMG model (WALDOR et al. 1983) but are not yet available for patients with MG (HOHLFELD and TOYKA 1985).

II. Congenital Myasthenia

1. Characteristics

For all congenital myasthenias a considerable effort has been expended to try to determine the underlying defect, but the very low incidence of the disorder and the limited amount of available human material impose restrictions. A number of cases of congenital myasthenia have been reported in which there is a presumptive postsynaptic abnormality. These have been divided into two types (ENGEL 1984), types B and D. A few other cases may have a different basic defect (VINCENT et al. 1981).

2. Congenital Myasthenia Type B

Clinical signs of muscle weakness, including extraocular muscles, appear after birth or early in childhood. Fatiguability is similar to that in aquired MG. Most cases are inherited in an autosomal dominant mode. AChE inhibitors are of no definite benefit. Physiologically, repetitive nerve stimulation of motor nerves leads a decremental response; on single stimuli a repetitive CMAP has been noted similar to that seen in type A congenital myasthenia (see Sect. B.IV.3). *In vitro* studies on nerve-muscle biopsy samples revealed decreased mEPP amplitudes with markedly prolonged decay times which could be further enhanced by inhibition of AChE. The EPP quantum content and mEPP frequency were not changed nor was the stock of readily releasable quanta. Structurally, the size of the nerve terminals was reduced to 65% of normal and focal degeneration occurred at or near the end-plate region including the muscle fibre. Cytochemical estimates of available nAChR by peroxidase-labelled α-BTX binding showed a marked reduction in areas with muscle fibre degeneration in the vicinity of the end-plate. This type of congenital myasthenia combines the features of a transmission disorder with those of a myopathy but functionally the defect in transmission seems to be dominant. The syndrome appears to be a complex genetic abnormality with an nAChR which probably has an abnormally prolonged ion channel opening time (ENGEL 1984).

3. Congenital Myasthenia Type D

The clinical features of this congenital myasthenia are similar to those of acquired MG including the positive response after administration of AChE inhibitors. In all cases of type D (ENGEL et al. 1981; VINCENT et al. 1981) the physiological defect revealed by *in vivo* and *in vitro* testing was similar to that of MG except that the ACh content of the nerve terminals was normal. Single channel conductance was measured in two familial cases with normal results. Structurally, endplates were elongated as in other cases of congenital myasthenia and LEMS. The exact character of the receptor defect is not known (ENGEL 1984).

4. Other Congenital Myasthenias

Recently another type of a human congenital myasthenia with a presumptive postsynaptic abnormality has been described with elongated and multiply branched axon terminals and a 3-fold increase in nAChR per muscle fibre (CORNBLATH et al. 1985). In dogs a variety of inherited congenital myasthenias have been described, some of which resemble the presynaptic and postsynaptic abnormalities described in human congenital myasthenia. These have to be distinguished from acquired MG in dogs which is the canine counterpart of human autoimmune MG (LENNON et al. 1981).

III. Inhibition of AChE

Two groups of clinical conditions associated with inhibition of AChE are common; one occurs with overdose of reversible, medically used inhibitors, for example neostigmine. The other occurs following intoxication with organophosphorus compounds. The latter bind to and block the esteratic site of the active centre of the enzyme by transferring a dialkylphosphoryl group to it. All these compounds have profound effects on muscarinic and nicotinic cholinergic synapses in the peripheral nervous system and the clinical signs depend on the relative contribution of each. In addition, the hydrophobic organophosphorus compounds, but not neostigmine and pyridostigmine, readily cross the blood-brain barrier and cause central nervous system pathology (GROB 1963a; HOBBIGER 1976). (For a detailed description of poisoning with organophosphorus compounds used as insecticides or in chemical warfare the reader is referred to NAMBA et al. 1971). Apart from the effects of acute intoxication there are chronic neurotoxic actions after injection of some organophosphorus compounds (e.g. triorthocresyl phosphate) which are independent of inhibition of AChE; this may include axonal degeneration with secondary demyelination (see Sect. D; DAVIS and RICHARDSON 1980; LOTTI et al. 1984). These peripheral nerve disorders are rare with the usual compounds used as insecticides. In acute intoxications symptoms of cholinergic hyperactivity in all cholinergic synapses occur which present clinically as fasciculations and cramps.

Finally, paralysis is due to irreversible blockade of neuromuscular transmission. Neurophysiologically, the effects at the level of the NMJ have been studied *in vitro* using nerve-muscle preparations. mEPP frequency and amplitude are initially increased and duration is prolonged. At a later stage mEPP and EPP amplitudes fall until blockade ensues. After a single nerve impulse multiple ACh-induced muscle fibre contractions can be seen; these have been explained by multiple interactions with nAChR. In addition, antidromic spiking in the nerve, probably mediated by increasing postsynaptic potassium efflux (GROB 1963a; HOHLFELD et al. 1981), may lead to the excitation of entire motor units via axon reflexes. This is thought to be the cause of the clinical fasciculations caused by AChE inhibitors.

Chronic intoxication by both organophosphorus compounds and reversible AChE-inhibitors leads to trophic changes in the postsynaptic membrane with loss of nAChR and in the muscle fibres; these resemble some of the changes seen in myasthenic syndromes (HUDSON et al. 1978). A single dose of paraoxon can induce grouped muscle fibre degeneration which has been attributed to the increased concentrations of ACh because it could be prevented by hemicholinium-3, a false transmitter, and by pralidoxim. Treatment of the clinical syndrome and the discussion of autonomic cholinergic dysfunction are covered in Sect. D. The widely used oxime antidotes appear to reactivate AChE at the NMJ more readily than at autonomic synapses (GROB 1963a).

IV. Snake Neurotoxins

The neurotoxins from cobras and α-BTX from the krait *Bungarus multicinctus* must be mentioned here because they have been crucial for our current knowledge concerning many of the disorders described above. The clinical disorder produced by snake bite is dominated by neuromuscular blockade although several other toxic polypeptides and proteins are contained in the crude venom causing systemic disease if the neuromuscular blockade is survived (CHANG 1979; CAMPBELL 1979). Cobra toxin and α-BTX are potent non-depolarizing blockers of nAChR with essentially the same physiological effects as d-tubocurarine. A myasthenia-like state can be maintained by cobra toxin for hours (SATYAMURTI et al. 1975).

Type I cobra toxins are used for the affinity purification of nAChR (KARLSSON et al. 1972) because of the reversible nature of their binding. By contrast, α-BTX binds irreversibly and is therefore unsuitable for affinity purification but highly suitable for labelling nAChR. Recent studies using nAChR produced by cloned cDNA in *Xenopus* oöcytes favour the view that the α-subunit is the most important one for binding α-BTX (MISHINA et al. 1984).

D. Disorders of the Peripheral Somatic Nervous System

I. Neuropathies with Predominant Motor Fibre Dysfunction and Motoneuron Disorders

1. Characteristics

In most human disorders of the peripheral nervous system the disease process is not restricted to cholinergic nerve cells and their respective axons; rather, diseases of neurons forming cholinergic synapses contribute to the disorder in variable proportions. In Table 2 the disorders with significant dysfunction of the peripheral cholinergic system have been listed. The aim of this section is not exhaustively to review all peripheral neuropathies but simply to outline examples where a significant cholinergic component is present.

The nerve cells of the peripheral motoneuron and their centrifugal processes including anterior root and spinal nerves end with their terminals on the cholinergic NMJ. Whenever any part of the lower motoneuron is affected by pathological conditions, dysfunction of the skeletal muscle fibre, referred to as 'denervation', will eventually result. This comprises (a) atrophy, (b) changes in contractile properties and (c) complete inexcitability to voluntary or reflex stimuli. Despite this uniform general principle the clinical appearance of various syndromes or diseases is extremely variable depending on the severity, duration, and site of the pathology. The clinical hallmarks are muscle atrophy and weakness; abnormal fatiguability is not a major feature, as it is in disorders of the NMJ.

In disorders affecting the motor nerve fibres an important subgroup is characterized by predominant pathology of the Schwann cell and the myelin sheath. There is an intimate relationship between the axon and the Schwann cell (AGUAYO and BRAY 1984) such that there is rarely isolated involvement of one or the other. Nevertheless, there is much to be said clinically for distinguishing 'Schwannopathies' from axonopathies and neuronopathies, because of the different sensitivities of Schwann cells, nerve cell bodies and axons to toxic, metabolic, physical, or inflammatory insults (DYCK et al. 1984).

2. Pathophysiological Aspects

The physiological signs of peripheral nerve involvement are detected *in vivo* by standard electrophysiological techniques including measurement of nerve conduction velocities and electromyography. Nerve conduction in motor nerves is usually only mildly impaired in axonopathies and neuronopathies but is markedly reduced in disorders of the myelin sheath. Conduction velocity is usually measured in fast-conducting large myelinated fibres, but estimates in medium-sized myelinated fibres are possible (KIMURA 1983). The events in single nerve fibres *in vivo* can be measured by refined microneurographic methods which are mainly used for myelinated sensory or autonomic fibres (FAGIUS and WALLIN 1980). The degree of degeneration and regeneration of peripheral nerve motor fibres can be estimated by standard needle electromyography and single fibre electromyography (STALBERG and TRONTELJ 1979; DAUBE 1981). More elaborate techniques including measurement of conduction over very short distances in single nerve fibres (BOSTOCK and SEARS 1978) can be used in animal models of peripheral nerve disorders.

Morphologically, a well-characterized sequence of events follows the degeneration of nerve fibres both in nerve and muscle (DYCK et al. 1984). More recently, some typical ultrastructural features of motor neurons and nerves exposed to various toxins have been described; heredodegenerative and inflammatory neuropathies have been defined (DYCK et al. 1984) and electrophysiological, biochemical, and biological changes have been described, including defects in fast and slow axoplasmic transport (OCHS 1984), abnormal storage or transport of substrates and enzymes (WEISS and GORIO 1984) and impulse blockade at nodal and paranodal regions of motor nerve fibres (BOSTOCK and SEARS 1978).

Examples of disorders affecting motoneurons and motor nerve fibres are given in Table 2. One disorder will be discussed as an example in which the cholinergic motor neurons are predominantly involved, amyotrophic lateral sclerosis (ALS).

II. Amyotrophic Lateral Sclerosis

1. Characteristics

Clinically, ALS is characterized by progressive weakness and atrophy of skeletal muscles with fasciculations and cramps due to loss of anterior horn cells in the spinal cord. In addition, signs of upper motoneuron involvement such as hyperreflexia and spasticity are present. Other systems including the sensory or cerebellar systems are involved but this rarely presents with clinical signs. Neuromuscular transmission may be mildly abnormal; this is secondary to degeneration of motor nerve fibres causing fatigue resembling that seen in myasthenic disorders (Denys and Norris 1979). Up to 10% of cases occur as a familial disorder with variable inheritance (Tandan and Bradley 1985a).

Physiological testing reveals normal or mildly reduced motor nerve conduction velocities and abnormal spontaneous activity in the electromyogram. Motor unit potentials are of prolonged duration and show increased amplitudes indicating reinnervated motor units in affected skeletal muscles. Morphologically, there is loss of motor neurons in the spinal cord and brain-stem motor nuclei and in upper motoneurons in the cortex with subsequent degeneration of the fibre tracts. Before complete cell death neurons may undergo a series of degenerative changes including loss of dendritic processes, accumulation of neurofilaments and swelling in the proximal axons (Tandan and Bradley 1985a). Similar morphological alterations have been seen in hereditary canine spinal muscular atrophy. One of the basis mechanisms in this model is a defect in slow axonal transport of neurofilaments similar to experimental intoxication with $\beta\beta$-iminodipropionitrile (Griffin et al. 1984). Another animal model is the mouse mutant 'wobbler' (Mitsumoto and Bradley 1982).

2. Aetiological and Pathogenic Aspects

The aetiology and pathogenesis of ALS is still obscure. Some clinical disorders resembling ALS in some aspects and with known aetiology have been thought to provide clues to the pathogenesis of ALS (Tandan and Bradley 1985b). These are heavy metal poisoning, e.g. lead or mercury; viral infections, such as those causing poliomyelitis and the 'late-polio syndrome' and Creutzfeldt-Jakob slow virus disease; metabolic defects, e.g. hexosaminidase deficiency; the ALS-dementia complex of Guam (possibly with a toxic aetiology); and chronic myeloradiculoneuropathy. But none of these pathogens or toxins have been proved to cause typical human ALS. Recently, four hypotheses have been proposed for the pathogenesis of ALS: (a) a defect in neuronal DNA repair mechanisms with a secondary defect in neuronal RNA synthesis or in RNA translation; some of these changes may be related to premature aging ('abiotrophy') of motoneurons; (b) a defect in neurotrophic hormones or neuronal hormone receptors including thyrotropin-releasing hormone; (c) serum factors or antibodies impairing neuronal outgrowth; (d) defective slow axonal transport with a neurofilamentous axonopathy (reviewed by Tandan and Bradley 1985b). In the familial form, a genetic factor directly linked to the aetiopathogenesis seems likely. Why the cholinergic motoneuron is the cell predominantly affected in ALS is unknown.

3. Treatment of ALS

Despite many controlled studies based on various putative aetiologic factors (TANDAN and BRADLEY 1985 a) with anti-viral agents, immunosuppressive drugs, plasma exchange, snake venom fractions, ganglioside preparations, vitamins and thyrotropin-releasing hormone, no curative therapy for ALS has been found.

E. Autonomic Neuropathies

I. Clinical Tests for Autonomic Involvement

Autonomic neuropathies and neuronopathies involve cells and fibres of the sympathetic and parasympathetic systems which usually utilize more than one neurotransmitter. A separation between cholinergic and non-cholinergic dysfunction can sometimes be made by tests of autonomic function such as the cholinergic sympathetic eccrine sudomotor reaction and the cholinergic cardiac reflexes induced by physical or pharmacological stimuli (LOW 1984). The anatomy and physiology of the autonomic nervous system has been recently reviewed (APPENZELLER 1982; BANNISTER 1983). Briefly, the sympathetic preganglionic rami communicantes albi (white, i.e. myelinated), the sympathetic trunk, and the predominantly cholinergic preganglionic splanchnic nerve contain largely myelinated fibres with a mean diameter of 4 μm and conduction velocities of more than 12 $m\ s^{-1}$ depending on size. By contrast, the parasympathetic vagus nerve contains largely unmyelinated fibres, as in the efferent fibres to the heart. Afferent fibres are more often of larger diameter (up to 12 μm).

Within the group of autonomic neuropathies two examples will be given: diabetic autonomic neuropathy and the acute pandysautonomia syndrome.

II. Cholinergic Autonomic Dysfunction in Diabetes Mellitus

1. Characteristics

In patients with long-standing diabetes mellitus up to 60% of patients show some deficit in peripheral nervous system function and in approximately 10% it is noted on clinical examination. Within these 10% a small proportion has fully developed autonomic neuropathy with dysfunction of pre- and postganglionic autonomic fibres including the parasympathetic vagus and major splanchnic nerves.

Clinically, the most important disturbance is postural hypotension. Up to 20% of patients with diabetic neuropathy may have abnormal blood pressure and heart rate changes upon rising from a sitting position or following deep inspiration. Only a few of these patients will present with symptoms such as dizziness or fainting immediately after standing up. In this case, the compensatory reflex rise in heart rate and blood pressure (in particular that caused by vasoconstriction of the splanchnic vascular bed) does not follow the physiological stimulus. Impotence and anhidrosis often occur in addition. Occasionally, acute cardiorespiratory arrest, possibly due to the absence of cholinergic vagal outflow to the smooth muscles of the airway, has occurred. Physiological findings are obtained by testing autonomic reflexes (LOW 1984). With modern microneurographic recording techniques, absence of spontaneous sympathetic fibre activity has been demonstrated in over half the diabetic subjects with signs of defective distal vasomotor control (FAGIUS 1982).

Morphologically, various changes in nerve fibres and ganglion cells have been described, including enlargement of cell bodies, loss of dendritic processes, vacuolation and fragmentation of axons, inflammatory infiltrates, segmental demyelination of preganglionic white (myelinated) rami or complete degeneration with fibre loss and terminal sprouting (THOMAS and ELIASSON 1984). The particular importance of the major splanchnic nerve for cholinergic outflow on to the coeliac ganglion has been shown in patients with postural hypotension (LOW 1984).

Usually, the longest nerve fibres show signs of degeneration first i.e. in the long nerves of the foot and in the vagus nerve.

2. Pathogenic Aspects

The pathogenesis of diabetic neuropathy in general and of diabetic autonomic neuropathy in particular is most likely due to a multiplicity of factors; these have recently been reviewed (BROWN and ASBURY 1984; THOMAS and ELIASSON 1984). A number of metabolic alterations in diabetics have been thought to cause generalized neuropathy including abnormal lipid metabolism, high glucose levels leading to increased non-enzymatic glycosylation, decreased protein synthesis and turnover, reduction in slow axonal transport and in fast retrograde transport, accumulation of sorbitol in Schwann cells and an accompanying decrease in myoinositol content and transport across the axon membrane. Myoinositol is not only a constituent of phosphatidylinositol in myelin; it also influences Na, K-ATPase activity and therefore nerve membrane excitability. Recently the pathogenic role of chronic nerve hypoxia secondary to the diabetic microangiopathy and to metabolic derangement has been shown experimentally (TUCK et al. 1984). It seems likely that the reduction in peripheral nerve blood flow with secondary hypoxia is the single most important noxious factor.

The presence of inflammatory nerve infiltrates and other disturbances of the immune system suggests that an autoimmune component may contribute to the condition either as a primary factor or as a secondary phenomenon (DUCHEN et al. 1980).

3. Treatment of Diabetic Neuropathy

The therapy of diabetic neuropathy is based on achieving optimal control of hyperglycemia; this apparently is an important factor for improving peripheral nerve function (BROWN and ASBURY 1984). The observation of disturbed myoinositol and sorbitol metabolism led to therapy with myoinositol supplementation or with the aldose reductase inhibiting compound sorbinil in an effort to block the formation of sorbitol from glucose. Although both treatments have not proved to be clinically effective, they are partly successful in correcting peripheral nerve dysfunction in the streptozotocin-induced animal model and in spontaneous diabetes in BB rats (THOMAS and ELLIASSON 1984). These animal models unfortunately do not develop a polyneuropathy of the severity seen in human diabetes so that the results are of doubtful relevance to the human disorder. Following the suggestion that autoimmune mechanisms might be involved in juvenile diabetes (type I, EISENBARTH 1986) and in autonomic neuropathy (DUCHEN et al. 1980) immunosuppressive measures have been introduced, but their efficacy in improving diabetic neuropathy has not yet been proved.

Symptomatic treatment of autonomic neuropathy is purely empirical, the therapeutic effects resting on the net balance of the predominant fibre populations involved. Drugs include 9a-fluorhydrocortisone, ergotamine derivatives, β-adrenergic blockers for postural hypotension, and cholinomimetics such as prostigmine, betanechol or carbachol for gastrointestinal and bladder atony. Such therapy is often of limited success (THOMAS and ELIASSON 1984; JOHNSON and LAMBIE 1986).

III. Acute Pandysautonomia

1. Characteristics

Acute or subacute autonomic disturbances are associated with a variety of acute polyneuropathies including the Guillain-Barré-Strohl syndrome, toxic neuropathies, the acute porphyrias and botulism and may occur as a remote non-metastatic effect of cancer. A pure pandysautonomia has been described in a small number of cases (YOUNG et al. 1975). Clinical symptoms and signs include blurred vision, pupillary defects, decreased tear formation, postural hypotension, loss of eccrine sweating, abdominal pain or colics, diarrhoea or constipation, impotence and sometimes incontinence. Progression occurs over days or weeks, with subsequent slow recovery to normal or with a residual deficit (APPENZELLER 1982). A subtype of this syndrome with the absence of postural hypotension has been named cholinergic dysautonomia and appears to be more common in children (HOPKINS et al. 1974).

Physiological signs and abnormal autonomic reflex tests are in principle the same as in other autonomic disturbances. In cases of pure cholinergic dysautonomia pharmacological testing of pupillary responses are consistent with cholinergic postganglionic denervation (HARIK et al. 1977). Morphologically, abnormalities have been described in human sural nerve biopsy samples containing somatic and autonomic nerve fibres with demyelination and axonal damage of small myelinated fibres and increased numbers of small unmyelinated fibres suggestive of regeneration (APPENZELLER 1982).

2. Aetiological and Pathogenic Aspects

The aetiology and pathogenesis of the condition is uncertain. It has been suggested that acute pandysautonomia may be a variant of the Guillain-Barré-Strohl syndrome because of its occasional association with this disease and its occurrence after viral infections (JOHNSON and LAMBIE 1986). In this respect, the finding of IgG deposits in the skin of a patient with pure dysautonomia is of interest because it suggests an autoimmune mechanism involving autoantibodies to postganglionic cholinergic sympathetic fibres (HARIK et al. 1977).

3. Treatment of Acute Pandysautonomia

Symptomatic treatment of this disorder is as in diabetic autonomic neuropathy (see Sect. D.II.2). Based on the autoimmune hypothesis, corticosteroids and immunosuppressive drugs might be helpful but no confirmative data are available.

IV. Animal Models of Autonomic Neuropathy

Several animal models have been used for the study of acute and subacute autonomic neuropathy; these include selective degeneration of peripheral adrenergic nerve terminals ('chemical sympathectomy', TRANZER and THOENEN 1968) induced by *in vivo* application of 6-hydroxydopamine; toxic neuropathies induced by acrylamide and vincristine, i.e. models with somatic and autonomic nerve fibre involvement (APPENZELLER 1982; SPENCER and SCHAUMBURG 1984); and experimental allergic neuritis induced by immunization with myelin proteins (RAINE 1985). A subtype of experimental allergic neuritis is experimental autonomic neuropathy induced in rabbits by immunization with antigens derived from sympathetic nerves and ganglion cells (APPENZELLER 1982). These animals are not clinically ill but display abnormal autonomic reflexes. Morphologically, perivascular infiltration with basophilic white cells and signs of degeneration and regeneration of small unmyelinated nerve fibres have been observed. No specific antigen or autoantibody has yet been identified.

V. Generalized Smooth Muscle Disease with Intestinal Pseudo-obstruction

This rare condition is of interest because in some cases it has been thought to be a disorder to the peripheral muscarinic cholinergic receptor (mAChR) (BANNISTER and HOYES 1981). Clinical signs in the first patient described included symptoms and signs of intestinal obstruction, failure of eccrine sweating and dilated and unresponsive pupils. Autonomic reflex testing revealed a slight to absent local sweating response to intradermal methacholine chloride, an absent constriction of the pupil to carbachol with scanty reaction to pilocarpine and eserine and an atonic bladder with some response to systemic carbachol. Bowel motility was not appreciably improved by carbachol and physostigmine (eserine).

Morphological investigations showed dense bodies and vacuolation in some axons and muscle cells but not degeneration of perikarya and dendrites in the myenteric plexus as seen in some autonomic neuropathies. The pathogenic mechanism is not clear. It has been speculated that the condition might be the counterpart of myasthenia (see Sect. C.I.) for the mAChR but supportive evidence is lacking.

F. Glossary

Compound muscle action potential (CMAP) denotes the potential of the entire skeletal muscle evoked by stimulation of the motor nerve which is recorded by surface (skin) electrodes.

Decrement denotes a fall in CMAP-amplitudes which reflects drop-out of individual muscle fibres due to decreasing amplitudes (or currents) of successive end-plate potentials (EPP) until they no longer reach that threshold for initiation of muscle action potentials.

Increment means a rise in successive CMAPs which reflects temporary increase of sub-threshold EPP amplitudes (or currents) such that more muscle fibres are activated and may contribute to the CMAP.

Repetitive nerve stimulation is a clinical neurophysiological test to study neuromuscular transmission *in vivo*. Successive CMAPs are evoked by repetitive and supramaximal stimulation of the motor nerve.

Acknowledgements. The author's laboratory has been supported in part by the Deutsche Forschungsgemeinschaft, the Ministerium für Wissenschaft und Forschung, Nordrhein-Westfalen, and the Hertie-Foundation. The critical review of the manuscript by Dr. E. LOGIGIAN and the continuing support by Professor H.-J. FREUND are gratefully acknowledged. Mrs. B. KOSEMETZKY provided expert secretarial aid.

References

Acuto O, Reinherz EL (1985) The human T-cell receptor: structure and function. N Engl J Med 312:1100–1111

Aguayo AJ, Bray GM (1984) Cell interactions studied in the peripheral nerves of experimental animals. In: Dyck PJ, Thomas PK, Lambert EH, Bunge R (eds) Peripheral neuropathy. Saunders, Philadelphia, pp 360–377

Aharonov A, Abramsky O, Tarrab-Hazdai R, Fuchs S (1975) Humoral antibodies to acetylcholine receptor in patients with myasthenia gravis. Lancet 2:340–342

Albuquerque EX, Rash JE, Mayer RF, Satterfield JR (1976) An electrophysiological and morphological study of the neuromuscular junction in patients with myasthenia gravis. Exp Neurol 51:536–563

Almon RR, Andrew CG, Appel SH (1974) Serum globulin in myasthenia gravis: inhibition of α-bungarotoxin binding to acetylcholine receptors. Science 186:55–57

Appenzeller O (1982) The autonomic nervous system: an introduction to basic and clinical concepts, 3rd edn. Elsevier, Amsterdam

Bannister R (1983) Autonomic failure. Oxford University Press, Oxford

Bannister R, Hoyes AD (1981) Generalized smooth-muscle disease with defective muscarinic receptor function. Br Med J 282:1015–1018

Besinger UA, Toyka KV, Hömberg M, Heininger K, Hohlfeld R, Fateh-Moghadam A (1983) Myasthenia gravis: long-term correlation of binding and bungarotoxin blocking antibodies against acetylcholine receptors with changes in disease severity. Neurology (Cleve) 33:1316–1321

Boroff DA, del Castillo J, Eroy WH, Steinhardt RA (1974) Observations on the action of type A botulinum toxin on frog neuromuscular junctions. J Physiol (Lond) 240:227–253

Bostock H, Sears TA (1978) The internodal axon membrane: electrical excitability and continuous conduction in segmental demyelination. J Physiol (Lond) 280:273–301
Brown MJ, Asbury AK (1984) Diabetic neuropathy. Ann Neurol 15:2–12
Campbell CH (1979) Symptomatology, pathology, and treatment of the bites of elapid snakes. In: Lee CY (ed) Snake venoms. (Handbook of experimental pharmacology, vol 52) Springer, Berlin Heidelberg New York, pp 898–921
Chang CC (1979) The action of snake venoms on nerve and muscle. In: Lee CY (ed) Snake venoms. (Handbook of experimental pharmacology, vol 52) Springer, Berlin Heidelberg New York, pp 309–376
Cherington M (1974) Botulism: ten years experience. Arch Neurol 30:432–437
Cleveland WL, Wassermann NH, Sarangarajan R, Penn AS, Erlanger BF (1983) Monoclonal antibodies to the acetylcholine receptor by a normally functioning autoantiidiotypic mechanism. Nature 305:56–57
Conti-Tronconi B, Brigonzi A, Fumagalli G, Sher M, Cosi V, Picolo G, Clementi F (1981) Antibody-induced degradation of acetylcholine receptor in myasthenia gravis: clinical correlates and pathogenic significance. Neurology (NY) 31:1440–1444
Cornblath DR, Sladky JT, Sumner AJ (1983) Clinical electrophysiology of infantile botulism. Muscle Nerve 6:448–452
Cornblath DR, Pestronk A, Yee WC, Kuncl RW (1985) Congenital myasthenia with increased numbers of acetylcholine receptors at neuromuscular junctions. Muscle Nerve 8:611
Cull-Candy SG, Lundh H, Thesleff S (1976) Effects of botulinum toxin on neuromuscular transmission in the rat. J Physiol (Lond) 260:177–203
Cull-Candy SG, Miledi R, Trautmann A (1979) End-plate currents and acetylcholine noise at normal and myasthenic human end-plates. J Physiol (Lond) 287:247–265
Dau PC (1984) Plasmapheresis: therapeutic or experimental procedure? Arch Neurol 41:647–653
Daube JR (1981) Quantitative EMG in nerve-muscle disorders. In: Stålberg E, Young RR (eds) Clinical neurophysiology. Butterworth, London, pp 33–65
Davis CS, Richardson RJ (1980) Organophosphorous compounds. In: Spencer PS, Schaumburg HH (eds) Experimental and clinical neurotoxicology. Williams and Wilkins, Baltimore, pp 527–544
Denys E, Norris FH (1979) Amyotrophic lateral sclerosis – impairment of neuromuscular transmission. Arch Neurol 36:202–205
Drachman DB (1981) The biology of myasthenia gravis. Am Rev Neurosci 4:195–225
Drachman DB, Angus CW, Adams RN, Michelson JD, Hoffmann GJ (1978) Myasthenic antibodies cross-link acetylcholine receptors to accelerate degradation. N Engl J Med 298:1116–1122
Drachman DB, Adams RN, Josifek LF, Self SG (1982) Functional activities of autoantibodies to acetylcholine receptors and the clinical severity of myasthenia gravis. N Engl J Med 307:769–775
Duchen LW, Anjorin A, Watkins PJ, Mackay JD (1980) Pathology of autonomic neuropathy in diabetes. Ann Intern Med 92:301–303
Dudel J, Birnberger KL, Toyka KV, Schlegel C, Besinger UA (1979) Effects of myasthenic immunoglobulins and of prednisolone on spontaneous miniature end-plate potentials in mouse diaphragms. Exp Neurol 66:365–380
Dyck PJ, Karnes J, Lais A, Lofgren EP, Stevens JC (1984) Pathologic alterations of the peripheral nervous system of humans. In: Dyck PJ, Thomas PK, Lambert EH, Bunge R (eds) Peripheral neuropathy. Saunders, Philadelphia, pp 760–870
Eisenbarth GS (1986) Type I diabetes mellitus; a chronic autoimmune disease. N Engl J Med 314:1350–1368
Engel AG (1984) Myasthenia gravis and myasthenic syndromes. Ann Neurol 16:519–534
Engel AG, Santa T (1971) Histometric analysis of the ultrastructure of the neuromuscular junction in myasthenia gravis and the myasthenic syndrome. Ann NY Acad Sci 183:46–63
Engel AG, Tsujihata M, Lindstrom JM, Lennon VA (1976) The motor end plate in myasthenia gravis and in experimental autoimmune myasthenia gravis: a quantitative ultrastructural study. Ann NY Acad Sci 274:60–79

Engel AG, Lambert EH, Mulder DM, Gomez M, Whittaker JN, Hart Z, Sahashi K (1981) Recently recognized congenital myasthenic syndromes: (a) end-plate acetylcholine esterase deficiency, (b) putative abnormality of the ACh-induced ion channel, (c) putative defect of ACh resynthesis or mobilization – clinical features, ultrastructure, and cytochemistry. Ann NY Acad Sci 377:614–639

Fagius J (1982) Microneurographic findings in diabetic polyneuropathy with special reference to sympathetic nerve activity. Diabetologia 23:415–420

Fagius J, Wallin BG (1980) Sympathetic reflex latencies and conduction velocities in patients with polyneuropathy. J Neurol Sci 47:449–461

Fambrough DM, Drachman DB, Satyamurti S (1973) Neuromuscular junction in myasthenia gravis: decreased acetylcholine receptors. Science 182:293–295

Fuchs S, Bartfeld D, Mochly-Rosen D, Souroujon M, Feingold C (1981) Acetylcholine receptor: molecular dissection and monoclonal antibodies in the study of experimental myasthenia. Ann NY Acad Sci 377:110–124

Fukunaga H, Engel AG, Osanie M, Lambert EH (1982) Paucity and disorganization of presynaptic membrane active zones in the Lambert-Eaton myasthenic syndrome. Muscle Nerve 5:686–697

Fukunaga H, Engel AG, Lang B, Newsom-Davis J, Vincent A (1983) Passive transfer of the Lambert-Eaton myasthenic syndrome with IgG from man to mouse depletes the presynaptic membrane active zones. Proc Natl Acad Sci USA 80:7636–7640

Fumagalli G, Engel AG, Lindstrom J (1982) Ultrastructural aspects of acetylcholine receptor turnover at the normal end-plate and in autoimmune myasthenia gravis. J Neuropathol Exp Neurol 41:567–579

Gomez CM, Richman DP (1983) Anti-acetylcholine receptor antibodies directed against the α-bungarotoxin binding site induce a unique form of experimental myasthenia. Proc Natl Acad Sci USA 80:4089–4093

Griffin JW, Cork LC, Hoffmann PN, Price DL (1984) Experimental models of motor neuron degeneration. In: Dyck PJ, Thomas PK, Lambert EH, Bunge R (eds) Peripheral neuropathy. Saunders, Philadelphia, pp 621–635

Grob D (1963a) Anticholinesterase intoxication in man and its treatment. In: Koelle GB (ed) Cholinesterases and anticholinesterase agents. (Handbook of experimental pharmacology, vol 15) Springer, Berlin Heidelberg New York, pp 990–1026

Grob D (1963b) Therapy of myasthenia gravis. In: Koelle GB (ed) Cholinesterases and anticholinesterase agents. (Handbook of experimental pharmacology, vol 15) Springer, Berlin Heidelberg New York, pp 1029–1050

Grob D, Namba T (1976) Characteristics and mechanisms of neuromuscular block in myasthenia gravis. Ann NY Acad Sci 274:143–173

Gutmann L, Pratt L (1976) Pathophysiologic aspects of human botulism. Arch Neurol 33:175–179

Harik SI, Ghandour MH, Farah FS, Afifi AK (1977) Postganglionic cholinergic dysautonomia. Ann Neurol 1:393–396

Heilbronn E, Mattsson C (1974) The nicotinic cholinergic receptor protein: improved purification method, preliminary aminoacid composition, and observed autoimmune response. J Neurochem 22:315–317

Hobbiger F (1976) Pharmacology of anticholinesterase drugs. In: Zaimis E (ed) Neuromuscular junction. (Handbook of experimental pharmacology, vol 42) Springer, Berlin Heidelberg New York, pp 487–582

Hohlfeld R, Toyka KV (1985) Strategies for the modulation of neuroimmunological disease at the level of autoreactive T lymphocytes. J Neuroimmunol 9:193–204

Hohlfeld R, Sterz R, Kalies I, Peper K, Wekerle H (1981) Neuromuscular transmission in experimental autoimmune myasthenia gravis (EAMG): quantitative ionophoresis and current fluctuation analysis at the normal and myasthenic rat end-plates. Pfluegers Arch 390:156–160

Hohlfeld R, Toyka KV, Heininger K, Grosse-Wilde H, Kalies I (1984) Autoimmune human T lymphocytes specific for acetylcholine receptor. Nature 310:244–246

Hohlfeld R, Toyka KV, Besinger UA, Gerhold B, Heininger K (1985) Myasthenia gravis: Reactivation of clinical disease and of autoimmune factors after discontinuation of long-term azathioprine. Ann Neurol 17:238–242

Hudson CS, Rash JE, Tiedt TN, Albuquerque EX (1978) Neostigmine-induced alterations at the mammalian neuromuscular junction. II. Ultrastructure. J Pharmacol Exp Ther 205:340–356

Ishikawa K, Engelhardt JK, Fujisawa T, Okamoto T, Katsuki H (1977) A neuromuscular transmission block produced by a cancer tissue extract derived from a patient with the myasthenic syndrome. Neurology (Minneap) 27:140–143

Johnson RH, Lambie DG (1986) Autonomic function and dysfunction. In: Asbury AK, McKhann GM, McDonald WI (eds) Diseases of the nervous system – clinical neurobiology. Ardmore, Philadelphia, pp 665–678

Kao I, Drachman DB (1977) Myasthenic immunoglobulin accelerates acetylcholine receptor degradation. Science 196:527–529

Kao I, Drachman DB, Price DL (1976) Botulinum toxin: mechanism of presynaptic blockade. Science 193:1256–1258

Karlsson E, Heilbronn E, Widlund L (1972) Isolation of the nicotinic acetylcholine receptor by biospecific chromatography on insolubilized *Naja naja* neurotoxin. FEBS Lett 28:107–111

Keesey J, Lindstrom J, Cokely H (1977) Anti-acetylcholine receptor antibody in neonatal myasthenia gravis. N Engl J Med 296:55

Kimura J (1983) Electrodiagnosis in diseases of nerve and muscle; principles and practice. Davis, Philadelphia

Lambert EH, Elmquist D (1981) Quantal components of end-plate potentials in the myasthenic syndrome. Ann NY Acad Sci 183:183–199

Lang B, Newsom-Davis J, Prior C, Wray D (1983) Antibodies to motor nerve terminals: an electrophysiological study of a human myasthenic syndrome transferred to mouse. J Physiol (Lond) 344:335–345

Lee CY (1972) Chemistry and pharmacology of polypeptide toxins in snake venoms. Annu Rev Pharmacol 12:265–286

Lennon VA, Lambert EH (1981) Monoclonal antibodies to acetylcholine receptors: evidence for a dominant idiotype and requirement of complement for pathogenicity. Ann NY Acad Sci 377:77–96

Lennon VA, Seybold ME, Lindstrom JM, Cochraine C, Ulevitch R (1978) Role of complement in the pathogenesis of experimental autoimmune myasthenia gravis. J Exp Med 146:973–983

Lennon VA, Lambert EH, Palmer AC, Cunningham JG, Christie TR (1981) Aquired and congenital myasthenia gravis in dogs – a study of 20 cases. In: Satoyoshi E (ed) Myasthenia gravis: pathogenesis and treatment. University of Tokyo Press, Tokyo, pp 41–54

Lerrick AJ, Wray D, Vincent A, Newsom-Davis J (1983) Electrophysiological effects of myasthenic serum factors studied in mouse muscle. Ann Neurol 13:186–191

Lindstrom JM, Seybold ME, Lennon VA, Whittingham S, Duane DD (1976) Antibody to acetylcholine receptor in myasthenia gravis: prevalence, clinical correlates, and diagnostic values. Neurology (Minneap) 26:1054–1059

Lotti M, Becker CE, Aminoff MJ (1984) Organophosphate polyneuropathy: pathogenesis and prevention. Neurology (NY) 34:658–662

Low PA (1984) Quantitation of autonomic responses. In: Dyck PJ, Thomas PK, Lambert EH, Bunge R (eds) Peripheral neuropathy. Saunders, Philadelphia, pp 1139–1165

Lundh H, Nilsson O, Rosen I (1984) Treatment of Lambert-Eaton myasthenic syndrome: 3,4-diaminopyridine and pyridostigmine. Neurology (NY) 34:1324–1330

Matthews G, Wickelgren WO (1977) Effects of guanidine on transmitter release and neuronal excitability. J Physiol (Lond) 266:69–89

Mishina M, Kurosaki T, Tobimatsu T, Morimoto Y, Noda M, Yamamoto T, Terao M, Lindstrom J, Takahashi T, Kuno M, Numa S (1984) Expression of functional acetylcholine receptor from cloned DNAs. Nature 307:604–608

Mitsumoto H, Bradley WG (1982) Murine motor neuron disease (the Wobbler mouse). Brain 105:811–834

Molenaar PC, Newsom-Davis J, Polak RL, Vincent A (1982) Eaton-Lambert syndrome: acetylcholine and choline acetyltransferase in skeletal muscle. Neurology (NY) 32:1062–1065

Molgo J, Lundh H, Thesleft S (1980) Potency of 3,4-diaminopyridine and 4-aminopyridine on mammalian neuromuscular transmission and the effect of pH changes. Eur J Pharmacol 61:25–34

Namba T, Nolte CT, Jackrel J, Grob D (1971) Poisoning due to organophosphate insecticides. Am J Med 50:475–492

Newsom-Davis J (1982) Autoimmune diseases of neuromuscular transmission. Clin Immunol Allergy 2:405–424

Newsom-Davis J (1985) Lambert-Eaton myasthenic syndrome. Springer Semin Immunopathol 8:129–140

Ochs S (1984) Basic properties of axoplasmic transport. In: Dyck PJ, Thomas PK, Lambert EH, Bunge R (eds) Peripheral neuropathy. Saunders, Philadelphia, pp 453–476

Oosterhuis HIGH (1984) Myasthenia gravis. Churchill Livingstone, Edinburgh

Patrick J, Lindstrom JM (1973) Autoimmune response to acetylcholine receptor. Science 180:871–872

Pennefather P, Quastel DMJ (1980) The effects of myasthenic IgG on miniature end-plate currents in mouse diaphragm. Life Sci 27:2047–2054

Pickett J, Berg B, Chaplin E, Brunstetter-Schafer M (1976) Syndrome of botulism in infancy: clinical and electrophysiological study. N Engl J Med 295:770–772

Raine CS (1985) Experimental allergic encephalomyelitis and experimental allergic neuritis. In: Vinken PJ, Bruyn GW, Klawans HL (eds) Handbook of clinical neurology, vol 27. Elsevier, Amsterdam

Rash JE, Hudson CS, Graham WF, Mayer RF, Warnick JE, Albuquerque EX (1981) Freeze fracture studies of human neuromuscular junctions. Membrane alterations observed in myasthenia gravis. Lab Invest 44:519–530

Richman DP, Gomez CM, Berman PW, Burres SA, Fitch FW, Arnason BGW (1980) Monoclonal anti-acetylcholine receptor antibodies can cause experimental myasthenia. Nature 286:738–739

Roberts A, Perera S, Lang B, Vincent A, Newsom-Davis (1985) Paraneoblastic myasthenia syndrome IgG inhibits $^{45}Ca^{2+}$ flux in a human small cell carcinoma line. Nature 317:737–739

Satyamurti S, Drachman DB, Sloane F (1975) Blockade of acetylcholine receptors: a model of myasthenia gravis. Science 187:955–957

Spencer PS, Schaumburg HH (1984) Experimental models of primary axonal disease by toxic chemicals. In: Dyck PJ, Thomas PK, Lambert EH, Bunge R (eds) Peripheral neuropathy. Saunders, Philadelphia, pp 636–649

Spitzer N (1972) Miniature end-plate potentials at mammalian neuromuscular junction poisoned by botulism toxin. Nature [New Biol] 237:26–27

Stalberg E, Trontelj J (1979) Single fibre electromyography. Miravelle, Old Woking

Stanley EF, Drachman DB (1978) Effect of myasthenic immunoglobulin acetylcholine receptors on intact neuromuscular junctions. Science 200:1285–1287

Sterz R, Hohlfeld R, Rajki K, Kaul M, Heininger K, Peper K, Toyka KV (1986) Effector mechanisms in myasthenia gravis: endplate function after passive transfer of IgG, Fab, and F(ab')2-hybrids. Muscle Nerve 9:306–312

Tandan R, Bradley WG (1985a) Amyotrophic lateral sclerosis. Part 1. Clinical features, pathology, and ethical issues in management. Ann Neurol 18:271–280

Tandan R, Bradley WG (1985b) Amyotrophic lateral sclerosis. Part 2. Etiopathogenesis. Ann Neurol 18:419–431

Thomas PK, Eliasson SG (1984) Diabetic neuropathy. In: Dyck PJ, Thomas PK, Lambert EH, Bunge R (eds) Peripheral neuropathy. Saunders, Philadelphia, pp 1773–1810

Toyka KV, Heininger K (1986) Acetylcholine-Rezeptor-Antikörper in der Diagnostik der Myasthenia gravis. Dtsch Med Wochenschr 111:1435–1439

Toyka KV, Drachman DB, Griffin DE, Pestronk A, Winkelstein JA, Fischbeck KH, Kao I (1977) Myasthenia gravis: study of humoral immune mechanisms by passive transfer to mice. N Engl J Med 296:125–131

Toyka KV, Birnberger KL, Anzil AP, Schlegel C, Besinger UA, Struppler A (1978) Myasthenia gravis: further electrophysiological and ultrastructural analysis of the transmission failure in the mouse passive transfer model. J Neurol Neurosurg Psychiatry 41:746–753

Tranzer JP, Thoenen H (1968) An electron microscopic study of selective degeneration of sympathetic nerve terminals after administration of 6-hydroxydopamine. Experientia 24:155–156

Tuck RR, Schmelzer JD, Low PA (1984) Endoneurial blood flow and oxygen tension in the sciatic nerve of rats with experimental diabetic neuropathy. Brain 107:935–950

Tzartos SJ, Lindstrom J (1980) Monoclonal antibodies used to probe acetylcholine receptor structure: localisation of the main immunogenic region and detection of similarities between subunits. Proc Natl Acad Sci USA 77:755–759

Tzartos SJ, Seybold ME, Lindstrom JM (1982) Specificities of antibodies to acetylcholine receptors in sera from myasthenia gravis patients measured by monoclonal antibodies. Proc Natl Acad Sci USA 79:188–192

Vincent A (1980) Immunology of acetylcholine receptors in relation to myasthenia gravis. Physiol Rev 60:757–824

Vincent A, Cull-Candy SG, Newsom-Davis J, Trautmann A, Molenaars PC, Polak RI (1981) Congenital myasthenia: end-plate acetylcholine receptors and electrophysiology in five cases. Muscle Nerve 4:306–318

Waldor M, Sriram S, McDevitt HO, Steinman L (1983) In vivo therapy with monoclonal anti-Ia antibody suppresses immune response to acetylcholine receptors. Proc Natl Acad Sci USA 80:2713–2717

Wapen BD, Gutmann L (1974) Wound botulism. JAMA 227:1416–1417

Watters D, Maelicke A (1983) Organization of ligand binding sites at the acetylcholine receptor: a study with monoclonal antibodies. Biochemistry 22:1811–1819

Weiss DG, Gorio A (eds) (1982) Axoplasmic transport in physiology and pathology. Springer, Berlin Heidelberg New York

Wekerle H, Ketelsen UP (1977) Intrathymic pathogenesis and dual genetic control of myasthenia gravis. Lancet 1:678–680

Wekerle H, Hohlfeld R, Ketelsen UP, Kalden JR, Kalies I (1981) Thymic myogenesis, T-lymphocytes and the pathogenesis of myasthenia gravis. Ann NY Acad Sci 377:455–475

Wilson S, Vincent A, Newsom-Davis J (1983) Acetylcholine receptor turnover in mice with passively transferred myasthenia gravis. II. Receptor synthesis. J Neurol Neurosurg Psychiatry 46:383–387

Young RR, Asbury AK, Corbett JL, Adams RD (1975) Pure pandysautonomia with recovery. Brain 98:613–639

CHAPTER 26

Central Cholinergic Neuropathologies

A. N. DAVISON

A. Introduction

Although acetylcholine (ACh) is not the predominant transmitter in the central nervous system, cholinergic innervation is of special importance in intellectual processing within the association cortex. Experimental studies in animals and pharmacological investigations in human subjects indicate the key role of ACh in memory. However, cholinergic neurons have other functions. For example, pyramidal motor neurons within the cortex and spinal cord are involved in motor activity. Loss of cholinergic nerve cells or interference with their function can lead to serious neurological disease, as is described in this chapter.

The cholinergic innervation of the human brain is similar to that of other species, as is shown by the distribution of choline acetyltransferase (ChAT), acetylcholinesterase (AChE) and muscarinic acetylcholine receptor (mAChR) activities (Table 1; Chap. 21). These are high in the basal ganglia (caudate nucleus and putamen) and in certain nuclei, for example those formed by the neurons in the nucleus basalis and diagonal band of Broca which project to the neocortex and hippocampus (Fig. 1; Chap. 22). The concentration of ChAT-rich cell bodies

Table 1. Distribution of ChAT and mAChR in human *post mortem* brain

Brain region	ChAT (nmol h^{-1} mg^{-1} protein	mAChR (pmol mg^{-1} protein)
Frontal cortex	3.67 ± 0.54 (16)	0.490 ± 0.196 (5)
Occipital cortex	3.11 ± 0.44 (16)	0.558 ± 0.059 (8)
Temporal cortex	3.80 ± 0.44 (16)	0.648 ± 0.064 (12)
Parietal cortex	3.44 ± 0.48 (16)	0.553 ± 0.048 (14)
Hippocampus	7.33 ± 1.12 (16)	0.378 ± 0.055 (12)
Amygdaloid nucleus	14.05 ± 1.57 (5)	0.340 ± 0.045 (5)
Caudate nucleus	80.50 ± 7.08 (16)	0.886 ± 0.097 (13)
Substantia nigra	4.67 ± 1.06 (11)	0.205 ± 0.085 (3)
Entorhinal cortex	4.75 ± 1.28 (5)	–
Mammillary body	3.84 ± 0.68 (9)	–
Cerebellum	4.28 ± 1.11 (5)	–
Putamen	346.5 ± 12.0 (76)	–
Nucleus accumbens	226.3 ± 10.3 (50)	–

ChAT activity was measured by the rate of ACh synthesis and mAChR by the binding of [^{3}H] QNB. Data shown are mean values ± SEM, with the number of cases in parentheses. (From PERRY and PERRY 1980; SPOKES 1979)

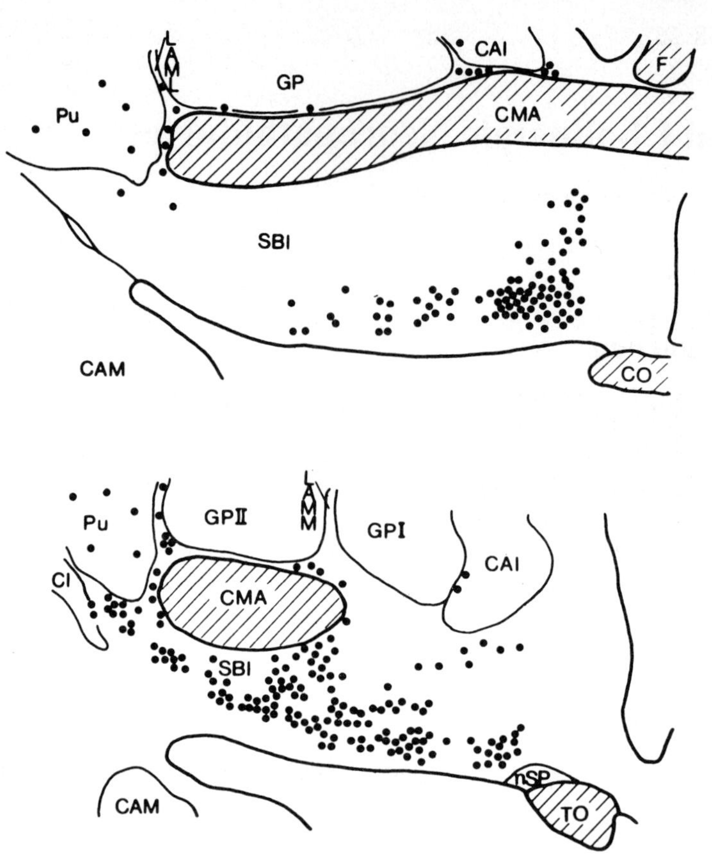

Fig. 1. Distribution of ChAT-positive cells in the human basal forebrain; ChAT-positive cells are indicated by *dots*. *CAI*, capsula interna; *CAM*, corpus amygdaloideum; *CMA*, commissura anterior; *CO*, chiasma opticus; *Cl*, claustrum; *F*, fornix; *GP*, globus pallidus; *GPI*, globus pallidus, crus I; *GPII*, globus pallidus, crus II; *LAML*, lamina medullaris lateralis; *LAMM*, lamina medullaris medialis; *nSP*, nucleus supraopticus hypothalami; *Pu*, putamen; *SBI*, substantia innominata; *TO*, tractus opticus. [After NAGAI et al. (1983) with permission of Dr. E. G. McGEER and Brain Research, Elsevier]

in the basal human forebrain has been described by NAGAI et al. (1983). Positive cells were located dorsal to the basal cortex and ventral to the putamen and globus pallidus throughout the entire rostro-caudal length of the pallidum.

B. Pathological Processes

I. Disposing Factors

Histological examination of the brain in some cholinergic neuropathies shows that there is extensive loss or shrinkage of nerve cells from some regions. This results in dysfunction or reduction in the number of nerve terminals projecting from the affected cholinergic cell bodies. Individuals who have inherited smaller numbers of nerve cells in a particular brain region or who have sustained neuronal loss as a result of disease (e.g. poliomyelitis) are clearly particularly vulnerable to pathological change. In addition the cholinergic system may be affected by developmental abnormalities as may happen in some types of mental retardation.

II. Metabolic

Some types of central cholinergic dysfunction have been shown to be secondary to systemic metabolic disorder (BLASS et al. 1981). These may produce some loss of intellectual function in severe cases with coma or by neurological dysfunction (e.g. ataxia). The cholinergic system is frequently involved; thus it is known that impaired cerebral carbohydrate catabolism leads to reduced ACh synthesis and that this is particularly sensitive to changes in cerebral oxidative metabolism.

III. Genetic

Studies in inbred mice show that genotype can influence development of cell number in the brain. For example CBA mice have 25% – 50% fewer nigral dopaminergic neurons (and dopamine receptors) than other strains of mice. It is interesting that near relatives of patients with phenothiazine-induced Parkinsonism have a significantly higher incidence of Parkinson's disease than relatives of similarly treated patients with no drug-induced symptoms. As pointed out by FINCH (1980), this implies that there are silent dopaminergic deficits and is consistent with genetic influence in Parkinson's disease. Mature individuals would be particularly susceptible to loss of neurons from the basal ganglia whether it be due to aging or disease. This hypothesis may explain the age-related incidence of some neurological diseases affecting the cholinergic system – such as motor neuron disease where the patients may have inherited a lower than normal population of cholinergic motor neurons, or in Alzheimer's disease where certain subcortical nuclei are affected.

IV. Neuronal Loss of Unknown Cause

In 1902 Sir William GOWERS realized that many degenerative brain conditions were due to an inexplicable progressive damage leading to loss of neurons. This state of cell death was termed abiotrophy. Although the cause of many cases of neurological degenerative disorder is still unknown experimental studies indicate possible mechanisms of cell loss and damage. It has been postulated that cell loss may be due to 'wear and tear' processes or accumulated errors within the perikaryon. Thus, one possibility is that late-onset neurodegenerative diseases are due to a defect in the capacity to repair DNA although this may be a secondary phenomenon in an already damaged or dying cell (ROBISON et al. 1985; LI and KAMINSKAS 1985). Changes in nucleic acid metabolism, in nucleolar volume and in protein synthesis have been described in the brains of subjects with Down's syndrome and Parkinson's and Alzheimer's diseases. Other data suggest that endogenous neurotoxins may accumulate in the mature CNS. It has been shown that excess glutamate, and kainate – the rigid glutamate analogue – are neurotoxic. Naturally occurring indolamine metabolites are also highly cytotoxic. Examples are the β-carbolines or tryptolines derived from the condensation of tryptamine and aldehydes (formaldehyde or acetaldehyde). Tetrahydro*nor*harmane can be shown to be a powerful inhibitor of different neurotransmitter systems including the cholinergic system (SKUP et al. 1983). Quinolinic acid (2,3-pyridine dicarboxylic acid) is another naturally occurring excitotoxin. This is also a metabolite of tryptophan with high convulsive and neuroexcitatory potency (SCHWARCZ et al. 1983). Injection of quinolinic acid into the hippocampus and ventricles causes pyramidal and granule cell damage. The metabolite is present naturally throughout the brain and concentrations are high (2 nmol g^{-1} wet weight) in the striatum and cortex compared to the cerebellum and other regions. There are indications that quinolinic acid accumulates with age and that loss of neurons in circumscribed brain nuclei (e.g. the nucleus basalis) may be due to selective neurotoxicity in these regions (SCHWARCZ and KOHLER 1983). Thus the striatum, pallidal formation (globus pallidus, nucleus basalis and thalamic nuclei) and hippocampus are all vulnerable to infusion of quinolinic acid (120 nmol). Regions such as the cerebellum, substantia nigra, amygdala, medial septum and hypothalamus are most resistant to the neurotoxic effect.

V. Environmental Poisons and Infective Agents

A number of anti-ChEs of the organophosphorus type, e.g. triorthocresylphosphate and diisopropylphosphorofluoridate (DFP), cause delayed neurotoxicity in man, domestic fowl and several other animals. Ataxia and paralysis of the legs occur in fowls 8–21 days after dosing. The neurotoxic lesions involve the long motor and sensory axons of the peripheral and central nervous system with some loss of nerve cell bodies. This effect is not related to acute cholinergic effects, and is thought to be initiated by the inhibition of another esterase, the so-called neurotoxic esterase.

Heavy metals are another class of environmental poisons which can interfere with brain development and in some instances have a neurotoxic action on the more mature brain. Thus it has frequently been considered that lead has a toxic action on motor neurons and accumulation of lead could be a causative factor in motor neuron disease. Neurotrophic viruses (e.g. poliomyelitis) are involved in some degenerative brain diseases. Similarly encephalitis produced by herpes simplex virus type 1 can result in selective cell loss, and the virus has been shown to spread from the olfactory pathway suggesting a possible infective mechanism. A neuronal cell line (PC 12) originally cloned from a rat pheochromocytoma has been used by RUBENSTEIN and PRICE (1984) to study the effects of herpes infection. AChE and ChAT activities are reduced by more than half within a few hours after infection. Both botulism and rabies have toxic effects on the peripheral cholinergic nervous system.

C. Central Cholinergic Neuropathies

I. Alzheimer's Disease

1. Types of Dementia

In dementia there is an acquired global impairment of higher cortical functions. Short-term memory is particularly affected. The natural history is frequently irreversible and progressive. There is still debate concerning the status of senile dementia as a syndrome of disease which is distinct from an exaggerated form of brain senescence. However, there are a number of conditions in which degenerative changes can be attributed to a definite cause. About a quarter of cases – defined as multi-infarct dementia – can be attributed to cerebrovascular disease. Other examples are the transmissible agent of Creutzfeldt-Jakob disease or the aluminium-induced dialysis dementias. However, the aetiology in the commonest form of dementia – nonvascular senile dementia – is unknown. The pathology of presenile dementia appears to be indistinguishable from that of senile dementia in those over 65 years of age. For this reason, when referring to Alzheimer's disease, it is convenient if both groups are included together (BOWEN 1981).

2. Pathology of Alzheimer's Disease

In Alzheimer's disease brain weight is usually reduced and there is relatively greater loss or shrinkage (PEARSON et al. 1983) of neurons in regions such as the nucleus basalis (Fig. 2; PRICE et al. 1982; PERRY et al. 1982), locus coeruleus (TOMLINSON et al. 1981), hippocampus (BALL 1977) and amygdala (HERZOG and KEMPER 1980). It has been proposed (ROSSOR 1981) that neuronal loss from the isodendritic core of the reticular formation leads to reduction in ascending projections from the subcortex and loss of the corresponding transmitter-containing terminals. As judged by the Golgi-staining method, the dendritic branching of the remaining neurons is less than in age-matched controls, and failure in sprouting by reactive synaptogenesis has been suggested to be due to a missing trophic factor (APPEL 1981). Currently available data led BUELL and COLEMAN (1979) to suggest a model in which there are two populations of neurons in normal aging cortex, one a group of dying neurons with shrinking dendritic trees, the other a group of surviving neurons with expanding dendritic trees. In normal aging the latter population prevails. With the passage of time there must be a shift of individual neurons from the surviving population to the dying population. The rate at which this shift takes place may be a function of genetic and non-genetic or extrinsic (toxicological, behavioural, infectious) factors. Infective agents or other environmental factors (e.g. aluminium) have been suspected to be involved. It has been suggested that nerve cell loss and damage (BALL 1982) may be due to reactivation of latent herpes virus (persisting in lymphocytes or even in ganglia), and the retrograde transport of the virus into the limbic system. Virus in the trigeminal ganglion may cross-infect nerve cells in the adjacent nuclei within the substantia innominata.

Histological changes are much more marked in Alzheimer's disease than in the aged brain. Granulovacuolar degeneration is seen in hippocampal cells and in-

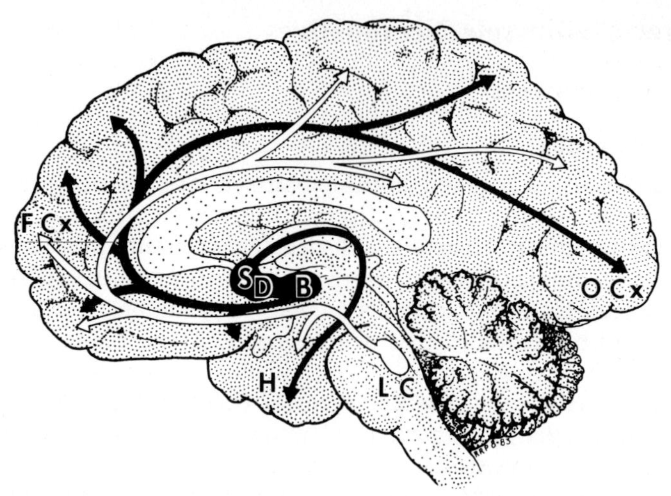

Fig. 2. Some of the subcortical nuclei and their projections affected in Alzheimer's disease. In *black* are the cholinergic ascending fibres from the medial septal nucleus (*S*) leading to the hippocampus (*H*) and projections from the diagonal band (*D*) and basal nucleus (*B*) leading to the cortex (*F Cx,* frontal cortex; *O Cx,* occipital cortex). The noradrenergic pathway from the locus coeruleus (*LC*) is shown in *white*. (Courtesy of Dr. Peter WHITEHOUSE)

creased numbers of plaques and especially tangles correlate with reduced intellectual ability. Neuritic plaques containing degenerating neurites, and synapses with an inner core of amyloid are frequently found close to capillaries. There is evidence that amyloid is derived from a β-protein precursor expressed by a gene locus on chromosome 21. 4.2 Kd subunits of an A4 peptide which are formed by proteolysis are thought to be assembled into amyloid and possibly tangle protein. Some degree of homology between the primary sequence of amyloid and serine proteinase has been shown. There is substantial loss of large neurons from nuclei of septal medialis, diagonal band of Broca and basal nucleus of Meynert in the substantia innominata (ARENDT et al. 1983). Surviving cells show several kinds of cell degeneration including the presence of numerous tangles (ROGERS et al. 1985). ARENDT et al. (1985) found a strong correlation between cerebral cortical plaque density and loss of neurons in the nucleus basalis. Similarly, loss of cells from the nucleus basalis has been shown to be related to formation of senile plaques and tangles in the temporal cortex in Alzheimer's disease (MANN et al. 1985). It was suggested that the first abnormalities were in distal intracortical axons and resulted from an unidentified pathological process such as a toxin, transmissible agent, genetic factor or aging. Studies (STRUBLE et al. 1982) on elderly monkeys show how the neuritic plaque may develop. The dystrophic process (abnormal deposition due to defective nutrition) causes cytoplasmic organelles, especialy neurofilaments, to accumulate, so increasing the diameter of neurites. As amyloid deposits (shown by Congo Red stain) increase, so there is a loss of AChE activity, indicating probable loss of cholinergic neurites and terminals (KITT et al. 1984). However, plaques have been found to be associated with other neurotransmitter systems (MORRISON et al. 1985), and not all plaques originate from the basal forebrain since neuritic plaques in the cerebellar cortex have no projections from these nuclei. Furthermore, experimental studies with scrapie show that plaque formation is closely associated with the distribution of the infective agent and not with the location of cholinergic terminals (WISNIEWSKI and MERZ 1984).

A second feature of Alzheimer's disease is the presence of neurofibrillary tangles which accumulate within the nerve cell perikarya and axonal terminals and thereby probably interfere with metabolism and axonal flow of material from the endoplasmic reticulum. The tangles have been identified as neurofilamentary in origin (ANDERTON et al. 1982). SELKOE et al. (1982) reported that the paired helical filaments were exceedingly insoluble, and postulated that cross-linked glutamyl-lysine bands are formed involving transglutaminase action. It is interesting that in aging brain, lipid peroxides (formed in autophagic vacuoles) can cross-link proteins such as neurofilaments, as well as induce lipofuscin formation. There may be a relationship between lipofuscin accumulation and the increase found in long-chain polyisoprenols (dolichols) in the brain with senescence and particularly with Alzheimer's disease (NG YING KIN 1983). In elderly patients with Alzheimer's disease, WILCOCK et al. (1982) have found that mental score is more closely correlated with tangle density than with the incidence of neuritic plaques.

3. Selective Involvement of the Cholinergic System

In 1971 BOWEN and his associates began a systematic search for biochemical changes in the brains of histologically confirmed cases of Alzheimer's disease at autopsy. It was found that meaningful biochemical data could be obtained, provided allowance was made for *pre-mortem* state (e.g. pneumonia, BOWEN et al. 1976) and other variables (SPOKES 1979). Analyses of homogenates of whole temporal lobe indicated that, in these cases, there was cell loss or shrinkage, loss of myelinated axons and reduction in enzyme activity connected with energy metabolism and transmitter synthesis (BOWEN et al. 1979). ChAT and AChE activities were markedly reduced in proportion to the degree of neuropathological change (BOWEN et al. 1976) but postsynaptic mAChR concentration was not significantly affected.

Biochemical data (BOWEN et al. 1979; ROSSOR et al. 1981; WILCOCK et al. 1982), and measurements of cerebral atrophy (HUBBARD and ANDERSON 1981) provide evidence that younger patients with Alzheimer's disease have a more severe form of pathological change. In elderly patients biochemical and pathological differences are much less marked possibly because older patients succumb more easily that younger healthy individuals (BOWEN et al. 1979).

An effect on markers of the cholinergic system was independently confirmed by other workers (Table 2). It was found that ChAT activity was particularly reduced in areas concerned with memory processes – the amygdala, hippocampus and temporal cortex. This research has been extended (Fig. 3) to a study of the enzymes (ChAT and AChE) in different layers of the cortex and indicates very substantial loss of enzyme activity. Most of the AChE activity lost is in the 10-S form which ATACK et al. (1983) postulate is specifically associated with the cho-

Table 2. Changes in cholinergic markers in the *post-mortem* brain of old and elderly patients with Alzheimer's disease. Values are percentages of controls

Brain region	ChAT		AChE	mAChR
	Patients		All patients	All patients
	Under 79	Over 79		
Amygdala	35	51	13	71
Hippocampus	42	51	11	88
Frontal cortex	42	78 (ns)	17	87
Temporal cortex	30	47	–	107
Parietal cortex	50	69	5	92
Occipital cortex	51	65	16	116
Putamen	96 (ns)	–	3	–
Nucleus accumbens	48	–	32	–
Substantia innominata	36	–	–	–

Data, obtained from various sources; ROSSOR et al. (1982b, 1984); PERRY and PERRY (1980); DAVIES (1979); DAVIES and MALONEY (1976). Figures are expressed as percentages of age-matched control values. All results are significantly different from controls except where indicated and for mAChR concentrations (ns, not significant)

linergic system. Earlier PERRY et al. (1978) showed a linear correlation between loss of ChAT and AChE activity and density of plaques. There are several reports that mAChR concentration is not significantly reduced. However, MASH et al. (1984) have reported a 50% reduction in M2 receptor concentration in the frontal cortex (Brodmann's area 10) relative to controls. Thus while postsynaptic (M1) receptors may be intact, the autoreceptors on presynaptic terminals may be lost. Since *post-mortem* tissue is subject to various factors which are difficult to control and as ChAT is not rate-controlling for the synthesis of ACh it was necessary to confirm the effects on the cholinergic system using fresh biopsy tissue (Fig. 4). [^{14}C]ACh synthesis from [U-^{14}C]glucose by tissue prisms was measured for 60 min at 37 °C after preincubation. Incubation mixtures contained 2.5 m*M* [U-^{14}C]glucose, 2 m*M* choline and 40 μ*M* paraoxon in modified Krebs-Ringer phosphate buffer. Incubations were performed in the presence of 5 or 31 m*M* K^+ as an indication of response under resting and stimulated conditions. [^{14}C]ACh was isolated from the incubation by precipitation with ammonium reineckate and extracted for scintillation counting. There was a considerable reduction in synthesis in tissue from Alzheimer's disease in comparison to control biopsy values. Moreover, FRANCIS et al. (1985) have reported a significant inverse correlation between ACh synthesis in the cortex from the medial temporal gyrus and the rating of cognitive impairment. In addition to a reduced formation of ACh a highly significant decrease in ChAT activity in samples of cortical grey matter was observed but only where there was histological evidence of Alzheimer's disease. Potassium-stimulated choline uptake into the brain prisms was also significantly reduced. Since stimulated choline uptake, ACh synthesis and ChAT activities are comparably reduced (Table 3) the most plausible explanation is that there is a loss of functionally active cholinergic nerve terminals in the cortex. The loss of such

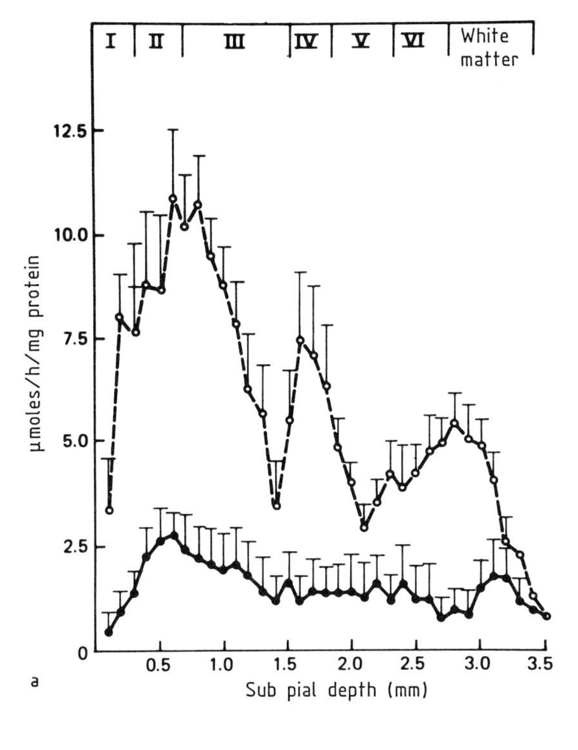

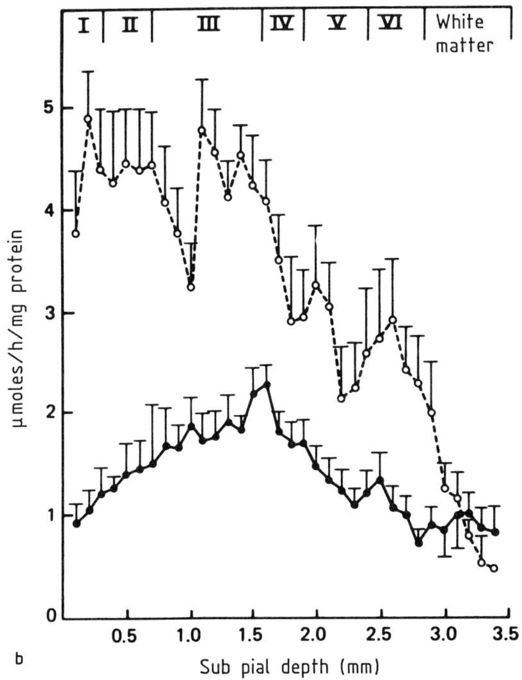

Fig. 3a, b. (Legend see p. 734)

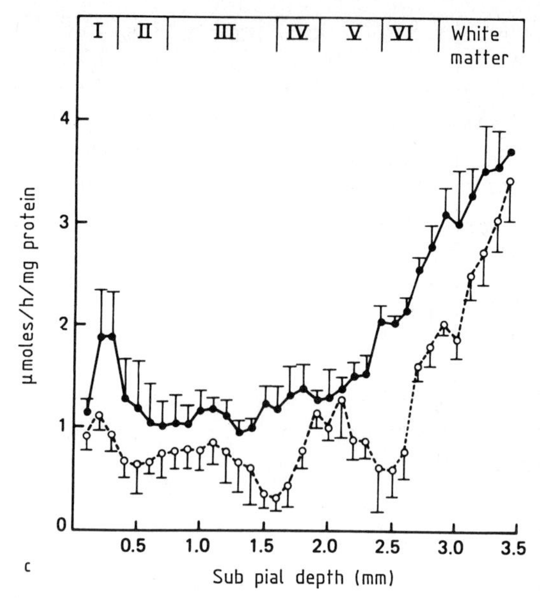

Fig. 3a–c. Comparison of laminar distribution of cholinergic enzyme activities in the human temporal cortex in controls and cases of Alzheimer's disease; data for normal samples are shown by *broken lines* and *open circles,* those for Alzheimer cases by *continuous lines* and *filled circles.* **a** ChAT, **b** AChE, **c** butyrocholinesterase activity. *Points* represent means ± SEM for four cases in each of the normal and Alzheimer disease groups except for **b**, where there are five in each group. (Reproduced with permission from PERRY et al. 1984 and the Journal of Neurochemistry)

terminals could be attributed to reduction in the population of cholinergic cell bodies in the subcortex with projections to the cortex (WHITEHOUSE et al. 1981), for there are only a small number of cholinergic interneurons in the cortex. Nevertheless the close relationship between loss of ChAT activity and cell loss in the nucleus basalis has been questioned. In one study between 75% and 90% of enzyme activity was lost from the nucleus and from the temporal cortex but the loss of neurons was only 33% (PERRY et al. 1982). This suggested that down-regulation of transmitter-specific enzyme production in cholinergic neurons occurs first and that loss of neurons may be a secondary feature of the disease. Evidence from animal studies indicates that cholinergic nerve endings constitute less than 10% of the total nerve terminals in the neocortex and probably even less in humans (RICHARDSON 1981; SIMS et al. 1981). The overall density of serotonin and noradrenergic innervation seems to account for less than 1% of all cortical nerve endings (BEAUDET and DECARRIES 1978). In addition there is a significant loss in the content of ganglioside *N*-acetylneuraminic acid (thought to be localized on neuronal cell body and dendrite membranes) of the entire temporal lobe and in different cortical layers (DEKOSKY and BASS 1982). This suggests that there is loss of neuronal structures other than those associated with cholinergic projections from the basal forebrain and the brain stem or from the small number of cholinergic neurons in the cortex. Thus the significant loss of β-galactosidase ac-

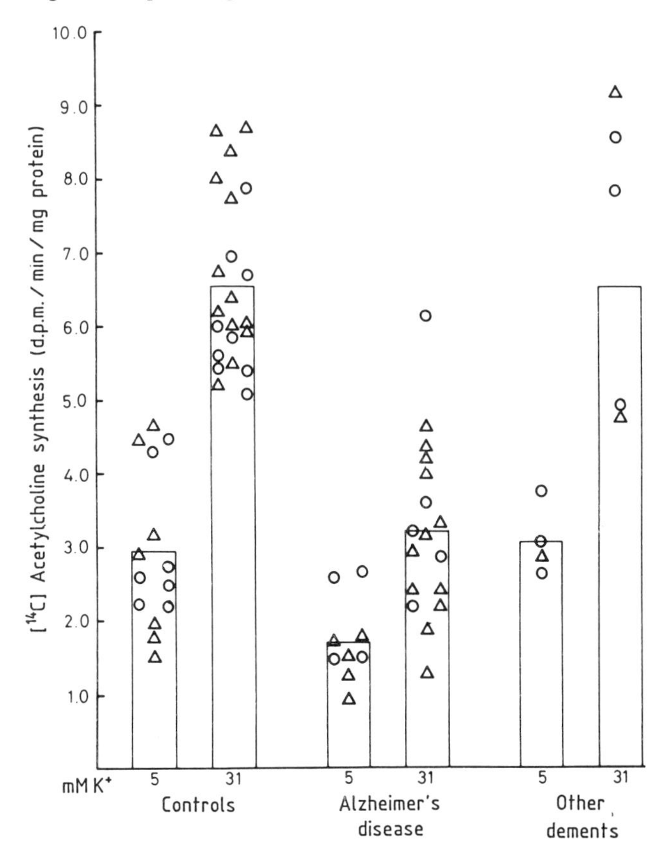

Fig. 4. [^{14}C]ACh synthesis measured in the presence of either 5 or 31 m*M* K^+ in samples of temporal neocortex (*triangles*) and frontal neocortex (*circles*) from neurosurgical controls, patients with histologically confirmed Alzheimer's disease and other demented patients. Samples from patients with Alzheimer's disease were significantly different ($P < 0.01$; Wilcoxon summed ranks test) from control samples measured under the same conditions. (Reproduced with permission from SIMS et al. 1983 and the Journal of Neurochemistry)

tivity (a marker of neuronal cytoplasm) from the temporal lobe in the late-onset Alzheimer samples would be consistent with the histologically observed reduction or shrinkage in large nerve cell populations in the neocortex (TERRY et al. 1981; COLON 1973; HUBBARD and ANDERSON 1985), and reduction of the somatostatin immunoreactivity probably associated with intrinsic cortical neurones (MCKINNEY et al. 1982). There is already evidence that the loss is selective, because the number of cholecystokinin-, vasoactive intestinal polypeptide- (ROSSOR et al. 1982a) glutamate- and γ-aminobutyrate- (GABA-)containing neurons remain constant (Smith et al. 1983). There is some evidence however that the major descending projections form the neocortex (e.g. glutamate neurons projecting to the thalamus and corpus striatum) are primarily affected (PEARCE et al. 1984; PEARSON et al. 1985).

Table 3. Comparison of values for different synaptic cholinergic markers in Alzheimer's disease and control fresh biopsy tissue. Data from SIMS et al. (1983)

	Stimulated choline uptake (pmol min^{-1} mg^{-1} protein)	Stimulated ACh synthesis (disintegrations min^{-1} mg^{-1} protein)	ChAT activity (pmol min^{-1} mg^{-1} protein)
Controls	3.04 ± 0.33 (6)	6.6 ± 1.2 (22)	86 ± 21 (11)
Alzheimer disease	1.72 ± 0.58 (6)	3.9 ± 1.2 (6)	37 ± 25 (6)
Other non-Alzheimer dements	2.71, 2.73	4.9, 8.5	70, 74

Neurosurgical samples were taken for diagnostic purposes and histologically assessed. Values are expressed as mean $\pm$ SD, with number of samples in parentheses. Single observations on two patients with non-Alzheimer dementia are included

4. Treatment

Studies with aged rats suggest that combination of a nootropic agent (piracetam) with choline enhances 24-h retention latency in a passive avoidance test (BARTUS et al. 1981). However, in general, attempts to improve cognition in patients with Alzheimer's disease have had little success (ROSENBERG et al. 1983). An alternative strategy to transmitter replacement therapy is to try to ameliorate the degenerative process within the CNS. Thus there is experimental evidence that choline stimulates sprouting of dendrites (MERVIS and BARTUS 1981), and a suggestion that long-term treatment with lecithin may modify the progressive deterioration seen in a subgroup of patients with dementia (LITTLE et al. 1985).

Administration of the anti-muscarinic drug, scopolamine, to young subjects produces cognitive deficits similar to those found in healthy elderly individuals (DRACHMAN and LEAVITT 1974). Conversely dosage with arecholine, a cholinergic agonist, and choline significantly increased serial learning in normal human subjects (SITARAM et al. 1978). Cholinomimetic drugs such as eserine have facilitated memory particularly in the old but only within a narrow dose range (BARTUS et al. 1982). An important new treatment of potential value is the use of continuous intracerebral injection of cholinergic agonists using an implanted pump (HARBAUGH et al. 1984). Improvement in cognition in Alzheimer patients has been reported following treatment with tetrahydroaminoacridine. This drug acts as a long term central anticholinesterase.

II. Down's Syndrome

Down's syndrome is a developmental disease in which there is trisomy of chromosome 21. In most cases the brain weighs less than normal and in subjects over the age of 35 years senile plaques, and tangles are frequently found. Some of these older patients appear to be demented. There are reports of a reduction in the cerebral cortical ChAT activity and of noradrenaline in the hypothalamus (Table 4). These changes may be related to loss of neurons from the nucleus basalis and locus coeruleus. MANN et al. (1984a, b) also reported a reduction in nucleolar volume in the temporal cortex and hippocampal neurons like that found

Table 4. Changes (percentage loss compared to controls) in Down's syndrome patients. Data from MANN et al. (1984a) and YATES et al. (1980)

	ChAT activity	Nerve cell number	Nucleolar volume
Nucleus basalis	–	24	20*
Locus coeruleus	–	78*	30*
Temporal cortex	83	30	42*
Hippocampus	90*	52*	58*

ChAT activities were estimated in eight areas from three cases of Down's syndrome. Significant reductions ($P<0.001$) were found in the caudate nucleus, amygdala, parahippocampal gyrus and anterior perforated substance. Morphological observations are calculated on values for six control and six Down's syndrome patients. Significant differences ($P<0.001$) are indicated with an asterisk

in Alzheimer's disease. Thus there are similarities in the pathology of Down's syndrome and Alzheimer's disease which lead to defects in cortical cholinergic activity.

III. Parkinson's Disease

Parkinson's disease is an extrapyramidal disease in which there is degeneration of the nigro-striatal dopaminergic system. This system is responsible for some of the motor disorders associated with the disease. However, the syndrome is also frequently accompanied by depression, cognitive defects and frontal type symptoms, the origins of which are unknown although a deficiency of the mesocorticolimbic dopaminergic pathways may be involved (RUBERG et al. 1982).

Intellectual impairment is prevalent among Parkinsonian subjects. The incidence of dementia is estimated to be five times higher in Parkinsonian subjects than in the general population, and the suggestion has been made on the basis of neuropathological similarities (presence of neurofibrillary tangles and senile plaques) that Parkinsonism and Alzheimer's dementia are related degenerative diseases. Nevertheless the majority of individuals with Parkinson-type dementia do not have Alzheimer-type pathology, tangles are absent and neuritic plaques are little more in number than in aged controls (PERRY et al. 1983).

It is of interest that RUBERG et al. (1982) and PERRY et al. (1983) have reported a reduction in ChAT activity in Parkinsonian patients with dementia in the absence of Alzheimer pathology (Table 5). These changes presumably relate to the extensive loss of neurons from the nucleus basalis (PERRY et al. 1985). Increased mAChR density has been observed in the frontal cortex and basal ganglia – possibly as the result of drug induction or, more plausibly, due to loss of presynaptic cholinergic terminals. However, unlike Alzheimer's disease, the apparent affinity of the ligand 3-quinuclidinyl benzylate (3-QNB) using the binding assay for mAChR was higher in the caudate nucleus and frontal cortex (RUBERG et al. 1982).

IV. Huntington's Chorea

Huntington's disease is a dominantly inherited movement disorder characterized by degeneration of cells in the basal ganglia. There is marked loss of neurons from

Table 5. Cholinergic activity in the brain of patients with Parkinson's disease (data from two separate laboratories). After RUBERG et al. (1982)
A. Regional distribution in mentally normal patients

Brain area	3-QNB binding (pmol mg^{-1} protein)		ChAT activity (nmol ACh h^{-1} mg^{-1} protein)	
	Control	Parkinson's disease	Control	Parkinson's disease
Caudate nucleus	1097 ± 41 (16)	1125 ± 38 (15)	180 ± 7	149 ± 13
Nucleus accumbens	826 ± 58 (13)	918 ± 59 (12)	91 ± 8	41 ± 12
Frontal cortex	700 ± 15 (17)	797 ± 31* (14)	4.9 ± 0.4	2.6 ± 0.3*
Hippocampus	334 ± 22 (13)	363 ± 29 (7)	13 ± 3	4.9 ± 0.8*

Results are means ± SEM with number of samples in parentheses. Asterisks denote significant differences ($P<0.001$)

B. Mental state and cortical ChAT activity. After PERRY et al. (1983)

	Control	Mentally normal	Cognitive impairment	With Alzheimer's disease
Frontal cortex	8.6 ± 2.6 (8)	5.78 ± 0.5 (3)	2.4 ± 0.6* (6)	2.4 (1)
Temporal cortex	4.6 ± 1.9 (8)	3.5 ± 1.7 (3)	0.7 ± 0.4* (6)	0.6 (1)

Results are mean amounts of ACh synthesized (in nmol h^{-1} mg^{-1} protein) ± SD with number of samples in parentheses. Asterisks denote significant differences ($P<0.001$)

the caudate nucleus and putamen. Some loss of cells occurs in the deeper layers of the cortex. The major biochemical change is in loss of GABA-ergic neurons (containing glutamate decarboxylase, GAD) in the caudate nucleus. Marked reduction of GAD activity and GABA concentration is found in both the striatum and substantia nigra. Large striatal cells that are probably cholinergic are reduced in number only when there is severe atrophy. However, in a small subgroup of Huntington's cases normal ChAT activity has been found. Unlike Alzheimer's disease no change in ChAT is found in the cortex and plaques and tangles are not present.

Intra-striatal injection of kainic acid has a potent neuroexcitatory effect producing structural and biochemical changes similar to those seen in Huntington's disease. Intrinsic intrastriatal neurons are destroyed including AChE-containing cells and their projected terminals in the substantia nigra (BIRD and SPOKES 1982).

V. Motoneuron Disease

The term motoneuron disease is the name given to a group of neuromuscular disorders including progressive muscular dystrophy, progressive bulbar palsy and amyotrophic lateral sclerosis which there occurs progressive destruction of upper and lower motoneurons in the adult cortex, cranial motor centres and spinal cord.

This topic has been discussed in Chap. 25, but insofar as there is often considerable of involvement of the, CNS it is, for the sake of completeness, mentioned again here. When lower motoneurons are predominantly affected, the condition is termed progressive muscular atrophy. In this case the large anterior horn cells of the spinal cord which supply the cholinergic innervation to peripheral skeletal muscle fibres are particularly affected. The cell bodies of spinal motoneurons are subject to extrinsic trophic or supportive influences within the CNS deriving from their glial micro-environment and are dependent on descending or afferent inputs from other nerve cells. In addition target muscle fibres themselves may, via retrograde transport, be vital in supplying neurotrophic support to motoneurons. Recent work by GURNEY and his associates (1984) has suggested that patients with motoneuron disease may have neutralizing serum autoantibodies to a muscle-derived factor which induces sprouting or peripheral motor nerve fibres. Whether these antibodies play a primary or secondary role in the aetiology of this disease remains, however, to be clearly defined.

D. Conclusion

It is clear that basic research on cholinergic neurotransmission is important to the understanding of the pathology of neurological and psychiatric disorder. A striking example has been the unexpected contribution from experimental studies on the nicotinic cholinergic recepetor to the problem of myasthenia gravis. The cholinergic system is clearly implicated in central neuropathy. Cognitive disorder can be related to impaired cortical innervation by ACh-containing neurons although the relationship of such studies to non-organic psychiatric disorder is less obvious.

However there is no doubt about the importance of all aspect of cholinergic research as described in this book, to neurological disease as well as to the wider field of contemporary neurobiology.

References

Anderton BH, Brainburg D, Downes MJ, Green PJ, Tomlinson BE, Ulrich J, Wood JN, Kahn J (1982) Monoclonal antibodies show that neurofibrillary tangles and neurofilaments share antigenic determinants. Nature 298:84–86

Appel SH (1981) A unifying hypothesis for the cause of amyotrophic lateral sclerosis, Parkinsonism and Alzheimer's disease. Ann Neurol 10:499–505

Arendt T, Bigl V, Arendt A, Tennstedt A (1983) Loss of neurons in the nucleus basalis of Meynert and Alzheimer's disease, paralysis agetans and Kossakoff's disease. Acta Neuropathol (Berl) 61:101–108

Arendt T, Bigl V, Tennstedt A, Arendt A (1985) Neuronal loss in different parts of the nucleus basalis is related to neuritic plaque formation in cortical target areas in Alzheimer's Disease. Neuroscience 14:1–15

Atack JR, Perry EK, Bonham JR, Perry RH, Tomlinson BE, Blessed G, Fairburn A (1983) Molecular forms of acetylcholinesterase in senile dementia of Alzheimer type: selective loss of the intermediate (IOS) form. Neurosci Lett 40:199–204

Ball MJ (1977) Neuronal loss, neuro-fibrillary tangles and granulovascular degeneration in the hippocampus with ageing and dementia. Acta Neuropathol (Berl) 37:111–118

Ball MJ (1982) Limbic predilection in Alzheimer dementia: is reactivated Herpes virus involved? Can J Neurol Sci 9:303–306

Bartus RT, Dean RL, Sherman KA, Friedman E, Beer B (1981) Profound effects of combining choline and piracetam and memory enhancement and cholinergic function in aged rats. Neurobiol Aging 2:105–111

Bartus RT, Dean RL, Beer B, Lippa AS (1982) The cholinergic hypothesis of geriatric memory dysfunction. Science 217:408–417

Beaudet A, Descarries L (1978) The monoamine innervation of rat cerebral cortex: synaptic and non-synaptic axon terminals. Neuroscience 3:851–860

Bird ED, Spokes EGS (1982) Huntington's chorea. In: Crow TJ (ed) Disorders of neurohumoural transmission. Academic, London, pp 144–182

Blass JP, Gibson GE, Duffy TE, Plum F (1981) Cholinergic dysfunction: a common denomination in metabolic encephalopathies. In: Peper G, Ladinsky H (eds) Cholinergic mechanisms. Plenum, New York, pp 921–928

Bowen DM (1981) Alzheimer's disease. In: Davison AN, Thompson RHS (eds) The molecular basis of neuropathology. Arnold, London p 649

Bowen DM, Smith CB, White P, Davison AN (1976) Neurotransmitter-related enzymes and indices of hypoxia in senile dementia and other abiotrophies. Brain 99:459–496

Bowen DM, White P, Spillane JA, Goodhardt MJ, Curzon G, Iwangoff P, Meier-Ruge W, Davison AN (1979) Accelerated ageing or selective neuronal loss as an important cause of dementia? Lancet 1:11–14

Buell SJ, Coleman PD (1979) Dendritic growth in the aged human brain and failure of growth in senile dementia. Science 206:854–855

Carrell RW (1988) Alzheimer's disease: enter a protease inhibitor. Nature 331:478–479

Colon EJ (1973) The cerebral cortex in presenile dementia. Acta Neuropathol (Berl) 23:281–290

Davies P (1979) Neurotransmitter-related enzymes in senile dementia of the Alzheimer type. Brain Res 171:319–328

Davies P, Maloney AJF (1976) Selective loss of cholinergic neurons in Alzheimer's disease. Lancet ii:1403

Dekosky St, Bass NH (1982) Aging, senile dementia and the intralaminar cytochemistry of cerebral cortex. Neurology (NY) 32:1227–1234

Drachman DA, Leavitt J (1974) Human memory and the cholinergic system, a relationship to ageing? Arch Neurol 30:113–121

Finch CE (1980) The relationships of aging changes in the basal ganglia to manifestations of Huntingtons chorea. Ann Neurol 7:406–411

Francis PT, Palmer AM, Sims NR, Bowen DM, Davison AN, Esiri MM, Neary D, Snowden JS, Wilcock GK (1985) Neurochemical studies of early-onset Alzheimer's disease: possible influence on treatment. N Engl J Med 313:7–11

Glenner GG (1988) Alzheimer's disease: its proteins and genes. Cell Vol 52:307–308

Gurney ME, Belton AC, Cashman N, Antel JP (1984) Inhibition of terminal axonal sprouting by serum from patients with amyotrophic lateral sclerosis. N Engl J Med 311:933–939

Harbaugh RE, Roberts DW, Coombs DW, Saunders RL, Reeder TM (1984) Preliminary Report: Intracranial cholinergic drug infusion in patients with Alzheimer's disease. Neurosurgery 15:514–518

Herzog AG, Kemper TL (1980) Amygdaloid changes in aging and dementia. Arch Neurol 37:625–629

Hubbard BM, Anderson JM (1981) A quantitative study of cerebral atrophy in old age and senile dementia. J Neurol Sci 50:135–145

Hubbard BM, Anderson JM (1985) Age-related variations in the neurone content of the cerebral cortex in senile dementia of Alzheimer type. Neuropathol Appl Neurobiol 11:369–382

Kitt CA, Price DL, Struble RG, Cork LC, Walker LC, Mobley WC, Becher MW, Joh TH, Wainer BH (1984) Transmitter specificity of neurites in senile plaques of aged monkeys. Soc Neurosci Abstr 10:271

Li JC, Kaminskas E (1985) Deficient repair of DNA lesions in Alzheimer's disease fibroblasts. Biochem Biophys Res Commun 129:733–738

Little A, Levy R, Chuaquikidd P, Hand D (1985) A double-blind, placebo controlled trial of high-dose lecithin in Alzheimer's disease. J Neurol Neurosurg Psychiatry 48:736–743

Mann DMA, Yates PO, Marcyniuk B (1984a) Alzheimer's presenile dementia, senile dementia of Alzheimer type and Down's syndrome in middle age from an age related continuum of pathological changes. Neuropathol Appl Neurobiol 10:185–200

Mann DMA, Yates PO, Marcyniuk B (1984b) Changes in nerve cells of the nucleus basalis of Meynert in Alzheimer's disease and their relationship to ageing and to the accumulation of lipofuscin pigment. Mech Ageing Dev 25:189–204

Mann DMA, Yates PO, Marcyniuk B (1985) Some morphometric observations on the cerebral cortex and hippocampus in presenile Alzheimer's disease, senile dementia of Alzheimer type and Down's syndrome in middle age. J Neurol Sci 69:139–159

Mash DC, Sevush S, Flynn DD, Norenberg MD, Potter LT (1984) Loss of M2-receptors in Alzheimer's disease. Neurology (NY) 34:120

McKinney M, Davies P, Coyles JT (1982) Somatostatin is not colocalised in cholinergic neurons innervating the rat cerebral cortex-hippocampal formation. Brain Res 243:169–172

Mervis RF, Bartus RT (1981) Modulation of pyramidal cell dendritic spine population in aging mouse neocortex: role of dietary choline. J Neuropathol Exp Neurol 40:313

Morrison JH, Rogers J, Scherr S, Benoit R, Bloom FE (1985) Somatostatin immunoreactivity in neuritic plaques of Alzheimer's patients. Nature 314:90–92

Nagai T, McGeer PL, Peng JH, McGeer EG, Dolman CE (1983) Choline acetyltransferase immunohistochemistry in brain of Alzheimer's disease patients and controls. Neurosci Lett 36:194–199

Ng Ying Kin NMK, Palo J, Haltia M, Wolfe LS (1983) High levels of brain dolichols in neuronal ceroid-lipafusionosis and senescence. J Neurochem 40:1465–1473

Pearce BR, Palmer AM, Bowen DM, Wilcock GK, Esiri MM, Davison AN (1984) Neurotransmitter dysfunction and atrophy of the caudate nucleus in Alzheimer's disease. Neurochem Pathol 2:221–233

Pearson RCA, Sofroniew MC, Cuello AC, Powell TPS, Eckenstein F, Esiri MM, Wilcock GK (1983) Persistence of cholinergic neurons in the basal nucleus in a brain with senile dementia of the Alzheimer's type demonstrated by immunohistochemical staining for choline acetyltransferase. Brain Res 289:375–379

Pearson RCA, Esiri MM, Hiorns RW, Wilcock GK, Powell TPS (1985) Anatomical correlates of the distribution of the pathological changes in the neocortex in Alzheimer disease. Proc Natl Acad Sci USA 82:4531–4535

Perry EK, Perry RH (1980) The cholinergic system in Alzheimer's disease. In: Roberts PJ (ed) Biochemistry of dementia. Wiley, Chichester, pp 135–183

Perry EK, Tomlinson BE, Blessed Y, Bergmann K, Gibson PH, Perry RH (1978) Correlation of cholinergic abnormalities with senile plaques and mental test scores in senile dementia. Br Med J 2:1457–1459

Perry RH, Candy JM, Perry EK, Irving D, Blessed D, Fairbairn AF, Tomlinson BE (1982) Extensive loss of choline acetyltransferase activity is not reflected by neuronal loss in the nucleus of Meynert in Alzheimer's disease. Neurosci Lett 33:311–315

Perry RH, Tomlinson BE, Candy JM, Blessed G, Foster JF, Bloxham CA, Perry EK (1983) Cortical cholinergic deficit in mentally impaired Parkinsonian patients. Lancet 2:789–790

Perry EK, Atack JR, Perry RH, Hardy JA, Dodd PR, Edwardson JA, Blessed G, Tomlinson BE, Fairbairn AF (1984) Intralaminar neurochemical distributions in human midtemporal cortex: comparison between Alzheimer's disease and the normal. J Neurochem 42:1402–1410

Perry EK, Curtis M, Dick DJ, Candy JM, Atack JR, Bloxham CA, Blessed G, Fairbairn A, Tomlinson BE, Perry RH (1985) Cholinergic correlates of cognitive impairment in Parkinson's disease: comparisons with Alzheimer's disease. J Neurol Neurosurg Psychiatry 48:413–421

Price DL, Whitehouse PJ, Struble RG, Clarke AW, Coyle JT, de Long MR, Hedreen JC (1982) Basal forebrain cholinergic systems in Alzheimer's disease and related dementias. Neurosci Commentar 1:84–92

Richardson PJ (1981) Quantitation of cholinergic synaptosomes form guinea pig brain. J Neurochem 37:258–260
Robison SH, Munzer JS, Tandan R, Bradley RS, Bradley WG (1985) DNA repair replication of alkylated DNA is reduced in Alzheimer's disease cells. J Neurol [Suppl] 232:63
Rogers JD, Brogan D, Mirra SS (1985) The nucleus basalis of Meynert in neurological disease: a quantitative morphological study. Ann Neurol 17:163–170
Rosenberg GS, Greenwald B, Davis KL (1983) In: Reisberg B (ed) Alzheimer's disease. Free Press, New York, pp 329–340
Rossor MN (1981) Parkinson's disease and Alzheimer's disease as disorders of the isodendritic core. Br Med J 283:1588–1590
Rossor MN, Iversen LL, Johnson AJ, Mountjoy CQ, Roth M (1981) Cholinergic deficit in frontal cerebral cortex in Alzheimer's disease is age dependent. Lancet ii:1422
Rossor MN, Emson PC, Mountjoy CQ, Roth M, Iversen LL (1982a) Neurotransmitter of the cerebral cortex in senile dementia of Alzheimer type. Exp Brain Res [Suppl] 5:153–157
Rossor MN, Garrett NJ, Johnson AL, Mountjoy CQ, Roth M, Iversen LL (1982b) A postmortem study of the cholinergic and GABA systems in senile dementia. Brain 105:313–330
Rossor MN, Iversen LL, Reynolds GP, Mountjoy CQ, Roth M (1984) Neurochemical characteristics of early and late onset types of Alzheimer's disease. Br Med J 288:961–964
Rubenstein R, Price RW (1984) Early inhibition of acetylcholinesterase and choline acetyltransferase activity in Herpes simplex virus type 1 infection of PC12 cells. J Neurochem 42:142–150
Ruberg M, Ploska A, Javoy-Agid F, Agid Y (1982) Muscarinic binding with choline acetyltransferase activity in Parkinsonian subjects with reference to dementia. Brain Res 232:129–139
Schwarcz R, Kohler C (1983) Differential vulnerability of central neurons of the rat to quinolinic acid. Neurosci Lett 38:85–90
Schwarcz R, Whetsell WO, Foster AC (1983) The neurodegenerative properties of intracerebral quinolinic acid and its structural analogisc-2,3-piperidine dicarboxylic acid. In: Fuxe K, Roberts P, Schwartz R (eds) Excitotoxins. Macmillan, London, pp 122–137
Selkoe DJ, Ihara Y, Salazar FJ (1982) Alzheimer's disease: insolubility of partially purified paired helical filaments in sodium dodecyl sulfate and urea. Science 215:1243–1245
Sims NR, Bowen DM, Davison AN (1981) [^{14}C]-acetylcholine synthesis and 14 C-carbon dioxide production from U-[^{14}C]-glucose by tissue prisms from human neocortex. Biochem J 196:867–876
Sims NR, Bowen DM, Allen SJ, Smith CCT, Neary D, Thomas DJ, Davison AN (1983) Presynaptic cholinergic dysfunction in patients with dementia. J Neurochem 40:503–509
Sitaram N, Weingartner H, Gillin JC (1978) Human serial learning: enhancement with arecholine and choline impairment with scopolamine. Science 201:274–276
Skup M, Oderfeld-Nowak B, Rommelspacher J (1983) In vitro studies on the effect of β-carbolines on the activities of acetylcholinesterase and choline acetyltransferase and on the muscarinic receptor binding of the rat brain. J Neurochem 41:62–68
Smith CCT, Bowen DM, Sims NR, Neary D, Davison AN (1983) Amino acid release from biopsy samples of temporal neocortex from patients with Alzheimer's disease. Brain Res 264:138–141
Spokes GS (1979) An analysis of factors influencing measurements of dopamine, noradrenaline, glutamate decarboxylase and choline acetylase in human post-mortem brain tissues. Brain 102:333–346
Struble RG, Cork LC, Whitehouse PJ, Price DL (1982) Cholinergic innervation in neuritic plaques. Science 216:413–415
Terry RD, Peck A, DeTeresa R, Schechter R, Horoupian DS (1981) Some morphometric aspects of the brain in senile dementia of the Alzheimer's type. Ann Neurol 10:184–192

Tomlinson BE, Irving B, Blessed G (1981) Cell loss in the locus coeruleus in senile dementia of Alzheimer type. J Neurol Sci 49:419–428
Whitehouse PJ, Price DL, Clark AW, Coyle JT, Delong MR (1981) Alzheimer disease: evidence for selective loss of cholinergic neurons in the nucleus basalis. Ann Neurol 10:122–126
Wilcock GK, Esiri MM, Bowen DM, Smith CCT (1982) Alzheimer's disease: correlation of cortical choline acetyltransferase activity with the severity of dementia and histological abnormalities. J Neurol Sci 57:407–417
Wisniewski HM, Merz GS (1984) Neuropathology of the aging brain and dementia of the Alzheimer type. In: Gaitz C, Samorajski T (eds) Aging 2000: our health care denstiny. Springer, Berlin Heidelberg New York, pp 231–243 (Biomedical issues, vol 1)
Yates CM, Simpson J, Maloney AFJ, Gordon A, Reid AM (1980) Alzheimer-like cholinergic deficiency in Down syndrome. Lancet 2:979

Subject Index

Handbook of Experimental Pharmacology

Springer-Verlag
Berlin Heidelberg
New York London
Paris Tokyo Hong Kong

Handbook of Experimental Pharmacology

Springer-Verlag
Berlin Heidelberg
New York London
Paris Tokyo Hong Kong

Volume 62
Aminoglycoside Antibiotics

Volume 63
Allergic Reactions to Drugs

Volume 64
Inhibition of Folate Metabolism in Chemotherapy

Volume 65
Teratogenesis and Reproductive Toxicology

Volume 66
Part 1: **Glucagon I**
Part 2: **Glucagon II**

Volume 67
Part 1
Antiobiotics Containing the Beta-Lactam Structure I

Part 2
Antibiotics Containing the Beta-Lactam Structure II

Volume 68, Parts 1 + 2
Antimalarial Drugs

Volume 69
Pharmacology of the Eye

Volume 70
Part 1
Pharmacology of Intestinal Permeation I

Part 2
Pharmacology of Intestinal Permeation II

Volume 71
Interferons and Their Applications

Volume 72
Antitumor Drug Resistance

Volume 73
Radiocontrast Agents

Volume 74
Antieliptic Drugs

Volume 75
Toxicology of Inhaled Materials

Volume 76
Clinical Pharmacology of Antiangial Drugs

Volume 77
Chemotherapy of Gastrointestinal Helminths

Volume 78
The Tetracycline

Volume 79
New Neuromuscular Blocking Agents

Volume 80
Cadmium

Volume 81
Local Anesthetics

Volume 82
Radioimmunoassay in Basic and Clinical Pharmacology

Volume 83
Calcium in Drug Actions

Volume 84
Antituberculosis Drugs

Volume 85
The Pharmacology of Lymphocytes